Mastering

AutoCAD® Civil 3D® 2010

Mastering

AutoCAD® Civil 3D® 2010

James Wedding

Scott McEachron

Wiley Publishing, Inc.

Senior Acquisitions Editor: Willem Knibbe
Development Editor: Kathryn Duggan
Technical Editor: Jonathan Stewart
Production Editor: Liz Britten
Copy Editor: Kathy Grider-Carlyle
Editorial Manager: Pete Gaughan
Production Manager: Tim Tate
Vice President and Executive Group Publisher: Richard Swadley
Vice President and Publisher: Neil Edde
Book Designer: Judy Fung and Bill Gibson
Proofreader: Jen Larsen, Word One
Indexer: Ted Laux
Project Coordinator, Cover: Lynsey Stanford
Cover Designer: Ryan Sneed
Cover Image: © Pete Gardner / Digital Vision / Getty Images

Published simultaneously in Canada

ISBN: 978-0-470-47353-5

Library of Congress Cataloging-in-Publication Data

Wedding, James, 1974-
Mastering Autocad Civil 3D 2010 / James Wedding, Scott McEachron. – 1st ed.
p. cm.
ISBN 978-0-470-47353-5 (paper/website)
1. Civil engineering — Computer programs. 2. Surveying — Computer programs. 3. Three-dimensional display systems. 4. AutoCAD Civil 3D (Electronic resource) I. McEachron, Scott, 1965- II. Title.
TA345.W44752 2009
624.0285'836–dc22

2009019191

10 9 8 7 6 5 4 3 2 1

Dear Reader,

Thank you for choosing *Mastering AutoCAD Civil 3D 2010*. This book is part of a family of premium-quality Sybex books, all of which are written by outstanding authors who combine practical experience with a gift for teaching.

Sybex was founded in 1976. More than 30 years later, we're still committed to producing consistently exceptional books. With each of our titles, we're working hard to set a new standard for the industry. From the paper we print on, to the authors we work with, our goal is to bring you the best books available.

I hope you see all that reflected in these pages. I'd be very interested to hear your comments and get your feedback on how we're doing. Feel free to let me know what you think about this or any other Sybex book by sending me an email at `nedde@wiley.com`. If you think you've found a technical error in this book, please visit `http://sybex.custhelp.com`. Customer feedback is critical to our efforts at Sybex.

Best regards,

Neil Edde
Vice President and Publisher
Sybex, an Imprint of Wiley

For Mom.
— JW

For Grandma.
— SM

Acknowledgments

This book is a team effort, with more authors than on are the front cover. We both have people to thank and acknowledge. Thank you to our clients and peers. Their generous sharing of data, time, and energy made many of the exercises and lessons in this book possible. We learn as much from our clients as we ever teach them.

Thank you to the team at Wiley: Willem Knibbe, Pete Gaughan, Jonathan Stewart, Kathryn Duggan, and Liz Britten. This team of editors performed an incredible job of guiding us through a faster production process than many thought possible.

Thank you to our friends in Manchester, New Hampshire. Autodesk has some truly great people working there, delivering the best product they can to users worldwide. Helping us with our questions the whole time they were preparing this release, the development team at Autodesk is an invaluable resource for this authoring team.

From The Authors

I have to first thank Scott. He took on the lion's share of the effort this year, and without him, some other team would have created the 2010 edition you have in your hands. Thanks to my partners at EE for understanding my love for this odd pet project, and for keeping the engine running while I was authoring. Thanks to Dan, Dave, Pete, Nick, Jessica, Dana, and Dana for being part of my sounding board and solution team this year. Your generous sharing of time and knowledge makes a task like this possible. Thank you Melinda and the girls for humoring all the "Hemingway" days and for still being excited about Daddy's books. And thanks Willem — you know why.

— *James Wedding, P.E.*

First and foremost I want to thank James for giving me the opportunity to work on this. This also couldn't have been possible without the trust of the EE team, and for that I'm grateful. I have to thank Nick for having the ability to help me keep things in perspective, and Travis for keeping me thinking outside the box. I need to thank my family and friends, and especially my grandmother for having enough faith in me 20 years ago to help me get started. I want to thank Greg and the team at Sherrill for giving me the opportunity to learn from the best, and Ross for having the courage to give me the push I'll always be grateful for. Finally, I want to thank Billy W. for being the cheerleader I needed, when I needed it.

— *Scott McEachron*

About the Authors

This book was written as a team effort from day one. Scott and I have covered the country and parts of the world training and teaching Civil 3D. Here's a bit more about each of us.

James Wedding, P.E., spent nearly a decade in the Dallas/Fort Worth land development industry before partnering with Engineered Efficiency (EE) in February 2006. A graduate of Texas Tech with a BSCE in 1997, he worked as a design engineer focused on private development. His design experience includes small commercial to multiphase single-family and master planned communities. James has served as president of the Preston Trail Chapter of the Texas Society of Professional Engineers, and he was selected their Young Engineer of the Year in 2003.

One of the earliest gunslingers for the Civil 3D product, James has worked extensively with the Autodesk product team to shape and guide the software's development. James is a highly rated repeat presenter at Autodesk University and a presenter on the Friday Civil 3D webcasts.

Scott McEachron, an Iowa native, received his Associate of Technology in Engineering Technology from the Morrison Institute of Technology in Morrison, Illinois in 1993. Prior to that he had been using AutoCAD and DCA in practice, but it wasn't until the spring of 1993 that he began a career in the civil/survey world in Edwardsville, IL. Scott began working in the Reseller community in 1998 and found his love for consulting then. Scott has been a speaker at Autodesk University for the past six years and is well known for his real-world approach to solving common civil/survey technology–related issues. Scott is known for his work in implementing Civil 3D under difficult circumstances and had one of the first documented success stories nearly five years ago.

Contents at a Glance

Contents

Foreword

When we began the development of AutoCAD® Civil 3D® software we had three key goals in mind:

- Provide automation tools for creating coordinated, reliable design information for a range of project types including land development, transportation, and environmental.
- Enable project teams to use the design information to accurately visualize, simulate, and analyze the performance of the project to come up with the best solutions.
- Facilitate delivery of higher quality construction documentation.

Our approach for achieving these goals was to develop a 3D information model that could accurately represent the civil engineering workflow and design process. The model, which dynamically connects design and construction documentation, has facilitated new ways of working and has helped civil engineers complete projects faster and with improved accuracy. Engineers are able to make design changes quickly and evaluate more alternatives, identify design issues and conflicts earlier in the process, and deliver higher quality designs faster.

These concepts have revolutionized the Civil Engineering process such that the plan production phase does not have to wait for the design to be completed. As a result, AutoCAD Civil 3D provides efficiencies in both the design automation and plan production stages of a typical design.

This is a departure from traditional 2D drafting based design software and it has a great potential to enhance design productivity and quality of design work. With AutoCAD Civil 3D, the entire design team can work from one model so that all phases of the project, from survey to construction documentation, remain coordinated.

The authors of *Mastering AutoCAD Civil 3D 2010* have embraced this vision from the start of its development. As the product has matured, I have had many conversations with James Wedding and the team at Engineered Efficiency to discuss product decisions and direction. I can recall conversations where James Wedding and I would discuss the reasons *why* we designed the product to work the way it does. As a result, this book contains much more than "picks and clicks." It has insightful tips, workflow recommendations, and best practices for using AutoCAD Civil 3D in a coordinated team environment.

On behalf of the entire AutoCAD Civil 3D product development team, I hope that AutoCAD Civil 3D enables you to work in ways that allow for creativity and profitability. This book is a great way to expand your understanding of the product and will help you gain the most out of the software.

Daniel A. Philbrick
Software Development Manager
Autodesk, Inc.

Introduction

Civil 3D was introduced in 2004 as a trial product. Designed to give the then Land Development desktop user a glimpse of the civil engineering software future, it was a sea change for AutoCAD-based design packages. Although there was need for a dynamic design package, many seasoned Land Desktop users wondered how they'd ever make the transition.

Over the past few years, Civil 3D has evolved from the wobbly baby introduced on those first trial discs to a mature platform used worldwide to handle the most complex engineering designs. With this change, many engineers still struggle with how to make the transition. The civil engineering industry as a whole is an old dog learning new tricks.

We hope this book will help you make the transition easier. As the user base grows and users get beyond the absolute basics, more materials are needed, offering a multitude of learning opportunities. Designed to help you get past the steepest part of the learning curve and teach you some guru-level tricks along the way, *Mastering AutoCAD Civil 3D 2010* should be a good addition to any Civil 3D user's bookshelf.

Who Should Read This Book

The *Mastering* book series is designed with specific users in mind. In the case of *Mastering AutoCAD Civil 3D 2010*, we expect you'll have some knowledge of AutoCAD in general and some basic engineering knowledge as well. We expect this book should appeal to a large number of Civil 3D users, but we envision a few primary users:

- Beginning users looking to make the move into using Civil 3D. These people understand AutoCAD and some basics of engineering, but they are looking to learn Civil 3D on their own, broadening their skillset to make themselves more valuable in their firms and in the market.
- Experienced Land Desktop users looking to transition to Civil 3D. These users understand design practice, and they need to learn how to do the familiar tasks in Civil 3D. They'll be able to jump to specific chapters and learn how to accomplish the task at hand.
- Civil 3D users looking for a desktop reference. With the digitization of the official help files, many users still long for a book they can flip open and keep beside them as they work. These people should be able to jump to the information they need for the task at hand, such as further information about a confusing dialog or troublesome design issue.
- Classroom instructors looking for better materials. This book was written with real data from real design firms. We've worked hard to make many of the examples match the real-world problems we have run into as engineers. This book also goes into greater depth than many basic texts, allowing short classes to teach the basics and leave the in-depth material for self-discovery, while longer classes can cover the full material presented.

This book can be used front-to-back as a self-teaching or instructor-based instruction manual. Each chapter has a number of exercises and most (but not all) build on the previous exercise. You can also skip to almost any exercise in any chapter and jump right in. We've created a large number of drawing files that you can download from `www.sybex.com/masteringcivil3d2010` to make picking and choosing your exercises a simpler task.

What You Will Learn

This book isn't a replacement for training. There are too many design options and parameters to make any book a good replacement for training from a professional. This book teaches you to use the tools available, explore a large number of the options available, and leave you with an idea of how to use each tool. At the end of the book, you should be able to look at any design task you run across, consider a number of ways to approach it, and have some idea of how to accomplish the task. To use one of our common analogies, reading this book is like walking around your local home-improvement warehouse. You see a lot of tools and use some of them, but that doesn't mean you're ready to build a house.

What You Need

Before you begin learning Civil 3D, you should make sure your hardware is up to snuff. Visit the Autodesk website and review graphic requirements, memory requirements, and so on. One of the most frustrating things that can happen is to be ready to learn, only to be stymied by hardware-related crashes. Civil 3D is a hardware-intensive program, testing the limits of every computer on which it runs.

We also really recommend using a dual-monitor setup. The number of dialogs, palettes, and so on make Civil 3D a real-estate hog. By having the extra space to spread out, you'll be able to see more of your design along with the feedback provided by the program itself.

You need to visit `www.sybex.com/go/masteringcivil3d2010` to download all of the data and sample files. Finally, please be sure to visit the Autodesk website at `www.autodesk.com` to download any service packs that might be available.

The Mastering Series

The *Mastering* series from Sybex provides outstanding instruction for readers with intermediate and advanced skills, in the form of top-notch training and development for those already working in their field and clear, serious education for those aspiring to become pros. Every *Mastering* book includes:

- Real-world scenarios ranging from case studies to interviews that show how the tool, technique, or knowledge presented is applied in actual practice.
- Skill-based instruction, with chapters organized around real tasks rather than abstract concepts or subjects.
- Self-review test questions, so you can be certain you're equipped to do the job right.

What Is Covered in This Book

Chapter 1: Getting Dirty: The Basics of Civil 3D introduces you to the interface and many of the common dialogs in Civil 3D. This chapter looks at the Toolbox and some underused Inquiry tools as well.

Chapter 2: Back to Basics: Lines and Curves examines various tools for creating linework. These tools include new best-fit tools that will let you interpolate a line or curve between known points.

Chapter 3: Lay of the Land: Survey looks at the Survey Toolspace and the unique toolset it contains for handling field surveying and fieldbook data handling. We also look at various surface and surveying relationships.

Chapter 4: X Marks the Spot: Points introduces Civil 3D points and the various methods of creating them. We also spend some time discussing the control of Civil 3D points with description keys and groups.

Chapter 5: The Ground Up: Surfaces in Civil 3D introduces the various methods of creating surfaces, using free and low-cost data to perform preliminary surface creation. Then we look at the various surface edits and analysis methods.

Chapter 6: Don't Fence Me In: Parcels describes the best practices for keeping your parcel topology tight and your labeling neat. It examines the various editing methods for achieving the desired results for the most complicated plats.

Chapter 7: Laying a Path: Alignments introduces the basic Civil 3D horizontal control element. This chapter also examines using layout tools that maintain the relationships between the tangents, curves, and spiral elements that create alignments.

Chapter 8: Cut to the Chase: Profiles looks at the sampling and creation methods for the vertical control element. We also examine the editing and element level control.

Chapter 9: Slice and Dice: Profile Views in Civil 3D examines all the various creation methods for building up profile views to reflect the required format for your design and plans. We also check out the new wizards used for creating split profile views.

Chapter 10: Templates Plus: Assemblies and Subassemblies looks at the building blocks of Civil 3D cross-sectional design. We look at the available tool catalogs and at building up full design sections for use in any design environment.

Chapter 11: Easy Does It: Basic Corridors introduces the basics of corridors — building full designs from horizontal, vertical, and cross-sectional design elements. We look at the various components to understand them better before moving to a more complex design set.

Chapter 12: The Road Ahead: Advanced Corridors looks at using corridors in unusual situations. We look at building surfaces, intersections, and some other areas of corridors that make them powerful in any design situation.

Chapter 13: Stacking Up: Cross Sections looks at slicing sections from surfaces, corridors, and pipe networks using alignments and the mysterious sample-line group. Working with the wizards and tools, we show you how to make your sections to order.

Chapter 14: The Tool Chest: Parts Lists and Part Builder gets into the building blocks of the pipe network tools. We look at modifying an existing part to add new sizes and then building up parts lists for various design situations.

Chapter 15: Running Downhill: Pipe Networks works with the creation tools for creating pipe networks. We look at both plan and profile views to get your plans looking like they should.

Chapter 16: Working the Land: Grading examines both feature lines and grading objects. We look at creating feature lines to describe critical areas and then using grading objects to describe mass grading. We also explore using the basic tools to calculate some simple volumes.

Chapter 17: Sharing the Model: Data Shortcuts looks at the data-shortcut mechanism for sharing data between Civil 3D users. We also look at updating and modifying the data behind the shortcuts and repairing broken references.

Chapter 18: Behind the Scenes: Autodesk Data Management Server walks you through installing and managing your own server for using Autodesk Vault as your project-management system. We also look at creating vaults and users for your design teams.

Chapter 19: Teamwork: Vault Client and Civil 3D walks you through bringing data into the ADMS created in Chapter 18, creating references between drawings and the update mechanism. We also look at the security features that allow team management and access control to individual files.

Chapter 20: Out the Door: Plan Production walks through the basics of creating view frame groups and creating sheets, and then it looks at some of the styles, templates, and editing techniques involved.

Chapter 21: Playing Nice with Others: LDT and LandXML looks at getting data back and forth with other software users. We look at importing data from your existing LDT projects to Civil 3D. We also examine the format of LandXML files to help you better understand what you can expect when you receive or send one out for sharing.

Chapter 22: Get the Picture: Visualization completes the main part of the book by taking all of the design elements and making presentation graphics from the design already modeled. We look at using the various rendering methods built into AutoCAD as well as some of the Civil 3D–specific tools.

Chapter 23: Projecting the Cost: Quantity Takeoff puts the Civil 3D model to use in the construction and contracting phase of the project. We examine pay items lists, tagging items for tabulation, and making the pipe network and corridor part of the inventory process. We also look at some basic reports for totaling up the bill.

The Appendix gathers together all the Master It problems from the chapters and provides a solution for each.

How to Contact the Authors

We welcome feedback from you about this book and/or about books you'd like to see from us in the future. You can reach us by writing to `Mastering@eng-eff.com`. For more information about our work, please visit our website at `www.eng-eff.com`.

Sybex strives to keep you supplied with the latest tools and information you need for your work. Please check their website at `www.sybex.com`, where we'll post additional content and updates that supplement this book if the need arises. Enter **Civil 3D** in the Search box (or type the book's ISBN — **9780470473535**) and click Go to get to the book's update page. You can also find updates and more information at `www.civil3d.com/errata`.

Thanks for purchasing *Mastering AutoCAD Civil 3D 2010*. We appreciate it, and look forward to exploring Civil 3D with you!

— *James Wedding, P.E.*

Chapter 1

Getting Dirty: The Basics of Civil 3D

Understanding Civil 3D's controls and operation is critical to mastering it. With its dizzying array of options and settings, getting Civil 3D to look and feel comfortable can take some effort. Learning how to use its numerous dialogs and tool palettes, as well as the Ribbon, is critical to driving Civil 3D and getting feedback about your design. This chapter explores the look and feel of Civil 3D as a CAD program, the unique components that make up the Civil 3D interface, and the creation of a working environment that matches the way you design.

By the end of this chapter, you'll learn to:

- Find any Civil 3D object with just a few clicks
- Modify the drawing scale and default object layers
- Modify the display of Civil 3D tooltips
- Add a new tool to the Toolbox
- Create a basic label style
- Create a new object style
- Navigate the Ribbon's contextual tabs

Windows on the Model

The most obvious change to the Civil 3D interface over its predecessors is the context-sensitive Ribbon. Many of Civil 3D's design tools can now be accessed via the Ribbon. A facelift to the Toolspace and enhancements to the general look and feel of the Civil 3D workspace combine to make this release easier to navigate than any of its predecessors. Figure 1.1 shows the Civil 3D palette sets along with the AutoCAD Tool Palettes and context-sensitive Ribbon displayed in a typical environment.

Toolspace

Toolspace is one of the unique Civil 3D palette sets. Toolspace can have as many as four tabs to manage user data. These tabs are as follows:

- Prospector
- Settings
- Survey
- Toolbox

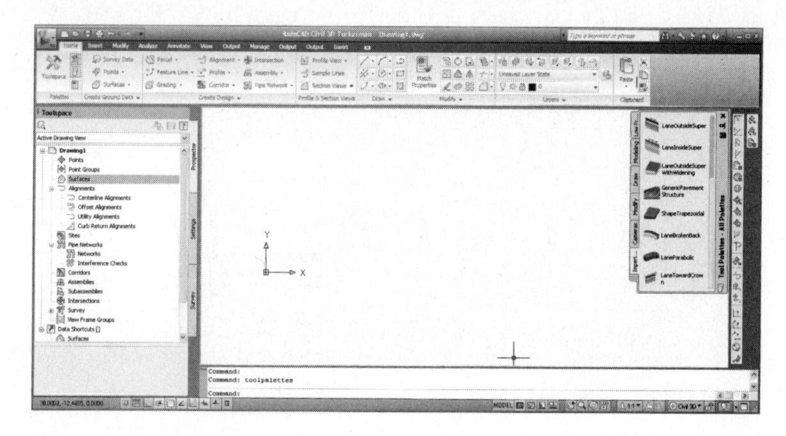

FIGURE 1.1
Civil 3D in a typical environment. Toolspace is docked on the left, and Panorama and Tool Palettes float over the drawing window. The Ribbon is at the top of the workspace

Using a Microsoft Windows Explorer–like interface within each, these tabs drive a large portion of the user control and data management of Civil 3D.

PROSPECTOR

Prospector is the main window into the Civil 3D object model. This palette or tab is where you go mining for data; it also shows points, alignments, parcels, corridors, and other objects as one concise, expandable list. In addition, in a project environment, this window is where you control access to your project data, create references to shared project data, and observe the check-in and check-out status of a drawing. Finally, you can also use Prospector to create a new drawing from the templates defined in the Drawing Template File Location branch in your AutoCAD Options dialog. Prospector has the following branches:

- Open Drawings
- Projects (only if the Vault client is installed)
- Data Shortcuts
- Drawing Templates

MASTER AND ACTIVE DRAWING VIEWS

If you can't see the Projects or Drawing Templates branch in Figure 1.1, look at the top of the Prospector pane. There is a drop-down menu for operating in Active Drawing View or Master View mode. Selecting Active Drawing View displays only the active drawing and Data Shortcuts. Master View mode, however, displays the Projects, the Drawing Templates, and the Data Shortcuts, as well as the branches of all drawings that are currently open.

In addition to the branches, Prospector has a series of icons across the top that toggle various settings on and off. Some of the Civil 3D icons from previous versions have been removed, and their functionality has been universally enabled for Civil 3D 2010. Those icons are noted here.

Item Preview Toggle Turns on and off the display of the Toolspace item preview within Prospector. These previews can be helpful when you're navigating drawings in projects (you can select one to check out) or when you're attempting to locate a parcel on the basis of its visual shape. In general, however, you can turn off this toggle — it's purely a user preference.

Preview Area Display Toggle When Toolspace is undocked, this button moves the Preview Area from the right of the tree view to beneath the tree view area.

Panorama Display Toggle Turns on and off the display of the Panorama window (which is discussed in a bit). To be honest, there doesn't seem to be a point to this button, but it's here nonetheless.

Help This should be obvious, but it's amazing how many people overlook it.

HAVE YOU LOOKED IN THE HELP FILE LATELY?

The AutoCAD Civil 3D development team in Manchester, New Hampshire, has worked hard to make the Help files in Civil 3D top notch and user friendly. The Help files should be your first line of support!

Open Drawings

This branch of Prospector contains the drawings currently open in Civil 3D. Each drawing is subdivided into groups by major object type, such as points, point groups, surfaces, and so forth. These object groups then allow you to view all the objects in the collection. Some of these groups are empty until objects are created. You can learn details about an individual object by expanding the tree and selecting an object.

Within each drawing, the breakdown is similar. If a collection isn't empty, a plus sign appears next to it, as in a typical Windows Explorer interface. Selecting any of these top-level collection names displays a list of members in the preview area. Right-clicking the collection name allows you to select various commands that apply to all the members of that collection. For example, right-clicking the Point Groups collection brings up the menu shown in Figure 1.2.

FIGURE 1.2
Context-sensitive menus in Prospector

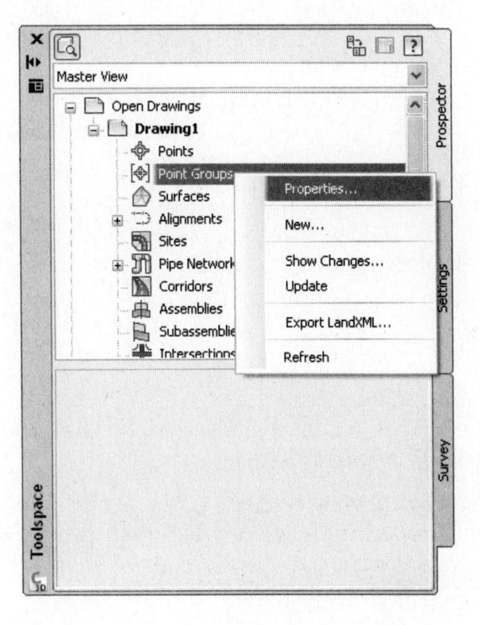

In addition, right-clicking the individual object in the list view offers many commands unique to Civil 3D: Zoom to Object and Pan to Object are typically included. By using these commands, you can find any parcel, point, cross section, or other Civil 3D object in your drawing almost instantly.

Many longtime users of AutoCAD have resisted right-clicking menus for their daily tasks since AutoCAD 14. In other AutoCAD products this may be possible, but in Civil 3D you'll miss half the commands! This book focuses on the specific options and commands for each object type during discussions of the particular objects.

Projects

The Projects branch of Prospector is the starting point for real team collaboration. This branch allows you to sign in and out of Vault, review what projects are available, manage the projects you sort through for information, check out drawings for editing, and review the status of drawings as well as that of individual project-based objects.

Data Shortcuts

Simply put, a data shortcut identifies the path to a specific object, in a specific drawing. Many users have found data shortcuts to be ideal in terms of project collaboration for two reasons: flexibility and simplicity.

Drawing Templates

The Drawing Templates branch is added more as a convenience than anything else. You can still create new drawings via the standard File ➤ New option, but by using the Drawing Templates branch, you can do the same thing without leaving Prospector. The Drawing Templates branch searches the file path specified in your AutoCAD Options dialog and displays a list of all the `.dwt` files it finds. You can customize this path to point to a server or other folder, but by default it's a local user-settings path. Right-clicking the name of a template presents you with the options shown in Figure 1.3.

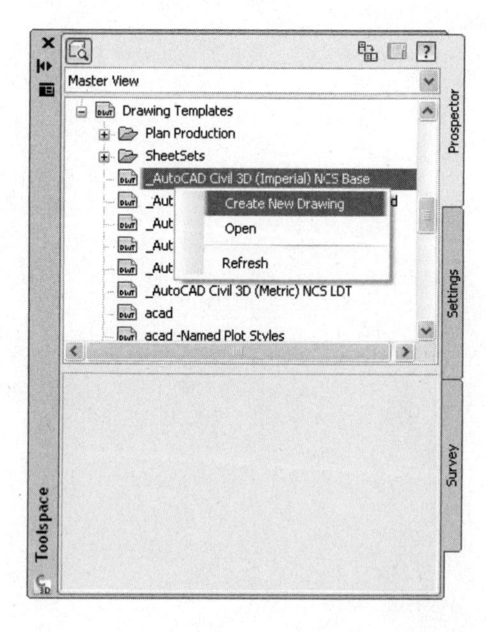

FIGURE 1.3 Creating a new drawing from within the Drawing Templates branch of Prospector. The templates shown here are located in the folder set in your AutoCAD Options window

Civil 3D is built on both AutoCAD and AutoCAD Map, so Civil 3D 2010 comes with a variety of templates. However, most users will want to select one of the top few, which start with `_Autodesk Civil 3D` and then have some descriptive text. These templates have been built on the basis of customer feedback to provide Civil 3D with a varying collection of object styles. These templates give you a good starting point for creating a template that meets your needs or the needs of your firm.

SETTINGS

The Settings tab of Toolspace is the proverbial rabbit hole. Here you can adjust how Civil 3D objects look and how the Civil 3D commands work. You use this tab to control styles, labels, and command settings for each component of Civil 3D. This book starts by looking at the top level of drawing settings and a few command settings to get you familiar, and then covers the specifics for each object's styles and settings in their respective chapters.

Drawing Settings

Starting at the drawing level, Civil 3D has a number of settings that you must understand before you can use the program efficiently. Civil 3D understands that the end goal of most users is to prepare construction documents on paper. To that end, most labeling and display settings are displayed in inches for imperial users and millimeters for metric users instead of nominal units like many other AutoCAD objects. Because much of this is based on an assumed working scale, let's look at how to change that setting, along with some other drawing options:

1. Open the file `Sample Site.dwg` from the installed tutorial drawings.
2. Switch to the Settings tab.
3. Right-click the filename, and select Edit Drawing Settings to display the dialog shown in Figure 1.4.

FIGURE 1.4
The Drawing Settings dialog

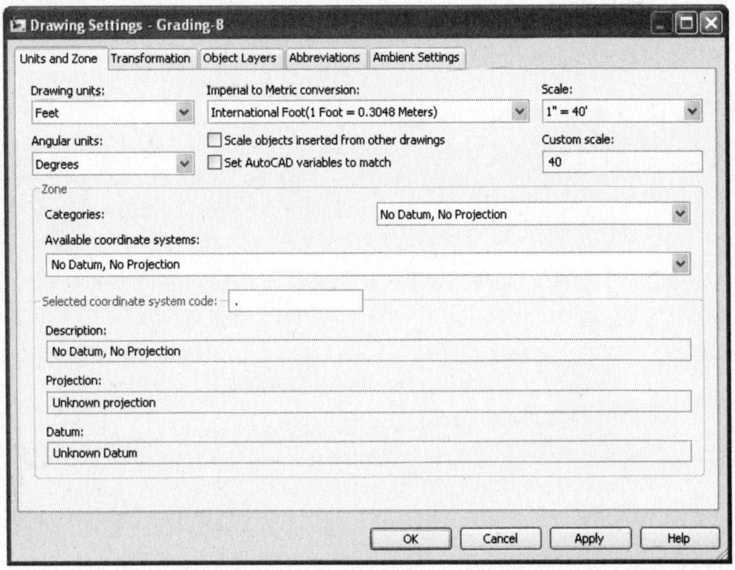

Each tab in this dialog controls a different aspect of the drawing. Most of the time, you'll pick up the Object Layers, Abbreviations, and Ambient Settings from a companywide template. But the drawing scale and coordinate information change for every job, so you'll visit the Units and Zone and the Transformation tabs frequently.

Units and Zone Tab

The Units and Zone tab lets you specify metric or imperial units for your drawing. You can also specify the conversion factor between systems. In addition, you can control the assumed plotting scale of the drawing. The drawing units typically come from a template, but the options for scaling blocks and setting AutoCAD variables depend on your working environment. Many engineers continue to work in an arbitrary coordinate system using the settings as shown earlier, but using a real coordinate system is easy! For example, setting up a drawing for a the Dallas, Texas, area, you'd follow this procedure:

1. Select USA, Texas from the Categories drop-down menu on the Units and Zone tab.
2. Select NAD83 Texas State Planes, North Central Zone, US Foot from the Available Coordinate Systems drop-down menu.

There are literally hundreds, if not thousands, of available coordinate systems. These are established by international agreement; because Civil 3D is a worldwide product, almost any recognized surveying coordinate system can be found in the options. Once your coordinate system has been established, you can change it on the Transformation tab if desired.

This tab also includes the options Scale Objects Inserted from Other Drawings and Set AutoCAD Variables to Match. In Figure 1.4, both are unchecked to move forward. The scaling option has been problematic in the past because many firms work with drawings that have no units assigned and therefore scale incorrectly; but you can experiment with this setting as you'd like. The Set AutoCAD Variables to Match option attempts to set the AutoCAD variables AUNITS, DIMUNITS, INSUNITS, and MEASUREMENT to the values placed in this dialog. You can learn about the nature of these variables via the Help system. Because of some inconsistencies between coordinate-based systems and the AutoCAD engine, sometimes these variables must be approximated. Again, you won't typically set this flag to True; you should experiment in your own office to see if it can help you.

Transformation Tab

With a base coordinate system selected, you can now do any further refinement you'd like using the Transformation tab. The coordinate systems on the Units and Zone tab can be refined to meet local ordinances, tie in with historical data, complete a grid to ground transformation, or account for minor changes in coordinate system methodology. These changes can include the following:

Apply Sea Level Scale Factor — Takes into account the mean elevation of the site and the spheroid radius that is currently being applied as a function of the selected zone ellipsoid.

Grid Scale Factor — Based on a 1:1 value, a user-defined uniform scale factor, a reference point scaling, or a prismoidal transformation in which every point in the grid is adjusted by a unique amount.

Reference Point — Can be used to set a singular point in the drawing field via pick or via point number, local northing and easting, or grid northing and easting values.

Rotation Point — Can be used to set the reference point for rotation via the same methods as the Reference Point.

Specify Grid Rotation Angle — Enter an amount or set a line to North by picking an angle or deflection in the drawing. You can use this same method to set the azimuth if desired.

Most engineering firms work on either a defined coordinate system or an arbitrary system, so none of these changes are necessary. Given that, this tab will be your only method of achieving the necessary transformation for certain surveying and Geographic Information System (GIS)–based, and Land Surveying–based tasks.

Object Layers Tab

Setting object layers to your company standard is a major part of creating the feel you're after when using Civil 3D in your office. The nearly 50 objects described here make up the entirety of the Civil 3D modeling components and the objects you and other users will deal with daily.

The layers listed in this dialog by default reflect a modified AIA CAD Layer Guideline as part of the National CAD Standard (NCS). This layering standard is built into many places in Civil 3D's templates and is becoming more widely adopted in the land-development industry. In addition to being fairly comprehensive and well known among engineering firms, the NCS has the benefit of being the roadmap for the future in terms of out-of-the-box content from Autodesk. Adopting this standard means you'll have fewer things to change with every release of the software. Nevertheless, it is important that every user know how to modify these defaults.

One common issue with the shipping templates is that the templates assume road design is the primary use of alignments. Use the following procedure to change the Alignment setting to the NCS for laying out a sanitary sewer:

1. Click the Layer column in the Alignment row, as shown in Figure 1.5.

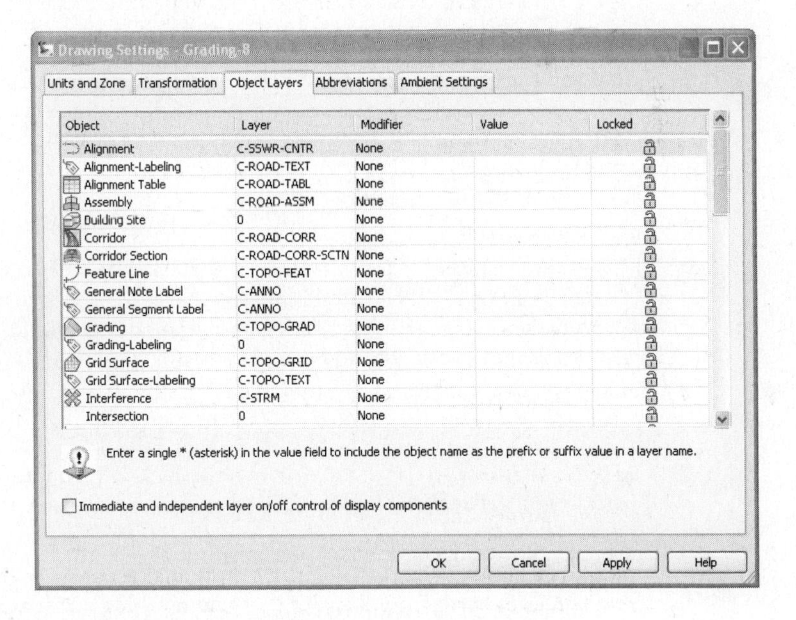

FIGURE 1.5
Changing the Layer setting for the Alignment object

2. In the Layer Selection dialog list, select C-SSWR-CNTR and click OK.

ONE OBJECT AT A TIME

Note that this procedure only changes the Alignment object. If you want to change the standard of all the objects, you need to adjust the Alignment Labeling, Alignment Table, Profile, Profile View, Profile View Labeling, and so on. To do this, it's a good idea to right-click in the grid view and select Copy All. You can then paste the contents of this matrix into Microsoft Excel for easy formatting and reviewing.

One common question that surrounds the Object Layers tab is the check box at lower left: Immediate and Independent Layer On/Off Control of Display Components. What the heck does that mean? Relax — it's not as complicated as it sounds.

Many objects in Civil 3D are built from underlying components. Take an alignment, for example. It's built from tangents, curves, spirals, extension lines, and so on. Each of these components can be assigned its own layer — in other words, the lines could be assigned to the LINES layer, curves to the CURVES layer, and so on. When this check box is selected, the *component's* layer exerts some control. In the example given, if the alignment is assigned to the ALIGN layer and the box is selected, turning off (not freezing) the LINES layer will make the line components of that alignment disappear. Deselect this control, and the LINES layer's status won't have any effect on the visibility of the alignment line components.

Finally, it's important to note that this layer control determines the object's parent layer *at creation*. Civil 3D objects can be moved to other layers at any time. Changing this setting doesn't change any objects already in place in the drawing.

Abbreviations Tab

One could work for years without noticing the Abbreviations tab. The options on this tab allow you to set the abbreviations Civil 3D uses when labeling items as part of its automated routines. The prebuilt settings are based on user feedback, and many of them are the same as the settings from Land Desktop, the last-generation civil engineering product from Autodesk.

Changing an abbreviation is as simple as clicking in the Value field and typing a new one. Notice that the Alignment Geometry Point Entity Data section has a larger set of values and some formulas attached. These are more representative of other label styles, and we'll visit the label editor a little later in this chapter.

THERE'S ALWAYS MORE TO LEARN

Until December 2006, James was still advising users to add "t." to their labels to get "Rt." or "Lt." in the final label. He'd forgotten that the abbreviations being used were set here! By changing the Left and Right abbreviation from "L" and "R" to "Lt." and "Rt.", respectively, you can skip that step in the label setup. Sometimes there are just too many options to remember them all!

Ambient Settings Tab

The Ambient Settings tab can be daunting at first. The term *ambient* means "surround" or "surrounding," and these settings control many of the math, labeling, and display features, as well as the user interaction surrounding the use of Civil 3D. Being familiar with the way this tab works will help you further down the line, because almost every other setting dialog in the program works like the one shown in Figure 1.6.

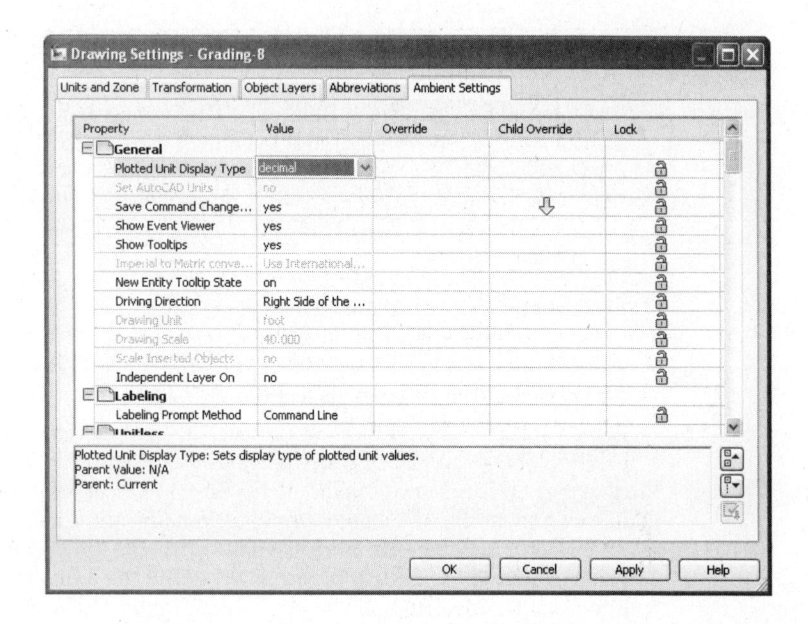

FIGURE 1.6
The Ambient Settings tab with the General branch expanded

You can approach this tab in the following ways:

- *Top to bottom* — Expand one branch, handle the settings in that branch, and then close it and move to the next.
- *Print and conquer* — Expand all the branches using the Expand All Categories button found at lower right.

After you have expanded the branches, right-click in the middle of the displayed options and select Copy to Clipboard. Then paste the settings to Excel for review, as you did with the Object Layers tab.

SHARING THE WORKLOAD

The Print and Conquer approach makes it easy to distribute multiple copies to surveyors, land planners, engineers, and so on and let them fill in the changes. Then, creating a template for each group is a matter of making their changes. If you're asking end users who aren't familiar with the product to make these changes, it's easy to miss one. Working line by line is fairly foolproof.

After you decide how to approach these settings, get to work. The settings are either drop-down menus or text boxes (in the case of numeric entries). Many of them are self-explanatory and common to land-development design. Let's look at these settings in more detail (see Figure 1.6).

Plotted Unit Display Type Remember, Civil 3D knows you want to plot at the end of the day. In this case, it's asking you how you would like your plotted units measured. For example, would you like that bit of text to be 0.25″ tall or $\frac{1}{4}$″ high? Most engineers are comfortable with the Leroy method of text heights (L80, L100, L140, and so on), so the decimal option is the default.

Set AutoCAD Units This displays whether or not Civil 3D should attempt to match AutoCAD drawing units, as specified on the Units and Zone tab.

Save Command Changes to Settings This setting is incredibly powerful but a secret to almost everyone. By setting it to Yes, your changes to commands will be remembered from use to use. This means if you make changes to a command during use, the next time you call that Civil 3D command, you won't have to make the same changes. It's frustrating to do work over because you forgot to change one out of the five things that needed changing, so this setting is invaluable.

Show Event Viewer Event Viewer is Civil 3D's main feedback mechanism, especially when things go wrong. It can get annoying, however, and it takes up valuable screen real estate (especially if you're stuck with one monitor!), so many people turn it off. We recommend leaving it on and pushing it to the side if needed.

Show Tooltips One of the cool features that people remark on when they first use Civil 3D is the small pop-up that displays relevant design information when the cursor is paused on the screen. This includes things such as Station-Offset information, Surface Elevation, Section information, and so on. Once a drawing contains numerous bits of information, this display can be overwhelming; therefore, Civil 3D offers the option to turn off these tooltips universally with this setting. A better approach is to control the tooltips at the object type by editing the individual feature settings. You can also control the tooltips by pulling up the properties for any individual object and looking at the Information tab.

Imperial to Metric Conversion This displays the conversion method specified on the Units and Zone tab. The two options currently available are US Survey Foot and International Foot.

New Entity Tooltip State You can also control tooltips on an individual object level. For instance, you might want tooltip feedback on your proposed surface but not on the existing surface. This setting controls whether the tooltip is turned on at the object level for new Civil 3D objects.

Driving Direction Specifies the side of the road that forward-moving vehicles use for travel. This setting is important in terms of curb returns and intersection design.

Drawing Unit, Drawing Scale, and Scale Inserted Objects These settings were specified on the Units and Zone tab but are displayed here for reference and so that you can lock them if desired.

Independent Layer On This is the same control that was set on the Object Layers tab.

The settings that are applied here can also be applied at the object levels. For example, you may typically want elevation to be shown to two decimal places; but when looking at surface elevations, you might want just one. The Override and Child Override columns give you feedback about these types of changes. See Figure 1.7.

The Override column shows whether the current setting is overriding something higher up. Because you're at the Drawing Settings level, these are clear. However, the Child Override column displays a down arrow, indicating that one of the objects in the drawing has overridden this setting. After a little investigation through the objects, you'll find the override is in the Edit Feature Settings of the Profile View as shown in Figure 1.8.

Notice that in this dialog, the box is checked in the Override column. This indicates that you're overriding the settings mentioned earlier, and it's a good alert that things have changed from the general Drawing Settings to this Object Level setting.

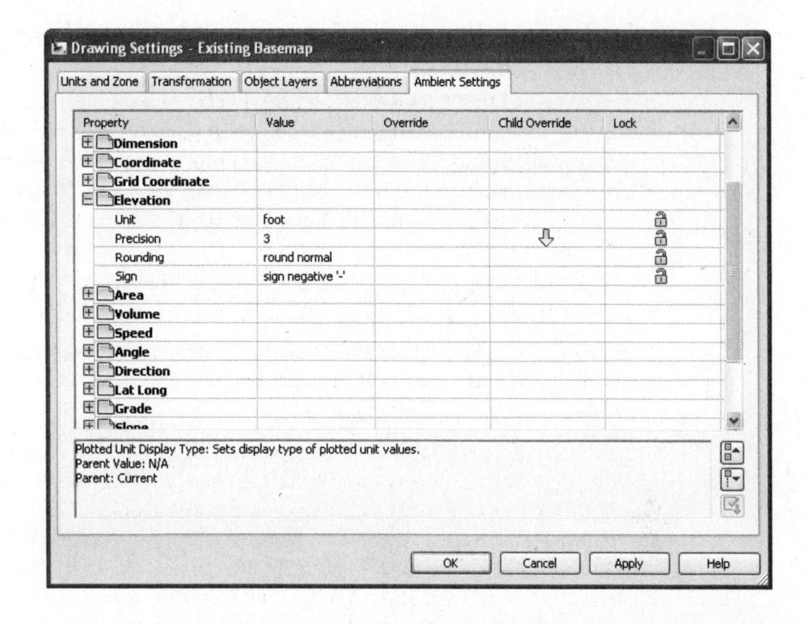

FIGURE 1.7
The Child Override indicator in the Elevation values

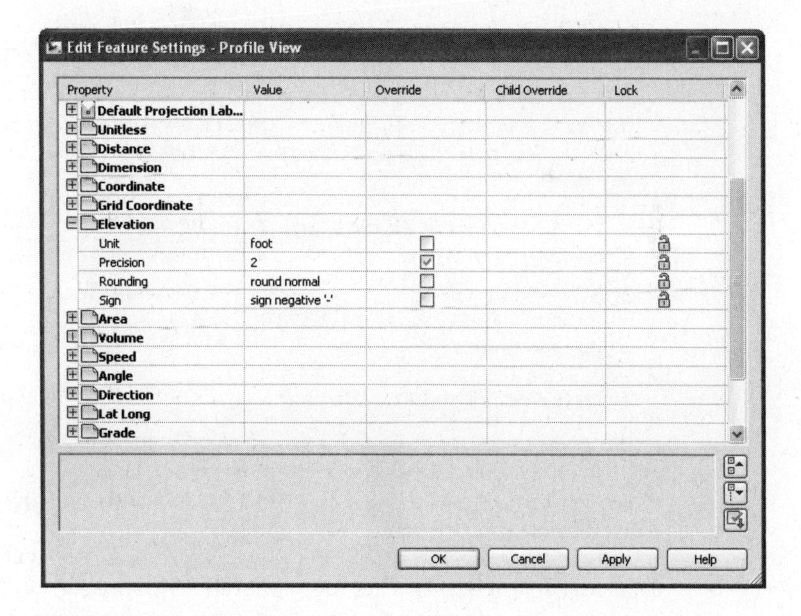

FIGURE 1.8
The Profile Elevation Settings and the Override indicator

But what if you don't want to allow those changes? Each Settings dialog includes one more column: Lock. At any level, you can lock a setting, graying it out for lower levels. This can be handy for keeping users from changing settings at the lower level that perhaps should be changed at a drawing level, such as sign or rounding methods.

Object Settings

If you click the Expand button next to the drawing name, you see the full array of objects that Civil 3D uses to build its design model. Each of these has special features unique to the object being described, but there are some common features as well. Additionally, the General collection contains settings and styles that are applied to various objects across the entire product.

The General collection serves as the catchall for styles that apply to multiple objects and for settings that apply to *no* objects. For instance, the Civil 3D General Note object doesn't really belong with the Surface or Pipe collections. It can be used to relate information about those objects, but because it can also relate to something like "Don't Dig Here!" it falls into the general category. The General collection has three components (or branches):

Multipurpose Styles These styles are used in many objects to control the display of component objects. The Marker Styles and Link Styles collections are typically used in cross-section views, whereas the Feature Line Styles collection is used in grading and other commands. Figure 1.9 shows the full collection of multipurpose styles and some of the marker styles that ship with the product.

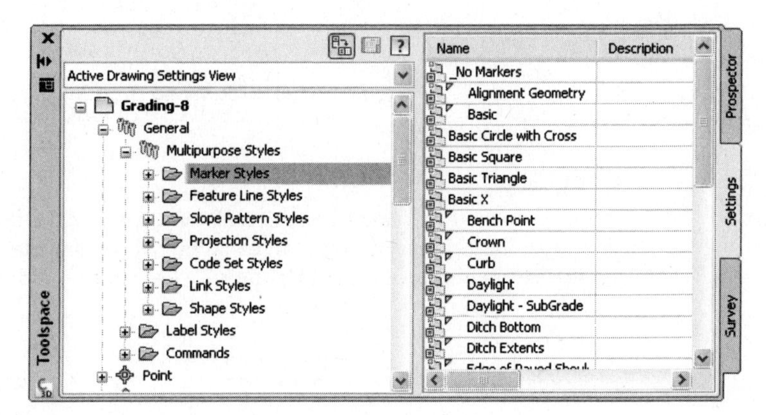

FIGURE 1.9
General multipurpose styles and some marker styles

Label Styles The Label Styles collection allows Civil 3D users to place general text notes or label single entities outside the parcel network while still taking advantage of Civil 3D's flexibility and scaling properties. With the various label styles shown in Figure 1.10, you can get some idea of their usage.

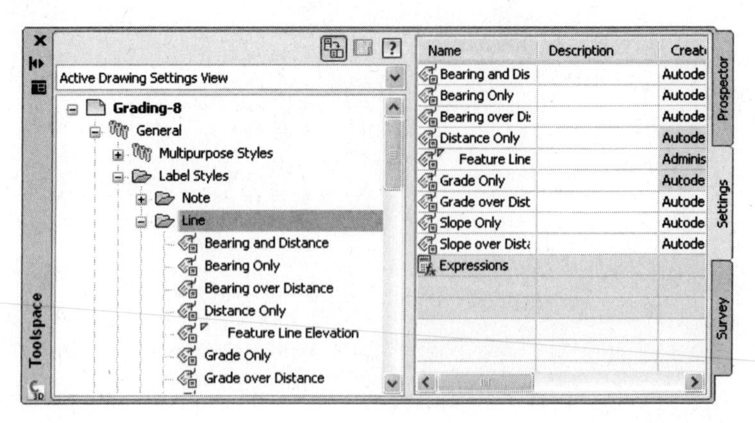

FIGURE 1.10
Line label styles

Because building label styles is a critical part of producing plans with Civil 3D, a later section of this chapter looks at how to build a new basic label and some of the common components that appear in every label style throughout the product.

Commands Almost every branch in the Settings tree contains a `Commands` folder. Expanding this folder, as shown in Figure 1.11, shows you the typical long, unspaced command names that refer to the parent object.

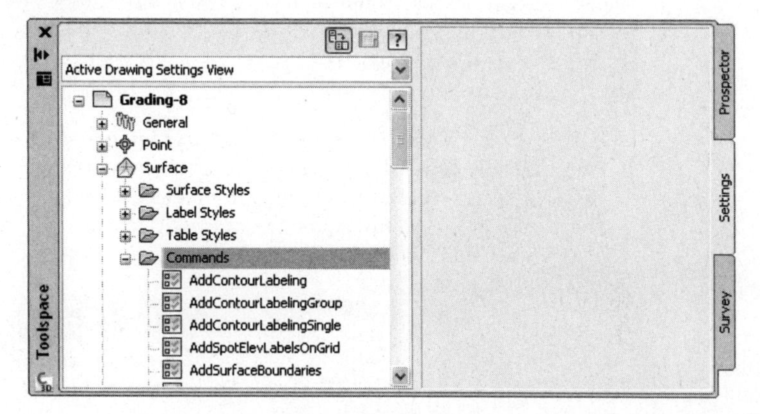

FIGURE 1.11 Surface command settings in Toolspace

SURVEY

The Survey palette is displayed optionally and controls the use of the survey, equipment, and figure prefix databases. Survey is an essential part of land-development projects. Because of the complex nature of this tab, all of Chapter 3, "Lay of the Land: Survey," is devoted to it.

TOOLBOX

The Toolbox is a launching point for add-ons and reporting functions. To access the Toolbox, from the Home tab in the Ribbon, select Toolspace ➢ Palettes ➢ Toolbox. Out of the box, the Toolbox contains reports created by Autodesk, but you can expand its functionality to include your own macros or reports. The buttons on the top of the Toolbox, shown in Figure 1.12, allow you to customize the report settings and add new content.

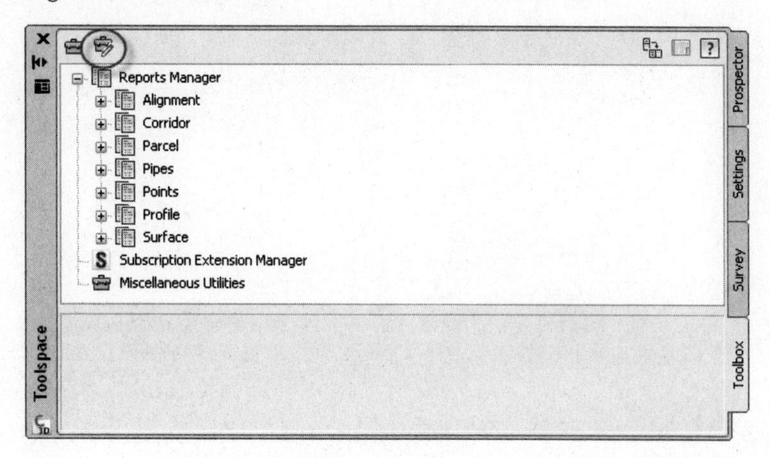

FIGURE 1.12 The Toolbox palette with the Edit Toolbox Content button circled

A Toolbox Built Just for You

You can edit the Toolbox content and the Report Settings by selecting the desired tool, right-clicking, and then executing. Don't limit yourself to the default reports that ship in the Toolbox, though. Many firms find that adding in-house customizations to the Toolbox gives them better results and is more easily managed at a central level than by customizing via the AutoCAD custom user interface (CUI) and workspace functionality.

Let's add one of the sample Civil 3D Visual Basic Application (VBA) macros to a new Toolbox:

1. Click the Edit Toolbox Content button shown in Figure 1.12 to open the Toolbox Editor in Panorama.
2. Click the button shown here to add a new root category.

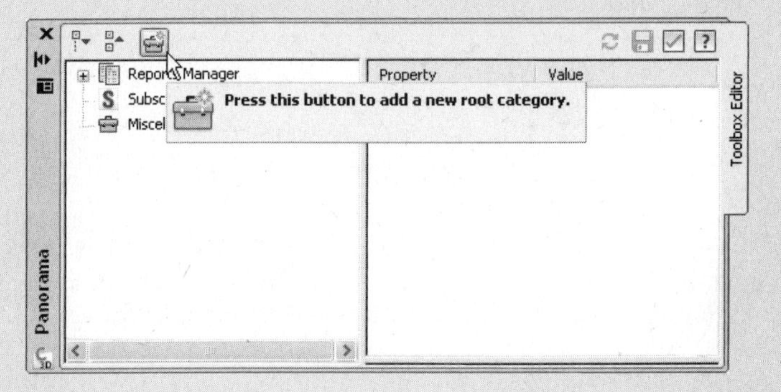

3. Click the Root Category1 toolbox that appears. The name will appear in the preview area, where you can edit it. Change the name to **Sample Files**, and press ↵.
4. Right-click the Sample Files toolbox, and select New Category as shown here.

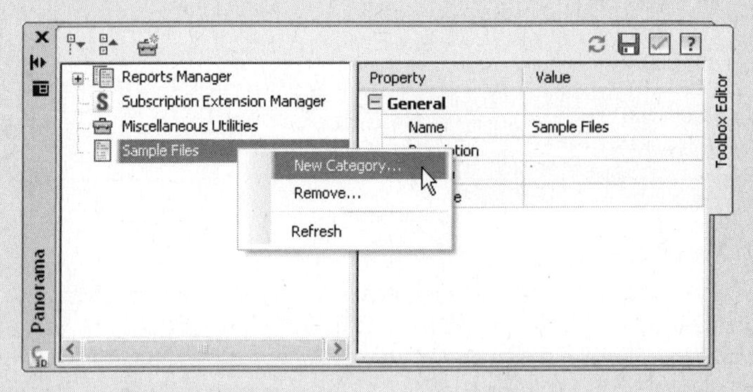

5. Expand the Sample Files toolbox to view the new category, and then click the name to edit it in the preview area. Change the name to **VBA**, and press ↵.

6. Right-click the VBA category, and select New Tool.

7. Expand the VBA category to view the new tool, and then click the name to edit it in the preview area. Change its name to **Pipe Sample**.

8. Change the Description to **Sample VBA**.

9. Working down through the properties in the preview area, select VBA in the drop-down menu in the Execute Type field.

10. Click in the Execute File field, and then click the More button.

11. Browse to `C:\Program Files\Autocad Civil 3D 2010\Sample\Civil 3D API\COM\Vba\Pipe\`, and select the file `PipeSample.dvb`.

12. Click Open.

13. Click in the Macro Name text field, and type **PipeSample** as shown here.

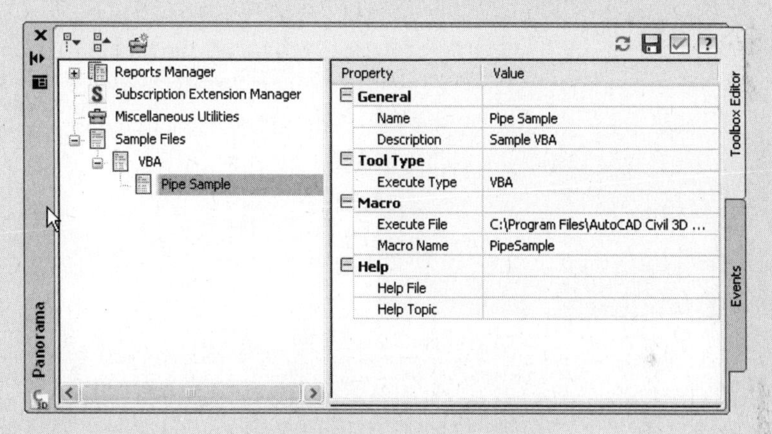

14. Click the green check box at upper right to dismiss the editor.

15. You will be asked "Would you like to apply those changes now?" Select Yes.

You've now added that sample VBA macro to your Toolbox. By adding commonly used macros and custom reports to your Toolbox, you can keep them handy without modifying the rest of your Civil 3D interface or programming buttons. It's just one more way to create an interface and toolset for the way you work.

Panorama

The Panorama window is Civil 3D's feedback and tabular editing mechanism. Designed to be a common interface for a number of different Civil 3D–related tasks, you can use it to provide information about the creation of profile views, to edit pipe or structure information, or to run basic volume analysis between two surfaces. For an example of Panorama in action, change to the View tab, and then select Palettes ➢ Event Viewer. You'll explore and use Panorama more during this book's discussion of specific objects and tasks.

RUNNING OUT OF SCREEN REAL ESTATE?

It's a good idea to turn on Panorama using this technique and then drag it to the side so you always see any new information. Although it's possible to turn it off, doing so isn't recommended — you won't know when Civil 3D is trying to tell you something! Place Panorama on your second monitor (now you see why you need to have a second monitor, don't you?), and you'll always be up to date with your Civil 3D model.

And in case you missed it, you were using Panorama when you added the sample VBA macro in the previous exercise.

Ribbon

As with AutoCAD, the Ribbon is the primary interface for accessing Civil 3D commands and features. When you select an AutoCAD Civil 3D object, the Ribbon displays commands and features related to that object. If several object types are selected, the Multiple contextual tab is displayed. Use the following procedure to familiarize yourself with the Ribbon:

1. Select one of the line labels in the northwest portion of the Sample Site drawing.
2. Notice that both the General Tools and Modify tabs are displayed as shown in Figure 1.13.

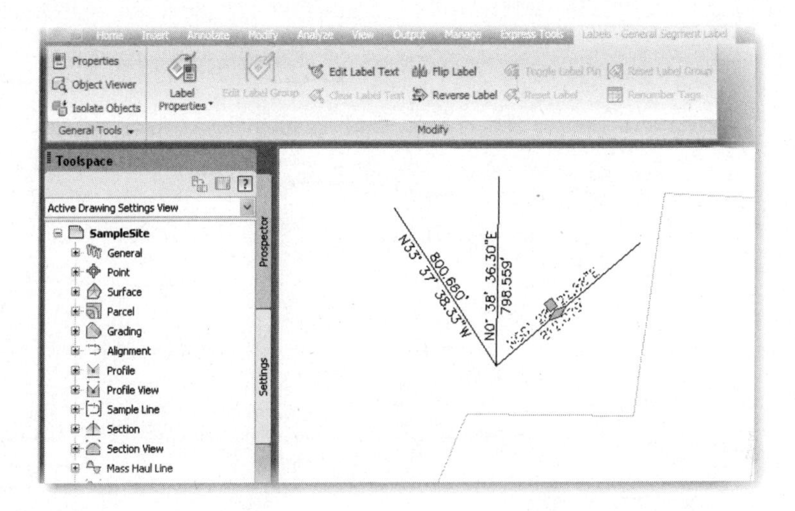

FIGURE 1.13
The context-sensitive Ribbon

3. Select a parcel label (the labels in the middle of the lot areas) and notice the display of the Multiple contextual tab.
4. Use the Esc key to cancel all selections.
5. Navigate to the Prospector and expand the Alignments ➤ Centerline Alignments collection.
6. Select the Avery Drive alignment, right-click, and choose the Select option on the menu. Notice the change in the Ribbon.

7. Select the down arrow next to the Modify panel. Using the pin at the bottom-left corner of the panel, pin the panel open.

8. Select the Properties command in the General Tools panel to open the AutoCAD Properties palette. Notice that the Modify panel remains opened and pinned.

It's All About Style

Before you get into the program itself, it's important to understand one bit of vocabulary and how it relates to Civil 3D: *style*. To put it simply, styles control the display properties of Civil 3D objects and labels. Styles control everything from the color of your point markers to the interval of your surface contours, and from your profile-view grid spacing to the text height in the Station-Offset label of your road alignment. Styles truly are where the power lies in Civil 3D. Label styles and object styles are the two major categories.

The difficult thing about styles is that it's hard to talk about them without being specific. Later chapters spend a fair amount of time talking about the specifics of the styles for each object, and this chapter looks at the common aspects of style manipulation; but styles may remain a mystery until you get your hands dirty later in the book.

Label Styles

To get started, look at the styles in the Spot Elevation branch by expanding the Surface branch and then the Label Styles branch on the Settings tab, as shown in Figure 1.14.

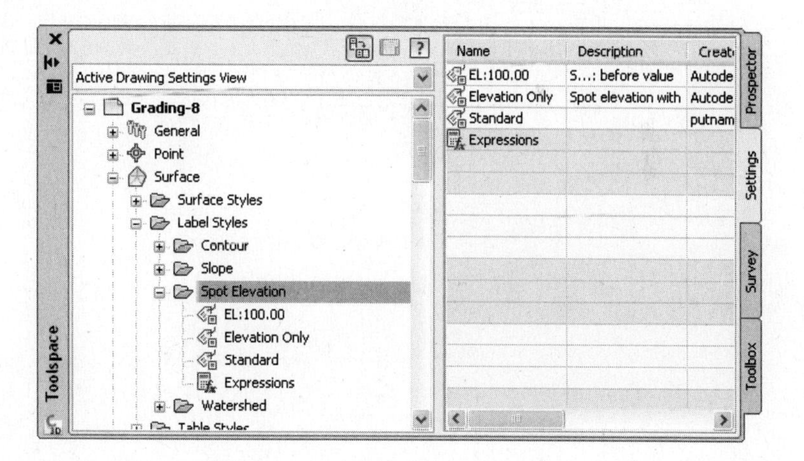

FIGURE 1.14
Spot Elevation label styles

There are two basic label styles in the Spot Elevation branch. Let's create a new one and explore the options for making labels. Remember, almost all of these options are present in other, object-specific label styles.

1. Right-click the Spot Elevation folder, and select New in the pop-up menu to open the Label Style Composer, as shown in Figure 1.15.

2. On the Information tab, change the style name to something appropriate. For this example, use **JW-EG**.

FIGURE 1.15
The Label Style Composer

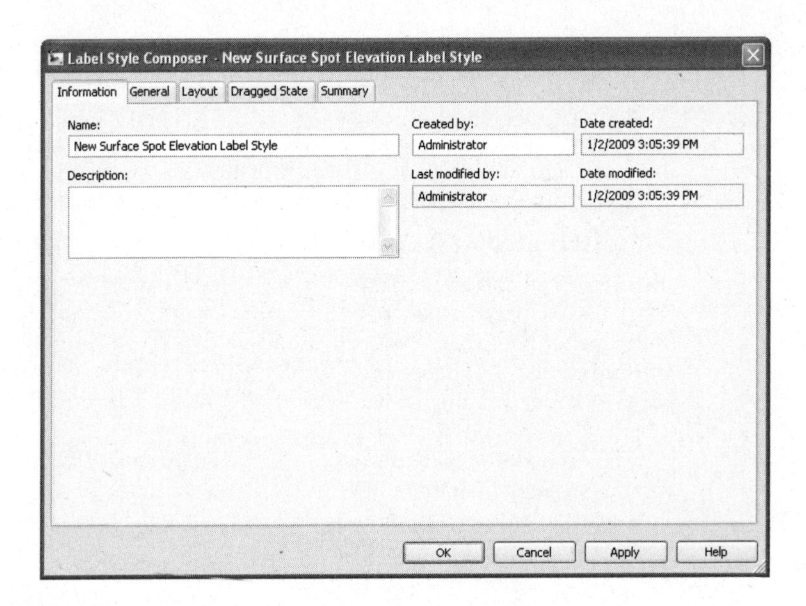

WHO BUILT THAT STYLE?

It's a good idea to always put something in the style name to indicate it wasn't in the box. Putting your initials or firm name at the beginning of the style is one way to make it easy to differentiate your styles from the prebuilt ones. Here, JW stands for James Wedding (EG stands for Existing Ground).

3. Switch to the General tab. Change the layer to **C-TOPO-TEXT** by clicking the layer cell and then the More button to the right of that cell.

   There are a fair number of options here, so let's pause the exercise, and look at them further:

   Text Style is the default style for text components that are created on the Layout tab. It's a good practice to use a zero-height text style with the appropriate font, because you'll set the plotted heights in the style anyway.

   Layer is the layer on which the *components* of a label are inserted, not the layer on which the label *itself* is inserted. Think of labels as nested blocks. The label (the block) gets inserted on the layer on the basis of the object layers you saw earlier. The components of the label get inserted on the layer that is set here. This means a change to the specified layer can control or change the appearance of the components if you like.

   Orientation Reference sets an object to act as the up direction in terms of readability. Civil 3D understands viewpoint rotation and offers the option to rotate or flip labels to keep them plan-readable. Most users set this to View to maintain the most plan-readable labels with the smallest amount of editing later.

Forced Insertion makes more sense in other objects and will be explored further. This feature essentially allows you to dictate the insertion point of a label on the basis of the object being labeled.

Plan Readable text maintains the up direction in spite of view rotation. This tends to be the "Ooooh, nice" feature that makes users smile. Rotating 100 labels is a tedious, thankless task, and this option handles it with one click.

Readability Bias is the angle at which readability kicks in. This angle is measured from the 0 degree of the x-axis that is common to AutoCAD angle measurements. When a piece of text goes past the readable bias angle, the text spins to maintain vertical orientation, as shown in Figure 1.16. Note how the label on the far left has rotated to accommodate the rotation past 110 degrees, the default bias angle. If you set the readability bias to 90.01, which is a typical setting, the text flips at a near-vertical angle.

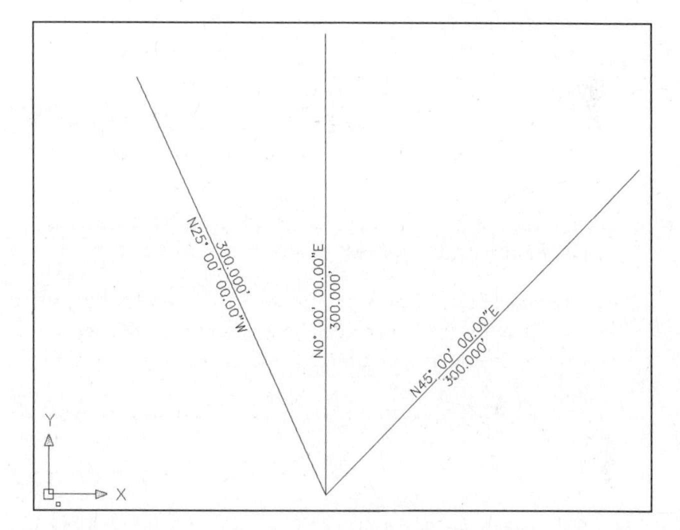

FIGURE 1.16
Examples of plan-readable text

Flip Anchors with Text determines how the text flips. Most users find that setting this to False gives the best results, but sometimes flipping an anchor point positions text as needed. You'll learn more about anchor points on the Layout tab.

4. Switch to the Layout tab. Again, a lot is going on here, so you'll work through the options and then make changes. As shown in Figure 1.17, each component of the label has a host of options. On the right is a preview of the label you're creating or editing. You can pan or zoom this view as needed to give you a better feel for the label style's appearance as you make changes.

A FULL THREE-DIMENSIONAL LABEL PREVIEW?

This preview defaults to a 3D Orbit control. Don't ask why; we're as confused as you are. Inevitably, you'll rotate the view out of a plan-top view, making the plan harder to understand. When this happens, right-click and select Preset Views ← Top to reorient yourself, or use the Viewcube function to pull to the top.

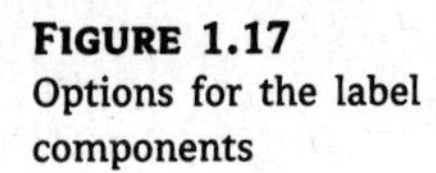

FIGURE 1.17
Options for the label components

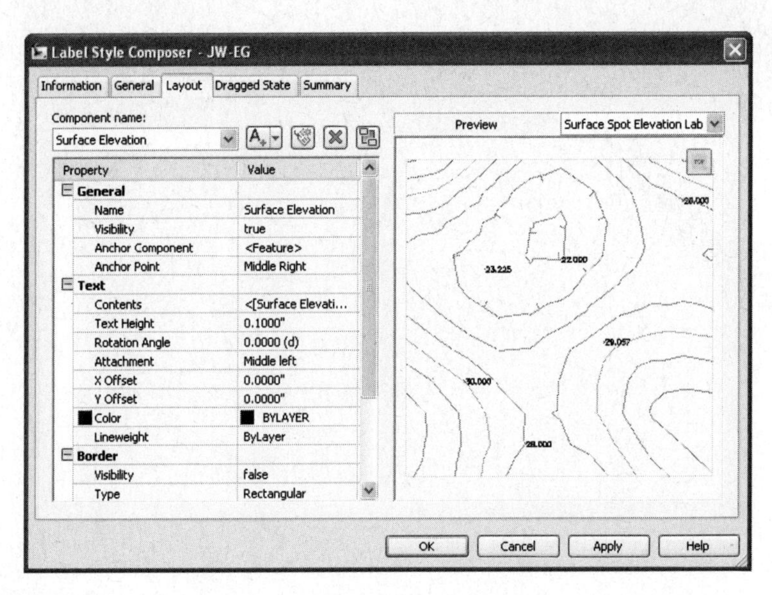

Again, pause and review some of the other options on this tab. Labels are made of individual components. A component can be text, a block, or a line, and the top row of buttons controls the selection, creation, and deletion of these components:

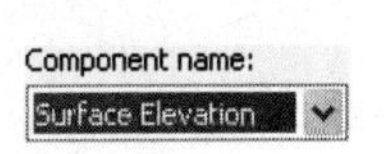

The Component drop-down menu activates which component is being modified in the options below. These components are listed in the order in which they were created.

The Create Text Component button lets you create new components. These components can be Text, Lines, Blocks, Reference Text, or Ticks. Some options aren't available for every label style.

The ability to label one object while referencing another (reference text) is one of the most powerful labeling features of Civil 3D. This is what allows you to label a spot elevation for both an existing and a proposed surface at the same time, using the same label. Alignments, COGO points, parcels, profiles, and surfaces can all be used as reference text.

The Copy Component button does just that. It copies the component currently selected in the Component drop-down menu.

The Delete Component button deletes components. Elements that act as the basis for other components can't be deleted.

The Component Draw Order button lets you shuffle components up and down within the label. This feature is especially important when you're using masks or borders as part of the label.

You can work your way down the component properties and adjust them as needed for a label:

Name is self-explanatory. It's the name used in the Component drop-down menu and when selecting other components. When you're building complicated labels, a little name description goes a long way.

Visibility set to True means this component shows on screen. Invisible components can be invaluable when you're creating complicated labels, as you'll see in later chapters.

Anchor Component and **Anchor Point** are straightforward, but many users have issues when first using these options. Every component of the label has an anchor component, anchor point, and attachment. The Anchor Component is how you tell Civil 3D where you want to hang the label component. This component is bounded by a box with nine anchor points, as shown in Figure 1.18. In this illustration, the nine possible anchor points are represented with Xs and the nine possible attachments with Os.

FIGURE 1.18
Anchor and attachment points

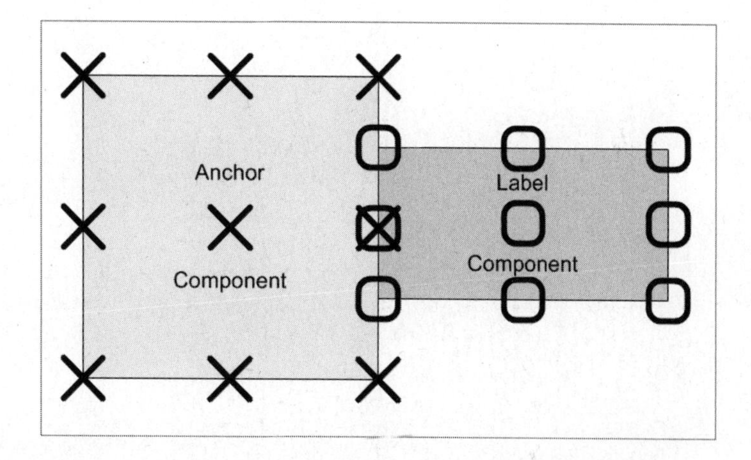

Now let's continue with the exercise and add a reference text component to this label.

5. Click the arrow on the right side of the component drop-down menu and select Reference Text to open the Select Type dialog as shown in Figure 1.19.
6. Select Surface as the type of reference object, and then click the OK button to exit the dialog.
7. The name of the text component is Reference Text.1 by default. Change the name to **Reference Text: Proposed Surface** and click the Apply button in the lower-right of the dialog. The middle portion of the dialog changes depending on the type of component, but the concepts are similar. In the case shown here, the middle portion is the Text property. Under Text, the first option is Contents, which determines the actual content of the text:
8. Click the Contents Value cell, and then click the ellipsis button that appears to the right to open the Text Component Editor.
9. Click in the preview window of the Text Component Editor. This is a simple text editor, and you can type anything you'd like in a label. You can also insert object information from Civil 3D objects, as you'll do now.
10. Highlight and delete the text in the preview window.
11. At left, select Surface Elevation from the drop-down list in the Properties text box if necessary.

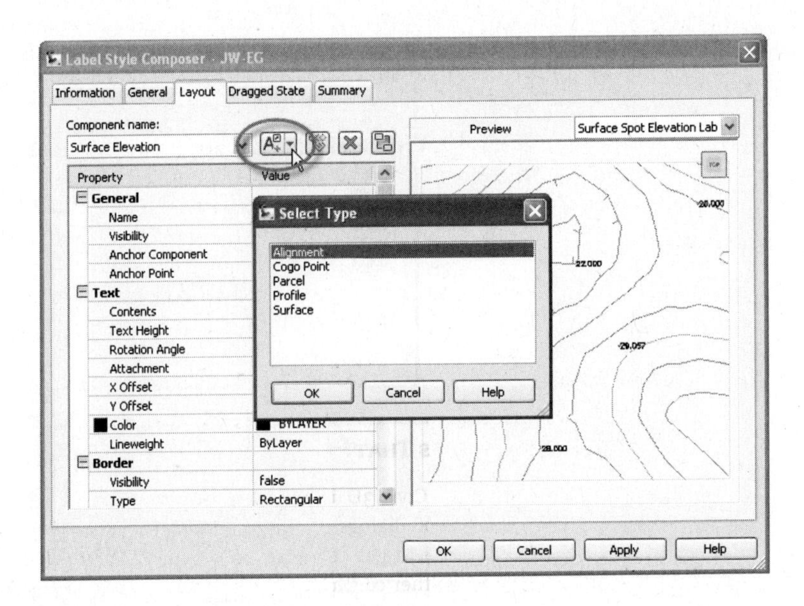

FIGURE 1.19
Reference text object selection

12. Change the precision to one decimal place by clicking in the column next to Precision and selecting 0.1, as shown in Figure 1.20.

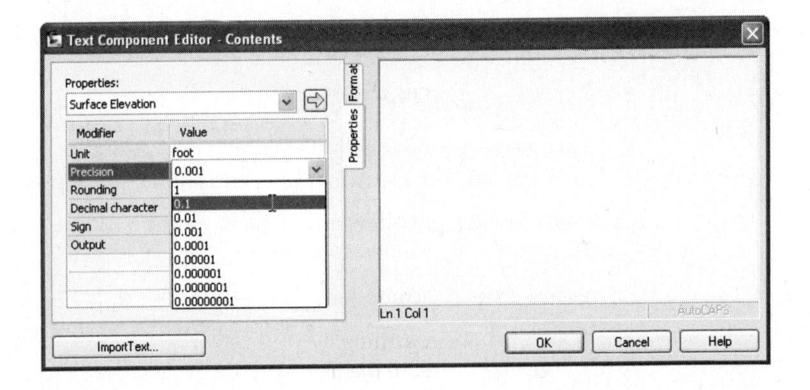

FIGURE 1.20
Setting label precision

13. Click the arrow circled in Figure 1.21 to insert your label text and elevation code into the preview area.
14. Click OK to exit this dialog, and you'll be back at the Label Style Composer.
15. Your label is complete. You can click OK to exit, but you might want to leave the label open as we discuss the Dragged State and Summary tabs next.

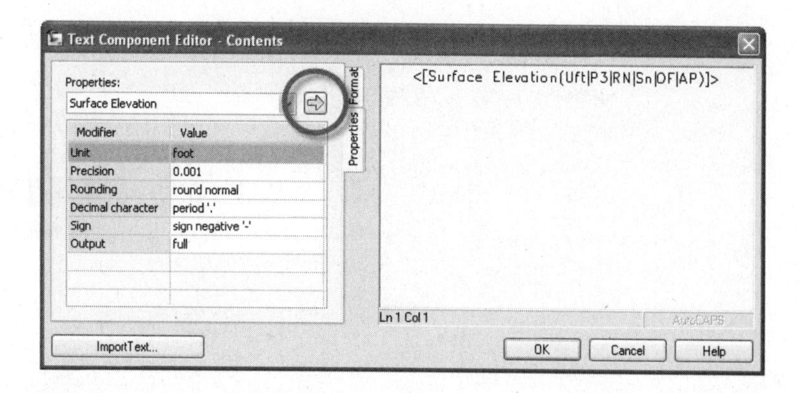

FIGURE 1.21
Don't forget the arrow!

HOW MANY DIALOGS IS THAT?

You can see why many Civil 3D instructors refer to label creation as "heading down the rabbit hole." You're a couple of dialogs deep just making the simplest of label styles, with one static text component. It's easy to get confused, but don't worry — it becomes second nature! The Text Component Editor is another common dialog that appears in every label-style creation exercise.

Let's look at the rest of the options, even though you won't be making any changes:

Text Height determines the plotted height of the label. Remember, Civil 3D knows you're going to print and will attempt to give you inches or millimeters.

Rotation Angle, **X Offset**, and **Y Offset** give you the ability to refine the placement of this component by rotating or displacing the text in an *x* or *y* direction.

Attachment determines which of the nine points on the label components bounding box are attached to the anchor point. See Figure 1.18 for an illustration. Change the attachment of the reference text component to Top Left. This will attach the upper-left corner of this proposed elevation reference text to the bottom left of the Surface Elevation component.

Color and **Lineweight** allow you to hard-code a color if desired. It's a good idea to leave these values set to ByLayer unless you have a good reason to change them.

The final piece of the component puzzle is a Border option. These options are as follows:

Visibility is obvious, turning the border on and off for this component. Remember that component borders shrink to the individual component: if you're using multiple components in a label, they all have their own borders.

Type allows you to select a rectangle, a rounded rectangle (slot), or a circle border.

Background Mask lets you determine whether linework and text behind this component are masked. This can be handy for construction notes in place of the usual wipeout tools.

Gap determines the offset from the component bounding box to the outer points on the border. Setting this to half of the text size usually creates a visually pleasing border.

Linetype and **Lineweight** give you the usual control of the border lines.

After working through all the options for the default label placement, you need to set the options that come into play when a label is dragged. Switch to the Dragged State tab. When a label is dragged in Civil 3D, it typically creates a leader, and text rearranges. The settings that control these two actions are on this tab. Unique options are explained here:

Arrow Head Style and **Size** control the tip of the leader. Note that Arrow Head Size also controls the tail size leading to the text object.

Type controls the leader type. Options are Straight Leader and Spline Leader. At the time of this writing, the AutoCAD multiple leader object can't be used.

Display controls whether components rearrange their placement to a stacked set of components (Stacked Text) or maintain their arrangement as originally composed (As Composed). Most users expect this to be set to As Composed for the most predictable behavior.

Every label has a Summary tab, and clicking the Expand All button circled in Figure 1.22 will present you with a full array of details about the label. Working down the Summary tab, you can review all of the options that have been selected for an individual label, as well as look for overrides, just like you did on the Settings tab. Click OK to exit the dialog; your new style will appear on the Settings tab.

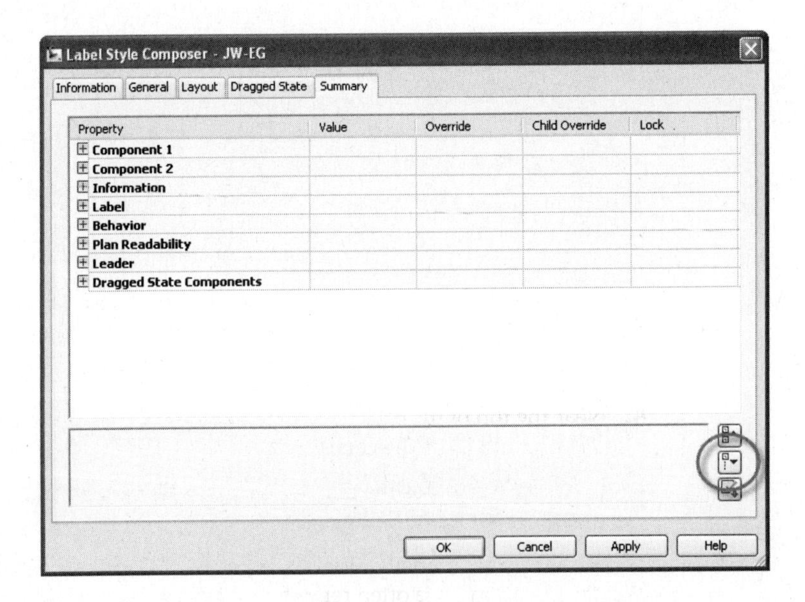

FIGURE 1.22 Summary tab with the Expand All button circled

The purpose of this exercise wasn't to build a Surface Spot Elevation label style; it was to familiarize you with the common elements of creating a label: the Label Style Composer and the Text Component Editor. However, you can try out the new label to check your work!

Object Styles

Beyond the styles used to label objects, Civil 3D also depends on styles to control the display of the native objects, including points, surface, alignments, and so on. Just as in label styles, certain components of the object styles are common to almost all objects, so let's create a new alignment style to introduce these common elements:

1. Expand the Alignment branch on the Settings tab, and then right-click the Alignment Styles folder. Select New, as shown in Figure 1.23.

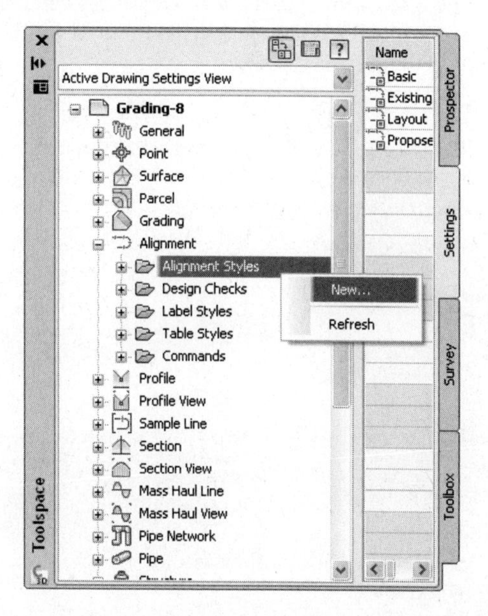

FIGURE 1.23 Creating a new alignment style via Prospector

2. Type a new name for your style on the Information tab, and enter a description if desired.
3. Switch to the Display tab. The other two tabs are unique for alignment objects, but the Display tab is part of every Civil 3D object style.
4. Turn off the Arrow component by clicking the lightbulb in the column next to it.
5. Near the top of the dialog, change the View Direction setting in the drop-down menu to Model, and notice the change.
6. Click OK to dismiss the dialog. Your new alignment style appears under the Alignment Styles branch in the Settings tab.

Objects can have distinctly different appearances when viewed in a plan view versus a 3D view. For example, surfaces are often represented by contours in plan view, but by triangular faces or a grid in 3D.

Object styles are a major component of efficient Civil 3D object modeling. Objects appear differently in varying plans. Having a full set of object styles to handle all of these uses can help make plan production as painless as possible.

A good way to start creating object styles is to pull out a set of existing plans that accurately represent your firm's standards. Pick an object, such as alignments or surfaces, and then begin working your way through the plan set, creating a new object style for each use case. Once you complete one object, pick another and repeat the exercise.

The Underlying Engine

Civil 3D is part of a larger product family from Autodesk. During its earliest creation, various features and functions from other products were recognized as important to the civil engineering community. These included the obvious things such as the entire suite of AutoCAD drafting, design, modeling, and rendering tools as well as more esoteric options such as Map's GIS capabilities. An early decision was made to build Civil 3D on top of the AutoCAD Map product, which in turn is built on top of AutoCAD.

This underlying engine provides a host of options and powerful tools for the Civil 3D user. AutoCAD and Map add features with every release that change the fundamental makeup of how Civil 3D works. With the introduction of workspaces in 2006, users can now set up Civil 3D to display various tools and palettes depending on the task at hand. Creating a workspace is like having a quick-fix bag of tools ready for the job at hand: preliminary design calls for one set of tools, and final plan production calls for another.

Workspaces are part of a larger feature set called the *custom user interface* (referred to as CUI in the help documentation and online). As you grow familiar with Civil 3D and the various tool palettes, menus, and toolbars, be sure to explore the CUI options that are available from the Workspace toolbar.

The Bottom Line

Find any Civil 3D object with just a few clicks. By using Prospector to view object data collections, you can minimize the panning and zooming that are part of working in a CAD program. When common subdivisions can have hundreds of parcels or a complex corridor can have dozens of alignments, jumping to the desired one nearly instantly shaves time off everyday tasks.

Master It Open `Sample Site.dwg` from the tutorials, and find parcel number five without using any AutoCAD commands.

Modify the drawing scale and default object layers. Civil 3D understands that the end goal of most drawings is to create hard-copy construction documents. By setting a drawing scale and then setting many sizes in terms of plotted inches or millimeters, Civil 3D removes much of the mental gymnastics that other programs require when you're sizing text and symbols. By setting object layers at a drawing scale, Civil 3D makes uniformity of drawing files easier than ever to accomplish.

Master It Change `Sample Site.dwg` from a 200-scale drawing to a 40-scale drawing.

Modify the display of Civil 3D tooltips. The interactive display of object tooltips makes it easy to keep your focus on the drawing instead of an inquiry or report tools. When too many objects fill up a drawing, it can be information overload, so Civil 3D gives you granular control over the heads-up display tooltips.

Master It Within the same Sample Site drawing, turn off the tooltips for the Avery Drive alignment.

Add a new tool to the Toolbox. The Toolbox provides a convenient way to access macros and reports. Many third-party developers exploit this convenient interface as an easier way to add functionality without disturbing users' workspaces.

Master It Add the Sample Pipe macro from `C:\Program Files\Autocad Civil 3D 2010\Sample\Civil 3D API\COM\Vba\Pipe`, and select `PipeSample.dvb`.

Create a basic label style. Label styles determine the appearance of Civil 3D annotation. The creation of label styles will constitute a major part of the effort in making the transition to Civil 3D as a primary platform for plan production. Your skills will grow with the job requirements if you start with basic labels and then make more complicated labels as needed.

Master It Create a copy of the Elevation Only Point label style, name it Elevation With Border, and add a border to the text component.

Create a new object style. Object styles in Civil 3D let you quit managing display through layer modification and move to a more streamlined style-based control. Creating enough object styles to meet the demands of plan production work will be your other major task in preparing to move to Civil 3D.

Master It Create a new Surface style named Contours_Grid, and set it to show contours in plan views but a grid display in any 3D view.

Navigate the Ribbon's contextual tabs. As with AutoCAD, the Ribbon is the primary interface for accessing Civil 3D commands and features. When you select an AutoCAD Civil 3D object, the Ribbon displays commands and features related to that object. If several object types are selected, the Multiple contextual tab is displayed.

Master It Using the Ribbon interface, access the Alignment Style Editor for the Proposed Alignment style. (Hint: it's used by the Avery Drive alignment.)

Chapter 2

Back to Basics: Lines and Curves

Engineers and surveyors are constantly creating lines and curves. Whether the task at hand involves re-creating and checking existing geometry from deeds or record plats, or designing a new land plan, it's important to have tools to assist in the accurate creation of this linework.

The lines and curves tools in Civil 3D, along with the complementary tools on the Transparent Commands and Inquiry toolbars, provide robust methods for creating and checking lines and curves. This linework can then be used as is, or as a foundation for creating sound parcels, alignments, and other Civil 3D object geometry.

It's important to note that creating Civil 3D objects such as parcels and alignments from this foundation geometry is necessary for robust reporting and labeling, such as legal descriptions, segment tables, and more detailed design and analysis.

By the end of this chapter, you'll learn to:

- Create a series of lines by bearing and distance
- Use the Inquiry commands to confirm that lines are drawn correctly
- Create a curve tangent to the end of a line
- Create a best-fit line for a series of Civil 3D points
- Label lines and curves

Labeling Lines and Curves

You can draw lines many ways in an AutoCAD-based environment. The tools found on the Draw panel of the Home tab create lines that are no more intelligent than those created by the standard AutoCAD Line command. How the Civil 3D lines differ from those created by the regular Line command isn't in the resulting entity, but in the process of creating them. In other words, in Civil 3D you provide directions to these line commands in survey terminology rather than in generic Cartesian parameters (see Figure 2.1).

Note that you can switch between any of the line commands without exiting the command. For example, if your first location is a point object, use Line by Point Object; then, without leaving the command, go back to the Lines/Curves menu and choose any Line or Curve command to continue creating your linework. You can also press the Esc key once, while in a Lines/Curves menu command, to resume the regular Line command.

FIGURE 2.1
Line-creation tools

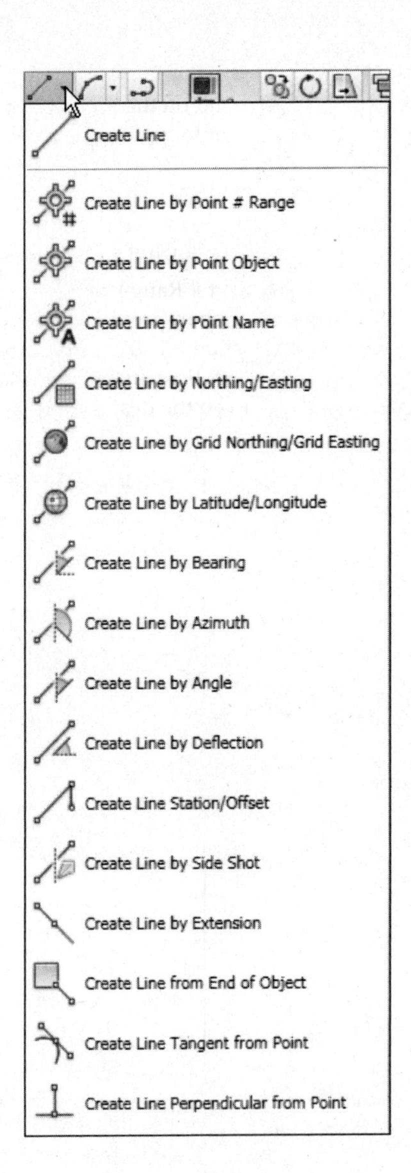

Coordinate Line Commands

The next few commands discussed in this section help you create a line using Civil 3D points and/or coordinate inputs. Each command requires you to specify a Civil 3D point, a location in space, or a typed coordinate input. These line tools are useful when your drawing includes Civil 3D points that will serve as a foundation for linework, such as the edge of pavement shots, wetlands lines, or any other points you'd like to connect with a line.

LINE COMMAND

The Create Line command on the Draw panel of the Home tab issues the standard AutoCAD Line command. It's equivalent to typing **line** on the command line or clicking the Line tool on the Draw toolbar.

CREATE LINE BY POINT # RANGE COMMAND

The Create Line by Point # Range command prompts you for a point number. You can type in an individual point number, press ↵, and then type in another point number. A line is drawn connecting those two points. You can also type in a range of points, such as **640-644**. Civil 3D draws a line that connects those lines in numerical order — from 640 to 641, and so on (see Figure 2.2). This order won't give you the desired linework for edge of asphalt, for example.

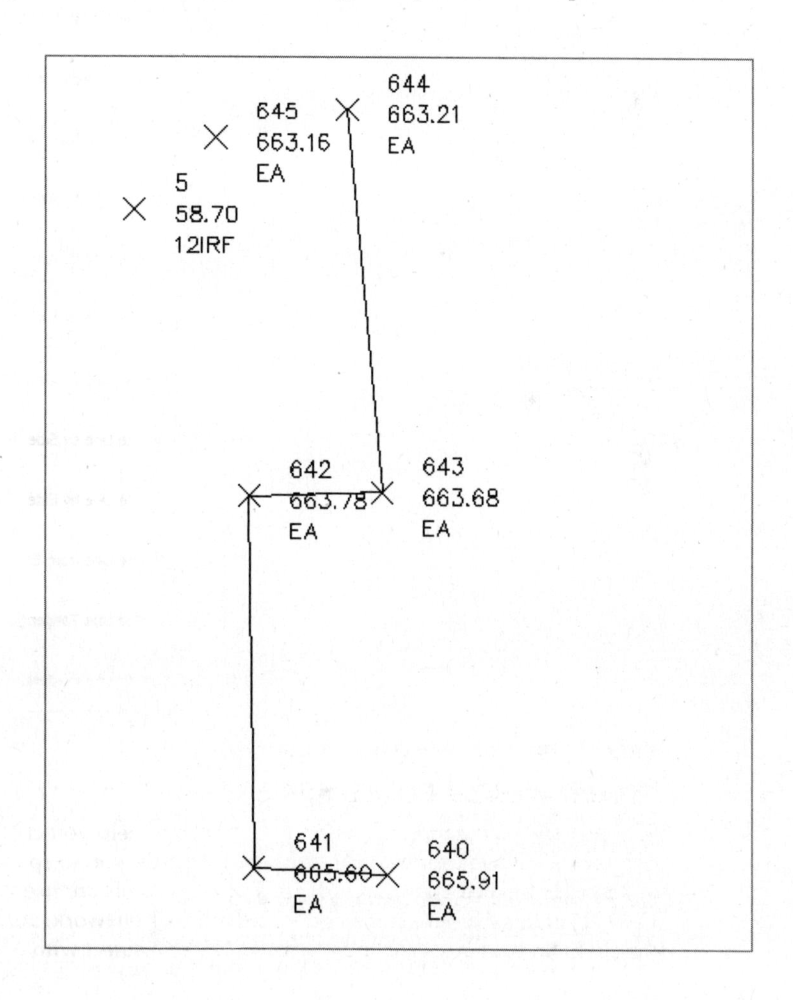

FIGURE 2.2
A line created using 640-644 as input

Alternatively, you can enter a list of points such as **640, 643, 644** (Figure 2.3). Civil 3D draws a line that connects the point numbers in the order of input. This is useful when your points were taken in a zigzag pattern (as is commonly the case when cross sectioning pavement), or when your points appear so far apart in the AutoCAD display that they can't be readily identified.

FIGURE 2.3
A line created using 640, 643, 644 as input

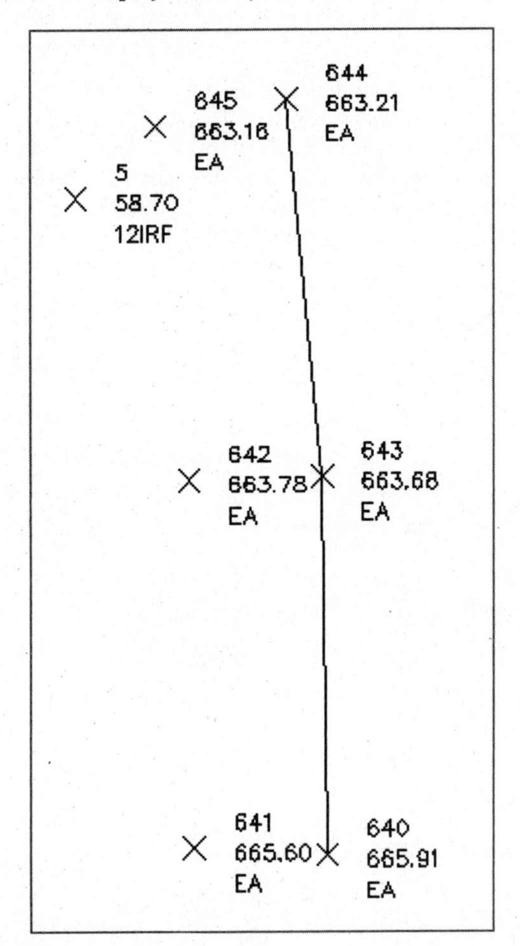

CREATE LINE BY POINT OBJECT COMMAND

The Create Line by Point Object command prompts you to select a point object. To select a point object, locate the desired start point and click any part of the point. This tool is similar to using the regular Line command and a Node osnap.

CREATE LINE BY POINT NAME COMMAND

The Create Line by Point Name command prompts you for a point name. A *point name* is a field in Point properties, not unlike the point number or description. The difference between a point name and a *point description* is that a point name must be unique. It is important to note that some survey instruments name points rather than number points as is the norm.

To use this command, enter the names of the points you want to connect with linework.

CREATE LINE BY NORTHING/EASTING AND CREATE LINE BY GRID NORTHING/EASTING COMMANDS

The Create Line by Northing/Easting and Create Line by Grid Northing/Easting commands let you input northing (*y*) and easting (*x*) coordinates as endpoints for your linework. The Create Line by Grid Northing/Easting command requires that the drawing have an assigned coordinate system. This command can be useful when working with known monumentation in a State Plane Coordinate System (SPCS).

CREATE LINE BY LATITUDE/LONGITUDE COMMAND

The Create Line by Latitude/Longitude command prompts you for geographic coordinates to use as endpoints for your linework. This command also requires that the drawing have an assigned coordinate system. For example, if your drawing has been assigned Delaware State Plane NAD83 US Feet and you execute this command, your Latitude/Longitude inputs are translated into the appropriate location in your state plane drawing. This command can be useful when drawing lines between waypoints collected with a standard handheld GPS unit.

Direction-Based Line Commands

The next few commands help you specify the direction of a line. Each of these commands requires you to choose a start point for your line before you can specify the line direction. You can specify your start point by physically choosing a location, using an osnap, or using one of the point-related line commands discussed earlier.

CREATE LINE BY BEARING COMMAND

The Create Line by Bearing command will likely be one of your most frequently used line commands.

This command prompts you for a start point, followed by prompts to input the Quadrant, Bearing, and Distance values. You can enter values on the command line for each input, or you can graphically choose inputs by picking them on screen. The glyphs at each stage of input guide you in any graphical selections. After creating one line, you can continue drawing lines by bearing, or you can switch to any other method by clicking one of the other Line By commands on the Draw panel (see Figure 2.4).

CREATE LINE BY AZIMUTH COMMAND

The Create Line by Azimuth command prompts you for a start point, followed by a north azimuth, and then a distance (Figure 2.5).

CREATE LINE BY ANGLE COMMAND

The Create Line by Angle command prompts you for a turned angle and then a distance (Figure 2.6). This command is useful when you're creating linework from angles right (in lieu of angles left) and distances recorded in a traditional handwritten field book (required by law in many states).

CREATE LINE BY DEFLECTION COMMAND

By definition, a *deflection angle* is the angle turned from the extension of a line from the backsight extending through an instrument. Although this isn't the most frequently used surveying tool in this day of data collectors and GPSs, on some occasions you may need to create this type of line. When you use the Create Line by Deflection command, the command line and tooltips prompt you for a deflection angle followed by a distance (Figure 2.7). In some cases, deflection angles are recorded in the field in lieu of angles right.

FIGURE 2.4
The tooltips for a quadrant (top), a bearing (middle), and a distance (bottom)

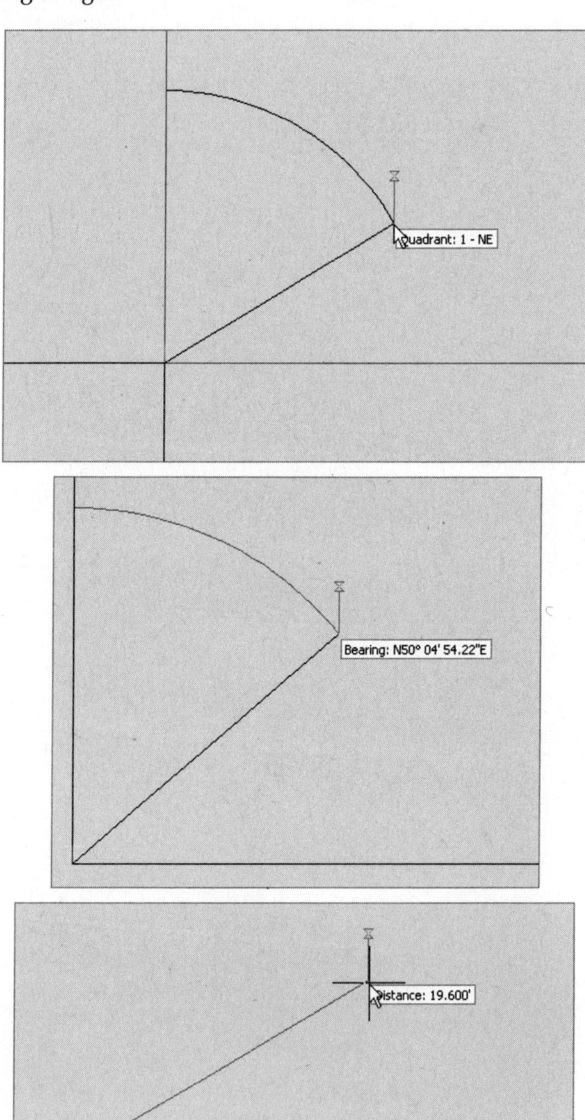

CREATE LINE BY STATION/OFFSET COMMAND

To use the Create Line by Station/Offset command, you must have a Civil 3D Alignment object in your drawing. The line created from this command allows you to start and/or end a line on the basis of a station and offset from an alignment.

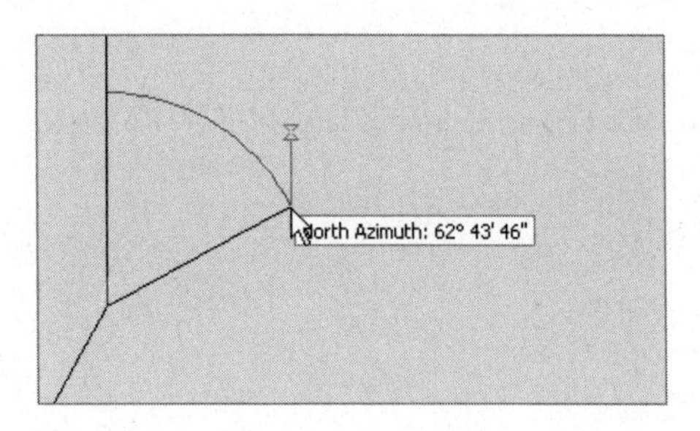

FIGURE 2.5
The tooltip for the Create Line by Azimuth command

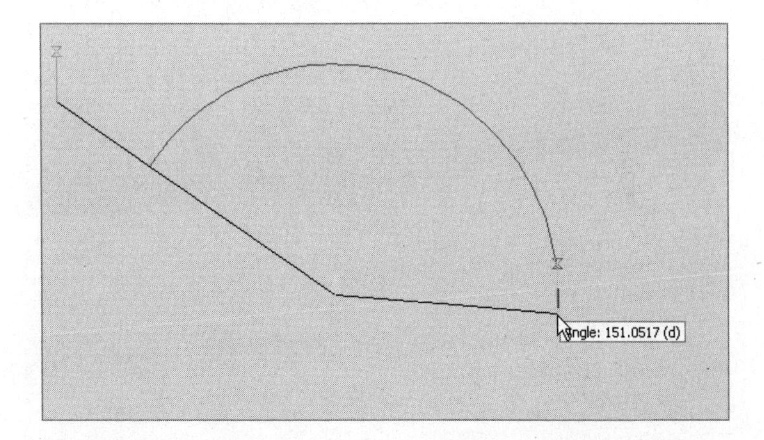

FIGURE 2.6
The tooltip for the Create Line by Angle command

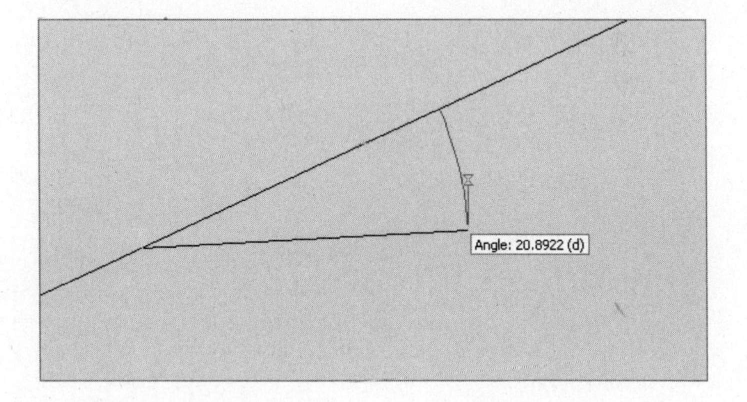

FIGURE 2.7
The tooltips for the Create Line by Deflection command

You're prompted to choose the alignment and then input a station and offset value. The line *begins* at the station and offset value. On the basis of the tooltips, you might expect the line to be drawn from the alignment station at offset zero and out to the alignment station at the input offset. This isn't the case.

When prompted for the station, you're given a tooltip that tracks your position along the alignment, as shown in Figure 2.8. You can graphically choose a station location by picking in the drawing (including using your osnaps to assist you in locking down the station of a specific feature). Alternatively, you can enter a station value on the command line.

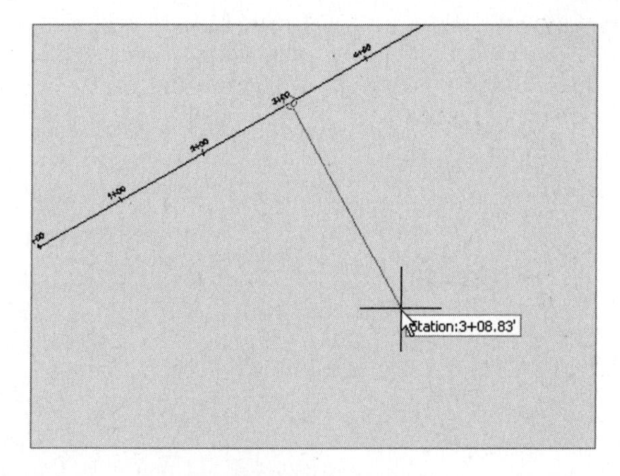

FIGURE 2.8
The Create Line by Station/Offset command provides a tooltip to track stationing along the alignment

Once you've selected the station, you're given a tooltip that is locked on that particular station and tracks your offset from the alignment (see Figure 2.9). You can graphically choose an offset by picking in the drawing, or you can type an offset value on the command line.

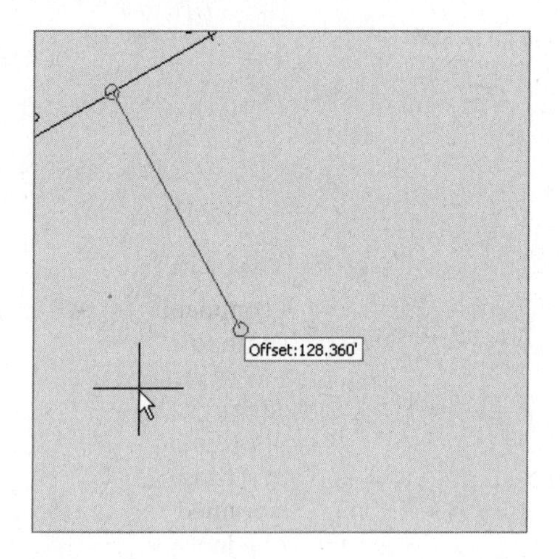

FIGURE 2.9
The Create Line by Station/Offset command gives a tooltip to track the offset from the alignment

CREATE LINE BY SIDE SHOT COMMAND

The Create Line by Side Shot command lets you occupy one point, designate a backsight, and draw a line that has endpoints relative to that point. The occupied point represents the setup of your surveying station, whereas the second point represents your surveying backsight. This tool may be most useful when you're creating stakeout information or re-creating data from field notes. If you know where your crew set up, and you have their side-shot angle measurements but you don't have electronic information to download, this tool can help. To specify locations relative to your occupied point, you can specify the angle, bearing, deflection, or azimuth on the command line or pick locations in your drawing. In some cases, it is more appropriate to supply a survey crew with handwritten notes regarding backsights, foresights, angles right, and distances rather than upload the same information to a data collector.

While the command is active, you can toggle between angle, bearing, deflection, and azimuth by following the command-line prompts.

When you're using the Create Line by Side Shot command, you're given a setup glyph at your occupied point, a backsight glyph, and a tooltip to track the angle, bearing, deflection, or azimuth of the side shot (see Figure 2.10). You can toggle between these options by following the command-line prompts.

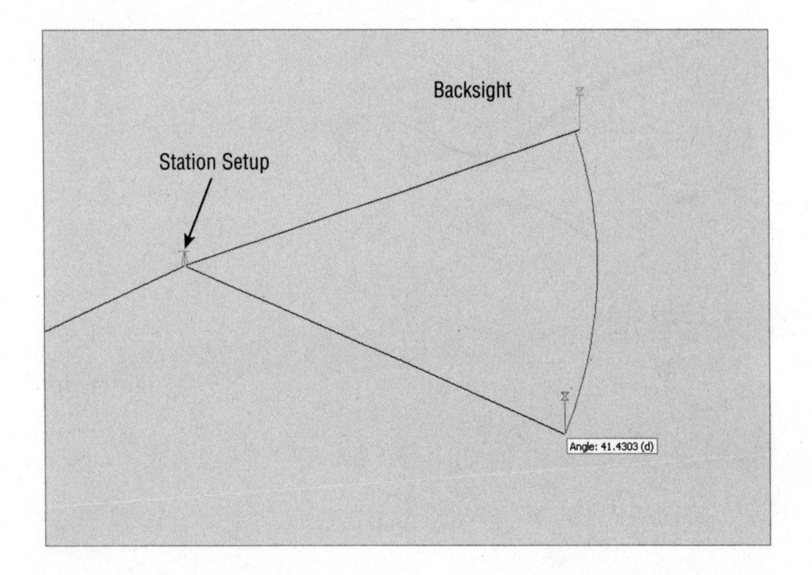

FIGURE 2.10
The tooltip for the Create Line by Side Shot command tracks the angle, bearing, deflection, or azimuth of the side shot

CREATE LINE EXTENSION COMMAND

The Create Line Extension command is similar to the AutoCAD Lengthen command. This command allows you to add length to a line or specify a desired total length of the line.

You are first prompted to choose a line. The command line then prompts you to `Specify distance to change, or [Total]`. This distance is added to the existing length of the line. The command draws the line appropriately and provides a short summary report on that line. The summary report in Figure 2.11 indicates the beginning line length was 100′ and that an additional distance of 50′ was specified with the Line Extension command. It is important to note that in some cases, it may be more desirable to create a line by a turned angle or deflection of 180 degrees so as not to disturb linework originally created from existing legally recorded documents.

FIGURE 2.11
The Create Line Extension command provides a summary of the changes to the line

```
Select line object:
Specify distance to change, or [Total]: 50
-------------------------------------------------------------------------------
                               LINE DATA
-------------------------------------------------------------------------------
Begin . . . . .  North: 6935078.5251'          East: 2450248.0169'
End . . . . . .  North: 6935199.4098'          East: 2450336.8251'
            Distance: 150.000'          Course: N36° 18' 11"E
```

If, instead, you specify a total distance on the command line, then the length of the line is changed to the distance you specify. The summary report shown in Figure 2.12 indicates that the beginning of the line was the same as in Figure 2.11 but with a total length of only 100′.

FIGURE 2.12
The summary report on a line where the command specified a total distance

```
LINEEXTENSION
Select line object:
Specify distance to change, or [Total]: t
Specify total distance, or [Change]: 100
--------------------------------------------------------------------------------
                                   LINE DATA
--------------------------------------------------------------------------------
Begin . . . . .  North: 6935078.5251'           East: 2450248.0169'
End . . . . . .  North: 6935159.1149'           East: 2450307.2223'
              Distance: 100.000'              Course: N36° 18' 11"E
```

CREATE LINE FROM END OF OBJECT COMMAND

The Create Line from End of Object command lets you draw a line tangent to the end of a line or arc of your choosing. Most commonly, you'll use this tool when re-creating deeds or other survey work where you have to specify a line that continues tangent from an arc (see Figure 2.13).

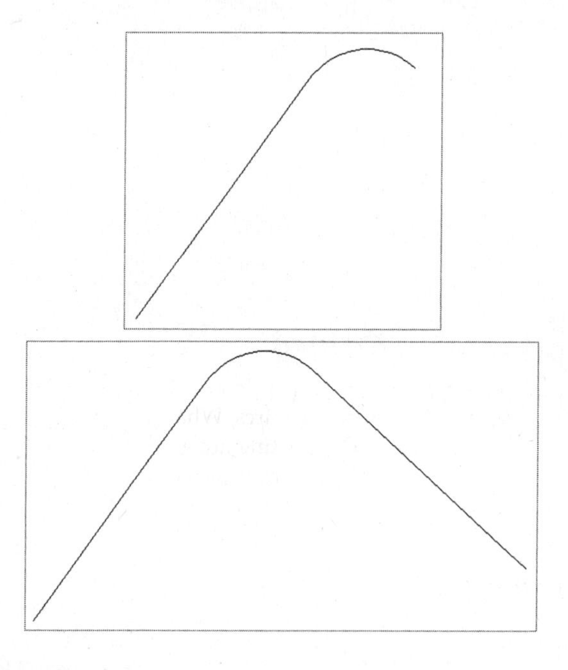

FIGURE 2.13
The Create Line from End of Object command can add a tangent line to the end of the arc

CREATE LINE TANGENT FROM POINT COMMAND

The Create Line Tangent from Point command is similar to the Create Line from End of Object command, but Create Line Tangent from Point allows you to choose a point of tangency that isn't the endpoint of the line or arc (see Figure 2.14).

CREATE LINE PERPENDICULAR FROM POINT COMMAND

Using the Create Line Perpendicular from Point command, you can specify that you'd like a line drawn perpendicular to any point of your choosing. In the example shown in Figure 2.15, a line is drawn perpendicular to the endpoint of the arc. This command can be useful when the distance from a known monument perpendicular to a legally platted line must be labeled in a drawing.

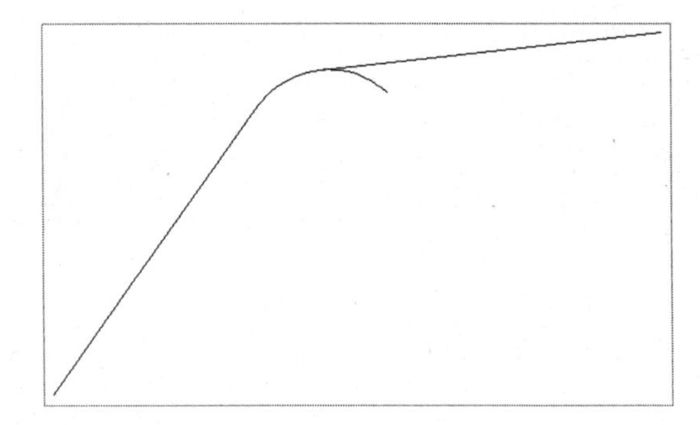

FIGURE 2.14
The Create Line Tangent from Point command can place a line tangent to the midpoint of an arc (or line)

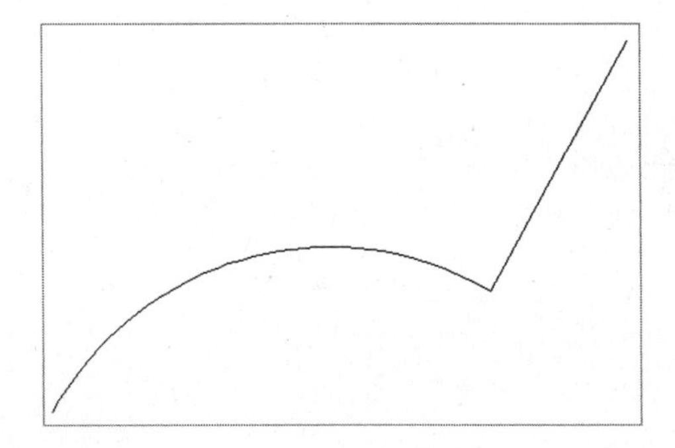

FIGURE 2.15
A perpendicular line is drawn from the endpoint of an arc, using the Create Line Perpendicular from Point command

Creating Curves

Curves are an important part of surveying and engineering geometry. In truth, curves are no different from AutoCAD arcs. What makes the curve commands unique from the basic AutoCAD commands isn't the resulting arc entity but the inputs used to draw that arc. Civil 3D wants you to provide directions to these arc commands using land surveying terminology rather than with generic Cartesian parameters.

Figure 2.16 shows the Create Curves menu options.

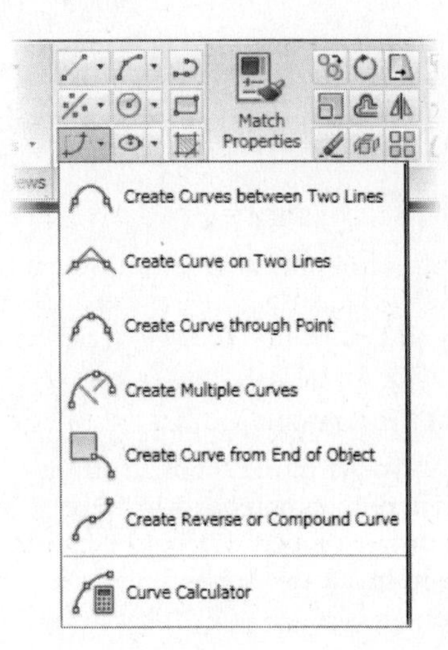

FIGURE 2.16
Create Curves commands

Standard Curves

When re-creating legal descriptions for roads, easements, and properties, engineers, surveyors, and mappers often encounter a variety of curves. Although standard AutoCAD arc commands could draw these arcs, the AutoCAD arc inputs are designed to be generic to all industries. The following curve commands have been designed to provide an interface that more closely matches land surveying, mapping, and engineering language.

Create Curve between Two Lines Command

The Create Curve between Two Lines command is much like the standard AutoCAD Fillet command, except you aren't limited to a radius parameter. The command draws a curve that is tangent to two lines of your choosing. This command also trims or extends the original tangents so their endpoints coincide with the curve endpoints. In other words, the lines are trimmed or extended to the resulting PC (point of curve; in other words, the beginning of a curve) and PT (point of tangency; in other words, the end of a curve) of the curve. You may find this command most useful when you're creating foundation geometry for road alignments, parcel boundary curves, and similar situations.

The command prompts you to choose the first tangent and then the second tangent. The command line gives the following prompt:

```
Select entry [Tangent/External/Degree/Chord/↵
Length/Mid-Ordinate/miN-dist/Radius]<Radius>:
```

Pressing ↵ at this prompt lets you input your desired radius. As with standard AutoCAD commands, pressing T changes the input parameter to tangent, pressing C changes the input parameter to chord, and so on.

As with the Fillet command, your inputs must be geometrically possible. For example, your two lines must allow for a curve of your specifications to be drawn while remaining tangent to both. Figure 2.17 shows two lines with a 25′ radius curve drawn between them. Note that the tangents have been trimmed so their endpoints coincide with the endpoints of the curve. If either line had been too short to meet the endpoint of the curve, then that line would have been extended.

Create Curve on Two Lines Command

The Create Curve on Two Lines command is identical to the Curve between Two Lines command, except that the Create Curve on Two Lines command leaves the chosen tangents intact. The lines aren't trimmed or extended to the resulting PC and PT of the curve.

Figure 2.18, for example, shows two lines with a 25′ radius curve drawn on them. The tangents haven't been trimmed and instead remain exactly as they were drawn before the Create Curve on Two Lines command was executed.

Create Curve through Point Command

The Create Curve through Point command lets you two choose two tangents for your curve followed by a pass-through point. This tool is most useful when you don't know the radius, length, or other curve parameters but you have two tangents and a target location. It isn't necessary that the pass-through location be a true point object; it can be any location of your choosing.

This command also trims or extends the original tangents so their endpoints coincide with the curve endpoints. The lines are trimmed or extended to the resulting PC and PT of the curve.

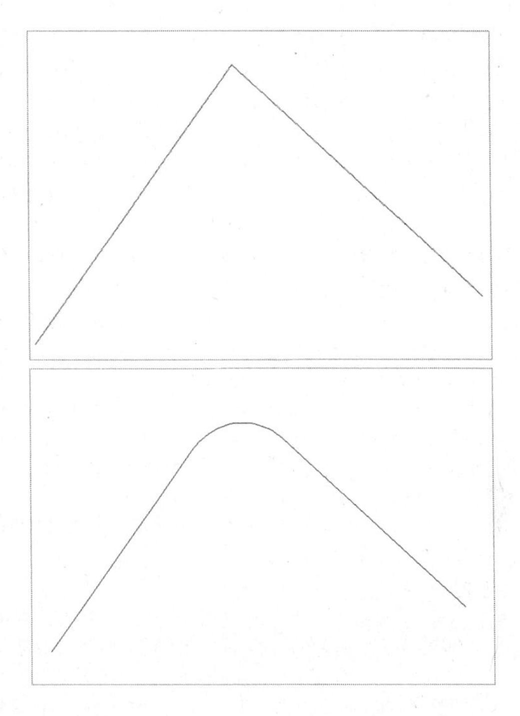

FIGURE 2.17
Two lines before (top) and after (bottom) using the Create Curve between Two Lines command

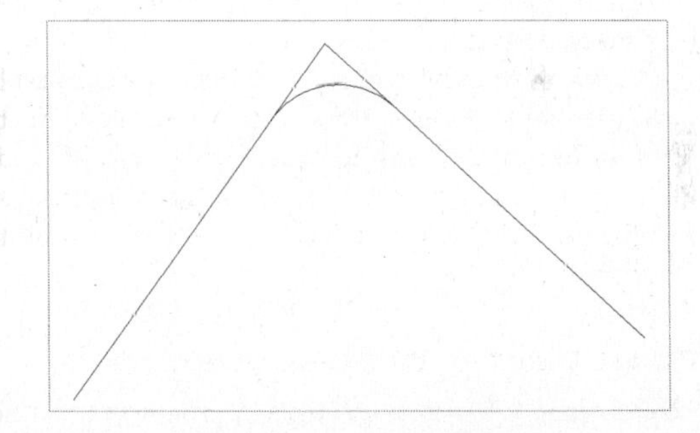

FIGURE 2.18
The original lines stay the same after you execute the Create Curve on Two Lines command

Figure 2.19, for example, shows two lines and a desired pass-through point. Using the Create Curve through Point command allows you to draw a curve that is tangent to both lines and that passes through the desired point. In this case, the tangents have been trimmed to the PC and PT of the curve.

CREATE MULTIPLE CURVES COMMAND

The Create Multiple Curves command lets you create several curves that are tangentially connected. The resulting curves have an effect similar to an alignment spiral section. This command can be useful when re-creating railway track geometry based upon field survey data.

The command prompts you for the two tangents. Then, the command-line prompts as follows:

```
Enter Number of Curves:
```

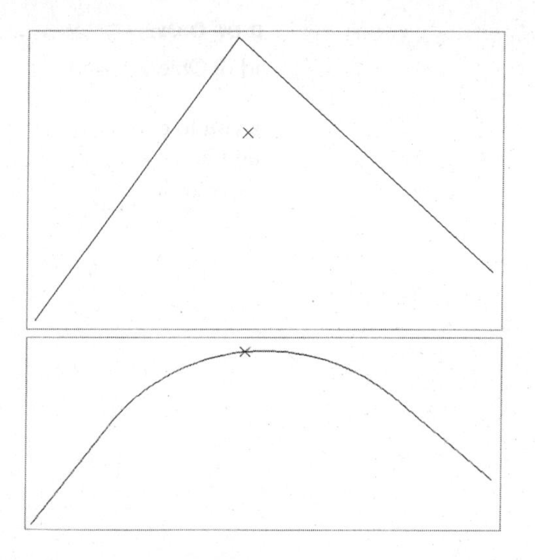

FIGURE 2.19
The first image shows two lines with a desired pass-through point. In the second image, the Create Curve through Point command draws a curve that is tangent to both lines and passes through the chosen point

The command allows for up to 10 curves between tangents.

One of your curves must have a flexible length that's determined on the basis of the lengths, radii, and geometric constraints of the other curves. Curves are counted clockwise, so enter the number of your flexible curve:

```
Enter Floating Curve #:
```

Enter the length and radii for all your curves:

```
Enter curve 1 Radius:
Enter curve 1 Length:
```

The floating curve number will prompt you for a radius but not a length.

As with all other curve commands, the specified geometry must be possible. If the command can't find a solution on the basis of your length and radius inputs, it returns no solution (see Figure 2.20).

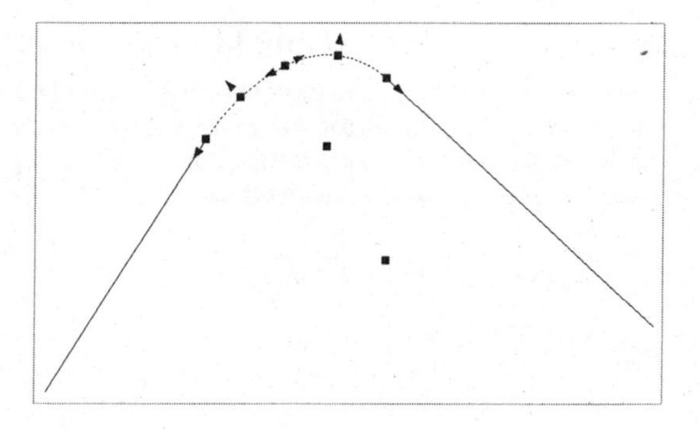

FIGURE 2.20
Two curves were specified with the #2 curve designated as the floating curve

Create Curve from End of Object Command

The Create Curve from End of Object command enables you to draw a curve tangent to the end of your chosen line or arc.

The command prompts you to choose an object to serve as the beginning of your curve. You can then specify a radius and an additional parameter (such as delta, length, and so on) for the curve or the endpoint of the resulting curve chord (see Figure 2.21).

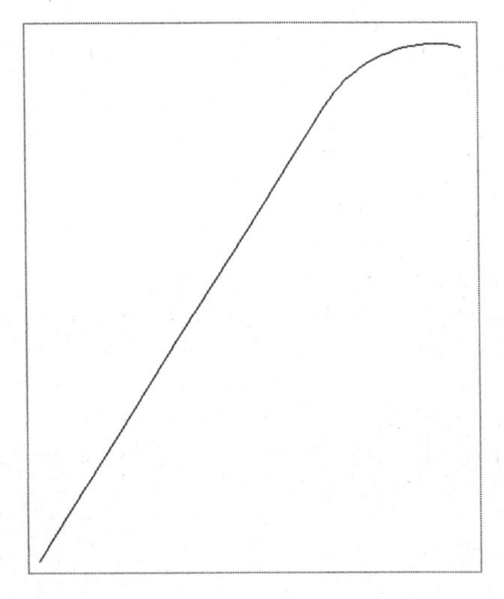

Figure 2.21
A curve, with a 25′ radius and a 30′ length, drawn from the end of a line

Create Reverse or Compound Curves Command

The Create Reverse or Compound Curves command allows you to add additional curves to the end of an existing curve. Reverse curves are drawn in the opposite direction (i.e., a curve to the right tangent to a curve to the left) from the original curve to form an S shape. In contrast, compound curves are drawn in the same direction as the original curve (see Figure 2.22). This tool can be useful when re-creating a legal description of a road alignment that contains reverse and/or compound curves.

Re-creating a Deed Using Line and Curve Tools

This exercise will help you apply some of the lines and curve tools you've learned so far to reconstruct a parcel similar to the property deed shown in Figure 2.23.

For ease of reading and clarity, the following is a summary of that description (note that like many real-world deeds, this deed will have a gap in closure of about 5′):

```
From Point of Beginning
Lines:
South 12 degrees 15 minutes 00 seconds West 828.23 feet to a point
North 86 degrees 18 minutes 25 seconds West 1039.50 feet to a point
```

```
North 18 degrees 40 minutes 07 seconds East 442.98 feet to a point
North 60 degrees 08 minutes 48 seconds East 107.43 feet to a point

Curve to the RIGHT:
Radius 761.35 feet
Arc length 204.70
Chord Length 204.08 feet
Chord Bearing North 67 degrees 50 minutes 56 seconds East to a point

Lines:
North 75 degrees 33 minutes 05 seconds East 671.23 feet to a point
North 77 degrees 10 minutes 37 seconds East 78.66 feet to a point

Curve to the RIGHT:
Radius 937.094
Arc length 62.94 feet
Chord Length 62.82
Chord Bearing North 79 degrees 52 seconds, 29 minutes East

Returning to Point of Beginning
The resulting enclosure should be: 15.11 acres (more or less)
```

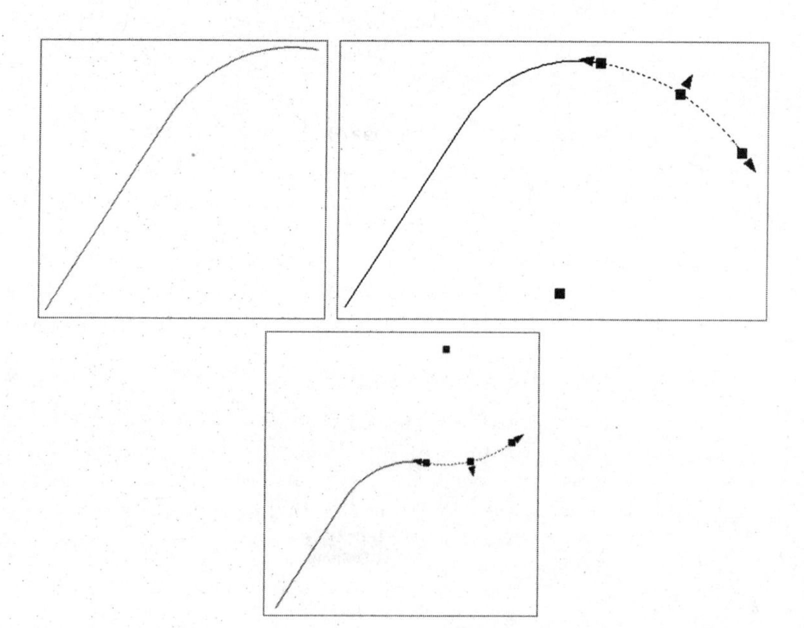

FIGURE 2.22
A tangent and curve before adding a reverse or compound curve (left); a compound curve drawn from the end of the original curve (right); and a reverse curve drawn from the end of the original curve (bottom)

FIGURE 2.23
A parcel deed

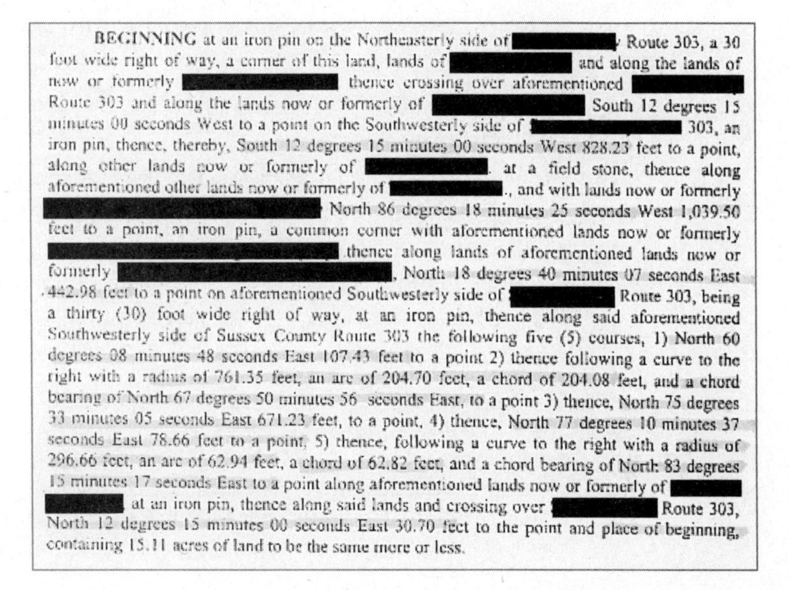

BEGINNING at an iron pin on the Northeasterly side of Route 303, a 30 foot wide right of way, a corner of this land, lands of and along the lands of now or formerly thence crossing over aforementioned Route 303 and along the lands now or formerly of South 12 degrees 15 minutes 00 seconds West to a point on the Southwesterly side of 303, an iron pin, thence, thereby, South 12 degrees 15 minutes 00 seconds West 828.23 feet to a point, along other lands now or formerly of . at a field stone, thence along aforementioned other lands now or formerly of ., and with lands now or formerly North 86 degrees 18 minutes 25 seconds West 1,039.50 feet to a point, an iron pin, a common corner with aforementioned lands now or formerly thence along lands of aforementioned lands now or formerly , North 18 degrees 40 minutes 07 seconds East 442.98 feet to a point on aforementioned Southwesterly side of Route 303, being a thirty (30) foot wide right of way, at an iron pin, thence along said aforementioned Southwesterly side of Sussex County Route 303 the following five (5) courses, 1) North 60 degrees 08 minutes 48 seconds East 107.43 feet to a point 2) thence following a curve to the right with a radius of 761.35 feet, an arc of 204.70 feet, a chord of 204.08 feet, and a chord bearing of North 67 degrees 50 minutes 56 seconds East, to a point 3) thence, North 75 degrees 33 minutes 05 seconds East 671.23 feet, to a point, 4) thence, North 77 degrees 10 minutes 37 seconds East 78.66 feet to a point, 5) thence, following a curve to the right with a radius of 296.66 feet, an arc of 62.94 feet, a chord of 62.82 feet, and a chord bearing of North 83 degrees 15 minutes 17 seconds East to a point along aforementioned lands now or formerly of at an iron pin, thence along said lands and crossing over Route 303, North 12 degrees 15 minutes 00 seconds East 30.70 feet to the point and place of beginning, containing 15.11 acres of land to be the same more or less.

Follow these steps:

1. Open the `Deed Create Start.dwg` file, which you can download from `www.sybex.com/masteringcivil3d2010`.
2. From the Draw panel on the Home tab, select the Line drop-down and select the Create Line by Bearing command.
3. At the `Select first point:` prompt, select any location in the drawing to begin the first line.
4. At the `>>Specify quadrant (1-4):` prompt, enter **3** to specify the SW quadrant, and then press ↵.
5. At the `>>Specify bearing:` prompt, enter **12.1500**, and press ↵.
6. At the `>>Specify distance:` prompt, enter **828.23**, and press ↵.
7. Repeat steps 4 through 6 for the next three courses.
8. Press Esc to exit the Create Line by Bearing command.
9. From the Draw panel on the Home tab, select the Curves ➤ Create Curve from End of Object and select the Create Curve from End of Object command.
10. At the `Select arc or line object:` prompt, select the northeast end of the last line drawn.

11. At the `Select entry [Radius/Point] <Radius>:` prompt, press ↵ to confirm the radius selection.
12. At the `Specify Radius, or [degreeArc/degreeChord]:` prompt, enter **761.35**, and then press ↵.
13. At the `Select entry [Tangent/Chord/Delta/Length/External/Mid-Ordinate] <Length>:` prompt, press ↵ to confirm the length selection.
14. At the `Specify length of curve:` prompt, enter **204.70**. The curve appears in the drawing, and a short report about the curve appears on the command line.
15. Continue using the Create Line by Bearing and the Create Curve from End of Object commands to complete the deed as listed at the beginning of this exercise.
16. The finished linework should look like Figure 2.24. There will be an error of closure of 4.55′. Typically, rounding errors, especially with the different curve parameters, can cause an error in closure. Perhaps reworking the deed holding a different curve parameter would improve your results. Consult your office survey expert about how this would be handled in house, and refer to Chapter 3, "Lay of the Land: Survey," for more information about traverse adjustment and similar tools.

FIGURE 2.24
The finished linework

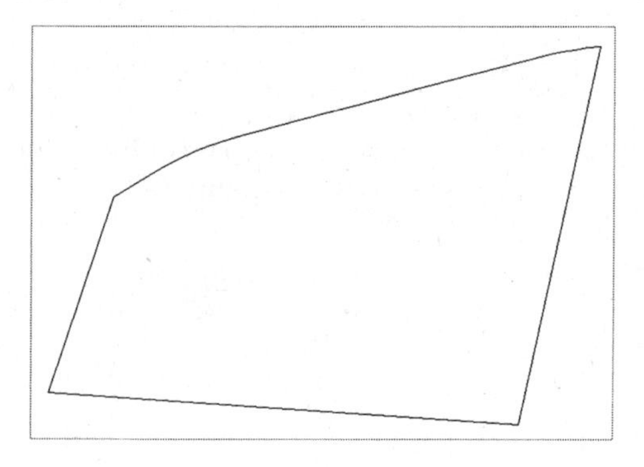

17. Save your drawing. You'll need it for the next exercise.

Best Fit Entities

Although engineers and surveyors do their best to make their work an exact science, sometimes tools like the Best Fit Entities are required.

Roads in many parts of the world have no defined alignment. They may have been old carriage roads or cart paths from hundreds of years ago that evolved into automobile roads. Surveyors and engineers are often called to help establish official alignments, vertical alignments, and right-of-way lines for such roads on the basis of a best fit of surveyed centerline data.

Other examples for using Best Fit Entities include property lines of agreement, road rehabilitation projects, and other cases where existing survey information must be approximated into "real" engineering geometry (see Figure 2.25).

CREATE BEST FIT LINE COMMAND

The Create Best Fit Line command under the Best Fit drop-down on the Draw panel takes a series of Civil 3D points, AutoCAD points, entities, or drawing locations and draws a single best-fit line segment from this information. In Figure 2.26, for example, the Create Best Fit Line command draws a best-fit line through a series of points that aren't quite collinear. Note that the best-fit line will change as more points are picked.

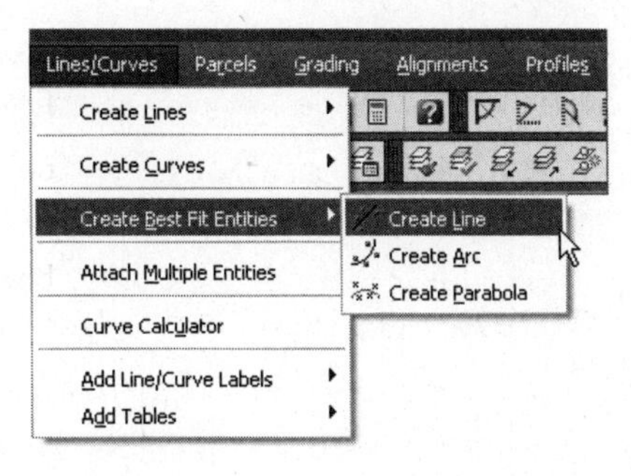

FIGURE 2.25
The Create Best Fit Entities menu options

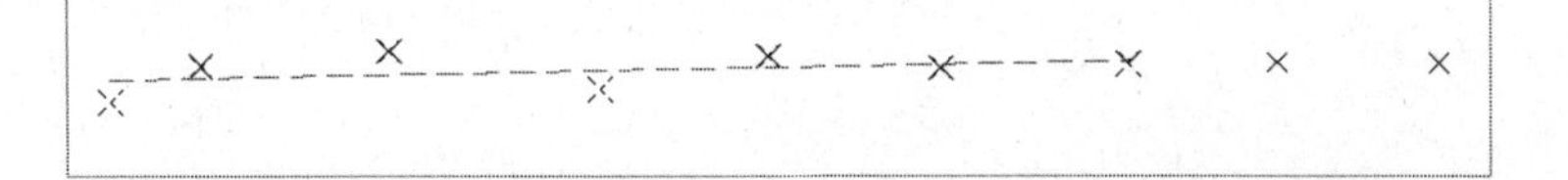

FIGURE 2.26
A preview line drawn through points that aren't quite collinear

Once you've selected your points, a Panorama window appears with information about each point you chose, as shown in Figure 2.27.

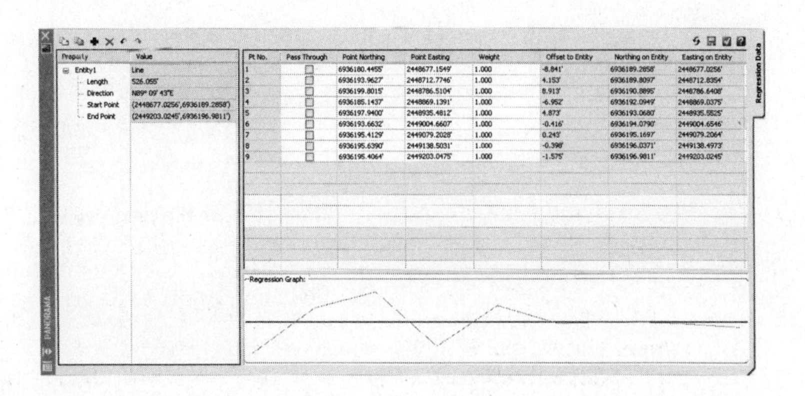

FIGURE 2.27
The Panorama window lets you optimize your best fit

This interface allows you to optimize your best fit by adding more points, selecting the check box in the Pass Through column to force one of your points on the line, or adjusting the value under the Weight column. Figure 2.28 shows a line drawn by best fit.

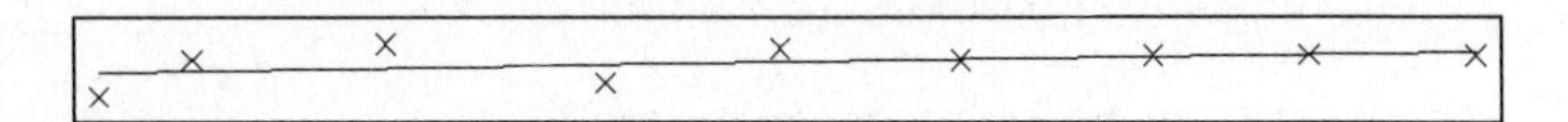

FIGURE 2.28
The resulting best-fit line through a series of points that aren't collinear

CREATE BEST FIT ARC COMMAND

The Create Best Fit Arc command under the Best Fit drop-down works identically to the Create Best Fit Line command, except the resulting entity is a single arc segment as opposed to a single line segment (see Figure 2.29).

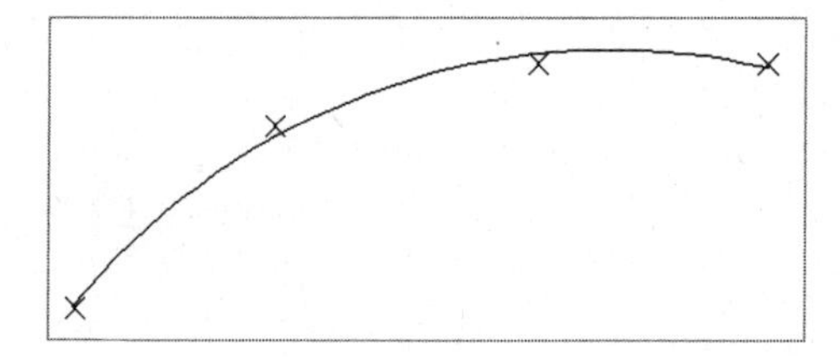

FIGURE 2.29
A curve created by best fit

CREATE BEST FIT PARABOLA COMMAND

The Create parabola command under the Create Best Fit Entities option works in a similar way to the line and arc commands just described. This command is most useful when you have TIN sampled or surveyed road information and you'd like to replicate true vertical curves for your design information.

After you select this command, the Parabola by Best Fit dialog appears (see Figure 2.30).

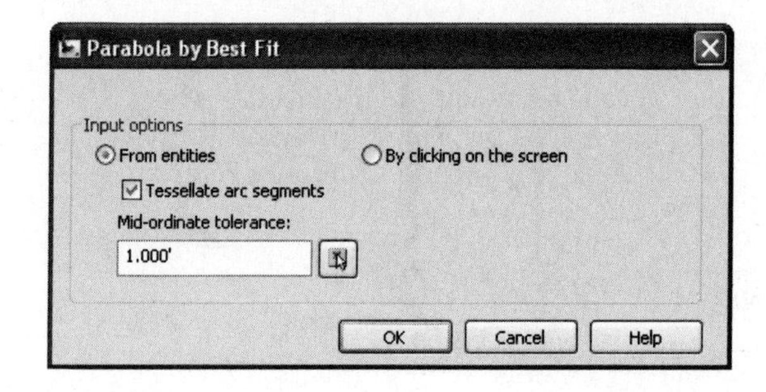

FIGURE 2.30
The Parabola by Best Fit dialog

You can select inputs from entities (such as lines, arcs, polylines, or profile objects) or by picking on screen. The command then draws a best-fit parabola on the basis of this information. In Figure 2.31, the shots were represented by AutoCAD points; more points were added by selecting the By Clicking on the Screen option and using the Node osnap to pick each point.

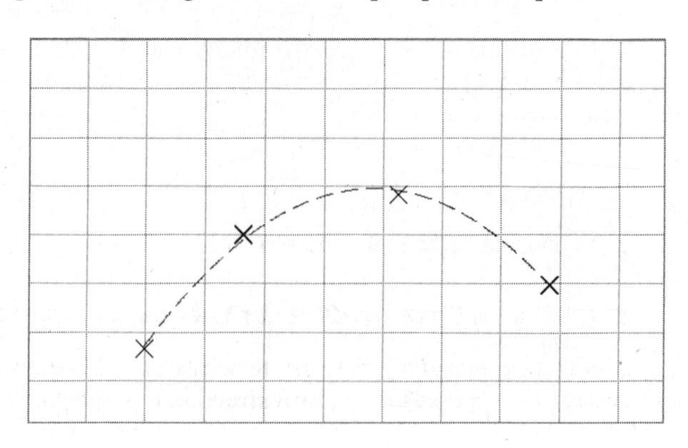

FIGURE 2.31
The best-fit preview line changes as more points are picked

Once you've selected your points, a Panorama window appears, showing information about each point you chose. Also note the information in the right pane regarding K-value, curve length, grades, and so forth.

In this interface (shown in Figure 2.32), you can optimize your K-value, length, and other values by adding more points, selecting the check box in the Pass Through column to force one of your points on the line, or adjusting the value under the Weight column.

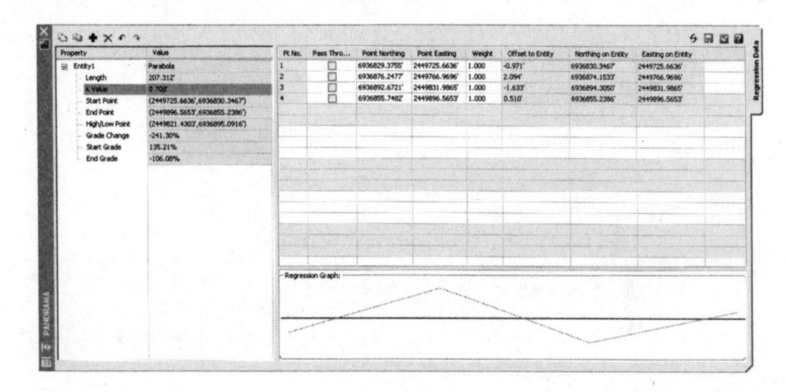

FIGURE 2.32
The Panorama window lets you make adjustments to your best-fit parabola

Attach Multiple Entities

The Attach Multiple Entities command is a combination of the Line from End of Object command and the Curve from End of Object command. This command is most useful for reconstructing deeds or road alignments from legal descriptions when each entity is tangent to the previous entity. Using this command saves you time because you don't have to constantly switch between the Line from End of Object command and the Curve from End of Object command (see Figure 2.33).

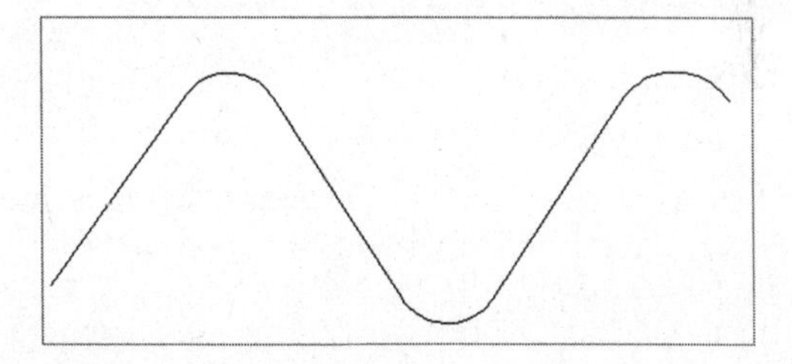

FIGURE 2.33
The Attach Multiple Entities command draws a series of lines and arcs so that each segment is tangent to the previous one

In previous releases, this command was part of the Lines and Curves menu. In Civil 3D 2010, the Attach Multiple Entities command isn't in a panel, but you can access it by typing **attachmultiple** at the command line.

Real World Scenario

CREATING AN EDGE-OF-ASPHALT LINE USING BEST FIT ENTITIES

It's common for surveyors to locate points along the edge of a road. Although road plans may call for perfectly straight edges, road construction is a different matter. Imperfections, branches, and other obstacles often make surveying the edge of asphalt difficult. When you're working with these types of points, you may find the Best Fit tools to be helpful for creating an optimized edge-of-asphalt linework.

Try this exercise:

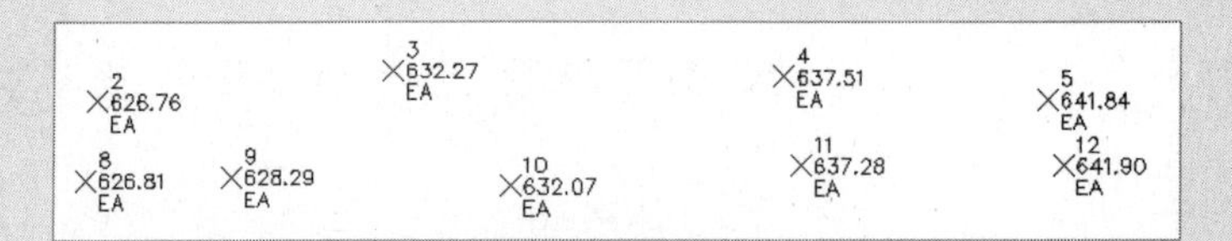

1. Open the Best Fit.dwg file, which you can download from `www.sybex.com/masteringcivil3d2010`.
2. From the Draw panel of the Home tab, select Best Fit ➤ Best Fit Line. The Line by Best Fit dialog appears.
3. Confirm that From Civil 3D Points is selected. Click OK.
4. At the `Select point objects or [Numbers/Groups]:` prompt, type **N** for numbers. Press ↵.
5. At the `Enter point numbers or [Select/Groups]:` prompt, type **2-7** and press ↵ to indicate that you'd like to connect point numbers 2 through 7 with a line of best fit. A red, dashed preview line appears.

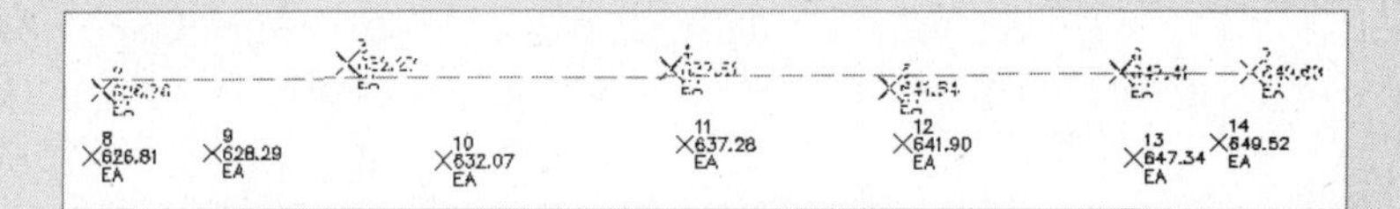

6. At the `Enter point numbers or [Select/Groups/Undo]:` prompt, press ↵. The Best Fit Panorama window appears. Click the green check box to dismiss Panorama and accept the best fit.

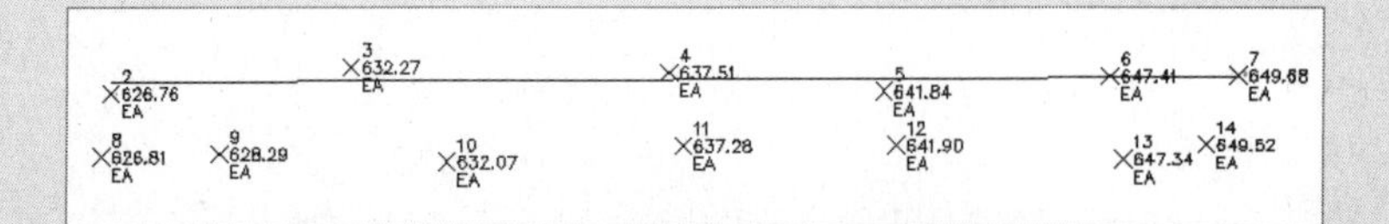

7. The best-fit line is drawn. Repeat the process for the other side of the road, noting what happens if you exclude certain points or force the line to a pass through them.

The Curve Calculator

Sometimes you may not have enough information to draw a curve properly. Although many of the curve-creation tools assist you in calculating the curve parameters, you may find an occasion where the deed you're working with is incomplete.

The Curve Calculator found in the Curves drop-down on the Draw panel helps you calculate a full collection of curve parameters on the basis of your known values and constraints. The units used in the Curve Calculator match the units assigned in your Drawing Settings.

The Curve Calculator can remain open on your screen while you're working through commands. You can send any value in the Calculator to the command line by clicking the button next to that value (see Figure 2.34).

FIGURE 2.34
The Curve Calculator

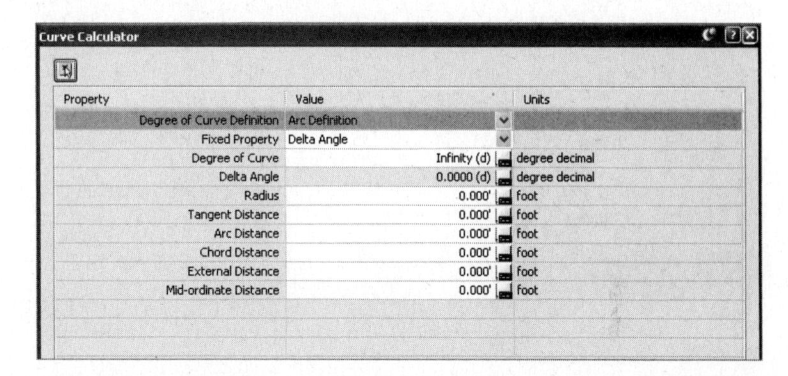

The button at upper left in the Curve Calculator inherits the arc properties from an existing arc in the drawing, and the drop-down menu in the Degree of Curve Definition selection field allows you to choose whether to calculate parameters for an arc or a chord definition.

The drop-down menu in the Fixed Property selection field also gives you the choice of fixing your radius or delta value when calculating the values for an arc or a chord, respectively (see Figure 2.35). Whichever parameter is chosen as the fixed value is held constant as additional parameters are calculated.

FIGURE 2.35
The Fixed Property drop-down menu gives you the choice of fixing your radius or delta value

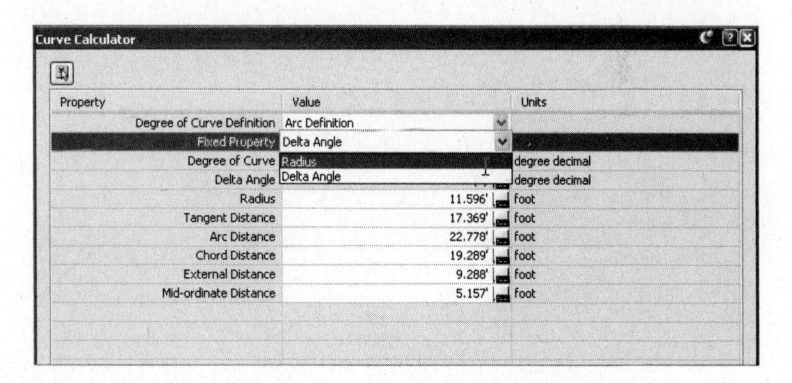

As explained previously, you can send any value in the Curve Calculator to the command line using the button next to that value (see Figure 2.36). This is most useful while you're active in a curve command and would like to use a certain parameter value to complete the command.

Adding Line and Curve Labels

Although most robust labeling of site geometry is handled using Parcel or Alignment labels, limited line- and curve-annotation tools are available in Civil 3D. The line and curve labels are composed much the same way as other Civil 3D labels, with marked similarities to Parcel and Alignment Segment labels.

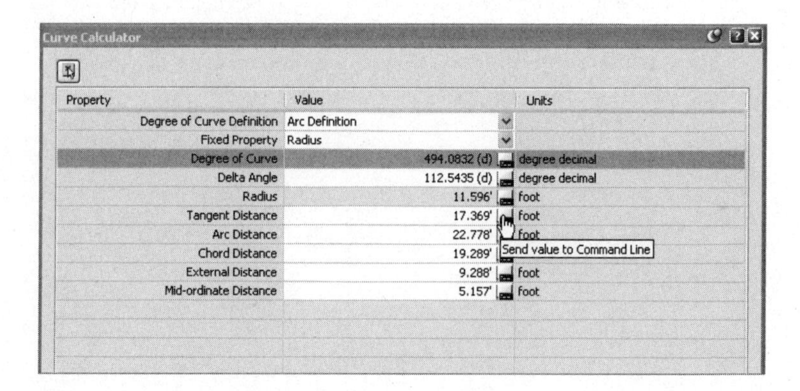

FIGURE 2.36
Click the button next to any value to send it to the command line

WHERE IS DELTA?

In the Text Component Editor for a curve label, the value that most people would refer to as a delta angle is called the General Segment Total Angle. To insert the Delta symbol in a label, simply type **\U+0394** in the Text Editor window on the right side of the Text Component Editor dialog box as shown here.

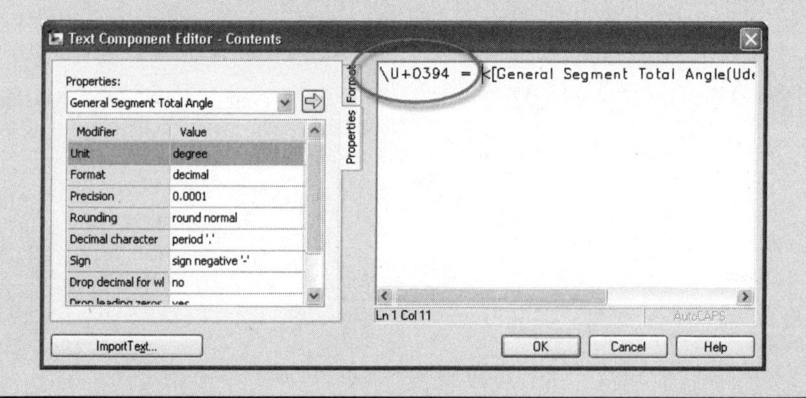

The next exercise leads you through labeling the deed you re-created earlier in this chapter:

1. Continue working in the `Deed Create Start.dwg` file.
2. Click the Labels button in the Labels & Tables panel on the Annotate tab. The Add Labels dialog appears, as shown in Figure 2.37.
3. Choose Line and Curve from the Feature drop-down menu.

FIGURE 2.37
The Add Labels dialog, set to Multiple Segment Labels

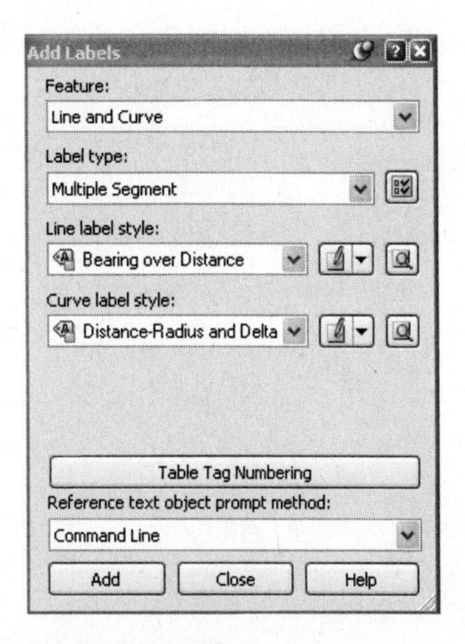

4. Choose Multiple Segment from the Label Type drop-down menu. The Multiple Segment option places the label at the midpoint of each selected line or arc.
5. Confirm that the Line Label Style is Bearing over Distance and that the Curve Label Style is Distance-Radius and Delta.
6. Click the Add button.
7. At the `Select Entity:` prompt, select each line and arc that you drew in the previous exercise. A label appears on each entity at its midpoint, as shown in Figure 2.38.

FIGURE 2.38
The labeled linework

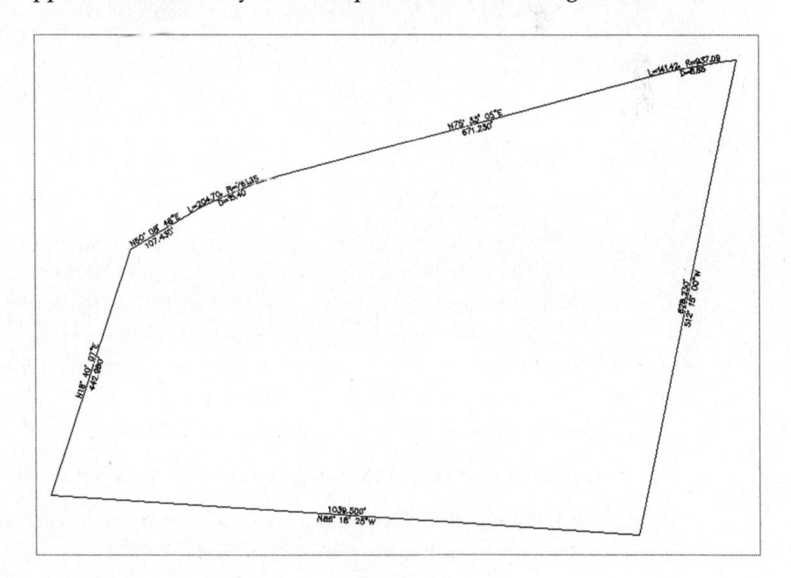

8. Save the drawing — you'll need it for the next exercise.

Converting Curve Labels to Tags and Making a Curve Table

Civil 3D 2010 added the ability to convert line and curve labels into tags for table creation. It's common to use tags or sequentially numbered short labels to identify lines and/or curves on the plan and then list the detailed information in a table. The following exercise shows you how to convert the curve labels that you added in the previous exercise into tags and make a table with the corresponding curve data:

1. Continue working in `Deed Create Start.dwg`.
2. Change to the Annotate tab in the Ribbon.
3. From the Labels & Tables panel, select Add Tables menu ➢ Line and Curve ➢ Add Curve Tables. The Table Creation dialog appears.
4. Select the Apply check box for General Curve: Distance-Radius and Delta, as shown in Figure 2.39. Click OK.

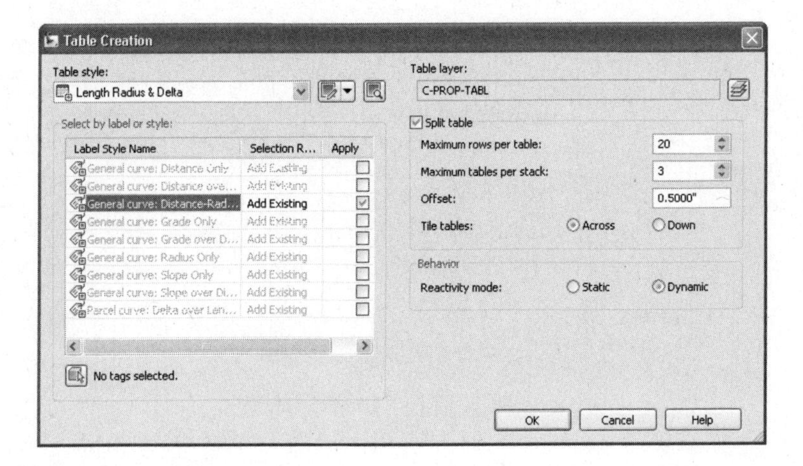

FIGURE 2.39
Select the Apply check box for General Curve: Distance-Radius and Delta

5. At the `Select upper left corner:` prompt, select any location. A table appears, as shown in Figure 2.40. Note that the curves that were formerly labeled with Distance-Radius and Delta are now labeled with sequentially numbered tags.

FIGURE 2.40
The curve table

Curve Table					
Curve #	Length	Radius	Delta	Chord Direction	Chord Length
C1	141.42	937.09	8.65	S79° 52' 29"W	141.28
C2	204.70	761.35	15.40	S67° 50' 57"W	204.08

6. Save your drawing.

Using Transparent Commands

In many cases, the ''Create Line by . . . '' commands in the Draw panel are the standard AutoCAD Line commands combined with the appropriate transparent commands.

A transparent command behaves somewhat similarly to an osnap command. You can't click the Endpoint button and expect anything to happen — you must be active inside another command, such as a line, an arc, or a circle command.

The same principle works for transparent commands. Once you're active in the Line command (or any AutoCAD or Civil 3D drawing command), you can choose the Bearing Distance transparent command and complete your drawing task using a bearing and distance.

As stated earlier, the transparent commands can be used in any AutoCAD or Civil 3D drawing command, much like an osnap. For example, you can be actively drawing an alignment and use the Northing/Easting transparent command to snap to a particular coordinate, and then press Esc once and continue drawing your alignment as usual.

While a transparent command is active, you can press Esc once to leave the transparent mode but stay active in your current command. You can then choose another transparent command if you'd like. For example, you can start a line using the Endpoint osnap, activate the Angle Distance transparent command, draw a line-by-angle distance, and then press Esc, which takes you out of angle-distance mode but keeps you in the Line command. You can then draw a few more segments using the Point Object transparent command, press Esc, and finish your line with a Perpendicular osnap.

You can activate the transparent commands using keyboard shortcuts (see Chapter 37 of the *Civil 3D Users Guide* PDF for more information) or using the Transparent Commands toolbar. Be sure you include the Transparent Commands toolbar (shown in Figure 2.41) in all your Civil 3D and survey-oriented workspaces.

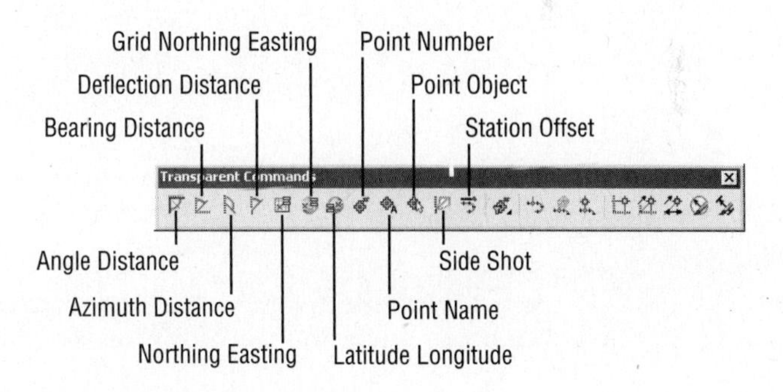

FIGURE 2.41
The Transparent Commands toolbar

The six profile-related transparent commands will be covered in Chapter 8, ''Cut to the Chase: Profiles.''

Standard Transparent Commands

The transparent commands shown in Table 2.1 behave identically to their like-named counterparts from the Draw panel (discussed earlier in this chapter). The difference is that you can call up these transparent commands in any appropriate AutoCAD or Civil 3D draw command, such as a line, polyline, alignment, parcel segment, feature line, or pipe-creation command.

TABLE 2.1: The Transparent Commands

Tool Icon	Menu Command
	Angle Distance
	Bearing Distance
	Azimuth Distance
	Deflection Distance
	Northing Easting
	Grid Northing Easting
	Latitude Longitude
	Point Number
	Point Name
	Point Object
	Side Shot
	Station Offset

Matching Transparent Commands

You may have construction or other geometry in your drawing that you'd like to match with new lines, arcs, circles, alignments, parcel segments, or other entities.

While actively drawing an object that has a radius parameter, such as a circle, an arc, an alignment curve, or a similar object, you can choose the Match Radius transparent command and then select an object in your drawing that has your desired radius. Civil 3D draws the resulting entity with a radius identical to that of the object you chose during the command. You'll save time using this tool because you don't have to first list the radius of the original object and then manually type in that radius when prompted by your circle, arc, or alignment tool.

The Match Length transparent command works identically to the Match Radius transparent command except that it matches the length parameter of your chosen object.

Using Inquiry Commands

A large part of a surveyor's work involves querying lines and curves for their length, direction, and other parameters.

The Inquiry Commands panel (Figure 2.42) is on the Analyze tab, and it makes a valuable addition to your Civil 3D and survey-related workspaces. Remember, panels can be dragged away from the Ribbon and set in the graphics environment much like a toolbar.

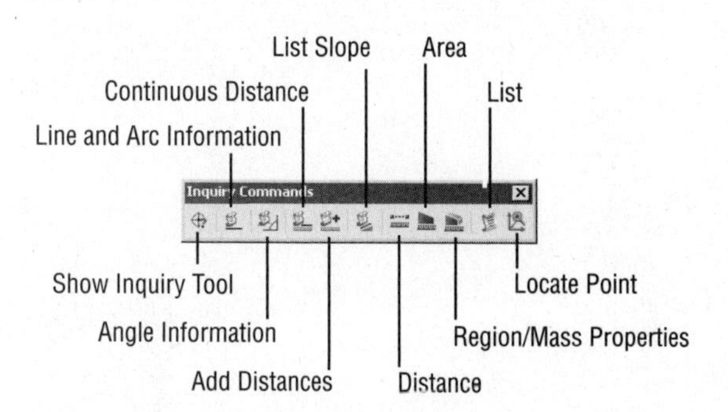

FIGURE 2.42
The Inquiry Commands panel

The Inquiry tool (shown in Figure 2.43) provides a diverse collection of commands that assist you in studying Civil 3D objects. You can access the Inquiry tool by clicking the Inquiry Tool button on the Inquiry panel.

Because the focus of this chapter is linework as opposed to Civil 3D objects, it doesn't include a detailed discussion of this tool.

The Station Tracker (shown in Figure 2.44) provides visual cues to help you locate stations that match the position of your cursor when working with alignments, profiles, and sections. The Station Tracker was an undocumented function in AutoCAD Civil 3D 2009, but the two related commands (ShowDrawingTips, used to track in the current viewport, and ShowDrawingTipsFull, used to track in all viewports) are now incorporated into the Station Tracker drop-down menu on the Inquiry panel of the Analyze tab. When the cursor is above a horizontal alignment, and one or more profile views for the same horizontal alignment are in the drawing, a temporary line is drawn in each profile view at the station defined by the cursor location. As the cursor moves, so will the temporary line. When the cursor is within a profile view, a temporary line is drawn perpendicular to the station defined by the cursor location along the parent horizontal alignment. As the cursor's position changes in the profile view, the position of the temporary line will move as well. When the cursor is within a section view, a temporary line is drawn perpendicular to the station defined by the cursor location along the parent horizontal alignment. As the cursor's position changes in the profile view, the position of the temporary line will move as well. Additionally, a vertical line is drawn in each associated profile view.

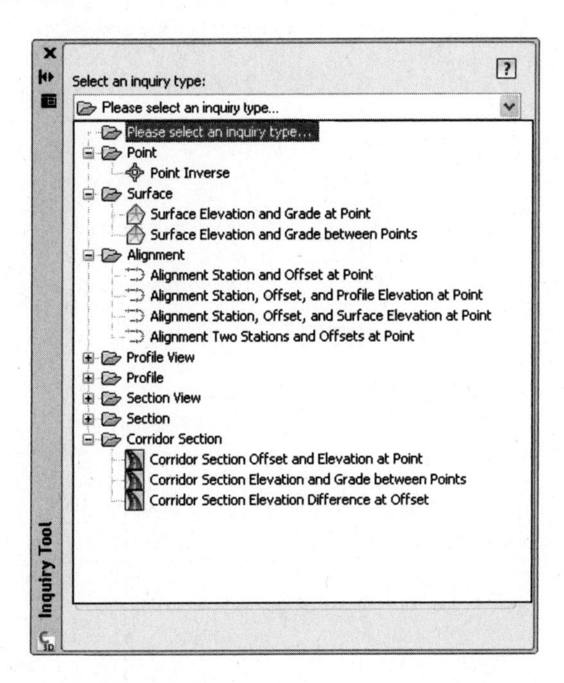

FIGURE 2.43
The Inquiry panel

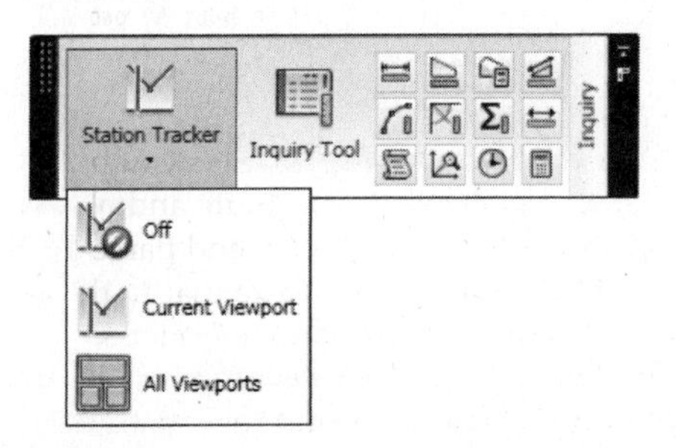

FIGURE 2.44
The Station Tracker tool

The Distance tool is the standard AutoCAD Distance command with which you may already be familiar. This command measures the distance between two points on your screen.

The Area tool is the standard AutoCAD Area command. This tool allows you to calculate the area and perimeter of several points on your screen.

The Region/Mass Properties tool lets you learn about the properties of a region or solid. Because it's rare that you'll create such an object while working with survey-type data, you may not use this very often.

The List Slope tool provides a short command-line report (like the one in Figure 2.45) that lists the elevations and slope of an entity (or two points) that you choose, such as a line or feature line.

FIGURE 2.45
The List Slope command-line report

```
Select object or [Points]:
First elev: 5.000', Second elev: 10.000', Elev diff: 5.000'
Grade: 2.37%, Slope: 42.28:1, Horiz dist: 211.389'

Select object or [Points]:
```

The Line and Arc Information tool provides a short report about the line or arc of your choosing (see Figure 2.46). This tool also works on parcel segments and alignment segments. Alternatively, you can type **P** for points at the command line to get information about the apparent line that would connect two points on screen.

FIGURE 2.46
The results of a line inquiry and an arc inquiry

```
Select object or [Points]:
-----------------------------------------------------------------
                         LINE DATA
-----------------------------------------------------------------
Begin . . . . .  X: 2449247.950'         Y: 6935520.120'
End . . . . . .  X: 2449743.105'         Y: 6935520.120'
              Distance: 495.155'          Course: N90° 00' 00"E

Select object or [Points]:
-----------------------------------------------------------------------
                           ARC DATA
-----------------------------------------------------------------------
Begin . . . . .  X: 2449247.950'         Y: 6935520.120'
Radial Point. .  X: 2449247.950'         Y: 6935495.120'
End . . . . . .  X: 2449223.048'         Y: 6935497.327'
PI  . . . . . .  X: 2449225.068'         Y: 6935520.120'
   Tangent: 22.882'        Chord: 33.759'       Course: S47° 31' 57"W
Arc Length: 37.060'       Radius: 25.000'        Delta: 84.9350 (d)
```

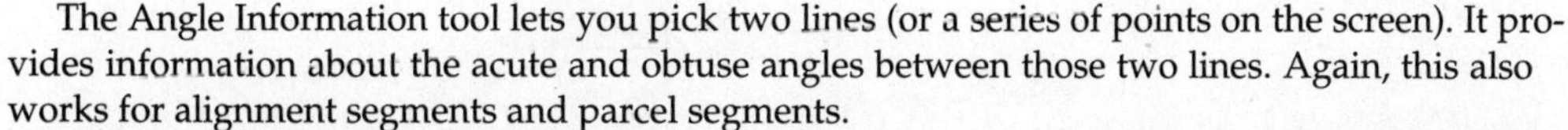

The Angle Information tool lets you pick two lines (or a series of points on the screen). It provides information about the acute and obtuse angles between those two lines. Again, this also works for alignment segments and parcel segments.

The Add Distances tool is similar to the Continuous Distance command, except the points on your screen don't have to be continuous.

The Continuous Distance tool provides a sum of distances between several points on your screen, or one base point and several points.

The List tool is the standard AutoCAD List command. This tool provides an AutoCAD text-window report of the entity type and some properties.

The ID Point tool is the standard AutoCAD ID command. This tool provides a short command-line report of the x-, y-, and z-coordinates of any location you select on your screen. You may most commonly use this tool to study part of a surface to identify its z elevation.

The Time tool is the standard AutoCAD Time command. This tool provides information regarding the date of creation of the drawing, the last save time, and the current time spent editing as shown in Figure 2.47.

FIGURE 2.47
The Time report

```
Current time:              Wednesday, February 11, 2009  5:27:29:219 PM
Times for this drawing:
  Created:                 Sunday, April 15, 2007  10:55:46:687 AM
  Last updated:            Wednesday, February 11, 2009  2:17:36:890 PM
  Total editing time:      0 days 02:02:05:860
  Elapsed timer (on):      0 days 02:02:05:751
  Next automatic save in:  <disabled>
```

The Quick Calculator performs a full range of mathematical functions, creates and uses variables, and converts units of measurement (see Figure 2.48).

FIGURE 2.48
The Quick Calculator

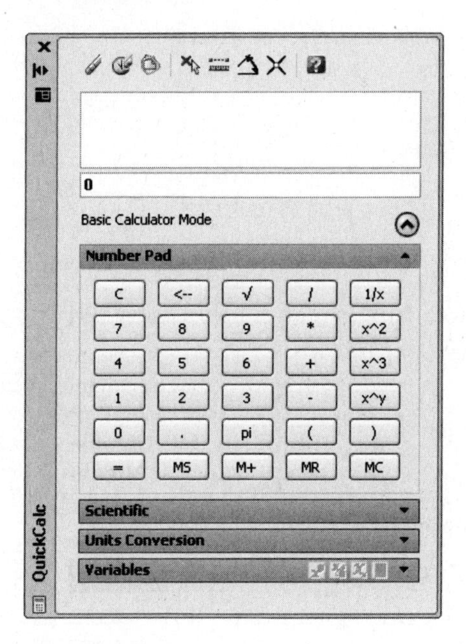

Establishing Drawing Settings

As you've worked through this chapter, your Drawing Settings have come into play several times. Let's look at a few of the locations where you should establish settings such as the coordinate system, precision, and units.

You access the Drawing Settings dialog by right-clicking the drawing name on the Settings tab of Toolspace.

Drawing Settings: Units and Zone

This chapter has noted several tools that require you to assign a coordinate system to your drawing. The coordinate system, among other settings, is assigned on the Units and Zone tab of the Drawing Settings dialog (see Figure 2.49). (We discussed setting a coordinate system in Chapter 1, "Getting Dirty: The Basics of Civil 3D," if you need to review.)

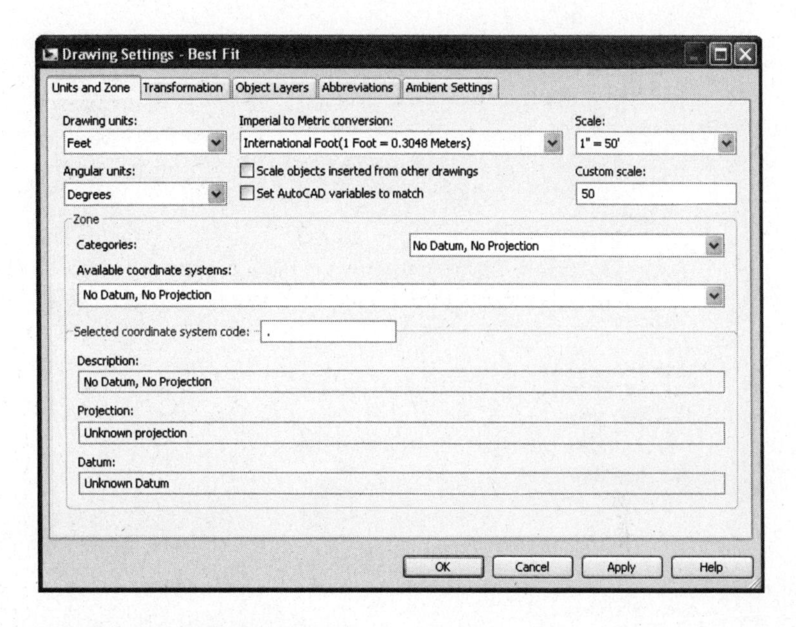

FIGURE 2.49
The Units and Zone tab of the Drawing Settings dialog

Drawing Settings: Ambient Settings

The Ambient Settings establish the default settings for all commands in Civil 3D. You can override these settings at the individual command settings level.

For example, perhaps your default precision for distance is three decimal places as established in the ambient settings, but for Alignment Layout you'd like to track eight decimal places. You can make that change under the Alignment Commands tree on the Settings tab of Toolspace (see Figure 2.50).

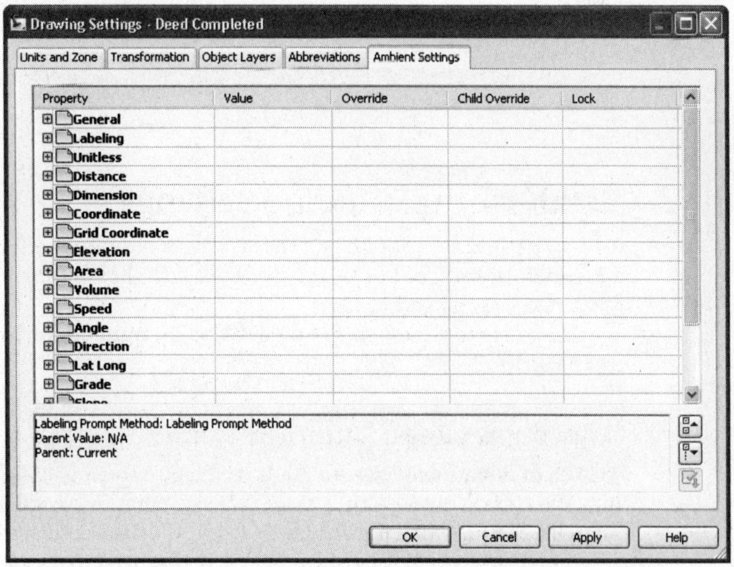

FIGURE 2.50
The Ambient Settings tab of the Drawing Settings dialog

Although many of the ambient settings apply to the creation of coordinate geometry, two categories frequently need adjusting (see Figure 2.51), ambient settings for direction and ambient settings for transparent commands.

> **DRAWING PRECISION VERSUS LABEL PRECISION**
>
> **You can create label styles to annotate objects using different precision, units, or specifications than those set in the Ambient or Command Settings dialogs. Establish settings to reflect how you'd like to input and track your data, not necessarily how you'd like to label your data.**

FIGURE 2.51
The Ambient Settings for Direction

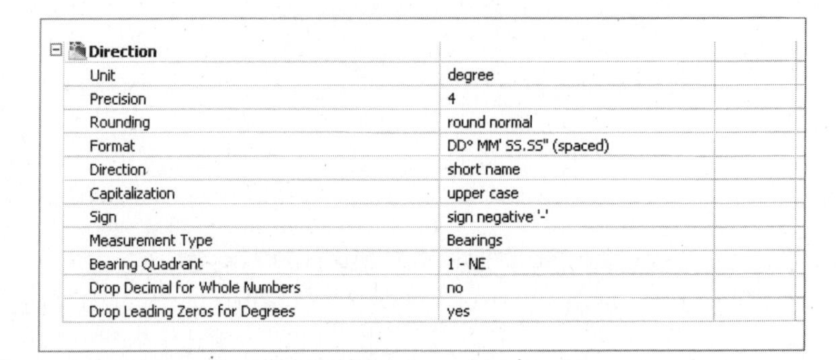

Direction		
Unit	degree	
Precision	4	
Rounding	round normal	
Format	DD° MM' SS.SS" (spaced)	
Direction	short name	
Capitalization	upper case	
Sign	sign negative '-'	
Measurement Type	Bearings	
Bearing Quadrant	1 - NE	
Drop Decimal for Whole Numbers	no	
Drop Leading Zeros for Degrees	yes	

The Ambient Settings for Direction offer the following choices:

- **Unit:** Degree, Radian, and Grad
- **Precision:** 0 through 8 decimal places
- **Rounding:** Round Normal, Round Up, and Truncate
- **Format:** Decimal, two types of DDMMSS, and Decimal DMS
- **Direction:** Short Name (spaced or unspaced) and Long Name (spaced or unspaced)
- **Capitalization**
- **Sign**
- **Measurement Type:** Bearings, North Azimuth, and South Azimuth
- **Bearing Quadrant**

From this list, it becomes clear where these settings apply to the tools discussed in this chapter. When you're using the Bearing Distance transparent command, for example, these settings control how you input your quadrant, your bearing, and the number of decimal places in your distance.

Explore the other categories, such as Angle, Lat Long, and Coordinate, and customize the settings to how you work.

At the bottom of the Ambient Settings tab is a Transparent Commands category, as shown in Figure 2.52.

FIGURE 2.52
The Transparent Commands area of the Ambient Settings tab

Transparent Commands	
Prompt for 3D Points	false
Prompt for Y before X	false
Prompt for Easting then Northing	false
Prompt for Longitude then Latitude	false

These settings control how (or if) you're prompted for the following information:

- **Prompt for 3D Points:** Controls whether you're asked to provide a *z* elevation after *x* and *y* have been located.
- **Prompt for Y before X:** For transparent commands that require *x* and *y* values, this setting controls whether you're prompted for the y-coordinate before the x-coordinate. Most users prefer this value set to False so they're prompted for an x-coordinate and then a y-coordinate.
- **Prompt for Easting then Northing:** For transparent commands that require Northing and Easting values, this setting controls whether you're prompted for the Easting first and the Northing second. Most users prefer this value set to False, so they're prompted for Northing first and then Easting.
- **Prompt for Longitude then Latitude:** For transparent commands that require longitude and latitude values, this setting controls whether you're prompted for Longitude first and Latitude second. Most users prefer this set to False, so they're prompted for Latitude and then Longitude.

Checking Your Work: The Mapcheck Analysis

After learning to use the line and curve tools in a production environment, it's time to check both the accuracy of your work and the accuracy of your labels.

Civil 3D 2010 includes a powerful Mapcheck Analysis tool located in the Survey menu on the Ground Data panel of the Analyze tab. The Mapcheck Analysis command computes closure based upon the labels you've created and subsequently modified along lines, curves, or parcel objects. The following exercise teaches you how to use the Mapcheck Analysis tool to check the accuracy of the deed you reestablished earlier:

1. Open the `Mapcheck.dwg` file, which you can download from `sybex.com/masteringcivil3d2010`.
2. Change to the Analyze tab, and select Survey ➢ Mapcheck from the Ground Data panel to display the Mapcheck Analysis palette.
3. Click the New Mapcheck button at the top of the menu bar, as shown in Figure 2.53.
4. At the `Enter name of mapcheck:` prompt, type **Record Deed**.

FIGURE 2.53
The New Mapcheck button in the Mapcheck Analysis palette

5. At the `Specify point of beginning (POB)` prompt, choose the north endpoint of the line representing the east line of the parcel.
6. Working clockwise, select each of the first four line labels as prompted.
7. Select the first tag encountered on the first curve along the north line of the parcel.
8. At the `Select a label or [Clear/Flip/New/Reverse]` prompt, type **R** and then press ↵ to reverse direction.
9. Select the next two line labels along the north line of the parcel.
10. Select the last curve label along the north line of the parcel and reverse the direction as needed.
11. Press ↵ to complete the parcel closure. The completed parcel should have eight sides.
12. Select the output view as shown in Figure 2.54 to verify closure.

NEW LABEL GRIPS

Select a label and then select the Drag Label grip to reveal two new grips. The Add Label Vertex and Move Leader Vertex grips are welcome additions. While a label is in a drag state, right-click to reveal the new Toggle Leader Tail option.

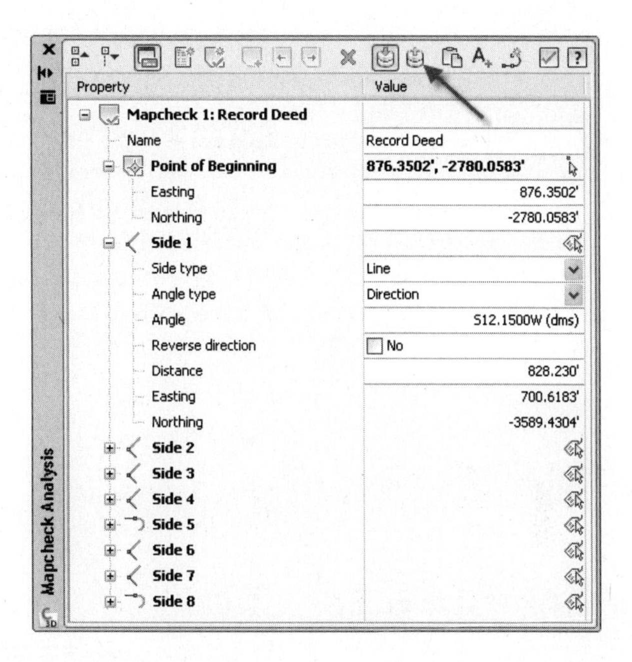

FIGURE 2.54
The completed deed in the Mapcheck Analysis palette

The Bottom Line

Create a series of lines by bearing and distance. By far the most commonly used command when re-creating a deed is Line by Bearing and Distance.

Master It Open the `Mastering Lines and Curves.dwg` file, which you can download from `www.sybex.com/masteringcivil3d2010`. Start from the Civil 3D point labeled START, and use any appropriate tool to create lines with the following bearings and distances (note that the Direction input format for this drawing has been set to DD.MMSSSS):

- N 57°06′56.75″ E; 135.441′
- S 41°57′03.67″ E; 118.754′
- S 27°44′41.63″ W; 112.426′
- N 50°55′57.00″ W; 181.333′

Use the Inquiry commands to confirm that lines are drawn correctly.

Master It Continue working in your drawing. Use any appropriate Inquiry command to confirm that each line has been drawn correctly.

Create a curve tangent to the end of a line. It's rare that a property stands alone. Often, you must create adjacent properties, easements, or alignments from their legal descriptions.

Master It Create a curve tangent to the end of the first line drawn in the first exercise that meets the following specifications:

- Radius: 200.00′
- Arc Length: 66.580′

Create a best-fit line for a series of Civil 3D points. Surveyed point data is rarely perfect. When you're creating drawing linework, it's often necessary to create a best-fit line to make up for irregular shots.

Master It Locate the Edge of Pavement (EOP) points in the drawing. Use the Create Line by Best Fit command to create a line using these points.

Label lines and curves. Although converting linework to parcels or alignments offers you the most robust labeling and analysis options, basic line- and curve-labeling tools are available when conversion isn't appropriate.

Master It Add line and curve labels to each entity created in the exercises. Choose a label that specifies the bearing and distance for your lines and length, radius, and delta of your curve.

Chapter 3

Lay of the Land: Survey

All civil-engineering projects start with a survey. Base maps provide engineers with data, which normally contain existing conditions. These maps can be used to develop an engineering-design model. Civil 3D 2010 provides an integrated solution that surveyors can use to create base maps that will reside in the same native format the engineers will use, thereby reducing potential (and costly) errors that result from translating data from one design software to another. In this chapter, you'll learn about tools and techniques that will link your survey equipment directly into the software, automate your drafting procedures from fieldwork, and provide a secure and independent database for storing and manipulating your survey data.

In this chapter, you'll learn to:

- Properly collect field data and import it into AutoCAD Civil 3D 2010
- Set up styles that will correctly display your linework
- Create and edit field book files
- Manipulate your survey data

Understanding the Concepts

Before you start working with the survey portion of Civil 3D, you first need to understand some basic concepts. When the majority of people think about surveying in any software, they generally think about going out into the field with a survey instrument and some form of data collector and returning to the office with a group of points — text entities with unique identifiers, northings, eastings, elevations, and some sort of descriptors. That point file, whether it be in ASCII format, text format, CSV format, or otherwise, is imported into a survey program that displays those points in some way, allowing drafters to essentially play a game of "connect the dots" to create a base plan. However, with the survey crew and the office staff working together, much of the "connect-the-dots" game can be played in the field. For example, in Figure 3.1, parking stripes, curb and gutter, asphalt, and concrete features have been connected correctly in the field with figures. This is important because it has the potential to reduce liability — always an important topic of conversation with surveyors. Because the field crew is on site and have actually witnessed existing conditions, they are in a better position to create the linework than a drafter who may have never seen the site.

Aside from having to get the survey field crew and the office staff working in harmony, there are a few other things you need to know. The first thing to know is that the survey functionality in Civil 3D doesn't use the old Point, Northing, Easting, Zed (elevation), Description (PNEZD) text file, but uses the raw data from your data collector to process an Autodesk field book (FBK) file, or a LandXML file. Granted, the option to use the PNEZD file is still there if you want "dumb" points. However, if you use an FBK file for your surveys, you have much more than just points on a screen — you actually have a record of how those points were collected. You will have the information you need to edit this file if needed. Instead of calculating new coordinates for a bad point, you will actually be able to navigate to the setup from which that point was collected and edit the rod height, instrument height, vertical angle, horizontal angle, or any other information that can be input directly into a data collector. This information is imported, stored, and manipulated in the survey database. The survey database is simply a Microsoft Access–formatted database file with all the information required to create the survey network, as shown in Figure 3.2.

IF IT'S JUST A DATABASE . . .

You may be tempted to think that Microsoft Access would give you an easy way to externally edit this database file. However, we advise you to *not* edit this information unless you are using the Civil 3D editing functions. These database records are not exactly named with intuitive and user-friendly names, and units may not be quite what you would expect. If edit this information in an external database program, you are doing so at your own risk!

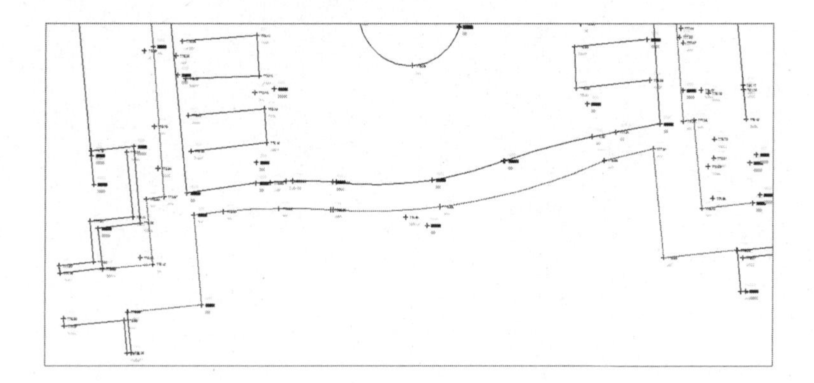

FIGURE 3.1
A portion of an as-built survey created with Civil 3D

Observation

Id	BacksightPo	StationPoin	Point	Data0	Data1	Data2	Data3	Std0	Std1	Std2	Std3	Mask	Observation	ScaleF
2	0	0	1	1524.00304801	1524.00304801	30.4800609601	-1E+20	-1E+20	-1E+20	-1E+20	-1E+20	15	56	
4	0	0	2	1584.96316993	1524.00304801	30.4800609601	-1E+20	-1E+20	-1E+20	-1E+20	-1E+20	13	56	
6	0	1	2	0	-1E+20	-1E+20	-1E+20	4.8481368E-05	-1E+20	-1E+20	-1E+20	1024	4	
9	2	1	3	0.62684469713	37.5605791212	1.57079632679	1.52400304801	-1E+20	0.00500352575	-1E+20	0.001	0	49984	
11	2	1	4	0.77263301917	42.4831089662	1.57079632679	1.52400304801	-1E+20	0.00500451030	-1E+20	0.001	0	49984	
13	2	1	5	0.89336132204	44.9733299467	1.57079632679	1.52400304801	-1E+20	0.00500505395	-1E+20	0.001	0	49984	
15	2	1	6	1.02181270685	43.6931675863	1.57079632679	1.52400304801	-1E+20	0.00500477046	-1E+20	0.001	0	49984	
17	2	1	7	1.37068463178	38.0330200660	1.57079632679	1.52400304801	-1E+20	0.00500361497	-1E+20	0.001	0	49984	
19	2	1	8	1.50889531599	35.10998222	1.57079632679	1.52400304801	-1E+20	0.00500308083	-1E+20	0.001	0	49984	
21	2	1	9	1.57281799985	29.6540513081	1.57079632679	1.52400304801	-1E+20	0.00500219792	-1E+20	0.001	0	49984	
23	2	1	10	1.57352097968	22.0340360681	1.57079632679	1.52400304801	-1E+20	0.0050012136	-1E+20	0.001	0	49984	
26	2	1	11	1.86719201888	44.0589281179	1.57079632679	1.52400304801	-1E+20	0.00500485062	-1E+20	0.001	0	49984	
29	2	1	12	2.15907409559	50.6517653035	1.57079632679	1.52400304801	-1E+20	0.00500640989	-1E+20	0.001	0	49984	
32	2	1	13	2.37035589782	60.4541529083	1.57079632679	1.52400304801	-1E+20	0.00500912843	-1E+20	0.001	0	49984	
34	2	1	14	2.38623858188	51.014478029	1.57079632679	1.52400304801	-1E+20	0.00500650196	-1E+20	0.001	0	49984	
36	2	1	15	2.37820503131	38.9047498095	1.57079632679	1.52400304801	-1E+20	0.00500378252	-1E+20	0.001	0	49984	
38	2	1	16	2.0170333833	29.8186436373	1.57079632679	1.52400304801	-1E+20	0.00500222238	-1E+20	0.001	0	49984	
40	2	1	17	2.40551943411	17.364490729	1.57079632679	1.52400304801	-1E+20	0.00500075376	-1E+20	0.001	0	49984	
42	2	1	18	2.74844269517	30.4312928626	1.57079632679	1.52400304801	-1E+20	0.00500231462	-1E+20	0.001	0	49984	
44	2	1	19	2.87887696793	44.887985776	1.57079632679	1.52400304801	-1E+20	0.00500503479	-1E+20	0.001	0	49984	

Record: 1 of 20 No Filter Search

FIGURE 3.2
A sample of the data stored in a typical survey database

Because this survey database file is located external to the drawing, it can be used simultaneously in multiple drawings, even if those drawings have different coordinate systems. The coordinate system information is set in the survey database settings and will automatically translate to any coordinate system set up in the drawing settings. This external database requires you to treat the survey database a bit differently than you would other aspects of Civil 3D. For example, many settings will reside in the survey database settings and not in the drawing template as is common for other Civil 3D settings. Other settings will reside in the Survey User Settings dialog, as shown in Figure 3.3.

I'VE CREATED A DATABASE FOR PRACTICE, NOW HOW DO I DELETE IT?

One small issue with the external database is that there is no way to delete it from within the program. This database is stored in `C:\Civil 3D Projects` by default (or your working folder if you're using Vault). The database is stored in a subfolder inside that working folder that has the same name as the database. To delete a database, you will be required to use Windows Explorer to delete the folder. Refreshing the database listing on the Survey tab will update the view and remove the deleted database.

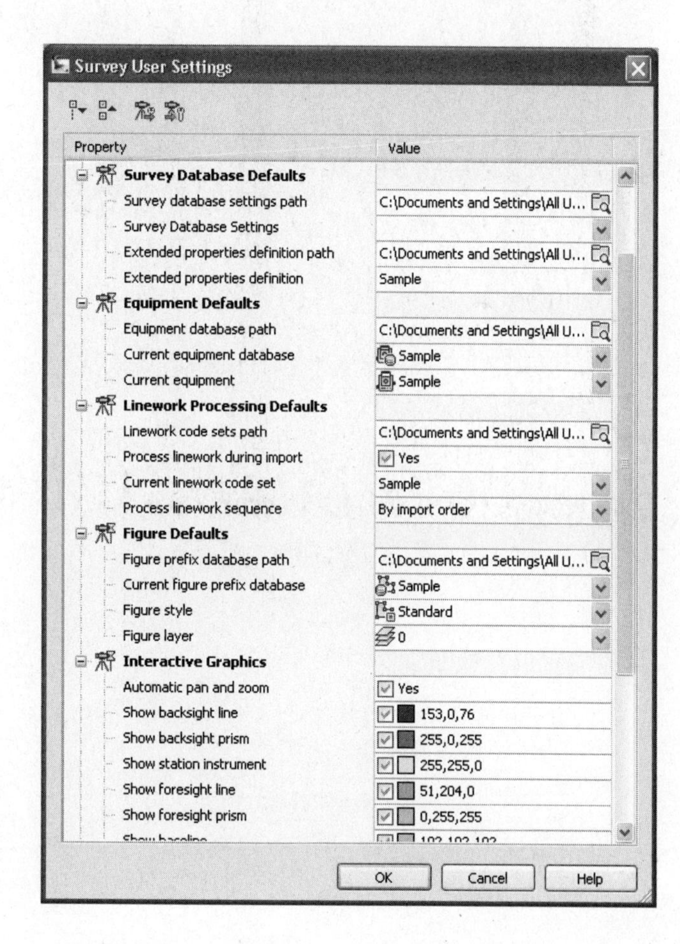

FIGURE 3.3
The Survey User Settings dialog

The settings in this dialog control many of the default choices for creating survey objects, much as Command settings do on the Settings tab of Civil 3D. You'll look at some in this exercise:

1. Create a new drawing by selecting the Application Menu ➤ New and picking the `_AutoCAD Civil 3D (Imperial) NCS.dwt` template file.
2. On the Survey tab in Toolspace, click the Edit Survey Settings icon in the upper-left corner, as shown in Figure 3.4. If the Survey tab is not available, change to the Home tab and click the Survey Toolspace button on the Palettes panel.
3. The Survey User Settings dialog opens. Look through the options and settings and observe all the defaults that can be chosen.

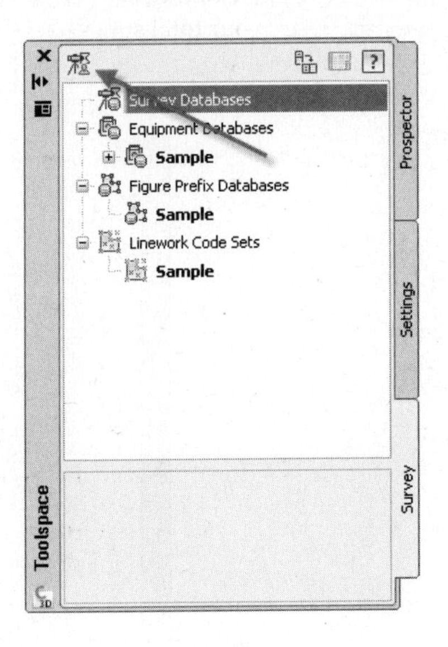

FIGURE 3.4
The Edit Survey Settings button in the upper left of the Survey tab in Toolspace

4. Click Cancel to dismiss the dialog without saving any changes.

Now that you've looked at the settings, let's get into the databases behind the scenes.

Databases Everywhere!

The Survey tab of Toolspace contains four different types of databases. The first is the survey database, where your survey networks are created and stored. This database contains the survey data that you can import into the program from the field. The equipment database is where you create replicas of your existing survey equipment for use when performing traverse analyses. The figure prefix database is where you enter your figure prefixes for use when you import a field book with linework into the program. The linework code sets database stores line and curve segment codes as well as coding methods and any special codes used in data collection. In this section, you'll explore the equipment database and the figure prefix database. Later you'll create and work with a survey database.

The Equipment Database

The equipment database is where you set up the different types of survey equipment that you are using in the field. This will allow you to apply the proper correction factors to your traverse analyses when it comes time to balance your traverse. Civil 3D comes with a sample piece of equipment for you to inspect to see what information you will need when it comes time to create your equipment. The Equipment Properties dialog (see Figure 3.5) provides all the default settings for the sample equipment in the equipment database. Expand the Equipment Databases ➤ Sample branches, right-click Sample, and select Properties to access this dialog in Toolspace. You will want to create your own equipment entries and enter the specifications for your particular total station. If you are unsure of the settings to enter, refer to the user documentation that you received when you purchased your total station.

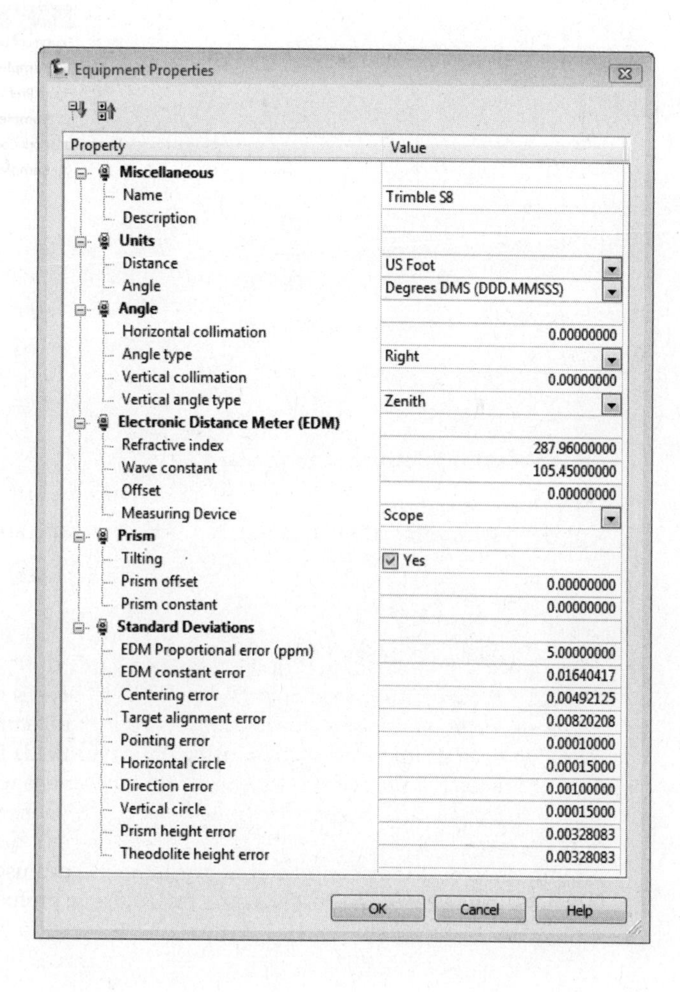

FIGURE 3.5
The Equipment Properties dialog

The Figure Prefix Database

Figures are created by codes entered into the data collector during field collection. Figures can have line segments, curve segments, elevations, and linetypes or colors assigned to them. For example, your field crew may locate a water line in the field and have it come in as line segments colored blue with a waterline linetype, or they could locate the back of a curb and display it with a solid line and a concrete style. They could set the back of this curb to come in as a breakline for surface creation. Figure 3.6 shows the Figure Prefixes Editor in Panorama.

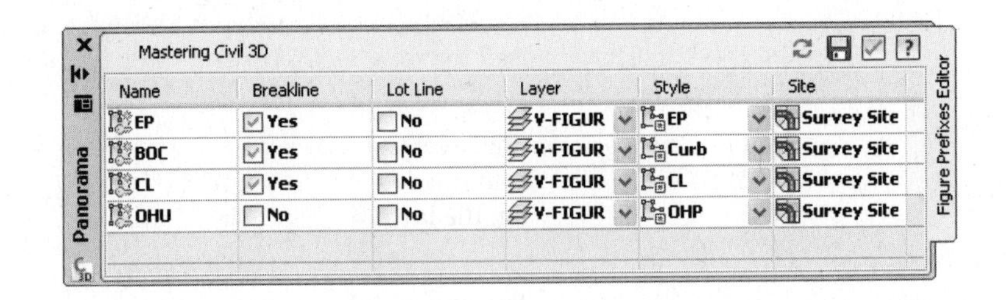

FIGURE 3.6
The Figure Prefixes Editor in Panorama

You'll need to have some styles ready to go when you create your figures, so you'll build a few now:

1. Open the drawing `figure styles.dwg`. (All drawings in this chapter can be downloaded from `www.sybex.com/go/masteringcivil3d2010`.)
2. In Toolspace, click the Settings tab.
3. Expand the Survey branch.
4. Right-click Figure Styles and select New. The Figure Style dialog opens.
5. On the Information tab, enter **CL** (for Centerline) in the Name text box. This is the name of the new style you are creating.
6. Switch to the Display tab, and make sure that Figure Lines is the only Component Type that is visible. Turn off any other ones that are visible.
7. Set the Layer for the Figure Lines component to V-FIGURE by clicking in the Layer cell. The Layer Selection dialog appears, with the available layers on a drop-down list. Because V-FIGURE is not part of the template, you will have to create it. Click New to open the Create Layer dialog. Enter **V-FIGURE** as the name of your new layer. Layer settings will be unimportant for this layer, because the properties will be assigned by the style. Click OK to dismiss the Create Layer dialog, and click OK again to dismiss the Layer Selection dialog.
8. Set the Color for the Figure Lines component to Red and the Linetype to CENTER2.
9. Create additional figure styles with the following names: **PROP**, **CONC**, **WATER**, **CATV**, **GAS**, **EP**, **BREAK**, **BLDG**, and **OHP**. Apply different colors and linetypes to each figure style as desired. Save the drawing — you will use it in the next exercise.
10. Click OK to dismiss the Figure Style dialog.

FIGURE SETTINGS

The following six settings can be specified for each figure prefix:

Name This specifies the name of the figure prefix. The figure prefix is used when you import a field book into a survey database network.

Breakline This specifies whether the figure should be used as a breakline during surface creation. These figures will have elevation data, allowing them to be selected manually in the breakline section of the surface definition.

Lot Line This specifies whether the figure should behave as a parcel segment. When toggled on, these figures are used to create parcels when inserted into a drawing and can be labeled as parcel segments.

Layer This specifies the layer that the figure will reside on when inserted into the drawing. If the layer already exists in the drawing, the figure will be placed on that layer. If the layer does not exist in the drawing, the layer will be created and the figure placed on the newly created layer.

Style This specifies the style to be used for each figure. The style contains the linetype and color of the figure. If the style exists in the current drawing, the figure will be inserted using that style. If the style does not exist in the drawing, the style will be created with the standard settings and the figure will be placed in the drawing according to the settings specified with the newly created style.

Site This specifies which site the figures should reside on when inserted into the drawing. As with previous settings, if the site exists in the drawing, the figure will be inserted into that site. If the site does not exist in the drawing, a site will be created with that name and the figure will be inserted into the newly created site. Because a figure can create parcels and surface breaklines, this functionality can be used to place those objects on the correct site.

Next you'll look at these settings in a practical exercise. You'll use the styles created in the previous exercise:

1. Open the drawing `figure prefix library.dwg`.
2. In the Survey tab of Toolspace, right-click Figure Prefix Databases and select New. The New Figure Prefix Database dialog opens.
3. Enter **Mastering Civil 3D** in the Name text box, and click OK to dismiss the dialog. Mastering Civil 3D will now be listed under the Figure Prefix Databases branch in the Survey tab on Toolspace.
4. Right-click the newly created Mastering Civil 3D figure prefix database and select Edit. The Figure Prefixes Editor in Panorama will appear.
5. Right-click the grid in the Figure Prefixes Editor and select New.
6. Click the SAMPLE name and rename the figure prefix **EP** (for Edge of Pavement).
7. Check the No box in the Breakline column to change it to Yes so the figure will be treated as a breakline. Leave the box in the Lot Line column unchecked, so the figure will not be treated as a parcel segment.

8. Under the Layer column, select V-FIGURE.
9. Under the Style column, select EP.
10. Under the Site column, leave the name of the site set to Survey Site.
11. Complete the Figure Prefixes table with the values shown in Table 3.1.

TABLE 3.1: Figure Settings

NAME	BREAKLINE	LOT LINE	LAYER	STYLE	SITE
PROP	No	Yes	V-FIGURE	PROP	PROPERTY
WALK	Yes	No	V-FIGURE	CONC	SURVEY SITE
CL	Yes	No	V-FIGURE	CL	SURVEY SITE
WATER	No	No	V-FIGURE	WATER	SURVEY SITE
CATV	No	No	V-FIGURE	CATV	SURVEY SITE
GAS	No	No	V-FIGURE	GAS	SURVEY SITE
BOC	Yes	No	V-FIGURE	CONC	SURVEY SITE
GUT	Yes	No	V-FIGURE	BREAK	SURVEY SITE
BLDG	No	No	V-FIGURE	BLDG	SURVEY SITE
OHP	No	No	V-FIGURE	OHP	SURVEY SITE

12. Click the green check mark in the upper-right corner to dismiss Panorama. You will receive a message that you have made changes but not yet applied them. Click Yes to apply the changes.

You've finished setting up the figure prefix database — now it's time to look at actually getting some data.

The Survey Database

Once you have all the front-end settings configured, you are ready to create your survey database. This database is stored in your working folder, which can be set by your network administrator or CAD manager if you are a Vault user, or it can simply be set to `C:\Civil 3D Projects`. When a survey database is created, it creates a folder with that database name in your working folder. Inside that folder, you will find one folder for each network in your database: a `survey.sdb` file, a `survey.sdx` file, and a `survey.bak` file. The SDB file is the Microsoft Access database format file that was mentioned earlier.

THE SURVEY DATABASE SETTINGS

Each survey database has numerous settings that control the database, as shown in Figure 3.7.

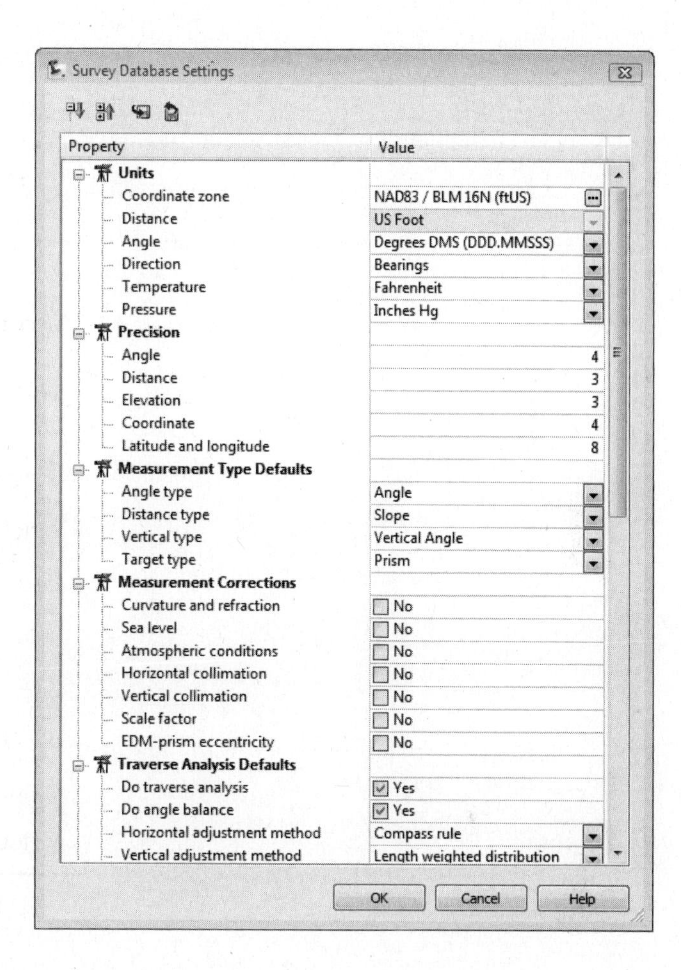

FIGURE 3.7
The Survey Database Settings dialog

The settings are as follows:

Units This is where you set your coordinate zone. It is the master coordinate zone for the database. If you insert any information in the database into a drawing with a different coordinate zone set, the program will translate that data to the drawing coordinate zone. Your coordinate zone units will lock the distance units in the Units branch. You can also set the angle, direction, temperature, and pressure units under this branch.

Precision This is where the precision information of angles, distance, elevation, coordinates, and latitude and longitude is defined and stored.

Measurement Type Defaults This is where you define the defaults for measurement types, such as angle type, distance type, vertical type, and target types.

Measurement Corrections This is used to define the methods (if any) for correcting measurements. Some data collectors allow you to make measurement corrections as you collect the data, so that needs to be verified, because double correction applications could lead to incorrect data.

Traverse Analysis Defaults This is where you choose how you perform traverse analyses and define the required precision and tolerances for each. There are four types of 2D-traverse analyses: Compass Rule, Transit Rule, Crandall Rule, and Least Squares Analysis. You can find more information and definitions of these analyses in the Civil 3D Help file.

There are also two types of 3D-traverse analyses: Length Weighted Distribution and Equal Distribution. You can also find more information and definitions of these analyses in the Civil 3D Help file. It should be noted that a 3D Least Squares Analysis can only be performed if the 2D Least Squares Analysis is also selected.

Least Squares Analysis Defaults This is where you set the defaults for a least squares analysis. You only need to change settings here if least squares analysis is the method you will use for your horizontal and/or vertical adjustments.

Survey Command Window This is the interface for manual survey tasks and for running survey batch files. This is where the default settings are defined for the Survey Command window interface.

Error Tolerance This is where tolerances are set for the survey database. If you perform an observation more than one time and the tolerances set within are not met, an error will appear in the Survey Command window and ask you what action you want to take.

Extended Properties This defines settings for adding extended properties to a survey LandXML file. This is useful for certain types of surveys, including Federal Aviation Administration (FAA)–certified surveys.

Next, you'll create a survey database and explore the database options before taking a look at the components of the database:

1. Open the drawing `survey database.dwg`.
2. On the Survey tab of Toolspace, right-click Survey Databases and select New Local Survey Database. The New Local Survey Database dialog opens.
3. Enter **Mastering Survey** in the text box and click the OK button. The Mastering Survey database will be created under the Survey Databases branch in the Survey tab of Toolspace.
4. Right-click your newly created survey database and select Edit Survey Database Settings. The Survey Database Settings dialog opens.
5. Browse through the various settings.
6. Click Cancel to dismiss the Survey Database Settings dialog.

SURVEY DATABASE COMPONENTS

Once you create a survey database and open it in the Survey tab of Toolspace, you will find that it contains seven components:

Import Events Import Events provide a framework for viewing and editing specific survey data, and they are created each time you import data into a survey database. The default name for the import event is the same as the imported filename (***`filename.ext`***). The Import Event collection contains the Networks, Figures, and Survey Points that are referenced from a specific import command, and provides an easy way to remove, re-import, and reprocess survey data in the current drawing.

Networks A survey network is a collection of connected data that is collected in the field. The network consists of setups (or stations), control points, non-control points, known directions, observations, setups, and traverses. This data is typically imported through a field book (also known as an FBK file) that is converted from raw data from the data collector, although it can also be imported through a LandXML file, a point file, or points from the drawing. A network must be created in a survey database before any data can be imported. A survey database can have multiple networks. For example, you can use different networks for different phases of a project. If working within the same coordinate zone, some users have used one survey database with many networks for every survey that they perform. This is possible because individual networks can be inserted into the drawing simply by dragging and dropping the network from the Survey tab in Toolspace into the drawing. You can hover your cursor over any Survey Network component in the drawing to see information about that component and the survey network. You can also right-click any component of the network and browse to the observation entry for that component.

Network Groups Network groups are collections of various survey networks within a survey database. These groups can be created to facilitate inserting multiple networks into a drawing at once simply by dragging and dropping.

Figures As discussed earlier, figures are linework created by codes and commands entered into the raw data file during data collection. These figures typically come from the descriptor or description of a point. A typical way of creating figures is illustrated in the following entry from an FBK file:

```
BEG EC
NE SS 3202 10040.90000 10899.21000 793.75000 "EC"
CONT EC
C3
NE SS 3203 10056.53000 10899.45000 793.47000 "EC"
NE SS 3204 10058.52000 10897.42000 793.53000 "EC"
NE SS 3205 10055.94000 10895.19000 793.61500 "EC"
CONT EC
C3
NE SS 3206 10040.43000 10895.11000 793.61500 "EC"
NE SS 3207 10038.18000 10896.59000 793.58000 "EC"
NE SS 3208 10040.47000 10899.34000 793.50500 "EC"
END
```

This FBK entry is going to create a figure named EC, which comes from the point description. This particular figure will have one line segment, and then a compound curve will be created from two three-point curves. The `BEG EC`, `CONT EC`, `C3`, and `END` commands tell the program how to draw the figure. Figures can be 2D or 3D (that is, they can be used simply for horizontal geometry or as breaklines for vertical geometry), can be used as parcel segments, and can be displayed in any of the same ways that a regular polyline can. The behavior of these figures is controlled by the Figure Prefix Library, which matches point descriptors with figure styles and displays them appropriately. Figures can be added to a drawing by dragging and dropping them into the drawing from the Survey tab in Toolspace.

Figure Groups Similar to network groups, figure groups are collections of individual figures. These groups can be created to facilitate quick insertion of multiple figures into a drawing.

Survey Points One of the most basic components of a survey database, points form the basis for each and every survey. Survey points look just like regular Civil 3D point objects, and their visibility can be controlled just as easily. However, one major difference is that a survey point cannot be edited within a drawing. Survey points are locked by the survey database, and the only way of editing is to edit the observation that collected the data for the point. This provides the surveyor with the confidence that points will not be accidentally erased or edited. Like figures, survey points can be inserted into a drawing by either dragging and dropping from the Survey tab of Toolspace or by right-clicking Surveying Points and selecting the Points ➤ Insert Into Drawing option.

Survey Point Groups Just like network groups and figure groups, survey point groups are collections of points that can be easily inserted into a drawing. When these survey point groups are inserted into the drawing, a Civil 3D point group is created with the same name as the survey point group. This point group can be used to control the visibility or display properties of each point in the group.

The Linework Code Sets Database

A linework code set (see Figure 3.8) interprets the meaning of field codes entered into a data collector by a survey crew. For example, the `B` code that is typically used to begin a line can be replaced by a code of your choosing, a decimal (`.`) can be used for a right-turn value, and a minus sign (`-`) can be used for a left-turn value. Linework code sets allow a survey crew to customize their data collection techniques based upon methods used by various types of software not related to Civil 3D.

This manner of data collection is new as of Civil 3D 2010 — it was not available in previous releases of Civil 3D, Land Desktop, Softdesk, or DCA. It allows for an incredible amount of flexibility in the field. For example, two consecutive points described in the field as follows could not have been interpreted properly previously (a figure would not be drawn):

```
FC1 VA 313 279.571080 61.851 89.414283 "BLDG9 B"
FC1 VA 314 228.405718 51.821 89.381046 "BLDG9 RT 100 50 25 -25 50 25 CLS"
```

However the same information interpreted by Civil 3D 2010 will yield a closed figure, as shown in Figure 3.9.

In the following example, you'll create an import event, and import an ASCII file with survey data. The survey data includes linework.

1. Create a new drawing from the `_AutoCAD Civil 3D (Imperial) NCS.dwt` template file.
2. Navigate to the Survey tab of Toolspace.
3. Right-click Survey Databases and select New Local Survey Database. The New Local Survey Database dialog opens.
4. Enter **Roadway** as the name of the folder in which your new database will be stored.
5. Click OK to dismiss the dialog. The Roadway survey database is created as a branch under the Survey Databases branch.
6. Expand the Roadway survey database, right-click Import Events, and select Import survey data to open the Import Survey Data dialog.
7. Select the Roadway survey database and select Next.

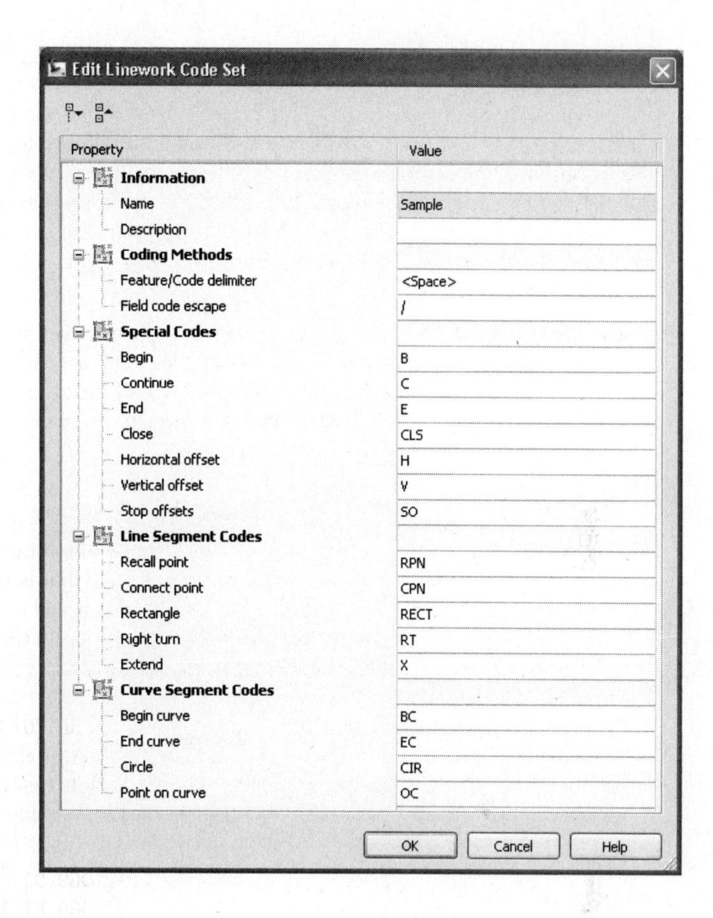

FIGURE 3.8
The Edit Linework Code Set dialog

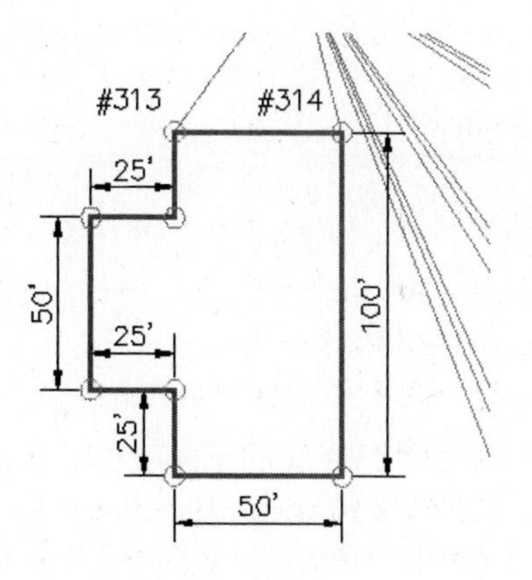

FIGURE 3.9
A closed figure interpreted properly

8. Select the Point File radio button, and specify the PNEZD (comma delimited) point file format as shown in Figure 3.10.

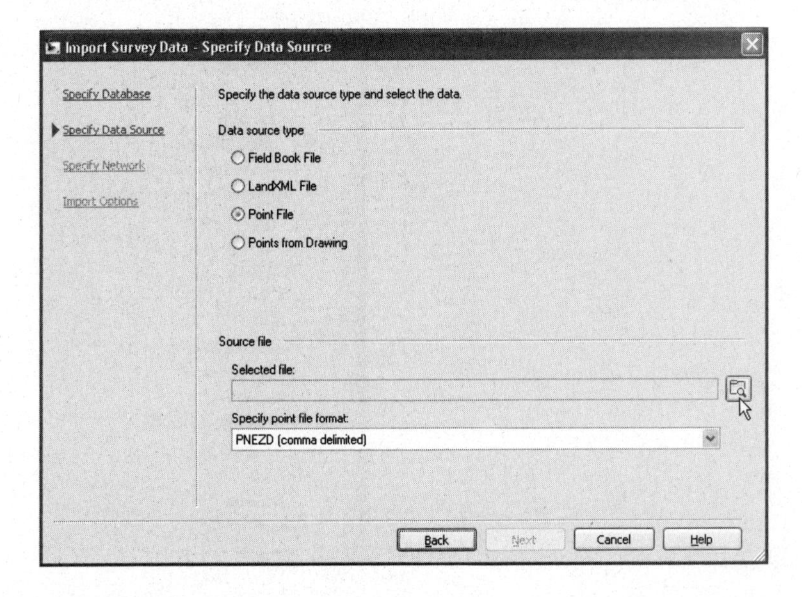

FIGURE 3.10
Selecting the appropriate file and file type from the Import Survey Data Wizard

9. Click the Browse button on the right side of the Selected File text box, and browse to the `import points.txt` file.
10. Click OK and the name will be listed in the dialog as shown in Figure 3.10. Select Next to open the Import Survey Data – Specify Network dialog.
11. Click the Create New Network button to open the New Network dialog.
12. Expand the Network property dialog, and enter **Roadway** as the Name of the new network. Click OK to exit the dialog, and click Next to open the Import Survey Data – Import Options dialog as shown in Figure 3.11.
13. Set the options as shown in Figure 3.11 and click the Finish button to import the data.

The data is imported and the linework is drawn; however, the building is missing the left side. The following steps will resolve the issue:

1. In the Survey tab, select Survey Databases ➢ Roadway ➢ Networks ➢ Roadway ➢ Non-Control points.
2. Right-click and select Edit to bring up the Non-Control Points Editor palette in Panorama.
3. Scroll to the bottom of the point list and notice the last line in the file describing point number 34.
4. Move your cursor to the left of the description and type **CLS** as shown in Figure 3.12. This is the default figure command "close."
5. Click the check box in the upper right of the palette to apply changes and save your edits. A warning dialog will appear. Click Yes to apply your changes.

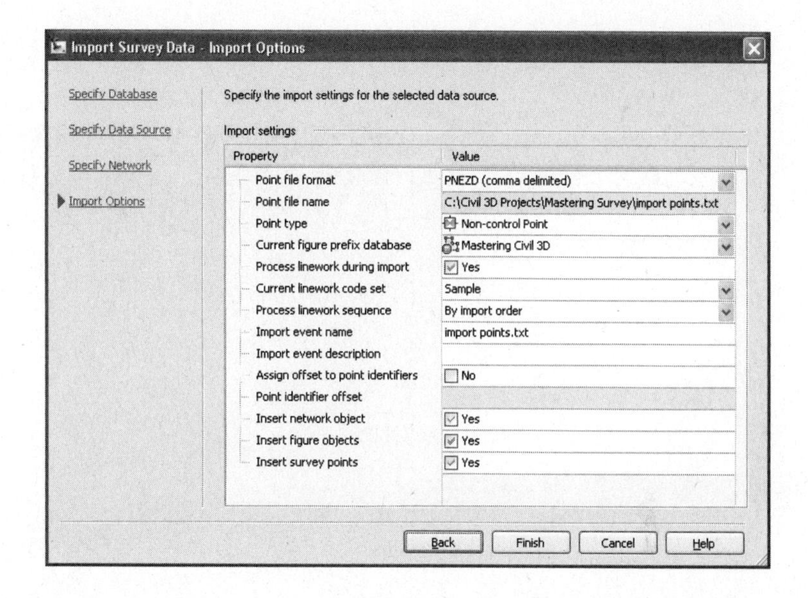

FIGURE 3.11
The Import Options dialog in the Import Survey Data Wizard

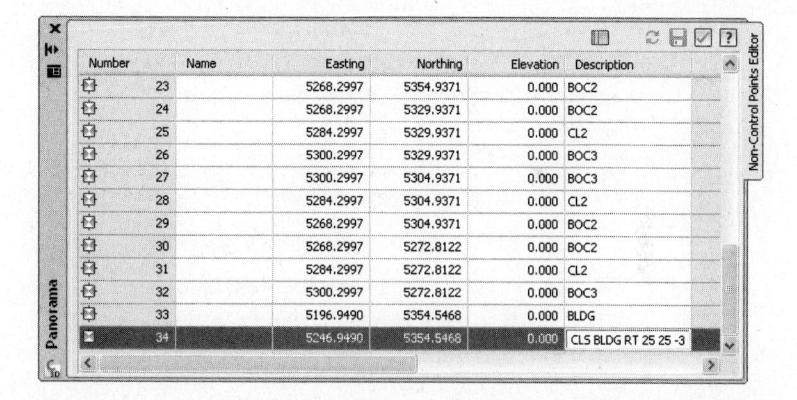

FIGURE 3.12
Editing `import points.txt` to add the CLS command and close the building geometry

6. Select Survey Databases ➤ Roadway ➤ Import Events ➤ `import points.txt`. Right-click `import points.txt`, and select Process Linework to bring up the Process Linework dialog.

7. Click OK to reprocess the linework with your updated point description. The building figure line and your drawing should look something like Figure 3.13.

EDITING THE SURVEYED POINT?

Many surveyors will cringe at this new ability to easily modify the drawing data without updating the source files from the field survey. We think this is an improvement in that you can create drawings that accurately reflect personal field observations without modifying the legal record that is the original survey data.

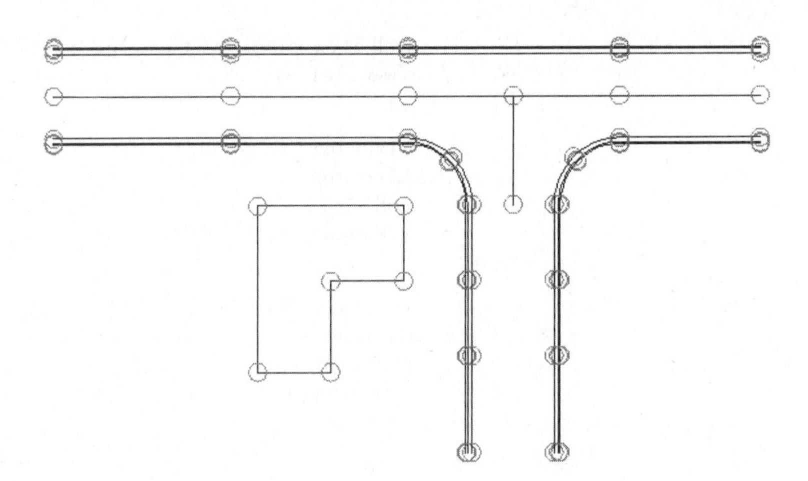

FIGURE 3.13 After editing and reprocessing the linework

Creating a Field Book

Once the data is collected and in its native format, you have to get the data into a format that can be imported into a survey database. Most users are accustomed to utilizing a simple ASCII or TXT file and importing the points from that file format, but the survey database requires a field book (FBK) file to be imported. This field book file is a simple text file that contains the survey information and the coding commands from the fieldwork, and it is most often converted from the raw data file in a data collector. How this data is converted really depends on the data collector being used — it is a common file type, but each data collector manufacturer writes their files a different way. The following paragraphs will help explain some of the more common data collectors and the methods used to convert their data into an FBK file.

Trimble Often considered the survey industry leader, Trimble data collectors are becoming more popular every day. Trimble has two major offerings in the data collector market: the Trimble TCS Survey Controller and the Trimble SCS900 Site Controller. Both of these data collectors interact directly with Civil 3D via a freely available download from the Trimble website called Trimble Link. This download will add a Trimble menu to Civil 3D and allow uploading and downloading of data directly using both data collectors as well as Trimble Site Vision software. Trimble Link will take a JOB file (the default file format for Trimble,) convert it to an FBK transparently, and create a new survey network in an existing database (or create a new database as well) and import the FBK into that newly created network, all in one step. You can download the software from `http://trimble.com/link.shtml`.

TDS Now a division of Trimble, Survey Pro is one of the most popular data collection software packages on the market today. Having been around for a long time, it seems as if everyone has seen some version of this software being used. The TDS RAW or RW5 file format is the format on which other data-collector manufacturers base their RAW data files. To convert from a TDS RAW data file to an FBK, users must either purchase TDS ForeSight DXM or use TDS Survey Link, which is included in your Civil 3D install. Survey Link can be initiated by typing **STARTSURVEYLINK** on the command line, or selecting Survey Data Collection Data Link from the Create Ground Data panel on the Home tab.

Leica Leica has been around for quite a while, and a lot of its older equipment is still very serviceable today. The Leica System 1200 data collectors can be integrated with Civil 3D via the use of Leica X-Change. Leica X-Change allows for the import of Leica data and conversion to the FBK format, as well as export options for Leica System 1200 collectors. You can download the software from www.leica-geosystems.com/corporate/en/downloads/lgs_page_catalog.htm?cid=239.

Microsurvey Microsurvey has two advantages not shared by any other data collector in this list: it does not need code commands to create linework, and it has the ability to directly export an FBK from the data collector. However, if users demand a conversion process, the Microsurvey RAW data file is based on TDS RAW data, and it can be converted using the included TDS Survey Link.

Carlson Like Trimble, Carlson offers a freely downloadable plugin for Civil 3D. Carlson Connect allows for direct import and export to Carlson SurvCE data collectors as well as a conversion option for drawings containing Carlson point blocks. You can download the software by going to http://update.carlsonsw.com/updates.php and selecting Carlson Connect as the product you are seeking.

Other If none of these tools appeal to you, you still have other options. The group over at CADApps in Australia makes two programs that can convert almost any raw data file into an FBK file. The first is Stringer Connect, a free download. The second is a much more full-functioned program called Stringer. This software will also work with a simple ASCII file of points to convert to an FBK. You can download the software from www.civil3dtools.com/catalog24.html.

A NOTE ABOUT DATA COLLECTORS AND ADD-ON SOFTWARE

The Trimble, Leica, and Carlson options were available with Civil 3D 2008, and they have historically been released soon after the yearly release of Civil 3D. However, as this is being written, none of these updates have been posted or even announced. In spite of this, we feel relatively confident that history will repeat itself again this year.

Working with Field Books

Once you have a survey database created, you have to create a network within that database to enable you to import survey data. As mentioned earlier, individual survey databases can be created for each survey job that you do, or you can create one overall database per coordinate zone and create individual networks for each survey job. Once you create a network, you can import either an FBK file or a survey LandXML file. This data can then be used to bring survey figures and survey points into a drawing, or it can be analyzed and adjusted. Figure 3.14 illustrates a typical network, with control points, directions, setups, and a traverse. This network also includes figures and survey points that have been derived from the network data.

CONTROL POINTS

Control points are typically points in your data that have a high confidence factor. Figure 3.15 illustrates the data related to a typical control point shown in the Preview pane of Toolspace.

NON-CONTROL POINTS

Non-control points are also stored in the survey database. Manually created points (hand-entered) or GPS-collected points, such as those located by real-time kinematic (RTK) GPS, are the only types of points that can be non-control points.

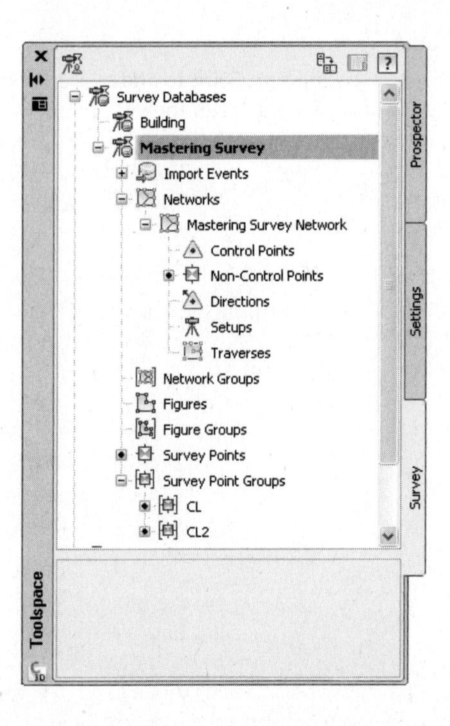

FIGURE 3.14
A typical survey database network and its data

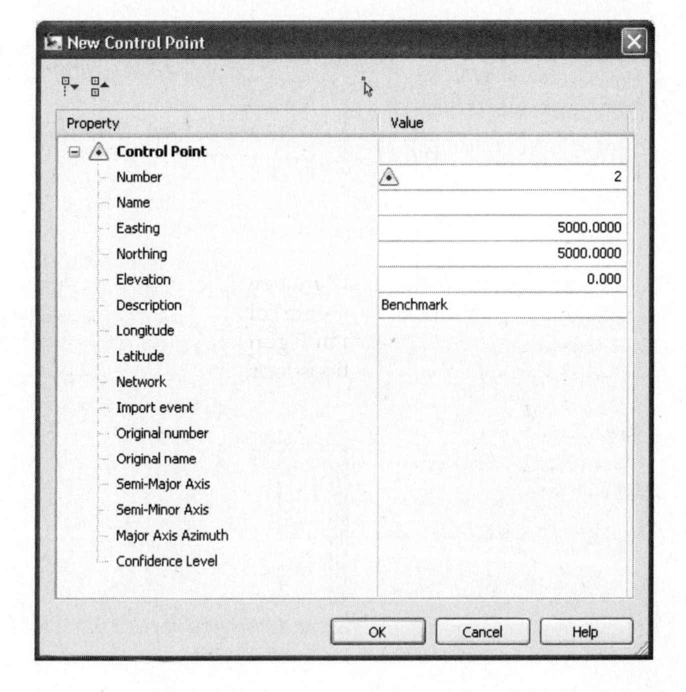

FIGURE 3.15
Control point data as shown in the Preview pane of Toolspace

Directions

The direction from one point to another must be manually entered into the data collector for the direction to show up later in the survey network for editing. The direction can be as simple as a compass shot between two initial traverse points that serves as a rough basis of bearings for a survey job. In the past, changing this direction was the easiest way to rotate a survey network, but it only worked if the information was provided in the raw data. An example of a direction is shown in Figure 3.16.

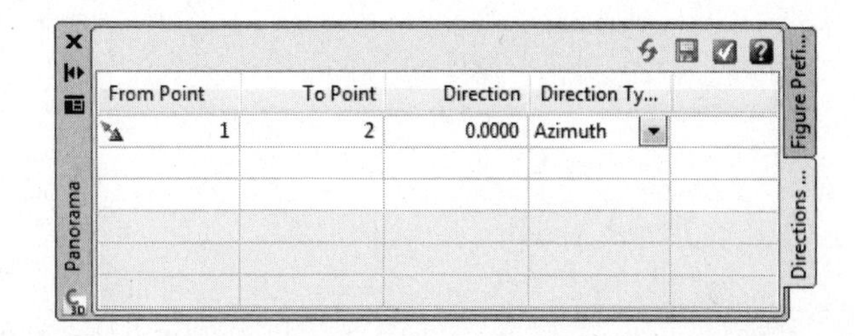

FIGURE 3.16 The direction from point 2 to point 1 is 00° 00′ 00″

Setups

The setup is typically where the meat of the data is found, especially when working with conventional survey equipment. Every setup, as well as the points (sideshots) located from that setup, can be found. Setups will contain two components: the station (or occupy point) and the backsight. Setups, as well as the observations located from the setup, can be edited. The interface for editing setups is shown in Figure 3.17. Angles and instrument heights can also be changed in this dialog.

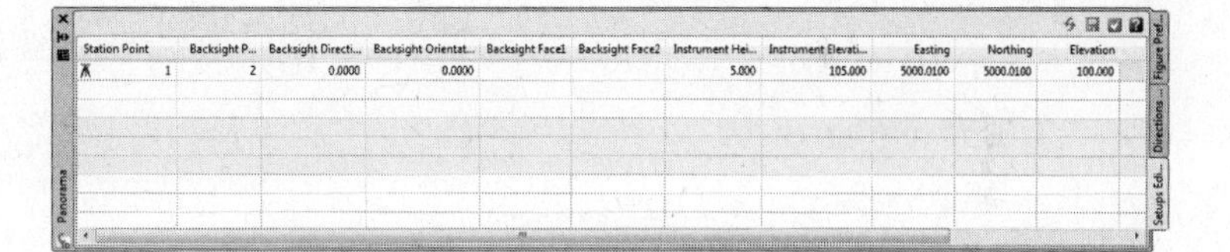

FIGURE 3.17 Setups and observations can be changed in the Setups Editor of Panorama

Traverses

The Traverses section is where new traverses are created or existing ones are edited. These traverses can come from your data collector, or they can be manually entered from field notes via the Traverse Editor, as shown in Figure 3.18. You can view or edit each setup in the Traverse Editor, as well as the traverse stations located from that setup.

Station: 1, Backsight: 2

Number	Name	Angle	Angle Type	Distance	Distance ...	Vertical	Vertical T...	Target Height	Target Type	Scale Fa...	Description
3		35.5456	Face1 Angle	123.230	Slope	90.0000	Vertical A	5.000	Prism	000000000	BOC
4		44.1607	Face1 Angle	139.380	Slope	90.0000	Vertical A	5.000	Prism	000000000	BOC
5		51.1109	Face1 Angle	147.550	Slope	90.0000	Vertical A	5.000	Prism	000000000	BOC
6		58.3244	Face1 Angle	143.350	Slope	90.0000	Vertical A	5.000	Prism	000000000	BOC
7		78.3204	Face1 Angle	124.780	Slope	90.0000	Vertical A	5.000	Prism	000000000	BOC
8		86.2712	Face1 Angle	115.190	Slope	90.0000	Vertical A	5.000	Prism	000000000	BOC
9		90.0657	Face1 Angle	97.290	Slope	90.0000	Vertical A	5.000	Prism	000000000	BOC
10		90.0922	Face1 Angle	72.290	Slope	90.0000	Vertical A	5.000	Prism	000000000	BOC
11		106.5856	Face1 Angle	144.550	Slope	90.0000	Vertical A	5.000	Prism	000000000	EP1
12		123.4221	Face1 Angle	166.180	Slope	90.0000	Vertical A	5.000	Prism	000000000	CL
13		135.4841	Face1 Angle	198.340	Slope	90.0000	Vertical A	5.000	Prism	000000000	EP2
14		148.1050	Face1 Angle	167.370	Slope	90.0000	Vertical A	5.000	Prism	000000000	EP2
15		136.1540	Face1 Angle	127.640	Slope	90.0000	Vertical A	5.000	Prism	000000000	CL

FIGURE 3.18 The Traverse Editor in Panorama

In this exercise, you'll create a traverse from an FBK file and save it for adjusting later:

1. Create a new drawing from the `_AutoCAD Civil 3D (Imperial) NCS.dwt` template file.
2. Navigate to the Survey tab of Toolspace.
3. Right-click Survey Databases and select New Local Survey Database. The New Local Survey Database dialog opens.
4. Enter **Shopping Center** as the name of the folder in which your new database will be stored.
5. Click OK to dismiss the dialog. The Shopping Center survey database is created as a branch under the Survey Databases branch.
6. Expand the Shopping Center survey database, right-click Networks, and select New. The New Network dialog opens.
7. Expand the Network branch in the dialog if needed. Name your new network **As-Built**. Enter your initials and today's date as the description. This will serve as an indication of who created the network and when.
8. Click OK. The As-Built network is now listed as a branch under the Networks branch in Prospector.
9. Right-click the As-Built network and select Import ➤ Import field book.
10. Open `shopping_center.fbk` and click OK. The Import Field Book dialog opens.
11. Make sure you have checked the boxes shown in Figure 3.19. The important settings are Show Interactive Graphics, which will allow you to see the network as it is being created, and Insert Figure Objects. Do not select Insert Survey Points at this time.

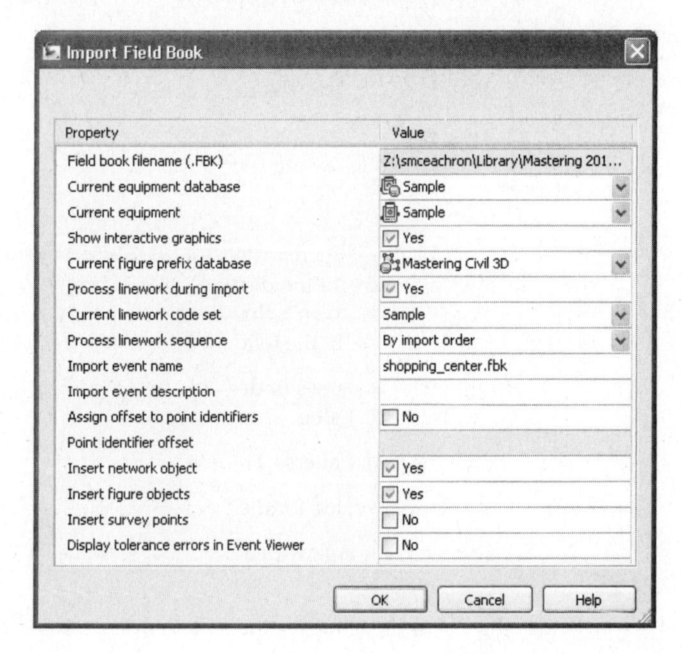

FIGURE 3.19
The Import Field Book dialog

12. Click OK to dismiss the dialog and watch your network as it is created. It may take a few minutes to import the entire network, but once it's imported, you should see figure objects in the drawing. Look at the network data within the survey database. All points in the network are listed as non-control points because the entire job was located using RTK GPS.

13. Save the drawing to your desktop as `Shopping Center.dwg` and close it.

Once you have defined a traverse, you can adjust it by right-clicking its name and selecting Traverse Analysis. You can adjust the traverse either horizontally or vertically, using a variety of methods. The traverse analysis can be written to text files to be stored, and the entire network can be adjusted on the basis of the new values of the traverse, as you'll do in the following exercise:

1. Create a new drawing from the `_AutoCAD Civil 3D (Imperial) NCS.dwt` template file.

2. Navigate to the Survey tab of Toolspace.

3. Right-click Survey Databases and select New Local Survey Database. The New Local Survey Database dialog opens.

4. Enter **Traverse** as the name of the folder in which your new database will be stored.

5. Click OK to dismiss the dialog. The Traverse survey database is created as a branch under the Survey Databases branch.

6. Expand the Traverse branch, right-click Networks, and select New. The New Network dialog opens.

7. Expand the Network branch in the dialog if needed. Name your new network **Traverse Practice**.

8. Click OK. The Traverse Practice network is now listed as a branch under the Networks branch of the Traverse survey database in Prospector.

9. Right-click the Traverse Practice network and select Import ➢ Import field book.

10. Select the `traverse.fbk` file and click OK. The Import Field Book dialog opens.

11. Make sure you have checked the boxes shown in Figure 3.20. You will be inserting the points into the drawing this time.

12. Click OK.

Inspect the data contained within the network. You have one control point — point 2 — that was manually entered into the data collector. There is one direction, and there are four setups. Each setup combines to form a closed polygonal shape that defines the traverse. Notice that there is no traverse definition. In the following exercise, you'll create that traverse definition for analysis:

1. Right-click Traverses under the Traverse Practice network and select New to open the New Traverse dialog.

2. Name the new traverse **Traverse 1**.

3. Enter **2** as the value for the Initial Station and **1** for the Initial Backsight.

   The traverse will now pick up the rest of the stations in the traverse and enter them into the next box.

4. Enter **2** as the value for the Final Foresight (the closing point for the traverse). Click OK.

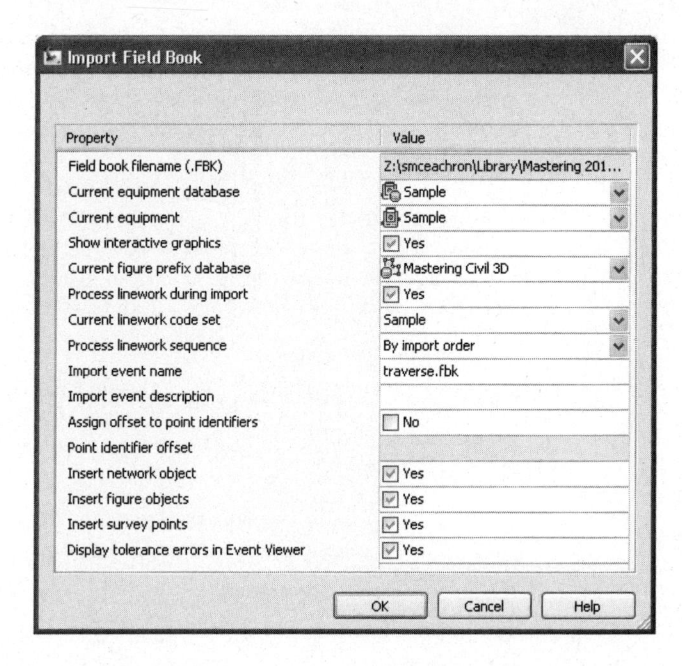

FIGURE 3.20
The Import Field Book dialog

5. Expand the Traverse branch and right-click Traverse 1. Select Traverse Analysis.
6. In the Traverse Analysis dialog, ensure that Yes is selected for Do Traverse Analysis and Do Angle Balance.
7. Select Least Squares for the Horizontal and Vertical Adjustment Method.
8. Select 30,000 for the Horizontal and Vertical Closure Limit 1:X.
9. Make sure the option Update Survey Database is set to Yes.
10. Click OK.

The analysis is performed, and four text files are displayed that show the results of the adjustment. Note that if you look back at your survey network, all points are now control points, because the analysis has upgraded all the points to control point status.

Figure 3.21 shows the results of the analysis and adjustment. Here, you can see the Elevation Error, Error North, Error East, Absolute Error, Error Direction, Perimeter (of the traverse), Number of Sides, and Area (of the traverse). You also can see that your new Precision is well within the tolerances set in step 8.

Figure 3.22 displays the results of the vertical analysis, displayed automatically upon completion of the analysis. You can see the individual points and their initial elevation, along with the adjusted elevations.

The third text file is shown in three separate portions. The first portion is shown in Figure 3.23. This portion of the file displays the various observations along with their initial measurements, standard deviations, adjusted values, and residuals. You can view other statistical data at the beginning of the file.

Figure 3.24 shows the second portion of this text file and displays the adjusted coordinates, the standard deviation of the adjusted coordinates, and information related to error ellipses displayed in the drawing. If the deviations are too high for your acceptable tolerances, you will need to redo the work or edit the field book.

FIGURE 3.21
Traverse analysis results

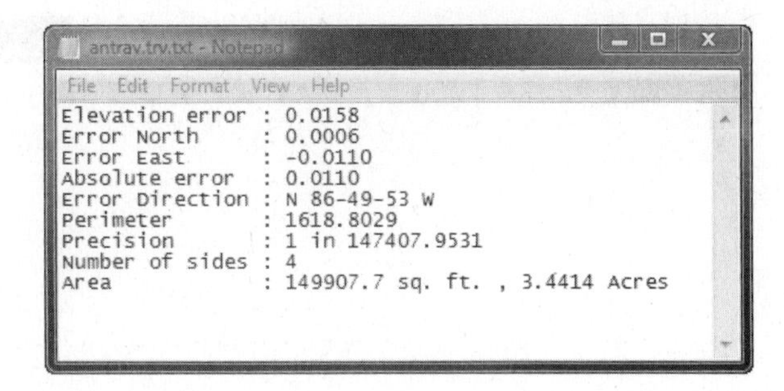

antrav.trv.txt - Notepad

File Edit Format View Help

```
Elevation error : 0.0158
Error North     : 0.0006
Error East      : -0.0110
Absolute error  : 0.0110
Error Direction : N 86-49-53 W
Perimeter       : 1618.8029
Precision       : 1 in 147407.9531
Number of sides : 4
Area            : 149907.7 sq. ft. , 3.4414 Acres
```

FIGURE 3.22
Vertical analysis results

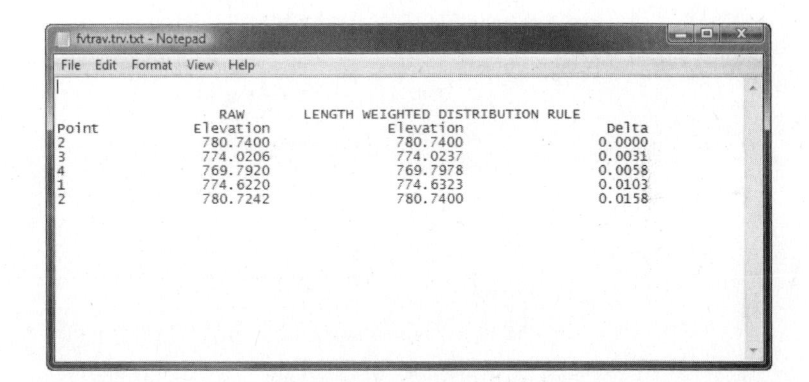

fvtrav.trv.txt - Notepad

File Edit Format View Help

Point	RAW Elevation	LENGTH WEIGHTED DISTRIBUTION RULE Elevation	Delta
2	780.7400	780.7400	0.0000
3	774.0206	774.0237	0.0031
4	769.7920	769.7978	0.0058
1	774.6220	774.6323	0.0103
2	780.7242	780.7400	0.0158

FIGURE 3.23
Statistical and observation data

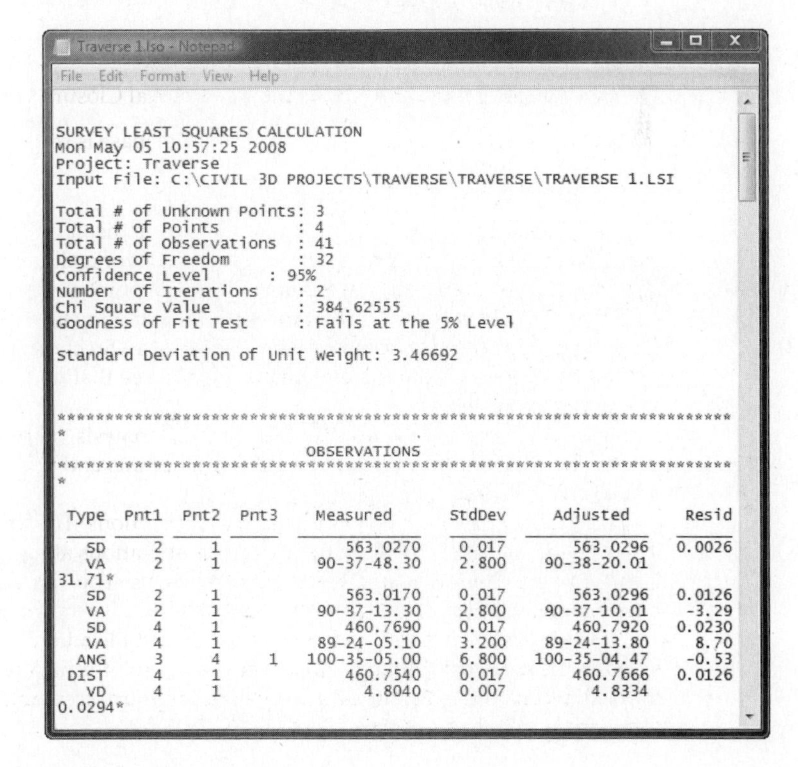

Traverse 1.lso - Notepad

File Edit Format View Help

```
SURVEY LEAST SQUARES CALCULATION
Mon May 05 10:57:25 2008
Project: Traverse
Input File: C:\CIVIL 3D PROJECTS\TRAVERSE\TRAVERSE\TRAVERSE 1.LSI

Total # of Unknown Points: 3
Total # of Points        : 4
Total # of Observations  : 41
Degrees of Freedom       : 32
Confidence Level       : 95%
Number  of Iterations    : 2
Chi Square Value         : 384.62555
Goodness of Fit Test     : Fails at the 5% Level

Standard Deviation of Unit Weight: 3.46692


*************************************************************************
*
                          OBSERVATIONS
*************************************************************************
*

 Type  Pnt1  Pnt2  Pnt3    Measured     StdDev    Adjusted      Resid
 ____  ____  ____  ____  ____________   ______  ____________   ______
   SD     2     1             563.0270   0.017       563.0296   0.0026
   VA     2     1          90-37-48.30   2.800    90-38-20.01
31.71*
   SD     2     1             563.0170   0.017       563.0296   0.0126
   VA     2     1          90-37-13.30   2.800    90-37-10.01    -3.29
   SD     4     1             460.7690   0.017       460.7920   0.0230
   VA     4     1          89-24-05.10   3.200    89-24-13.80     8.70
  ANG     3     4     1   100-35-05.00   6.800   100-35-04.47    -0.53
 DIST     4     1             460.7540   0.017       460.7666   0.0126
   VD     4     1               4.8040   0.007         4.8334
0.0294*
```

FIGURE 3.24
Adjusted coordinate information

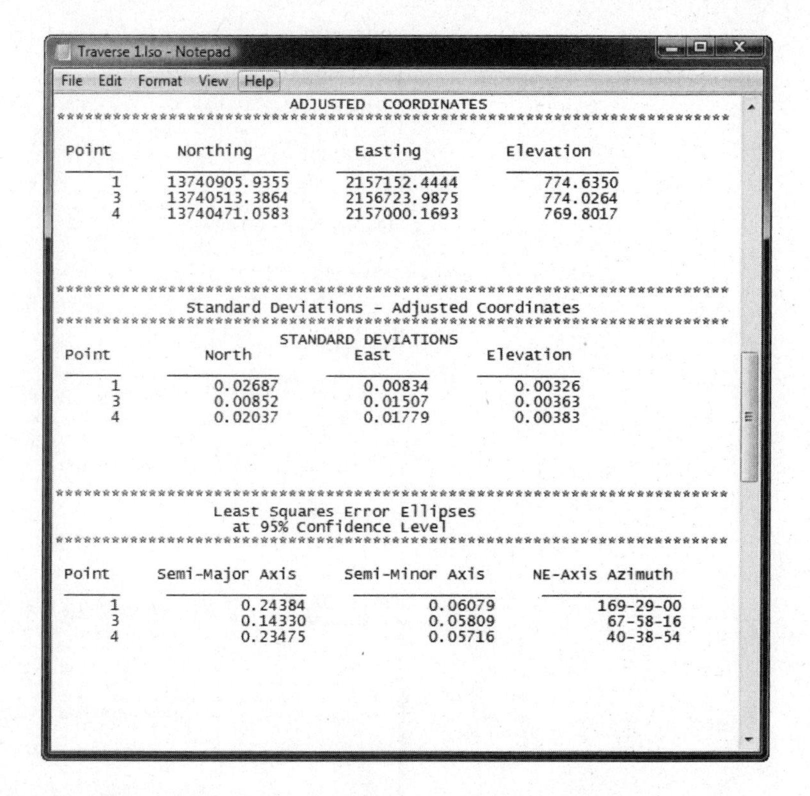
Traverse 1.lso - Notepad

File Edit Format View Help

ADJUSTED COORDINATES

Point	Northing	Easting	Elevation
1	13740905.9355	2157152.4444	774.6350
3	13740513.3864	2156723.9875	774.0264
4	13740471.0583	2157000.1693	769.8017

Standard Deviations - Adjusted Coordinates

STANDARD DEVIATIONS

Point	North	East	Elevation
1	0.02687	0.00834	0.00326
3	0.00852	0.01507	0.00363
4	0.02037	0.01779	0.00383

Least Squares Error Ellipses
at 95% Confidence Level

Point	Semi-Major Axis	Semi-Minor Axis	NE-Axis Azimuth
1	0.24384	0.06079	169-29-00
3	0.14330	0.05809	67-58-16
4	0.23475	0.05716	40-38-54

Figure 3.25 displays the final portion of this text file — the Blunder Detection and Analysis. Civil 3D will look for and analyze data in the network that is obviously wrong and choose to keep it or throw it out of the analysis if it doesn't meet your criteria. If a blunder (or bad shot) is detected, the program will not fix it. You will have to edit the data manually, whether by going out in the field and collecting the correct data or by editing the FBK file.

One more method of creating traverses is to create a traverse from figures. Originally introduced to Civil 3D 2008 as a program extension, this command allows a user to connect points with a polyline, create a figure from that polyline, and then define a traverse from the newly created figure.

1. Open the `Traverse Analysis.dwg` file. This drawing shows four points, numbered 1 through 4.

2. Select the Polyline tool to draw a polyline from point 2 to point 3 to point 4 to point 1, and then point 2. It is best to use the Point Number transparent command when creating this polyline. To do this, select the Polyline tool, click the Point Number transparent command, and enter the point number. Press ↵, and enter the next point number. Continue until your closed polyline is complete, and press Esc to exit from the transparent command. Press Esc again to exit from the Polyline command.

3. Right-click on the current survey database from the Toolspace and choose Create Figure from Object.

4. Pick the polyline connecting the points. The Create Figure from Object dialog opens.

5. Name the figure **TRAV** and be sure the Associate Survey Points to Vertices property is set to Yes. Click OK to dismiss the dialog. Press ↵ to end the command.

6. Expand the Figures branch. You will see one figure named TRAV. Drag and drop it into the current drawing.

7. Choose Survey ➤ Define Traverse Stations from Figure.

8. In the dialog that appears, set the Survey Network to Traverse from Figure, and then click Next.

9. At the bottom of the dialog, select New Traverse.

10. In the next dialog, select the TRAV figure and click Next.

11. Your Initial Station should already be filled in as Station 2. The stations should be entered into the gray area as 3,4,1,2. Enter **1** as the Initial Backsight and **2** as the Final Foresight.

12. Save the drawing and close it.

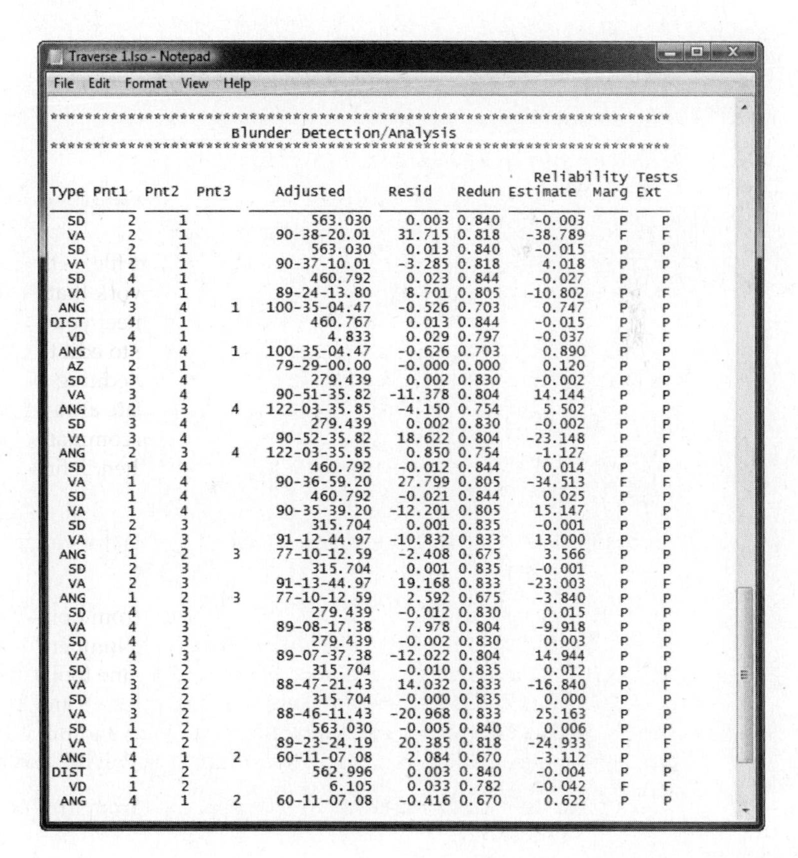

Traverse 1.lso - Notepad

File Edit Format View Help

Blunder Detection/Analysis

Type	Pnt1	Pnt2	Pnt3	Adjusted	Resid	Redun	Estimate	Reliability Tests Marg	Ext
SD	2	1		563.030	0.003	0.840	-0.003	P	P
VA	2	1		90-38-20.01	31.715	0.818	-38.789	F	F
SD	2	1		563.030	0.013	0.840	-0.015	P	P
VA	2	1		90-37-10.01	-3.285	0.818	4.018	P	P
SD	4	1		460.792	0.023	0.844	-0.027	P	P
VA	4	1		89-24-13.80	8.701	0.805	-10.802	P	F
ANG	3	4	1	100-35-04.47	-0.526	0.703	0.747	P	P
DIST	4	1		460.767	0.013	0.844	-0.015	P	P
VD	4	1		4.833	0.029	0.797	-0.037	F	F
ANG	3	4	1	100-35-04.47	-0.626	0.703	0.890	P	P
AZ	2	1		79-29-00.00	-0.000	0.000	0.120	P	P
SD	3	4		279.439	0.002	0.830	-0.002	P	P
VA	3	4		90-51-35.82	-11.378	0.804	14.144	P	P
ANG	2	3	4	122-03-35.85	-4.150	0.754	5.502	P	P
SD	3	4		279.439	0.002	0.830	-0.002	P	P
VA	3	4		90-52-35.82	18.622	0.804	-23.148	P	F
ANG	2	3	4	122-03-35.85	0.850	0.754	-1.127	P	P
SD	1	4		460.792	-0.012	0.844	0.014	P	P
VA	1	4		90-36-59.20	27.799	0.805	-34.513	F	F
SD	1	4		460.792	-0.021	0.844	0.025	P	P
VA	1	4		90-35-39.20	-12.201	0.805	15.147	P	P
SD	2	3		315.704	0.001	0.835	-0.001	P	P
VA	2	3		91-12-44.97	-10.832	0.833	13.000	P	P
ANG	1	2	3	77-10-12.59	-2.408	0.675	3.566	P	P
SD	2	3		315.704	0.001	0.835	-0.001	P	P
VA	2	3		91-13-44.97	19.168	0.833	-23.003	P	F
ANG	1	2	3	77-10-12.59	2.592	0.675	-3.840	P	P
SD	4	3		279.439	-0.012	0.830	0.015	P	P
VA	4	3		89-08-17.38	7.978	0.804	-9.918	P	P
SD	4	3		279.439	-0.002	0.830	0.003	P	P
VA	4	3		89-07-37.38	-12.022	0.804	14.944	P	P
SD	3	2		315.704	-0.010	0.835	0.012	P	P
VA	3	2		88-47-21.43	14.032	0.833	-16.840	P	F
SD	3	2		315.704	-0.000	0.835	0.000	P	P
VA	3	2		88-46-11.43	-20.968	0.833	25.163	P	P
SD	1	2		563.030	-0.005	0.840	0.006	P	P
VA	1	2		89-23-24.19	20.385	0.818	-24.933	F	F
ANG	4	1	2	60-11-07.08	2.084	0.670	-3.112	P	P
DIST	1	2		562.996	0.003	0.840	-0.004	P	P
VD	1	2		6.105	0.033	0.782	-0.042	F	F
ANG	4	1	2	60-11-07.08	-0.416	0.670	0.622	P	P

FIGURE 3.25 Blunder analysis

QUESTIONING THE RESULTS

A traverse completed using the aforementioned process will create a `Traverse 1.lsi` file. This is known as the *input file*. The input file can be edited using Windows Notepad. Opening this file reveals something like this snippet:

```
?NEZ    1 13740905.915851 2157152.441122    774.599442
NEZ    2 13740803.176500 2156598.905200   780.740000
?NEZ    3 13740513.387064 2156723.987321    774.020102
?NEZ    4 13740471.057054 2157000.167405    769.790530
```

If you remove the question mark from the front of any line, Civil 3D will then use that point as another monument when balancing the survey. This may be necessary because the initial point and the first foresight or backsight may be known local monuments with known coordinate values.

Use the following steps as a guideline to balancing a traverse via this method:

1. Create a network and import a field book using normal methods.
2. Right-click the name of the network and select Least Squares Analysis ➢ Perform Analysis.
3. Right-click the name of the network and select Least Squares Analysis ➢ Edit Input File.
4. Remove the necessary question marks in the `.lsi` file and save the file.
5. Right-click the name of the network and select Least Squares Analysis ➢ Process Input File.
6. Right-click the name of the network and select Least Squares Analysis ➢ Update Survey Database.
7. Right-click the name of the network and select Points ➢ Update.

As you can see, this process is longer than the simple traverse adjustment of the prior exercise, but it allows greater flexibility. It's interesting to note that when you use this methodology, a traverse does not need to be defined.

OTHER METHODS OF MANIPULATING SURVEY DATA

Often, it is necessary to edit the entire survey network at one time. For example, rotating a network to a known bearing or azimuth from an assumed one happens quite frequently. However, unless directions are defined in the field book file, changing that rotation is difficult. Previously, you would have to calculate several coordinates to make a rotation and/or translation. However, if you change to the Modify tab and choose Survey from the Ground Data panel, you'll open a Survey tab. On the Survey tab, choose Translate Database from the Modify panel to manipulate the location of a network.

MANIPULATING THE NETWORK

Translating a survey network can move a network from an assumed coordinate system to a known coordinate system, it can rotate a network, and it can adjust a network from assumed elevations to a known datum.

1. Create a new drawing from the NCS `Extended Imperial template` file.
2. In the Survey tab of Toolspace, right-click and select New Local Survey Database. The New Local Survey Database dialog opens. Enter **Translate** in the text box. This is the name of the folder for the new database.
3. Click OK, and the Translate database will now be listed under the Survey Databases branch on the Survey tab.
4. Select Networks under the new Translate branch. Right-click and select New to open the New Network dialog. Enter **Translate** as the name of this new network. Click OK to dismiss the dialog.
5. Right-click the Translate network, and select Import ➤ Import Field Book.
6. Navigate to the `traverse.fbk` file, and click Open. The Import Field Book dialog opens. Select OK to accept the default options.
7. Update the Point Groups collection in Prospector so that the points show up in the drawing if needed.
8. Draw an orthogonal polyline directly to the north of point 3. It can be any length, but be sure to use the object snap "node."
9. Change to the Modify tab and choose Survey from the Ground Data panel to open the Survey tab.
10. Change to the Survey tab and choose Translate Database from the Modify panel drop-down menu.
11. For the purposes of this exercise, leave the points on their same coordinate system, but change the point 3-point 2 bearing to due north. Elevations will remain unchanged.
12. In the first window, type **3** as the Number. This is the Base Point number (the number that you will be rotating the points around). Click Next.
13. In the next window, click the Pick in Drawing button in the lower-left corner to specify the new angle.
14. Using osnaps, pick point 3 and then point 2 for your Reference Angle.
15. Pick point 3 and then somewhere along the orthogonal polyline for the second point to define the new angle. The new angle should be just over 23 degrees. Click Next.
16. In the next window, click the Pick in Drawing button on the lower left to pick point 3 as the Destination Point. This will essentially negate any translation features and provide you with only a rotation.
17. Leave the Elevation Change box empty. If you were raising or lowering the elevations of the network, this box is where you would enter the change value.
18. Click Next to review your results, and Finish to complete the translation.
19. Go back to the drawing and inspect your points. Point 2 should now be due north of point 3.
20. Close the drawing without saving. Even though you did not save the drawing, the changes were made directly to the database and the network can be imported into any drawing.

Other Survey Features

Two other components of the survey functionality included with Civil 3D 2010 are the Astronomic Direction Calculator and the Geodetic Calculator. The Astronomic Direction Calculator, shown in Figure 3.26, is used to calculate sun shots or star shots.

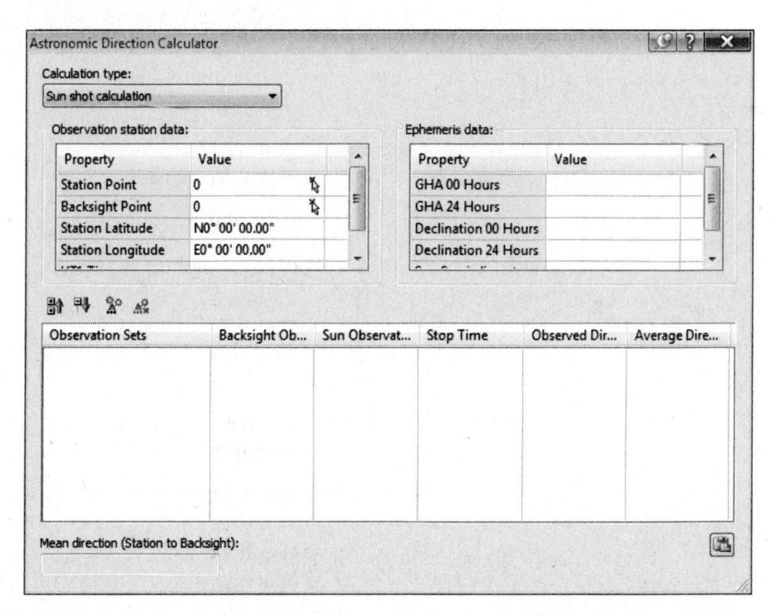

FIGURE 3.26
The Astronomic Direction Calculator

The Geodetic Calculator is used to calculate and display the latitude and longitude of a selected point, as well as their local and grid coordinates. It can also be used to calculate unknown points. If you know the grid coordinates, the local coordinates, or the latitude and longitude of a point, you can enter it into the Geodetic Calculator and create a point at that location. Note that the Geodetic Calculator only works if a coordinate system is assigned to the drawing in the Drawing Settings dialog. In addition, any transformation settings specified in this dialog will be reflected in the Geodetic Calculator, shown in Figure 3.27.

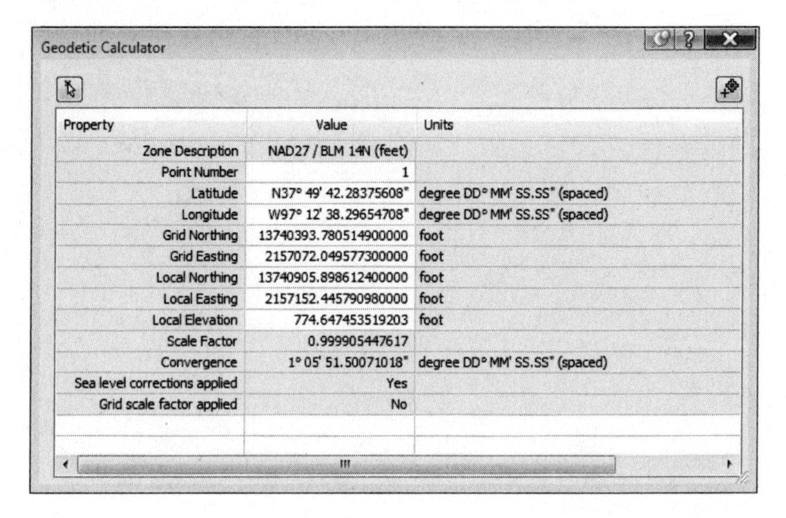

FIGURE 3.27
The Geodetic Calculator

The Bottom Line

Properly collect field data and import it into AutoCAD Civil 3D 2010. You learned best practices for collecting data, how the data is translated into a usable format for the survey database, and how to import that data into a survey database. You learned what commands draw linework in a raw data file, and how to include those commands into your data collection techniques so that the linework is created correctly when the field book is imported into the program.

Master It In this exercise, you'll create a new drawing and a new survey database and import the `Shopping_Center.fbk` file into the drawing.

Set up styles that will correctly display your linework. You learned how to set up styles for figures. You also learned that figures can be set as breaklines for surface creation and lot lines and that they can go on their own layer and be displayed in many different ways.

Master It In this exercise, you'll use the `Mastering1.dwg` file and survey database from the previous exercise and create figure styles and a figure prefix database for the various figures in the database.

Create and edit field book files. You learned how to create field book files using various data collection techniques and how to import the data into a survey database.

Master It In this exercise, you'll create a new drawing and survey database. Open the `field notes.pdf` file and use a data collector (data collector emulators can be downloaded from the websites of the data collector manufacturers) to input the field notes. Export the raw data, convert it into an FBK file, and import it into the new survey database.

Manipulate your survey data. You learned how to use the traverse analysis and adjustments to create data with a higher precision.

Master It In this exercise, you'll use the survey database and network from the previous exercises in this chapter. You'll analyze and adjust the traverse using the following criteria:

- Use the Compass Rule for Horizontal Adjustment.
- Use the Length Weighted Distribution Method for Vertical Adjustment.
- Use a Horizontal Closure Limit of 1:25,000.
- Use a Vertical Closure Limit of 1:25,000.

Chapter 4

X Marks the Spot: Points

The foundation of any civil engineering project is the simple point. Most commonly, points are used to identify the location of existing features, such as trees and property corners; topography, such as ground shots; or stakeout information, such as road geometry points. However, points can be used for much more. This chapter will both focus on traditional point uses and introduce ideas to apply the dynamic power of point editing, labeling, and grouping to other applications.

By the end of this chapter, you'll learn to:

- Import points from a text file using description key matching
- Create a point group
- Export points to LandXML and ASCII format
- Create a point table

Anatomy of a Point

Civil 3D *points* (see Figure 4.1) are intelligent objects that represent *x*, *y*, and *z* locations in space. Each point has a unique number and, optionally, a unique name that can be used for additional identification and labeling.

You can view and change point properties in the AutoCAD Properties palette, as shown in Figure 4.2.

Creating Basic Points

You can create points many ways, using the Points menu from the Create Ground Data panel on the Home tab. Points can also be imported from text files or external databases or converted from AutoCAD, Land Desktop, or Softdesk point objects.

Point Settings

Before you begin creating points, it's important to investigate which settings may make the task easier. Individual point objects are placed on an object layer that controls their visibility, which you can manage in both the Command settings and the Create Points dialogs. You can also establish default style, default label style, and default elevations, names, and descriptions in your Civil 3D template, which will make it easy to follow your company standard when creating points.

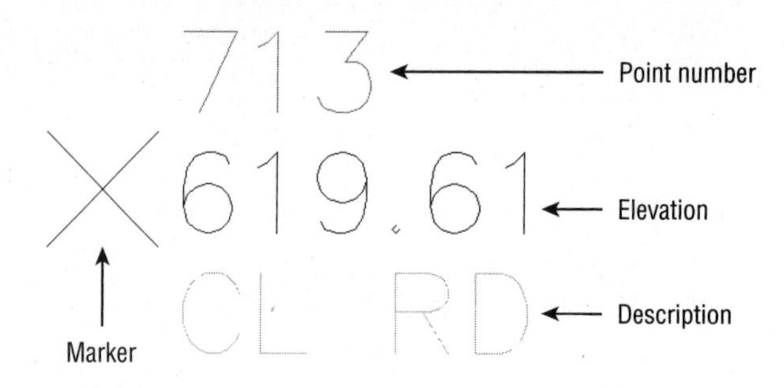

FIGURE 4.1
A typical point object showing a marker, a point number, an elevation, and a description

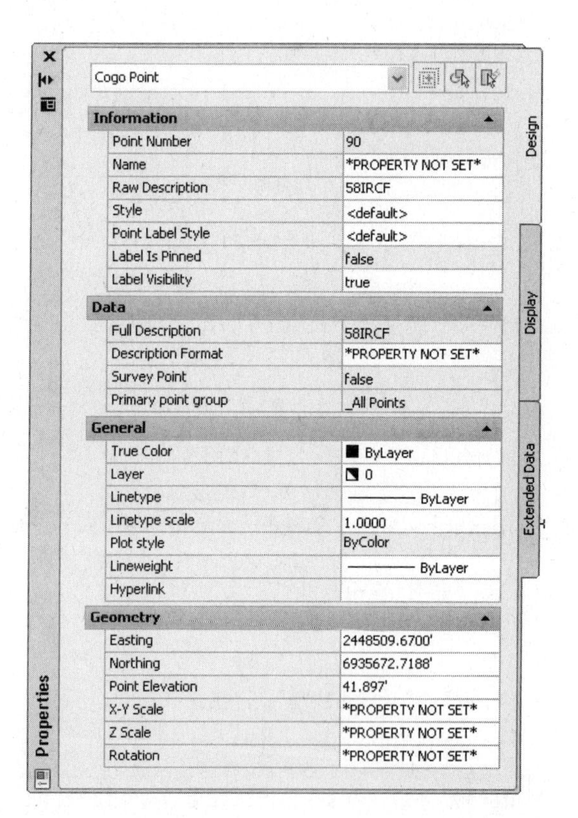

FIGURE 4.2
The AutoCAD Properties palette

DEFAULT LAYER

For most Civil 3D objects, the object layer is established in the Drawing Settings. In the case of points, the default object layer is set in the Command Settings for point creation and can be overridden in the Create Points dialog (see Figure 4.3).

PROMPT FOR ELEVATIONS, NAMES, AND DESCRIPTIONS

When creating points in your drawing, you have the option of being prompted for elevations, names, and descriptions (see Figure 4.4). In many cases, you'll want to leave these options set to Manual. The command line will ask you to assign an elevation, name, and description for every point you create.

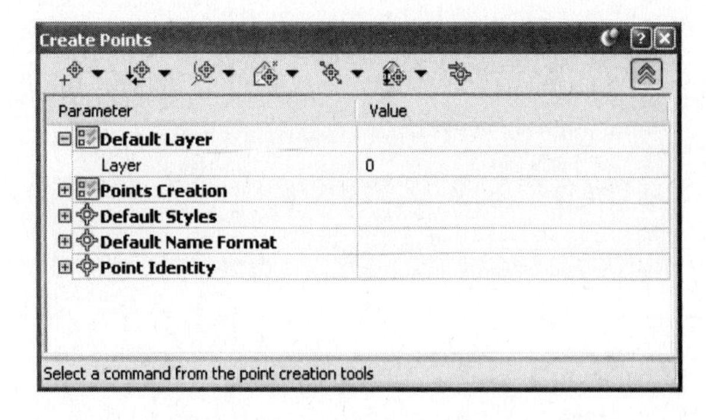

FIGURE 4.3
You should set the point object layer before creating points

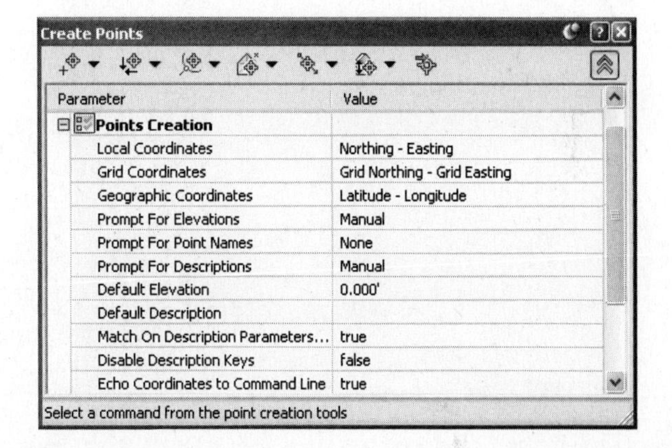

FIGURE 4.4
You can change the Elevations, Point Names, and Descriptions settings from Manual to Automatic

If you're creating a batch of points that have the same description or elevation, you can change the Prompt toggle from Manual to Automatic and then provide the description and elevation in the default cells. For example, if you're setting a series of trees at an elevation of 10′, you can establish settings as shown in Figure 4.5.

Be sure to change these settings back to Manual before you import points from external sources that provide a Z elevation. If not, all imported points will be assigned the default elevation regardless of the Z provided by the imported file.

Importing Points from a Text File

One of the most common means of creating points in your drawing is to import an external text file. This file may be the result of surveyed information or an export from another program (see Figure 4.6).

The import process supports these point file formats: TXT, PRN, CSV, XYZ, AUF, NEZ, and PNT. The following are the most common formats for imported point lists:

- Autodesk Uploadable File
- External Project Point Database
- ENZ, space delimited or comma delimited
- NEZ, space delimited or comma delimited
- PENZ, space delimited or comma delimited

- PENZD, space delimited or comma delimited
- PNE, space delimited or comma delimited
- PNEZ, space delimited or comma delimited
- PNEZD, space delimited or comma delimited

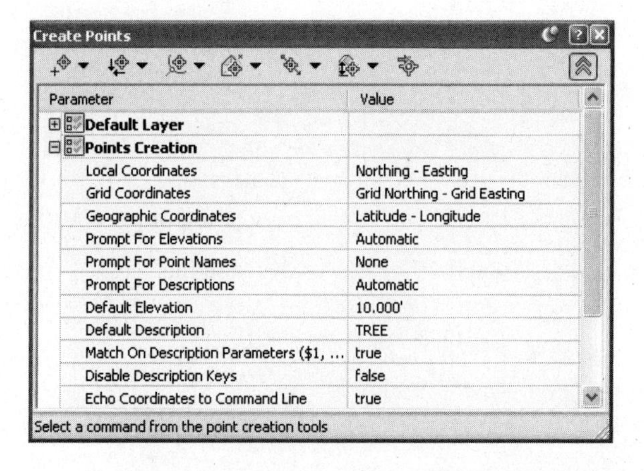

FIGURE 4.5
Default settings for placing tree points at an elevation of 10′

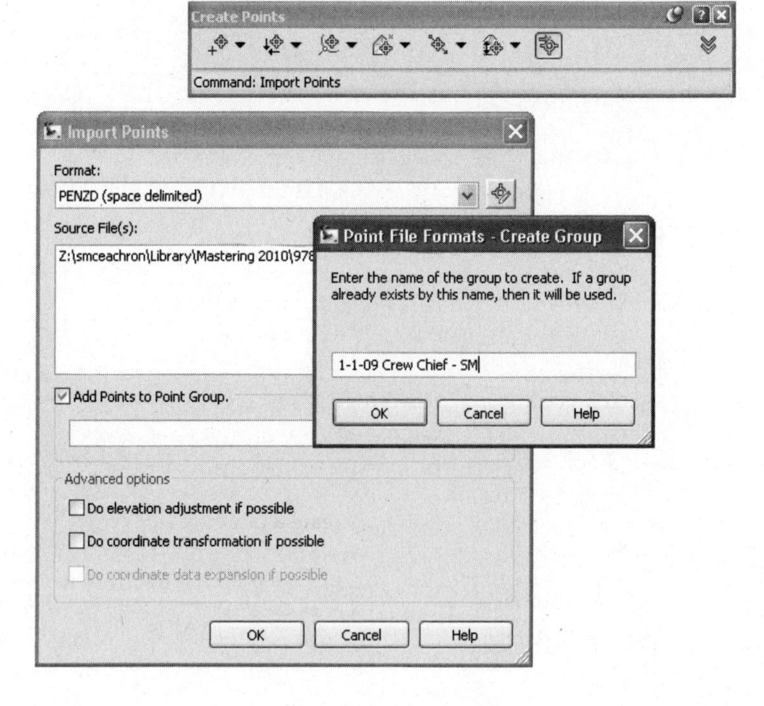

FIGURE 4.6
The Import Points tool in the Create Points dialog (top), and creating a point group for the newly imported points (bottom)

You can also perform an elevation adjustment, a coordinate system transformation, or a coordinate data expansion on import.

An elevation adjustment can be performed if the point file contains additional columns for thickness, Z+, or Z-. You can add these columns as part of a custom format. See the *Civil 3D 2009 Users Guide* section "Using Point File Format Properties to Perform Calculations" for more details.

You can perform a coordinate system transformation if a coordinate system has been assigned both to your drawing (under the Drawing Settings) and as part of a custom point format. In this case, the program can also do a coordinate data expansion, which calculates the latitude and longitude for each point.

IMPORTING A TEXT FILE OF POINTS

In this exercise, you'll learn how to import a `.txt` file of points into Civil 3D:

1. Open the `Import Points.dwg` file, which you can download from `www.sybex.com/go/masteringcivil3d2010`.
2. Open the Import Points dialog by selecting Points from File on the Import tab of the Insert tab.
3. Change the Format to PENZD Space Delimited.
4. Click the file folder's plus (+) button to the right of the Source File field, and navigate out to locate the `PENZD space.txt` file.
5. Leave all other boxes unchecked.
6. Click OK. You may have to zoom extents to see the imported points.

Converting Points from Land Desktop, Softdesk, and Other Sources

Civil 3D contains several tools for migrating legacy point objects to the current version. The best results are often obtained from an external point list, such as a text file, LandXML, or an external database. However, if you come across a drawing that contains the original Land Desktop, Softdesk, AutoCAD, or other types of point objects, tools and techniques are available to convert those objects into Civil 3D points.

A Land Desktop point database (the `Points.mdb` file found in the `COGO` folder in a Land Desktop project) can be directly imported into Civil 3D in the same interface in which you'd import a text file.

Land Desktop point objects, which appear as `AECC_POINT`s in the AutoCAD Properties palette, can also be converted to Civil 3D points (see Figure 4.7). Upon conversion, this tool gives you the opportunity to assign styles, create a point group, and more.

Occasionally, you'll receive AutoCAD point objects drawn at elevation from aerial topography information or other sources. It's also not uncommon to receive Softdesk point blocks from outside surveyors. Both of these can be converted to or replaced by Civil 3D points under the Points pull-down on the Create Ground Data panel of the Home tab (see Figure 4.8).

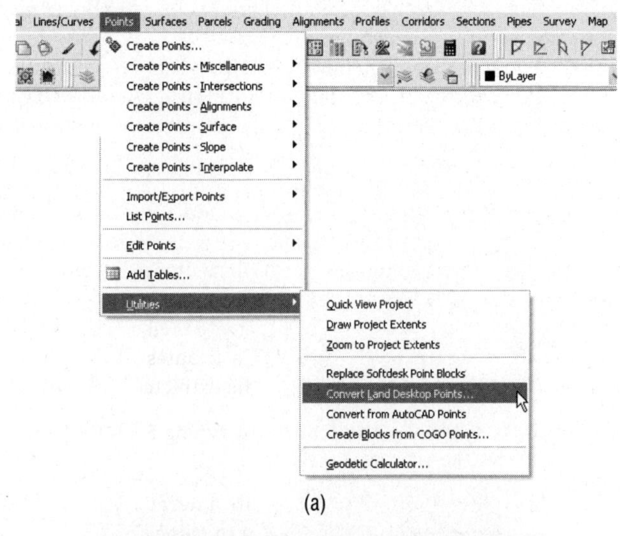

(a)

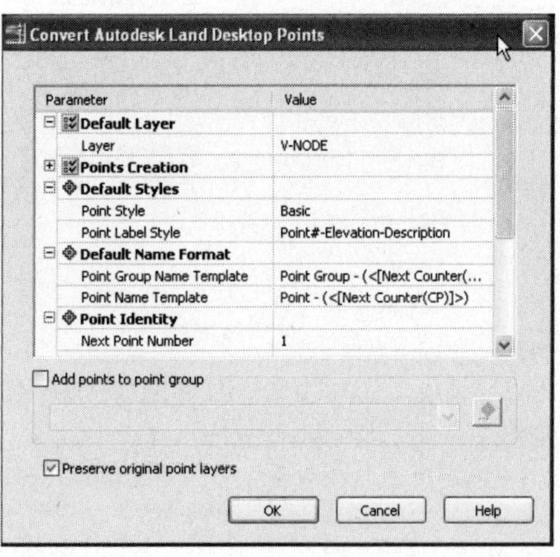

(b)

FIGURE 4.7 The Convert Land Desktop Points option (a) opens the Convert Autodesk Land Desktop Points dialog (b)

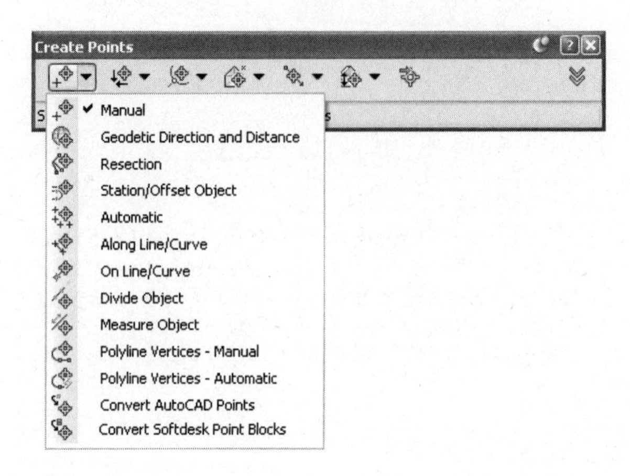

FIGURE 4.8 Use the Create Points dialog to convert AutoCAD point entities or Softdesk point blocks.

Using AutoCAD Attribute Extraction to Convert Outside Program Point Blocks

Occasionally, you may receive a drawing that contains point blocks from a third-party program. These point blocks may look similar to Softdesk point blocks, but the block attributes may have been rearranged and you can't convert them directly to Civil 3D points using Civil 3D tools.

The best way to handle this would be to ask the source of the drawing (the outside surveyor, for example) to provide a text file or LandXML file of their points database. However, often this isn't possible.

Because these objects are essentially AutoCAD blocks with special attributes, you can use the Data Extraction tools from AutoCAD to harvest their attributes and make a text file. You can then import this text file back into Civil 3D to create Civil 3D points. The points should store number, description, and elevation information in their attributes. As AutoCAD objects, they also understand their *x* and *y* position. These properties can all be extracted and reimported using the following procedure:

1. Use the Data Extraction tool by typing **EATTEXT** in the command line to launch the Data Extraction Wizard.
2. Select the radio button to create a new data extraction, and click Next. The Save Extraction As dialog appears, prompting you to name and save this extraction. Give the extraction a meaningful name, and save it in the appropriate folder. Click Next.
3. Confirm that the drawings to be scanned for attributed blocks are on the list:

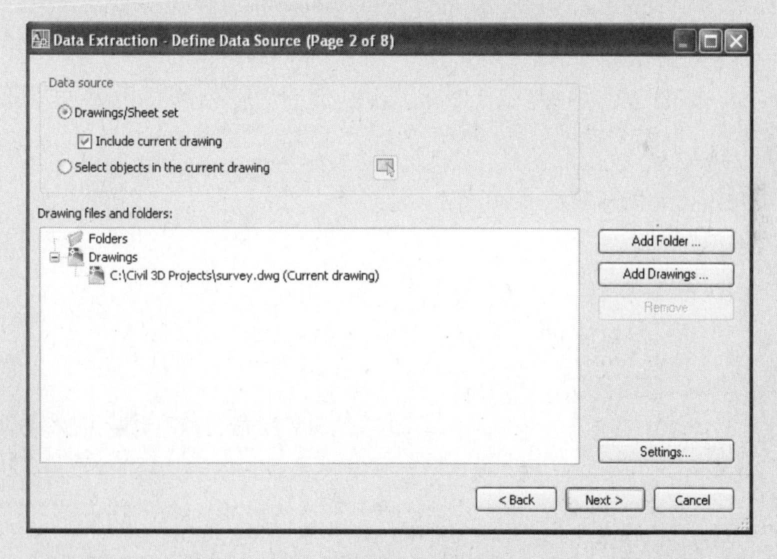

4. In the Select Objects screen of the Data Extraction dialog, select the Display Blocks Only radio button, and check Display Blocks With Attributes Only; this will include your point blocks and a few others. Eliminate the other types of attributed blocks by unchecking their boxes. (If you're unsure which block is your survey point block, exit the wizard and investigate one of your point blocks. Depending on what software package they came from, they may have a different name and different attributes.) Click Next.

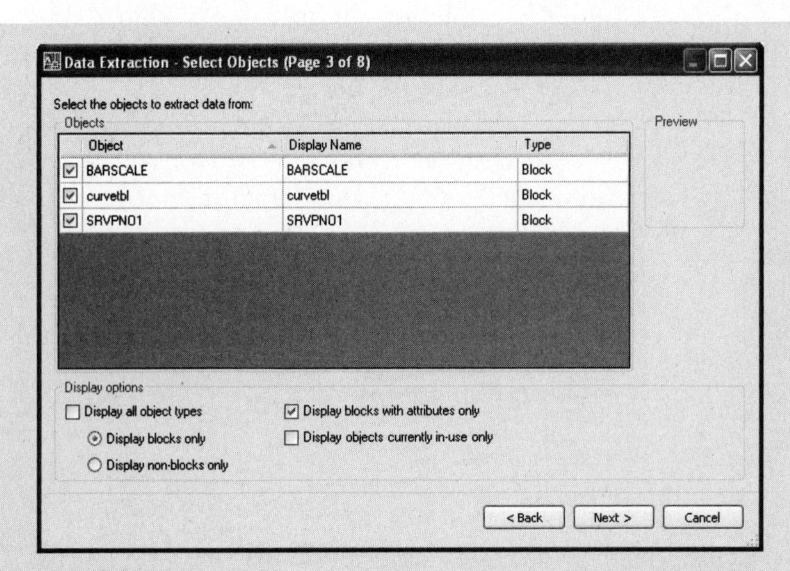

5. In the Select Properties dialog, locate the Point Number, Elevation, and Description, as well as X and Y. Click Next.

6. In the Refine Data screen of the Data Extraction dialog, rearrange the columns into a PNEZD format and remove any extra columns, such as Count and Name. Then, click Next.

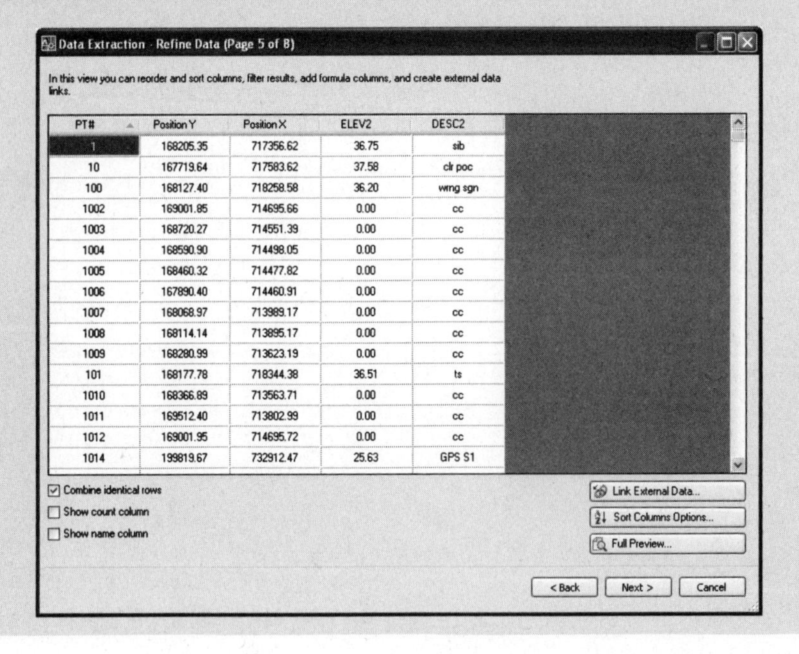

7. In the Choose Output screen of the Data Extraction dialog, choose Output Data to External File, and save your extraction as an `.xls` (Microsoft Excel) file in a logical place. If you save it as a `.txt` file initially, the file will have extra spaces in the wrong places.

8. Open this file in Microsoft Excel, and remove the first line of text (the header information). Use Save As to create a `.txt` file.

A1 ▾ fx PT#

	A	B	C	D	E	F	G	H
1	PT#	Position Y	Position X	ELEV2	DESC2			
2	1	168205.4	717356.62	36.75	sib			
3	10	167719.6	717583.62	37.58	clr poc			
4	100	168127.4	718258.58	36.2	wrng sgn			
5	1002	169001.9	714695.66	0	cc			
6	1003	168720.3	714551.39	0	cc			
7	1004	168590.9	714498.05	0	cc			
8	1005	168460.3	714477.82	0	cc			

9. In Civil 3D, use the Import Points tool in the Create Points dialog to import the `.txt` file.

Converting Points

In this exercise, you'll learn how to convert Land Desktop point objects and AutoCAD point entities into Civil 3D points:

1. Open the `Convert Points.dwg` file, which you can download from `www.sybex.com/go/masteringcivil3d2010`.
2. Use the List command or the AutoCAD Properties palette to confirm that most of the objects in this drawing are `AECC_POINTs`, which are points from Land Desktop. Also note a cluster of cyan-colored AutoCAD point objects in the western portion of the site.
3. Run the Land Desktop Point Conversion tool by choosing Points ➢ Utilities ➢ Convert Land Desktop Points. Note that the Convert Autodesk Land Desktop Points dialog allows you to choose a default layer, point creation settings, styles, and so on and also add the points to a point group. Leave the defaults, and click OK.
4. Civil 3D scans the drawing looking for Land Desktop point objects.
5. Once Civil 3D has finished the conversion, zoom in on any of the former Land Desktop points. The points should now be `AECC_COGO_POINTs` in both the List command and in the AutoCAD Properties palette, confirming that the conversion has taken place. The Land Desktop points have been replaced with Civil 3D points, and the original Land Desktop points are no longer in the drawing.
6. Zoom in on the cyan AutoCAD point objects.
7. Run the AutoCAD Point Conversion tool by choosing Points ➢ Utilities ➢ Convert From AutoCAD points.
8. The command line prompts you to `Select AutoCAD Points`. Use a crossing window to select all of the cyan-colored AutoCAD points.

9. At the command-line prompt, enter a description of **GS** (Ground Shot) for each point.

10. Zoom in on one of the converted points, and confirm that it has been converted to a Civil 3D point. Also note that the original AutoCAD points have been erased from the drawing.

Getting to Know the Create Points Dialog

In Civil 3D 2010, you can find point-creation tools directly under the Points pull-down on the Create Ground Data panel of the Home tab as well as in the Create Points dialog. The dialog is *modeless*, which means it stays on your screen even when you switch between tasks:

Miscellaneous Point-Creation Options The options in the Miscellaneous category are based on manually selecting a location or on an AutoCAD entity, such as a line, pline, and so on. Some common examples include placing points at intervals along a line or polyline, as well as converting Softdesk points or AutoCAD entities (see Figure 4.8 earlier in this chapter).

Intersection Point-Creation Options The options in the Intersection category allow you to place points at a certain location without having to draw construction linework. For example, if you needed a point at the intersection of two bearings, you could draw two construction lines using the Bearing Distance transparent command, manually place a point where they intersect, and then erase the construction lines. Alternatively, you could use the Direction/Direction tool in the Intersection category (see Figure 4.9).

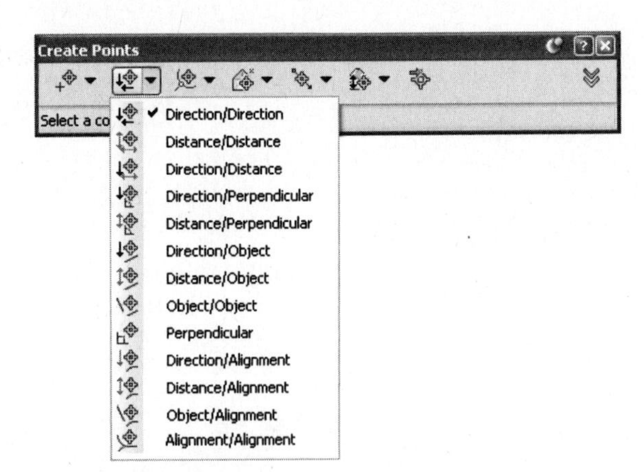

FIGURE 4.9 Intersection point-creation options

Alignment Point-Creation Options The options in the Alignment category are designed for creating stakeout points based on a road centerline or other alignments. You can also set Profile Geometry points along the alignment using a tool from this menu. See Figure 4.10.

Surface Point-Creation Options The options in the Surface category let you set points that harvest their elevation data from a surface. Note that these are points, not labels, and therefore aren't dynamic to the surface. You can set points manually, along a contour or a polyline, or in a grid. See Figure 4.11.

Interpolation Point-Creation Options The Interpolation category lets you fill in missing information from survey data or establish intermediate points for your design tasks. For example, suppose your survey crew picked up centerline road shots every 100 feet, and you'd like to interpolate intermediate points every 25 feet. Instead of doing a manual slope calculation, you could use the Incremental Distance tool to create additional points (see Figure 4.12).

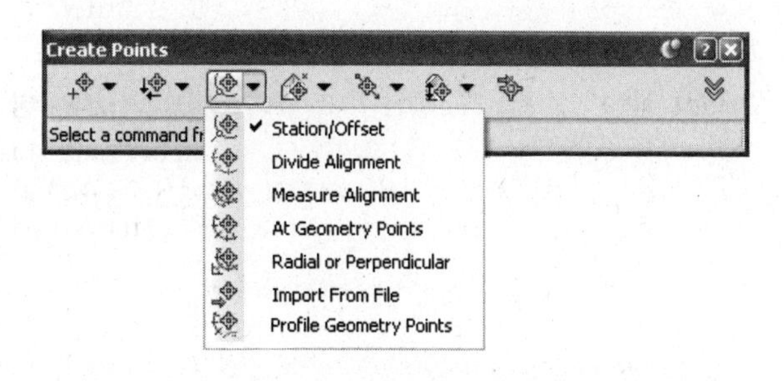

FIGURE 4.10
Alignment point-creation options

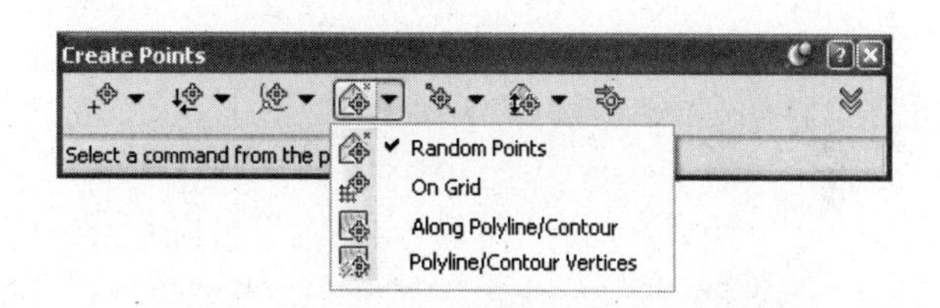

FIGURE 4.11
Surface point-creation options

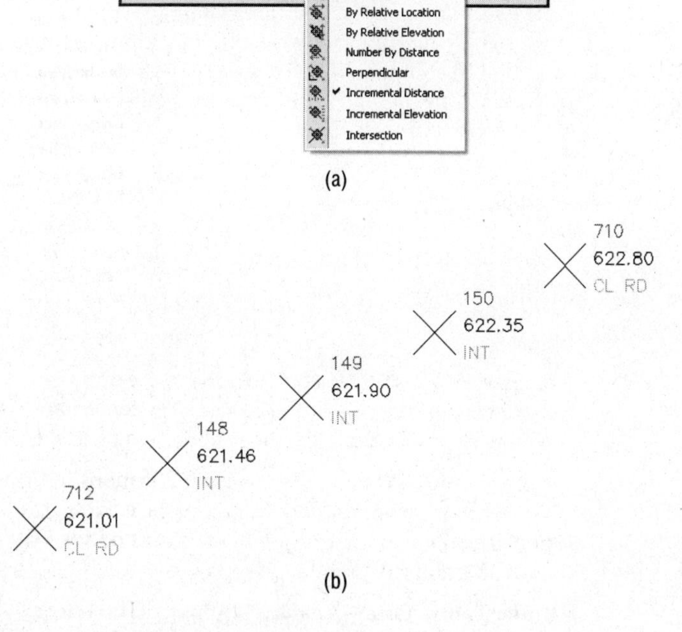

FIGURE 4.12
The interpolation point-creation options (a) and intermediate points created using the Incremental Distance tool (b)

Another use would be to set intermediate points along a pipe stakeout. You could set a point for the starting and ending invert, and then set intermediate points along the pipe to assist the field crew.

Slope Point-Creation Options The Slope category allows you to set points between two known elevations by setting a slope or grade. Similar to the options in the Interpolate and Intersect categories, these tools save you time by eliminating construction geometry and hand calculations (see Figure 4.13).

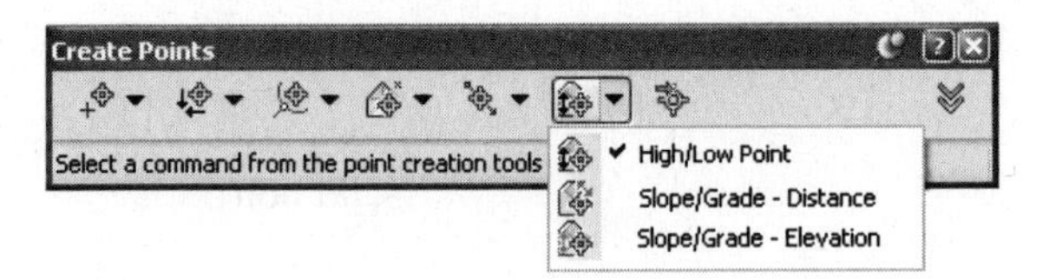

FIGURE 4.13
Slope point-creation options

CREATING POINTS

In this exercise, you'll learn how to create points along a parcel segment and along a surface contour:

1. Open the `Create Points Exercise.dwg` file, which you can download from `www.sybex.com/go/masteringcivil3d2010`. Note that the drawing includes an alignment, a series of parcels, and an existing ground surface.
2. Open the Create Points dialog by selecting Points ➤ Point Creation Tools on the Create Ground Data panel.
3. Click the down arrows at far right to expand the dialog.
4. Expand the Points Creation option. Change the Prompt for Elevations value to None and the Prompt for Descriptions value to Automatic by clicking in the respective cell, clicking the down arrow, and selecting the appropriate option. Enter **LOT** for the Default Description (see Figure 4.14). This will save you from having to enter a description and elevation each time. Because you're setting stakeout points for rear lot corners, elevation isn't important in this case.

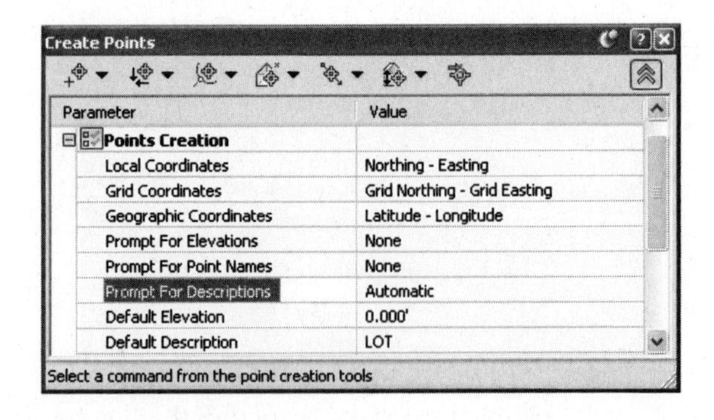

FIGURE 4.14
Point-creation settings in the Create Points dialog

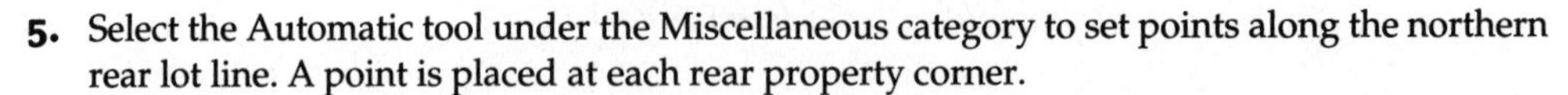

5. Select the Automatic tool under the Miscellaneous category to set points along the northern rear lot line. A point is placed at each rear property corner.

6. Select the Measure Object tool under the Miscellaneous category to set points along the southern rear lot line. After selecting the object, this tool prompts you for starting and ending stations (press ↵ twice to accept the measurements), and offset (press ↵ to accept 0), and an interval (enter **25**).

7. Experiment with other point-creation tools as desired.

8. Go back into the Point Settings options, and change Prompt for Elevations to Manual and the Default Description to EG. The next round of points you'll set will be based on the existing ground elevation.

9. Select the Along Polyline/Contour tool in the Surface category to create points every 100 feet along any contour. The command line prompts you to physically choose a surface object, ask for a point spacing interval, and then pick a polyline or contour. Pick any contour on your surface, and note that points are placed at surface elevation along that contour.

Basic Point Editing

Despite your best efforts, points will often be placed in the wrong location or need additional editing after their initial creation. It's common for property-corner points to be rotated to match a different assumed benchmark or for points used in a grading design to need their elevations adjusted.

Physical Point Edits

Points can be moved, copied, rotated, deleted and more using standard AutoCAD commands and grip edits. In Civil 3D 2010, you can now rotate a point with the special Point-Rotation grip, as shown in Figure 4.15.

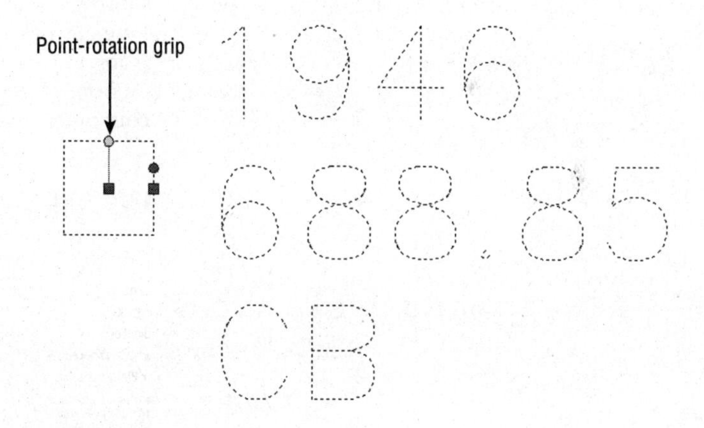

FIGURE 4.15
Civil 3D 2010 adds a special grip for point rotation

Properties Box Point Edits

You can access many point properties through the AutoCAD Properties palette. Pick a point, right-click, and choose Properties (see Figure 4.2).

Panorama and Prospector Point Edits

You can access many point properties through the Point Editor in Panorama. Choose a point (or points), right-click, and choose Edit Points. Panorama brings up information for the selected point(s) (see Figure 4.16).

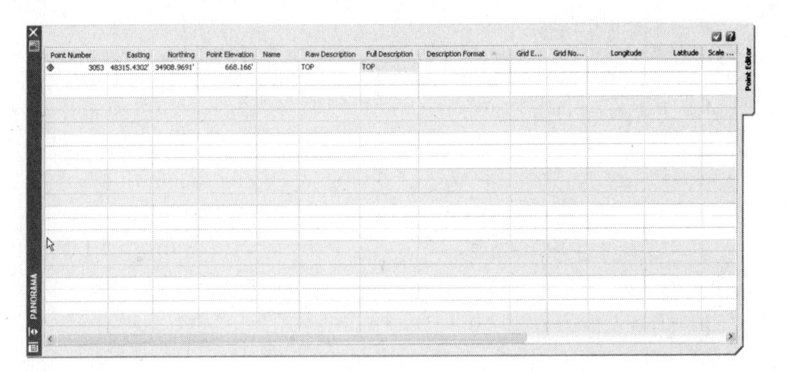

FIGURE 4.16
Edit points in Panorama

You can access a similar interface in the Prospector tab of Toolspace by highlighting the Points collection (see Figure 4.17).

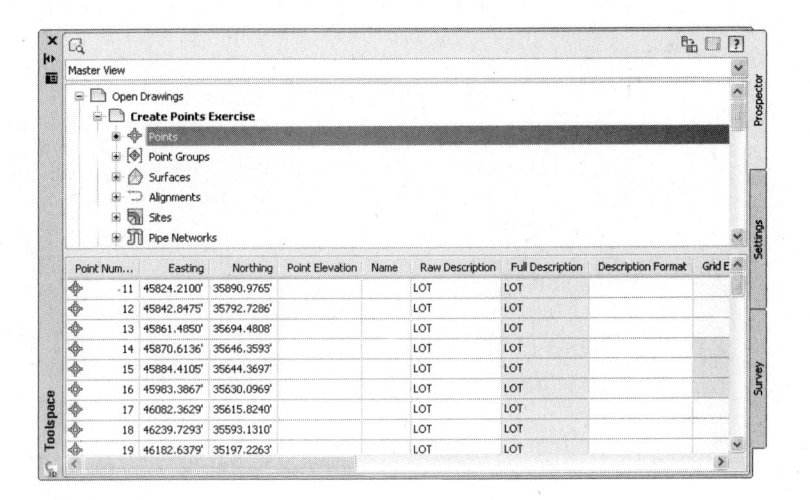

FIGURE 4.17
Prospector lets you view your entire Points collection at once

Changing Point Elevations

In addition to the basic point-editing functions, advanced tools are available for manipulating points (see Figure 4.18), in the context-sensitive panel that opens when a point object is selected.

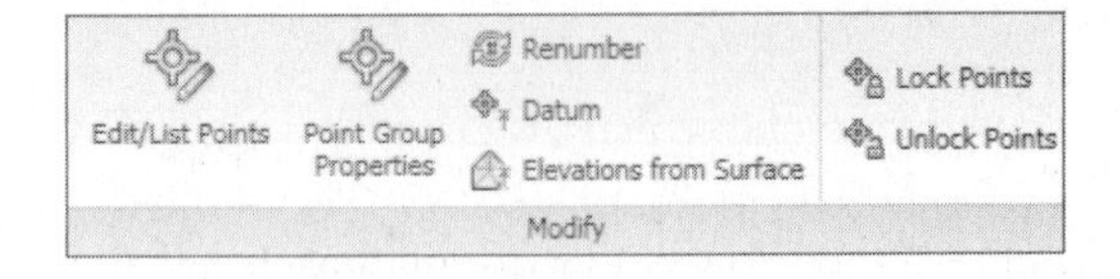

FIGURE 4.18
Advanced point-editing commands in the ribbon

As noted earlier, you can create points from a surface elevation. But if you already have a batch of points in your drawing that you'd like to move up to a surface, you can choose Datum from the Modify panel after choosing Points from the Ground Data panel on the Modify tab (see Figure 4.19).

FIGURE 4.19
Tree points that were moved up to surface elevation

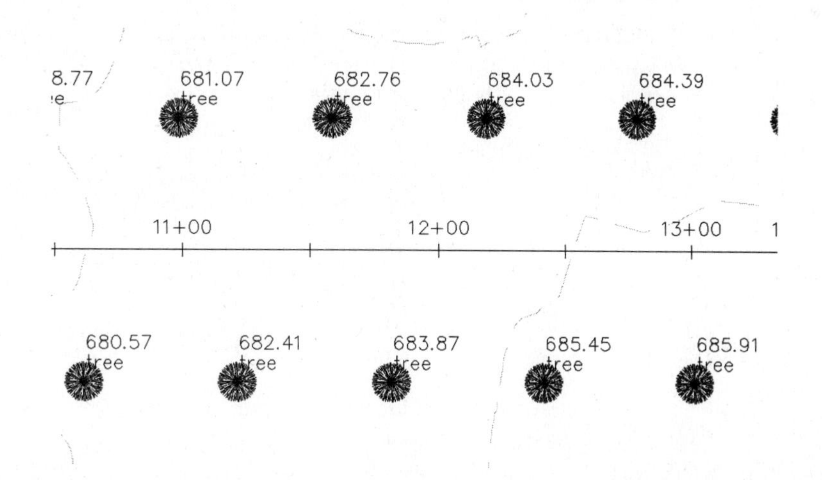

Selecting a point object, and selecting Elevations from Surface from the Modify panel allows you to raise or lower all the points in a drawing.

If you'd like to change the datum of only a certain selection of points, pick the points, and select Datum from the Modify panel. In Panorama, highlight the points, right-click, and choose Datum. This technique can be used for any point in Panorama or in the Preview pane of Prospector.

Point Styles

As with all other Civil 3D objects, the way the point object appears is controlled by an object style. Point styles control the shape, size, location, and visibility of the point marker as well as the visibility of the point label (see Figure 4.20).

FIGURE 4.20
The components of a point object

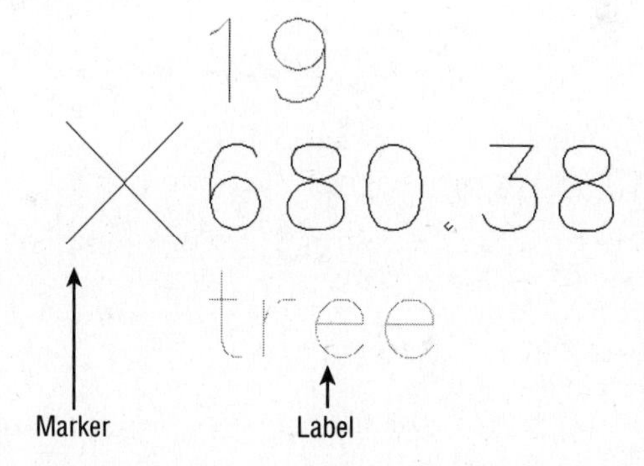

Point styles appear on the Settings tab of Toolspace under the Point branch (see Figure 4.21).

You can edit a point style by double-clicking it. The Point Style dialog, which contains five tabs, appears. Each of these tabs is used to customize the point style:

Information Tab The Information tab provides a place to name and describe your style (see Figure 4.22).

Marker Tab The Marker tab (see Figure 4.23) lets you customize the point marker. You can use an AutoCAD point, a custom marker (such as an X, +, or tick), or any AutoCAD block. The marker can have a fixed rotation.

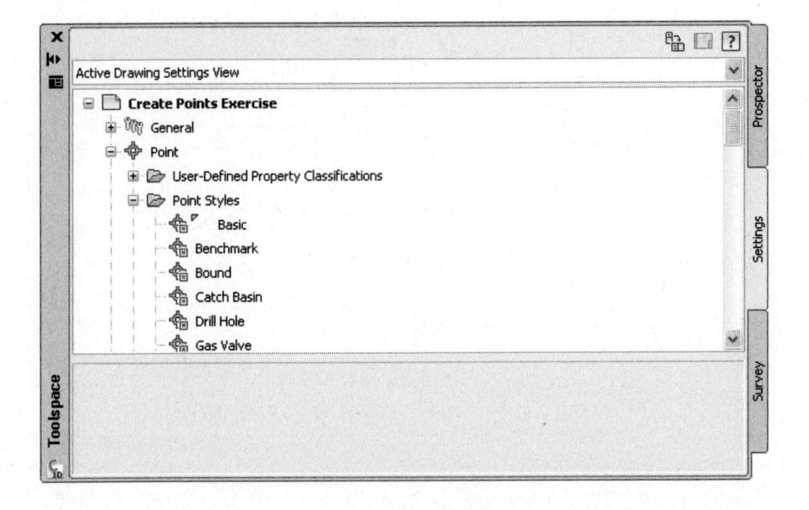

FIGURE 4.21
Point styles on the Settings tab

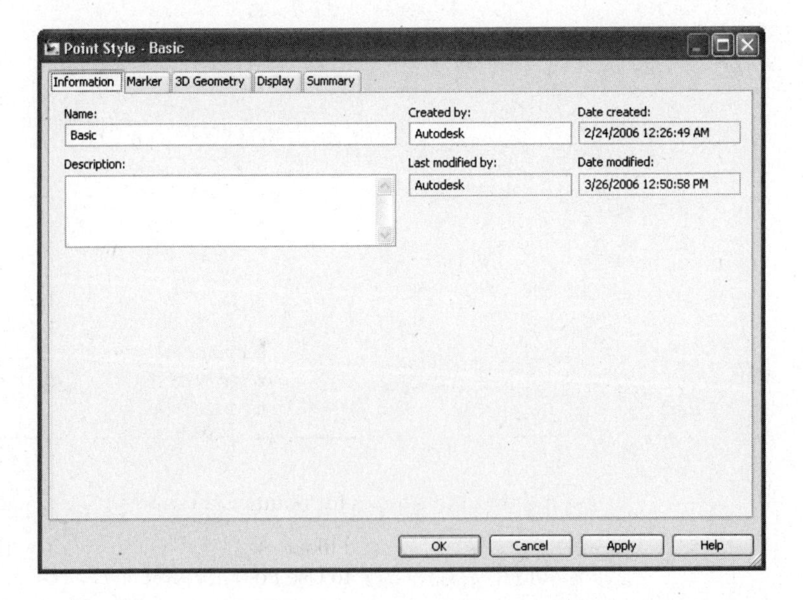

FIGURE 4.22
The Information tab in the Point Style dialog

Table 4.1 shows the options for point-object size and the effects of each option.

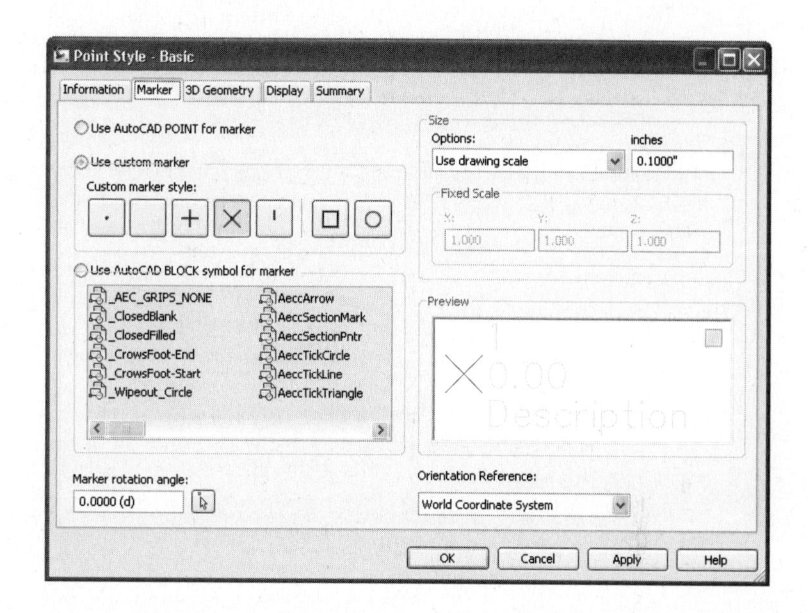

FIGURE 4.23
The Marker tab in the Point Style dialog

TABLE 4.1: Size Options and Effects

SIZE OPTION	EFFECT
Use Drawing Scale	Requires that you specify the plotted marker size in inches. Civil 3D then scales the marker on the basis of your drawing or viewport scale.
Use Fixed Scale	Requires that you specify the actual size of the marker and then specify additional scale factors in the Fixed Scale area.
Use Size in Absolute Units	Requires that you specify an actual size for the marker.
Size Relative to Screen	Requires that you specify a percentage of the screen. As you zoom in or out, the points resize when the screen is regenerated.

Table 4.2 shows the options for point-orientation reference and the effect of each option.

3D Geometry Tab If you'd like to see your points drawn at the elevations they represent, you can set Point Display Mode to Use Point Elevation. Most often, the best practice is to use Flatten Points to Elevation so that points are drawn at elevation zero with the balance of your normal linework.

TABLE 4.2: Orientation References and Effects

ORIENTATION REFERENCE OPTION	EFFECT
World Coordinate System	The point marker always relates to the World Coordinate System, regardless of custom UCS changes or viewport orientation.
View	The point marker responds to changes in viewport orientation.
Object	The point marker is oriented with respect to the point location.

This doesn't affect their intelligence, only where the points are physically drawn. For example, building points from surfaces uses the elevation value stored within the point intelligence, not the physical *z* location of the point object.

This dialog also allows you to apply an exaggeration on the point object's location via the Scale Factor property, if desired (see Figure 4.24).

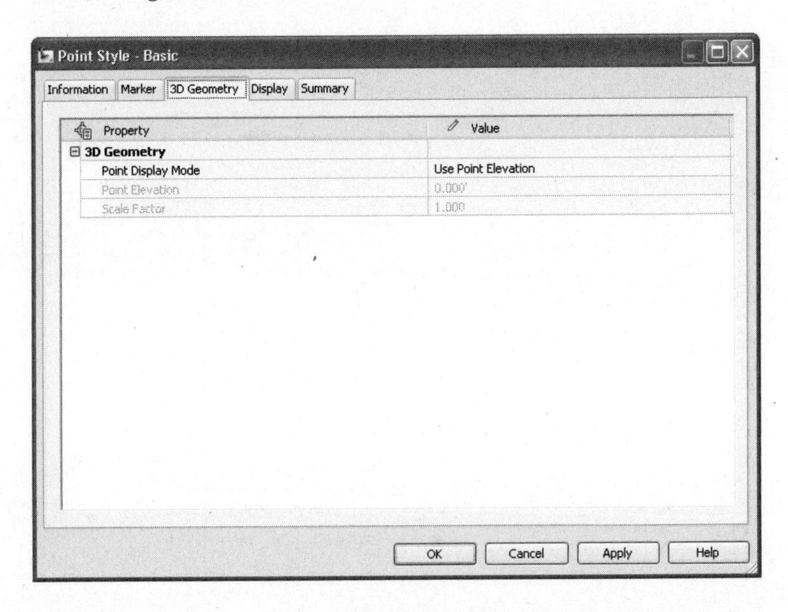

FIGURE 4.24
The 3D Geometry tab in the Point Style dialog

Display Tab The Display tab lets you control visibility, layer mapping, and other properties in both plan (2D) and model (3D) views of the Marker and Label components (see Figure 4.25).

Summary Tab The Summary tab provides a list of all the point style properties (see Figure 4.26).

Point Label Styles

As with all other Civil 3D objects, the way a point is labeled is controlled by the label style. Point label styles control which information is extracted from the point object for labeling purposes and how it's presented. This can be as simple as the point description in a standard font or as elaborate as several user-defined properties, coordinates, and more information in several colors and text styles.

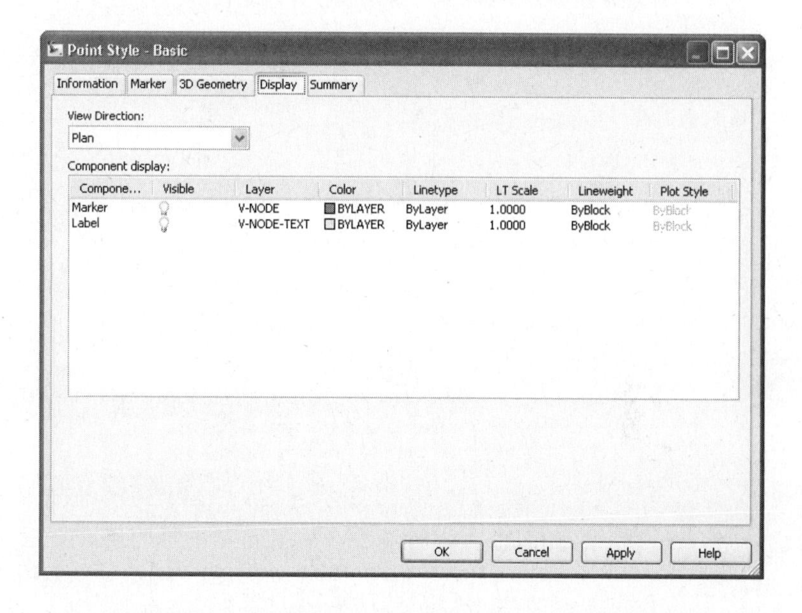

FIGURE 4.25
The Display tab in the Point Style dialog

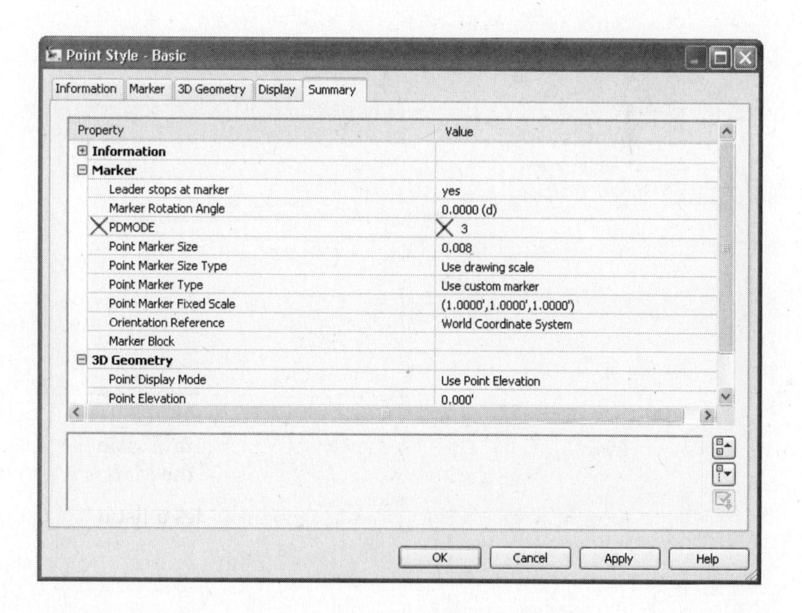

FIGURE 4.26
The Summary tab in the Point Style dialog

Point label styles are composed much the same as other labels in Civil 3D. For additional information about composing label styles, see Chapter 1 or the *Civil 3D Users Guide*.

Label styles appear on the Settings tab of Toolspace under the Point branch (see Figure 4.27).

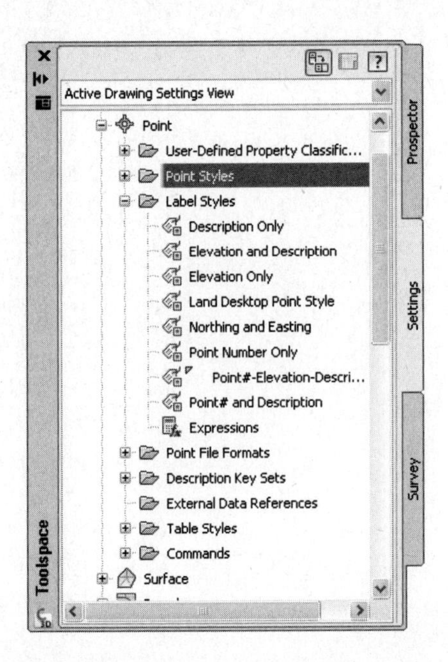

FIGURE 4.27
Label styles on the Settings tab

You edit a label style by double-clicking it. The Label Style Composer dialog, which contains five tabs, appears. Each tab provides different options for customizing the label styles of the points in your drawing:

Information Tab The Information tab provides a place to name and describe your label style (see Figure 4.28).

FIGURE 4.28
The Information tab in the Label Style Composer dialog

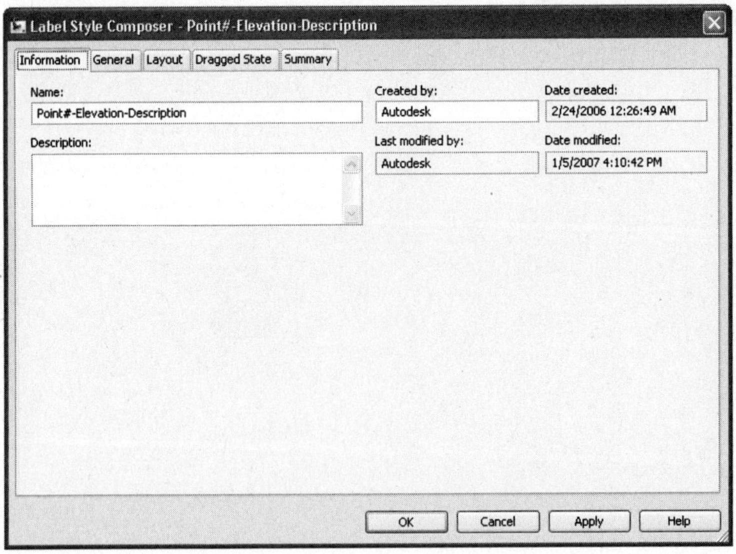

General Tab The General tab provides options for setting text style, label visibility, layer, orientation, and readability (Figure 4.29).

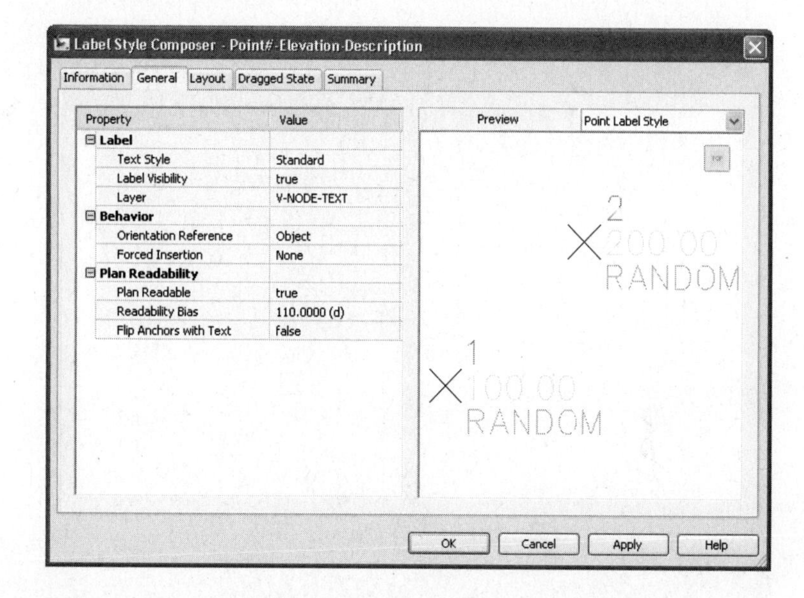

FIGURE 4.29
The General tab in the Label Style Composer dialog

Layout Tab The Layout tab in the Label Style Composer dialog is identical to the Layout tab of other label styles. Note that you can add text, blocks, and lines to the labels of points, but points have no reference text options (see Figure 4.30).

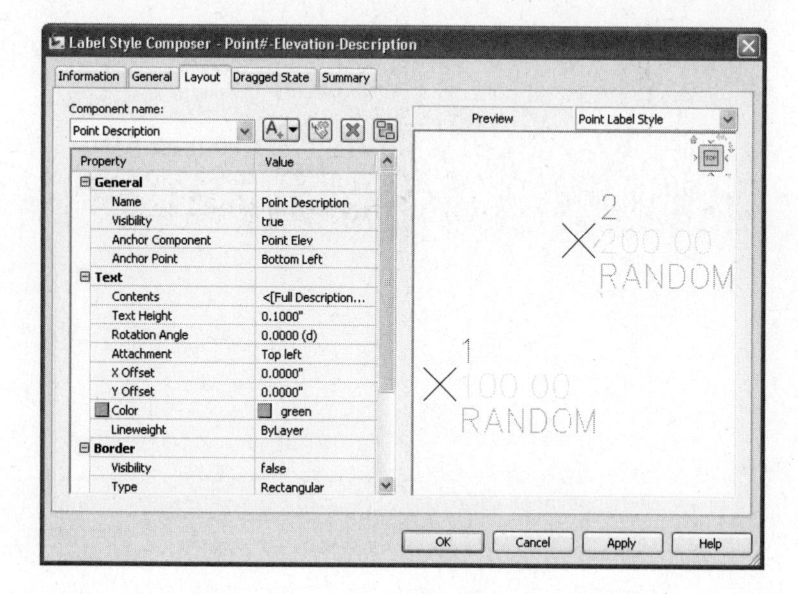

FIGURE 4.30
The Layout tab in the Label Style Composer dialog

Dragged State Tab The Dragged State tab is identical to the Dragged State tab of other label styles (see Figure 4.31).

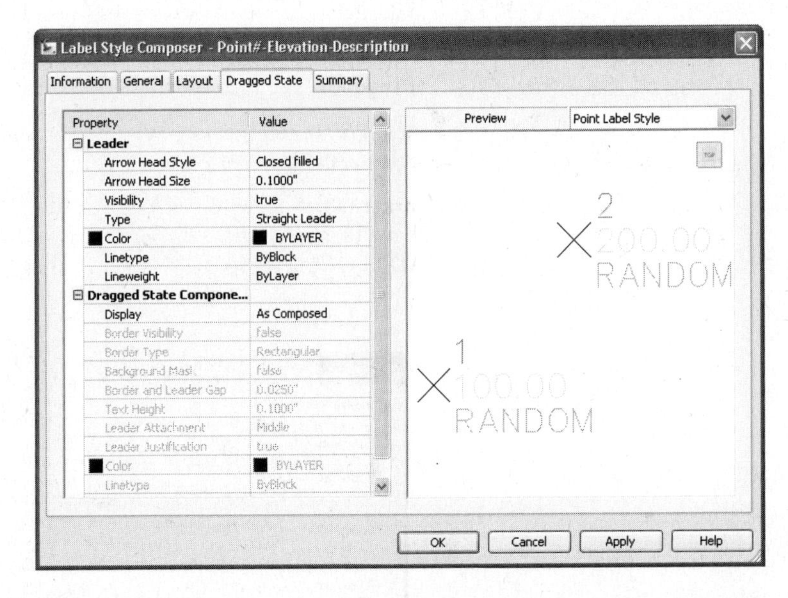

FIGURE 4.31
The Dragged State tab in the Label Style Composer dialog

Summary Tab This tab provides a list of all the point label-style properties (see Figure 4.32).

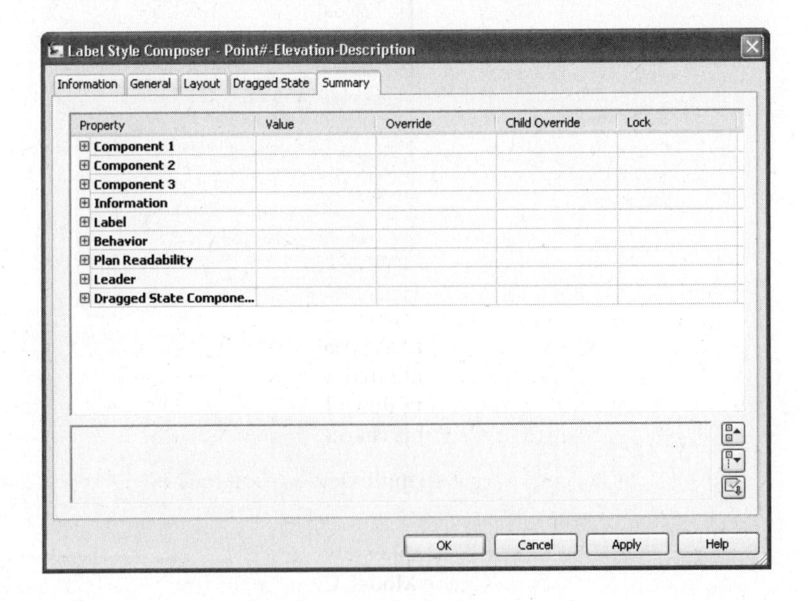

FIGURE 4.32
The Summary tab in the Label Style Composer

Creating More Complex Point and Point-Label Styles

In this lesson, you'll learn how to customize a point style to include a multiview block and how to customize a point-label style to include elevation, northing, and easting. Consider using points for objects you may have previously represented with AutoCAD blocks. Points are easier to label and edit, and you also gain the power of dynamic tables. Add the use of a 3D or multiview block, and you gain the ability to make 3D visualizations with little effort. Follow these steps:

1. Open the file `Point Style Exercise.dwg`, which you can download from `www.sybex.com/go/masteringcivil3d2010`.
2. Expand the Point ➤ Point Styles branches of the Settings tab of Toolspace. Double-click the Tree point style to open the Point Style dialog.
3. Switch to the Marker tab. A 2D tree block called Tree 6 is currently the marker for this style (you may have to scroll over to see the marker). Right-click in the whitespace of the Use AutoCAD BLOCK Symbol for Marker list box, and click Browse, as shown in Figure 4.33.

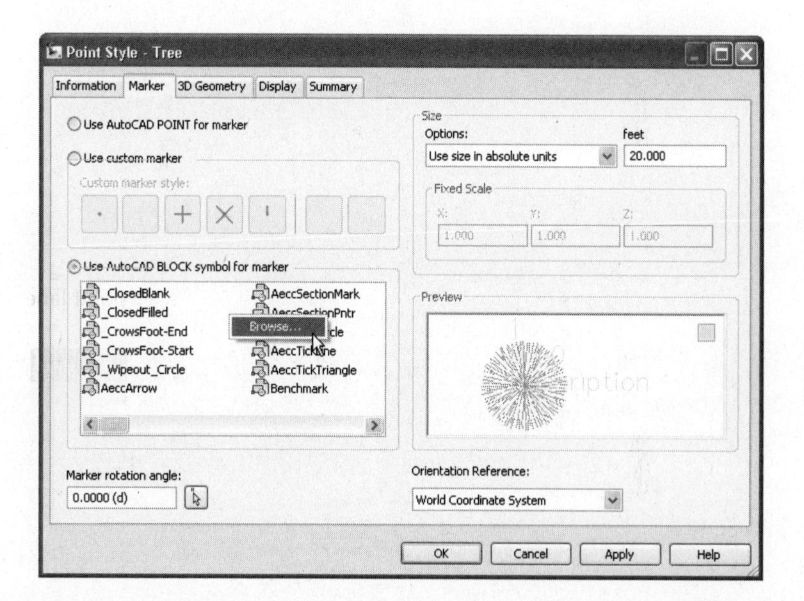

FIGURE 4.33
Right-click in the block area, and click Browse

4. Browse to `C:\Documents and Settings\All Users\Application Data\Autodesk\C3D2010\enu\Data\Symbols\MvblocksAutoCAD Civil 3D 2010\Data\Symbols\Mvblocks`. (Note that the path may vary depending on your OS or installation settings.) If you have difficulty locating this folder, you can download a multiview block called `American Beech.dwg` for this chapter from `www.sybex.com/go/masteringcivil3d2010`.
5. Choose any tree multiview block in this directory, such as American Beech. Click Open to select this block as your point marker.
6. Switch to the Display tab of the Point Style dialog. Change the View Direction drop-down list from Plan to Model. Confirm that the Label component is turned off in the Model view direction. Only the multiview tree block is visible, not the label text, when the drawing is rotated into isometric view.
7. Click OK to dismiss the Point Style dialog. Type **REGEN**↵ at the command line to see the changes in your point marker.

8. Change to the Views tab and select Unsaved Views ➤ SW Isometric from the Views panel. Zoom in on the site, and note that the trees appear as 3D trees in isometric.
9. Using the process outlined in the previous step, change the view to Top.
10. Expand the Point ➤ Label Styles branches on the Settings tab of Toolspace. Right-click the Elevation and Description label style, and select Copy.
11. On the Information tab of the Label Style Composer, enter **Elevation Description Northing and Easting** in the Name text box.
12. Switch to the Layout tab. Click the down arrow next to Create Text Component, and choose Text from the drop-down menu, as shown in Figure 4.34.

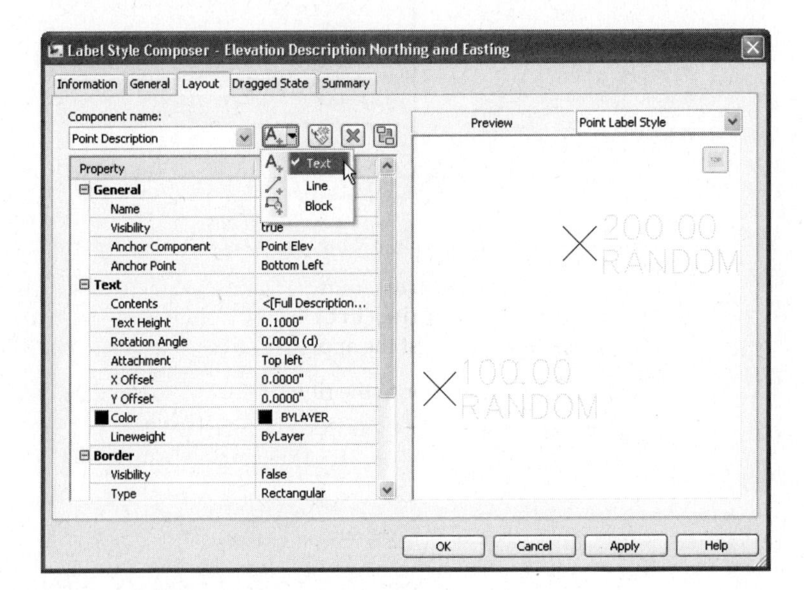

FIGURE 4.34 Adding a text component

13. In the Name field, replace Text.1 with **Northing Easting**.
14. Change the Anchor Component to **Point Description**. This option will anchor the Northing Easting text to the Description text.
15. Change the Anchor Point to **Bottom Left**. This option will make sure the label uses the bottom-left corner of the Point Description component as an attachment point.
16. Under the Text category, change the Attachment point to **Top Left**. This will ensure that the Northing Easting text uses the top-left corner to attach to the Point Description component. The Preview window should show the phrase *Label Text* lined up under the word *RANDOM*.
17. Click in the field for the Contents value to activate the ellipsis. Click the ellipsis button to open the Text Component Editor, as shown in Figure 4.35.
18. Erase the text that is currently in the Text Component Editor (*Label Text*).
19. Under the Properties pull-down menu, select Northing.

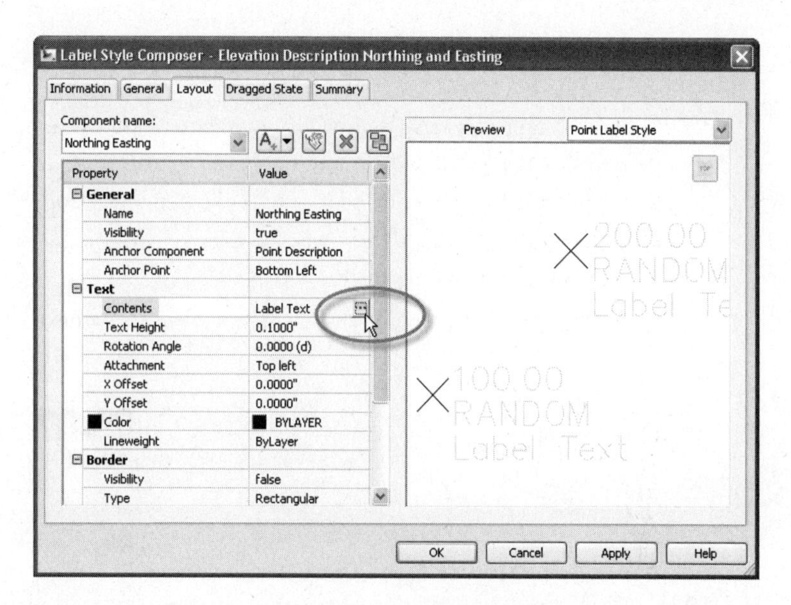

FIGURE 4.35
Click the ellipsis button in the Contents field

20. Change the precision value to .01 using the Precision pull-down menu.

21. Click the small arrow button. The following text string appears on the right side of the dialog: "`<[Northing(Uft|P2|RN|AP|Sn|OF)]>`". This is a special formula that Civil 3D will use to harvest the appropriate point northing from the point data.

22. Move your cursor to the right of this text string, and add a comma, followed by a space. This tells Civil 3D to separate your Northing value and your Easting value with a comma and a space.

23. Repeat steps 19–22 using Easting instead of Northing.

24. Click OK to exit the Text Component Editor and again to exit the Label Style Composer.

25. Switch to the Prospector tab of Toolspace, and locate the TREE point group under the Point Groups branch. Right-click, and select Properties.

26. Select Elevation Description Northing and Easting from the drop-down menu in the Point Label Style selection box, and click OK to exit the Point Group Properties.

27. Zoom into one of your points, and note the new label style. These labels can be dragged off to the side, or the label style can be adjusted to display smaller text or different attachment points to ease readability.

Point Tables

You've seen some of the power of dynamic point editing; now let's look at how those dynamic edits can be used to your advantage in point tables.

Most commonly, you may need to create a point table for survey or stakeout data; it could be as simple as a list of point numbers, northing, easting, and elevation. These types of tables are easy to create using the standard point-table styles and the tools located in the Points menu under the Add Tables option.

Also consider that many plans require schedules and tables listing the locations and specifications of things like trees, signs, and light posts. Instead of representing these items with regular AutoCAD blocks labeled with quick leaders, consider building those blocks into your point-object style and setting them as point objects instead. In the next exercise, you'll see how to create a tree schedule from a customized point-table style.

User-Defined Properties

Standard point properties include items such as number, easting, northing, elevation, name, description, and the other entries you see when examining points in Prospector or Panorama. But what if you'd like a point to know more about itself?

It's common to receive points from a soil scientist that list additional information such as groundwater elevation or infiltration rate. Surveyed manhole points often include invert elevations or flow data. Tree points may also contain information about species or caliber measurements. All this additional information can be added as user-defined properties to your point objects. You can then use user-defined properties in point labeling, analysis, point tables, and more.

Creating a Point Table and User-Defined Properties for Tree Points

In this exercise, you'll learn how to create and customize a point table for tree location points, including the addition of user-defined properties for tree species:

1. Open the `Point Table.dwg` file, which you can download from `www.sybex.com/go/masteringcivil3d2010`. Note the series of points, which represents trees along an alignment.
2. Change to the Annotate tab and select Add Tables ➢ Add Point Table on the Labels & Tables panel. The Point Table Creation dialog appears.
3. Click the Select Point Groups button on the right side of the middle of the dialog, next to the text that says No Point Group Selected, to open the Point Groups dialog. Select Tree, and click OK.
4. Leave all other options at their defaults, and click OK.
5. Place your table somewhere in the drawing off to the right side of the surface. Note that the table splits into three columns for readability. This table may be fine for many purposes, but let's customize it to be more suitable for a Tree schedule.
6. Expand the Point ➢ Table Styles branches on the Settings tab of the Toolspace.
7. Right-click the PNEZD format table style, and select Copy. The Table Style dialog appears.
8. On the Information tab, enter **Tree Table** in the Name text box.
9. Switch to the Data Properties tab.
10. Double-click the cell that says Point Table to display the Text Component Editor. Delete the words *Point Table*, and enter **Tree Schedule**. Click OK.
11. Click to activate the Description column, and click the red X to delete the Description column.
12. Click OK to exit the Table Style dialog.

13. Pan over to your table(s). Pick a table and select Table Properties from the Modify panel on the ribbon. The Table Properties dialog appears.
14. Select Tree Table (the style you just created) from the Table Style drop-down menu. Click OK.

Next, you'll create a user-defined property so that you can add a column for tree species:

1. Select the User-Defined Property Classifications entry under the Point branch on the Settings tab of Toolspace. Right-click, and choose New. The User-Defined Property Classifications dialog appears.
2. Enter **Tree Properties** in the Classification Name text box. Click OK.
3. Right-click the new **Tree Properties** entry under User-Defined Property Classification, and select New. The New User-Defined Property dialog appears.
4. Enter **Species** in the Name text box, and make sure the Property Field Type selection box is set to String. Click OK. Note that you could continue to create additional properties such as diameter, canopy, or other extra tree values.
5. Expand the Point ➤ Table Styles branches on the Settings tab. Double-click the Tree Table style to open the Table Style dialog.
6. Switch to the Data Properties tab (if you aren't already there), and click in the Easting column to activate it. Click the white + to add a new column.
7. Double-click the empty column header (you may need to scroll to the right) to open the Text Component Editor. Enter **Species** in the Preview screen to label the column. Click OK.
8. In the Column Value row, double-click under the word *Automatic* in your Species column to open the Text Component Editor. Choose the user-defined property Species from the Properties drop-down menu, change the Capitalization Modifier to a Value of **Preserve Case** and click the white arrow to move the name to the right side of the screen, as shown in Figure 4.36. (Sometimes it's necessary to save the drawing, exit Civil 3D, and reenter the drawing to see the User-Defined Properties as choices in the Text Component Editor and other places. If you don't see Species as a choice, click OK to exit the Text Component Editor, and click OK again to exit the Table Style dialog. Save your drawing, close and reopen it, and come back to this point.)
9. Click OK to exit the Text Component Editor, and click OK again to exit the Table Style dialog. You should now see an empty Species column on the right side of your table in the drawing.

THINKING AHEAD: CAPITALIZATION MATTERS

In the previous exercise, the Capitalization Modifier was changed to a value of **Preserve Case.** As a general rule of thumb, when you're working with text values in Civil 3D, it is important to think ahead and remain consistent in terms of capitalization. This is especially true when working with points, because at some point, the data in the drawing may be exported to a surveyor's data collector (or vice versa). In many data collectors, there is a distinct difference between a Pine Tree, a pine tree, and a PINE TREE. As a result, some data collection software will misinterpret the data and see three species of tree, when in fact there is only one. This same principle holds true in Description Keys, as discussed later in this chapter.

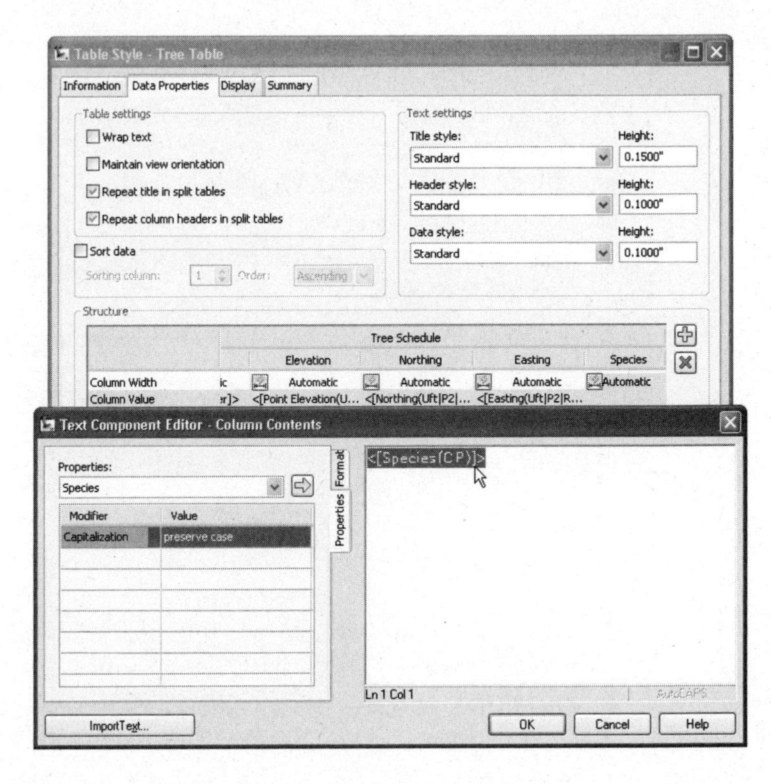

FIGURE 4.36
The Text Component Editor

Next, you need to assign the Species property to your Tree point group:

1. Highlight the Point Groups entry on the Prospector tab of Toolspace, and look in the Preview pane to find the Classification column, as shown in Figure 4.37. (You may have to scroll to the right to see the Classification column.)

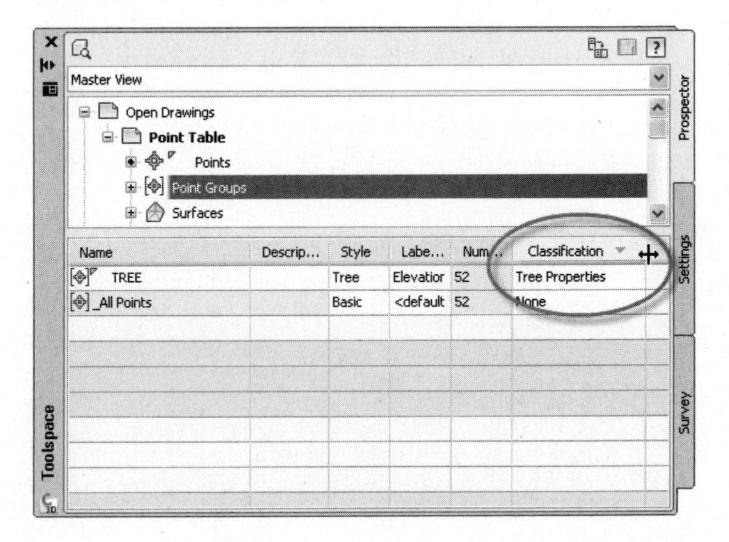

FIGURE 4.37
The Classification column in Prospector

2. Click in the TREE Classification column, and select the Tree Properties option from the drop-down menu. This tells the point group to add fields for any additional tree properties, such as your Species entry.

3. Select TREE under the Point Group branch in Prospector. Species should now be the last column in the Preview pane.

4. Reduce the number of visible columns by right-clicking in a column header and unchecking the entries for latitude, longitude, and other columns you aren't using right now, as shown in Figure 4.38. This will give you more room to work.

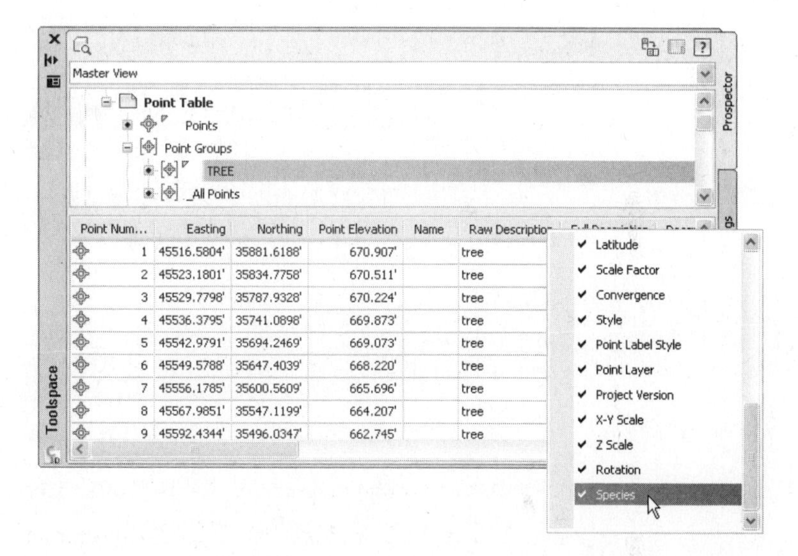

FIGURE 4.38 Remove unnecessary columns from the Preview pane

5. Enlarge the Species column by holding your mouse on the right side of the column header and dragging the column wider. Make a few entries in the Species column, such as **maple**, **oak**, **pine**, and so on.

6. Go out to the Point Table, and note the addition of these entries for user-defined properties, such as light-post descriptions, hydrant specifications, and other proposed and existing features, as well as design elements.

7. The completed exercise can be seen in `Point Table Finished.dwg`.

Creating a Point Group to Control Visibility and Moving a Point Group to Surface

Earlier, you saw that default point styles and label styles can be assigned in the Command settings, but what happens when the points are already in your drawing or are accidentally placed with the wrong style? It would be tedious to select each point individually to make the change. Even making batch changes in Prospector would take a great deal of time.

Points often fit into categories. For example, topographic points can be grouped together for surface building, traverse points can be grouped to mark the path of surveyors, tree points can be separated into groups by specific species, and so on. These different categories can often be sorted and grouped for analysis, labeling, exporting, and other tasks. *Point groups* provide a tool for building sets of points that can be used not only to control visibility but also to organize points for more advanced applications, such as table creation.

You were introduced briefly to point groups in the section on point tables. This section, however, will more closely examine the options for creating and using point groups.

Point groups can organize points on the basis of many different properties including, but not limited to, number (or range of numbers), elevations, descriptions, and names. Advanced query-building functions, overrides, exclusions, and other tools can also help fine-tune your point group.

In this exercise, you'll learn how to use the Divide Alignment tool to create a series of tree points, create a point group for them, and then use the Elevations from Surface tool to bring the tree points up to the existing ground surface elevation:

1. Open the drawing `Tree exercise.dwg`, which you can download from `www.sybex.com/go/masteringcivil3d2010`. Note that it includes a surface and an alignment.
2. On the Home tab, select Points ➤ Choose Point Creation Tools on the Create Ground Data panel. The Create Points dialog appears.
3. Click the down arrow on the Alignment category, and select the Divide Alignment tool.
4. At the command prompt, pick the alignment; enter **25**↵ at the segment prompt and **40**↵ at the offset prompt. This places 25 points at equal intervals along the alignment at a 40′ offset. (When Civil 3D prompts for an offset, a positive number is a right offset.)
5. Enter **TREE**↵ as the point description, and press ↵ at the elevation prompt. This assigns an initial elevation of zero to the points. Press ↵ to accept each point placement; note that red Xs, designating the points, are created at even intervals, as shown in Figure 4.39. Your drawing may show an elevation of zero or, alternatively, no elevation listing. Both cases are OK at this point. (Note that instead of pressing ↵ each time to accept the description and elevation, you could also have changed the default point settings as described in the "Creating Basic Points" section of this chapter.)

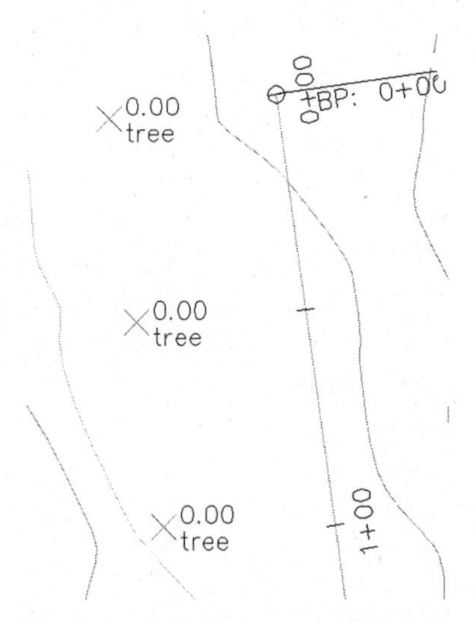

FIGURE 4.39
Using the Divide Alignment tool to place points

6. Repeat the process for the other side of the road. You can use the Divide Alignment tool, the Measure Alignment tool, or any other tool that appeals to you. Just remember that at this point, elevation isn't important.

7. Right-click the Point Groups collection in Prospector, and select New. The Point Group Properties dialog appears, and you can create a point group for your tree points.
8. Enter **Tree** in the Name text box. Select Tree from the drop-down menu in the Point Style selection box, and Elevation and Description in the Point Label Style selection box.
9. Switch to the Include tab. Select the With Raw Descriptions Matching check box, and type **TREE** into the text box, as shown in Figure 4.40.

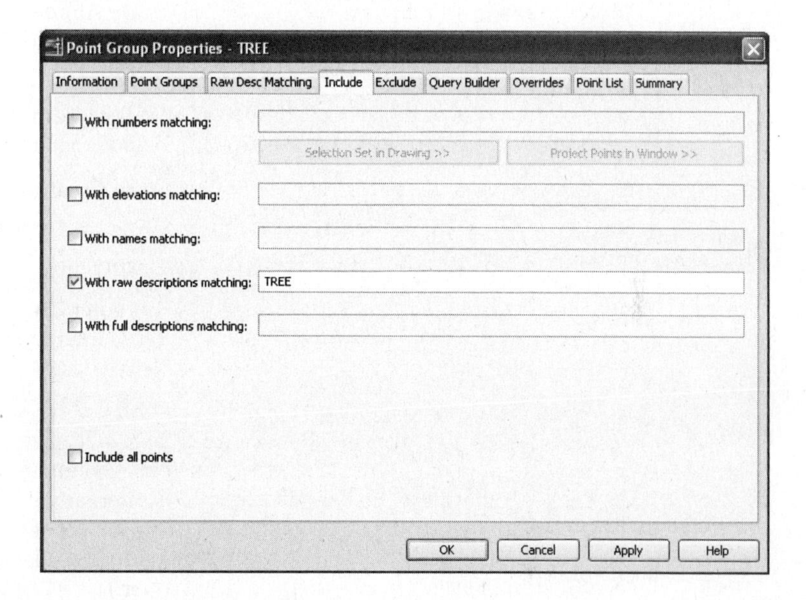

FIGURE 4.40
The Include tab of the Point Group Properties dialog

10. Click OK, and the point group will be created.
11. Zoom in on your points in the drawing. You should see the points as a tree block, as shown in Figure 4.41, with a label that reads *tree*. There are no elevations yet, but your drawing may list them as elevation zero.

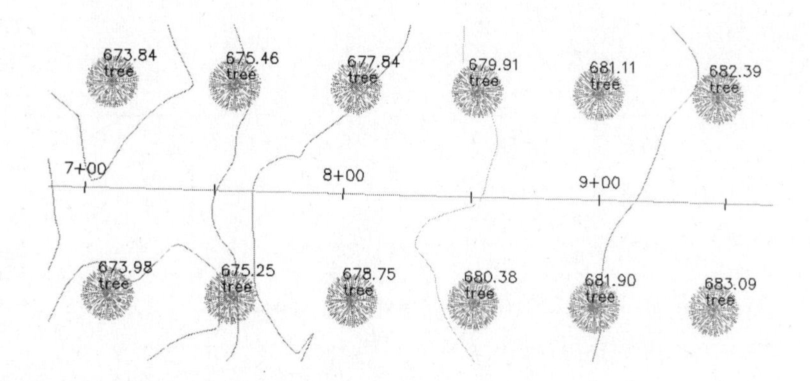

FIGURE 4.41
The completed exercise

12. Select a point and then select Elevations from Surface on the Modify panel. The Select Surface dialog appears.

13. Choose EG from the drop-down menu, and click OK.

14. The command line prompts for Points [All/Numbers/Group/Selection]<All>. Type **G** and press ↵ for Group. The Point Groups dialog appears.

15. Select Tree, and click OK. Note that the tree points now have elevations reflecting the existing ground (EG) surface, as shown in Figure 4.41. If your EG changes, reapply the Elevations from Surface tool.

Working with Description Keys

Description keys are a tool you can use to automatically control the visibility of points that meet certain criteria. Point groups control the style and labeling of the entire group of points, but description keys are applied to a single point as an override. In other words, a description key is an automated equivalent of manually choosing a point and changing its point style in Panorama.

If a new point is created when a description key set is active, the description key matches a point's raw description with a predefined set of style, label, format, and layer parameters. The example shown in Figure 4.42 indicates that all points created with the raw description of XTREE will be immediately identified by the description key as needing a tree style, a standard label style, and the existing tree format, and assigned to the layer V-NODE-TREE.

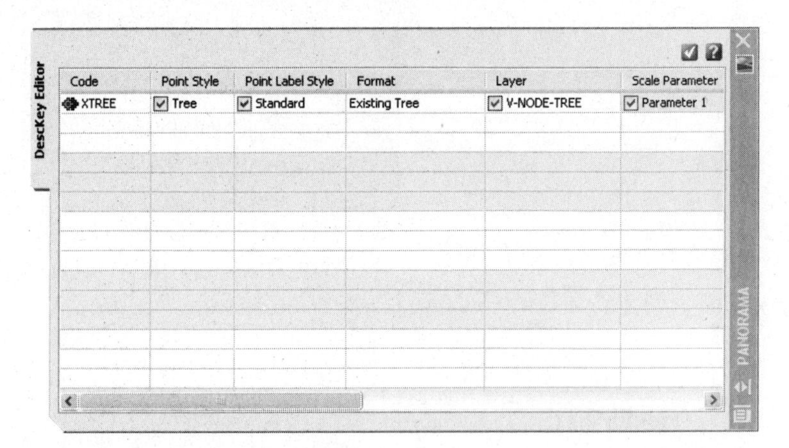

FIGURE 4.42
A tree description key

RAW DESCRIPTION VERSUS FULL DESCRIPTION

In Civil 3D vocabulary, the raw description as listed in Prospector and the code as listed in the DescKey Editor are identical. You can think of raw description or code as a field or machine-friendly string of characters that a surveyor would input while collecting points in an effort to save time by using minimal keystrokes. The description key set uses these codes to identify which description key to apply to a certain point. Typically, the codes are standardized by your company.

Along the same vein, full description and format are identical. Full description can "humanize" the raw description for use in labeling and identification purposes. For example, a code of TOB might be difficult to understand if used in the point label. In that case, you can apply a format of Top of Bank in the TOB description key and use the full description in the point-label style.

In cases where you'd like the format/full description to match the code/raw description, type **$*** in the Format column of the DescKey Editor. In description key speak, each word followed by a space is called a parameter, and each parameter is given the $ sign plus a number (beginning with 0) designator. When the asterisk is used directly after the dollar sign, you are simply telling Civil 3D to accept every parameter as described.

The important thing to remember about description keys is that unlike point groups, they only work for points that are newly created, imported, or converted in the drawing when the description key set is active. If you already have points in your drawing, the description key won't scan the drawing to make changes to existing points. This functionality is different than that of Land Desktop.

CREATING A DESCRIPTION KEY SET

Description key sets appear on the Settings tab of Toolspace under the Point branch. You can create a new description key set by right-clicking the Description Key Sets collection and choosing New, as shown in Figure 4.43.

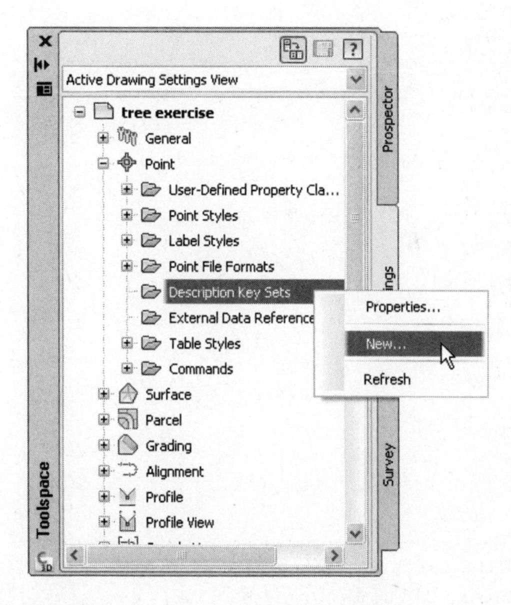

FIGURE 4.43
Creating a description key set

In the resulting Description Key Set dialog, give your description key set a meaningful name, and click OK. You'll create the actual description keys in another dialog.

CREATING DESCRIPTION KEYS

To enter the individual description key codes and parameters, right-click your description key set, as illustrated in Figure 4.44, and select Edit Keys. The DescKey Editor in Panorama appears.

To enter new codes, right-click a row with an existing key in the DescKey Editor, and choose New or Copy from the shortcut menu, as shown in Figure 4.45.

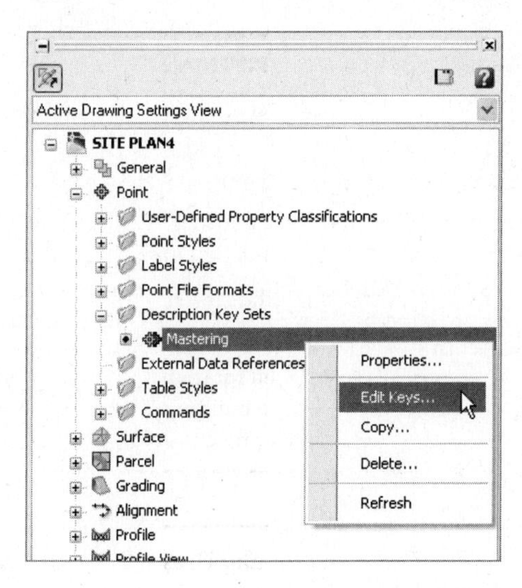

FIGURE 4.44
Editing a description key set

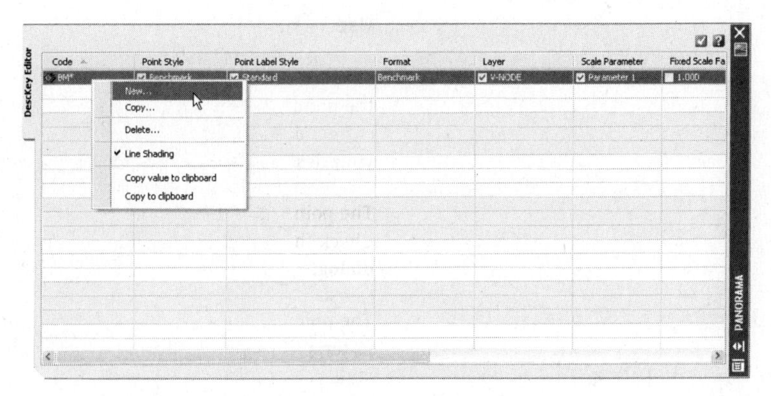

FIGURE 4.45
Creating or copying a description key

Table 4.3 describes the first five columns in the DescKey Editor, which are the most commonly used.

Most firms will want to create a complete description key set for their standard field code list. Figure 4.46 shows an example of a short description key set using some common field codes, styles, and formats.

USING WILDCARDS

The asterisk (*) acts as a wildcard in many places in Civil 3D. Two of the most common places to use a wildcard are the DescKey Editor and the Point Group Properties dialog. Whereas a DescKey code of TREE flags any points created with that raw description, a DescKey code of TREE* also picks up raw descriptions of TREE1, TREE2, TREEMAPLE, TREEOAK, and so on. You can use the wildcard the same way in the Point Group Properties dialog when specifying items to include or exclude.

More wildcard characters are available for use in the DescKey Editor. See page 533 of the "Civil 3D Users Guide" PDF (found under the Help menu) for more information.

POINT GROUPS OR DESCRIPTION KEYS?

After reading the last two sections, you're probably wondering which method is better for controlling the look of your points. This question has no absolute answer, but there are some things to take into consideration when making your decision.

Point groups are useful for both visibility control and sorting. They're dynamic and can be used to control the visibility of points that already exist in your drawing.

Both can be standardized and stored in your Civil 3D template.

Your best bet is probably a combination of the two methods. For large batches of imported points or points that require advanced rotation and scaling parameters, description keys are the better tool. For preparing points for surface building, exporting, and changing the visibility of points already in your drawing, point groups will prove most useful.

TABLE 4.3: Columns and Uses

COLUMN	USE
Code	The raw description or field code entered by the person collecting or creating the points, which works as an identifier for matching the point with the correct description key. Click inside this field to activate it, and then type your desired code. Wildcards are allowed.
Point Style	The point style that will be applied to points that meet the code criteria. Check the box, and then click inside the field to activate a style-selection dialog.
Point Label Style	The point-label style that will be applied to points that meet the code criteria. Check the box, and then click inside the field to activate a style-selection dialog.
Format	The format or full description that will be applied to points that meet the code criteria. You can type a value in this field or use **$*** to match code.
Layer	The layer that will be applied to points that meet the code criteria. Click inside this field to activate a layer-selection dialog.

ADVANCED DESCRIPTION KEY PARAMETERS

After the first five columns, you'll see additional columns containing advanced parameters. You can use these parameters to automatically scale and rotate your points based on information collected in the field. For more information and complete documentation on these features, see the *Civil 3D Users Guide* under the Help menu.

ACTIVATING A DESCRIPTION KEY SET

Once you've created a description key set, you must change the settings for your commands so that Civil 3D knows to match your newly created points with the appropriate key.

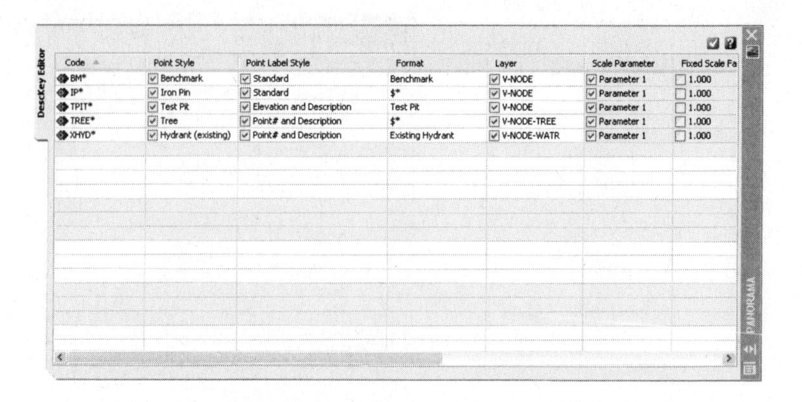

Code	Point Style	Point Label Style	Format	Layer	Scale Parameter	Fixed Scale Fa
BM*	Benchmark	Standard	Benchmark	V-NODE	Parameter 1	1.000
IP*	Iron Pin	Standard	$*	V-NODE	Parameter 1	1.000
TPIT*	Test Pit	Elevation and Description	Test Pit	V-NODE	Parameter 1	1.000
TREE*	Tree	Point# and Description	$*	V-NODE-TREE	Parameter 1	1.000
XHYD*	Hydrant (existing)	Point# and Description	Existing Hydrant	V-NODE-WATR	Parameter 1	1.000

FIGURE 4.46
A simple description key set containing five description keys

The Commands ➤ CreatePoints branches are stored on the Settings tab of Toolspace under the Point branch. Edit these command settings by right-clicking, as shown in Figure 4.47.

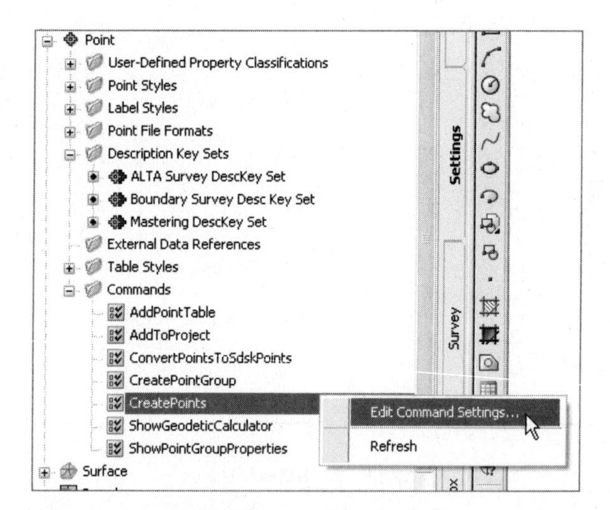

FIGURE 4.47
The CreatePoints command settings

In the Edit Command Settings dialog, ensure that Match On Description Parameters is set to True and Disable Description Keys is set to False, as shown in Figure 4.48.

If you have multiple description key sets in your drawing, you must specify a matching search order for the sets. When a new point is created, Civil 3D will search through the first set on the list and work its way down through all the description keys in your drawing.

To access this search order, locate the Description Key Sets collection on the Settings tab of Toolspace, right-click, and then choose Properties. The dialog shown in Figure 4.49 appears. Use the arrows to move the desired description key set to adjust the search order.

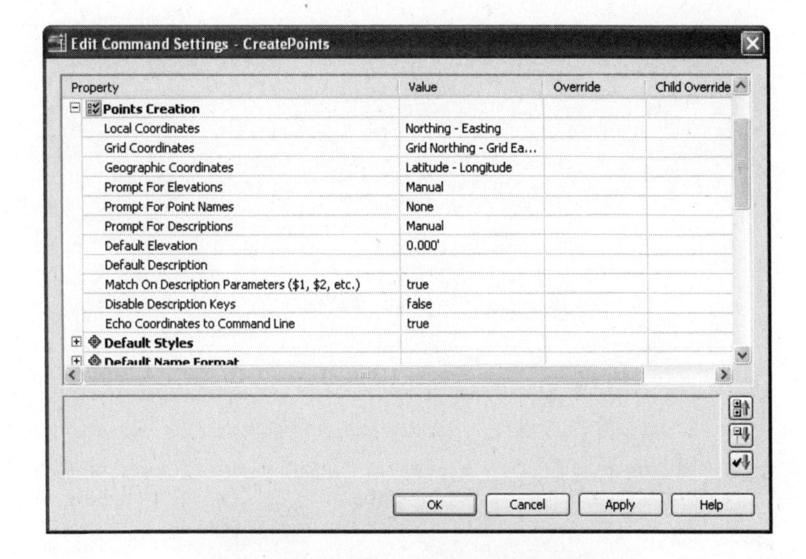

FIGURE 4.48
Set Disable Description Keys to False

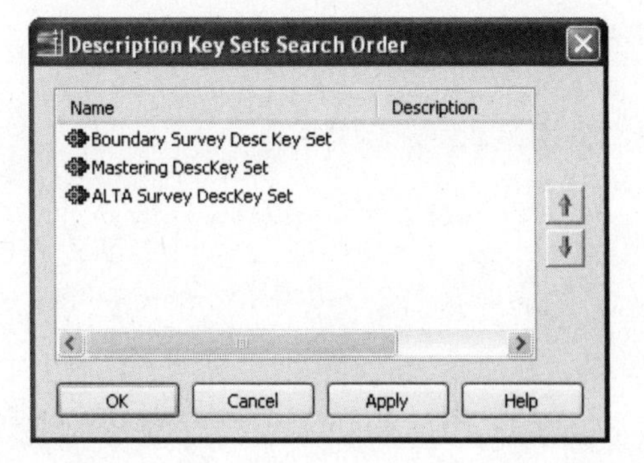

FIGURE 4.49
The Description Key Sets Search Order dialog

Real World Scenario

WORKING WITH LAYERS AND DESCRIPTION KEYS TOGETHER

It's common for surveyors to import points, apply description keys, and use the LAYISO command to isolate a group of points (or "shots" as they are commonly referred to) and create two-dimensional linework or breaklines. The following exercise walks you through the steps to apply this concept effectively.

1. Open the `Description Keys.dwg` file, which you can download from `www.sybex.com/masteringcivil3d2010`.

2. Choose Points from File from the Import panel of the Insert tab. The Import Points dialog appears.
3. Be sure the Format is PNEZD (comma-delimited), and then click the white + sign and select `Survey.txt` (which you can download from `www.sybex.com/masteringcivil3d2010`).
4. Select Open and then click OK to exit the dialogs and review the results.
5. At the `Command:` prompt, type **LAYISO** and press ←. Select one of the points labeled Top of Bank and press ← and notice that the layer is isolated. (Note: You may have to REGEN if you are still seeing other points not on the Top of Bank layer.)
6. Open the Description Key Sets branch on the Settings tab of the Toolspace.
7. Right-click on Civil 3D and select Edit Keys. The DescKey Editor will open in Panorama.
8. Review the layers found on the Display tab for the TOPB* and BOTB* Style.
9. Review the layer selected on the General tab in the Label Style Composer for both the TOPB* and BOTB* Point Label Style.
10. Review the Layer settings for both the BOTB* and TOPB* codes as selected in the DescKey Editor.

The Bottom Line

Import points from a text file using description-key matching. Most engineering offices receive text files containing point data at some point during a project. Description keys provide a way to automatically assign the appropriate styles, layers, and labels to newly imported points.

Master It Create a new drawing from `_AutoCAD Civil 3D (Imperial) NCS.dwt`. Revise the Civil 3D description key set to use the parameters listed in Table 4.4.

TABLE 4.4: Civil 3D Description Key Set Parameters

CODE	POINT STYLE	POINT LABEL STYLE	FORMAT	LAYER
GS	Standard	Elevation Only	Ground Shot	V-NODE
GUY	Guy Pole	Elevation and Description	Guy Pole	V-NODE
HYD	Hydrant (existing)	Elevation and Description	Existing Hydrant	V-NODE-WATR
TOP	Standard	Point#-Elevation-Description	Top of Curb	V-NODE
TREE	Tree	Elevation and Description	Existing Tree	V-NODE-TREE

Import the `MasteringPointsPNEZDspace.txt` file from the data location, and confirm that the description keys made the appropriate matches by looking at a handful of points of each type. Do the trees look like trees? Do the hydrants look like hydrants?

Create a point group. Building a surface using a point group is a common task. Among other criteria, you may want to filter out any points with zero or negative elevations from your Topo point group.

Master It Create a new point group called Topo that includes all points *except* those with elevations of zero or less.

Export points to LandXML and ASCII format. It's often necessary to export a LandXML or an ASCII file of points for stakeout or data-sharing purposes. Unless you want to export every point from your drawing, it's best to create a point group that isolates the desired point collection.

Master It Create a new point group that includes all the points with a raw description of TOP. Export this point group via LandXML and to a PNEZD Comma Delimited text file.

Create a point table. Point tables provide an opportunity to list and study point properties. In addition to basic point tables that list number, elevation, description, and similar options, you can customize point-table formats to include user-defined property fields.

Master It Create a point table for the Topo point group using the PNEZD format table style.

Chapter 5

The Ground Up: Surfaces in Civil 3D

Although it's fun to play in fantasy land, designing in a void, at some point you have to get real to get things built. Once the survey has come in, the boundaries have been laid out, and the project has been defined, you have to start building a real model. One of the most primitive elements in a 3D model of any design is the surface. This chapter looks at various methods of surface creation and editing. Then it moves into discussing ways to view, analyze, and label surfaces, and explores how they interact with other parts of your project.

By the end of this chapter, you'll be able to

- Create a preliminary surface using freely available data
- Modify and update a TIN surface
- Prepare a slope analysis
- Label surface contours and spot elevations

Digging In

A surface in Civil 3D is built on the basis of mathematical principles of planar geometry. Each face of a surface is based on three points defining a plane. Each of these triangular planes shares an edge with another, and a continuous surface is made. This methodology is typically referred to as a triangulated irregular network (TIN). On the basis of Delaunay triangulation (c. 1934), this means that for any given (x,y) point, there can be only one unique z value within the surface (as slope is equal to rise over run, when the run is equal to zero, the result is "undefined"). What does this mean to you? It means surfaces in Civil 3D have two major limitations:

No Thickness Operations on the basis of solid modeling are not possible. You cannot add or subtract surfaces or look for their unions as you can with a solid that has thickness in the vertical direction.

No Vertical Faces Vertical faces cannot exist in a TIN because two points on the surface cannot have the same (x,y) coordinate pair. At a theoretical level, this limits the ability of Civil 3D to handle true vertical surfaces, such as walls or curb structures. This must be considered when modeling corridors as discussed in Chapter 11.

Beyond these basic limitations, surfaces are flexible and can describe any object's face in astonishing detail. The surfaces can range in size from a few square feet to square miles and generally process quickly.

There are two main categories of surfaces in Civil 3D: standard surfaces and volume surfaces. A standard surface is based on a single set of points, whereas a volume surface builds a surface by measuring vertical distances between surfaces. Each of these surfaces can also be a grid or TIN surface. The grid version is still a TIN upon calculation of planar faces, but the data points are arranged in a regularly spaced grid of information. The TIN surface definition is made from randomly located points that may or may not follow any pattern to their location.

Creating Surfaces

Before you can analyze or do any other fun things with a surface, you have to make one! To the land development company today, this can mean pulling information from a large number of sources, including Internet sources, old drawings, and fieldwork. Working with each requires some level of knowledge about the reliability of the information and how to handle it in Civil 3D. In this section, you look at obtaining data from a couple of free sources and bringing it into your drawing, creating new surfaces, and making a volume surface.

Before creating surfaces, you need to know a bit about the components that can be used as part of a surface definition:

LandXML Files These typically come from an outside source or are exported from another project. LandXML has become the *lingua franca* of the land development industry. These files include information about points and triangulation, making replication of the original surface a snap.

TIN Files Typically, a TIN file will come from a land development project on which you or a peer has worked. These files contain the baseline TIN information from the original surface and can be used to replicate it easily.

DEM Files Digital Elevation Model (DEM) files are the standard format files from governmental agencies and GIS systems. These files are typically very large in scale, but can be great for planning purposes.

Point Files Point files work well when you're working with large data sets where the points themselves don't necessarily contain extra information. Examples include laser scanning or aerial surveys.

Point Groups Civil 3D point groups or survey point groups can be used to build a surface from their respective members and maintain the link between the membership in the point group and being part of the surface. In other words, if a point is removed from a group used in the creation of a surface, it is also removed from the surface.

Boundaries Boundaries are closed polylines that determine the visibility of the TIN inside the polyline. Outer boundaries are often used to eliminate stray triangulation, whereas others are used to indicate areas that could perhaps not be surveyed, such as a building pad.

Breaklines Breaklines are used for creating hard-coded triangulation paths, even when those paths violate the Delaunay algorithms for normal TIN creation. These can describe anything from the top of a ridge to the flowline of a curb section. A TIN line may not cross the path of a breakline.

Drawing Objects AutoCAD objects that have an insertion point at an elevation (e.g., text, blocks, etc.) can be used to populate a surface with points. It's important to remember that the objects themselves are not connected to the surface in any way.

Edits Any manipulation after the surface is completed, such as adding or removing triangles or changing the datum, will be part of the edit history. These changes can be viewed in the Properties of a surface and can be toggled on and off individually to make reviewing changes simple.

Working with all of these elements, you can model and render almost anything you'd find in the world — and many things you wouldn't. In the next section, you start actually building some surfaces.

Free Surface Information

You can find almost anything on the Internet, including information about your project site that probably includes level information you can use to build a surface. For most users, free surface information can be gathered from government entities or Google. You look at both in this section.

SURFACES FROM GOVERNMENT DIGITAL ELEVATION MODELS

One of the most common forms of free data is the Digital Elevation Model (DEM). These files have been used by the U.S. Department of the Interior's United States Geological Survey (USGS) for years and are commonly produced by government organizations for their GIS systems. The DEM format can be read directly by Civil 3D, but the USGS typically distributes the data in a complex format called Spatial Data Transfer Standard (SDTS). The files can be converted using a freely available program called `sdts2dem`. This DOS-based program converts the files from the SDTS format to the DEM format you need. Once you are in possession of a DEM file, creating a surface from it is relatively simple, as you'll see in this exercise:

1. Start a new blank drawing from the NCS Imperial Extended template that ships with Civil 3D.
2. Switch to the Settings tab of Toolspace, right-click the drawing name, and select Edit Drawing Settings. Set the coordinate system as shown in Figure 5.1 via the Drawing Settings dialog and click OK. The coordinate system of the DEM file that you will import will be set to match the coordinate system of the drawing.

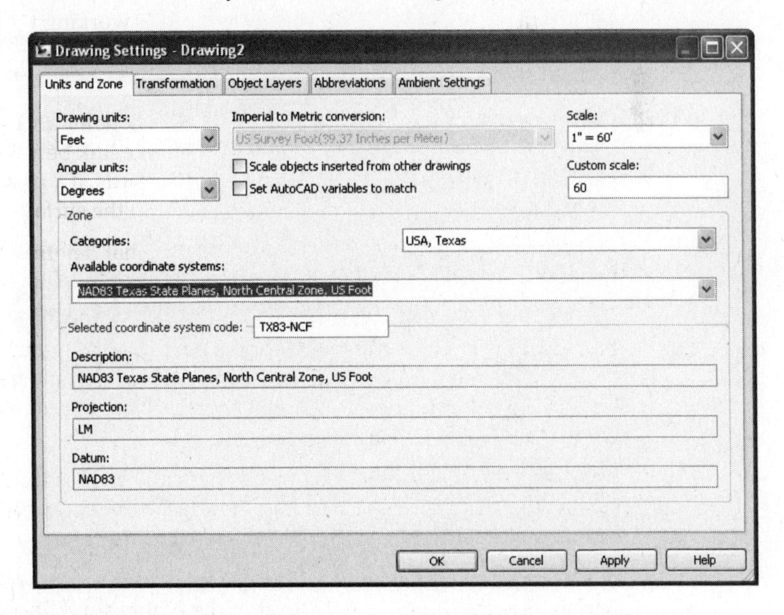

FIGURE 5.1 Civil 3D coordinate settings for DEM import

3. In Prospector, right-click the Surfaces collection and select the Create Surface option. The Create Surface dialog appears.

4. Accept the options in the dialog, and click OK to create the surface. This surface is added as Surface 1 to the Surfaces collection.

5. Expand the Surfaces ➤ Surface 1 ➤ Definition branch, as shown in Figure 5.2.

FIGURE 5.2
Adding DEM data to a surface

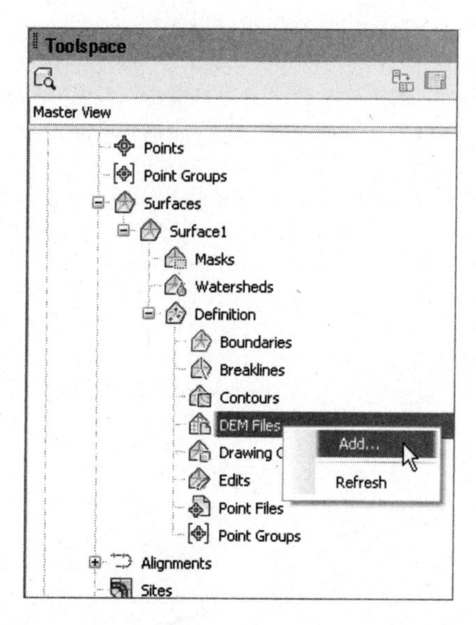

6. Right-click DEM Files and select the Add option (see Figure 5.2). The Add DEM File dialog appears.

7. Navigate to the `McKinneyWest.DEM` file and click Open. (Remember, all data and drawing files for this book can be downloaded from `www.sybex.com/go/masteringcivil3d2010`.)

8. Set the values in the DEM File Properties dialog as shown in Figure 5.3 and click OK. This translates the DEM's coordinate system to the drawing's coordinate system.

9. Right-click Surface 1 in Prospector and select Surface Properties. (Alternatively, you could right-click Surface 1 in Prospector and select Zoom To to bring the surface into view, and then right-click the surface in your drawing and select Surface Properties.) The Surface Properties dialog appears.

10. On the Information tab, change the Name field to **McKinney W**.

11. Change the Surface Style drop-down list to Border Only, and then click OK to dismiss the Surface Properties dialog.

Once you have the DEM data imported, you can pause over any portion of the surface and see that Civil 3D is providing feedback through a Tooltip. This surface can be used for preliminary planning purposes but isn't accurate enough for construction purposes. The main drawback to DEM data is the sheer bulk of the surface size and point count. The McKinney West DEM file just imported contains 1.6 million points and covers more than 62 square miles. This much data can be overwhelming, and it covers an area much larger than the typical site. You'll look at some data reduction methods later in this chapter.

In addition to making a DEM a part of a TIN surface, you can build a surface directly from the DEM (Select the Surfaces branch, right-click, and select Create Surface from DEM). The drawback to this is that no coordinate transformation is possible. Because one of the real benefits of using georectified data is pulling in information from differing coordinate systems, you're skipping this method to focus on the more flexible method shown here.

WHERE TO FIND FREE INFORMATION

Numerous websites contain free GIS information, but it can be hard to keep up with them. Scott McEachron has been a fan of free GIS data for years. His Autodesk University presentations on getting free data and using it are a favorite — he always keeps his list of sites up to date. Now you can find that list at his Civil 3D–related blog "Paving the Way" at `http://blog.121pcs.com`.

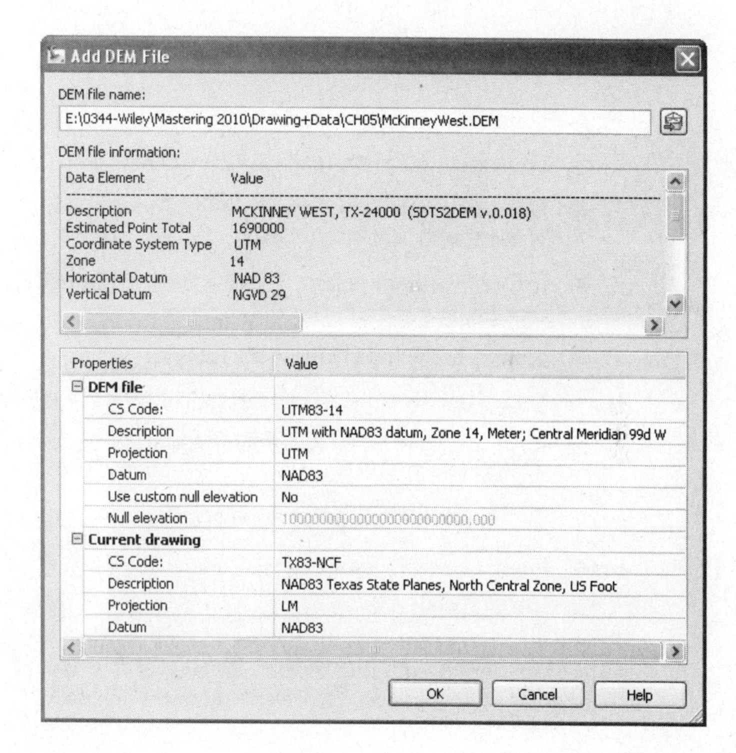

FIGURE 5.3
Setting the McKinney `West.DEM` file properties

SURFACES FROM GOOGLE EARTH

Civil 3D also includes an importing function that brings in surface and image information directly from Google Earth. The Digital Elevation Models (DEMs) used by Google Earth were collected over a 10-day span in February 2000 by the Space Shuttle Endeavor. The data, known as SRTM (Shuttle Radar Topography Mission) data, is typically not updated on a large scale. Ground control was not used during the collection of data, and the mission sought to achieve a vertical accuracy of just 16 meters. Because most freely available DEMs have been gathered by digitizing USGS QUADs, you can generally assume that SRTM data is the best freely available information out there. In this exercise, you'll look at importing a Google Earth location as a Civil 3D surface.

1. Download the latest version of Google Earth from `http://earth.google.com` and install it.

GOOGLE EARTH AND VERSIONS

This data and exercise were tested with Version 5.0.11337.1968. Due to a programming change on the Google Earth side, some later versions are picky about the amount of data Civil 3D pulls. Depending on the version installed on your machine, you might want to search the Web for information regarding the Civil 3D and Google Earth interactions. Obviously, we would suggest you first stop at `www.civil3d.com` and search for more information.

2. Launch Google Earth and get connected.
3. From the main menu, choose File ➤ Open.
4. Navigate to the Data directory and select the `Efficiency Acres.kmz` file to restore a view of a site in McKinney, Texas.
5. In Civil 3D, create a new drawing from the Imperial NCS Extended template and set the coordinate system as you did in the prior exercise.
6. Change to the Insert tab on the Ribbon.
7. On the Import panel, select Google Earth ➤ Google Earth Surface.
8. Press ↵ to accept the coordinate system as shown and the Surface Creation dialog will appear.
9. Accept the defaults in the Surface Creation dialog and click OK to dismiss the dialog.
10. From the main menu, choose View ➤ Zoom ➤ Extents to see something like Figure 5.4.

The interesting thing about surfaces built from Google Earth is that their accuracy is zoom-level dependent. This means that the tighter you are zoomed into a site in Google Earth, the better the surface you derive from that picture. Because of this dependence, you should attempt to zoom in as tightly as possible on the area of interest when using Google Earth for preliminary surface information.

Inexpensive Surface Approximations

Inexpensive is a relative term, but compared with on-the-ground surveying, aerial and laser-scanning services are inexpensive, especially in difficult terrain or over large tracts. In this section, you'll work with elevated polylines and a large point cloud delivered as a text file. These are quite common, and historically it can be difficult making an acceptable surface from them.

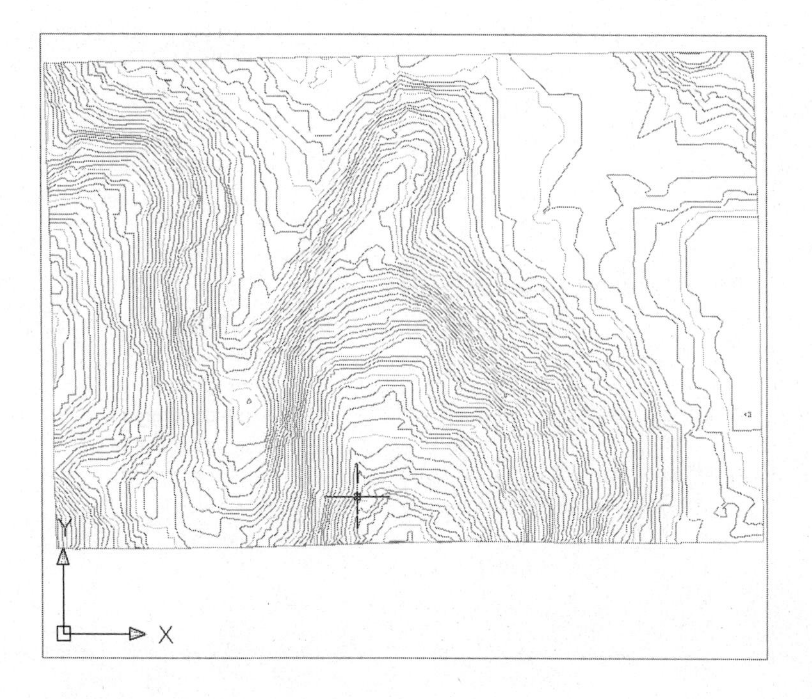

FIGURE 5.4
Completed Google Earth surface import

SURFACES FROM AERIAL CONTOUR INFORMATION

One of the common complaints of converting a drawing full of contours at elevation into a working digital surface is that the resulting contours don't accurately reflect the original data. Civil 3D includes a series of surface algorithms that work very well at matching the resulting surface to the original contour data. You'll look at those surface edits in this series of exercises.

1. Open the `Aerial Contours.dwg` file. Note that the contours in this file are composed of polylines.
2. In Prospector, right-click the Surfaces branch, and select the Create Surface option. The Create Surface dialog appears.
3. Leave the Type field as TIN Surface but change the Name value to **EG-Aerial**.
4. Change the Description to something appropriate.
5. Change the Style drop-down list to Aerial and click OK to close the dialog.
6. Expand the Surfaces ➤ EG-Aerial ➤ Definition branches.
7. Right-click Contours and select the Add option. The Add Contour Data dialog appears.
8. Set the options as shown in Figure 5.5 and click OK. (You will return to the Minimize Flat Areas By options in a bit.)
9. Enter **ALL** at the command line to select all of the entities in the drawing. You can dismiss Panorama if it appears and covers your screen.

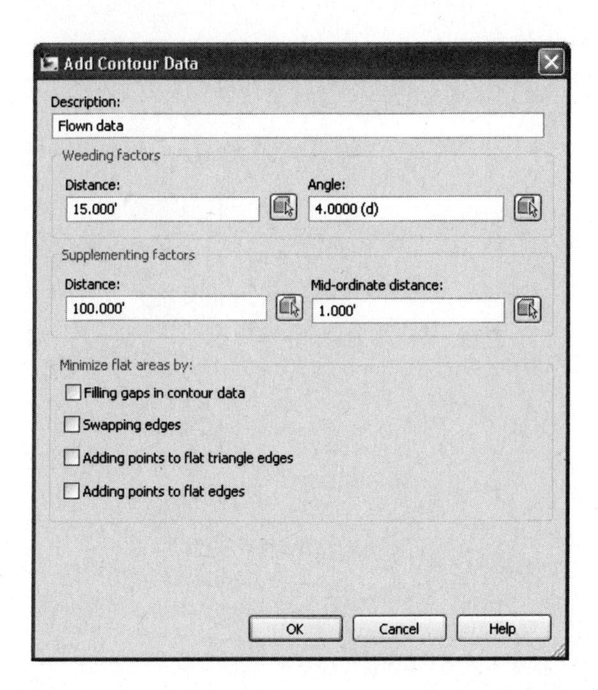

FIGURE 5.5
The Add Contour Data dialog

The contour data has some tight curves and flat spots where the basic contouring algorithms simply fail. Zoom in to any portion of the site, and you can see these areas by looking for the red original contour not matching the new Civil 3D-generated contour. You'll fix that now:

1. Expand the Definition branch and right-click Edits. Select the Minimize Flat Areas option to open the Minimize Flat Areas dialog. Note that the dialog has the same options found in that portion of your original Add Contour Data dialog. You just did it as two steps to illustrate the power of these changes!
2. Click OK.

Now the contours displayed more closely match the original contour information. There might be a few instances where there are gaps between old and new contour lines, but in a cursory analysis, none was off by more than 0.4′ in the horizontal direction — not bad when you're dealing with almost a square mile of contour information. You'll see how this was done in this quick exercise:

1. Zoom into an area with a dense contour spacing and select the surface.
2. Click the Surface Properties button from the Modify panel.
3. Change the Surface Style field to Contours with Points and click OK to see a drawing similar to Figure 5.6.

In Figure 5.6, you're seeing the points the TIN is derived from, with some styling applied to help understand the creation source of the points. Each point in red is a point picked up from the contour data itself. The blue points are all added data on the basis of the Minimize Flat Areas edits. These points make it possible for the Civil 3D surface to match almost exactly the input contour data.

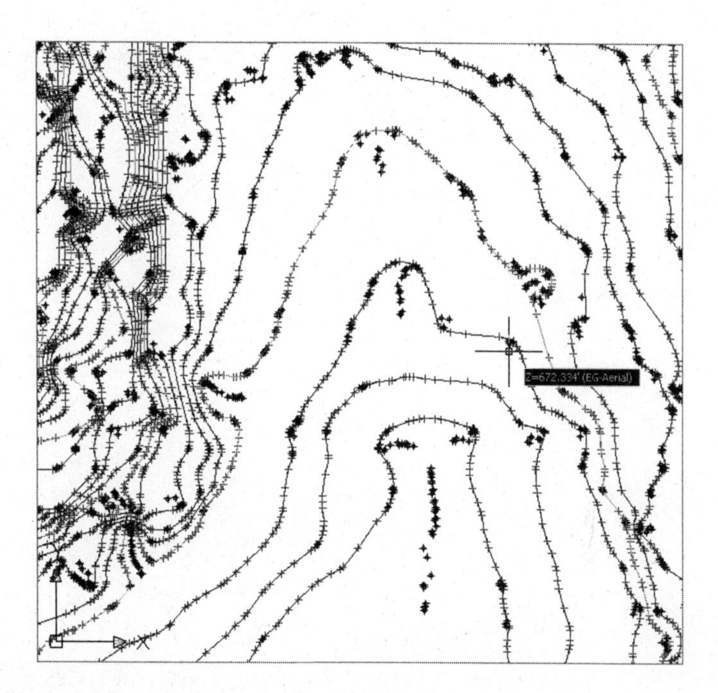

FIGURE 5.6
Surface data points and derived data points

SURFACES FROM POINT CLOUDS OR TEXT FILES

Besides receiving polylines, it is common for an aerial surveying company to also send a simple text file with points. This isn't an ideal situation because you have no information about breaklines or other surface features, but it is better than nothing or using a Google Earth–derived surface. Because you have the same aerial surface described as a series of points, you'll add them to a surface in this exercise:

1. Create a new drawing using the NCS Extended Template.
2. From the Create Ground Data panel on the Home tab, choose Surfaces ➤ Create Surface. The Create Surface dialog appears.
3. Change the Name value to **Aerial Points**, and click OK to close the dialog.
4. In Prospector, expand the Surfaces ➤ Aerial Points ➤ Definition branches.
5. Right-click Point Files and select the Add option. The Add Point File dialog shown in Figure 5.7 appears.
6. Set the Format field to ENZ (Comma Delimited).
7. Click the Browse button shown in Figure 5.7. The Select Source File dialog opens.
8. Navigate to the Data folder, and select the `Point Cloud.txt` file. Click OK.
9. Click OK to exit the Add Point File dialog and build the surface. Panorama will appear, but you can dismiss it.
10. Right-click Aerial Points in Prospector and select the Zoom To option to view the new surface created.

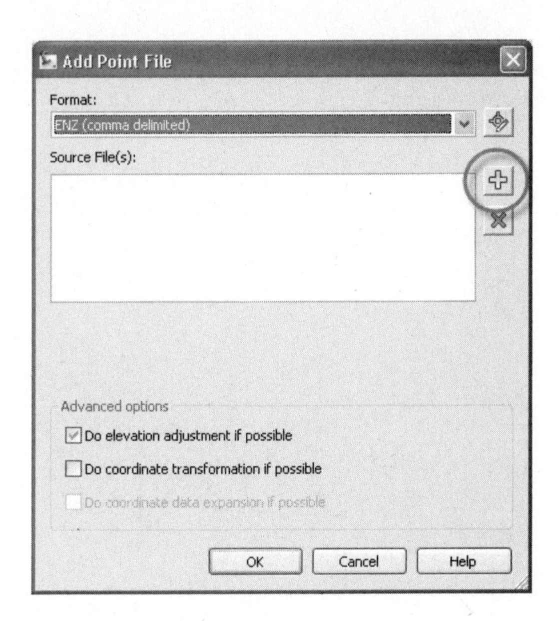

FIGURE 5.7
Adding a point file to the surface definition

This surface looks much like the one created from polylines, as it should. In both cases, you're making surfaces from the best information available. When doing preliminary work or large-scale planning, these types of surfaces are great. For more accurate and design-based surfaces, you typically have to get into field-surveyed information. You'll look at that next.

On-the-Ground Surveying

DEM and Google Earth are good starting points, and aerial or laser-scanned data can be a solid addition, but most land development projects get built after the ground topographic (topo for short) work is performed. In this exercise, you'll look at building a surface from a point group created by surveyed points. Once you've completed the basic surface building, it will be time to edit and refine it further.

1. Open the `Surface Points.dwg` file.
2. Right-click the Surfaces branch in Prospector and select the Create Surface option. The Create Surface dialog appears.
3. Change the Name to **EG**.
4. Click the Style value field and then click the ellipsis button to open the Select Surface Style dialog. Select the Contours 2′ and 10′ (Design) option, and then click OK to close the dialog.
5. Click OK again to close the Create Surface dialog.
6. In Prospector, expand the Surfaces ➤ EG ➤ Definition branches.
7. Right-click Point Groups and select the Add option. The Point Groups dialog appears.
8. Select the Field Work point group and click OK. The dialog closes and your screen should look like Figure 5.8.

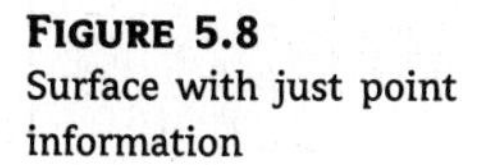

FIGURE 5.8
Surface with just point information

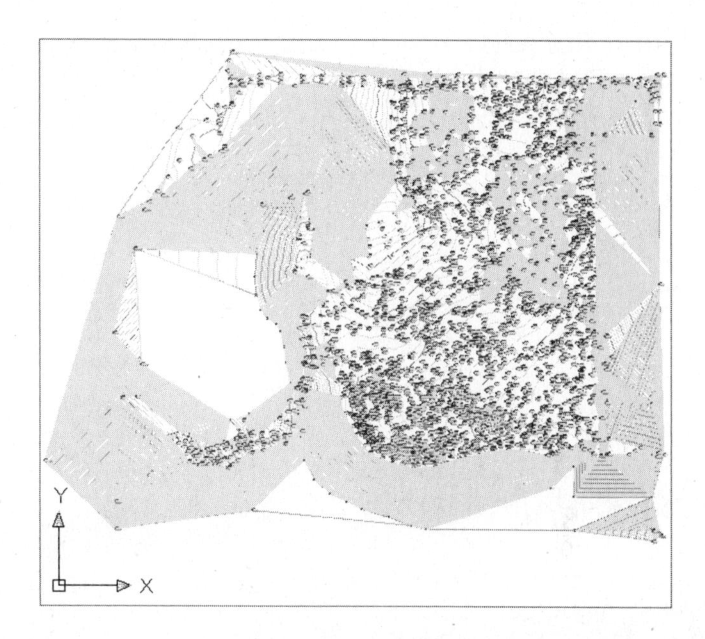

Simply adding surface information to a TIN definition isn't enough. To get beyond the basics, you need to look at the edits and other types of information that can be part of a surface.

Refining and Editing Surfaces

Once a basic surface is built, and, in some cases, even before it is built, you can do some cleanup and modification to the TIN construction that make it much more usable and realistic. Some of these edits include limiting the input data, tweaking the triangulation, adding in breakline information, or hiding areas from view. In this section, you look at a number of ways to refine surfaces to end up with the best possible model from which to build.

Surface Properties

The most basic steps you can perform in making a better model are right in the Surface Properties dialog. The surface object contains information about the build and edit operations, along with some values used in surface calculations. These values can be used to tweak your surface to a semi-acceptable state before more manual operations are needed.

In this exercise, you go through a couple of the basic surface-building controls that are available. You'll do them one at a time in order to measure their effects on the final surface display.

1. Open the `Surface Properties.dwg` file.
2. Expand the Surfaces branch.
3. Right-click EG and select Surface Properties. The Surface Properties dialog appears.
4. Select the Definition tab. Note the list at the bottom of the dialog.
5. Under the Definition Options at the top of the dialog, expand the Build option.

The Build options of the Definition tab allow you to tweak the way the triangulation occurs. The basic options are listed here:

Copy Deleted Dependent Objects When you select Yes and an object that is part of the surface definition (such as the polylines you used in your aerial surface, for instance) is deleted, the information derived from that object is copied into the surface definition. Setting this option to True in the Aerial Surface properties would let you erase the polylines from the drawing file while still maintaining the surface information.

Exclude Elevations Less Than Setting this to Yes puts a floor on the surface. Any point that would be built into the surface, but is lower than the floor, is ignored. In the EG surface, there are calculated boundary points with zero elevations, causing real problems that can be solved with this simple click. The floor elevation is controlled by the user.

Exclude Elevations Greater Than The idea is the same as with the preceding option, but a ceiling value is used.

Use Maximum Triangle Length This setting attempts to limit the number of narrow "sliver" triangles that typically border a site. By not drawing any triangle with a length greater than the user input value, you can greatly refine the TIN.

Convert Proximity Breaklines to Standard Toggling this to Yes will create breaklines out of the lines and entities used as proximity breaklines. You'll look at this more later.

Allow Crossing Breaklines Determines what Civil 3D should do if two breaklines in a surface definition cross each other. As mentioned, an (x,y) coordinate pair cannot have two z values, so some decision must be made about crossing breaklines. If you set this to Yes, you can then select whether to use the elevation from the first or the second breakline or to average these elevations.

In this next portion of the exercise, you limit the build options in order to create a better model:

1. Set the Exclude Elevations Less Than value to Yes.
2. Set the Value to **200** and click OK to exit the dialog. Elevations less than 200′ will be excluded.

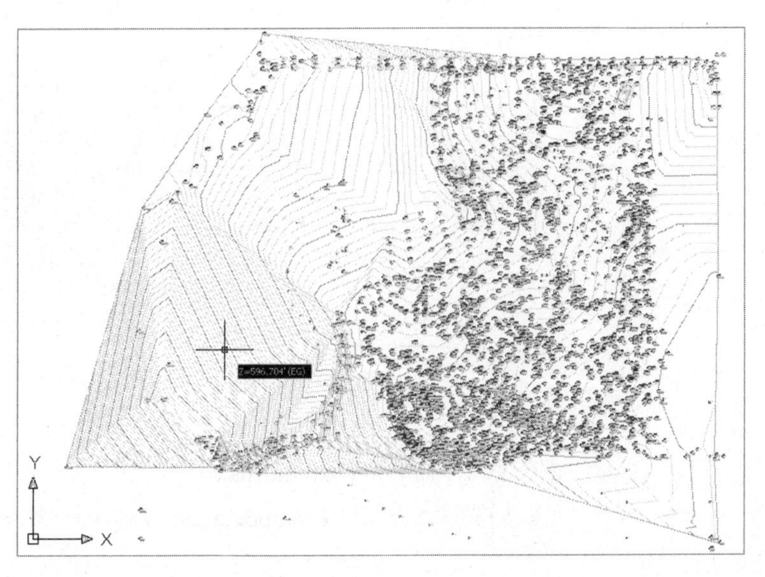

FIGURE 5.9
EG surface after ignoring low elevations

3. A warning message will appear. Civil 3D is simply warning you that your surface definition has changed. Click Rebuild the Surface to rebuild the surface. When it's done, it should look like Figure 5.9.

Although this surface is better than the original, there are still huge areas being contoured that probably shouldn't be. By changing the style to review the surface, you can see where you still have some issues.

1. Bring up the Surface Properties dialog again, and switch to the Information tab.
2. Change the Surface Style field to Contours and Triangles.
3. Click Apply. This makes the changes without exiting the dialog.
4. Drag the dialog to the side so you can see the site. On the west side of the site, you can see some long triangles formed in areas where a survey was taken to pick up an offsite easement and tie into existing survey monuments.
5. Switch to the Definition tab.
6. Expand the Build option.
7. Set the Use Maximum Triangle Length value to Yes.

8. In the Maximum Triangle Length value field, there is a Pick on Screen button. Click it and the dialog disappears.
9. Pick points as shown in Figure 5.10. The value should return around 850′. After picking the second point, the Surface Properties dialog should reappear.

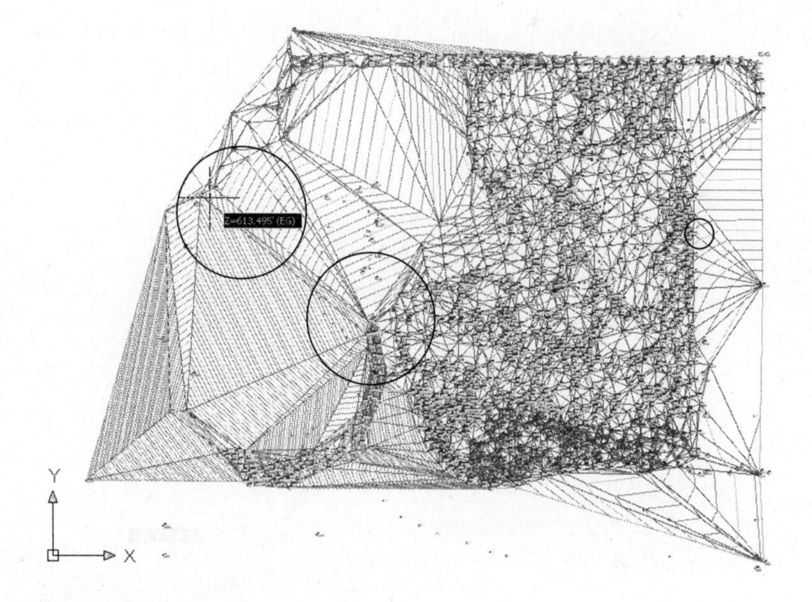

FIGURE 5.10
Pick points for the Max Triangle Length

10. Click OK to apply and exit the dialog.
11. Click Rebuild the Surface to update and dismiss the warning message.

The value is a bit high, but it is a good practice to start with a high value and work down to avoid losing any pertinent data. Setting this value to 600′ will result in a surface that is acceptable because it doesn't lose a bunch of important points. Beyond this, you'll need to look at making some edits to the definition itself instead of modifying the build options.

Surface Additions

Beyond the simple changes to the way the surface is built, you can look at modifying the pieces that make up the surface. In the case of your drawing so far, you have merely been building from points. While this is OK for small surfaces, you need to go further in the case of this surface. In this section, you add a few breaklines and a border and finally perform some manual edits to your site.

You Can't Always Get What You Want

But sometimes you get what you need. Autodesk has included the ability to reorder the build operations within the Surface Definition tab. If you look at the lower left of the Surface Properties Definition tab shown here, you'll find that there are arrows to the left of the list box showing all the data, edits, and changes you've made to the surface.

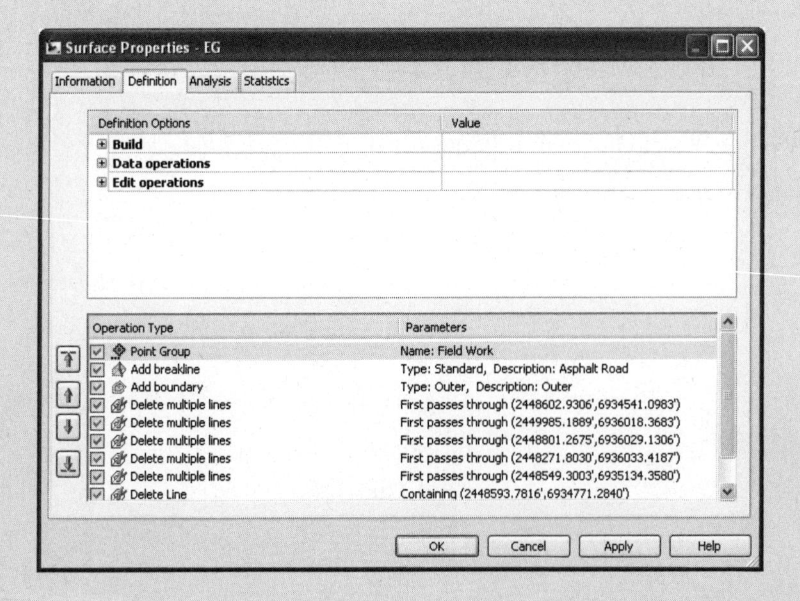

As Civil 3D builds a surface, it processes this data and information from top to bottom — in this case, adding points, and then the breaklines, and then a boundary, and so on. If a later operation modifies one of these additions or edits, the later operation takes priority. To change the processing order, select an operation, and then use the arrows at left to push it up or down within the process. One common example of this will be to place a boundary as the last operation to ensure accurate triangulation. You'll look at boundaries in the next section.

Adding Breakline Information

Breaklines can come from any number of sources. They can be approximated on the basis of aerial photos of the site that help define surface features or can be directly input from fieldbook files and the Civil 3D survey functionality. Five types of breaklines are available for use:

- **Standard breaklines** — Built on the basis of 3D lines, feature lines, or polylines. They typically connect points already included but can contain their own elevation data. Simple-use cases for connecting the dots include linework from a survey or drawing a building pad to ensure that a flat area is included in the surface. Feature lines and 3D polylines are often used as the mechanism for grading design and include their own vertical information. This might be the description of a parking lot area or a drainage swale behind a building, for instance.
- **Proximity breaklines** — Allow you to force triangulation without picking precise points. These lines force triangulation but will not add vertical information to the surface.
- **Wall breaklines** — Define walls in surfaces. Because of the limitation of true vertical surfaces, a wall breakline will let you approximate a wall without having to create an offset. These are defined on the basis of an elevation at a vertex, and then an elevation difference at each vertex.
- **Nondestructive breaklines** — Designed to maintain the integrity of the original surface while updating triangulation.
- **From File** — Can be selected if a text file contains breakline information. This can be the output of another program and can be used to modify the surface without the creation of additional drawing objects.

In most cases, you'll build your surfaces from standard, proximity, and wall breaklines. In this example, you add in some breaklines that describe road and surface features:

1. Open the `Surface Additions.dwg` file.
2. In Prospector, expand the Surfaces ➢ EG ➢ Definition branches.
3. Right-click Breaklines and select the Add option. The Add Breaklines dialog appears.
4. Enter a description and the settings as shown in Figure 5.11. Click OK to accept the settings and close the dialog.
5. Pick the two polylines along the north portion of the site and press ↵ to finish.
6. Right-click Breaklines and select the Add option again.
7. In the Description field, enter **Gravel Road**, and click OK.
8. Zoom to the southwest portion of the site and pick the two longer polylines, as shown in Figure 5.12.
9. Press ↵ to finish the command.
10. Add one more set of breaklines and enter **Slopes** in the Description field.
11. Pick the other polylines on the site. The surveyors tagged these features with Toe and Top point descriptions, so you want to make sure the surface reflects the grade breaks that were found.
12. Press ↵ to complete the command.

FIGURE 5.11
The Add Breaklines dialog

Add Breaklines
Description:
Asphalt Road
Type:
Standard
File link options:
Break link to file
Weeding factors
Distance:
15.000'
Angle:
4.0000 (d)
Supplementing factors
Distance:
100.000'
Mid-ordinate distance:
1.000'
OK Cancel Help

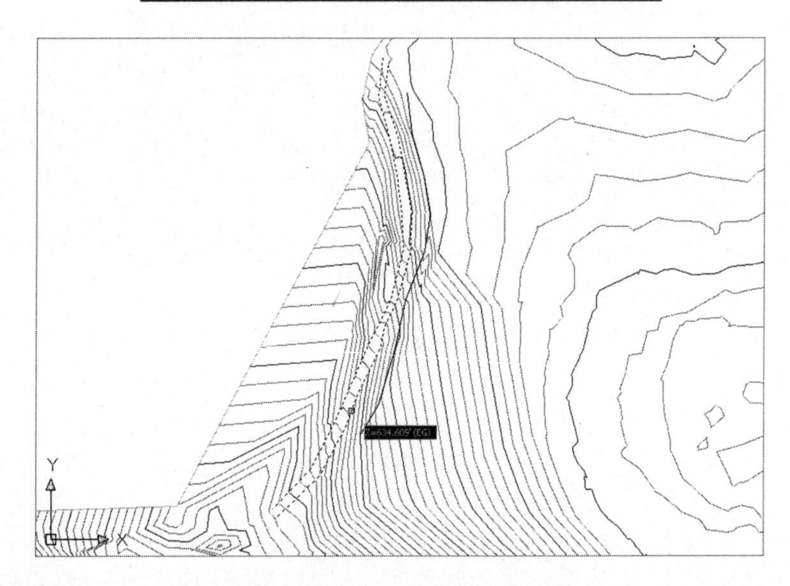

FIGURE 5.12
Selecting the gravel road breaklines

The surface changes in this case are fairly subtle but are still visible. On sites with more extreme grade breaks, such as those that might follow a channel or a site grading, breaklines are invaluable in building the correct surface.

ADDING A SURFACE BORDER

In the previous exercise, you fixed some minor breakline issues. However, in the data presented, the bigger issue is still the number of inappropriate triangles that are being drawn along the edge of the site. It is often a good idea to leave these triangles untouched during the initial build of a surface, because they serve as pointers to topographical data (such as monumentation, control, utility information, and so on) that may otherwise go unnoticed without a visit to the site. This is a common problem and can be solved by using a surface border. You can sketch in a polyline to approximate a border, but the Extract Objects from Surface utility gives you the ability to use the surface itself as a starting point.

The Extract Objects from Surface utility allows you to re-create any displayed surface element as an independent AutoCAD entity. This can be the contours, grid, 3D faces, and so forth. In this exercise, you extract the existing surface boundary as a starting point for creating a more refined boundary that will limit triangulation:

1. Select Extents from the Navigate panel on the View tab to view the whole surface on screen.
2. Select the surface and select Extract Objects on the Surface Tools panel to open the Extract Objects from Surface dialog.
3. Deselect the Major and Minor Contour options, as shown in Figure 5.13.

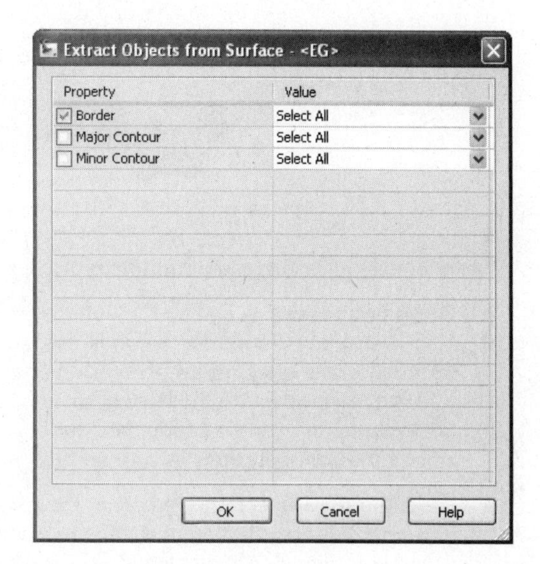

FIGURE 5.13
Extracting the border from the surface object

4. Click OK to finish the process.
5. Pick the green border line, and notice from the grips displayed you are no longer selecting the surface but a 3D polyline.
6. This polyline will form the basis for your final surface boundary. By extracting the polyline from the existing surface, you save a lot of time playing connect the dots along the points that are valid. Next, you refine this polyline and add it to the surface as a boundary.
7. In Prospector, right-click Point Groups and select the Properties option. The Point Groups dialog appears.
8. Move Field Work to the top of the list using the up and down arrows on the right.
9. Click OK to display all of your points on the screen.
10. Working your way around the site, grip-edit the polyline you made in step 4 to exclude some of the large triangles such as those on the eastern border. When complete, your polyline might look something like Figure 5.14.

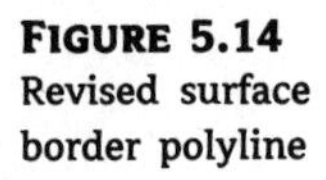

FIGURE 5.14
Revised surface border polyline

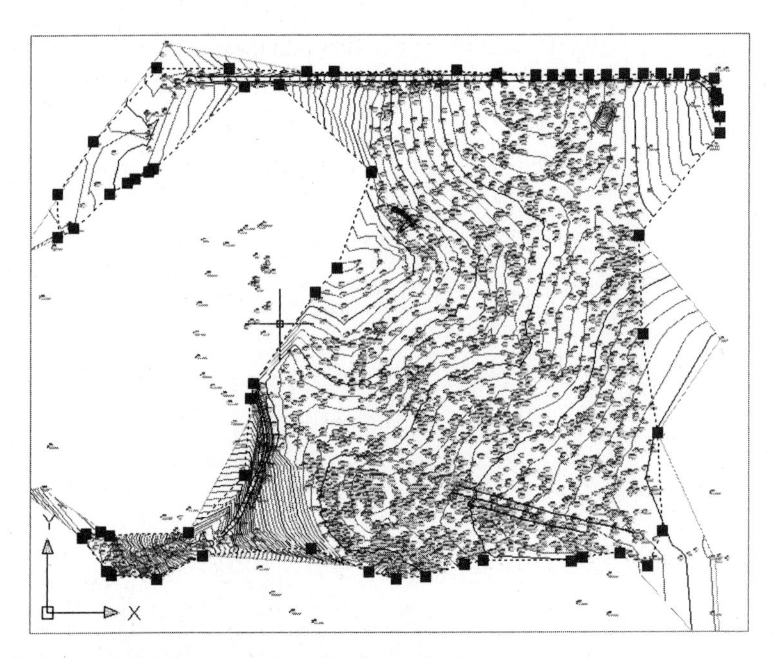

Just like breaklines, there are multiple types of surface boundaries:

- **Outer boundaries** — Define the outer edge of the shown boundary. When the Non-destructive Breakline option is used, the points outside the boundary are still included in the calculations; then additional points are created along the boundary line where it intersects with the triangles it crosses. This trims the surface for display but does not exclude the points outside the boundary. You'll want to have your outer boundary among the last operations in your surface building process.
- **Hide boundaries** — Punch a hole in the surface display for things like building footprints or a wetlands area that are not to be touched by design. Hidden surface areas are *not* deleted but merely not displayed.
- **Show boundaries** — Show the surface inside a hide boundary, essentially creating a donut effect in the surface display.
- **Data clip boundaries** — Place limits on data that will be considered part of the surface from that point going *forward*. This is different from an Outer boundary in that the data clip boundary will keep the data from ever being built into the surface as opposed to limiting it after the build. This is handy when attempting to build Civil 3D surfaces from large data sources such as LIDAR or DEM files. Because they limit data being placed into the surface definition, you'll want to have data clips as among the first operations in your surface.

The addition of every boundary is considered a separate part of the building operations. This means that the order in which the boundaries are applied controls their final appearance. For example, a show boundary selected before a hide boundary will be overridden by that hide operation. To finish the exercise, add the outer boundary twice, once as a nondestructive breakline and once with a standard breakline, and observe the difference.

1. In Prospector, expand the Surfaces branch.
2. Right-click EG and select the Surface Properties option. The Surface Properties dialog appears.
3. Change the Surface Style to Contours and Triangles.
4. In Prospector, expand the Surfaces ➤ EG ➤ Definition branches.
5. Right-click Boundaries and select the Add option. The Add Boundaries dialog opens.
6. Enter a name if you like and check the Non-destructive Breakline option.
7. Pick the polyline and notice the immediate change.
8. Zoom in on the southeast portion of your site, as shown in Figure 5.15.

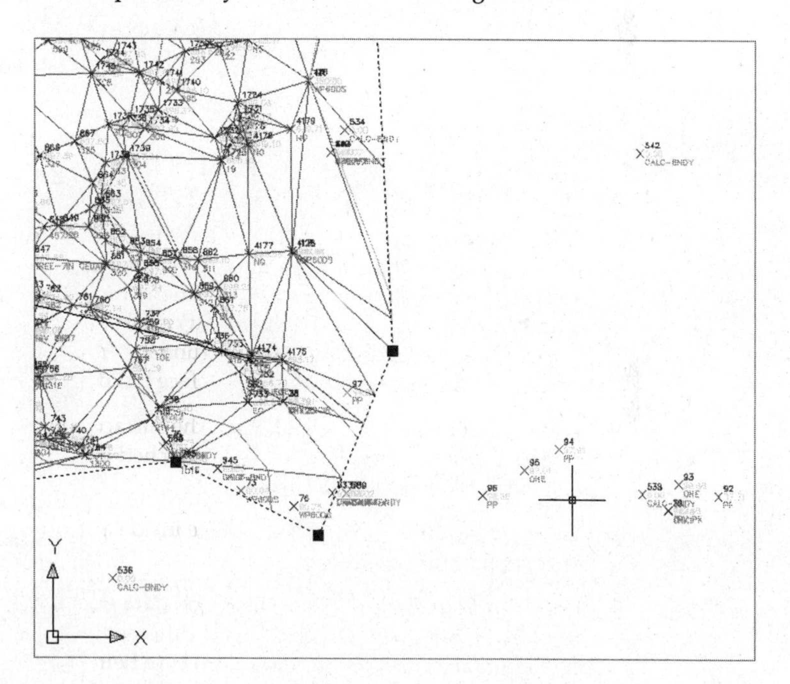

FIGURE 5.15
A nondestructive border in action

Notice how the triangulation appears to include lines to nowhere. This is the nature of the nondestructive breakline. The points you attempted to exclude from the surface are still being included in the calculation; they are just excluded from the display. This isn't the result you were after, so fix it now:

1. In Prospector, expand the Surfaces ➤ EG ➤ Definition branches and select Boundaries.
2. A listing of the boundaries appears in the preview area.

3. Right-click the border you just created and select Delete, as shown in Figure 5.16. Click OK in the warning dialog that says that the selected definition items will be permanently removed from the surface.

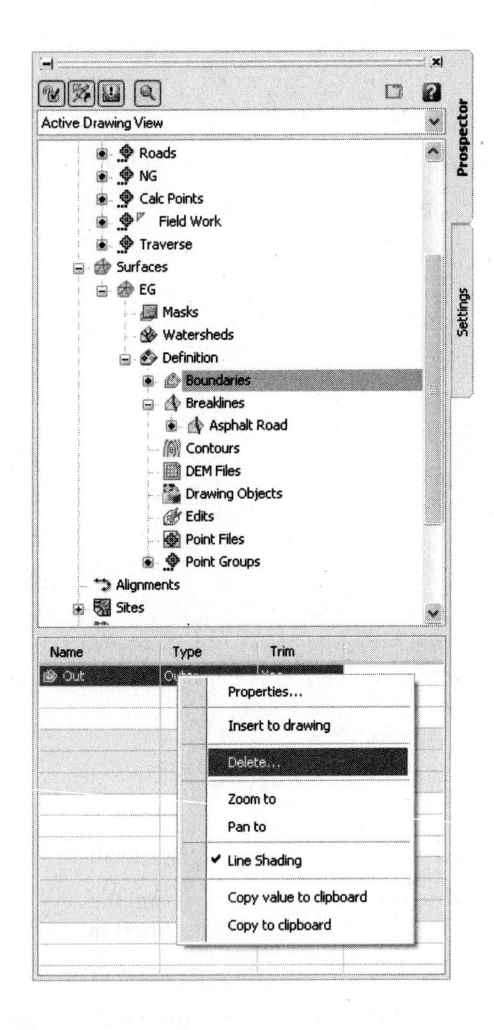

FIGURE 5.16
Deleting a surface boundary

4. In Prospector, expand the Surfaces branch and right-click EG. Select the Rebuild option to return to the prior version of the surface.
5. Right-click Boundaries and select the Add option again. The Add Boundaries dialog appears.
6. This time, leave the Non-destructive Breakline option deselected and click OK.
7. Pick the border polyline on your screen. Notice that no triangles intersect your boundary now where it does not connect points.
8. On the main menu, choose View ➢ Zoom ➢ Extents to see the result of the border addition.

In spite of adding breaklines and a border, you still have some areas that need further correction or changes.

Surface Masks

Surface masking is useful when you want to hide a portion of the surface or to create a rendering area. In this example, you use a closed polyline to create a rendering area for the surface:

1. Open the `Surface Masks.dwg` file.
2. Expand the Surfaces ➤ EG branches.
3. Right-click Masks and select the Create Mask option.
4. Pick the magenta polyline on the southern half of the site and right-click or press ↵ to complete the selection. The Create Mask dialog appears.
5. Change the settings in the dialog as shown in Figure 5.17.

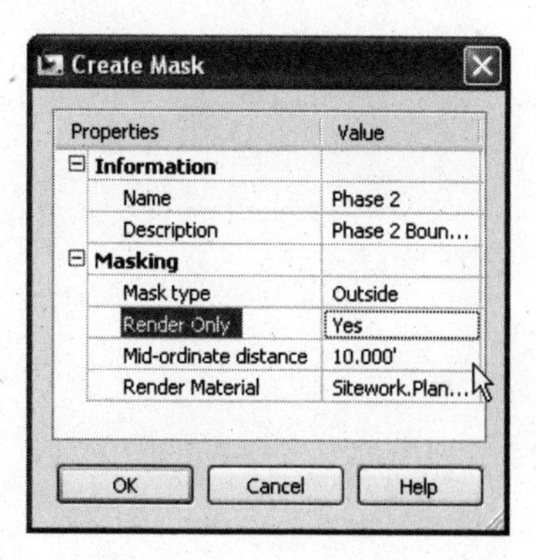

Figure 5.17
Setting the options in the Create Mask dialog

6. Click in the Render Material value field to activate the ellipsis button.
7. Click the ellipsis button, and the Select Render Material Style dialog appears. Select the Sitework.Planting Grass.Short material option.
8. Click OK to dismiss the Select Render Material Style dialog.
9. Click OK again to dismiss the Create Mask dialog.
10. In Prospector, right-click EG and select the Surface Properties option to bring up the Surface Properties dialog.
11. Change the Surface Style field to Contours and Triangles. Click OK.
12. Type **shademode** on the command line and type **r** for the Realistic option. The main site will be displayed with a generic soil pattern, whereas your site will be rendered with a grass textural pattern, as shown in Figure 5.18.

Surface masks can be used to render surfaces with textures, to color large areas with solid colors for marketing purposes, or to hide information during surface presentation.

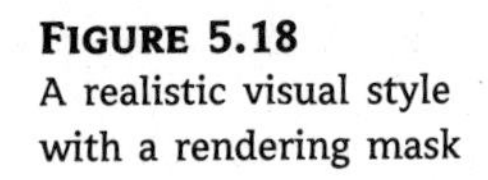

FIGURE 5.18
A realistic visual style with a rendering mask

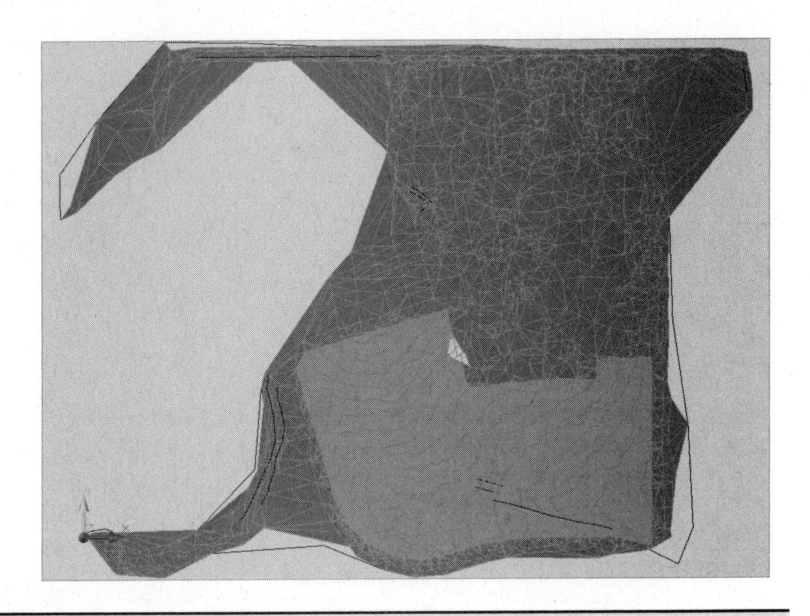

> **WHY CHANGE THE STYLE?**
>
> A Civil 3D surface must display triangles in order for rendering materials to be calculated and shown. You will inevitably forget this; it's just one of those frustrating anomalies in the program.

MANUAL SURFACE EDITS

In your surface, you have a few ''finger'' surface areas where the surveyors went out along narrow paths from the main area of topographic data. The nature of TIN surfaces is to connect dots, and so these fingers often wind up as webbed areas of surface information that's not really accurate or pertinent. A number of manual edits can be performed on a surface. These edit options are part of the definition of the surface and include the following:

- **Add Line** — Connects two points where a triangle did not exist before. This essentially adds a breakline to the surface, so adding a breakline would generally be a better solution.
- **Delete Line** — Removes the connection between two points. This is used frequently to clean up the edge of a surface or to remove internal data where a surface should have no triangulation at all. This can be an area such as a building pad or water surface.
- **Swap Edge** — Changes the direction of the triangulation methodology. For any four points, there are two solutions to the internal triangulation, and the Swap Edge edit alternates from one solution to the other.
- **Add Point** — Allows for the manual addition of surface data. This function is often used to add a peak to a digitized set of contours that might have a flat spot at the top of a hill or mountain.
- **Delete Point** — Allows for the manual removal of a data point from the surface definition. Generally, it's better to fix the source of the bad data, but this can be a fix if the original data is not editable (in the case of a LandXML file, for example).

- **Modify Point** and **Move Point** — Variations on the same idea. Modify Point moves a surface point in the *z* direction, whereas a Move Point is limited to horizontal movement. In both cases, the original data input is not modified but merely the TIN point.
- **Minimize Flat Areas** — Performs the edits you saw earlier in this chapter to add supplemental information to the TIN and to create a more accurate surface, forcing triangulation to work in the *z* direction instead of creating flat planes.
- **Raise/Lower Surface** — A simple arithmetic operation that moves the surface in the *z* direction. This is useful for testing rough grading schemes for balancing dirt or for adjusting entire surfaces after a new benchmark has been observed.
- **Smooth Surface** — Presents a pair of methods for supplementing the surface TIN data. Both of these work by extrapolating more information from the current TIN data, but they are distinctly different in their methodology:
    - **Natural Neighbor Interpolation (NNI)** — Adds points to a surface on the basis of the weighted average of nearby points. This data generally works well to refine contouring that is sharply angular because of limited information or long TIN connections. NNI works only within the bounds of a surface; it cannot extend beyond the original data.
    - **Kriging** — Adds points to a surface based on one of five distinct algorithms to predict the elevations at additional surface points. These algorithms create a trending for the surface beyond the known information and can therefore be used to extend a surface beyond even the available data. Kriging is very volatile, and you should understand the full methodology before applying this information to your surface. Kriging is frequently used in subsurface exploration industries such as mining, where surface (or strata) information is difficult to come by and the distance between points can be higher than desired.
- **Paste Surface** — Pulls in the TIN information from the selected surface and replaces the TIN information in the host surface with this new information. This is helpful in creating composite surfaces that reflect both the original ground and the design intent. You look more at pasting in Chapter 16 in the discussion on grading.
- **Simplify Surface** — Allows you to reduce the amount of TIN data being processed via one of two methods. These are Edge Contraction, wherein Civil 3D tries to collapse two points connected by a line to one point; and Point Removal, which removes selected surface points based on some algorithms designed to reduce data points that are similar.

Manual editing should always be the last step in updating a surface. Fixing the surface is a poor substitution for fixing the underlying data the TIN is built from, but in some cases, it is the quickest and easiest way to make a more accurate surface.

Point and Triangle Editing

In this section, you remove triangles manually, and then finish your surface by correcting what appears to be a blown survey shot.

1. Open the `Surface Edits.dwg` file.
2. In Prospector, expand the Surfaces ➢ EG ➢ Definition branches.

3. Right-click Edits and select the Delete Line option.

4. Enter **C** as the command line to enter a crossing selection mode.

5. Start at the lower right of the pick area shown in Figure 5.19, and move to the upper-left corner as shown. Right-click or press ↵ to finish the selection.

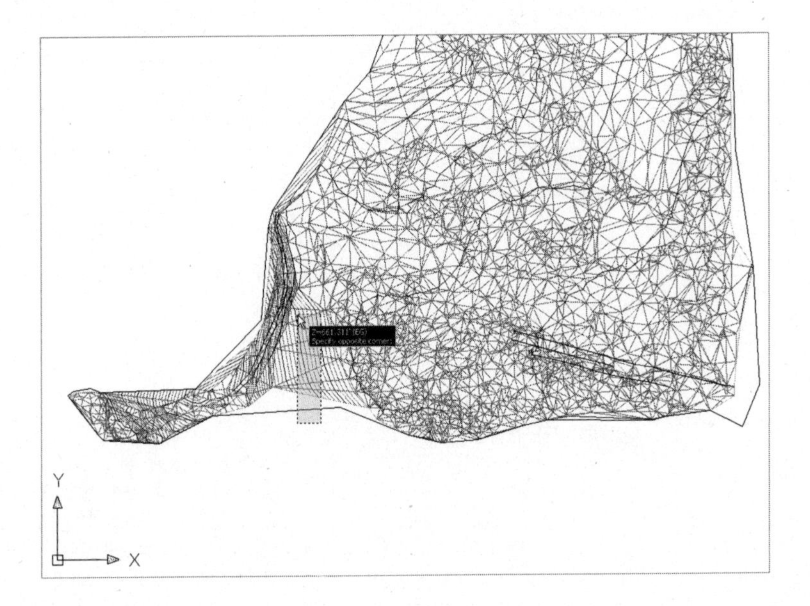

FIGURE 5.19
Crossing the window selection to delete TIN lines

6. Repeat this process in the upper right, and then on the upper left, removing triangles until your site resembles Figure 5.20.

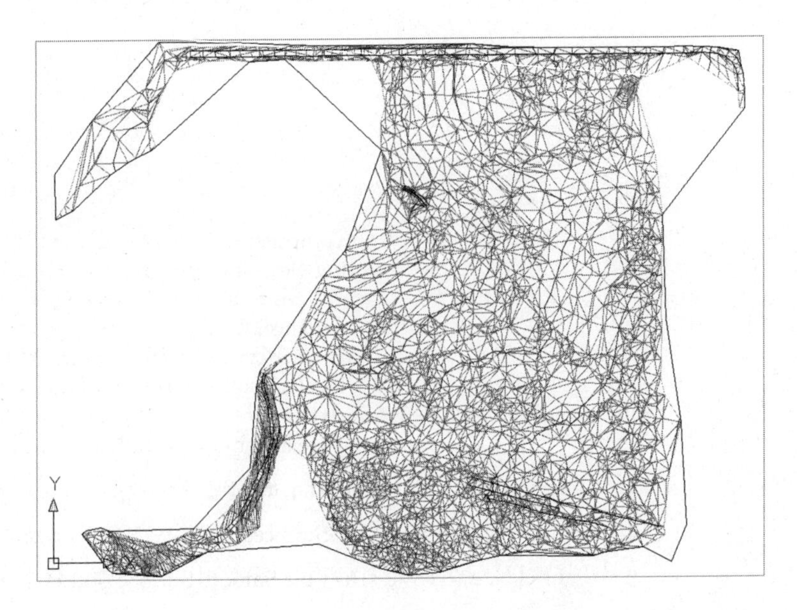

FIGURE 5.20
Surface after removal of extraneous triangles

7. Zoom to the northeast corner of your site, and you'll notice a collection of contours that seems out of place.
8. Change the Surface Style to Contours and Points.
9. Right-click Edits again and select the Delete Point option.
10. Select the red + marker in the middle of the contours, as shown in Figure 5.21.

FIGURE 5.21
Blown survey shot to be removed

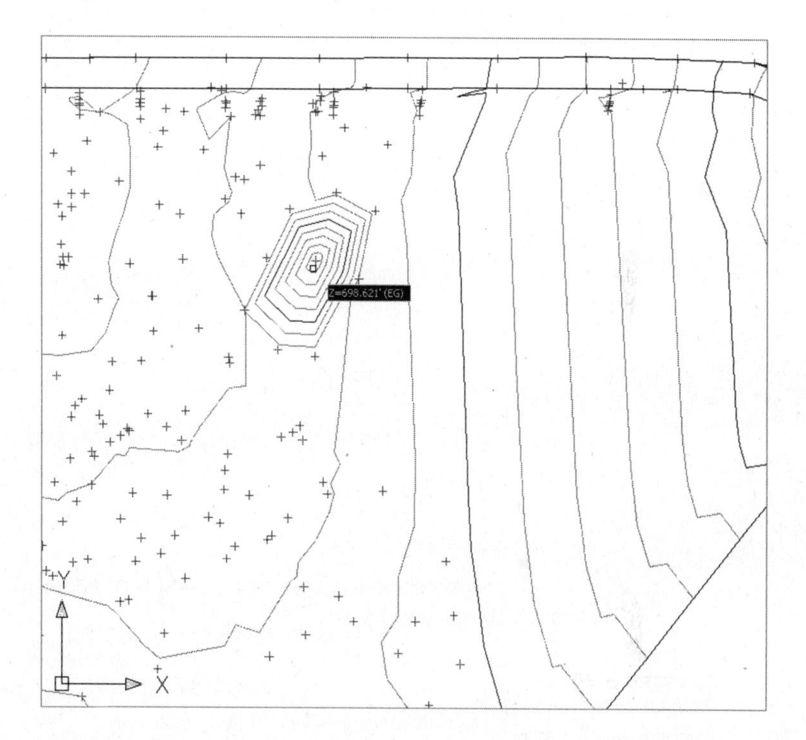

11. Right-click to complete the edit, and notice the immediate change in the contouring.

Surface Smoothing

One common complaint about computer-generated contours is that they're simply too precise. The level of calculations in setting elevations on the basis of linear interpolation along a triangle leg makes it possible for contour lines to be overly exact, ignoring contour line trends in place of small anomalies of point information. Under the eye of a board drafter, these small anomalies were averaged out, and contours were created with smooth flowing lines.

While you can apply object level smoothing as part of the contouring process, this smoothes the end result but not the underlying data. In this section, you use the NNI smoothing algorithm to reduce surface anomalies and create a more visually pleasing contour set:

1. Open the `Surface Smoothing.dwg` file. The area to be smoothed is shown in Figure 5.22.
2. In Prospector, expand the Surfaces ➤ EG ➤ Definition branches.
3. Right-click Edits and select the Smooth Surface option. The Smooth Surface dialog opens.

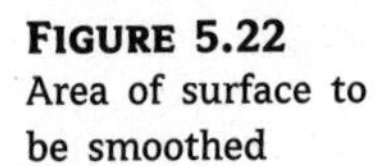

FIGURE 5.22
Area of surface to be smoothed

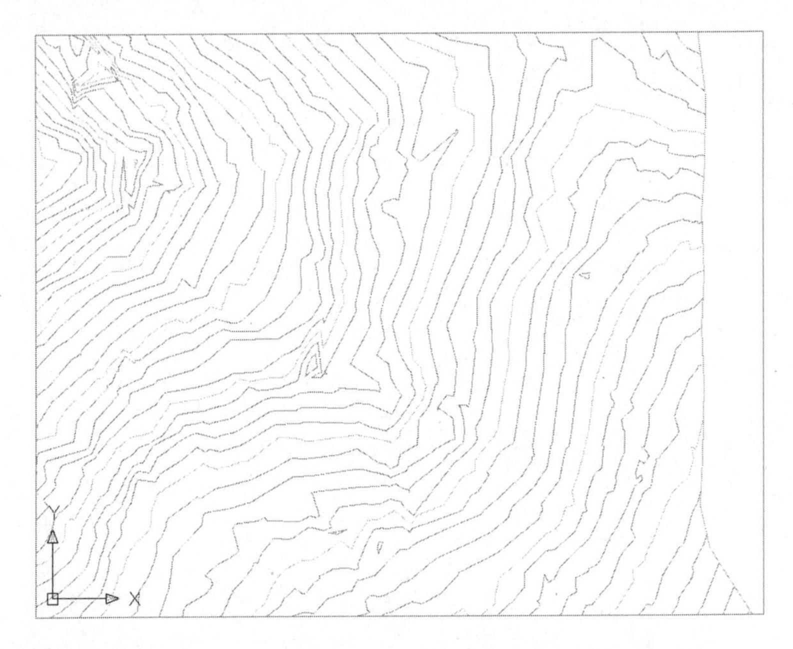

4. Expand the Smoothing Methods branch, and verify that Natural Neighbor Interpolation is the Select Method value.
5. Expand the Point Interpolation/Extrapolation branch, and click in the Select Output Region value field. Click the ellipsis button.
6. Select the rectangle drawn on screen, and press ↵ to return to the Smooth Surface dialog.
7. Enter **20** for the Grid X-Spacing and Grid Y-Spacing values, and then press ↵. Note that Civil 3D will tell you how many points you are adding to the surface immediately below this input area by the value given in the Number of Output Points field. It's grayed out, but it does change on the basis of your input values.
8. Click OK and the surface will be smoothed as shown in Figure 5.23.

Note that you said for the *surface* to be smoothed — not the contours. To see the difference, change the Surface Style to Contours and Points to display your image as shown in Figure 5.24.

Note all of the points with a circle cross symbol. These points are all new, created by the NNI surface-smoothing operation. These are part of your surface, and the contours reflect the updated surface information.

Surface Simplifying

Because of the increasing use in land development projects of GIS and other data-heavy inputs, it's critical that Civil 3D users know how to simplify the surfaces produced from these sources. In this exercise, you simplify the surface created from a point cloud earlier in this chapter.

1. Open the `Surface Simplifying.dwg` file. For reference, the surface statistics for the Aerial Points surface are shown in Figure 5.25.
2. Within Prospector, expand Surfaces ➤ Aerial Points ➤ Definition.

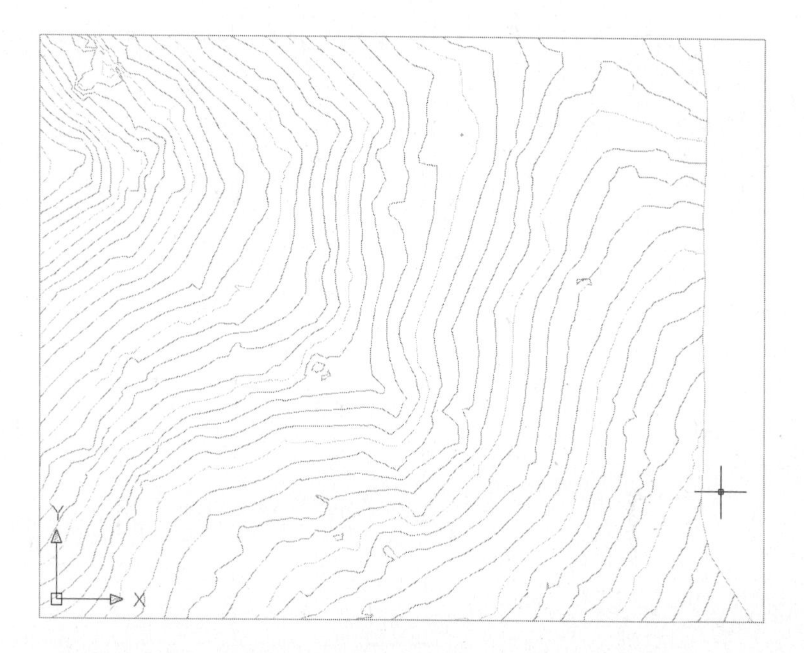

FIGURE 5.23
Using NNI to smooth the surface

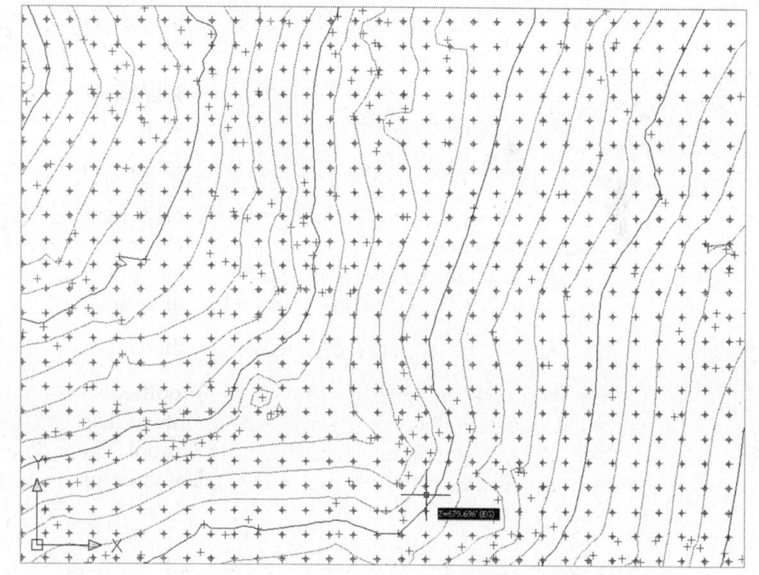

FIGURE 5.24
Points added via NNI surface smoothing

3. Right-click Edits and select Simplify Surface to display the Simplify Surface Wizard.
4. Select the Point Removal radio button as shown in Figure 5.26 and click Next to move to the Region Options.
5. Leave the Region Option set to Use Existing Surface Border. There are also options for selecting areas with a window or polygon, as well as selecting based on an existing entity. Click Next to move to the Reduction Options.

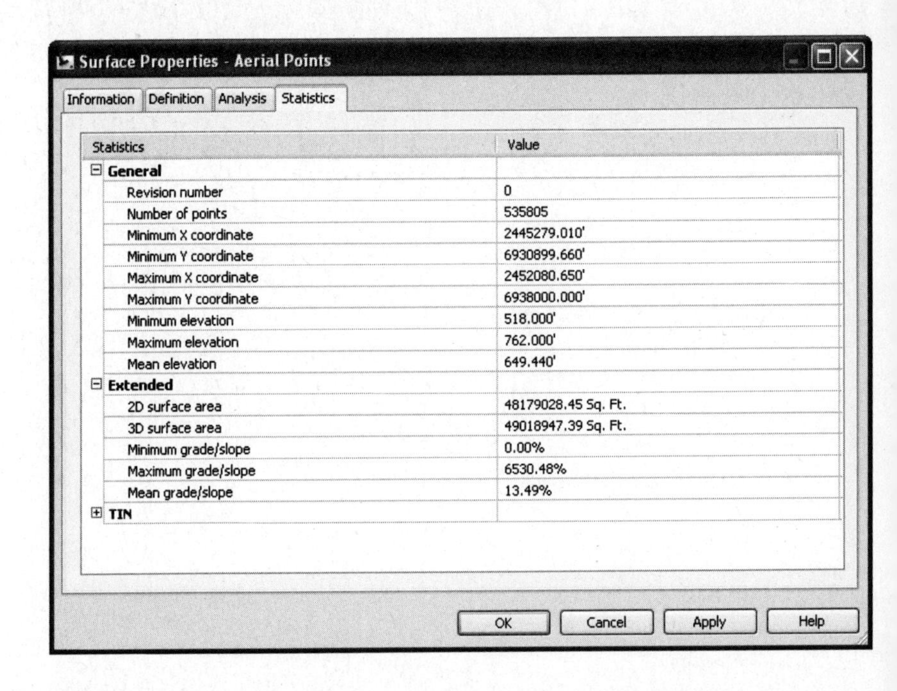

FIGURE 5.25 Aerial Points surface statistics before simplification

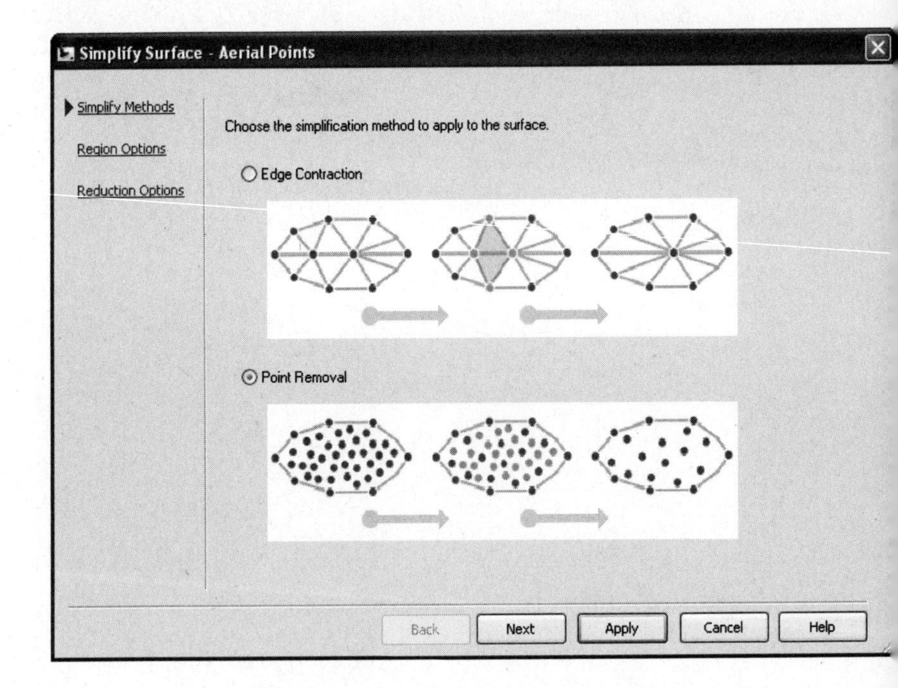

FIGURE 5.26 The Simplify Surface dialog

6. Set the Percentage of points to remove to 10 percent and then uncheck the Maximum change in elevation option. This value is the maximum allowed change between the surface elevation at any point before and after the simplify process has run.
7. Click Apply. The program will process this calculation and display a Total Points Removed number as shown in Figure 5.27. You can adjust the slider or toggle on the Maximum Change in Elevation button to experiment with different values.

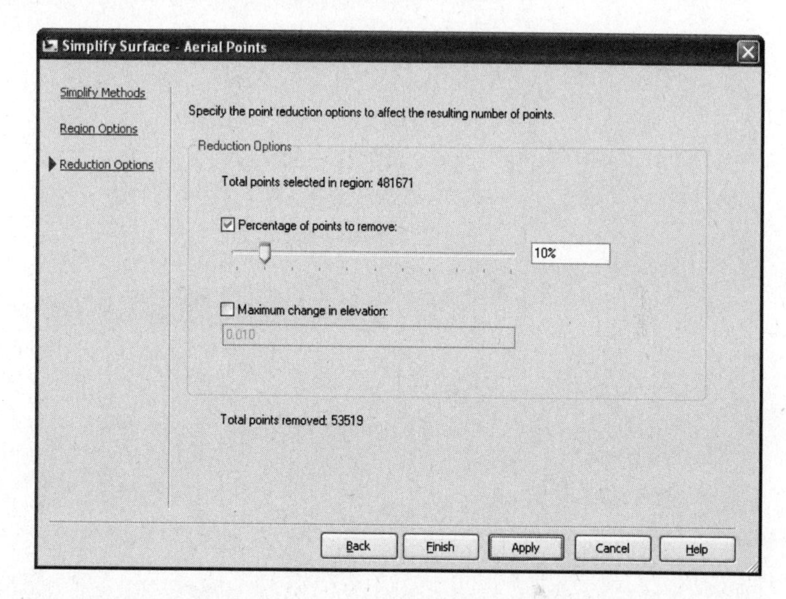

FIGURE 5.27
Reduction Options in the Simplify Surface Wizard

8. Click Finish to dismiss the wizard and fully commit to the Simplify edit.

A quick visit to the Surface Properties Statistics tab shows that the number of points has been reduced to 434,000. On something like an aerial topography or DEM, reducing the point count probably will not reduce the usability of the surface, but this simple 10 percent point reduction actually takes almost 20 percent off the file size. Remember, you can always remove the edit or uncheck the operation in the Definition tab of the Surface Properties dialog.

The creation of a surface is merely the starting point. Once you have a TIN to work with, you have a number of ways to view the data using analysis tools and varying styles.

Surface Styling and Analysis

Once a surface is created, you can display information in a large number of ways. The most common so far has been contours and triangles, but these are the basics. By using varying styles, you can show a large amount of data with one single surface. Not only can you do simple things such as adjust the contour interval, but Civil 3D can apply a number of analysis tools to any surface:

- **Contours** — Allows the user to specify a more specific color scheme or linetype as opposed to the typical minor-major scheme. Commonly used in cut-fill maps to color negative colors one way, positive contours another, and the balance or zero contours yet another color.
- **Elevations** — Creates bands of color to differentiate various elevations. This can be a simple weighted distribution to help in creation of marketing materials, hard-coded elevations

to differentiate floodplain and other elevation-driven site concerns, or ranges to help a designer understand the earthwork involved in creating a finished surface.

- **Direction Analysis** — Draws arrows showing the normal direction of the surface face. This is typically used for aspect analysis, helping site planners review the way a site slopes with regard to cardinal directions and the sun.
- **Slopes Analysis** — Colors the face of each triangle on the basis of the assigned slope values. While a distributed method is the normal setup, a common use is to check site slopes for compliance with Americans with Disabilities Act (ADA) requirements or other site slope limitations, including vertical faces (where slopes are abnormally high).
- **Slope Arrows** — Displays the same information as a slope analysis, but instead of coloring the entire face of the TIN, this option places an arrow pointing in the downhill direction and colors that arrow on the basis of the specified slope ranges.
- **User-Defined Contours** — Refers to contours that typically fall outside the normal intervals. These user-defined contours are useful to draw lines on a surface that are especially relevant but don't fall on one of the standard levels. A typical use is to show the normal pool elevation on a site containing a pond or lake.

In the following exercises, you'll look at the basic style manipulations to get various contour and color schemes. Then you'll work through an elevation analysis using standard value distribution methods, a custom elevation analysis, and a slope analysis.

Surface Styles

Just like every other Civil 3D object, surfaces are displayed on the basis of styles. Like most other styles, the basic color and linetype controls are part of the style, but so are more specific surface components such as contour interval, the use of depression ticks, the colors used in elevation banding — and the list goes on. In this section, you'll look at the way the surface-specific styles are built and some of their unique tricks. You'll start with adjusting surface contouring and then move into styles that are primarily focused on analysis.

The Contouring Basics

Contouring is the standard surface representation on which land development plans are built. But in past programs such as Autodesk's Land Development, changing the contouring interval was akin to pulling teeth. With the use of styles in Civil 3D, you can have any number of styles prebuilt to allow you to quickly and painlessly change how contours are displayed. In this example, you'll copy an existing surface contouring style and modify the interval to a setting more suitable for commercial site design review:

1. Open the `Surface Styles.dwg` file. This surface is currently displayed with a 5′ minor contour and 25′ major contour.
2. Select the surface by picking any contour or the boundary, and then click the Surface Properties button on the Modify panel. The Surface Properties dialog appears.
3. On the Information tab, click the down arrow next to the Style Editor button. Select the Copy Current Selection option to display the Surface Style dialog.
4. On the Information tab, change the Name field to **Contours 0.25′ and 1′** and remove the description in place.

5. Switch to the Contours tab and expand the Contour Intervals property, as shown in Figure 5.28.

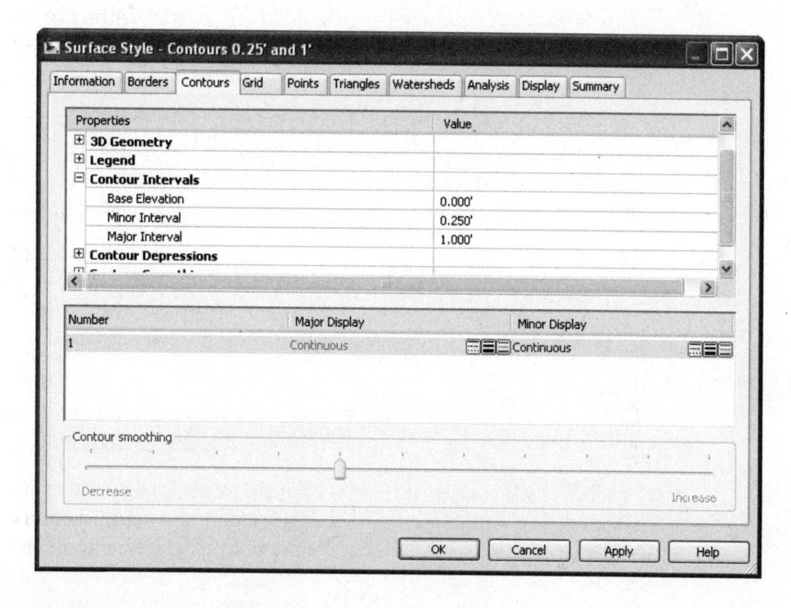

FIGURE 5.28
The expanded Contour Intervals setting

6. Change the Minor Interval value to **0.25**′, and press ↵. The Major Interval value will jump to 1.25′, maintaining the ratio that was previously in place.
7. Change the Major Interval value to 1.0′, and press ↵.
8. Expand the Contour Smoothing property (you may have to scroll down). Select a Smooth Contours value of True, which activates the Contour Smoothing slider bar near the bottom. Don't change this Smoothing value, but keep in mind that this gives you a level of control over how much Civil 3D modifies the contours it draws.
9. Click OK to close this dialog and then click OK again to close the Surface Properties dialog.

SURFACE VERSUS CONTOUR SMOOTHING

Remember, contour smoothing is *not* surface smoothing. Contour smoothing applies smoothing at the individual contour level but not at the surface level. If you want to make your surface contouring look fluid, you should be smoothing the surface.

The surface should be rendered faster than you can read this sentence even with the incredibly tight contour interval you've selected. This style doesn't make much sense on a site like this one, but it can be used effectively on something like a commercial site or highway entrance ramp where the low surface slope values make 1′ contours close to meaningless in terms of seeing what is going on with the surface.

You skipped over one portion of the surface contours that many people consider a great benefit of using Civil 3D: depression contours. If this option is turned on via the Contours tab, ticks will be added to the downhill side of any closed contours leading to a low point. This is a stylistic option, and usage varies widely.

Now let's look at a few of the other options and areas you ignored in creating this style. There are some interesting changes from many other Civil 3D objects, as you can see in the component listing in Figure 5.29.

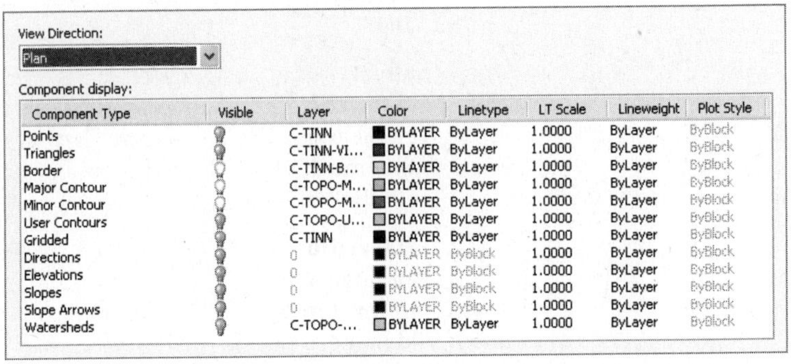

FIGURE 5.29
Listing of surface style components in the Plan direction

View Direction:
Plan

Component display:

Component Type	Visible	Layer	Color	Linetype	LT Scale	Lineweight	Plot Style
Points		C-TINN	BYLAYER	ByLayer	1.0000	ByLayer	ByBlock
Triangles		C-TINN-VI...	BYLAYER	ByLayer	1.0000	ByLayer	ByBlock
Border		C-TINN-B...	BYLAYER	ByLayer	1.0000	ByLayer	ByBlock
Major Contour		C-TOPO-M...	BYLAYER	ByLayer	1.0000	ByLayer	ByBlock
Minor Contour		C-TOPO-M...	BYLAYER	ByLayer	1.0000	ByLayer	ByBlock
User Contours		C-TOPO-U...	BYLAYER	ByLayer	1.0000	ByLayer	ByBlock
Gridded		C-TINN	BYLAYER	ByLayer	1.0000	ByLayer	ByBlock
Directions		0	BYLAYER	ByBlock	1.0000	ByLayer	ByBlock
Elevations		0	BYLAYER	ByBlock	1.0000	ByLayer	ByBlock
Slopes		0	BYLAYER	ByBlock	1.0000	ByLayer	ByBlock
Slope Arrows		0	BYLAYER	ByBlock	1.0000	ByLayer	ByBlock
Watersheds		C-TOPO-...	BYLAYER	ByLayer	1.0000	ByLayer	ByBlock

Under the Component Type column, Points, Triangles, Border, Major Contour, Minor Contour, User Contours, and Gridded are standard components and are controlled like any other object component. The Plan (aka 2D), Model (aka 3D), and Section views are independent, and surfaces are one of the objects where different plan and model views are common. The Directions, Elevations, Slopes, and Slope Arrows components are unique to surface styles. Note that the Layer, Color, and Linetype fields are grayed out for these components. Each of these components has its own special coloring schemes, which you look at in the next section.

ELEVATION BANDING

Displaying surface information as bands of color is one of the most common display methods for engineers looking to make a high-impact view of the site. Elevations are a critical part of the site design process, and understanding how a site flows in terms of elevation is an important part of making the best design. Elevation analysis typically falls into two categories: showing bands of information on the basis of pure distribution of linear scales or showing a lesser number of bands to show some critical information about the site. In this first exercise, you'll use a pretty standard style to illustrate elevation distribution along with a prebuilt color scheme that works well for presentations:

1. Open the `Surface Analysis.dwg` file.
2. On the Settings tab of Toolspace, expand the Surface ➤ Surface Styles branches.
3. Right-click Elevation Banding (2D) and select the Copy option. The Surface Style dialog appears.
4. On the Information tab, change the Name field to **Elevation Banding (3D)** and switch to the Analysis tab.
5. Expand the Elevations property to review the settings built into the style.
6. Set the Group by Value field to Equal Interval.

These distribution methods show up in nearly all of the surface analysis methods. Here's what they mean:

- **Quantile** — Often referred to as an equal count distribution and will create ranges that are equal in sample size. These ranges will not be equal in linear size but in distribution across a surface. This method is best used when the values are relatively equally spaced throughout the total range, with no extremes to throw off the group sizing.
- **Equal Interval** — A stepped scale, created by taking the minimum and maximum values and then dividing the delta into the number of selected ranges. This method can create real anomalies when extremely large or small values skew the total range so that much of the data falls into one or two intervals, with almost no sampled data in the other ranges.
- **Standard Deviation** — The bell curve that most engineers are familiar with and is well suited for when the data follows the bell distribution pattern. It generally works well for slope analysis, where very flat and very steep slopes are common and would make another distribution setting unwieldy.

For the elevation analysis, you'll use Equal Interval because your data is constrained to a relatively small range.

7. Change the Display Type drop-down list to 3D Faces to facilitate the isometric view you'll want to create later.
8. Change the Scheme drop-down list to Land.
9. Change the Elevations Display Mode drop-down list to Exaggerate Elevations. This will make the elevation differences more apparent when you select an isometric viewpoint.
10. Change the value of the Exaggerate Elevations by Scale Factor to **5**. Your dialog should look like Figure 5.30.

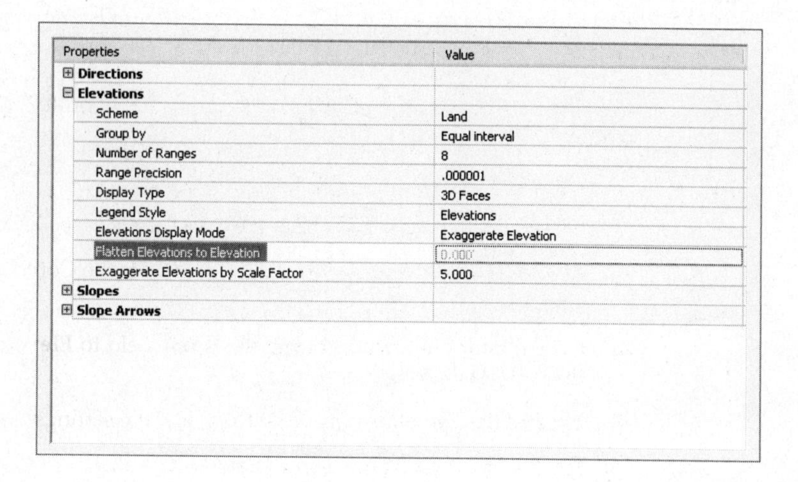

FIGURE 5.30
Changes to the Elevation fields on the Analysis tab for better isometric views

11. Switch to the Display tab. The only component turned on is Elevations.
12. Change the View Direction field to Model.
13. Turn off the Triangles and turn on Elevations by clicking the lightbulb in the Visible field. Click OK.
14. Pick the surface on your screen.
15. Right-click and select Surface Properties. The Surface Properties dialog appears.
16. On the Information tab, change the Surface Style field to your new Elevation Banding (3D) style.
17. Switch to the Analysis tab.
18. Click the blue Run Analysis arrow in the middle of the dialog to populate the Range Details area.
19. Click OK to close the Surface Properties dialog.
20. On the View tab, choose the drop-down arrow to the right of the Unsaved View option on the Views panel, and select SW Isometric.
21. Zoom in if necessary to get a better view.
22. Type **shademode** on the command line and type **r** for the Realistic option to see a semi-rendered view that should look something like Figure 5.31.

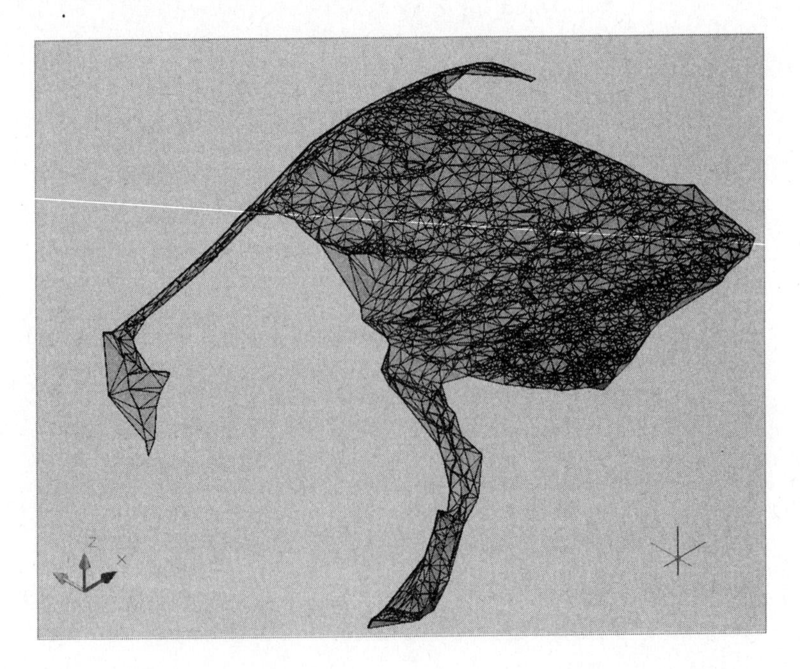

FIGURE 5.31
Conceptual view of the site with the Elevation Banding style

AutoCAD Visual Styles

The triangles seen are part of the view style and can be modified via the Visual Styles Manager. Turning the edge mode off will leave you with a nicely gradated view of your site. You can edit the visual style by clicking View ➢ Visual Styles ➢ Visual Style Manager on the main menu.

You look at more of the visualization techniques in Chapter 22. For now, you'll use a 2D elevation to clearly illustrate portions of the site that cannot be developed. In this exercise, you manually tweak the colors and elevation ranges on the basis of design constraints from outside the program:

1. Open the View tab. In the Views panel, click the drop-down arrow to the right of the Unsaved View option and select Top.
2. Type **shademode** on the command line and type **2** for the 2D Wireframe option.
3. On the View tab, choose Extents from the Navigate panel to return to a triangle view of your site.
4. On the Settings tab, right-click the Elevation Banding (2D) style and select the Copy option. The Surface Style dialog appears.
5. On the Information tab, change the Name field to **Zoning**.
6. Switch to the Analysis tab and expand the Elevations property.
7. Change the Number field in the Ranges area to **3** and click OK to close the Surface Style editor.

   This site has a limitation placed in that no development can go below the elevation of 664. Your analysis will show you the areas that are below 664, a buffer zone to 665, and then everything above that.

8. Select the surface and right-click to select the Surface Properties option. The Surface Properties dialog appears.
9. On the Information tab, change the Surface Style field to Zoning.
10. On the Analysis tab, change the Number field in the Ranges area to **3**.
11. Click the Run Analysis arrow in the Ranges area to populate the Range Details area.
12. Double-clicking in the Minimum and Maximum Elevations fields allows for direct editing. Double-clicking the Color Swatch field allows for manual picking. Modify your surface properties to match Figure 5.32. (The colors are red, yellow, and green from top to bottom, respectively.)
13. Click OK to exit the dialog.

Understanding surfaces from a vertical direction is helpful, but many times, the slopes are just as important. In the next section, you'll take a look at using the slope analysis tools in Civil 3D.

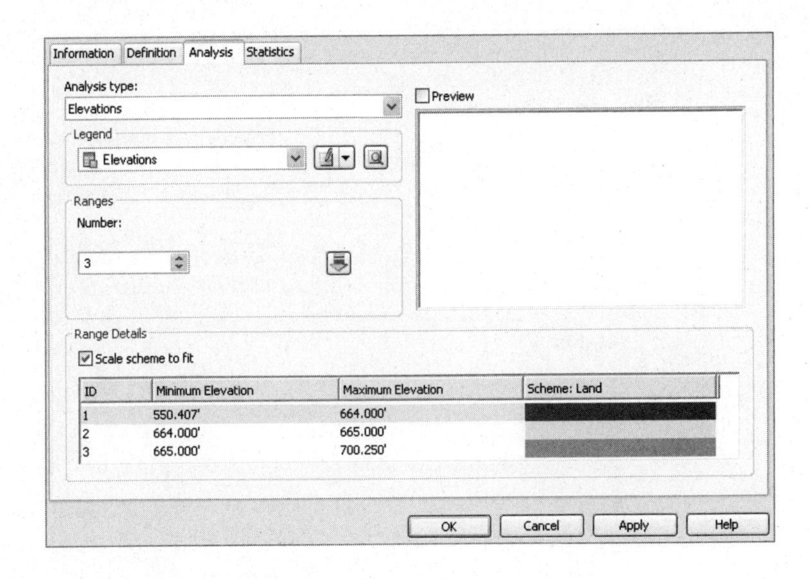

FIGURE 5.32
The Surface Properties dialog after manual editing

Slopes and Slope Arrows

Beyond the bands of color that show elevation differences in your models, you also have tools that display slope information about your surfaces. This analysis can be useful in checking for drainage concerns, meeting accessibility requirements, or adhering to zoning constraints. Slope is typically shown as areas of color as the elevations were or as colored arrows that indicate the downhill direction and slope. In this exercise, you look at a proposed site grading surface and run the two slope analysis tools:

1. Open the `Surface Slopes.dwg` file.
2. Select the surface in the drawing and click the Surface Properties button on the Modify panel. The Surface Properties dialog appears.
3. On the Information tab, change the Surface Style field to Slope Banding (2D).
4. Change to the Analysis tab.
5. Change the Analysis Type drop-down list to Slopes.
6. Change the Number field in the Ranges area to **5** and click the Run Analysis button. The Range Details area will populate.
7. Click OK to close the dialog. Your screen should look like Figure 5.33.

The colors are nice to look at, but they don't mean much, and slopes don't have any inherent information that can be portrayed by color association. To make more sense of this analysis, add a table:

1. Select the surface again and select the Add Legend button on the Labels & Tables panel.
2. Type **S** at the command line to select Slopes, and press ↵.
3. Press ↵ again to accept the default value of a Dynamic legend.

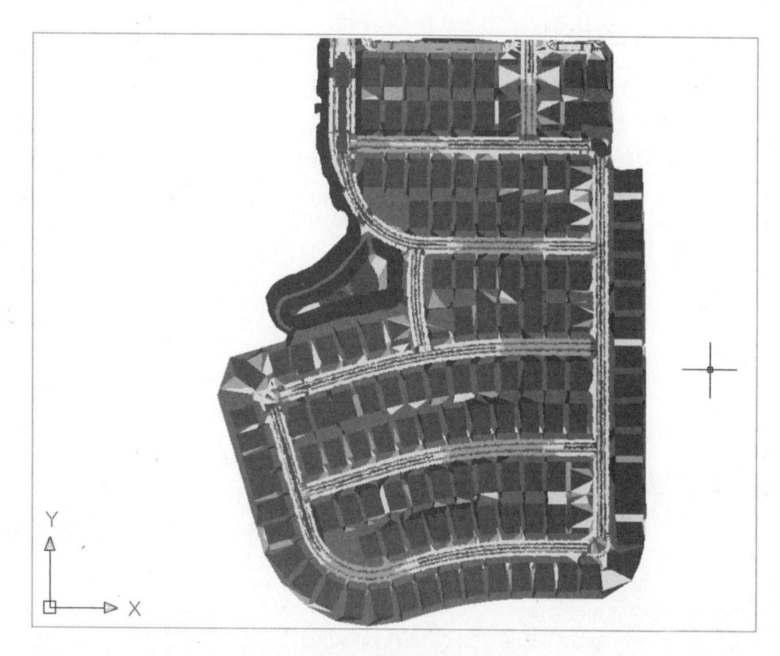

FIGURE 5.33
Slope color banding analysis

4. Pick a point on screen to draw the legend, as shown in Figure 5.34.

FIGURE 5.34
The Slopes legend table

Slopes Table

Number	Minimum Slope	Maximum Slope	Area	Color
1	0.00%	2.03%	829313.87	
2	2.03%	3.27%	224037.88	
3	3.27%	4.58%	173537.18	
4	4.58%	20.70%	575088.48	
5	20.70%	176874.45%	175598.72	

By including a legend, you can actually make sense of the information presented in this view. Because you know what the slopes are, you can also see which way they go.

1. On the Settings tab of Toolspace, expand the Surface ➤ Surface Styles branches.
2. Right-click Slope Banding (2D) and select Copy. The Surface Style dialog appears.

3. On the Information tab, change the Name field to **Slope Arrows**.
4. Switch to the Display tab and turn off the Slopes component by clicking on the lightbulb in the Visible field.
5. Turn on the Slope Arrows component by clicking the lightbulb in the Visible field.
6. Click OK to close the dialog.
7. Select the surface and click the Surface Properties button on the Modify panel. The Surface Properties dialog appears.
8. On the Information tab, change the Surface Style drop-down list to Slope Arrows.
9. Change to the Analysis tab.
10. Change the Analysis Type drop-down list to Slope Arrows.
11. Change the Number field in the Ranges area to **5** and click the Run Analysis button.
12. Click OK to close the dialog.

The benefit of arrows is in looking for "birdbath" areas that will collect water. These arrows can also verify that inlets are in the right location as in Figure 5.35. Look for arrows pointing to the proposed drainage locations and you'll have a simple design-verification tool.

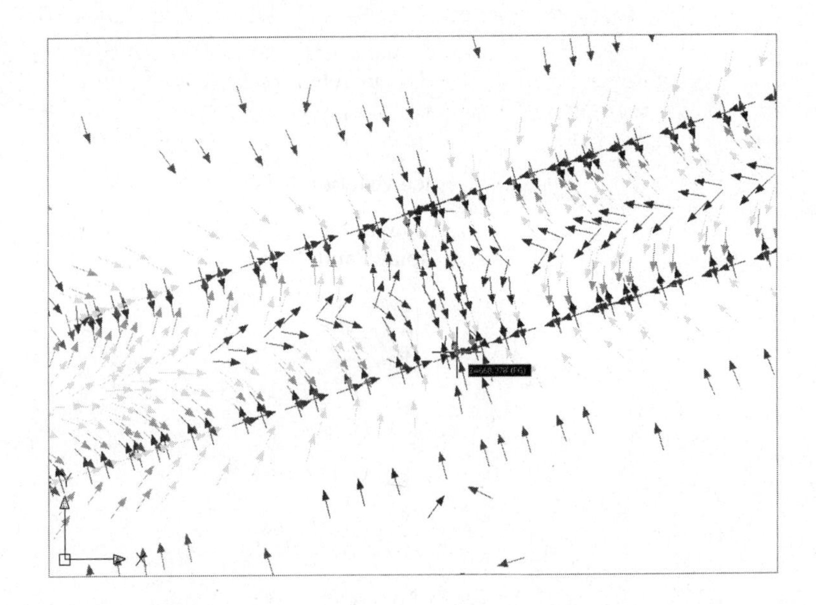

FIGURE 5.35
Slope arrows pointing to a proposed inlet location

With these simple analysis tools, you can show a client the areas of their site that meet their constraints. Visually strong and simple to produce, this is the kind of information that a 3D model makes available. Beyond the basic information that can be represented in a single surface, Civil 3D also contains a number of tools for comparing surfaces. You compare this existing ground surface to a proposed grading plan in the next section.

Comparing Surfaces

Earthwork is a major part of almost every land development project. The money involved with earthmoving is a large part of the budget, and for this reason, minimizing this impact is a critical part of the final design. Civil 3D contains a number of surface analysis tools designed to help in this effort, and you'll look at them in this section. First, a simple comparison provides feedback about the volumetric difference, and then a more detailed approach enables you to perform an analysis on this difference.

For years, civil engineers have performed earthwork using a section methodology. Sections were taken at some interval, and a plot was made of both the original surface and the proposed surface. Comparing adjacent sections and multiplying by the distance between them yields an end-area method of volumes that is generally considered acceptable. The main problem with this methodology is that it ignores the surfaces in the areas between sections. These areas could include areas of major change, introducing some level of error. In spite of this limitation, this method worked well with hand calculations, trading some accuracy for ease and speed.

With the advent of full-surface modeling, more precise methods became available. By analyzing both the existing and proposed surfaces, a volume calculation can be performed that is as good as the two surfaces. At every TIN vertex in both surfaces, a distance is measured vertically to the other surface. These delta amounts can then be used to create a third volume surface representing the difference between the surfaces. Civil 3D uses this methodology to perform its calculations, but the end-area method can still be used if desired.

Simple Volumes

When performing rough analysis, the total volume is the most important part. Once an acceptable volume has been created, more refined analysis and comparison can be performed. In this exercise, you compare two surfaces to simply pull a basic volume number, and then modify the proposed grade to illustrate how quickly changes can be reviewed:

1. Open the `Surface Volumes.dwg` file.
2. Change to the Analyze tab, and then click the Volumes button on the Volumes and Materials panel to display Panorama with the Composite Volumes tab, as shown in Figure 5.36.

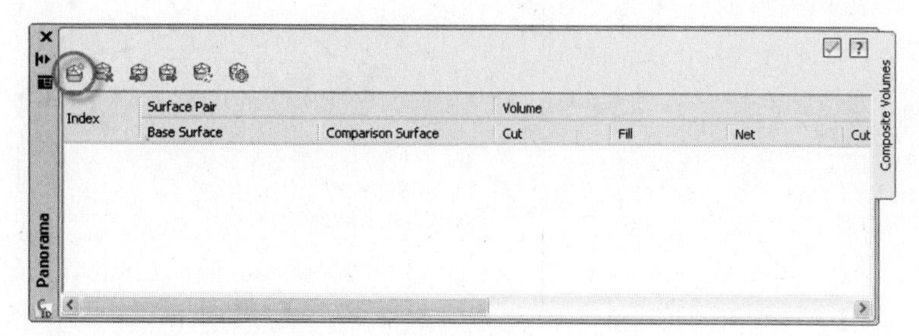

FIGURE 5.36
The Composite Volumes tab in Panorama

3. Click the Create New Volume Entry button on the far left, as indicated in Figure 5.36, to create a new volume entry.
4. Click the <select surface> field under the Base Surface heading and select EG.
5. Click the <select surface> field under the Comparison Surface heading and select FG. Civil 3D will calculate the volume (Figure 5.37). Note that you can apply a cut or fill factor by typing directly into the cells for these values.

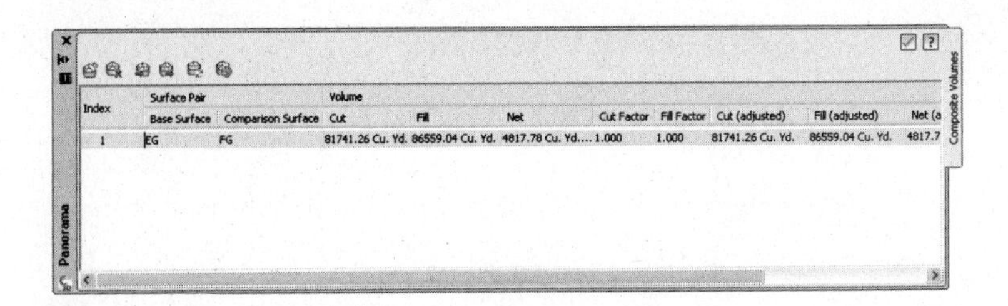

FIGURE 5.37
Composite volume calculated

DON'T TOUCH THAT CLOSE BUTTON!

This utility's calculations will disappear if you close Panorama. This information can be exported to XML or copied and pasted into another document if a record is required.

---

6. Without closing Panorama, move to Prospector, and expand the Surfaces ➤ FG ➤ Definition branches.
7. Right-click Edits and select the Raise/Lower Surface option.
8. Enter **−0.25** at the command line to drop the site 3″.
9. In Panorama, click the Recompute Volumes button (as shown in Figure 5.38) to update the calculations.

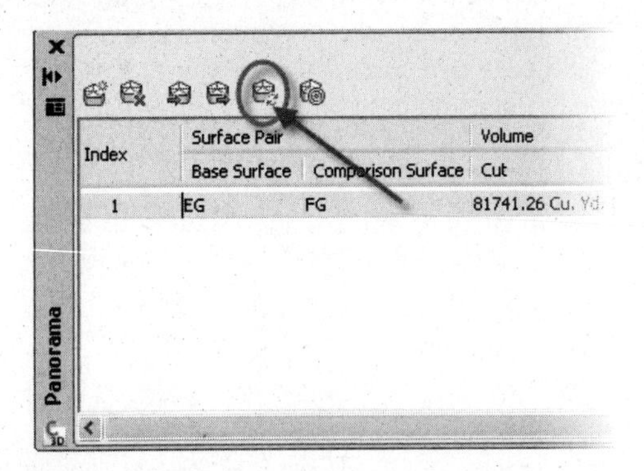

FIGURE 5.38
Recomputing the composite volume

10. Right-click the Edits list to remove the lowering edit and select Delete.
11. Return to Panorama and recompute to return to the original volume calculation.

The original design was quite good in terms of cut and fill, so you will look at a more detailed analysis of the earthwork by using a TIN volume surface in the next section.

Volume Surfaces

Using the volume utility for initial design checking is helpful, but quite often, contractors and other outside users want to see more information about the grading and earthwork for their own uses. This requirement typically falls into two categories: a cut-fill analysis showing colors or contours or a grid of cut-fill tick marks.

Color cut-fill maps are helpful when reviewing your site for the locations of movement. Some sites have areas of better material or can have areas where the cost of cut is prohibitive (such as rock). In this exercise, you use two of the surface analysis methods to look at the areas for cut-fill on your site:

1. Open the `Surface Volumes.dwg` file if it is not already open.
2. In Prospector, right-click the Surfaces branch and select Create Surface. The Create Surface dialog appears.
3. Change the Type field to TIN Volume Surface.
4. Expand the Information property, and change the Name to **Volume**.
5. In the Style value field, click the ellipsis button to open the Select Surface Style dialog. Select Elevation Banding (2D) and click OK.
6. Expand the Volume Surfaces property, and click in the Base Surface value field. Click the ellipsis button to open the Select Base Surface dialog. Select EG and click OK.
7. Click in the Comparison Surface value field. Click the ellipsis button to open the Select Comparison Surface dialog. Select FG and click OK. The dialog should look like Figure 5.39. Note that you can apply cut and fill factors to your calculations by filling them in here.

FIGURE 5.39
Creating a volume surface

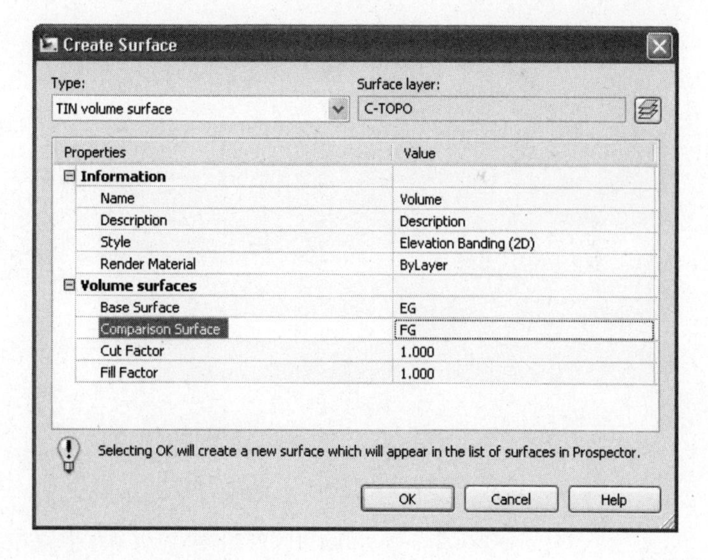

8. Click OK to complete the surface creation.

   This new Volume surface appears in Prospector's Surfaces collection, but notice that the icon is slightly different, showing two surfaces stacked on each other. The color mapping currently shown is just a default set, though, and does not indicate much.

9. Right-click Volume in the Surfaces branch of the Prospector and select the Surface Properties option. The Surface Properties dialog appears.
10. Switch to the Statistics tab and expand the Volume branch.

The value shown for the Net Volume (Unadjusted) is what was calculated in the Surface Volume utility in the previous exercise. This information can be cut and pasted into other programs for saving or other analysis if needed.

11. Switch to the Analysis tab.
12. Change the Number field in the Ranges area to **3**, and click the Run Analysis arrow.
13. Change the values in the cells by double-clicking and editing to match Figure 5.40. Pick any colors you like.

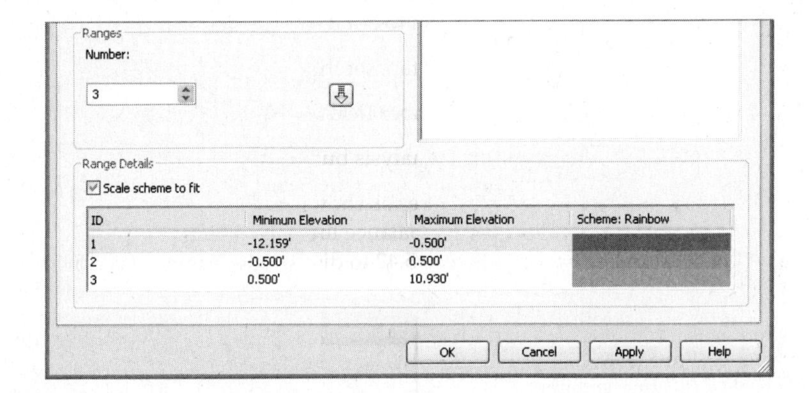

FIGURE 5.40
Elevation analysis settings for earthworks

14. Click OK to close the dialog.

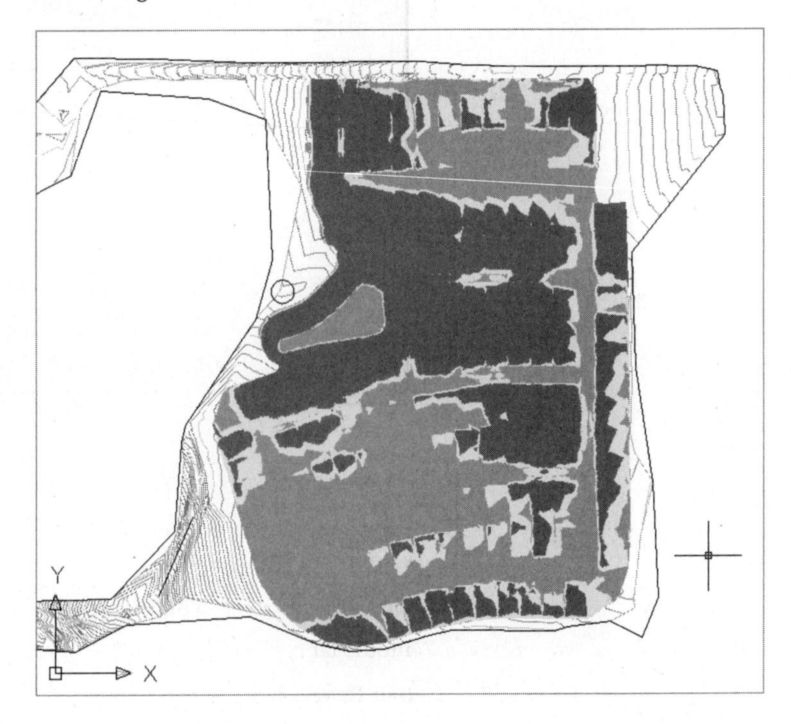

FIGURE 5.41
Completed elevation analysis

The volume surface now indicates areas of cut, fill, and areas near balancing, similar to Figure 5.41. If you leave a small range near the balance line, it's clearer to see the areas that are being left nearly undisturbed.

To show where large amounts of cut or fill could incur additional cost (such as compaction, or excavation protection), you would simply modify the analysis range as required.

The Elevation Banding surface is great for onscreen analysis, but the color fills make it hard to plot or use in many applications. In this next exercise, you use the Contour Analysis tool to prepare cut-fill contours in these same colors:

1. Right-click Volume in the Surfaces collection of the Prospector and select the Surface Properties option to open the Surface Properties dialog again.
2. On the Analysis tab, set the Analysis Type field to Contours.
3. Change the Number field in the Ranges area to **3**.
4. Click the Run Analysis button.
5. Change the ranges as shown in Figure 5.42. The contour colors are shades of red for cut, a yellow for the balance line, and shades of green for the fill areas. Click the small button shown in Figure 5.42 to display the AutoCAD Select Color dialog.

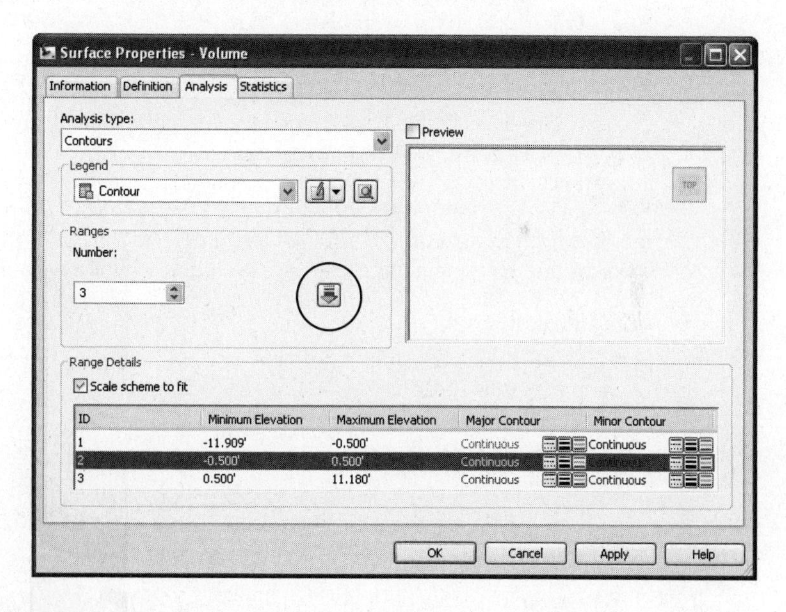

FIGURE 5.42
Earthworks contour analysis

6. Switch to the Information tab on the Surface Properties dialog, and change the Surface Style to Contours 1′ and 5′ (Design).
7. Click the down arrow next to the Style field and select the Copy Current Selection option. The Surface Style Editor appears.
8. On the Information tab, change the Name field to **Contours 1′ and 5′ (Earthworks)**.
9. Switch to the Contours tab.
10. Expand the Contour Ranges branch.

11. Change the value of the Use Color Scheme property to True. It's safe to ignore the values here because you hard-coded the values in your surface properties.
12. Click OK to close the Surface Style Editor and click OK again to close the Surface Properties dialog.

The volume surface can now be labeled using the surface-labeling functions, which you look at in the next section.

Labeling the Surface

Once you've created the surface model, it is time to communicate the model's information in various formats. This includes labeling contours, creating legends for the analysis you've created, or adding spot labels. These exercises work through these main labeling requirements and building styles for each.

Contour Labeling

The most common requirement is to place labels on surface-generated contours. In Land Desktop, this was one of the last steps because a change to a surface required erasing and replacing all the labels. Once labels have been placed, their styles can be modified.

Placing Contour Labels

Contour labels in Civil 3D are created by special lines that understand their relationship with the surface. Everywhere one of these lines crosses a contour line, a label is applied. This label's appearance is based on the style applied and can be a major, minor, or user-defined contour label. Each label can have styles selected independently, so using some AutoCAD selection techniques can be crucial to maintaining uniformity across a surface. In this exercise, you'll add labels to your surface and explore the interaction of contour label lines and the labels themselves.

1. Open the `Surface Labeling.dwg` file.
2. Select the surface in the drawing to display the Tin Surface tab. On the Labels and Tables panel, select Add Labels ➢ Contour – Single.
3. Pick any spot on a blue major contour to add a label.
4. On the Labels and Tables panel, select Add Labels ➢ Contour – Multiple.
5. Pick a point on the west of the road contours (to the north) and then a second point to the east, crossing a number of contours in the process.
6. On the Labels and Tables panel, select Add Labels ➢ Contour – Multiple – At Interval.
7. Pick a point near the middle left of the site and a second point across the site to the east.
8. Enter **400** at the command line for an interval value.

You've now labeled your site in three ways to get contour labels in a number of different locations. You would need additional labels in the northeast and southwest to complete the labeling, because you did not cross these contour objects with your contour label line. You could add more labels by clicking Add, but you can also use the labels created already to fill in these missing areas. By modifying the contour line labels, you can manipulate the label locations and add new labels. In this exercise, you fill in the labeling to the northeast:

1. Zoom to the northeast portion of the site, and notice that some of the contours are labeled only along the boundary or not at all, as shown in Figure 5.43.

FIGURE 5.43
Contour labels applied

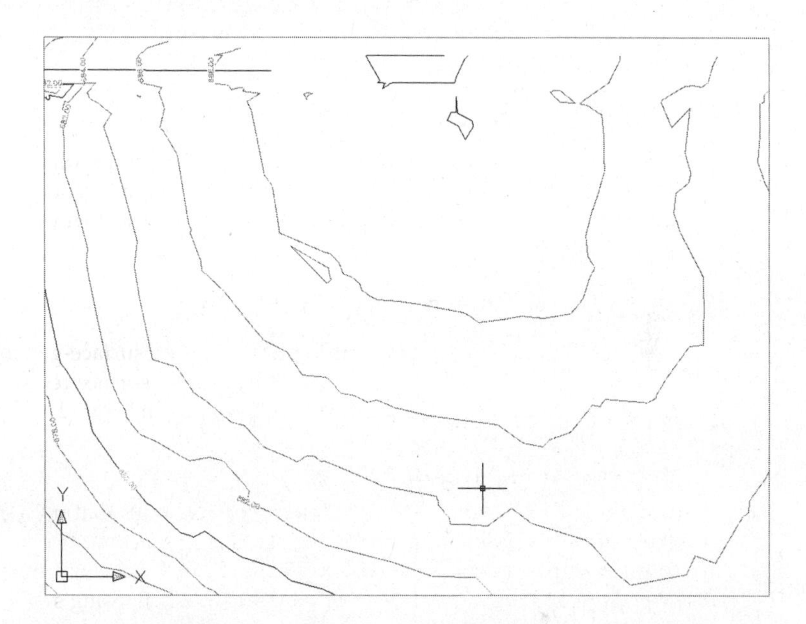

2. Zoom in to any contour label placed using the Contour – Single button, and pick the text. Three grips will appear. The original contour label lines are quite apparent, but in reality, every label has a hidden label line beneath it.
3. Grab the northernmost grip and drag across an adjacent contour, as in Figure 5.44. New labels will appear everywhere your dragged line now crosses a contour.

FIGURE 5.44
Grip-editing a contour label line

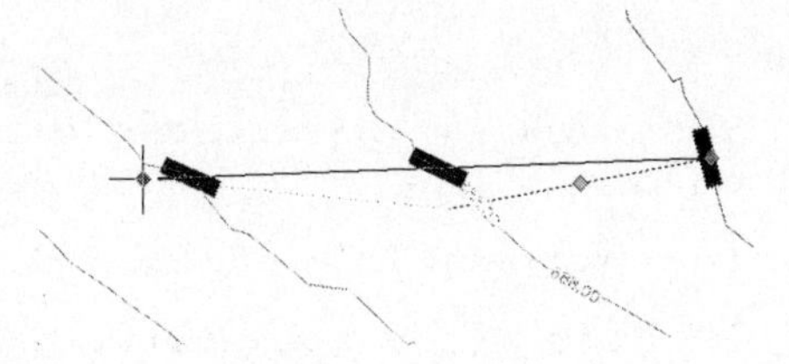

4. Drop the grip somewhere to create labels as desired.

By using the created label lines instead of adding new ones, you'll find it easier to manage the layout of your labels.

CONTOUR LABELING STYLES

The fewer label lines produced, the easier is it is to manage or modify them. Before you perform a change, build a new contour label style that uses a boundary around the text and has no decimal places:

1. Switch to the Settings tab of Toolspace, and expand Surface ➤ Label Styles ➤ Contour.
2. Right-click Existing Major Labels and select Copy. The Label Style Composer dialog appears.
3. On the Information tab, change the Name field to **Existing With Box** and switch to the Layout tab.
4. Under the Border property, set the Visibility to True.
5. Click in the Contents value, and then click the ellipsis button to bring up the Text Component Editor.
6. Click in the preview area to select the text and delete it.
7. On the Properties tab, change the value of the Precision Modifier to 1 and click the insert arrow.
8. Click OK to close and zoom in on the preview, as shown in Figure 5.45, to verify your changes have stuck.

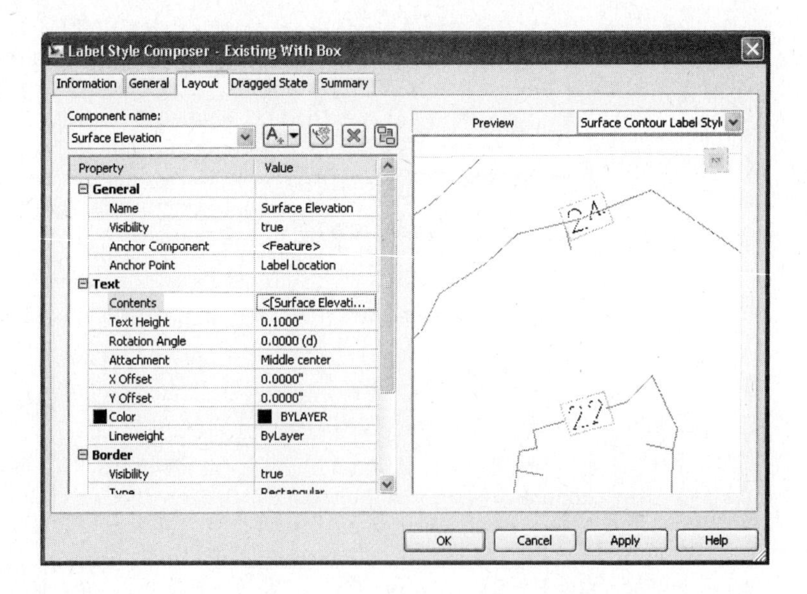

FIGURE 5.45 Completed Existing With Box contour label style

9. Click OK to close the Label Style Composer.

Now that you have a contour label style you're happy with, you can update all the label lines you've already created. In this exercise, you'll change the style used on your entire site:

1. Change to the View tab and perform a zoom extents on your labeled surface by clicking the Extents button on the Navigate panel.

2. Pick one of the visible contour label lines.

3. Right-click and choose Select Similar from the menu to pick all of the surface contour label lines. This selection is based on type and layer, so be careful that you don't pick up extraneous objects.

4. Click the Properties button on the General Tools panel to open the AutoCAD Object Properties Manager dialog, shown in Figure 5.46.

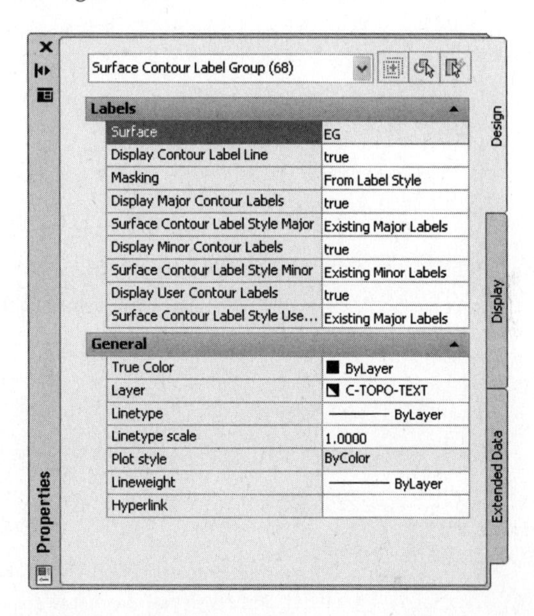

FIGURE 5.46
Contour label group in the Object Properties Manager dialog box

5. Change the Display Contour Label Line value to True if you would like to see all the label lines created. Be sure to set it to False before continuing.

6. Change the Surface Contour Label Style Major value to Existing With Box.

7. Change the Display Minor Contour Labels value to False to turn off the minor contour labels.

8. Close the AutoCAD Object Properties Manager. Press the Esc key to dismiss the selection, and perform a zoom extents to see the results.

Even when the contour label lines are set to not display, selecting any label will activate the grips, allowing for the manipulation or selection of other contour label lines. Your site is generally labeled, so you can now move on to more specific point labels.

Surface Point Labels

In every site, there are points that fall off the contour line but are critical. In an existing surface, this can be the low point in a pond or a driveway that has to be matched. When you're working with commercial sites, the spot grade is the most common review element. One of the most time-consuming issues in land development is the preparation of grading plans with hundreds of individual spot grades. Every time a site grading scheme changes, these are typically updated manually, leaving lots of opportunities for error.

With Civil 3D's surface modeling, spot labels are dynamic and react to changes in the underlying surface. By using surface labels instead of points or text callouts, you can generate a grading plan early on in the design process and begin the process of creating sheets. In this section, you label surface slopes in a couple of ways, create a single spot label for critical information, and conclude by creating a grid of labels similar to many estimation software packages.

LABELING SLOPES

Beyond the specific grade at any single point, most grading plans use slope labels to indicate some level of trend across a site or drainage area. Civil 3D can generate the following two slope labels:

- One-point slope labels indicate the slope of an underlying surface triangle. These work well when the surface has large triangles, typically in pad or mass grading areas.
- Two-point slope labels indicate the slope trend on the basis of two points selected and their locations on the surface. A two-point slope label works by dividing the surface elevation distance between the points by the planar distance between the pick points. This works well in existing ground surface models to indicate a general slope direction but can be deceiving in that it does not consider the terrain between the points.

In this exercise, you'll apply both types of slope labels, and then look at a minor style modification that is commonly requested:

1. Open the `Surface Slope Labeling.dwg` file.
2. Select the surface to display the Tin Surface tab. On the Labels and Tables panel, select Add Labels ➤ Slope.
3. At the command line, press ↵ to select a one-point label style.
4. Zoom in on the circle drawn on the western portion of the site and use a Center snap to place a label at its center, as shown in Figure 5.47.

FIGURE 5.47
A one-point slope label

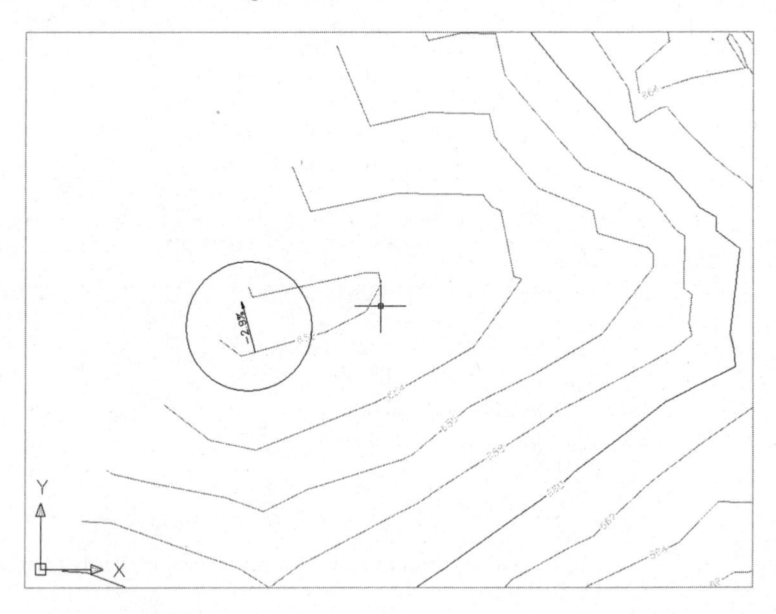

5. Press Esc or ↵ to exit the command.

6. Select the surface to display the Tin Surface tab. On the Labels and Tables panel, select Add Labels ➤ Slope.

7. At the command line, press T to switch to a two-point label style.

8. Pan to the southwest portion of the site, and use an Endpoint snap to pick the northern end of the line shown in Figure 5.48.

FIGURE 5.48
First point in a two-point slope label

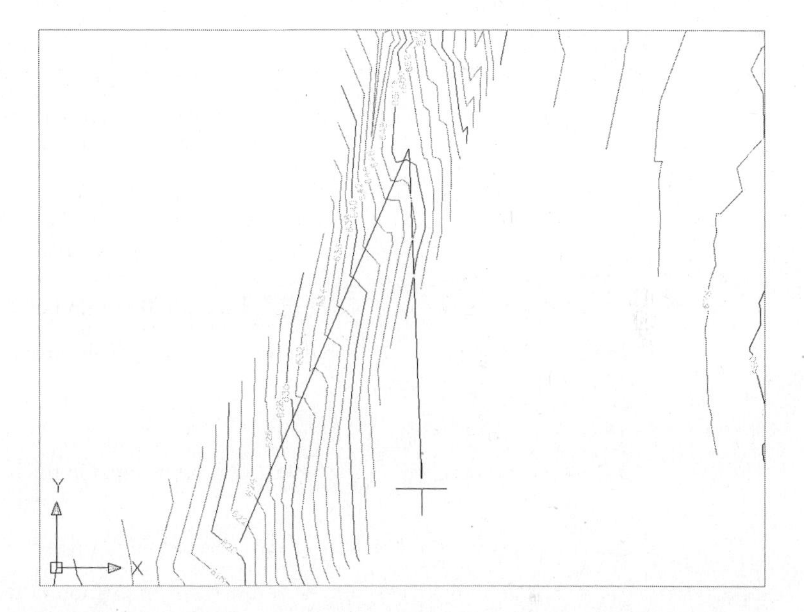

9. Use an Endpoint snap to select the other end of the line to complete the label, and press Esc or ↵ to exit the command.

This second label indicates the average slope of a dirt road that is cut into the side of the site. By using a two-point label, you get a better understanding of the trend, as opposed to a specific point.

One concern for many users is the sign on the spot label. Because the arrow on the two-point label is always drawn from point one to point two, the arrow can point in both an upslope and a downslope direction, so the sign is important. On a one-point label, however, the arrow always points downhill, making the sign redundant. In this exercise, you create a new style to drop the sign:

1. Select the one-point slope label created earlier.

2. Click the Label Properties button on the Modify panel to display the AutoCAD Object Properties Manager palette.

3. Click the drop-down arrow for Surface Slope Label Style and select Create/Edit at the bottom of the list to display the Style Selection dialog.

4. Click the drop-down arrow on the right of the dialog and select Create Child Of Current Selection to bring up the Label Style Composer.

5. On the Information tab, change the Name field to **Percent-No Sign** and change to the Layout tab.

6. Change the Component Name field at the top of the Layout tab to Surface Slope.

7. Click in the Contents value, and then click the ellipsis button to bring up the Text Component Editor.
8. Delete the text in the preview area.
9. Change the Sign property to Drop Sign and click the arrow to insert the data field.
10. Click OK to close the Text Component Editor dialog. The preview should look like Figure 5.49.

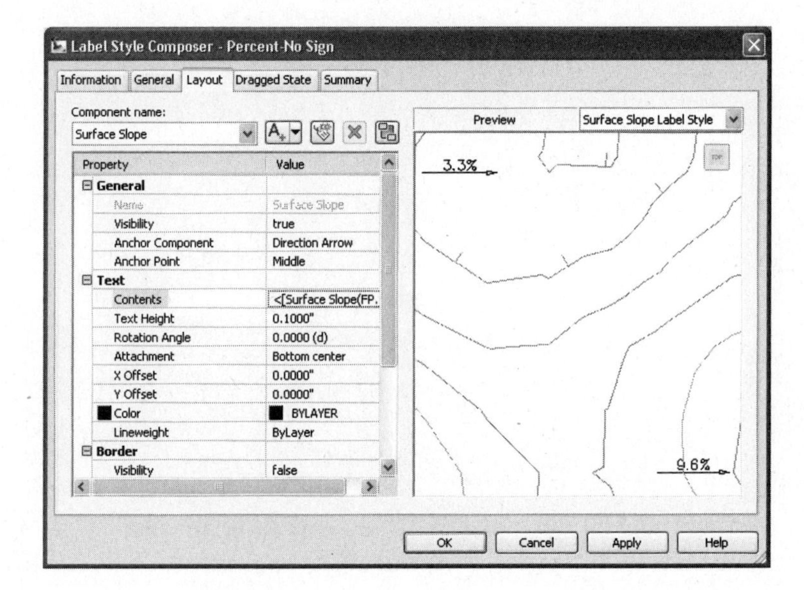

FIGURE 5.49
Label Style Composer for the Percent-No Sign label

11. Click OK to close the Style Selection dialog.
12. Click OK again to close the Surface Slope Label Style dialog.

Note that the style is selected and the screen has updated already. Just a reminder: by creating a child style, you've built in a relationship between your Percent-No Sign label and the Percent label. A change in the Percent label style to layer, color, size, and so on will be reflected in your child style.

CRITICAL POINTS

A typical grading plan is a sea of critical points that drive the site topography. In the past, much of this labeling and point work was done by creating COGO points and simply displaying their properties. Although this is effective, it has two distinct disadvantages. First, these points are not reflective of the design but part of the design. This makes the sheet creation a part of the grading process, not a parallel process. Second, the addition of COGO points to any drawing and project when they're not truly needed just weighs down the design model. Point management is a mentally intensive task, and anything that can limit extraneous data is worth investigating.

Surface labels react dynamically to the surface and to the point of insertion. Moving any of these labels would update the information to reflect the surface underneath. This relationship makes it possible for one user to place labels on a grading plan while the final surface is still in flux. A change in the proposed surface is reflected in an update from the project, and an updated sheet can be on the plotter in minutes.

Surface Grid Labels

Sometimes, more than a few points are requested. Estimation software typically creates a grid of point labels that can be easily reviewed or passed to a contractor for field work. In this exercise, you'll use the volume surface you generated earlier in this chapter to create a set of surface labels that reflect this requirement:

1. Open the `Surface Volume Grid Labels.dwg` file.
2. Change to the Annotate tab, and select Labels ➤ Surface ➤ Spot Elevations On Grid.
3. Click one of the colored contours to pick the Volume surface.
4. Pick a point in the southwest of the surface to set a base point for the grid.
5. Press ↵ to set the grid rotation to zero.
6. Enter **25** at the command line to set the *x* spacing.
7. Enter **25** at the command line to set the *y* spacing.
8. Click to the northeast of the surface to set the area for the labels.
9. Verify the preview box contains the Volume surface and press ↵ at the command line to continue.
10. Wait a few moments as Civil 3D generates all the labels just specified. Your drawing should look similar to Figure 5.50.

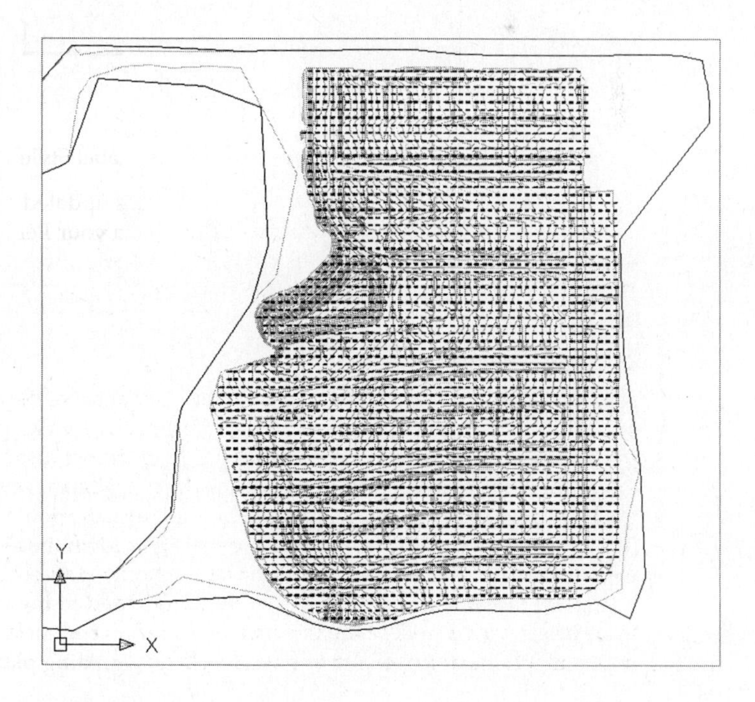

Figure 5.50
Volume surface with grid labels

USING SURFACES TO DO MORE

In this exercise, you'll use a surface label to make a building pad label. By using surface information, you can also take advantage of Civil 3D's Expressions to include extra elevation information.

1. Open the `Surface Spot Labeling.dwg` file.
2. In the Settings tab of Toolspace, expand the Surfaces ➢ Label Styles ➢ Spot Elevation branches.
3. Right-click Expressions and select the New option. The New Expression dialog appears.
4. Change the Name field to **FF**.
5. Change the Description to **Finished Floor Elevation**.
6. Click the Insert Property button and select Surface Elevation, as shown here.

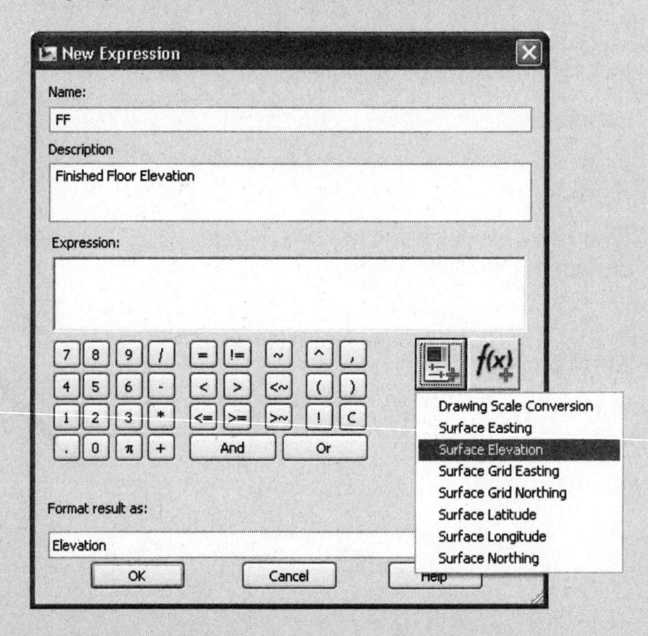

7. Click next to the Surface Elevation entry and type **+0.7**.
8. Click OK to close the dialog.

You use expressions throughout Civil 3D to label or modify labels with information that can be derived mathematically from a surface. These expressions can include some level of logic, but in this case, it's simple math to make 2 bits of information from 1 bit of data.

1. Right-click the Spot Elevation branch and select the New option. The Label Style Composer dialog appears.

2. On the Information tab, change the Name field to **Pad Label**.
3. Switch to the Layout tab.
4. Click in the Contents value of the Surface Elevation component and click the ellipsis button to bring up the Text Component Editor.
5. Erase the text in the preview area, and then type **FF**:.
6. Select FF from the Properties drop-down list.
7. Change the Precision value to 0.1 and click the insert arrow.
8. Click to enter the text area, and press ↵ at the end of the first line of text to create a line break.
9. Type **FP**:.
10. Select Surface Elevation from the Properties drop-down list.
11. Change the Precision value to 0.1 and click the insert arrow. Your label should look like what's shown here.

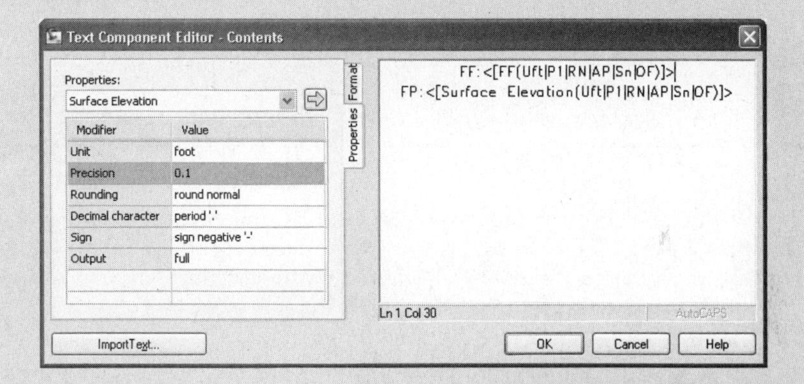

12. Click OK to close the Text Component Editor.
13. Change the Anchor Point value to Middle Center.
14. Change the Attachment value to Middle Center.
15. Click OK to close the dialog.
16. On the Annotate tab, select the Labels button from the Labels & Tables panel.
17. Change the Feature field to Surface.
18. Change the Label Type field to Spot Elevation.
19. Change the Spot Elevation Label Style field to Pad Label.
20. Change the Marker Style field to <none>.
21. Click the Add button.

22. Click in the center of the circle surrounding the building pad to insert a label. Your result will be similar to what's shown here.

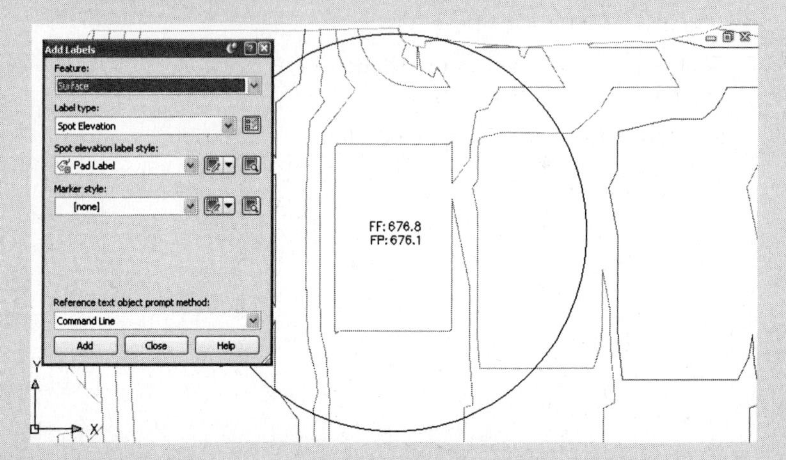

If you would like to color these labels based on cut or fill, check out the Autodesk Civil Community at `http://civilcommunity.autodesk.com`, where Peter Funk was kind enough to walk through the expressions and label styles necessary to create a label that's one color in cut, another in fill. His example was for points, but the same technique works for surface labels.

Labeling the grid is imprecise at best. Grid labeling ignores anything that might happen between the grid points, but it presents the surface data in a familiar way for engineers and contractors. By using the tools available and the underlying surface model, you can present information from one source in an almost infinite number of ways.

The Bottom Line

Create a preliminary surface using freely available data. Almost every land development project involves a surface at some point. During the planning stages, freely available data can give you a good feel for the lay of the land, allowing design exploration before money is spent on fieldwork or aerial topography. Imprecise at best, this free data should never be used as a replacement for final design topography, but it's a great starting point.

Master It Create a new drawing from the Civil 3D Extended template and bring in a Google Earth surface for your home or office location. Be sure to set a proper coordinate system to get this surface in the right place.

Modify and update a TIN surface. TIN surface creation is mathematically precise, but sometimes the assumptions behind the equations leave something to be desired. By using the editing tools built into Civil 3D, you can create a more realistic surface model.

Master It Modify your Google Earth surface to show only an area immediately around your home or office. Create an irregular shaped boundary and apply it to the Google Earth surface.

Prepare a slope analysis. Surface analysis tools allow users to view more than contours and triangles in Civil 3D. Engineers working with nontechnical team members can create strong meaningful analysis displays to convey important site information using the built-in analysis methods in Civil 3D.

Master It Create an Elevation Banding analysis of your home or office surface and insert a legend to help clarify the image.

Label surface contours and spot elevations. Showing a stack of contours is useless without context. Using the automated labeling tools in Civil 3D, you can create dynamic labels that update and reflect changes to your surface as your design evolves.

Master It Label the contours on your Google Earth surface at 1′ and 5′ (Design).

Chapter 6

Don't Fence Me In: Parcels

Land-development projects often involve the subdivision of large pieces of land into smaller lots. Even if your projects don't directly involve subdivisions, you're often required to show the legal boundaries of your site and the adjoining sites.

In previous CAD systems, a few tools were available for parcel management. You could create AutoCAD entities, such as lines and arcs, to represent the lot boundaries and then create a closed polyline to assist in determining the parcel area. You could also create static text labels for area, bearing, and distance. Even if you took advantage of some of the parcel-management tools in Land Desktop, the most minor change to the project, such as a road widening or a horizontal alignment adjustment, required days of editing, adjustment, and relabeling.

Civil 3D parcels give you a dynamic way to create, edit, manage, and annotate these legal land divisions. If you edit a parcel segment to make a lot larger, all of the affected labels will update — including areas, bearings, distances, curve information, and table information.

By the end of this chapter, you'll learn to:

- Create a boundary parcel from objects
- Create a right-of-way parcel using the right-of-way tool
- Create subdivision lots automatically by layout
- Add multiple parcel-segment labels

Creating and Managing Sites

In Civil 3D, a *site* is a collection of parcels, alignments, grading objects, and feature lines that share a common topology. In other words, Civil 3D objects that are on the same site are related to, as well as interact with, each other. These objects that react to each other are called *site geometry objects*.

Best Practices for Site Topology Interaction

At first glance, it may seem that the only uses for parcels are subdivision lots and, therefore, you may think that you need only one site for your drawing.

However, once you begin working with parcels, you'll find features like dynamic area labels to be useful for delineating and analyzing soil boundaries; paving, open-space, and wetlands areas; and any other region enclosed with a boundary. The automatic layer enforcement of parcel object styles also adds to the appeal of using parcels. Using additional types of parcels will require you to come up with a site-management strategy to keep everything straight.

It's important to understand how site geometry objects react to one another. Figure 6.1 shows a typical parcel that might represent a property boundary.

When an alignment is drawn and placed on the same site as the property boundary, the parcel splits into two parcels, as shown in Figure 6.2.

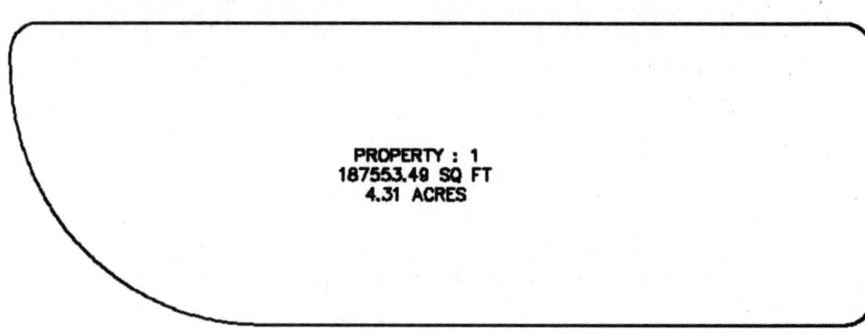

FIGURE 6.1
A typical property boundary

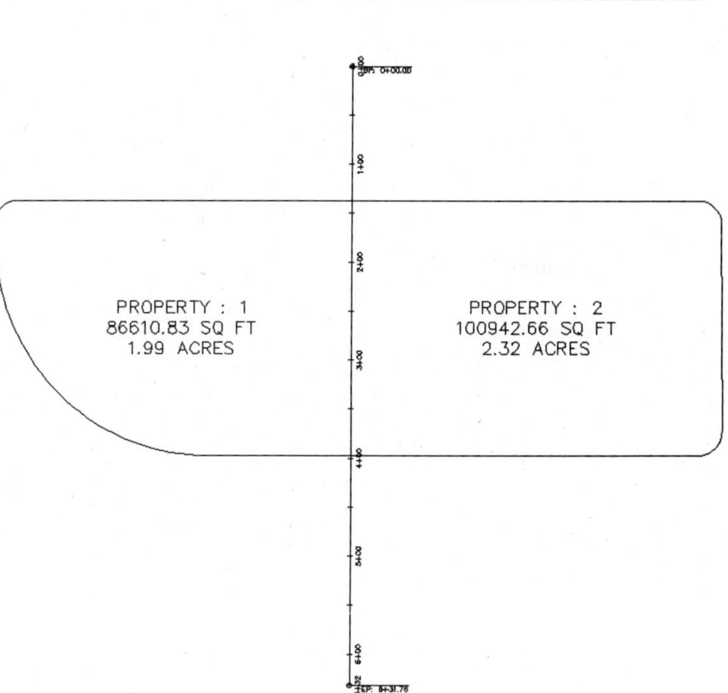

FIGURE 6.2
An alignment that crosses a parcel divides the parcel in two if the alignment and parcel exist on the same site

You must plan ahead to create meaningful sites based on interactions between the desired objects. For example, if you want a road centerline, a road right-of-way (ROW) parcel, and the lots in a subdivision to react to each other, they need to be on the same site (see Figure 6.3).

The alignment (or road centerline), ROW parcel, and lots all relate to one another. A change in the centerline of the road should prompt a change in the ROW parcel and the subdivision lots.

If you'd like to avoid the interaction between site geometry objects, place them on different sites. Figure 6.4 shows an alignment that has been placed on a different site from the boundary parcel. Notice that the alignment doesn't split the boundary parcel.

It's important that only objects that are intended to react to each other be placed on the same site. For example, in Figure 6.5, you can see parcels representing both subdivision lots and soil boundaries. Because it wouldn't be meaningful for a soil-boundary parcel segment to interrupt the area or react to a subdivision-lot parcel, the subdivision-lot parcels have been placed on a Subdivision Lots site, and the soil boundaries have been placed on a Soil Boundaries site.

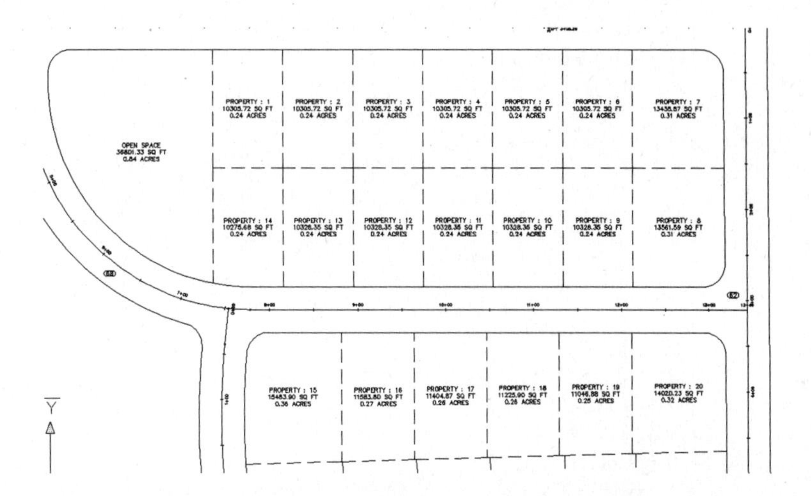

FIGURE 6.3
Alignments, ROW parcels, open-space parcels, and subdivision lots react to one another when drawn on the same site

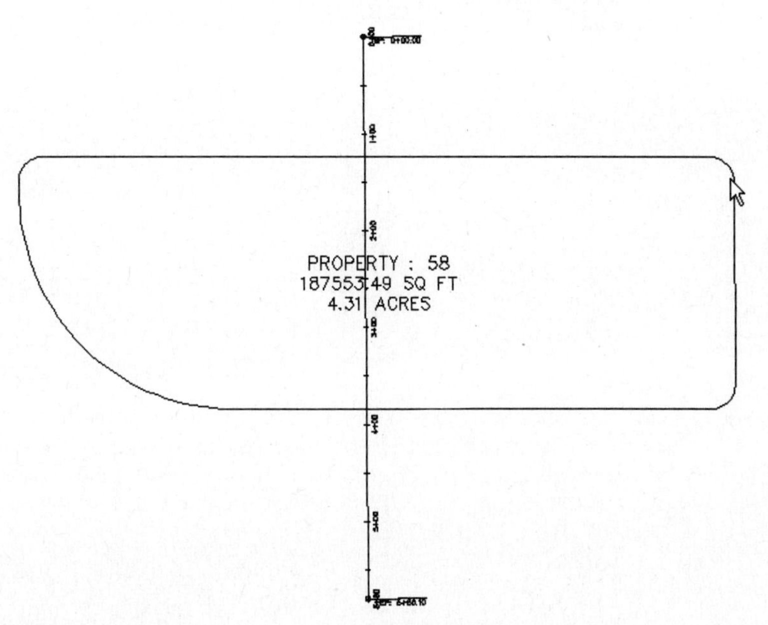

FIGURE 6.4
An alignment that crosses a parcel won't interact with the parcel if they exist on different sites

If you didn't realize the importance of site topology, you might create both your subdivision-lot parcels and your soil-boundary parcels on the same site and find that your drawing looks similar to Figure 6.6. This figure shows the soil-boundary segments dividing and interacting with subdivision-lot parcel segments, which doesn't make any sense.

Another way to avoid site-geometry problems is to do site-specific tasks in different drawings and use a combination of external references and data references to share information.

For example, you could have an existing base drawing that housed the soil-boundaries site, XRefed into a subdivision plat drawing that housed the subdivision-lots site instead of separating the two drawings onto two different sites.

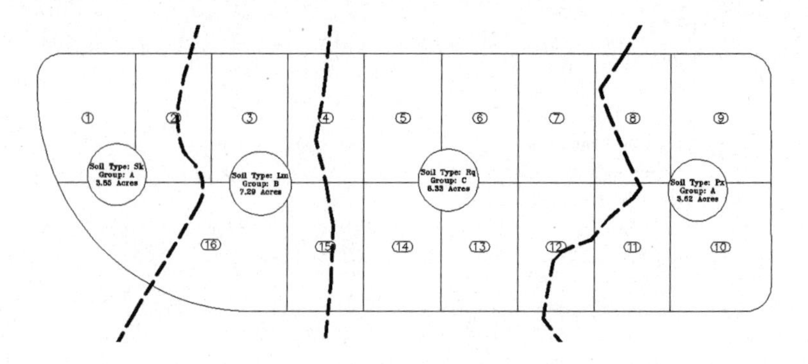

FIGURE 6.5
Parcels can be used for subdivision lots and soil boundaries as long as they're kept on separate sites

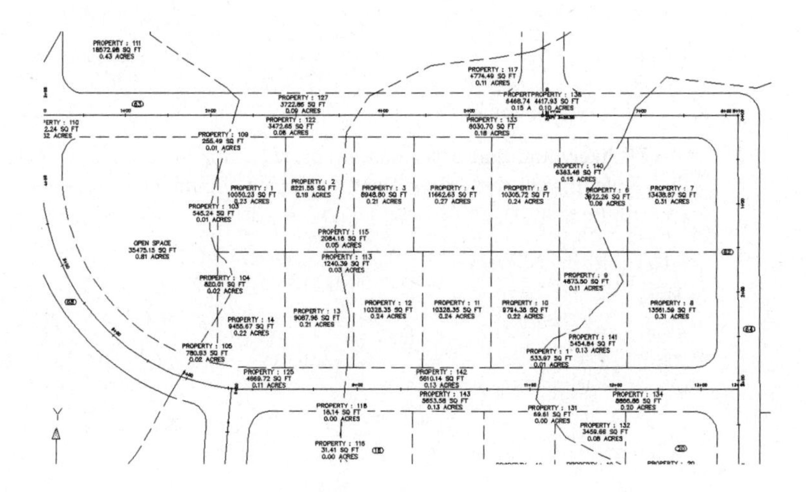

FIGURE 6.6
Subdivision lots and soil boundaries react inappropriately when placed on the same site

You should consider keeping your legal site plan in its own drawing. Because of the interactive and dynamic nature of Civil 3D parcels, it might be easy to accidentally grab a parcel segment when you meant to grab a manhole, and unintentionally edit a portion of your plat.

You'll see other workflow examples and drawing divisions later in this chapter, as well as in the chapters on data shortcuts (Chapter 17, "Sharing the Model: Data Shortcuts") and Vault (Chapter 19, "Teamwork: Vault Client and Civil 3D").

If you decide to have sites in the same drawing, here are some sites you may want to create. These suggestions are meant to be used as a starting point. Use them to help find a combination of sites that works for your projects:

Roads and Lots This site could contain road centerlines, ROW, platted subdivision lots, open space, adjoining parcels, utility lots, and other aspects of the final legal site plan.

Grading Feature lines and grading objects are considered part of site geometry. If you're using these tools, you must make at least one site for them. You may even find it useful to have several grading sites.

Easements If you'd like to use parcels to manage, analyze, and annotate your easements, you may consider creating a separate site for easements.

Stormwater Management If you'd like to use parcels to manage, analyze, and annotate your stormwater subcatchment boundaries, you may consider creating a separate site for stormwater management.

As you learn new ways to take advantage of alignments, parcels, and grading objects, you may find additional sites that you'd like to create at the beginning of a new project.

WHAT ABOUT THE "SITELESS" ALIGNMENT?

The previous section mentioned that alignments are considered site geometry objects. Civil 3D 2008 introduced the concept of the "siteless" alignment: an alignment that is placed on the <none> site. An alignment that is created on the <none> site doesn't react with other site geometry objects or with other alignments created on the <none> site.

However, you can still create alignments on traditional sites, if you desire, and they will react to other site geometry objects. This may be desirable if you want your road centerline alignment to bisect a ROW parcel, for example.

You'll likely find that best practices for most alignments are to place them on the <none> site. For example, if road centerlines, road-transition alignments, swale centerlines, and pipe-network alignments are placed on the <none> site, you'll save yourself quite a bit of site geometry management.

It is important to note that although <none> sites cannot be seen or selected in a drawing, they still exist in the drawing database. For example, if you've used the <none> site option 12 times, you'll have 12 uniquely numbered <none> site definitions in the drawing database.

See Chapter 7, "Laying a Path: Alignments," for more information about alignments and sites.

Creating a New Site

You can create a new site in Prospector. You'll find the process easier if you brainstorm potentially needed sites at the beginning of your project and create those sites right away — or, better yet, save them as part of your standard Civil 3D template. You can always add or delete sites later in the project.

The Sites collection is stored in Prospector, along with the other Civil 3D objects in your drawing.

The following exercise will lead you through creating a new site that you can use for creating subdivision lots:

1. Open the `Create Site.dwg` file, which you can download from `www.sybex.com/go/masteringcivil3d2010`. Note that the drawing contains alignments and soil-boundary parcels, as shown in Figure 6.7.
2. Locate the Sites collection on the Prospector tab of Toolspace.
3. Right-click the Sites collection, and select New to open the Site Properties dialog (see Figure 6.8).
4. On the Information tab of the Site Properties dialog, enter **Subdivision Lots** for the name of your site.
5. Confirm that the settings on the 3D Geometry tab match what is shown on Figure 6.9.

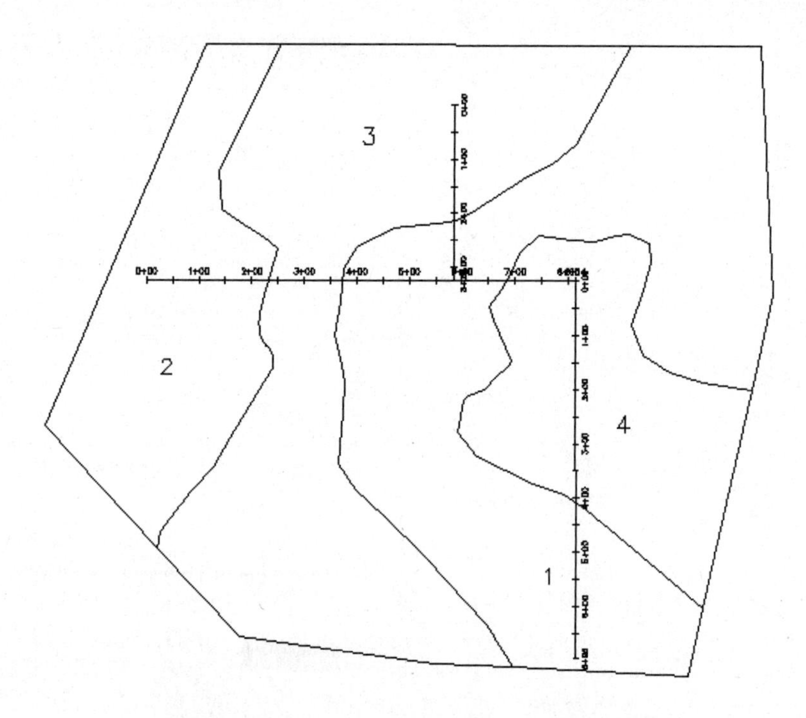

FIGURE 6.7
The Create Site drawing contains alignments and soil-boundary parcels

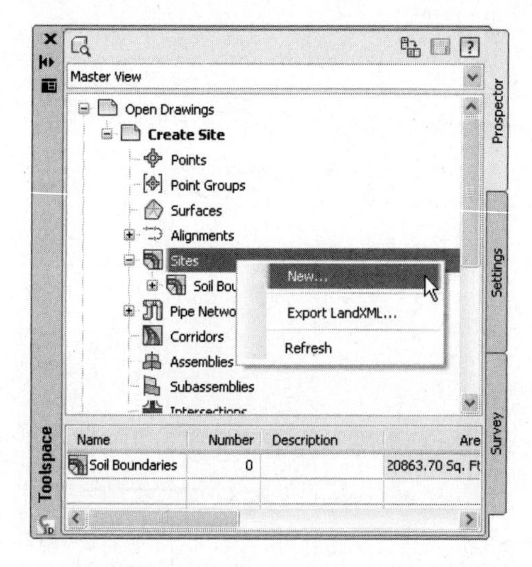

FIGURE 6.8
Right-click the Sites collection, and select New

6. Confirm that the settings on the Numbering tab match Figure 6.10. Everything should be set to 1.
7. Locate the Sites collection on the Prospector tab of Toolspace, and note that your Subdivision Lots site appears on the list.

You can repeat the process for all the sites you anticipate needing over the course of the project.

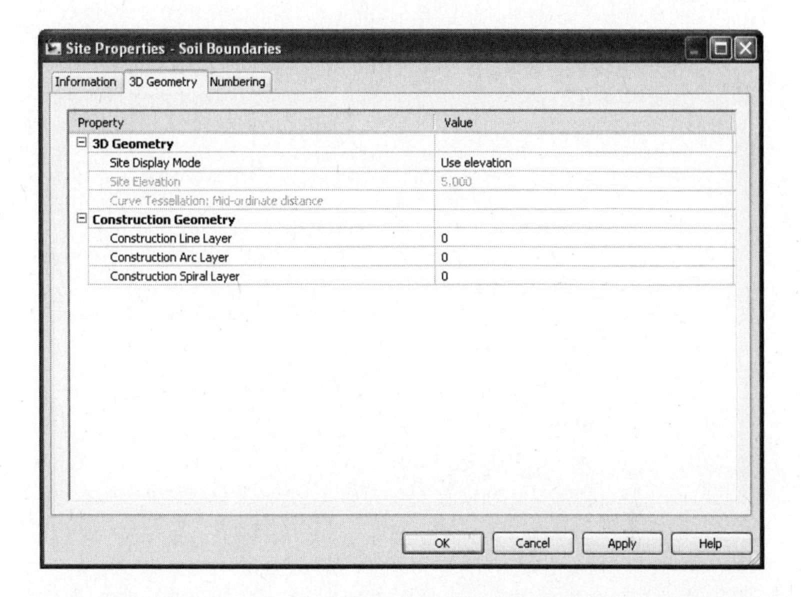

FIGURE 6.9
Confirm the settings on the 3D Geometry tab

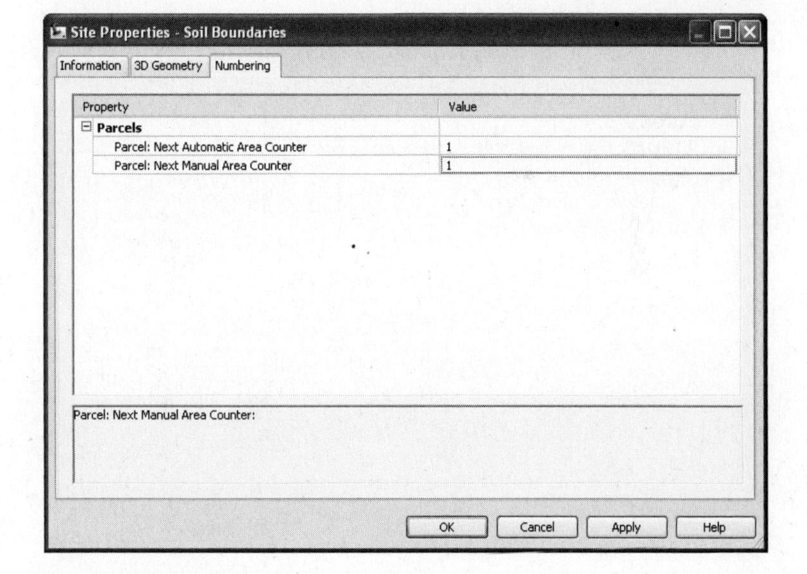

FIGURE 6.10
Confirm the settings on the Numbering tab

Creating a Boundary Parcel

The Create Parcel from Objects tool allows you to create parcels by choosing AutoCAD entities in your drawing or in an XRefed drawing. In a typical workflow, it's common to encounter a boundary created by AutoCAD entities, such as polylines, lines, and arcs.

When you're using AutoCAD geometry to create parcels, it's important that the geometry be created carefully and meets certain requirements. The AutoCAD geometry must be lines, arcs, polylines, 3D polylines, or polygons. It can't include blocks, ellipses, circles, or other entities.

Civil 3D may allow you to pick objects with an elevation other than zero, but you'll find you get better results if you flatten the objects so that all objects have an elevation of zero. Sometimes the geometry appears sound when elevation is applied, but you may notice that this isn't the case once the objects are flattened. Flattening all objects before creating parcels can help you prevent frustration when creating parcels.

This exercise will teach you how to create a parcel from Civil 3D objects:

1. Open the `Create Boundary Parcel.dwg` file, which you can download from `www.sybex.com/go/masteringcivil3d2010`. This drawing has several alignments, which were created on the Subdivision Lots site, and some AutoCAD linework representing a boundary. In addition, parcels were created when the alignments formed closed areas on the Subdivision Lots site.
2. On the Home tab, select Parcel ➤ Choose Create Parcel from Objects on the Create Design panel.
3. At the `Select lines, arcs, or polylines to convert into parcels or [Xref]:` prompt, pick the red polyline that represents the site boundary. Press ↵.
4. The Create Parcels – From Objects dialog appears. Select Subdivision Lots; Property; and Name Square Foot & Acres from the drop-down menus in the Site, Parcel Style, and Area Label Style selection boxes, respectively. Leave everything else set to the defaults. Click OK to dismiss the dialog.
5. The boundary polyline forms parcel segments that react with the alignments. Area labels are placed at the newly created parcel centroids, as shown in Figure 6.11.

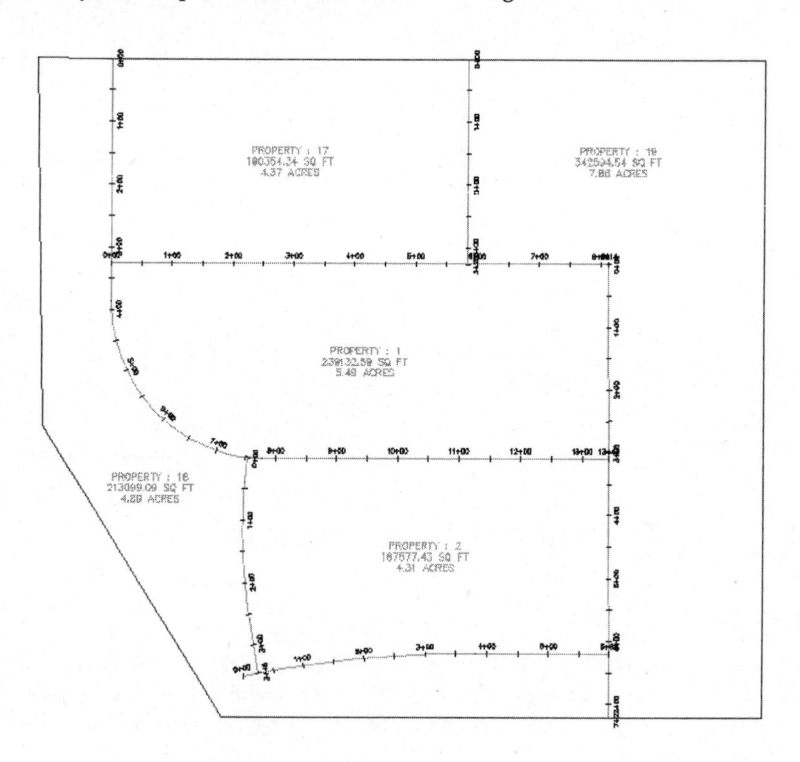

FIGURE 6.11 The boundary parcel segments, alignments, and area labels

Creating a Wetlands Parcel

Although you may never have thought of things like wetlands areas, easements, and stormwater-management facilities as parcels in the past, you can take advantage of the parcel tools to assist in labeling, stylizing, and analyzing these features for your plans.

This exercise will teach you how to create a parcel representing wetlands using the transparent commands and Draw Tangent-Tangent with No Curves tool from the Parcel Layout Tools toolbox:

1. Open the `Create Wetlands Parcel.dwg` file, which you can download from `www.sybex.com/go/masteringcivil3d2010`. Note that this drawing has several alignments, parcels, and a series of points that represent a wetlands delineation.
2. Choose Parcel ➤ Parcel Creation Tools on the Create Design panel. The Parcel Layout Tools toolbar appears.

3. Click the Draw Tangent-Tangent with No Curves tool on the Parcel Layout Tools toolbar. The Create Parcels – Layout dialog appears.
4. In the dialog, select Subdivision Lots, Property, and Name Square Foot & Acres from the drop-down menus in the Site, Parcel Style, and Area Label Style selection boxes, respectively. Keep the default settings for all other options. Click OK.

5. At the `Specify start point:` prompt, click the Point Object transparent command on the Transparent Commands toolbar, and then pick point 1. Continue picking the wetlands points in numerical order, as shown in Figure 6.12.

FIGURE 6.12
Pick each wetlands point in numerical order

6. Once you've reached point 9, be sure to pick point 1 again to close the loop. Press ↵ to exit the Point Object transparent command and press ↵ again. Type **X**, and then press ↵ again to dismiss the Create Parcels – Layout dialog. Your result should look similar to Figure 6.13.

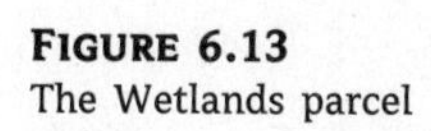

FIGURE 6.13
The Wetlands parcel

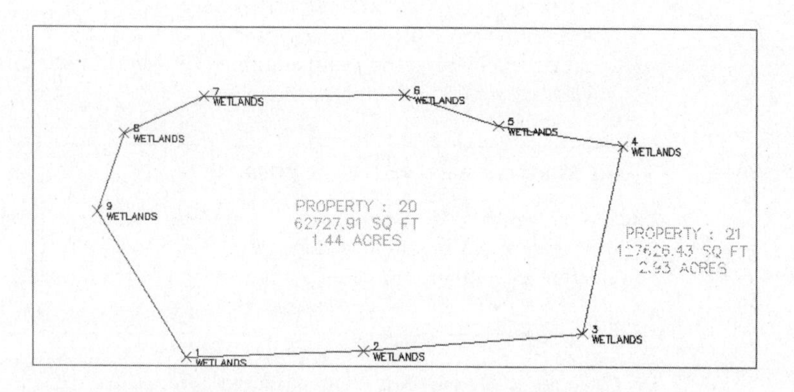

7. It's usually easier to change the appearance of the parcel and its area label after the parcel has been created. Change the style of the parcel by picking the parcel area label and picking Parcel Properties from the Modify panel. The Parcel Properties dialog appears.

8. Select Wetlands from the drop-down menu in the Object Style selection box on the Information tab, and then click OK to dismiss the dialog. The parcel segments turn green, and a swamp hatch pattern appears inside the parcel to match the Wetlands style.
9. To change the style of the parcel area label, first select the Wetlands parcel area label, and then right-click and select Edit Area Selection Label Style. The Parcel Area Label Style dialog appears.
10. Select the Wetlands Area Label style from the drop-down menu in the Parcel Area Label Style selection box. Click OK to dismiss the dialog. A label appears, labeling the wetlands as shown in Figure 6.14. Later sections will discuss parcel style and parcel area label style in more detail.

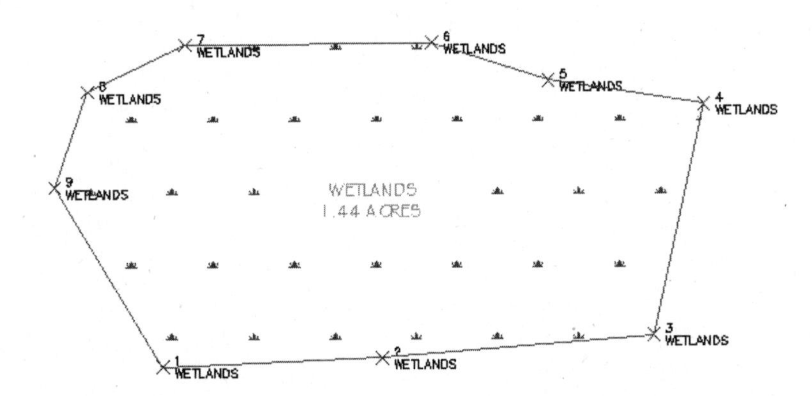

FIGURE 6.14
The Wetlands parcel with the appropriate label styles applied

Creating a Right-of-Way Parcel

The Create ROW tool creates ROW parcels on either side of an alignment based on your specifications. The Create ROW tool can be used only when alignments are placed on the same site as the boundary parcel, as in Figure 6.15.

The resulting ROW parcel will look similar to Figure 6.16.

Options for the Create ROW tool include offset distance from alignment, fillet or chamfer cleanup at parcel boundaries, and alignment intersections. Figure 6.17 shows an example of chamfered cleanup at alignment intersections.

MAKE SURE YOUR GEOMETRY IS POSSIBLE

Make sure you provide parameters that are possible. If the program can't achieve your filleting requirements at any one intersection, a ROW parcel won't be created. For example, if you specify a 25′ filleting radius, but the roads come together at a tight angle that would only allow a 15′ radius, then a ROW parcel won't be created.

Once the ROW parcel is created, it's no different from any other parcel. For example, it doesn't maintain a dynamic relationship with the alignment that created it. A change to the alignment will require the ROW parcel to be edited or, more likely, re-created.

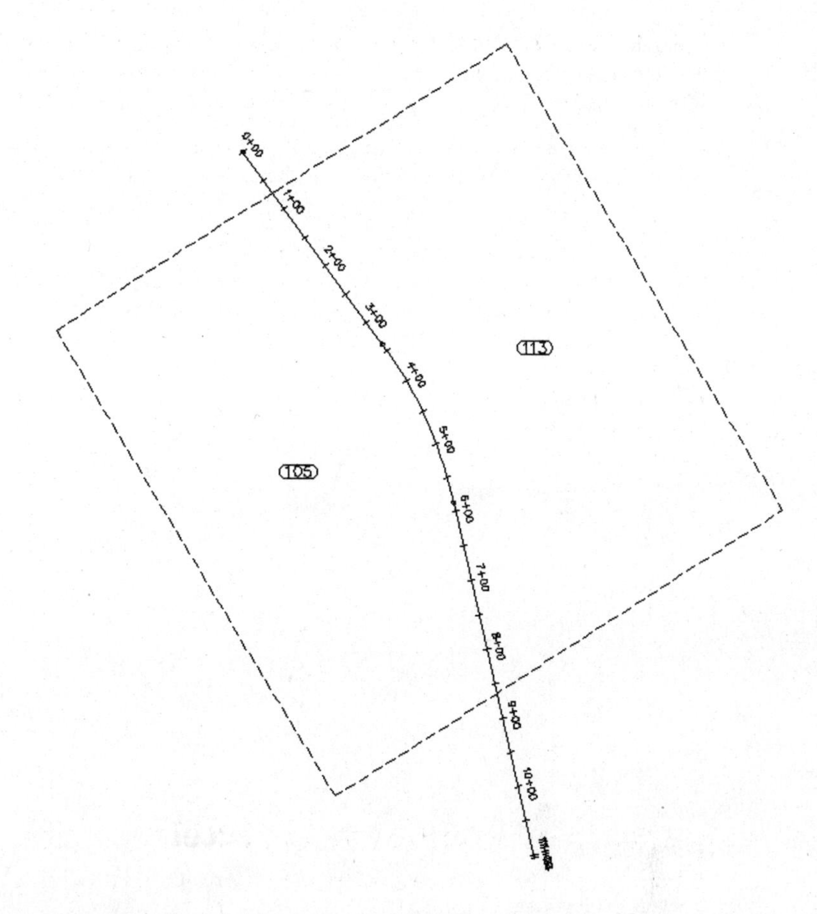

FIGURE 6.15
An alignment on the same site as parcels

This exercise will teach you how to use the Create ROW tool to automatically place a ROW parcel for each alignment on your site:

1. Open the `Create Right of Way Parcel.dwg` file, which you can download from `www.sybex.com/go/masteringcivil3d2010`. Note that this drawing has several alignments on the same site as the boundary parcel, resulting in several smaller parcels between the alignments and boundary.
2. Choose Parcel ➤ Create Right of Way on the Create Design panel.
3. At the `Select parcels:` prompt, pick Property: 1, Property: 2, Property: 3, Property: 4, and Property: 5 on the screen. Press ↵ to stop picking parcels. The Create Right Of Way dialog appears, as shown in Figure 6.18.
4. Expand the Create Parcel Right of Way parameter, and enter **20′** as the value for Offset from Alignment.
5. Expand the Cleanup at Parcel Boundaries parameter. Enter **20′** as the value for Fillet Radius at Parcel Boundary Intersections. Select Fillet from the drop-down menu in the Cleanup Method selection box.

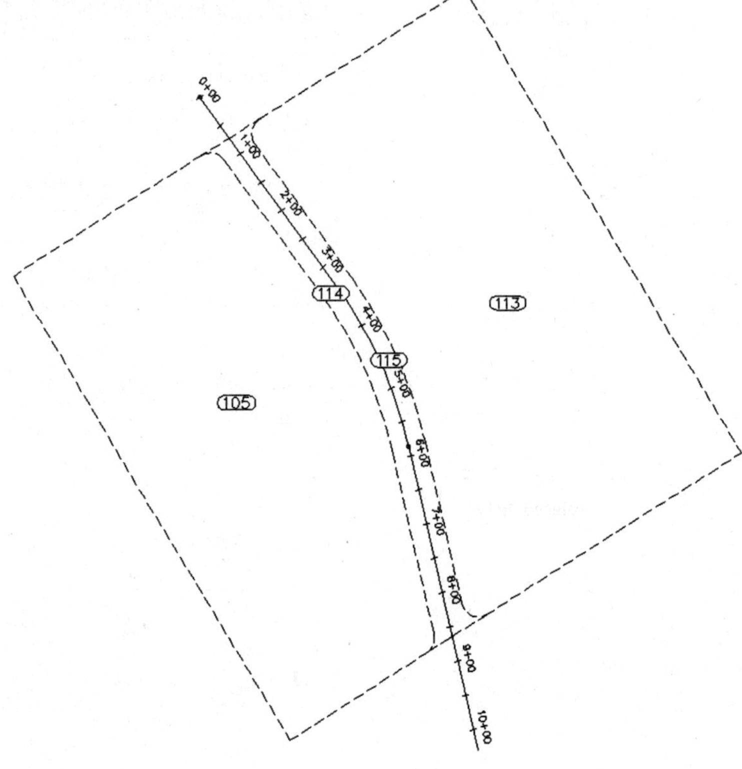

FIGURE 6.16
The resulting parcels after application of the Create ROW tool

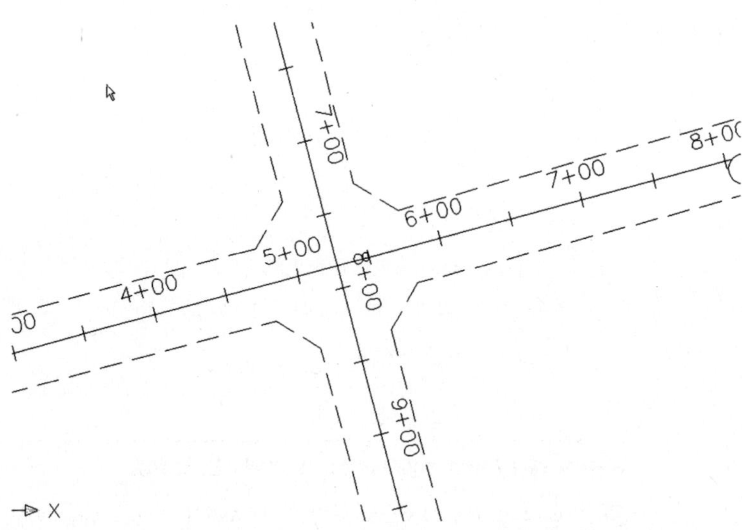

FIGURE 6.17
A ROW with chamfer cleanup at alignment intersections

6. Expand the Cleanup at Alignment Intersections parameter. Enter **20'** as the value for Fillet Radius at Alignment Intersections. Select Fillet from the drop-down menu in the Cleanup Method selection box.

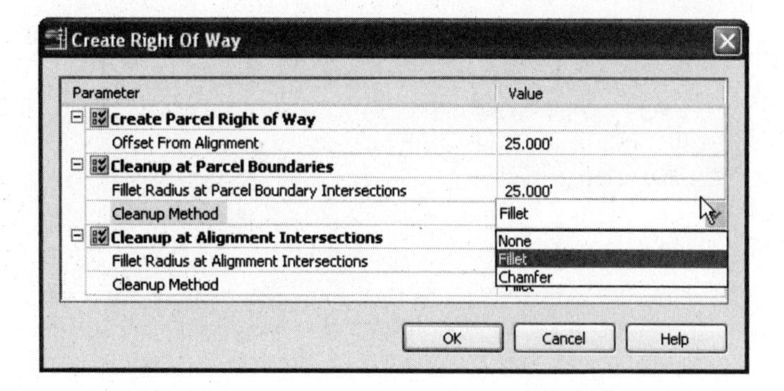

FIGURE 6.18
The Create Right Of Way dialog

7. Click OK to dismiss the dialog and create the ROW parcels. Your drawing should look similar to Figure 6.19.

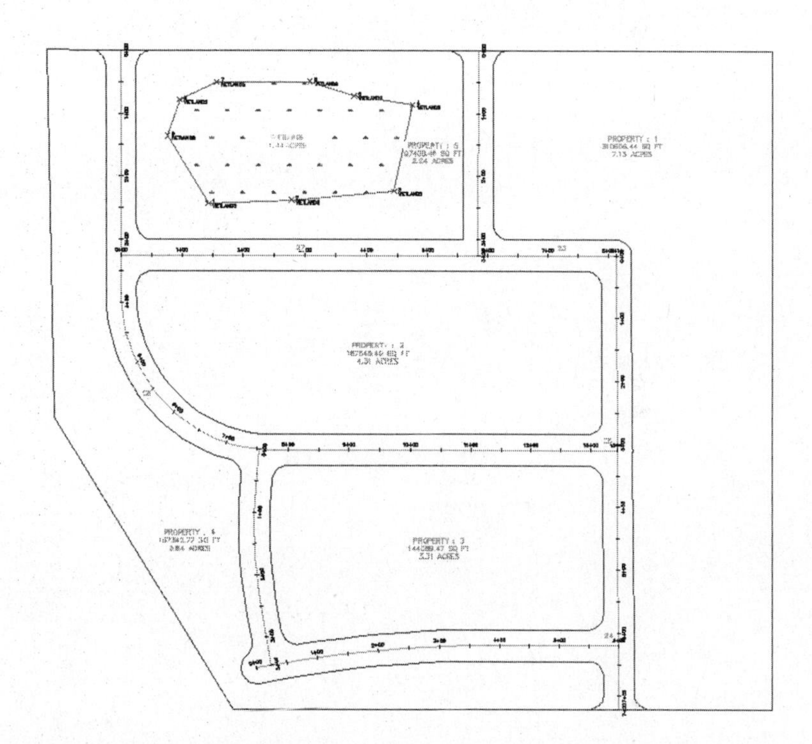

FIGURE 6.19
The completed ROW parcels

WHEN THE CREATE ROW TOOL ISN'T ENOUGH

The Create ROW tool works well for straightforward road plans that have even widths. If you need something a little more intricate for design elements, such as cul-de-sacs, width changes, or knuckles, you can do the same thing the Create ROW tool does using the AutoCAD lines and curves commands and the Parcel Creation Tools from the Parcel menu on the Create Design panel.

Creating Subdivision Lot Parcels Using Precise Sizing Tools

The precise sizing tools allow you to create parcels to your exact specifications. You'll find these tools most useful when you have your roadways established and understand your lot-depth requirements. These tools provide automatic, semiautomatic, and freeform ways to control frontage, parcel area, and segment direction.

Attached Parcel Segments

Parcel segments created with the precise sizing tools are called *attached segments*. Attached parcel segments have a start point that is attached to a frontage segment and an endpoint that is defined by the next parcel segment they encounter. Attached segments can be identified by their distinctive diamond-shaped grip at their start point and no grip at their endpoint (see Figure 6.20).

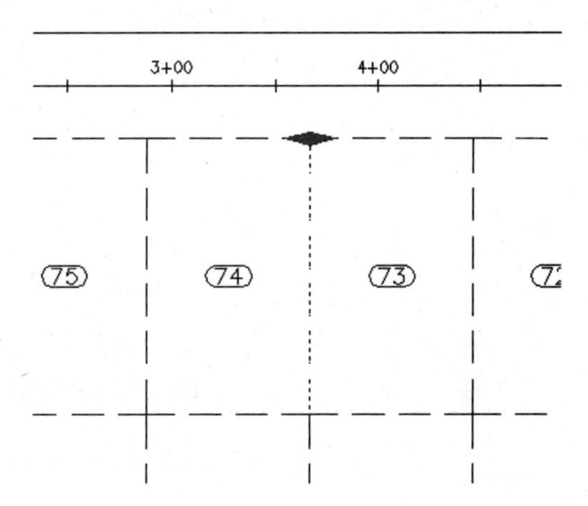

FIGURE 6.20
An attached parcel segment

In other words, you establish their start point and their direction, but they seek another parcel segment to establish their endpoint. Figure 6.21 shows a series of attached parcel segments. You can tell the difference between their start and endpoints because the start points have the diamond-shaped grips.

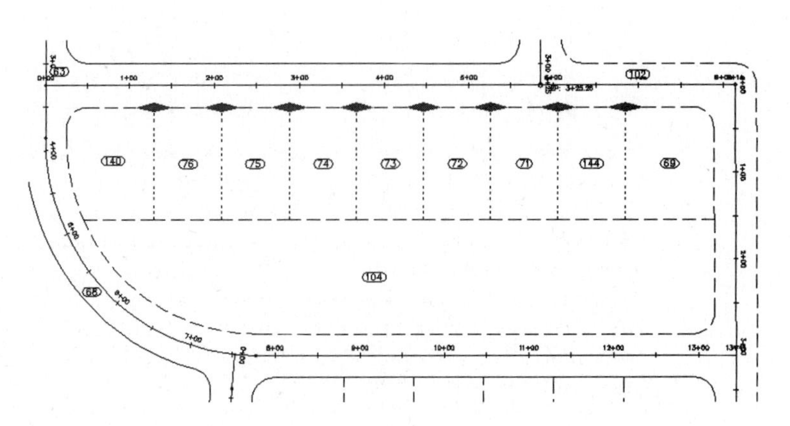

FIGURE 6.21
A series of attached parcel segments, with their endpoint at the rear lot line

You can drag the diamond-shaped grip along the frontage to a new location, and the parcel segment will maintain its angle from the frontage. If the rear lot line is moved or erased, the attached parcel segments find a new endpoint (see Figure 6.22) at the next available parcel segment.

FIGURE 6.22
The endpoints of attached parcel segments extend to the next available parcel segment if the initial parcel segment is erased

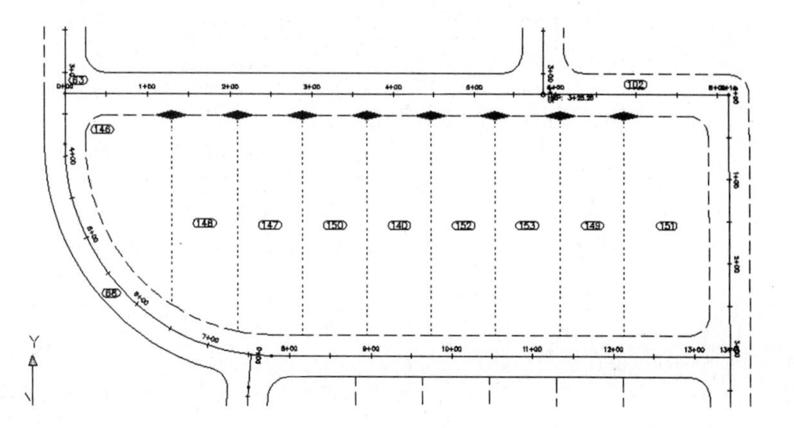

Precise Sizing Settings

The precise sizing tools consist of the Slide Angle, Slide Direction, and Swing Line tools (see Figure 6.23).

FIGURE 6.23
The precise sizing tools on the Parcel Layout Tools toolbar

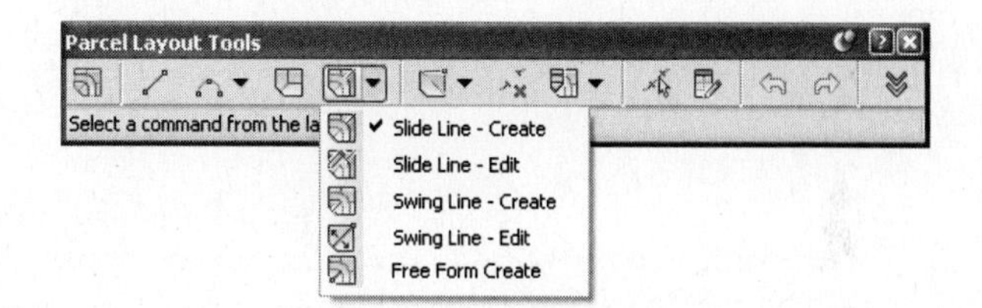

The Parcel Layout Tools toolbar can be expanded so that you can establish settings for each of the precise sizing tools (see Figure 6.24). Each of these settings is discussed in detail in the following sections.

NEW PARCEL SIZING

When you create new parcels, the tools respect your default area and minimum frontage (measured from either a ROW or a building setback line). The program always uses these numbers as a minimum; it bases the actual lot size on a combination of the geometry constraints (lot depth, frontage curves, and so on) and the additional settings that follow. Keep in mind that the numbers you establish under the New Parcel Sizing option must make geometric sense. For example, if you'd like a series of 7,500-square-foot lots that have 100′ of frontage, you must make sure that your rear parcel segment allows for at least 75′ of depth; otherwise, you may wind up with much larger frontage values than you desire or a situation where the software can't return a meaningful result.

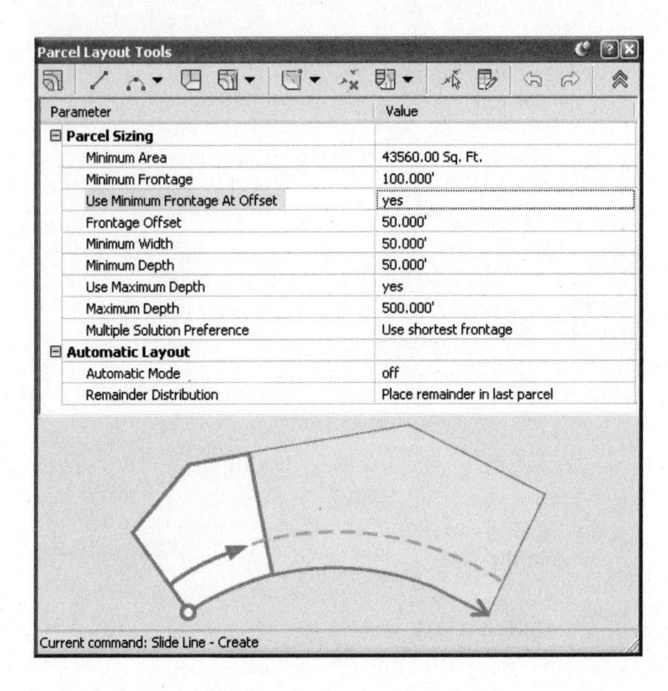

FIGURE 6.24
The settings on the Parcel Layout Tools toolbar

AUTOMATIC LAYOUT

Automatic Layout has two parameters when the list is expanded: Automatic Mode and Remainder Distribution. The Automatic Mode parameter can have the following values:

On Automatically follows your settings and puts in all the parcels, without prompting you to confirm each one.

Off Allows you to confirm each parcel as it's created. In other words, this option provides you with a way to semi-automatically create parcels.

The Remainder Distribution parameter tells Civil 3D how you'd like ''extra'' land handled. This parameter has the following options:

Create Parcel from Remainder Makes a last parcel with the leftovers once the tool has made as many parcels as it can to your specifications on the basis of the settings in this dialog. This parcel is usually smaller than the other parcels.

Place Remainder in Last Parcel Adds the leftover area to the last parcel once the tool has made as many parcels as it can to your specifications on the basis of the settings in this dialog.

Redistribute Remainder Takes the leftover area and pushes it back through the default-sized parcels once the tool has made as many parcels as it can to your specifications on the basis of the settings in this dialog. The resulting lots aren't always evenly sized because of differences in geometry around curves and other variables, but the leftover area is absorbed.

There aren't any rules per se in a typical subdivision workflow. Typically the goal is to create as many parcels as possible within the limits of available land. To that end, you'll use a combination of AutoCAD tools and Civil 3D tools to divide and conquer the particular tract of land with which you are working.

Slide Line – Create Tool

The Slide Line – Create tool creates an attached parcel segment based on an angle from frontage. You may find this tool most useful when your jurisdiction requires a uniform lot-line angle from the right of way.

This exercise will lead you through using the Slide Line – Create tool to create a series of subdivision lots:

1. Open the `Create Subdivision Lots.dwg` file, which you can download from `www.sybex.com/go/masteringcivil3d2010`. Note that this drawing has several alignments on the same site as the boundary parcel, resulting in several smaller parcels between the alignments and boundary.
2. Choose Parcel ➤ Parcel Creation Tools on the Create Design panel. The Parcel Layout Tools toolbar appears.
3. Expand the toolbar by clicking the Expand the Toolbar button, as shown in Figure 6.25.

FIGURE 6.25
The Expand the Toolbar button

4. Change the value of the following parameters by clicking in the Value column and typing in the new values if they aren't already set. Notice how the preview window changes to accommodate your preferences:
   - Default Area: **7500.00 Sq. Ft.**
   - Minimum Frontage: **75.000′**
   - Use Minimum Frontage at Offset: **yes**
   - Frontage Offset: **25.000′**
   - Minimum Width: **75.000′**
   - Minimum Depth: **50.000′**
   - Use Maximum Depth: **no**
   - Maximum Depth: **500.000′**
   - Multiple Solution Preference: **Use shortest frontage**
5. Change the following parameters by clicking in the Value column and selecting the appropriate option from the drop-down menu, if they aren't already set:
   - Automatic Mode: **on**
   - Remainder Distribution: **Redistribute remainder**
6. Click the Slide Line – Create tool (see Figure 6.26). The Create Parcels – Layout dialog appears.

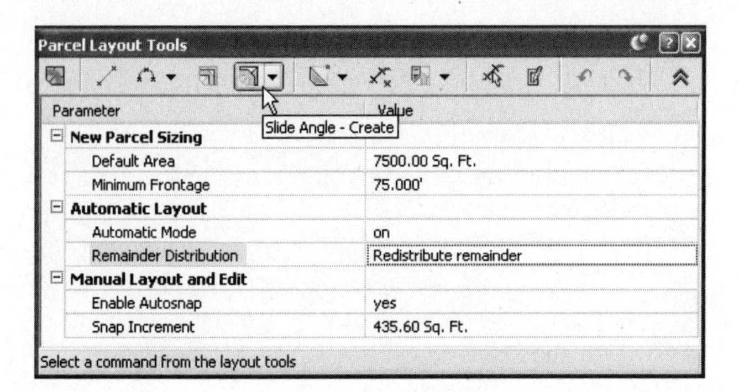

FIGURE 6.26
The Slide Line – Create tool

7. Select Subdivision Lots, Single Family, and Name Square Foot & Acres from the drop-down menus in the Site, Parcel Style, and Area Label Style selection boxes, respectively. Leave the rest of the options set to the default. Click OK to dismiss the dialog.

8. At the `Select parcel to be subdivided or [Pick]:` prompt, type **P** and press ↵. Pick a point on the screen inside Property: 29.

9. At the `Select start point on frontage:` prompt, use your Endpoint osnap to pick the point of curvature along the ROW parcel segment for Property: 29 (see Figure 6.27).

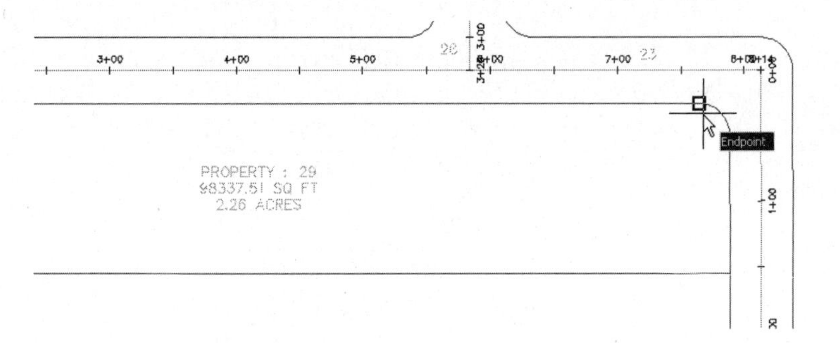

FIGURE 6.27
Pick the point of curvature along the ROW parcel segment

10. The parcel jig appears. Move your mouse slowly along the ROW parcel segment, and notice that the parcel jig follows the parcel segment. At the `Select end point on frontage:` prompt, use your Endpoint osnap to pick the point of curvature along the ROW parcel segment for Property: 29 (see Figure 6.28).

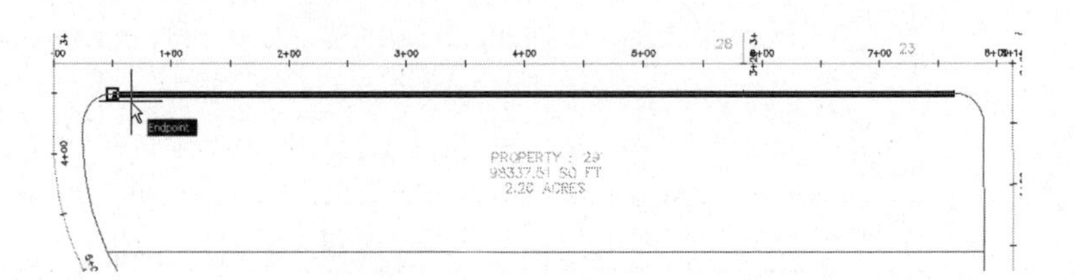

FIGURE 6.28
Allow the parcel-creation jig to follow the parcel segment, and then pick the point of curvature along the ROW parcel segment

11. At the `Specify angle or [Bearing/aZimuth]:` prompt, enter **90**↵. Notice the preview (see Figure 6.29).

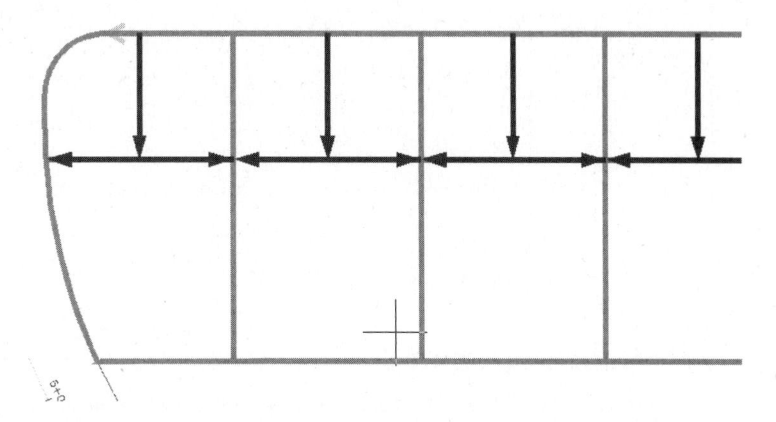

FIGURE 6.29
A preview of the results of the automatic parcel layout

12. At the `Accept result? [Yes/No] <Yes>:` prompt, press ↵ to accept the default Yes.
13. At the `Select parcel to be subdivided or [Pick]:` prompt, press ↵, and then type **X** and press ↵ to exit the command.
14. Your drawing should look similar to Figure 6.30. Note that Property: 29 still exists among the newly defined parcels and has kept its original parcel style and area label style.

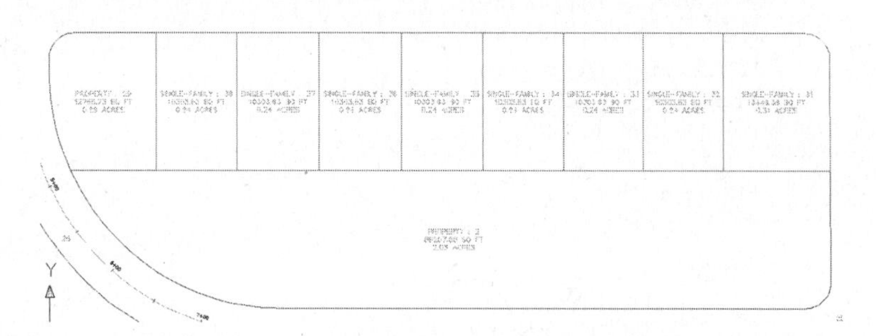

FIGURE 6.30
The automatically created lots

15. Repeat steps 6 through 13 for Property: 2, if desired. Try changing the options in the Parcel Layout Tools dialog and review the preview window to inspect the results of your selections.

CURVES AND THE FRONTAGE OFFSET

In most cases, the frontage along a building setback is graphically represented as a straight line drawn tangent or parallel to, and behind, the setback. When you specify a Minimum Width along a Frontage Offset (the building setback line) in the Parcel Layout Tools dialog, and when the lot frontage is curved, the distance you enter is measured along the curve. In most cases, this result may be insignificant, but in a large development, the error could be the defining factor in your decision to add or subtract a parcel from the development.

You may find this tool most useful when you're re-creating existing lots or when you'd like to create a series of parallel lot lines with a known bearing.

Swing Line – Create Tool

The Swing Line – Create tool creates a "backward" attached parcel segment where the diamond-shaped grip appears not at the frontage but at a different location that you specify. The tool respects your minimum frontage, and it adjusts the frontage larger if necessary in order to respect your default area.

The Swing Line – Create tool is semiautomatic because it requires your input of the swing-point location.

You may find this tool most useful around a cul-de-sac or in odd-shaped corners where you must hold frontage but have a lot of flexibility in the rear of the lot.

Creating Open Space Parcels Using the Free Form Create Tool

A site plan is more than just single-family lots. Areas are usually dedicated for open space, stormwater-management facilities, parks, and public-utility lots. The Free Form Create tool can be useful when you're creating these types of parcels. This tool, like the precise sizing tools, creates an attached parcel segment with the special diamond-shaped grip.

In the following exercise, you'll use the Free Form Create tool to create an open space parcel:

1. Open the `Create Open Space.dwg`. Note that this drawing contains a series of subdivision lots.
2. Pan over to Property: 1. Property: 1 is currently 7.13 acres. Note the small marker point noting the desired location of the open space boundary.
3. Select Parcel ➤ Parcel Creation Tools on the Create Design Panel. Select the Free Form Create tool (see Figure 6.31). The Create Parcels – Layout dialog appears.

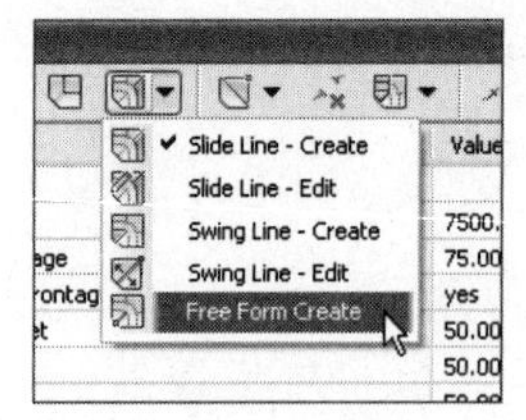

FIGURE 6.31
The Free Form Create tool

4. Select Subdivision Lots, Open Space, and Name Square Foot & Acres from the drop-down menus in the Site, Parcel Style, and Area Label Style selection boxes, respectively. Keep the default values for the remaining options. Click OK to dismiss the dialog.
5. Slide the Free Form Create attachment point around the Property: 1 frontage (see Figure 6.32). At the `Select attachment point:` prompt, use your Node osnap to pick the point labeled Open Space Limit.
6. At the `Specify lot line direction`: prompt, press ↵ to specify a perpendicular lot-line direction.

7. A new parcel segment is created from your Open Space Limit point, perpendicular to the ROW parcel segment, as shown in Figure 6.33.

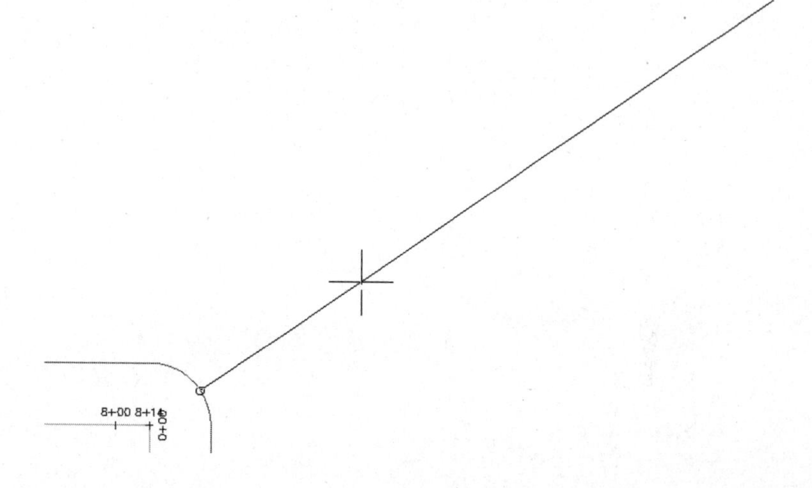

FIGURE 6.32
Use the Free Form Create tool to select an attachment point

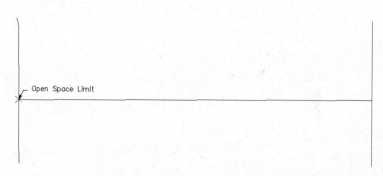

FIGURE 6.33
Attach the parcel segment to the marker point provided

8. Note that a new Open Space parcel has formed, and Property: 1 has been reduced from 7.13 acres to 3.31 acres.
9. Press ↵ to exit the Free Form Create command. Enter **X**, and then press ↵ to exit the toolbar.
10. Pick the new parcel segment so that you see its diamond-shaped grip. Grab the grip, and slide the segment along the ROW parcel segment (see Figure 6.34).
11. Notice that when you place the parcel segment at a new location the segment endpoint snaps back to the rear parcel segment (see Figure 6.35). This is typical behavior for an attached parcel segment.

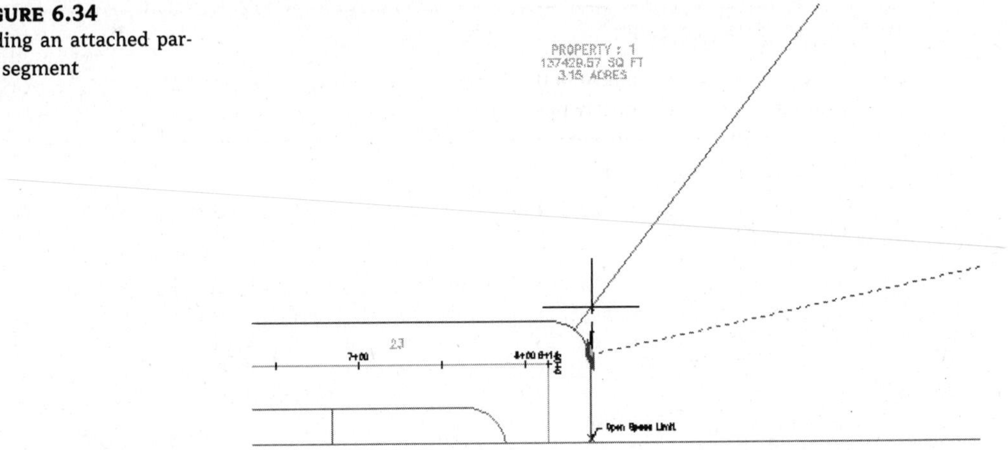

FIGURE 6.34
Sliding an attached parcel segment

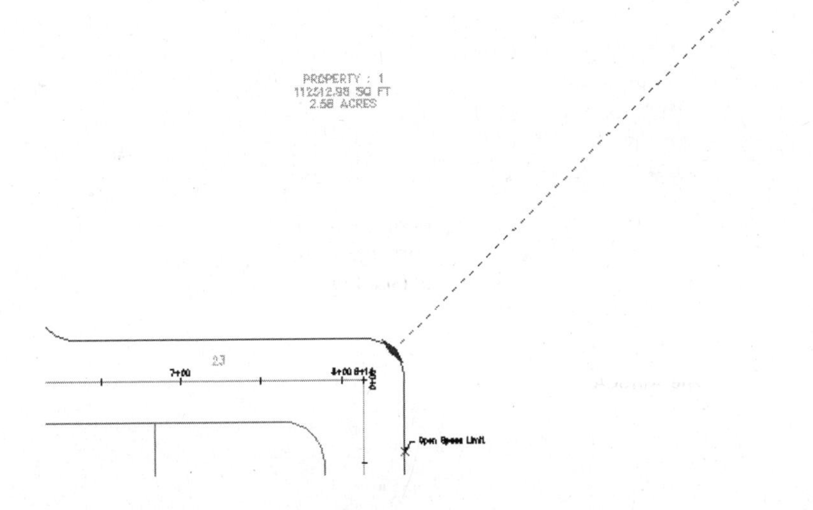

FIGURE 6.35
Attached parcel segments snap back to the rear parcel segment

Editing Parcels by Deleting Parcel Segments

One of the most powerful aspects of Civil 3D parcels is the ability to perform many iterations of a site-plan design. Typically, this design process involves creating a series of parcels and then deleting them to make room for iteration with different parameters, or deleting certain segments to make room for easements, public utility lots, and more.

You can delete parcel segments using the AutoCAD Erase tool or the Delete Sub-Entity tool on the Parcel Layout Tools toolbar.

BREAK THE UNDO HABIT

You'll find that parcels behave better if you use one of the segment-deletion methods described in this section to erase improperly placed parcels rather than using the Undo command.

It's important to understand the difference between these two methods. The AutoCAD Erase tool behaves as follows:

- If the parcel segment was originally created from a polyline (or similar parcel-layout tools, such as the Tangent-Tangent No Curves tool), the AutoCAD Erase tool erases the entire segment (see Figure 6.36).

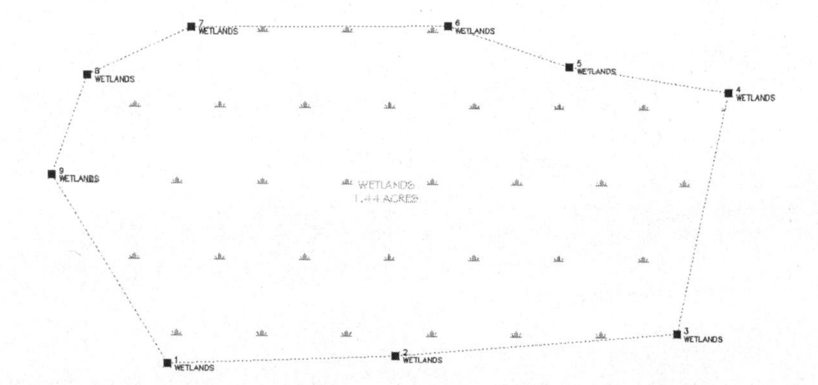

FIGURE 6.36 The segments indicated by the blue grips will be erased after using the AutoCAD Erase tool

- If the parcel segment was originally created from a line or arc (or similar parcel-layout tools, such as the precise sizing tools), then AutoCAD Erase erases the entire length of the original line or arc (see Figure 6.37).

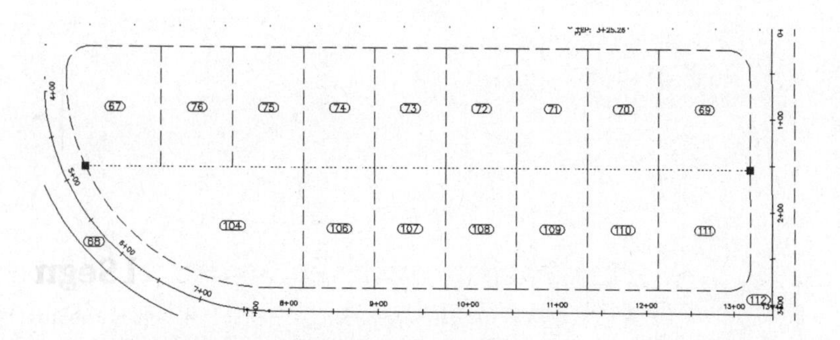

FIGURE 6.37 The AutoCAD Erase tool will erase the entire segment indicated by the blue grips

The Delete Sub-Entity tool acts more like the AutoCAD Trim tool. The Delete Sub-Entity tool only erases the parcel segments between parcel vertices. For example, if Parcel 76, as shown in Figure 6.38, must be absorbed into Parcel 104 to create a public utility lot with dual road access, you'd want to only erase the segment at the rear of Parcel 76 and not the entire segment shown previously in Figure 6.37.

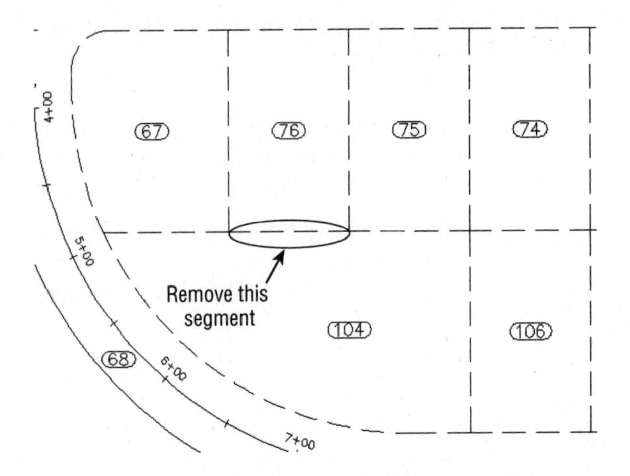

FIGURE 6.38
Use the Delete Sub-Entity tool to erase the rear parcel segment for Parcel 76

Selecting the Parcel 76 label and then picking Parcel Layout Tools on the Modify panel brings up the Parcel Layout Tools toolbar. Selecting the Delete Sub-Entity tool allows you to pick only the small rear parcel segment for Parcel 76. Figure 6.39 shows the result of this deletion.

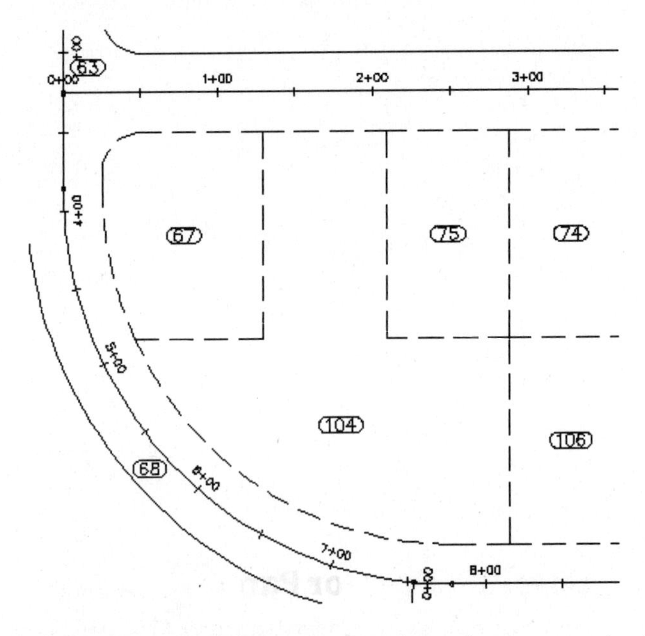

FIGURE 6.39
The rear lot line for Parcel 76 was erased using the Delete Sub-Entity tool, creating a larger Parcel 104

The following exercise will lead you through deleting a series of parcel segments using both the AutoCAD Erase tool and the Delete Sub-Entity tool:

1. Open the `Delete Segments.dwg` file. Note that this drawing contains a series of subdivision lots, along with a wetlands boundary.
2. Let's say you just received word that there was a mistake with the wetlands delineation and you need to erase the entire wetlands area. Use the AutoCAD Erase tool to erase the

parcel segments that define the wetlands parcel. Note that the entire parcel disappears in one shot, because it was created with the Tangent-Tangent No Curves tool (which behaves similarly to creating a polyline).

3. Next, you discover that parcel Single-Family: 29 needs to be removed and absorbed into parcel Property: 2 to enlarge a stormwater-management area. Choose Parcel Layout Tools from the Parcel pull-down on the Create Design panel of the Home tab.

4. Click the Delete Sub-Entity tool (see Figure 6.40).

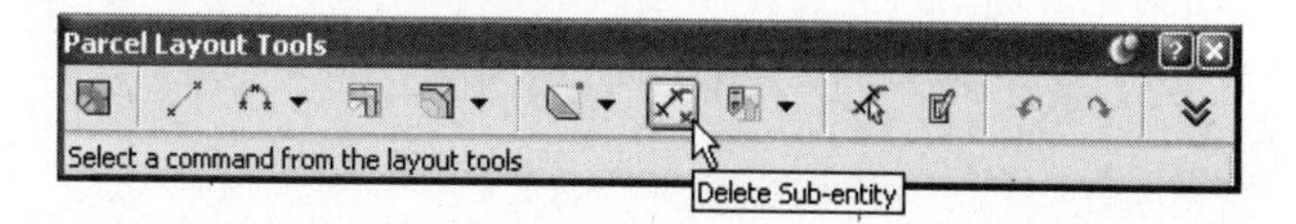

FIGURE 6.40
The Delete Sub-Entity tool

5. At the `Select subentity to remove:` prompt, pick the rear lot line between Single-Family: 29 and Property: 2. Press ↵ to exit the command, enter **X**, and then press ↵ to exit the Parcel Layout Tools toolbar.

6. The parcel segment immediately disappears. The resulting parcel has a combined area from Single-Family: 29 and Property: 2, as shown in Figure 6.41.

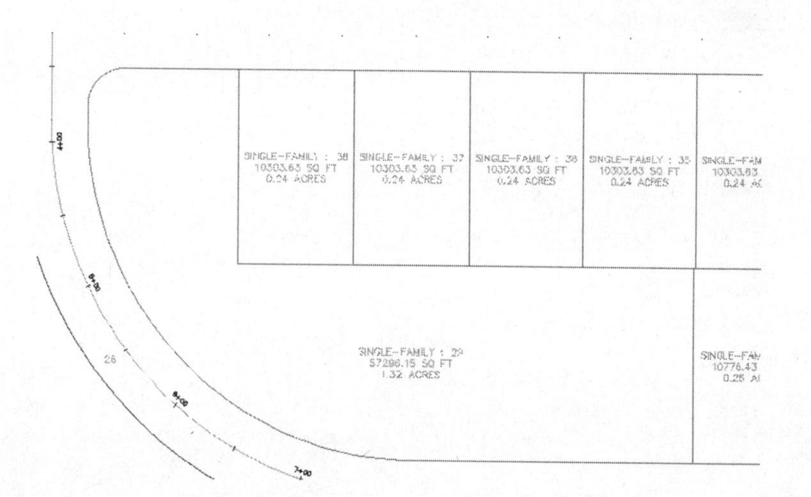

FIGURE 6.41
The parcel after erasing the rear lot line

Best Practices for Parcel Creation

Now that you have an understanding of how objects on a site interact and you've had some practice creating and editing parcels in a variety of ways, you'll take a deeper look at how parcels must be constructed to achieve topology stability, predictable labeling, and desired parcel interaction.

Forming Parcels from Segments

In the earlier sections of this chapter, you saw that parcels are created only when parcel segments form a closed area (see Figure 6.42).

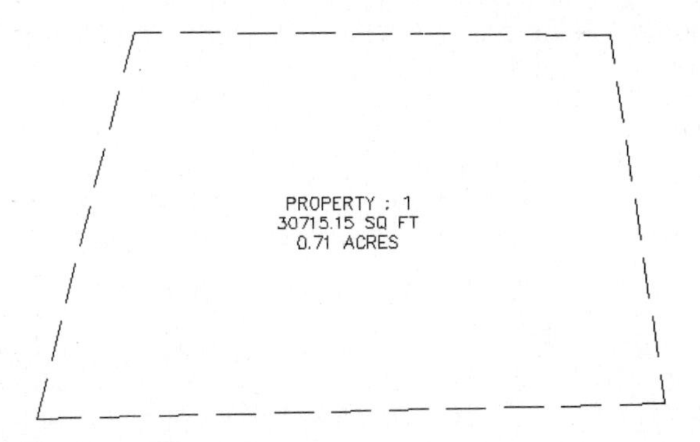

FIGURE 6.42
A parcel is created when parcel segments form a closed area

Parcels must always close. Whether you draw AutoCAD lines and use the Create Parcel from Objects menu command or use the parcel segment creation tools, a parcel won't form until there is an enclosed polygon. Figure 6.43 shows four parcel segments that don't close; therefore, no parcel has been formed.

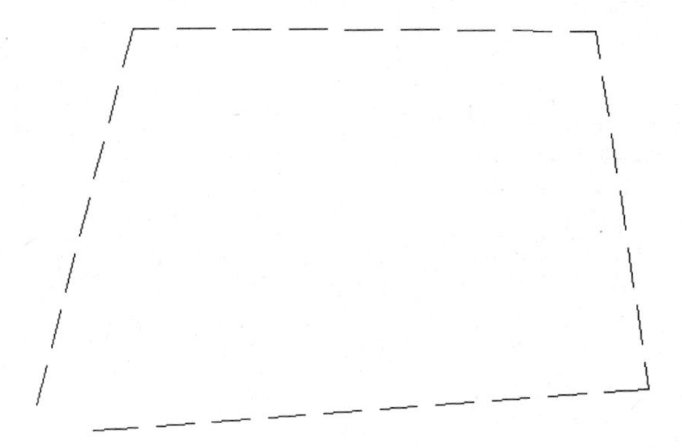

FIGURE 6.43
No parcel will be formed if parcel segments don't completely enclose an area

There are times in surveying and engineering when parcels of land don't necessarily close when created from legal descriptions. In this case, you must work with your surveyor to perform an adjustment or find some other solution to create a closed polygon.

You also saw that even though parcels can't be erased, if you erase the appropriate parcel segments, the area contained within a parcel is assimilated into neighboring parcels.

Parcels Reacting to Site Objects

Parcels require only one parcel segment to divide them from their neighbor (see Figure 6.44). This behavior eliminates the need for duplicate segments between parcels, and duplicate segments must be avoided.

As you saw in the section on site interaction, parcels understand their relationships to one another. When you create a single parcel segment between two subdivision lots, you have the ability to move one line and affect two parcels. Figure 6.45 shows the parcels from Figure 6.44 once the parcel segment between them has been shifted to the left. Note that both areas change in response.

A mistake that many people new to Civil 3D make is to create parcels from closed polylines, which results in a duplicate segment between parcels. Figure 6.46 shows two parcels created from two closed polylines. These two parcels may appear identical to the two seen in the previous example, because they were both created from a closed polyline rectangle; however, the segment between them is actually two segments.

FIGURE 6.44
Two parcels, with one parcel segment between them

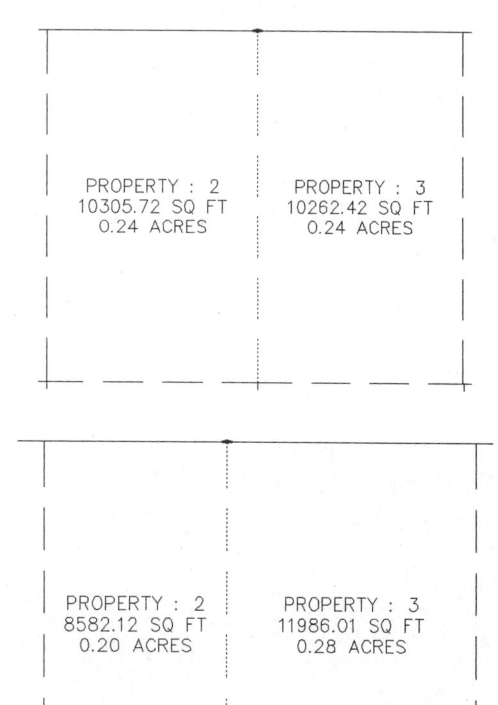

FIGURE 6.45
Moving one parcel segment affects the area of two parcels

PROPERTY : 2
8582.12 SQ FT
0.20 ACRES

PROPERTY : 3
11986.01 SQ FT
0.28 ACRES

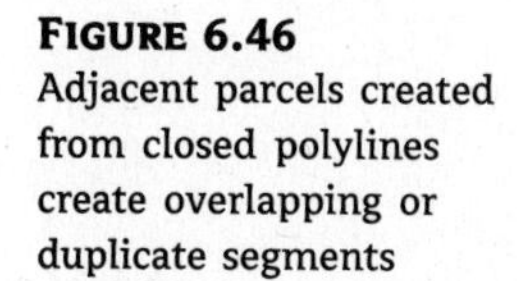

FIGURE 6.46
Adjacent parcels created from closed polylines create overlapping or duplicate segments

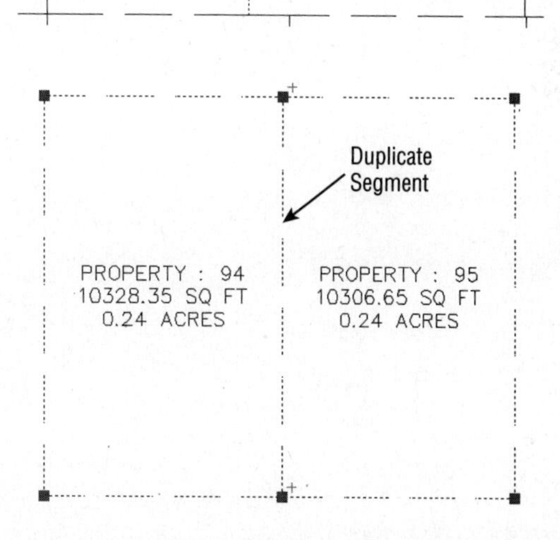

The duplicate segment becomes apparent when you attempt to grip-edit the parcel segments. Moving one vertex from the common lot line, as seen in Figure 6.47, reveals the second segment. Also note that a sliver parcel is formed. Duplicate site-geometry objects and sliver parcels make it difficult for Civil 3D to solve the site topology and can cause drawing stability problems and unexpected parcel behavior. You must avoid this at all costs. Creating a subdivision plat of parcels this way almost guarantees that your labeling won't perform properly and could potentially lead to data loss and drawing corruption.

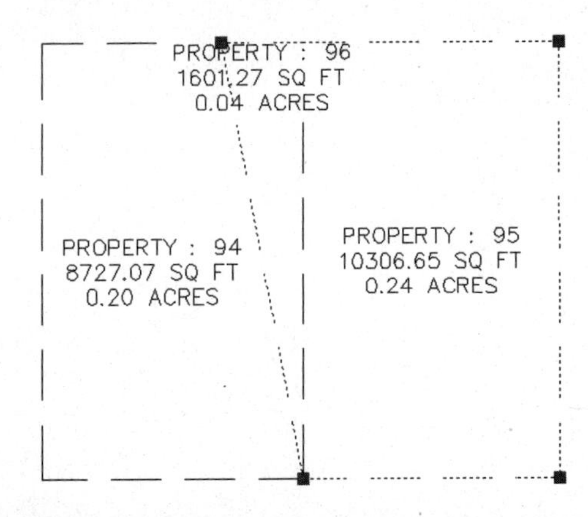

FIGURE 6.47
Duplicate segments become apparent when they're grip-edited and a sliver parcel is formed

> **MIGRATE PARCELS FROM LAND DESKTOP? JUST DON'T DO IT!**
>
> The Land Desktop Parcel Manager essentially created Land Desktop parcels from closed polylines. If you migrate Land Desktop parcels into Civil 3D, your resulting Civil 3D parcels will behave poorly and will almost universally result in drawing corruption.

Parcels form to fill the space contained by the original outer boundary. You should always begin a parcel-division project with an outer boundary of some sort (see Figure 6.48).

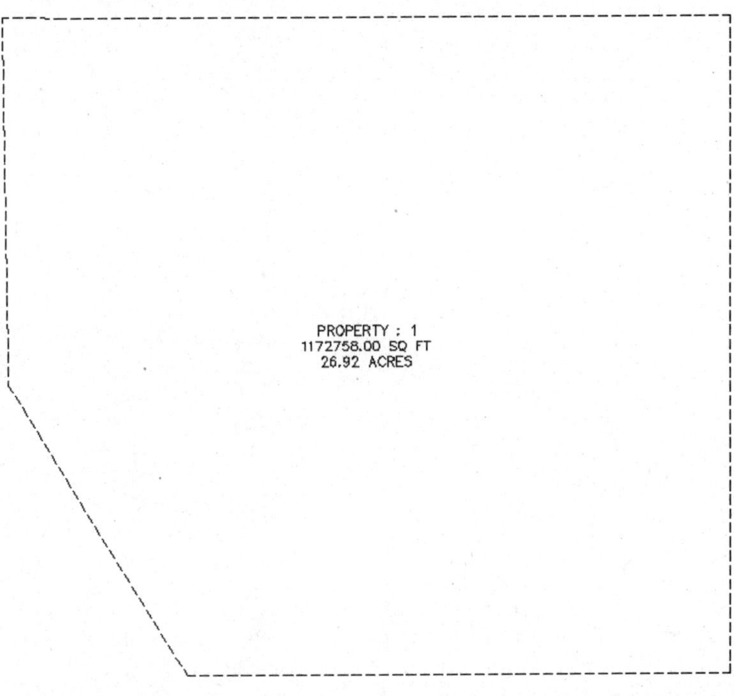

FIGURE 6.48
An outer boundary parcel

You can then add road centerline alignments to the site, which divides the outer boundary as shown in Figure 6.49.

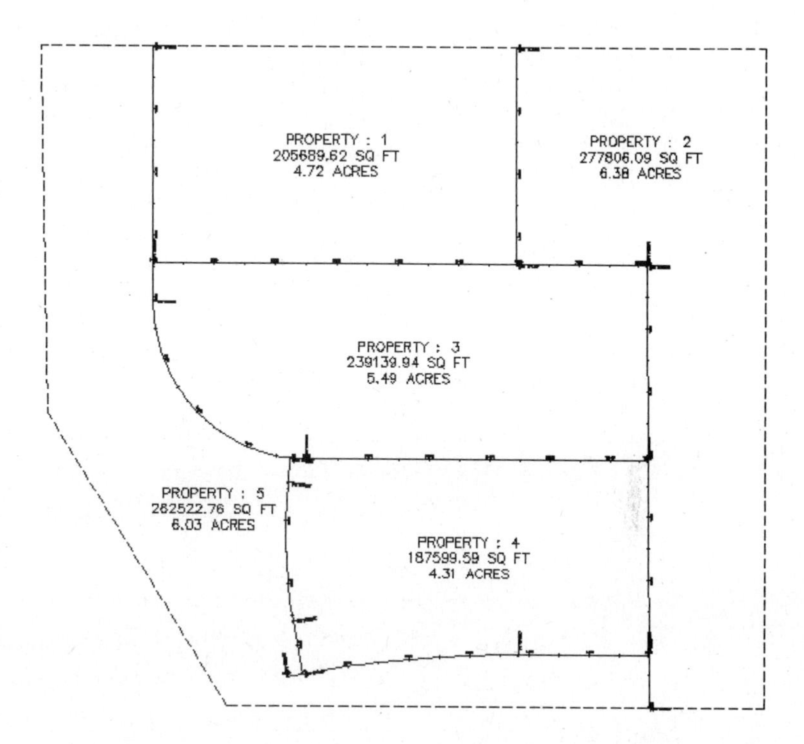

FIGURE 6.49
Alignments added to the same site as the boundary parcel divide the boundary parcel

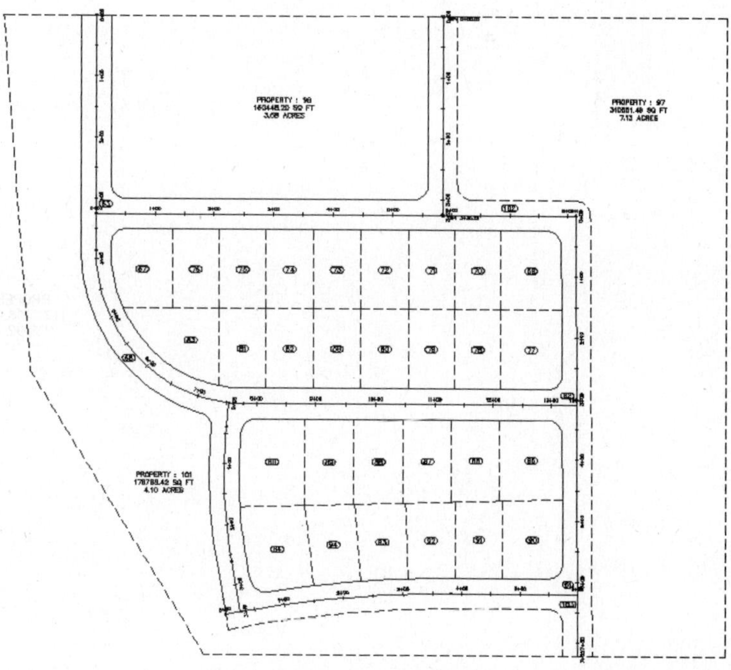

FIGURE 6.50
The total area of parcels contained within the original boundary sums to equal the original boundary area

It's important to note that the boundary parcel no longer exists intact. As you subdivide this site, parcel 1 is continually reallocated with every division. As road ROW and subdivision lots are formed from parcel segments, more parcels are created. Every bit of space that was contained in the original outer boundary is accounted for in the mesh of newly formed parcels (see Figure 6.50).

From now on, you'll consider ROW, wetlands, parkland, and open-space areas as parcels, even if you didn't before. You can make custom label styles to annotate these parcels however you like, including a "no show" or none label.

Real World Scenario

IF I CAN'T USE CLOSED POLYLINES, HOW DO I CREATE MY PARCELS?

How do you create your parcels if parcels must always close, but you aren't supposed to use closed polylines to create them?

In the earlier exercises in this chapter, you learned several techniques for creating parcels. These techniques included using AutoCAD objects and a variety of parcel-layout tools. A summary of some of the best practices for creating parcels are listed here:

- Create closed polylines for boundaries and islands, and then use the Create Parcel from Objects menu command. Closed polylines are suitable foundation geometry in cases where they won't be subject to possible duplicate segments. The following graphic shows a boundary parcel and a designated open-space parcel that were both created from closed polylines. Other examples of island parcels include isolated wetlands, ponds, or similar features that don't share a common segment with the boundary parcel.

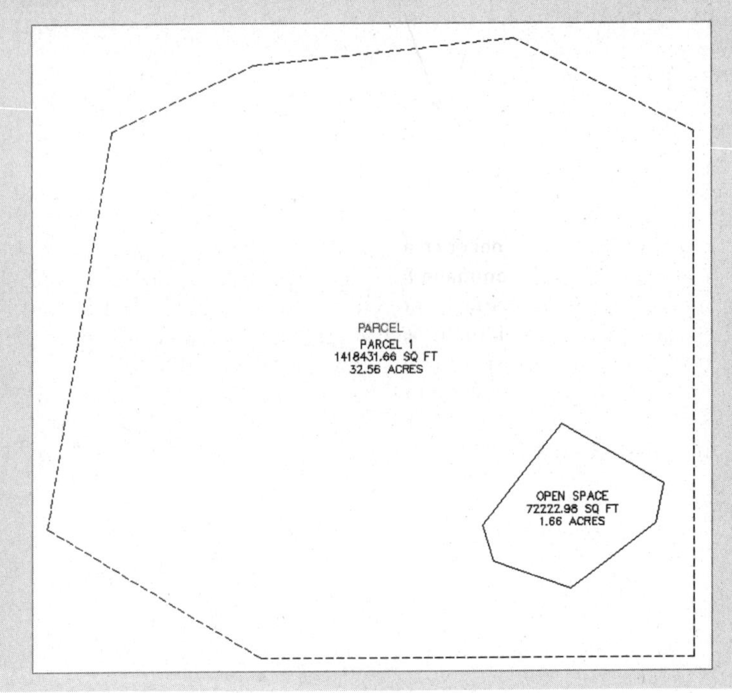

- Create trimmed/extended polylines for internal features, and then use the Create Parcel from Objects menu command. For internal features such as easements, buffers, open space, or wetlands that share a segment with the outer boundary, draw a polyline that intersects the outer boundary, but be careful not to trace over any segments of the outer boundary. Use the Create Parcel from Objects menu command to convert the polyline into a parcel segment. The following graphic shows the technique used for an easement.

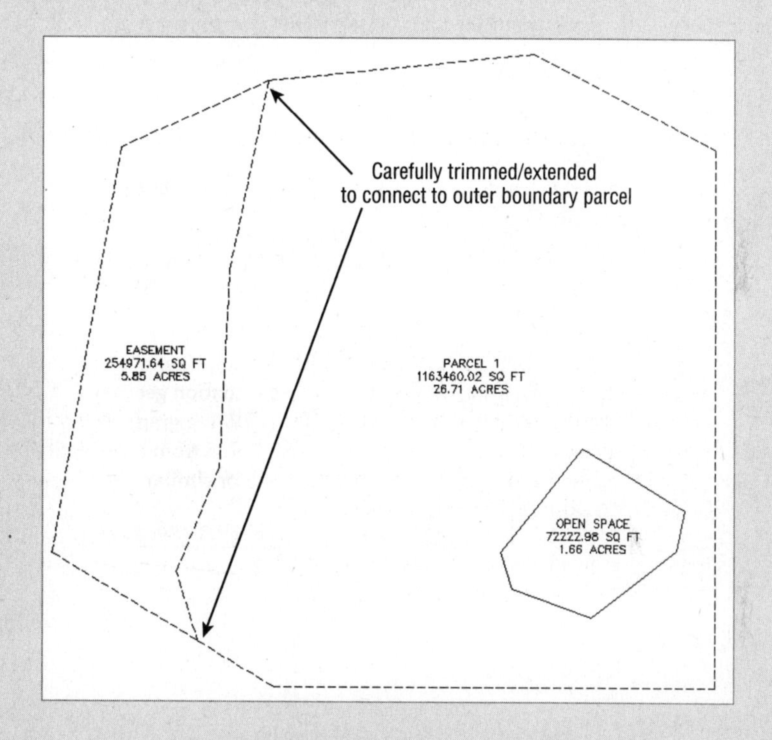

- Use the Create ROW tool or create trimmed/extended polylines, and then use the Create Parcel from Objects menu command for ROW segments. In a previous exercise, you used the Create ROW tool. You also learned that even though this tool can be useful, it can't create cul-de-sacs or changes in ROW width. In cases where you need a more intricate ROW parcel, use the AutoCAD Offset tool to offset your alignment. The resulting offsets are polylines. Use circles, arcs, fillet, trim, extend, and other tools to create a joined polyline to use as foundation geometry for your ROW parcel. Use the Create Parcel from Objects menu command to convert this linework into parcel segments, as shown here.

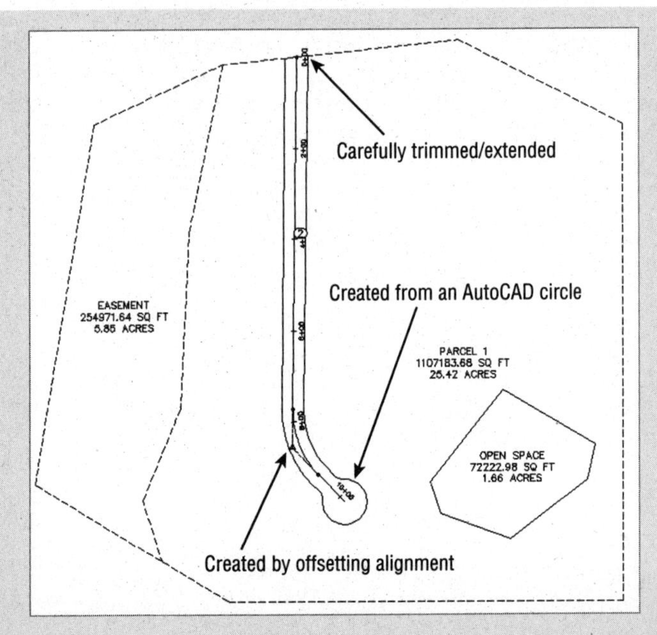

- Create trimmed or extended polylines for rear lot lines, and then use the Create Parcel from Objects menu command. The precise sizing tools tend to work best when given a rear lot line as a target endpoint. Create this rear target by offsetting your ROW parcel to your desired lot depth. The resulting offset is a polyline. Use Trim, Extend, and other tools to create a joined polyline to use as foundation geometry for your rear lot parcel segment. Use the Create Parcel from Objects menu command to convert this linework into parcel segments, as shown here.

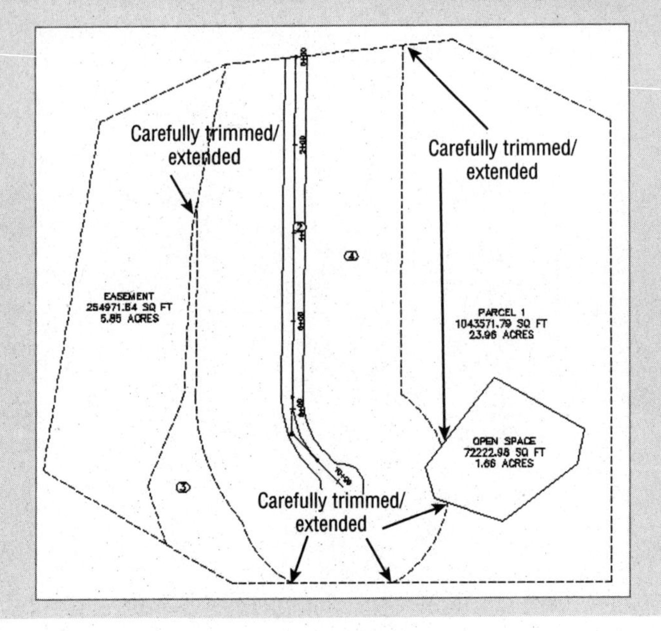

- Use the Parcel Creation Tools menu command with the attached segment tools for internal parcels. The precise sizing tools and Free Form Create automatically create only one attached segment between parcels, so using them to create your subdivision lot boundaries ensures proper parcel geometry, as shown in Properties 1 through 19 in the following graphic.

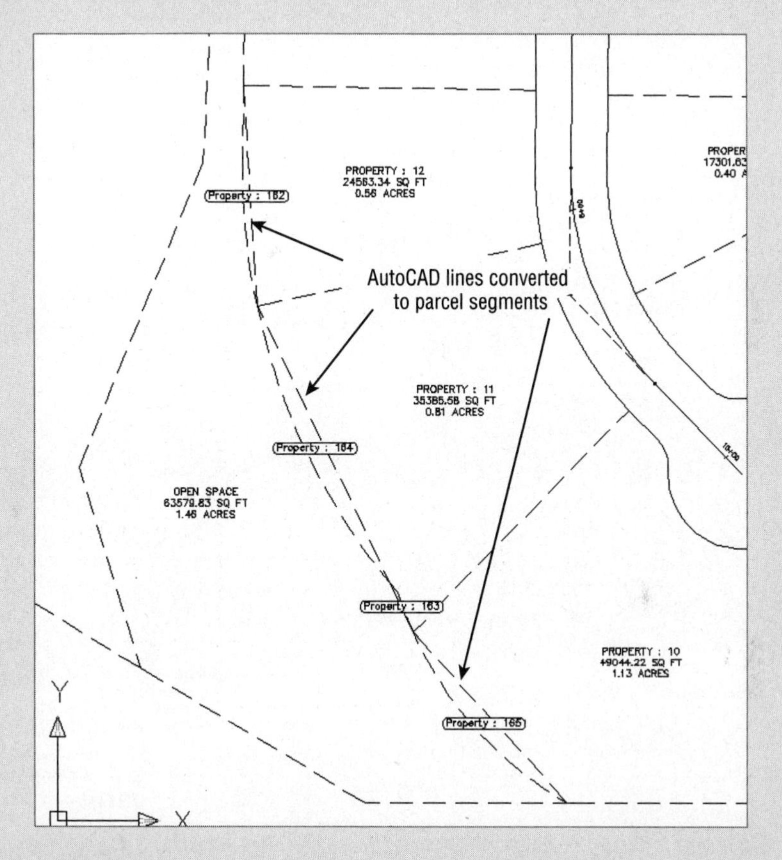

- Use line segments and the Create Parcel from Objects menu command for final detailed segment work. Surveyors often prefer to lay out straight-line segments rather than curves, so for the final rear-lot line cleanup, create AutoCAD lines across the back of each lot and then use Create Parcel from Objects to turn those lines into parcel segments, as shown here.

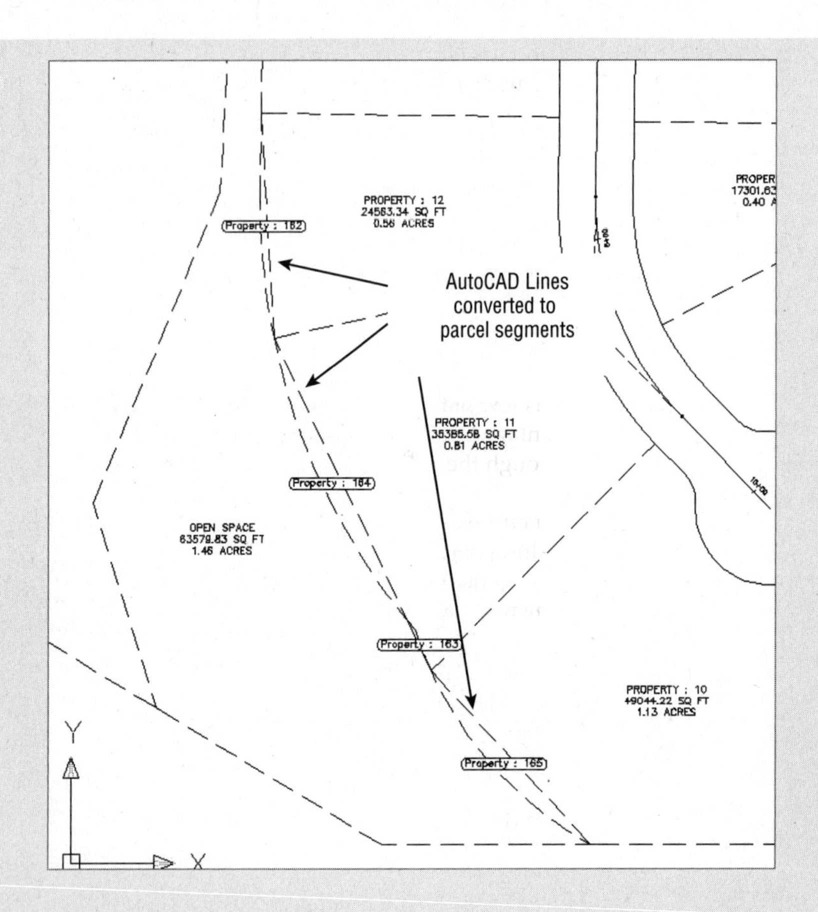

This final cleanup is best saved for the very end of the project. Parcel iterations and refinement work much better with a continuous rear lot line.

Constructing Parcel Segments with the Appropriate Vertices

Parcel segments should have natural vertices only where necessary and split-created vertices at all other intersections. A natural vertex, or point of intersection (PI), can be identified by picking a line, polyline, or parcel segment and noting the location of the grips (see Figure 6.51).

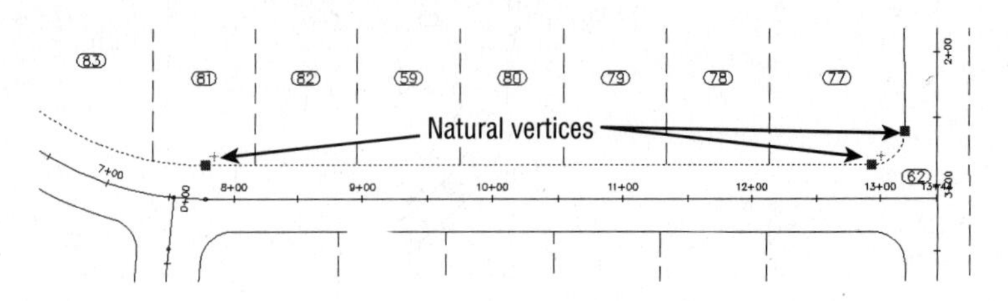

FIGURE 6.51
Natural vertices on a parcel segment

A split-created vertex occurs when two parcel segments touch or cross each other. Note that in Figure 6.52, the parcel segment doesn't show a grip even where each individual lot line touches the ROW parcel.

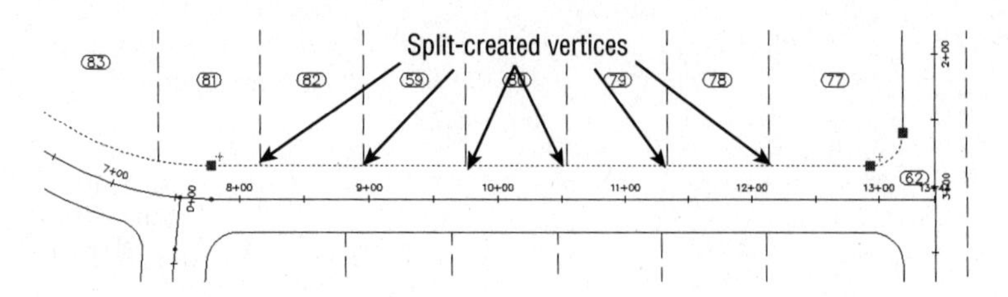

FIGURE 6.52
Split-created vertices on a parcel segment

It's desirable to have as few natural vertices as possible. In the example shown previously in Figure 6.51, the ROW frontage line can be expressed as a single bearing and length from the end of the arc in parcel 81 through the beginning of the arc in parcel 77, as opposed to having seven smaller line segments.

If the foundation geometry was drawn with a natural vertex at each lot line intersection as in Figure 6.53, then the resulting parcel segment won't label properly and may cause complications with editing and other functions. This subject will be discussed in more detail later in the section "Labeling Spanning Segments," later in this chapter.

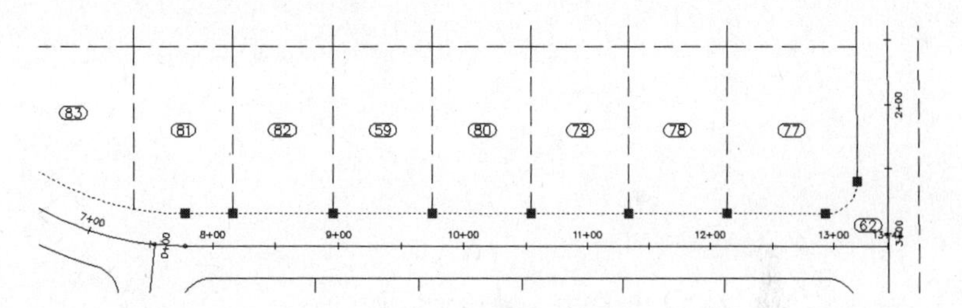

FIGURE 6.53
Unnecessary natural vertices on a parcel segment create problems for labeling and editing

Parcel segments must not overlap. Overlapping segments create redundant vertices, sliver parcels, and other problems that complicate editing parcel segments and labeling. Figure 6.54 shows a segment created to form parcel 104 that overlaps the rear parcel segment for the entire block. This segment should be edited to remove the redundant parcel segment across the rear of parcel 76 to ensure good parcel topology.

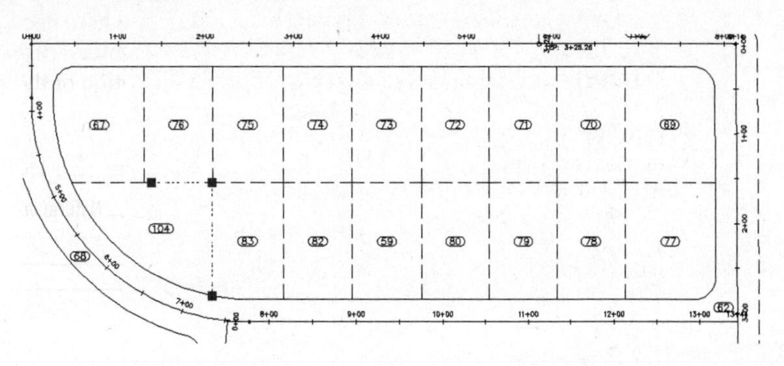

FIGURE 6.54
Avoid creating overlapping parcel segments

Parcel segments must not overhang. Spanning labels are designed to overlook the location of intersection formed (or T-shaped) split-created vertices. However, these labels won't span a crossing formed (X- or + [plus]-shaped) split-created vertex. Even a very small parcel segment overhang will prevent a spanning label from working and may even affect the area computation for adjacent parcels. The overhanging segment in Figure 6.55 would prevent a label from returning the full spanning length of the ROW segment it crosses.

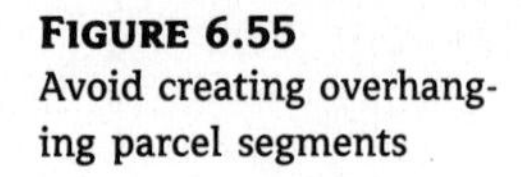

FIGURE 6.55
Avoid creating overhanging parcel segments

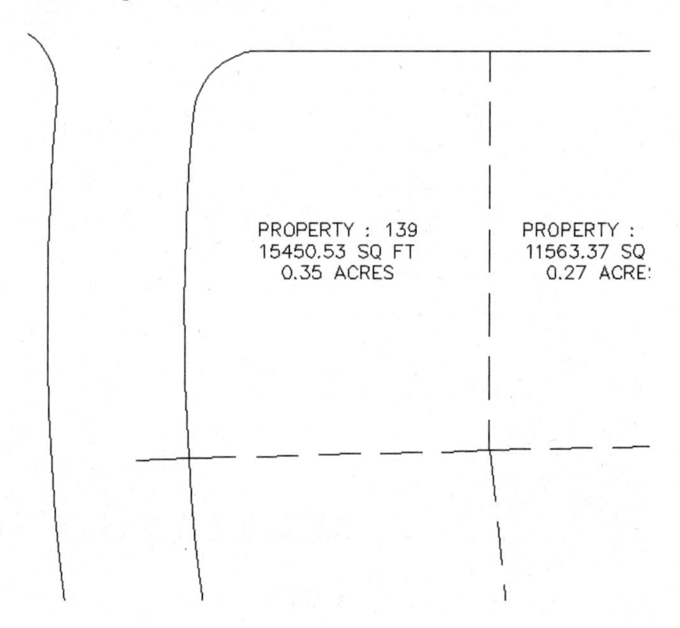

Labeling Parcel Areas

A parcel area label is placed at the parcel centroid by default, and it refers to the parcel in its entirety. When asked to pick a parcel, you pick the area label. An area label doesn't necessarily have to include the actual area of the parcel.

Area labels can be customized to suit your fancy. Figure 6.56 shows a variety of customized area labels.

Parcel area labels are composed like all other labels in Civil 3D. You can select the following default parcel properties for text components of the label from the drop-down menu in the Properties selection box in the Text Component Editor dialog (see Figure 6.57):

- Name
- Description
- Parcel Area
- Parcel Number
- Parcel Perimeter
- Parcel Address
- Parcel Site Name

- Parcel Style Name
- Parcel Tax ID

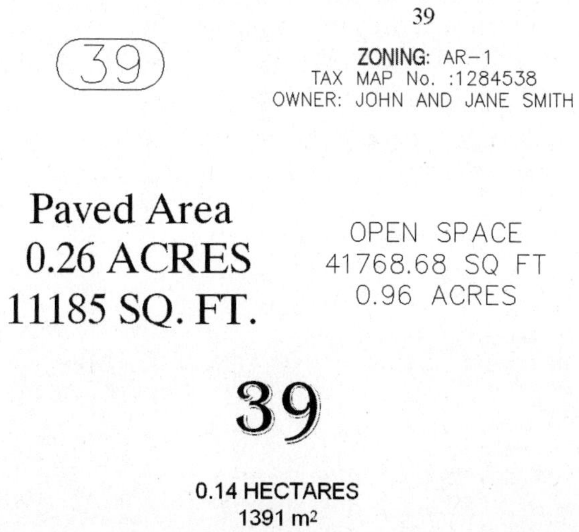

FIGURE 6.56
Sample area labels

39

ZONING: AR-1
TAX MAP No. :1284538
OWNER: JOHN AND JANE SMITH

Paved Area
0.26 ACRES
11185 SQ. FT.

OPEN SPACE
41768.68 SQ FT
0.96 ACRES

39

0.14 HECTARES
1391 m^2

FIGURE 6.57
Text Component Editor showing the various properties that the text components of an area label can have

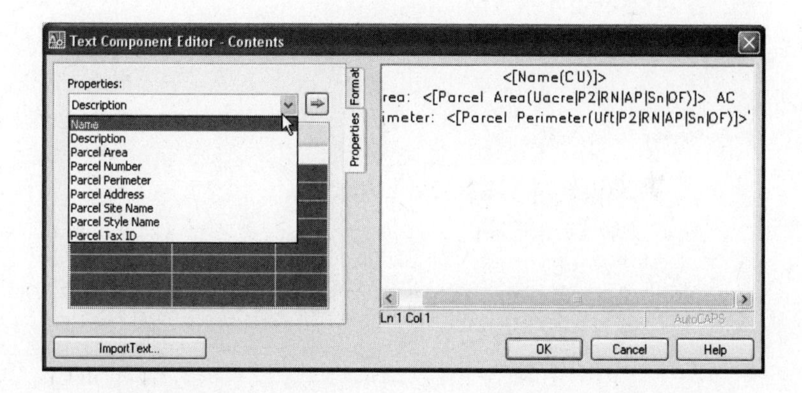

Area labels often include the parcel name or number. You can rename or renumber parcels using Renumber/Rename from the Modify panel after selecting a parcel.

The following exercise will teach you how to renumber a series of parcels:

1. Open `Change Area Label.dwg`. Note that this drawing contains many subdivision lot parcels.
2. Select parcel Single-Family: 31 and select Renumber/Rename from the Modify panel. The Renumber/Rename Parcels dialog appears.
3. In the Renumber/Rename Parcels dialog, make sure Subdivision Lots is selected from the drop-down menu in the Site selection box. Change the value of the Starting Number selection box to **1**. Click OK.

4. At the `Specify start point or [Polylines/Site]:` prompt, pick a point on the screen anywhere inside the Single-Family: 31 parcel, which will become your new Single-Family: 1 parcel at the end of the command.

5. At the `End point or [Undo]:` prompt, pick a point on the screen anywhere inside the Single-Family: 29 parcel, almost as if you were drawing a line; then, pick a point anywhere inside Property: 2 (be sure not to cross other parcel lines); and then, pick a point inside Single-Family: 39. Press ↵ to stop choosing parcels. Press ↵ again to end the command.

Note that your parcels have been renumbered from 1 through 16. Repeat the exercise with other parcels in the drawing for additional practice if desired.

The next exercise will lead you through one method of changing an area label using the Edit Parcel Properties dialog:

1. Continue working in the `Change Area Label.dwg` file.

2. Select parcel Single-Family:1 and select Multiple Parcel Properties from the Modify panel. At the `Specify start point or [Polylines/All/Site]:` prompt, pick a point on the screen anywhere inside the Single-Family: 1 parcel.

3. At the `End point or [Undo]:` prompt, pick a point on the screen anywhere inside Single-Family: 2, almost as if you were drawing a line. Press ↵ to stop choosing parcels. Press ↵ again to open the Edit Parcel Properties dialog (see Figure 6.58).

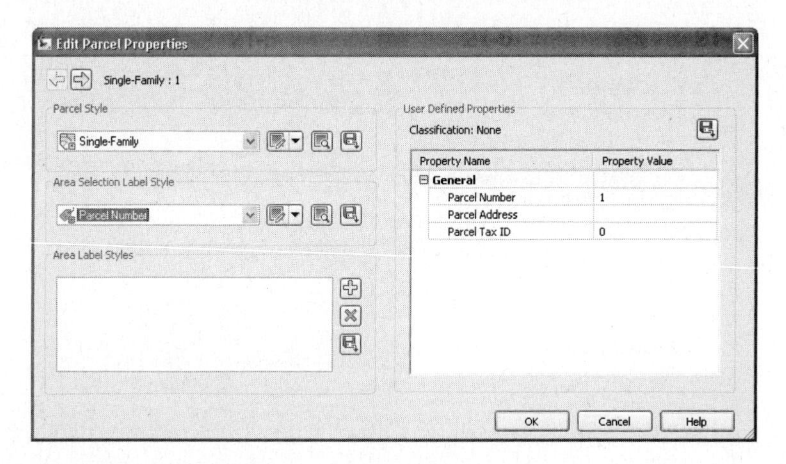

FIGURE 6.58
The Edit Parcel Properties dialog

4. In the Area Selection Label Styles portion of the Edit Parcel Properties dialog, use the drop-down menu to choose the Parcel Number area label style.

5. Click the Apply to All Parcels button.

6. Click Yes in the dialog displaying the question "Apply the area selection label style to the 2 selected parcels?"

7. Click OK to exit the Edit Parcel Properties dialog.

The two parcels now have parcel area labels that call out numbers only. Note that you could also use this interface to add a second area label to certain parcels if required.

This final exercise will show you how to use Prospector to change a group of parcel area labels at the same time:

1. Continue working in the `Change Area Label.dwg` file.
2. In Prospector, expand the Sites ➤ Subdivision Lots ➤ Parcels collection (see Figure 6.59).

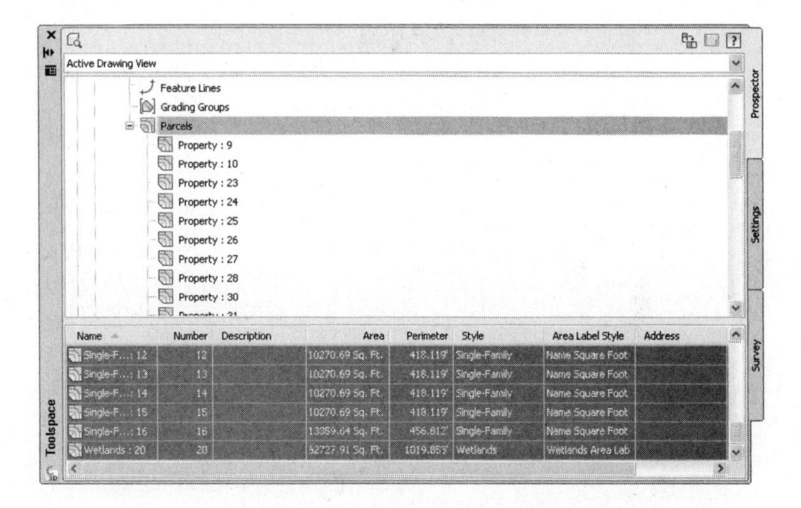

FIGURE 6.59
The Parcels collection in Prospector

3. In the Preview pane, click the Name column to sort the Parcels collection by name.
4. Hold down the Shift key, and click each Single-Family parcel to select them all. Release the Shift key, and your parcels should remain selected.
5. Slide over to the Area Label Style column. Right-click the column header, and select Edit (see Figure 6.60).

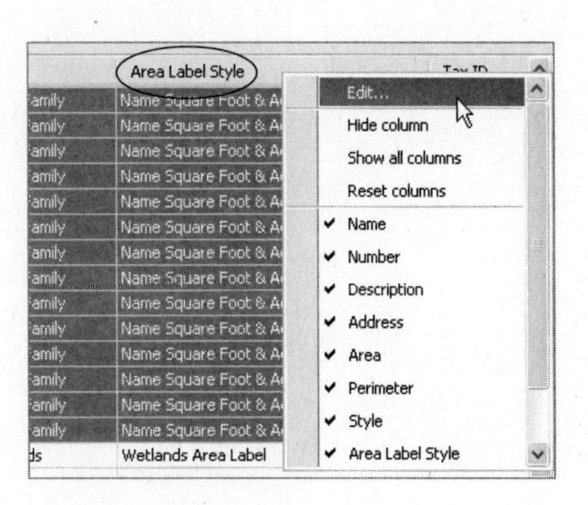

FIGURE 6.60
Right-click the Area Label Style column header, and select Edit

6. In the Select Label Style dialog, select Parcel Number from the drop-down menu in the Label Style selection box. Click OK to dismiss the dialog.

7. The drawing will process for a moment. Once the processing is finished, minimize Prospector and inspect your parcels. All the Single-Family parcels should now have the Parcel Number area label style.

What If the Area Label Needs to Be Split onto Two Layers?

You may have a few different types of plans that show parcels. Because it would be awkward to have to change the parcel area label style before you plot each sheet, it would be best to find a way make a second label on a second layer so that you can freeze the area component in sheets or viewports when it isn't needed. Here's an example where the square footage has been placed on a different layer so it can be frozen in certain viewports:

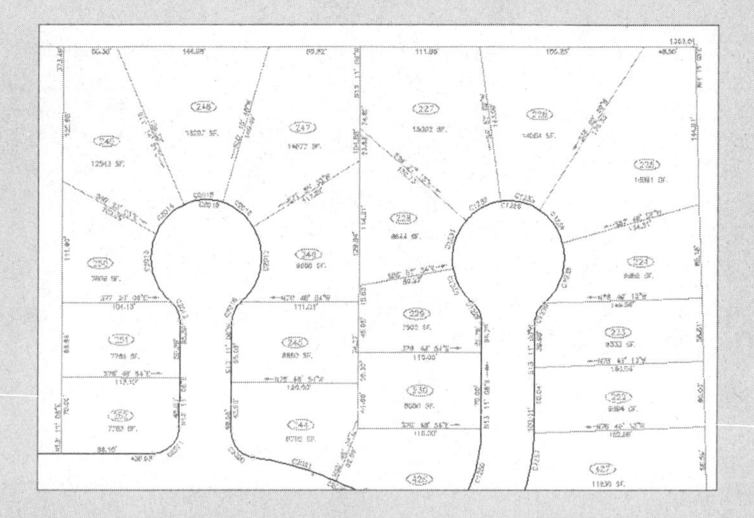

You can accomplish this by either creating a second parcel area label that calls out the area only, or creating a General Note label that contains parcel reference text.

Here's one way to add a second parcel area label:

1. Change to the Annotate tab. From the Labels & Tables panel, select Add Labels ➤ Parcel ➤ Add Parcel Labels.

2. Select Area from the drop-down menu in the Label Type selection box, and then select an area style label that will be the second area label.

3. Click Add, and then pick your parcel on screen.

A second option is to create a General Note label that has reference text, which calls out the parcel area as shown in the following graphic.

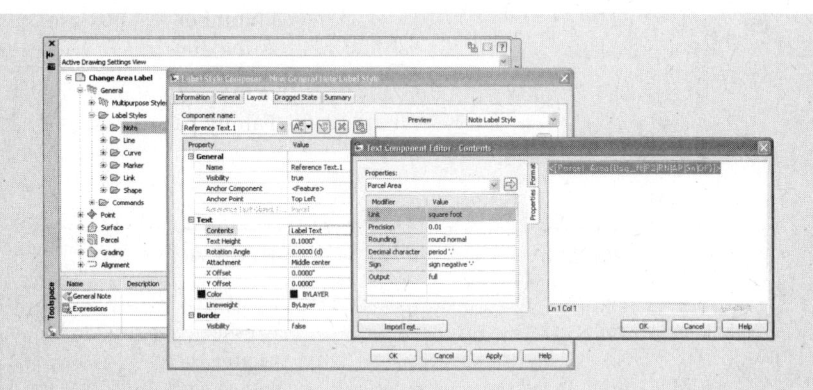

Here's an example of how you can compose this label:

1. Still in the Annotate tab, select Add Labels ➤ Notes from the Labels and Tables.
2. In the Add Labels dialog, select Note from the drop-down menu in the Feature selection box, and then select the Parcel Note label style.
3. Click Add, and pick an insertion point for the label. You're prompted to pick the parcel you'd like to reference.

You'll find a second parcel area label to be a little more automatic when you place it (it already knows what parcel to reference), but the General Note label is more flexible about location, easier to pin, and easier to erase.

You can also use the Edit Parcel Properties dialog, as shown in an exercise earlier in the chapter, to add a second label.

Labeling Parcel Segments

Although parcels are used for much more than just subdivision lots, most parcels you create will probably be used for concept plans, record plats, and other legal subdivision plans. These plans, such as the one shown in Figure 6.61, almost always require segment labels for bearing, distance, direction, crow's feet, and more.

Labeling Multiple Parcel Segments

The following exercise will teach you how to add labels to multiple parcel segments:

1. Open the `Segment Labels.dwg` file, which you can download from `www.sybex.com/go/masteringcivil3d2010`. Note that this drawing contains many subdivision lot parcels.
2. Switch to the Annotate tab, and select Add Labels from the Labels & Tables panel on the Annotate tab.
3. In the Add Labels dialog, select Parcel, Multiple Segment, Bearing over Distance, and Delta over Length and Radius from the drop-down menus in the Feature, Label Type, Line Label Style, and Curve Label Style selection boxes, respectively.

4. Click Add.
5. At the `Select parcel to be labeled by clicking on area label:` prompt, pick the area label for parcel 1.
6. At the `Label direction [CLockwise/COunterclockwise]<CLockwise>:` prompt, press ↵ to accept the default and again to exit the command.
7. Each parcel segment for parcel 1 should now be labeled. Continue picking parcels 2 through 15 in the same manner. Note that segments are never given a duplicate label, even along shared lot lines.
8. Press ↵ to exit the command.

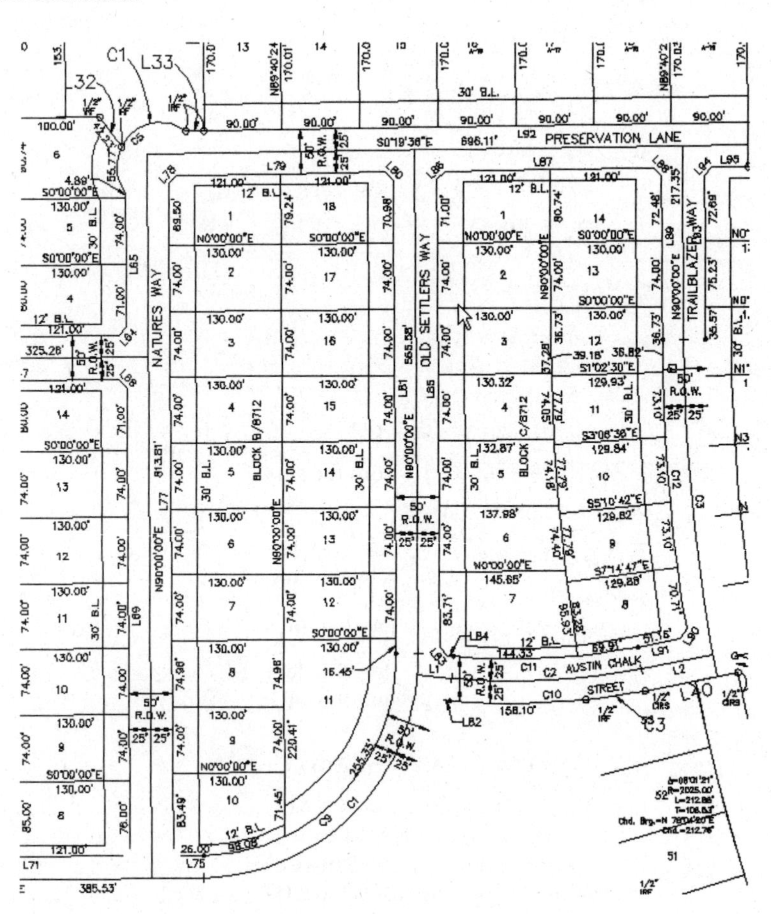

FIGURE 6.61
A fully labeled site plan

The following exercise will teach you how to edit and delete parcel segment labels:

1. Continue working in the `Segment Labels.dwg` file.
2. Zoom in on the label along the frontage of parcel 8 (see Figure 6.62).

82.51'
N90° 00' 00.00"E
129.86'
N0° 00' 00.00"E
8
129.86'
N0° 00' 00.00"E

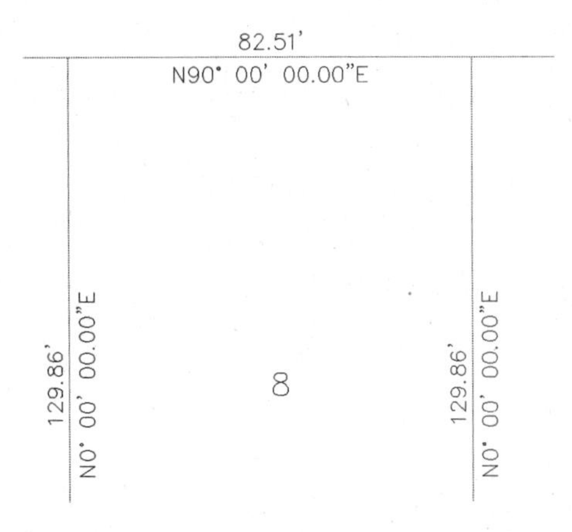

FIGURE 6.62
The label along the frontage of parcel 8

3. Select the label. You'll know your label has been picked when you see a diamond-shaped grip at the label midpoint (see Figure 6.63).

82.51'
N90° 00' 00.00"E

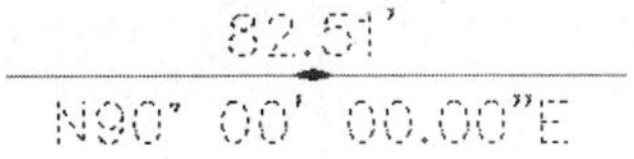

FIGURE 6.63
A diamond-shaped grip appears when the label has been picked

4. Once your label is picked, right-click over the label to bring up the shortcut menu.
5. Select Flip Label from the shortcut menu. The label flips so that the bearing component is on top of the line and the distance component is underneath the line.
6. Select the label again, right-click, and select Reverse Label. The label reverses so that the bearing now reads NW instead of NE.
7. Repeat steps 3 through 6 for several other segment labels, and note their reactions.
8. Select any label. Once the label is picked, execute the AutoCAD Erase tool or press the Delete key. Note that the label disappears.

Labeling Spanning Segments

Spanning labels are used where you need a label that spans the overall length of an outside segment, such as the example in Figure 6.64.

Spanning labels require that you use the appropriate vertices as discussed in detail in a previous section. Spanning labels have the following requirements:

- Spanning labels can only span across split-created vertices. Natural vertices will interrupt a spanning length.
- Spanning label styles must be composed to span the outside segment (see Figure 6.65).

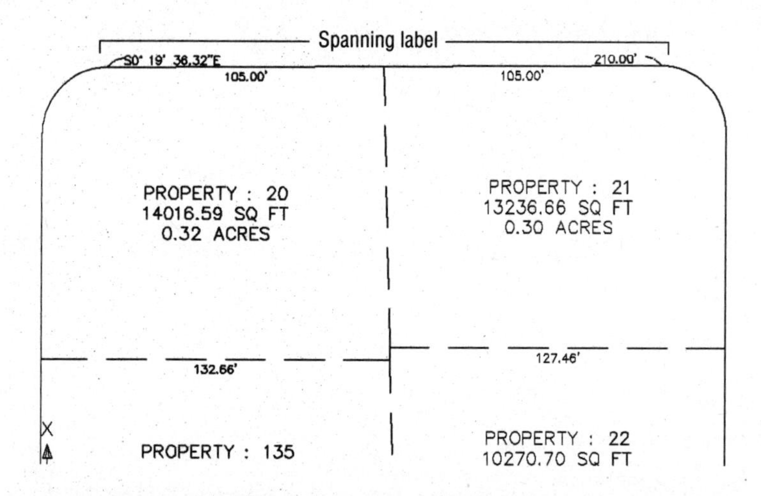

FIGURE 6.64
A spanning label

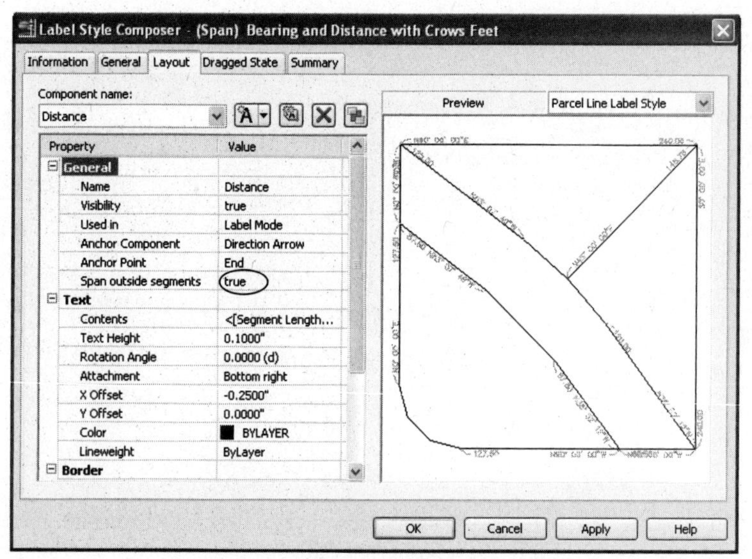

FIGURE 6.65
Set all components of a spanning label to span outside segments

- Spanning label styles must be composed to attach the desired spanning components (such as length and direction arrow) on the outside segment (as shown previously in Figure 6.64), with perhaps a small offset (see Figure 6.66).

THINKING AHEAD: CROW'S FEET

The small arc placed at the endpoint of parcel segment is commonly referred to as a "crow's foot." Crow's feet are traditionally used in lieu of standard extension and dimension lines because they take up very little space.

At some point during construction, property boundary markers (such as iron pipes) must be placed at each lot corner. In Civil 3D, this is typically represented by placing point objects at those locations.

The crow's feet have an Endpoint object snap, so care must be taken to select the lot corner, and not the end of the crow's foot, when creating points. Depending upon your offset value as shown in Figure 6.66, it is possible to set a point in the wrong location and incorrectly identify a property boundary marker.

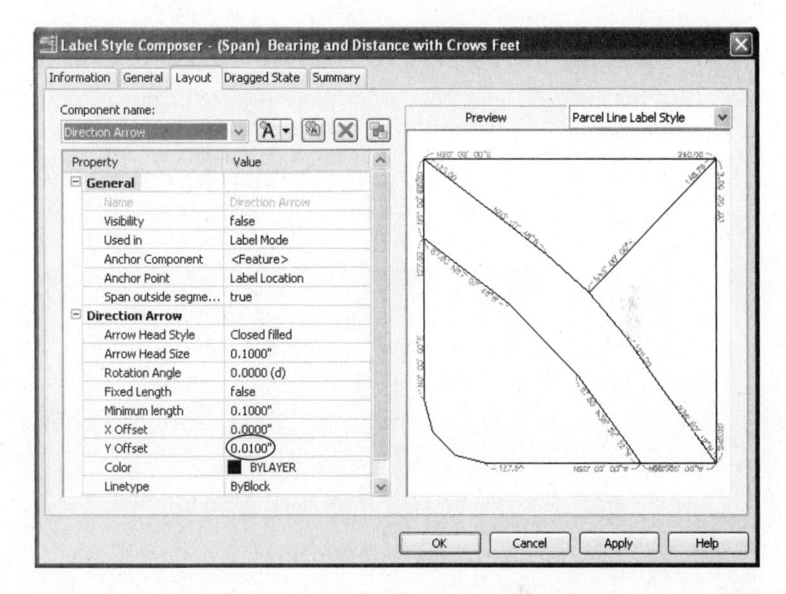

FIGURE 6.66
Give each component a small offset in the *y* direction

Once you've confirmed that your geometry is sound and your label is properly composed, you're set to span. The following exercise will teach you how to add spanning labels to single parcel segments:

1. Continue working in the `Segment Labels.dwg` file.
2. Zoom in on the ROW parcel segment that runs from parcel 1 through parcel 9.
3. Change to the Annotate tab and select Add Labels ➤ Parcel ➤ Add Parcel Labels from the Labels & Tables panel.
4. In the Add Labels dialog, select Single Segment, (Span) Bearing and Distance with Crow's Feet, and Delta over Length and Radius from the drop-down menus in the Label Type, Line Label Style, and Curve Label Style selection boxes, respectively.
5. Click Add.
6. At the `Select label location:` prompt, pick somewhere near the middle of the ROW parcel segment that runs from parcel 10 through parcel 16.
7. A label that spans the full length between natural vertices appears (see Figure 6.67).

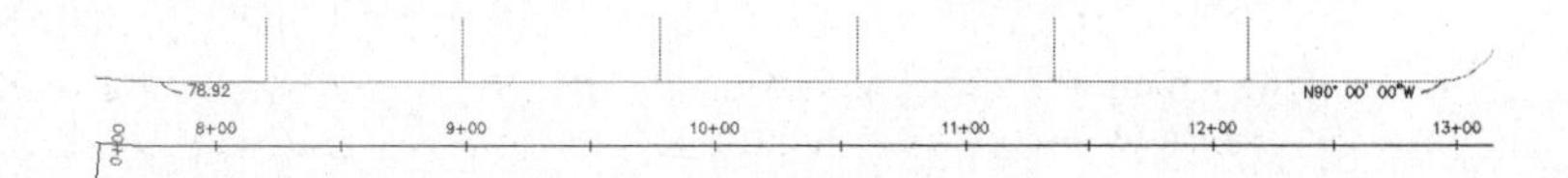

FIGURE 6.67
The spanning label

FLIP IT, REVERSE IT

If your spanning label doesn't seem to work on your first try and you've followed all the spanning-label guidelines, try flipping your label to the other side of the parcel segment, reversing the label, or using a combination of both flipping and reversing.

Adding Curve Tags to Prepare for Table Creation

Surveyors and engineers often make segment tables to simplify plan labeling, produce reports, and facilitate stakeout. Civil 3D parcels provide tools for creating dynamic line and curve tables, as well as a combination of line and curve tables.

Parcel segments must be labeled before they can be used to create a table. They can be labeled with any type of label, but you'll likely find it to be best practice to create a tag-only style for segments that will be placed in a table.

The following exercise will teach you how to replace curve labels with tag-only labels, and then renumber the tags:

1. Continue working in the `Segment Labels.dwg` file. Note that the labels along tight curves, such as parcels 9 and 15, would be better represented as curve tags.
2. Change to the Annotate tab. Select Add Labels ➤ Parcel ➤ Add Parcel Labels from the Labels & Tables panel.
3. In the Add Labels dialog, select Replace Multiple Segment, Bearing over Distance, and Spanning Curve Tag Only from the drop-down menus in the Label Type, Line Label Style, and Curve Label Style selection boxes, respectively.
4. At the `Select parcel to be labeled by clicking on area label or [CLockwise/COunterclockwise]<CLockwise>:` prompt, pick the area label for parcel 1. Note that the line labels for parcel 1 are reset and the curve labels convert to tags.
5. Repeat step 4 for parcels 2 through 15. Press ↵ to exit the command.

Now that each curve label has been replaced with a tag, it's desirable to have the tag numbers be sequential. The following exercise will teach you how to renumber tags:

1. Continue working in the `Segment Labels.dwg` file.
2. Zoom into the curve on the upper-left side of parcel 9 (see Figure 6.68). Your curve may have a different number from the figure.
3. Select a parcel and select Renumber/Rename from the Modify panel.
4. At the `Select label to renumber tag or [Settings]:` prompt, type **S**, and then press ↵.
5. The Table Tag Numbering dialog appears (see Figure 6.69). Change the value in the Curves Starting Number selection box to **1**. Click OK.
6. Click each curve tag in the drawing at the `Select label to renumber tag or [Settings]:` prompt. The command line may say `Current tag number is being used, press return to skip to next available or [Create duplicate]`, in which case, press ↵ to skip the used number. When you're finished, press ↵ to exit the command.

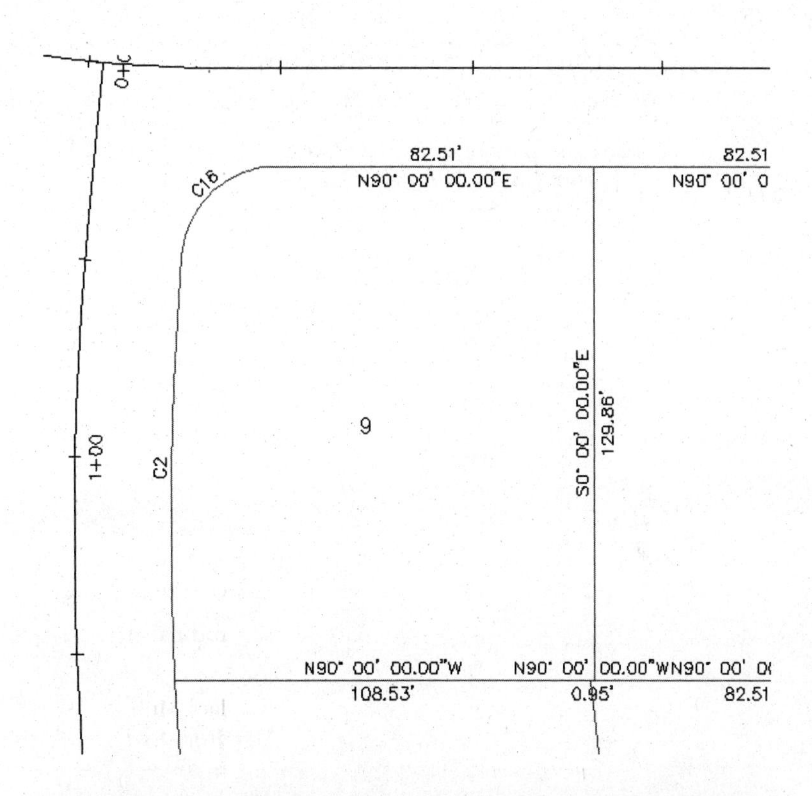

FIGURE 6.68
Curve tags on parcel 9

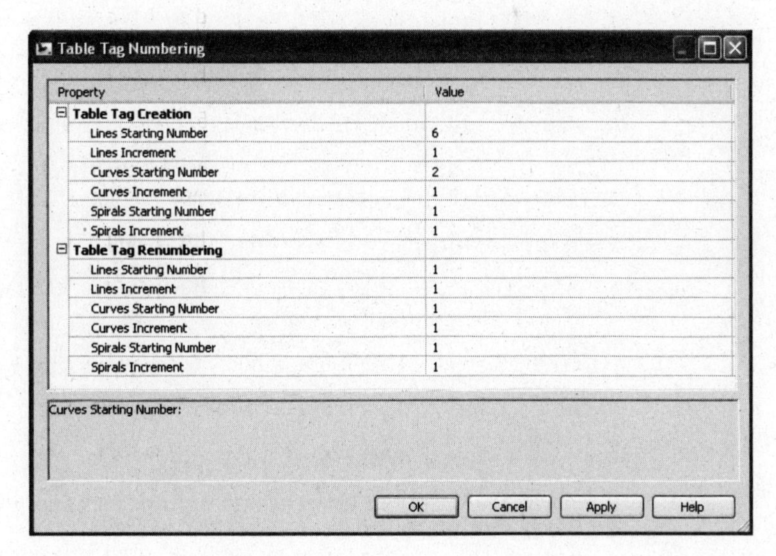

FIGURE 6.69
The Table Tag Numbering dialog

Creating a Table for Parcel Segments

The following exercise will teach you how to create a table from curve tags:

1. Continue working in the `Segment Labels.dwg` file. You should have several curves labeled with the Curve Tag Only label.
2. Select a parcel and choose Add Curve under the Add Tables from the Labels & Tables panel.
3. In the Table Creation dialog, select Length Radius & Delta from the drop-down menu in the Table Style selection box. In the Selection area of the dialog, select the Apply check box for the Spanning Curve Tag Only entry under Label Style Name. Keep the default values for the remaining options. The dialog should look like Figure 6.70. Click OK.

FIGURE 6.70
The Table Creation dialog

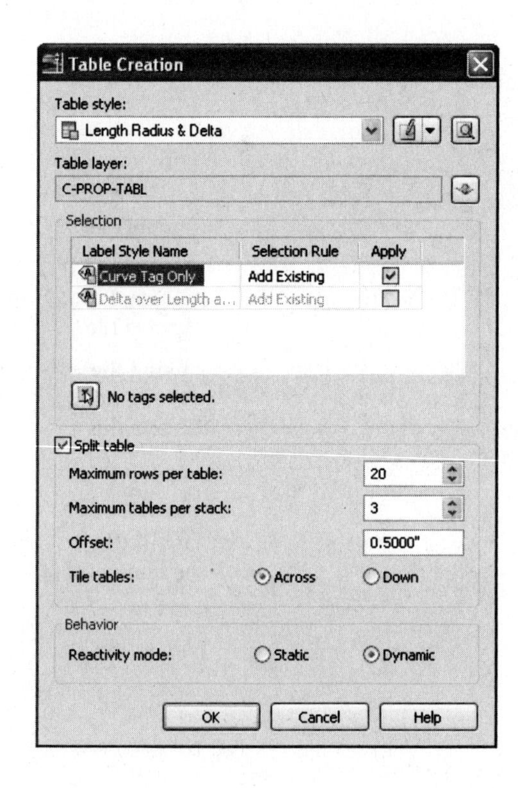

4. At the `Select upper left corner:` prompt, pick a location in your drawing for the table. A curve table appears, as shown in Figure 6.71.

FIGURE 6.71
A curve table

Curve Table					
Curve #	Length	Radius	Delta	Chord Direction	Chord Length
C1	37.06	25.00	84.93	N47° 31' 58"E	33.76
C15	39.13	25.00	89.68	S45° 09' 28"E	35.26
C14	39.41	25.00	90.31	S44° 50' 33"W	35.45
C13	311.43	2025.00	8.81	S85° 35' 39"W	311.12
C12	311.43	2025.00	8.81	S85° 35' 39"W	311.12
C11	311.43	2025.00	8.81	S85° 35' 39"W	311.12
C8	311.43	2025.00	8.81	S85° 35' 39"W	311.12
C7	0.00	1.13	0.04	S81° 11' 18"W	0.00
C6	38.66	25.00	88.60	N54° 30' 36"W	34.92
C3	206.60	775.00	15.27	N2° 34' 19"W	205.99
C2	206.60	775.00	15.27	N2° 34' 19"W	205.99

The Bottom Line

Create a boundary parcel from objects. The first step to any parceling project is to create an outer boundary for the site.

Master It Open the `Mastering Parcels.dwg` file, which you can download from `www.sybex.com/go/masteringcivil3d2010`. Convert the polyline in the drawing to a parcel.

Create a right-of-way parcel using the right-of-way tool. For many projects, the ROW parcel serves as frontage for subdivision parcels. For straightforward sites, the automatic Create ROW tool provides a quick way to create this parcel.

Master It Continue working in the `Mastering Parcels.dwg` file. Create a ROW parcel that is offset by 25′ on either side of the road centerline with 25′ fillets at the parcel boundary.

Create subdivision lots automatically by layout. The biggest challenge when creating a subdivision plan is optimizing the number of lots. The precise sizing parcel tools provide a means to automate this process.

Master It Continue working in the `Mastering Parcels.dwg` file. Create a series of lots with a minimum of 10,000 square feet and 100′ frontage.

Add multiple parcel segment labels. Every subdivision plat must be appropriately labeled. You can quickly label parcels with their bearings, distances, direction, and more using the segment labeling tools.

Master It Continue working in the `Mastering Parcels.dwg` file. Place Bearing over Distance labels on every parcel line segment and Delta over Length and Radius labels on every parcel curve segment using the Multiple Segment Labeling tool.

Chapter 7

Laying a Path: Alignments

The world is 3D, but almost every design starts as a concept: a flat line on a flat piece of paper. Cutting a way through the trees, the hills, and the forests, you can design around a basic layout to get some idea of horizontal placement. This horizontal placement is the alignment and drives much of the design. This chapter shows you how alignments can be created, how they interact with the rest of the design, how to edit and analyze them, how styles are involved with display and labeling, and finally, how they work with the overall project.

By the end of this chapter, you'll learn to:

- Create an alignment from a polyline
- Create a reverse curve that never loses tangency
- Replace a component of an alignment with another component type
- Create a new label set
- Override individual labels with other styles

Alignments, Pickles, and Freedom

Before you can efficiently work with alignments, you must understand two major concepts: the interaction of alignments and sites, and the idea of geometry that is fixed, floating, or free.

Alignments and Sites

Prior to Civil 3D 2008, alignments were always a part of a site and interacted with the topology contained in that site. This interaction led to the pickle analogy: alignments are like pickles in a mason jar. You don't put pickles and pepper in the same jar unless you want hot pickles, and you don't put lots and alignments in the same site unless you want subdivided lots.

Since the 2008 release, Civil 3D now has two ways of handling alignments in terms of sites: they can be contained in a site as before, or they can be independent of a site. Notice how Figure 7.1 shows Parker Place as a member of the Alignments collection directly under the drawing and Carson Court as a member of the Alignments collection that is part of the Rose Acres Sites collection, which is directly under the drawing.

Both the Parker Place and the Carson Court alignments can be used to cut profiles or control corridors, but only the Carson Court alignment will react with and create parcels as a member of a site topology.

Many users of versions prior to 2008 had issues keeping alignments and sites straight. Unless you have good reason for them to interact (as in the case of an intersection), it makes sense to create alignments outside of any site object. They can be moved later if necessary. For the purpose of the exercises in this chapter, you won't place any alignments in a site.

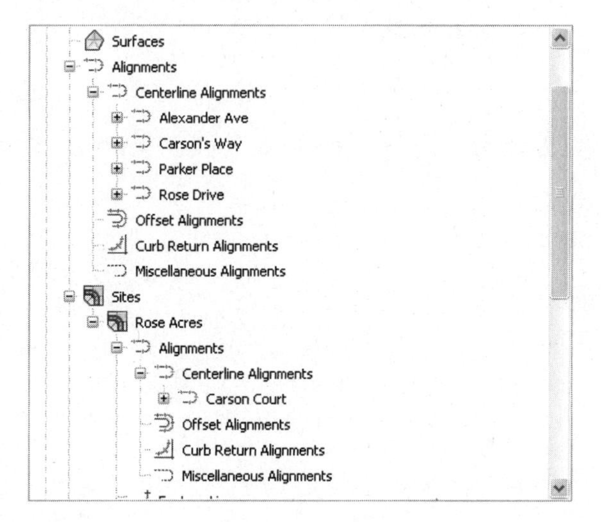

FIGURE 7.1
Alignments in and out of the site collection

Alignment Entities and Freedom

Civil 3D recognizes four types of alignments: centerline alignments, offset alignments, curb return alignments, and miscellaneous alignments. Each alignment type can consist of three types of entities or segments: lines, arcs, and spirals. These segments control the horizontal alignment of your design. Their relationship to one another is described by the following terminology:

- Fixed segments are fixed in space. They're defined by connecting points in the coordinate plane and are independent of the segments that occur either before or after them in the alignment. Fixed segments may be created as tangent to other components, but their independence from those objects lets you move them out of tangency during editing operations. This can be helpful when you're trying to match existing field conditions.
- Floating segments float in space but are attached to a point in the plane and to some segment to which they maintain tangency. Floating segments work well in situations where you have a critical point but the other points of the horizontal alignment are flexible.
- Free segments are functions of the entities that come before and after them in the alignment structure. Unlike fixed or floating segments, a free segment must have segments that come before and after it. Free segments maintain tangency to the segments that come before and after them and move as required to make that happen. Although some geometry constraints can be put in place, these constraints can be edited and are user dependent.

During the exercises in this chapter, you'll use a mix of these entity types to understand them better. Autodesk has also published a drawing called Playground that you can find by searching on the Web. This drawing contains examples of most of the types of entities that you can create.

Creating an Alignment

Alignments in Civil 3D can be created from AutoCAD objects (lines, arcs, or polylines,) or by layout. This section looks at both ways to create an alignment and discusses the advantages and disadvantages of each. The exercise will use the street layout shown in Figure 7.2 as well as the different methods to achieve your designs.

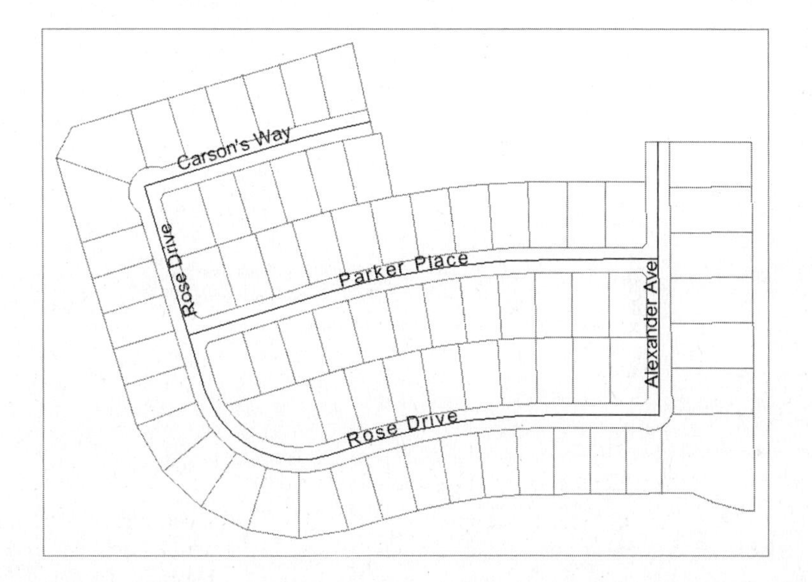

FIGURE 7.2
Proposed street layout

Creating from a Polyline

Most designers have used either polylines or lines and arcs to generate the horizontal control of their projects. It's common for surveyors to generate polylines to describe the center of a right of way or for an environmental engineer to draw a polyline to show where a new channel should be constructed. These team members may or may not have Civil 3D, so they use their familiar friend — the polyline — to describe their design intent.

Although polylines are good at showing where something should go, they don't have much data behind them. To make full use of these objects, you should convert them to native Civil 3D alignments that can then be shared and used for myriad purposes. Once an alignment has been created from a polyline, offsets can be created to represent rights of way, building lines, and so on. In this exercise, you'll convert a polyline to an alignment and create offsets:

1. Open the `Alignments from Polylines.dwg` file.
2. Change to the Home tab and choose Alignment ➤ Create Alignment from Objects.
3. Pick the polyline labeled Parker Place, shown previously in Figure 7.2 and press ↵. Press ↵ again to accept the default direction; the Create Alignment from Objects dialog appears.
4. Change the Name field to **Parker Place**, and select the Centerline Type as shown in Figure 7.3.
5. Accept the other settings, and click OK.

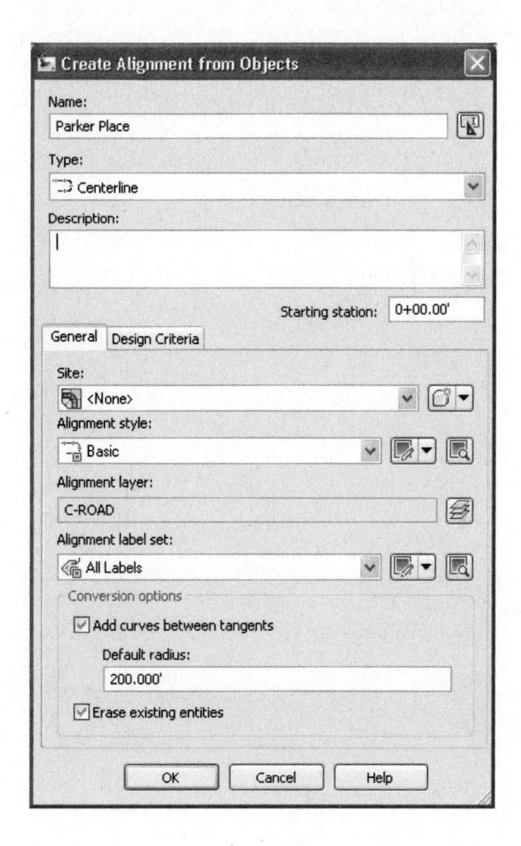

FIGURE 7.3
The settings used to create the Parker Place alignment

You've created your first alignment and attached stationing and geometry point labels. It is common to create offset alignments from a centerline alignment to begin to model rights of way. In the following exercise, you'll create offset alignments and mask them where you don't want them to be seen:

1. Change to the Home tab and choose Alignment ➤ Create Offset Alignment.
2. Pick the Parker Place alignment to open the Create Offset Alignments dialog as shown in Figure 7.4.
3. Click OK to accept the defaults as shown in Figure 7.4.
4. Select the offset alignment just created along the northerly right of way of Parker Place.
5. Choose Alignment Properties from the Modify panel to open the Alignment Properties dialog.
6. Change to the Masking tab and click the Add Masking Region button (circled in Figure 7.5). Type **9+60.16′** for the first station and **9+94.15′** for the second station when prompted. Click OK. Notice that the alignment is now masked at the west end.
7. Repeat the process for the right of way alignment along the south side of Parker Place.

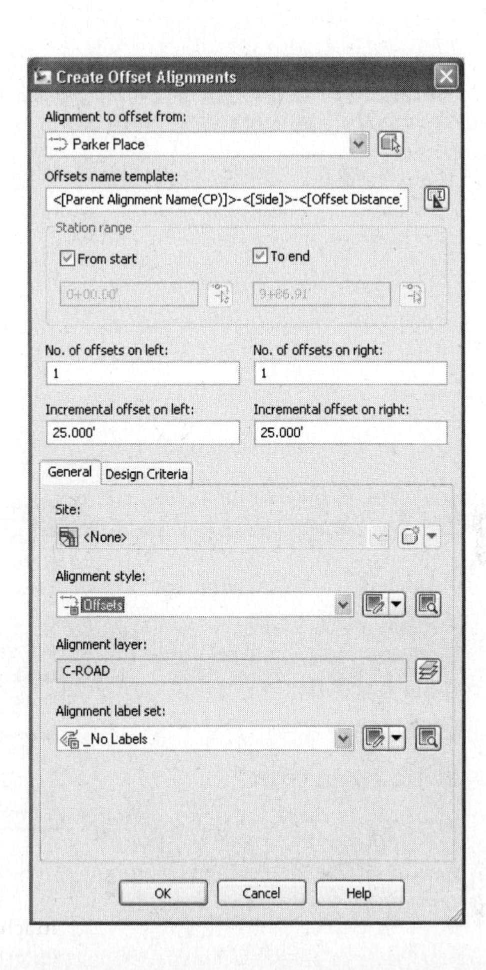

FIGURE 7.4
The Create Offset Alignments dialog

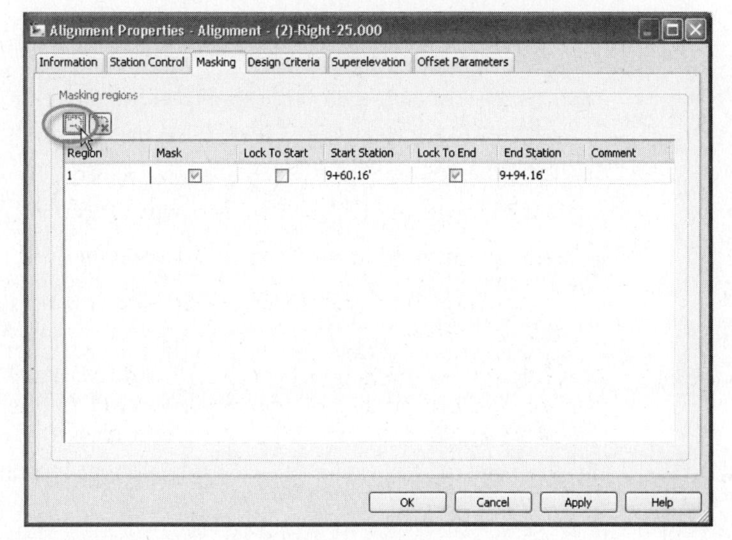

FIGURE 7.5
Creating an alignment mask

Offset alignments are simple to create, and they are dynamically linked to a centerline alignment. To test this, simply grip the centerline alignment, select the endpoint grip to make it hot, and stretch the alignment to the west. Notice the change.

Just like every other Civil 3D object, alignments and their labels are controlled by styles. In this case, setting the Alignment Label Set to Major and Minor Only means that only the labels for major and minor stations are displayed. A later section looks at label sets; for now, let's see what other options you have for labeling alignments:

1. Change to the Home tab and choose Alignment ➤ Create Alignment from Objects.
2. Pick the shorter polyline label that will define Carson's Way on the northerly end of your site.
3. In the Create Alignment from Objects dialog, do the following:
   - Change the Name field to **Carson's Way**.
   - Set the Alignment Style field to Select Proposed.
   - Set the Alignment Label Set field to _No Labels.
4. Click OK.

No labels are displayed when the Alignment Label Set is set to No Labels.

You created two street alignments from polylines and two offsets. They're ready for use in corridors, in profiling, or for any number of other uses.

Creating by Layout

Now that you've made a series of alignments from polylines, let's look at the other creation option: Create by Layout. You'll use the same street layout (Figure 7.2) that was provided by a planner, but instead of converting from polylines, you'll trace the alignments. Although this seems like duplicate work, it will pay dividends in the relationships created between segments:

1. Open the `Alignments by Layout.dwg` file.
2. Change to the Home tab and choose Alignment ➤ Alignment Creation Tools from the Create Design tab. The Create Alignment – Layout dialog appears, as shown in Figure 7.6.
3. Change the Name field to **Parker Place**, and then click OK to accept the other settings. The Alignment Layout Tools toolbar appears, as shown in Figure 7.7.
4. Click the down arrow next to the Draw Tangent – Tangent without Curve tool at far left, and select the Tangent – Tangent (With Curves) option. The tool places a curve automatically; you'll adjust the curve, watching the tangents extend as needed.
5. Pick the far left end of Parker Place using an Endpoint snap.
6. Pick just above the arc in the middle of the street, and then pick the endpoint at far right (see Figure 7.8) to finish creating this alignment.
7. Right-click or press ↵, and you'll be back at the command line, but the toolbar will still be open.
8. Click the red X button at the upper-right on the toolbar to close it.

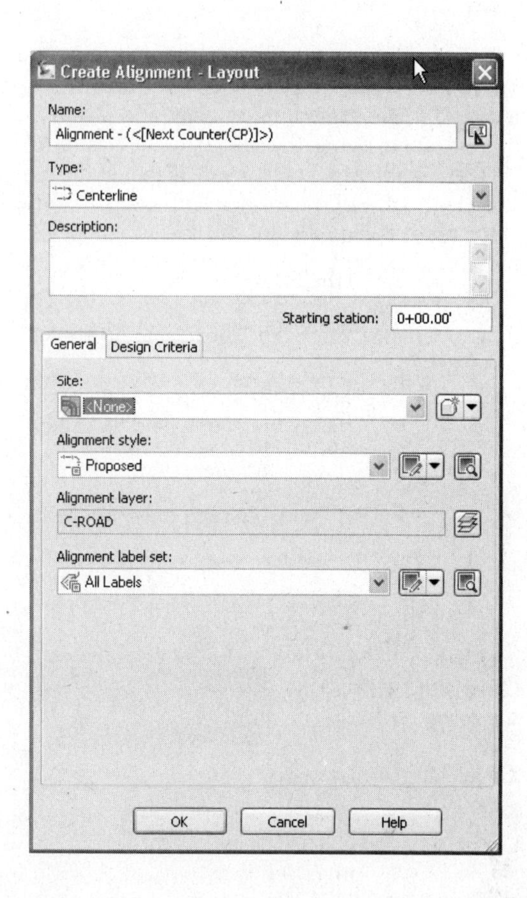

FIGURE 7.6
Creating an alignment by layout

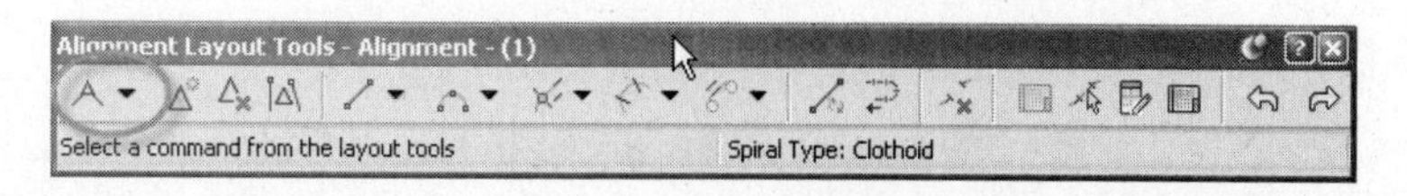

FIGURE 7.7
The Alignment Layout Tools toolbar

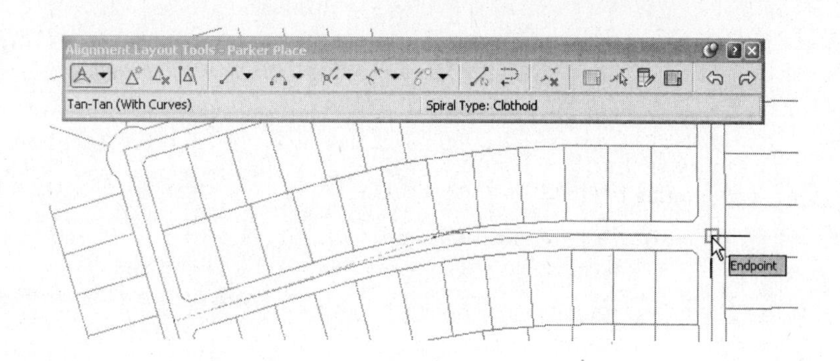

FIGURE 7.8
Completing the Parker Place alignment

Zoom in on the arc, and notice that it doesn't match the arc the planner put in for you to follow. That's OK — you'll fix it in a few minutes. It bears repeating that in dealing with Civil 3D objects, it's good to get something in place and *then* refine. With Land Desktop or other packages, you didn't want to define the object until it was fully designed. In Civil 3D, you design and then refine.

The alignment you just made is one of the most basic. Let's move on to some of the others and use a few of the other tools to complete your initial layout. In this exercise, you build the alignment at the north end of the site, but this time you use a floating curve to make sure the two segments you create maintain their relationship:

1. Change to the Home tab and choose Alignment ➢ Alignment Creation Tools from the Create Design tab. The Create Alignment – Layout dialog appears
2. In the Create Alignment – Layout dialog, do the following:
   - Change the Name field to **Carson's Way**.
   - Set the Alignment Style field to Layout.
   - Set the Alignment Label Set field to Major and Minor Only.
3. Click OK, and the Alignment Layout Tools toolbar appears.
4. Select the Draw Fixed Line – Two Points tool (see Figure 7.9).

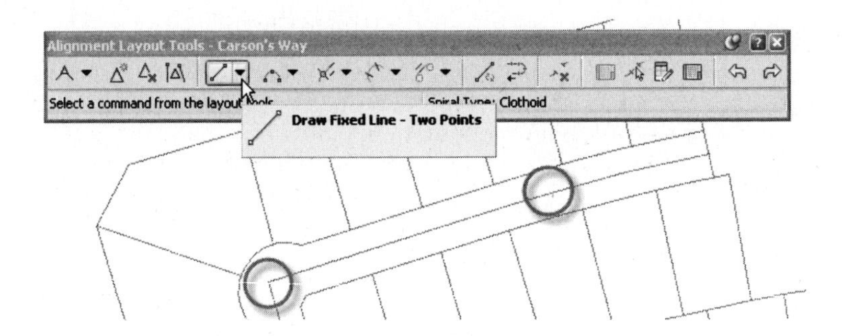

FIGURE 7.9
The Draw Fixed Line – Two Points tool

5. Pick the two points circled in Figure 7.9, using Endpoint snaps and working left to right to draw the fixed line. When you've finished, the command line will state `Specify start point:`.
6. Click the down arrow next to the Add Fixed Curve (Three Points) tool on the toolbar, and select the More Floating Curves ➢ Floating Curve (From Entity End, through Point) option, as shown in Figure 7.10.

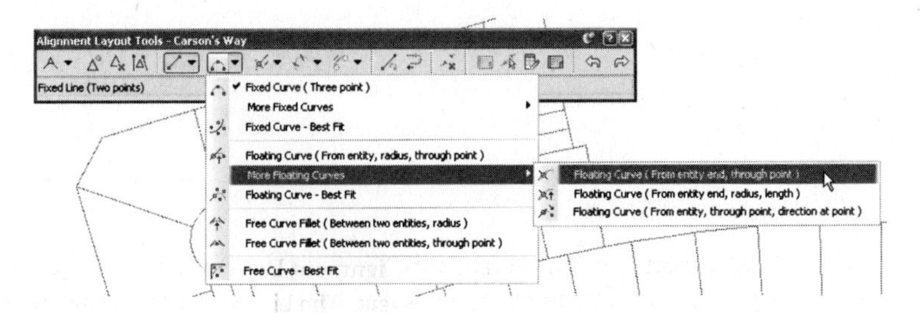

FIGURE 7.10
Selecting the Floating Curve tool

7. Ctrl+click the fixed-line segment you drew in steps 4 and 5. A blue rubber band should appear, indicating that the alignment of the curve segment is being floated off the end-point of the fixed segment (see Figure 7.11). The Ctrl+click is required to make the pick activate the alignment segment and not the polyline entity.

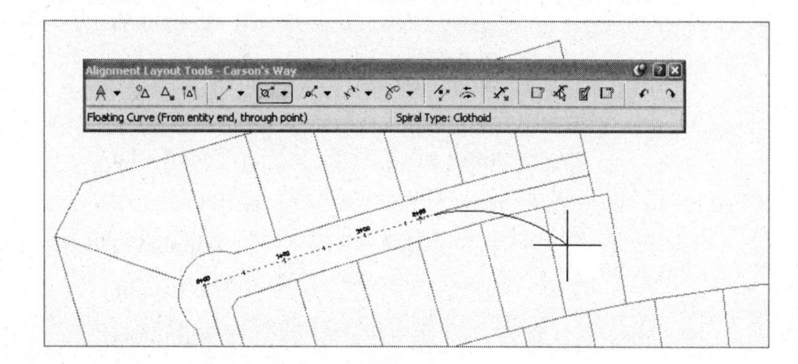

FIGURE 7.11
Adding a floating curve to the Carson's Way alignment

8. Pick the east end of the Carson's Way polyline arc segment.
9. Right-click or press ↵ to exit the command.
10. Close the toolbar to return to Civil 3D.

In your previous drawing, you picked a point on the Carson's Way alignment and pulled it away to illustrate the lack of connection between the two segments. This time, pick the grip near the western end and pull it away from its location in the cul-de-sac. Notice that the line and the arc move in sync, and tangency is maintained (see Figure 7.12).

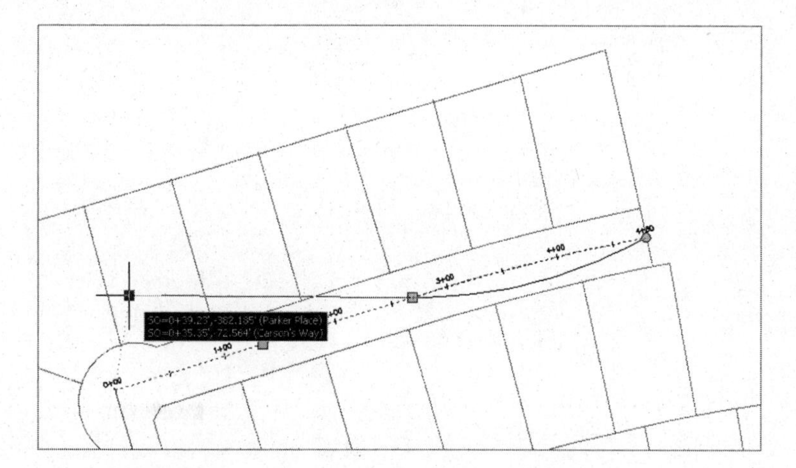

FIGURE 7.12
Floating curves maintain their tangency.

MAKING IT FIT

One aspect of using component alignment layout that this chapter doesn't discuss is the ability to use Best Fit entities within an alignment. The Line and Curve drop-down menus on the Alignment Layout

toolbar include options for Floating and Fixed Lines by Best Fit, as well as Best Fit curves in all three flavors: Fixed, Float, and Free. These options are of great assistance when you're doing rehab work or other jobs where some form of the data already exists. Explore the Best Fit algorithms in Chapter 2, "Back to Basics: Lines and Curves."

Next, let's look at a more complicated alignment construction — building a reverse curve where the planner left a short segment connecting two curves:

1. Change to the Home tab and choose Alignment ➤ Alignment Creation Tools from the Create Design tab. The Create Alignment – Layout dialog appears.
2. In the Create Alignment – Layout dialog, do the following:
   - Change the Name field to **Rose Drive**.
   - Set the Alignment Style field to Layout.
   - Set the Alignment Label Set field to _No Labels.
3. Click OK, and the Alignment Layout Tools toolbar appears.
4. Start by drawing a fixed line from the north end of the western portion to its endpoint using the same Draw Fixed Line (Two Points) tool as before.
5. Use the same Floating Curve (From Entity End, through Point) tool to draw a curve from the end of this segment to the midpoint of the small tangent on the south end (see Figure 7.13).

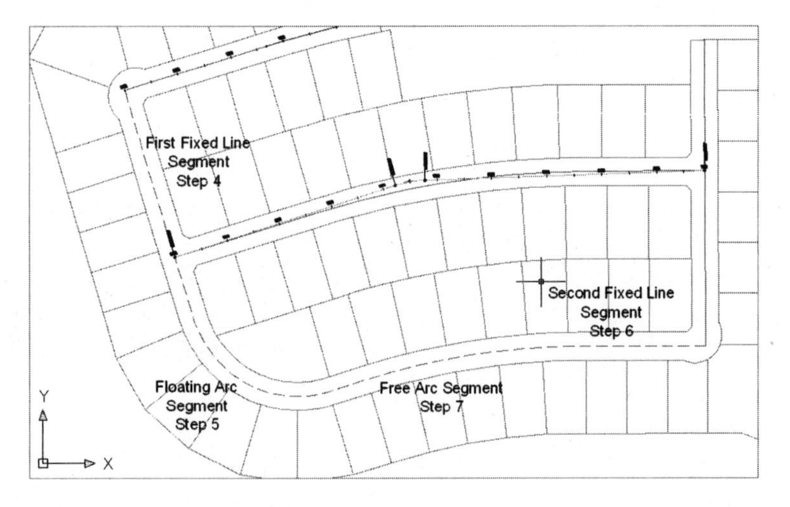

FIGURE 7.13
Segment layout for Rose Drive

6. Return to the toolbar, and select the Draw Fixed Line (Two Points) tool again. Draw the line on the east end of this proposed street. This segment is still part of the alignment, in spite of not being connected!

   To finish your reverse curve, you need a free curve to tie the floating curve to the segment you just drew.

7. From the drop-down menu next to the Add Fixed Curve (Three Points) tool, select the Free Curve Fillet (Between Two Entities, Radius) option.
8. Pick the first arc you drew, as shown in Figure 7.14. The free curve attaches to this entity.

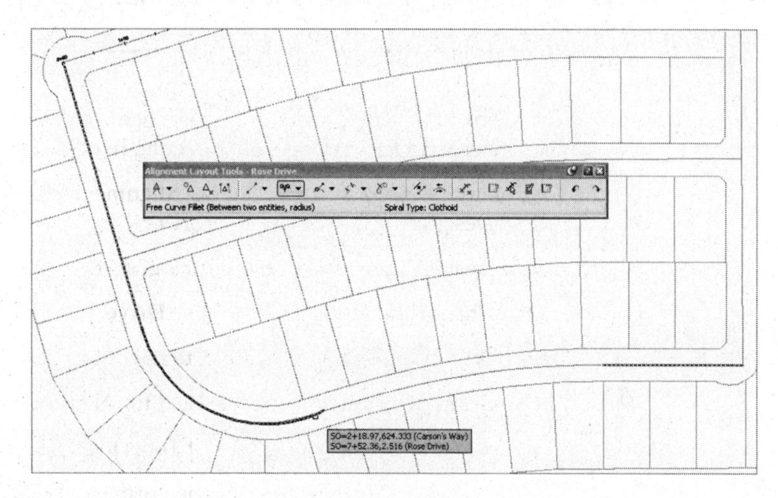

FIGURE 7.14
Adding the floating curve segment

9. Pick the detached segment you made a moment ago.
10. Press ↵ at the command line for a solution of less than 180. Draw a curve as opposed to a cloverleaf, which is the only other solution that will solve this geometry.
11. Enter **R** at the command line to select a reverse curve.
12. Use a Center snap to pick the center of the sketched polyline arc, and then enter **1380** at the command line.
13. Right-click or press ↵, and close the toolbar.

The alignment now contains a perfect reverse curve. Move any of the pieces, and you'll see the other segments react to maintain the relationships shown in Figure 7.15. This flexibility in design isn't possible with the converted polylines you used previously. Additionally, the flexibility of the Civil 3D tools allows you to explore an alternative solution (the reverse curve) as opposed to the basic solution (two curves with a short tangent). Flexibility is one of Civil 3D's strengths.

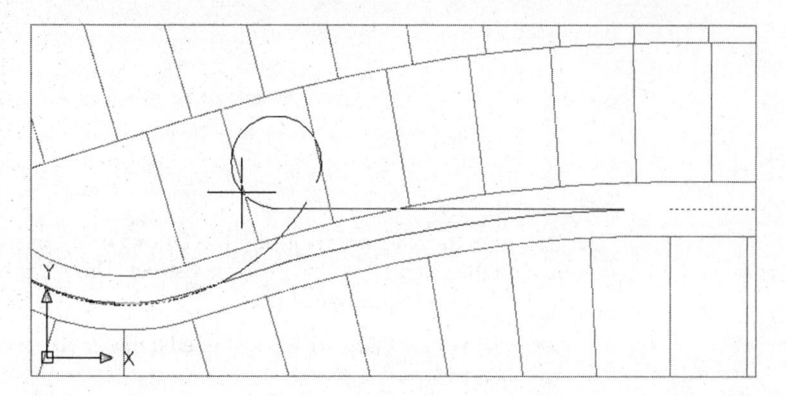

FIGURE 7.15
Curve relationships during a grip-edit

Let's do one more exercise and draw a straight line from south to north on the east portion of the site using the Alignment Layout Tools toolbar:

1. Change to the Home tab's Create Design panel and choose Alignment ➤ Alignment Creation Tools. The Create Alignment – Layout dialog appears.
2. In the Create Alignment – Layout dialog, do the following:
   - Change the Name field to **Alexander Ave**.
   - Set the Alignment Style field to Layout.
   - Set the Alignment Label Set field to All Labels.
3. Click OK, and the Alignment Layout Tools toolbar appears.
4. Select the Draw Fixed Line tool, and then pick the south and north ends of Alexander Ave to create a straight line.
5. Right-click or press ↵, and close the toolbar.

You've completed your initial layout (see Figure 7.16). There are some issues with curve sizes, and the reverse curve may not be acceptable to the designer, but you'll look at those changes in the later section on editing alignments.

FIGURE 7.16
Completed alignment layout

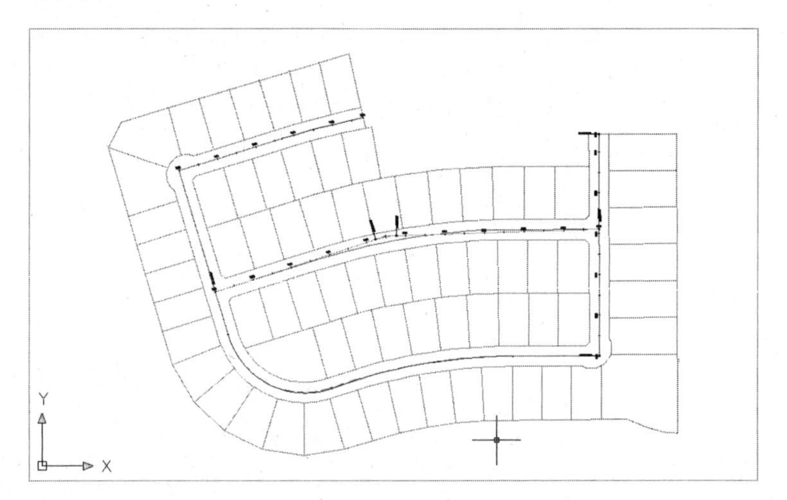

Creating with Design Constraints and Check Sets

Starting in Civil 3D 2009, users have the ability to create and use design constraints and design check sets during the process of aligning and creating design profiles. Typically, these constraints check for things like curve radius, length of tangents, and so on. Design constraints use information from the American Association of State Highway and Transportation Officials (AASHTO) or other design manuals to set curve requirements. Check sets allow users to create their own criteria to match local requirements, such as subdivision or county road design. First, you'll make one quick set of design checks:

1. Open the `Creating Checks.dwg` file.
2. Change to the Settings tab in Toolspace, and expand the Alignment\Design Checks branch.

3. Right-click the Line Folder, and select New to display the New Design Check dialog shown in Figure 7.17.

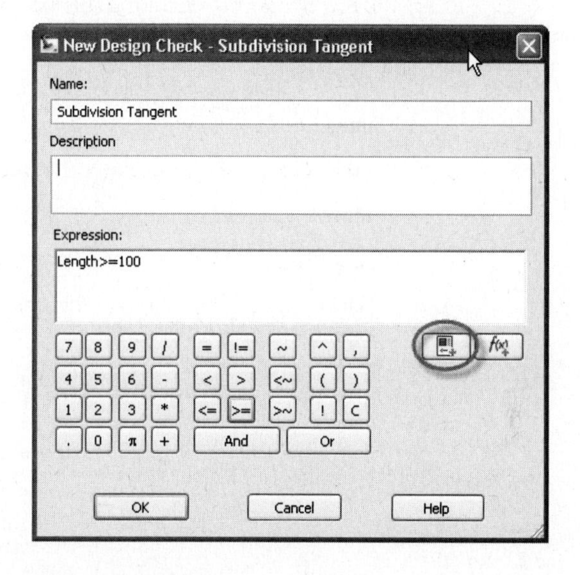

FIGURE 7.17 Completed Subdivision Tangent design check. The result of a design check is true or false; in this case, it tells you whether the alignment line segment is longer than 100′.

4. Change the name to **Subdivision Tangent**.
5. Click the Insert Property drop-down menu circled in Figure 7.17, and select Length.
6. Click the greater-than symbol (>), and then type **100** in the Design Check field as shown. When complete, your dialog should look like Figure 7.17. Click OK to close the dialog.
7. Right-click the Curve folder, and create the Subdivision Curve design check as shown in Figure 7.18.

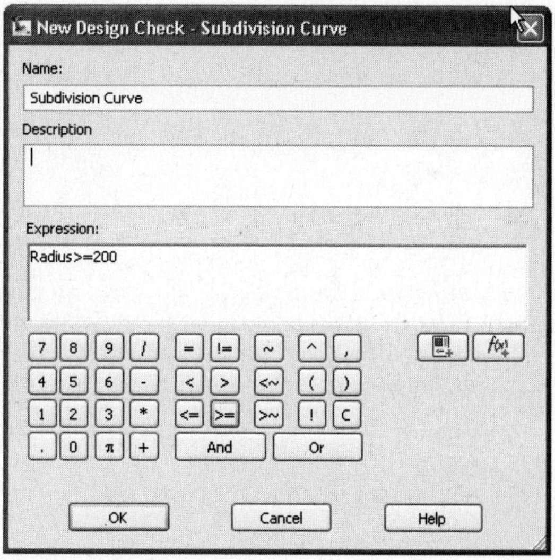

FIGURE 7.18 The Completed Subdivision Curve design check. The >= indicates that an acceptable curve is equal to or greater than 200′. Without the equal portion, a curve would require a radius of 200.01′ to pass the check.

8. Right-click the Design Check Sets folder, and select New to display the Alignment Design Check Set dialog.

9. Change the name to **Subdivision Streets** (as shown in the header of Figure 7.19), and then switch to the Design Checks tab.

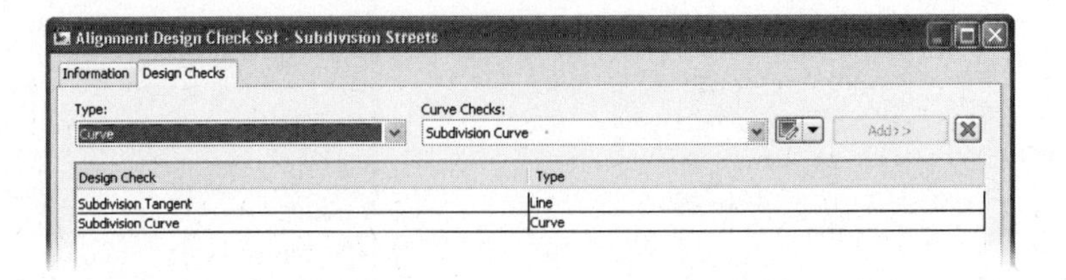

FIGURE 7.19
The Completed Subdivision Streets design check set

10. Click the Add button to add the Subdivision Tangent check to the set.

11. Choose Curve from the Type drop-down list, and click Add again to complete the set as shown in Figure 7.19.

Once you've created a number of design checks and design check sets, you can apply them as needed during the design and layout stage of your projects.

DESIGN CHECKS VERSUS DESIGN CRITERIA

In typical fashion, the language used for this feature isn't clear. What's the difference? A design check uses basic properties such as radius, length, grade, and so on to check a particular portion of an alignment or profile. These constraints are generally dictated by a governing agency based on the type of road involved. Design criteria use speed and related values from design manuals such as AASHTO to establish these geometry constraints.

Think of having a suite of check sets, with different sets for each city and type of street or each county, or for design speed. The options are open. In the next exercise, you'll see the results of your Subdivision Streets check set in action:

1. Open the `Checking Alignments.dwg` file.

2. Change to the Home tab and choose Alignment ➤ Alignment Creation Tools from the Create Design tab. The Create Alignment – Layout dialog appears.

3. Click the Design Criteria tab, and set the design speed to **30** as shown in Figure 7.20. Set the check boxes as shown in the figure. (Note that the Use Criteria-Based Design check box must be selected to activate the other two.) Click OK to close the dialog.

4. Select the Tangent-Tangent with Curves option at left side on the Alignment Layout toolbar. The stock curve radius of 150′ is left in place to illustrate the design check failure indicators.

5. Connect the center points of the circles on the screen to create the alignment shown in Figure 7.21. Notice the two exclamation-point symbols, which indicate that a design check has been violated.

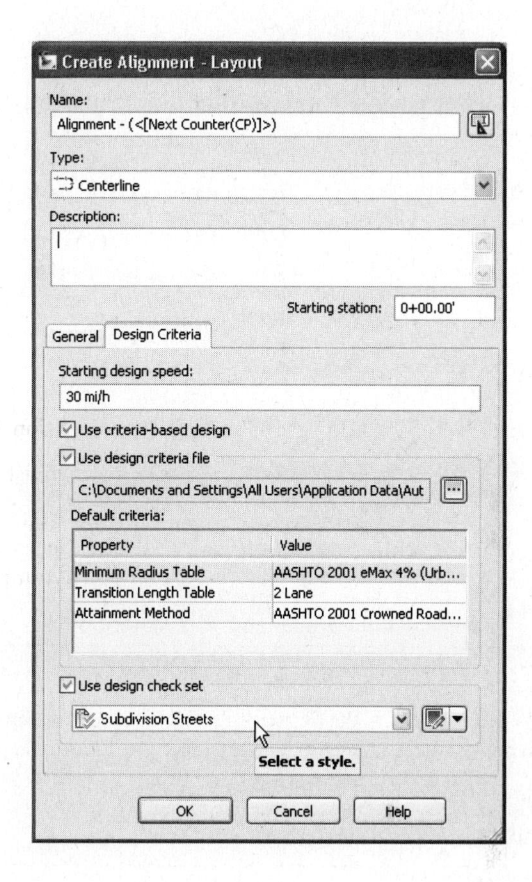

FIGURE 7.20
Setting up design checks during the creation of alignments

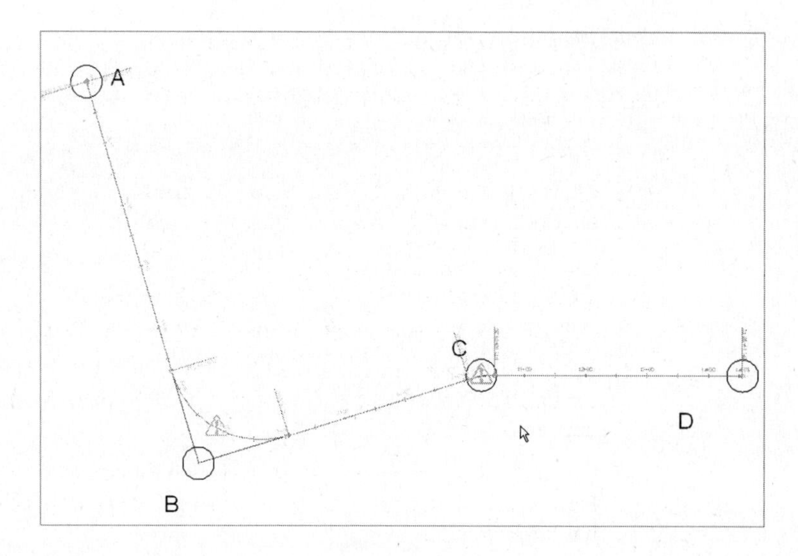

FIGURE 7.21
Completed alignment layout

Now that you know how to create an alignment that doesn't pass the design checks, let's look at different ways of modifying alignment geometry. As you correct and fix alignments that violate the assigned design checks, those symbols will disappear.

Editing Alignment Geometry

The general power of Civil 3D lies in its flexibility. The documentation process is tied directly to the objects involved, so making edits to those objects doesn't create hours of work in updating the documentation. With alignments, there are three major ways to edit the object's horizontal geometry without modifying the underlying construction:

Graphical Select the object, and use the various grips to move critical points. This method works well for realignment, but precise editing for things like a radius or direction can be difficult without construction elements.

Tabular Use Panorama to view all the alignment segments and their properties, typing in values to make changes. This approach works well for modifying lengths or radius values, but setting a tangent perpendicular to a screen element or placing a control point in a specific location is better done graphically.

Segment Use the Alignment Layout Parameters dialog to view the properties of an individual piece of the alignment. This method makes it easy to modify one piece of an alignment that is complicated and that consists of numerous segments, whereas picking the correct field in a Panorama view can be difficult.

In addition to these methods, you can use the Alignment Layout Tools toolbar to make edits that involve removing components or adding to the underlying component count. The following sections look at the three simple edits and then explain how to remove components from and add them to an alignment without redefining it.

Grip-Editing

You already used graphical editing techniques when you created alignments from polylines, but those techniques can also be used with considerably more precision than shown previously. The alignment object has a number of grips that reveal important information about the elements' creation (see Figure 7.22).

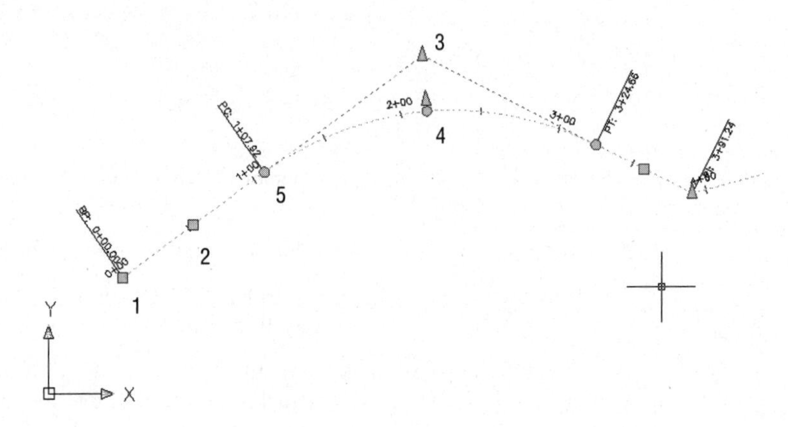

FIGURE 7.22
Alignment grips

You can use the grips in Figure 7.22 to do the following actions:

- The square grip at the beginning of the alignment, Grip 1, indicates a segment point that can be moved at will. This grip doesn't attach to any other components.
- The square grip in the middle of the tangents, Grip 2, allows the element to be translated. Other components attempt to hold their respective relationships, but moving the grip to a location that would break the alignment isn't allowed.
- The triangular grip at the intersection of tangents, Grip 3, indicates a PI relationship. The curve shown is a function of these two tangents and is free to move on the basis of incoming and outgoing tangents, while still holding a radius.
- The triangular grip near the middle of the curve, Grip 4, lets the user modify the radius directly. The tangents must be maintained, so any selection that would break the alignment geometry isn't allowed.
- Circular grips on the end of the curve, Grip 5, allow the radius of the curve to be indirectly changed by changing the point of the PC of the alignment. You make this change by changing the curve length, which in effect changes the radius.

OFFSETS, THAT'S A PLUS

Offset alignments have two special grips: the arrow and the plus sign. The arrow is used to change the offset value, and the plus sign is used to create a transition such as a turning lane. Even offset alignments with turning lanes remain dynamic to their host alignment. Offset alignment objects can be found in the Prospector in the Alignments collection.

In the following exercise, you'll use grip-edits to make one of your alignments match the planner's intent more closely:

1. Open the `Editing Alignments.dwg` file.
2. Expand the Alignments branch in Prospector, right-click Parker Place, and select Zoom To.
3. Zoom in on the curve in the middle of the alignment. This curve was inserted using the default settings and doesn't match the guiding polyline well.
4. Select the alignment to activate the grips.
5. Select the triangular grip that appears near the PI as shown in Figure 7.23, and use your scroll wheel to zoom out.
6. Use an Extended Intersection snap to place the PI at the intersection of the two straight polyline segments.
7. Zoom in again on the curve. Notice that the curve still doesn't follow the polyline.
8. Select the circular grip shown in Figure 7.23, and use a Nearest snap to place it on the dashed polyline. This changes the radius without changing the PI.

Your alignment now follows the planned layout. With no knowledge of the curve properties or other driving information, you've quickly reproduced the design's intent.

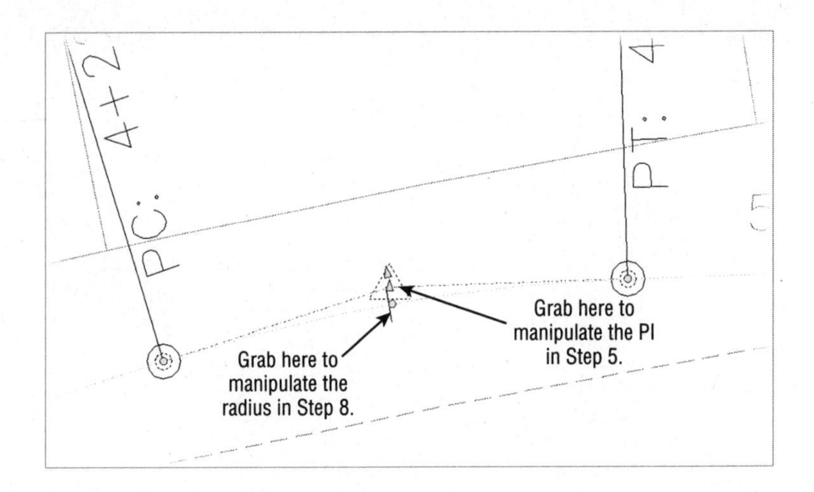

FIGURE 7.23
Grip-editing the Parker Place curve

Tabular Design

When you're designing on the basis of governing requirements, one of the most important elements is meeting curve radius requirements. It's easy to work along an alignment in a tabular view, verifying that the design meets the criteria. In this exercise, you'll verify that your curves are suitable for the design:

1. If necessary, open the `Editing Alignments.dwg` file.
2. Zoom to the Carson's Way alignment, and select it in the drawing window to activate the grips.
3. Select Geometry Editor from the Modify panel. The Alignment Layout Tools toolbar opens.
4. Select the Alignment Grid View tool, as shown in Figure 7.24.

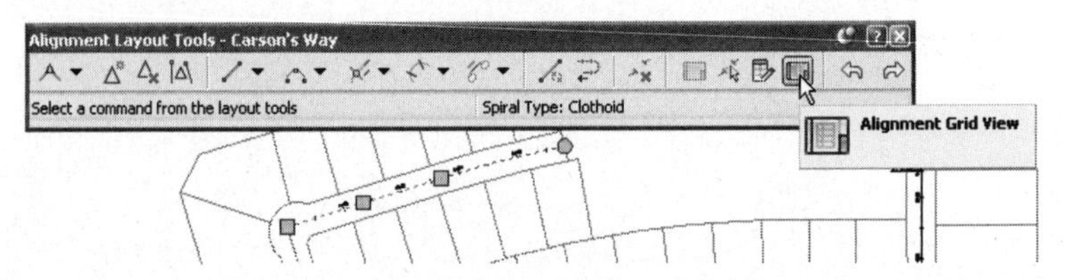

FIGURE 7.24
Selecting the Alignment Grid View tool

5. Panorama appears, with the two elements of the alignment listed along the left. You can use the scroll bar along the bottom to review the properties of the alignment if necessary. Note that the columns can be resized as well as toggled off by right-clicking the column headers.

CREATING AND SAVING CUSTOM PANORAMA VIEWS

If you right-click a column heading and select Customize near the bottom of the menu, you're presented with a Customize Columns dialog. This dialog allows you to set up any number of column views, such as Road Design or Stakeout, that show different columns. These views can be saved, allowing you to switch between views easily. This feature is a great change from previous versions where the column view changes weren't held or saved between viewings.

6. The radius for the first curve can't be edited. Remember that the location of the curve was based on the curve being tangent to the line before and passing through a point.

7. Click the check box to dismiss Panorama, and then close the toolbar.

Panorama allows for quick and easy review of designs and for precise data entry, if required. Grip-editing is commonly used to place the line and curve of an alignment in an approximate working location, but then you use the tabular view in Panorama to make the values more reasonable — for example, to change a radius of 292.56 to 300.00.

Component-Level Editing

Once an alignment gets more complicated, the tabular view in Panorama can be hard to navigate, and deciphering which element is which can be difficult. In this case, reviewing individual elements by picking them on screen can be easier:

1. Continue with the `Editing Alignments.dwg` file.

2. Zoom to Rose Drive, and select it to activate the grips.

3. Select Geometry Editor from the Modify panel. The Alignment Layout Tools toolbar appears.

4. Select the Sub-Entity Editor tool, as shown in Figure 7.25, to open the Alignment Layout Parameters dialog.

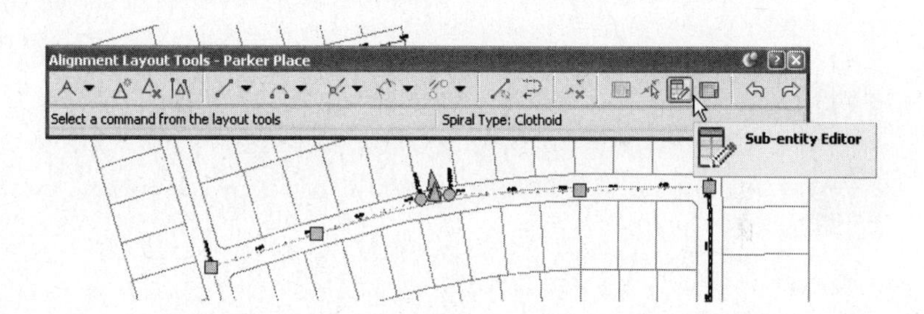

FIGURE 7.25
The Sub-Entity Editor tool

5. Select the Pick Sub-Entity tool (to the left of the Sub-Entity Editor tool) on the Alignment Layout Tools toolbar.

6. Pick the first curve on the southwest corner of the site to display its properties in the Alignment Layout Parameters dialog (see Figure 7.26). The properties are mostly grayed out, which indicates that the values for this curve are being derived from other parameters. This curve was drawn so that it would be tangent to a line and would pass through a point (Pass Through Point3), which controls every other aspect of the curve.

7. Zoom in, and pick the second curve in the reverse curve. Notice that the Radius field is now black (see Figure 7.26) and is available for editing.

8. Change the value in the Radius field to **2000**, and watch the screen update. This value is too far from the original design intent to be a valid alternative.

9. Change the value in the Radius field to **1400**, and again watch the update. This value is closer to the design and is acceptable.

10. Close the Alignment Layout Parameters dialog and the Alignment Layout Tools toolbar.

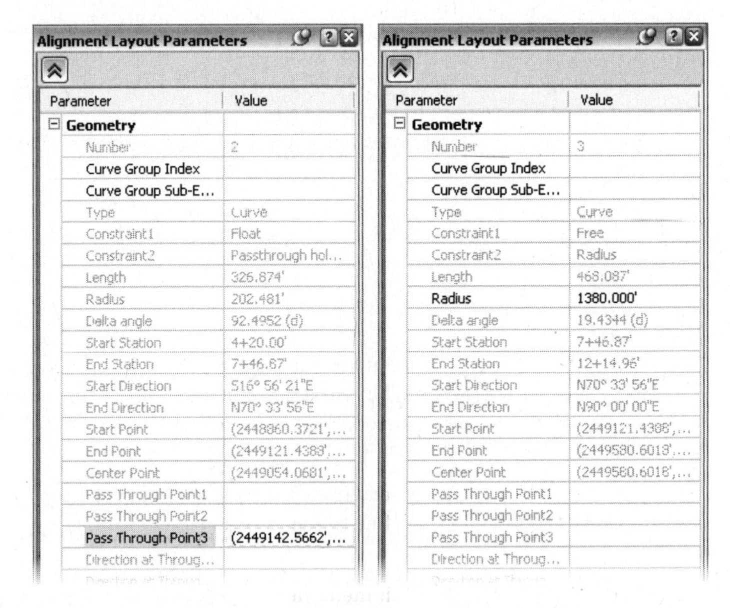

FIGURE 7.26
The Alignment Layout Parameters dialogs for the first curve (on the left) and the second curve (on the right) on Rose Drive

By using the Alignment Layout Parameters dialog, you can concisely review all the individual parameters of a component. In each of the editing methods discussed so far, you've modified the elements that were already in place. Now, let's look at changing the makeup of the alignment itself, not just the values driving it.

Changing Alignment Components

One of the most common changes is adding a curve where there was none before or changing the makeup of the curves and tangents already in place in an alignment. Other design changes can include swapping out curves for tangents or adding a second curve to smooth a transition area.

Real World Scenario

SOMETIMES THE PLANNER IS RIGHT

It turns out that your perfect reverse curve isn't allowed by the current ordinances for subdivision design! In this example, you'll go back to the design the planner gave you for the southwest corner of the site (Rose Drive) and place a tangent between the curves:

1. Open the `Editing Alignments.dwg` file if necessary.
2. Zoom to and select Rose Drive to activate the grips.
3. Select Geometry Editor from the Modify panel. The Alignment Layout Tools toolbar appears.
4. Select the Delete Sub-Entity tool.

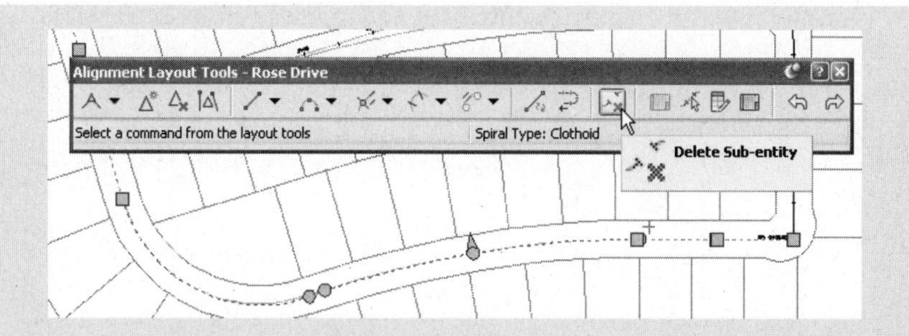

5. Pick the two curves in Rose Drive to remove them. You have to pick the eastern-most curve first, because the other curve (the floating curve) has dependencies that must be removed before it can be deleted. Note that the last tangent is still part of the alignment — it just isn't connected.
6. Select the Draw Fixed Line – Two Points tool, and snap to the endpoints of the short tangent in the red polyline to the southeast. Be sure to pick from left to right to get the direction correct.
7. From the drop-down menu next to Add Fixed Curve – Three Points, select the Free Curve Fillet (Between Two Entities, through Point) option.
8. Ctrl+click the line on the western tangent, and then Ctrl+click the short line you just created. The blue arc shown indicates the placement of the proposed fillet. In the following image, you haven't yet picked a point for the arc to pass through.

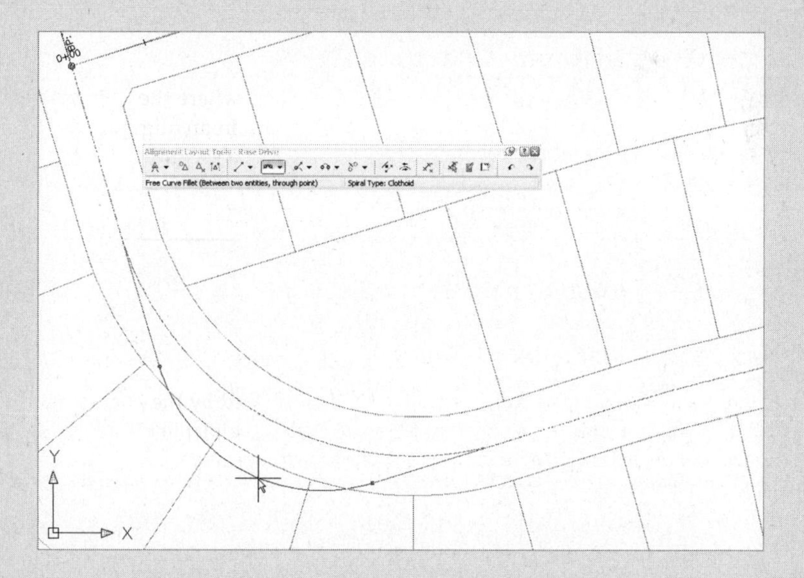

9. Use a Nearest snap, and pick a point along the arc.
10. Repeat this process to complete the other curve and connect the full alignment.
11. When you've finished, close the Alignment Layout Tools toolbar.

So far in this chapter, you've created and modified the horizontal alignments, adjusted them on screen to look like what your planner delivered, and tweaked the design using a number of different methods. Now let's look beyond the lines and arcs and get into the design properties of the alignment.

Alignments as Objects

Beyond the simple nature of lines and arcs, alignments represent other things such as highways, streams, sidewalks, or even flight patterns. All these items have properties that help define them, and many of these properties can also be part of your alignments. In addition to obvious properties like names and descriptions, you can include functionality such as superelevation, station equations, reference points, and station control. This section will look at other properties that can be associated with an alignment and how to edit them.

Renaming Objects

The default naming convention for alignments is flexible (and configurable) but not descriptive. In previous sections, you ignored the descriptions and left the default names in place, but now let's modify them. In addition, you'll learn the easy way to change the object style and how to add a description.

Most of an alignment's basic properties can be modified in Prospector. In this exercise, you'll change the name in a couple of ways:

1. Open the `Alignment Properties.dwg` file, and make sure Prospector is open.
2. Expand the Alignments/Centerline Alignments collection, and note that Alignment - (1) through Alignment - (4) are listed as members.

DIDN'T YOU ALREADY DO THIS?

Yes. You named the alignments in earlier exercises to make referencing them in the text simpler and easier to understand. Hope you'll forgive the rewind!

3. Click the Centerline Alignments branch, and the individual alignments appear in a preview area (see Figure 7.27).
4. Down in the grid-view area, click in the Name field for Alignment - (1), and pause briefly before clicking again. The text highlights for editing.
5. Change the name to **Parker Place**, and press ↵. The field updates, as does Prospector.
6. Click in the Description field, and enter a description. Press ↵.
7. Click the Style field, and the Select Label Style dialog appears. Select Layout from the drop-down menu, and click OK to dismiss. The screen updates.

That's one method. The next is to use the AutoCAD Object Properties Manager (OPM) palette:

1. Open the OPM palette by using the Ctrl+1 keyboard shortcut or some other method.
2. Select Alignment - (3) in the drawing. The OPM looks like Figure 7.28.

3. Click in the Name field, and change the name to **Rose Drive**.
4. Click in the Description field, and enter a description. Click OK.
5. Notice that the Style field for the alignment can't be changed, which somewhat limits this method.
6. Press Esc on your keyboard to deselect all objects, and close the OPM dialog if you'd like.

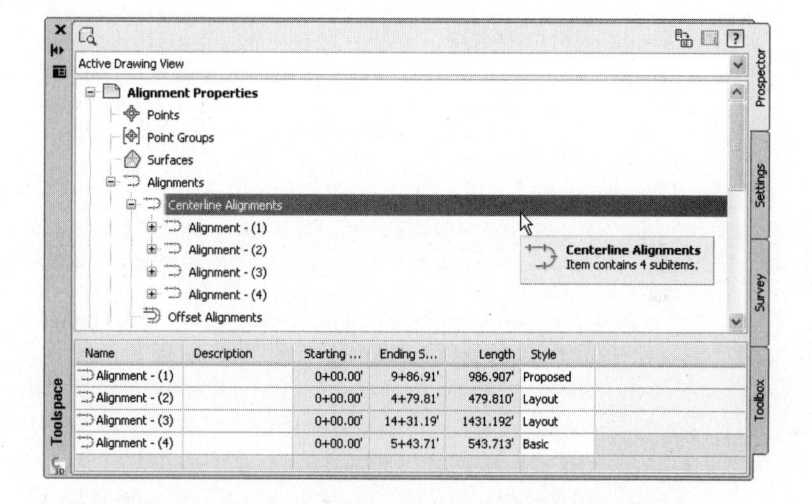

FIGURE 7.27
The Alignments collection listed in the preview area of Prospector

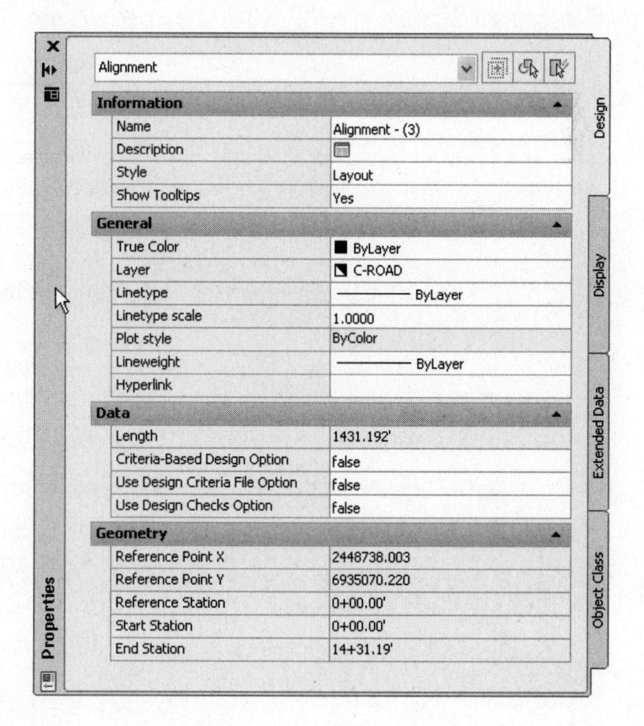

FIGURE 7.28
Alignment - (3) in the AutoCAD Object Properties Manager palette

The final method involves getting into the Alignment Properties dialog, your access point to information beyond the basics:

1. In the main Prospector window, right-click Alignment - (2), and select Properties. The Alignment Properties dialog for Alignment - (2) opens.
2. Change to the Information tab if it isn't selected.
3. Change the name to **Carson's Way**, and enter a description in the Description field.
4. Set Object Style to Existing.
5. Click Apply. Notice that the dialog header updates immediately, as does the display style in the drawing.
6. Click OK to exit the dialog.

Now that you've updated your alignments, let's make them all the same style for ease of viewing. The best way to do this is in the Prospector preview window:

1. Pick the Alignments branch, and highlight one of the alignments in the preview area.
2. Press Ctrl+A to select them all, or pick the top and then Shift+click the bottom item. The idea is to pick *all* of the alignments.
3. Right-click the Style column header and select Edit (see Figure 7.29).

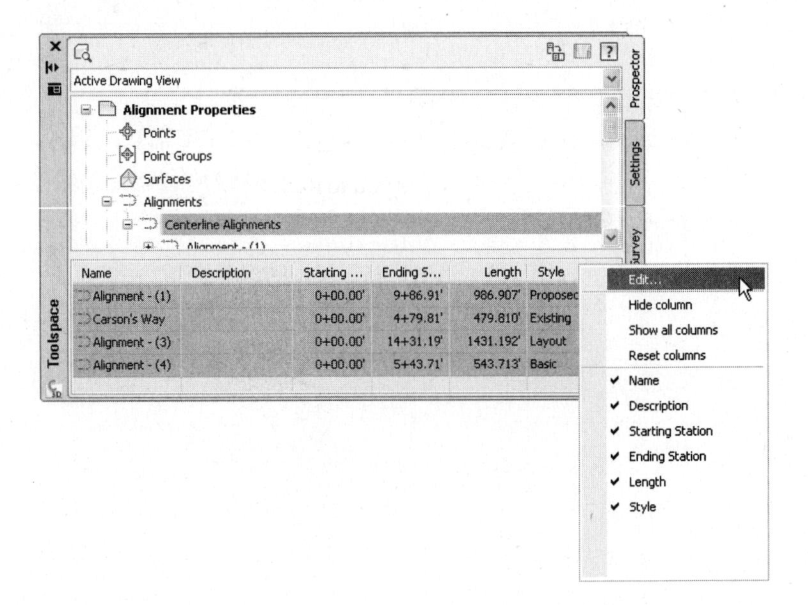

FIGURE 7.29
Editing alignment styles en masse via Prospector

4. Select Layout from the drop-down list in the Select Label Style dialog that appears, and click OK. Notice that all alignments pick up this style. Although the dialog is named Edit Label Style, you are actually editing the Object Style.
5. While you're here, change the name of Alignment - (4) to **Alexander Ave**.

DON'T FORGET THIS TECHNIQUE

This technique works on every object that displays in the List Style preview: parcels, pipes, corridors, assemblies, and so on. It can be painfully tedious to change a large number of objects from one style to another using any other method.

The alignments now look the same, and they all have a name and description. Let's look beyond these basics at the other properties you can modify and update.

The Right Station

At the end of the process, every alignment has stationing applied to help locate design information. This stationing often starts at zero, but it can also tie to an existing object and may start at some arbitrary value. Stationing can also be fixed in both directions, requiring station equations that help translate between two disparate points that are the basis for the stationing in the drawing.

One common problem is an alignment that was drawn in the wrong direction. Thankfully, Civil 3D has a quick edit command to fix that:

1. Open the `Alignment Properties.dwg file`, and make sure Prospector is open.
2. Pick Alexander Ave to the east and choose Reverse Direction from the Modify drop-down panel.
3. A warning message appears, reminding you of the consequences of such a change. Click OK to dismiss it.
4. The stationing reverses, with 0+00 now at the north end of the street.

This technique allows you to reverse an alignment almost instantly. The warning that appears is critical, though! When an alignment is reversed, the information that was derived from its original direction may not translate correctly, if at all. One prime example of this is design profiles: they don't reverse themselves when the alignment is reversed, and this can lead to serious design issues if you aren't paying attention.

Beyond reversing, it's common for alignments to not start with zero. For example, the Alexander Ave alignment is a continuation of an existing street, and it makes sense to make the starting station for this alignment the end station from the existing street. In this exercise, you'll set the beginning station:

1. Select the Alexander Ave alignment.
2. Select Alignment Properties from the Modify panel.
3. Switch to the Station Control tab. This tab controls the base stationing and lets you create station equations.
4. Enter **456.79** in the Station field in the Reference Point area (see Figure 7.30), and click Apply.
5. Dismiss the warning message that appears, and click Apply again. The Station Information area to the top right updates. This area can't be edited, but it provides a convenient way to review the alignment's length and station values.

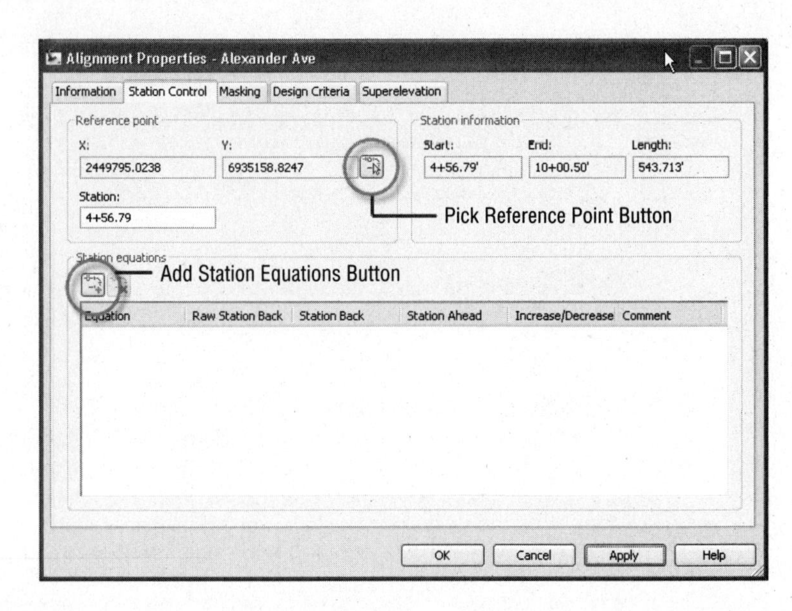

FIGURE 7.30
Setting a new starting station on the Alexander Ave alignment

In addition to changing the value for the start of the alignment, you could also use the Pick Reference Point button, circled in Figure 7.30, to select another point as the stationing reference point.

Station equations can occur multiple times along an alignment. They typically come into play when plans must match existing conditions or when the stationing has to match other plans, but the lengths in the new alignment would make that impossible without some translation. In this exercise, you'll add a station equation about halfway down Alexander Ave for illustrative purposes:

1. On the Station Control tab of the Alignment Properties dialog, click the Add Station Equations button (see the circled button in Figure 7.30).
2. Use an End snap to pick the intersection of the two alignments about halfway down Alexander Ave.
3. Change the Station Ahead value to **1000**. (Again, you're going for illustration, not reality!)
4. Click Apply, and notice the change in the Station Information area (see Figure 7.31).

Click OK to close the dialog, and review the stationing that has been applied to the alignment.

Stationing constantly changes as alignments are modified during the initial stages of a development or as late design changes are pushed back into the plans. With the flexibility shown here, you can reduce the time you spend dealing with minor changes that seem to ripple across an entire plan set.

Assigning Design Speeds

One driving part of transportation design is the design speed. Civil 3D considers the design speed a property of the alignment, which can be used in labels or calculations as needed. In this exercise, you'll add a series of design speeds to Rose Drive. Later in the chapter, you'll label these sections of the road:

1. Bring up the Alignment Properties dialog for Rose Drive to the south and east using any of the methods discussed.
2. Switch to the Design Criteria tab.
3. Click the Add Design Speed button on the top row.
4. Click in the Design Speed field for Number 1, and enter **30**. This speed is typical for a subdivision street.
5. Click the Add Design Speed button again.
6. Click in the Start Station field for Number 2. A small Pick On Screen button appears to the right of the Start Station value, as shown in Figure 7.32.
7. Click the Pick On Screen button, and then use a snap to pick the PC on the southwest portion of the site, near station 4+20.
8. Enter a value of **20** in the Design Speed field for Number 2.
9. Click the Pick On Screen button again to add one more design speed portion, and snap to the end of the short tangent.
10. Enter a value of **30** for this design speed. When complete, the tab should look like Figure 7.33.

FIGURE 7.31
Alexander Ave station equation in place

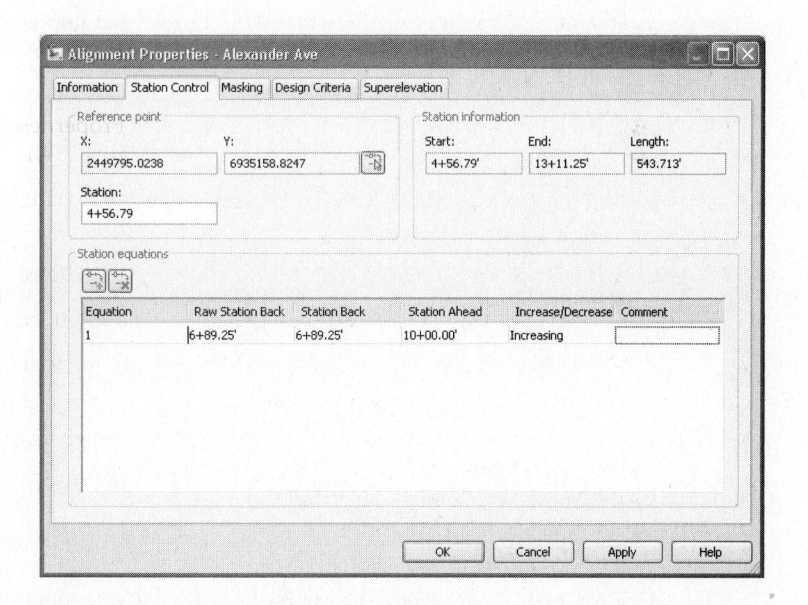

In a subdivision, these values can be inserted for labeling purposes. In a highway design, they can be used to drive the superelevation calculations that are critical to a working design. The next section looks at this subject.

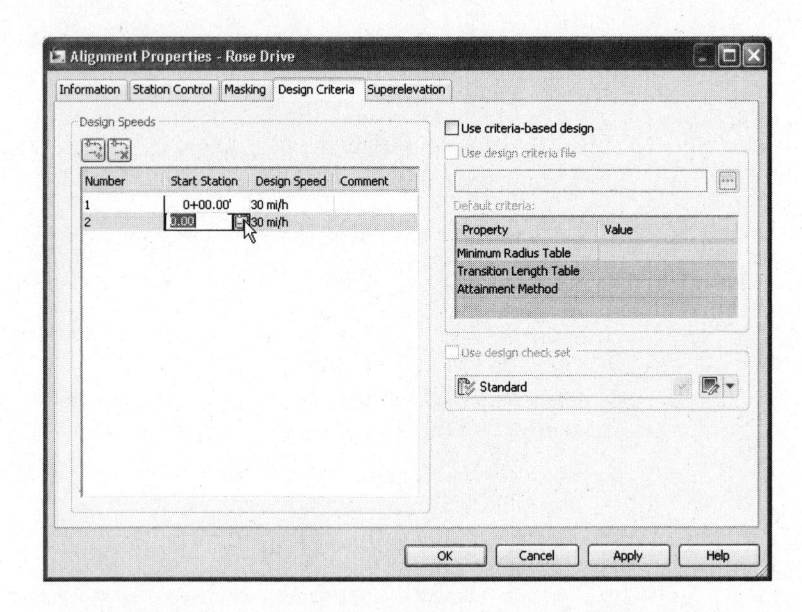

FIGURE 7.32
Setting the design speed for a Start Station field

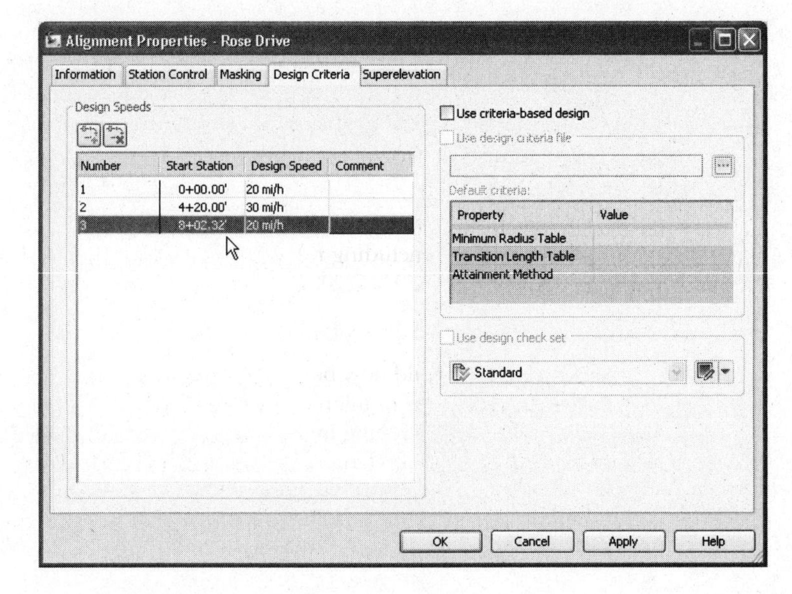

FIGURE 7.33
The design speeds assigned to Rose Drive

Banking Turn Two

Once you move beyond the basic subdivision collector street, you enter the realm of thoroughfare and highway design. In the United States, this design is governed by AASHTO manuals that dictate equations for inserting superelevation and transition zones. In this exercise, you'll create superelevation tables for an alignment:

1. Open the `Alignments with Superelevation.dwg` file. This drawing contains an alignment with a design speed of 65 mph, which is more typical of highway design.

2. Pull up the Route 66 Alignment Properties dialog, and switch to the Superelevation tab (see Figure 7.34).

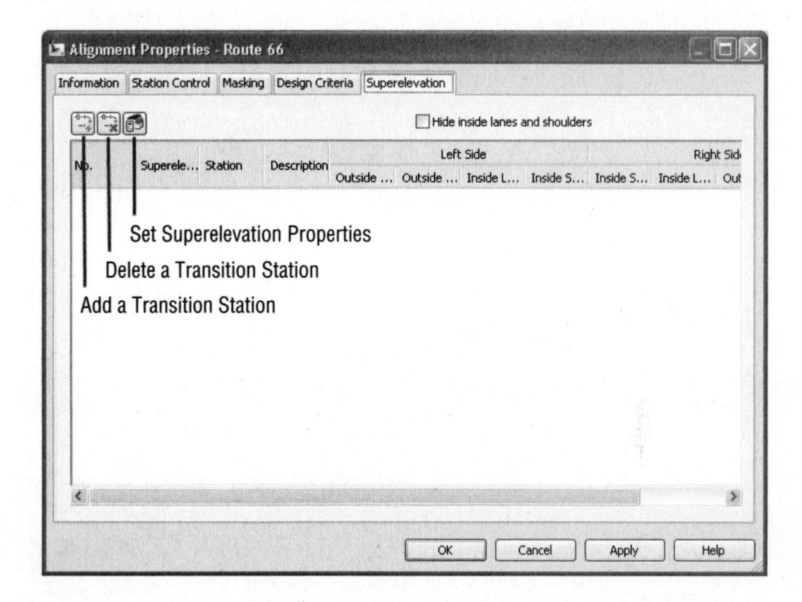

FIGURE 7.34
Assigning superelevation to the Route 66 alignment

3. Click the Set Superelevation Properties button to display the full range of options. The 2001 AASHTO manual is selected by default, but you can select other criteria from this dialog.
4. Change the Superelevation Rate Table to AASHTO 2001 eMax 8%, and click OK. Superelevation ranges, including runout areas, are created on the basis of the curves in the alignment and the AASHTO 2001 specification.
5. Click OK to close the dialog.

This information could now be used by corridor assemblies to determine cross slope for lanes, shoulders, or other areas of interest. Additionally, in simpler design cases, you can accomplish transitions manually by typing in slope amounts at critical stations. Finally, if you need to create your own criteria, a Design Criteria Editor is available under the Alignments menu.

All the information applied to an alignment is great, but if you don't put that information on paper, it's useless. Let's focus now on labeling this information and making labels efficiently.

SUPERELEVATION: IT'S NOT JUST FOR HIGHWAYS

Don't get trapped into thinking that just because you don't do highway design, you can't use some of the same tools. Some firms use superelevation to help design waterslides. The fun in learning Civil 3D is finding new ways to use the tools given to solve one problem to solve another problem entirely. You'll know you've mastered the software when you find yourself abusing the tools in this manner.

Styling Alignments

You'll deal with three major areas in setting up alignment styles:

- The alignment
- Alignment labels, including
  - Label sets
  - Station
  - Station offsets
  - Lines
  - Curves
  - Spirals
  - Tangent intersections
  - Vertical data (high and low points)
- Table styles with options for
  - Line
  - Curve
  - Spiral
  - Segment

This breakdown follows exactly the same format as the Settings tab of Toolspace, so let's look at them the same way. You can manipulate each of these label styles in similar ways, so this section covers most of them but not every single one.

The Alignment

The alignment style controls how the actual alignment appears. This is outside the scope of the labeling or layering that may be in play. As you've seen in the previous portions of this chapter, an alignment style can dramatically alter the appearance of the simplest alignments, such as when you used the Layout style to show lines in red and arcs in blue.

Civil 3D ships with a number of alignment styles, but they're pretty generic. In this exercise, you'll copy the existing style and modify it to create a plot style that you might use in your construction documents:

1. Open the `Alignment Styles.dwg` file. All of the alignments currently have the Layout style applied.
2. On the Settings tab, expand the Alignment ➢ Alignment Styles branch.
3. Right-click Existing, and select Copy.
4. On the Information tab, change the name in the Name field to **Offsets**, and enter a description. You'll work your way through the tabs to build your style.

5. Switch to the Markers tab, where you have the option to display marker objects at all the major critical points. You'll add the Begin of Alignment and End of Alignment points to the Alignment Geometry style.

6. Click the small Style icon to the right of the Marker Style field for Begin of Alignment (see Figure 7.35), and select Alignment Geometry from the drop-down list in the Pick Marker Style dialog. These Marker Styles are established in the General branch of the Settings tab.

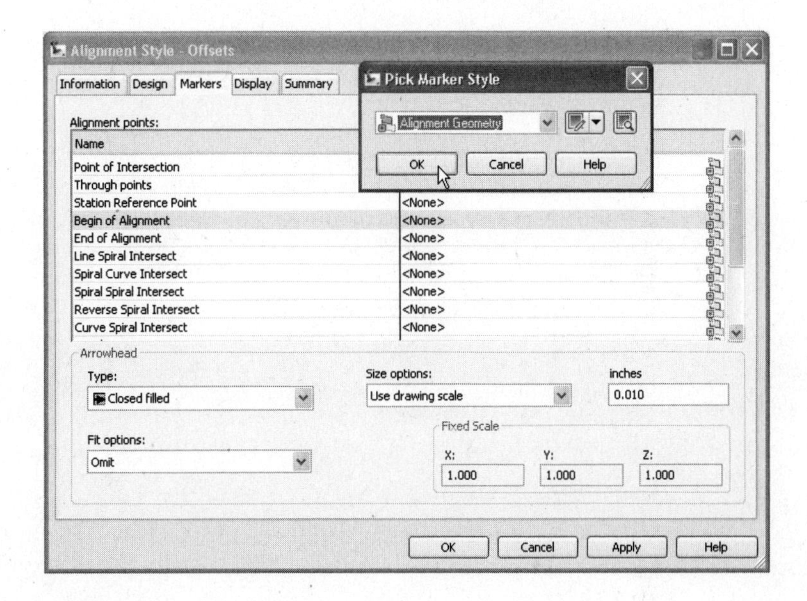

FIGURE 7.35
Selecting the marker style

7. Click OK, and close the dialog.

8. Repeat these three steps for End of Alignment.

9. Click OK to close the Alignment Style dialog.

Your new style appears in the list of alignment styles. Go ahead and change all the alignments to use this style, using the techniques covered earlier in this chapter. Changing the alignment style is fairly straightforward. As you think about your office, you might build alignment styles for various types of roads, pipe networks, sidewalks, or trails — any number of things.

MATCHMAKER

The AutoCAD MATCHPROP (Match Properties) command can also be used to match alignment styles. Although this command doesn't currently work on every object type, it works on several. The Match Properties tool can be found on the Modify panel of the Home tab. Paint the town!

Labeling Alignments

Labeling in Civil 3D is one of the program's strengths, but it's also an easy place to get lost. There are myriad options for every type of labeling situation under the sun, and keeping them straight can be difficult. In this section, you'll begin by building label styles for stationing along an alignment, culminating in a label set. Then, you'll create styles for station and offset labels, using reference text to describe alignment intersections. Finally, you'll add a street-name label that makes it easier to keep track of things.

The Power of Label Sets

When you think about it, any number of items can be labeled on an alignment, before getting into any of the adjoining objects. These include major and minor stations, geometry points, design speeds, and profile information. Each of these objects can have its own style. Keeping track of all these individual labeling styles and options would be burdensome and uniformity would be difficult, so Civil 3D features the concept of *label sets*.

A label set lets you build up the labeling options for an alignment, picking styles for the labels of interest, or even multiple labels on a point of interest, and then save them as a set. These sets are available during the creation and labeling process, making the application of individual labels less tedious. Out of the box, a number of sets are available, primarily designed for combinations of major and minor station styles along with geometry information.

You'll create individual label styles over the next couple of exercises and then pull them together with a label set. At the end of this section, you'll apply your new label set to the alignments.

Major Station

Major station labels typically include a tick mark and a station callout. In this exercise, you'll build a style to show only the station increment and run it parallel to the alignment:

1. Open `Alignment Labels.dwg` file.
2. Switch to the Settings tab, and expand the Alignment ➢ Label Styles ➢ Station ➢ Major Station branch.
3. Right-click the Parallel with Tick style, and select Copy. The Label Style Composer dialog appears.
4. On the Information tab, change the Name field to **Station Index Only**.
5. Switch to the Layout tab.
6. Click in the Contents Value field, under the Text property, and then click the ellipsis button to open the Text Component Editor dialog.
7. Click in the preview area, and delete the text that's already there.
8. Click in the Output Value field, and click the down arrow to open the drop-down list.
9. Select the Left of Station Character option, as shown in Figure 7.36 (you may have to scroll down). Click the insert arrow circled in the figure.

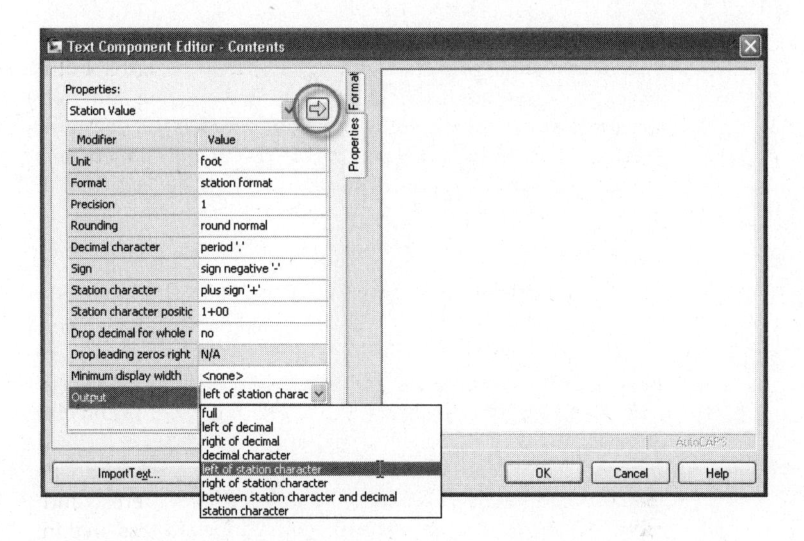

FIGURE 7.36
Modifying the Station Value Output value in the Text Component Editor dialog

10. Click OK to close the Text Component Editor dialog.
11. Click OK to close the Label Style Composer dialog.

The label style now shows in your label styles, but it's not applied to any alignments yet.

Geometry Points

Geometry points reflect the PC, PT, and other points along the alignment that define the geometric properties. The existing label style doesn't reflect a plan-readable format, so you'll copy it and make a minor change in this exercise:

1. Expand the Alignment ➤ Label Styles ➤ Station ➤ Geometry Point branch.
2. Right-click the Perpendicular with Tick and Line style, and select Copy to open the Label Style Composer dialog.
3. On the Information tab, change the name to **Perpendicular with Line**, and change the description, removing the circle portion.
4. Switch to the General tab.
5. Change the Readability Bias setting to **90**. This value will force the labels to flip at a much earlier point.
6. Switch to the Layout tab.
7. Set the Component Name field to the Tick option.
8. Click the Delete Component button (the red X button).
9. Click OK to close the Label Style Composer dialog.

This new style flips the plan-readable labels sooner and removes the circle tick mark.

Label Set

In contrast to the prebuilt styles, you can build your label set on the basis of its designed use. As a result, it's easier to pick it from a list than if it were being picked on the basis of a combination of its components. This exercise builds a Paving label set from scratch, but you could copy a similar label set and modify it for future sets:

1. Expand the Alignment ➤ Label Styles ➤ Label Sets branch.
2. Right-click Label Sets, and select New to open the Alignment Label Set dialog.
3. On the Information tab, change the name to **Paving**.
4. Switch to the Labels tab.
5. Set the Type field to the Major Stations option and the Major Station Label Style field to the Station Index Only style that you just created, and click the Add button.
6. Set the Type field to the Minor Stations option and the Minor Station Label Style field to the Tick option, and click the Add button.
7. Set the Type field to Geometry Points and the Geometry Point Label Style field to Perpendicular with Line, and click the Add button to open the Geometry Points dialog as shown Figure 7.37. Deselect the Alignment Beginning and Alignment End options as shown. Click OK to dismiss the dialog.

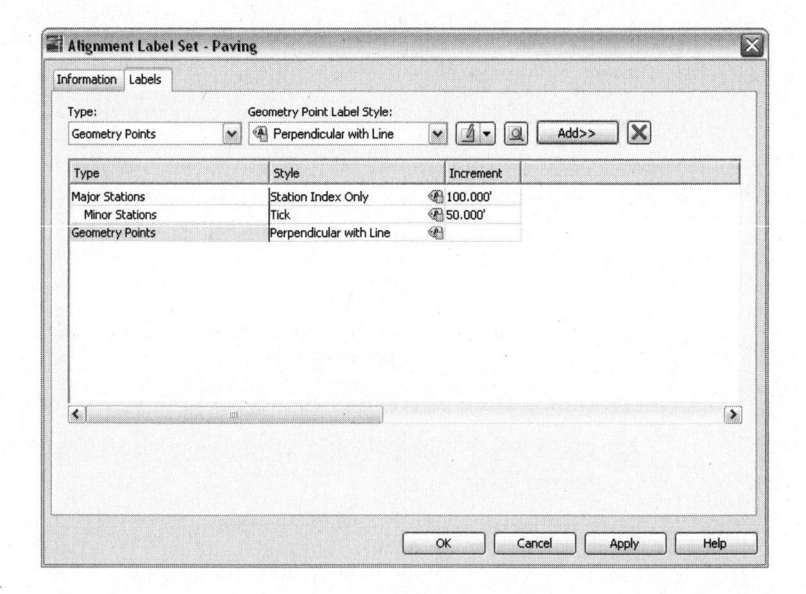

FIGURE 7.37
Deselecting the Label Alignment Beginning and Label Alignment End Geometry Point Options

8. Review the settings to make sure they match Figure 7.38, and then click OK to dismiss the Alignment Label Set dialog.

You've built a new label set that you can apply to paving alignment labels. In early versions of Civil 3D, the labels were part of the alignment, which sometimes made it difficult to get labels the way you wanted. Now, the label set is a different object. Enter the LIST command, and pick a label. You'll see a reference to a label group instead of an alignment object. What does this mean to you as an end user? A couple of things: first, you can use the AutoCAD properties to set label styles for individual groups; and second, the Labeling tab in the Alignment Properties dialog is no more.

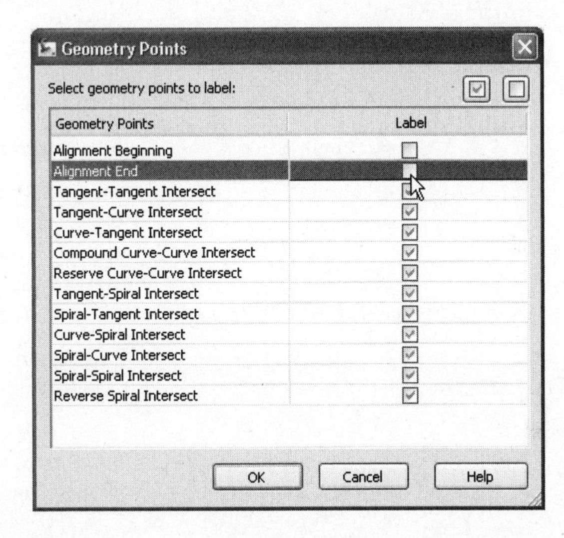

FIGURE 7.38
The completed Paving alignment label set

In this exercise, you'll apply your label set to all of your alignments and then see how an individual label can be changed from the set:

1. Select the Rose Drive alignment on screen. To find it easily, simply hover your cursor over each of the alignments until the tooltip is displayed and note the alignment name, style, layer, and station values.
2. Right-click, and select Edit Alignment Labels to display the Alignment Labels dialog shown in Figure 7.39.

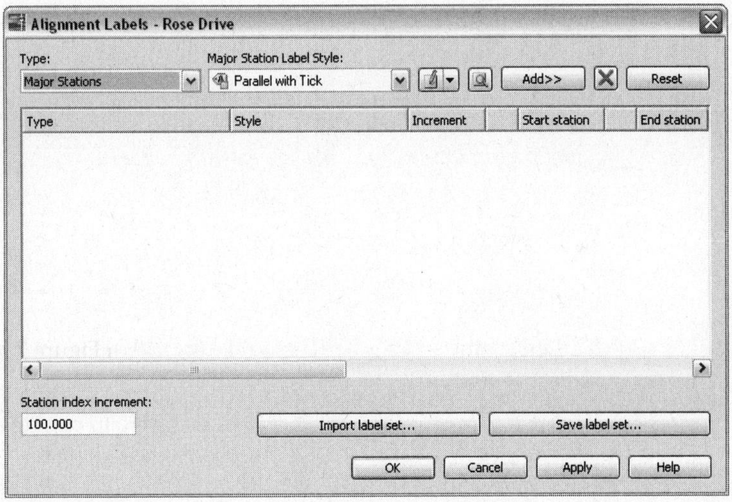

FIGURE 7.39
The Alignment Labels dialog for Rose Drive

3. Click the Import Label Set button near the bottom of this dialog.
4. In the Select Style Set drop-down list, select the Paving Label Set and click OK.
5. The Style field for the alignment labels populates with the option you selected.
6. Click OK to dismiss the dialog.
7. Repeat this process across the rest of the alignments.
8. When you've finished, zoom in on any of the major station labels.
9. Hold down the Ctrl key, and select the label. Notice that a single label is selected, not the label set group.
10. Right-click and select Label Properties.
11. The Label Properties dialog appears, allowing you to pick another label style from the drop-down list.
12. Change the Label Style value to Parallel with Tick, and change the Flip Label value to True.
13. Press Esc to deselect the label item and exit this dialog.

By using alignment label sets, you'll find it easy to standardize the appearance of labeling and stationing across alignments. Building label sets can take some time, but it's one of the easy, effective ways to enforce standards.

CTRL+CLICK? WHAT IS THAT ABOUT?

Prior to AutoCAD Civil 3D 2008, clicking an individual label picked the label and the alignment. Because labels are part of a label set object now, Ctrl+click is the *only* way to access the Flip Label and Reverse Label functions!

STATION OFFSET LABELING

Beyond labeling an alignment's basic stationing and geometry points, you may want to label points of interest in reference to the alignment. Station offset labeling is designed to do just that. In addition to labeling the alignment's properties, you can include references to other object types in your station offset labels. The objects available for referencing are as follows:

- Alignments
- COGO points
- Parcels
- Profiles
- Surfaces

In this exercise, you'll use an alignment reference to create a label suitable for labeling the intersection of two alignments. It will pick up the stationing information from both:

1. Open the `Alignment Styles.dwg` file.
2. On the Settings tab, expand Alignment ➢ Label Styles ➢ Station Offset.

3. Right-click Station and Offset, and select Copy to open the Label Style Composer dialog.
4. On the Information tab, change the name of your new style to **Alignment Intersection**.
5. Switch to the Layout tab. In the Component Name field, delete the Marker component.
6. In the Component Name field, select the Station Offset component.
7. Change the Name field to **Main Alignment**.
8. In the Contents Value field, click the ellipsis button to bring up the Text Component Editor.
9. Select the text in the preview area, and delete it all.
10. Type **Sta.** in the preview area; be sure to leave a space after the period.
11. In the Properties drop-down field, select Station Value.
12. Click the insert arrow, press the right arrow to move your cursor to the end of the line, and type one space.
13. In the Properties drop-down field, select Alignment Name.
14. Click the insert arrow to add this bit of code to the preview.
15. Click your mouse in the preview area, or press the right arrow or End key. Move to the end of the line, and type an equal sign (=).
16. Click OK to return to the Label Style Composer dialog, shown in Figure 7.40.

FIGURE 7.40
Alignment text changed to the new values

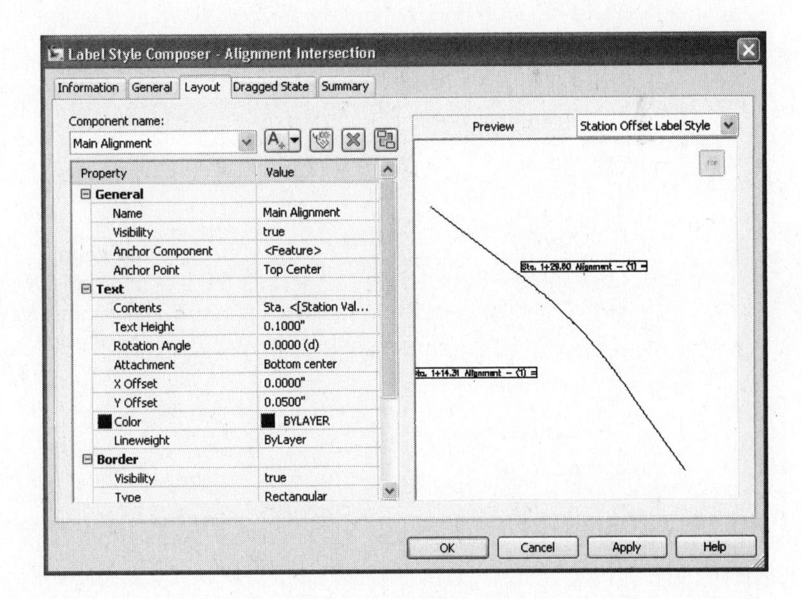

17. Under the Border Property, set the Visibility field to False.
18. Select Reference Text from the drop-down list next to the Add Component tool (see Figure 7.41).

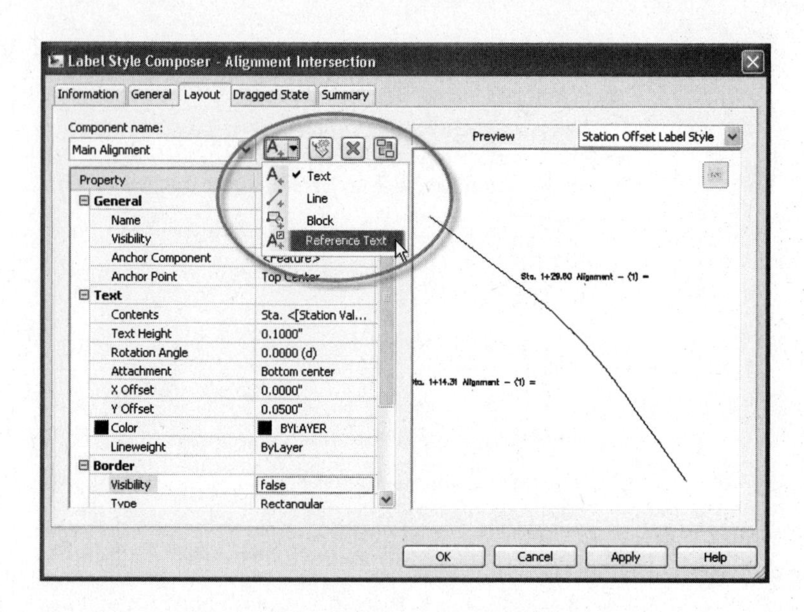

FIGURE 7.41 Adding a Reference Text component to a label

19. In the Select Type dialog that appears, select Alignment and click OK.
20. Change the Name field to **Intersecting Alignment**.
21. In the Anchor Component field, select Main Alignment.
22. In the Anchor Point field, select Bottom Left.
23. In the Attachment field, select Top Left. By choosing the anchor point and attachment point in this fashion, the bottom left of the Main Alignment text is linked to the top left of the Intersection alignment text.
24. Click in the Contents field, and click the ellipsis button to open the Text Component Editor.
25. Move your cursor over the ''Label Text'' that appears. Pick three times with the left mouse button to highlight both words and then delete them.
26. Type **Sta.** in the preview area; be sure to leave a space after the period.
27. In the Properties drop-down list, select Station Value.
28. Click the insert arrow, move your cursor to the right with the arrow or End key, and add a space.
29. In the Properties drop-down list, select Alignment Name.
30. Click the insert arrow to add this bit of code to the preview.
31. Click OK to exit the Text Component Editor, and click OK again to exit the Label Style Composer dialog.

Wow, that seemed like a lot of work for one label! But if you never have to rebuild that label again, then it's worth it, right? With the Station Offset label type, you have two options:

- Station Offset is used for labeling points that are important because of where they are along an alignment, such as a buffer area or setback. In these cases, the station along the alignment is the driving force in where the label occurs.
- Station Offset-Fixed Point is used for labeling points that are fixed in space, such as for a fire hydrant or curb return. In the case of the Fixed Point option, the point being labeled isn't dependent on the alignment station for relevance but for location.

With those out of the way, you can test your new label and pick up all the intersections of the streets. You'll use Rose Drive as your main alignment and work your way along it:

1. Go to the Annotate tab of the ribbon. From the Add Labels pulldown, select Alignment ➤ Add Alignment Labels ➤ Add Alignment Labels. The Add Labels dialog appears.
2. In the Label Type drop-down list, select Station Offset.
3. In the Station Offset Label Style drop-down list, select Alignment Intersection.
4. Leave the Marker Style field alone, but remember that you could use any of these styles to mark the selected point.
5. Click the Add button.
6. Pick the Rose Drive alignment.
7. Snap to the endpoint at the far northwest end.
8. Enter **0** for the offset amount, and press ↵.
9. The command line prompts you to `Select Alignment for Label Style Component Intersecting Alignment`. Pick the Carson's Way alignment.
10. Click the Add button again, and repeat the process at the other two alignment intersections.

There are two things to note in this process: first, you click Add between adding labels because Civil 3D otherwise assumes you want to use the same reference object for every instance of the label; second, the labels are sitting right on the point of interest. Drag them to a convenient location, and you're set to go. When you do this, your label should look something like the one shown in Figure 7.42.

Using station offset labels and their reference object ability, you can label most site plans quickly with information that dynamically updates. Because of the flexibility of labels in terms of style, you can create "design labels" that are used to aid in modeling yet never plot and aren't seen in the final deliverables. This is ideal in corridor modeling as discussed in Chapter 12, "The Road Ahead: Advanced Corridors."

SEGMENT LABELING

Every land-development professional has a story about the developer who named an entire subdivision after their kids, grandkids, dogs, golf buddies, favorite bars, and so on. As these plans work their way through reviewing agencies, there are inevitably changes, and the tedium of changing a street name on 45 pages of construction documents can't be described.

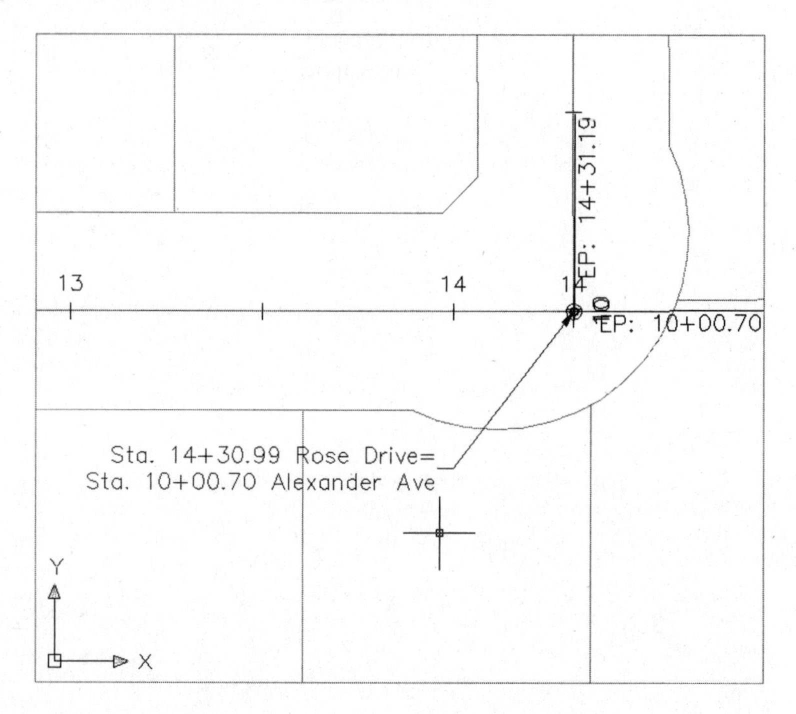

FIGURE 7.42
The Alignment Intersection label style in use

Thankfully, as you've already seen in the station offset label, you can access the properties of the alignment to generate a label. In this exercise, you'll use that same set of properties to create street-name labels that are applied and always up to date:

1. Open the `Segment Labeling.dwg` file.
2. In the Settings tab, expand the Alignment ➢ Label Styles ➢ Line branch.
3. Right-click Line, and select New. The Label Style Composer dialog appears.
4. On the Information tab, change the Name field to **Street Names**. Switch to the General tab.
5. Click in the Layer Value field, and click the ellipsis button. The Layer Selection dialog appears.
6. Select the layer C-ROAD-LABL, and click OK to close the Layer Selection dialog.
7. Change the Readability Bias Value field to 90(d). Switch to the Layout tab.
8. In the Component Name drop-down list, delete the Direction Arrow and Distance components by clicking the red X button.
9. In the Component Name drop-down list, select the Bearing component, and change its name in the Value field to **Street Name**.
10. Click the Contents field, and click the ellipsis button. The Text Component Editor appears.

11. Delete the entire preview contents.
12. In the Properties drop-down list, select Alignment Name.
13. Set the Capitalization field to Upper Case. By forcing capitalization, you can standardize the way street names appear without having to double-check every bit of user input.
14. Click the insert arrow, and then click OK to exit the Text Component Editor.
15. Click OK again to exit the Label Style Composer dialog.
16. Change to the Annotate tab and choose the Add Labels button on the Labels & Tables panel to display the Add Labels dialog shown in Figure 7.43.

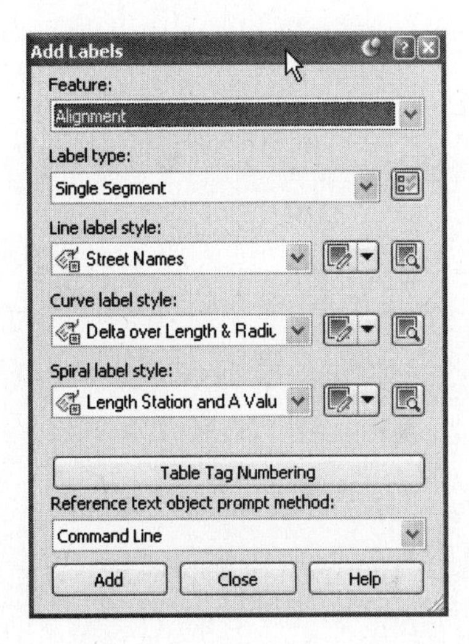

FIGURE 7.43
The Add Labels dialog

17. In the Label Type field, select Single Segment from the drop-down list. Then, in the Line Label Style field, select your Street Names style. Note that curves and spirals have their own set of styles. If you click a curve during the labeling process, you won't get a street name — you'll get something else.
18. Click the Add button.
19. Pick various line segments around the drawing. Each street is labeled with the appropriate name.
20. Click the Close button to close the Add Labels dialog.

Days of work averted! The object properties of an alignment can be invaluable in documenting your design. Creating a collection of styles for all the various components and types takes time but pays you back in hours of work saved on every job.

Alignment Tables

There isn't always room to label alignment objects directly on top of them. Sometimes doing so doesn't make sense, or a reviewing agency wants to see a table showing the radius of every curve in the design. Documentation requirements are endlessly amazing in their disparity and seeming randomness. Beyond labels that can be applied directly to alignment objects, you can also create tables to meet your requirements and get plans out the door.

You can create four types of tables:

- Lines
- Curves
- Spirals
- Segments

Each of these is self-explanatory except perhaps the Segments table. That table generates a mix of all the lines, curves, and spirals that make up an alignment, essentially re-creating the alignment in a tabular format. In this section, you'll generate a new line table and draw the segment table that ships with the product.

All the tables work in a similar fashion. Go to the Annotate tab of the Ribbon. From the Add Tables pull-down, select Alignment, and then pick a table type that is relevant to your work. The Table Creation dialog appears (see Figure 7.44).

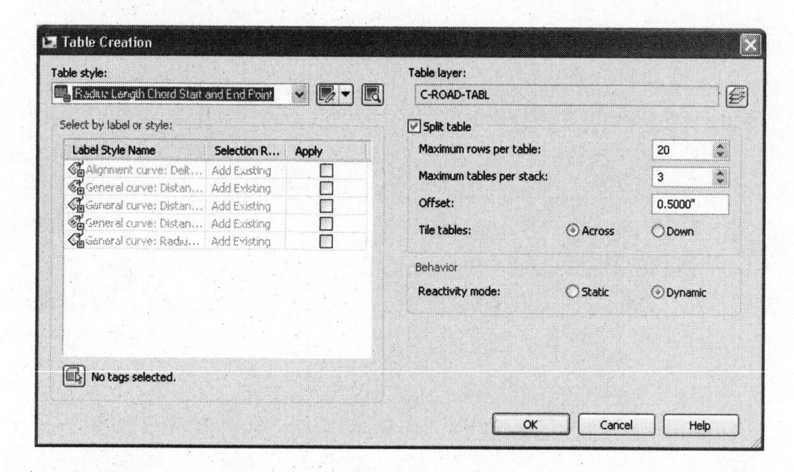

FIGURE 7.44
The Table Creation dialog

You can select a table style from the drop-down list or create a new one. Select a table layer by clicking the blue arrow. The selection area determines how the table is populated. All the label-style names for the selected type of component are presented, with a check box to the right of each one. Applying one of these styles enables the Selection Rule, which has the following two options:

Add Existing Any label using this style that currently exists in the drawing is converted to a tag format, substituting a key number such as L1 or C27, and added to the table. Any labels using this style created in the future will *not* be added to the table.

Add Existing and New Any label using this style that currently exists in the drawing is converted to a tag format and added to the table. In addition, any labels using this style created in the future will be added to the table.

To the right of the Select area is the Split Table area, which determines how the table is stacked up in Model space once it's populated. You can modify these values after a table is generated, so it's often easier to leave them alone during the creation process.

Finally, the Behavior area provides two selections for the Reactivity Mode: Static and Dynamic. These selections determine how the table reacts to changes in the driving geometry. In some cases in surveying, this disconnect is used as a safeguard to the platted data; but in general, the point of a 3D model is to have live labels that dynamically react to changes in the object.

Before you draw any tables, you need to apply labels so the tables will have data to populate. In this exercise, you'll throw some labels on your alignments, and then you'll move on to drawing tables in the next sections:

1. Open the `Segment Labeling.dwg` file.
2. Change to the Annotate tab and choose Add Labels from the Labels & Tables panel to open the Add Labels dialog.
3. In the Feature field, select Alignment, and then in the Label Type field, select Multiple Segment from the drop-down list. With this option, you'll click each alignment one time, and every subcomponent will be labeled with the style selected here.
4. Verify that the Line Label Style field (not the General Line Label Style) is set to Bearing Over Distance. You won't be left with these labels — you just want them for selecting elements later.
5. Click Add, and select all four alignments.
6. Click Close to close the Add Labels dialog.

Now that you've got labels to play with, let's build some tables.

CREATING A LINE TABLE

Most line tables are simple: a line tag, a bearing, and a distance. You'll also see how Civil 3D can translate units without having to change anything at the drawing level:

1. On the Settings tab, expand the Alignment ➤ Alignment Styles ➤ Table Styles branch.
2. Right-click Line, and select New to open the Table Style dialog.
3. On the Information tab, change the Name field to **Bearing & Distance (ft + m)**. Switch to the Data Properties tab shown in Figure 7.45.
4. Click in the Start Point column header. Your cursor will change temporarily to look like a hand, and the Start Point column button will highlight. Click the red X button at right to delete the column.
5. Repeat this step for the End Point column.
6. Double-click the Length column header to bring up the Text Component Editor.
7. Add a space and then **(ft.)** to the end of the text already in place.
8. Click OK to close the editor.
9. Click the + button at right in the Table Style dialog to add an additional column.
10. Double-click the header to bring up the Text Component Editor.
11. Title it **Length (m)** in the preview area. Click OK to close the editor.

FIGURE 7.45
Before the table edits

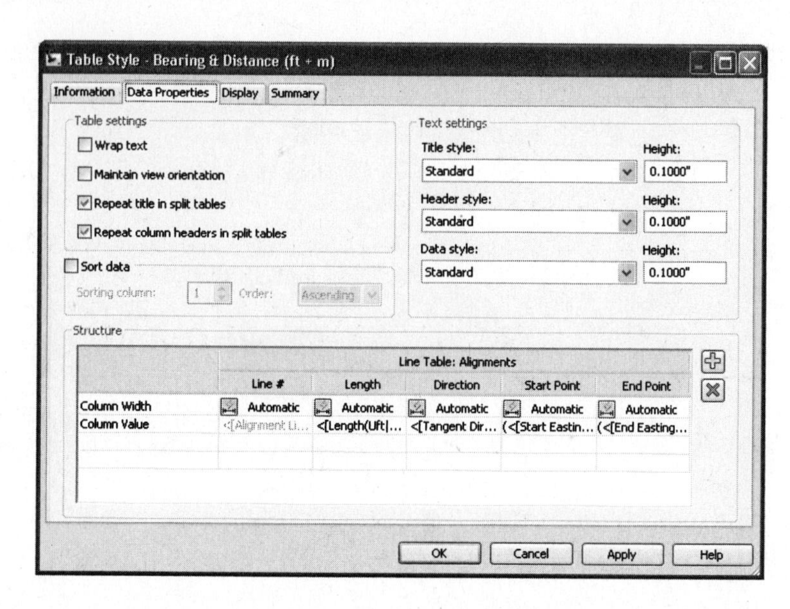

12. Double-click the Column Value field under the new column to bring up the Text Component Editor.
13. Set the Properties drop-down list to Length, and change the Units Value field to Meter.
14. Click the insert arrow, and click OK to close the editor.
15. Click and drag the Direction column header to the left until a small table icon appears. This indicates rearranging columns. Place the Direction column third. Your table should look like Figure 7.46.

FIGURE 7.46
Completed table edits

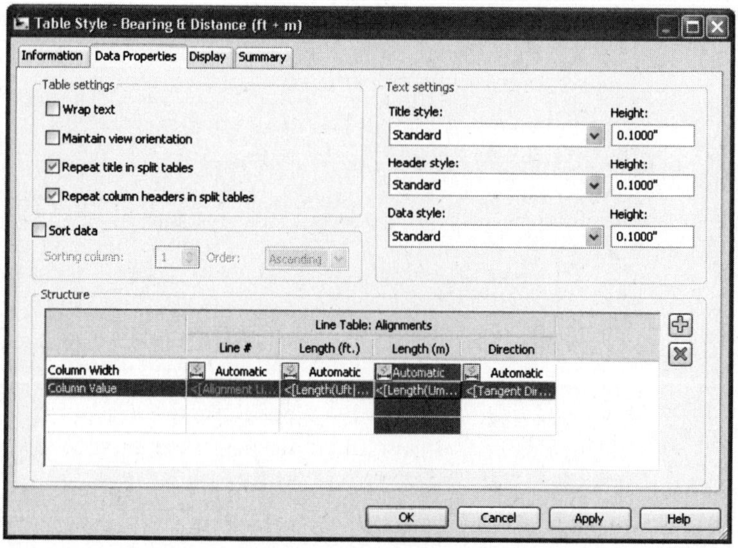

16. Switch to the Display tab, and turn off the display of the three fill components by clicking the lightbulbs.

17. Click OK to close the dialog.

18. Choose the Add Tables drop-down arrow from the Labels & Tables panel, and choose Alignment ➤ Add Line to open the Table Creation dialog.

19. Set the dialog options as shown in Figure 7.47, click the Pick On-Screen button in the lower right of the dialog, and click OK.

FIGURE 7.47
Creating a line table

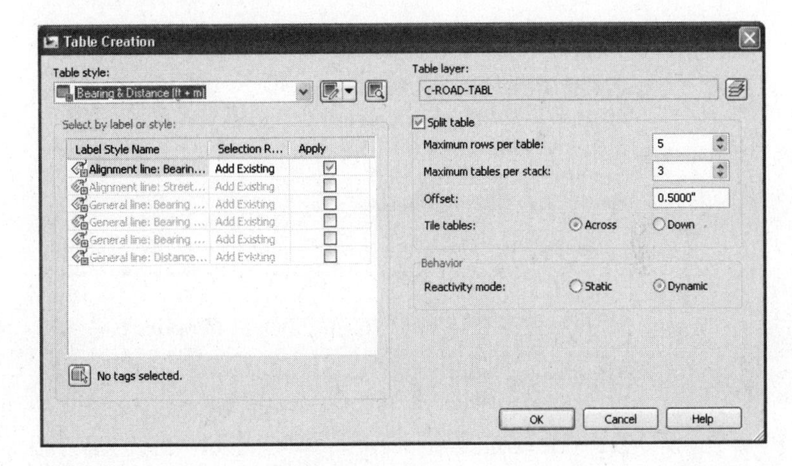

20. Pick a point on screen, and the table will generate.

Pan back to your drawing, and you'll notice that the line labels have turned into tags on the line segments. After you've made one table, the rest are similar. Be patient as you create tables — a lot of values must be tweaked to make them look just right. By drawing one on screen and then editing the style, you can quickly achieve the results you're after.

AN ALIGNMENT SEGMENT TABLE

An individual segment table allows a reviewer to see all the components of an alignment. In this exercise, you'll draw the segment table for Rose Drive:

1. Choose the Add Tables drop-down arrow from the Labels & Tables panel, and choose Alignment ➤ Add Segment to open the Table Creation dialog.

2. In the Select Alignment field, choose the Rose Drive alignment from the drop-down list, and click OK.

3. Pick a point on the screen, and the table will be drawn.

Note that a number of segments seem to be exactly the same! Because you previously made a table that changed the existing label styles and then built this table on the basis of the Rose Drive alignment, the lines have duplicate tags. If you erase some of these tags, you can update this table to be correct. You'll also renumber for ease of use:

1. Move along Rose Drive, making sure each segment has only one tag. Remove any duplicated tags by erasing them.

2. Select one of the tag labels and choose the Renumber Tags tool from the Labels & Tables panel.
3. Pick each of the line tags along the alignments.
4. Right-click to exit the command.

Both tables update to reflect the new numbering scheme and still reflect the properties of each segment.

The Bottom Line

Create an alignment from a polyline. Creating alignments based on polylines is a traditional method of building engineering models. With Civil 3D's built-in tools for conversion, correction, and alignment reversal, it's easy to use the linework prepared by others to start your design model. These alignments lack the intelligence of crafted alignments, however, and you should use them sparingly.

Master It Open the `Mastering Alignments.dwg` file, and create alignments from the linework found there.

Create a reverse curve that never loses tangency. Using the alignment layout tools, you can build intelligence into the objects you design. One of the most common errors introduced to engineering designs is curves and lines that aren't tangent, requiring expensive revisions and resubmittals. The free, floating, and fixed components can make smart alignments in a large number of combinations available to solve almost any design problem.

Master It Open the `Mastering Alignments.dwg` file, and create an alignment from the linework on the right. Create a reverse curve with both radii equal to 200 and with a passthrough point in the center of the displayed circle.

Replace a component of an alignment with another component type. One of the goals in using a dynamic modeling solution is to find better solutions, not just a solution. In the layout of alignments, this can mean changing components out along the design path, or changing the way they're defined. Civil 3D's ability to modify alignments' geometric construction without destroying the object or forcing a new definition lets you experiment without destroying the data already based on an alignment.

Master It Convert the arc indicated in the `Mastering Alignments.dwg` file to a free arc that is a function of the two adjoining segments. The curve radius is 150′.

Create a new label set. Label sets let you determine the appearance of an alignment's labels and quickly standardize that appearance across all objects of the same nature. By creating sets that reflect their intended use, you can make it easy for a designer to quickly label alignments according to specifications with little understanding of the requirement.

Master It Within the `Mastering Alignments.dwg` file, create a new label set containing only major station labels, and apply it to all the alignments in that drawing.

Override individual labels with other styles. In spite of the desire to have uniform labeling styles and appearances between alignments within a single drawing, project, or firm, there are always exceptions. Using AutoCAD's Ctrl+click method for element selection, you can access commands that let you modify your labels and even change their styles.

Master It Create a copy of the Perpendicular with Tick Major Station style called Major with Marker. Change the Tick Block Name to Marker Pnt. Replace some (but not all) of your major station labels with this new style.

Chapter 8

Cut to the Chase: Profiles

Profile information is the backbone of vertical design. Civil 3D takes advantage of sampled data, design data, and external input files to create profiles for a number of uses. Even the most basic designs require profiles. In this chapter you'll look at creation tools, editing profiles, and display styles, and you'll learn about ways to get your labels just so. Profile views are a different subject and will be covered in more detail in the next chapter.

In this chapter, you'll learn to:

- Sample a surface profile with offset samples
- Lay out a design profile on the basis of a table of data
- Add and modify individual components in a design profile
- Apply a standard label set

Elevate Me

The whole point of a three-dimensional model is to include the elevation element that's been missing for years. But to get there, designers and engineers still depend on a flat 2D representation of the vertical dimension as shown in a profile view (see Figure 8.1).

A profile is nothing more than a series of data pairs in a station, elevation format. There are basic curve and tangent components, but these are purely the mathematical basis for the paired data sets. In Civil 3D, you can generate profile information in one of the following three ways:

- Sampling from a surface involves taking vertical information from a surface object every time the sampled alignment crosses a TIN line of the surface.
- Using a layout to create allows you to input design information, setting critical station and elevation points, calculating curves to connect linear segments, and typically working within requirements laid out by a reviewing agency.
- Creating from a file lets you point to a specially formatted text file to pull in the station and elevation pairs. This can be helpful in dealing with other analysis packages or spreadsheet tabular data.

This section looks at all three methods of creating profiles.

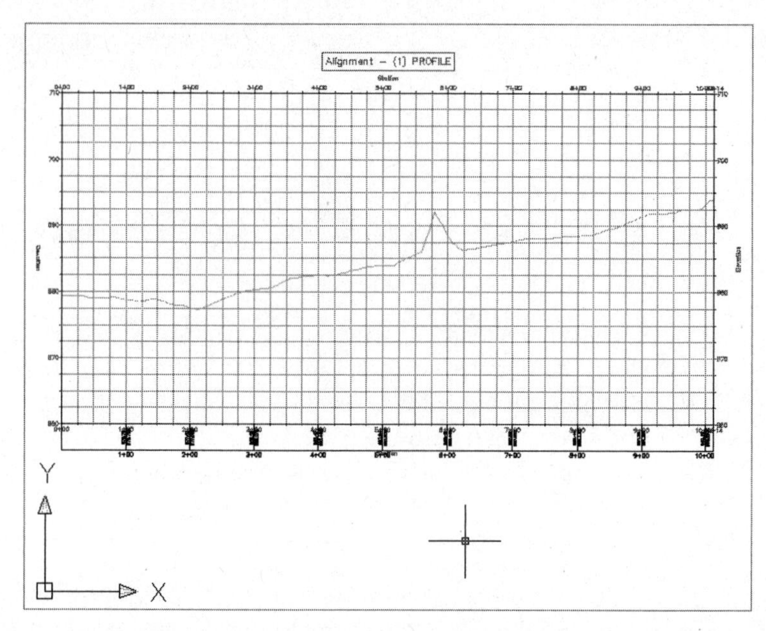

FIGURE 8.1
A typical profile view of the surface elevation along an alignment

Surface Sampling

Working with surface information is the most elemental method of creating a profile. This information can represent a simple existing or proposed surface, a river flood elevation, or any number of other surface-derived data sets. Within Civil 3D, surfaces can also be sampled at offsets, as you'll see in the next series of exercises. Follow these steps:

1. Open the `Profile Sampling.dwg` file shown in Figure 8.2. (Remember, all data files can be downloaded from `www.sybex.com/go/masteringcivil3d2010`.)

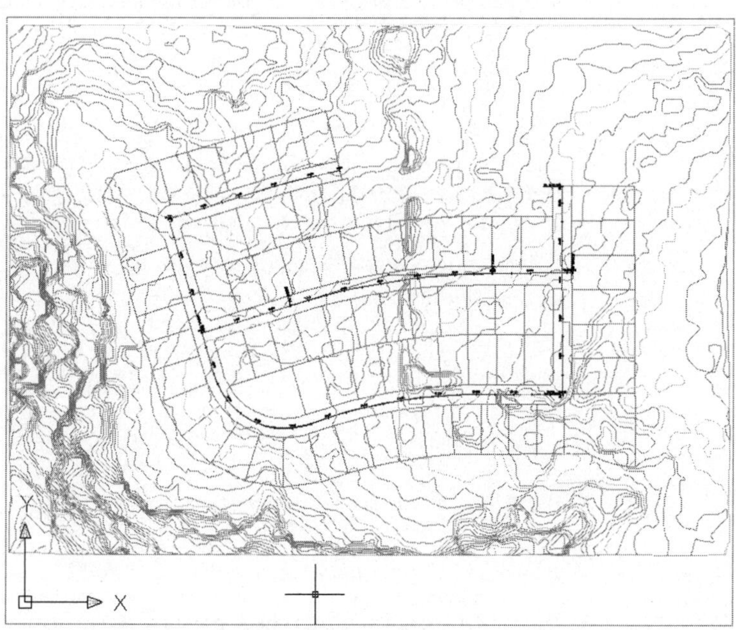

FIGURE 8.2
The drawing used for the exercise on profile sampling

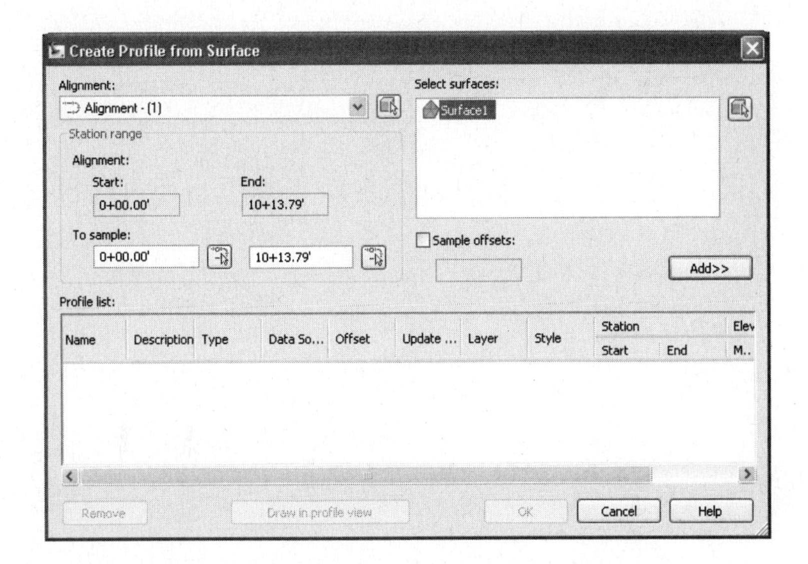

FIGURE 8.3
The Create Profile from Surface dialog

2. Select Create Surface Profile from the profile drop-down on the Create Design panel to open the Create Profile from Surface dialog box as shown in Figure 8.3.

   This dialog has a number of important features, so take a moment to see how it breaks down:

   - The upper-left quadrant is dedicated to information about the alignment. You can select the alignment from a drop-down list, or you can click the Pick On Screen button. The Station Range area is automatically set to run from the beginning to the end of the alignment, but you can control it manually by entering the station ranges in the To Sample text boxes.
   - The upper-right quadrant controls the selection of the surface and the offsets. You can select a surface from the list, or you can click the Pick On Screen button. Beneath the Select Surfaces list area is a Sample Offsets check box. The offsets aren't applied in the left and right direction uniformly. You must enter a negative value to sample to the left of the alignment or a positive value to sample to the right. In all cases, the profile isn't generated until you click the Add button.
   - The Profile List box displays all profiles associated with the alignment currently selected in the Alignment drop-down menu. This area is generally static (it won't change), but you can modify the Update Mode, Layer, and Style columns by clicking the appropriate cells in this table. You can stretch and rearrange the columns to customize the view.

3. Select the Alignment - (1) option from the Alignment drop-down menu if it isn't already selected.
4. In the Select Surfaces list area, select Surface1.
5. Click the Add button, and the Profile List area is populated with profile information for Surface1 on Alignment - (1).
6. Select the Sample Offsets check box to make the entry box active, and enter **-25,25**.

7. Click Add again. The dialog should look like Figure 8.4.

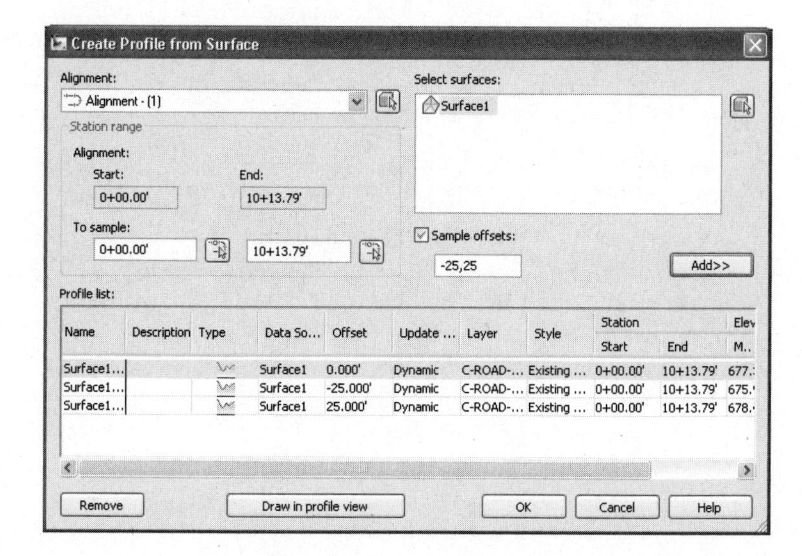

FIGURE 8.4
The Create Profile from Surface dialog showing the profiles sampled on Surface1

8. In the Profile List area, select the cell in the Style column that corresponds to the –25.000′ value in the Offset column (see Figure 8.4) to activate the Pick Profile Style dialog.
9. Select the Left Sample Profile option, and click OK. The style changes from the Existing Ground Profile to the Left Sample Profile in the table.
10. Select the cell in the Style column that corresponds to the 25.000′ value in the Offset column.
11. Select the Right Sample Profile option, and click OK. The dialog should look like Figure 8.5.

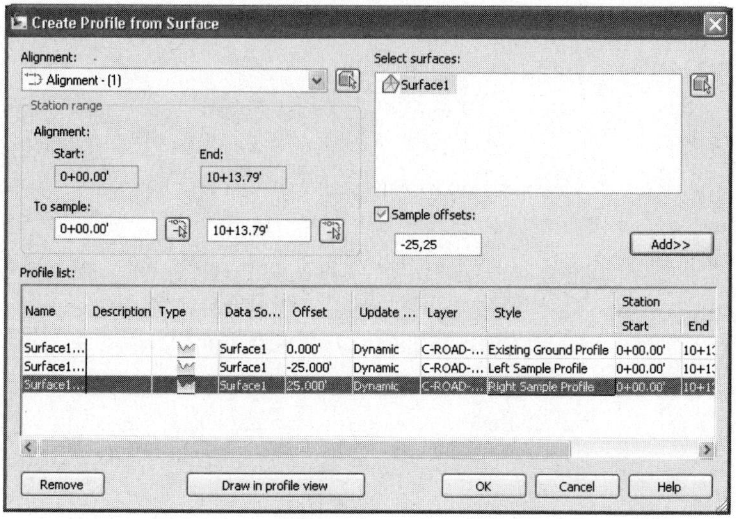

FIGURE 8.5
The Create Profile from Surface dialog with styles assigned on the basis of the Offset value

12. Click OK to dismiss this dialog.

13. The Events tab in Panorama appears, telling you that you've sampled data or if an error in the sampling needs to be fixed. Click the green check mark or the X to dismiss Panorama.

Profiles are dependent on the alignment they're derived from, so they're stored as profile branches under their parent alignment on the Prospector tab, as shown in Figure 8.6.

> **AREN'T YOU GOING TO DRAW THE PROFILE VIEW?**
>
> Under normal conditions, you would click the Draw in Profile View button to go through the process of creating the grid, labels, and other components that are part of a completed profile view. You'll skip that step for now because you're focusing on the profiles themselves.

FIGURE 8.6
Alignment profiles on the Prospector tab

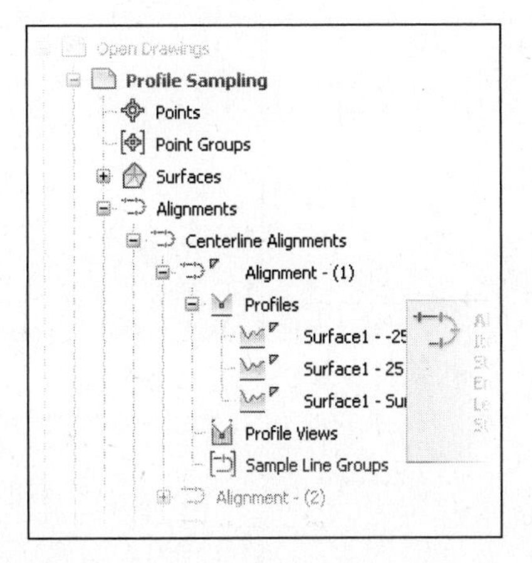

By maintaining the profiles under the alignments, it becomes simpler to review what has been sampled and modified for each alignment. Note that the profiles are dynamic and continuously update, as you'll see in this next exercise:

1. Open the `Dynamic Profiles.dwg` file. This drawing has profiles that were created in the first exercise from sampling the surface along Alignment - (1) and along offsets that were 25′ to the right and left of Alignment - (1).

2. On the View tab's Viewports panel, select Viewport Configurations ➢ Two: Horizontal.

3. Click in the top viewport to activate it.

4. On the Prospector tab, expand the Alignments branch to view the alignment types, expand the Centerline Alignments branch, and right-click Alignment - (1). Select the Zoom To option, as shown in Figure 8.7.

5. Click in the bottom viewport to activate it.

6. Expand the Alignments ➢ Centerline Alignments ➢ Alignment - (1) ➢ Profile Views branches.

7. Right-click Alignment - (1)3, and select Zoom To. Your screen should now look like Figure 8.8.

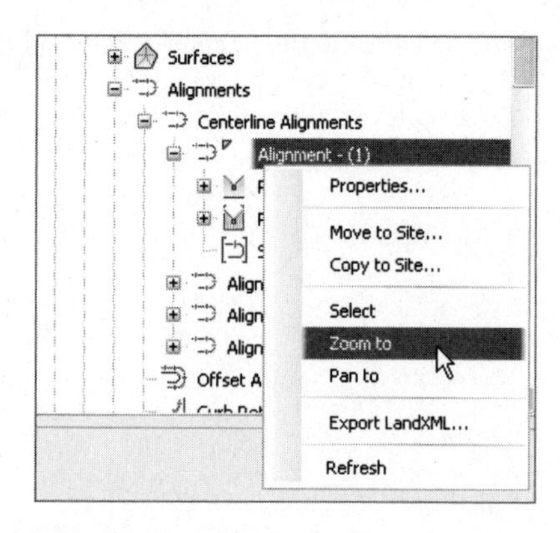

FIGURE 8.7
The Zoom To option on Alignment - (1)

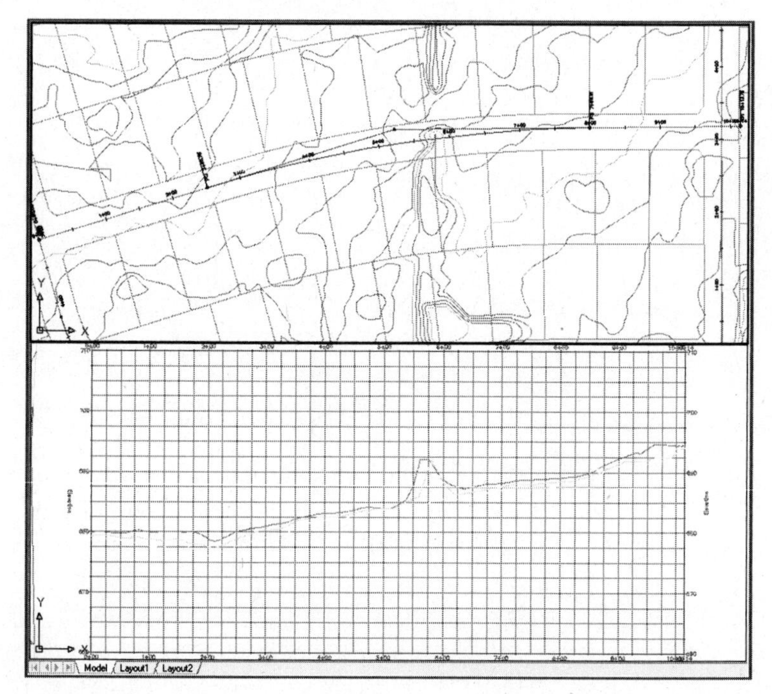

FIGURE 8.8
Splitting the screen for plan and profile editing

8. Click in the top viewport.
9. Zoom out until a circle appears on the left side of the drawing, beyond the lot layout.
10. Pick the alignment to activate the grips, and stretch the beginning grip to the center of the circle, as shown in Figure 8.9.
11. Click to complete the edit. The alignment profile (the green line) automatically adjusts to reflect the change in the starting point of the alignment. Note that the offset profiles (the yellow and red lines) move dynamically as well.

By maintaining the relationships between the alignment, the surface, the sampled information, and the offsets, Civil 3D creates a much more dynamic feedback system for designers. This can be useful when you're analyzing a situation with a number of possible solutions, where the surface information will be a deciding factor in the final location of the alignment. Once you've selected a location, you can use this profile view to create a vertical design, as you'll see in the next section.

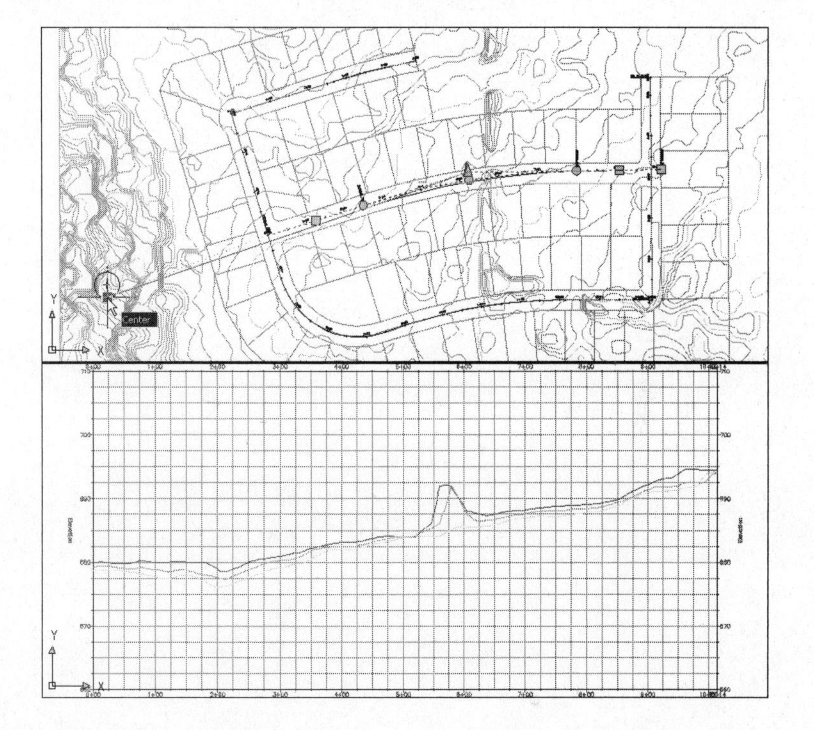

FIGURE 8.9
Grip-editing the alignment

Layout Profiles

Working with sampled surface information is dynamic, and the improvement over previous generations of Autodesk Civil design software is profound. Moving into the design stage, you'll see how these improvements continue as you look at the nature of creating design profiles. By working with layout profiles as a collection of components that understand their relationships with each other as opposed to independent finite elements, you can continue to use the program as a design tool instead of just a drafting tool.

You can create layout profiles in two basic ways:

- PVI-based layouts are the most common, using tangents between points of vertical intersection (PVIs) and then applying curve parameters to connect them. PVI-based editing allows editing in a more conventional tabular format.
- Entity-based layouts operate like horizontal alignments in the use of free, floating, and fixed entities. The PVI points are derived from pass-through points and other parameters that are used to create the entities. Entity-based editing allows for the selection of individual entities and editing in an individual component dialog.

You'll work with both methods in the next series of exercises to illustrate a variety of creation and editing techniques. First, you'll focus on the initial layout, and then you'll edit the various layouts.

Layout by PVI

PVI layout is the most common methodology in transportation design. Using long tangents that connect PVIs by derived parabolic curves is a method most engineers are familiar with, and it's the method you'll use in the first exercise:

1. Open the `Layout Profiles 1.dwg` file.
2. On the Home tab's Create Design panel, select Profiles ➢ Profile Creation Tools.
3. Pick the Alignment - (1) profile view by clicking one of the grid lines. The Create Profile dialog appears, as shown in Figure 8.10.

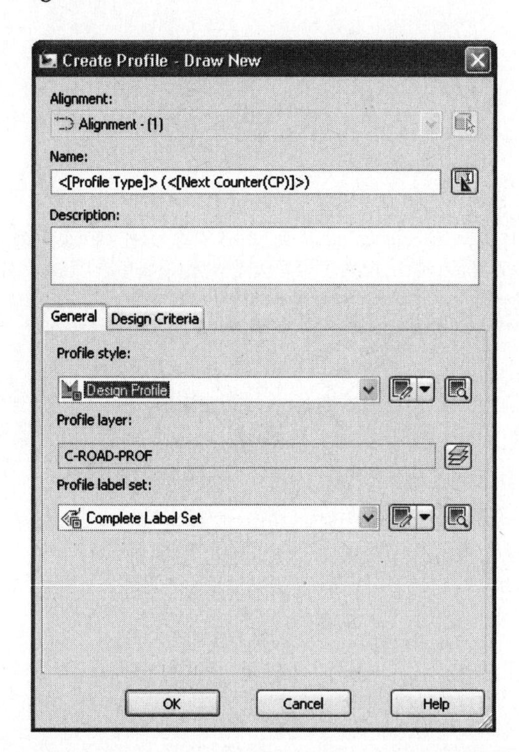

Figure 8.10
The Create Profile dialog

4. Click OK to accept the default settings. The Profile Layout Tools toolbar appears. Notice that the toolbar (see Figure 8.11) is *modeless*, meaning it stays open even if you do other AutoCAD operations such as Pan or Zoom.

   At this point, you're ready to begin laying in your vertical design. Before you do, however, note that this exercise skips two options. The first is the idea of Profile Label Sets. You'll explore them in a later section of this chapter. The other option is Criteria-Based Design on the design tab. Criteria-based design operates in profiles similar to Alignments in that the software compares the design speed to a selected design table (typically AASHTO 2001 in the North America releases) and sets minimum values for curve K values. This can be helpful when you're laying out long highway design projects, but most site and subdivision designers have other criteria to design against.

5. On the toolbar, click the arrow by the Draw Tangents without Curves tool on the far left. Select the Curve Settings option, as shown in Figure 8.11. The Vertical Curve Settings dialog opens.

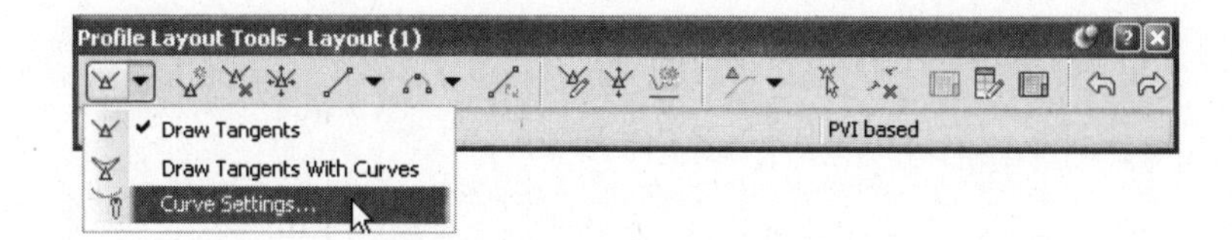

FIGURE 8.11
The Curve Settings option on the Profile Layout Tools toolbar

6. The Select Curve Type drop-down menu should be set to Parabolic, and the Length values in both the Crest Curves and Sag Curves areas should be 150.000′. Selecting a Circular or Asymmetric curve type activates the other options in this dialog.

TO K OR NOT TO K

You don't have to choose. Realizing that users need to be able to design using both, Civil 3D 2010 lets you modify your design based on what's important. You can enter a K value to see the required length, and then enter a nice even length that satisfies the K. The choice is up to you.

7. Click OK to close the Vertical Curve Settings dialog.
8. On the Profile Layout Tools toolbar, click the arrow next to the Draw Tangents without Curves tool again. This time, select the Draw Tangents with Curves option.
9. Use a Center osnap to pick the center of the circle at far left in the profile view. A jig line, which will be your layout profile, appears.
10. Continue working your way across the profile view, picking the center of each circle with a Center osnap.
11. Right-click or press ↵ after you select the center of the last circle; your drawing should look like Figure 8.12.

A JIG? BUT I DON'T DANCE!

A *jig* is a temporary line shown onscreen to help you locate your pick point. Jigs work in a similar way to osnaps in that they give feedback during command use but then disappear when the selection is complete. Civil 3D uses jigs to help you locate information on the screen for alignments, profiles, profile views, and a number of other places where a little feedback goes a long way.

The layout profile is labeled with the Complete Label Set you selected in the Create Profile dialog. As you'd expect, this labeling and the layout profile are dynamic. If you click and then zoom in on this profile line, not the labels or the profile view, you'll see something like Figure 8.13.

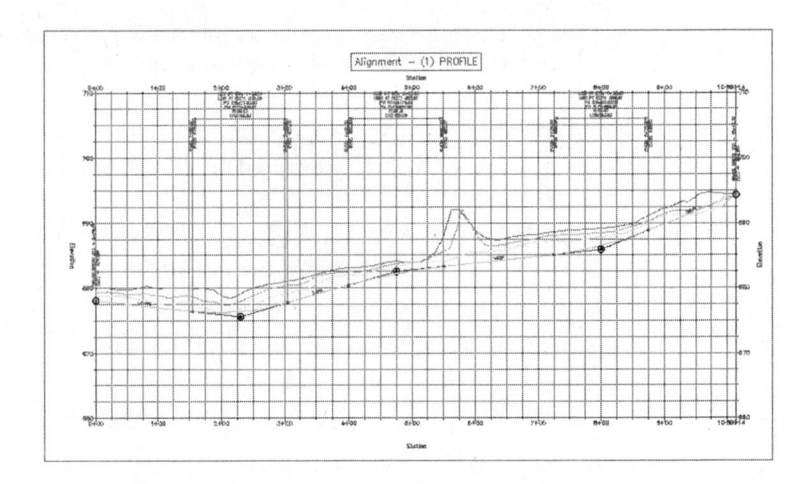

FIGURE 8.12
A completed layout profile with labels

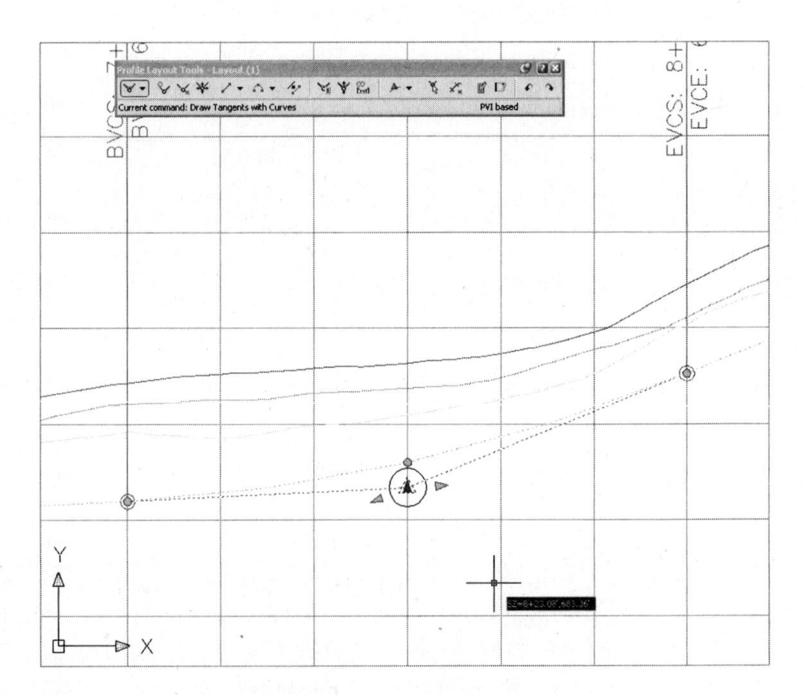

FIGURE 8.13
The types of grips on a layout profile

The PVI-based layout profiles include the following unique grips:

- The red triangle at the PVI point is the PVI grip. Moving this alters the inbound and outbound tangents, but the curve remains in place with the same design parameters of length and type.
- The triangular grips on either side of the PVI are sliding PVI grips. Selecting and moving either moves the PVI, but movement is limited to along the tangent of the selected grip. The curve length isn't affected by moving these grips.

- The circular grips near the PVI and at each end of the curve are curve grips. Moving any of these grips makes the curve longer or shorter without adjusting the inbound or outbound tangents or the PVI point.

Although this simple pick-and-go methodology works for preliminary layout, it lacks a certain amount of control typically required for final design. For that, you'll use another method of creating PVIs:

1. Open the `Layout Profiles 2.dwg` file. Make sure the Transparent Commands toolbar (Figure 8.14) is displayed somewhere on your screen.

FIGURE 8.14
The Transparent Commands toolbar

2. On the Home tab's Create Design panel, select Profiles ➤ Profile Creation Tools.
3. Pick a grid line on the Alignment - (1) profile view to display the Create Profile – Draw New dialog.
4. Click OK to accept the default settings.
5. On the Profile Layout Tools toolbar, click the drop-down menu by the Draw Tangents without Curves tool, and select the Draw Tangents with Curves tool, as in the previous exercise. Use a Center osnap to snap to the center of the circle at the left edge of the profile view.

6. On the Transparent Commands toolbar, select the Profile Station Elevation command.
7. Pick a grid line on the profile view. If you move your cursor within the profile grid area, a vertical red line, or *jig*, appears; it moves up and down and from side to side. Notice the tooltips.
8. Enter **245**↵ at the command line for the station value. If you move your cursor within the profile grid area, a horizontal and vertical jig appears (see Figure 8.15), but it can only move vertically along station 245. If you've turned on Dynamic Input, the text in the white box shows the elevation along the 245 jig, and the text in the black box shows the horizontal and vertical location of the cursor.
9. Enter **675**↵ at the command line to set the elevation for the second PVI.
10. Press Esc only once. The Profile Station Elevation command is no longer active, but the Draw Tangents with Curves tool that you previously selected on the Profile Layout Tools toolbar continues to be active.

11. On the Transparent Commands toolbar, select the Profile Grade Station command.
12. Enter **3**↵ at the command line for the profile grade.
13. Enter **650**↵ for the station value at the command line. Press Esc only once to deactivate the Profile Grade Station command.

14. On the Transparent Commands toolbar, select the Profile Grade Length command.
15. Enter **2**↵ at the command line for the profile grade.
16. Enter **300**↵ for the profile grade length.

17. Press Esc only once to deactivate the Profile Grade Length command and to continue using the Draw Tangent with Curves tool.

18. Use a Nearest osnap to select a point along the far-right side of the profile view. Be careful to select a point on the grid or within the grid. Note that a curve may not be inserted between the last two tangents. If the PVI at 9+50 doesn't leave room for a 150′ vertical curve to fit before the last PVI at 10+13.79, the curve will not be drawn.

19. Press ↵ to complete the profile. Your profile should look like Figure 8.16.

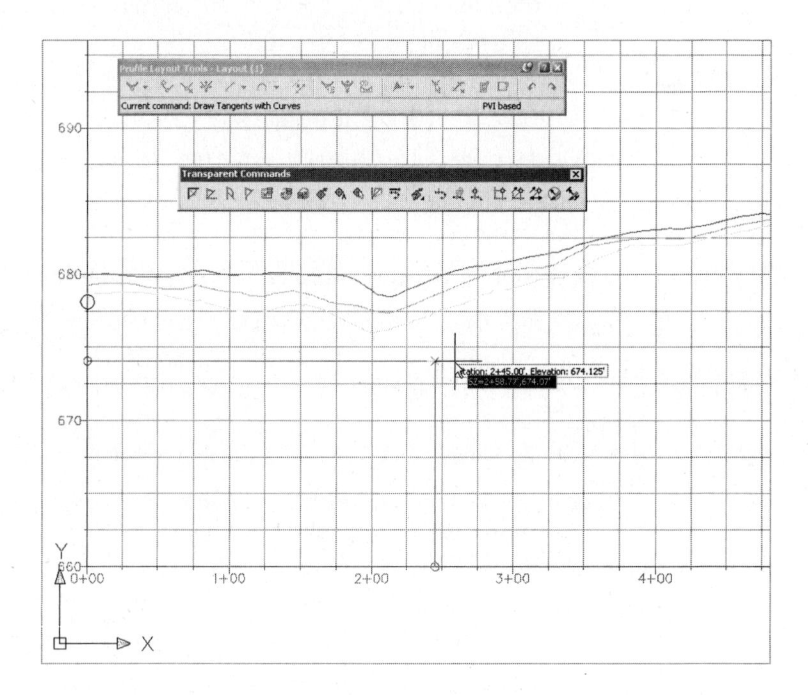

FIGURE 8.15 A jig appears when you use the Profile Station Elevation transparent command

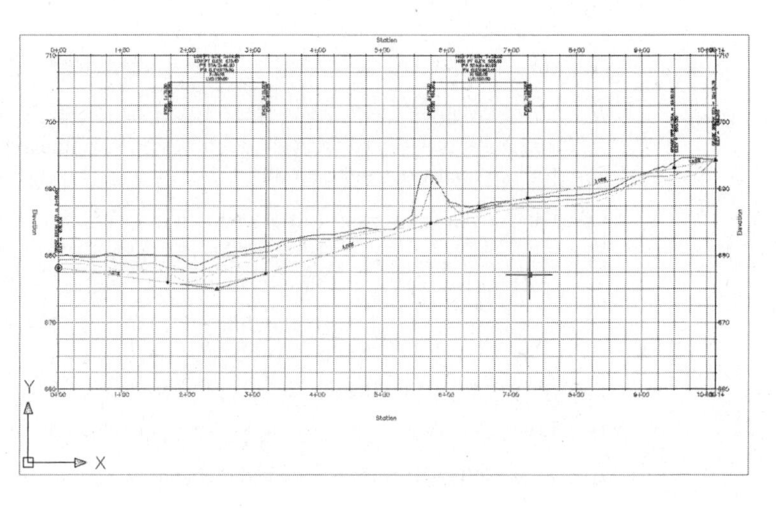

FIGURE 8.16 Using the Transparent Commands toolbar to create a layout profile

Using PVIs to define tangents and fitting curves between them is the most common approach to create a layout profile, but you'll look at an entity-based design in the next section.

LAYOUT BY ENTITY

Working with the concepts of fixed, floating, and free entities as you did in Chapter 7, ''Laying a Path: Alignments,''you'll lay out a design profile in this exercise:

1. Open the `Layout Profiles 3.dwg` file.
2. On the Home tab's Create Design panel, select Profiles ➢ Profile Creation Tools.
3. Pick a grid line on the Alignment - (1) profile view to display the Create Profile dialog.
4. Click OK to accept the default settings and open the Profile Layout Tools toolbar.
5. Click the arrow by the Draw Fixed Tangent by Two Points tool, and select the Fixed Tangent (Two Points) option, as shown in Figure 8.17.

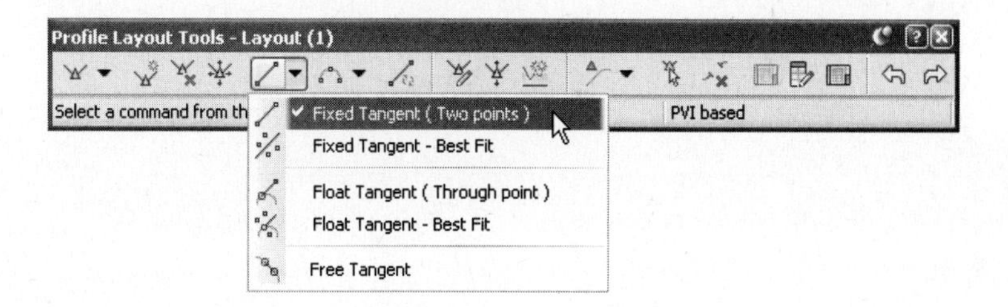

FIGURE 8.17 Selecting the Fixed Tangent (Two Points) tool on the Profile Layout Tools toolbar

6. Using a Center osnap, pick the circle at the left edge of the profile view. A rubberbanding line appears.
7. Using a Center osnap, pick the circle located at approximately station 2+30. A tangent is drawn between these two circles.
8. Using a Center osnap, pick the circle located at approximately station 8+00. Another rubberbanding line appears.
9. Using a Center osnap, pick the circle located at the right edge of the profile view. A second tangent is drawn. Right-click to exit the Fixed Line (Two Points) command; your drawing should look like Figure 8.18. Note that there are no station labels on the second tangent, because it isn't yet tied to the first segment. The labeling begins at station 0 +00 and continues until there is a break, as there was at the end of the first tangent.
10. Click the arrow next to the Draw Fixed Parabola by Three Points tool on the Profile Layout Tools toolbar. Choose the More Fixed Vertical Curves ➢ Fixed Vertical Curve (Entity End, Through Point) option, as shown in Figure 8.19.
11. Pick the left tangent to attach the fixed vertical curve. Remember to pick the tangent line and not the end circle. A rubberbanding line appears.
12. Using a Center osnap, select the circle located at approximately station 4+75.
13. Right-click to exit the Fixed Vertical Curve (Entity End, Through Point) command. Your drawing should look like Figure 8.20.

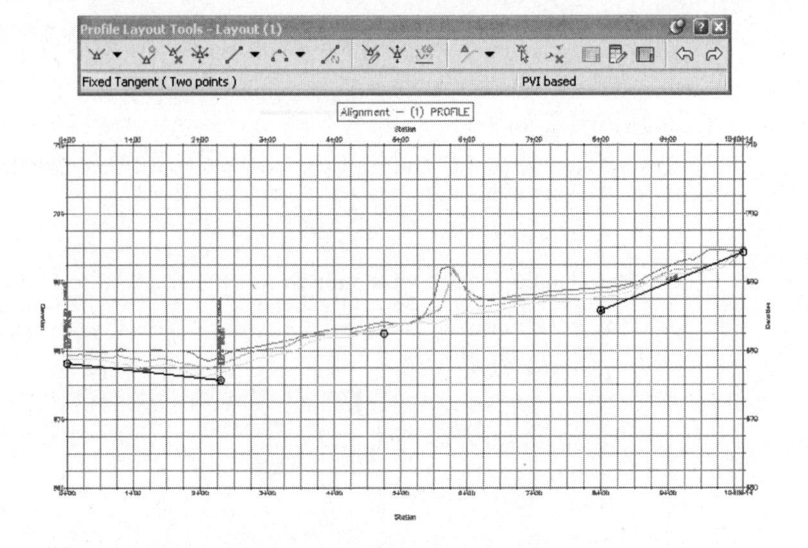

FIGURE 8.18
Layout profile with two tangents drawn

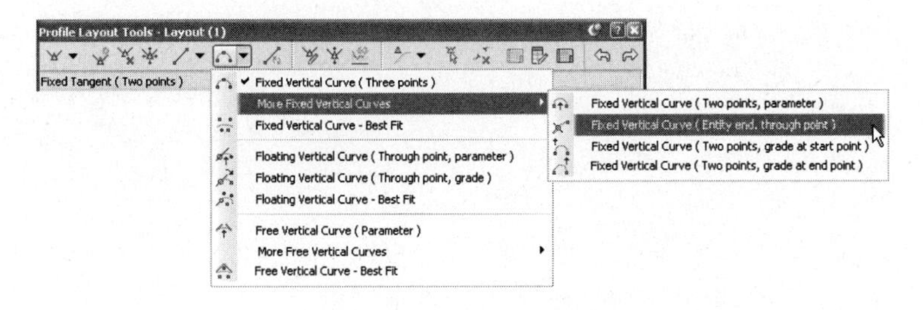

FIGURE 8.19
The Fixed Vertical Curve (Entity End, Through Point) tool on the Profile Layout Tools toolbar

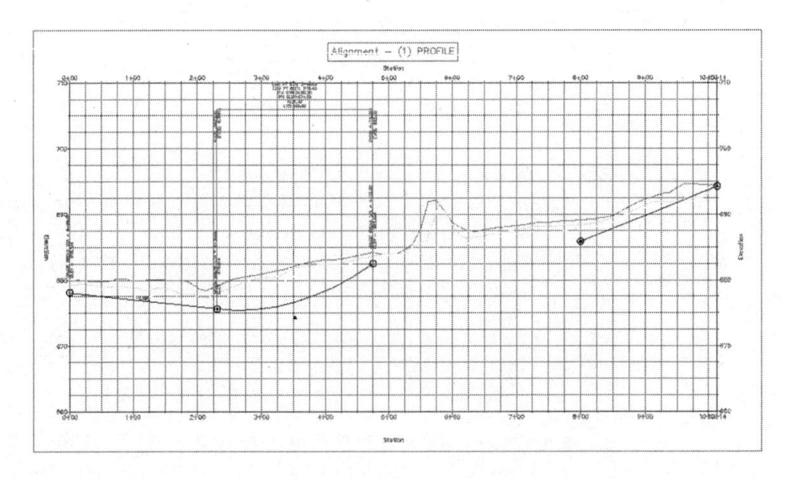

FIGURE 8.20
Completed curve from the entity end

14. Click the arrow next to the Draw Fixed Tangent by Two Points tool, and select the Float Tangent (Through Point) option, as shown in Figure 8.21.

15. Pick the curve you just created, and then pick the beginning of the tangent you created at far right.

FIGURE 8.21
Selecting the Float Tangent (Through Point) tool on the Profile Layout Tools toolbar

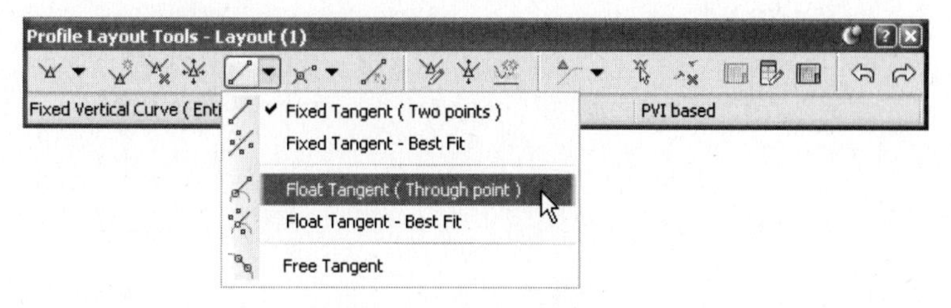

16. Click the arrow next to the Draw Fixed Parabola Tangential to End of an Entity and Passing through a Point tool, and select the Free Vertical Curve (Parameter) option as shown in Figure 8.22.

FIGURE 8.22
Selecting the Free Vertical Curve (Parameter) tool on the Profile Layout Tools toolbar

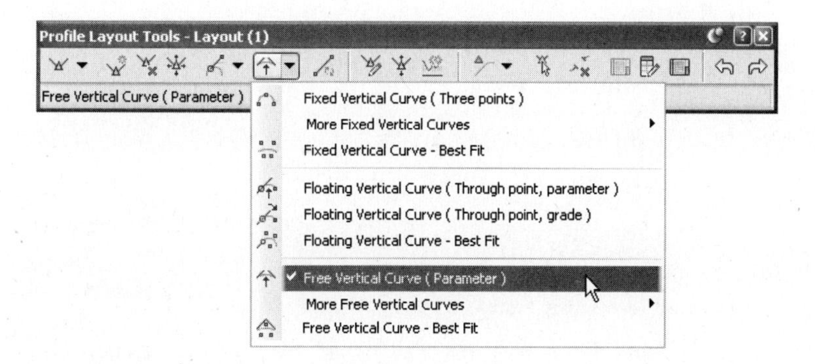

17. Pick the tangent just created, and then pick the tangent that ends your layout profile.
18. Enter **150↵** at the command line for a curve length.
19. Right-click or press ↵ to complete the profile, and then close the Profile Layout Tools toolbar by clicking the red X button. Your drawing should look like Figure 8.23.

FIGURE 8.23
Completed layout profile created with entity tools

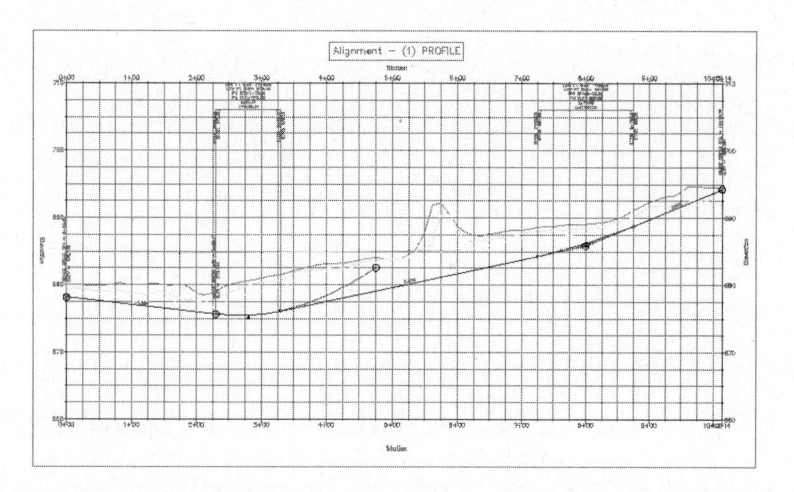

With the entity-creation method, grip-editing works in a similar way to other layout methods. You'll look at more editing methods after trying the final creation method.

Creating a Profile from a File

Working with profile information in Civil 3D is nice, but it's not the only place where you can create or manipulate this sort of information. Many programs or analysis packages generate profile information. One common case is the plotting of a hydraulic grade line against a stormwater network profile of the pipes. When information comes from outside the Civil 3D program, it's often output in myriad formats. If you convert this format to the required format for Civil 3D, the profile information can be input directly.

Civil 3D has a specific format. Each line is a PVI definition (station and elevation). Curve information is an optional third bit of data on any line. Here's one example:

```
0 550.76
127.5 552.24
200.8 554 100
256.8 557.78 50
310.75 561
```

In this example, the third and fourth lines include the curve length as the optional third piece of information. The only inconvenience of using this input method is that the information in Civil 3D doesn't directly reference the text file. Once the profile data is imported, no dynamic relationship exists with the text file.

In this exercise, you'll import a small text file to see how the function works:

1. Open the `Profile from File.dwg` file.
2. On the Home tab's Create Design panel, select Profiles ➤ Profile Creation Tools.
3. Select the `TextProfile.txt` file, and click Open. The Create Profile dialog appears.
4. In the dialog, choose Complete Label Set from the Profile Label Set drop-down menu.
5. Click OK. Your drawing should look like Figure 8.24.

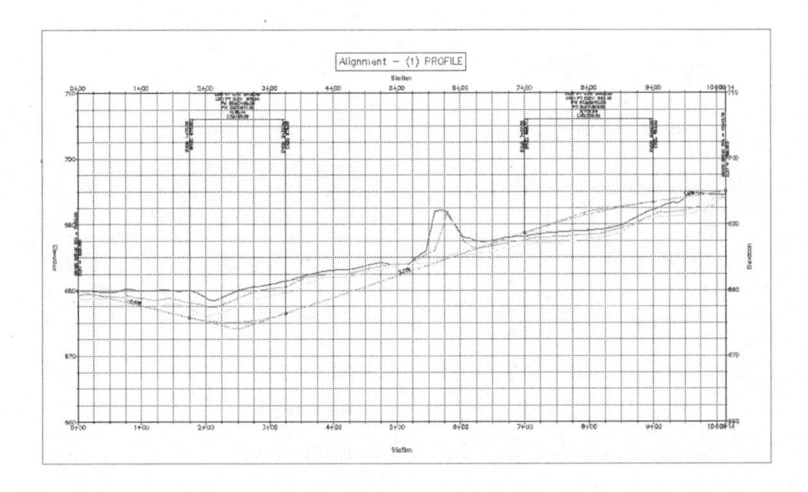

Figure 8.24
A completed profile created from a file

Now that you've tried the three main ways of creating profile information, you'll edit a profile in the next section.

Editing Profiles

The three methods just reviewed let you quickly create profiles. You saw how sampled profiles reflect changes in the parent alignment and how some grips are available on layout profiles, and you also imported a text file that could easily be modified. In all these cases, the editing methods left something to be desired, from either a precision or a dynamic relationship viewpoint.

This section looks at profile editing methods. The most basic is a more precise grip-editing methodology, which you'll learn about first. Then you'll see how to modify the PVI-based layout profile, how to change out the components that make up a layout profile, and how to use some editing functions that don't fit into a nice category.

GRIP PROFILE EDITING

Once a layout is in place, sometimes a simple grip edit will suffice. But for precision editing, you can use a combination of the grips and the tools on the Transparent Commands toolbar, as in this short exercise:

1. Open the `Grip Editing Profiles.dwg` file.
2. Zoom in, and pick the layout profile (the only profile with labels) to activate its grips.
3. Pick the triangular grip pointing upward on the left vertical curve to begin a grip stretch of the PVI, as shown in Figure 8.25.

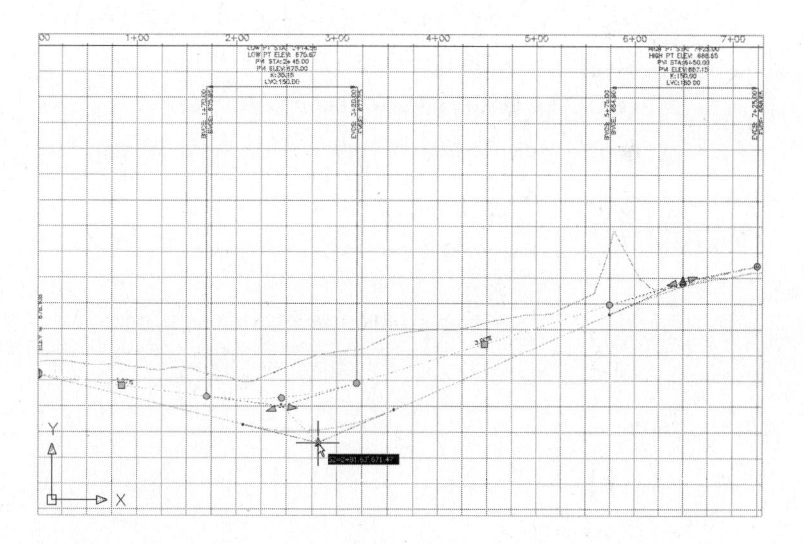

FIGURE 8.25
Grip-editing a PVI

4. On the Transparent Commands toolbar, select the Profile Station Elevation command.
5. Pick a grid line on the profile view.
6. Enter **275**↵ at the command line to set the profile station.
7. Enter **677**↵ to set the profile elevation.

8. Press Esc to deselect the layout profile object you selected in step 2, and regenerate your view to complete the changes, as shown in Figure 8.26.

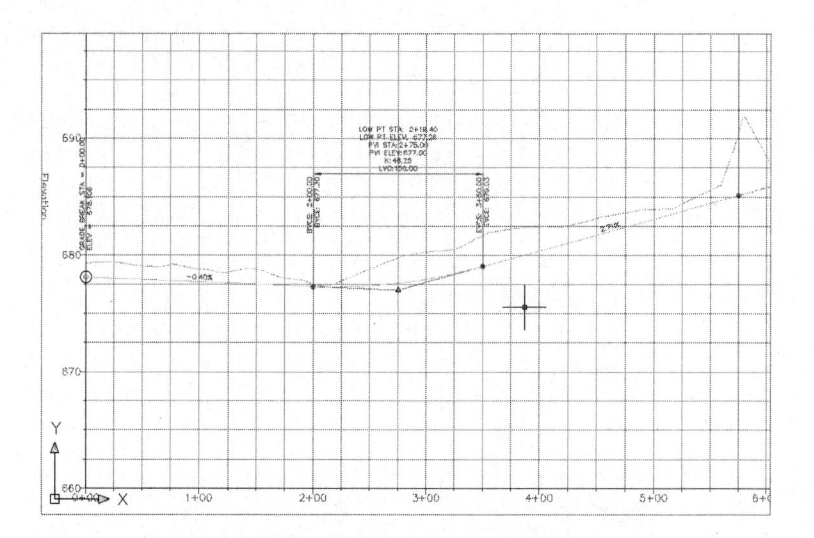

FIGURE 8.26
Completed grip edit using the transparent commands for precision

The grips can go from quick-and-dirty editing tools to precise editing tools when you use them in conjunction with the transparent commands in the profile view. They lack the ability to precisely control a curve length, though, so you'll look at editing a curve next.

PARAMETER AND PANORAMA PROFILE EDITING

Beyond the simple grip edits, but before changing out the components of a typical profile, you can modify the values that drive an individual component. In this exercise, you'll use the Profile Layout Parameter dialog and the Panorama palette set to modify the curve properties on your design profile:

1. Open the `Parameter Editing Profiles.dwg` file.
2. Pick the layout profile (the only profile with labels) to activate its grips.
3. Select Geometry Editor from the Modify Profile panel, as shown in Figure 8.27.

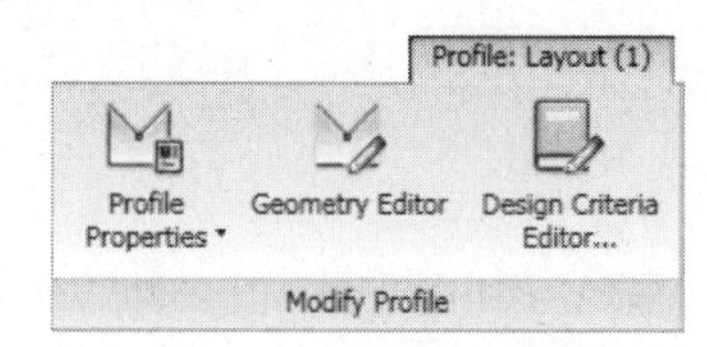

FIGURE 8.27
The Edit Profile Geometry option on the Modify Profile panel

4. On the Profile Layout Tools toolbar, click the Profile Layout Parameters tool to open the Profile Layout Parameters dialog.
5. Click the Select PVI tool, as shown in Figure 8.28, and zoom in to click near the PVI at station 6+50 to populate the Profile Layout Parameters dialog.

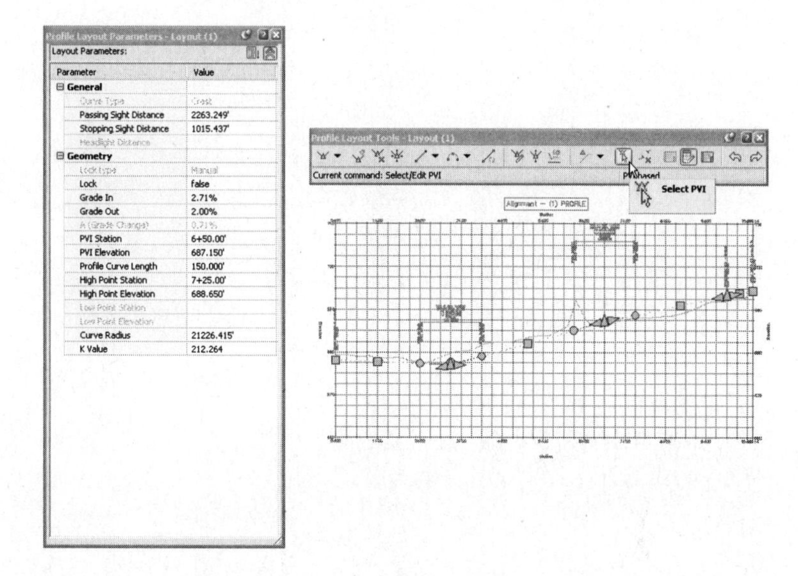

FIGURE 8.28
The Profile Layout Parameters dialog and the PVI tool shown on the Profile Layout Tools toolbar

Values that can be edited are in black; the rest are mathematically derived and can be of some design value but can't be directly modified. The two buttons at the top of the dialog adjust how much information is displayed.

6. Change the Profile Curve Length in the Profile Layout Parameters dialog to **250.000′** (see Figure 8.29).

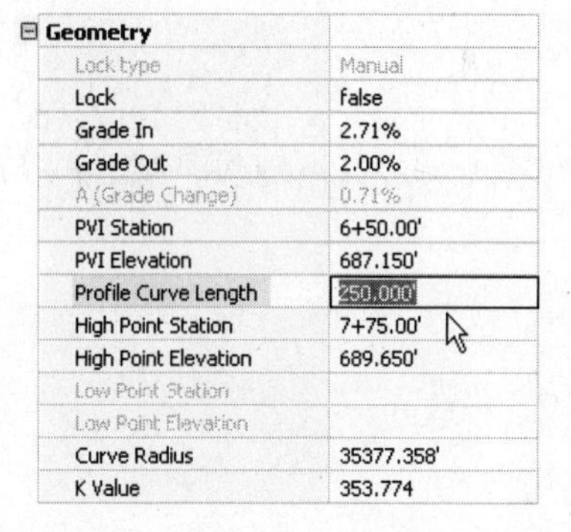

Geometry	
Lock type	Manual
Lock	false
Grade In	2.71%
Grade Out	2.00%
A (Grade Change)	0.71%
PVI Station	6+50.00'
PVI Elevation	687.150'
Profile Curve Length	250.000'
High Point Station	7+75.00'
High Point Elevation	689.650'
Low Point Station	
Low Point Elevation	
Curve Radius	35377.358'
K Value	353.774

FIGURE 8.29
Direct editing of the curve layout parameters

7. Close the Profile Layout Parameters dialog by clicking the red X in the upper-right corner. Then, press the Esc key to deactivate the Select PVI tool.

8. On the Profile Layout Tools toolbar, click the Profile Grid View tool to activate the Profile Entities tab in Panorama. Panorama allows you to view all the profile components at once, in a compact form.
9. Scroll right in Panorama until you see the Profile Curve Length column.
10. Double-click in the cell for the Entity 2 value in the Profile Curve Length column (see Figure 8.30), and change the value from 150.000′ to **250.000′**.

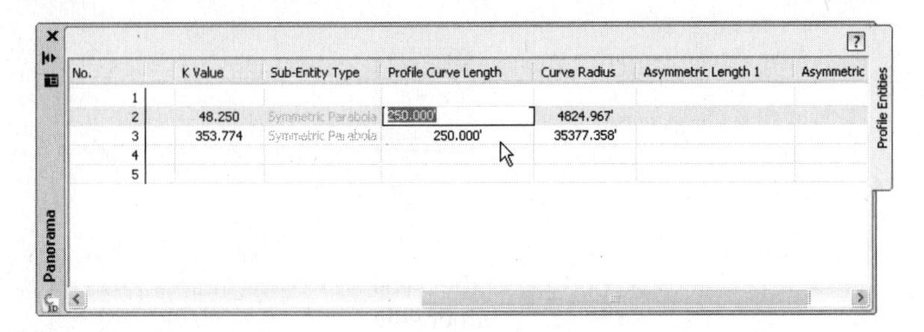

FIGURE 8.30 Direct editing of the curve length in Panorama

11. Close the Profile Layout Tools toolbar, and zoom out to review your edits. Your complete profile should now look like Figure 8.31.

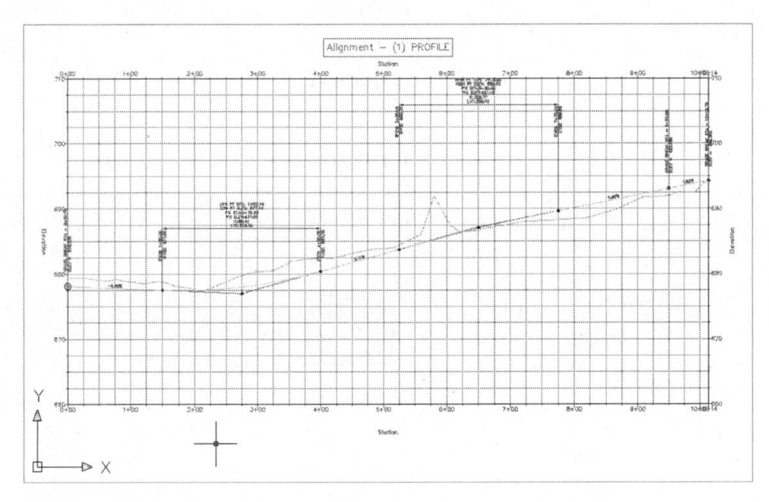

FIGURE 8.31 The completed editing of the curve length in the layout profile

You can use these tools to modify the PVI points or tangent parameters, but they won't let you add or remove an entire component. You'll do that in the next section.

> **PVIs IN LOCKDOWN**
>
> You can lock a PVI at a specific station and elevation in the Profile Layout Parameters dialog. PVIs that are locked cannot be moved with edits to adjacent entities. However, it's important to note that a PVI can be unlocked with editing grips.

COMPONENT-LEVEL EDITING

In addition to editing basic parameters and locations, sometimes you have to add or remove entire components. In this exercise, you'll add a PVI, remove a curve from an area that no longer requires one, and insert a new curve into the layout profile:

1. Open the `Component Editing Profiles.dwg` file.
2. Select the layout profile to activate its grips.
3. Select Geometry Editor from the Modify Profile panel.
4. On the Profile Layout Tools toolbar, click the Insert PVI tool, as shown in Figure 8.32.

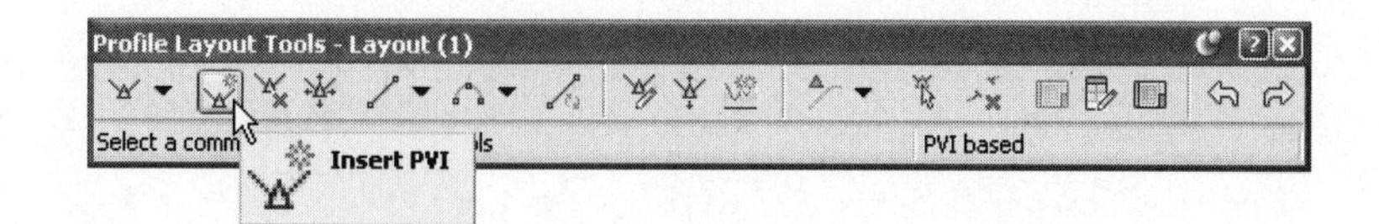

FIGURE 8.32
Click the Insert PVI tool

5. Pick a point near the 3+50 station, with an approximate elevation of 680. The position of the tangents on either side of this new PVI is affected, and the profile is adjusted accordingly. The vertical curves that were in place are also modified to accommodate this new geometry.

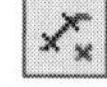

6. On the Profile Layout Tools toolbar, click the Delete Entity tool. Notice the grips disappear and the labels update.

7. Zoom in, and pick the curve entity near station 2+00 to delete it. Then, right-click to update the display.

8. On the Profile Layout Tools toolbar, click the arrow by the Draw Fixed Parabola by Three Points tool. Select the More Free Vertical Curves ➤ Free Vertical Parabola (PVI Based) option, as shown in Figure 8.33.

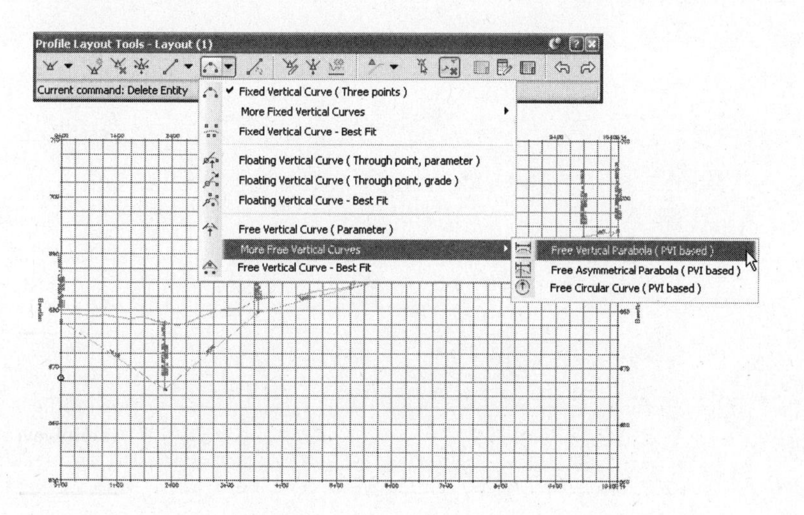

FIGURE 8.33
The layout profile after curve deletion, and selecting the Free Vertical Parabola tool on the Profile Layout Tools toolbar

9. Pick near the PVI you just inserted (near the 3 + 50 station in step 5) to add a curve.

10. Enter **100↵** at the command line to set the curve length.

11. Right-click to exit the command and update the profile display.

12. Close the Profile Layout Tools toolbar.

Editing profiles using any of these methods gives you precise control over the creation and layout of your vertical design. In addition to these tools, some of the tools on the Profile Layout Tools toolbar are worth investigating and somewhat defy these categories. You'll look at them next.

OTHER PROFILE EDITS

Some handy tools exist on the Profile Layout Tools toolbar for performing specific actions. These tools aren't normally used during the preliminary design stage, but they come into play as you're working to create a final design for grading or corridor design. They include raising or lowering a whole layout in one shot, as well as copying profiles. Try this exercise:

1. Open the `Other Profile Edits.dwg` file.
2. Pick the layout profile to activate its grips.
3. Select Geometry Editor from the Modify Profile panel. The Profile Layout Tools toolbar appears.

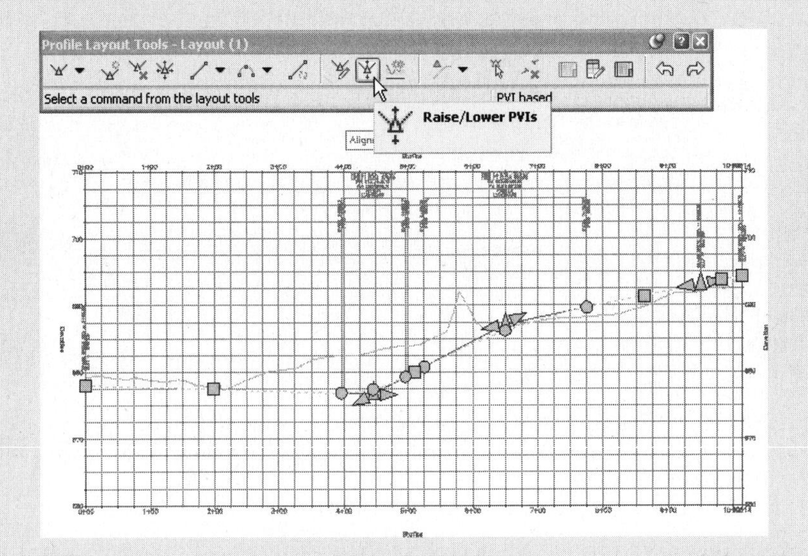

4. Click the Raise/Lower PVIs tool. The Raise/Lower PVI Elevation dialog shown here appears:

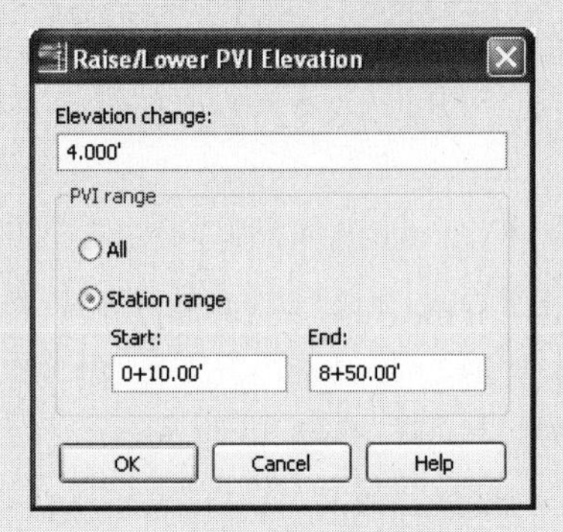

5. Set Elevation Change to **4.000′**. Click the Station Range radio button, and set the Start value to 0+ 10′ and the End value to 8+50.00′. You're moving the internal PVIs 4′ vertically while maintaining the endpoints of the design. Click OK.
6. Click the Copy Profile tool just to the right of the Raise/Lower PVIs tool to display the Copy Profile Data dialog.

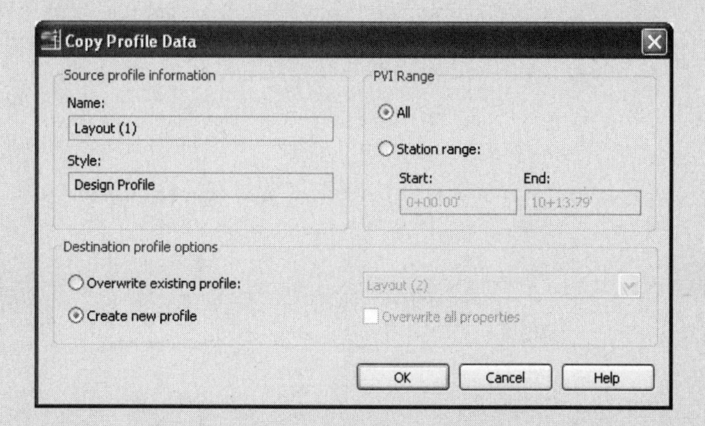

7. Click OK to create a new layout profile directly on top of Layout 1. Note that after picking the newly created layout profile, the Profile Layout Tools toolbar now references this newly created Layout (2) profile.
8. Use the Raise/Lower tool to drop Layout (2) by 0.5′ to simulate an edge-of-pavement design.

Using the layout and editing tools in these sections, you should be able to design and draw a combination of profile information presented to you as a Civil 3D user.

Profile Display and Stylization

No matter how profiles are created, they need to be shown and labeled to make the information more understandable. In this section, you'll learn about the style options for the profile linework as well as options for the profile labels.

Profile Styles

Like every other object in Civil 3D, the display of profiles is controlled through styles. In this series of exercises, you'll look at the components that make up a profile display and prepare the display for plotting:

1. Open the `Profile Styles.dwg` file.
2. Zoom in on any BVC or EVC point, and notice the double circles around the PVI points.

The style used for the layout profile and the style used to label the vertical curve both have small circles at the BVC and EVC points, creating the double circles seen in the drawing. These circles on the profile style are called *markers*; you can apply them at selected profile data points to illustrate various features on a profile, such as high or low points. Just like most objects in Civil 3D, the markers have their own style.

The double circles aren't needed, so you'll modify the Design Profile style to remove them from the vertical curve points on the layout profile.

3. In Toolspace, switch to the Settings tab.
4. Expand the Profile ➢ Profile Styles branches to expose the styles already in this drawing.
5. Right-click Design Profile, and select the Copy option to open the Profile Style dialog.
6. On the Information tab, change the Name field to **Road Plot**.
7. Change to the Design tab (see Figure 8.34).

FIGURE 8.34
The Profile Style Design tab

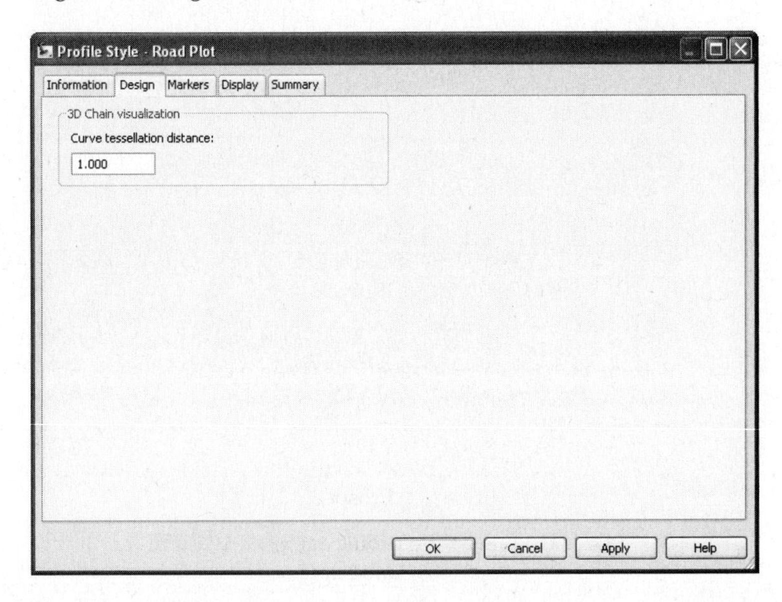

3D Chain Visualization is Civil 3D's way of displaying profile information in an isometric view. Profile information is displayed as a 3D polyline with *x* and *y* information from the alignment and *z* information from the profile data at that location. Because 3D polylines can't accurately reflect a curve in space, the coordinate values are tessellated by using the distance in this dialog.

8. Switch to the Markers tab, as shown in Figure 8.35.

Note that the Alignment Geometry marker style, which places a small circle around each of the vertical curve points, is used in your new Road Plot profile style.

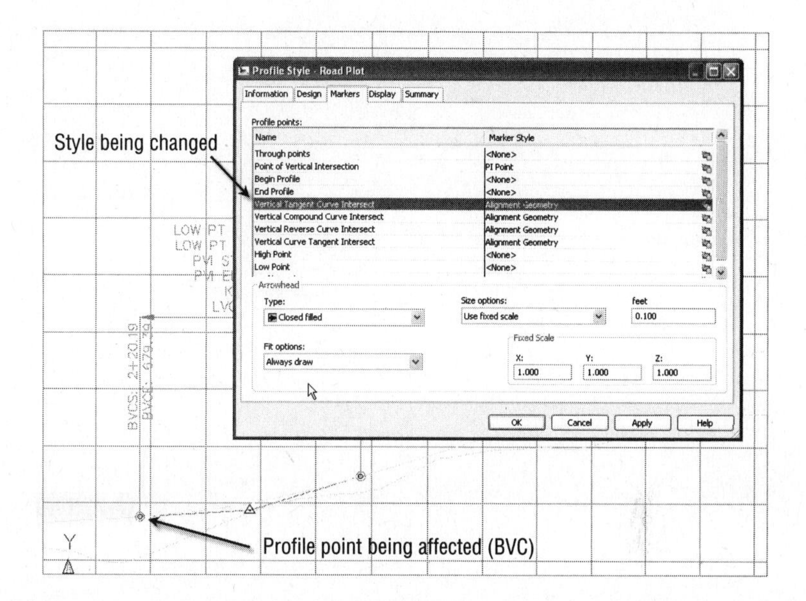

FIGURE 8.35
The Markers tab and profile style circles

9. Double-click the small icon on the right of the Marker field for the Vertical Tangent Curve Intersect to display the Pick Marker Style dialog.
10. Select <none> from the drop-down menu to remove the Alignment Geometry marker style, and click OK to close this dialog.
11. Repeat steps 9 and 10 for the Vertical Compound Curve Intersect, the Vertical Reverse Curve Intersect, and the Vertical Curve Tangent Intersect profile points. Your Profile Style dialog will look like Figure 8.36.
12. The Display tab works as it does in other dialogs you've explored, so you'll skip it in this exercise. Click OK to close the dialog. The new profile style is listed under the Profile Styles branch in Prospector.
13. Pick the layout profile again, right-click, and select the Profile Properties option. The Profile Properties dialog appears.
14. On the Information tab, set the Object Style field to Road Plot, and click OK. The double circles are gone, and your profile is ready for plotting.

Take a few moments to check out the Layout Profile style in this exercise's drawing. Remember, you need to pick the layout profile, select Profile Properties from the Modify Profile panel, and then set the Object Style field on the Information tab to Layout Profile. This style uses different colors for the components that form a layout profile, making it easy to visually discern a curve versus a tangent onscreen. This type of style is tailored for the design phase, when information is primarily displayed onscreen as opposed to plotted on paper. No matter how a profile is colored or marked with symbols, the labels tell the story.

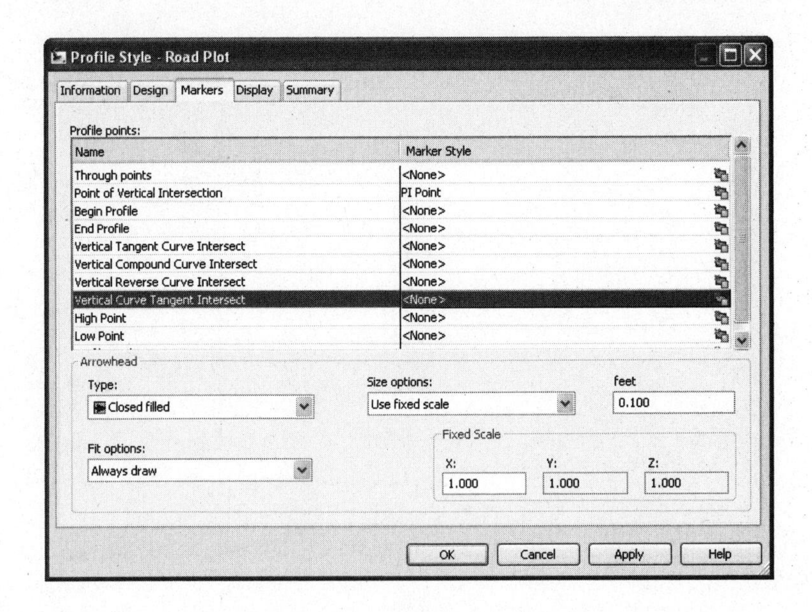

FIGURE 8.36
Marker styles set to <none> for all vertical profile intersect points

PROFILE LABELS

It's important to remember that the profile and the profile view aren't the same thing. The labels discussed in this section are those that relate directly to the profile. This usually means station-based labels, individual tangent and curve labels, or grade breaks. You'll look at individual label styles for these components and then at the concept of the label set.

APPLYING LABELS

Like alignments, you apply labels as a group of objects separate from the profile. In this exercise, you'll learn how to add labels along a profile object:

1. Open the `Applying Profile Labels.dwg` file.
2. Pick the cyan layout profile (the profile with two vertical curves) to activate the profile object.
3. Select Edit Profile Labels from the Labels panel to display the Profile Labels dialog (see Figure 8.37).

   Selecting the type of label from the Type drop-down menu changes the Style drop-down menu to include styles that are available for that label type. Next to the Style drop-down menu are the usual Style Edit/Copy button and a preview button. Once you've selected a style from the Style drop-down menu, clicking the Add button places it on the profile. The middle portion of this dialog displays information about the labels that are being applied to the profile selected; you'll look at that in a moment.
4. Choose the Major Stations option from the Type drop-down menu. The name of the second drop-down menu changes to Profile Major Station Label Style to reflect this option. Verify that Perpendicular with Tick is selected in this menu.

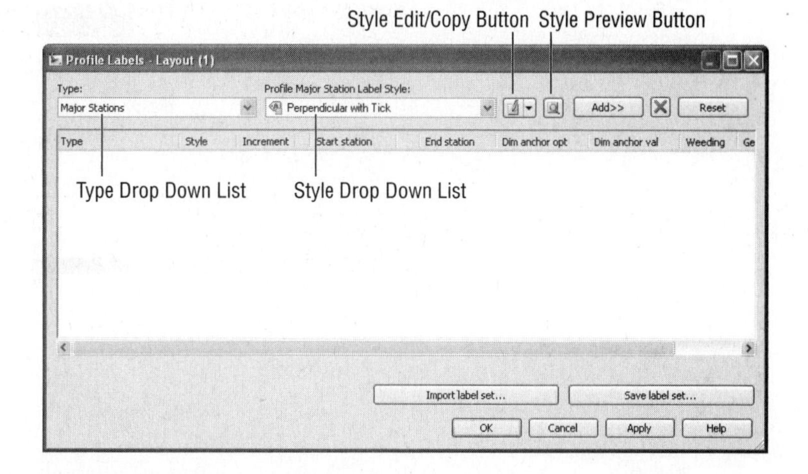

FIGURE 8.37
An empty Profile Labels dialog

5. Click Add to apply this label to the profile.
6. Choose Horizontal Geometry Points from the Type drop-down menu.
7. The name of the Style drop-down menu changes to Profile Horizontal Geometry Point. Select the Standard option, and click Add again to display the Geometry Points dialog shown in Figure 8.38. This dialog lets you apply different label styles to different geometry points if necessary.

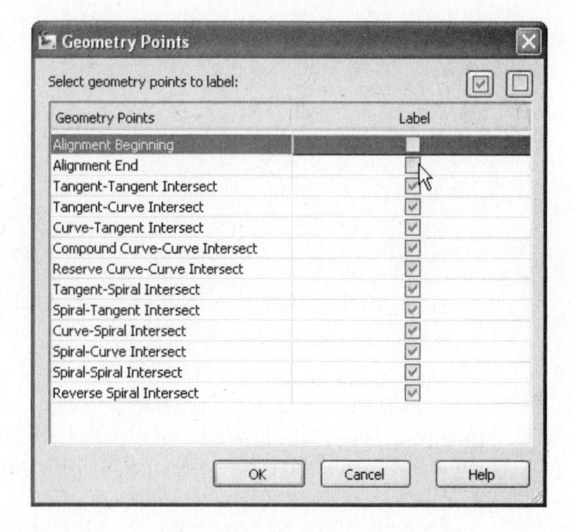

FIGURE 8.38
The Geometry Points dialog appears when you apply labels to horizontal geometry points

8. Deselect the Alignment Beginning and Alignment End rows, as shown in Figure 8.38, and click OK to close the dialog.
9. Click the Apply button. Drag the dialog out of the way to view the changes to the profile (see Figure 8.39).

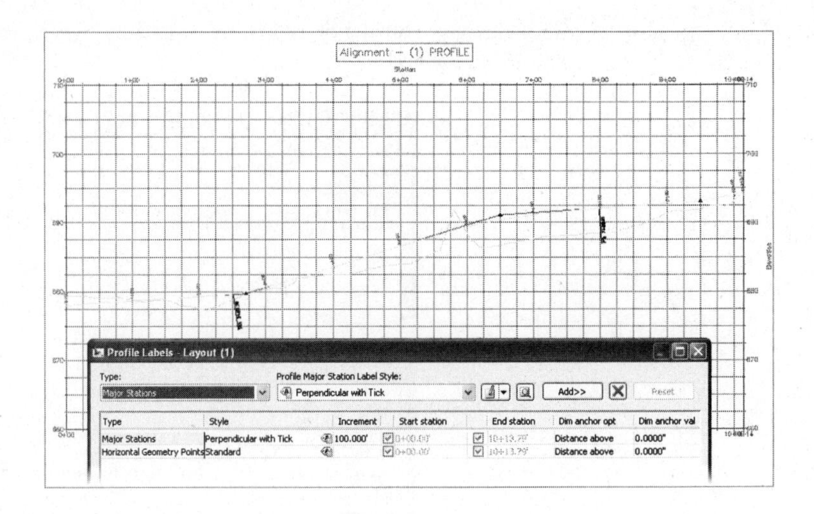

FIGURE 8.39
Labels applied to major stations and alignment geometry points

10. In the middle of the Profile Labels dialog, change the Increment value in the Major Stations row to **50**, as shown in Figure 8.40. This modifies the labeling increment only, not the grid or other values.

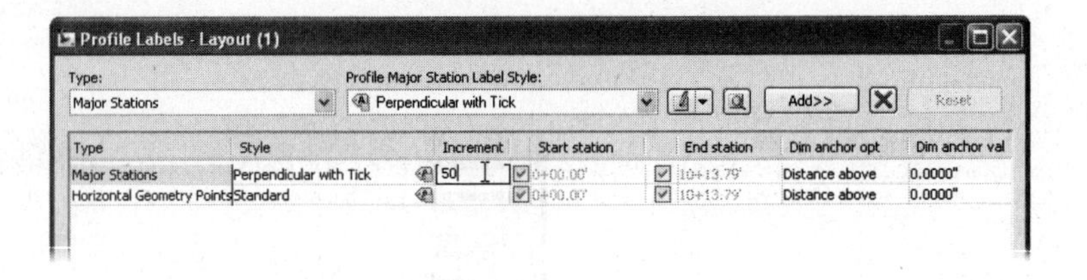

FIGURE 8.40
Modifying the major station labeling Increment

11. Click OK to close the Profile Labels dialog.

As you can see, applying labels one at a time could turn into a tedious task. After you learn about the types of labels available, you'll revisit this dialog and look at the two buttons at the bottom for dealing with label sets.

STATION LABELS

Labeling along the profile at major, minor, and alignment geometry points lets you insert labels similar to a horizontal alignment. In this exercise, you'll modify a style to reflect a plan-readable approach and remove the stationing from the first and last points along the profile:

1. Open the `Station Profile Labels.dwg` file.
2. Pick the layout profile and select Edit Profile Labels from the Labels panel to display the Profile Labels dialog.
3. Deselect the Start Station and End Station check boxes for the Major Stations label.

4. Change the value for the Start Station to **50** and the value of the End Station to **1012**, as shown in Figure 8.41.

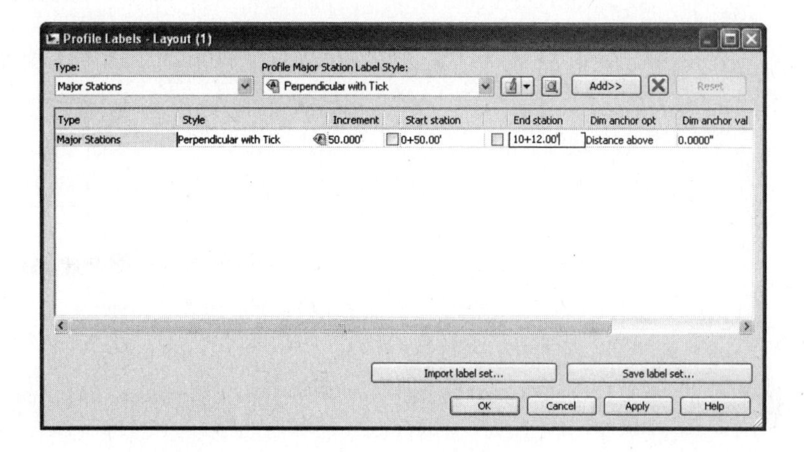

FIGURE 8.41
Modifying the values of the starting and ending stations for the major labels

5. Click the icon in the Style field to display the Pick Label Style dialog.
6. Select the Edit Current Selection option from the Style drop-down menu. The Label Style Composer dialog appears.
7. On the General tab, change the value of Orientation Reference to View, as shown in Figure 8.42.

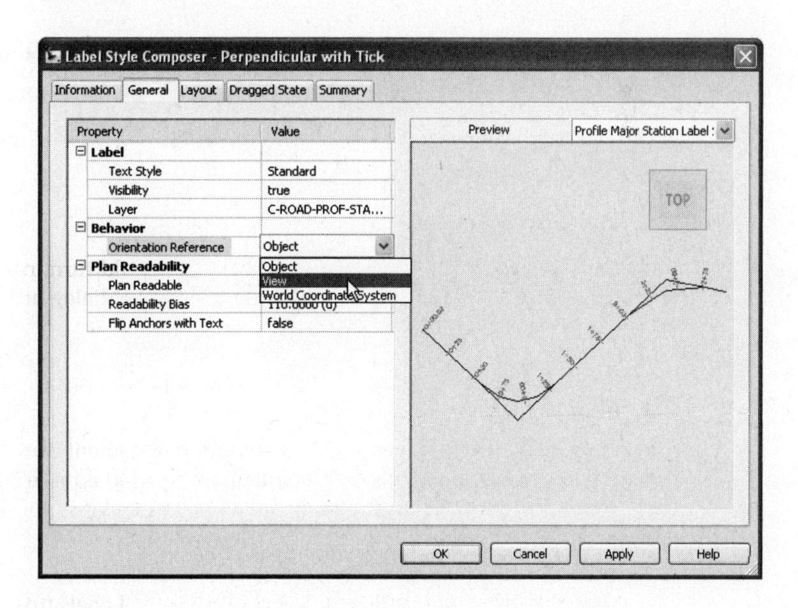

FIGURE 8.42
Changing the orientation reference of a label

8. Click OK to close the Label Style Composer dialog. Click OK again to close the Pick Label Style dialog.

9. Click OK to close the Profile Labels dialog. Instead of each station label being oriented so that it's perpendicular to the profile at the station, all station labels are now oriented vertically along the top of the profile at the station.

By controlling the frequency, starting and ending station, and label style, you can create labels for stationing or for conveying profile information along a layout profile.

LINE LABELS

Line labels in profiles are typically used to convey the slope or length of a tangent segment. In this exercise, you'll add a length and slope to the layout profile.

1. Open the `Line Profile Labels.dwg` file.
2. Switch to the Settings tab of Toolspace.
3. Expand the Profile ➤ Label Styles ➤ Line branches.
4. Right-click Percent Grade, and select the New option to open the Label Style Composer dialog and create a child style.
5. On the Information tab, change the Name field to Length and Percent Grade.
6. Change to the Layout tab.
7. Click in the cell for the Contents value, and click the ellipsis button to display the Text Component Editor.
8. Change the Properties drop-down menu to the Tangent Slope Length option and the Precision value to **0.01**, as shown in Figure 8.43.

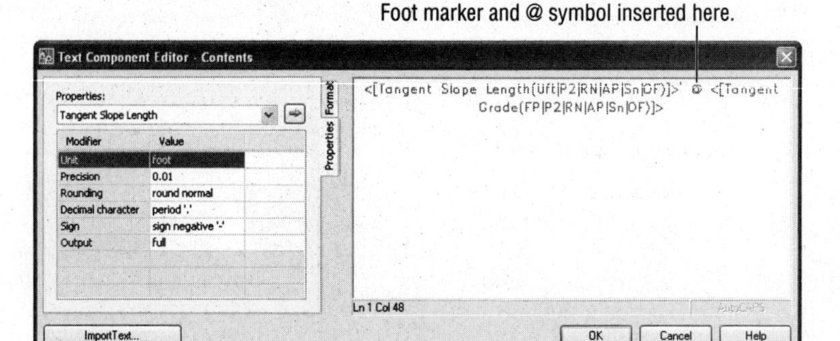

FIGURE 8.43
The Text Component Editor with the values for the Tangent Slope Length entered

9. Click the insert arrow, and then add a foot symbol, a space, an @ symbol, and another space in the editor's preview pane so that it looks like Figure 8.43.
10. Click OK to close the Text Component Editor, and click OK again to close the Label Style Composer dialog.
11. Pick the layout profile and select Edit Profile Labels from the Labels panel to display the Profile Labels dialog.
12. Change the Type field to the Lines option. The name of the Style drop-down menu changes to Profile Tangent Label Style. Select the Length and Percent Grade option.

13. Click the Add button, and then click OK to exit the dialog. The profile view should look like Figure 8.44.

> **WHERE IS THAT DISTANCE BEING MEASURED?**
>
> The *tangent slope length* is the distance along the horizontal geometry between vertical curves. This value doesn't include the tangent extensions. There are a number of ways to label this length; be sure to look in the Text Component Editor if you want a different measurement.

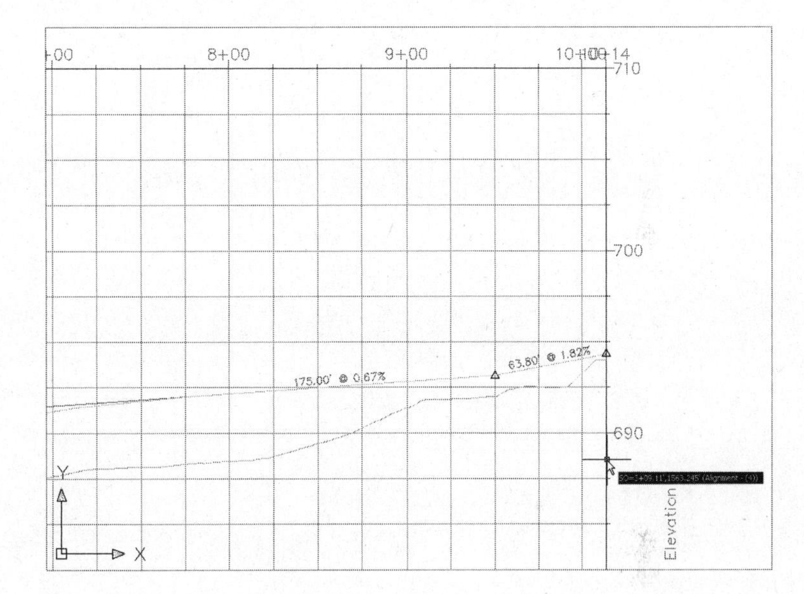

FIGURE 8.44
A new line label applied to the layout profile

CURVE LABELS

Vertical curve labels are one of the most confusing aspects of profile labeling. Many people become overwhelmed rapidly because there's so much that can be labeled and there are so many ways to get all the right information in the right place. In this quick exercise, you'll look at some of the special label anchor points that are unique to curve labels and how they can be helpful:

1. Open the `Curve Profile Labels.dwg` file.
2. Pick the layout profile and select Edit Profile Labels from the Labels panel to display the Profile Labels dialog.
3. Choose the Crest Curves option from the Type drop-down menu. The name of the Style drop-down menu changes. Select the Crest and Sag option.
4. Click the Add button to apply the label.
5. Choose the Sag Curves option from the Type drop-down menu. The name of the Style drop-down menu changes to Profile Sag Curve Label Style. Click Add to apply the same label style.
6. Click OK to close the dialog; your profile should look like Figure 8.45.

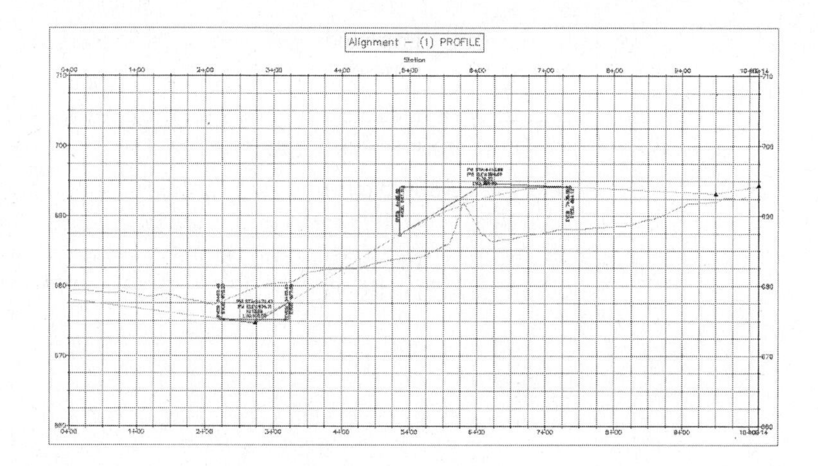

FIGURE 8.45
Curve labels applied with default values

Most labels are applied directly on top of the object being referenced. Because typical curve labels contain a large amount of information, putting the label right on the object can yield undesired results. In the following exercise, you'll modify the label settings to review the options available for curve labels:

1. Pick the layout profile and select Edit Profile Labels from the Labels panel to display the Profile Labels dialog.
2. Scroll to the right in the middle of the dialog, and locate the Dim Anchor Opt column.
3. Change the Dim Anchor Opt value for the Sag Curves to Distance Below.
4. Change the Dim Anchor Val to **2″**.
5. Change the Dim Anchor Val value for the Crest Curves to **2″** as well. Your dialog should look like the one shown in Figure 8.46.

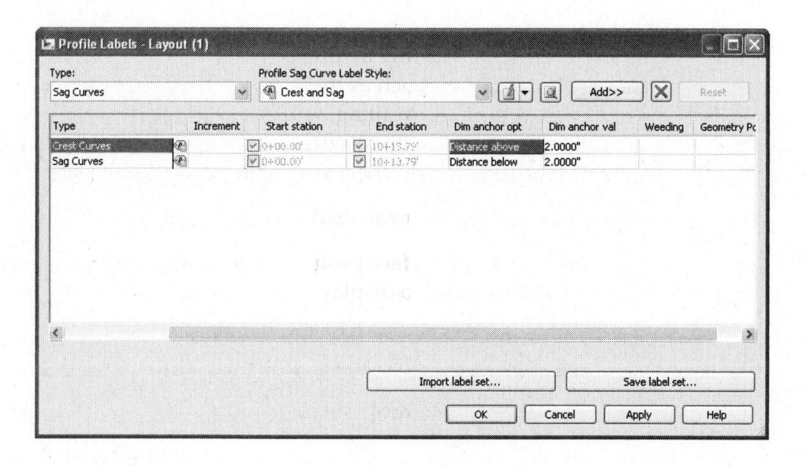

FIGURE 8.46
Curve labels with distance values inserted

6. Click OK to close the dialog.

The labels can also be grip-modified to move higher or lower as needed, but you'll try one more option in the following exercise:

1. Pick the layout profile and select Edit Profile Labels from the Labels panel to display the Profile Labels dialog.
2. Scroll to the right, and change both Dim Anchor Opt values for the Crest and Sag Curves to Graph View Top.
3. Change the Dim Anchor Val for both curves to **-1″**, and click OK to close the dialog. Your drawing should look like Figure 8.47.

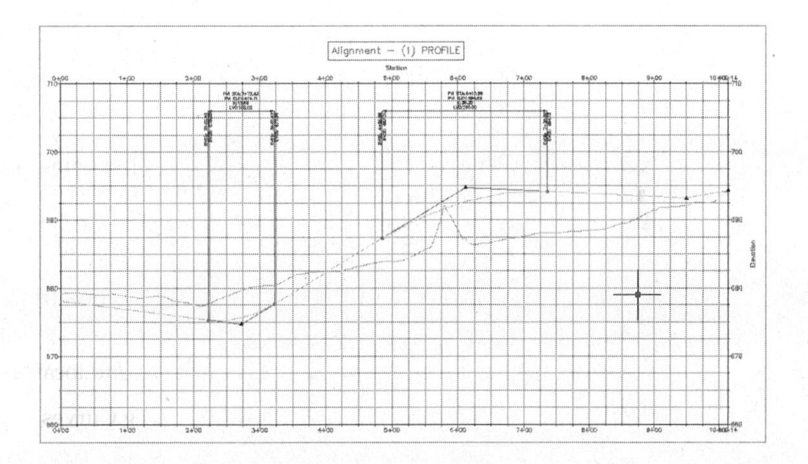

FIGURE 8.47
Curve labels anchored to the top of the graph

By using the top or bottom of the graph as the anchor point, you can apply consistent and easy labeling to the curve, regardless of the curve location or size.

GRADE BREAKS

The last label style typically involved in a profile is a grade-break label at PVI points that don't fall inside a vertical curve, such as the beginning or end of the layout profile. Additional uses include things like water-level profiling, where vertical curves aren't part of the profile information or existing surface labeling. In this exercise, you'll add a grade-break label and look at another option for controlling how often labels are applied to profile data:

1. Open the `Grade Break Profile Labels.dwg` file.
2. Pick the green surface profile (the irregular profile), and then select Edit Profile Labels from the Labels panel to display the Profile Labels dialog.
3. Choose Grade Breaks from the Type drop-down menu. The name of the Style drop-down menu changes. Select the Station over Elevation style and click the Add button.
4. Click Apply, and drag the dialog out of the way to review the change. It should appear as in Figure 8.48.

   A sampled surface profile has grade breaks every time the alignment crosses a surface TIN line. Why wasn't your view coated with labels?

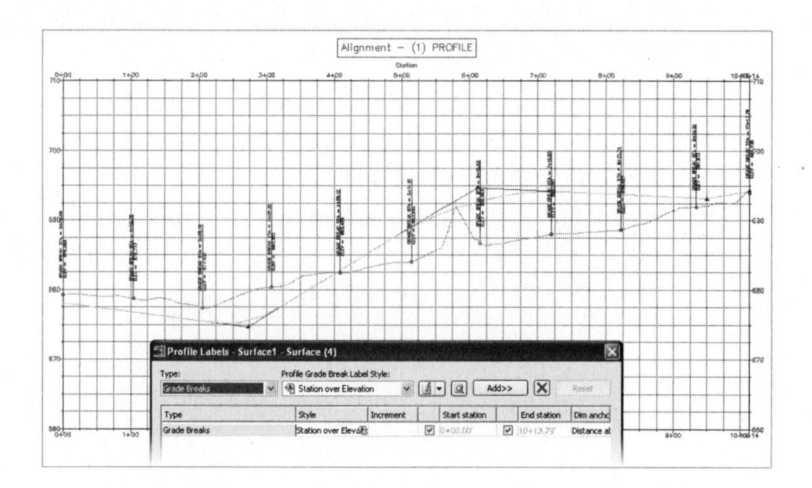

FIGURE 8.48
Grade-break labels on a sampled surface

5. Scroll to the right, and change the Weeding value to **50′**.
6. Click OK to dismiss the dialog. Your profile should look like the one in Figure 8.49.

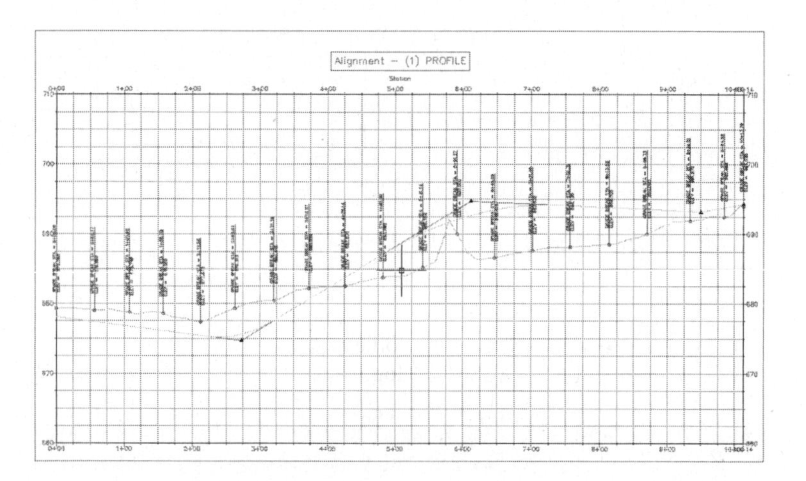

FIGURE 8.49
Grade-break labels with a 50′ Weeding value

Weeding lets you control how frequently grade-break labels are applied. This makes it possible to label dense profiles, such as a surface sampling, without being overwhelmed or cluttering the view beyond usefulness.

As you've seen, there are many ways to apply labeling to profiles, and applying these labels to each profile individually could be tedious. In the next section, you'll build a label set to make this process more efficient.

PROFILE LABEL SETS

Applying labels to both crest and sag curves, tangents, grade breaks, and geometry with the label style selection and various options can be monotonous. Thankfully, Civil 3D gives you the ability to use label sets, as in alignments, to make the process quick and easy. In this exercise, you'll apply a label set, make a few changes, and export a new label set that can be shared with team members or imported to the Civil 3D template. Follow these steps:

1. Open the `Profile Label Sets.dwg` file.
2. Pick the layout profile, and then select Edit Profile Labels from the Labels panel to display the Profile Labels dialog.
3. Click the Import Label Set button near the bottom of the dialog to display the Select Style Set dialog.
4. Select the Standard option from the drop-down menu, and click OK.
5. Click OK again to close the Profile Labels dialog and see the profile view, as in Figure 8.50.

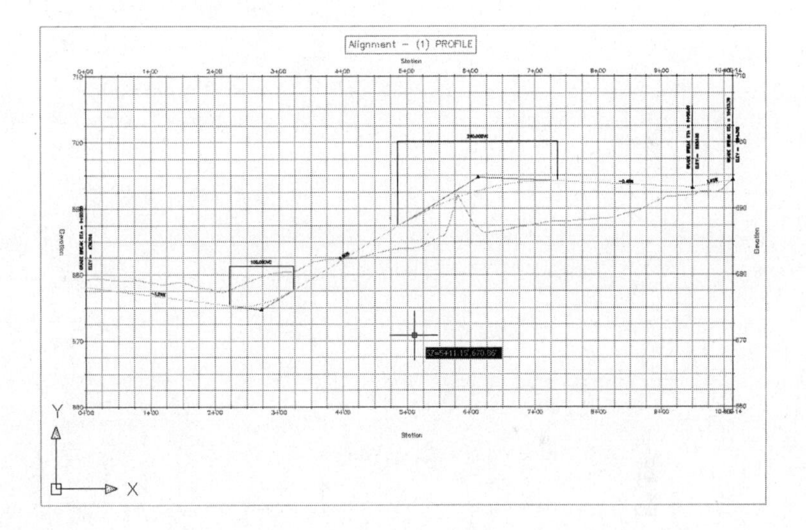

FIGURE 8.50
Profile with the Standard label set applied

6. Pick the layout profile and select Edit Profile Labels from the Labels panel to display the Profile Labels dialog.
7. Click Import Label Set to display the Select Style Set dialog.
8. Select the Complete Label Set option from the drop-down menu, and click OK.
9. Double-click the icon in the Style cell for the Lines label. The Pick Label Style dialog opens. Select the Length and Percent Grade option from the drop-down menu, and click OK.
10. Double-click the icon in the Style cell for both the Crest and Sag Curves labels. The Pick Label Style dialog opens. Select the Crest and Sag option for both curves from the drop-down menu, and click OK.
11. Click the Apply button, and drag the dialog out of the way, as shown in Figure 8.51, to see the changes reflected.
12. Click the Save Label Set button to open the Profile Label Set dialog and create a new profile label set.

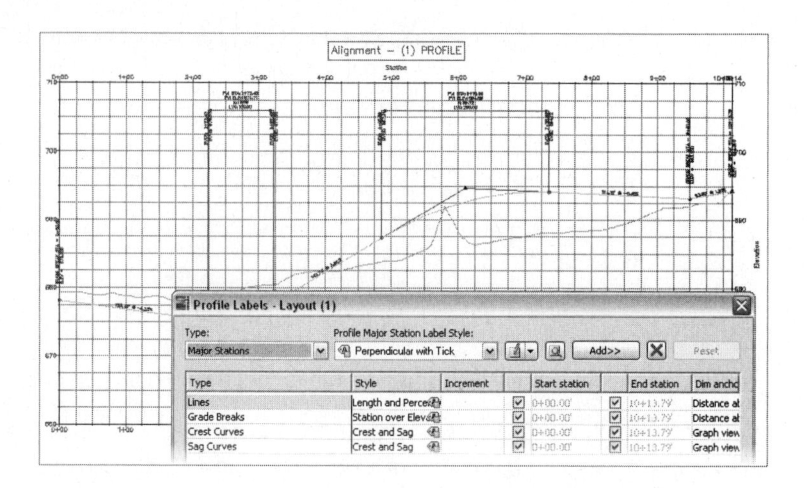

FIGURE 8.51
Establishing the Road Profile Labels label set

13. On the Information tab, change the Name to **Road Profile Labels**. Click OK to close the Profile Label Set dialog.
14. Click OK to close the Profile Labels dialog.
15. On the Settings tab of Toolspace, select Profile ➤ Label Styles ➤ Label Sets. Note that the Road Profile Labels set is now available for sharing or importing to other profile label dialogs.

SOMETIMES YOU DON'T WANT TO SET EVERYTHING

Resist the urge to modify the beginning or ending station values in a label set. If you save a specific value, that value will be applied when the label set is imported. For example, if you set a station label to end at 15+00 because the alignment is 15+15 long, that label will stop at 15+00, even if the target profile is 5,000 feet long!

Label sets are the best way to apply profile labeling uniformly. When you're working with a well-developed set of styles and label sets, it's quick and easy to go from sketched profile layout to plan-ready output.

The Bottom Line

Sample a surface profile with offset samples. Using surface data to create dynamic sampled profiles is an important advantage in working with a three-dimensional model. Quick viewing of various surface slices and grip-editing alignments makes for an effective preliminary planning tool. Combined with offset data to meet review agency requirements, profiles are robust design tools in Civil 3D.

Master It Open the `Mastering Profile.dwg` file and sample the ground surface along Alignment - (2), along with offset values at 15′ left and 25′ right of the alignment.

Lay out a design profile on the basis of a table of data. Many programs and designers work by creating pairs of station and elevation data. The tools built into Civil 3D let you input this data precisely and quickly.

Master It In the `Mastering Profiles.dwg` file, create a layout profile on Alignment (4) with the following information:

STATION	PVI ELEVATION	CURVE LENGTH
0+00	694	
2+90	696.50	250′
5+43.16	688	

Add and modify individual components in a design profile. The ability to delete, modify, and edit the individual components of a design profile while maintaining the relationships is an important concept in the 3D modeling world. Tweaking the design allows you to pursue a better solution, not just a working solution.

Master It In the `Mastering Profile.dwg` file, move the third PVI (currently at 9+65, 687) to 9+50, 690. Then, add a 175′ parabolic vertical curve at this point.

Apply a standard label set. Standardization of appearance is one of the major benefits of using Civil 3D styles in labeling. By applying label sets, you can quickly create plot-ready profile views that have the required information for review.

Master It In the `Mastering Profile.dwg` file, apply the Road Profiles label set to all layout profiles.

Chapter 9

Slice and Dice: Profile Views in Civil 3D

Although you work with profiles of all lengths while you're designing, that design profile eventually has to be printed for reviewers, contractors, and other project members. Profile views come into action when you convert the long profiles into workable chunks of data that fit nicely on the printed page with the labels and important information placed in the right places.

By the end of this chapter, you'll learn to:

- Create a simple view as part of the sampling process
- Change profile views and band sets as needed
- Split profile views into smaller views

A Better Point of View

Working with vertical data is an integral part of building the Civil 3D model. Once profile information has been created in any number of ways, displaying it to make sense is another whole task. It can't be stated enough that profiles and profile views are not the same thing in Civil 3D. The profile view is the method that Civil 3D uses to display profile data. A single profile can be shown in an infinite number of views, with different grids, exaggeration factors, labels, or linetypes. In this first part of the chapter, you'll look at the various methods available for creating profile views.

Creating During Sampling

The easiest way to create a profile view is to draw it as an extended part of the surface sampling procedure. In this brief exercise, you'll sample a surface and then create the view in one series of steps:

1. Open the `Profile Views 1.dwg` file. (Remember, all data files can be downloaded from `www.sybex.com/go/masteringcivil3d2010`.)
2. Change to the Home tab and select Profile ➢ Create Surface Profile from the Create Design panel to display the Create Profile from Surface dialog.
3. In the Alignment text box, select Parker Place. In the Select Surfaces list box, select the EG surface. Click the Add button.
4. Click the Draw in Profile View button to move into the Create Profile View Wizard, shown in Figure 9.1.

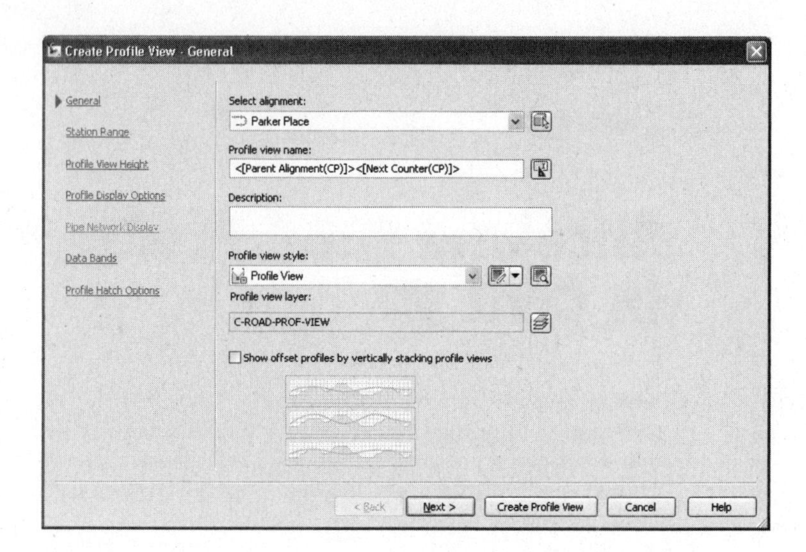

FIGURE 9.1
The Create Profile View Wizard

Profile views are created with the help of a wizard. The wizard offers the advantage of stepping through all the options involved in creating a view or simply accepting the command settings and creating the profile view quickly and simply.

5. Verify that Parker Place is selected in the Alignment drop-down list, and click Next.
6. Verify that the Station Range area has the Automatic option selected and click Next.
7. Verify that the Profile View Height field has the Automatic option selected and click Next.
8. Click the Create Profile View button in the Profile Display Options window.
9. Pick a point on screen somewhere to the right of the site and surface to draw the profile view, as shown in Figure 9.2.

By combining the profile sampling step with the creation of the profile view, you have avoided one more trip to the menus. This is the most common method of creating a profile view, but you look at a manual creation in the next section.

Creating Manually

Once an alignment has profile information associated with it, any number of profile views might be needed to display the proper information in the right format. To create a second, third, or tenth profile view once the sampling is done, it's necessary to use a manual creation method. In this exercise, you'll create a profile view manually for an alignment that already has a surface-sampled profile associated with it:

1. Open the `Profile Views 1.dwg` file if you have not already done so.
2. Change to the Home tab and select Profile View ➢ Create Profile View from the Profile & Section Views panel.
3. In the Select Alignment text box, select Carson's Way from the drop-down list. The profile was already sampled from the surface.

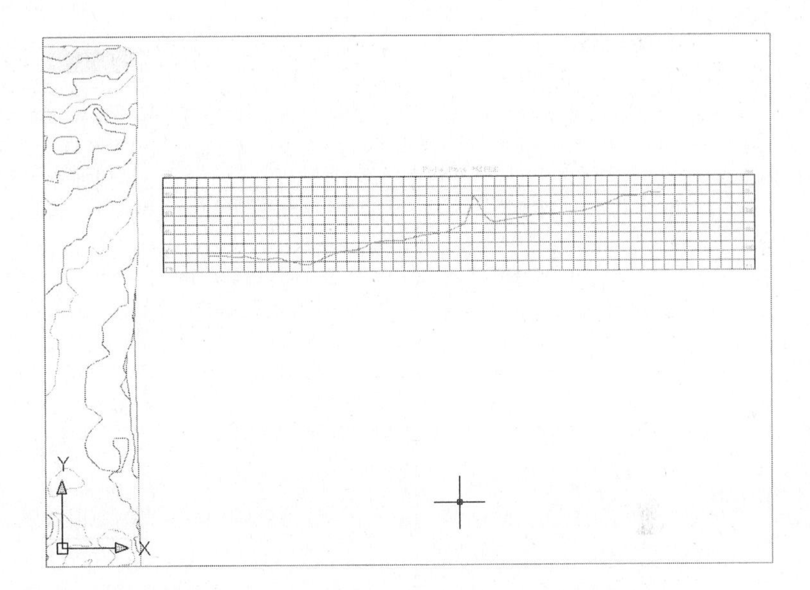

FIGURE 9.2
The completed profile view for Parker Place

4. In the Profile View Style drop-down list, select the Full Grid style.
5. Click the Create Profile View button and pick a point on screen to draw the profile view, as shown in Figure 9.3.

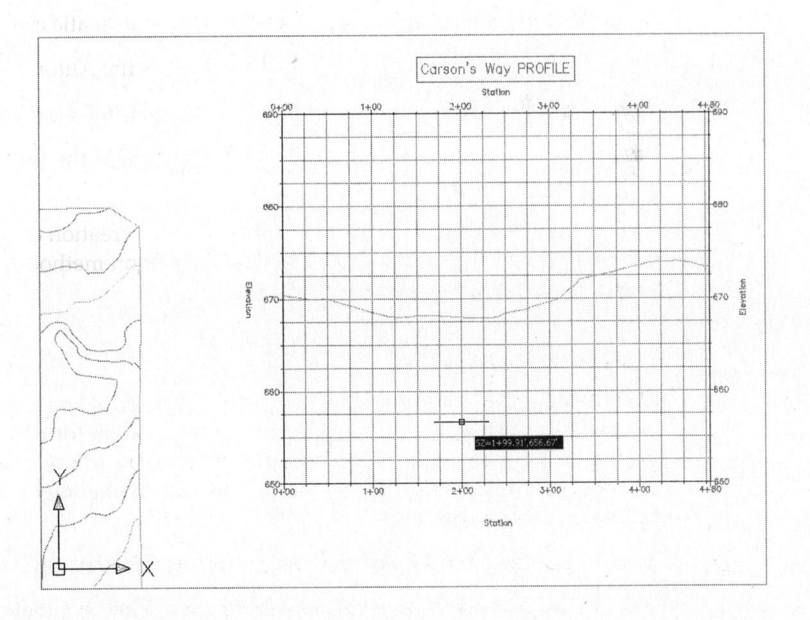

FIGURE 9.3
The completed profile view of Carson's Way

Using these two creation methods, you've made simple views, but you look at a longer alignment in the next exercise, and some more of the options available in the Create Profile View Wizard.

Splitting Views

Dividing up the data shown in a profile view can be time consuming. Civil 3D's Profile View Wizard is used for simple profile view creation, but the wizard can also be used to create manually limited profile views, staggered (or stepped) profile views, multiple profile views with gaps between the views, and stacked profiles (aka three-line profiles). You'll look at these different variations on profile view creation in this section.

Creating Manually Limited Profile Views

Continuous profile views like you made in the first two exercises work well for design purposes, but they are often unusable for plotting or exhibiting purposes. In this exercise, you'll sample a surface, and then use the wizard to create a manually limited profile view. This variation will allow you to control how long and how high each profile view will be, thereby making the views easier to plot or use for other purposes.

1. Open the `Profile Views 2.dwg` file.
2. Change to the Home tab and select Profile ➢ Create Surface Profile from the Create Design panel to display the Create Profile from Surface dialog.
3. In the Alignment text box, select Rose Drive; in the Select Surface list box, select the EG surface and click the Add button.
4. Click the Draw in Profile View button to enter the wizard.
5. In the Profile View Style drop-down list, select the Full Grid style and click Next.
6. In the Station Range area, select the User Specified Range radio button. Enter **0** for the Start station and **10+00** for the End station, as shown in Figure 9.4. Notice the preview picture shows a clipped portion of the total profile. Click Next.

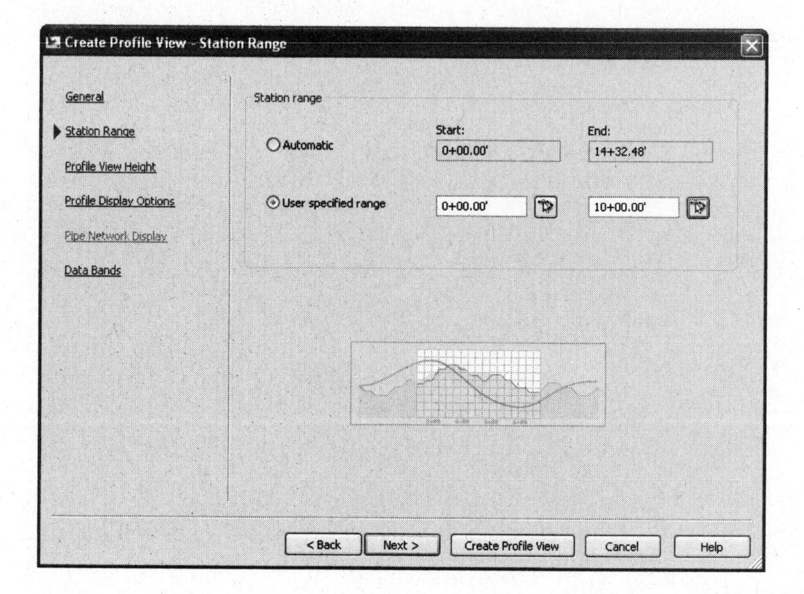

FIGURE 9.4 The start and end stations for the user-specified profile view

7. In the Profile View Height area, select the User Specified radio button. Set the Minimum height to **665** and the Maximum height to **705**.

8. Click the Create Profile View button and pick a point on screen to draw the profile view. Your screen should look similar to Figure 9.5.

FIGURE 9.5
Applying user-specified station and height values to a profile view

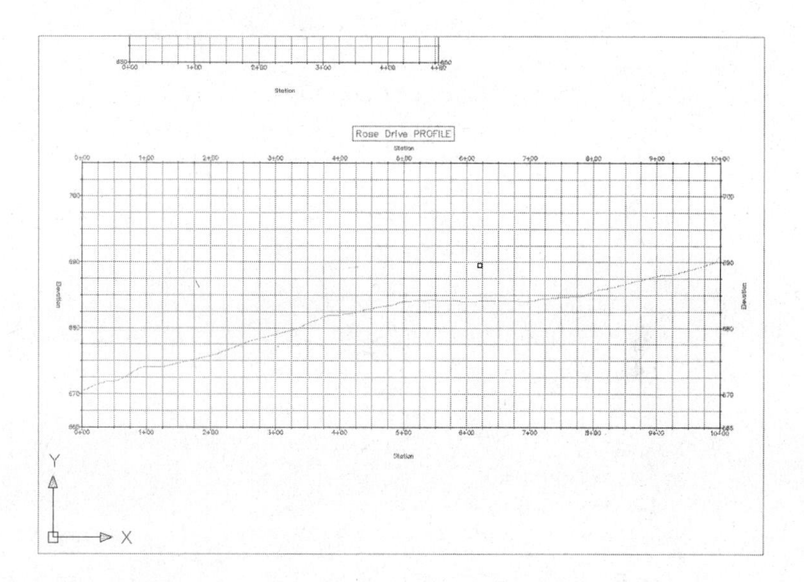

CREATING STAGGERED PROFILE VIEWS

When large variations occur in profile height, the graph must often be split just to keep from wasting much of the page with empty gridlines. In this exercise, you use the wizard to create a staggered, or stepped, view:

1. Open the `Profile Views 2.dwg` file if you haven't already.

2. Change to the Home tab and select Profile ➢ Create Surface Profile from the Create Design panel to display the Create Profile from Surface dialog.

3. In the Alignment text box, select Escarpment; in the Select Surface list box, select the EG surface and click the Add button.

4. Click the Draw in Profile View button to enter the wizard. Click Next to move to the Station Range options.

5. Click Next in the Station Range window to allow the view to show the full length.

6. In the Profile View Height field, select the User Specified option and set the values to **628.00′** and **668.00′** as shown in Figure 9.6.

7. Check the Split Profile View options and set the view styles, as shown in Figure 9.6.

8. Click the Create Profile View button and pick a point on screen to draw the staggered display, as shown in Figure 9.7.

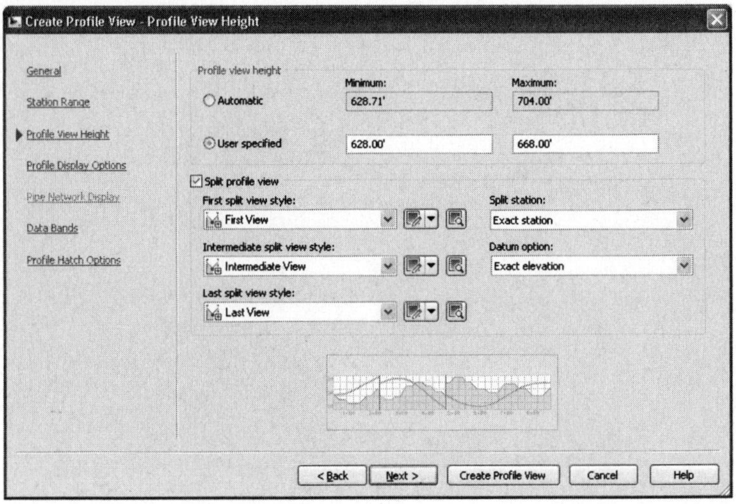

FIGURE 9.6
Split Profile View settings

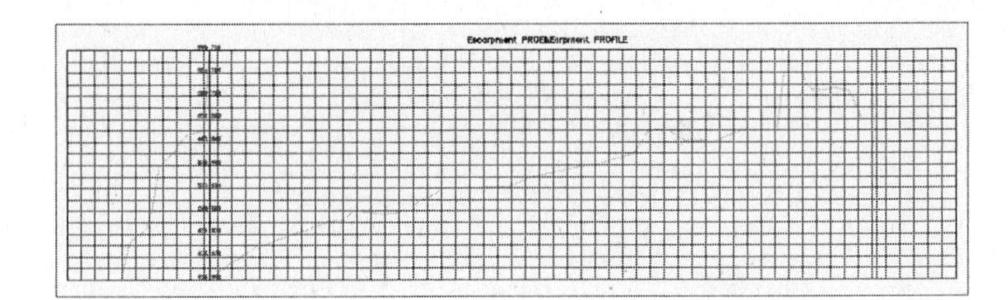

FIGURE 9.7
A staggered (stepped) profile view created via the wizard

The profile view is split into views according to the settings that were selected in the Create Profile View Wizard in step 7. The first section shows the profile from 0 to the station where the elevation change of the profile exceeds the limit for height. The second section displays the rest of the profile. Each of these sections is part of the same profile view and can be adjusted via the Profile View Properties dialog.

CREATING GAPPED PROFILE VIEWS

Profile views must often be limited in length and height to fit a given sheet size. Gapped views are a way to show the entire length and height of the profile, by breaking the profile into different sections with "gaps" or spaces between each view. In this exercise, you use a variation of the Create Profile View Wizard to create gapped views automatically:

1. Open the `Profile View 2.dwg` file if you haven't already. *If you did the previous exercise, skip steps 2 and 3!*
2. Change to the Home tab and select Profile ➢ Create Surface Profile from the Create Design panel to display the Create Profile from Surface dialog.
3. In the Alignment text box, select Escarpment; in the Select Surface list box, select the EG surface and click the Add button. Click OK to exit the dialog.

4. Change to the Home tab and select Profile View ➤ Create Multiple Profile Views to display the Create Multiple Profile Views Wizard (see Figure 9.8).

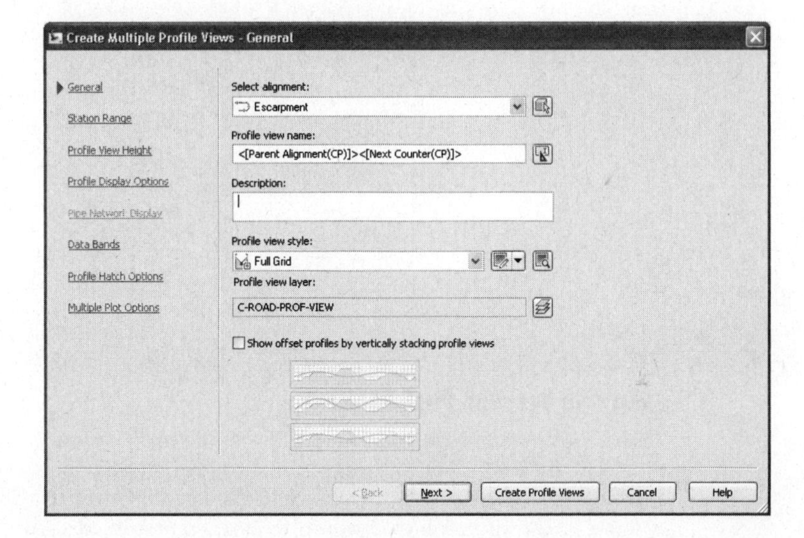

FIGURE 9.8
The Create Multiple Profile Views Wizard

5. In the Select Alignment drop-down list, select Escarpment, and in the Profile View Style drop-down list, select the Full Grid option, as shown in Figure 9.8. Click Next.
6. In the Station Range area, make sure the Automatic option is selected. This area is also where the Length of Each View is set. Click Next.
7. In the Profile View Height area, make sure the Automatic option is selected. Note that you could use the Split Profile View options from the previous exercise here as well.
8. Click the Multiple Plot Options text on the left side of the dialog to jump to that step in the wizard. This step controls whether the gapped profile views will be arranged in a column, row, or a grid. The Escarpment alignment is fairly short, so the gapped views will be aligned in a row. However, it could be prudent with longer alignments to stack the profile views in a column or a compact grid, thereby saving screen space.
9. Click the Create Profile Views button and pick a point on screen to create a view similar to Figure 9.9.

The gapped profile views are the two profile views on the bottom of the screen and, just like the staggered profile view, show the entire alignment from start to finish. Unlike the staggered view, however, the gapped view is separated by a "gap" into two views. In addition, the gapped views are independent of each other so they have their own styles, properties, and labeling associated with them, making them useful when you don't want a view to show information that is not needed on a particular section. This is also the primary way to create divided profile views for sheet production.

Note that when using the Create Multiple Views option, every profile view is the full length as defined in the wizard, even if the alignment is not that long.

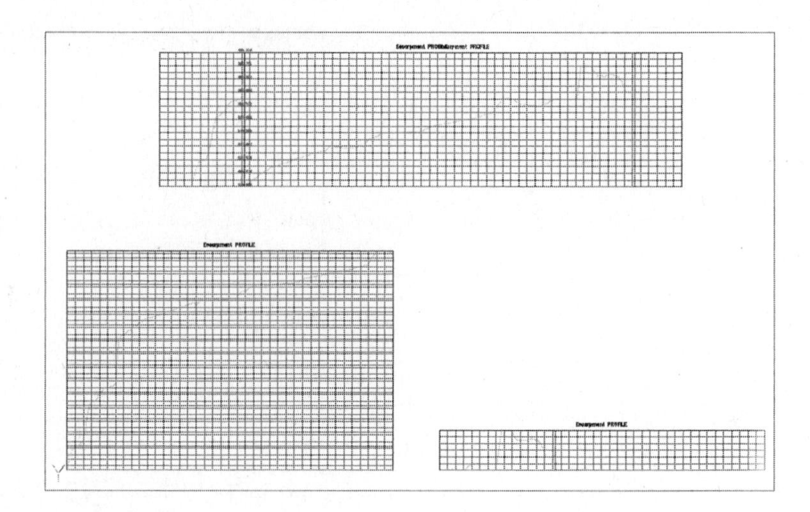

FIGURE 9.9
The staggered and gapped profile views of the Escarpment alignment

CREATING STACKED PROFILE VIEWS

In the southwestern United States, a three-line profile view is a common requirement. In this situation, the centerline is displayed in a central profile view, with left and right offsets shown in profile views above and below the centerline profile view. These are then typically used to show top-of-curb design profiles in addition to the centerline design. In this exercise, you look at how the Profile View Wizard makes generating these views a simple process:

1. Open the `Stacked Profiles.dwg` file. This drawing has sampled profiles for the Rose Drive alignment at center as well as left and right offsets.
2. Change to the Home tab and select Profile View ➢ Create Profile View to display the Create Profile View Wizard.
3. Select Rose Drive from the Select Alignment drop-down list.
4. Check the Show Offset Profiles by Vertically Stacking Profile Views option on the General page of the wizard.
5. Click the Stacked Profile text on the left side of the wizard (it's a hyperlink) to jump to the Stacked Profile step and the dialog will appear as shown in Figure 9.10.
6. Set the style for each view as shown in Figure 9.10 and click Next.
7. Toggle the Draw option for the first profile (EG – Surface (12)) as shown in Figure 9.11. Note that Middle View - [1] is currently selected in the Select Stacked View to Specify Options For list box.
8. Click Top View in the Select Stacked View to Specify Options For list box, and then toggle on the EG - - 25.000 profile (the second profile listed). Note the two "-" symbols because of the naming template that ships in Civil 3D. This is the left-hand offset.
9. Click Bottom View in the Select Stacked View to Specify Options For list box, and then toggle on the EG - 25.000 profile (the third profile listed). This is the right-hand offset.
10. Click the Create Profile View button, and snap to the center of the circle to the southeast of the site as a pick point. Your result should look similar to Figure 9.12.

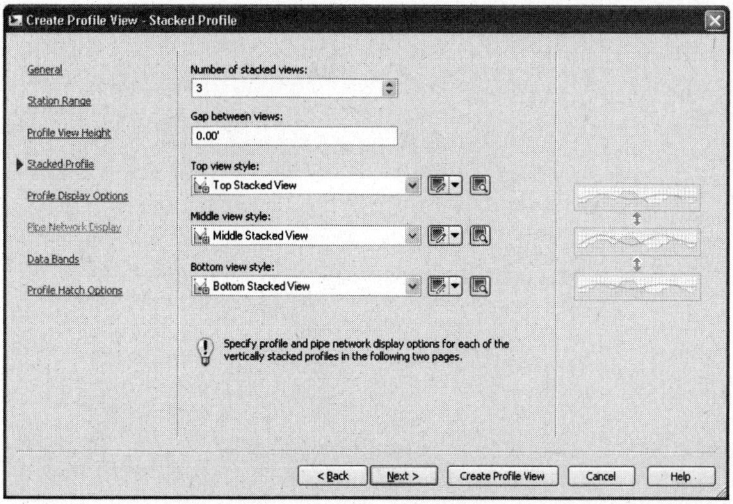

FIGURE 9.10
Setting up stacked profile views

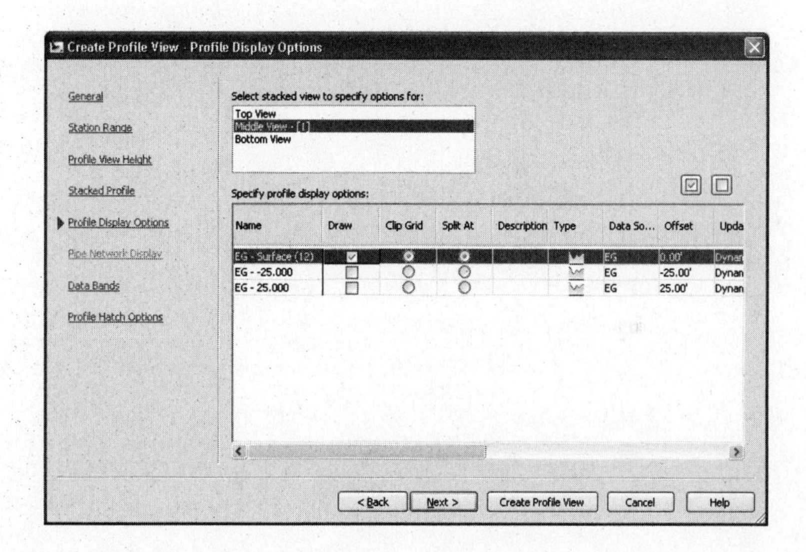

FIGURE 9.11
Setting the stacked view options for each view

As you saw in Chapter 8, "Cut to the Chase: Profiles," the styles for each profile can be adjusted, as can the styles for each profile view. The stacking here simply automates a process that many users found tedious. Once you have created layout profiles, they can also be added to these views by editing the Profile View Properties.

Profile Utilities

One common requirement is to compare profile data for objects that are aligned similarly but not parallel. Another is the ability to project objects from a plan view into a profile view. The abilities to superimpose profiles and project objects are both discussed in this section.

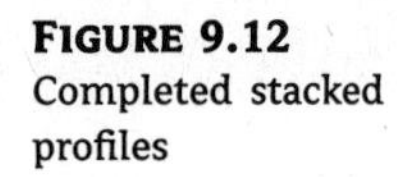

FIGURE 9.12
Completed stacked profiles

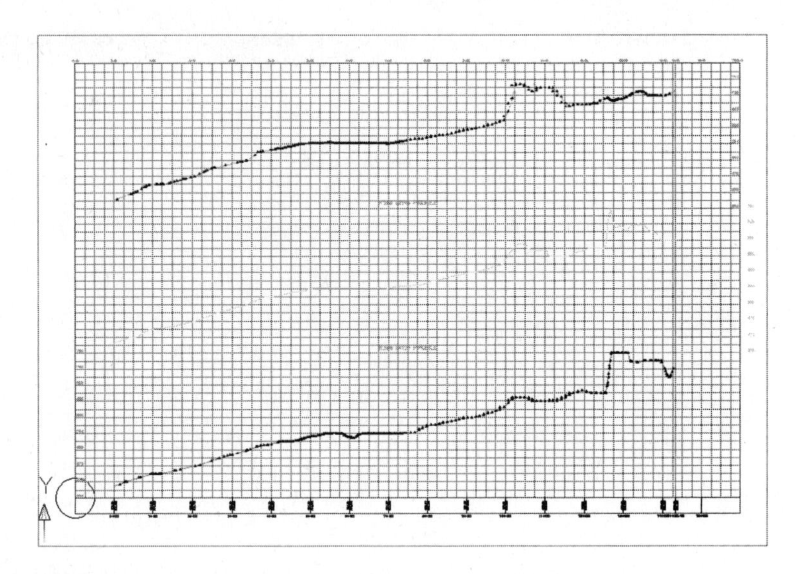

Superimposing Profiles

In a profile view, a profile is sometimes superimposed to show one profile adjacent to another (e.g., a ditch adjacent to a road centerline). In this brief exercise, you'll superimpose one of your street designs onto the other to see how they compare over a certain portion of their length:

1. Open the `Superimpose Profiles.dwg` file. This drawing has two profile views created, one with a layout profile.
2. Change to the Home tab and select Profile ➤ Create Superimposed Profile. Civil 3D will prompt you to select a source profile.
3. Zoom into the Carson's Way profile view and pick the cyan layout profile. Civil 3D will prompt you to select a destination profile view for display.
4. Pick the Parker Place profile view to display the Superimpose Profile Options dialog shown in Figure 9.13.
5. Click OK to dismiss the dialog, accepting the default settings.
6. Zoom in on the left side of the Parker Place profile view to see the superimposed data, as shown in Figure 9.14.

Note that the vertical curve in the Carson's Way layout profile has been approximated on the Parker Place profile view, using a series of PVIs. Superimposing works by projecting a line from the target alignment (Parker Place) to an intersection with the other source alignment (Carson's Way).

The target alignment is queried for an elevation at the intersecting station and a PVI is added to the superimposed profile. See Figure 9.15 for a bit of clarification. Note that this superimposed profile is still dynamic! A change in the Carson's Way layout profile will be reflected on the Parker Place profile view.

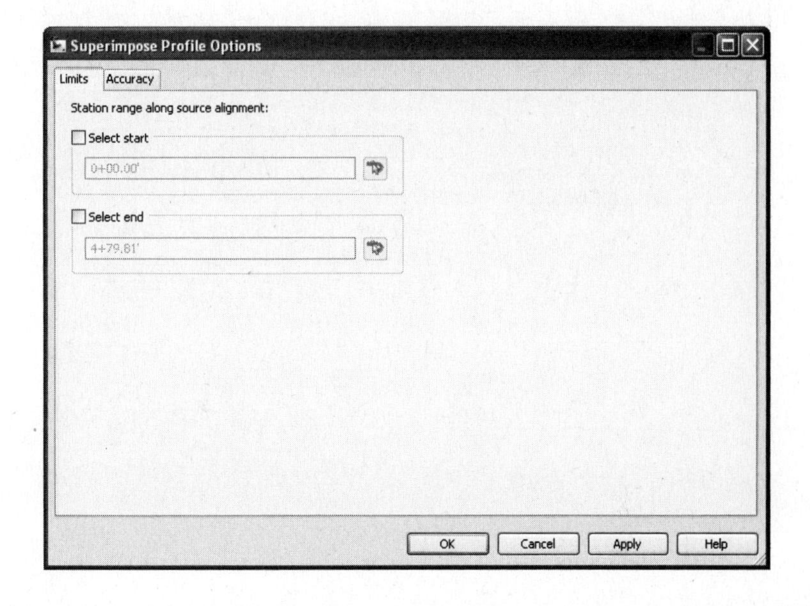

FIGURE 9.13
The Superimpose Profile Options dialog

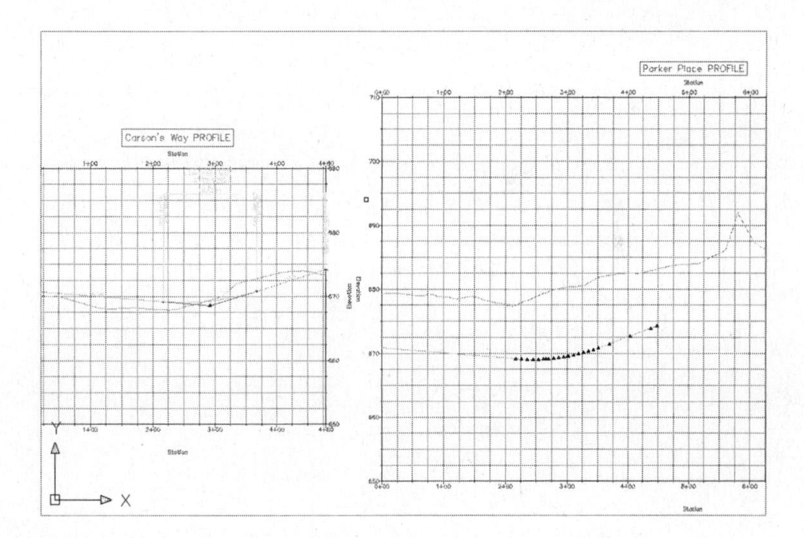

FIGURE 9.14
The Carson's Way layout profile superimposed on the Parker Place profile view

Object Projection

Some AutoCAD and some AutoCAD Civil 3D objects can be projected from a plan view into a profile view. The list of available AutoCAD objects includes points, blocks, 3D solids, and 3D polylines. The list of available AutoCAD Civil 3D objects includes COGO points, feature lines, and survey figures. In the following exercise, you'll project a COGO point located at the intersection of two alignments into a profile view:

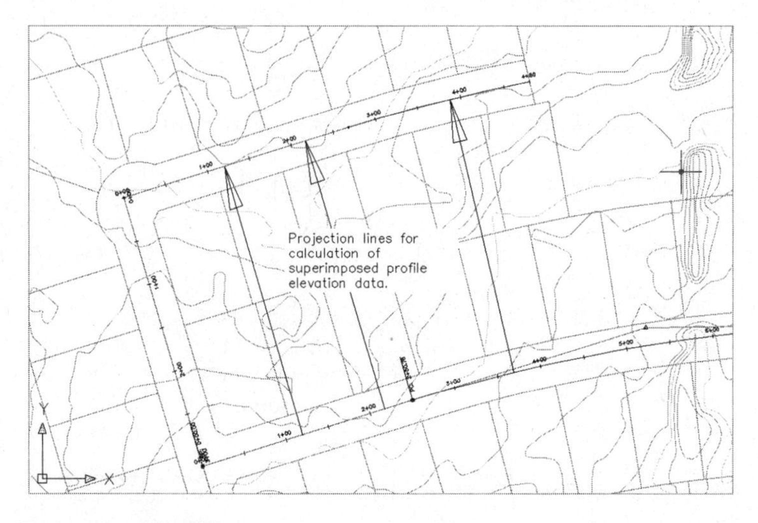

FIGURE 9.15
An example of profile sample lines that have been superimposed

1. Open the `Object Projection.dwg` file.
2. Change to the Home tab and select Profile View ➢ Project Objects to Profile View from the Profile & Section Views panel. Select the CL-CL point object located in the center of the screen and press ↵. Civil 3D will prompt you to select a profile view.
3. Pan to the Rose Drive profile view located in the bottom right of the drawing and select a grid line. The Project Objects To Profile View dialog opens.
4. Select the Layout (2) Elevation Option and verify that the other options match those in Figure 9.16. Click OK to dismiss the dialog, and review your results as shown in Figure 9.17

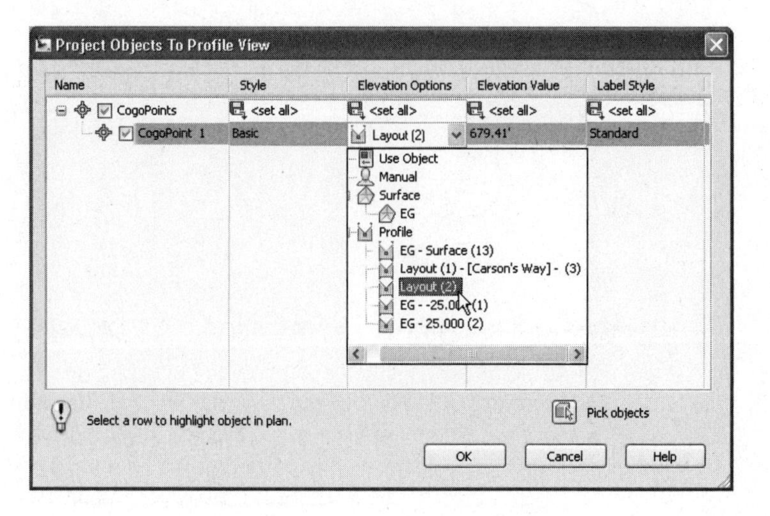

FIGURE 9.16
A completed Project Objects To Profile View dialog

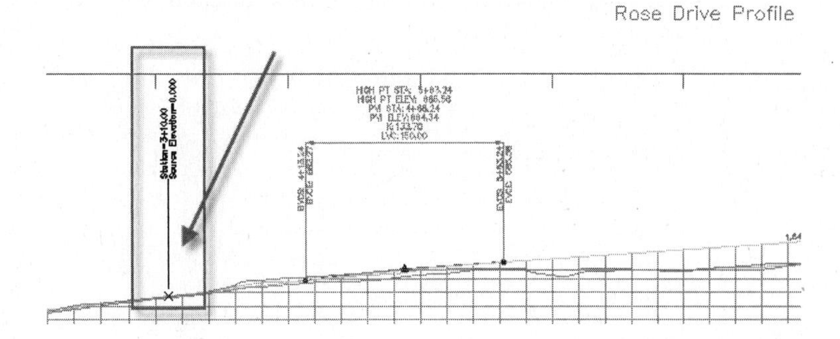

FIGURE 9.17
The COGO point object projected into a profile view.

Once an object has been projected into a profile view, the Profile View Properties dialog will display a new Projections tab. Projected objects will remain dynamically linked with respect to their plan placement. Because profile views and section views are similar in nature, objects can be projected into section views in the same fashion.

Editing Profile Views

Once profile views have been created, things gets interesting. The number of modifications to the view itself that can be applied, even before editing the styles, makes profile views one of the most flexible pieces of the AutoCAD Civil 3D package. In this series of exercises, you look at a number of changes that can be applied to any profile view in place.

Profile View Properties

Picking a profile view and selecting Profile View Properties from the Modify View panel yields the dialog shown in Figure 9.18. The properties of a profile include the style applied, station and elevation limits, the number of profiles displayed, and the bands associated with the profile view. If a pipe network is displayed, a tab labeled Pipe Networks will appear.

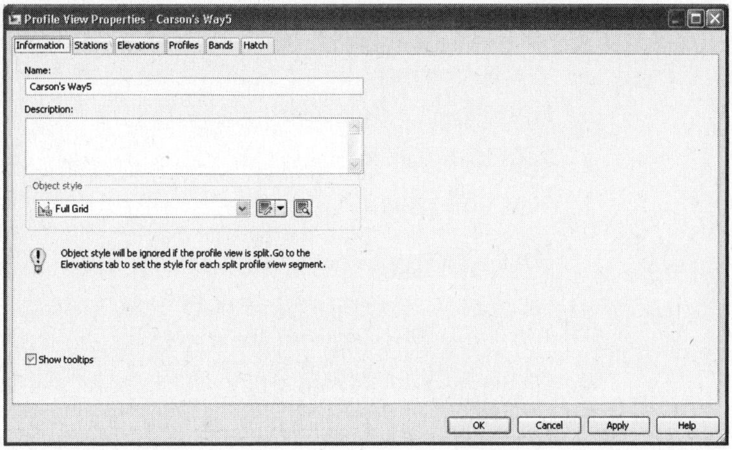

FIGURE 9.18
Typical Profile View Properties dialog

ADJUSTING THE PROFILE VIEW STATION LIMITS

In spite of the wizard, there are often times when a profile view needs to be manually adjusted. For example, the most common change is to limit the length or height (or both) of the alignment that is being shown so it fits on a specific size of paper or viewport. You can make some of these changes during the initial creation of a profile view (as shown in a previous exercise), but you can also make changes after the profile view has been created.

One way to do this is to use the Profile View Properties dialog to make changes to the profile view. The profile view is a Civil 3D object, so it has properties and styles that can be adjusted through this dialog to make the profile view look like you need it to.

1. Open the `Profile View Properties.dwg` file.
2. Zoom to the Rose Drive profile view.
3. Pick a grid line, and select Profile View Properties from the Modify View panel.
4. On the Stations tab, click the User Specified Range radio button, and set the value of the End station to 10+00, as shown in Figure 9.19.

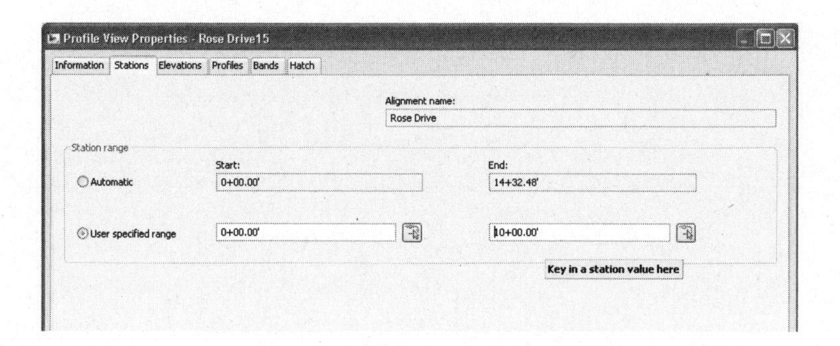

FIGURE 9.19
Adjusting the end station values for Rose Drive

5. Click OK to close the dialog. The profile view will now reflect the updated end station value.

   One of the niceties in Civil 3D is that copies of a profile view retain the properties of that view, making a gapped view easy to create manually if they were not created with the wizard.

6. Enter **Copy** ↵ on the command line. Pick the Rose Drive profile view you just modified.
7. Press F8 on your keyboard to toggle on the orthogonal mode, and then press ↵.
8. Pick a base point and move the crosshairs to the right. When the crosshairs reach a point where the two profile views do not overlap, pick that as your second point, and press ↵ to end the Copy command.
9. Pick the copy just created and select Profile View Properties from the Modify View panel. The Profile View Properties dialog appears.
10. On the Stations tab, change the stations again. This time, set the Start field to **10+00** and the End field to **14+32.48**. The total length of the alignment will now be displayed on the two profile views, with a gap between the two views at station 10+00. Click OK, and your drawing will look like Figure 9.20.

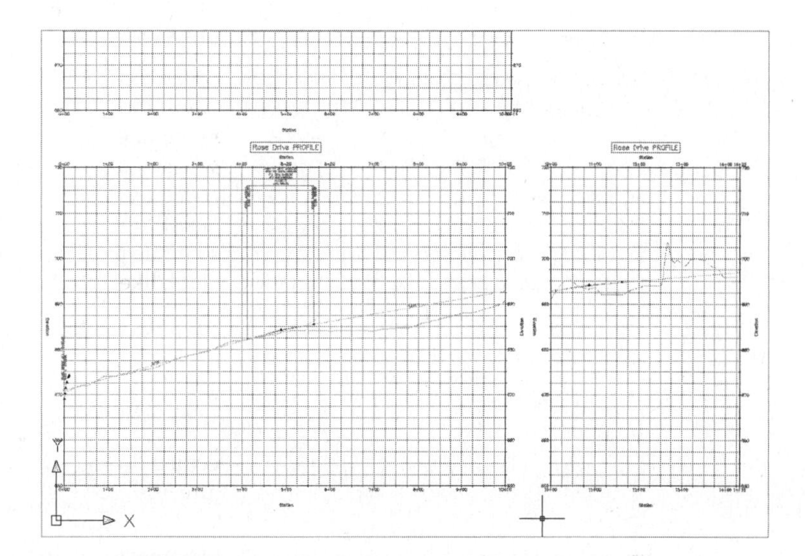

FIGURE 9.20
A manually created gap between profile views

In addition to creating gapped profile views by changing the profile properties, you can also show phase limits by applying a different style to the profile in the second view.

ADJUSTING THE PROFILE VIEW ELEVATIONS

Another common issue is the need to control the height of the profile view. Civil 3D automatically sets the datum and the top elevation of profile views on the basis of the data to be displayed. In most cases this is adequate, but in others, this simply creates a view too large for the space allocated on the sheet or wastes a large amount of that space.

1. Open the `Profile View Properties.dwg` file if you have not already done so.
2. Pan or zoom to the Parker Place profile view.
3. Select Profile View Properties from the Modify View panel. The Profile View Properties dialog appears.
4. Change to the Elevations tab.
5. In the Elevation Range area, check the User Specified Height radio button and enter the Minimum and Maximum heights, as shown in Figure 9.21.

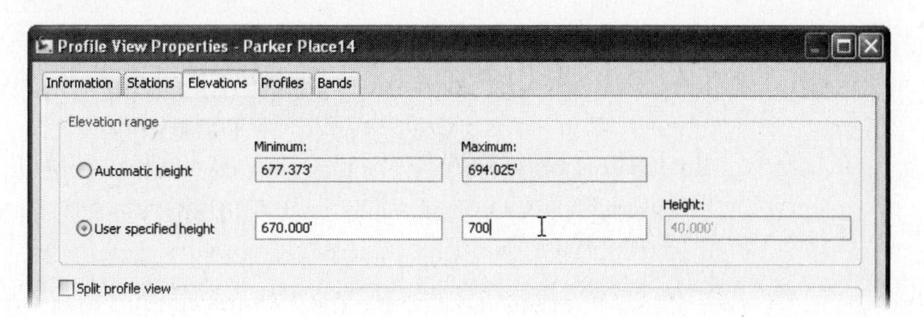

FIGURE 9.21
Modifying the height of the profile view

6. Click OK to close the dialog. The profile view of Parker Place should reflect the updated elevations as in Figure 9.22.

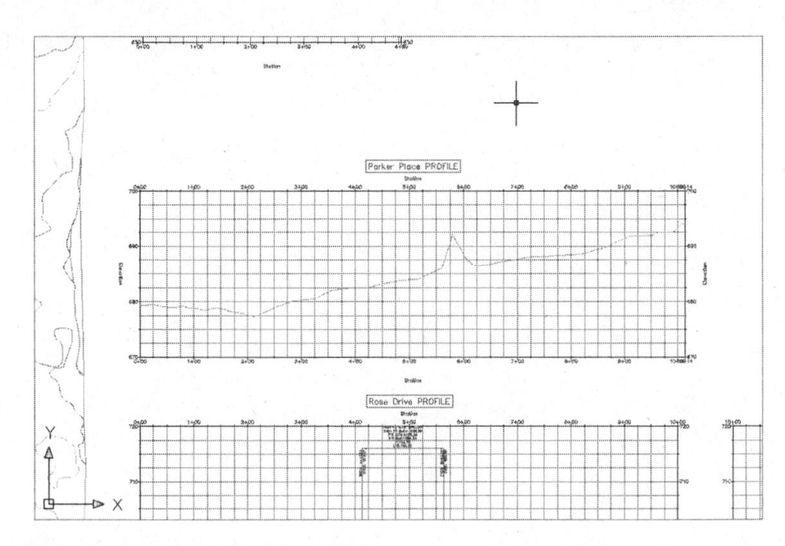

FIGURE 9.22
The updated profile view with the heights manually adjusted

The Elevations tab can also be used to split the profile view and create the staggered view that you previously created with the wizard.

1. Pick the Escarpment profile view, right-click a grid line, and select the Profile View Properties option to open the Profile View Properties dialog.
2. Switch to the Elevations tab. In the Elevations Range area, click the User Specified Height radio button.
3. Check the Split Profile View option.
4. Notice that the Height field is now active. Set the height to **40**.
5. Click OK to exit the dialog. The profile view should look like Figure 9.23.

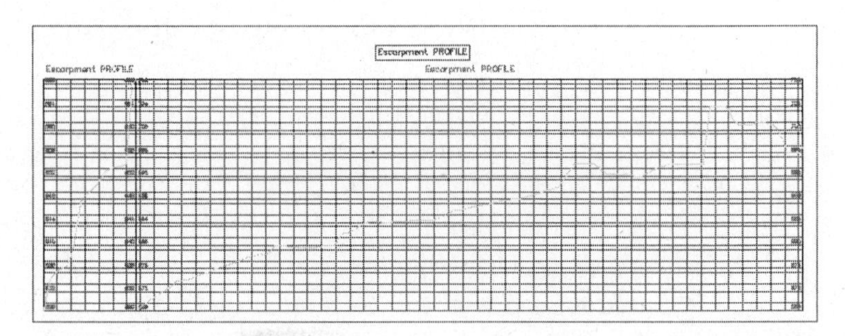

FIGURE 9.23
A split profile view for the Escarpment alignment

The automation gives you a rough idea of what your profile looks like, but it simply doesn't work or look as good as you'd like it. In this exercise, you tweak this even further by manually splitting the view:

1. Select the Escarpment profile view, and select Profile View Properties from the Modify View panel. The Profile View Properties dialog appears.
2. In the Split Profile View area, click the Manual radio button to turn on the Split Profile View Data table.

3. In the Split Profile View Data table, change the values of the Split Station, the Adjusted Datum, and the Profile View Style for profile views 1 and 2 so that they match Figure 9.24. Notice that you also had to change the Height field in the Elevation Range area.

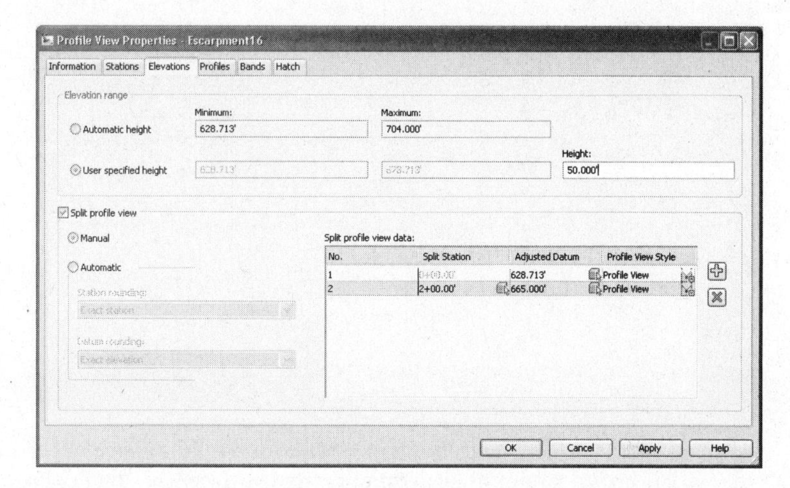

FIGURE 9.24
Updating the Elevations tab

4. Click OK to close the dialog, and your screen should now be similar to Figure 9.25.

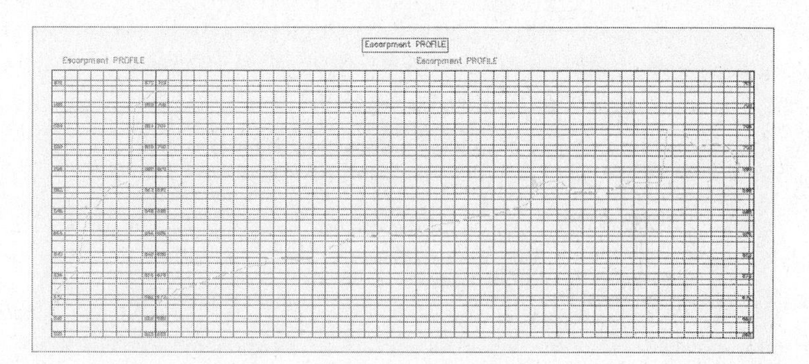

FIGURE 9.25
Completed split profile edits

Automatically creating split views is a good starting point, but you'll often have to tweak them as you've done here. The selection of the proper profile view styles is an important part of the Split Profile View process. You look at styles in a later section of this chapter.

PROFILE DISPLAY OPTIONS

Civil 3D allows the creation of literally hundreds of profiles for any given alignment. This makes it easy to evaluate multiple design solutions, but it can also mean that profile views get very crowded. In this exercise, you'll look at some profile display options that allow the toggling of various profiles within a profile view:

1. Open the `Profile View Properties.dwg` file if you have not done so already.
2. Pick the Carson's Way profile view, and select Profile View Properties from the Modify View panel. The Profile View Properties dialog appears.
3. Switch to the Profiles tab.

4. Uncheck the Draw option in the EG Surface row and click the Apply button.
5. Drag your dialog out of the way and your profile view should look similar to Figure 9.26.

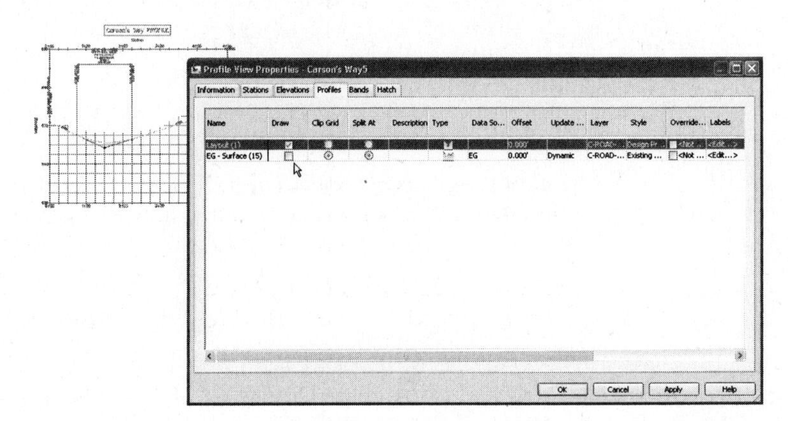

FIGURE 9.26
The Carson's Way profile view with the Draw option toggled off

Toggling off the Draw option for the EG surface has created a profile view style in which a profile of the existing ground surface will not be drawn on the profile view. In addition, this style includes an option that removes, or "clips," the grid lines above the EG surface profile. In effect, the EG profile line acts as the grid-clipping line. Clipping is typically used to satisfy a reviewer's request more than to help the Civil 3D user. Clipping is generally a reviewer requirement more than a user one.

6. Click the radio button in the Layout (1) row for the Clip Grid option.
7. Click OK to have a profile view similar to Figure 9.27.

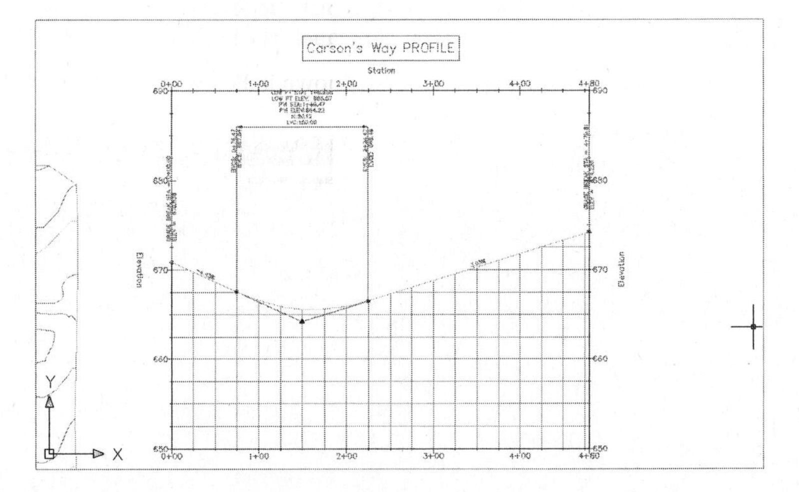

FIGURE 9.27
The Carson's Way profile view limited to one profile line

The sampled profile from the EG surface still exists under the Carson's Way alignment; it simply isn't shown in the current profile view. Now that you've modified a number of styles, you'll look at another option that is available on the Profile View Properties dialog: bands.

Profile View Bands

Data bands are horizontal elements that display additional information about the profile or alignment that is referenced in a profile view. Bands can be applied to both the top and bottom of a profile view, and there are six different band types:

- **Profile Data bands** — Display information about the selected profile. This information can include simple elements such as elevation, or more complicated information such as the cut-fill between two profiles at the given station.
- **Vertical Geometry bands** — Create an iconic view of the elements making up a profile. Typically used in reference to a design profile, vertical data bands make it easy for a designer to see where vertical curves are located along the alignment.
- **Horizontal Geometry bands** — Create a simplified view of the horizontal alignment elements, giving the designer or reviewer information about line, curve, and spiral segments and their relative location to the profile data being displayed.
- **Superelevation bands** — Display the various options for Superelevation values at the critical points along the alignment.
- **Sectional Data bands** — Can display information about the sample line locations, distance between them, and other sectional-related information.
- **Pipe Data bands** — Can show specific information about each pipe or structure being shown in the profile view.

In this exercise, you add bands to give feedback on the EG and layout profiles, as well as horizontal and vertical geometry:

1. Open the `Profile View Bands.dwg` file.
2. Zoom out and down to pick the Rose Drive profile view, select Profile View Properties from the Modify View panel. The Profile View Properties dialog appears.
3. Click the Bands tab, as shown in Figure 9.28.

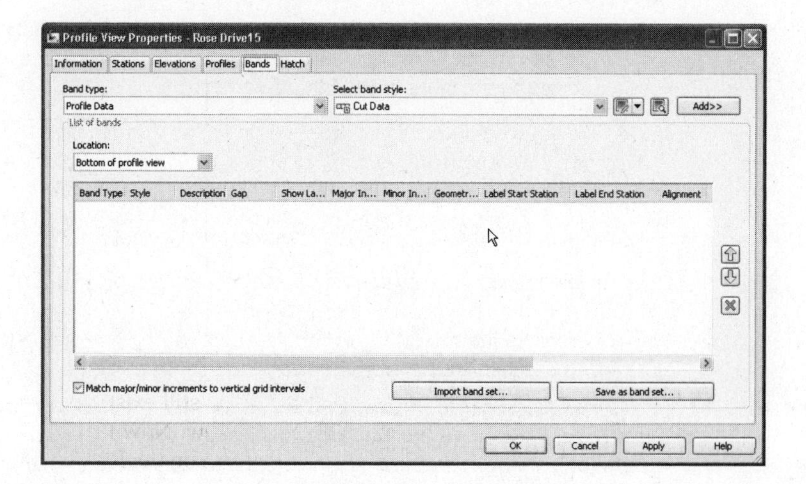

FIGURE 9.28
The Bands tab of the Profile View Properties dialog

4. Verify that the Band Type drop-down list is set to Profile Data and set the Select Band Style drop-down list to Elevations and Stations. Click the Add button to display the Geometry Points to Label in Band dialog shown in Figure 9.29.

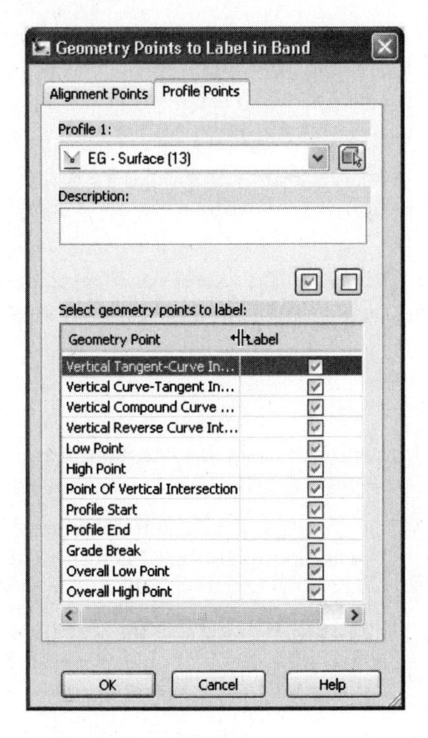

FIGURE 9.29
The Geometry Points to Label in Band dialog

5. Click OK to close the dialog and return to the Profile View Properties dialog. The Profile Data band should now be listed in the middle of the dialog.
6. Set the Location drop-down list to Top of Profile View.
7. Change the Band Type list box to the Horizontal Geometry option and the Select Band Style list box to the Geometry option. Click Add. The Horizontal Geometry band will now be added to the table in the List of Bands area.
8. Change the Band Type drop-down list to Vertical Geometry. Do not change the Select Band Style list box from its current option of Geometry. Click Add. The Vertical Geometry band will also be added to the table in the List of Bands area.
9. Click OK to exit the dialog. Your profile view should look like Figure 9.30.

However, there are obviously problems with the bands. The Vertical Geometry band is a mess and is located above the title of the profile view, whereas the Horizontal Geometry band actually overwrites the title. In addition, the elevation information has two numbers with different rounding applied. In this exercise, you'll fix those issues:

1. Pick the Rose Drive profile view, and then select Profile View Properties from the Modify View panel. The Profile View Properties dialog appears.
2. Switch to the Bands tab.

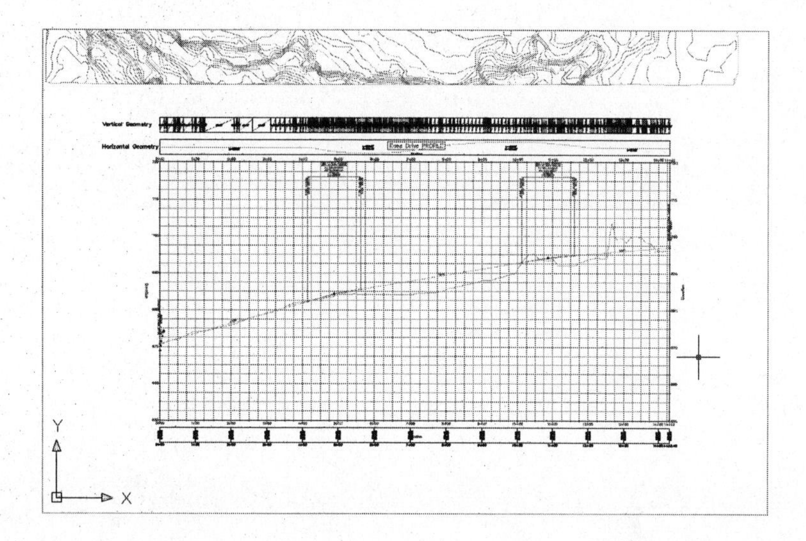

FIGURE 9.30
Applying bands to a profile view

3. Verify that the Location list box in the List of Bands area is set to the Bottom of Profile View option.

4. Verify that the option at the bottom of the screen that says "Match major/minor increments to vertical grid intervals" is turned on. This will ensure the major/minor intervals of the profile data band match the major/minor profile view style's major/minor grid spacing.

5. The Profile Data band is the only band currently listed in the table in the List of Bands area. Scroll right in the Profile Data row and notice the two columns labeled Profile 1 and Profile 2. Change the value of Profile 2 to Layout (2) as shown in Figure 9.31.

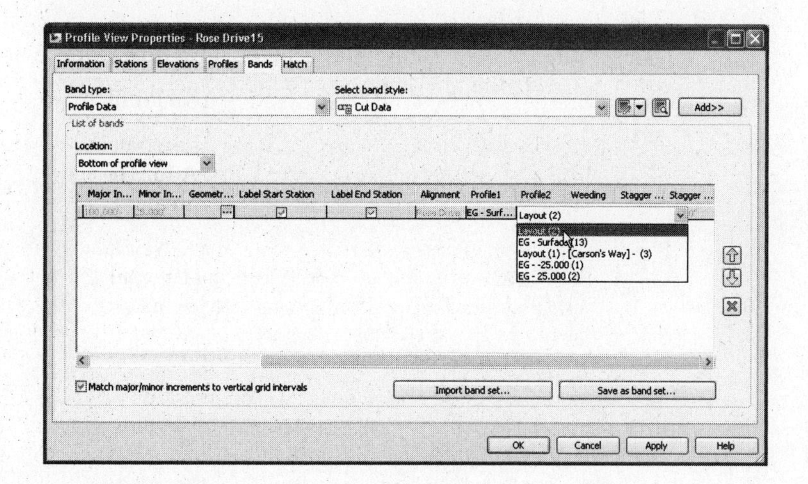

FIGURE 9.31
Setting the Profile View Bands to reference the Layout (2) profile

6. Change the Location drop-down selection to Top of Profile View.

7. The Horizontal and the Vertical Geometry bands are now listed in the table as well. Scroll to the right again, and set the value of Profile 1 in the Vertical Geometry band to Layout (2).

8. Scroll back to the left and set the Gap for the Horizontal Geometry band to **1.5″**. This value controls the distance from one band to the next or to the edge of the profile view itself.
9. Click OK to close the dialog. Your profile view should now look like Figure 9.32.

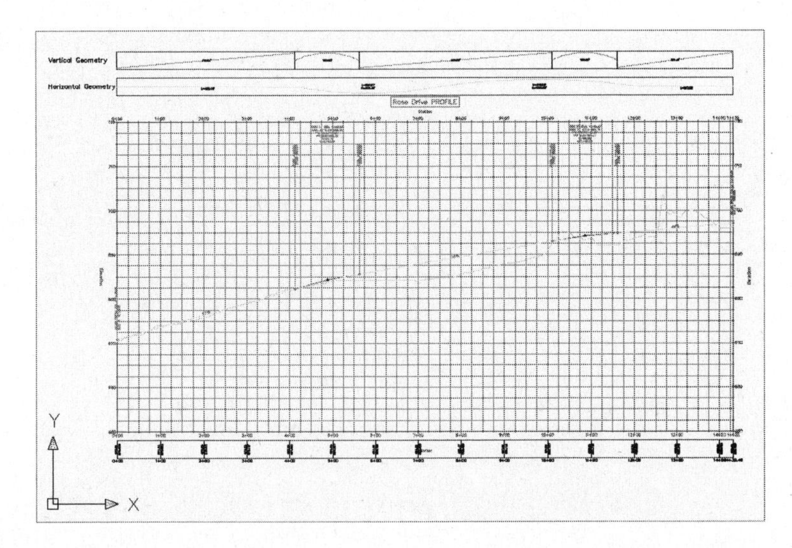

FIGURE 9.32
Completed profile view with the Bands set appropriately

Bands use the Profile 1 and Profile 2 designation as part of their style construction. By changing what profile is referenced as Profile 1 or 2, you change the values that are calculated and displayed (e.g., existing versus proposed elevations). These bands are just more items that are driven by styles, so the next section describes all the various styles in play with a profile view.

PROFILE VIEW HATCH

Many times it is necessary to shade cut/fill areas in a profile view. The settings on the Hatch tab are used to specify upper and lower cut/fill boundary limits for associated profiles (see Figure 9.33). Shape styles from the General Multipurpose Styles collection found on the Settings tab of the Toolspace can also be selected here. These settings include the following:

- **Cut Area** — Click this button to add hatching to a profile view in areas of the cut.
- **Fill Area** — Click this button to add hatching to a profile view in areas of the fill.
- **Multiple boundaries** — Click this button to add hatching to a profile view in areas of a cut/fill where the area must be averaged between two existing profiles (for example, finished ground at the centerline vs. the left and right top of curb).
- **From criteria** — Click this button to import Quantity Takeoff Criteria.

Profile View Styles

Profile view styles are among the most complicated to establish in Civil 3D, matched only by cross-sectional views. The style controls so many things, including annotation along all four axes, grid-and-tick spacing and clipping, and horizontal alignment information. The nice thing is that just as with every other stylized object, you have to go through the process only once and you can then apply the style to other views and share it.

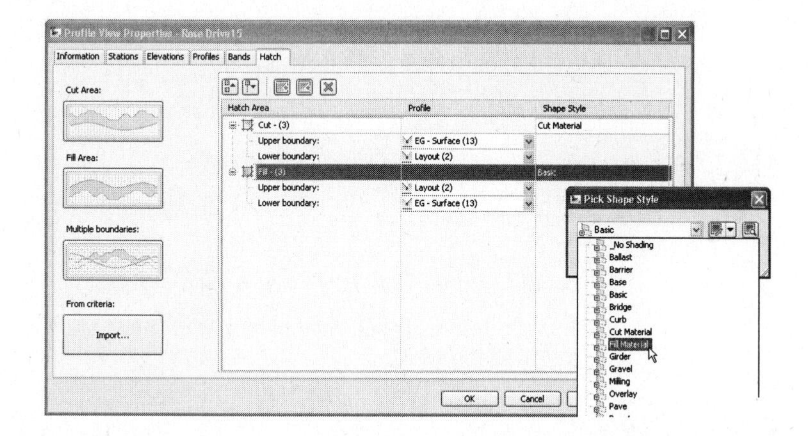

FIGURE 9.33
Shape style selection on the Hatch tab of the Profile View Style dialog

MASTERING PROFILES AND PROFILE VIEWS

One of the most difficult concepts to master in AutoCAD Civil 3D is the notion of which settings control which display property. Although the following two rules may sound overly simplistic, they are easily forgotten in times of frustration:

- Every object has a label and a style.
- Every label has a style.

Furthermore, if you can remember that there is a distinct difference between a profile object and a profile view object you place it in, you'll be well on your way to mastering profiles and profile views. When in doubt, select an object, right-click, and pay attention to the Civil 3D commands available between two horizontal lines, as shown here.

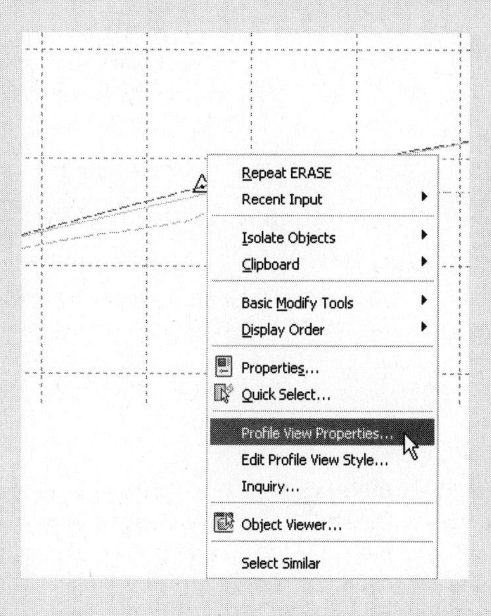

Profile View Style Selection

Selection of a profile view style is straightforward, but because of the large number of settings in play with a profile view style, the changes can be dramatic. In the following quick exercise, you'll change the style and see how much a profile view can change in appearance:

1. Open the `Profile View Styles.dwg` file.
2. Pick a grid line in the Carson's Way profile view and right-click. Notice the available commands.
3. Select the Profile View Properties option. The Profile View Properties dialog opens.
4. On the Information tab, change the Object Style list box to Major Grids and click OK to arrive at Figure 9.34.

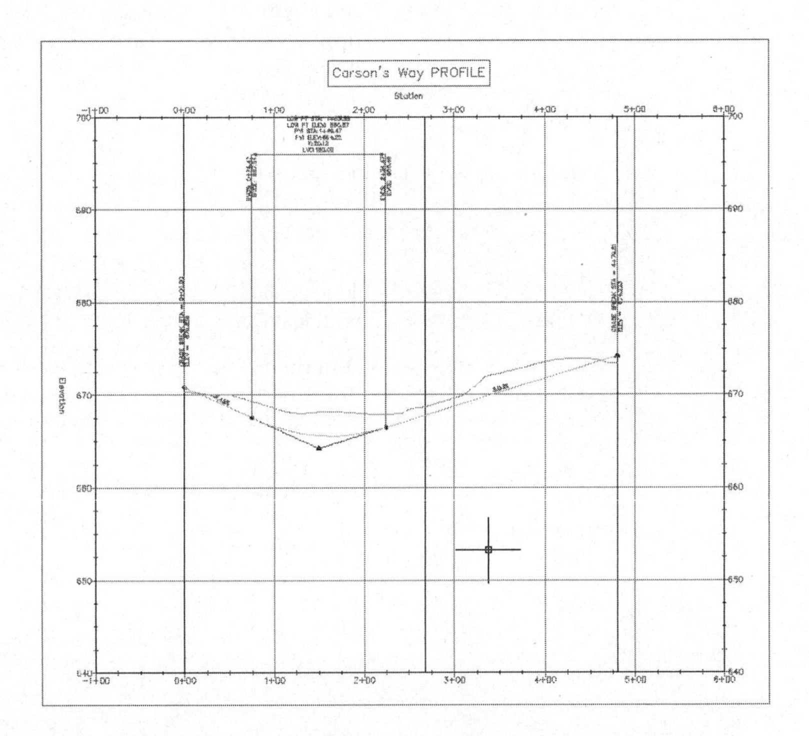

Figure 9.34
The Carson's Way profile view with the Major Grids style applied

A profile view style includes information such as labeling on the axis, vertical scale factors, grid clipping, and component coloring. Using various styles lets you make changes to the view to meet requirements without changing any of the design information. Changing the style is a straightforward exercise, so next you look at what's happening behind the scenes when a profile view style is modified.

Profile View Style Editing

Like every other object, profile view styles control every aspect of how a profile view looks. In this series of exercises, you go in and out of the Profile View Style Editor quite frequently so you can see each change individually:

1. Open the `Profile View Styles.dwg` file if you haven't already.
2. In the Settings tab on Toolspace, expand the Profile View ➤ Profile View Styles branches.
3. Right-click Full Grid and select the Copy option.
4. On the Information tab, change the Name field to **Mastering** and click OK to close the dialog.
5. Select the Rose Drive profile view, and select Profile View Properties from the Modify View panel. The Profile View Properties dialog appears.
6. On the Information tab, change the style name in the Object Style list box from the Full Grid profile view style to the new Mastering profile view style and click OK.

   You haven't truly changed anything, because the Mastering profile view style is still just a copy of the Full Grid profile view style. However, now that the Rose Drive profile view references your new style, you can step through a large number of changes and evaluate the results by simply clicking the Apply button in the dialog and reviewing the updated profile view.

7. In the Settings tab on Toolspace, right-click Mastering, and select the Edit option. The Profile View Style dialog opens.
8. Change to the Graph tab.
9. Change the Vertical Scale drop-down list to 1″= 5′. The Profile View Direction could also be modified here, but you'll leave it as Left to Right.
10. Click the Apply button, but do not close the dialog, to see the change in the Rose Drive profile view (see Figure 9.35).

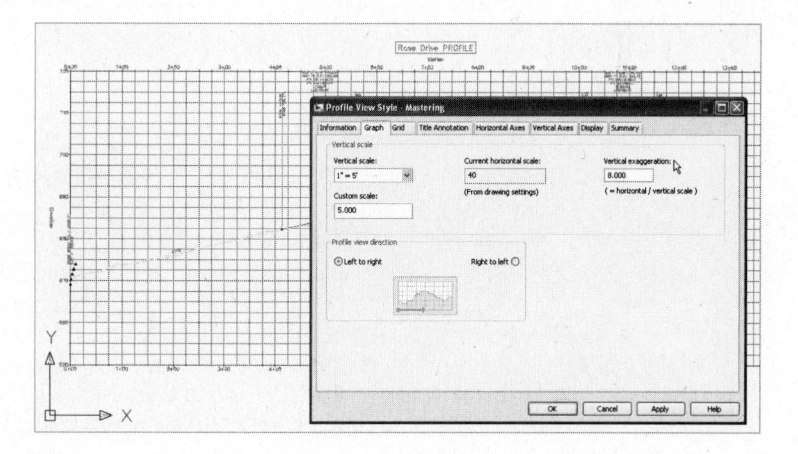

FIGURE 9.35
The Rose Drive profile view with an updated vertical exaggeration

11. Change to the Grid tab.
12. In the Grid Options area, check the Clip Vertical Grid option and then check the Clip to Highest Profile(s) option as shown in Figure 9.36.
13. Click the Apply button so your screen looks like Figure 9.36. The vertical grid lines have been removed, or "clipped."

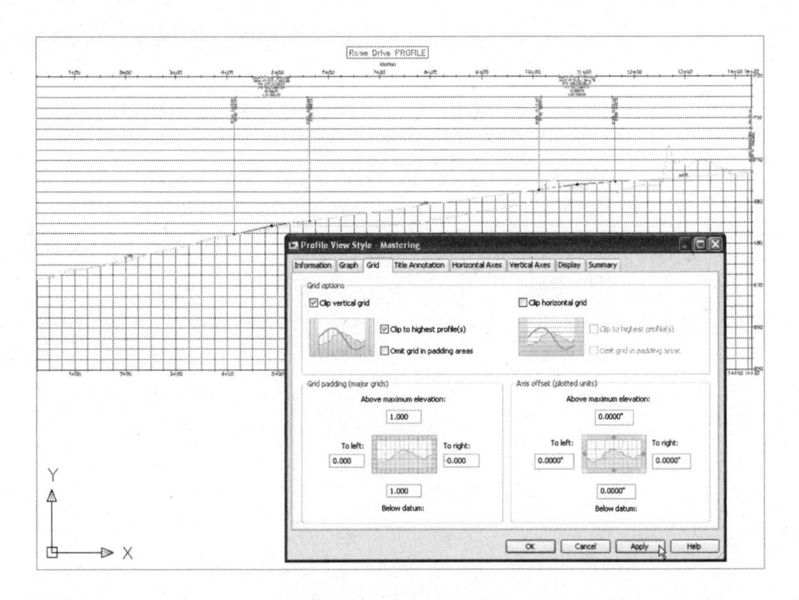

FIGURE 9.36
Clipping the vertical grid lines on the Rose Drive profile view

14. In the Grid Options area, check the Clip Horizontal grid option.
15. Click the Apply button and your view should look like Figure 9.37. By using the Apply button, you can verify changes without having to exit the dialog, and then reenter if you want to continue editing.

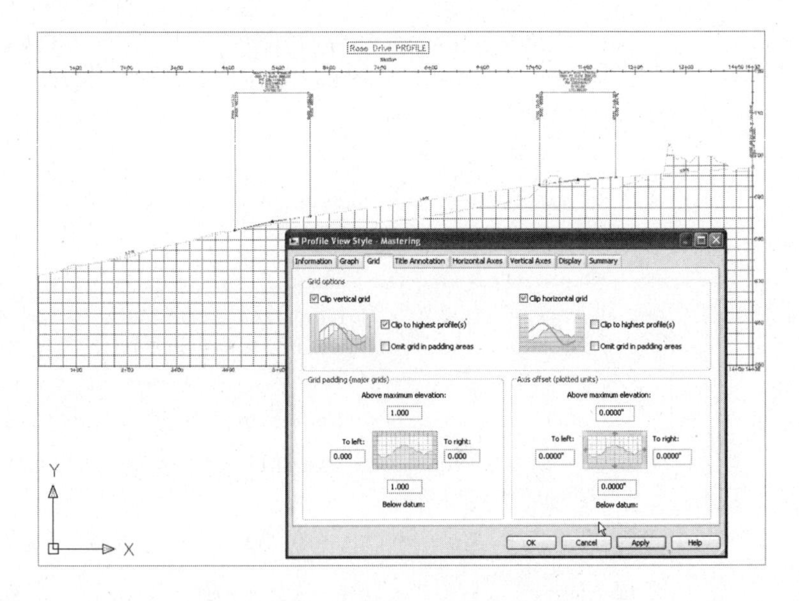

FIGURE 9.37
The Rose Drive profile view with both the vertical and horizontal grids clipped

16. If the profile looks like it should, click OK to close the dialog. If not, review the prior steps, making changes as needed, and use the Apply button to check your work.
17. Zoom in on the left-hand axis to make the next changes easier to view.

18. Right-click Mastering in the Settings tab again and select the Edit option. The Profile View Style dialog opens.

19. Still on the Grid tab, change the value of the To Left field in the Grid Padding (Major Grids) to 1 and click Apply to see the change in the profile view shown in Figure 9.38.

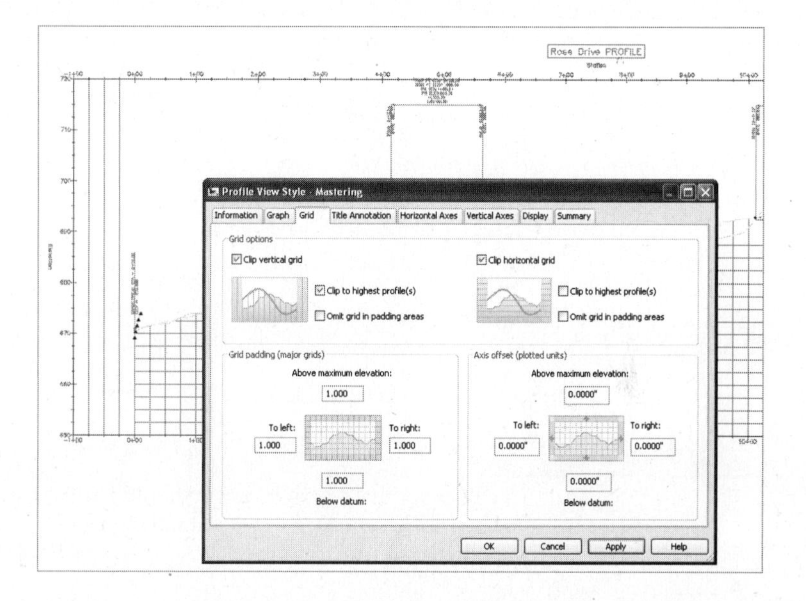

FIGURE 9.38
A grid padding applied to the left of the Rose Drive profile view

20. Change the value of the To Left field in the Grid Padding (Major Grids) area back to 0, and change the value of the To Left and To Right fields in the Axis Offset (Plotted Units) area to 0.5″.

21. Click OK to close the dialog.

The padding and offset values are used to add extra grid and buffer space around the main portion of the profile view. By using various values in conjunction, almost any spacing requirement can be accommodated.

It's important to remember that in addition to all of the configurations you're stepping through, you also have the ability to turn on and off individual components on the Display tab as you do with other objects. You'll get there in a few more tabs. Now you look at modifying the title above the profile:

1. Open the `Profile View Styles.dwg` file if you haven't already.

2. Zoom to the title of the Rose Drive profile view so you can more clearly see the changes about to be applied.

3. Right-click Mastering again in the Settings tab and select the Edit option. The Profile View Style dialog opens.

4. Change to the Title Annotation tab, as shown in Figure 9.39.

   The left portion of the Title Annotation tab in Figure 9.39 is devoted to the title of the profile view, and the right portion is set up to control the annotation placed on each axis. The right axis settings are visible, as indicated by both the radio button and the different-colored text in the small preview picture.

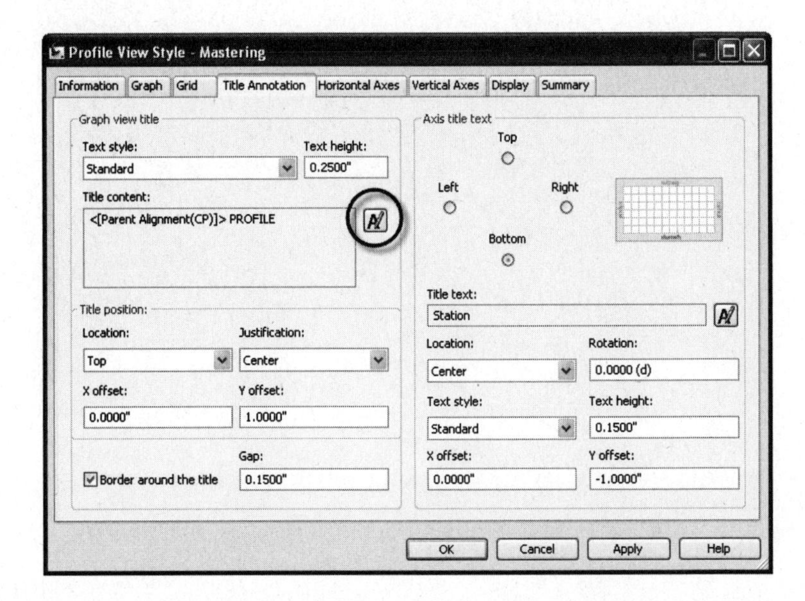

FIGURE 9.39
The Title Annotation tab in the Profile View Style dialog

5. Click the Edit Mtext button circled in Figure 9.39 to bring up the Text Component Editor for the title text.
6. Click in the preview area, change PROFILE to **Profile**, and press ↵ to create a line break.
7. In the Properties list box, select the Drawing Scale option and change the Precision value to 1.
8. Click the white insert arrow to add this property to the label. Press ↵ at the end of this new line to create a line break.
9. In the Properties list box, select the Graph View Vertical Scale option and change the Precision value to 1.
10. Click the white insert arrow, as shown in Figure 9.40, to add this property to the label.
11. Highlight the second and third line in the preview screen, as shown on Figure 9.41. Right-click and select the Cut option.
12. Click OK to close the Text Component Editor.
13. At the bottom of the Graph View Title area (on the left-hand side of the dialog), uncheck the Border Around The Title checkbox. Click Apply to see the change.

The Drawing Scale and the Graph View Vertical Scale properties, which were cut from the preview screen in step 10, aren't available for selection in the axis labels, only in the title label. But the program understands the field codes you just cut and will let you use them in the axis labels by pasting them in, even if they can't be selected directly. (This is one of our favorite hacks for getting around the limitations that are in Civil 3D!)

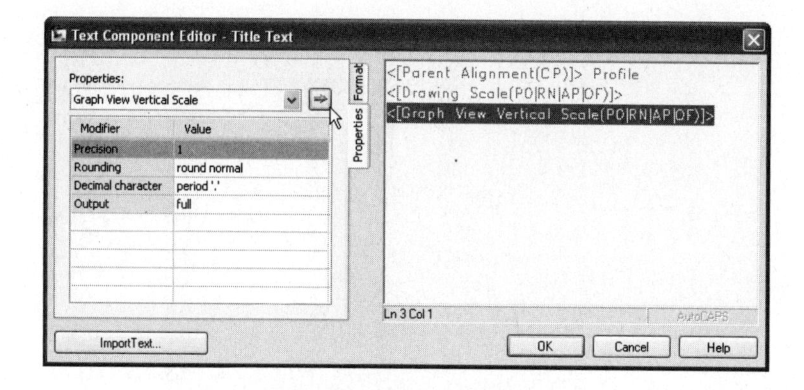

FIGURE 9.40
Inserting the label components for the title of the profile view

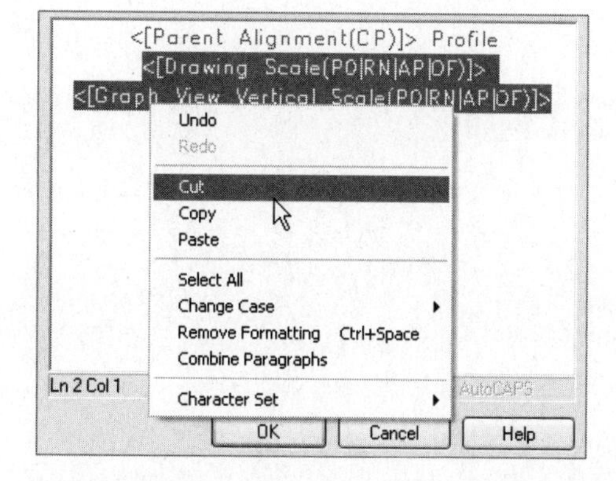

FIGURE 9.41
Cutting label components from the preview screen in the Text Component Editor

1. In the Axis Title Text (on the top-right side of the dialog) area, verify that the Bottom radio button is selected.
2. Click the Edit Mtext button just to the right of the Title Text text box to enter the Text Component Editor for the Axis Title Text. Note that the scale labels aren't available from the Properties drop-down list shown in Figure 9.42.

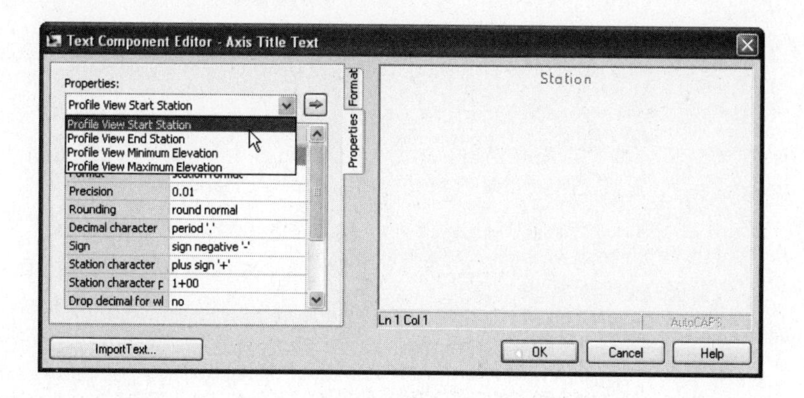

FIGURE 9.42
The Text Component Editor for the axis label text and its available properties

3. Highlight Station in the preview area, right-click, and select the Paste option.

4. Click in the preview area and add text to match Figure 9.43.

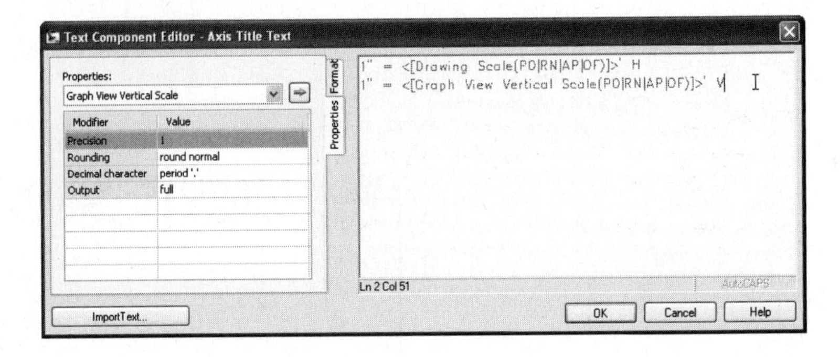

FIGURE 9.43
Hacked axis label with additional properties available

5. Click OK to close the Text Component Editor.
6. Set the Location list box to the Right option, the X Offset value to −**1.25**″, and the Y Offset value to **0.75**″.
7. Click OK to exit the dialog, and Pan to the lower right of the profile view to see the change, as shown in Figure 9.44.

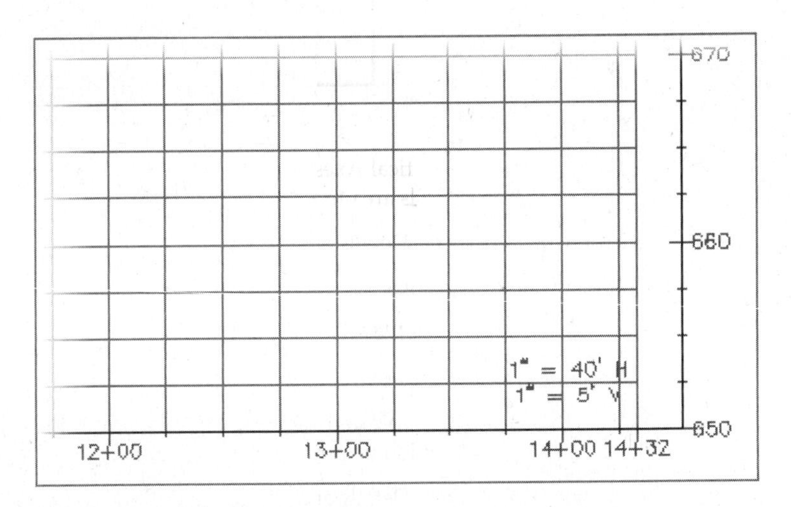

FIGURE 9.44
Applied bottom axis label with scale inserted and border offsets

8. Right-click the Mastering profile view style yet again on the Settings tab, and select the Edit option to open the Profile View Style dialog.
9. On the Horizontal Axes tab, set the options as follows:
   - In the Select Axis to Control area, select the Top radio button and notice how the highlighted area in the preview picture changes.
   - In the Minor Tick Details area, set the Interval value to **20**′.
   - In both the Major and Minor Tick Details area, set the Tick Justification field to the Top option.

10. Click Apply to see the changes to the drawing, as shown in Figure 9.45. You may need to drag your dialog out of the way to see these changes.

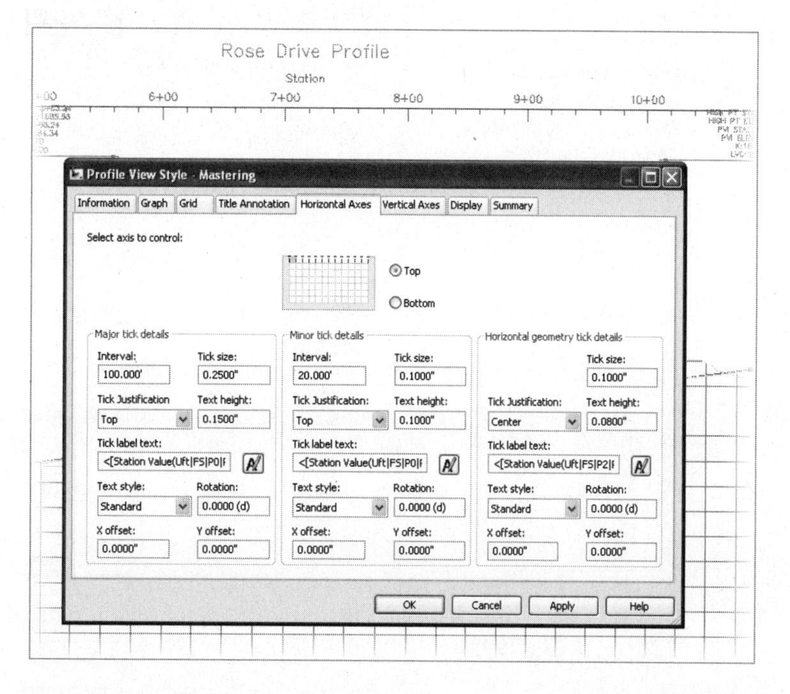

FIGURE 9.45
Horizontal ticks and vertical grid after modifications in step 9

11. Switch to the Vertical Axes tab, and set the Tick Justification field in both the Major and Minor Tick Details area to the Left option.
12. In the Major Tick Details area, set the following options to match the dialog in Figure 9.46:
    - Tick Size: **0.5″**
    - X Offset: **0.1200″**
    - Y Offset: **0.1000″**
13. In the Select Axis to Control area on the top of the screen, click the Right radio button. Set the Tick Justification field in both the Major and Minor Tick Details area to the Right option.
14. In the Major Tick Details area, set the following options:
    - Tick Size: **0.5″**
    - X Offset: **−0.1200″**
    - Y Offset: **0.1000″**
15. Switch to the Display tab and turn off the following additional components:
    - Left Axis Title
    - Right Axis Title
    - Top Axis Title

- Top Axis Annotation Major
- Top Axis Ticks Minor

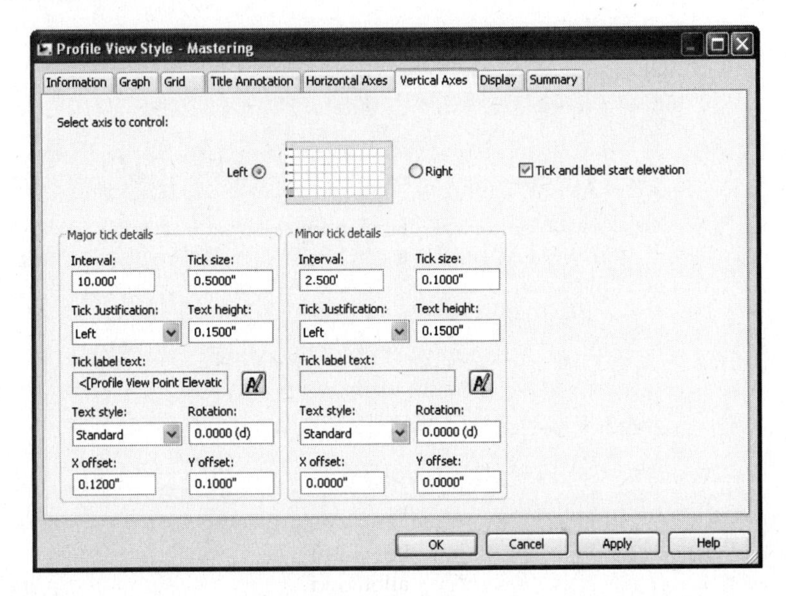

FIGURE 9.46 Labeling of the vertical ticks using the settings shown on the Vertical Axes tab

16. Click OK to close the Profile View Style dialog. Your profile view should look like Figure 9.47.

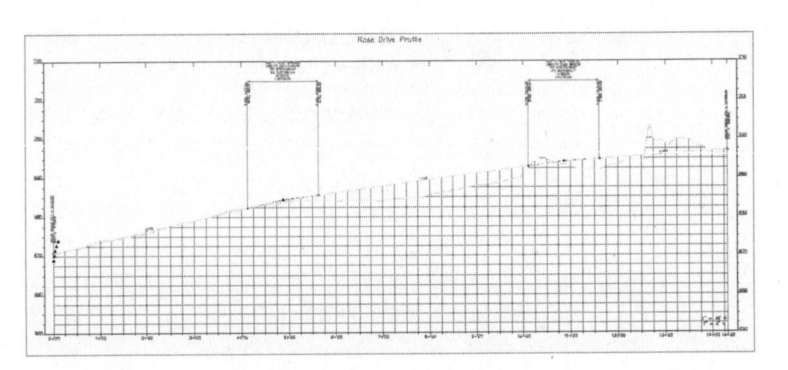

FIGURE 9.47 Completed style modifications applied to the Rose Drive profile view

WHAT'S DRIVING THE MINOR AXIS VALUES?

For a long time, it wasn't documented, but the Minor Tick Interval on the Left Vertical and Bottom Horizontal axes is what controls the grid spacing. Even if you don't turn on the ticks on these axes, the spacing increment will be reflected in the minor grid lines.

Although that style took a long time to create and modify, it's now ready for use on any other profile view in the drawing, or to be moved to a drawing template file for sharing with other team members. Now that you have the grid looking how you like, you'll apply some more labels and see how the data bands are built.

Labeling Styles

Now that the profile is labeled, the profile view grid spacing is set, and the titles all look good, it's time to add some specific callouts and detail information. Civil 3D uses profile view labels and bands for annotating.

View Annotation

Profile view annotations label individual points in a profile view, but they are not tied to a specific profile object. These labels can be used to label a single point or the depth between two points in a profile. We say "depth" because the label recognizes the vertical exaggeration of the profile view and applies the scaling factor to label the correct depth. Profile view labels can be either station elevation or depth labels. In this exercise, you'll use both:

1. Open the `Profile View Labels.dwg` file.
2. Zoom in on the Carson's Way profile view.
3. Switch to the Annotate tab and select Add Labels from the Labels & Tables panel. The Add Labels dialog opens.
4. In the Feature list box, select Profile View; in the Label Type list box, verify the selection of the Station Elevation option; and in the Station Elevation Label Style list box, make sure that the Station and Elevation style is selected.
5. Click the Add button.
6. Click a grid line in the Carson's Way profile view. Zoom in on the right side so that you can see the point where the EG and layout profiles cross over.
7. Pick this profile crossover point visually, and then pick the same point to set the elevation and press ↵. Your label should look like Figure 9.48.
8. In the Add Labels dialog, change the Label Type list box and the Depth Label Style list box to the Depth option. Click the Add button.
9. Click a grid line on the Carson's Way profile view.
10. Pick a point along the layout profile and then pick a point along the EG profile and press ↵. The depth between the two profiles will be measured as shown in Figure 9.49.
11. Close the Add Labels dialog.

Why Don't Snaps Work?

There's no good answer to this question. For a number of releases now, users have been asking for the ability to simply snap to the intersection of two profiles. We mention this because you'll try to snap, and wonder if you've lost your mind. You haven't — it just doesn't work. Maybe next year?

Depth labels can be handy in earthworks situations where cut and fill become critical, and individual spot labels are important to understanding points of interest, but most design documentation is accomplished with labels placed along the profile view axes in the form of data bands. The next section describes these band sets.

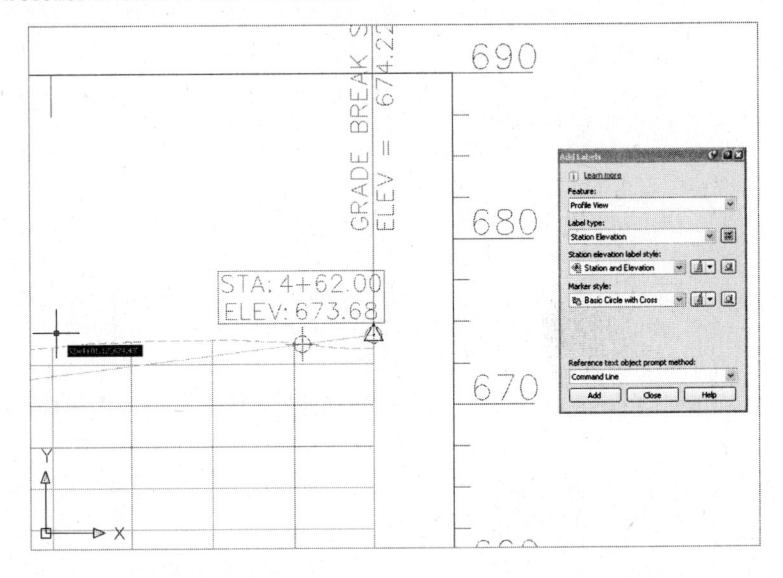

FIGURE 9.48 An elevation label for a profile station

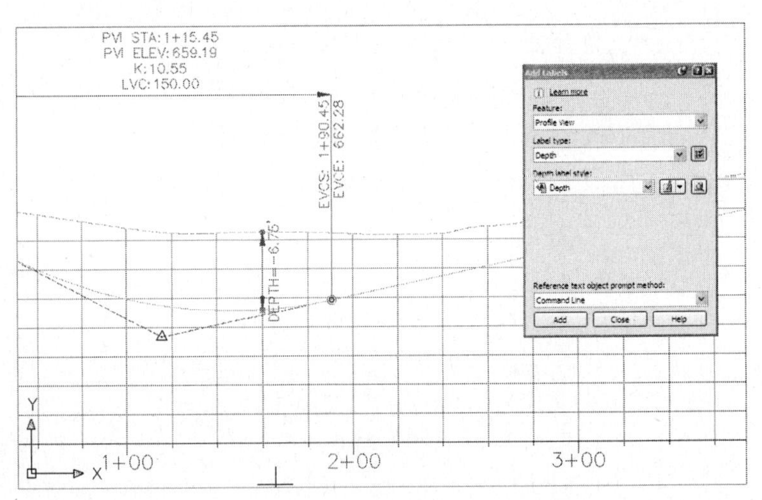

FIGURE 9.49 A depth label applied to the Carson's Way profile view

Real World Scenario

MAKING THE BAND

You looked at assigning bands to the profile view earlier in the chapter, but now you'll look at how bands are composed. This really gets into how you'll use Civil 3D in your office, as making profile views look just like the reviewer wants them is one of the most important tasks. Setting up the various bands for every agency can be time consuming, but the uniformity will pay dividends when you create the required views later. In this exercise, you'll modify an existing band style and apply it to your Carson's Way profile view:

1. Open the `Multiple Bands.dwg` file.
2. On the Settings tab, expand the Profile View ➢ Band Styles ➢ Profile Data branches.
3. Right-click the Elevations and Stations band style and select the Copy option. The Profile Data Band Style dialog opens.
4. On the Information tab, change the Name text box to **Elevations Only**.
5. Change to the Band Details tab as shown in the following image. The left side of this tab controls the various options for the title text for the band. These options were turned off on the Display tab in the style you copied, so you can ignore them. The right side controls the labeling of other critical points.

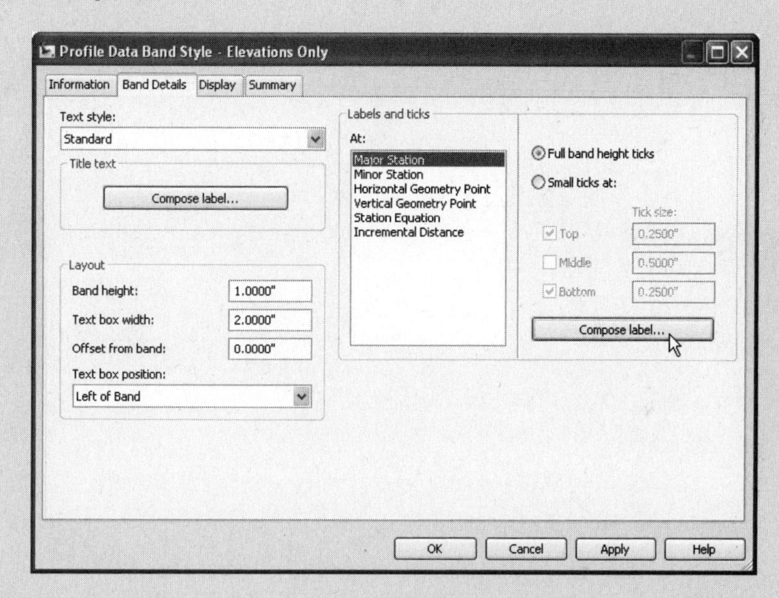

6. In the Labels and Ticks area, select the Major Station option in the At list box, and click the Compose Label button to bring up the Label Style Composer.
7. In the Component Name list box, verify the selection of the Station Value component and set the Visibility property to False.
8. Change to the EG Elevation component in the Component Name list box.
9. Click in the Contents value cell and click the More button to bring up the Text Component Editor.
10. Delete the text in the preview area, and type **EG**: (include a space after the colon), and then select the Profile 1 Elevation property in the Properties list box with a Precision value of 0.1. Click the white insert arrow to insert these properties into the preview, as shown in the image here.

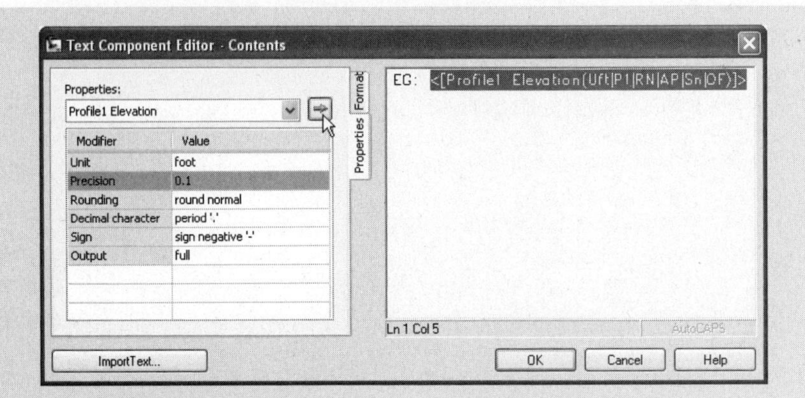

11. Click OK to close the Text Component Editor.
12. In the Label Style Composer, change the EG Elevation to the FG Elevation component in the Component Name list box. Click in the Contents value cell and click the More button to bring up the Text Component Editor.
13. Delete the text in the preview area, and type **FG**: (include a space after the colon), and then select the Profile 2 Elevation property in the Properties list box with a Precision value of 0.01. Click the white insert arrow to insert these properties into the preview area. Your dialog should look like the following image.

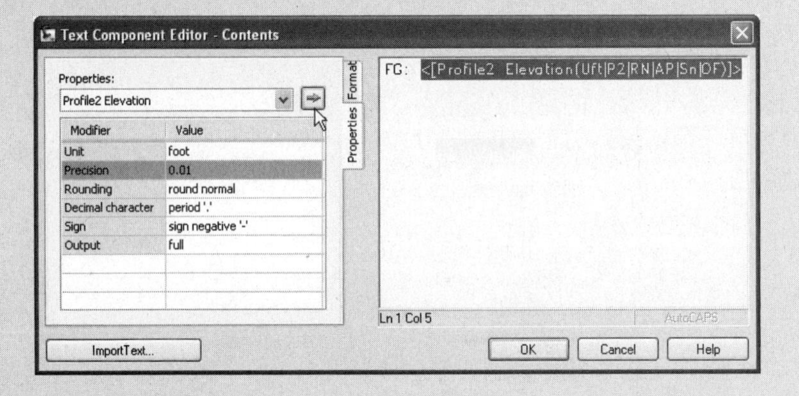

14. Click OK three times to exit the Text Component Editor, the Label Style Composer, and the Profile Data Band Style dialogs.
15. Select the Carson's Way profile view, right-click, and select the Profile View Properties option. The Profile View Properties dialog opens.
16. Change to the Bands tab, and set the Band Type list box to Profile Data and the Select Band Style list box to Elevations Only. Click Add to display the Geometry Points to Label in Band dialog. Click OK to dismiss it.

17. Add the Offsets band to the Bottom of Profile View as well. This band will label the offset surface elevations.

18. Click OK to close the dialog and update your profile view. Your screen should look like the following image.

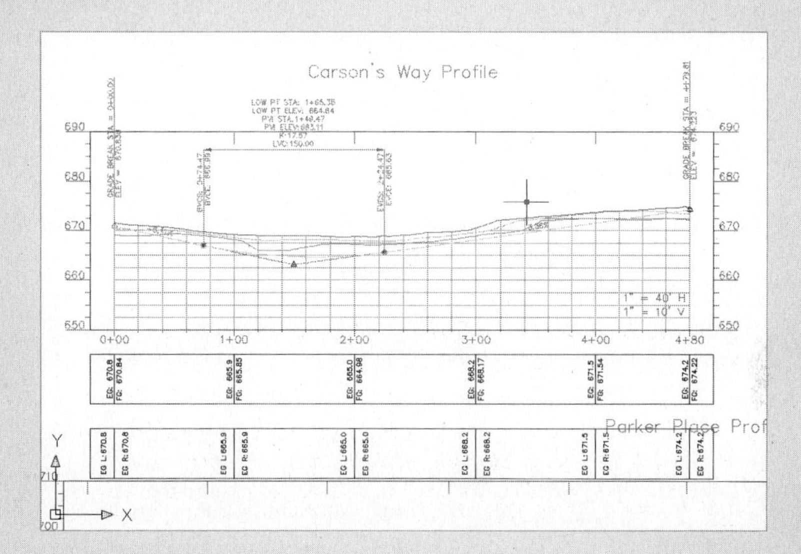

19. Select the profile view, right-click, and return to the Profile View Properties dialog.

20. Scroll to the right in the List of Bands area, and change the Profile assignments as shown here.

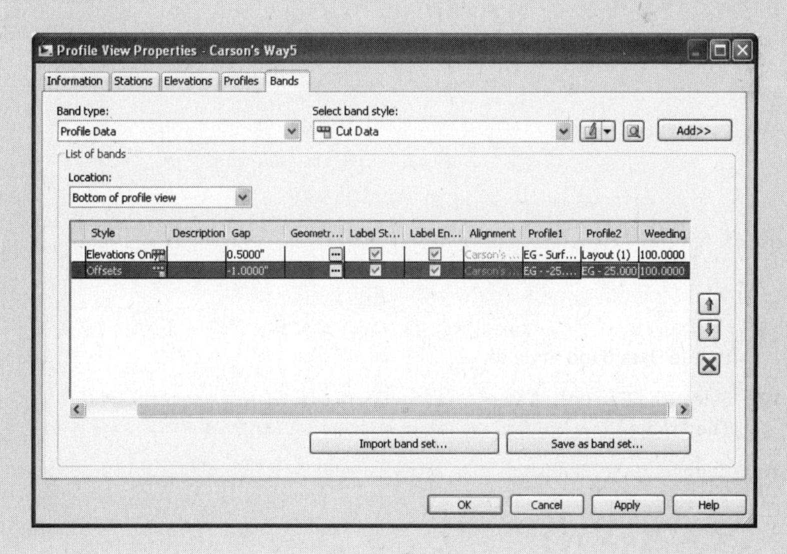

21. Change the Gap for the Offsets band to **−1″** as shown. This superimposes the Offsets band on top of the Elevations Only band.

22. Click OK and update your profile view as shown here.

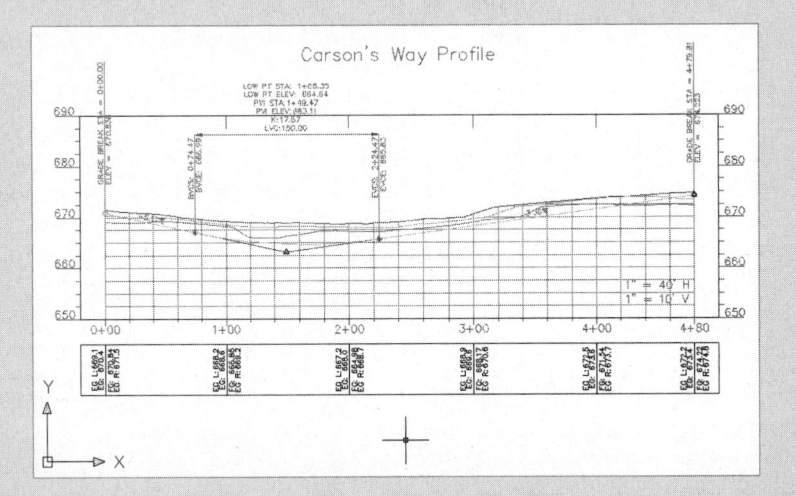

Band information can show any number of profiles, and by creatively using the offsets and profile assignments, you can apply labeling simply as well.

Band Sets

Band sets are simply collections of bands, much like the profile label sets or alignment label sets. In this brief exercise, you'll save a band set, and then apply it to a second profile view:

1. Open the `Profile View Band Sets.dwg` file.
2. Pick the Carson's Way profile view, and select Profile View Properties from the Modify View panel. The Profile View Properties dialog opens.
3. Switch to the Bands tab.
4. Click the Save as Band Set button to display the Band Set dialog in Figure 9.50.
5. Change the Name field to **EG+FG and Offsets**.
6. Click OK to close the Band Set dialog.
7. Click OK to close the Profile View Properties dialog.
8. Pick the Rose Drive profile view, and select Profile View Properties from the Modify View panel.
9. Switch to the Bands tab.
10. Click the Import Band Set button, and the Band Set dialog opens.

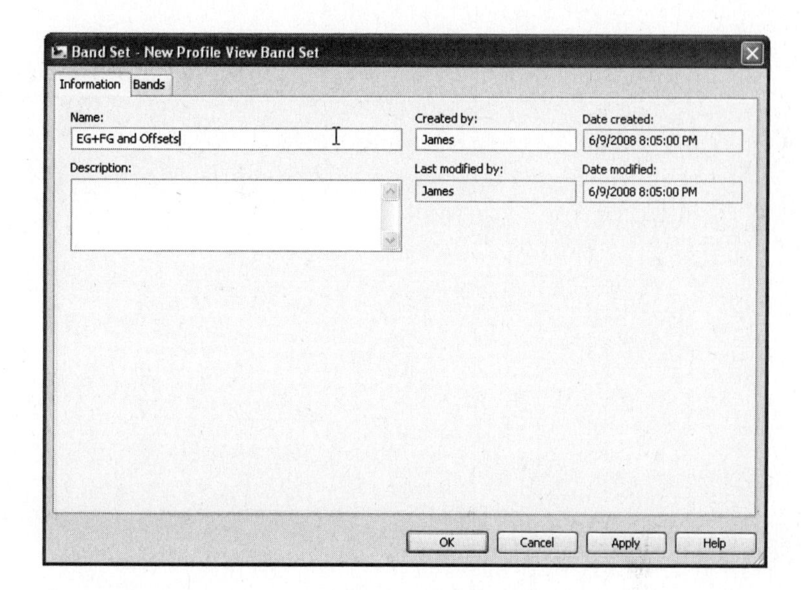

FIGURE 9.50
The Information tab for the Band Set dialog

11. Select the EG+FG and Offsets option from the drop-down list and click OK.
12. Change the Profile assignments for the Elevations Only and Offsets band styles, as shown in Figure 9.51.

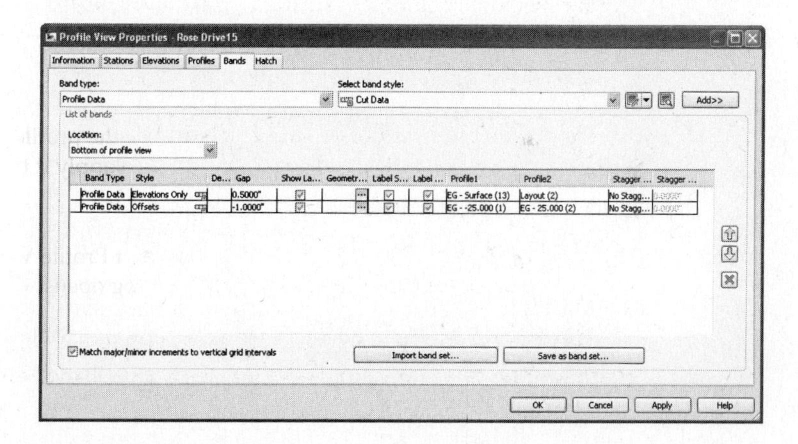

FIGURE 9.51
Assigning relevant alignments and profiles to the bands

13. Click OK to exit the Profile View Properties dialog. Your profile view should look like Figure 9.52.

Your Rose Drive profile view now looks like the Carson's Way profile view. Band sets allow you to create uniform labeling and callout information across a variety of profile views. By using a band set, you can apply myriad settings and styles that you've assigned to a single profile view to a number of profile views. The simplicity of enforcing standard profile view labels and styles makes using profiles and profile views simpler than ever.

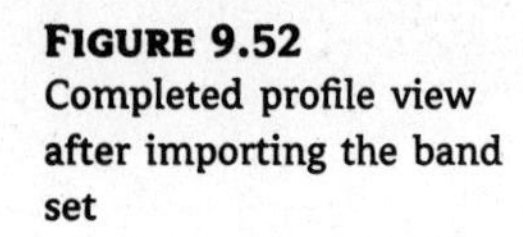

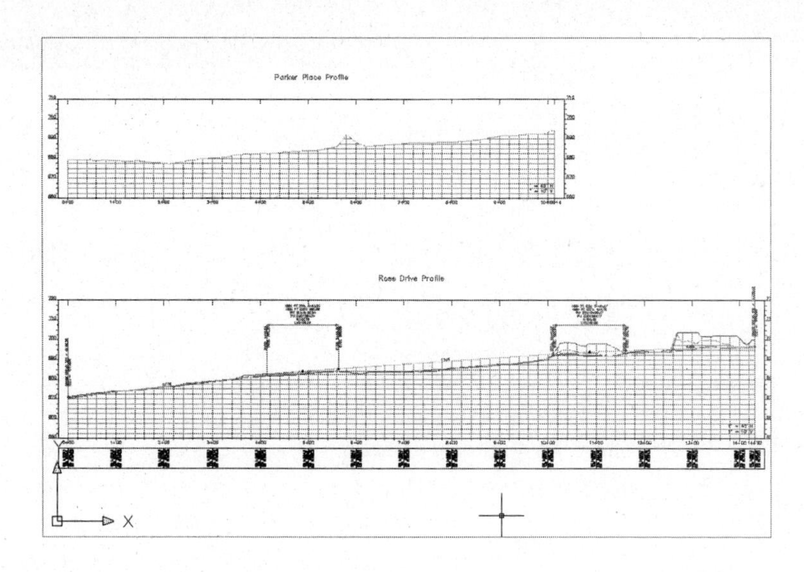

FIGURE 9.52
Completed profile view after importing the band set

The Bottom Line

Create a simple view as part of the sampling process. You will seldom want to sample a surface without creating a view of that data. By combining the steps into one quick process, you'll save time and effort as profile views are generated.

> **Master It** Open the `Mastering Profile Views.dwg` file and create a view using the Full Grid profile view style for the Alexander Ave alignment. Display only the layout profile, the EG and offsets at 15′ left and 25′ right.

Change profile views and band sets as needed. Using profile view styles and band sets allows for the quick customization and standardization of profile data. Because it's easy to change the styles and bands, many users design using one style and then change the style as required for submission.

> **Master It** Change the Alexander Ave profile view to the Mastering style and assign the EG+FG and Offsets band set. Assign appropriate profiles to the bands.

Split profile views into smaller views. Designing in one continuous profile view makes the designer's job easier, but plotting typically requires multiple views. Using the wizard or individual profile view properties makes it easy to split apart profile view information for presentation or submittal purposes.

> **Master It** Create a new pair of profile views for the Parker Place alignment, each 600′ long. Assign the Mastering profile view style but no bands.

Chapter 10

Templates Plus: Assemblies and Subassemblies

Roads, ditches, trenches, and berms usually follow a predictable pattern known as a *typical section*. Assemblies are how you tell Civil 3D what these typical sections look like. Assemblies are made up of smaller components called *subassemblies*. For example, a typical road section assembly contains subassemblies such as lanes, sidewalks, and curbs.

These typical sections, or assemblies, will be strung together into simple and complex corridor models in Chapter 11, "Easy Does It: Basic Corridors," and Chapter 12, "The Road Ahead: Advanced Corridors." In this chapter, the focus will be on understanding where these assemblies come from and how to build and manage them. Because it's difficult to understand the extensive applications of assemblies without seeing them in action in a corridor model, you may find it useful to work through the examples in this chapter and then come back and reread it after working through Chapters 11 and 12.

By the end of this chapter, you'll learn to:

- Create a typical road assembly with lanes, curbs, gutters, and sidewalks
- Edit an assembly
- Add daylighting to a typical road assembly

Subassemblies

A *subassembly* is a building block of a typical section, known as an *assembly*. Examples of subassemblies include lanes, curbs, sidewalks, channels, trenches, daylighting, and any other component required to complete a typical corridor section.

The Corridor Modeling Catalog

An extensive catalog of subassemblies has been created using the Microsoft .NET programming language for use in Civil 3D. Over 100 subassemblies are available in the standard Imperial catalog (there is also a Metric catalog), and each subassembly has a list of adjustable parameters. There are also about a dozen generic links you can use to further refine your most complex assembly needs. From ponds and berms, to swales and roads, the design possibilities are almost infinite.

It's possible to create additional subassemblies by programming in .NET or using Create Subassembly from Polyline in the Home tab's Create Design panel. Because the Create Subassembly from Polyline tool isn't intuitive, and it's rare to need a new subassembly, this chapter will focus on taking advantage of and customizing subassembly parts from the standard Imperial catalog.

If you've exhausted the possibilities in the standard catalog and feel you need to create your own custom subassembly, you can find more information about doing so in the Help file.

ACCESSING THE CORRIDOR MODELING CATALOG

The Corridor Modeling Catalog is installed by default on your local hard drive. Change to the Modify tab and then choose Corridor from the Design panel to open the Corridor contextual tab. On the Corridor contextual tab, click the Catalog button on the Launch Pad panel to open a content browser interface that allows you to explore the entire collection of subassemblies available in each category (see Figure 10.1).

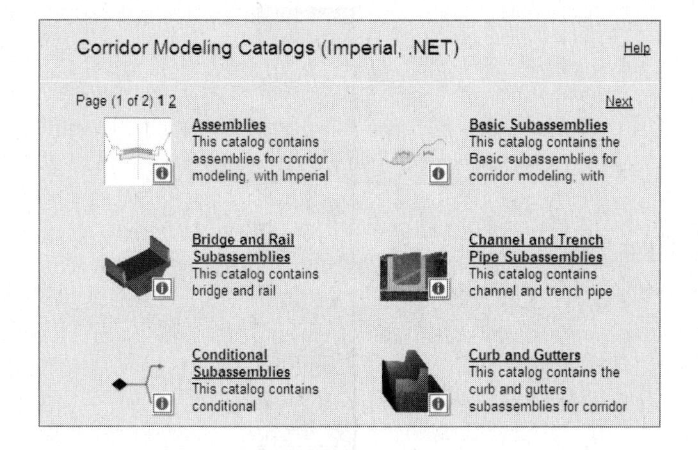

FIGURE 10.1
The front page of the Corridor Modeling Catalog

ACCESSING SUBASSEMBLY HELP

Later, this chapter will point out other shortcuts to access the extensive subassembly documentation. You can get quick access to information by right-clicking any subassembly entry on the Corridor Modeling Catalog page and selecting the Help option (see Figure 10.2).

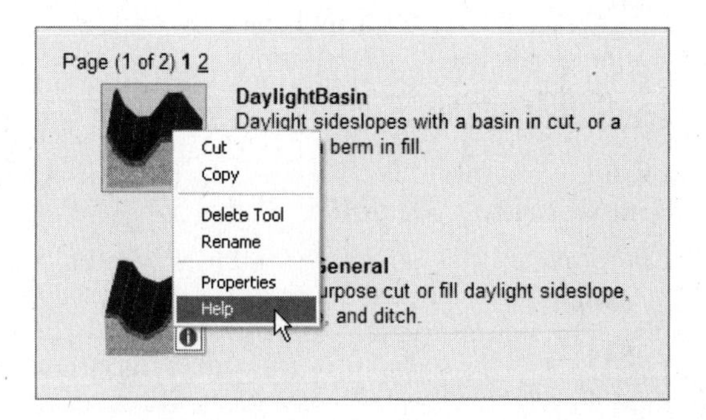

FIGURE 10.2
Accessing the Help file through the Corridor Modeling Catalog

The Subassembly Reference in the Help file provides a detailed breakdown of each subassembly, examples for its use, its parameters, a coding diagram, and more. While you're searching the catalog for the right parts to use, you'll find the Subassembly Reference infinitely useful.

ADDING SUBASSEMBLIES TO A TOOL PALETTE

The creation of assemblies relies heavily on the use of tool palettes, as you'll see later in this chapter. By default, Civil 3D has several tool palettes created for corridor modeling. You can access these tool palettes by changing to the Home tab and clicking the Tool Palettes button on the Palettes panel.

If you'd like to add additional subassemblies to your tool palettes, or for some reason your default palettes weren't installed, you can use the i-drop to grab subassemblies from the catalog and drop them onto a tool palette. To use the i-drop, click the small blue *i* next to any subassembly, and continue to hold down your left mouse button until you're over the desired tool palette. Release the button, and your subassembly should appear on the tool palette (see Figure 10.3).

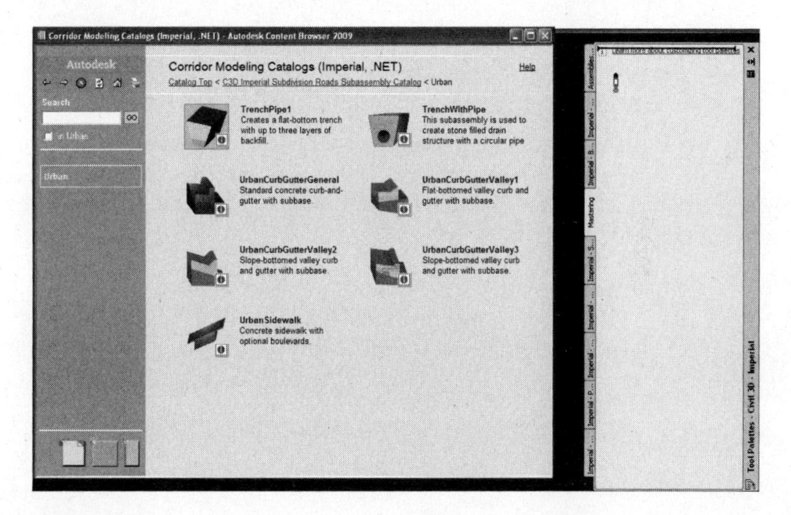

FIGURE 10.3
Using the i-drop to add a subassembly to a tool palette

Building Assemblies

You build an assembly by changing to the Home tab and choosing Assembly ➤ Create Assembly from the Create Design panel to create an assembly baseline. Once the assembly baseline has been created, you'll complete the process by adding subassembly components to that baseline from a tool palette. A typical assembly baseline is shown in Figure 10.4.

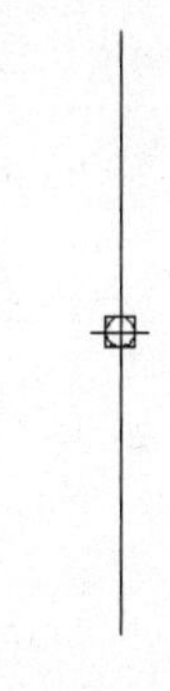

FIGURE 10.4
An assembly baseline

The process is extremely simple, much like building with interlocking blocks. Each component you add has an understanding of how it needs to connect to the assembly. Although creating assemblies is easy, it takes a bit of practice to get into the rhythm and to understand all the different subassembly parameters.

> **GETTING TO THE TOOL PALETTES**
>
> The exercises in the rest of this chapter depend heavily on the use of the Tool Palette feature of AutoCAD when pulling together assemblies from subassemblies. To avoid some redundancy, we're going to omit the step of opening the tool palette in every single exercise. If it's open, leave it open, if it's closed, open it. This is part of the Mastering series after all — you know how to handle AutoCAD basics! In case you need a reminder, the easiest way to open the Tool Palette feature is Ctrl+3.

Creating a Typical Road Assembly

The most common assembly used in corridor modeling is a typical road assembly. This assembly uses the road centerline alignment and profile as its baseline.

The process for building an assembly requires the use of the Tool Palette feature and the AutoCAD Properties palette, both of which can be docked. You'll quickly learn how to best orient these palettes with your limited screen real estate. If you run dual monitors, you may find it useful to place both of these palettes on your second monitor.

When you're creating your first few assemblies, it's common to miss a prompt or misplace a subassembly. To prevent these errors, proceed slowly, read the command line, and know that you can always erase misplaced subassemblies and replace them.

This exercise builds a typical assembly using the BasicLane, BasicCurbandGutter, and BasicSidewalk subassemblies (see Figure 10.5) to match a road section consisting of 10′ lanes, a curb and gutter, and a 5′ sidewalk with 2′ boulevard buffer strips on either side.

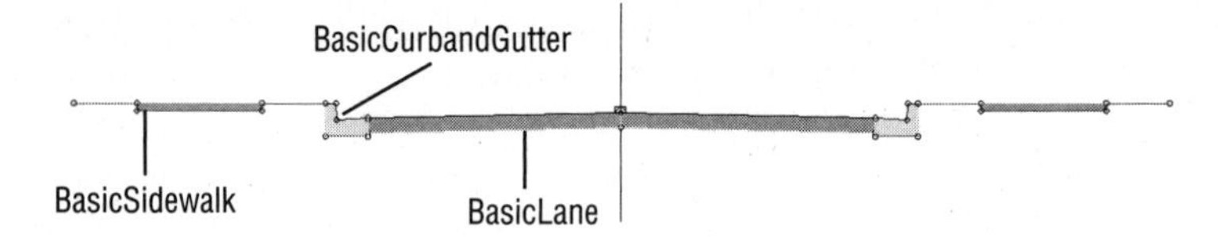

FIGURE 10.5
A typical road assembly

Let's have a more detailed look at each component you'll use in the following exercise. A quick peek into the subassembly Help will give you a breakdown of attachment options, input parameters, target parameters, output parameters, behavior, layout-mode operation, and the point, link, and shape codes. (For more information about points, links, and shapes, see Chapter 11.)

THE BASICLANE SUBASSEMBLY

The BasicLane subassembly creates a simple lane with only a few parameters. This is typically the best lane subassembly to use for your first attempts at corridor modeling as well as any situation where a straightforward lane is required. The BasicLane subassembly has parameters for customizing its side, width, and slope as well as depth of material (see Figure 10.6).

FIGURE 10.6
The BasicLane subassembly

Corridors built using the BasicLane subassembly are most commonly used for building top and datum surfaces, rendering paved areas, and creating cross sections. They can also return quantities of excavated material. Keep in mind that there is only one depth parameter, so this subassembly isn't useful for a detailed breakdown of placed material, such as a road section that has a layer of asphalt, a layer of gravel, and so on.

The BasicLane can't be superelevated, nor does it have targets for transitions (turning lanes, etc.).

THE BASICCURBANDGUTTER SUBASSEMBLY

The BasicCurbandGutter subassembly (Figure 10.7) is another simple component that creates an attached curb and gutter. Looking into the subassembly Help, you'll see a diagram of the BasicCurbandGutter with callouts for its seven parameters: side; insertion point; gutter width and slope; and curb height, width, and depth. You can adjust these parameters to match many standard curb-and-gutter configurations.

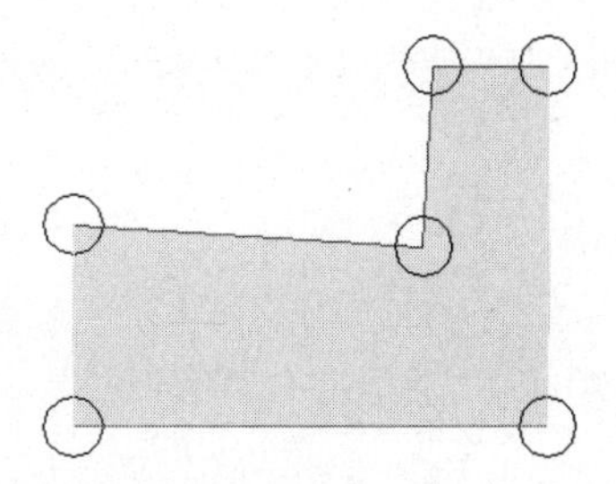

FIGURE 10.7
The BasicCurbandGutter subassembly

Corridors built using the BasicCurbandGutter subassembly are most commonly used for building top and datum surfaces, rendering curb-and-gutter areas, and creating cross sections. The BasicCurbandGutter subassembly can return quantities of concrete (or other curb-and-gutter construction material) but not gravel bedding or other advanced material layers.

BasicCurbandGutter doesn't have targets for transitions.

THE BASICSIDEWALK SUBASSEMBLY

The BasicSidewalk subassembly (Figure 10.8) creates a sidewalk and boulevard buffer strips. The Help file lists the following five parameters for the BasicSidewalk subassembly: side, width, depth, buffer width 1, and buffer width 2. These parameters let you adjust the sidewalk width, material depth, and buffer widths to match your design specification.

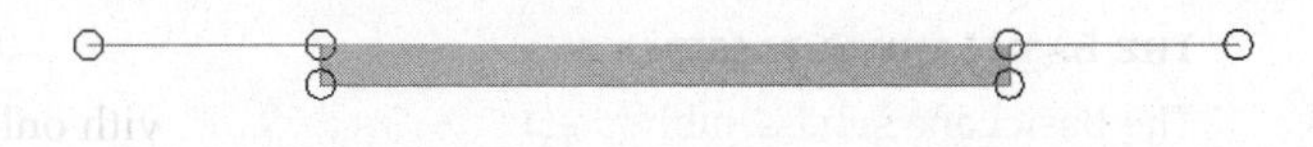

FIGURE 10.8
The BasicSidewalk subassembly

Corridors built using the BasicSidewalk subassembly are most commonly used for building top and datum surfaces and rendering concrete sidewalk areas. The BasicSidewalk subassembly can return quantities of concrete (or other sidewalk construction material) but not gravel bedding or other advanced material layers.

In the following exercise, you'll build a typical road assembly using these subassemblies. Note that the BasicSidewalk is a flat sidewalk section. If your standard sidewalk detail requires a cross slope, use the UrbanSidewalk subassembly (which is discussed later in this chapter). Follow these steps:

1. Create a new drawing from the `_AutoCAD Civil 3D (Imperial) NCS.dwt` template.
2. Change to the Home tab and choose Assembly ➢ Create Assembly from the Create Design panel. The Create Assembly dialog opens.
3. Enter **Typical Road** in the Name text box. Make sure the Assembly Style text box is set to Basic and the Code Set Style text box is set to All Codes. Click OK.
4. Pick a location in your drawing for the assembly — somewhere in the center of your screen is fine.
5. Locate the Imperial-Basic tab on the tool palette. Position the palette on your screen so that you can clearly see the assembly baseline.
6. Click the BasicLane button on the tool palette (see Figure 10.9). The AutoCAD Properties palette appears. Position the palette on your screen so that you can clearly see both the assembly baseline and the Imperial-Basic tool palette.

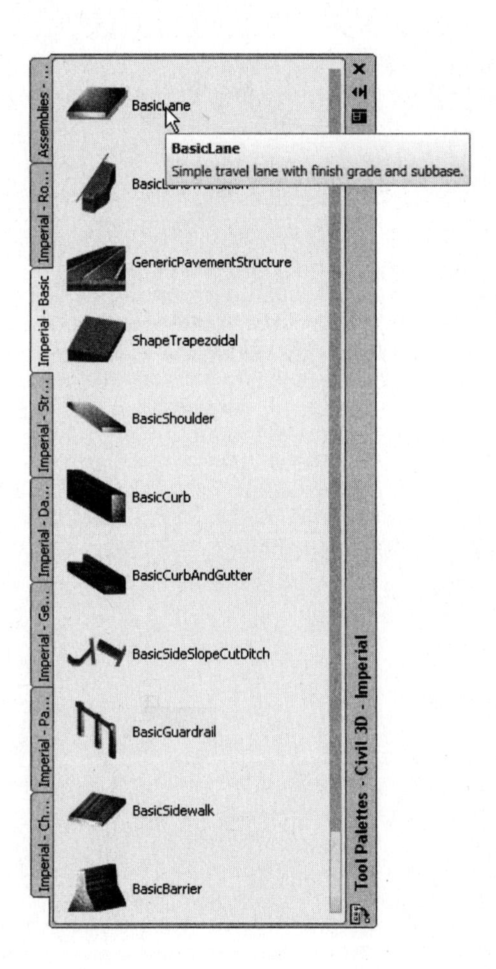

FIGURE 10.9
BasicLane button on the Imperial-Basic tool palette

7. Locate the Advanced section on the Design tab of the AutoCAD Properties palette. This section lists the BasicLane parameters. Make sure the Side parameter says Right, and change the Width parameter to 10′. This prepares you to place a 10′-wide lane on the right side of the assembly.

8. The command line states `Select marker point within assembly or [RETURN for Detached]`: Click the assembly on the right side of the center point marker to place a 10′-wide lane on the right side of the assembly.

9. Return to the AutoCAD Properties palette, and change the Side parameter to Left. Click the assembly on the left side of the center point marker to place a 10′ lane on the left side of the assembly. (Be sure to click the assembly baseline marker and not any part of the right BasicLane.) If you lost the AutoCAD Properties palette, you can resume the BasicLane subassembly placement by clicking the BasicLane button on your tool palette. Note that you'll have to change the Width parameter again.

MIRROR, MIRROR ON THE ROAD

You can mirror your subassemblies. To do so, select the assembly you'd like to mirror, right-click, choose Mirror from the shortcut menu, and choose the desired location for the mirrored subassembly.

10. Click the BasicCurbandGutter button on the tool palette. The Advanced section of the AutoCAD Properties palette Design tab lists the BasicCurbandGutter parameters. Change the Side parameter to Right. Note that the Insertion Point parameter has been established at the Gutter Edge, meaning the curb will attach to the lane at the desired gutter edge location. This is typically at the top edge of the pavement.

11. The command line states `Select marker point within assembly or [RETURN for Detached]`:. Click the circular point marker located at the top right of the BasicLane subassembly. This marker represents the top-right edge of pavement (see Figure 10.10). If you misplace your BasicCurbandGutter, use the AutoCAD Erase command to erase the misplaced subassembly and return to step 10.

FIGURE 10.10
The BasicCurbandGutter subassembly placed on the BasicLane subassembly

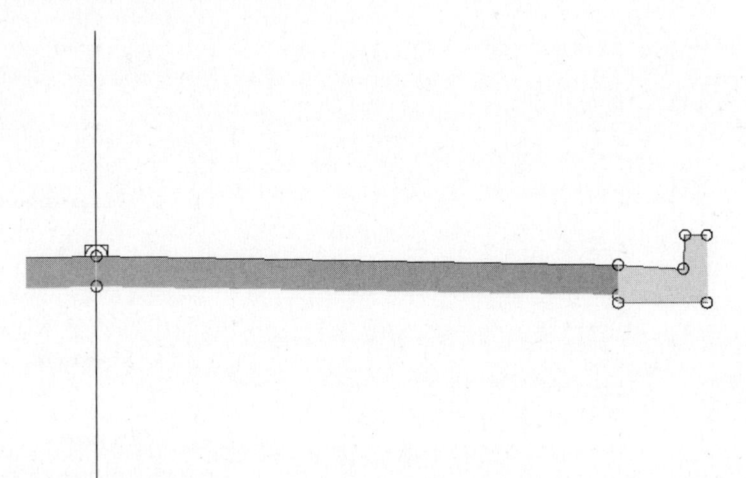

12. Change the Side parameter on the AutoCAD Properties palette to Left. Click the circular point marker located at the top left of the BasicLane subassembly. This marker represents the top-left edge of pavement.

13. Click the BasicSidewalk button on the tool palette. In the Advanced section of the Design tab on the AutoCAD Properties palette, change the Side parameter to Right, the Width parameter to 5′, and the Buffer Width 1 and Buffer Width 2 parameters to 2′. Doing so creates a sidewalk subassembly that has a 5′-wide sidewalk with 2′-wide boulevard strips on either side.

14. The command line states `Select marker point within assembly or [RETURN for Detached]:`. Click the circular point marker on the BasicCurbandGutter subassembly that represents the top rear of the curb to attach the BasicSidewalk subassembly (see Figure 10.11). If you misplace the subassembly, use the AutoCAD Erase command to erase the misplaced subassembly and return to step 13.

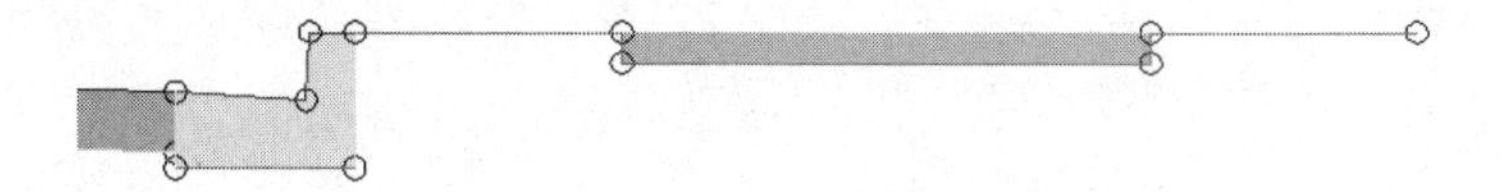

FIGURE 10.11 The BasicSidewalk subassembly placed on the BasicCurbandGutter subassembly

15. Change the Side parameter on the AutoCAD Properties palette to Left. Click the circular point marker on the right of the BasicCurbandGutter subassembly that represents the top rear of the curb. Press Esc to exit the command.

You have now completed a typical road assembly. Save your drawing if you'd like to use it in a future exercise.

Alternative Subassemblies

Once you gain some skills in building assemblies, you can explore the Corridor Modeling Catalog to find subassemblies that have more advanced parameters so that you can get more out of your corridor model. For example, if you must produce detailed schedules of road materials such as asphalt, coarse gravel, fine gravel, subgrade material, and so on, the catalog includes lane subassemblies that allow you to specify those thicknesses for automatic volume reports.

The following section includes some examples of different components you can use in a typical road assembly. Many more alternatives are available in the Corridor Modeling Catalog. The Help file provides a complete breakdown of each subassembly in the catalog; you'll find this useful as you search for your perfect subassembly.

Each of these subassemblies can be added to an assembly using exactly the same process specified in the first exercise in this chapter. Choose your alternative subassembly instead of the basic parts specified in the exercise, and adjust the parameters accordingly.

ALTERNATIVES TO THE BASICLANE SUBASSEMBLY

Although the BasicLane subassembly is suitable for many roads, you may need a more robust road lane that provides an opportunity for superelevation, additional materials, or transitioning.

BasicLaneTransition

You can use the BasicLaneTransition subassembly (Figure 10.12) instead of the BasicLane. The BasicLaneTransition is limited to a few parameters and builds a corridor model that can have a top surface and a datum surface. However, this subassembly provides an opportunity for the lane to be widened or narrowed, as you'll see in exercises in both Chapters 11 and 12. Refer to those exercises for more detailed examples.

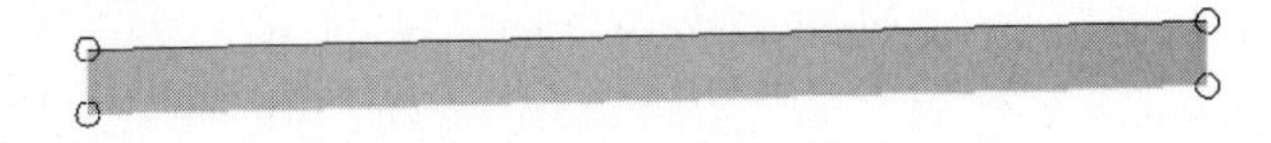

FIGURE 10.12
The BasicLaneTransition subassembly

The transition parameters for the BasicLaneTransition are as follows:

Hold Offset and Elevation behaves as a normal lane with no widening or narrowing.

Hold Elevation, Change Offset holds the design elevation at the edge of the pavement and calculates a new grade to accommodate a stretch on the basis of a target alignment.

Hold Grade, Change Offset holds the lane grade as specified in the parameters but calculates a new design elevation to accommodate a stretch on the basis of a target alignment.

Hold Offset, Change Elevation holds the lane width as specified in the parameters but uses a design elevation as specified by a target profile.

Change Offset and Elevation determines both the elevation and grade at the edge of the pavement by a target alignment and profile.

LaneParabolic

The LaneParabolic subassembly (Figure 10.13) is used for road sections that require a parabolic lane in contrast to the linear grade of the BasicLane. The LaneParabolic subassembly also adds options for two pavement depths and a base depth. This is useful in jurisdictions that require two lifts of asphalt and granular subbase material. Taking advantage of these additional parameters gives you an opportunity to build corridor models that can return more detailed quantity takeoffs and volume calculations.

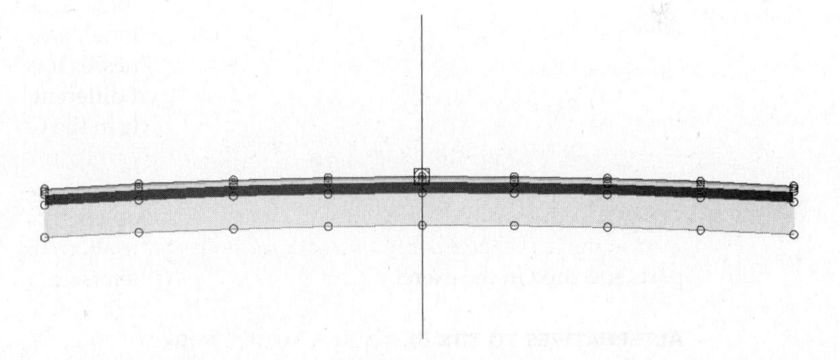

FIGURE 10.13
The LaneParabolic subassembly

Note that the LaneParabolic subassembly doesn't have a Side parameter. The parabolic nature of the component results in a single attachment point that would typically be the assembly centerline marker. Keep in mind that subassemblies by definition are AutoCAD objects and, therefore, can be moved to the right or to the left (as well as up and down) when specifications require design profiles along the top back of curb as opposed to the centerline of a travel way.

LaneBrokenBack

If your design calls for multiple lanes, and those lanes must each have a unique slope, investigate the LaneBrokenBack subassembly (Figure 10.14). This subassembly provides parameters to change the road-crown location and specify the width and slope for each lane. Like LaneParabolic, the LandBrokenBack subassembly provides parameters for additional material thicknesses.

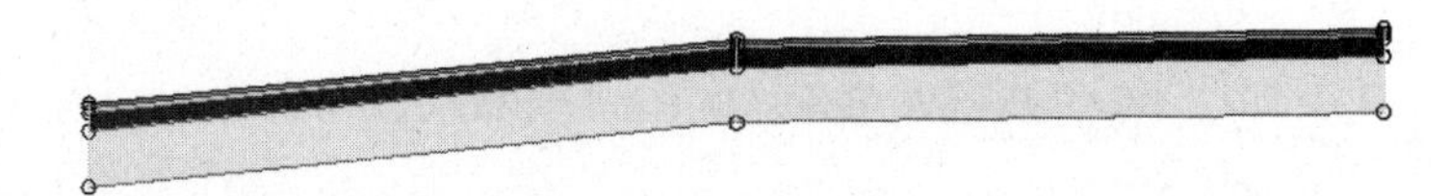

FIGURE 10.14
The LaneBrokenBack subassembly

The LaneBrokenBack subassembly, like BasicLaneTransition, allows for the use of target alignments and profiles to guide the subassembly horizontally and/or vertically.

ALTERNATIVES TO THE BASICCURBANDGUTTER

There are many types of curbs, and the BasicCurbandGutter subassembly can't model them all. Sometimes you may need to extract subbase quantities for your curbing or a more complicated set of curb dimensions, or perhaps you need a shoulder. In those cases, the Corridor Modeling Catalog provides many alternatives to BasicCurbandGutter.

BasicCurb

The BasicCurb subassembly (see Figure 10.15) is even simpler than the BasicCurbandGutter assembly. This subassembly is a straight-faced, gutterless curb that is typically attached to an outside edge of pavement. However, it can also be used on the inside edge of a median or anywhere else a straight-faced curb component is required.

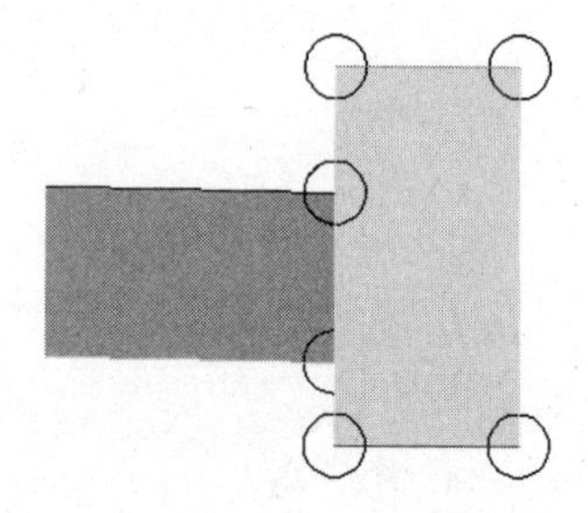

FIGURE 10.15
The BasicCurb subassembly

BasicShoulder

BasicShoulder (see Figure 10.16) is another simple yet effective subassembly for use with road sections that require a shoulder.

FIGURE 10.16
The BasicShoulder subassembly

UrbanCurbGutterGeneral

The UrbanCurbGutterGeneral subassembly (Figure 10.17) is similar to BasicCurbandGutter, except that it provides more dimension parameters and additional material parameters. If your jurisdiction specifies a curb that can't be replicated using the simple dimensions of BasicCurb-andGutter, investigate the Help file for this subassembly. Also, if your design requires detailed quantity takeoffs for the subbase used under your curb and gutter structures, this subassembly has parameters for subbase depth and slope.

FIGURE 10.17
The UrbanCurbGutter-General subassembly

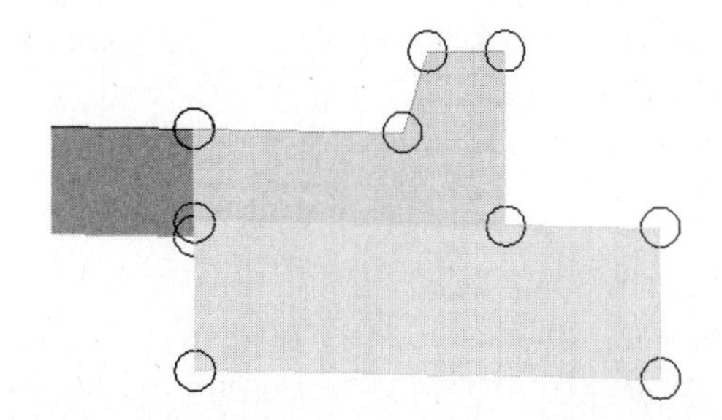

ALTERNATIVES TO THE BASICSIDEWALK SUBASSEMBLY

BasicSidewalk can reproduce many sidewalk designs, but it isn't as customizable as the other side-walk subassembly. In addition to the UrbanSidewalk assembly discussed next, consider generic links, guardrails, and other roadside structures to enhance your corridor model.

UrbanSidewalk

Whereas BasicSidewalk produces a flat sidewalk and boulevard area, the UrbanSidewalk subassembly (see Figure 10.18) can assign a slope to its sidewalk and boulevards. Additionally, you can assign the sidewalk alignment targets that are useful in cases where your sidewalk or boulevard must be widened to accommodate a bus stop, lane widening, or other pedestrian feature.

FIGURE 10.18
The UrbanSidewalk subassembly

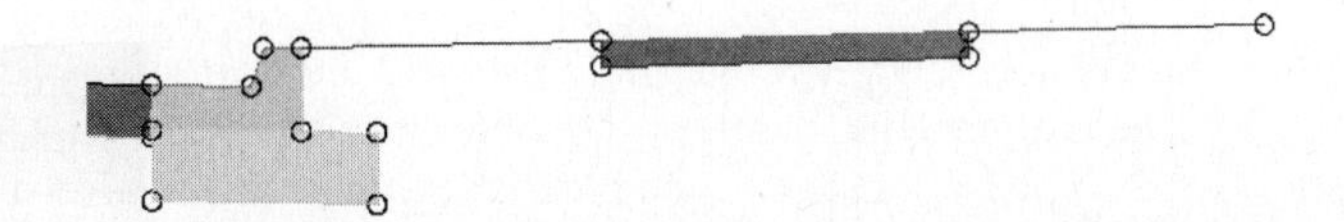

Editing an Assembly

When you first begin making assemblies, you'll be tempted to erase components and begin again when you make a mistake such as specifying an incorrect lane width. Although there is no harm in starting over, it's simple to change a subassembly parameter.

EDITING A SINGLE SUBASSEMBLY

Once your assembly is created, you can edit individual subassemblies as follows:

1. Pick the subassembly you'd like to edit.
2. Select the Subassembly Properties option from the Modify Subassembly panel.

3. The Subassembly Properties dialog appears. Click the Subassembly Help button at bottom right in the dialog if you want to shortcut to the Help page that gives detailed information about the use of this particular subassembly.
4. Switch to the Parameters tab to access the same parameters you saw in the AutoCAD Properties palette when you first placed the subassembly.
5. Click inside any field on the Parameters tab to make changes.

Editing the Entire Assembly

Sometimes it's more efficient to edit all the subassemblies in an assembly at once. To do so, pick the assembly baseline marker, or any subassembly that is connected to the assembly you'd like to edit. This time, select the Assembly Properties option from the Modify Assembly panel.

Renaming the Assembly

The Information tab on the Assembly Properties dialog gives you an opportunity to rename your assembly and provide an optional description. In many cases, it's ideal to rename your assemblies and include the station range in which they can be found along a baseline (for example, **Patricia Parkway Sta 1+34.76 to 3+74.21**).

Changing Parameters

The Construction tab on the Assembly Properties dialog houses each subassembly and its parameters. You can change the parameters for individual subassemblies by selecting the subassembly on the left side of the Construction tab and changing the desired parameter on the right side of the Construction tab.

Renaming Groups and Subassemblies

Note that the left side of the Construction tab displays a list of groups. Under each group is a list of the subassemblies in use in your assembly. A new group is formed every time a subassembly is connected directly to the assembly marker.

For example, in Figure 10.19, you see Group - (13). The first subassembly under Group - (13) is BasicLane - (74). If you dig into its parameters on the right side of the dialog, you'll learn that this lane is attached to the right side of the assembly marker, a BasicCurbandGutter is attached to right side of the BasicLane, and a BasicSidewalk is attached to the right side of the BasicCurbandGutter. The next group, Group - (14), is identical but attached to the left side of the assembly marker.

The automatic naming conventions aren't terribly self-explanatory, and it would be convenient not to have to dig into the subassembly parameters to determine which side of the assembly a certain group is on. Later, when you're making complex corridors, you'll be provided a list of subassemblies to choose from; it's certainly easier to figure out which BasicLane you need to choose when your choice is Basic Lane Right as opposed to Basic Lane - (74). Therefore, it's in your best interest to rename your subassemblies once you've built your assembly.

You can rename both groups and subassemblies on the Construction tab of the Assembly Properties dialog by selecting the entry you'd like to rename, right-clicking, and selecting Rename.

There is no official best practice on renaming your groups and subassemblies, but you may find it useful to designate what type of subassembly it is, what side of the assembly it falls on, and other distinguishing features (see Figure 10.20). For example, if a lane is to be designated as a transition lane or a generic link used as a ditch foreslope, it would be useful to name them descriptively.

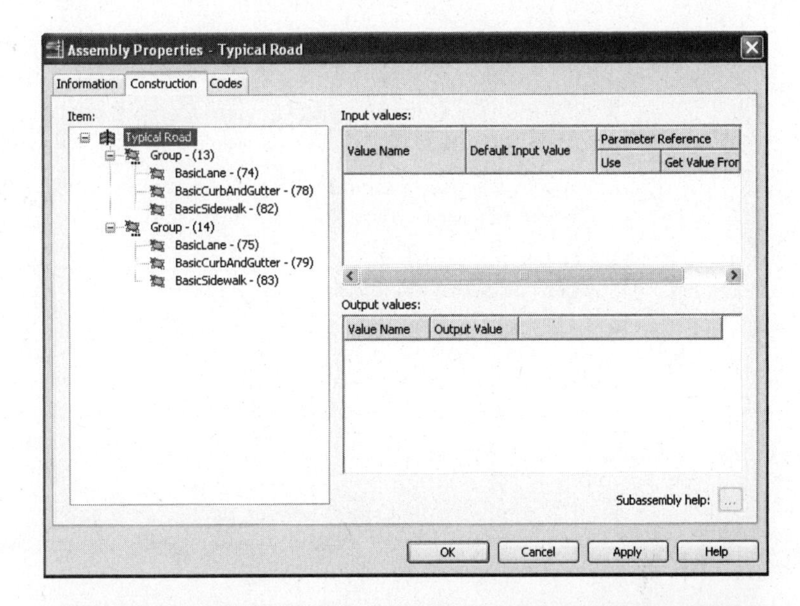

FIGURE 10.19 The Construction tab shows the default group and subassembly naming.

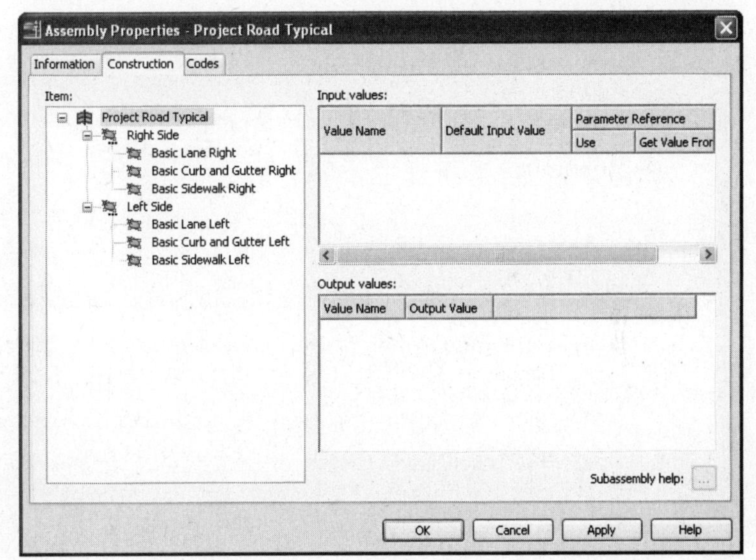

FIGURE 10.20 The Construction tab showing renamed groups and subassemblies

Creating Assemblies for Nonroad Uses

There are many uses for assemblies and their resulting corridor models aside from road sections. The Corridor Modeling Catalog also includes components for retaining walls, rail sections, bridges, channels, pipe trenches, and much more. In Chapter 11, you'll use a channel assembly and a pipe-trench assembly to build corridor models. Let's investigate how those assemblies are put together by building a channel assembly for a stream section:

1. Create a new drawing from the `_AutoCAD Civil 3D (Imperial) NCS.dwt` template, or continue working in your drawing from the first exercise in this chapter.

2. Change to the Home tab and choose Assembly ➤ Create Assembly from the Create Design panel. The Create Assembly dialog opens.
3. Enter **Channel** in the Name text box. Confirm that the Assembly Style text box is set to Basic and that Code Set Style is set to All Codes. Click OK.
4. Specify a location in your drawing for the assembly. Somewhere in the center of your screen where you have room to work is fine.
5. Locate the Trench Pipes tab on the tool palette. Position the palette on your screen so that you can clearly see the assembly baseline.
6. Click the Channel button on the tool palette. The AutoCAD Properties palette appears.
7. Locate the Advanced section of the Design tab on the AutoCAD Properties palette. You'll place the channel with its default parameters and make adjustments through the Assembly Properties dialog, so don't change anything for now. Note that there is no Side parameter. This subassembly will be centered on the assembly marker.
8. The command line states `Select marker point within assembly or [RETURN for Detached]:`. Pick the assembly center-point marker, and a channel is placed on the assembly (see Figure 10.21).

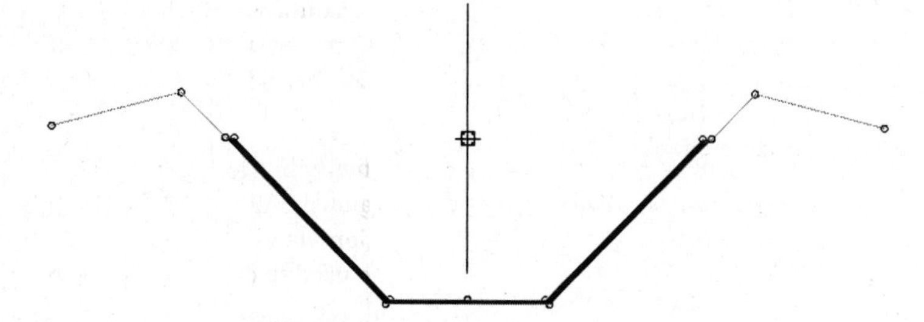

FIGURE 10.21
The Channel subassembly placed on the assembly center point marker

9. Press Esc to leave the assembly-creation command and dismiss the palette.
10. Select the assembly marker and select Assembly Properties from the Modify Assembly panel.
11. The Assembly Properties dialog appears. Switch to the Construction tab.
12. Select the Channel Assembly entry on the left side of the dialog. Click the Subassembly Help button located at bottom right in the dialog's Construction tab.
13. The Subassembly Reference portion of the AutoCAD Civil 3D 2010 Help file appears. Familiarize yourself with the diagram and input parameters for the Channel subassembly. Especially note the attachment point, bottom width, depth, and sideslope parameters. The attachment point indicates where your baseline alignment and profile will be applied.
14. Minimize the Help file.
15. To match the engineer's specified design, you need a stream section 6′ deep with a 6′-wide bottom, 1:1 sideslopes, and no backslopes. Change the following parameters in the Assembly Properties dialog:
    - Bottom Width: 6′
    - Depth: 6′

- Left and Right Backslope Width: 0′
- Sideslope: 1:1

16. Click OK, and confirm that your completed assembly looks like Figure 10.22.

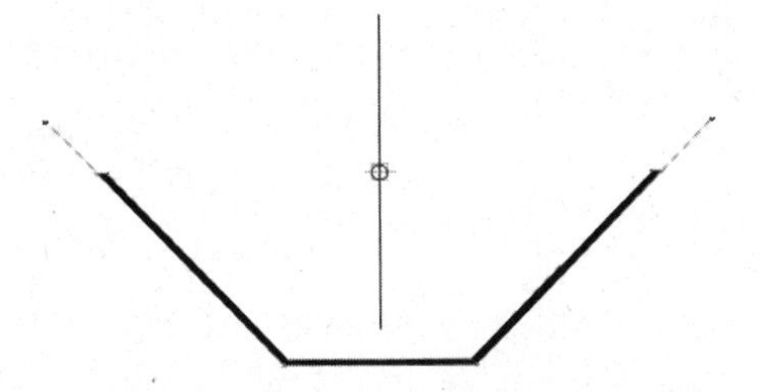

FIGURE 10.22
A completed channel assembly

Real World Scenario

A PIPE TRENCH ASSEMBLY

Projects that include piping, such as sanitary sewers, storm drainage, gas pipelines, or similar structures, almost always include trenching. The trench must be carefully prepared to ensure the safety of the workers placing the pipe, as well as provide structural stability for the pipe in the form of bedding and compacted fill.

The corridor is an ideal tool for modeling pipe trenching. With the appropriate assembly combined with a pipe-run alignment and profile, you can not only design a pipe trench but also use cross-section tools to generate section views (Graphic), materials tables, and quantity takeoffs. The resulting corridor model can also be used to create a surface for additional analysis and use.

The following exercise will lead you through building a pipe trench corridor based on an alignment and profile that follow a pipe run, and a typical trench assembly.

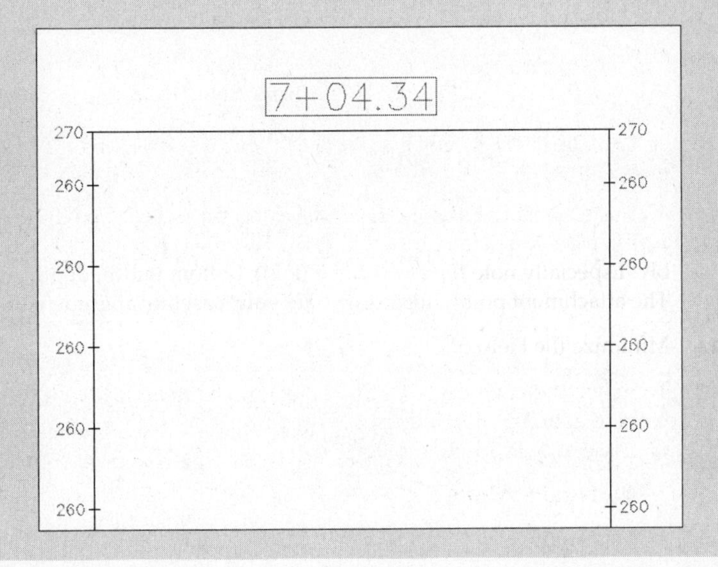

1. Create a new drawing using the _AutoCAD Civil 3D (Imperial) NCS.dwt template, or continue working in your drawing from the previous exercise.
2. Change to the Home tab and choose Assembly ➢ Create Assembly from the Create Design panel. The Create Assembly dialog opens.
3. Enter **Pipe Trench** in the Name text box to change the assembly's name. Confirm that the Assembly Style text box is set to Basic and Code Set Style is set to All Codes. Click OK.
4. Pick a location in your drawing for the assembly. Somewhere in the center of your screen where you have room to work is fine.
5. Locate the Trench Pipes tab on the tool palette. Position the palette on your screen so that you can clearly see the assembly baseline.
6. Click the TrenchPipe1 button on the tool palette. The AutoCAD Properties palette appears. Position the AutoCAD Properties palette on your screen so that you can clearly see both the assembly baseline and the tool palette.
7. Locate the Advanced section of the Design tab on the AutoCAD Properties palette. This section lists the TrenchPipe1 parameters. You'll place TrenchPipe1 with its default parameters and make adjustments through the Assembly Properties dialog, so don't change anything for now. Note that there is no Side parameter. This subassembly will be placed centered on the assembly marker.
8. The command line states `Select marker point within assembly or [RETURN for Detached]:`. Pick the assembly center-point marker. A TrenchPipe1 subassembly is placed on the assembly as shown.

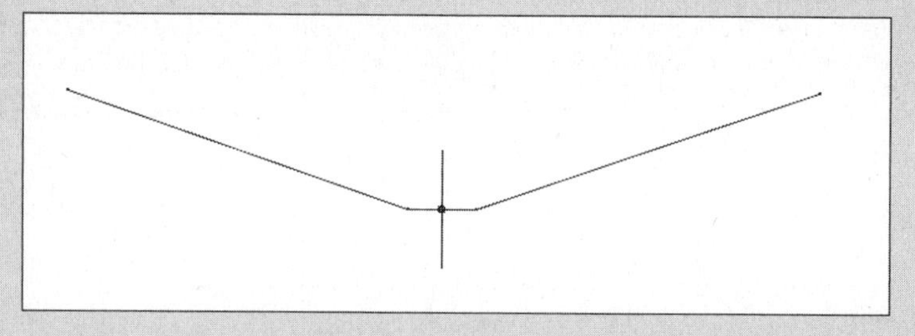

9. Press Esc to leave the assembly-creation command and dismiss the AutoCAD Properties palette.
10. Select the assembly marker and select Assembly Properties from the Modify Assembly panel.
11. The Assembly Properties dialog appears. Switch to the Construction tab.
12. Select the TrenchPipe1 assembly entry on the left side of the dialog. Click the Subassembly Help button located at bottom right.
13. The Subassembly Reference portion of the AutoCAD Civil 3D 2010 Help file appears. Familiarize yourself with the diagram and input parameters for the TrenchPipe1 subassembly. In this case, the profile grade line will attach to a profile drawn to represent the pipe invert. Because the trench will be excavated deeper than the pipe invert to accommodate gravel bedding, you'll use the bedding depth parameter in a moment. Also note under the Target

Parameters that this subassembly needs a surface target to determine where the sideslopes terminate.

14. Minimize the Help file.

15. To match the engineer's specified design, the pipe trench should be 3′ deep and 6′ wide with 3:1 sideslopes and 1′ of gravel bedding. Change the following parameters in the Assembly Properties dialog:

   - Bedding Depth: 1′
   - Offset to Bottom: -3′
   - Sideslope: 3:1

16. Click OK.

17. Confirm that your completed assembly looks like the graphic shown here, and save your drawing.

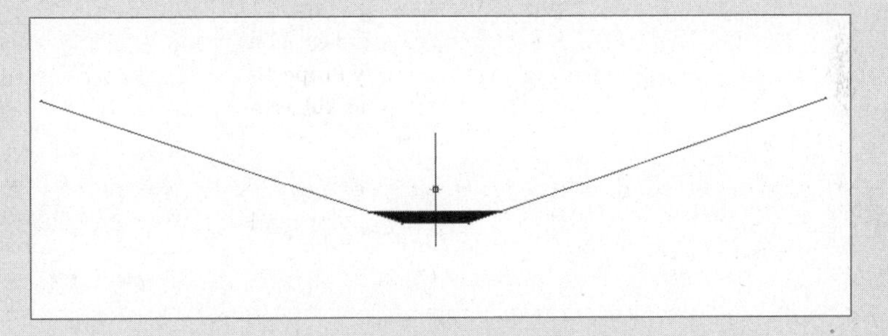

This assembly will be used to build a pipe-trench corridor in Chapter 11.

Working with Generic Subassemblies

Despite the more than 100 subassemblies available in the Corridor Modeling Catalog, sometimes you may not find the perfect component. Perhaps none of the channel assemblies exactly meet your design specifications, and you'd like to make a more customized assembly; or neither of the sidewalk subassemblies allow for the proper boulevard slopes. Maybe you'd like to try to do some preliminary lot grading using your corridor, or mark a certain point on your subassembly so that you can extract important features easily.

You can tackle all these items by programming your own custom subassemblies, creating a custom subassembly from a polyline — better yet, you can handle them using subassemblies from the Generic Subassembly Catalog (see Figure 10.23). These simple and flexible components can be used to build almost anything, although they lack the coded intelligence of some of the more intricate assemblies (such as knowing if they're paved, grass, or similar, and understanding things like subbase depth, and so on).

Enhancing Assemblies Using Generic Links

Let's look at two examples where you might take advantage of generic links.

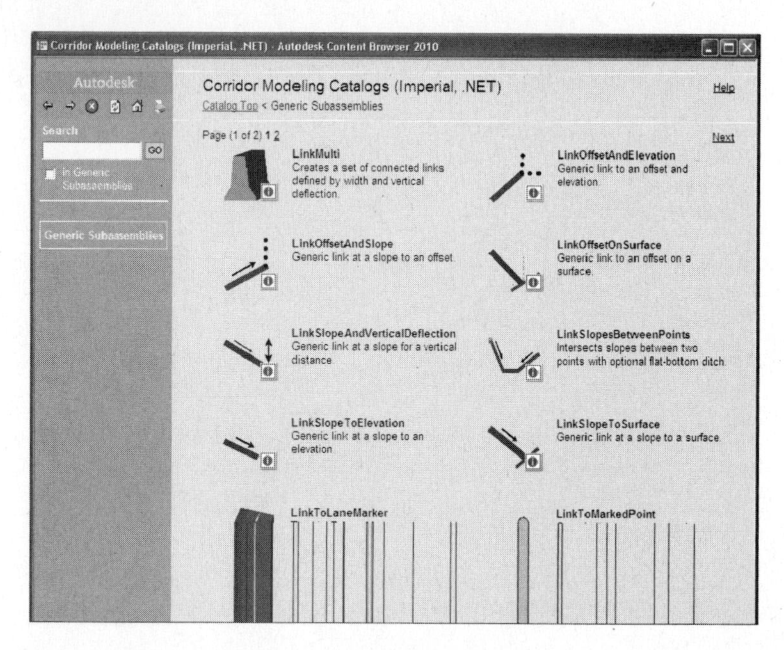

FIGURE 10.23
The Generic Subassembly Catalog

The first example involves the typical road section you built in the first exercise in this chapter. You saw that BasicSidewalk doesn't allow for a sloped sidewalk or sloped buffer strips. UrbanSidewalk does have a slope parameter, but each buffer strip has the same slope as the sidewalk itself. If you need a 6′-wide buffer strip with a 3 percent slope, and then a 5′ sidewalk with a 2 percent slope, followed by another buffer strip that is 10′ wide with a slope of 5 percent, you can use generic links to assist in the construction of the proper assembly:

1. Open the `Sidewalk Start.dwg` file (which you can download from `www.sybex.com/go/masteringcivil3d2010`), or continue working in your drawing from the first exercise in this chapter.
2. Zoom in on the road-section assembly. If you're working in your drawing from the first exercise in the chapter, erase the BasicSidewalk subassemblies from either side of your assembly.
3. Locate the Imperial-Generic tab on the tool palette. Position the palette on your screen so that you can clearly see the assembly baseline.
4. Click the LinkWidthandSlope button, and the AutoCAD Properties dialog appears. Position the dialog on your screen so that you can clearly see both the assembly baseline and the tool palette.
5. Locate the Advanced section of the Design tab in the AutoCAD Properties palette. This section lists the parameters for this subassembly. Change the parameters as follows to create the first buffer strip:
   - Side: Right
   - Width: 6′
   - Slope: 3%

6. The command line states `Select marker point within assembly or [RETURN for Detached]:`. Select the circular point marker on the right BasicCurbandGutter subassembly, which represents the top back of the curb. A Link subassembly appears, as shown in Figure 10.24.

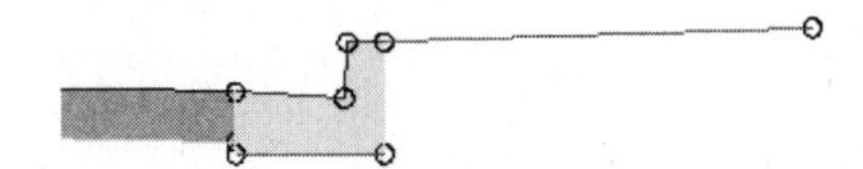

FIGURE 10.24
The Generic Link subassembly

7. Switch to the Imperial-Curbs tab of the tool palette. Click the UrbanSidewalk button, and the AutoCAD Properties palette appears. Position the palette on your screen so that you can clearly see it, the assembly baseline, and the tool palette.
8. Locate the Advanced section of the Design tab in the AutoCAD Properties palette. This section lists the parameters for the UrbanSidewalk subassembly. Change the parameters as follows to create the sidewalk:
   - Side: Right
   - Sidewalk Width: 5′
   - Slope: 2%
   - Inside Boulevard Width: 0′
   - Outside Boulevard Width: 0′
9. The command line states `Select marker point within assembly or [RETURN for Detached]:`. Select the circular point marker on the right LinkWidthandSlope subassembly. An UrbanSidewalk subassembly appears, as shown in Figure 10.25.
10. Switch to the Imperial-Generic tab of the tool palette. Click the LinkWidthandSlope button, and the AutoCAD Properties palette appears. Position the palette on your screen so that you can still see the assembly baseline and the tool palette.

FIGURE 10.25
The UrbanSidewalk subassembly

11. Locate the Advanced section of the Design tab on the AutoCAD Properties palette. This section lists the parameters for the LinkWidthandSlope subassembly you saw in step 5. Change the parameters as follows to create the second buffer strip:
    - Side: Right
    - Width: 10′
    - Slope: 5%

    Your drawing should now look like Figure 10.26.

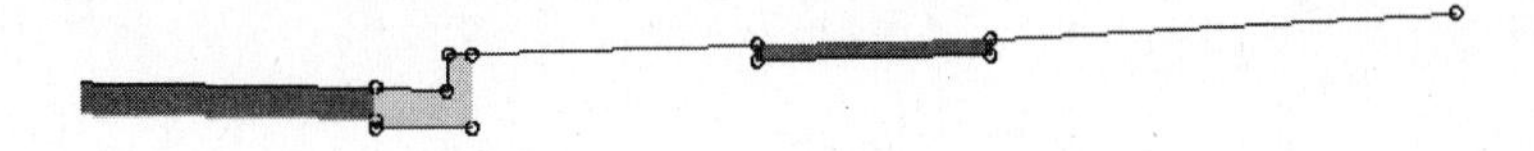

FIGURE 10.26
The sidewalk and buffer strips

12. Select the two generic links and the sidewalk assembly.
13. Select Mirror Subassemblies from the Modify Subassembly panel. The command line displays `Select marker point within assembly:`.
14. Select the marker point on the left back of curb. The completed assembly should look like Figure 10.27.

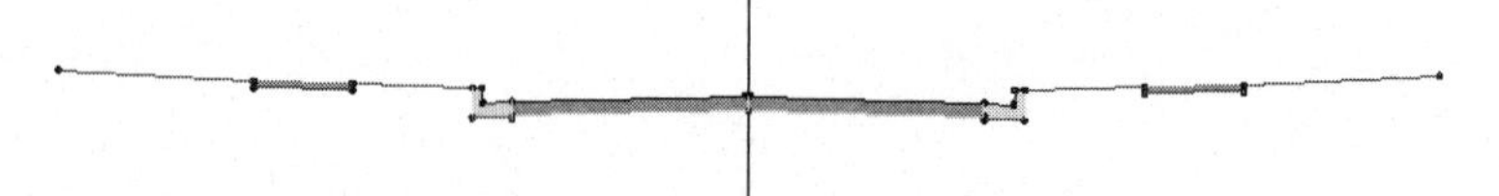

FIGURE 10.27
The completed assembly

15. Save your drawing if you'd like to use it in a future exercise.

You've now created a custom sidewalk boulevard for a typical road.

The second example involves the channel section you built earlier in this chapter. Although the TrenchPipe1 subassembly includes a surface target, the Channel assembly doesn't. This exercise will lead you through using the LinkSlopetoSurface generic subassembly, which will provide a surface target to the Channel assembly that will seek the target assembly at a 25 percent slope. For more information about surface targets, see Chapters 11 and 12. Follow these steps:

1. Open the `Channel Link Start.dwg` file (which you can download from `www.sybex.com/go/masteringcivil3d2010`), or continue working in your drawing from the channel exercise in this chapter.
2. Zoom in on the Channel assembly.
3. Locate the Imperial-Generic tab on the tool palette. Position the palette on your screen so that you can clearly see the assembly baseline.
4. Click the LinkSlopetoSurface button. The AutoCAD Properties palette appears. Position the palette on your screen so that you can still see both the assembly baseline and the tool palette.
5. Locate the Advanced section of the Design tab on the AutoCAD Properties palette. This section lists the parameters for the LinkWidthandSlope subassembly. Change the parameters as follows to create a surface target link:
    - Side: Right
    - Slope: 25%
6. The command line states `Select marker point within assembly or [RETURN for Detached]:`. Click the circular point marker at upper right on the Channel subassembly that is farthest away. A surface target link appears (see Figure 10.28).

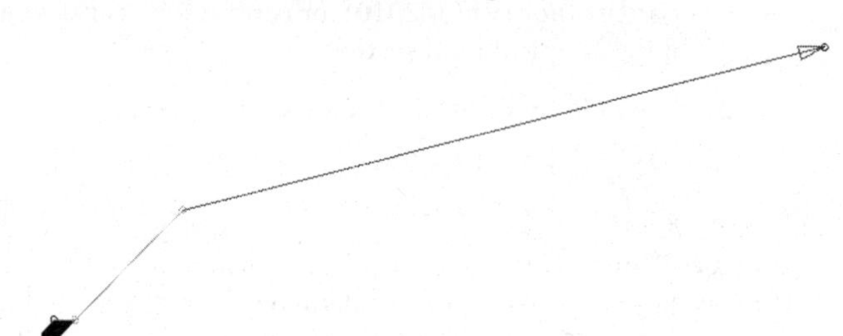

FIGURE 10.28
The attachment location for the LinkSlopetoSurface subassembly

7. To complete the left side of the assembly, repeat steps 4 through 6, and change the Side parameter to the Left option.

8. The completed assembly should look like Figure 10.29.

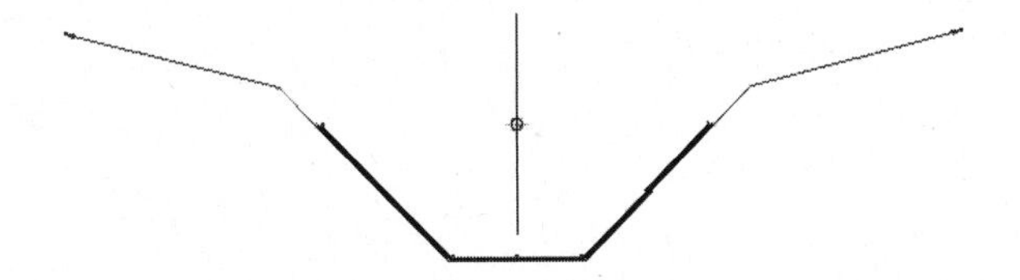

FIGURE 10.29
The completed Channel assembly

Adding a surface link to a Channel assembly provides a surface target for the assembly. When you're designing a channel, it's important to tie into existing ground. In its original form, the Channel subassembly doesn't include a target parameter that would allow you to choose an existing ground; therefore, you'd need to do quite a bit of hand grading between the top of the bank and existing ground. Now that you've added the LinkSlopetoSurface, you can specify your existing ground as the surface target, and the subassembly will grade between top of the bank and the surface for you. You can achieve additional flexibility for connecting to existing ground with the more complicated Daylight subassemblies, as discussed in the next section.

Working with Daylight Subassemblies

Most typical sections have many absolute requirements, such as a cross slope for a lane or the height of a curb. But from that last engineered point on the left and right of a typical section, some design decisions have flexibility.

In the example of the typical road section from the first part of this chapter, the engineer needs to design the grade from the last buffer strip until the section ties into existing ground. The location where the design meets existing ground is known as *daylighting*.

Daylight subassemblies provide tools to assist the engineer in meeting the design intent between existing ground and the typical section. Some Daylight subassemblies are shown in Figure 10.30.

Enhancing an Assembly with a Daylight Subassembly

Using the typical road section from the first exercise in this chapter, your subdivision layout allows for grading 25′ from the end of the sidewalk buffer strip. This grading has a 4:1 maximum for both cut-and-fill situations. In the following exercise, you'll use the DaylightMaxWidth subassembly, which contains parameters for specifying the grading width and the maximum cut-and-fill slopes:

1. Open the `Daylight Start.dwg` file (which you can download from `www.sybex.com/go/masteringcivil3d2010`), or continue working in any drawing from this chapter that contains a typical road section.

2. Zoom in on the typical road-section assembly.

3. Locate the Imperial-Daylight tab on the tool palette. Position the palette on your screen so that you can clearly see the assembly baseline.

4. Click the DaylightMaxWidth button on the tool palette. The AutoCAD Properties palette appears. Position the palette on your screen so that you can still clearly see both the assembly baseline and the tool palette.

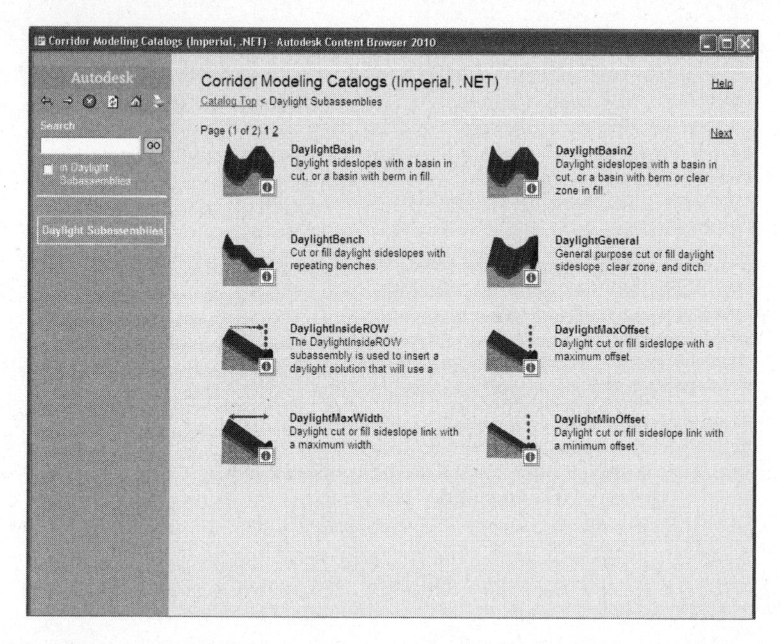

FIGURE 10.30
Some Daylight subassemblies in the Corridor Modeling Catalog

5. Locate the Advanced section of the Design tab on the AutoCAD Properties palette. This section lists the parameters for the DaylightMaxWidth subassembly. Change the following parameters to create the daylight as required:

   - Side: Right
   - Cut Slope: 4:1 (or 25%)
   - Fill Slope: 4:1 (or 25%)
   - Max Width: 25

6. The command line states `Select marker point within assembly or [RETURN for Detached]:`. Select the circular point marker on the farthest right link. The subassembly appears as in Figure 10.31.

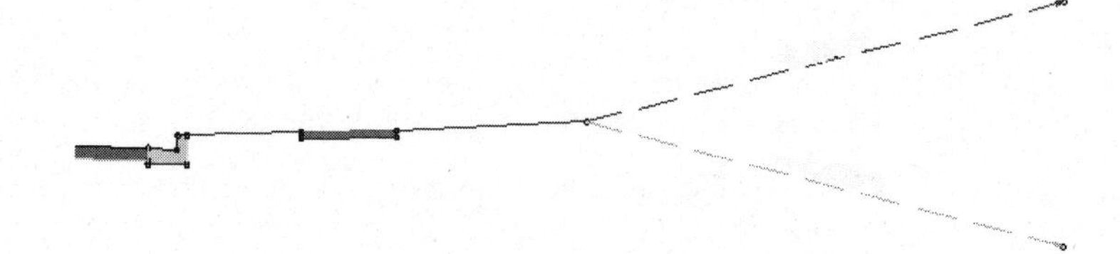

FIGURE 10.31
Placement of the DaylightMaxWidth subassembly

7. Press Esc to exit the assembly-creation command.

8. Pick the DaylightMaxWidth subassembly, and then choose Subassembly Properties from the Modify Subassembly panel.

9. Switch to the Parameters tab in the Subassembly Properties dialog.

10. Click the Subassembly Help button in the lower-right corner. The Subassembly Reference opens in a new window. Familiarize yourself with the options for the DaylightMaxWidth subassembly, especially noting the optional parameters for a lined material, a mandatory

daylight surface target, and an optional alignment target that can be used for the maximum width.

11. Minimize the Subassembly Reference window.
12. To complete the left side of the assembly, repeat steps 3 through 6, changing the Side parameter for each subassembly to the Left option. The completed assembly should look like Figure 10.32.

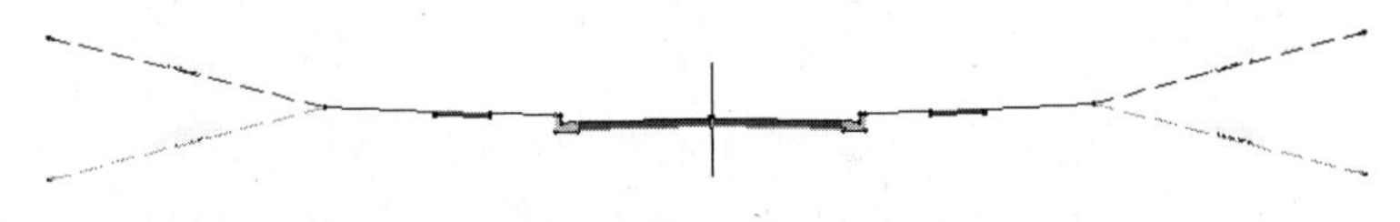

FIGURE 10.32
An assembly with the daylight subassembly properly attached

WHEN TO IGNORE PARAMETERS

The first time you attempt to use many Daylight subassemblies, you may become overwhelmed by the sheer number of parameters, as shown here.

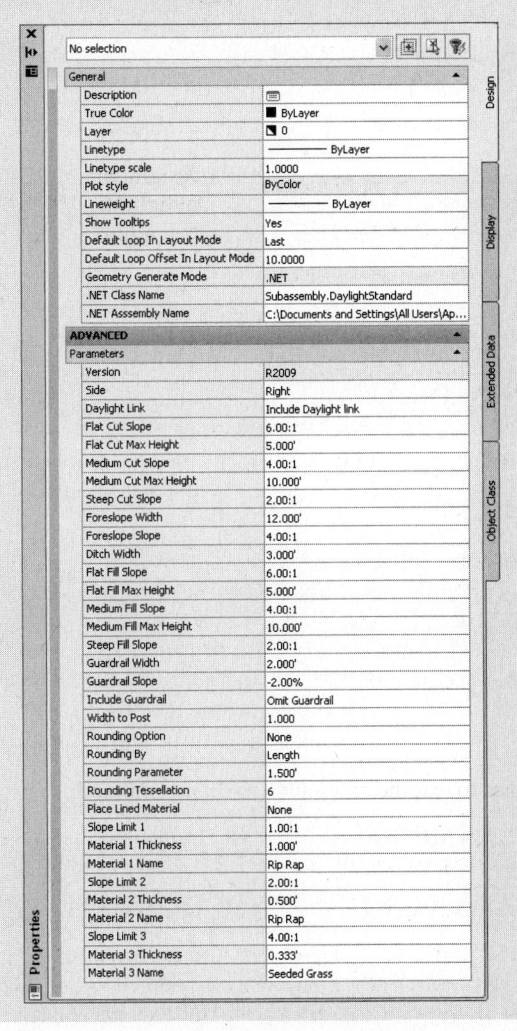

The good news is that many of these parameters are unnecessary for most uses. For example, many Daylight subassemblies, such as DaylightGeneral, include multiple cut-and-fill widths for complicated cases where the design may call for test scenarios. If your design doesn't require this level of detail, leave those parameters set to zero.

Some Daylight subassemblies include guardrail options. If your situation doesn't require a guardrail, leave the default parameter set to the Omit Guardrail option and ignore it from then on. Another common, confusing parameter is Place Lined Material, which can be used for riprap or erosion-control matting. If your design doesn't require this much detail, ensure that this parameter is set to None, and ignore the thickness, name, and slope parameters that follow.

One parameter you can't ignore is Target Surface. The entire function of daylighting is to tie into a target surface. Without that target, many of the daylight parameters have no point of reference, and often your Daylight subassembly won't work and you'll get errors upon building your corridor.

That being said, you can temporarily omit the Daylight link on subassemblies where ditch or bench construction occurs if your target surface isn't ready. An example of this might be if you're tying into an adjacent plot of land that is already under construction. When the construction is finished and you've obtained the final surface model, then you should change the parameter to include the Daylight link.

If you're ever in doubt about which parameters can be omitted, investigate the Help file for that subassembly.

Alternative Daylight Subassemblies

At least a dozen Daylight subassemblies are available, varying from a simple cut-fill parameter to a more complicated benching or basin design. Your engineering requirements may dictate something more challenging than the exercise in this section. Here are some alternative Daylight subassemblies and the situations where you might use them. For more information on any of these subassemblies and the many other daylighting choices, see the AutoCAD Civil 3D 2010 Subassembly Reference in the Help file.

DaylightToROW

The DaylightToROW subassembly (see Figure 10.33) forces a tie-in to the target surface using the controlling parameter of ROW Offset from Baseline. Because this value is calculated from the baseline location, you can place lanes, sidewalks, curbs, and more between the baseline and the Daylight subassembly and not worry about recalculating the width of the Daylight subassembly as you would with DaylightMaxWidth. This subassembly is most useful in design situations where you absolutely must not grade outside of the ROW.

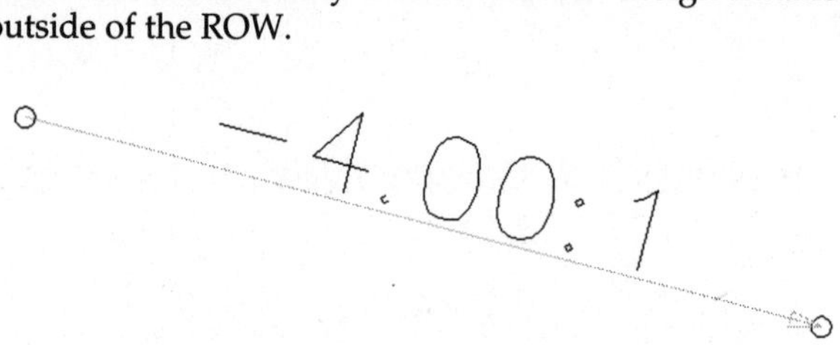

Figure 10.33
The DaylightToROW subassembly

BasicSideSlopeCutDitch

In addition to including cut-and-fill parameters, the BasicSideSlopeCutDitch subassembly (see Figure 10.34) is capable of creating a ditch when it detects a cut condition. This is most useful for road sections that require a roadside ditch through cut sections but omit it when passing through areas of fill. If your corridor model is revised in a way that changes the location of cut-and-fill boundaries, the ditch will automatically adjust.

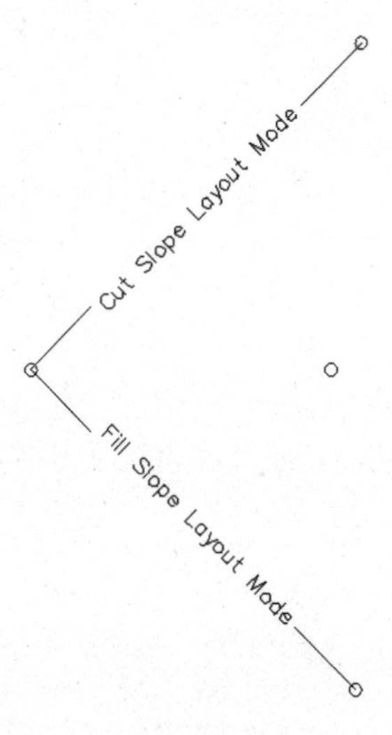

FIGURE 10.34
The BasicSideSlopeCut-Ditch subassembly

DaylightBasin

Many engineers must design berms to contain roadside swales when the road design is in the fill condition. The process for determining where these berms are required is often tedious. The DaylightBasin subassembly (see Figure 10.35) provides a tool for automatically creating these "false berms." The subassembly contains parameters for the specification of a basin (which can be easily adapted to most roadside ditch cross sections as well) and parameters for containment berms that appear only when the subassembly runs into areas of roadside cut.

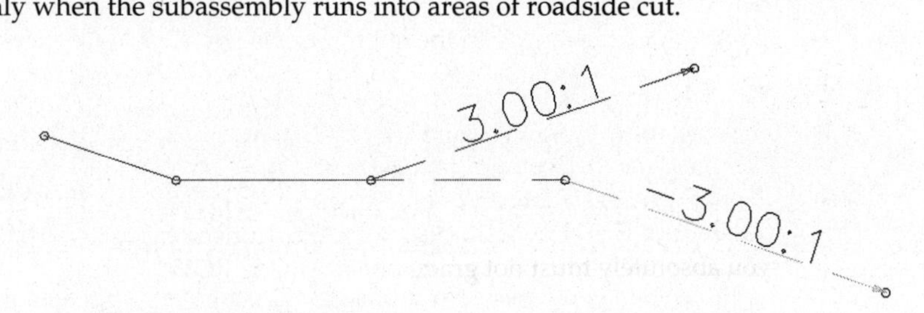

FIGURE 10.35
The DaylightBasin subassembly

Saving Subassemblies and Assemblies for Later Use

Customizing subassemblies and creating assemblies are both simple tasks. However, you'll save time in future projects if you store these assemblies for later use. In Civil 3D 2010, some common assemblies are built into the default tool palettes to get you started, as shown in Figure 10.36.

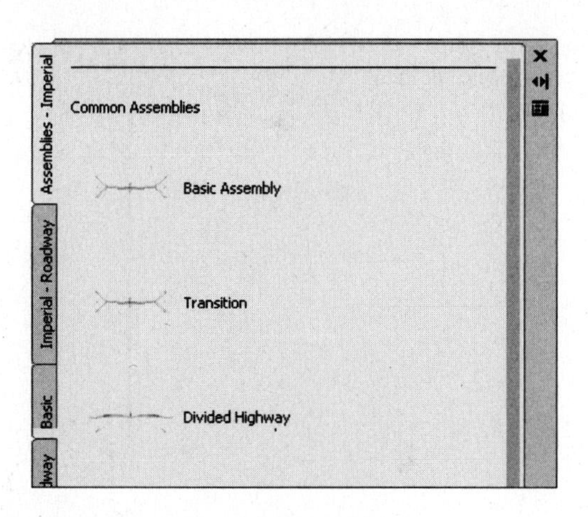

FIGURE 10.36
Some completed assemblies are provided on the default tool palettes.

Storing a Customized Subassembly on a Tool Palette

A typical jurisdiction usually has a finite number of allowable lane widths, curb types, and other components. It would be extremely beneficial to have the right subassemblies with the parameters already set available on your tool palette.

The following exercise will lead you through storing a customized subassembly on a tool palette:

1. Open the `Storing Subassemblies and Assemblies.dwg` file (which you can download from `www.sybex.com/go/masteringcivil3d2010`).
2. Locate the Imperial-Basic tab on the tool palette. Position the palette on your screen so that you can clearly see the assembly baseline.
3. Right-click in the Tool Palette area, and select New Palette to create a new tool palette. Enter **My Road Parts** in the Name text box.
4. Select the right lane from the assembly. You'll know it's selected when you can see it highlighted and the grip appears at the assembly baseline.
5. Click the lane with the left mouse button anywhere except the grip location until you see an arrow-shaped glyph appear. It takes a bit of practice to get this arrow to show up, and sometimes it's hard to see until you move your cursor a bit. If you have trouble, try clicking near the edges of the lane instead of in the hatched shape area. Once the arrow appears, continue to hold down the left mouse button, move your cursor to the tool palette, and release the mouse button once you're over the tool palette. It's easier to place the item if you hover your cursor toward the top of your new tool palette.
6. When you release the mouse button, an entry appears on your tool palette for BasicLane. Right-click this entry, and select the Properties option. The Tool Properties dialog appears (see Figure 10.37).
7. Enter **10-Foot Wide Basic Lane at 2%** in the Name text box. You can also change the image, description, and other parameters in this dialog. Click OK.
8. Repeat this process for each lane and each curb in the drawing, if desired. The resulting tool palette looks similar to Figure 10.38.

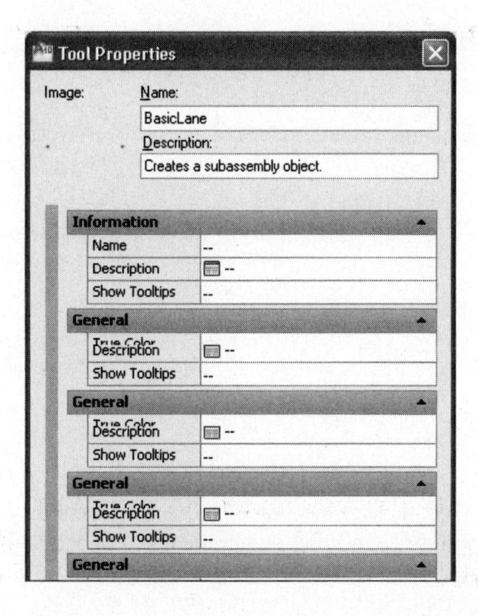

FIGURE 10.37
The Tool Properties dialog

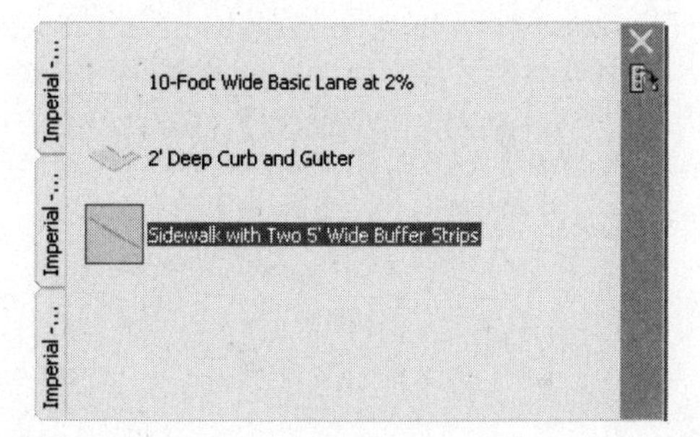

FIGURE 10.38
A tool palette with three customized subassemblies

Note the tool palette entries for each subassembly point to the location of the `Subassembly.NET` directory, and not to this drawing. If you share this tool palette, make sure the subassembly directory is either identical or accessible to the person with whom you're sharing.

Storing a Completed Assembly on a Tool Palette

In addition to storing individual subassemblies on a tool palette, it's often useful to warehouse entire completed assemblies. Many jurisdictions have several standard road cross sections; once each standard assembly has been built, you can save time on future similar projects by pulling in a prebuilt assembly.

The process for storing an assembly on a tool palette is nearly identical to the process of storing a subassembly. Simply select the assembly baseline, hover your cursor over the assembly baseline, left-click, and drag to a palette of your choosing.

It's usually a good idea to create a library or *sandbox* drawing in a shared network location for common completed assemblies and to create all assemblies in that drawing before dragging them onto the tool palette. By using this approach, you'll be able to test your assemblies for validity before they are rolled into production.

The Bottom Line

Create a typical road assembly with lanes, curb, gutter, and sidewalk. Most corridors are built to model roads. The most common assembly used in these road corridors is some variation of a typical road section consisting of lanes, curb, gutter, and sidewalk.

Master It Create a new drawing from the `_AutoCAD Civil 3D (Imperial) NCS.dwt` template. Build a symmetrical assembly using BasicLane, BasicCurbandGutter, and BasicSidewalk. Use widths and slopes of your choosing.

Edit an assembly. Once an assembly has been created, it can be easily edited to reflect a design change. Often, at the beginning of a project, you won't know the final lane width. You can build your assembly and corridor model with one lane width and then later change the width and rebuild the model immediately.

Master It Working in the same drawing, edit the width of each BasicLane to 14′, and change the cross slope of each BasicLane to -3.08%.

Add daylighting to a typical road assembly. Often, the most difficult part of a designer's job is figuring out how to grade the area between the last hard engineered point in the cross section (such as the back of a sidewalk) and existing ground. An extensive catalog of daylighting subassemblies can assist you with this task.

Master It Working in the same drawing, add the DaylightMinWidth subassembly to both sides of your typical road assembly. Establish a minimum width of 10′.

Chapter 11

Easy Does It: Basic Corridors

The corridor object is a three-dimensional road model that combines the horizontal geometry of an alignment, the vertical geometry of a profile, and the cross-sectional geometry of an assembly.

Corridors range from extremely simple roads to complicated highways and interchanges. This chapter focuses on building several simple corridors that can be used to model and design roads, channels, and trenches.

By the end of this chapter, you'll learn to:

- Build a single baseline corridor from an alignment, profile, and assembly
- Create a corridor surface
- Add an automatic boundary to a corridor surface

Understanding Corridors

It its simplest form, a corridor is a three-dimensional combination of an alignment, a profile, and an assembly (see Figure 11.1).

You can also build corridors with additional combinations of alignments, profiles, and assemblies to make complicated intersections, interchanges, and branching streams (see Figure 11.2).

The horizontal properties of the alignment, the vertical properties of the profile, and the cross-sectional properties of the assembly are merged together to form a dynamic model that can be used to build surfaces, sample cross sections, and much more.

Most commonly, corridors are thought of as being used to model roads, but they can also be adapted to model berms, streams, lagoons, trails, and even parking lots (see Figure 11.3).

Creating a Simple Road Corridor

The first ingredient in any corridor is an alignment. This alignment is referred to as a *baseline*. A baseline requires a corresponding profile and an assembly. A corridor can have multiple baselines, and a baseline can be divided into *regions*. You'll see how regions are used a little later in this chapter. (Corridors with multiple baselines are discussed in Chapter 12, "The Road Ahead: Advanced Corridors.")

When you create and iterate a design, you may be tempted to use its default name, such as Alignment-64 or Basic Lane-(3)(3), instead of a much more meaningful name. Before building a corridor, even a simple corridor, it is important to make sure your alignments, profiles, assemblies, and subassemblies have good names. If you get into the habit of giving your objects significant and meaningful names — even for the simplest corridor — you will be rewarded when you build larger corridors (see Figure 11.4).

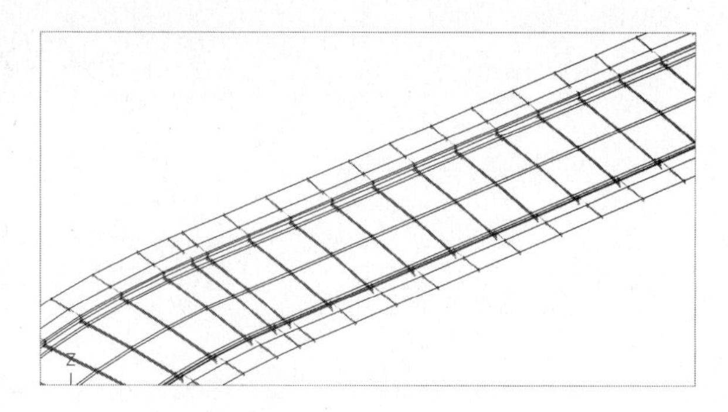

FIGURE 11.1
A simple corridor

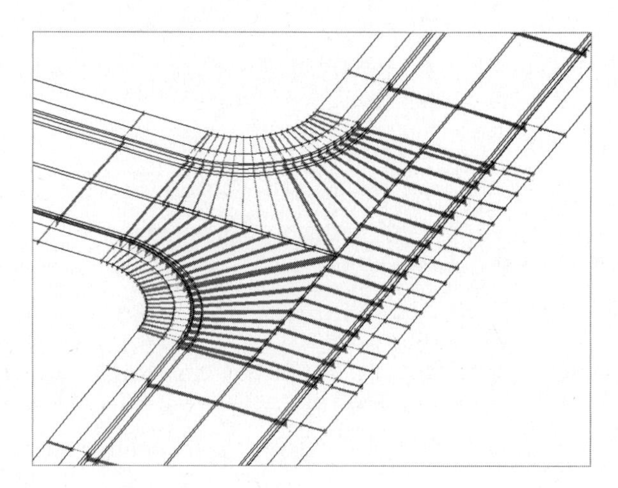

FIGURE 11.2
An intersection modeled with a corridor

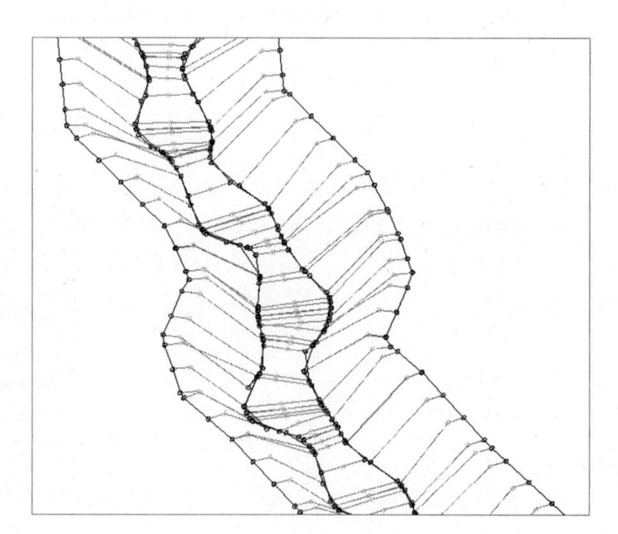

FIGURE 11.3
A complex stream modeled with a corridor

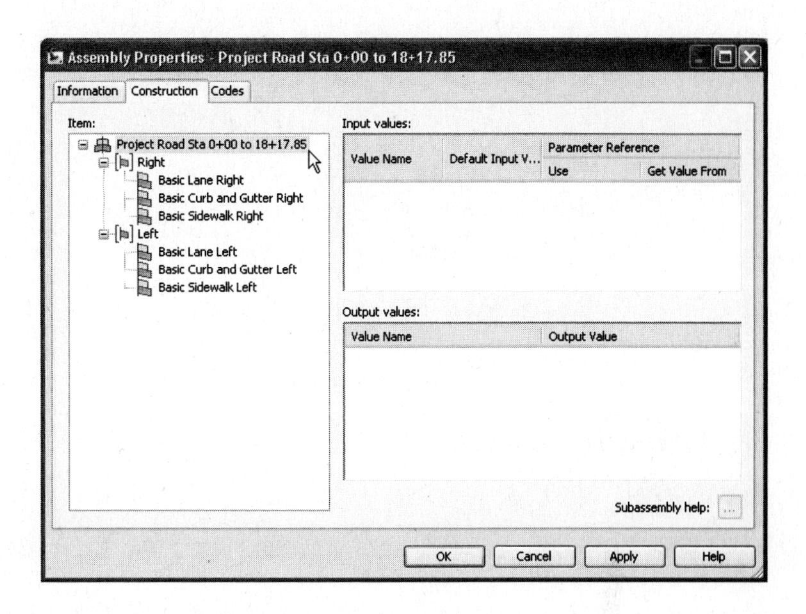

FIGURE 11.4
Check the names of your alignments, profiles, assemblies, and subassemblies

This exercise gives you hands-on experience in building a corridor model from an alignment, a profile, and an assembly:

1. Open the `Simple Corridor.dwg` file, which you can download from `www.sybex.com/go/masteringcivil3d2010`. Note that the drawing has an alignment, a profile view with two profiles, and an assembly, as well as an existing ground surface.
2. Change to the Home tab and select Corridor ➢ Create Simple Corridor from the Create Design panel. The Create Simple Corridor dialog opens.
3. In the Name text box, give your corridor a meaningful name, such as **Project Road**. Keep the default values for Corridor Style and Corridor Layer (see Figure 11.5).

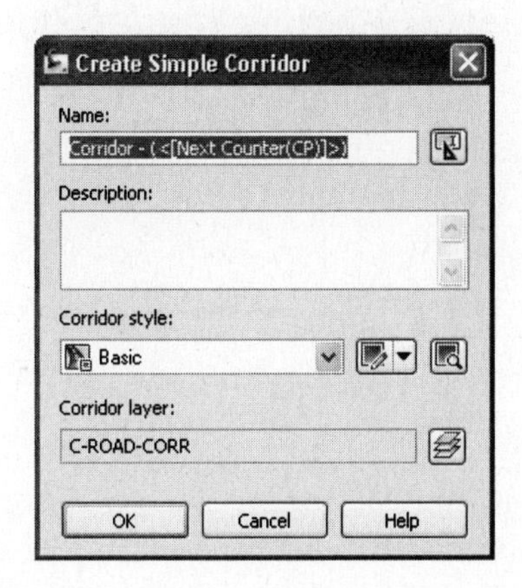

FIGURE 11.5
Change the corridor name to something meaningful for easy bookkeeping

4. Click OK to dismiss the dialog.

5. At the `Select baseline alignment <or press enter key to select from list>:` prompt, pick the alignment in the drawing. Alternatively, you could press ↵ and select your alignment from a list.

6. At the `Select a profile <or press enter key to select from list>:` prompt, pick the Finished Ground profile (the profile with labels) in the drawing. Alternatively, you could press ↵ and select your profile from a list.

7. At the `Select an assembly <or press enter key to select from list>:` prompt, pick the vertical line of the assembly in the drawing. Alternatively, you could press ↵ and select your assembly from a list.

8. The program will process and build the corridor. Dismiss the Panorama if it appears over the road centerline alignment as shown in Figure 11.6.

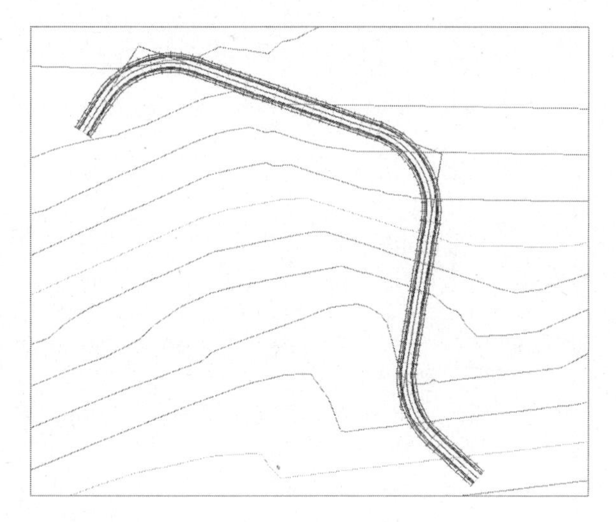

FIGURE 11.6
The completed simple corridor

Utilities for Viewing Your Corridor

Once your corridor is built, chances are you will want to examine the corridor in section view and use 3D to view the model and check for problems. For a station-by-station look at a corridor, pick the corridor and choose Corridor Section Editor from the Modify panel. The Section Editor contextual tab opens (see Figure 11.7).

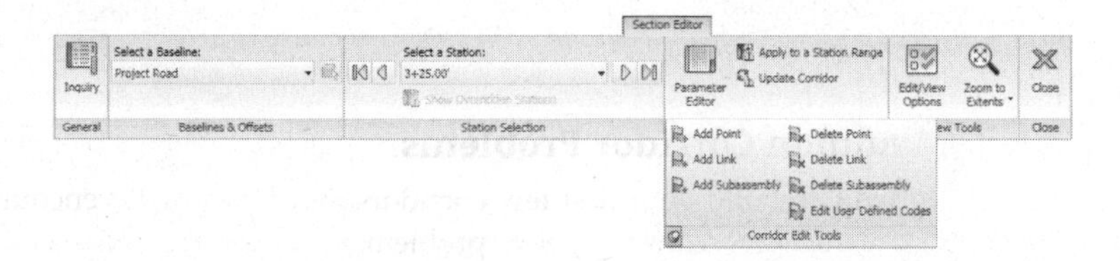

FIGURE 11.7
The Section Editor contextual tab with the Corridor Edit Tools panel pinned open

The Station Selection panel on the Section Editor contextual tab allows you to move forward and backward through your corridor to see what each section looks like. There are also options for advanced editing such as overriding specific stations, inserting and deleting subassemblies, points, links, and more on other panels. To exit the Section Editor, simply click the Close button on the Close panel.

To view your corridor in an isometric view, switch to the View tab and select any of the available options from the Views panel. To return to plan view, change to the View tab and select Top from the Views panel.

Rebuilding Your Corridor

A corridor is a *dynamic* model — which means if you modify any of the objects that were used to create the corridor, the corridor must be updated to reflect those changes. For example, if you make a change to the Finished Ground profile, the corridor needs to be rebuilt to reflect the new design. The same principle applies to changes to alignments, assemblies, target surfaces, and any other corridor ingredients or parameters.

You have three options for rebuilding corridors. The first is to manually rebuild your corridor by clicking the corridor object itself and choosing Rebuild Corridor from the shortcut menu. The second option is to right-click the corridor name in the Corridor collection in Prospector, and select Rebuild – Automatic (see Figure 11.8). The third option is to click the corridor and select Rebuild from the Modify panel on the Corridor contextual tab. Although Rebuild – Automatic is great for small corridors or while you are actively iterating a portion of your corridor and would like to see the results immediately, it is not a good idea to have this set as a general rule. Every time you make a change that even remotely affects your corridor, the corridor will go through a rebuilding process, during which you cannot work. If you have a large corridor or you need to make a series of changes, this can be disruptive.

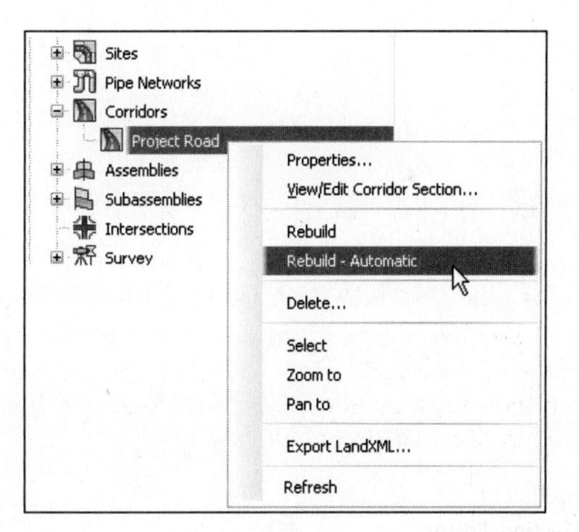

FIGURE 11.8
Right-click the corridor name in the Corridor collection in Prospector to rebuild it

Common Corridor Problems

When you build your first few corridors, many new users encounter several problems. Here is a list of some of the most typical problems and how to solve them:

Problem Your corridor seems to fall off a cliff, meaning the beginning or ending station of your corridor drops down to zero, as shown in Figure 11.9.

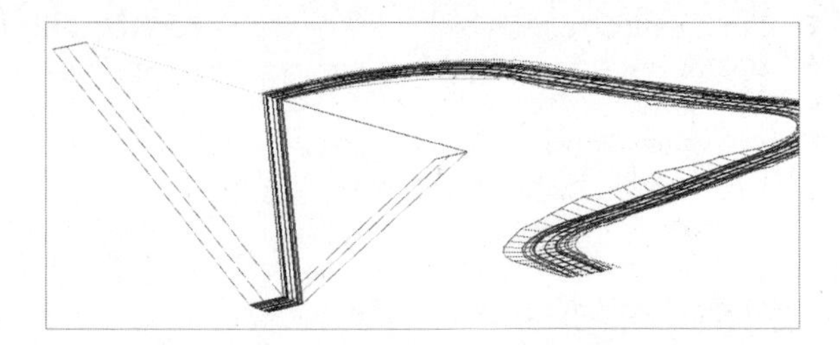

FIGURE 11.9
A corridor that drops down to zero

Typical Cause Your profile is not exactly the same length as the baseline alignment.

Fix Adjust your profile to be exactly as long as your alignment or edit your corridor region to begin/end before the trouble area. A good guide for determining if your profile is the same length as your alignment is by looking at the length of your Existing Ground profile. Unless your alignment goes off the surface, your Existing Ground line should be exactly the same length as your alignment. Using the Endpoint osnaps is a good way to check whether they are the same length. In Figure 11.10, you can see that the finished grade profile has been snapped to the endpoint of the Existing Ground profile.

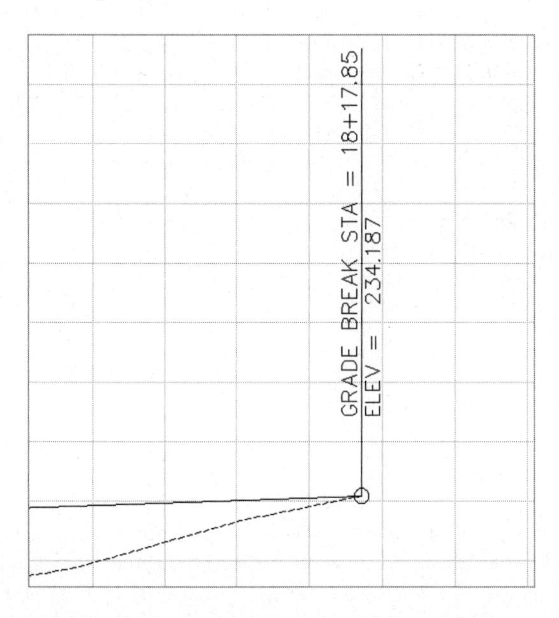

FIGURE 11.10
Your proposed profile and the Existing Ground profile must be the same length

Problem Your corridor seems to take longer to build and sample at a higher frequency than you intended. Also, your daylighting seems to be nonexistent (see Figure 11.11).

Typical Cause You accidentally chose the Existing Ground profile instead of the Finished Ground profile for your baseline profile. Most corridors are set up to sample at every vertical geometry point, and a sampled profile, such as the Existing Ground profile, has many more vertical geometry points than a layout profile (in this case, the Finished Ground profile). These additional points on the Existing Ground profile cause the unexpected sample lines and flags that something is wrong.

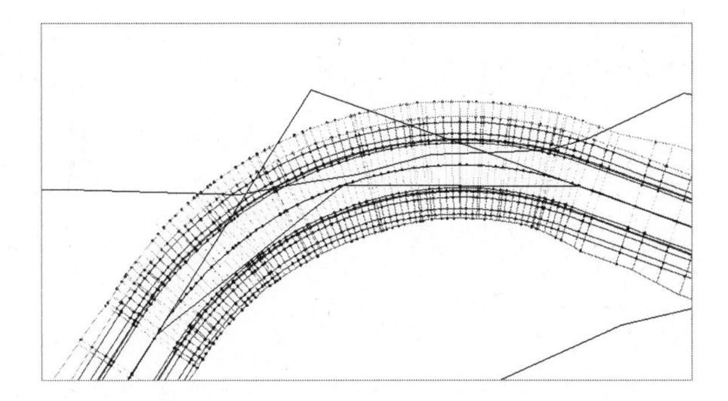

FIGURE 11.11
An example of unexpected corridor frequency

Fix Always use care to choose the correct profile. Either physically pick the profile on screen or make sure your naming conventions clearly define your finished grade as finished grade. If your corridor is already built, pick your corridor, right-click, and choose Corridor Properties. On the Parameters tab of the Corridor Properties dialog, change the baseline profile from Existing Ground to Finished Ground. Figure 11.12 shows the Parameters tab with Finished Ground properly listed as the baseline profile.

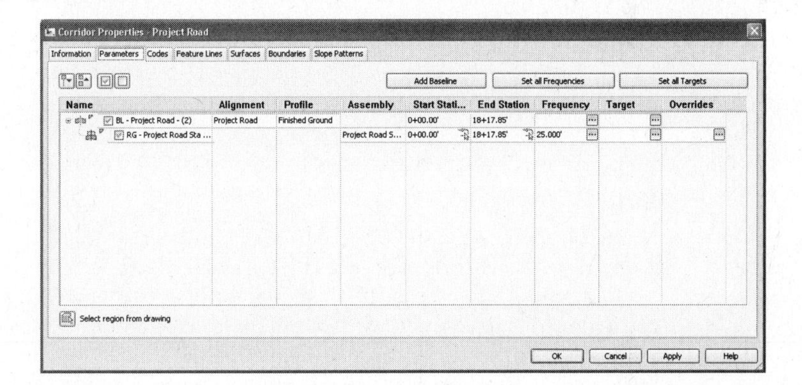

FIGURE 11.12
The Parameters tab of the Corridor Properties dialog

Corridor Anatomy

Corridors are made up of several components. If you explore the corridor you created in the previous section, you should see the four components that make up every corridor. *Points* and *links* are coded into the subassemblies that comprise the assembly. *Feature lines* glue the points and links together along the baseline. *Shapes*, which are not required for the actual model building, add an additional visual cue for material type.

A corridor is a collection of cross sections at a given frequency. The cross section comes from the assembly. In Figure 11.13, you can see that a corridor cross section looks very much like an assembly. Note the location of points, links, and shapes in the cross section. Links connect the points, and the shapes fill the areas created by links with color.

The cross sections are placed at intervals along the corridor baseline and then connected with corridor feature lines (see Figure 11.14). The corridor feature lines connect points from one cross section to the next.

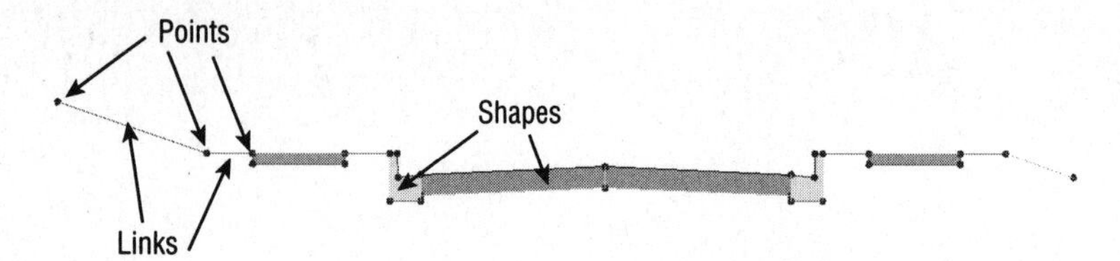

FIGURE 11.13
A single corridor section

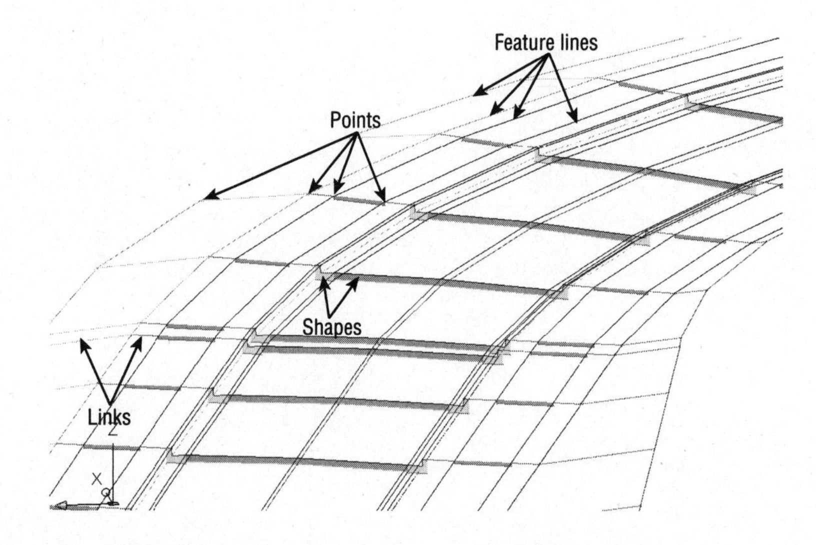

FIGURE 11.14
A corridor is a collection of cross sections connected with feature lines

Points

Not to be confused with Civil 3D COGO points or AutoCAD points, corridor points are at the foundation of a corridor section. They provide the first dimension of the corridor cross section.

As discussed in Chapter 10, "Templates Plus: Assemblies and Subassemblies," points are coded at the time of subassembly programming to have an understanding of their identity. A corridor point knows whether it is a Crown, an Edge Of Travel Way, an Edge Of Paved Shoulder, or one of more than 50 other standard point codes. These point codes indicate where the point can be found on the subassembly. For example, a point that has the Crown point code will appear at the crown of a road lane subassembly, and Crown-coded points will be placed in each cross section of the corridor model in the same location.

Figure 11.15 shows the same cross section from Figure 11.14, with only the points turned on.

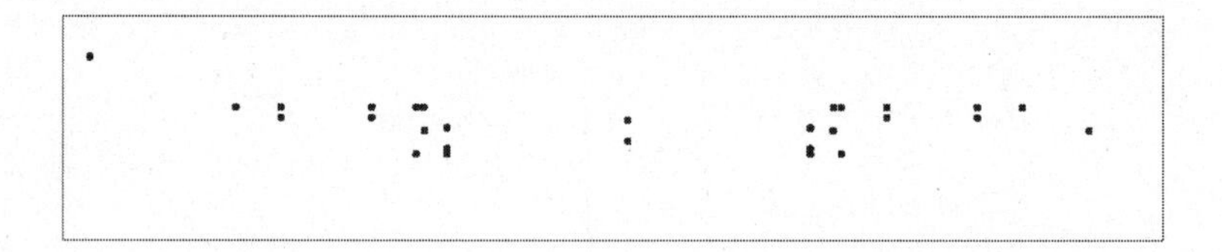

FIGURE 11.15
Corridor points in cross section

Links

Links provide the second dimension to the corridor cross section. Think of links as special, intelligent lines that connect the corridor points. Like points, links are coded at the time of subassembly programming to understand that they are a Top, Base, Pave, or one of the more than 15 other standard link codes. Similar to point codes, link codes indicate where the links are used. For example, links that are assigned to the Top link code connect points that are located at the finished grade surface regardless of whether they are paved or unpaved, whereas links assigned to the Pave link code are used to link points representing only the paved elements of the finished grade. A link can be assigned more than one code, if applicable. For example, a road lane would be assigned both Top and Pave, whereas a grassed buffer strip would be assigned only Top.

Figure 11.16 also shows the same cross section from Figure 11.14, except now both the points and links are shown.

FIGURE 11.16
A cross-sectional view of corridor points connected with links

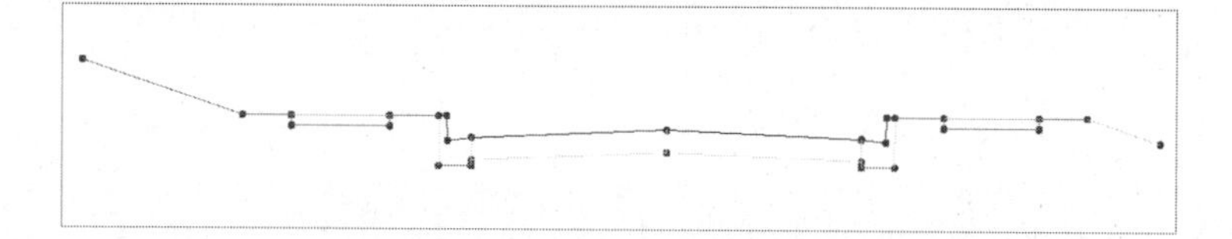

Shapes

Like points and links, shapes are also coded as part of the subassembly. Shapes are defined from links that form a closed polygon, such as a course of pavement, a gravel-base course, or a thickness of sidewalk. Figure 11.17 shows shapes that represent sidewalk, curb, and pavement materials.

FIGURE 11.17
Shapes can represent pavement, curb, and sidewalk materials

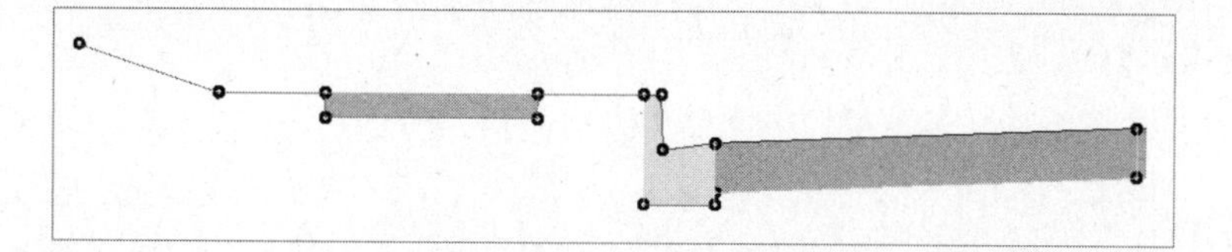

Corridor Feature Lines

Points and links come from subassemblies, but corridor feature lines are created when the corridor is built. Corridor feature lines, sometimes simply referred to as feature lines, should not be confused with grading feature lines as discussed in a later chapter. These special feature lines are the third dimension that takes a corridor from being simply a collection of cross sections to being a model with meaningful flow (see Figure 11.18).

FIGURE 11.18
A three-dimensional view showing corridor feature lines connecting each cross section

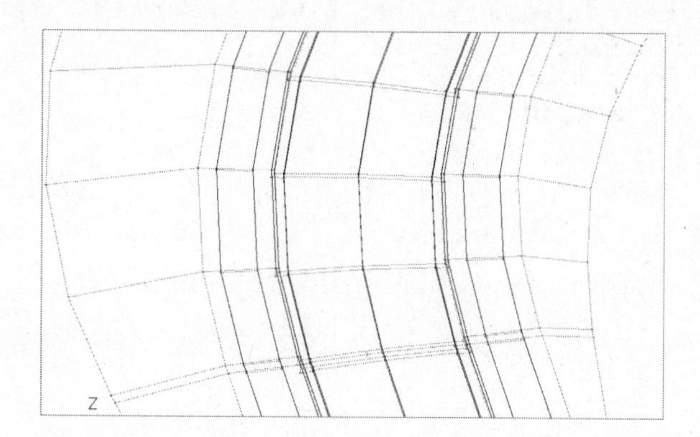

Like grading feature lines, corridor feature lines can also be used as breaklines when a corridor surface is built.

Corridor feature lines are first drawn connecting the same point codes. For example, a feature line will work its way down the corridor and connect all the TopCurb points. If there are TopCurb points on the entire length of your corridor, then the feature line does not have any decisions to make. If your corridor changes from having a curb to having a grassed buffer or ditch, the feature line needs to figure out where to go next.

The Feature Lines tab of the Corridor Properties dialog has a drop-down menu called Branching (see Figure 11.19), with two options — Inward and Outward. Inward branching forces the feature line to connect to the next point it finds toward the baseline. Outward branching forces the feature line to connect to the next point it finds away from the baseline.

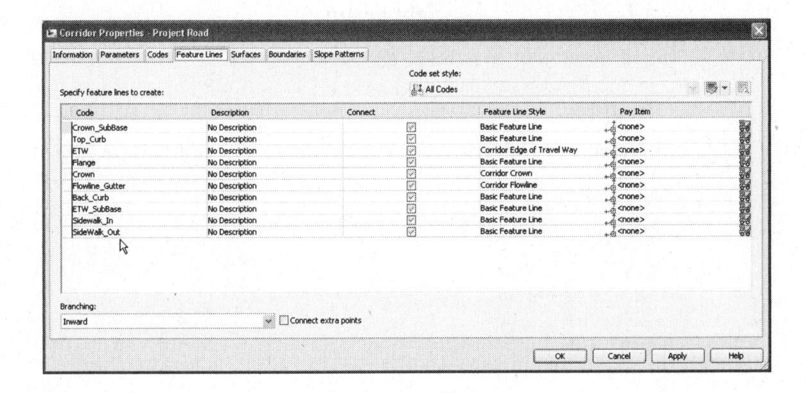

FIGURE 11.19
The Feature Lines tab of the Corridor Properties dialog

As mentioned earlier, a feature line will only connect the same point codes by default. However, the Feature Lines tab of the Corridor Properties dialog allows you to eliminate certain feature lines on the basis of the point code. For example, if for some reason you did not want your TopCurb points connected with a feature line, you could toggle that feature line off.

Feature lines can be extracted from a corridor to produce alignments, profiles, and grading feature lines that are dynamically linked to the corridor. In the following exercise, a corridor feature line is extracted to produce an alignment and profile:

1. Open the `Corridor Feature Line.dwg` file.
2. From the Home tab's Create Design panel, select Alignment ➤ Create Alignment from Corridor.
3. Select the top outside edge of the corridor near the center of the circle (the circle is for reference in the figure only). This opens the Select a Feature Line dialog as shown in Figure 11.20.

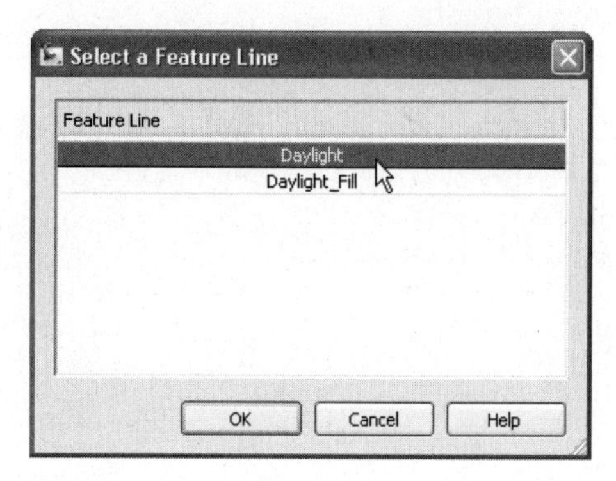

FIGURE 11.20
Selecting the Daylight feature line to be extracted as an alignment

4. Select the Daylight Feature Line as shown in Figure 11.20.
5. Select OK to close the Select a Feature Line dialog. The Create Alignment from Objects dialog is displayed as shown in Figure 11.21.

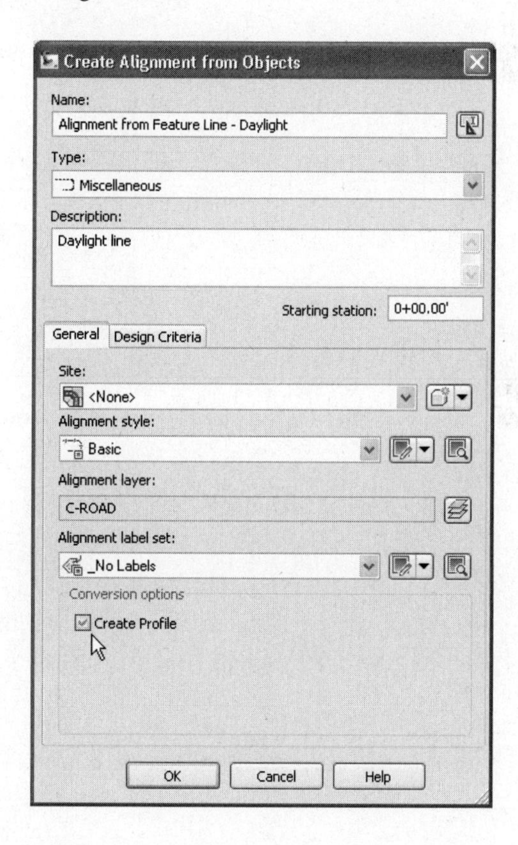

FIGURE 11.21
The completed Create Alignments from Objects dialog

6. Change your options to match those shown in Figure 11.21. Notice that the Create Profile option has been selected.
7. Click OK to dismiss the Create Alignment from Objects dialog. The Create Profile – Draw New dialog opens as shown in Figure 11.22.

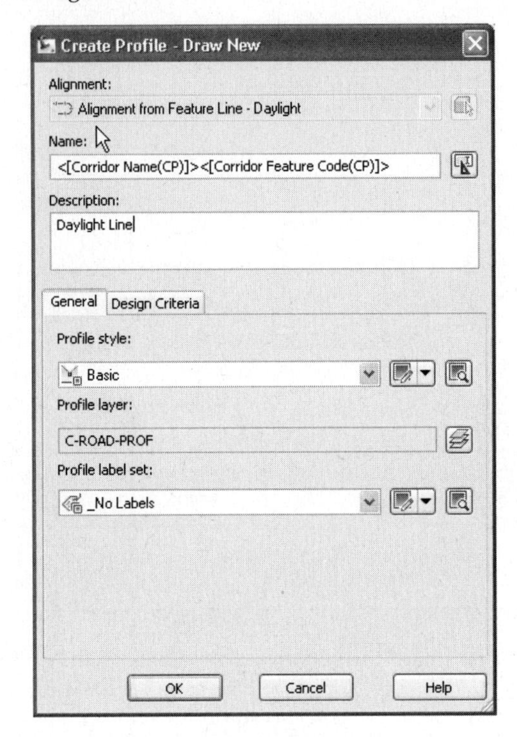

FIGURE 11.22
The Create Profile – Draw New dialog

8. Change your options to match those shown in Figure 11.22.
9. Click OK to dismiss the dialog and then press ↵ to exit the command.
10. Review the alignment and profile as shown on the Prospector tab of the Toolspace (see Figure 11.23).

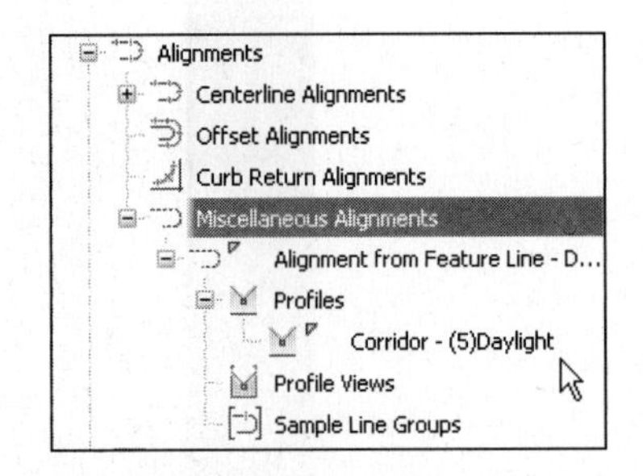

FIGURE 11.23
A completed alignment and profile shown on the Prospector tab of the Toolspace

Alignments and profiles extracted from corridor feature lines are not dynamically linked to the corridor. However, you have an option to create dynamically linked grading feature lines in the Create Feature Line from Corridor dialog as shown in Figure 11.24.

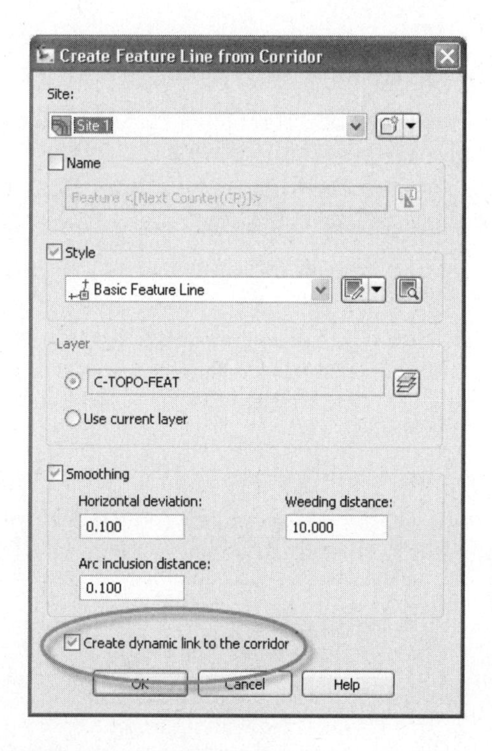

FIGURE 11.24
The Create Feature Line from Corridor dialog with the option to create a dynamic link turned on

Adding a Surface Target for Daylighting

A road cross section between centerline and right-of-way is usually clearly defined by the local road–design specifications. The area between the right-of-way and the existing ground surface, however, is not always so straightforward. Chapter 10 talked about daylighting subassemblies that can assist in grading this in-between area. Figure 11.25 shows a rendered corridor that uses a daylight subassembly to tie into existing ground.

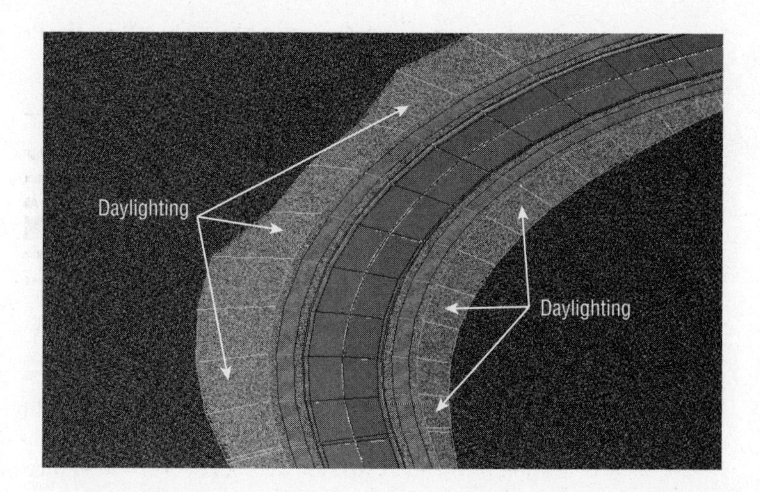

FIGURE 11.25
A rendered corridor showing daylighting between the sidewalk and existing ground

A daylighting subassembly needs to understand which surface it is targeting. This exercise teaches you how to assign the target surface to a corridor:

1. Open the `Corridor Daylight.dwg` file. Note that the drawing contains an alignment, a profile view with two profiles, and an assembly, as well as an Existing Ground surface.
2. Pan over to the assembly. In addition to the lanes, curbs, and sidewalk, there is a Daylight subassembly built into the assembly, as shown in Figure 11.25.
3. Change to the Home tab and select Corridor ➢ Create Corridor from the Create Design panel.
4. Create your corridor exactly as you did in the first exercise in this chapter. Follow the prompts to pick the alignment, profile, and assembly. Select Finished Ground as the profile to apply to the corridor. The Create Corridor dialog appears after all objects have been picked.
5. Click the Set All Targets button. The Target Mapping dialog appears.
6. Click in the Object Name column field in the Target Mapping dialog where it says "Click here to set all." The Pick a Surface dialog appears and prompts you to choose a surface for the daylighting subassembly to target.
7. Select the Existing Ground surface. Click OK to dismiss the dialog. Click OK again to dismiss the Target Mapping dialog.
8. Click OK to exit the Create Corridor dialog. The corridor will build and create daylighting points, links, and feature lines that show how it ties into the Existing Ground surface. If the Panorama appears, dismiss the dialog by clicking the green check mark in upper-right corner of the dialog. The result should be similar to Figure 11.26.

FIGURE 11.26
The completed corridor with daylighting

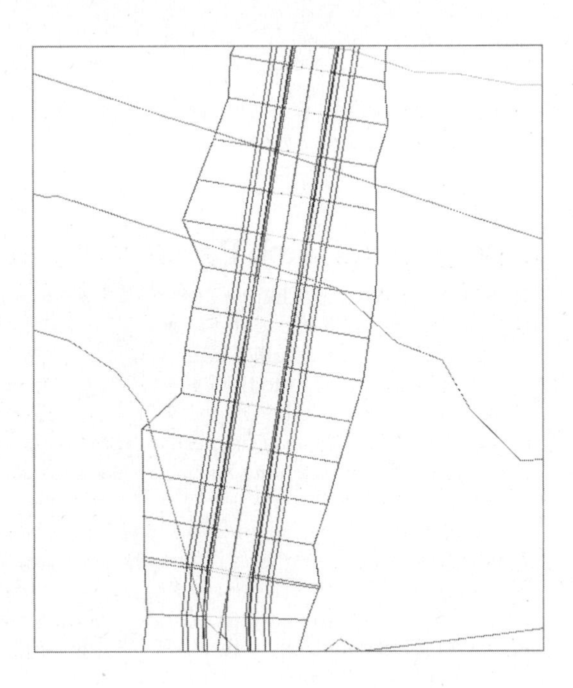

Common Daylighting Problems

Adding a surface target throws another variable into the mix. Here is a list of some of the most typical problems new users face and how to solve them:

Problem Your corridor doesn't show daylighting even though you have a Daylight subassembly on your assembly. You may get a Target Object Not Found or a similar error message in Event Viewer.

Typical Cause You forgot to set the surface target when you created your surface.

Fix If your corridor is already built, pick your corridor, right-click, and select Corridor Properties. On the Parameters tab of the Corridor Properties dialog, click Set All Targets. The Target Mapping dialog opens, and its first entry is Surfaces. Click in the Object Name column field. This will prompt you to choose a surface for the daylighting subassembly to target.

Problem Your corridor seems to be missing patches of daylighting. You may also get an error message in Event Viewer.

Typical Cause Your target surface doesn't fully extend the full length of your corridor or your target surface is too narrow at certain locations.

Fix Add more data to your target surface so that it is large enough to accommodate daylighting down the full length of the corridor. If this is not possible, omit daylighting through those specific stations, and once your corridor is built, do hand grading using feature lines or grading objects. You can also investigate other subassemblies such as Link Offset to Elevation that will meet your design intention without requiring a surface target.

Problem Your corridor daylighting falls short of a tie-in to the existing ground surface. You may get an error message in Event Viewer, such as No Intersection with Link Found.

Typical Cause Your Daylight subassembly parameters are too restrictive to grade all the way to your target surface. The Daylight link cannot find the target surface within the grade, width, or other parameters you've set in the subassembly properties.

Fix Revisit your Daylight subassembly settings to give the program a wider offset or steeper grade. If your settings cannot be adjusted, you'll have to adjust your horizontal and/or vertical design to properly grade.

Applying a Hatch Pattern to a Corridor

In this chapter and in Chapter 10, you learned that links have codes that give them intelligence about what part of the road they are on. These codes can be used to apply styles automatically. For example, if you wanted all of your paved areas to have a certain hatch pattern or render material, you could assign a style to the Pave link code, as in Figure 11.27.

A code set style can be created to enhance your corridor's appearance for things like exhibits at public hearings. Instead of spending time creating a series of hatch boundaries using polylines and then manually applying hatches to areas of paving, sidewalks, curbs, and so on, you can have the code set style automatically hatch those areas for you.

Another task that can be performed with a code set style is the application of render materials. This type of code set style is studied in more detail in Chapter 22, "Get the Picture: Visualization." Similar to a hatch-pattern code set style, the Render Material code set style will automatically apply render materials to your corridor on the basis of link codes. For example, the corridor in Figure 11.28 was stylized with a code set that automatically assigned an asphalt render material to the Pave code, a concrete render material to the Curb and Sidewalk codes, and a grass render material to the Daylight code.

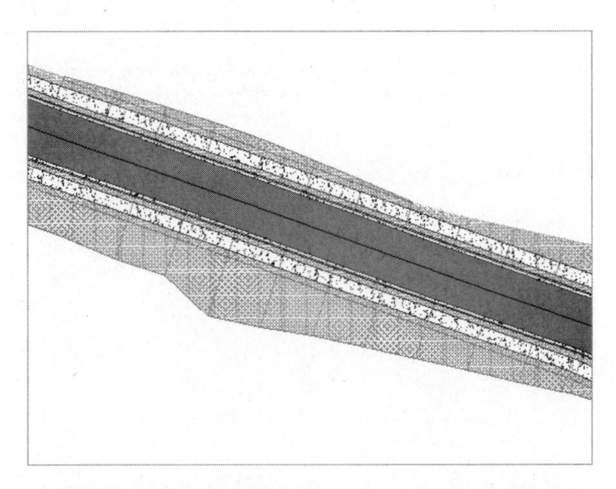

FIGURE 11.27
A corridor with a Hatching code set style

FIGURE 11.28
An image of a corridor with a Render Material code set style

You can view the Code Set Style collection on the Settings tab of Toolspace by expanding the General ➤ Multipurpose Styles ➤ Code Set Styles branches. As with any style listed on the Settings tab, you can edit code set styles by right-clicking the code set style and choosing Edit. The Code Set Style dialog is shown in Figure 11.29.

A code set style is a compilation of styles for links, points, shapes, and feature lines. A code set style could be thought of as similar to an Alignment label set. Since an alignment can have many different types of labels, the label set lets you collect and stylize them in one spot instead of having to assign each style individually.

Just as you might create an Alignment label set for local roads, highways, streams, and other special situations, you can create different code set styles for different desired corridor looks. Some examples might be a code set style that applies elaborate hatching for preliminary site plans, another code set style that applies render materials for rendering and drive-throughs, and maybe another that applies a different hatch to designate that a road is already constructed for use in future road plans.

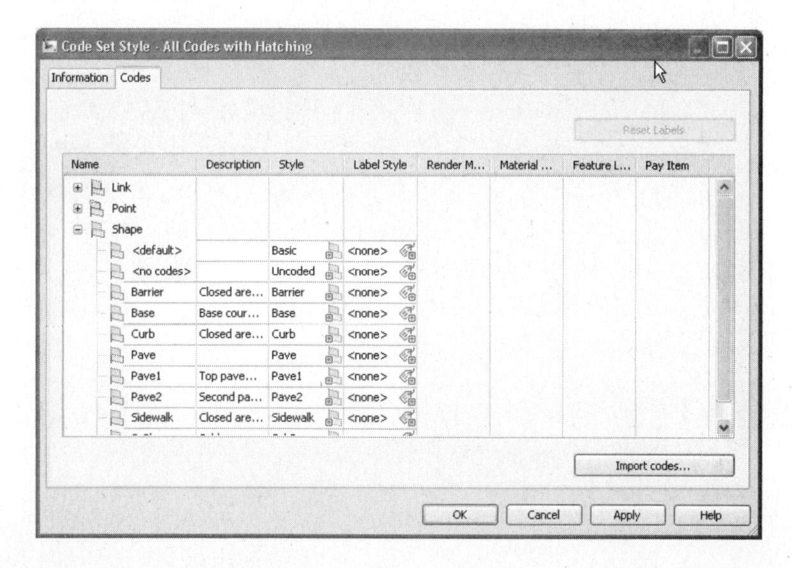

FIGURE 11.29
The Code Set Style dialog

In addition to assigning fill material (hatching) and render material to specific links, the code set controls the appearance of all corridor components. If you would like to customize the color, layer, linetype, and so on of links, points, shapes, or feature lines, this is where you would do that. Figure 11.30 shows an example of a corridor where the color of each point, link, feature line, and shape has been customized in the code set style.

WHAT ABOUT THE FEATURE LINES?

When you build your corridor, the default code set style in your Command settings is applied to links, points, shapes, render materials, fill materials, and feature lines. Once the corridor is built, changes to the code set style will update all of these items except the feature lines. Changes to feature lines once the corridor is built must be made in the Corridor Properties dialog.

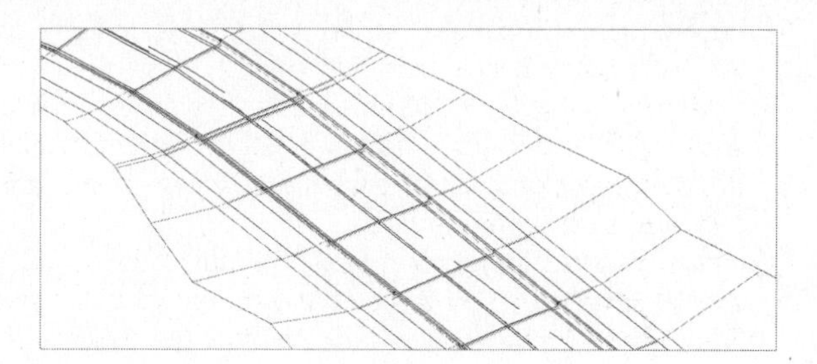

FIGURE 11.30
A corridor's appearance is controlled by link, point, shape, and feature line styles

In this next exercise, you are going to examine a default code set style and apply it to your corridor to see the hatch pattern:

1. Open the `Hatch Corridor.dwg` file. Note that this drawing contains a corridor.
2. Pick the corridor and select Corridor Properties from the Modify panel. The Corridor Properties dialog opens.
3. In the Corridor Properties dialog, switch to the Codes tab. Select All Codes with Hatching from the drop-down list in the Code Set Style selection box.
4. Click OK to dismiss the dialog.
5. Your corridor should now have hatching applied as per the code set style, similar to Figure 11.27.
6. Expand the General ➢ Multipurpose Styles ➢ Code Set Styles branches on the Settings tab of Toolspace.
7. Double-click the All Codes with Hatching code set style and the Code Set Style dialog opens. Switch to the Codes tab.
8. Scroll to the right until you see the Material Area Fill Style column. The Material Area Fill Style specifies the hatch pattern for each link code. You can customize these hatch patterns by clicking any entry in this column and modifying the style.

Creating a Corridor Surface

A corridor provides the raw material for surface creation. Just as you would use points and breaklines to make a surface, a corridor surface uses corridor points as point data and uses feature lines and links like breaklines.

The Corridor Surface

Because it needs more information about what you want to build, Civil 3D does not automatically build a corridor surface when you build a corridor. From examining subassemblies, assemblies, and the simple corridors you built in the previous exercises, you have probably noticed that there are many "layers" of points, links, and feature lines. Some represent the very top of the finished ground of your road design, some represent subsurface gravel or concrete thicknesses, and some represent subgrade, among other possibilities. You can choose to build a surface from any one of these layers or from all of them. Figure 11.31 shows an example of a TIN surface built from the links that are all coded Top, which would represent final finished ground.

When you first create a surface from a corridor, the surface is dependent on the corridor object. This means that if you change something that affects your corridor and then rebuild the corridor, the surface will also update. In Civil 3D 2007, and all subsequent versions since, this surface shows up as a true surface under the Surfaces branch in Prospector. After you create the initial corridor surface, you can create a static export of the surface by changing to the Home tab and using Surfaces ➢ Create Surface from Corridor on the Create Ground Data panel. A detached surface will not react to corridor changes and can be used to archive a version of your surface.

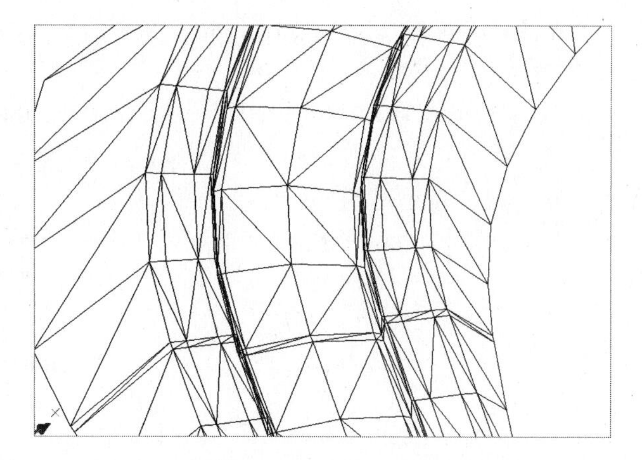

FIGURE 11.31
A surface built from Top code links

Creation Fundamentals

You create corridor surfaces on the Surfaces tab in the Corridor Properties dialog using the following two steps (which are examined in detail later in this section):

1. Click Create a Corridor Surface to add a surface entry (see Figure 11.32)

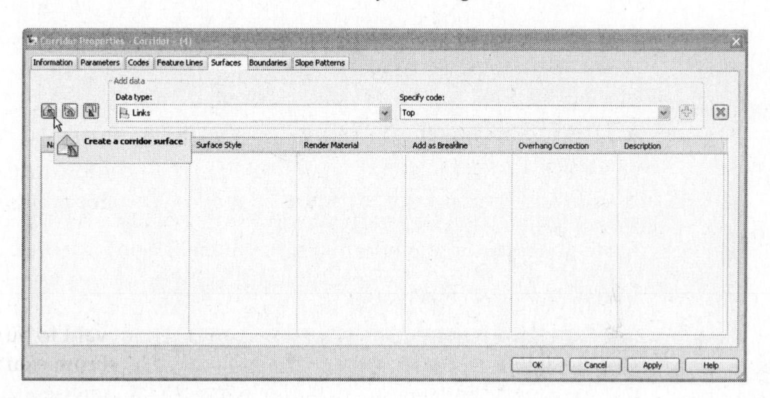

FIGURE 11.32
The Create a Corridor Surface button

2. Choose data to add, and then click the + sign.

DATA TYPES

You can choose to create your corridor surface on the basis of links, feature lines, or a combination of both.

Creating a Surface from Link Data

Most of the time, you will build your corridor surface from links. As discussed earlier, links understand which "layer" they fit into on your corridor. Choosing to build a surface from Top links will create a surface that triangulates between the points at the link vertices that represent the final finished grade. The most commonly built link-based surfaces are Top, Datum, and Subbase; however, you can build a surface from any link code in your corridor. Figure 11.33 shows a corridor and its surface, which was created from link data. You can see the triangulation lines connecting the link vertex points.

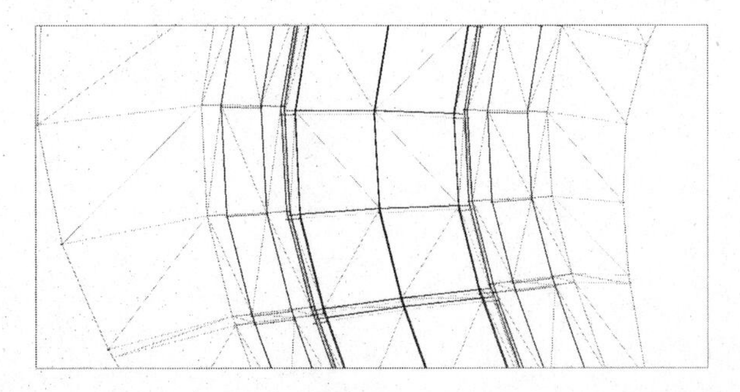

FIGURE 11.33
A corridor and its surface were created from link data

When building a surface from links, you have the option of checking a box in the Add as Breakline column. Checking this box will add the actual link lines themselves as additional breaklines to the surface. In most cases, especially in intersection design, checking this box forces better triangulation. It would be good practice to always check this box. If you find that you have an extremely large corridor and run into performance problems, consider leaving this box unchecked.

Creating a Surface from Feature Lines

There might be cases where you would like to build a simple surface from your corridor — for example, by using just the crown and edge-of-travel way. If you build a surface from feature lines only or a combination of links and feature lines, you have more control over what Civil 3D uses as breaklines for the surface.

If you added each Top feature-line code to your surface entry and built a surface, you would get a very similar result as if you had added the Top link codes. Each feature line usually has a vertex where the corridor points would normally fall; therefore, triangulation occurs almost identically to a link-based surface. However, you would have to choose and add each feature line individually, which would take more time than building a link-based surface. Also, if your corridor is complex and has transitions, a feature line may not be continuous along the length of your corridor and would cause unexpected triangulation. For the most part, you will probably find that you rarely build a surface from feature lines alone. Feature lines are most useful when added to link-based corridor surfaces to reinforce triangulation.

Figure 11.34 shows what the Surfaces tab of the Corridor Properties dialog would look like if you chose to build a surface from the Back_Curb, Crown, Daylight, and Edge of Travel Way (ETW) feature lines.

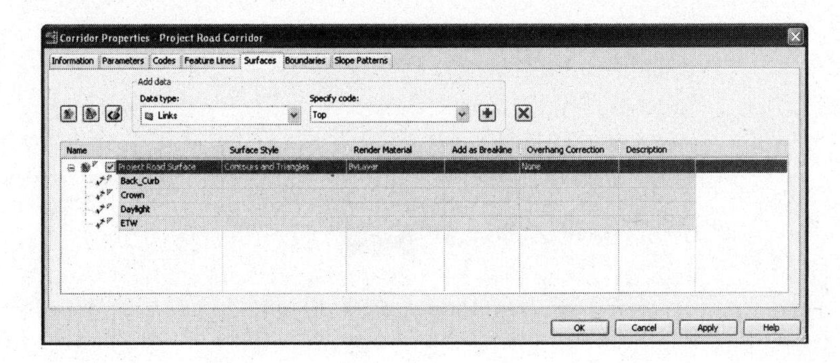

FIGURE 11.34
The Surfaces tab indicates that the surface will be built only from certain feature lines

The resulting surface is shown in Figure 11.35. Although there are few applications for a feature line–only corridor surface, it is useful to understand what happens when feature lines are added to a corridor surface.

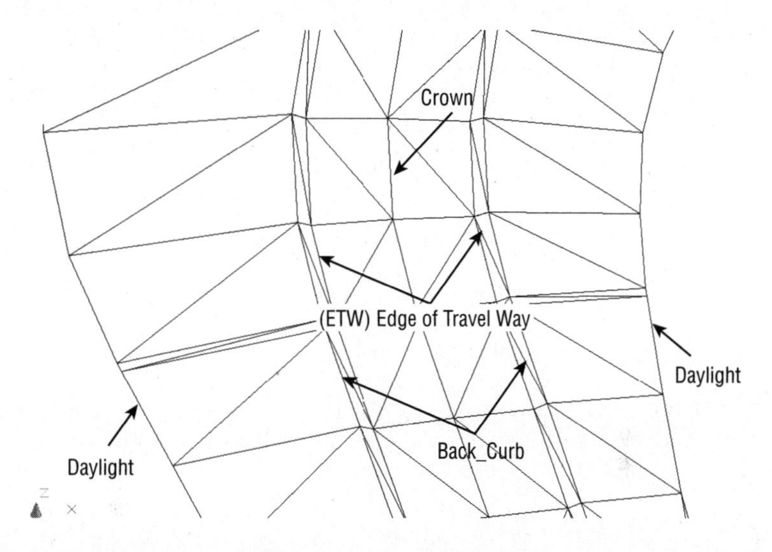

FIGURE 11.35
A surface built from only the Back_Curb, Crown, Daylight, and ETW feature lines

Creating a Surface from Both Link Data and Feature Lines

A link-based surface can be improved by the addition of feature lines. A link-based surface does not automatically include the corridor feature lines, but instead uses the link vertex points to create triangulation. Therefore, the addition of feature lines ensures that triangulation occurs where desired. This is especially important for intersection design, curves, and other corridor surfaces where triangulation around tight corners is critical. Figure 11.36 shows the Surfaces tab of the Corridor Properties dialog where a Top link surface will be improved by the addition of Back_Curb, ETW, and Top_Curb feature lines.

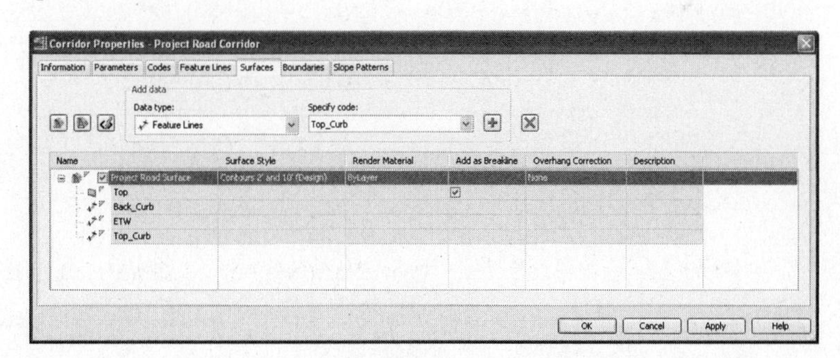

FIGURE 11.36
The Surfaces tab indicates that the surface will be built from Top links as well as from several feature lines

If you are having trouble with triangulation or contours not behaving as expected, experiment with adding a few feature lines to your corridor surface definition.

OTHER SURFACE TASKS

You can do several other tasks on the Surfaces tab. You can set a Surface Style, assign a meaningful name, and provide a description for your surface. Alternatively, you can do all those things once the surface appears in the drawing and through Prospector.

Adding a Surface Boundary

For accurate results in both volume and quantity takeoff reports, you should consider creating a boundary for every corridor surface. Tools that can automatically and interactively add surface boundaries, using the corridor intelligence, are available. Figure 11.37 shows a corridor surface before the addition of a boundary.

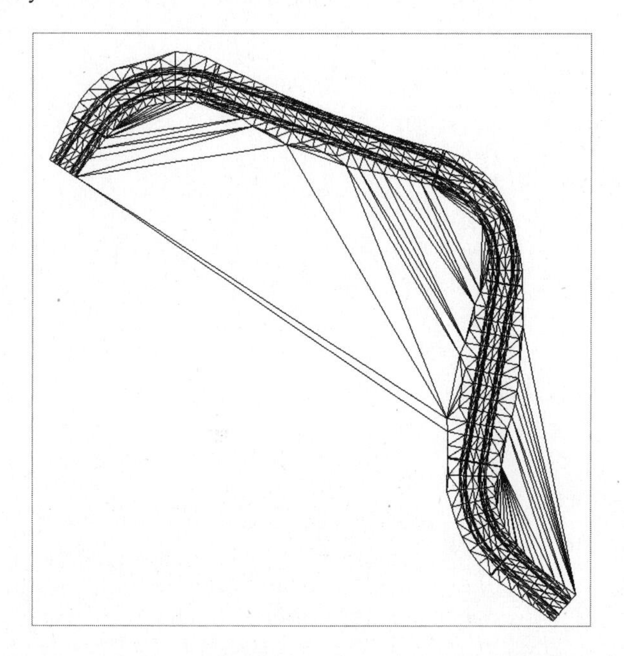

FIGURE 11.37
A corridor surface before the addition of a boundary

You can create corridor surface boundaries using the Boundaries tab of the Corridor Properties dialog. It stands to reason, if this page is blank, a corridor surface has not yet been created. Figure 11.38 shows a corridor surface after the application of an automatic boundary. Notice how the extraneous triangulation has been eliminated along the line of intersection between the existing ground and the proposed ground (the daylight line), thereby creating a much more accurate surface ready to be rendered.

BOUNDARY TYPES

There are several tools to assist in corridor surface boundary creation. They can be automatic, semiautomatic, or manual in nature depending on your needs and the complexity of the corridor.

You access these options on the Boundaries tab of the Corridor Properties dialog by right-clicking the name of your surface entry, as shown in Figure 11.39.

Corridor Extents as Outer Boundary This is available only when you have a corridor with multiple baselines. With this selection, Civil 3D will shrink-wrap the corridor, taking into account intersections and various daylight options on different alignments. This will probably be your most-used boundary option.

Add Automatically The Add Automatically boundary tool allows you to pick a Feature Line code to use as your corridor boundary. This tool is available only for single baseline corridors. Because this tool is the most automatic and easiest to apply, you will use it almost every time you build a single baseline corridor.

Add Interactively The Add Interactively boundary tool allows you to work your way around a multibaseline corridor and choose which corridor feature lines you would like to use as part of the boundary definition.

Add from Polygon The Add from Polygon tool allows you to choose a closed polyline or polygon in your drawing that you would like to add as a boundary for your corridor surface.

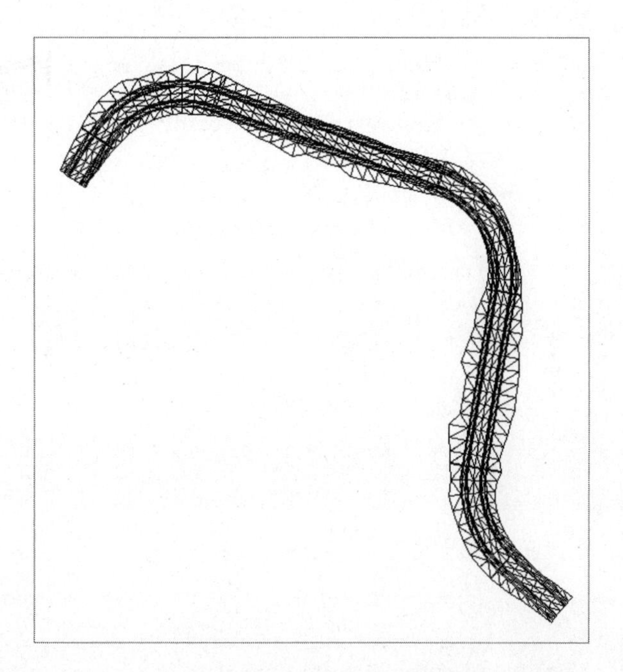

FIGURE 11.38
A corridor surface after the addition of an automatic daylight boundary

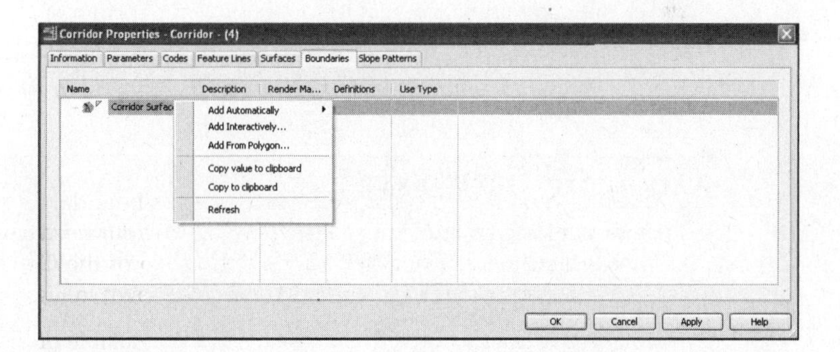

FIGURE 11.39
Corridor Surface Boundary options

Now that you have studied the components of how corridor surfaces are built and how to create corridor surface boundaries, this next exercise leads you through creating a corridor surface with an automatic boundary:

1. Open the `Corridor Surface.dwg` file. Note that there is a corridor in this drawing.
2. Pick the corridor and select Corridor Properties from the Modify panel. The Corridor Properties dialog opens.

3. Switch to the Surfaces tab.
4. Click the Create a Corridor Surface button on the far left side of the dialog. You should now have a surface entry in the bottom half of the dialog.
5. Click the surface entry under the Name column and change the default name of your surface to **Project Road Corridor Surface**. Notice the surface style can also be changed under the Surface Style column.
6. Confirm that Links has been selected from the drop-down menu in the Data Type selection box and that Top has been selected from the drop-down menu in the Specify Code selection box. Click the + button to add Top Links to the Surface Definition.
7. Click OK to leave this dialog, and examine your surface. The area inside the corridor model itself should look fine; however, because you have not yet added a boundary to this surface, undesirable triangulation is occurring outside your corridor area.
8. Expand the Surfaces branch in Prospector. Note that you now have a corridor surface listed.
9. Pick the corridor and select Corridor Properties from the Modify panel. The Corridor Properties dialog opens. If the Corridor Properties button is not available on the Modify panel, you may have accidently chosen the corridor surface.
10. In the Corridor Properties dialog, switch to the Boundaries tab.
11. Right-click the surface entry. Hover over the Add Automatically flyout, and select Sidewalk_Out as the feature line that will define the outer boundary of the surface. Note the Add Automatically option is available only on single baseline corridors.
12. Confirm that the Use Type column says Outside Boundary to ensure that the boundary definition will be used to define the desired extreme outer limits of the surface.
13. Click OK to dismiss the dialog. Examine your surface, and note that the triangulation terminates at the Sidewalk_Out point all along the corridor model.
14. Experiment with making changes to your finished grade profile, assembly, or alignment geometry and rebuilding both your corridor and finished ground surface.

Rebuild: Leave It On or Off?

Upon rebuilding your corridor, your surface will need to be updated. Typically, the best practice is to leave Rebuild – Automatic OFF for corridors and keep Rebuild – Automatic ON for surfaces. This practice is usually okay for your corridor-dependent surfaces. The surface will only want to rebuild when the corridor is rebuilt. For very large corridors, this may become a bit of a memory lag, so try it both ways and see what you like best.

Common Surface Creation Problems

Some common problems encountered when creating surfaces are as follows:

Problem Your corridor surface does not appear or seems to be empty.

Typical Cause You might have created the surface entry but no data.

Fix Open the Corridor Properties dialog and switch to the Surfaces tab. Select an entry from the drop-down menus in the Data Type and Specify Code selection boxes, and click the + sign. Make sure your dialog shows both a surface entry and a data type, as per Figure 11.40.

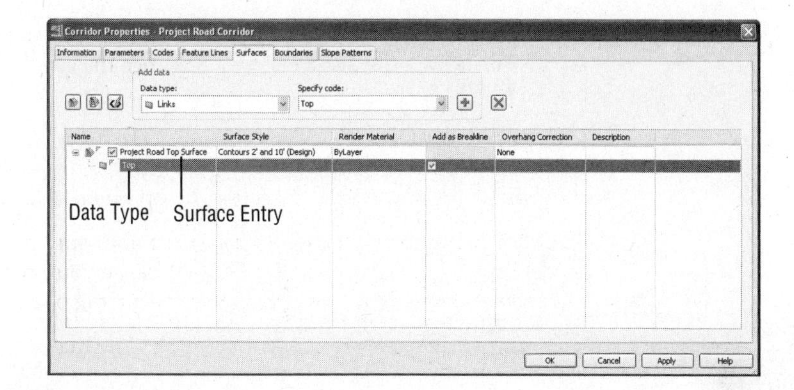

FIGURE 11.40
A surface cannot be created without both a surface entry and a data type

Problem Your corridor surface does not seem to respect its boundary after a change to the assembly or surface-building data type (in other words, you switched from link data to feature lines).

Typical Cause Automatic and interactive boundary definitions are dependent on the codes used in your corridor. If you remove or change the codes used in your corridor, the boundary needs to be defined.

Fix Open the Corridor Properties dialog and switch to the Boundaries tab. Erase any boundary definitions that are no longer valid (if any). Redefine your boundaries.

Problem Your corridor surface seems to have gaps at PCs and PTs near curb returns.

Typical Cause You may have encountered an error in rounding at these locations, and you may have inadvertently created gaps in your corridor. This is commonly the result of building a corridor using two-dimensional linework as a guide, but some segments of that linework do not touch.

Fix Be sure your corridor region definitions produce no gaps. You might consider using the `PEDIT` command to join lines and curves representing corridor elements that will need to be modeled later. You might also consider setting a COGO point at these locations (PCs, PTs, and so on) and using the Node object snap instead of the Endpoint object snap to select the same location each time you are required to do so.

Performing a Volume Calculation

One of the most powerful aspects of Civil 3D is having instant feedback on your design iterations. Once you create a preliminary road corridor, you can immediately compare a corridor surface to existing ground and get a good understanding of earthwork magnitude. When you make an adjustment to the finished grade profile and then rebuild your corridor, you can see the effect that this change had on your earthwork within a minute or two, if not sooner.

Even though volumes were covered in detail in Chapter 5, "The Ground Up: Surfaces in Civil 3D," it is worth revisiting the subject here in the context of corridors.

This exercise uses a TIN-to-TIN composite volume calculation; average end area and other section-based volume calculations are covered in Chapter 13, ''Stacking Up: Cross Sections.''

1. Open the `Corridor Surface Volume.dwg` file. Note that this drawing has a corridor and a corridor surface.
2. Change to the Analyze tab and choose Volumes ➤ Volumes from the Volumes and Materials panel.
3. The Composite Volume palette in Panorama appears.
4. Click the Create New Volume Entry button toward the top left of the Volume palette. A Volume entry with an Index of 1 should appear in the palette.
5. Click inside the cell in the Base Surface column and select Existing Ground for the Volume entry with an Index of 1.
6. Click inside the cell in the Comparison Surface column and select the Project Road Corridor Surface.
7. A Cut/Fill breakdown should appear in the remaining columns, as shown in Figure 11.41. Make a note of these numbers.

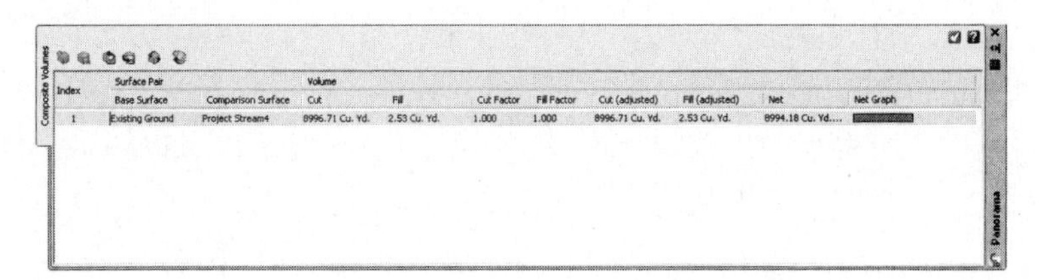

FIGURE 11.41 Panorama showing an example of a volume entry and the cut/fill results (the numbers shown here may vary from those entered in the exercise)

8. Leave Panorama open on your screen (make it smaller, if desired), and pan over to your Finished Ground profile.
9. Pick the Finished Ground profile and grip-edit a PVI so that the profile changes drastically — in other words, so there would suddenly be a great deal more cut or fill.
10. Pick the corridor, right-click, and choose Rebuild Corridor. Notice that the corridor changes, and therefore the corridor surface changes as well.
11. Click Recompute Volumes in Panorama and note the new values for cut and fill.

Common Volume Problem

A common volume problem is as follows:

Problem Your volume number does not update.

Typical Cause You might have forgotten to rebuild the corridor, rebuild the corridor surface, or click Recompute Volumes.

Fix Check Prospector to see if either your corridor or corridor surface is out of date. First, rebuild the corridor, and then rebuild the corridor surface.

Creating a Corridor with a Lane Widening

So far, all of the corridor examples you looked at have a constant cross section. In the next section, you take a look at what happens when a portion of your corridor needs to transition to a wider section and then transition back to normal.

Using Target Alignments

Earlier in this chapter, we discussed baselines and mentioned that baselines can be broken up into different regions. Regions are examined in detail in Chapter 12.

In this section, you apply another corridor parameter called a *target*. We mentioned the idea of targets when you added a surface target for a daylighting subassembly. In addition to surfaces, alignments and profiles can be used as targets.

Many subassemblies have been programmed to allow for not only a baseline attachment point but also additional attachment points on target alignments and/or profiles. Figure 11.42 shows a centerline alignment to be used as a baseline and an edge-of-travel way alignment to be used as a target.

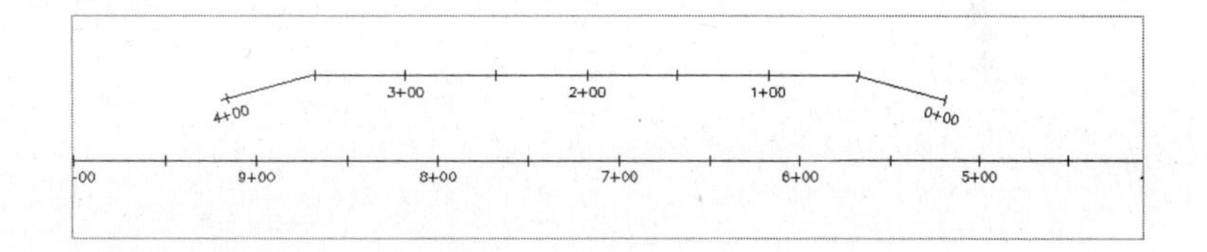

FIGURE 11.42
A centerline alignment used as a baseline and an edge-of-travel way alignment used as a target

The subassembly will be stretched, raised, lowered, and adjusted to reflect the location and elevation of the target. In this chapter, we discuss target alignments. In Chapter 12, we go into more detail about using profile targets.

For example, the BasicLaneTransition subassembly can be set up to hook onto an alignment and a profile. Think of the lane as a rubber band that is attached both to the baseline of the corridor (such as the road centerline) and the target alignment. As the target alignment, such as a lane widening, gets further from the baseline, the rubber band is stretched wider. As that target alignment transitions back toward the baseline, the rubber band changes to reflect a narrower cross section. Figure 11.43 shows a corridor built using the edge-of-travel way alignment previously shown in Figure 11.42 as a target.

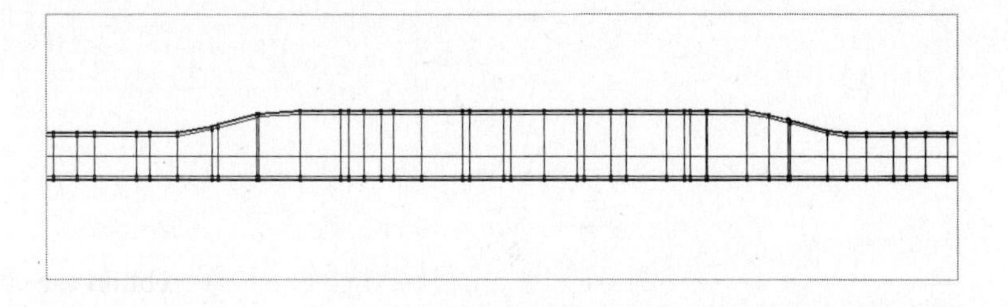

FIGURE 11.43
A corridor built using the centerline and the edge-of-travel way alignments

As discussed in Chapter 10, the BasicLaneTransition subassembly has several options for how it will transition. For the next example, you use Hold Grade, Change Offset. Hold Grade means the subassembly will hold the default grade of −2 percent as it is stretched to change offset with the target alignment. Using the Hold Grade, Change Offset setting eliminates the need for a target profile, because the elevation at the edge-of-travel way will be determined by the default grade (−2 percent, in this case) and lane width at a given sampling location.

The following exercise teaches you how to use a transition alignment as a corridor target for a lane widening:

1. Open the `Corridor Widening.dwg` file. Note that this drawing has a corridor.
2. Freeze the layer C-ROAD-CORR. Note that the Widening EOP alignment represents the edge-of-pavement for a street parking zone.
3. Thaw the layer C-ROAD-CORR.
4. Pan over to the assembly in the drawing. Select the right lane subassembly and choose Subassembly Properties from the Modify Subassembly palette. Switch to the Parameters tab on the Subassembly Properties dialog.
5. Note that the entry for the Transition field is Hold Grade, Change Offset, as shown in Figure 11.44. Click OK to dismiss the dialog.

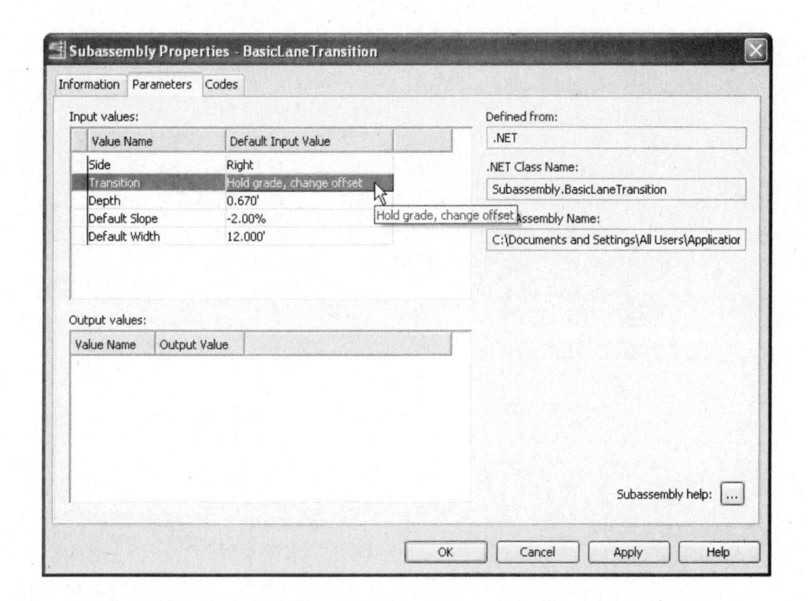

FIGURE 11.44 Set the Default Input Value for the Transition field to Hold Grade, Change Offset

6. Pan over to your corridor. Pick the corridor and choose Corridor Properties from the Modify panel. Switch to the Parameters tab on the Corridor Properties dialog.
7. Click Set All Targets. The Target Mapping dialog opens. Click in the field next to Width or Offset Targets to bring up the Set Width or Offset Target dialog.
8. In the Set Width or Offset Target Dialog, select the Widening EOP alignment, and then click the Add button. The Widening EOP alignment will appear in the lower portion of the dialog.
9. Click OK to dismiss the dialog. Click OK again to dismiss the Target Mapping dialog, and then click OK once more to dismiss the Corridor Properties dialog.

10. The corridor will rebuild and reflect the wider lane where appropriate. The finished corridor should look similar to Figure 11.43.

Common Transition Problems

A common problem encountered when creating transitions is as follows:

Problem Your corridor does not reflect your lane widening.

Typical Cause No. 1 You forgot to set the targets.

Fix Refer to steps 6, 7, and 8 in the previous exercise.

Typical Cause No. 2 Your subassembly isn't set to Change Offset.

Fix Examine your lane subassembly. Make sure it is a subassembly with a transition parameter, such as a BasicLaneTransition. Swap out the subassembly if necessary. Once you have confirmed the proper subassembly is in place, make sure that you have chosen a transition parameter that meets your design intent, such as Hold Grade, Change Offset.

Creating a Stream Corridor

Corridors can be used for far more than just road designs. You explore some more advanced corridor models in Chapter 12, but there are plenty of simple, single-baseline applications for alternative corridors such as channels, berms, streams, retaining walls, and more. You can take advantage of several specialized subassemblies or build your own custom assembly using a combination of generic links. Figure 11.45 shows an example of a stream corridor.

FIGURE 11.45
A simple stream corridor viewed in 3D built from the Channel subassembly and a generic link subassembly

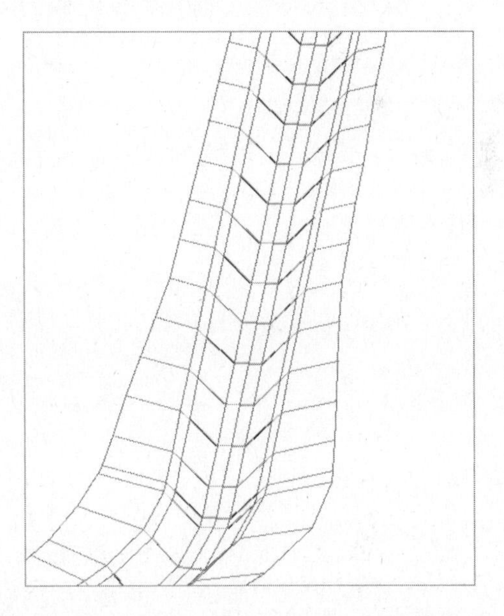

One of the subassemblies discussed in Chapter 10 is the Channel subassembly. The following exercise shows you how to apply this subassembly to design a simple stream:

1. Open the `Corridor Stream.dwg` file. Note that there is an alignment that represents a stream centerline, a profile that represents the stream normal water line, and an assembly created using the Channel and LinkSlopetoSurface subassemblies.

2. Change to the Home tab and choose Corridor ➤ Create Simple Corridor on the Create Design panel. The Create Simple Corridor dialog opens.
3. Name your corridor something appropriate, such as Project Stream. Click OK.
4. Follow the prompts, and pick the Stream CL alignment, the Stream NWL profile, and the Project Stream assembly. The Target Mapping dialog will appear after all objects have been picked.
5. In the Target Mapping dialog, choose the Existing Ground surface for all surface targets. Keep the default values for the additional targets. Click OK to dismiss the dialog.
6. The stream corridor will build itself and will look similar to Figure 11.46. Select the corridor and choose Corridor Section Editor from the Modify panel. Navigate through the stream cross sections. When you are finished viewing the sections, dismiss the dialog by clicking the X on the Close panel.

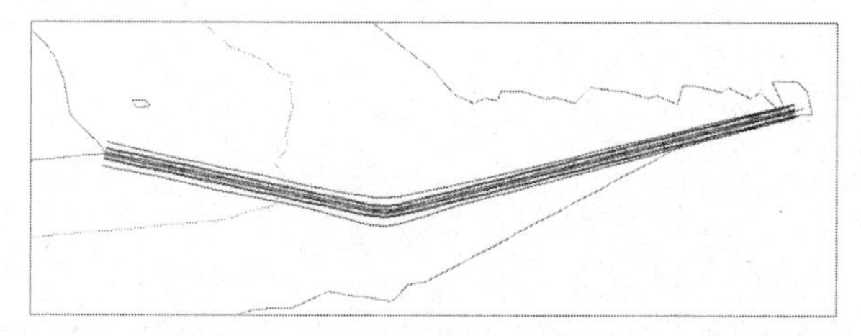

FIGURE 11.46
The completed stream corridor

This corridor can be used to build a surface for a TIN-to-TIN volume calculation or can be used to create sections and generate material quantities, cross-sectional views, and anything else that can be done with a more traditional road corridor.

Real World Scenario

CREATING A PIPE TRENCH CORRIDOR

Another alternative use for a corridor is a pipe trench. A pipe trench corridor is useful for determining quantities of excavated material, limits of disturbance, trench-safety specifications, and more. This graphic shows a completed pipe trench corridor:

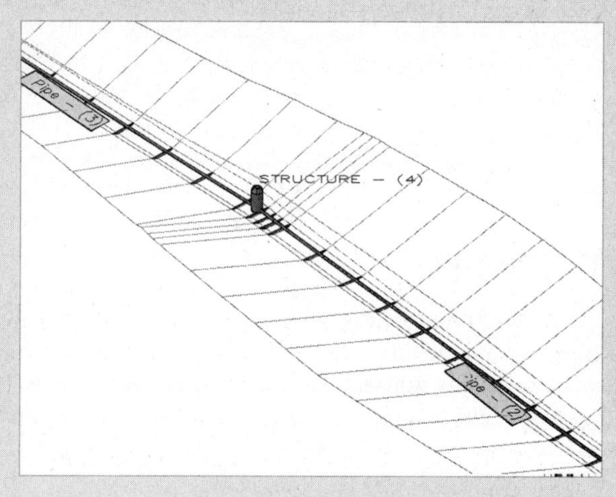

One of the subassemblies discussed in Chapter 10 is the TrenchPipe1 subassembly. The following exercise leads you through applying this subassembly to a pipe trench corridor:

1. Open the `Corridor Pipe Trench.dwg` file. Note that there is a pipe network, with a corresponding alignment, profile view, and pipe trench assembly. Also note that there is a profile drawn that corresponds with the inverts of the pipe network.
2. Change to the Home tab and choose Corridor ➤ Create Simple Corridor. Give your corridor a meaningful name, such as **Pipe Trench Corridor**. Click OK.
3. Follow the prompts, press ↵, and pick the Pipe Centerline alignment from the list, the Bottom of Pipe profile, and the Pipe Trench assembly as your corridor components. Once these selections are made, the Target Mapping dialog appears.
4. In the Target Mapping dialog, choose Existing Ground as the target surface. Click OK.
5. The corridor will build itself. Select the corridor and choose Corridor Section Editor from the Modify panel. Browse the cross sections through the trench. When you are finished viewing the sections, dismiss the dialog by clicking the X on the Close panel. This corridor can be used to build a surface for a TIN-to-TIN volume calculation or can be used to create sections and generate material quantities, cross-sectional views, and anything else that can be done with a more-traditional road corridor.

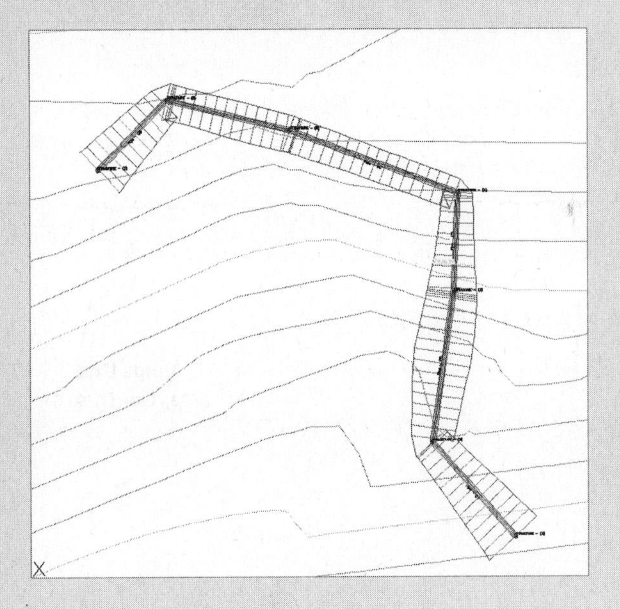

The Bottom Line

Build a single baseline corridor from an alignment, profile, and assembly. Corridors are created from the combination of alignments, profiles, and assemblies. Although corridors can be used to model many things, most corridors are used for road design.

Master It Open the `Mastering Corridors.dwg` file. Build a corridor on the basis of the Project Road alignment, the Project Road Finished Ground profile, and the Project Typical Road Assembly.

Create a corridor surface. The corridor model can be used to build a surface. This corridor surface can then be analyzed and annotated to produce finished road plans.

Master It Continue working in the `Mastering Corridors.dwg` file. Create a corridor surface from Top links.

Add an automatic boundary to a corridor surface. Surfaces can be improved with the addition of a boundary. Single baseline corridors can take advantage of automatic boundary creation.

Master It Continue working in the `Mastering Corridors.dwg` file. Use the Automatic Boundary Creation tool to add a boundary using the Daylight code.

Chapter 12

The Road Ahead: Advanced Corridors

In Chapter 11, ''Easy Does It: Basic Corridors,'' you built several simple corridors and began to see the dynamic power of the corridor model. The focus of that chapter was to get things started, but it's unrealistic to think that a project would have only one road in the middle of nowhere with no intersections, no adjustments, and no complications. You may be having trouble visualizing how you'll build a corridor to tackle your more complex design projects, such as the one pictured in Figure 12.1.

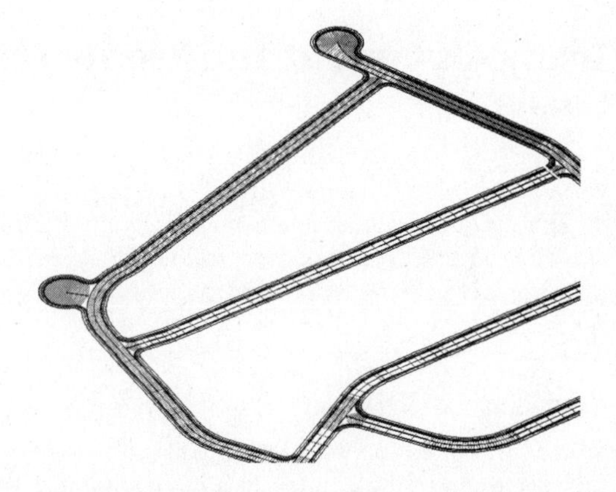

FIGURE 12.1
A corridor model for a medium-sized subdivision

This chapter focuses on taking your corridor-modeling skills to a new level by introducing more tools to your corridor-building toolbox, such as intersecting roads, cul-de-sacs, advanced techniques, and troubleshooting. Keep in mind that this is only the beginning. There are many ways to manipulate your assemblies, alignments, profiles, and the corridor itself to model anything you can imagine.

This chapter assumes that you've worked through the examples in the alignments, profiles, profile view, assemblies, and basic corridor chapters. Without a strong knowledge of the foundation skills, many of the tasks in this chapter may prove to be difficult.

By the end of this chapter, you'll learn to:

- Add a baseline to a corridor model for a cul-de-sac
- Add alignment and profile targets to a region for a cul-de-sac
- Use the Interactive Boundary tool to add a boundary to the corridor surface

Getting Creative with Corridor Models

New users often ask more-experienced users to teach them how to design an intersection (or a cul-de-sac, or a site, or anything) using Civil 3D. By the end of this chapter, you'll understand why this request is not only unrealistic, but probably impossible. There are as many ways to design an intersection as there are intersections in the world.

The best you can do is to learn how the corridor tools can be applied to a few typical scenarios. But don't take this chapter as gospel: use the skills you learn here to create a foundation for your own models in your own design situations. Users often dismiss an intersection from being applicable to their situation because it doesn't include a turn lane or perhaps their intersection comes together at an odd angle. This is unfortunate, because the same fundamental tools can be adapted to accommodate additional design constraints.

Another example you may consider is adapting the corridor model for use in a parking lot or in a commercial site. As you saw in Chapter 11 with stream and pipe-trench corridors, the corridor model is not a road-only tool. It can be used for ponds, berms, curbs, and gutters, and much more.

Civil 3D in general, and the corridor model specifically, won't be truly useful to you unless you can see them as limitless, flexible models that you control to your design constraints. Build something; try something. If it doesn't work, look back through the chapter for more ideas and keep refining, improving, and learning.

Using Alignment and Profile Targets to Model a Roadside Swale

The previous chapter included an example where the road lane used an alignment target to add a variable width to the lane without changing the vertical design.

This chapter deals with a roadside swale that follows a variable horizontal alignment, as well as a vertical profile that doesn't follow the centerline of the road. This happens frequently when existing culvert crossings must be met or when you have different slope requirements for the roadside swale.

Corridor Utilities

To create an alignment and profile for the swale, you'll take advantage of some of the corridor utilities found in the Launch Pad by changing to the Modify tab and choosing Corridor from the Design panel to open the Corridor tab. The utilities on this tab are as follows:

- **Feature Lines from Corridor** — This utility extracts a grading feature line from a corridor feature line. This grading feature line can remain dynamic to the corridor, or it can be a static extraction. Typically, this extracted feature line will be used as a foundation for some feature-line grading or projection grading. If you choose to extract a dynamic feature line, it can't be used as a corridor target due to possible circular references.
- **Alignments from Corridor** — This utility creates an alignment that follows the horizontal path of a corridor feature line. You can use this alignment to create target alignments, profile views, special labeling, or anything else for which a traditional alignment could be used. Extracted alignments aren't dynamic to the corridor.
- **Profile from Corridor** — This utility creates a profile that follows the vertical path of a corridor feature line. This profile appears in Prospector under the baseline alignment and is drawn on any profile view that is associated with that baseline alignment. This profile is

typically used to extract edge of pavement (EOP) or swale profiles for a finished profile view sheet or as a target profile for additional corridor design, as you'll see in this section's exercise. Extracted profiles aren't dynamic to the corridor.

- **Points from Corridor** — This utility creates Civil 3D points that are based on corridor point codes. You select which point codes to use, as well as a range of corridor stations. A Civil 3D point is placed at every point-code location in that range. These points are a static extraction and don't update if the corridor is edited. For example, if you extract COGO points from your corridor and then revise your baseline profile and rebuild your corridor, your COGO points won't update to match the new corridor elevations.
- **Polyline from Corridor** — This utility extracts a 3D polyline from a corridor feature line. The extracted 3D polyline isn't dynamic to the corridor. You can use this polyline as is or flatten it to create road linework.
- **Static Surface from Corridor** — This utility copies a dynamic corridor surface and converts it into a static corridor surface. This tool is most useful for creating an archive surface that won't react to future corridor revisions.

The following exercise takes you through revising a model from a symmetrical corridor with roadside swales to a corridor with a transitioning roadside swale centerline. You'll also take advantage of some of the static extraction corridor utilities discussed in this section:

1. Open the `Corridor Swale.dwg` file, which you can download from `www.sybex.com/go/masteringcivil3d2010`. Note that the drawing contains a symmetrical corridor (see Figure 12.2), which was built using an assembly that includes two roadside swales. You can view the corridor in 3D by picking it, and choosing Object Viewer from the General Tools panel. You can also change your view of the corridor by using the 3D Orbit tools.

FIGURE 12.2 The initial corridor with symmetrical roadside swales

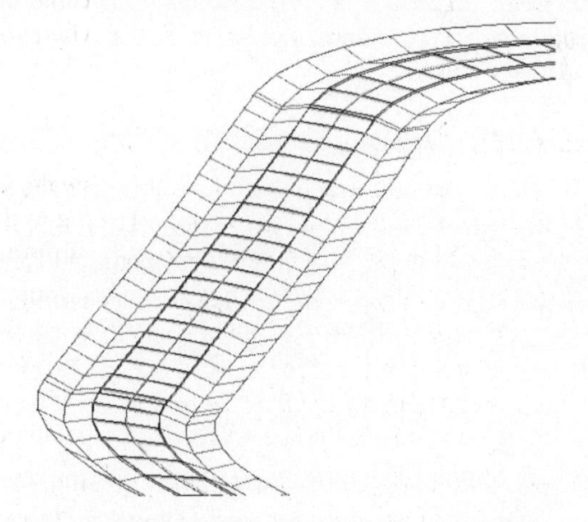

2. Select the corridor and choose the Alignments from Corridor tool from the Launch Pad panel. When prompted, pick the corridor feature line that represents the swale on the right side of the centerline, as shown in Figure 12.3, to create an alignment. Note that if you select one of the corridor feature lines to activate the Corridors tab, that feature line will be created by default. You can pick one of the section lines to avoid this problem!

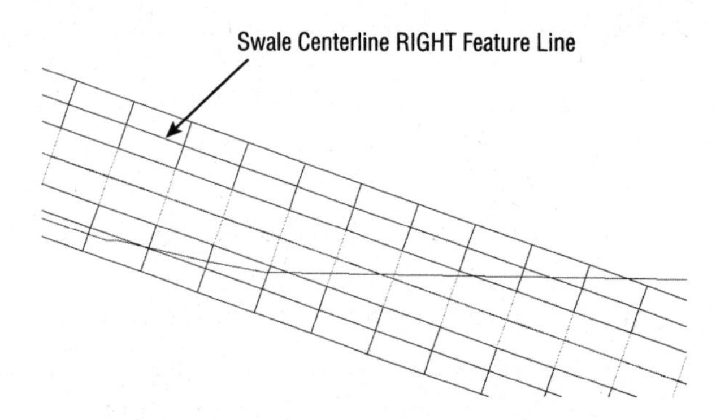

FIGURE 12.3
The corridor feature line that represents the swale on the right side of the centerline

3. In the Create Alignment dialog, name the alignment **Swale CL**. Keep the default values for the Style and Label options, and deselect the Create Profile check box. You'll add the profile another way. Click OK to dismiss the dialog, and ESC to exit the command. Notice an alignment has been created at the corridor feature line (see Figure 12.4).

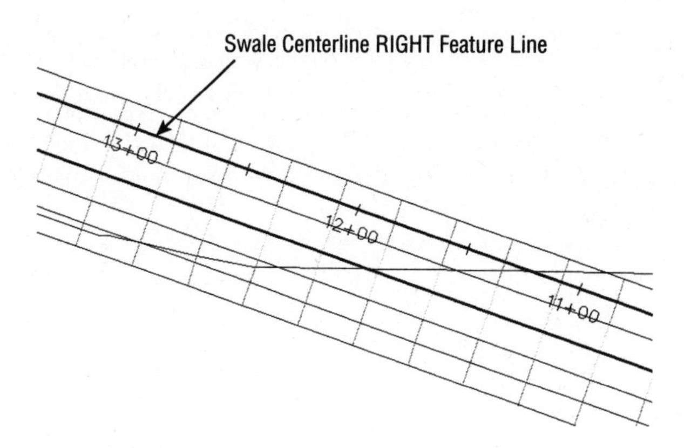

FIGURE 12.4
The resulting extracted alignment

4. Select the corridor and choose the Profile from Corridor tool on the Launch Pad panel. Pick the same corridor feature line from which to create the profile. (Note that you may have to send your alignment to the back using the **draworder** command so you can pick the corridor feature line.) In the Create Profile dialog, name the profile **Project Road-Swale CL Profile**, and keep the default values for the Style and Label options. Click OK to dismiss the dialog, and press Esc to exit the command.

5. Pan over to the Project Road profile view to see the profile you just created. The profile represents the vertical path of the feature line representing the swale centerline, as shown in Figure 12.5.

6. Pan over to your newly extracted swale centerline alignment. Add a PI around Station 13+00 in plan. You can use the transparent commands to snap the PI to Station 13+00 exactly or place it approximately at Station 13+00.

7. Grip-edit the PI, and stretch it approximately 10′ to the north. Use the Station Offset transparent command if you'd like to be precise.

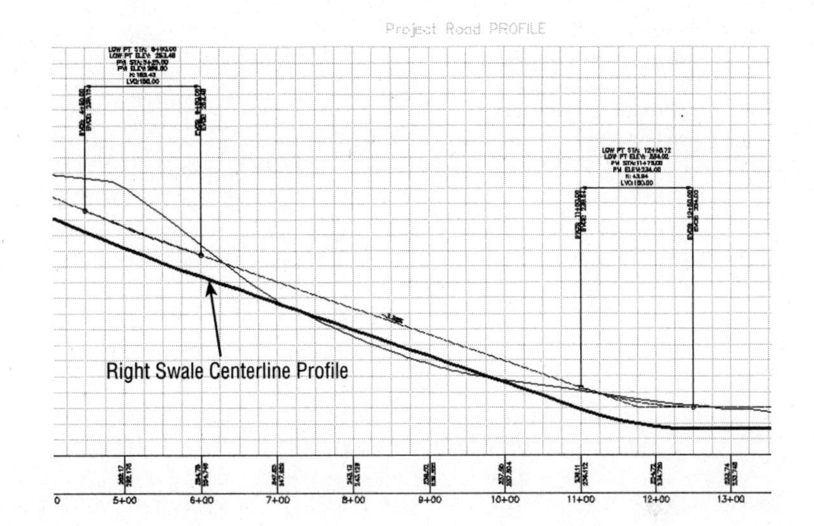

FIGURE 12.5
The resulting extracted profile

8. Pan over to the profile view. Grip-edit the swale centerline profile to provide an exaggerated low spot around station 13+00, as shown in Figure 12.6. Press Esc to exit the command.

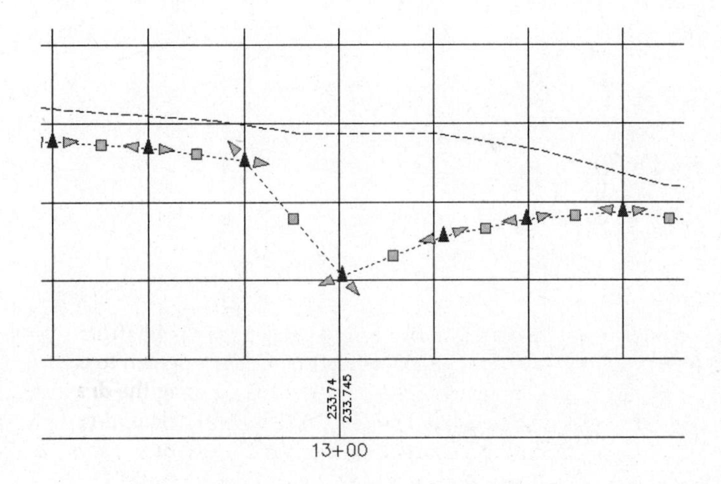

FIGURE 12.6
Stretch a PVI to provide an exaggerated low spot.

9. Select the corridor and select Corridor Properties from the Modify panel to open the Corridor Properties dialog. Switch to the Parameters tab. Click Set All Targets to open the Target Mapping dialog.

10. Expand the Width or Offset Targets option, and set the Ditch Foreslope RIGHT subassembly to target the Object Name Swale CL Alignment. Expand the Slope or Elevation Targets option, and set the Ditch Foreslope RIGHT subassembly to target the Swale CL Profile. Figure 12.7 shows the Target Mapping dialog with the alignments and profiles appropriately mapped.

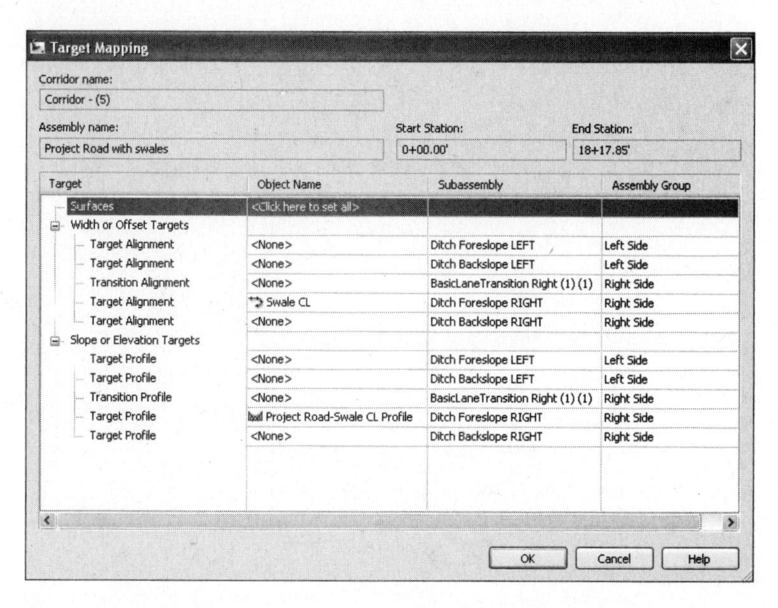

FIGURE 12.7 Set the Ditch Foreslope subassembly to follow the swale centerline alignment and profile.

11. Click OK to dismiss the Target Mapping dialog. Click OK again to dismiss the Corridor Properties dialog and rebuild the corridor. When viewed in 3D, the corridor should now look like Figure 12.8.

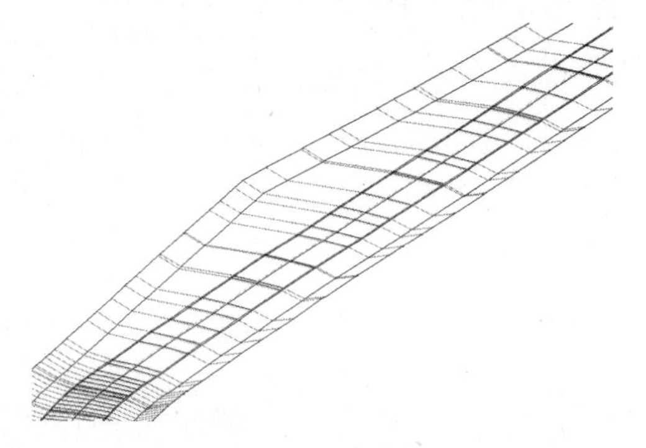

FIGURE 12.8 The adjusted corridor

Note that the corridor has been adjusted to reflect the new target alignment and profile. Also note that you may want to increase the sampling frequency. You can view the sections using the View/Edit Corridor Section tools. You can also view the corridor in 3D by picking it, right-clicking, and choosing Object Viewer. Use the 3D Orbit tools to change your view of the corridor.

Modeling a Peer-Road Intersection

When you're using the corridor model, an important distinction to make is that even though you're building a model, you aren't designing it. Technically speaking, you could build the model and design the intersection simultaneously; however, most users who attempt to ''figure it out'' as they're constructing the model find the task tedious and frustrating.

The first step is to figure out how your intersection works. Give yourself some modeling guidelines, trends, and design constraints (but *not* actual hard elevations just yet), either onscreen or on a small plotted schematic (see Figure 12.9).

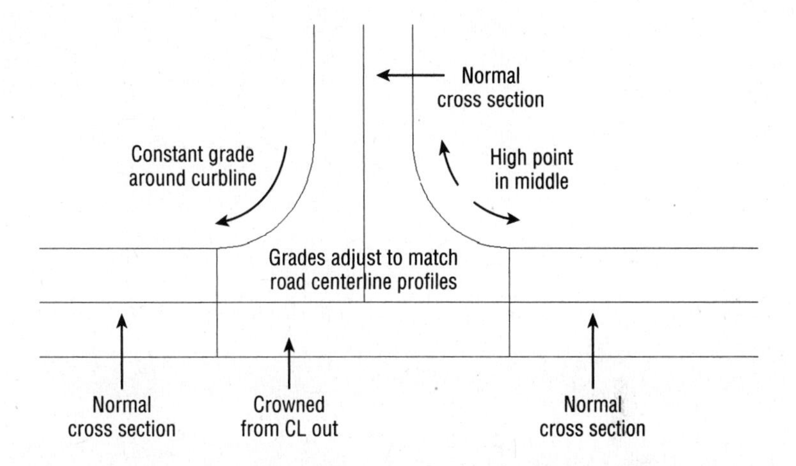

FIGURE 12.9
Plan your intersection model in sketch form.

Next, plan what alignments, profiles, and assemblies you'll need to create the right combination of baselines, regions, and targets to model an intersection that will interact the way you want.

Figure 12.10 shows a sketch of required baselines. *Baselines* are the horizontal and vertical foundation of a corridor, as you saw in Chapter 11. Each baseline consists of an alignment and its corresponding finished ground (FG) profile. You may never have thought of edge of pavement (EOP) in terms of profiles, but after building a few intersections, thinking that way will become second nature. The Intersection tool on the Create Design panel of the Home tab will create EOP baselines as curb return alignments for you, but it will rely on your input for curb return radii.

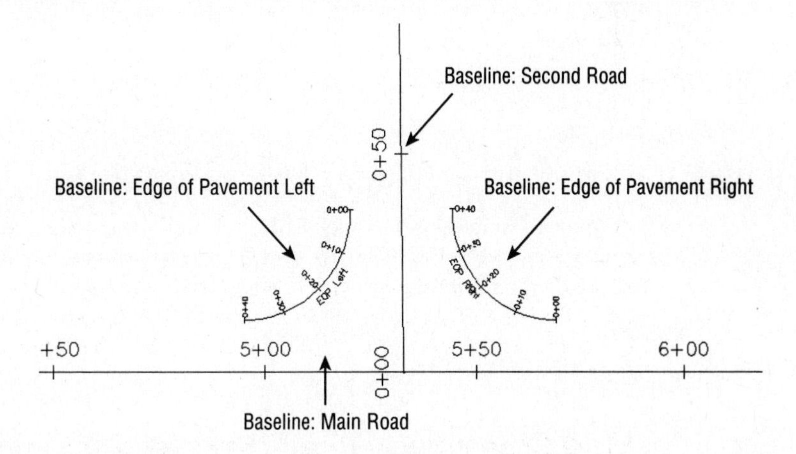

FIGURE 12.10
Required baselines for modeling a typical intersection

Figure 12.11 breaks each baseline into regions where a different assembly or different target will be applied. The next exercise will give you hands-on experience in splitting regions along a corridor baseline as a precursor to building an intersection automatically. Once the intersection has been created, target mapping as well as other particulars can be modified as needed.

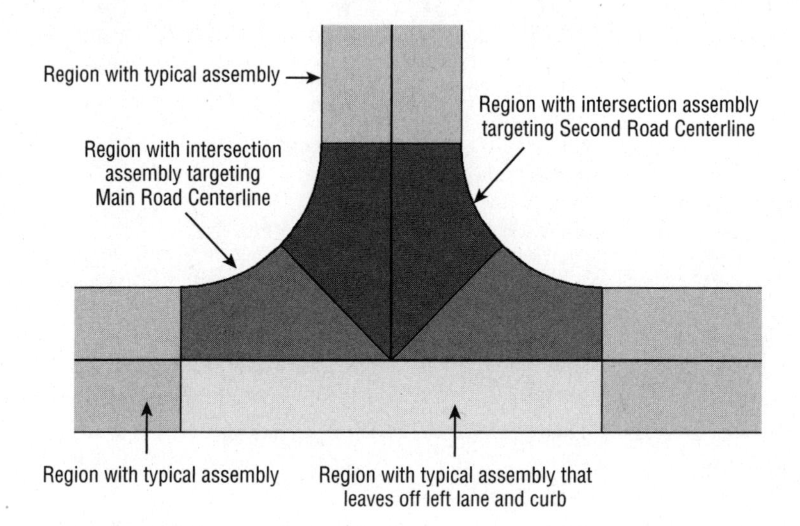

FIGURE 12.11
Required regions for modeling an intersection created by the Intersection tool

Using the Intersection Wizard

In Chapter 11, you learned briefly about regions. A baseline consists of a combination of an alignment and a profile, whereas assemblies are applied to specific regions. By default, every baseline has one region, which you created in Chapter 11 whenever you made a new corridor. If certain zones of a baseline require the application of a different assembly, you split a baseline into multiple regions. The Intersection tool will do this for you automatically.

On the basis of the schematic you drew of your intersection, your main road will need two assemblies to reflect two different road cross sections. The first assembly, as shown in Figure 12.12, is the typical or ''normal'' case. The bulk of your neighborhood will use this typical assembly along straight pieces that aren't intersections, widening areas, or similar structures.

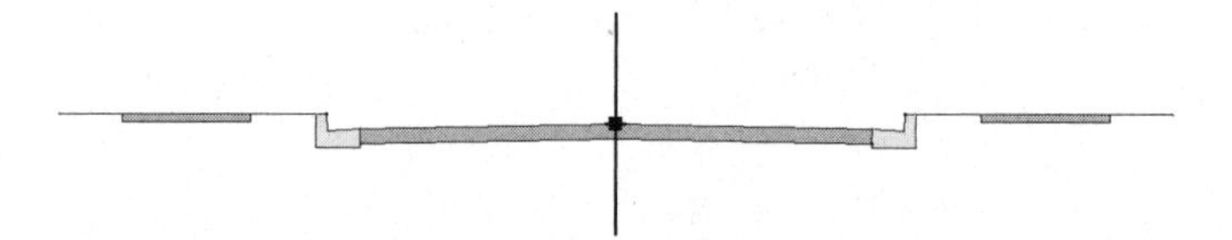

FIGURE 12.12
A typical assembly will be applied to all ''normal'' regions.

The next assembly (see Figure 12.13) is a right lane–only assembly that you'll apply through the intersection. On the basis of your sketch, this particular main road won't maintain a full crown through the intersection; however, the right half of the road will be normal.

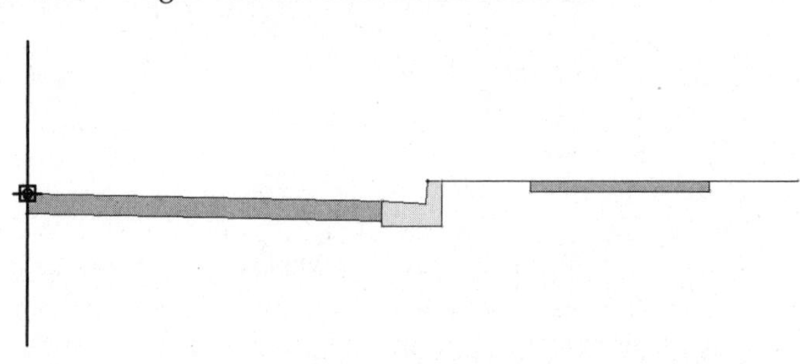

FIGURE 12.13
Right lane–only assembly

This exercise will take you through building a typical peer-road intersection using the Intersection wizard :

1. Open the `Corridor Peer Intersection.dwg` file, which you can download from `www.sybex.com/go/masteringcivil3d2010`. The drawing contains two centerline alignments (Project Road and Second Road). The drawing also contains a single baseline corridor running down Project Road.
2. Open the Corridor Properties dialog, and switch to the Parameters tab. Note that Baseline (1) uses the Project Road alignment and the Project Road FG profile. Also note that there is currently one region for the Project Road typical assembly baseline. This baseline will require more regions to match the schematic previously shown in Figure 12.11. You will create one region by splitting an existing region, but the Intersection tool will automatically create the rest.
3. Right-click Region (1) and select Split Region from the menu.
4. Select the intersection of the two alignments using the Intersection object snap.
5. Press Esc to return to the Corridor Properties dialog. The new region is displayed as shown in Figure 12.14.

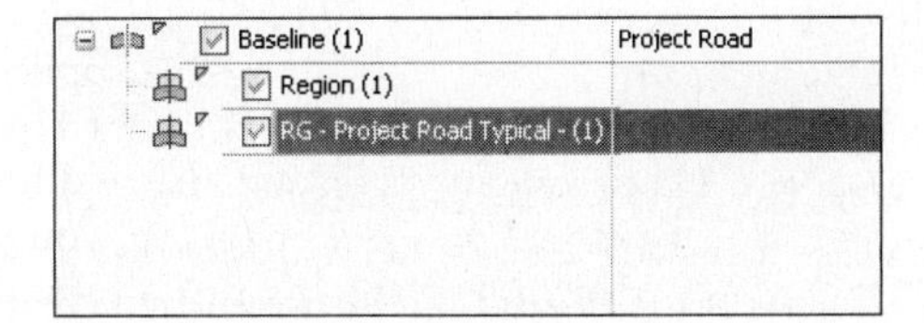

FIGURE 12.14
A new region created by splitting an existing region

6. Select OK to exit the Corridor Properties dialog.
7. Select the corridor to enable the region grips.
8. Using the triangle grips to the left and right of the intersection, spread the region apart as shown in Figure 12.15.

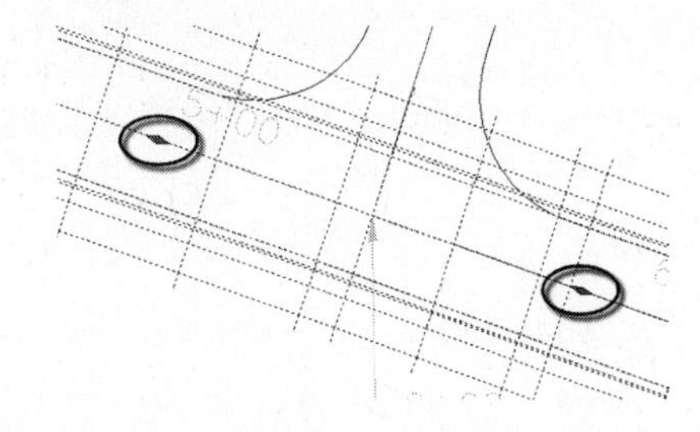

FIGURE 12.15
The region grips spread apart to make room for an intersection

9. Choose Rebuild from the Modify panel to rebuild the corridor and notice the split region appears.
10. Press Esc to cancel the grips.

11. Change to the Home tab and choose the Intersection tool on the Create Design panel. Using the Intersection object snap, choose the intersection of the two existing alignments. The Create Intersection – General dialog opens as shown in Figure 12.16.

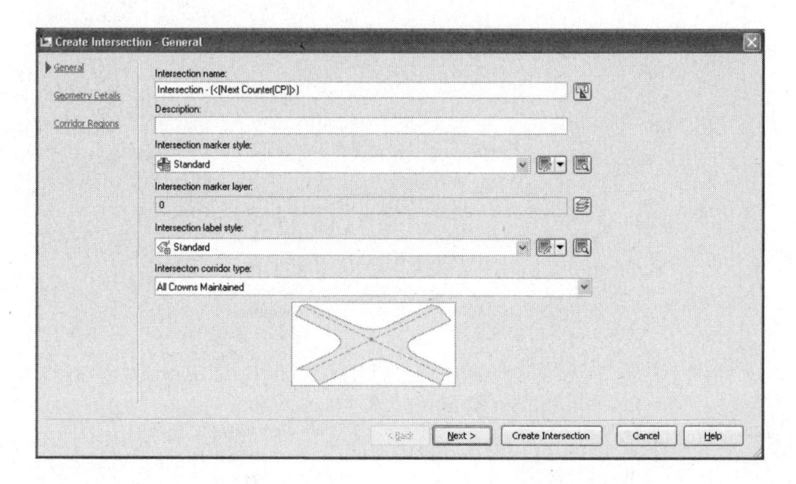

FIGURE 12.16
The Create Intersection – General dialog

Create Intersection - General
General
Geometry Details
Corridor Regions
Intersection name:
Intersection - (<[Next Counter(CP)]>)
Description:
Intersection marker style:
Standard
Intersection marker layer:
0
Intersection label style:
Standard
Intersecton corridor type:
All Crowns Maintained
< Back
Next >
Create Intersection
Cancel
Help

12. Accept the defaults as shown in Figure 12.16 and click Next to open the Create Intersection – Geometry Details dialog. Notice the priority of the intersecting alignments. The priority can be modified using the up and down arrows as shown in Figure 12.17.

FIGURE 12.17
The Create Intersection – Geometry Details dialog

Create Intersection - Geometry Details
General
Geometry Details
Corridor Regions
Intersecting alignments:
Priority Alignment Station Profile
1 Project Road 5+32.57 Project Road FG
2 Second Road 0+00.00 Second Road FG
Offset and curb returns
Create or specify offset alignments
Offset Parameters
Create curb return alignments
Curb Return Parameters
Offset and curb return profiles
Create offset and curb return profiles
Lane Slope Parameters
Curb Return Profile Parameters
< Back
Next >
Create Intersection
Cancel
Help

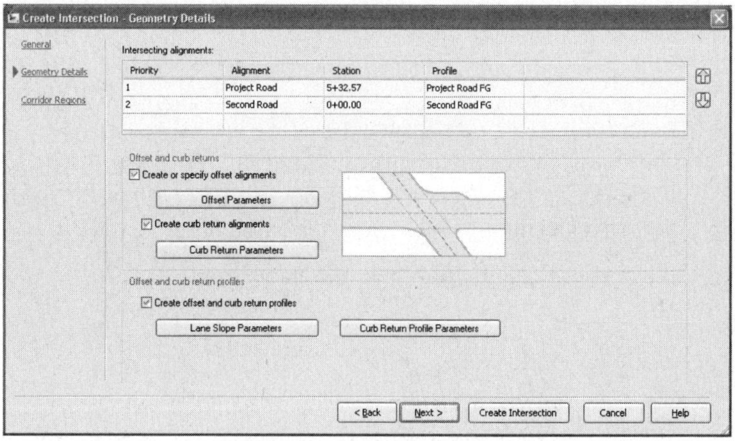

13. Click the Offset Parameters button to open the Intersection Offset Parameters dialog as shown in Figure 12.18. The options in this dialog are used to automatically generate dynamic offset alignments along the edges of pavement to target as necessary.

14. Click OK to accept the defaults and exit the Intersection Offset Parameters dialog.

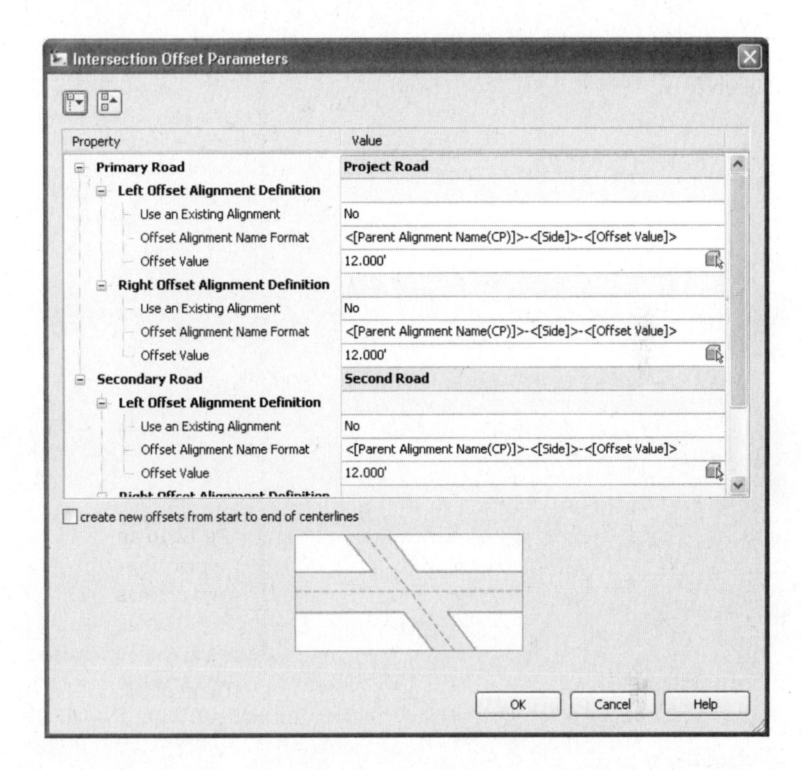

FIGURE 12.18
The Intersection Offset Parameters dialog

15. Click the Curb Return Parameters button to open the Intersection Curb Return Parameters dialog as shown in Figure 12.19. This dialog is used to specify radii along all four curb returns and to widen turn lanes for incoming and outgoing roads as necessary.

16. Click OK to accept the defaults and return to the Create Intersection – Geometry Details dialog.

17. Click the Lane Slope Parameters button to open the Intersection Lane Slope Parameters dialog as shown in Figure 12.20. This dialog is used to establish the lane cross fall from the centerline to the edge of pavement. An existing profile can be used as well. However, the slope cannot be picked from existing lane subassemblies.

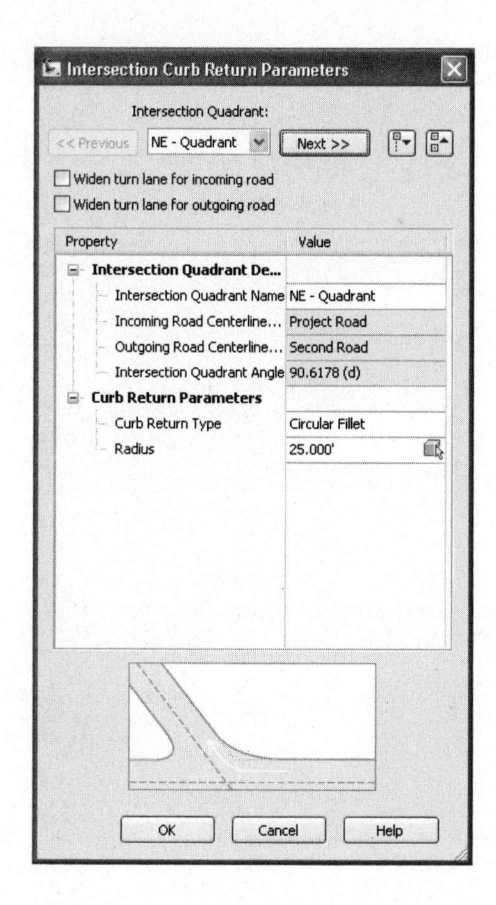

FIGURE 12.19
The Intersection Curb Return dialog

18. Click OK to accept the defaults and return to the Create Intersection – Geometry Details dialog. Click the Curb Return Profile Parameters button to open the Intersection Curb Return Profile Parameters dialog, as shown in Figure 12.21. The parameters in this dialog can be set accordingly should you need to extend the curb return profiles along the incoming and outgoing tangents before and after a curb return.

19. Click OK to accept the defaults and return to the Create Intersection – Geometry Details dialog. Click Next to advance to the Create Intersection – Corridor Regions dialog as shown in Figure 12.22. This dialog is used to specify whether or not to add to an existing corridor or simply create a new one. It is also used to specify which assemblies to use at specific locations.

20. Be sure your settings match those shown in Figure 12.22, and then click the Create Intersection button to review the results. Your intersection should look something like Figure 12.23.

Do not delete the intersection marker. This will sever the linking of the alignments within the intersection. Also notice that the curb return profiles get locked (as indicated by the diamond-shaped grips). If you manually edit the curb return profiles and update the intersection later, you will sever the link between the curb returns and other parts of the model.

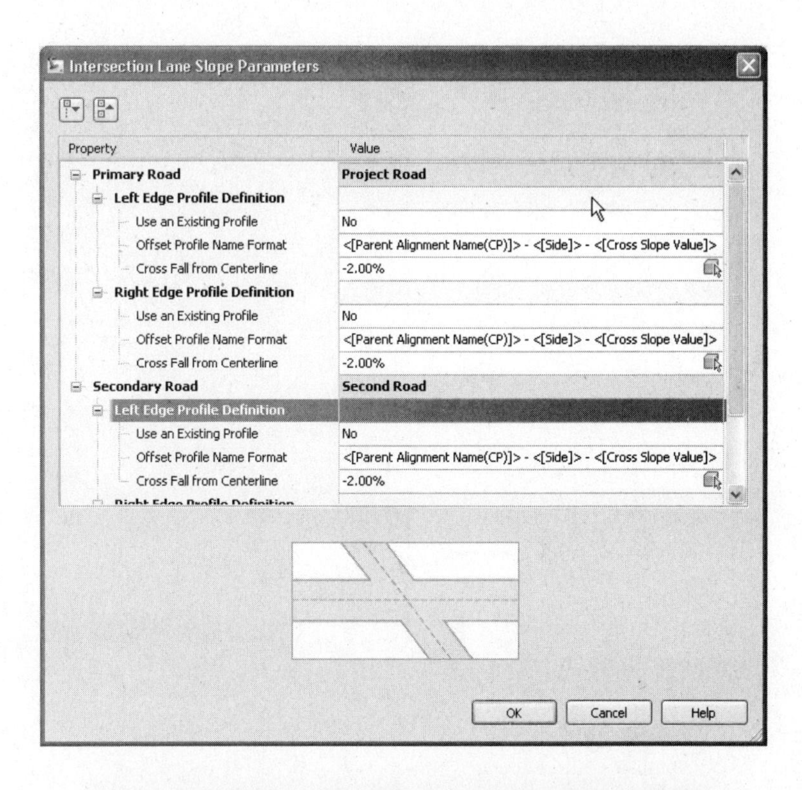

FIGURE 12.20
The Intersection Lane Slope Parameters dialog

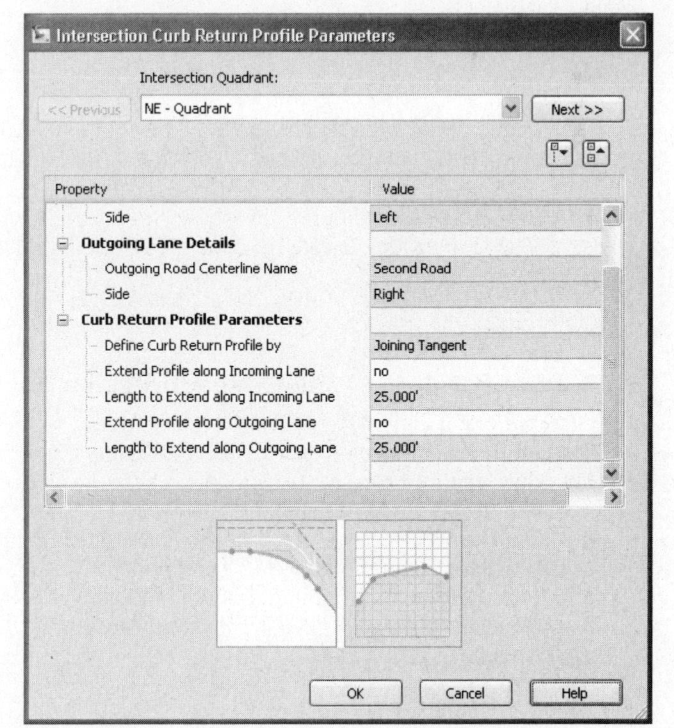

FIGURE 12.21
The Intersection Curb Return Profile Parameters dialog

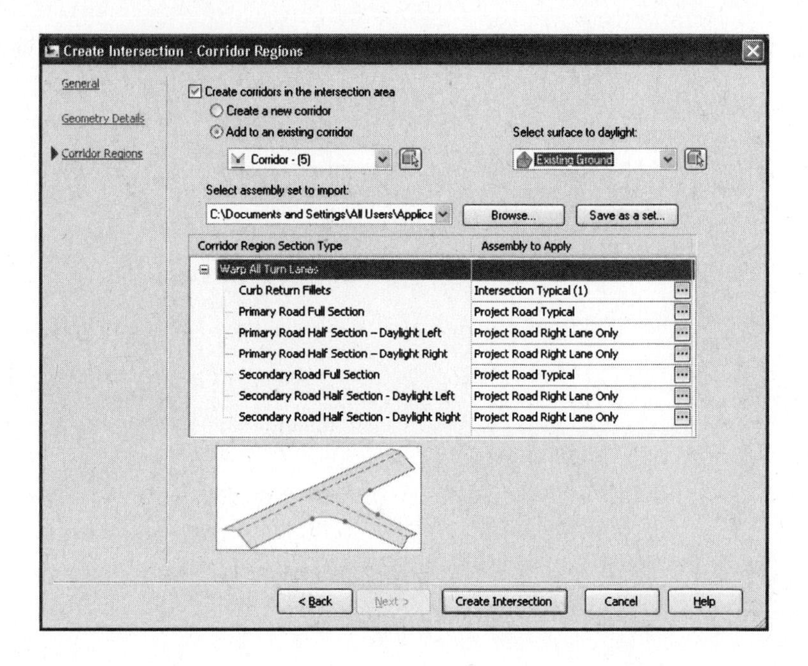

FIGURE 12.22
The Create Intersection – Corridor Regions dialog

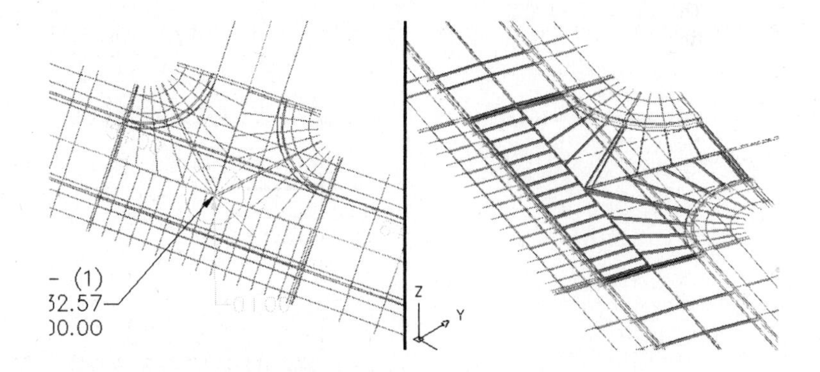

FIGURE 12.23
Completed intersection built using the Intersection Wizard

Manually Adding a Baseline and Region for an Intersecting Road

A corridor isn't limited to one baseline. Depending on the size of your project, you may build one corridor that includes many baselines. It isn't uncommon to build a corridor with 60 or more baselines that represent road centerlines, transitions, swales, and more. In the previous example, the intersection was modeled using the Intersection tool, and all the components are now dynamically linked. If the profile of one of the streets is modified, the profile of the intersecting street, as well as the curb returns and other related components will update. However, in some cases, you may find it necessary to model an intersection manually. That process typically begins with adding a baseline to an existing corridor. In this example, you'll add a baseline for an intersecting road to your corridor:

1. Open the `Corridor Peer Intersection 2.dwg` file, which you can download from `www.sybex.com/go/masteringcivil3d2010`.
2. Open the Corridor Properties dialog, and switch to the Parameters tab.
3. Click Add Baseline. The Pick Horizontal Alignment dialog opens.
4. Pick the Second Road alignment. Click OK to dismiss the dialog.
5. Click in the Profile field on the Parameters tab of the Corridor Properties dialog. The Select a Profile dialog opens.
6. Pick the Second Road FG. Click OK to dismiss the dialog.
7. In the Corridors Properties dialog, right-click BL - Second Road - (1) and select Add Region.
8. In the Pick an Assembly dialog, select the Project Road Typical Assembly. Click OK to dismiss the dialog.
9. Expand BL - Second Road - (1) by clicking the small + sign, and see the new region you just created.
10. Click OK to dismiss the Corridor Properties dialog; your corridor will automatically rebuild. Your corridor should now look like Figure 12.24.

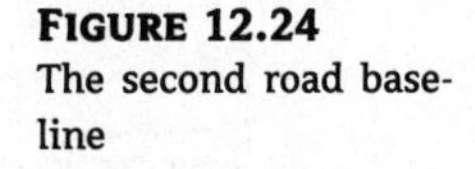

FIGURE 12.24
The second road baseline

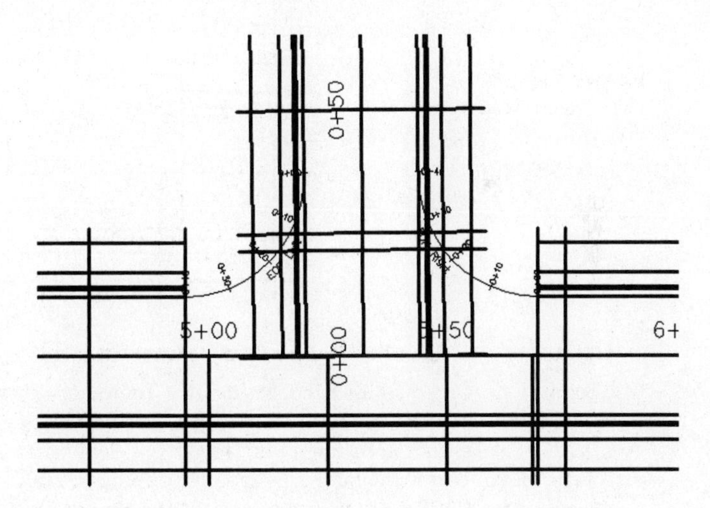

Notice that the region for the second baseline extends all the way to the end of the Second Road alignment. You must now adjust this to match the EOP points of curvature:

1. Open the Corridor Properties dialog, and switch to the Parameters tab. Using the techniques covered in the previous exercise, adjust the Start Station of the region to 0+37.40. You could also pick the corridor and use the triangular-shaped grip that represents the end of the region to adjust the station, as shown in Figure 12.25.
2. Once you move the region grip, you must rebuild your corridor. Pick your corridor, right-click, and choose Rebuild Corridor. Your corridor should now match Figure 12.26.

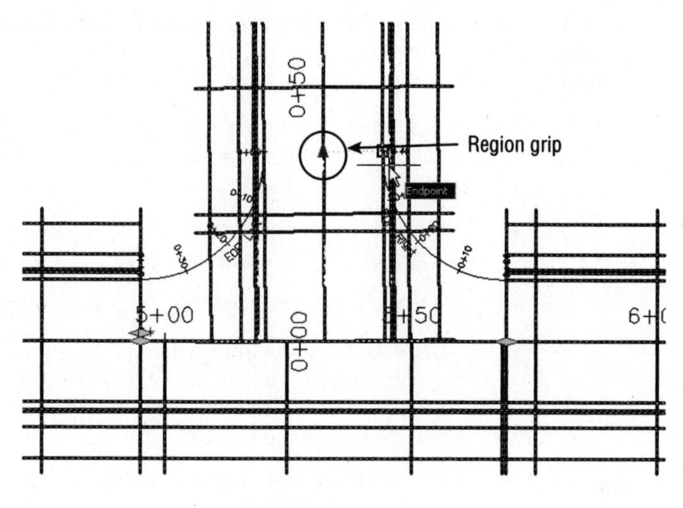

FIGURE 12.25
Using the region grip to move the region End Station

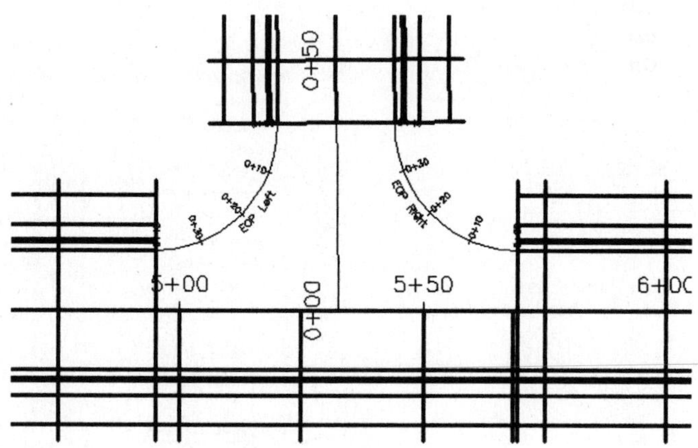

FIGURE 12.26
The finished corridor

Creating an Assembly for the Intersection

You built several assemblies in Chapter 10, ''Templates Plus: Assemblies and Subassemblies,'' but most of them were based on the paradigm of using the assembly marker along a centerline. This next exercise leads you through building an assembly that attaches at the EOP.

The exercise uses the BasicLaneTransition subassembly (see Figure 12.27), but you're by no means limited to this subassembly in practice. You'll find that BasicLaneTransition is ideal for your first few intersections because it doesn't have many options for targeting or materials, which can often be confusing to new users.

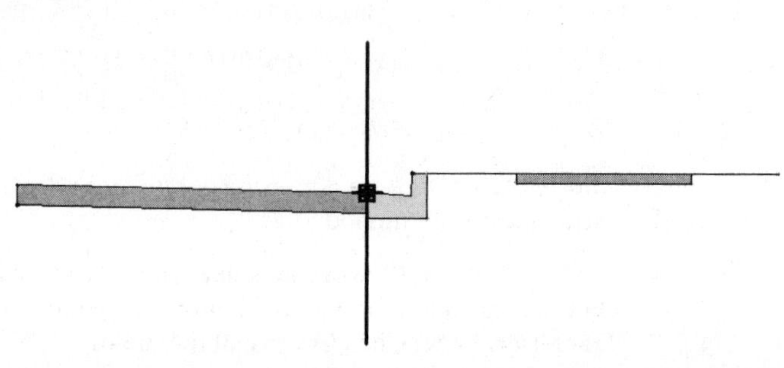

FIGURE 12.27
The BasicLaneTransition subassembly

You can also use the LaneOutsideSuper and LaneTowardCrown subassemblies, as well as any other lane with an alignment and profile target. Spend some time in the subassembly Help file to find your perfect subassembly. Then, follow these steps:

1. Open the `Corridor Peer Intersection Assembly.dwg` file (which you can download from `www.sybex.com/go/masteringcivil3d2010`), or continue working in your drawing from the previous exercise.
2. Zoom in on the assemblies, and locate the Intersection Typical assembly. Note that the assembly doesn't yet include a lane.
3. Bring in your tool palettes, and switch to the Imperial-Basic palette.
4. Click the BasicLaneTransition subassembly. On the Properties palette, change the Transition parameter to the Change Offset and Elevation option (see Figure 12.28), which allows the subassembly to react to both a target alignment and a target profile. Also change the value of the Default Slope to +2%.

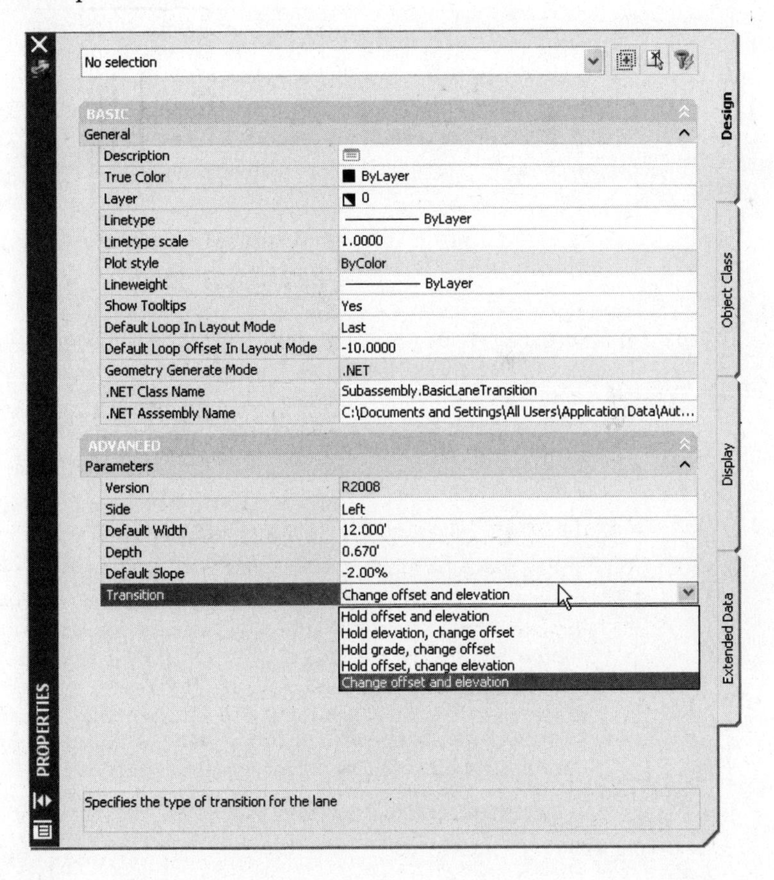

FIGURE 12.28
Change the Transition parameter to Change Offset and Elevation.

5. Add the BasicLaneTransition subassembly to the left side of the assembly, and then press Esc to exit the command.
6. Pick the BasicLaneTransition subassembly, right-click, and choose Subassembly Properties. On the Information tab of the dialog, change the name of the subassembly to **Intersection Transition Lane**. Click OK to exit the dialog.

Adding Baselines, Regions, and Targets for the Intersections

The Intersection assembly attaches to alignments created along the EOP. Because a baseline requires both horizontal and vertical information, you also need to make sure that every EOP alignment has a corresponding finished ground (FG) profile.

It may seem awkward at first to create alignment and profiles for things like the EOP, but after some practice you'll start to see things differently. If you've been designing intersections using 3D polylines or feature lines, think of the profile as a vertical representation of a feature line and the profile grid view as the feature-line elevation editor. If you've designed intersections by setting points, think of alignment PIs and profile PVIs as points. If you need a low point midway through the EOP, as indicated in the sketch at the beginning of this section, you'll add a PVI with the appropriate elevation to the EOP FG profile.

Here are some things to keep in mind:

Site Geometry Make sure any alignments you create are either *siteless* (placed on the <none> site) or placed on another appropriate site.

Naming Conventions Instead of allowing your alignments and profiles to be named Alignment-1, Alignment-2, and so on, give each one a meaningful name that will help you keep them straight, so that you can identify them on a list and locate them in plan. The same rule applies for the FG profiles. If the alignment is named EOP Right, then name the profile EOP Right FG or something similar. Explore alignment- and profile-name templates to see if you can help yourself by automating the naming.

Organization Figure out a way to keep your profile views organized. Perhaps line them up next to the appropriate road centerline, or use some other convention. If you just stick them anywhere, you'll have a difficult time navigating and finding what you need.

Styles Because you may not need to show these alignments and profiles on a plan sheet anywhere, consider making some design styles that give you the information you need. The style can also look different enough from your normal alignment and profile view styles that you can look at it and, in a glance, identify it as an EOP profile view versus another type. The examples in the next exercise include smaller lettering, alignment name labels, tighter labeling intervals, and other style features to assist the designer.

You'll build a corridor intersection in the next exercise. Along the way, you'll examine the corridor in various stages of completion so that you'll understand what each step accomplishes. In practice, you'll likely continue working until you build the entire model. Follow these steps:

1. Open the `Corridor Peer Intersection 3.dwg` file (which you can download from `www.sybex.com/go/masteringcivil3d2010`), or continue working in your drawing from the previous exercise.
2. Zoom to the area where the profile views are located. Notice that existing and proposed ground profiles are created for both EOP alignments (see Figure 12.29).
3. Open the Corridor Properties dialog, and switch to the Parameters tab.
4. Click Add Baseline. The Pick Horizontal Alignment dialog opens.
5. Select the EOP Left alignment. Click OK to dismiss the dialog.

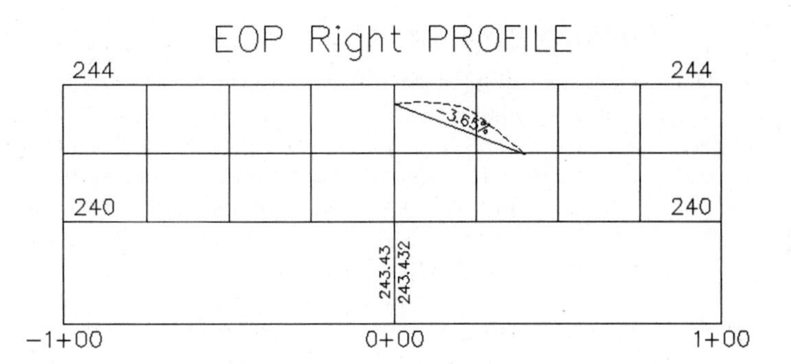

FIGURE 12.29
A profile view for the existing and proposed EOP profiles

6. Click in the Profile field on the Parameters tab of the Corridor Properties dialog.
7. In the Select a Profile dialog, select EOP Left FG. Click OK to dismiss the dialog.
8. Right-click BL – EOP Left – (1), and select Add Region.
9. In the Pick an Assembly dialog, select Intersection Typical Assembly. Click OK to dismiss the dialog.
10. Expand the baseline, and see the new region you just created.
11. Click OK to dismiss the Corridor Properties dialog and automatically rebuild your corridor. Your corridor should now look similar to Figure 12.30.

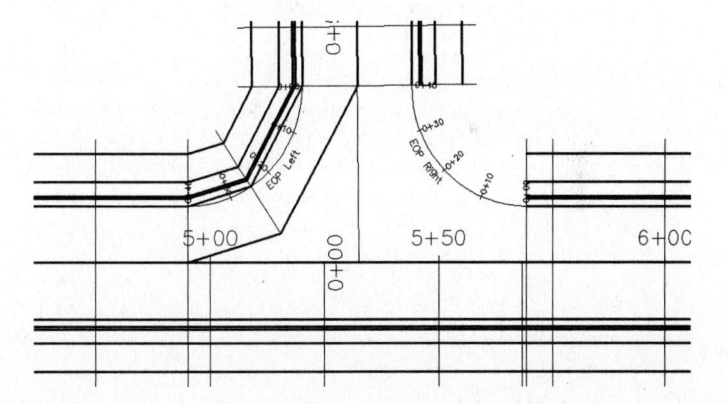

FIGURE 12.30
Your corridor after applying the Intersection Typical assembly

Your assembly has been applied; but because you haven't set the targets, the lane is only 12′ wide. You'll need to target the Second Road centerline from Station 0+00 through approximately Station 0+20 and the Project Road centerline from approximately Station 0+20 through 0+40:

1. Open the Corridor Properties dialog, and switch to the Parameters tab. Change the End Station of baseline BL – EOP Left – (1) region RG – Intersection Typical – (1) to **0+19.77**. The best way to do this is to click the Specify Station button and use your Intersection osnap to choose the intersection of Project Road and Second Road, as shown in Figure 12.31.
2. Click the ellipsis button in the Target column for the Baseline (3) Region (1) row, as shown in Figure 12.32.

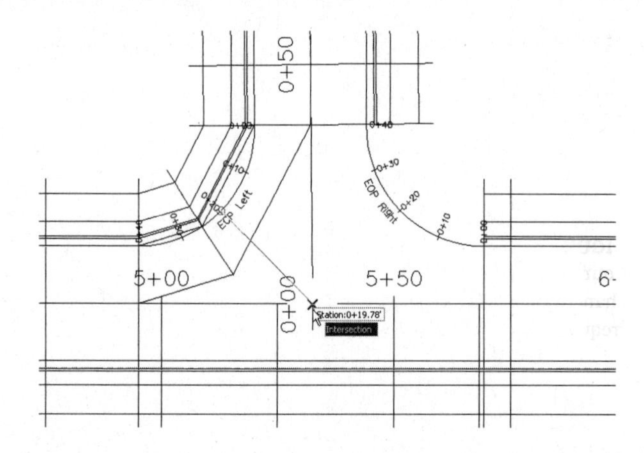

FIGURE 12.31
Use the Select Station button and the Intersection osnap to change the End Station.

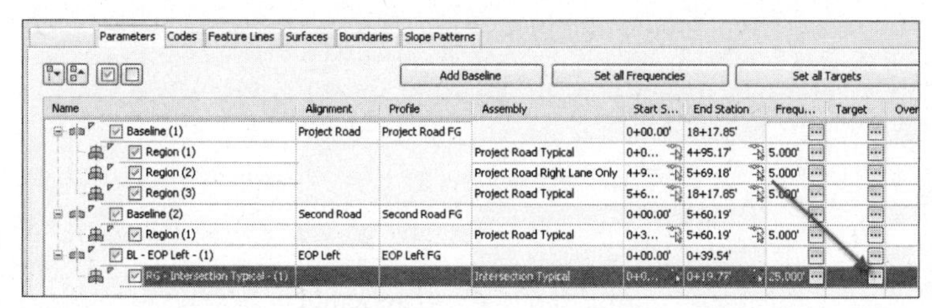

FIGURE 12.32
Click the ellipsis button.

3. In the Target Mapping dialog, choose Second Road for the Transition Alignment and Second Road FG for the Transition Profile. Click OK.
4. Click OK again to dismiss the Corridor Properties dialog and automatically rebuild your corridor. Your corridor should now look similar to Figure 12.33.

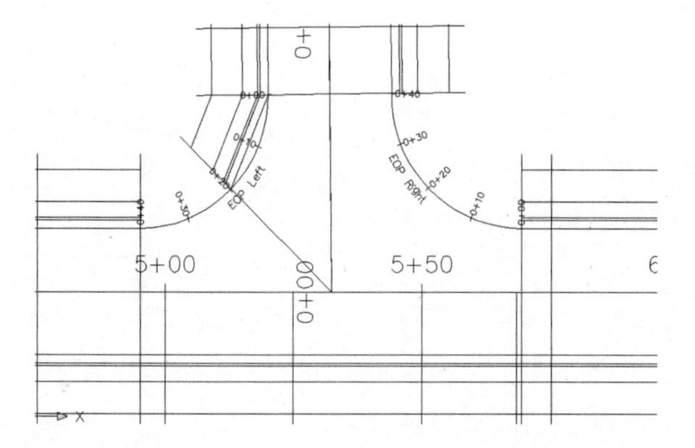

FIGURE 12.33
Your corridor after setting the targets

The target is set correctly for your first region, but the sampling frequency of the corridor is too far apart, resulting in a chunky appearance. You'll now adjust the sampling frequency:

1. Open the Corridor Properties dialog, and switch to the Parameters tab. Click the Frequency button for Baseline (3) Region (1).

2. In the Frequency to Apply to Assemblies dialog, change the Along Curves value to 2 and the At Profile High/Low Points value to Yes. Because your EOP alignments consist of a single curve, you can ignore the options for tangents and spirals. Click OK.

3. Click OK again to dismiss the Corridor Properties dialog and automatically rebuild your corridor. Your corridor should look like Figure 12.34.

FIGURE 12.34
Your corridor after changing the sampling frequency

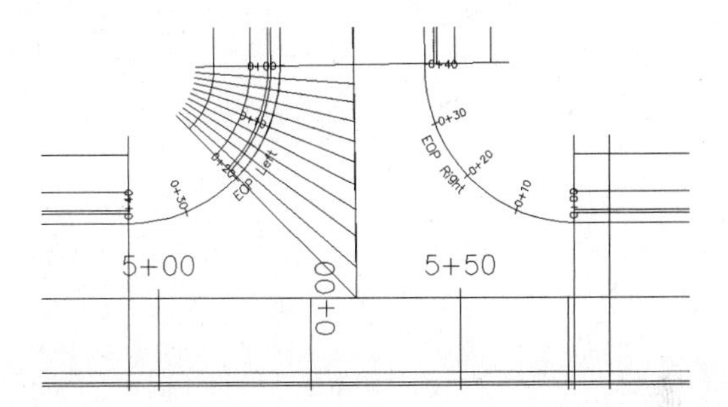

The first part of the EOP Left region is modeled properly. Now you can complete the left side of the corridor:

1. Open the Corridor Properties dialog, and switch to the Parameters tab. Click RG - Intersection Typical - (1) under BL - EOP Left - (1) and choose Insert Region – After ... Use the Intersection Typical Assembly. The new region automatically picks up from station 0+19.77 and continues to the end of the EOP Left alignment.

2. Click the ellipsis button in the Target column for this region. In the Target Mapping dialog, change the Transition Alignment to Project Road and the Transition Profile to Project Road FG. Click OK.

3. Click the Frequency button for this region. In the Frequency to Apply to Assemblies dialog, change the Along Curves value to 2 and the At Profile High/Low Points value to Yes. Click OK.

4. Click OK to dismiss the Corridor Properties dialog and automatically rebuild your corridor. Your corridor should now look similar to Figure 12.35.

FIGURE 12.35
Your corridor after mapping the targets

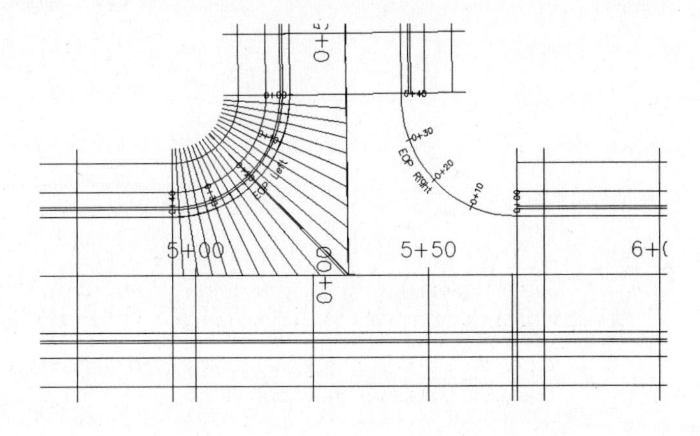

You must now repeat the process for the EOP Right baseline. To make it easier to work with your corridor, you can turn off baselines that you've finished building:

1. Open the Corridor Properties dialog again. Deselect the check boxes next to Baselines 1 through 3 on the Parameters tab, and click Apply.
2. Move the Corridor Properties dialog on your screen so that you can see the intersection alignments. Your corridor has temporarily disappeared because you've turned off all the baselines.
3. On the Parameters tab, click the Collapse All Categories button (see Figure 12.36) to "roll up" your baselines and regions and give you more room to work.

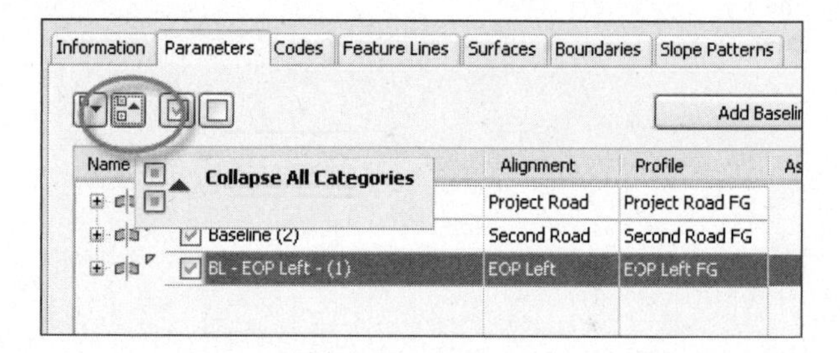

FIGURE 12.36
The Collapse All Categories button

4. Add a baseline and appropriate regions, targets, and frequencies. A subtle, but important, difference for this side of the intersection is that the EOP Right alignment stationing is reversed from the EOP Left, so you'll first target the Project Road. Also note that if you'd like to preview your work, you can click Apply and view the results without leaving the dialog. The right side of your intersection should look like Figure 12.37.

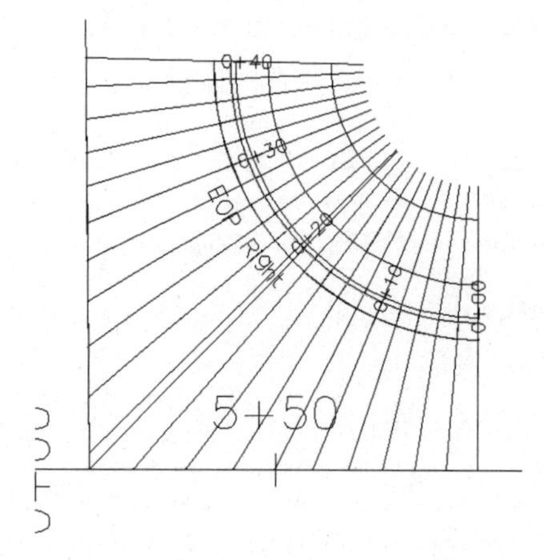

FIGURE 12.37
The right side of the intersection

5. On the Parameters tab, click the Turn On All the Baselines button to activate your corridor baselines (see Figure 12.38).

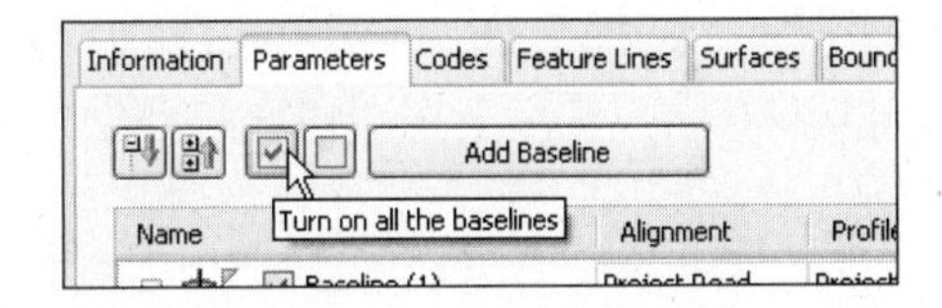

FIGURE 12.38
The Turn On All the Baselines button

6. Click OK to dismiss the dialog and automatically rebuild your corridor. Your corridor should now look like Figure 12.39

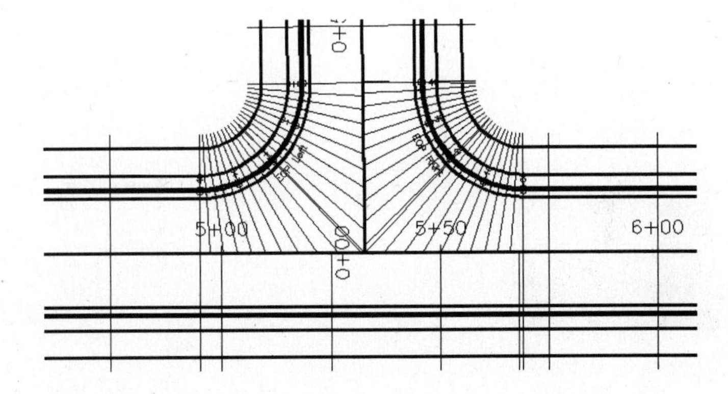

FIGURE 12.39
The properly modeled intersection

At this point, you have a properly modeled intersection, yet you haven't even begun the design. The next few sections will lead you through some techniques to assist you in refining the model and beginning the design process.

Troubleshooting Your Intersection

The best way to learn how to build advanced corridor components is to go ahead and build them, make mistakes, and try again. This section provides some guidelines on how to "read" your intersection to identify what steps you may have missed.

YOUR LANES APPEAR TO BE BACKWARD

Occasionally, you may find that your lanes wind up on the wrong side of the EOP alignment, as in Figure 12.40. The most common cause is that the direction of your alignment isn't compatible with your subassembly.

Fix this problem by reversing your alignment direction and rebuilding the corridor, or by editing your subassembly to swap the lane to the other side of the assembly. If you want to minimize the number of assemblies in your drawing, note which directions the intersection alignments should run to accommodate one intersection assembly.

In the exercise, you used an intersection assembly with the transition lane on the left side, so you made sure the left intersection EOP alignment ran clockwise and the right intersection EOP alignment ran counterclockwise. It's also easy to reverse an alignment and quickly rebuild the corridor if you catch the mistake after the corridor is built.

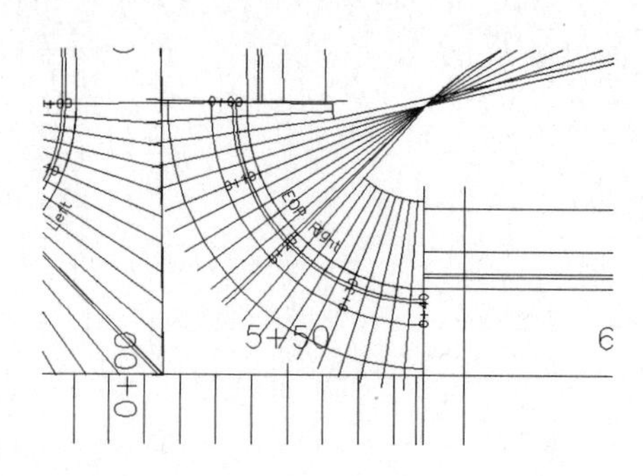

FIGURE 12.40
An intersection with the lanes modeled on the wrong side

YOUR INTERSECTION DROPS DOWN TO ZERO

A common problem when modeling corridors is the cliff effect, where a portion of your corridor drops down to zero. You probably won't notice in plan view, but if you rotate your corridor in 3D using the Object Viewer (see Figure 12.41), you'll see the problem. If your baseline profile isn't exactly as long as your baseline alignment, this will always occur.

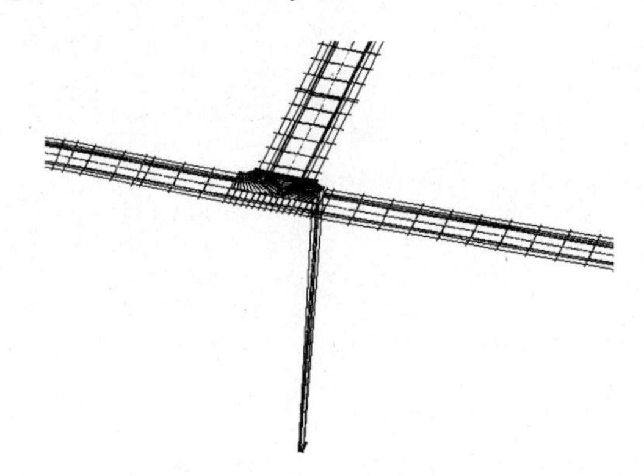

FIGURE 12.41
A corridor viewed in 3D, showing a drop down to zero

Fix this problem by making sure your baseline profile is exactly as long as your baseline alignment. The easiest way to ensure this is to use your osnaps to snap the proposed profile to the start points and endpoints of the existing ground (EG) profile and then use the profile grid view to refine the elevations. Alternatively, you can adjust your region limits to reflect the appropriate station range. You'll find that it's much more foolproof to make sure your baseline alignment and profile match exactly.

YOUR LANES EXTEND TOO FAR IN SOME DIRECTIONS

There are several variations on this problem, but they all appear similar to Figure 12.42. All or some of your lanes extend too far down a target alignment, or they may cross one another, and so on.

This occurs when the wrong target alignment and profile have been set for one or more regions. In the case of Figure 12.42, the first region of the EOP Left baseline was set to target the main road alignment instead of the secondary road alignment.

You can fix this problem by opening the Target Mapping dialog for the affected regions and confirming that the appropriate targets have been set. Once the regions are targeting appropriately, your corridor should look like Figure 12.43.

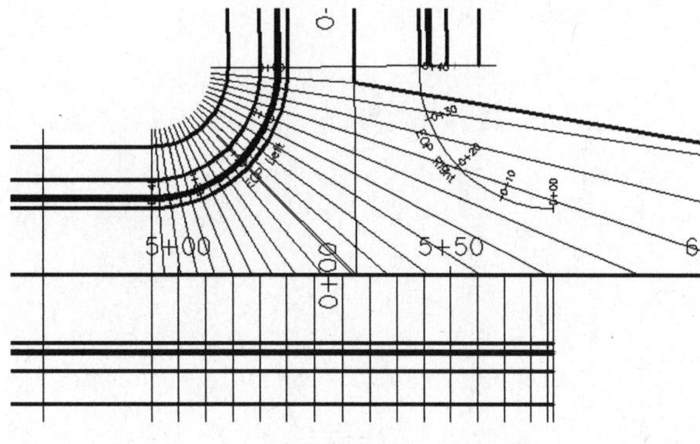

FIGURE 12.42
The intersection lanes extend too far down the main road alignment.

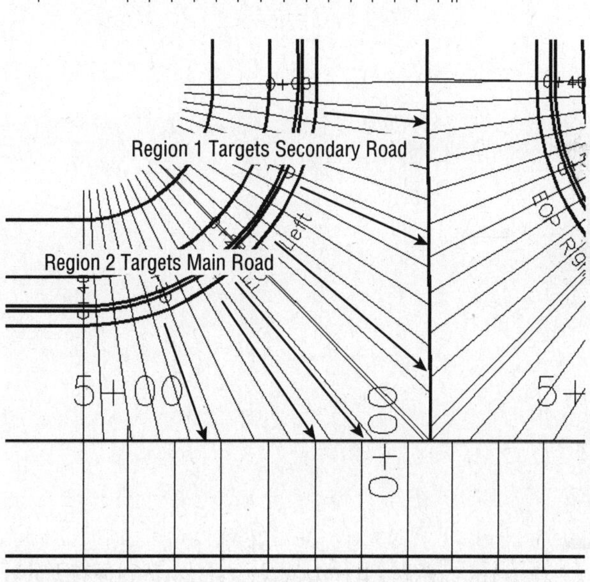

FIGURE 12.43
The intersection lanes have been repaired using the appropriate targets.

YOUR LANES DON'T EXTEND FAR ENOUGH

If your intersection or portions of your intersection look like Figure 12.44, you neglected to set the correct target alignment and profile.

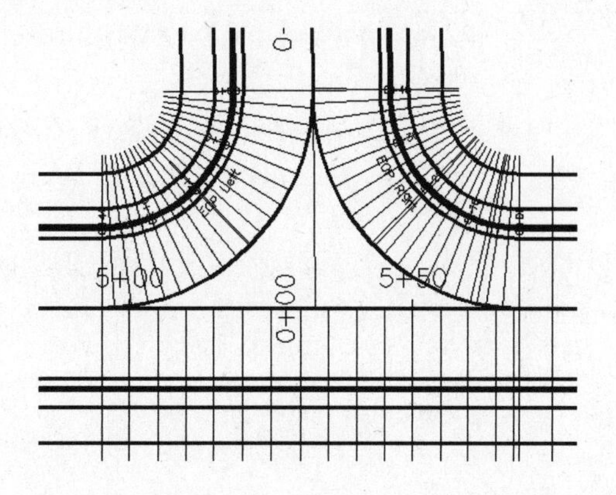

FIGURE 12.44
Intersection lanes don't extend out far enough.

You can fix this problem by opening the Target Mapping dialog for the appropriate regions and double-checking that you assigned targets to the right subassembly. It's also common to accidentally set the target for the wrong subassembly if you use Map All Targets or if you have poor naming conventions for your subassemblies.

OTHER TYPES OF INTERSECTIONS

The corridor model is a tool that can be adapted to almost any design situation. There are many different ways to design an intersection; through a different combination of alignments, profiles, assemblies, and targets, you can create a model that will assist in the design of any intersection.

Another common method for intersection design occurs when the crowned section of the main road is held through the intersection. This type of intersection can be modeled with the same baselines as the peer-road intersection from the exercise in this chapter. There are a few differences, however.

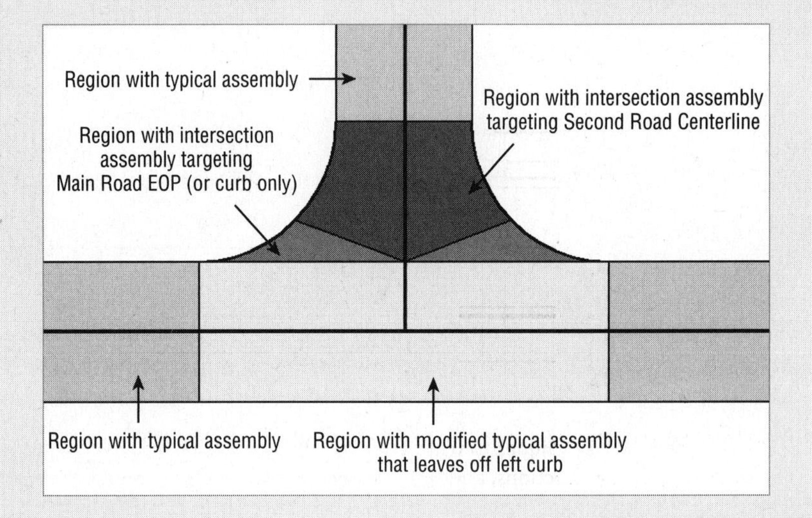

Instead of a right-lane-only assembly, you need an assembly that is fundamentally the typical road section minus the left curb, sidewalk, and so on, as in the following graphic.

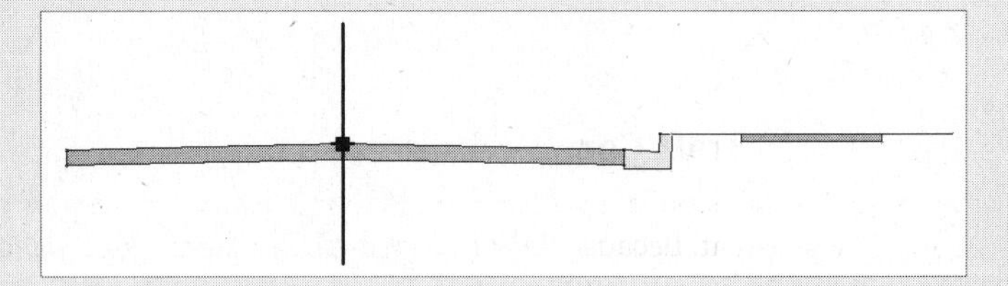

You can use the same intersection assembly for the EOP baselines as in the exercise, but the region that would normally target the Main Road centerline now targets an alignment and corresponding profile that follows the edge of the Main Road left lane. The following graphic shows the location of this target alignment:

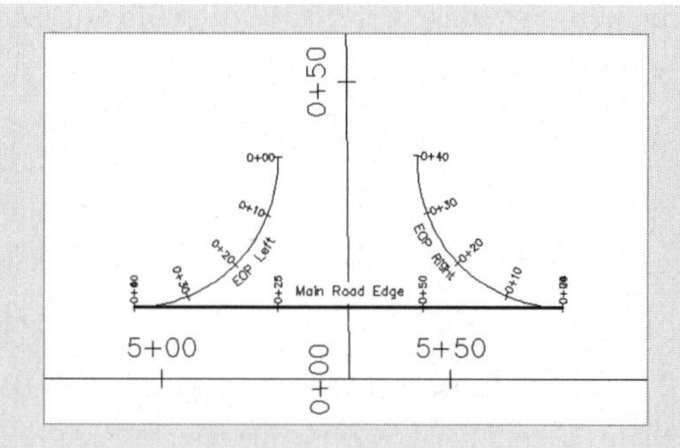

Once the region targets have been adjusted appropriately, the intersection corridor should look like this:

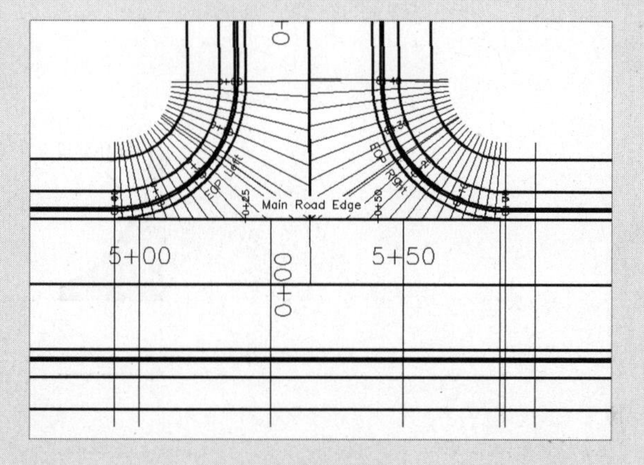

You can use similar techniques to model additional design elements such as left-turn lanes, widening, asymmetrical intersections, and interchanges. The best way to tackle an intersection is to sketch the intersection layout and figure out how the components are related. Map out baselines, regions, targets, and subassemblies. Once you have a plan for your intersection, build all the required pieces and use the corridor to join your model together.

Building a First-Draft Corridor Surface

Once the initial model is built, the time comes to make sure the elevations on the EOP profiles match your design intent. Because each end of the EOP alignment touches a portion of the normal cross section, it could be assumed that the elevations provided by a corridor surface at those locations are your design intent. This information may change as you iterate the design, but those elevations, whatever they may be, will drive the design of the EOP profiles.

One technique for determining the elevations at those locations is to build a first-draft corridor surface. This surface will be ugly, and it won't contain any good information within the intersection area, but it will return solid elevations throughout the normal portions of your corridor.

Before beginning this exercise, review the basic corridor surface building sections in Chapter 11. Follow these steps:

1. Open the `Corridor Peer Intersection Draft Surface.dwg` file (which you can download from `www.sybex.com/go/masteringcivil3d2010`), or continue working in your drawing from the previous exercise.

2. Open the Corridor Properties dialog, and switch to the Surfaces tab.
3. Click the Create a Corridor Surface button. A Corridor Surface entry appears.
4. Ensure that Links is selected as the Data Type and Top appears under Specify Code. Click the + button. An entry for Top appears under the Corridor Surface entry.
5. Click OK, and your corridor will automatically rebuild. You should have a corridor surface that appears similar to Figure 12.45.

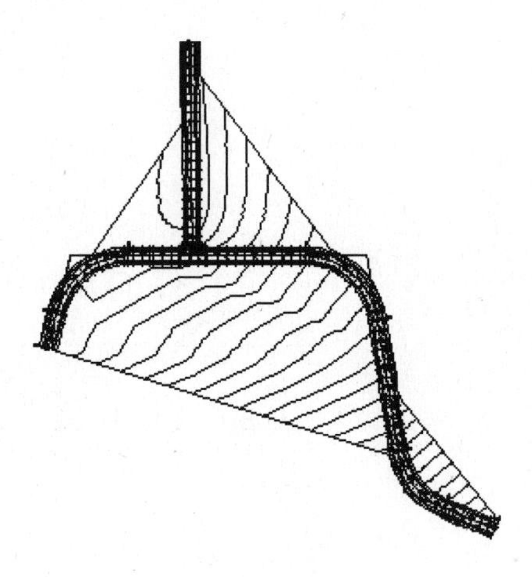

FIGURE 12.45
A first-draft corridor surface

6. Pick the surface, right-click, and choose Surface Properties.
7. On the Information tab of the Surface Properties dialog, change the Surface Style setting to No Display. This makes the surface invisible so you can use it for data but not be distracted by seeing it.

Perfecting Your Model to Optimize the Design

As stated in previous sections, your model can be built properly to reflect your desired trends and design constraints without having any real design information applied. Consider the example in Figure 12.46. The corridor has been built to properly link all the components together, but Civil 3D doesn't automatically assign elevations to those relationships. The program depends on you to assign appropriate elevations to perfect your model and begin design iterations.

Once the appropriate elevations have been assigned, your corridor is ready for design iterations (see Figure 12.47).

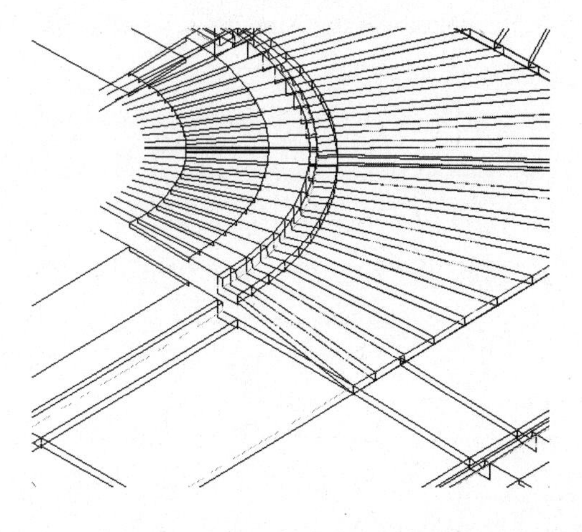

FIGURE 12.46
A properly built corridor showing a location where the model needs to be corrected

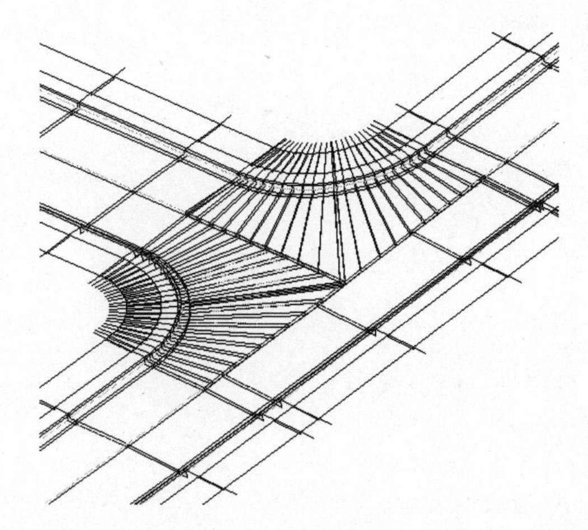

FIGURE 12.47
The same corridor after the model has been adjusted

This exercise will lead you through placing some design labels to assist in perfecting your model and then show you how to easily edit your EOP FG profiles to match the design intent on the basis of the draft corridor surface built in the previous section:

1. Open the `Corridor Peer Intersection 4.dwg` file (which you can download from `www.sybex.com/go/masteringcivil3d2010`), or continue working in your drawing from the previous exercise.

   If you continue working in your current drawing, you may want to change the corridor surface style to show contours from time to time. Also, you may find it easier to work if you freeze the C-ROAD-CORR layer. Thaw or freeze this layer as necessary to get a better view or to be able to osnap to your alignments more easily. If you've frozen your corridor, you can always access its properties by right-clicking the name of the corridor in

Prospector. You can also change the Corridor Surface style to No Display at any point during the exercise for the same reasons.

2. Open the Corridor Properties dialog, and switch to the Parameters tab.

3. Deselect the check boxes next to the two intersection baselines — BL - EOP Left - (1) and BL - EOP Right - (2). By building the corridor (and therefore corridor surface) without these baselines, you eliminate the possibility of accidentally grabbing a bogus elevation produced by one of these yet-undesigned baselines.

4. Click OK, and automatically rebuild your corridor. You should have a corridor that appears similar to Figure 12.48. Your model may also display contours. If desired, temporarily change the Surface Style to No Display to get a closer look at the corridor.

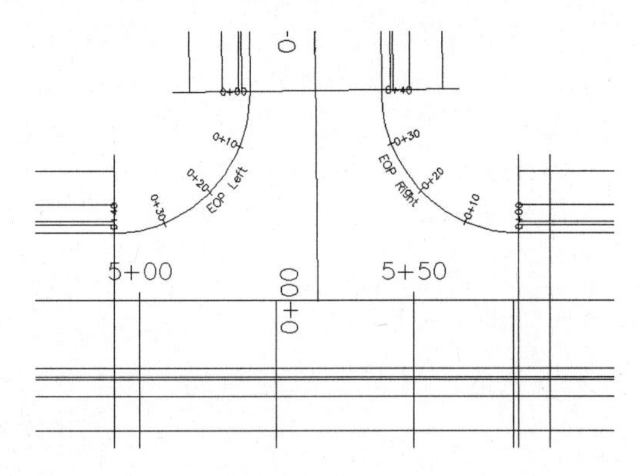

FIGURE 12.48
The corridor model with the intersection baselines turned off and the Surface Style set to No Display

Your rough surface also automatically rebuilds, ignoring the intersection, as shown in Figure 12.49.

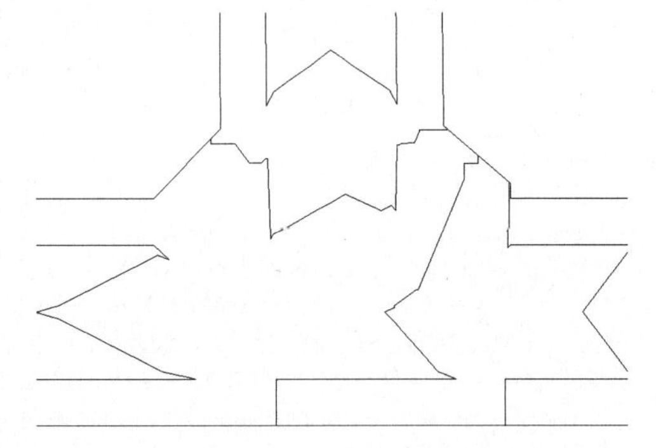

FIGURE 12.49
The corridor surface model with the intersection baselines turned off

5. Add a label by selecting the alignment and choosing Add Labels ➢ Add Alignment Labels from the Labels & Tables panel. The Add Labels dialog appears.

6. In the Alignment Label dialog, select Station Offset for the Label Type, Intersection Centerline Label for the Station Offset label style, and Basic X for the Marker Style. Click Add.

7. This label was composed to reference two alignments and two profiles. At the `Select Alignment` command-line prompt, pick the Project Road alignment.

8. At the `Specify Station` command-line prompt, use your Intersection osnap to choose the intersection of the two alignments.

9. At the `Specify Station Offset` command-line prompt, type **0** (zero).

10. At the `Select profile for label style component Profile` 1 command-line prompt, right-click to bring up a list of profiles, and choose Project Road FG. Click OK to dismiss the dialog.

11. At the `Select alignment for label style component Alignment` 2 command-line prompt, pick the Second Road alignment.

12. At the `Select profile for label style component Profile` 2 command-line prompt, right-click to bring up a list of profiles, and choose Second Road FG. Click OK to dismiss the dialog.

13. Press ↵, and then click Close to dismiss the Alignment Label dialog. Thaw the C-ROAD-CORR layer if it's frozen.

14. Pick the label, and use the square-shaped grip to drag the label somewhere out of the way. Your label should look like Figure 12.50.

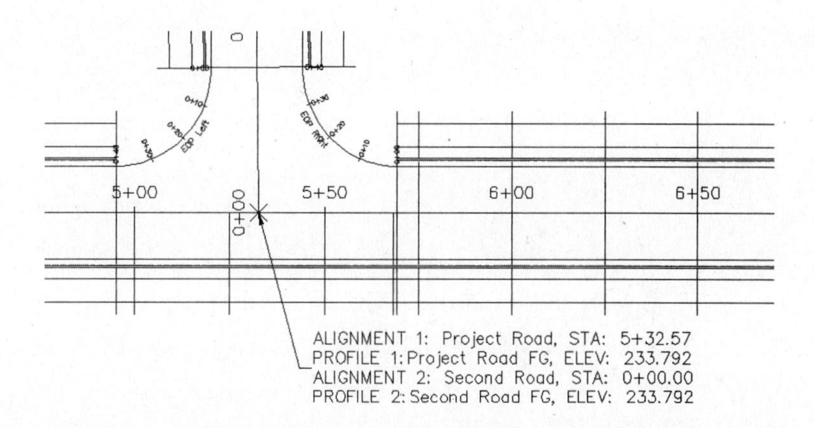

FIGURE 12.50
The intersection center-line design label

Note that the elevations of the Project Road FG and Second Road FG are both equal to 233.792. If the label indicated that these elevations weren't the same, you would need to edit your profile geometry to make sure they matched. This label will be useful during later stages of the design process. As you iterate your FG profiles, this label will alert you to any adjustments that are required to make the profiles match at their intersection.

The next part of the exercise guides you through the process of adding a label to help determine what elevations should be assigned to the Start and End Stations of the EOP Right and EOP Left alignments:

1. Pick the Project Road alignment and choose Add Labels ➤ Add Alignment Labels from the Labels & Tables panel. The Add Labels dialog appears.

2. In the Alignment Label dialog, select Station Offset for the Label Type, Intersection EOP Label for the Station Offset label style, and Basic X for the Marker Style. Click Add.

3. These labels have been composed to be a label of the road centerline that references a surface elevation. At the `Select Alignment` command-line prompt, pick the Project Road alignment.
4. At the `Specify Station` command-line prompt, use your Endpoint osnap to pick the Start Station of the EOP Right alignment.
5. At the `Specify Station Offset` command-line prompt, use your Endpoint osnap to pick the Start Station of the EOP Right alignment.
6. At the `Select surface for label style component Corridor Surface` command-line prompt, right-click to bring up a list of surfaces, and select Corridor Surface.
7. Press ↵, and then click Close to dismiss the Alignment Label dialog.
8. Pick the label, and use the square-shaped grip to drag the label somewhere that is out of the way. Your label should look like Figure 12.51.

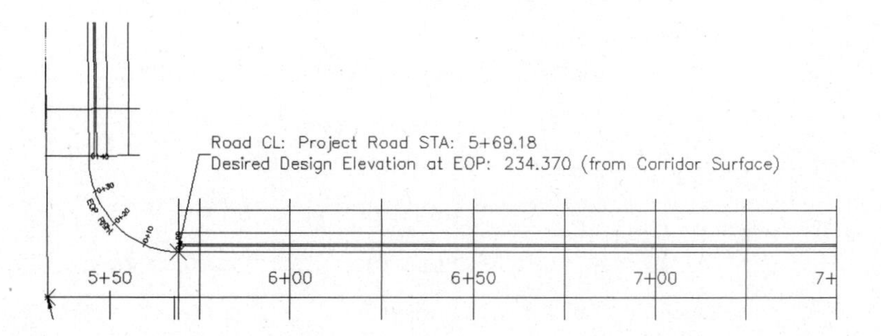

FIGURE 12.51
The intersection centerline design label

9. Repeat the previous steps to provide labels for the End Station of the EOP Right alignment and the Start and End Stations of the EOP Left alignment. Note that you'll have to exit the command when you need to switch the reference road centerline.
10. Once all the labels have been placed, your corridor should look like Figure 12.52.

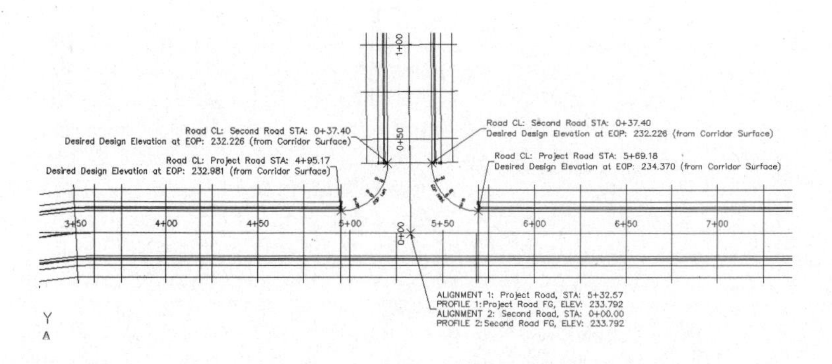

FIGURE 12.52
The corridor with all labels placed

Now that you've added labels to the Start and End Stations of the EOP Right and EOP Left alignments, you can edit the geometry of the EOP Right and Left FG profiles and then view the contours from the corridor surface:

1. Change to the View tab and choose Viewport Configurations List ➤ Two: Vertical from the Viewports tab to split your screen into two viewports so that you can easily see your EOP alignments and labels at the same time as the corresponding profiles. Press ↵ at the command line to specify a vertical split. Your screen should look similar to Figure 12.53.

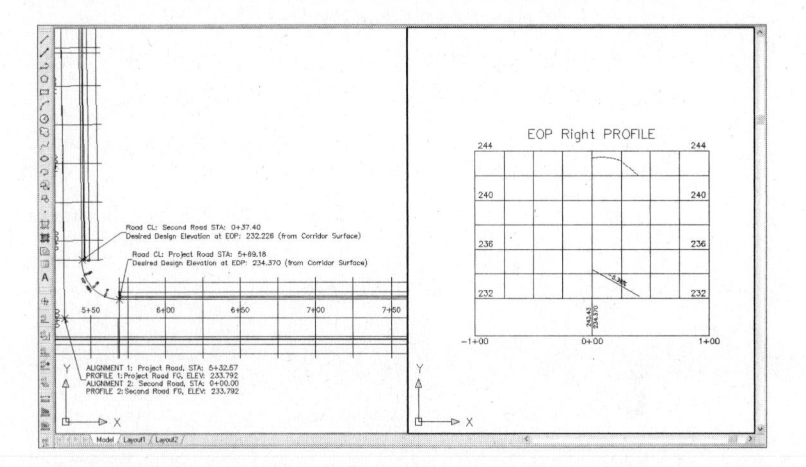

FIGURE 12.53
Use a split screen to see the plan and profile simultaneously.

2. Pick the EOP Right FG profile. Right-click, and choose Edit Profile Geometry.
3. On the Profile Layout Tools toolbar, click Profile Grid View.
4. In the Profile Entities palette in Panorama, change the PVI elevations to match the values from the plan labels, as shown in Figure 12.54.

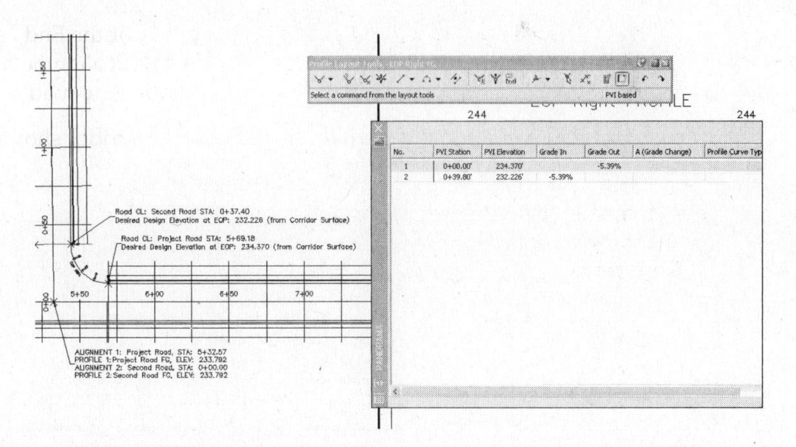

FIGURE 12.54
Edit the PVI elevations to match the desired EOP elevations.

5. Dismiss Panorama. Repeat the previous steps for the EOP Left FG profile.
6. Open the Corridor Properties dialog, and switch to the Parameters tab. Turn on all baselines. Click OK, and your corridor will automatically rebuild.
7. Pick your corridor. Right-click, and choose Object Viewer.

8. Navigate through Object Viewer to confirm that your corridor model is now appropriately tied together at the EOP. Note that any changes in the centerline won't automatically update your EOP alignments. Your corridor should look like Figure 12.55.

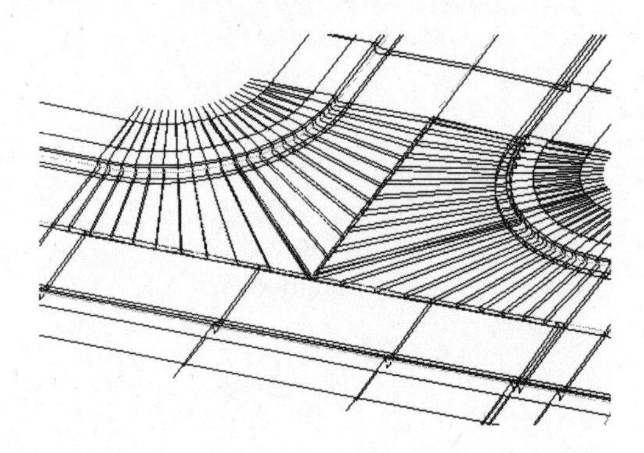

FIGURE 12.55
The properly modeled intersection

9. As you continue iterating your design and making edits to the centerline profiles, rebuild the corridor and use your updated labels to refine the desired elevations on the EOP profiles.
10. Exit Object Viewer.
11. Select your corridor surface under the Surfaces branch in Prospector. Right-click, and choose Surface Properties. The Surface Properties dialog opens. Change the Surface Style to Contours 1′ and 5′ (Design). Click OK to dismiss the dialog.
12. Pick the corridor surface, and use Object Viewer to study the TIN in the intersection area.

Study the contours in the intersection area. Even though the contours may not necessarily be optimally designed at this point, you should have decent contours within the intersection area, with no pits or holes in the corridor surface. Ignore the contours outside the corridor limits.

Refining a Corridor Surface

Once your model makes more sense, you can build a better corridor surface. Don't be confused into thinking this is your final design. You can continue to edit, refine, and optimize your corridor as you gain new information. From now on, however, it will take only a few minutes to edit the corridor and see the reaction of the corridor surface.

In this exercise, you'll use links as breaklines and add feature lines. Be sure to review the section on surface building from corridors in Chapter 11, which mentioned them.

When you're building a surface from links, you have the option of selecting a check box in the Add as Breakline column. Doing so adds the actual link lines as additional breaklines to the surface. In most cases, especially intersection design, selecting this check box forces better triangulation.

In the next exercise, you'll add a few meaningful corridor feature lines to the surface to force triangulation along important features like edge of travel way and top of curb.

The exercise also gives you hands-on experience in adding an interactive boundary to your corridor. In Chapter 11, you were able to use an automatic boundary. However, you no longer have the automatic boundary option for multiple baseline corridors. The Interactive Boundary tool provides an interface that allows you to choose a bounding feature line around your corridor.

While 2010 adds the ability to the corridor extents as the outer boundary, it's good to know the manual methods just in case.

The more appropriate data you add to the surface definition, the better your surface (and therefore your contours) will look — right from the beginning, which means fewer edits and less temptation to grade by hand. Follow these steps:

1. Open the `Corridor Peer Intersection 5.dwg` file (which you can download from `www.sybex.com/go/masteringcivil3d2010`), or continue working in your drawing from the previous exercise.

2. Pick your corridor surface in the drawing. Right-click and choose Surface Properties. The Surface Properties dialog opens. Change Surface Style of Corridor - 5(5) to No Display so you don't accidentally pick it when choosing a corridor boundary. Click OK to dismiss the dialog.

3. Select the corridor, right-click, and choose Corridor Properties. Switch to the Surfaces tab. In the row for the Top, under Corridor Surface, select the check box in the Add as Breakline column.

4. Select Feature Lines from the drop-down menu in the Data Type selection box.

5. Use the drop-down menu in the Specify Code selection box and the + button to add the Back_Curb, Crown, ETW, Flange, Flowline_Gutter, and Top_Curb feature lines to the Corridor Surface, as shown in Figure 12.56.

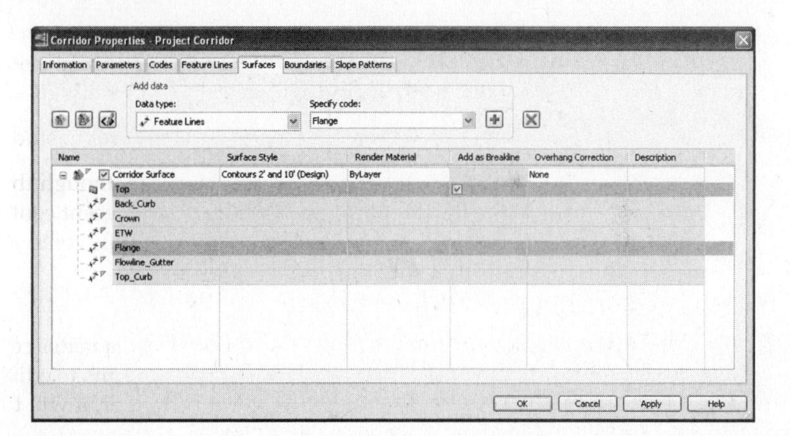

FIGURE 12.56
Adding feature lines to the corridor surface

6. Switch to the Boundaries tab.

7. Right-click the corridor surface entry, and select Add Interactively.

8. Zoom down to the Start Station of Project Road. The following command-line prompt appears: `To define boundary, select the first point on a corridor feature line.` Use your Endpoint osnap to pick the leftmost feature line on the corridor.

9. The command line prompts you to `Select next point on this feature line or click on another feature line or [Undo/Close]:`.

10. Move your mouse, and notice that a red jig follows your cursor along the chosen feature line. It continues to follow you until the end of a region.

11. Once you reach the next region, pick the leftmost feature line in that region. Continue the process around the entire corridor. As you progress, the jig follows your cursor and picks, as shown in Figure 12.57.

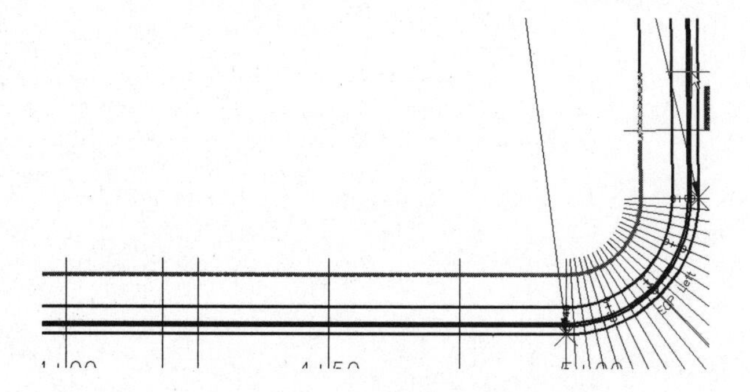

FIGURE 12.57
The corridor boundary jig

12. When you come back to the Start Station of Project Road, type **C** to close the boundary.
13. The Boundaries tab of the Corridor Properties dialog returns.
14. Click OK to dismiss the dialog, and your corridor will automatically rebuild, along with your corridor surface.
15. Select your corridor surface under the Surfaces branch in Prospector. Right-click and choose Surface Properties. The Surface Properties dialog opens. Change the Surface Style to Contours 1′ And 5′ (Design), and click OK. Click OK to dismiss the Surface Properties dialog.
16. Your surface is now limited to the area inside the interactive boundary. Use Object Viewer to examine your surface TIN. You may see some improvement in the triangulation because you added the corridor links and feature lines as breaklines. In plan view, your surface contours should look like Figure 12.58.

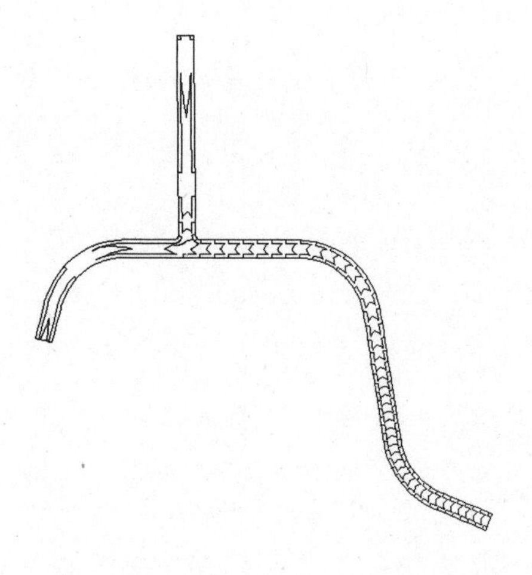

FIGURE 12.58
The finished corridor surface contours

In isometric view, your surface TIN at the intersection should appear as shown in Figure 12.59.

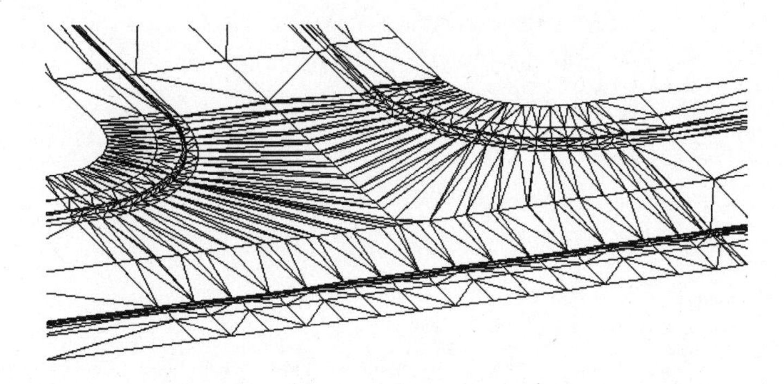

FIGURE 12.59
The surface TIN viewed in 3D

You can make additional, optional edits to the TIN if you're still unhappy with your road surface. Edits that may prove useful include increasing the corridor frequency in select regions, adjusting the Start and End stations in a region, or using surface-editing commands to swap edges or delete points.

Real World Scenario

TAKE ADVANTAGE OF YOUR TIN

After competing the corridor surface-creation section in Chapter 11 and the corridor surface refinements in this chapter, you should have a good understanding of how a corridor surface is built. There are many ways you can take advantage of surface creation that let the TIN fill in the blanks for you and reduce the number of target alignments and profiles you must create and maintain.

A previous section mentioned a method for creating an intersection where the crown of the main road is held through the intersection. That method requires the creation and maintenance of a fifth alignment and profile that follows the main road EOP. In this example, you can use a curb-only assembly, similar to the following graphic, in the region that would normally target the edge-of-lane alignment:

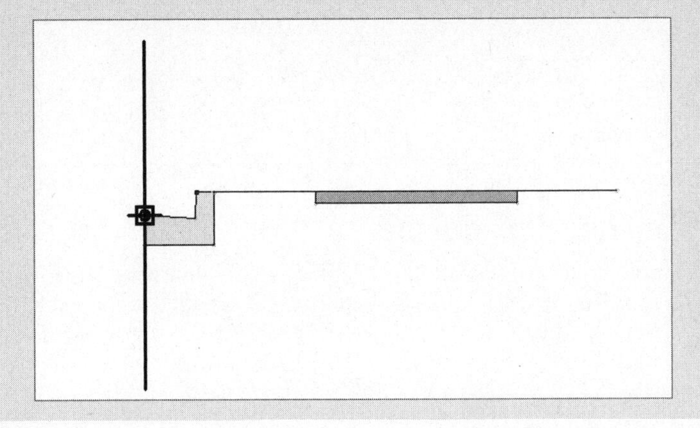

The resulting corridor doesn't have any links in this region, as in this graphic, which may seem strange until you build the corridor surface:

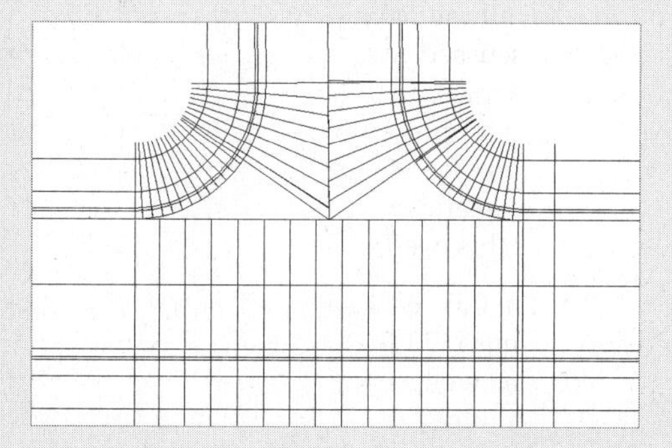

The resulting corridor surface uses the principles explained in the section "Creating a Corridor Surface" in Chapter 11 to create the TIN, and therefore corridor points are connected to complete the lanes with TIN lines. The next image clearly shows the triangulation along the crowned section and no gaps anywhere in the intersection:

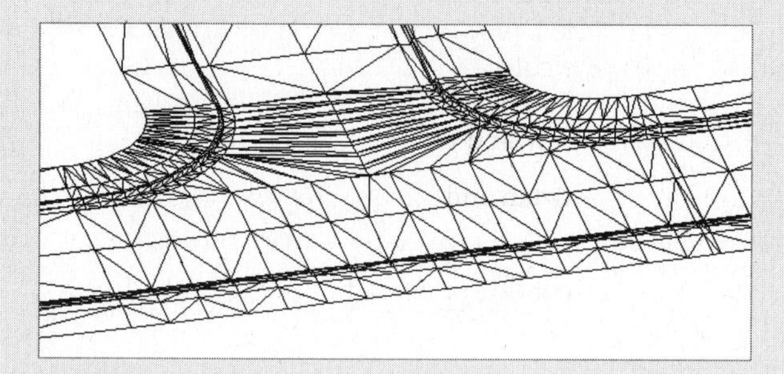

The resulting TIN occasionally needs more edge swapping than a corridor surface modeled with a complete set of links. Also keep in mind that because there are no links through part of the intersection, code-set styles designed for hatching and rendering may not work through the linkless region, and corridor-based material quantities may not be valid.

Also consider combining extracted dynamic feature lines with grading objects to enhance your surfaces.

All things considered, this technique can be a major timesaver when you're modeling intersections.

Modeling a Cul-de-sac

Another common corridor design roadblock is the cul-de-sac. Although cul-de-sacs come in all shapes and sizes, you can apply and adapt the principles explained in this section in many design scenarios, including off-center cul-de-sacs, asymmetrical cul-de-sacs, knuckles, turnarounds, and other designs.

Adding a Baseline, Region, and Targets for the Cul-de-sac

As you would with an intersection, you should sit down and plan your cul-de-sac before beginning the model. How does your cul-de-sac look? What is driving the elevations? Is there a crown in the cul-de-sac, or does all the pavement grade to one side? Answer these questions, and draw a quick sketch. Then, as discussed earlier in the section "Modeling a Peer-Road Intersection," plan your baselines, regions, targets, and subassemblies.

For this example, you'll apply an assembly to a single baseline that traces the cul-de-sac EOP all the way around the bulb. It's strongly recommended that you work through the intersection exercise before doing this exercise. Many of the techniques are identical and therefore won't be explained in the steps for this exercise:

1. Open the `Corridor Cul-de-sac.dwg` file, which you can download from `www.sybex.com/go/masteringcivil3d2010`. It includes an alignment that follows the EOP for a 38′ radius cul-de-sac, as well as a corresponding profile. It also includes a corridor, a corridor surface, a few assemblies, and some Intersection EOP labels.
2. Change to the View tab and choose Viewport Configurations List ➢ Two: Vertical from the Viewports panel. Use your Pan and Zoom commands so that you see the cul-de-sac plan on one side of your screen and the cul-de-sac EOP profile view on the other side.
3. Pick the cul-de-sac EOP FG profile, and select Geometry Editor from the Modify Profile panel. In the Profile Layout Tools toolbar, select the Profile Grid tool. Confirm the Start and End Stations of the cul-de-sac EOP FG profile match the desired elevations as listed in the plan-view design labels. If the elevations don't match, make the necessary adjustments. Dismiss the grid view and the profile toolbar.
4. Change to the View tab and choose Viewport Configurations List ➢ Single from the Viewports panel.
5. Zoom over to the cul-de-sac in plan view. Open the Corridor Properties dialog, and switch to the Parameters tab.
6. Click Add Baseline. Select the Cul-de-sac EOP as the baseline alignment in the Pick Horizontal Alignment dialog. Click OK.
7. Click in the Profile field. Select Cul-de-sac EOP FG as the baseline profile in the Select a Profile dialog. Click OK.
8. Right-click the baseline you just created, and select Add Region. Select the Intersection Typical Assembly in the Pick an Assembly dialog. Click OK.
9. Expand your baseline, and see the new region you just created.
10. Click the Frequency button for your region. In the Frequency to Apply to Assemblies dialog, set all the frequency intervals to 5′, and change At Profile High/Low Points to Yes. Click OK.
11. Click the Target button for your region. In the Target Mapping dialog, set Transition Alignment to Project Road and Transition Profile to Project Road FG. Click OK.

12. Click OK again to dismiss the Corridor Properties dialog and automatically rebuild your corridor, along with your corridor surface. Your corridor should look similar to Figure 12.60.

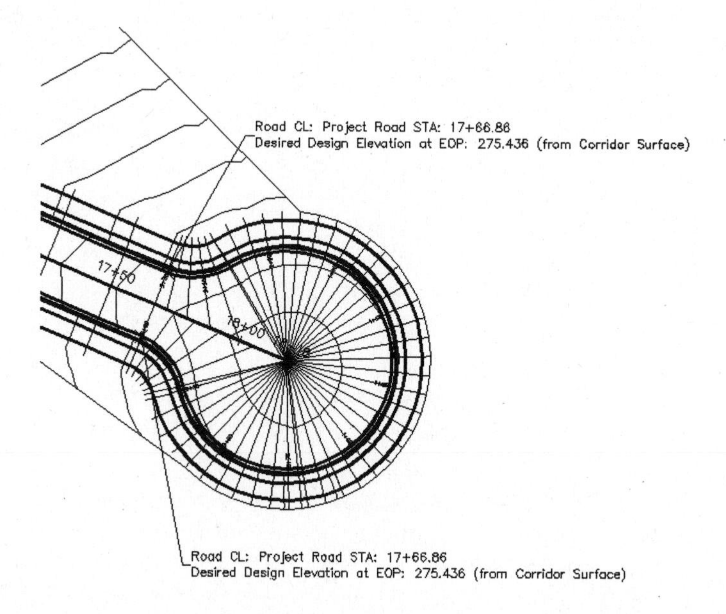

FIGURE 12.60
The modeled cul-de-sac

13. Add a boundary to your corridor, as detailed in the previous section. View your surface in Object Viewer or in a 3D view to check for triangulation problems, and study the surface result. The corridor should look similar to Figure 12.61a in Object Viewer, whereas the corridor-surface contours and the corridor-surface TIN should look similar to Figures 12.61b and 12.61c, respectively.

Additionally, you may want to improve the quality of your corridor surface using some of the techniques described in the "Refining a Corridor Surface" section.

Troubleshooting Your Cul-de-sac

People make several common mistakes when modeling their first few cul-de-sacs.

YOUR CUL-DE-SAC APPEARS WITH A LARGE GAP IN THE CENTER

If your curb line seems to be modeling correctly but your lanes are leaving a large empty area in the middle (see Figure 12.62), chances are pretty good that you neglected to assign targets or perhaps assigned the incorrect targets.

Fix this problem by opening the Target Mapping dialog for your region and checking to make sure you assigned the road centerline alignment and FG profile for your transition lane. If you have a more advanced lane subassembly, then you may have accidentally set the targets for another subassembly somewhere in your corridor instead of the lane for the cul-de-sac transition, especially if you have poor subassembly-naming conventions. Poor naming conventions become especially confusing if you used the Map All Targets button.

FIGURE 12.61
(a) The corridor model viewed in 3D, (b) the resulting corridor-surface contours, and (c) the resulting corridor-surface TIN

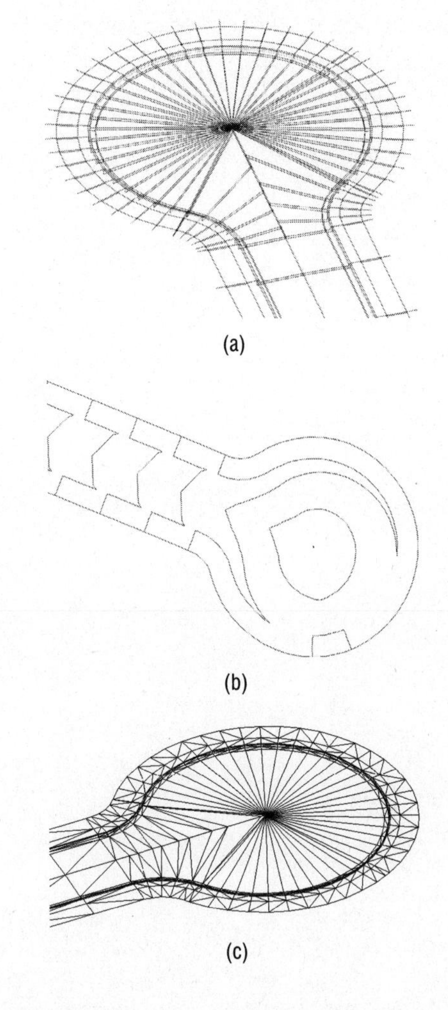

FIGURE 12.62
A cul-de-sac without targets

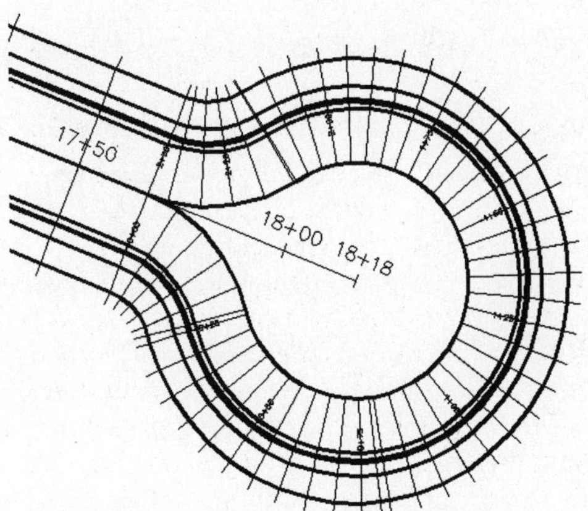

YOUR CUL-DE-SAC APPEARS TO BE BACKWARD

Occasionally, you may find that your lanes wind up on the wrong side of the EOP alignment, as shown in Figure 12.63. The most common cause is that the direction of your alignment isn't compatible with your subassembly.

FIGURE 12.63
A cul-de-sac with the lanes modeled on the wrong side

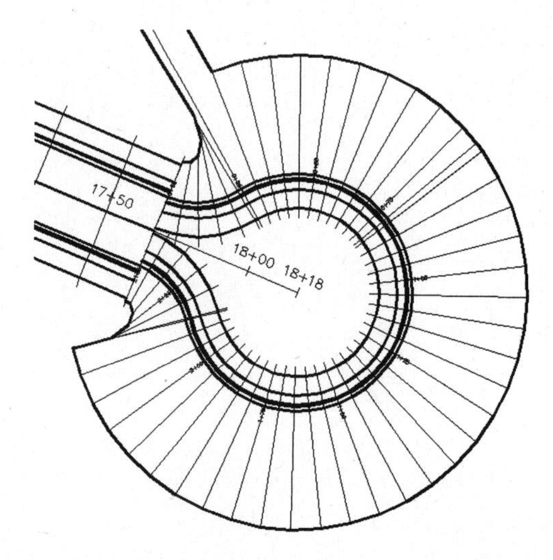

You can fix this problem by reversing your alignment direction and rebuilding the corridor or editing your subassembly to swap the lane to the other side of the assembly.

YOUR CUL-DE-SAC DROPS DOWN TO ZERO

A common problem when you first begin modeling cul-de-sacs, intersections, and other corridor components is that one end of your baseline drops down to zero. You probably won't notice the problem in plan view; but once you build your surface (see Figure 12.64a) or rotate your corridor in 3D (see Figure 12.64b), you'll see it. This problem will always occur if your baseline profile isn't the same length as your baseline alignment.

FIGURE 12.64
(a) Contours indicating that the corridor surface drops down to zero, and (b) a corridor viewed in 3D showing a drop down to zero

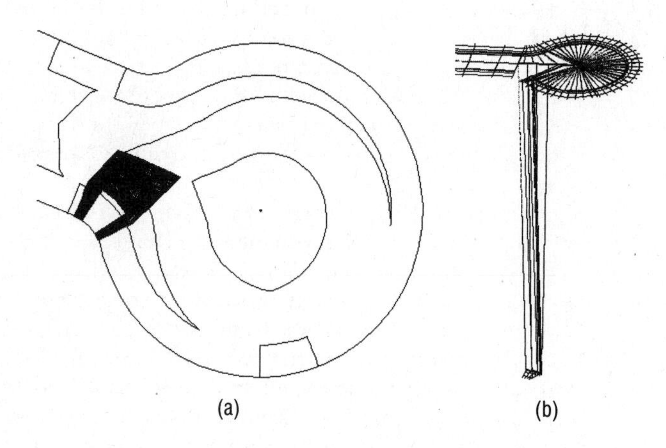

You can fix this problem by making sure your baseline profile is the same length as your baseline alignment. See the section "Troubleshooting Your Intersection" for more tips and information on fixing this problem.

YOUR CUL-DE-SAC SEEMS FLAT

When you're first learning the concept of targets, it's easy to mix up baseline alignments and target alignments. In the beginning, you may accidentally choose your EOP alignment as a target instead of the road centerline. If this happens, your cul-de-sac will look similar to Figure 12.65.

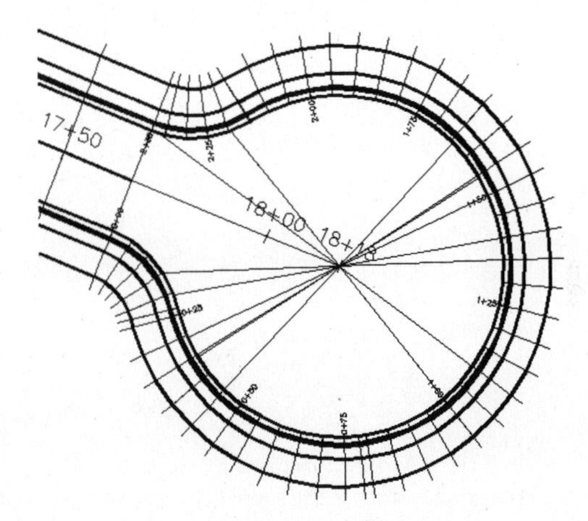

FIGURE 12.65
A cul-de-sac with the wrong targets set

You can fix this problem by opening the Target Mapping dialog for this region and making sure the target alignment is set to the road centerline and the target profile is set to the road centerline FG profile.

Modeling a Widening with an Assembly Offset

As you continue to improve your corridor-building skills, you'll want to investigate increasingly more advanced methods for refining your model to better meet your design intent.

In Chapter 11, you did a road-widening example with a simple lane transition. Earlier in this chapter, you worked with a roadside-ditch transition, intersections, and cul-de-sacs. These are just a few of the techniques for adjusting your corridor to accommodate a widening, narrowing, interchange, or similar circumstances. There is no one method for how to build a corridor model; every method discussed so far can be combined in a variety of ways to build a model that reflects your design intent.

Another tool in your corridor-building arsenal is the assembly offset. In the transition lane and transition ditch examples, you may have noticed that every corridor link is modeled perpendicular to the baseline, as shown in Figure 12.66.

You can use the assembly offset when the elements outside your transition area are better modeled perpendicular to the transition alignment, as shown in Figure 12.67.

Typical examples of when you'll use an assembly offset include transitioning ditches, widening roads, traffic-calming lanes, interchanges, and other applications. The assembly in Figure 12.68, for example, includes two assembly offsets. The assembly could be used for transitioning roadside swales, similar to the first exercise in this chapter.

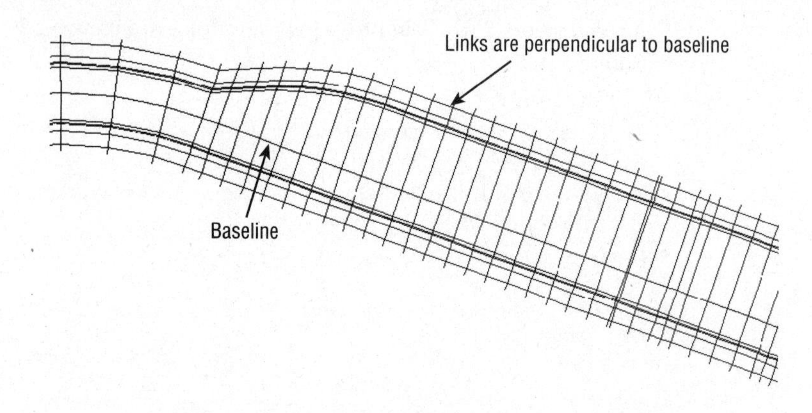

FIGURE 12.66
Modeling a road widening

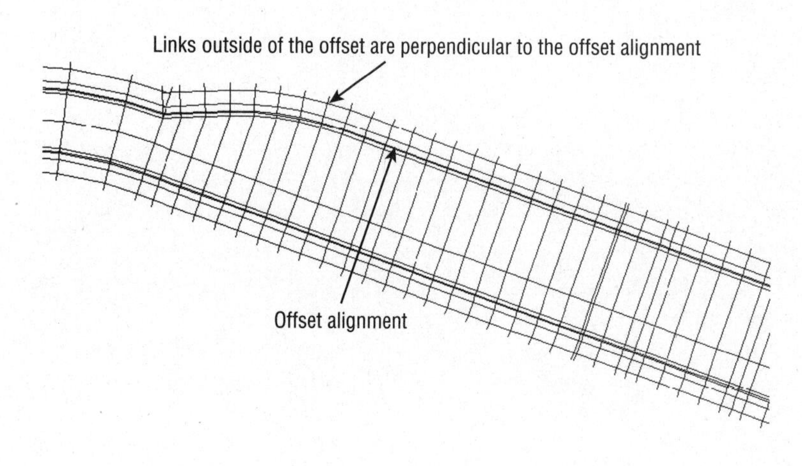

FIGURE 12.67
Modeling a road widening with an assembly offset

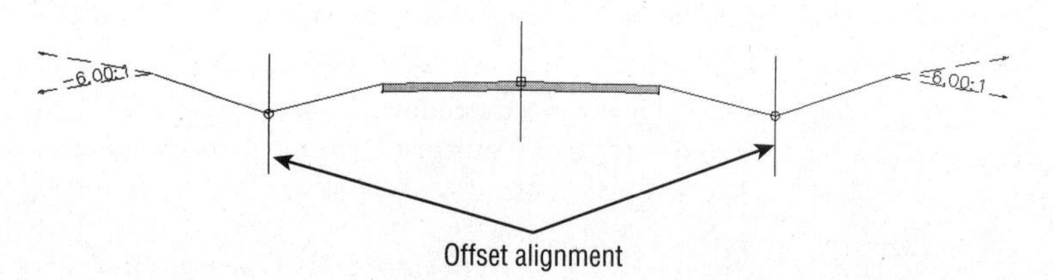

FIGURE 12.68
An assembly with two offsets representing roadside swale centerlines

In addition to adapting your assembly to include offsets, each offset requires a dedicated alignment and profile.

Even though each offset requires its own profile, it isn't always necessary to use Profile ➤ Profile Creation Tools from the Create Design panel and design a profile from scratch. Use some creativity to figure out additional methods to achieve your design intent. Extracting a profile from a corridor, sampling a profile from a first-draft surface, importing a profile from another source, copying a main-road profile and moving it to different elevation, and superimposing profiles are all valid methods for creating profiles for targeting and assembly offsets.

In this exercise, you'll improve the basic road-widening model from Chapter 11 with an assembly offset:

1. Open the `Assembly Offset Assembly Corridor.dwg` file, which you can download from `www.sybex.com/go/masteringcivil3d2010`. This is the same model you created in Chapter 11 using a simple transition with a target alignment. In addition to the road model, there is a profile view of the Widening EOP alignment.
2. Zoom to the area of the drawing where the assemblies are located. You'll see an incomplete assembly called Project Road Offset Assembly.
3. Change to the Home tab and choose Assembly ➤ Add Assembly Offset from the Create Design panel.
4. At the `Select an assembly <or press enter key to select from list>` command-line prompt, pick the Project Road Offset Assembly.
5. At the `Specify offset location` command-line prompt, pick the far-right point of the right lane on the Project Road Offset Assembly. Your result should look like Figure 12.69.

FIGURE 12.69
The Project Road Offset assembly

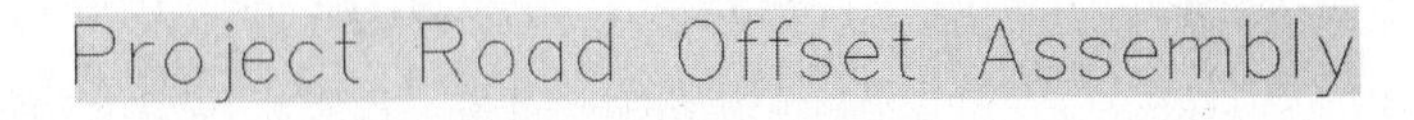

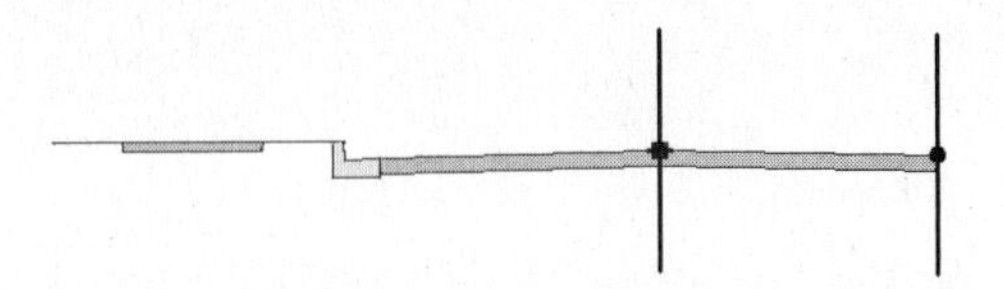

6. Use the AutoCAD Copy command to copy the Right Curb and Sidewalk subassemblies from one of the other assemblies. Place them near your Project Road Offset Assembly.
7. Pick the curb. Right-click, and choose Add to Assembly; then, pick the assembly offset you placed on the Project Road Offset assembly.
8. Repeat step 7 with the Sidewalk subassembly, except attach the sidewalk to the appropriate place on the curb. Your result should look like Figure 12.70.

FIGURE 12.70
The Project Road Offset assembly with the Right Curb and Sidewalk subassemblies

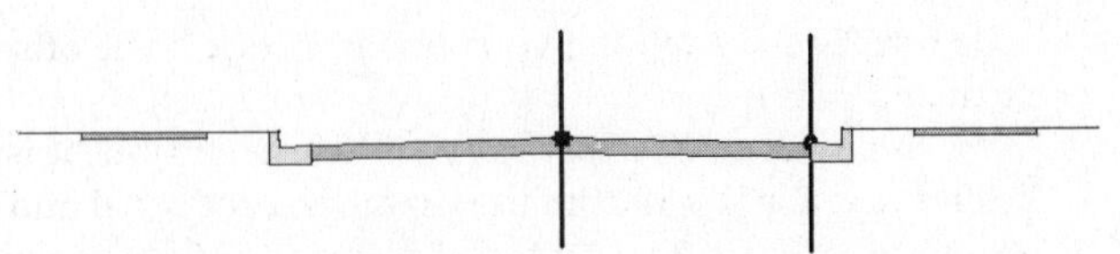

Now that the assembly is built, you need a profile representing EOP before you can adjust the corridor model. You have several options for creating this profile, such as using Profiles ➤ Create by Layout or any other valid profile-creation technique. In this case, the corridor surface in this drawing can provide a valid profile for your offset:

1. Change to the Home tab and choose Profile ➤ Create Surface Profile from the Create Design panel. In the Create Profile from Surface dialog, select Widening EOP Alignment and Project Road Corridor Surface. Click Add to sample this surface.
2. In the Profile list, change the Update Mode of the profile to Static, and change Style to Right Sample Profile. Click OK to dismiss the dialog. A surface profile should automatically appear in the Widening EOP profile view. If the Panorama window appears to notify you that a profile has been created, click the green check mark to dismiss it.
3. Now you must swap the more basic transition-lane assembly for the more robust offset assembly you created earlier in the exercise. Open the Corridor Properties dialog, and select the Project Road Corridor on the Parameters tab.
4. Change the Assembly for Region (2) from Project Road Transition Lane to Project Road Offset Assembly.
5. Expand Region (2) to list the offsets. Select Widening EOP under Alignment and Project Road Corridor Surface Profile under Profile.
6. Open the Target Mapping dialog for this region, and confirm that the transition alignment for the right lane is still set to Widening EOP. Set this value if necessary. Click OK to close the dialog.
7. Click OK again to dismiss the Corridor Properties dialog and to automatically rebuild your corridor, along with your corridor surface.

If you examine the corridor model and the resulting surface, you'll notice that there is a subtle, but potentially important, difference in the curb and sidewalk. In the previous model, the curb and sidewalk links were placed perpendicular to the baseline, which in this case was the Project Road centerline. In the offset assembly–based model, the links are perpendicular to the EOP alignment, which is often closer to your design intent. If you zoom in on the curb and sidewalk area, your corridor should look similar to Figure 12.71.

FIGURE 12.71
A close-up view of the completed corridor

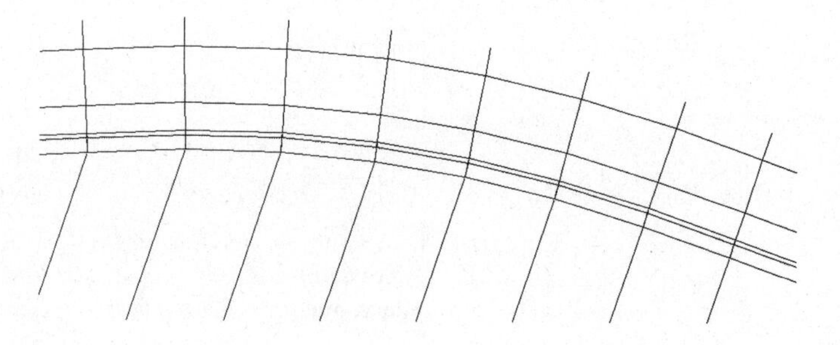

The Trouble with Bowties

In your adventures with corridors, chances are pretty good that you'll create an overlapping link or two. These overlapping links are known affectionately as *bowties*. A mild example can be seen in the following graphic:

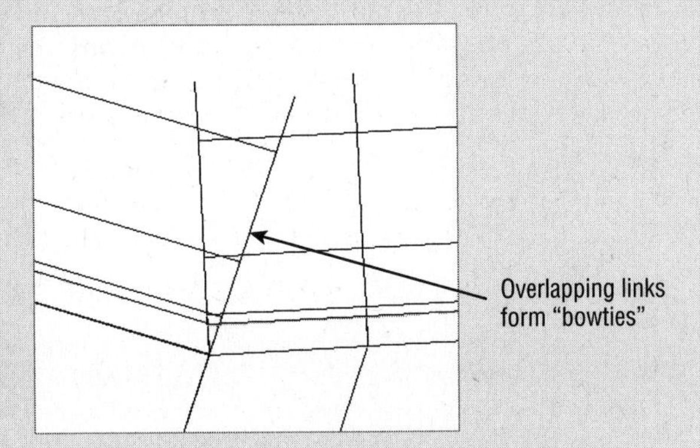

An even more pronounced example can be seen in the following river corridor image:

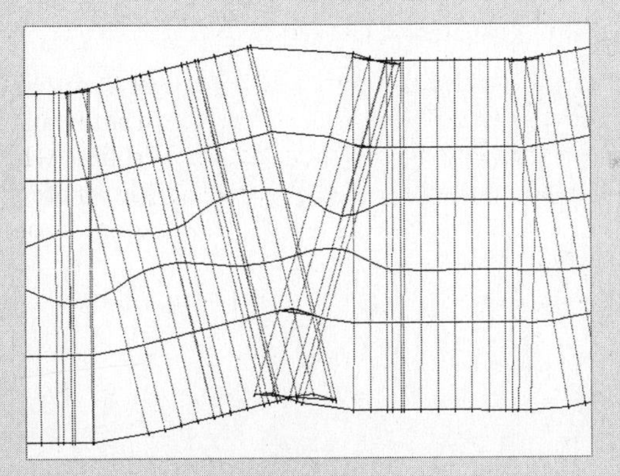

Bowties are problematic for several reasons. In essence, the corridor model has created two or more points at the same *x* and *y* locations with a different *z*, making it difficult to build surfaces, extract feature lines, create a boundary, and apply code-set styles that render or hatch.

When your corridor surface is created, the TIN has to make some assumptions about crossing breaklines that can lead to strange triangulation and incoherent contours, such as in the following graphic:

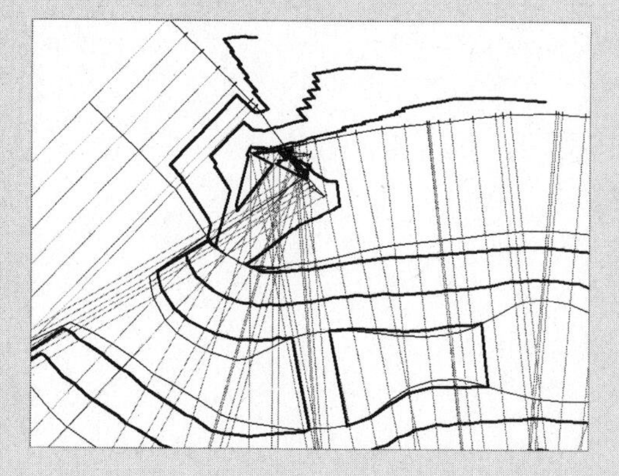

When you create a corridor that produces bowties, the corridor won't behave as expected. Choosing Corridors ➤ Utilities to extract polylines or feature lines from overlapping corridor areas yields an entity that is difficult to use for additional grading or manipulation because of extraneous, overlapping, and invalid vertices. If the corridor contains many overlaps, you may have trouble even executing the extraction tools. The same concept applies to extracted alignments, profiles, and COGO points.

If you try to add an automatic or interactive boundary to your corridor surface, either you'll get an error or the boundary jig will stop following the feature line altogether, making it impossible to create an interactive boundary.

Chapter 22, "Get The Picture: Visualization," will explore some methods for applying code-set styles to a corridor model for link-based rendering. The software goes through superficial surface modeling to apply the render materials. Because the corridor has no inherent valid boundary when there are overlapping links, you'll get undesirable results when rendering such a corridor, as in the following graphic, which shows a corridor similar to the one in the assembly-offset exercise when the realistic visual style is applied:

To prevent these problems, the best plan is to try to avoid link overlap. Be sure your baseline, offset, and target alignments don't have redundant or PI locations that are spaced excessively close.

If you initially build a corridor with simple transitions that produce a lot of overlap, try using an assembly offset and an alignment besides your centerline as a baseline. Another technique is to split your assembly into several smaller assemblies and to use your target assemblies as baselines, similar to using an assembly offset. This method was used to improve the river corridor shown in the previous graphics. The following graphic shows the two assemblies that were created to attach at the top of bank alignments instead of the river centerline:

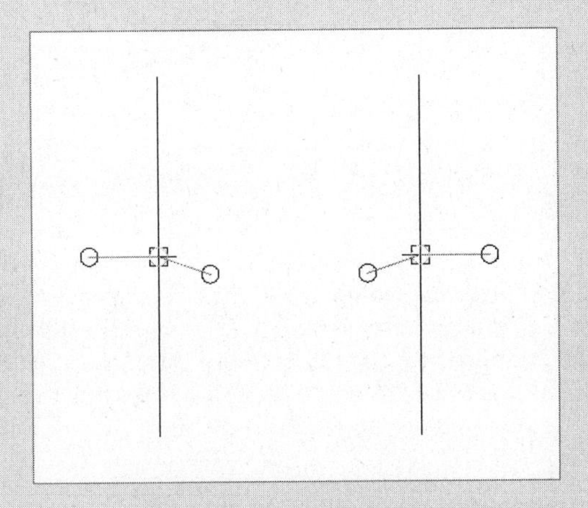

The resulting corridor is shown in the following graphic:

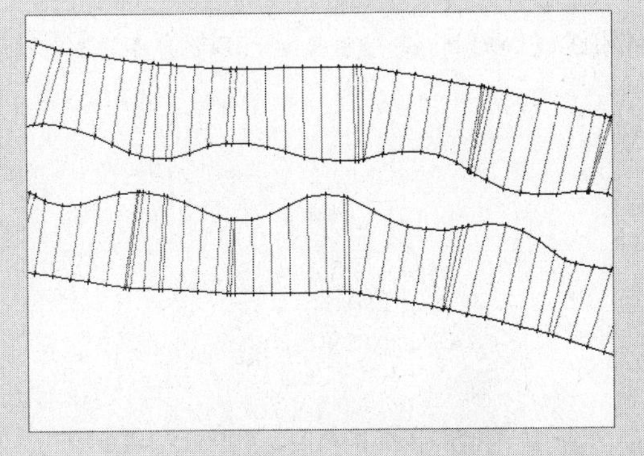

The TIN connected the points across the flat bottom and modeled the corridor perfectly, as you can see in the following image:

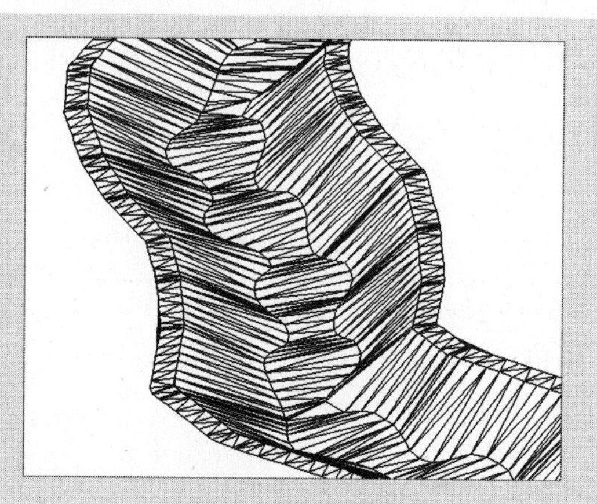

Another method for eliminating bowties is to notice the area where they seem to occur and adjust the regions. If your daylight links are overlapping, perhaps you can create an assembly that doesn't include daylighting and create a region to apply that new assembly.

If overlap can't be avoided in your corridor, don't panic. If your overlaps are minimal, you should still be able to extract a polyline or feature line — just be sure to weed vertices and clean up the extracted entity before using it for projection grading. You can create a boundary for your corridor surface by drawing a regular polyline around your corridor and adding it as a boundary to the corridor surface under the Surfaces branch in Prospector. The surface-editing tools, such as Swap Edge, Delete Line, and Delete Point, can also prove useful for the final cleanup and contour improvement of your final corridor surface.

As you gain more experience building corridors, you'll be able to prevent or fix most overlap situations, and you'll also gain an understanding of when they aren't having a detrimental effect on the quality of your corridor model and resulting surface.

Using a Feature Line as a Width and Elevation Target

You've gained some hands-on experience using alignments and profiles as targets for swale, intersection, and cul-de-sac design. Civil 3D 2010 adds options for corridor targets beyond alignments and profiles. You can now use grading feature lines, survey figures, or polylines to drive horizontal and/or vertical aspects of your corridor model.

Dynamic Feature Lines Cannot Be Used As Targets

It's important to note that dynamic feature lines extracted using the Feature Lines from Corridor tool can't be used as Targets. The possibility of circular references would be too difficult for the program to anticipate and resolve.

Imagine using an existing polyline that represents a curb for your lane-widening projects without duplicating it as an alignment, or grabbing a survey figure to assist with modeling an existing road for a rehabilitation project. The next exercise will lead you through an example where a lot-grading feature line is integrated with a corridor model:

1. Open the `Feature Line Target.dwg` file, which you can download from `www.sybex.com/go/masteringcivil3d2010`. This drawing includes a corridor as well as a yellow feature line that runs through a few lots.
2. Zoom to the area of the drawing where the assemblies are located. There is an assembly that includes a LinkWidthAndSlope subassembly attached to the sidewalk on the left side. You'll be using the yellow feature line as a target for this subassembly.
3. Zoom to the corridor. Select the corridor, right-click, and choose Corridor Properties.
4. Switch to the Parameters tab in the Corridor Properties dialog. Click the Set All Targets button. The Target Mapping dialog appears.
5. Click the <None> field next to Target Alignment for the LinkWidthAndSlope subassembly. The Set Width or Offset Target dialog appears.
6. Choose Feature Lines, Survey Figures and Polylines in the Select Object Type to Target drop-down menu (see Figure 12.72).

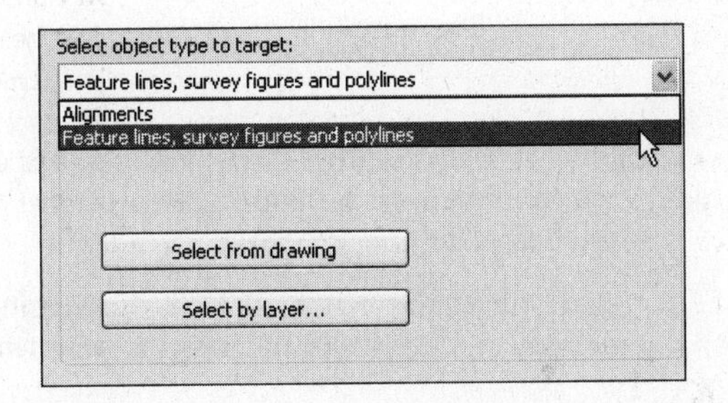

FIGURE 12.72
The Select Object Type to Target pull-down menu as seen in both the Set Width or Offset and Set Profile or Elevation Target dialogs

7. Click the Select from Drawing button. The command line prompts you to `Select feature lines, survey figures or polylines to target`. Select the yellow feature line, and then press ↵. The Set Width or Offset Target dialog is redisplayed, with an entry in the Selected Entries to Target area. Click OK to return to the Target Mapping dialog.

   If you stopped at this point, the horizontal location of the feature line would guide the LinkWidthAndSlope assembly, and the vertical information would be driven by the slope set in the subassembly properties. Although this has its applications, most of the time you'll want the feature-line elevations to direct the vertical information. The next few steps will teach you how to dynamically apply the vertical information from the feature line to the corridor model.
8. Click the <None> field next to Target Profile for the LinkWidthAndSlope subassembly. The Set Slope or Elevation Target dialog appears.
9. Make sure Feature Lines, Survey Figures and Polylines is selected in the Select Object Type to Target drop-down menu.
10. Click the Select from Drawing button. The command line prompts you to `Select feature lines, survey figures or polylines to target`. Select the yellow feature line, and then

press ↵. The Set Slope or Elevation Target dialog is redisplayed, with an entry in the Selected Entries to Target area. Click OK to return to the Target Mapping dialog.

11. Click OK to return to the Corridor Properties dialog.
12. Click OK to exit the Corridor Properties dialog. The corridor will rebuild to reflect the new target information and should look similar to Figure 12.73.

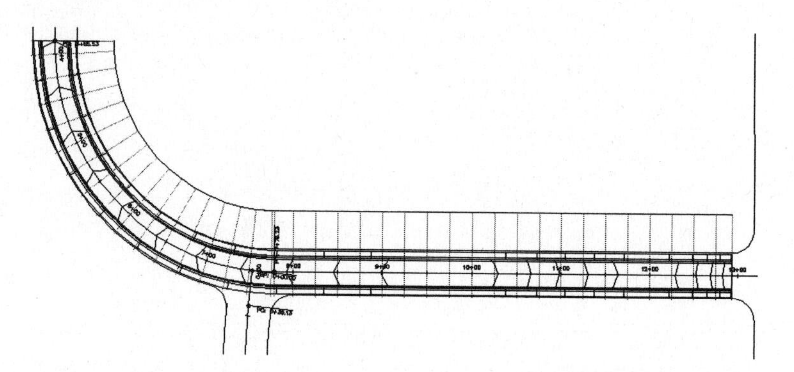

FIGURE 12.73
The corridor now uses the grading feature line as a width and elevation target.

Once you've linked the corridor to this feature line, any edits to this feature line will be incorporated into the corridor model. You can establish this feature line at the beginning of the project and then make horizontal edits and elevation changes to perfect your design. The next few steps will lead you through making some changes to this feature line and then rebuilding the corridor to see the adjustments.

13. Select the corridor, right-click, and choose Display Order ➢ Send to back. This sends the corridor model behind the target feature line.
14. Switch on your ORTHO setting. Use the AutoCAD Move command to move the feature line back approximately 10–20 feet, as in Figure 12.74. Note that you could also edit individual vertices or use any of the horizontal feature-line editing tools available under the Grading menu and on the Feature Line toolbar.

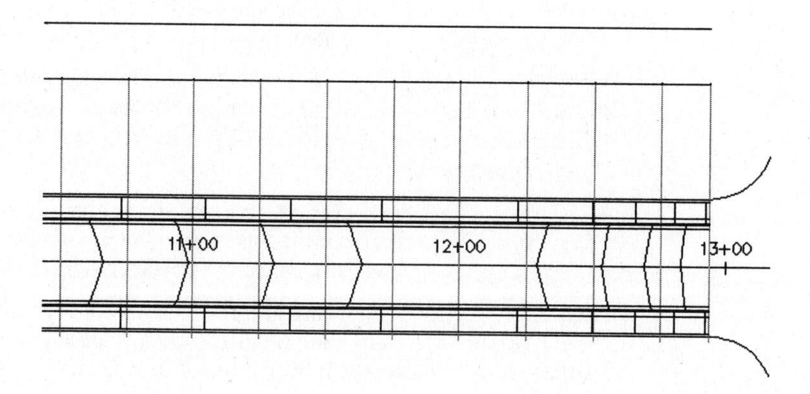

FIGURE 12.74
Move the feature line approximately 10–20 feet deeper into the lot.

15. Select the feature line. Right-click, and choose Raise/Lower from the shortcut menu.

16. The command line prompts you to `Specify elevation difference` <1.00>. Type **5**, and then press ↵. Each vertex of the feature line rises 5′ vertically. Note that you could also edit individual vertices or use any of the vertical feature-line editing tools available under the Grading menu and on the Feature Line toolbar.

17. Select the corridor. Right-click, and choose Rebuild Corridor. The corridor will rebuild to reflect the changes to the target feature line and should appear similar to Figure 12.75.

FIGURE 12.75
The completed corridor model

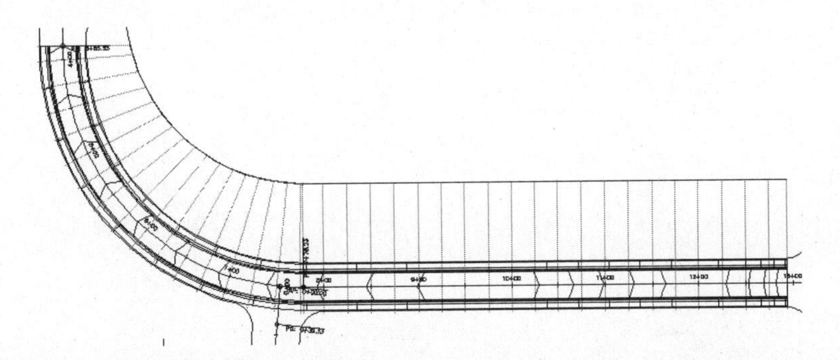

Edits to targets — whether they're feature lines, alignments, profiles, or other Civil 3D objects — drive changes to the corridor model, which then drives changes to any corridor surfaces, sections, section views, associated labels, and other objects that are dependent on the corridor model. Brainstorm ways that you can take advantage of this dynamic connection, such as making a corridor surface and then a quick profile or two, so that you can see your iterations of the feature line in immediate action as you work through your design.

The Bottom Line

Add a baseline to a corridor model for a cul-de-sac. Although for simple corridors you may think of a baseline as a road centerline, other elements of a road design can be used as a baseline. In the case of a cul-de-sac, the EOP, the top of curb, or any other appropriate feature can be converted to an alignment and profile and used as a baseline.

Master It Open the `Mastering Advanced Corridors.dwg` file, which you can download from `www.sybex.com/go/masteringcivil3d2010`. Add the cul-de-sac alignment and profile to the corridor as a baseline. Create a region under this baseline that applies the Typical Intersection assembly.

Add alignment and profile targets to a region for a cul-de-sac. Adding a baseline isn't always enough. Some corridor models require the use of targets. In the case of a cul-de-sac, the lane elevations are often driven by the cul-de-sac centerline alignment and profile.

Master It Continue working in the `Mastering Advanced Corridors.dwg` file. Add the Second Road alignment and Second Road FG profile as targets to the cul-de-sac region. Adjust the Assembly Application Frequency to 5′, and make sure the corridor samples are profile PVIs.

Use the Interactive Boundary tool to add a boundary to the corridor surface. Every good surface needs a boundary to prevent bad triangulation. Bad triangulation creates inaccurate and unsightly contours. Civil 3D provides several tools for creating corridor surface boundaries, including an Interactive Boundary tool.

Master It Continue working in the `Mastering Advanced Corridors.dwg` file. Create an interactive corridor surface boundary for the entire corridor model.

Chapter 13

Stacking Up: Cross Sections

Cross sections are used in Civil 3D to allow the user to have a graphic confirmation of design intent, as well as to calculate the quantities of materials used in a design. Sections must have at least two types of Civil 3D objects to be created: an alignment and a surface. Other objects, such as pipes, structures, and corridor components, can be sampled in a sample line group, which is used to create the graphical section that is displayed in a section view. These section views and sections remain dynamic throughout the design process, reflecting any changes made to the sampled information. This reduces potential errors in materials reports, keeping often costly mistakes from happening during the construction process.

In this chapter, you'll learn to:

- Create sample lines
- Create section views
- Define materials
- Generate volume reports

The Corridor

Before you create sample lines, you often start with a corridor. The corridor allows you to display the materials being used, as well as to show the new surface with cut-and-fill areas. In this chapter, the corridor is a relatively short roadway (1,340′) designed for a residential subdivision (see Figure 13.1).

This corridor has both a top surface and a datum surface created for inclusion in the sample line group, as shown in Figure 13.2. Creating surfaces from the different links and feature lines in a corridor allows you to use sections to calculate volumes between those surfaces. These volumes are calculated by specifying which surfaces to compare when you create a materials list.

Creating the Best Possible Surface for Sampling

Note that you can create corridor surfaces in two different ways — from links and from feature lines. Links will provide you with a total surface along the width of a corridor, such as the top of pavement, top of base, and top of subbase. Feature lines require selecting a few more objects to add into a corridor surface to accurately create the surface.

When you create your sample line group, you will have the option to sample any surface in your drawing, including corridor surfaces, the corridor assembly itself, and any pipes in your drawing. The sections are then sampled along the alignment with the left and right widths specified and at the intervals specified. Once the sample lines are created, you can then choose to create section views or to define materials.

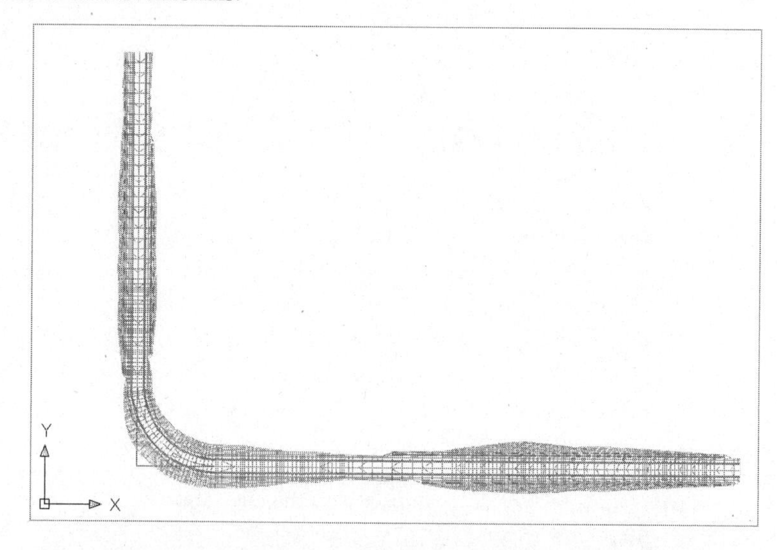

FIGURE 13.1
The Elizabeth Lane corridor

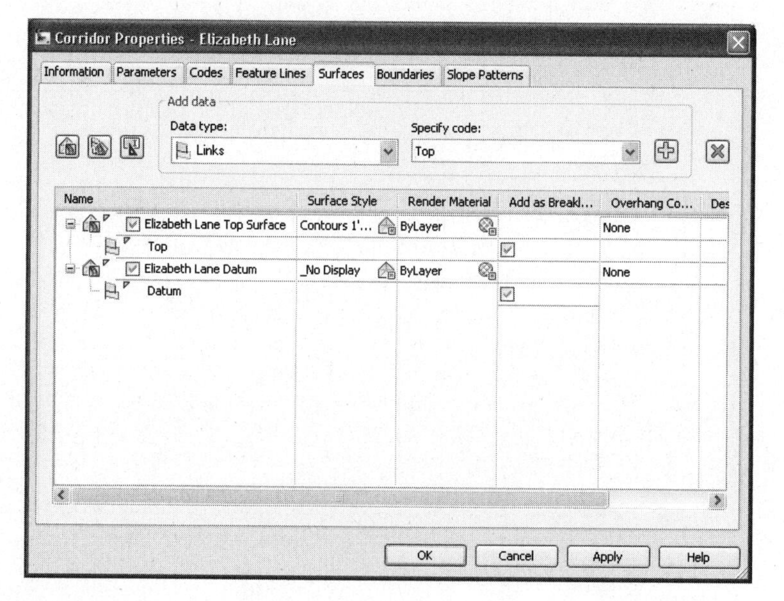

FIGURE 13.2
The Corridor Properties dialog

Lining Up for Samples

Sample lines are the engine underneath both sections and materials and are held in a collection called sample line groups. One alignment can have multiple sample line groups, but a sample line group can sample only one alignment. Sample lines typically consist of two components: the sample lines and the sample line labels, as shown in Figure 13.3.

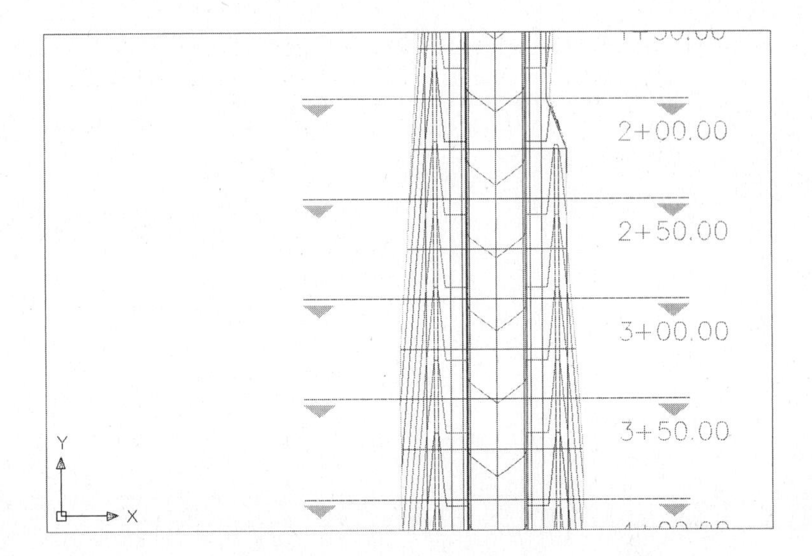

FIGURE 13.3
Sample lines consist of the lines and their labels.

If you pick a sample line, you will see it has three different types of grips, as shown in Figure 13.4. The diamond grip on the alignment allows you to move the sample line along the alignment. The triangular grip on the end of the sample line allows you to move the sample line along an extension of the line, either making it longer or shorter. The square grip on the end of the sample line allows you to not only move the sample line in or out, but also move it in any direction on the XY-plane.

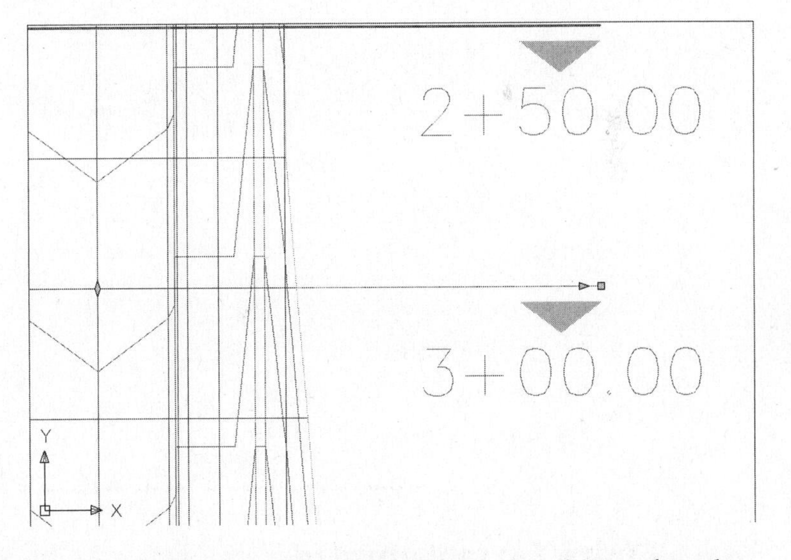

FIGURE 13.4
The three types of grips on a sample line

To create a sample line group, change to the Home tab and choose Sample Lines from the Profile & Section Views panel. After selecting the appropriate alignment, the Create Sample Line Group dialog as shown in Figure 13.5 will display. This dialog prompts you to name the sample line group, apply a sample line and label style, and choose the objects in your drawing that you would like to sample. Every object that is available will be displayed in this box, with an area to set the section style, whether to sample the data, what layer each sampled item would be applied to, and a setting to specify whether the data should be static or dynamic. For example, you would typically select your existing ground (EG) surface to be sampled, displayed with an EG style, and be static. Your finished grade (FG) surface would also be sampled, but would be displayed with an FG style and be dynamic.

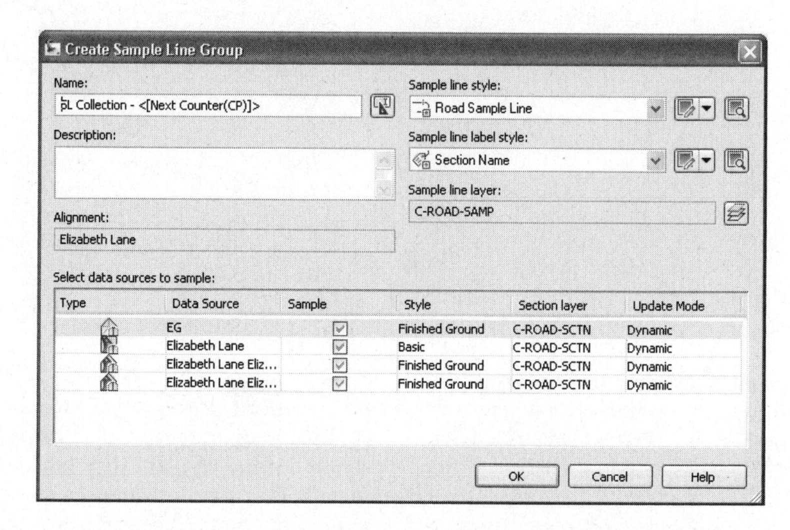

FIGURE 13.5
The Create Sample Line Group dialog

Once the sample data has been selected, the Sample Line Tools toolbar will appear, as shown in Figure 13.6. This toolbar is context-sensitive and is displayed only when you are creating sample lines.

FIGURE 13.6
The Sample Line Tools toolbar

Once you have completed the Sample Line Creation process, close the toolbar, and the command ends. Because most of the information is already set for you in this toolbar, the Sample Line Creation Methods button is the only one that is really needed. This gives you the following five options for creating sample lines:

- By Range of Stations
- At a Station
- From Corridor Stations
- Pick Points on Screen
- Select Existing Polylines

In Civil 3D 2010, these options are listed in order from most used to least used. Because the most common method of creating sample lines is from one station to another at set intervals, the By Range of Stations option is first. You can use At a Station to create one sample line at a specific station. From Corridor Stations allows you to insert a sample line at each corridor assembly insertion. Pick Points on Screen allows you to pick any two points to define a sample line. This option can be useful in special situations, such as sampling a pipe on a skew. The last option, Select Existing Polylines, lets you define sample lines from existing polylines.

A Warning About Using Polylines to Define Sample Lines

Be careful when picking existing polylines to define sample lines. Any osnaps used during polyline creation can throw off the Z-values of the section, giving sometimes undesirable results.

To define sample lines, you need to specify a few settings. Figure 13.7 shows the settings that need to be defined in the Create Sample Lines By Station Range dialog. The Right Swath Width is the width from the alignment that you sample. Most of the time this distance is greater than the ROW distance. You also select your Sampling Increments, and choose whether to include special stations, such as horizontal geometry (PC, PT, and so on), vertical geometry (PVC, high point, low point, and so on), and superelevation critical stations.

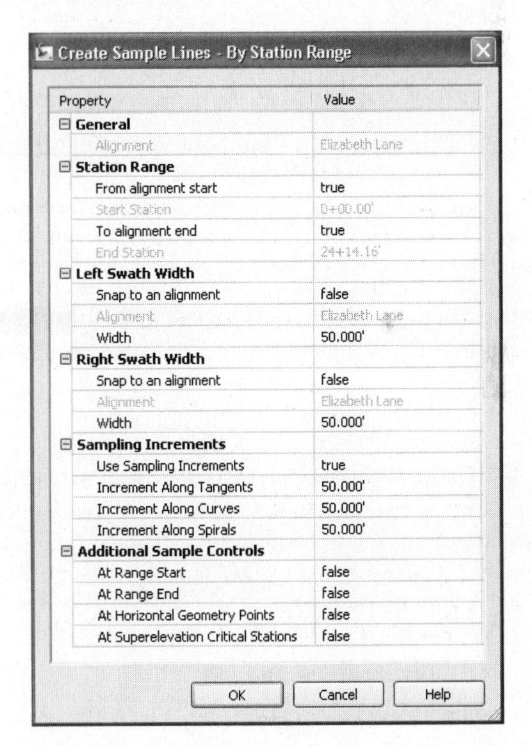

FIGURE 13.7
Sample Line settings

Creating Sample Lines along a Corridor

Before creating cross sections, you must sample the information that will be displayed. You do this by creating sample lines, which are part of a sample line group. Only one alignment can be sampled per sample line group. When creating sample lines, you will have to determine the frequency of your sections and the objects that you want included in the section views. In the following exercise, you create sample lines for Elizabeth Lane:

1. Open the `sections1.dwg` file, which you can download from `www.sybex.com/go/masteringcivil3d2010`.
2. Change to the Home tab and choose Sample Lines from the Profile & Section Views panel.
3. Press ↵ to display the Select Alignment dialog.
4. Select the Elizabeth Lane alignment and click OK. You can also pick the alignment in the drawing.
5. Make sure all the Sample boxes are checked in the data sources.
6. Set the styles for the sections as follows: EG – Existing Ground, Elizabeth Lane – Basic, Elizabeth Lane Top Surface – Finished Ground, and Elizabeth Lane Datum – _NULL.
7. Click OK.

8. On the Sample Line Tools toolbar, click the Sample Line Creation Methods drop-down arrow and then click the By Range of Stations button on the drop-down list. Observe the settings, but do not change anything.
9. Click OK, and press ↵ to end the command.
10. If you receive a Panorama view telling you that your corridor is out of date and may require rebuilding, dismiss it.
11. Close the drawing without saving.

The sample lines can be edited by using the grips as just described or by selecting a sample line and choosing Edit Sample Line from the Modify panel. This command displays both the Sample Line Tools toolbar and the Edit Sample Line dialog, allowing you to pick your alignment. Once the alignment is picked, the sample lines can be edited individually or as a group. The Edit Sample Lines dialog allows you to pick a sample line and edit the information on an individual basis, but it is much more efficient to edit the all of the sample lines at the same time.

Editing the Swath Width of a Sample Line Group

There may come a time when you will need to show information outside the limits of your section views or not show as much information. To edit the width of a section view, you will have to change the swath width of a sample line group. These sample lines can be edited manually on an individual basis, or you can edit the entire group at once. In this exercise, you edit the widths of an entire sample line group:

1. Open the `sections2.dwg` file.
2. Select a sample line and choose Edit Sample Line from the Modify panel. The Sample Line Tools toolbar and the Edit Sample Line dialog appear.
3. Click the Alignment Picker button on the toolbar, and then pick the Elizabeth Lane alignment.
4. Click OK.

5. Click the down arrow for the sample line editing tools on the toolbar and then click the Edit Swath Widths for Group button. The Edit Sample Line Widths dialog appears.
6. Type **100** in both the Left and Right Swath Width text boxes. Click OK.
7. Press ↵ to end the command.
8. If you receive a Panorama view telling you that your corridor is out of date and may require rebuilding, dismiss it.
9. Examine your sample lines, noting the wider sample lines.
10. Close the drawing without saving.

Creating the Views

Once the sample line group is created, it is time to create views. Views can be created in three ways: single view, all views, or all views-by-page. Creating views-by-page allows you to lay out your cross sections on a sheet-by-sheet basis. This is accomplished by creating a page setup for your proposed sheet size and defining that page setup in your sheet style. You can arrange the section views by either rows or columns and specify the space between each consecutive section view. This even allows you to put a predefined grid on your cross-section sheet. Although the setup for this is quite tedious, the payoff is incredible if you plot many sheets' worth of cross sections at a time. Figure 13.8 shows a layout containing section views arranged to plot by page.

FIGURE 13.8
Section views arranged to plot by page

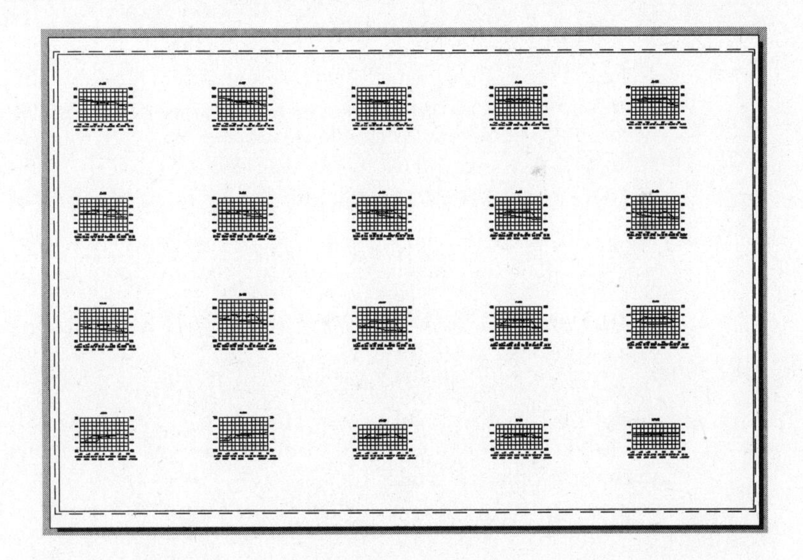

Section views are nothing more than a window showing the section. The view contains horizontal and vertical grids, tick marks for axis annotation, the axis annotation itself, and a title. Views can also be configured to show horizontal geometry, such as the centerline of the section, edges of pavement, and right-of-way. Tables displaying quantities or volumes can also be shown for individual sections. Figure 13.9 shows a typical section view with such a table.

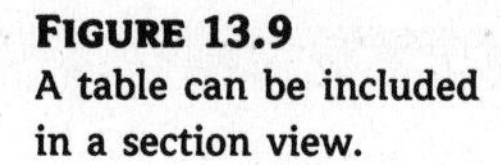
FIGURE 13.9
A table can be included in a section view.

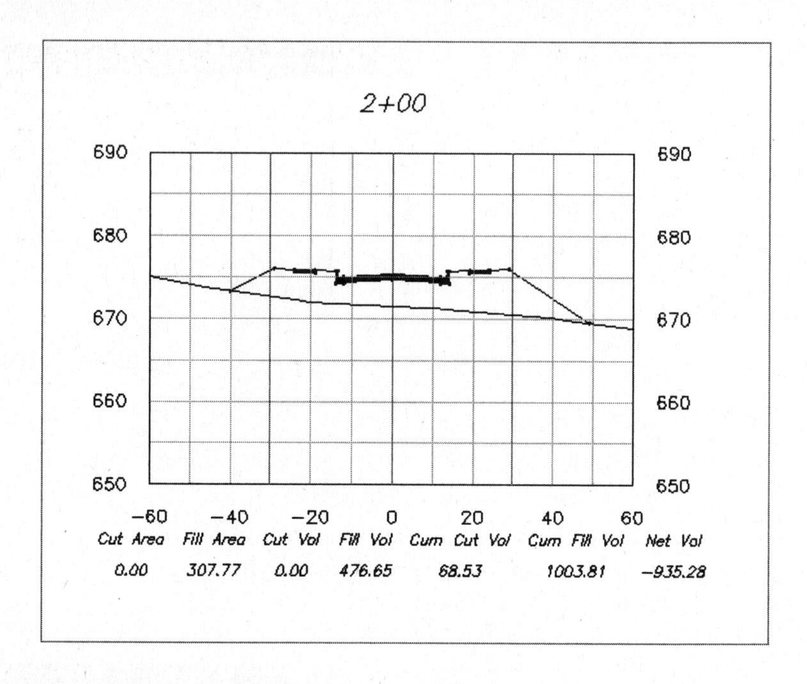

Creating a Single-Section View

There are occasions when all sections views are not needed. In these situations, a single-section view can be created. In this exercise, you create a single-section view of station 3+00 from the sample lines created in the previous exercise:

1. Open the `sections3.dwg` file.
2. Change to the Home tab and choose Section Views ➤ Create Section View from the Profile & Section Views panel. The Create Section View Wizard appears.
3. Make sure the settings in the Create Section View Wizard match the settings shown in the following figures. The first window of the wizard is the General window, shown in Figure 13.10. It allows you to select the alignment, sample line group name, desired sample line or station, and the section view style. You can navigate from one window to another by either selecting the Next button at the bottom of the wizard or by selecting the hyperlinks on the left side of the wizard.

   The second window is where the offset range is selected, as shown in Figure 13.11. This allows the section views to either be the same width by entering a user-specified offset or allows the width of the section view to be controlled by the width of the sample line.

   The third window, shown in Figure 13.12, is where the elevation range is selected. This allows the height of the section view to be set automatically based on the depth of cut or fill or to be a consistent height by entering a user-specified elevation range.

FIGURE 13.10
The General window of the Create Section View Wizard

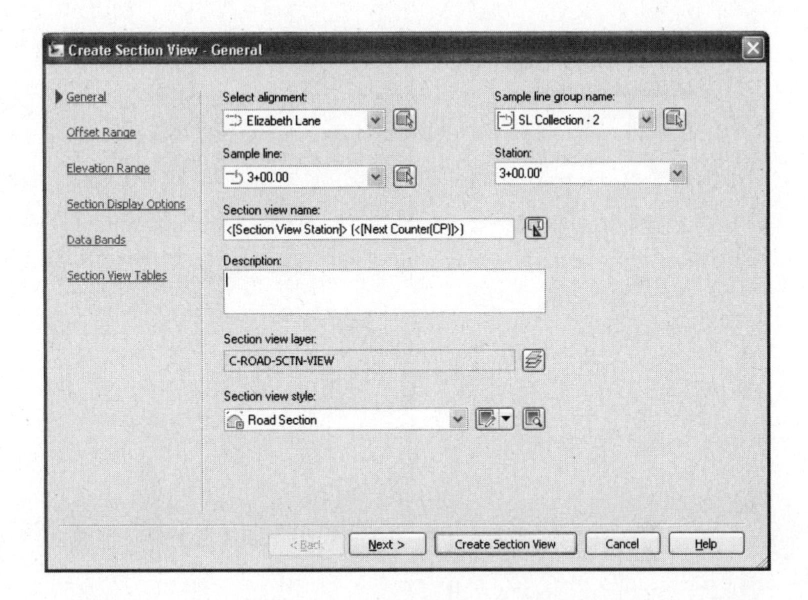

FIGURE 13.11
The Offset Range window of the Create Section View Wizard

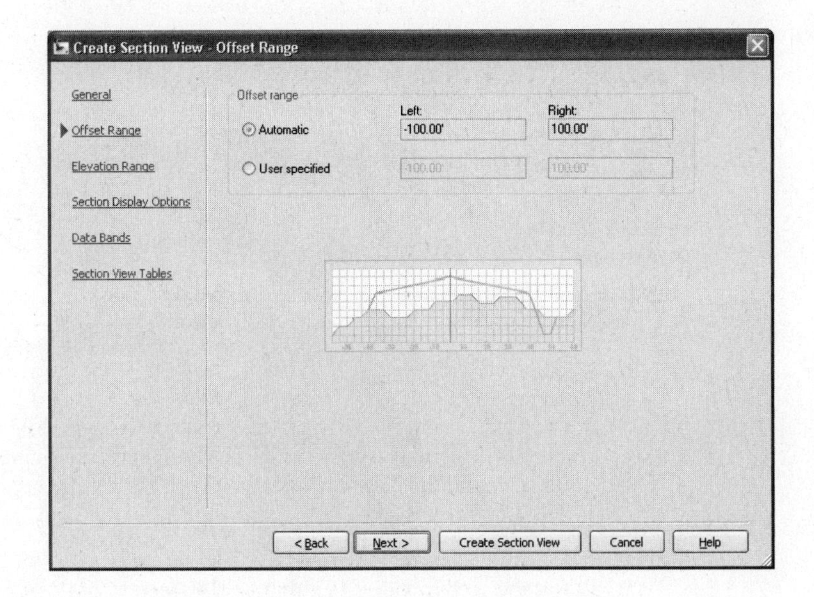

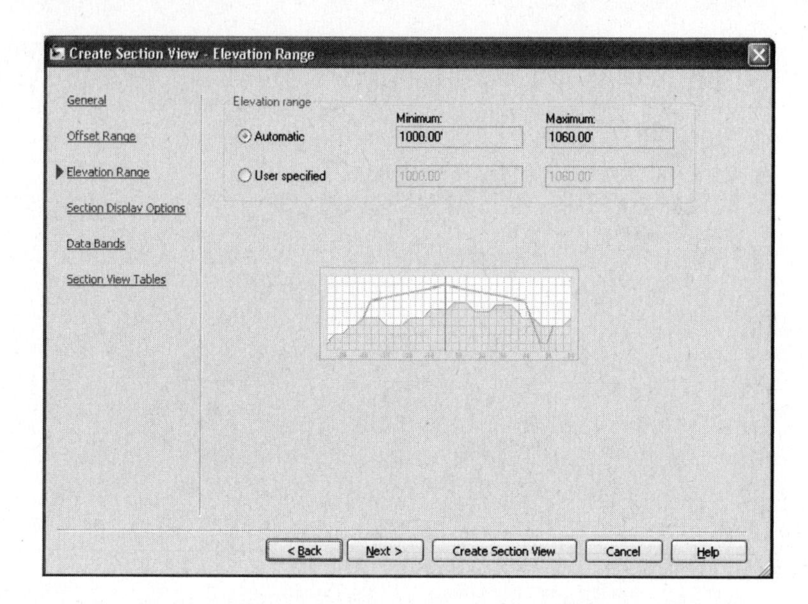

FIGURE 13.12 The Elevation Range window of the Create Section View Wizard

The fourth window is where the section display options are specified, as shown in Figure 13.13. In this window, you can pick which sections to draw in the view, if and how you want the grid clipped in the section view, and specify section view labels and section styles.

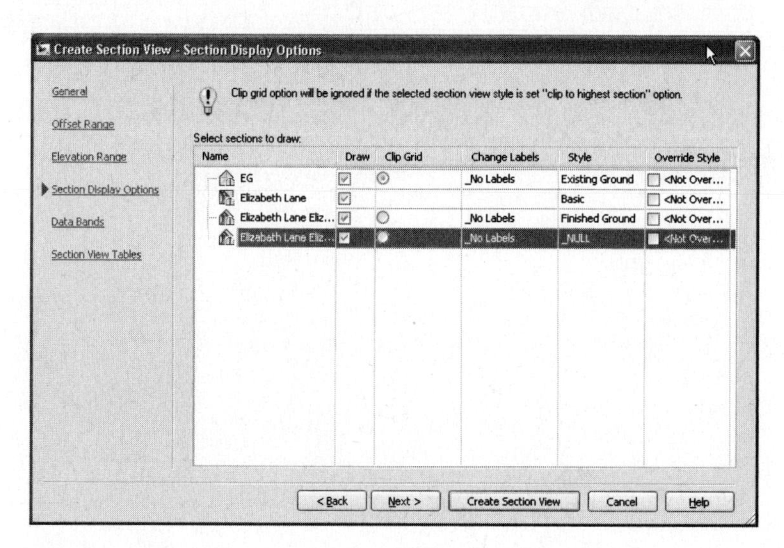

FIGURE 13.13 The Section Display Options window of the Create Section View Wizard

The fifth window, shown in Figure 13.14, is where the data band options are specified. In this window, you can select band sets to add to the section view, pick the location of the band, and choose the surfaces to be referenced in the bands.

The sixth and last window, shown in Figure 13.15, is where the section view tables are set up. It should be noted that this window will be available only if you have already computed materials for the sample line group. In this window, you can select the type of table to specify and the table style, and select the position of the table relative to the section view. Notice the graphic on the right side of the window that illustrates the current settings.

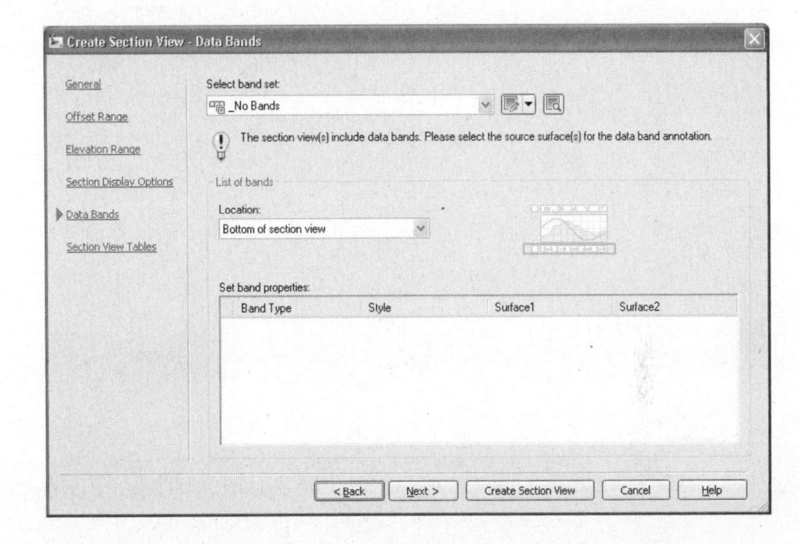

FIGURE 13.14
The Data Bands window of the Create Section View Wizard

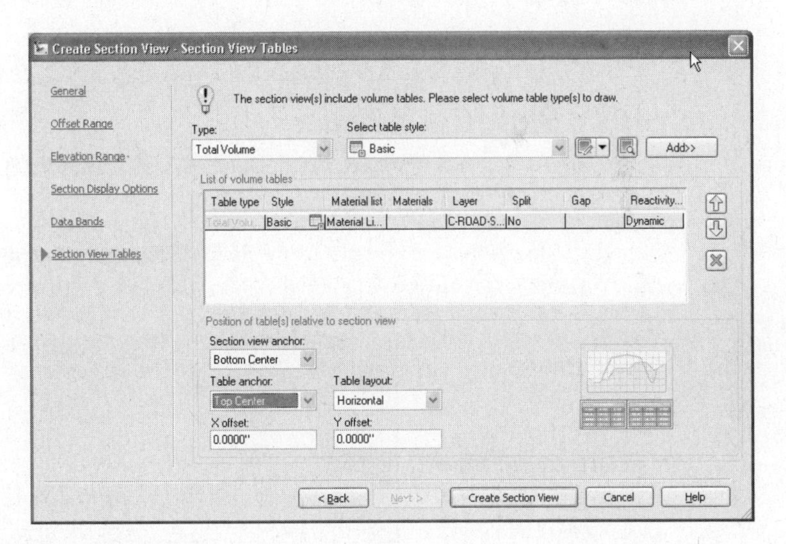

FIGURE 13.15
The Section View Tables window of the Create Section View Wizard

4. Select Create Section View.
5. Pick any point in the drawing area to place your section view.
6. Examine your section view. The display should match Figure 13.16.
7. Close the drawing without saving.

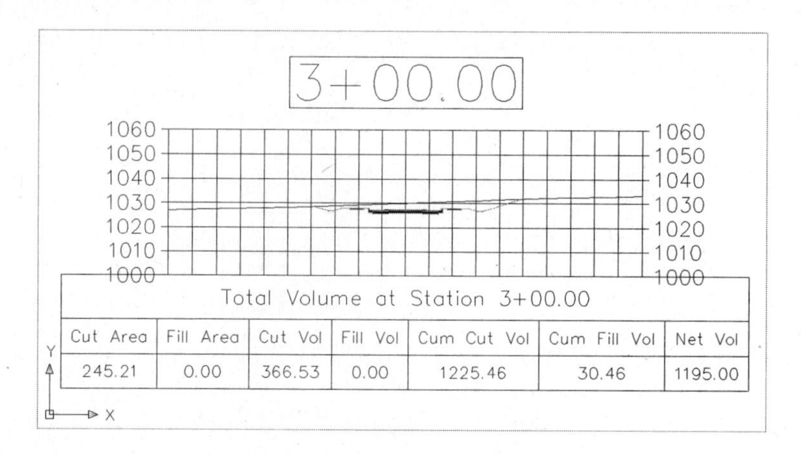

FIGURE 13.16
The finished section view

SECTION VIEW OBJECT PROJECTION

One of the new features in the 2010 release of Civil 3D is the ability to project AutoCAD points, blocks, 3D solids, 3D polylines, AutoCAD Civil 3D COGO points, feature lines, and survey figures into section and profile views. Each of the objects listed can be projected to a section view and labeled appropriately. This function is explored more in Chapter 9, "Slice and Dice: Profile Views in Civil 3D." You access the command by changing to the Home tab's Profile and Section Views panel and selecting Section Views ➤ Project Objects To Section.

It's a Material World

Once alignments are sampled, volumes can be calculated from the sampled surface or from the corridor section shape. These volumes are calculated in a materials list and can be displayed as a label on each section view or in an overall volume table, as shown in Figure 13.17.

The volumes can also be displayed in an XML report, as shown in Figure 13.18.

Once a materials list is created, it can be edited to include more materials or to make modifications to the existing materials. For example, soil expansion (fluff or swell) and shrinkage factors can be entered to make the volumes more accurately match the true field conditions. This can make cost estimates more accurate, which can cause fewer surprises during the construction phase of any given project.

Creating a Materials List

Materials can be created from surfaces or from corridor shapes. Surfaces are great for earthwork because you can add cut or fill factors to the materials, whereas corridor shapes are great for determining quantities of asphalt or concrete. In this exercise, you practice calculating earthwork quantities for the Elizabeth Lane corridor:

1. Open the `sections4.dwg` file.
2. Change to the Analyze tab and choose Compute Materials from the Volumes and Materials panel. The Select a Sample Line Group dialog appears.

FIGURE 13.17
A total volume table inserted into the drawing

Total Volume Table						
Station	Fill Area	Cut Area	Fill Volume	Cut Volume	Cumulative Fill Vol	Cumulative Cut Vol
0+50.00	97.77	38.27	0.00	0.00	0.00	0.00
1+00.00	132.28	17.87	213.01	51.99	213.01	51.99
1+50.00	207.01	0.00	314.15	16.55	527.16	68.53
2+00.00	307.77	0.00	476.65	0.00	1003.81	68.53
2+50.00	387.68	0.00	643.94	0.00	1647.75	68.53
3+00.00	446.32	0.00	772.22	0.00	2419.97	68.53
3+50.00	350.11	0.00	737.43	0.00	3157.41	68.53
4+00.00	383.68	0.00	679.43	0.00	3836.84	68.53
4+50.00	459.29	0.00	780.52	0.00	4617.36	68.53
5+00.00	431.67	0.00	824.96	0.00	5442.32	68.53
5+50.00	474.38	0.00	838.93	0.00	6281.25	68.53
6+00.00	543.49	0.00	942.47	0.00	7223.72	68.53
6+50.00	697.29	0.00	1148.87	0.00	8372.59	68.53
7+00.00	631.80	0.00	1230.65	0.00	9603.23	68.53
7+50.00	588.72	0.00	1130.12	0.00	10733.36	68.53
8+00.00	203.01	24.69	733.09	22.86	11466.44	91.39
8+50.00	116.07	49.97	295.44	69.13	11761.88	160.52
9+00.00	375.17	0.00	454.85	46.27	12216.73	206.79
9+50.00	293.14	0.00	618.81	0.00	12835.53	206.79
10+00.00	12.65	27.88	283.14	25.82	13118.68	232.60
10+50.00	0.00	157.91	11.71	172.03	13130.39	404.63
11+00.00	0.00	341.15	0.00	462.09	13130.39	866.72
11+50.00	0.00	447.02	0.00	729.78	13130.39	1596.50
12+00.00	0.00	443.95	0.00	824.97	13130.39	2421.47
12+50.00	0.00	316.31	0.00	703.95	13130.39	3125.42
13+00.00	0.00	127.75	0.00	411.17	13130.39	3536.59

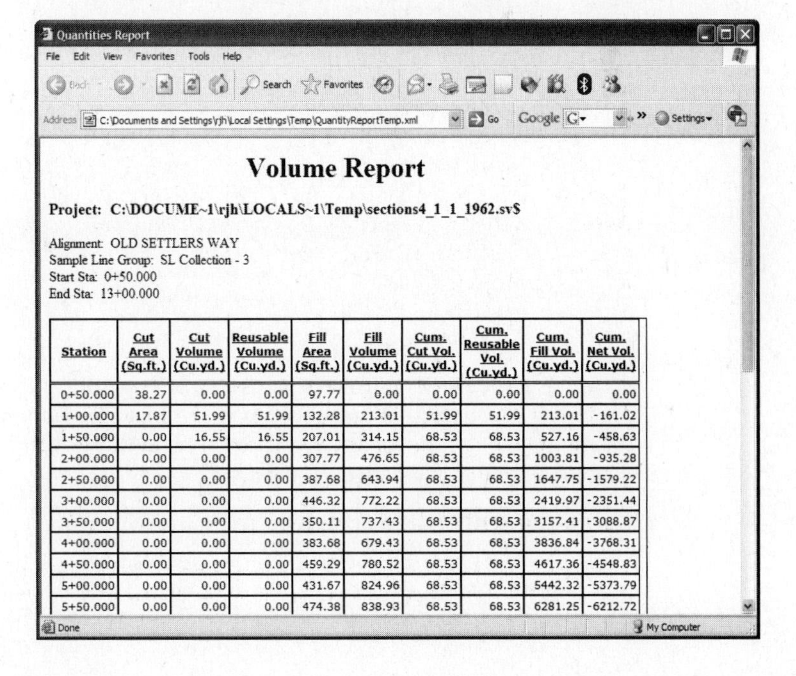
Quantities Report

Volume Report

Project: C:\DOCUME~1\rjh\LOCALS~1\Temp\sections4_1_1_1962.sv$

Alignment: OLD SETTLERS WAY
Sample Line Group: SL Collection - 3
Start Sta: 0+50.000
End Sta: 13+00.000

Station	Cut Area (Sq.ft.)	Cut Volume (Cu.yd.)	Reusable Volume (Cu.yd.)	Fill Area (Sq.ft.)	Fill Volume (Cu.yd.)	Cum. Cut Vol. (Cu.yd.)	Cum. Reusable Vol. (Cu.yd.)	Cum. Fill Vol. (Cu.yd.)	Cum. Net Vol. (Cu.yd.)
0+50.000	38.27	0.00	0.00	97.77	0.00	0.00	0.00	0.00	0.00
1+00.000	17.87	51.99	51.99	132.28	213.01	51.99	51.99	213.01	-161.02
1+50.000	0.00	16.55	16.55	207.01	314.15	68.53	68.53	527.16	-458.63
2+00.000	0.00	0.00	0.00	307.77	476.65	68.53	68.53	1003.81	-935.28
2+50.000	0.00	0.00	0.00	387.68	643.94	68.53	68.53	1647.75	-1579.22
3+00.000	0.00	0.00	0.00	446.32	772.22	68.53	68.53	2419.97	-2351.44
3+50.000	0.00	0.00	0.00	350.11	737.43	68.53	68.53	3157.41	-3088.87
4+00.000	0.00	0.00	0.00	383.68	679.43	68.53	68.53	3836.84	-3768.31
4+50.000	0.00	0.00	0.00	459.29	780.52	68.53	68.53	4617.36	-4548.83
5+00.000	0.00	0.00	0.00	431.67	824.96	68.53	68.53	5442.32	-5373.79
5+50.000	0.00	0.00	0.00	474.38	838.93	68.53	68.53	6281.25	-6212.72

FIGURE 13.18
A total volume XML report shown in Microsoft Internet Explorer

3. In the Select Alignment field, verify that the alignment is set to Elizabeth Lane, and in the Select Sample Line Group field, verify that the sample line group is set to SL Collection – 2.
4. Click OK. The Compute Materials dialog appears.
5. Make sure Material List – (1) is selected and click the button with the X in the upper-right of the dialog to delete the material list.
6. Click the Import Another Criteria button at the bottom of the dialog.
7. Select Earthworks from the drop-down menu in the Quantity Takeoff Criteria selection box and click OK.
8. Click the Object Name cell for the Existing Ground surface, and select EG from the drop-down menu.
9. Click the Object Name cell for the Datum surface, and select Elizabeth Lane Elizabeth Lane Datum (this is the name of the corridor followed by the name of the surface) from the drop-down menu.
10. Verify that your settings match those shown in Figure 13.19.

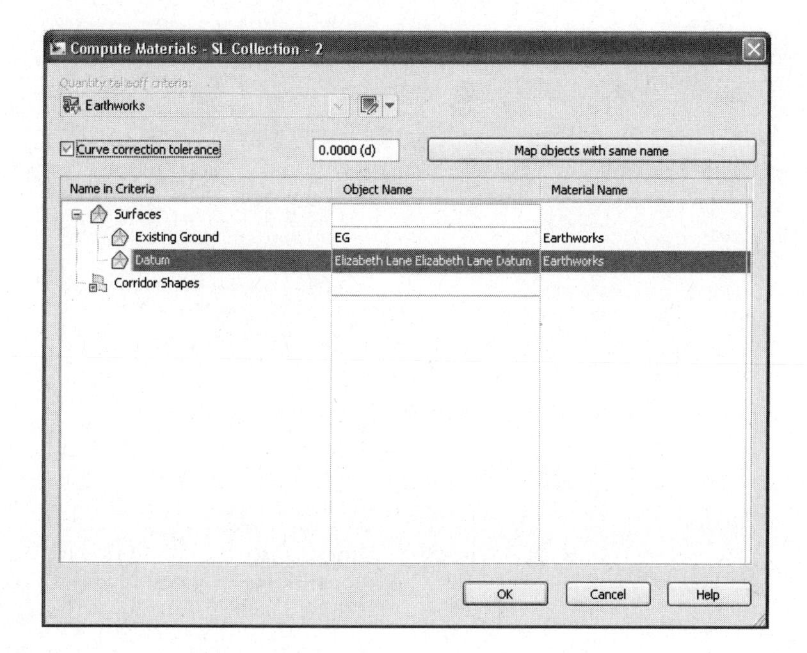

FIGURE 13.19
The settings for the Compute Materials dialog

11. Click OK, and then click OK again.
12. Save the drawing.

Creating a Volume Table in the Drawing

In the preceding exercise, materials were created that represent the total dirt to be moved or used in the sample line group. In the next exercise, you insert a table into the drawing so you can inspect the volumes:

1. Continue using the `sections4.dwg` file.
2. Change to the Analyze tab and choose Total Volume Table from the Volumes and Materials panel. The Create Total Volume Table appears.
3. Verify that your settings match those shown in Figure 13.20. Pay close attention and make sure that Reactivity Mode at the bottom of the dialog is set to Dynamic. This will cause the table to update if any changes are made to the sample line collection.

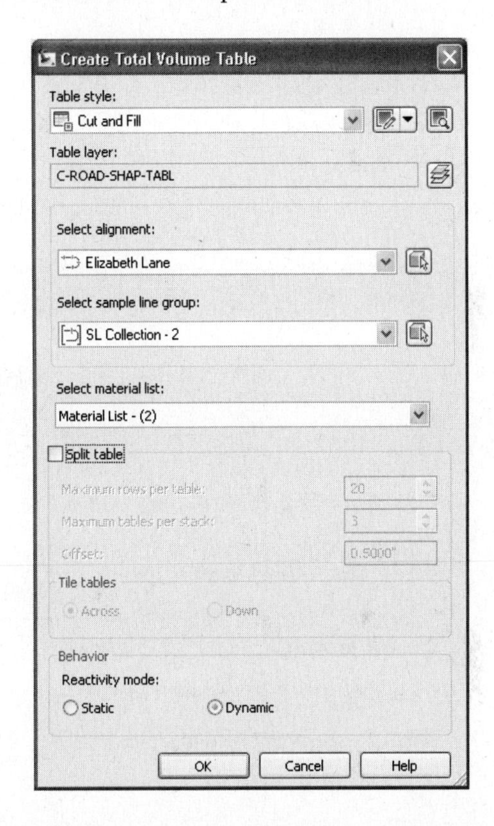

FIGURE 13.20
The Create Total Volume Table dialog settings

4. Click OK.
5. Pick a point in the drawing to place the volume table. The table indicates a Cumulative Fill Volume of 7,239.85 cubic yards and a Cumulative Cut Volume of 21,120.65 cubic yards, as shown in Figure 13.21. This means you will have to haul off just over 13,880 cubic yards of dirt from the site during road construction.
6. Save the drawing.

Adding Soil Factors to a Materials List

Because this design obviously has an excessive amount of fill, the materials need to be modified to bring them closer in line with true field numbers. For this exercise, the shrinkage factor will be assumed to be 0.80 and 1.20 for the expansion factor (20 percent shrink and swell). These numbers are arbitrary — numbers used during an actual design will be based on soil type and conditions.

In addition to these numbers (which Civil 3D represents as Cut Factor for swell and Fill Factor for shrinkage), a Refill Factor can also be set. This specifies how much cut can be reused for fill. For this exercise, assume a Refill Factor of 1.00:

FIGURE 13.21
The total volume table

13+00.00	108.82	0.00	349.84	0.00	6771.24	6872.55
13+50.00	110.80	0.00	203.35	0.00	6974.59	6872.55
14+00.00	39.06	9.24	138.75	8.55	7113.34	6881.10
14+50.00	12.76	35.39	47.98	41.32	7161.32	6922.42
15+00.00	17.76	26.69	28.26	57.48	7189.58	6979.90
15+50.00	15.28	46.86	30.60	68.11	7220.18	7048.00
16+00.00	1.18	89.80	15.24	126.54	7235.42	7174.55
16+50.00	0.00	200.82	1.09	269.10	7238.52	7443.64
17+00.00	0.00	400.61	0.00	556.88	7236.52	8000.53
17+50.00	0.00	593.80	0.00	920.75	7238.52	8921.28
18+00.00	0.00	704.81	0.00	1202.42	7238.52	10123.70
18+50.00	0.00	906.32	0.00	1491.79	7236.52	11615.50
19+00.00	0.00	907.96	0.00	1679.89	7236.52	13295.39
19+50.00	0.00	720.96	0.00	1508.26	7236.52	14803.65
20+00.00	0.00	529.20	0.00	1157.56	7238.52	15961.21
20+50.00	0.00	573.57	0.00	1021.08	7236.52	16982.29
21+00.00	0.00	529.76	0.00	1021.60	7238.52	18003.89
21+50.00	0.00	444.97	0.00	902.53	7236.52	18906.42
22+00.00	0.00	359.19	0.00	744.59	7236.52	19651.01
22+50.00	0.00	266.74	0.00	579.57	7238.52	20230.57
23+00.00	0.00	182.12	0.00	415.61	7236.52	20646.18
23+50.00	0.00	118.71	0.00	278.55	7236.52	20924.74
24+00.00	1.67	82.19	1.54	167.51	7238.06	21092.24
24+14.16	5.17	46.13	1.79	28.40	7239.85	21120.65

1. Continue using drawing `sections4.dwg`.
2. Change to the Analyze tab and choose Compute Materials from the Volumes and Materials panel. The Select a Sample Line Group dialog appears.
3. Select the Elizabeth Lane alignment and the SL Collection – 2 sample line group, if not already selected.
4. Click OK. The Edit Material List dialog appears.
5. Enter a Cut Factor of **1.20**, a Fill Factor of **0.80**, and verify that all other settings are the same as in Figure 13.22. Click OK.
6. Examine the Total Volume table again. Notice that the new Cumulative Fill Volume is 5,791.88 cubic yards and the new Cumulative Cut Volume is 25,344.54 cubic yards.
7. Save the drawing.

CAN YOU HAVE ACCURATE VOLUME NUMBERS WITHOUT SECTIONS?

Why yes, you can! Civil 3D has the capability to add cut and fill factors to both volume surfaces during creation and in the surface volumes panorama.

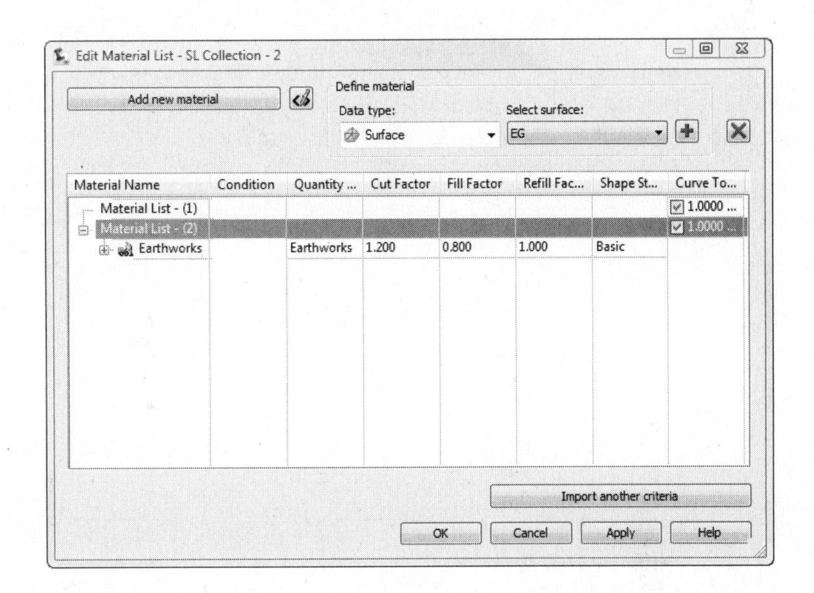

FIGURE 13.22
The Edit Material List dialog

Volumes can also be created in a format that can be printed and put into a project documentation folder. This is accomplished by creating a volume report, which is populated through LandXML. This report will open and display in your browser, but you can convert it to Word or Excel format with a simple copy and paste from the XML report. The XML report style sheet can be edited. The following is a sample of the default code in the style sheet:

```
<CrossSect name="1+00.00" number="2" sta="100" staEq="100"
areaCut="17.8692907698351" areaUsable="17.8692907698351"
areaFill="132.279872457422" volumeCut="62.3825333709404"
volumeUsable="51.9854444757837" volumeFill="170.406674689733"
cumVolumeCut="62.3825333709404" cumVolumeUsable="51.9854444757837"
cumVolumeFill="170.406674689733" massHaul="-108.024141318792">
- <MaterialCrossSects>
- <MaterialCrossSect name="Earthworks(Cut)" area="17.8692907698351"
volume="62.3825333709404" cumVolume="62.3825333709404">
- <MaterialCrossSectEnvelop area="0.136156978705912">
 <CrossSectPnt OE="-13.512209, 673.728273" />
 <CrossSectPnt OE="-12.263952, 673.582858" />
 <CrossSectPnt OE="-13.500000, 673.508695" />
 <CrossSectPnt OE="-13.512209, 673.728273" />
 </MaterialCrossSectEnvelop>
```

This produces a table that looks like Table 13.1.

TABLE 13.1: Output of the Volume Report

STATION	CUT AREA (SQ.FT.)	CUT VOLUME (CU.YD.)	REUSABLE VOLUME (CU.YD.)	FILL AREA (SQ.FT.)	FILL VOLUME (CU.YD.)	CUM. CUT VOL. (CU.YD.)	CUM. REUSABLE VOL. (CU.YD.)	CUM. FILL VOL. (CU.YD.)	CUM. NET VOL. (CU.YD.)
0+50.000	38.27	0.00	0.00	97.77	0.00	0.00	0.00	0.00	0.00
1+00.000	17.87	62.38	51.99	132.28	170.41	62.38	51.99	170.41	−108.02
1+50.000	0.00	19.85	16.55	207.01	251.32	82.24	68.53	421.73	−339.49
2+00.000	0.00	0.00	0.00	307.77	381.32	82.24	68.53	803.05	−720.81
2+50.000	0.00	0.00	0.00	387.68	515.15	82.24	68.53	1318.20	−1235.96
3+00.000	0.00	0.00	0.00	446.32	617.78	82.24	68.53	1935.98	−1853.74
3+50.000	0.00	0.00	0.00	350.11	589.95	82.24	68.53	2525.92	−2443.69
4+00.000	0.00	0.00	0.00	383.68	550.27	82.24	68.53	3076.20	−2993.96
4+50.000	0.00	0.00	0.00	459.29	646.11	82.24	68.53	3722.30	−3640.07
5+00.000	0.00	0.00	0.00	431.67	678.67	82.24	68.53	4400.97	−4318.73
5+50.000	0.00	0.00	0.00	474.38	687.00	82.24	68.53	5087.97	−5005.73
6+00.000	0.00	0.00	0.00	543.49	768.16	82.24	68.53	5856.13	−5773.89

Generating a Volume Report

Volume reports can be included on a drawing but normally aren't because of liability issues. However, it is often necessary to know what these volumes are and have some record of them. Civil 3D provides you with a way to create a report that is suitable for printing or for transferring to a word processing or spreadsheet program. In this exercise, you'll create a volume report for the Elizabeth Lane corridor:

1. Open the `sections5.dwg` file.
2. Change to the Analyze tab and choose Volume Report from the Volumes and Materials panel. The Report Quantities dialog appears.
3. Verify that Material List – (2) is selected in the dialog and click OK.
4. You may get a warning message that says, "Scripts are usually safe. Do you want to allow scripts to run?" Click Yes.
5. Note the cut-and-fill volumes and compare them to your volume table in the drawing. Close the report when you are done viewing it.
6. Close the drawing without saving.

A Little More Sampling

Although it is good practice to create your section views as one of the last steps in your project, occasionally data is added that needs to be shown in a section view after the section views are created. To accomplish this, you need to add that data to the sample line group. This is accomplished by navigating to the sample line group and editing the properties of the sample line group. The Sections tab shown in Figure 13.23 shows the Sample More Sources button required to add more data to the sample line group.

FIGURE 13.23
The Sections tab of the Sample Line Group Properties dialog

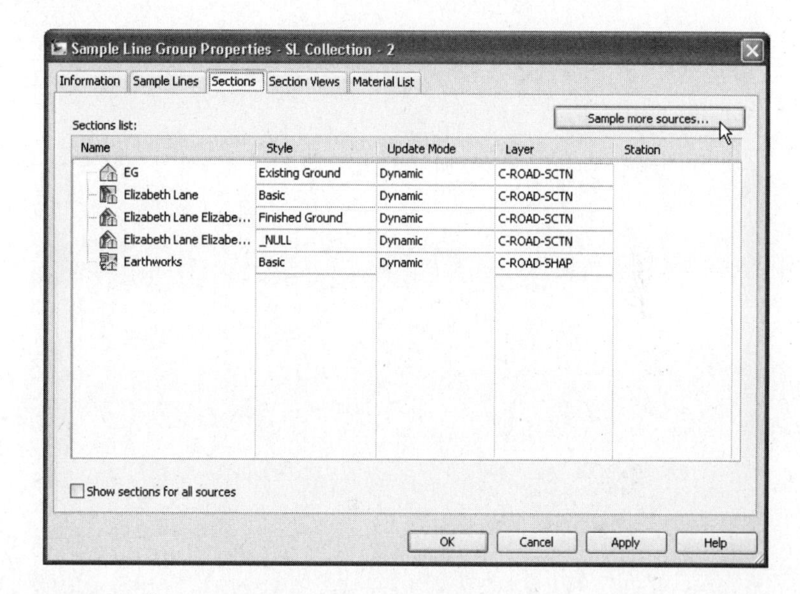

Once you click Sample More Sources, the Section Sources dialog opens. This dialog allows you to either add more sample sources to or delete sample sources from the sample line group. In this example, you have a sanitary sewer network that was added to the project. Because you need to show the locations of the sanitary pipes with respect to the designed road, you need to add the sanitary sewer network to the sample line group. To do this, simply select the sanitary sewer network and click Add, as shown in Figure 13.24.

Adding a Pipe Network to a Sample Line Group

In prior releases of Civil 3D, it was difficult to add more information to a section view once the sample lines were created. Quite often, it was easier to delete the sample line group and create a new one from scratch, including the information that you wanted to show. With the 2010 release, it is much easier to sample additional information. In this exercise, you'll add a pipe network to a sample line group and inspect the existing section views to ensure that the pipe network was added correctly:

1. Open the `sections6.dwg` file.
2. In Prospector, expand the Alignments ➢ Centerline Alignments ➢ Elizabeth Lane ➢ Sample Line Groups ➢ SL Collection – 2 branches, as shown in Figure 13.25.

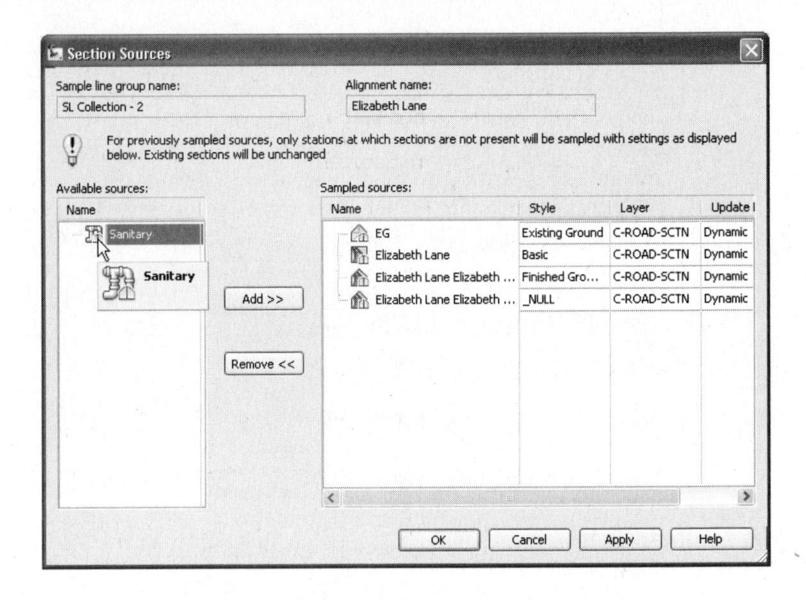

FIGURE 13.24 Adding a sanitary sewer network to the sample line group

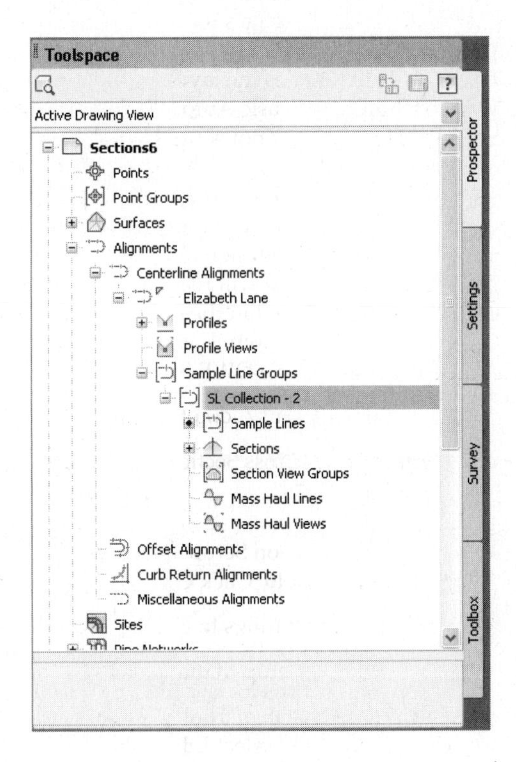

FIGURE 13.25 The location of the Sample Line Group, located under the Alignments branch

3. Right-click SL Collection – 2 and select Properties. The Sample Line Group Properties dialog appears.
4. Switch to the Sections tab if necessary.
5. Click Sample More Sources. The Section Sources dialog appears.
6. Select the Sanitary Network from the drop-down menu in the Available Sources selection box.
7. Click Add.
8. Click OK to dismiss the Section Sources dialog.
9. Click OK again to dismiss the Sample Line Group Properties dialog.
10. Close the drawing without saving.

Automating Plotting

By using automated plotting, you can make your cross sections arrange themselves on sheets. Creating multiple section views typically arranges the section views in rows and columns. This is fine for viewing the sections on screen, but local requirements often require that the section views be arranged and plotted on sheets for construction documents. Civil 3D has had the capability to arrange the sections on pages for a while, but it has always been a confusing and complicated task for users to grasp. The Plot by Page group plot style refers to a sheet style (a style within a style, if you will) that determines the layout and plot area for a given sheet size. This sheet style requires an understanding of a basic AutoCAD concept: page layouts. These page layouts are the basis for the entire process, and if not set up correctly they will render the Plot by Page group plot style useless.

In this exercise, you'll learn how to set up a sheet style, apply that sheet style to the Plot by Page group plot style, and then use the new styles to create section views on pages. For this exercise, you'll plot the section views on an arch D sheet size (24″×36″) with a scale of 1″ = 20′ horizontal and 1″ = 10′ vertical. This will require a vertical exaggeration of 2 for the section view style, which has already been specified in the drawing. Because page layouts are simpler to set up if you have something to start with, there is already a layout in the drawing with the correct parameters set to plot the arch D sheets to a DWF file.

1. Open the `sections7.dwg` file.
2. Right-click the Cross Sections Setup tab of the drawing and select Page Setup Manager.
3. Select New.
4. Select Cross Section Setup under Start With: and enter a new page setup name of **Cross Sections 20 and 10**. Click OK.
5. Verify the plot settings in the Page Setup dialog box as shown in Figure 13.26 and click OK.
6. Close the Page Setup Manager dialog.
7. On the Settings tab of Toolspace, expand the Section View ➢ Group Plot Styles, right-click Plot by Page, and select Edit.

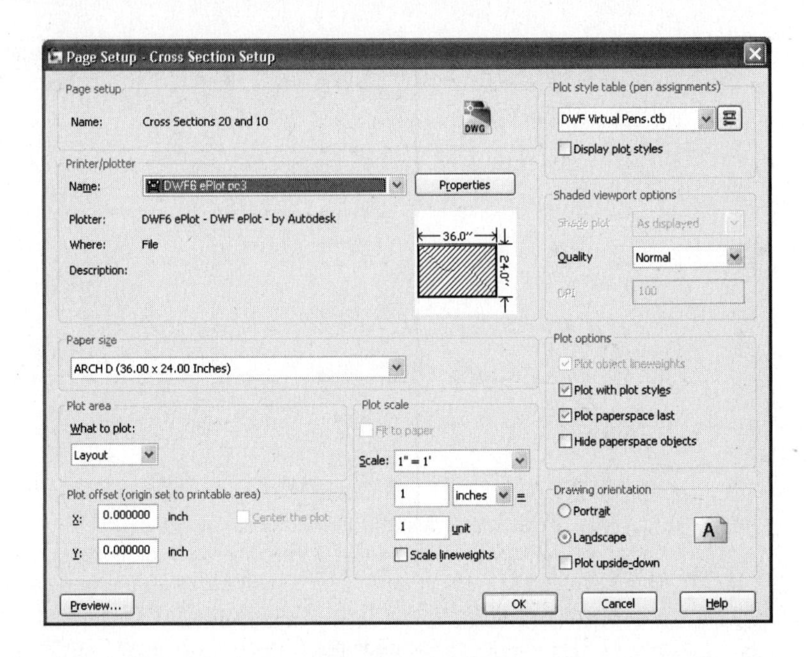

FIGURE 13.26
Settings for the Page Setup dialog

8. Switch to the Plot Area tab and make sure the Plot by Page radio button is selected. In the sheet style pull-down list, select Sheet Size – D (24x36). Then click the pull-down button to the right and select Edit Current Selection.

9. Switch to the Sheet tab and select the Cross Sections 20 and 10 page layout that you created earlier in the exercise. Set your page margins as follows:

   - Top: **0.25″**
   - Bottom: **0.25″**
   - Left: **0.25″**
   - Right: **1.00″**

   Click OK.

10. Back in the Plot By Page style dialog, switch to the Array tab. Because these section views are being plotted at a relatively large scale, not very many will fit on each sheet due to the width and height of your section views. Verify that your settings match the settings shown in Figure 13.27 and click OK.

11. Change to the Home tab and choose Section Views ➢ Create Multiple Views from the Profile & Section Views panel.

12. Because you have already used the Create Section View command in this chapter, the wizard remembers your default settings. Because of this, you only need to change one thing on the General page of the wizard — the group plot style. Change it to Plot by Page and select Create Section Views.

13. Pick a point in the drawing to insert your section views. It is a good idea to select an area to the right of the drawing and away from the design area.

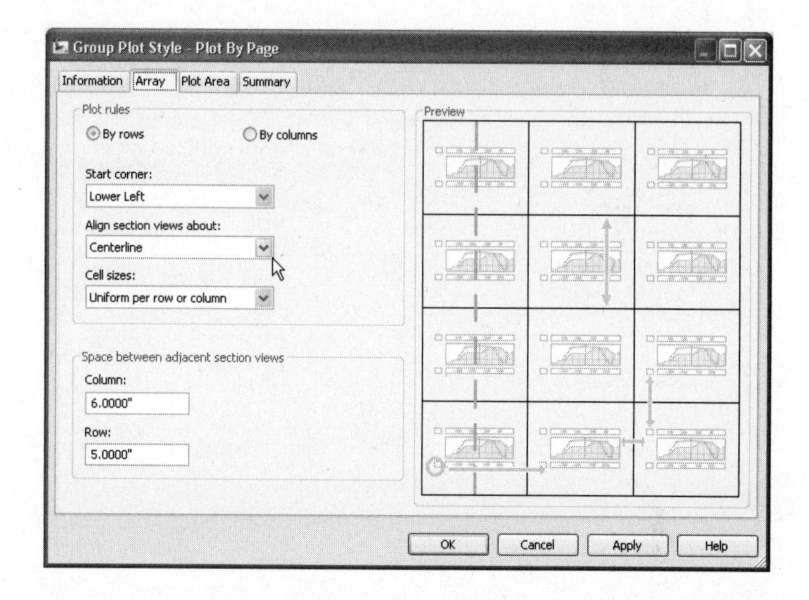

FIGURE 13.27
Array settings for plotting section views by page.

14. View the newly created section views. You will notice that they are arranged within two squares, one yellow and one green, as shown in Figure 13.28. The green square represents the edge of paper and the yellow square represents the plottable area of the paper.

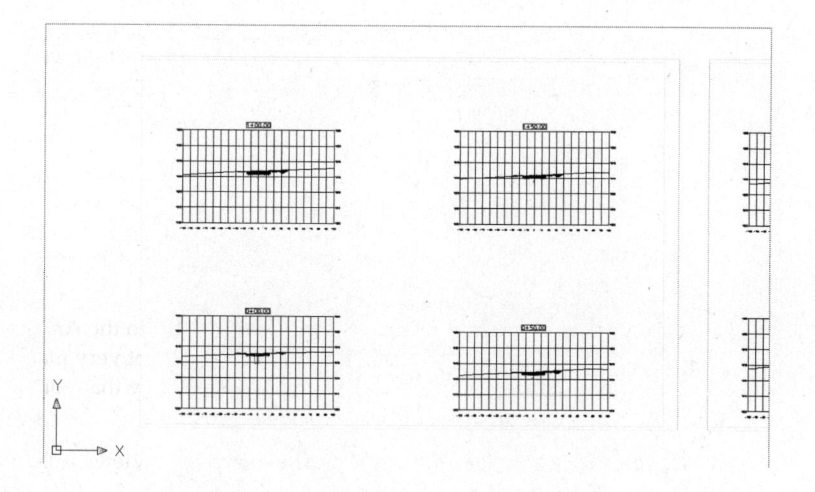

FIGURE 13.28
Section views arranged on sheets for plotting

15. To put the section views into layout views, activate the Cross Section Setup layout tab. Double-click within the viewport area to activate it and pan to the first set of section views. Take caution not to zoom, or you will change the viewport scale. Once you get to the area of the first sheet, you may have to pan it around to get it positioned exactly where you want it.

16. Close the drawing without saving.

Annotating the Sections

Now that the views are created, annotation is added to further explain design intent. Labels can be added through the code set style or by adding labels to the section view itself. These section labels can be used to label the section offset, elevation, or slope. You can see an example in Figure 13.29.

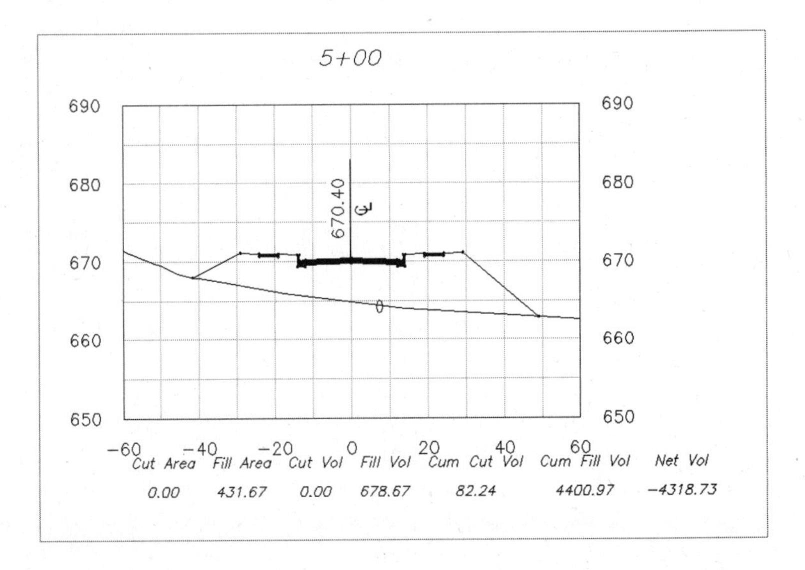

FIGURE 13.29 Adding labels to a section view will help explain design intent.

Real World Scenario

ADDING LABELS TO THE SECTION VIEW

There is often a need for labels to show what the various graphics mean in a section view. To show exact elevations, labels are much more efficient than scaling information from the grid. In this exercise, you'll label the section-view elevation at the designed centerline of the road:

1. Open the `sections8.dwg` file.
2. Change to the Annotate tab and choose Add Labels from the Labels & Tables panel. The Add Labels dialog appears.
3. Select Section View from the drop-down menu in the Feature selection box.
4. Select Offset Elevation from the drop-down menu in the Label Type selection box.
5. Select CL Elevation from the drop-down menu in the Label Style selection box.
6. Click Add.
7. Pick a grid on the section view to select the view.
8. Pick the top centerline of the corridor in the view. It may be easier to pick using the Intersection osnap.

9. Click Close in the Add Labels dialog. You now have labels reflecting the section-view elevation at the centerline.
10. Close the drawing without saving.

The Bottom Line

Create sample lines. Before any section views can be displayed, sections must be created from sample lines.

Master It Open `sections1.dwg` and create sample lines along the alignment every 50′.

Create section views. Just as profiles can only be shown in profile views, sections require section views to display. Section views can be plotted individually or all at once. You can even set them up to be broken up into sheets.

Master It In the previous drawing, you created sample lines. In that same drawing, create section views for all the sample lines.

Define materials. Materials are required to be defined before any quantities can be displayed. You learned that materials can be defined from surfaces or from corridor shapes. Corridors must exist for shape selection and surfaces must already be created for comparison in materials lists.

Master It Using `sections4.dwg`, create a materials list that compares EG with Elizabeth Lane Top Road Surface.

Generate volume reports. Volume reports give you numbers that can be used for cost estimating on any given project. Typically, construction companies calculate their own quantities, but developers often want to know approximate volumes for budgeting purposes.

Master It Continue using `sections4.dwg`. Use the materials list created earlier to generate a volume report. Create an XML report and a table that can be displayed on the drawing.

Chapter 14

The Tool Chest: Parts Lists and Part Builder

Before you can begin modeling and designing your pipe network (which will be discussed in Chapter 15, ''Running Downhill: Pipe Networks''), consider what components you'll need in your design. Civil 3D allows you to build a parts list that contains the specific pipes and structures you'll need at your fingertips while designing a pipe network. Civil 3D installs a large part catalog, as well as an interface for customizing pipe-network parts if necessary. In addition, the parts list enables you to specify how each pipe and structure will look and behave in a design situation. This chapter explores creating a parts list for a typical sanitary sewer network.

By the end of this chapter, you'll learn to:

- Add pipe and structure families to a new parts list
- Create rule sets that apply to pipes and structures in a parts list
- Apply styles to pipes and structures in a parts list

Planning a Typical Pipe Network: A Sanitary Sewer Example

Before you begin designing a pipe network, it's important to brainstorm all of the parts you'll need to construct the network, how these objects will be represented in plan and profile, and the behavior of these parts (which you'll specify using the slope, cover, rim, and sump parameters). Once you have a list of the elements you need, you can locate the appropriate parts in the part catalog, build the appropriate rule sets, create the proper styles to match your CAD standard, and tie it all together in a parts list.

Let's look at a typical sanitary sewer design. You typically start by going through the sewer specifications for the jurisdiction in which you're working. There is usually a published list of allowable pipe materials, manhole details, slope parameters, and cover guidelines. Perhaps you have concrete and PVC pipe manufacturer's catalogs that contain pages of details for different manholes, pipes, and junction boxes. There is usually a recommended symbology for your submitted drawings — and, of course, you have your own in-house CAD standards. Assemble this information, and make sure you address these issues:

- Recommended structures, including materials and dimensions (be sure to attach detail sheets)
- Structure behavior, such as required sump, drops across structures, and surface adjustment
- Structure symbology

- Recommended pipes, including materials and dimensions (again, be sure to attach detail sheets)
- Pipe behavior, such as cover requirements; minimum, maximum, and recommended slopes; velocity restrictions; and so on
- Pipe symbology

The following is an example of a completed checklist:

Sanitary Sewers in Sample County

Recommended Structures: Standard concentric manhole, small-diameter cleanout.

Structure Behavior: All structures have 1.5′ sump, rims, a 0.10′ invert drop across all structures. All structures designed at finished road grade.

Structure Symbology: Manholes are shown in plan view as a circle with an *S* inside and a diameter that corresponds to the actual diameter of the manhole. Cleanouts are shown as a solid, filled circle with a diameter that corresponds to the actual diameter of the cleanout. (See Figure 14.1.)

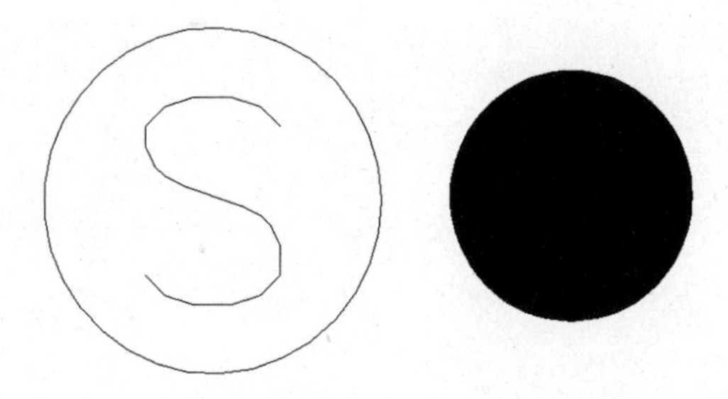

FIGURE 14.1 Sanitary sewer manhole in plan view (left) and a cleanout in plan view (right)

Manholes are shown in profile view with a coned top and rectangular bottom. Cleanouts are shown as a rectangle (see Figure 14.2).

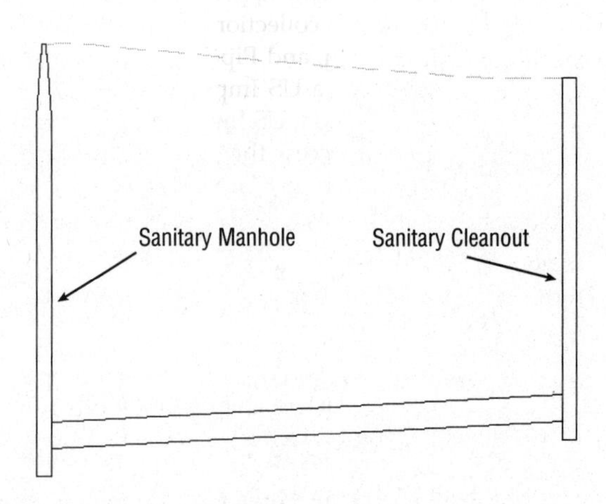

FIGURE 14.2 Profile view of a sanitary sewer manhole (left) and a cleanout (right)

Recommended Pipes: 8″, 10″, and 12″ PVC pipe per manufacturer specifications.

Pipe Behavior: Pipes must have cover of 4′ to the top of the pipe; the maximum slope for all pipes is 10 percent, although minimum slopes may be adjusted to optimize velocity as follows:

Sewer Size	Minimum Slope
8″	0.40%
10″	0.28%
12″	0.22%

Pipe Symbology: In plan view, pipes are shown with a CENTER2 linetype line that has a thickness corresponding to the inner diameter of the pipe. In profile view, pipes show both inner and outer walls, with a hatch between the walls to highlight the wall thickness (see Figure 14.3).

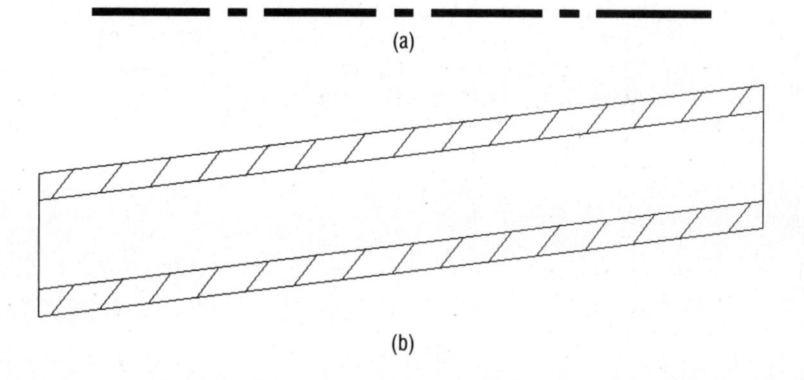

Figure 14.3
Sanitary pipe in plan view (a) and in profile view (b)

The Part Catalog

Once you know what parts you require, you need to investigate the part catalog to make sure these parts (or a reasonable approximation) are available.

The part catalog is a collection of two *domains* that contain two catalogs each. Structures are considered one domain, and Pipes are the second. The Structures domain consists of a Metric Structures catalog and a US Imperial Structures catalog; the Pipes domain consists of a Metric Pipes catalog and a US Imperial Pipes catalog.

Although you can access the parts from the catalogs while creating your parts lists in Civil 3D, you can't examine or explore the catalogs easily while in the Civil 3D interface. It's useful to understand where these catalogs reside and how they work.

The part catalog (see Figure 14.4) is installed locally by default at

```
C:\Documents and Settings\All Users\Application Data\Autodesk\C3D2010\enu\Pipes
Catalog\
```

Note that all paths in this chapter are the Windows XP install paths. If you're running Civil 3D on Windows Vista, please check the *Civil 3D Users Guide* for information on the `Pipes Catalog` folder install location.

If you can't locate the `Pipes Catalog` folder, it may be because your network administrator installed the catalogs at a network location when Civil 3D was deployed.

FIGURE 14.4
The Pipes Catalog folder

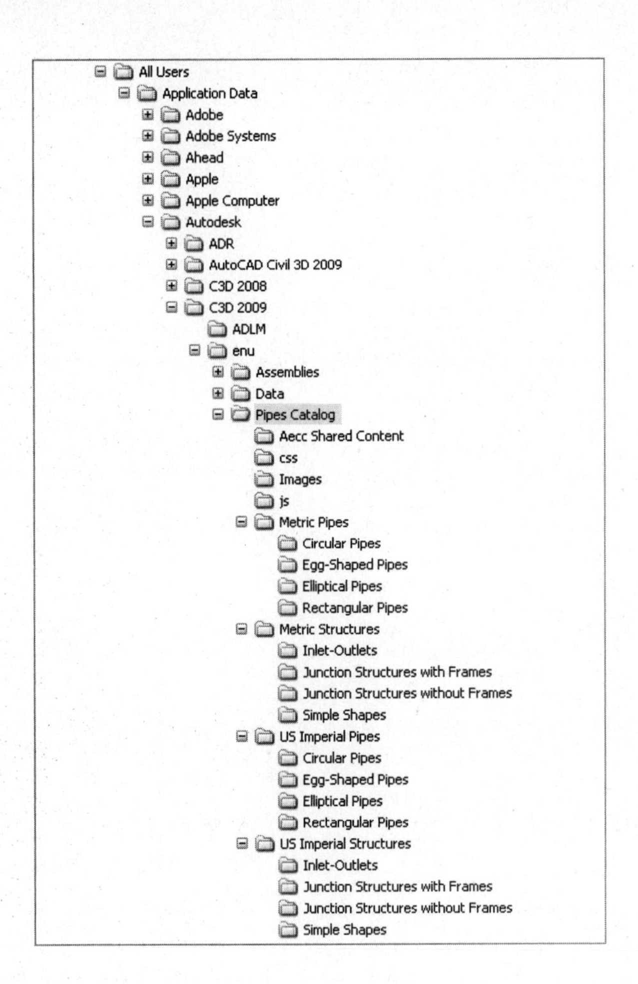

The Structures Domain

To learn more about how a catalog is organized, let's explore the `US Imperial Structures` folder.

The first file of interest in the `US Imperial Structures` folder is an HTML document called `US Imperial Structures` (see Figure 14.5).

FIGURE 14.5
The US Imperial Structures HTML document

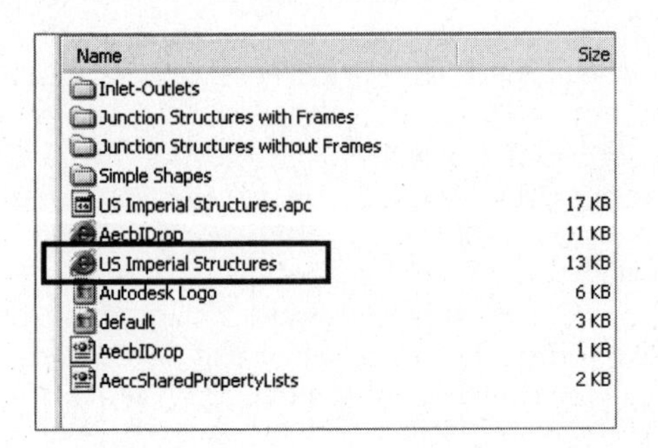

Double-click this file. Internet Explorer opens with a window so you can explore the US Imperial Structures catalog. A tree with different structure types on the left is under the Catalog tab. Expanding the tree allows you to explore the types of structures that are available. You may have to allow ActiveX controls to view the file.

Structures that fall into the same type have behavioral properties in common but may vary in shape and proportion. The four structure types in the default catalogs are as follows:

- Inlet-Outlets
- Junction Structures with Frames
- Junction Structures without Frames
- Simple Shapes

Under each structure type are several *shapes*. The shape spells out the details of how the structure is shaped and proportioned, and it shows what happens to each dimension when the size increases. If you drill into the Junction Structures with Frames type and highlight the AeccStructConcentricCylinder_Imperial shape, you can see this in action (see Figure 14.6).

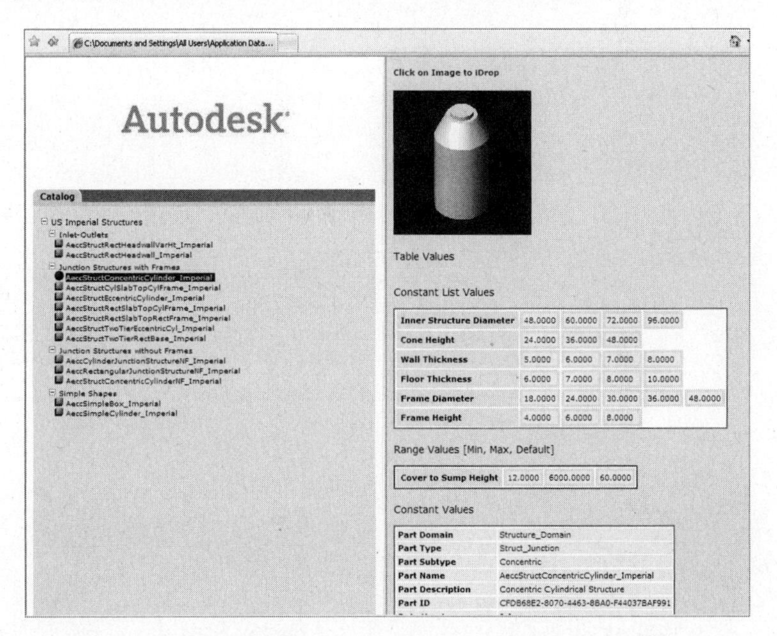

FIGURE 14.6
A closer look at the structure type and shape

In this view, you can see the available sizes of the Concentric Cylindrical structure, such as 48″, 60″, 72″, and 96″, but you may not edit them. Editing must be completed in the Part Builder as discussed later in this chapter. Follow the table to see what happens to the cone height, wall thickness, floor thickness, and frame diameter and height as the diameter dimension increases. The structure may get bigger or smaller, but its basic form remains the same and its behavior is predictable.

Explore the other structures of the Junction Structures with Frames type, and note why they're all considered to be the same type. Their shape changes, but their fundamental behavior and intent are similar. For example, each has a frame, each has a similar size range, and each is used as a junction for pipes.

Let's return to the example of a sanitary sewer network. You need a standard concentric manhole and a simple, small-diameter cleanout, and most likely you have specification sheets from the concrete products company handy. You know that you'll need to have a least a couple of Junction structures, and based on the required structures' spec sheets, you know you need a Junction Structure with Frame. Your manhole most closely resembles the Concentric Cylindrical structure, and the 48″ and 60″ match the allowable sizes. For the cleanout, the Cylindrical Slab Top structure is the appropriate shape and behavior, but you need a 6″ size. The smallest size available by default is 15″. Make notes on your checklist to add a part size for a 6″ Cylindrical Slab Top structure. (You'll take care of that in the next section.)

Something to keep in mind as you're searching for the appropriate structures to meet your standard is that the part you choose doesn't necessarily have to be a perfect match for your specified standard detail. The important things to look for are general shape, insertion behavior, and key dimensions.

Ask yourself the following questions:

- Will I be able to orient this structure in an appropriate way (is it round, rectangular, concentric, or eccentric)?
- Will I be able to label the insertion point (rim elevation) correctly?
- Will this structure look the way I need it to in the plan, profile, and section views? Does it look the same viewed from every angle, or would some views show a wider/narrower structure?
- Can I adjust my structure style in plan and profile views to display the structure the way I want to see it?

If you can find a standard shape that is conducive to all of these, it's worth your time to try it out before resorting to building custom parts with Part Builder.

For example, if a certain catch basin required in your town has a unique frame and is an elongated rectangle, try one of the standard rectangle frame structures first (see Figure 14.7). You probably already use a certain type of block in your CAD drawings to represent this type of catch basin, and that block can be applied to the structure style. It's not always necessary to build a custom part for small variations in shape. The important thing is that you model and label your rims and inverts properly. You'll learn more about this in the next few sections.

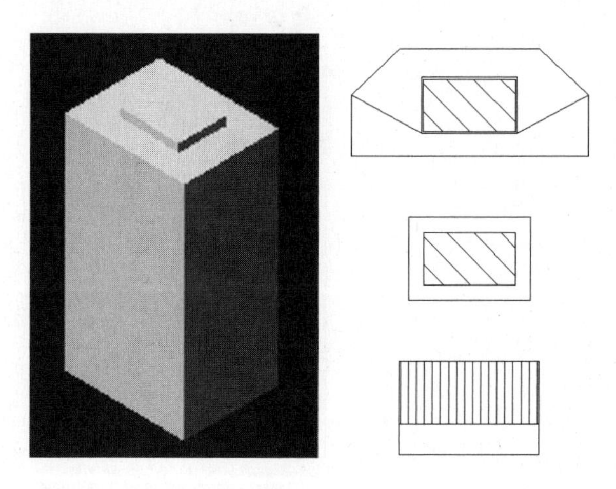

FIGURE 14.7
Each catch basin can be modeled with the same Rectangular Slab Top structure

The Pipes Domain

The second part domain is named Pipes. Pipes have one only type, which is also called Pipes.

Locate the catalog HTML file under the `US Imperial Pipes` folder (see Figure 14.8), and explore this catalog the same way you explored the US Imperial Structures catalog.

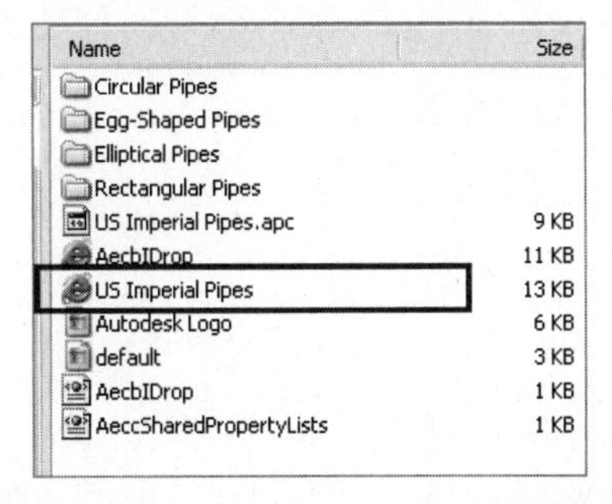

FIGURE 14.8
The catalog HTML file is in the US Imperial Pipes folder

Double-click the HTML file. Internet Explorer opens with a window that allows you to view the US Imperial Pipes catalog. A tree with different pipe shapes appears on the left under the Catalog tab. Expanding the tree allows you to explore the pipe shapes that are available. Pipe shapes are broken into smaller categories by material (in the case of a circular pipe) or orientation (for an elliptical pipe).

The four pipe shapes in the default catalogs are as follows:

- Circular Pipes (Concrete, Ductile Iron, and Polyvinyl Chloride)
- Egg-Shaped Pipes (Concrete)
- Elliptical Pipes (Concrete Vertical and Concrete Horizontal)
- Rectangular Pipes (Concrete)

Drill into the Circular Pipes shape, and highlight the AeccCircularConcretePipe_Imperial material (see Figure 14.9). Let's dig into this a bit more.

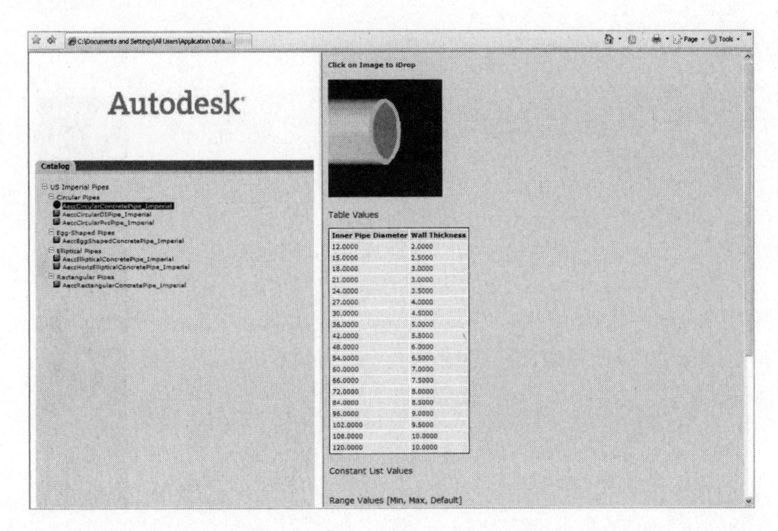

FIGURE 14.9
A closer look at the pipe shape and material

In this view, you can see the different inner pipe diameters and their corresponding wall thicknesses. Explore the other pipe shapes to get a feel for what is available.

You need 8″, 10″, and 12″ pipe for the sanitary sewer example. The spec sheet from the PVC pipe supplier indicates that the inner diameters and corresponding wall thicknesses are appropriate for your design.

Again, keep in mind as you're searching for the appropriate pipes to meet your standard that the pipe you choose doesn't necessarily have to be a perfect match for your specified standard detail. For example, your storm-drainage system might have the option to use high-density polyethylene (HDPE) pipe, and you might think you'd have to build a custom pipe type using Part Builder.

However, examine your symbology and labeling requirements a little further before jumping into building a custom part. The wall thickness of HDPE pipe is different from the same size PVC, but that may not be important for your drawing representation or labeling.

If you show HDPE as a single line or a double line representing inner diameter, wall thickness is irrelevant. You can create a custom label style to label the pipe HDPE instead of PVC.

If you show HDPE represented as its outer diameter or have particular requirements to show wall thickness in crossing and profile views, then building a custom pipe in Part Builder may be your best option (see Figure 14.10).

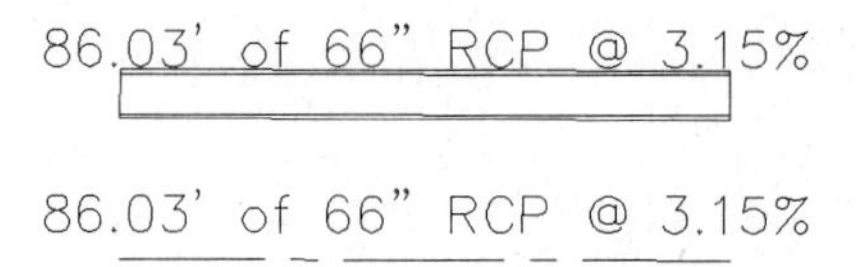

FIGURE 14.10
The top pipe must specify the outer diameter exactly. The bottom pipe, which uses a centerline to represent the inner diameter, doesn't have this requirement

The important things to consider are general shape and wall thickness. Ask yourself these questions:

- Does this pipe approximate the shape of my specified pipe?
- Does this pipe offer the correct inner diameter choice?
- How important is wall thickness to my model, plans, profiles, and sections?
- Can I adjust my pipe style in plan and profile views to display the pipe the way I need to see it?
- Can I create a pipe label that will label this pipe the way I need it labeled?

Again, if you can find a standard pipe that is conducive to all of these, it's worth your time to try it before you resort to building custom pipes with Part Builder.

The Supporting Files

Each part family in the catalog has three corresponding files located at

```
C:\Documents and Settings\All Users\Application Data\Autodesk\C3D2010\enu\Pipes
Catalog\
```

Two files are mandatory sources of data required for the part to function properly, and one is an optional bitmap that provides the preview image you see in the catalog browser and parts-list interface.

First, dig into the `US Imperial Pipes` folder as follows:

```
C:\Documents and Settings\All Users\Application Data\Autodesk\C3D2010\enu\Pipes
Catalog\US Imperial Pipes\Circular Pipes
```

This reveals the following files for the Imperial Circular Pipe:

- `*partname*.dwg`: This part drawing file contains the geometry that makes up the part as well as the definition of the parametric relationships.
- `*partname*.xml`: This part XML file is created as soon as a new part file is started in Part Builder. This file contains information on the parameter sizes, such as wall thickness, width, and diameter.
- `*partname*.bmp`: This part image file is used as a visual preview for that particular part in the HTML catalog (as you saw in the previous section). This file is optional.

You'll study some of these files in greater detail in the next section.

Mark on your checklist which parts you'd like to use from the standard catalog, and also note the parts that you may have to customize using Part Builder.

Part Builder

Part Builder is an interface that allows you to build and modify pipe network parts. Part Builder is accessed by selecting Create Design ➤ Partbuilder from the Home tab. At first, you may use Part Builder to add a few missing pipes or structure sizes. As you become more familiar with the environment, you may build your own custom parts from scratch.

This section is intended to be an introduction to Part Builder and a primer in some basic skills required to navigate the interface. It isn't intended to be a robust "how-to" for creating custom parts. For more details on building custom parts, please see the *Civil 3D Users Guide* — which you can access through the Civil 3D Help interface, as well as in `.pdf` format under Help ➤ Users Guide (pdf). Civil 3D 2010 also includes three detailed tutorials for creating three types of custom structures. The tutorials lead you through creating a Cylindrical Manhole structure, a Drop Inlet Manhole structure, and a Vault structure. You can find these tutorials by going to Help ➤ Tutorials and then navigating to AutoCAD Civil 3D Tutorials ➤ Part Builder Tutorials.

BACK UP THE PART CATALOGS

Here's a warning: before exploring Part Builder in any way, it's critical that you make a backup copy of the part catalogs. Doing so will protect you from accidentally removing or corrupting default parts as you're learning and will provide a means of restoring the original catalog.

The catalog (as discussed in the previous section) can be found by default at

```
C:\Documents and Settings\All Users\Application Data\Autodesk\C3D2010\enu\Pipes
Catalog\
```

To make a backup, copy this entire directory and then save that copy to a safe location, such as another folder on your hard drive or network, or to a CD.

Parametric Parts

The parts in the Civil 3D pipe network catalogs are *parametric*. Parametric parts are dynamically sized according to a set of variables, or parameters. In practice, this means you can create one part and use it in multiple situations.

For example, in the case of circular pipes, if you didn't have the option of using a parametric model, you'd have to create a separate part for each diameter of pipe you wanted, even if all other aspects of the pipe remained the same. If you had 10 pipe sizes to create, that would mean 10 sets of `partname.dwg`, `partname.xml`, and `partname.bmp` files, as well as an opportunity for mistakes and a great deal of redundant editing if one aspect of the pipe needed to change.

Fortunately, you can create one parametric model that understands how the different dimensions of the pipe are related to each other and what sizes are allowable. When a pipe is placed in a drawing, you can change its size. The pipe will understand how that change in size affects all the other pipe dimensions such as wall thickness, outer diameter, and more; you don't have to sort through a long list of individual pipe definitions.

Part Builder Orientation

The Civil 3D pipe network catalogs are drawing specific. If you're in a metric drawing, you need to make sure the catalog is mapped to metric pipes and structures, whereas if you're in an imperial drawing, you'll want the imperial. By default, the Civil 3D templates should be appropriately mapped, but it's worth the time to check. Set the catalog by changing to the Home tab and selecting Create Design ➤ Set Pipe Network Catalog. Verify the appropriate folder and catalog for your drawing units in the Pipe Network Catalog Settings dialog (see Figure 14.11), and you're ready to go.

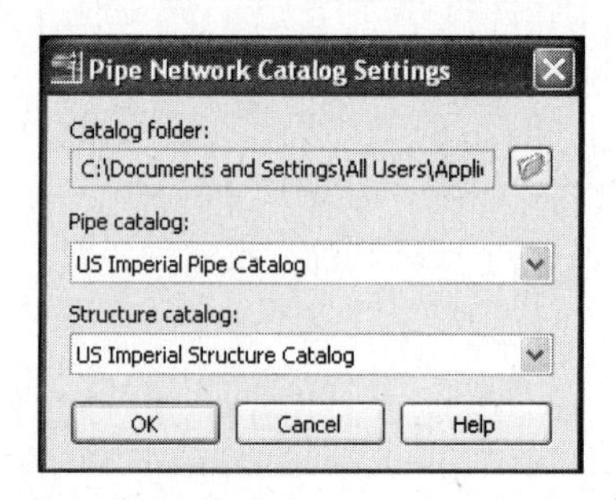

FIGURE 14.11
Choose the appropriate folder and catalog for your drawing units

UNDERSTANDING THE ORGANIZATION OF PART BUILDER

The vocabulary used in the Part Builder interface is related to the vocabulary in the HTML catalog interface that you examined in the previous section, but there are several differences that are sometimes confusing.

The first screen that appears when you start the Part Builder is Getting Started – Catalog Screen (see Figure 14.12).

At the top of this window is a drop-down menu for selecting the pipe catalog. The choices, in this case Pipe and Structure, are based on what has been set for the drawing (either Metric or Imperial).

Below the Part Catalog input box is a listing of chapters. (In terms of Part Builder vocabulary, a *pipe chapter* is roughly equivalent to the catalog interface term *shape*.) US Imperial Pipe Catalog has four default chapters: Circular Pipes, Egg-Shaped Pipes, Elliptical Pipes, and Rectangular Pipes. You can create new chapters for different-shaped pipes, such as Arch Pipe.

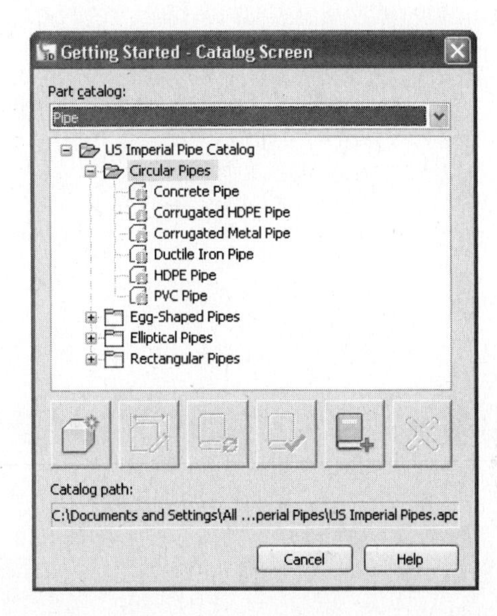

FIGURE 14.12
The Getting Started – Catalog Screen

US Imperial Structure Catalog also has four default chapters: Inlets-Outlets, Junction Structures With Frames, Junction Structures Without Frames, and Simple Shapes. You can create new chapters for custom structures. (In terms of Part Builder vocabulary, a *structure chapter* is roughly equivalent to the catalog interface term *type*.)

You can expand each chapter folder to reveal one or more part families. US Imperial Pipe Catalog has six default families (Concrete Pipe, Corrugated HDPE Pipe, Corrugated Metal Pipe, Ductile Iron Pipe, HDPE Pipe, and PVC Pipe) under the Circular Pipes chapter. Pipes that reside in the same family typically have the same parametric behavior, with only differences in size.

US Imperial Pipe Catalog has four default families (Inlet-Outlets, Concentric Cylindrical Structure NF, Cylindrical Junction Structure NF, and Rectangular Junction Structure NF) under the Junction Structures without Frames chapter. Like pipes, structures that reside in the same family typically have the same parametric behavior, with only differences in size.

As Table 14.1 shows, a series of buttons on the Getting Started – Catalog Screen lets you perform various edits to chapters, families, and the catalog as a whole.

Exploring Part Families

The best way to get oriented to the Part Builder interface is to explore one of the standard part families. In this case, examine the US Imperial Pipe Catalog ➢ Circular Pipe Chapter ➢ Concrete Pipe family by highlighting Concrete Pipe and clicking the Modify Part Sizes button. A Part Builder task pane appears with `AeccCircularConcretePipe_Imperial.dwg` on the screen, as shown in Figure 14.13.

The Part Builder task pane, or Content Builder (Figure 14.14), is well documented in the *Civil 3D Users Guide*. Please refer to the *Users Guide* for detailed information about each entry in Content Builder.

TABLE 14.1: The Part Builder Catalog Tools

ICON	FUNCTION
	The New Parametric Part button creates a new part family.
	The Modify Part Sizes button allows you to edit the parameters for a specific part family.
	The Catalog Regen button refreshes all the supporting files in the catalog when you've finished making edits to the catalog.
	The Catalog Test button validates the parts in the catalog when you've finished making edits to the catalog.
	The New Chapter button creates a new chapter.
	The Delete button deletes a part family. Use this button with caution, and remember that if you accidentally delete a part family, you can restore your backup catalog as mentioned in the beginning of this section.

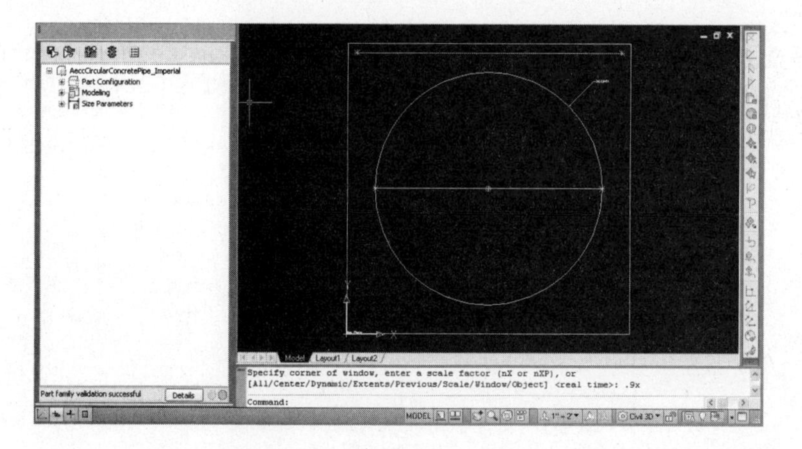

FIGURE 14.13
The Parametric Building environment

Adding a Part Size Using Part Builder

The hypothetical municipality requires a 12″ sanitary sewer cleanout. After studying the catalog, you decide that Concentric Cylindrical Structure with No Frames is the appropriate shape for your model, but the smallest inner diameter size in the catalog is 48″. The following tutorial gives you

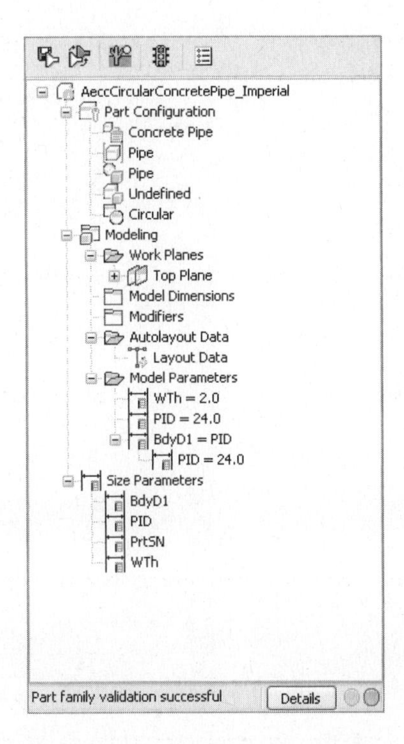

FIGURE 14.14
Content Builder

some practice in adding a structure size to the catalog — in this case, adding a 12″ structure to the US Imperial Structures catalog:

1. You can make changes to the US Imperial Structures catalog from any drawing that is mapped to that catalog, which is probably any imperial drawing you have open. For this exercise, start a new drawing from `_AutoCAD Civil 3D (Imperial) NCS Base.dwt`.
2. On the Home tab, select Create Design ➢ Part Builder.
3. Choose Structure from the drop-down list in the Part Catalog selection box.
4. Expand the Junction Structure without Frames chapter.
5. Highlight the Concentric Cylindrical Structure NF (no frames) part family.
6. Click the Modify Part Sizes button.
7. The Part Builder interface opens `AeccStructConcentricCylinderNF_Imperial.dwg` along with the Content Builder task pane.
8. Expand the Size Parameters tree.
9. Right-click the SID (Structure Inner Diameter) parameter, and choose Edit. The Edit Part Sizes dialog appears.
10. Locate the SID column (see Figure 14.15). Double-click inside the box, and note that a drop-down menu shows the available inner diameter sizes: 48, 60, 72, and 96.

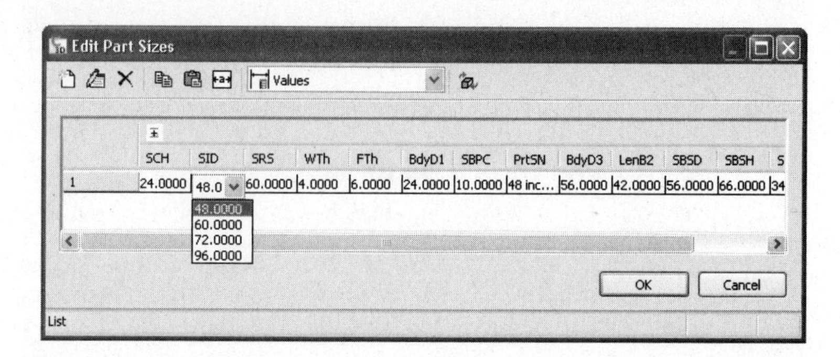

FIGURE 14.15
Choosing a part size

11. Locate the Edit button (see Figure 14.16). Make sure you're still active in the SID column cell, and then click Edit. The Edit Values dialog appears. Click Add, and type **12**. Click OK to close the Edit Values dialog, and click OK again to close the Edit Part Sizes dialog.

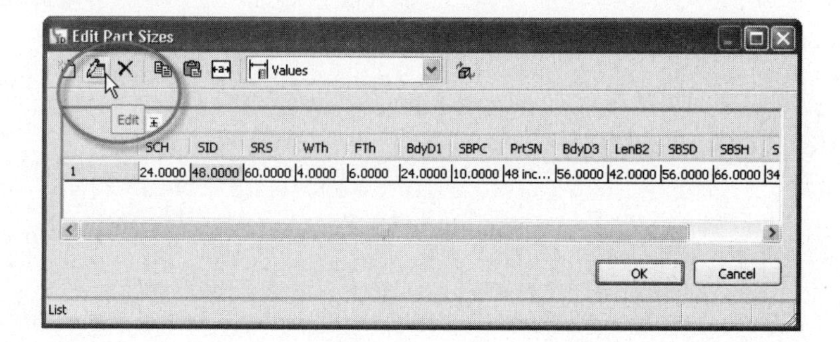

FIGURE 14.16
Click the Edit button to open the Edit Values dialog

12. Click the small X in the upper-right corner of the Content Builder task pane to exit Part Builder. When prompted with Save Changes to Concentric Cylindrical Structure NF, click Yes. (You could also click Save in Content Builder to save the part and remain active in the Part Builder interface.)

13. You're back in your original drawing. If you created a new parts list in any drawing that references the US Imperial Structures catalog, the 12″ structure would now be available for selection.

Sharing a Custom Part

You may find that you need to go beyond adding pipe and structure sizes to your catalog and build custom part families or even whole custom chapters. Perhaps instead of building them yourself, you're able to acquire them from an outside source.

The following section can be used as reference for adding a custom part to your catalog from an outside source, as well as sharing custom parts that you've created. The key to sharing a part is to locate the three files mentioned earlier.

Adding a custom part size to your catalog requires these steps:

1. Locate the *partname*.dwg, *partname*.xml, and (optionally) *partname*.bmp files of the part you'd like to obtain.

2. Make a copy of the *partname*.dwg, *partname*.xml, and (optionally) *partname*.bmp files.

3. Insert the *partname*.dwg, *partname*.xml, and (optionally) *partname*.bmp files in the correct folder of your catalog.

4. Run the **partcatalogregen** command in Civil 3D.

Adding an Arch Pipe to Your Part Catalog

This exercise will teach you how to add a premade custom part to your catalog:

1. You can make changes to the US Imperial Pipes catalog from any drawing that is mapped to that catalog, which is probably any imperial drawing you have open. For this exercise, start a new drawing from the _AutoCAD Civil 3D (Imperial) NCS Base.dwt file.

2. Create a new folder called **Arch pipe** in the following directory:

   ```
   C:\Documents and Settings\All Users\Application Data\Autodesk\C3D2010\
   enu\Pipes Catalog\US Imperial Pipes\
   ```

   This directory should now include five folders: Arch Pipe, Circular Pipes, Egg-Shaped Pipes, Elliptical Pipes, and Rectangular Pipes.

3. Copy the Concrete Arch Pipe.dwg and Concrete Arch Pipe.xml files into the Arch Pipe folder. (Note that there is no optional bitmap for this custom pipe.)

4. Return to your drawing, and enter **PARTCATALOGREGEN** in the command line. Press P to regenerate the Pipe catalog. Press ↵ to exit the command.

5. If you created a new parts list at this point in any drawing that references the US Imperial Pipes catalog, the arch pipe would be available for selection.

6. To confirm the addition of the new pipe shape to the catalog, locate the catalog HTML file at

   ```
   C:\Documents and Settings\All Users\Application Data\Autodesk\C3D2010\
   enu\Pipes Catalog\US Imperial Pipes\US Imperial Pipes.htm
   ```

   Explore the catalog as you did in the "The Part Catalog" section earlier.

Part Styles

The catalog defines how parts are modeled, and styles define how parts are represented in the drawing. Styles can be true reflections of the model — or, more commonly, they can be customized and enhanced with items such as AutoCAD blocks.

Your company or municipality should have a CAD standard that spells out the symbology you need to represent your pipes and structures. Prepare a list of blocks, layers, and other specifications that you need to compose the appropriate styles.

Creating Structure Styles

On the basis of the hypothetical sanitary sewer example, you need to create structure styles to reflect the conventions shown earlier in Figures 14.1 and 14.2. You need a different structure style for each type shown. With your list of specifications, explore the style options to plan how best to build your styles. You access the Structure Style dialog by expanding Structure Styles on the Settings tab of Toolspace and double-clicking an existing style or creating a new style.

The following tour through the structure-style interface can be used for reference as you create company standard styles.

THE MODEL TAB

This tab (Figure 14.17) controls what represents your structure when you're working in 3D. Typically, you want to leave this set to Use Catalog Defined 3D Part so that when you look at your structure, it looks like your concentric manhole or whatever you've chosen in the parts list.

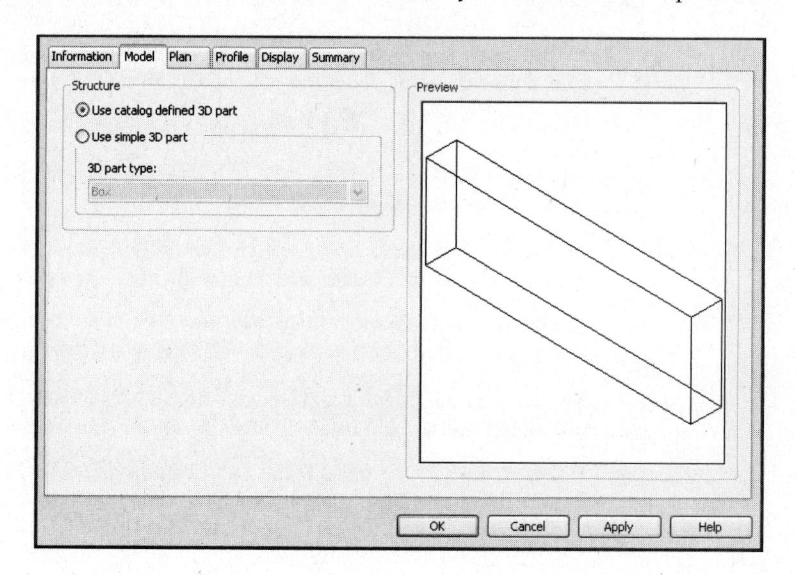

FIGURE 14.17
The Model tab in the Structure Style dialog

THE PLAN TAB

During your brainstorming session, you figured out how you need your structure represented in the plan. The Plan tab (Figure 14.18) enables you to compose your object style to match that specification.

Options on the Plan tab include the following:

Use Outer Part Boundary uses the actual limits of your structure from the parts list and shows you an outline of the structure as it would appear in the plan.

User Defined Part uses any block you specify. In the case of your sanitary manhole, you chose a symbol to match the CAD standard.

Size Options has several options. Click Help to learn more about the specifics of each option.

Enable Part Masking creates a wipeout or mask inside the limits of the structure. Any pipes that connect to the center of the pipe appear trimmed at the limits of the structure.

THE PROFILE TAB

Once you know what your structure must look like in the profile, you use the Profile tab (Figure 14.19) to create the style.

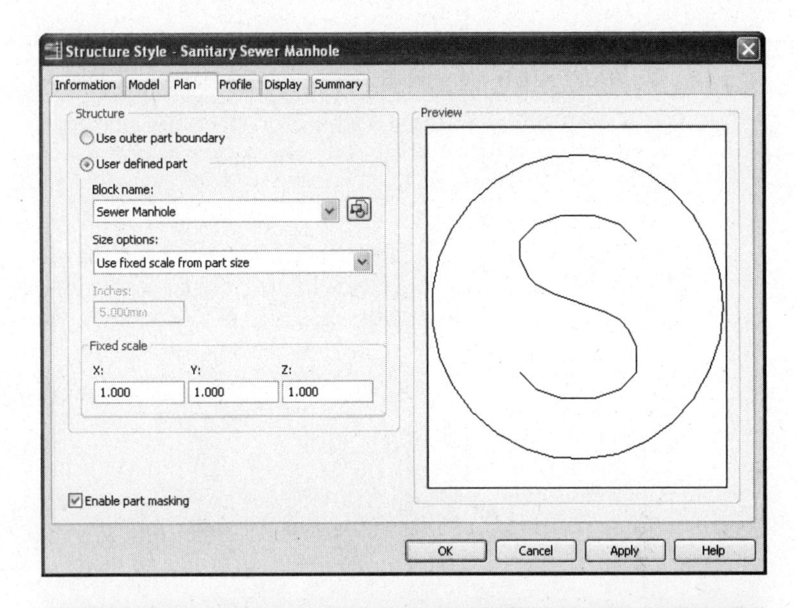

FIGURE 14.18
The Plan tab in the Structure Style dialog

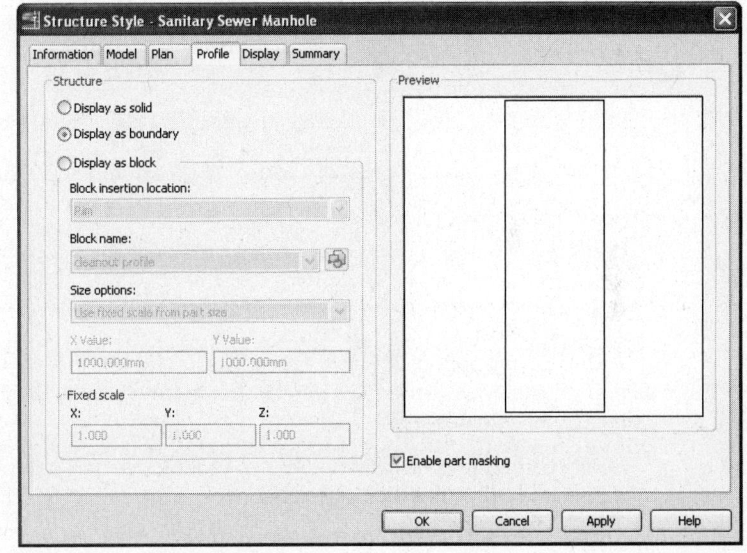

FIGURE 14.19
The Profile tab in the Structure Style dialog

Options on the Profile tab include the following:

Display as Solid uses the actual limits of your structure from the parts list and shows you the mesh of the structure as it would appear in profile view.

Display as Boundary uses the actual limits of your structure from the parts list and shows you an outline of the structure as it would appear in profile view. You'll use this option for the sanitary manhole.

Display as Block uses any block you specify. You'll use this option for the sanitary cleanout.

Size Options has several options. Click Help to learn more about the specifics of each option.

Enable Part Masking creates a wipeout or mask inside the limits of the structure. Any pipes that connect to the center of the pipe appear trimmed at the limits of the structure.

THE DISPLAY TAB

The Display tab (Figure 14.20) enables you to control the visibility and display properties of all possible structure style elements in Plan, Model, Profile, and Section. When you're creating a structure style, check all of the directions to be sure your structure will be shown as desired.

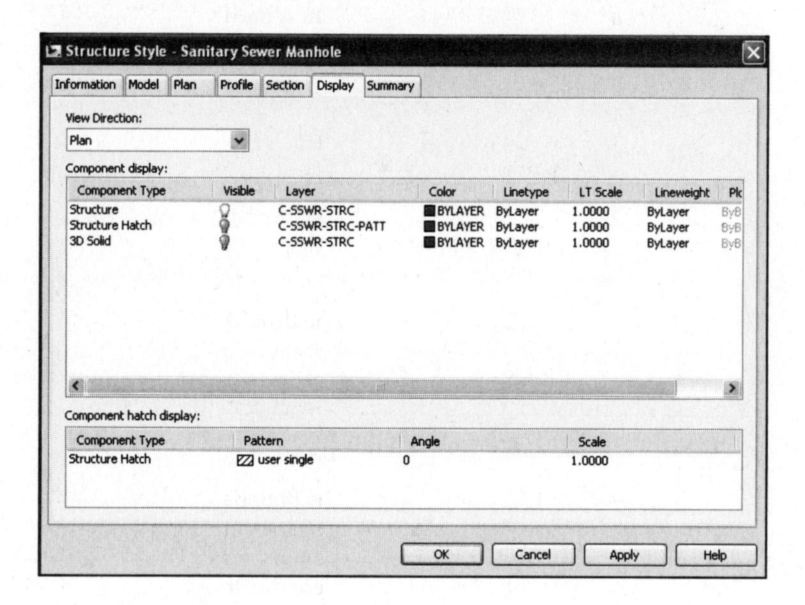

FIGURE 14.20
The Display tab in the Structure Style dialog

The display options for structure styles have a number of components that you need to review and understand. Use the View Direction pull-down menu to navigate between Plan, Model, Profile and Section components:

Plan Structure controls whether the structure is visible in plan view and allows for layer mapping and color/linetype overrides.

Plan Structure Hatch controls the visibility of any structure hatch in plan view and also allows for layer mapping and color/linetype overrides.

Model 3D Solid controls the visibility of the 3D model in plan view and allows for layer mapping and color/linetype overrides. In the 2D view direction, this is most commonly set to not visible.

Profile Structure controls whether the structure is visible in profile view and allows for layer mapping and color/linetype overrides.

Profile Structure Hatch controls the visibility of any structure hatch in profile view and allows for layer mapping and color/linetype overrides.

Section Structure controls whether the structure is visible in a section view and allows for layer mapping and color/linetype overrides.

Section Structure Hatch controls the visibility of any structure hatch in a section view and allows for layer mapping and color/linetype overrides.

Profile Structure Pipe Outlines controls the visibility of incoming connected pipes in profile view and allows for layer mapping and color/linetype overrides.

Section Structure Pipe Outlines controls the visibility of incoming connected pipes in a section view and allows for layer mapping and color/linetype overrides.

Creating Structure Styles for a Sanitary Manhole and Sanitary Cleanout

In this exercise, you'll create the structure styles for the sanitary sewer project:

1. Open `Parts List.dwg`.
2. Expand the Structures ➤ Structures Styles trees on the Settings tab of Toolspace. Right-click the Sanitary Sewer Manhole style, and choose Edit. The Structure Style dialog appears.
3. Switch to the Plan tab, and note that the User Defined Part option is selected. Note that the Block Name selection box has a drop-down menu you can use to choose any block contained in the drawing.
4. Under Size Options, use the drop-down menu to change the size from the Use Drawing Scale option to Use Fixed Scale. This automatically sizes the block to reflect the modeled structure size: a 48″ manhole is represented smaller than a 72″ manhole, and so on. Note that the Enable Part Masking check box is selected.
5. Switch to the Profile tab.
6. Confirm that the Display as Boundary option is selected. This makes an outline in profile view that reflects the true shape of the model (ignore the preview window at this point). You'll specify that your sanitary manholes are concentric cylindrical structures when you build the parts list; therefore, this boundary will be a conical-topped structure with a square bottom, as required by the CAD standard. Note that the Enable Part Masking check box is selected.
7. Switch to the Display tab. Use the View Direction drop-down to confirm that the structure will be visible in plan, model, profile, and section views. Confirm that all hatches are set to not visible.
8. Click OK.
9. Right-click the Sanitary Sewer Manhole style that you just edited, and choose Copy.
10. On the Information tab, rename the copied style **Sanitary Sewer Cleanout**.
11. Switch to the Plan tab, and change the selection to the Use Outer Part Boundary option (ignore the preview window at this point); this makes an outline around the outer limits of the model, as seen in plan view. Because you're using a circular structure to represent your cleanouts, this option returns a circle that corresponds with the diameter of your cleanout.
12. Switch to the Profile tab, and select the Display as Block option; select Rim from the drop-down menu in the Block Insertion Location selection box; and then click the button

next to the drop-down menu in the Block Name selection box to choose a block that is outside the drawing (see Figure 14.21). Navigate to the drawing `cleanout profile.dwg`. Select this drawing to be the block that represents your cleanout in profile.

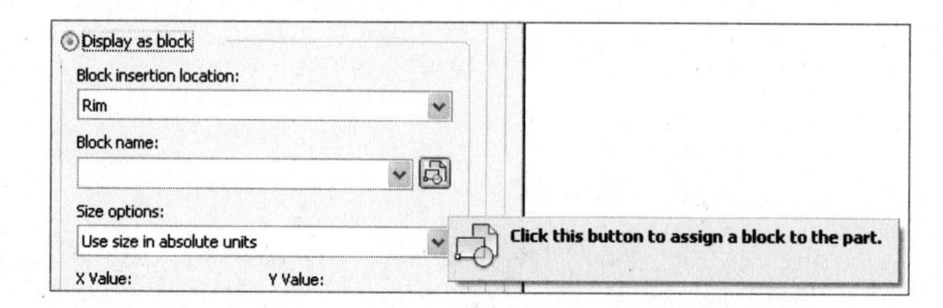

FIGURE 14.21
Choose a block from outside the drawing

13. Select the Use Fixed Scale from Part Size option from the drop-down menu in the Size Options selection box. This option stretches the block to accommodate the depth of the structure when your simple rectangular block is inserted at the structure rim location.

14. Switch to the Display tab. Leave everything the same, except make the Plan Structure Hatch visible. Click in the Plan Hatch pattern column at the bottom of the dialog; the Hatch Pattern dialog appears. Select Solid Fill (see Figure 14.22) from the Type drop-down menu; this fills your circle with a solid hatch to match the CAD standard. Click OK.

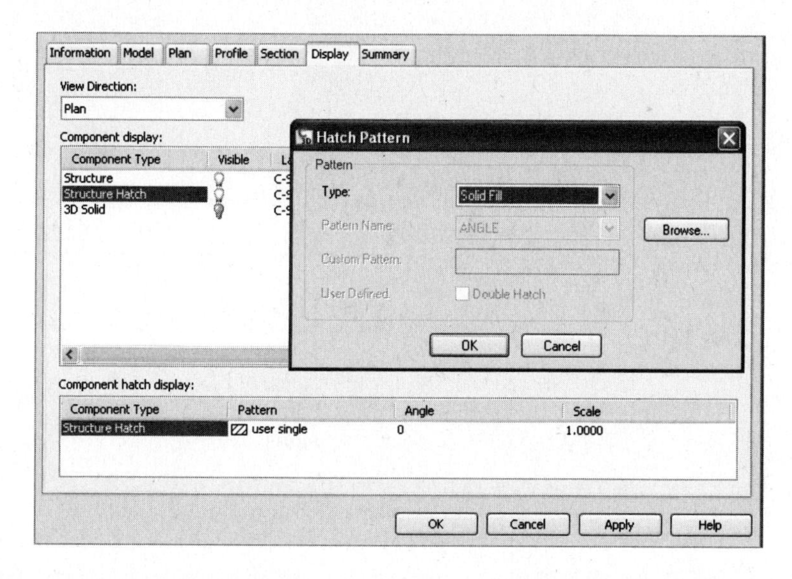

FIGURE 14.22
The Hatch Pattern dialog

15. Click OK to close the Structure Style dialog. You'll use these styles later when you define your parts list.

16. Save your drawing to use in the next exercise.

Creating Pipe Styles

On the basis of the hypothetical sanitary sewer example, you need to create pipe styles to reflect the pipes shown earlier in Figure 14.3.

You need one pipe style to handle your sanitary sewer pipe. In your real projects, you'll need to do similar brainstorming for storm drainage, water, and any other pipes you may be designing. With your list of specifications, explore the style options to plan how best to build your styles. You access the Pipe Style dialog by expanding Pipe Styles on the Settings tab of Toolspace and double-clicking an existing style or creating a new style. The following tour through the pipe-style interface can be used for reference as you create company standard styles.

THE PLAN TAB

This tab (see Figure 14.23) controls what represents your pipe when you're working in plan view.

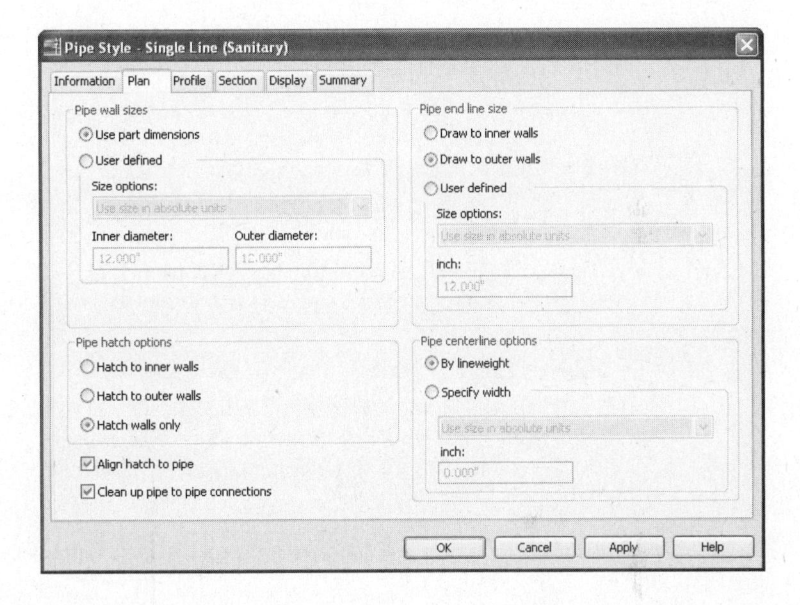

FIGURE 14.23
The Plan tab in the Pipe Style dialog

Options on the Plan tab include the following (click Help at any time for a detailed breakdown of each option):

Pipe Wall Sizes: You have a choice of having the program apply the part size directly from the catalog part (that is, the literal pipe dimensions as defined in the catalog) or specifying your own constant or scaled dimensions.

Pipe Hatch Options: If you choose to show pipe hatching (see the Display tab), this part of the dialog gives you options to control that hatch. You can hatch the entire pipe to the inner or outer walls, or you can hatch the space between the inner and outer walls only, as shown in Figure 14.24.

Pipe End Line Size: If you choose to show an end line (see the Display tab), you can control its length with these options. An end line can be drawn connecting the outer walls (see Figure 14.25) or the inner walls, or you can specify your own constant or scaled dimensions.

Pipe Centerline Options: If you choose to show a centerline (see the Display tab), you can display it by the lineweight established in the Display tab, or you can specify your own part-driven, constant, or scaled dimensions. Use this option for your sanitary pipes in places where the width of the centerline widens or narrows on the basis of the pipe diameter.

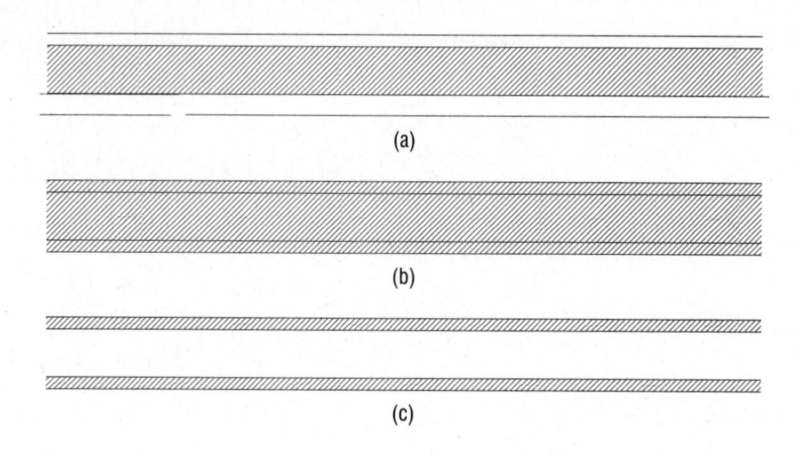

FIGURE 14.24
Pipe hatch to inner walls (a), outer walls (b), and hatch walls only (c)

FIGURE 14.25
Pipe end line shown to the outer walls

THE PROFILE TAB

The Profile tab (see Figure 14.26) is almost identical to the Plan tab, except the controls here determine what your pipe looks like in profile view. The only additional settings on this tab are the crossing-pipe hatch options. If you choose to display crossing pipe with a hatch, these settings control the location of that hatch.

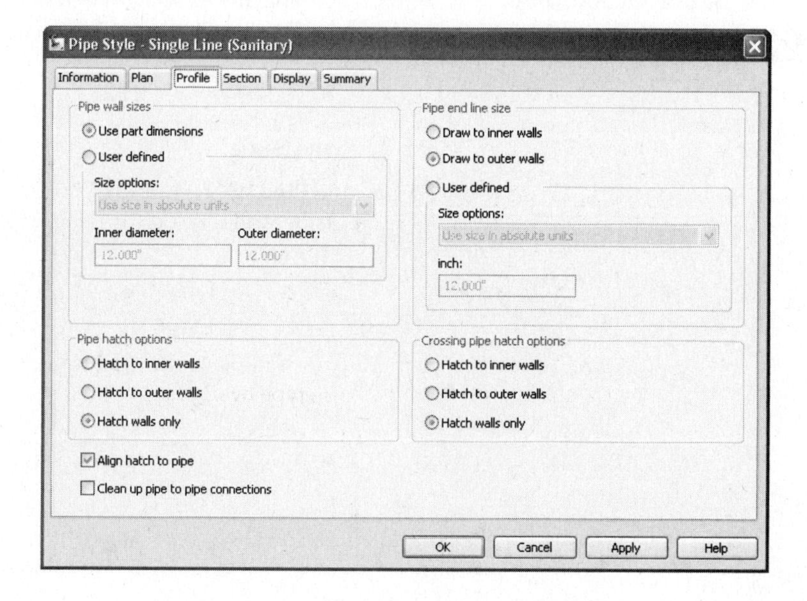

FIGURE 14.26
The Profile tab in the Pipe Style dialog

THE SECTION TAB

If you choose to show a hatch on your pipes in section, you control the hatch location on this tab (see Figure 14.27).

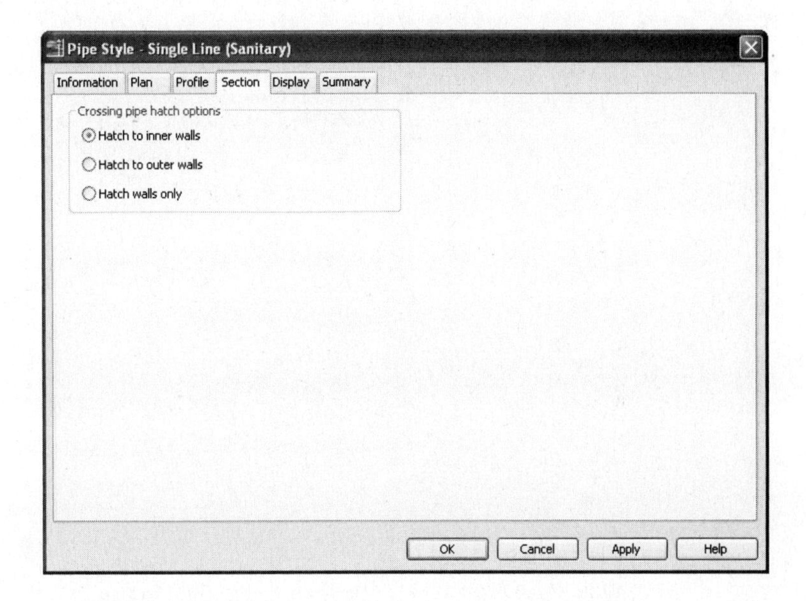

FIGURE 14.27
The Section tab in the Pipe Style dialog

THE DISPLAY TAB

The Display tab (see Figure 14.28) enables you to control the visibility and display properties of all possible pipe-style elements. Note that the same settings apply for Plan, Model, Profile, and Section View Direction settings. Check all the directions to be sure your pipe will be shown as desired.

Just like the structure styles, the pipe styles have a long list of display components that you can adjust and modify to suit your needs. Use the View Direction pull-down menu to navigate between Plan, Model, Profile, and Section components:

Plan Pipe Centerline/Inside Pipe Walls/Outside Pipe Walls controls whether the pipe centerline/inside walls/outside walls are visible in plan view and allows for layer mapping and color/linetype overrides.

Plan Pipe End Line controls the visibility of the end line specified on the Plan tab and allows for layer mapping and color/linetype overrides.

Plan Pipe Hatch controls the visibility of the hatch specified on the Plan tab and allows for layer mapping and color/linetype overrides.

Model 3D Solid controls the visibility of the 3D model in plan view and allows for layer mapping and color/linetype overrides. In the 2D View Direction, this is most commonly set to not visible.

Profile Pipe Centerline/Inside Wall/Outside Wall controls whether the pipe centerline/inside walls/outside walls are visible in profile view and allows for layer mapping and color/linetype overrides.

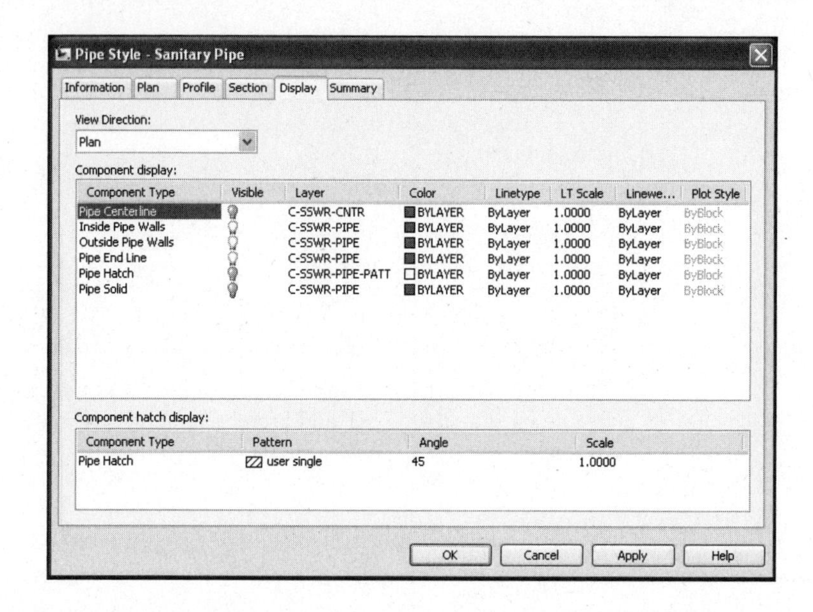

FIGURE 14.28
The Display tab in the Pipe Style dialog

Profile Pipe End Line controls the visibility of the end line specified on the Profile tab and allows for layer mapping and color/linetype overrides.

Profile Pipe Hatch controls the visibility of the hatch specified on the Profile tab and allows for layer mapping and color/linetype overrides.

Profile Crossing Pipe Inside Wall is a control you can use to view the inside walls of pipes that cross profile views and allows for layer mapping and color/linetype overrides. Most commonly this isn't visible except in specific pipe-crossing styles.

Profile Crossing Pipe Outside Wall is a control you can use to view the outside walls of pipes that cross profile views and allows for layer mapping and color/linetype overrides. Most commonly this isn't visible except in specific pipe-crossing styles.

Profile Crossing Pipe Hatch is a control you can use to view the hatch of pipes that cross profile views and allows for layer mapping and color/linetype overrides. Most commonly this isn't visible except in specific pipe-crossing styles.

Section Pipe Inside Walls controls the visibility of the inside pipe walls in section view and allows for layer mapping and color/linetype overrides.

Section Crossing Pipe Hatch controls the visibility of the outside pipe walls in section view and allows for layer mapping and color/linetype overrides.

CREATING A PIPE STYLE FOR A SANITARY SEWER

In this exercise, you'll create the pipe styles for the sanitary sewer:

1. Continue working in your drawing from the last exercise.
2. Expand the Pipes ➢ Pipe Styles tree on the Settings tab in Toolspace. Right-click the Single Line Sanitary style, and choose Edit. The Pipe Styles dialog appears.

3. Switch to the Plan tab, and locate the Pipe Centerline Options area in the lower-right corner of the dialog. Select the Specify Width option, and then use the drop-down list to change the selection to the Draw to Inner Walls option. This stretches the linetype of the specified layer to be the same width as the pipe inner diameter. For example, an 8″ pipe has a thinner linetype than a 12″ pipe.
4. Switch to the Profile tab. Locate Pipe Hatch Options in the lower-left corner of the dialog. Confirm that the Hatch Walls Only option is selected.
5. Switch to the Display tab.
6. Note that Plan Pipe Centerline is mapped to the C-SSWR-CNTR layer. You'll change its linetype in the Layer Properties Manager dialog in step 9.
7. Make sure the following options are visible: Plan Pipe Centerline, Profile Pipe Inside Wall, Profile Pipe Outside Wall, and Profile Pipe Hatch. The rest should be set to not visible. Use the View Direction pull-down menu to switch between Plan and Profile options.
8. Click OK.
9. Open the Layer Properties Manager dialog (see Figure 14.29), and change the linetype on C-SSWR-CNTR to CENTER2. Close the Layer Properties Manager.

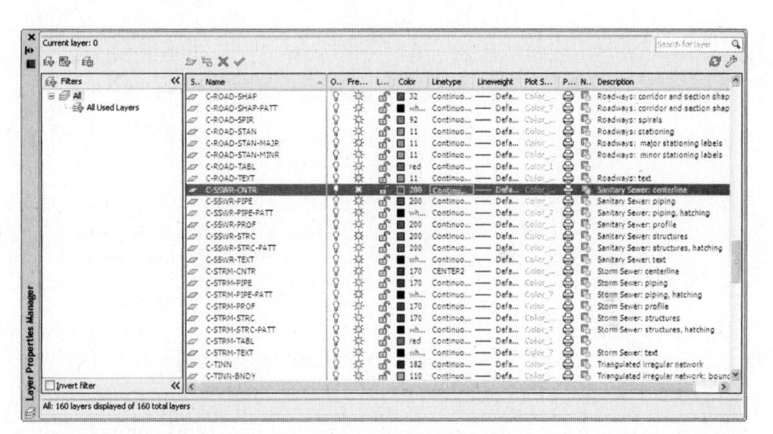

FIGURE 14.29 Change the linetype in the Layer Manager dialog

10. Save your drawing — you'll use it in the next exercise.

Part Rules

At the beginning of this chapter, you made notes about your hypothetical municipality having certain requirements for how structures and pipes behave — things like minimum slope, sump depths, and pipe-invert drops across structure. Depending on the type of network and the complexity of your design, there may be many different constraints on your design. Civil 3D allows you to establish structure and pipe rules that will assist in respecting these constraints during initial layout and edits. Some rules don't change the pipes or structures during layout but provide a "violation only" check that you can view in Prospector.

Rules are separated into two categories — structure rules and pipe rules — and are collected in rule sets. You can then add these rule sets to specific parts in your parts list, which you'll build at the end of this chapter.

Structure Rules

Structure rule sets are located on the Settings tab of Toolspace, under the Structure tree.

For a detailed breakdown of structure rules and how they're applied, including images and illustrations, please see the *Civil 3D Users Guide*. This section will serve as reference when you're creating rules for your company standards.

Under the Structure Rule Set tree, right-click Basic and click Edit. Click the Add Rule button on the Rules tab in the Structure Rule Set dialog. The Add Rule dialog appears, which allows you to access all the various structure rules (see Figure 14.30).

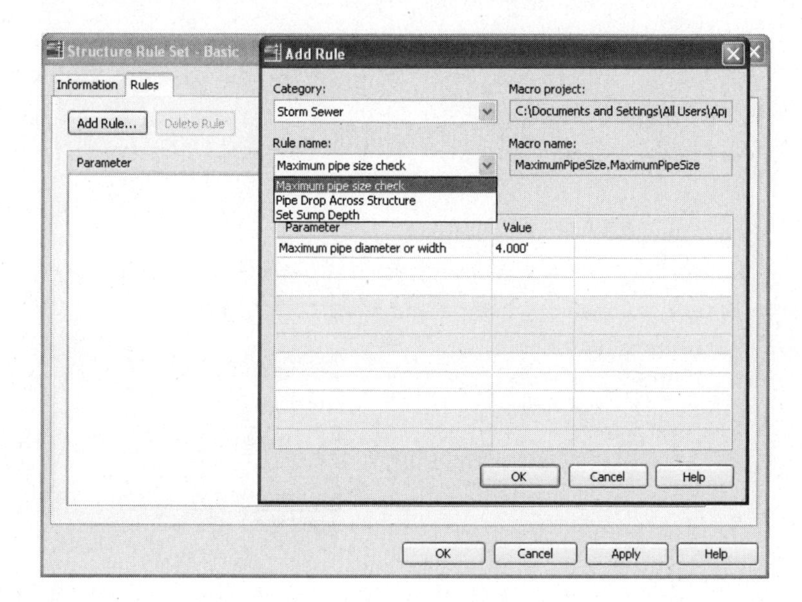

FIGURE 14.30
The Add Rule dialog

MAXIMUM PIPE SIZE CHECK

The Maximum Pipe Size Check rule (see Figure 14.31) examines all pipes connected to a structure and flags a violation in Prospector if any pipe is larger than your rule. This is a violation-only rule — it won't change your pipe size automatically.

PIPE DROP ACROSS STRUCTURE

The Pipe Drop Across Structure rule (see Figure 14.32) places a piece of intelligence onto the structure that tells any connected pipes how their inverts (or, alternatively, their crowns or centerlines) must relate to one another.

When a new pipe is connected to a structure that has the Pipe Drop Across Structure rule applied, the following checks take place:

- A pipe drawn to be exiting a structure has an invert equal to or lower than the lowest pipe entering the structure.
- A pipe drawn to be entering a structure has an invert equal to or higher than the highest pipe exiting the structure.
- Any minimum specified drop distance is respected between the lowest entering pipe and the highest exiting pipe.

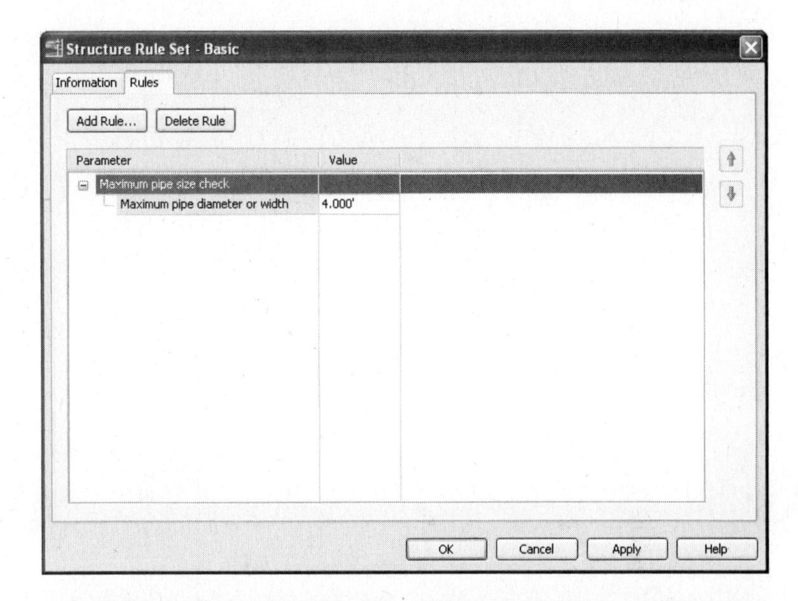

FIGURE 14.31
The Maximum Pipe Size Check rule option

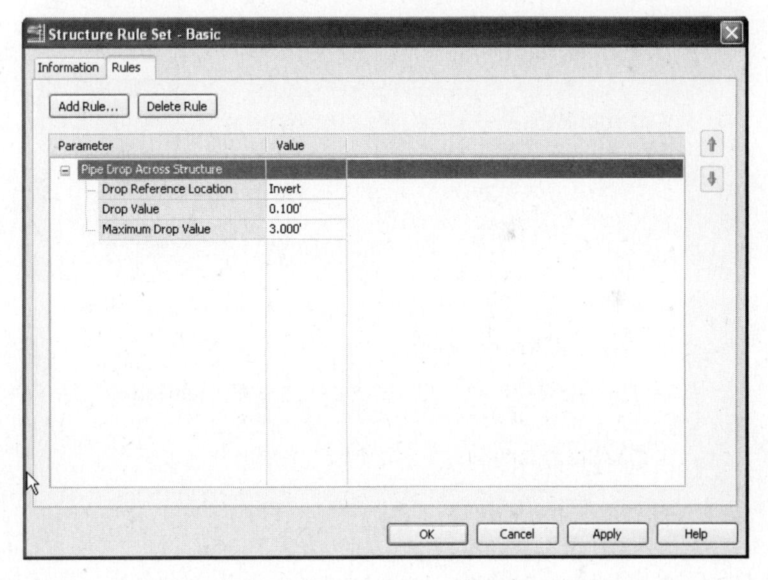

FIGURE 14.32
The Pipe Drop Across Structure rule options

In the hypothetical sanitary sewer example, you're required to maintain a 0.10′ invert drop across all structures. You'll use this rule in your structure rule set in the next exercise.

SET SUMP DEPTH

The Set Sump Depth rule (Figure 14.33) establishes a desired sump depth for structures. It's important to add a sump-depth rule to all your structure rule sets; otherwise, Civil 3D will assume a sump that is most often undesirable and is difficult to modify once your structures have been drawn.

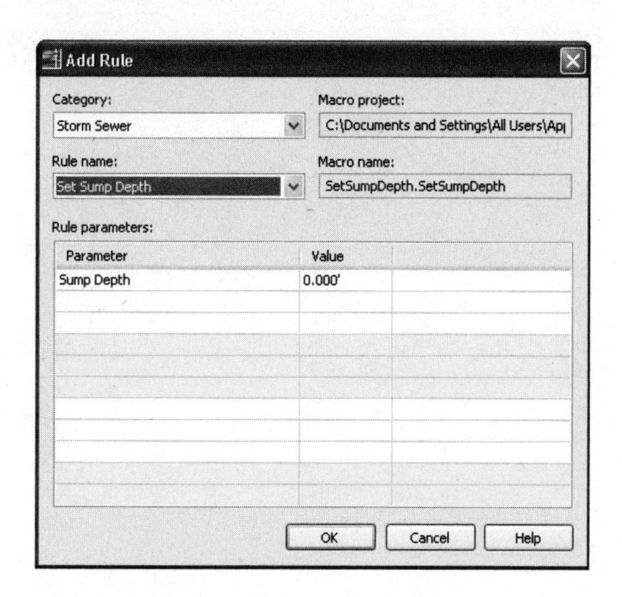

FIGURE 14.33
The Set Sump Depth rule options

In the hypothetical sanitary sewer example, all the structures have a 1.5′ sump depth. You'll use this rule in your structure rule set in the next exercise.

Pipe Rules

Pipe rule sets are located on the Settings tab of Toolspace, under the Pipe tree. For a detailed breakdown of pipe rules and how they're applied, including images and illustrations, please see the *Civil 3D Users Guide*.

After you right-click on a Pipe Rule Set and click Edit, you can access all the pipe rules by clicking the Add Rule button on the Rules tab of the Pipe Rule Set dialog.

Cover And Slope Rule

The Cover And Slope rule (Figure 14.34) allows you to specify your desired slope range and cover range. You'll create one Cover And Slope rule for each size pipe in the hypothetical sanitary sewer example.

Cover Only Rule

The Cover Only rule (Figure 14.35) is designed for use with pressure-type pipe systems where slope can vary or isn't a critical factor.

Pipe to Pipe Match Rule

The Pipe to Pipe Match rule (Figure 14.36) is also designed for use with pressure-type pipe systems where there are no true structures (only null structures), including situations where pipe is placed to break into an existing pipe. This rule determines how pipe inverts are assigned when two pipes come together, similar to the Pipe Drop Across Structure rule.

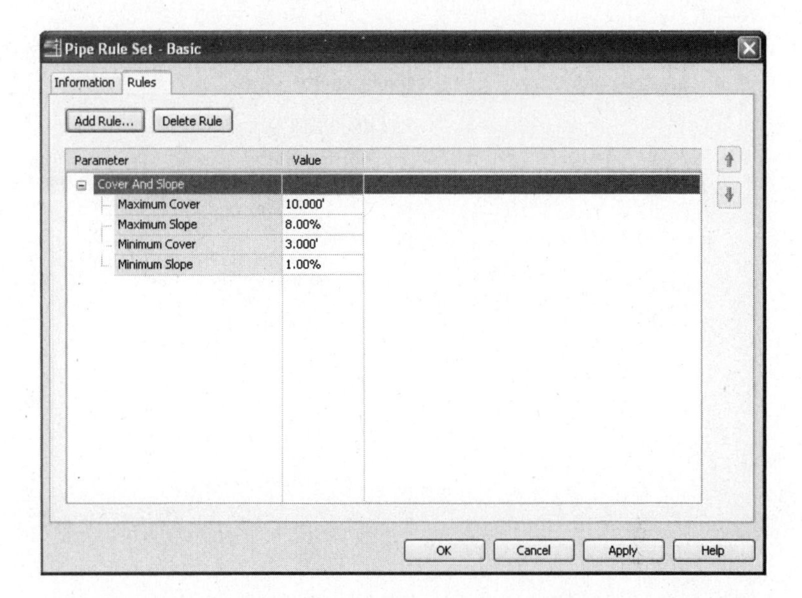

FIGURE 14.34
The Cover And Slope rule options

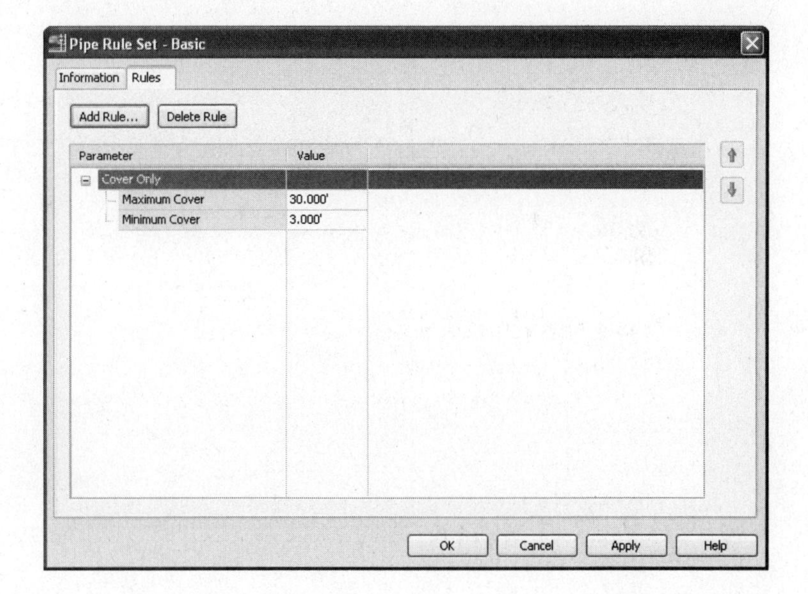

FIGURE 14.35
The Cover Only rule options

LENGTH CHECK

Length Check is a violation-only rule; it won't change your pipe length size automatically. The Length Check options (see Figure 14.37) allow you to specify a minimum and maximum pipe length.

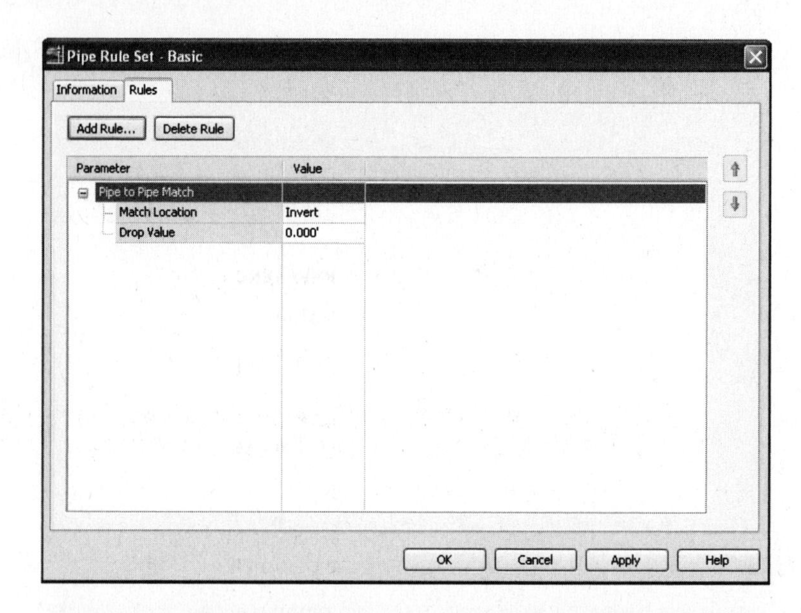

FIGURE 14.36
The Pipe to Pipe Match rule options

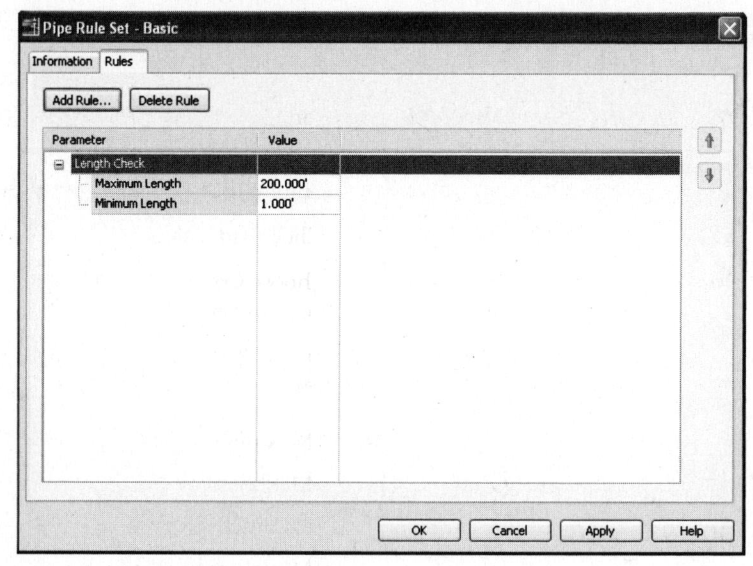

FIGURE 14.37
The Length Check rule options

Creating Structure and Pipe Rule Sets

In this exercise, you'll create one structure rule set and three pipe rule sets for a hypothetical sanitary sewer project:

1. Continue working in your drawing from the previous exercise.
2. Locate the Structure Rule Set on the Settings tab of Toolspace under the Structure tree. Right-click the Structure rule set, and choose New.

3. On the Information Tab, enter **Sanitary Structure Rules** in the Name text box.
4. Switch to the Rules tab. Click the Add Rule button.
5. In the Add Rule dialog, choose Pipe Drop Across Structure in the Rule Name drop-down list. Click OK. (You can't change the parameters.)
6. Confirm that the parameters in the Structure Rule Set dialog are the following:

DROP REFERENCE LOCATION	**INVERT**
Drop Value	0.1′
Maximum Drop Value	3′

   These parameters establish a rule that will match your hypothetical municipality's standard for the drop across sanitary sewer structures.
7. Click the Add Rule button.
8. In the Add Rule dialog, choose Set Sump Depth in the Rule Name drop-down list. Click OK. (You can't change the parameters.)
9. Change the Sump Depth parameter to 1.5′ in the Structure Rule Set dialog to meet the hypothetical municipality's standard for sump in sanitary sewer structures.
10. Click OK.
11. Locate the Pipe Rule Set on the Settings tab of Toolspace under the Pipe tree. Right-click the Pipe Rule Set, and choose New.
12. On the Information tab, enter **8 Inch Sanitary Pipe Rules** for the name.
13. Switch to the Rules tab. Click Add Rule.
14. In the Add Rule dialog, choose Cover And Slope in the Rule Name drop-down list. Click OK. (You can't change the parameters.)
15. Modify the parameters to match the constraints established by your hypothetical municipality for 8″ pipe, as follows:

Maximum Cover	10′
Maximum Slope	10%
Minimum Cover	4′
Minimum Slope	0.40%

16. Click OK.
17. Select the rule set you just created (8 Inch Sanitary Pipe Rules). Right-click, and choose Copy.
18. On the Information tab, enter **10 Inch Sanitary Pipe Rules** in the Name text box.
19. Modify the parameters to match the constraints established by your hypothetical municipality for a 10″ pipe, as follows:

Maximum Cover	10′
Maximum Slope	10%
Minimum Cover	4′
Minimum Slope	0.28%

20. Repeat the process to create a rule set for the 12″ pipe using the following parameters:

Maximum Cover	10′
Maximum Slope	10%
Minimum Cover	4′
Minimum Slope	0.22%

21. You should now have one structure rule set and three pipe rule sets.
22. Save your drawing — you'll use it in the next exercise.

Parts List

When you know what parts you need, what they need to look like, and how you want them to behave, you can standardize your needs in the form of a parts list. If you think of the part catalog, the styles in your template, and the rule sets as being your well-stocked workshop, the parts list is the toolbox that you fill with only the equipment you need to get the job done. Parts lists are stored in your standard Civil 3D template so they're at your fingertips when new jobs are created.

For example, when you're designing a sanitary sewer system, you may need only a small spectrum of PVC pipe sizes and manhole types that follow a few sets of rules and require only a style or two. You wouldn't want to have to sort through your entire collection of parts, rules, and styles every time you created a sanitary sewer network. You can make a parts list called Sanitary Sewers (or something similar) and stock it with the pipes, structures, styles, and rules you'll need to get the job done.

Similarly, depending on the type of work you do, you'll want at least a Storm Drainage parts list with concrete pipe, catch basins, storm manholes, applicable rule sets and styles, and a Water Network parts list containing PVC pipe and null structures as well as some cover-only rule sets. As you begin your first few pilot projects, you'll begin to see which parts lists are most useful, and you can continue to build them as part of your standard template.

The parts lists are located on the Settings tab of Toolspace under the Pipe Network tree. You can create parts lists by right-clicking the Parts Lists entry and choosing Create Parts List. You can edit parts lists by choosing a specific parts list, right-clicking, and choosing Edit.

Adding Part Families on the Pipes Tab

When you create a parts list, the Pipes tab is initially blank. You can add pipe families by right-clicking the entry on the Pipes tab (it may say New Parts List, or the name of your newly created parts list) and choosing Add Part Family (Figure 14.38).

In the Part Catalog dialog (see Figure 14.39), you're given the opportunity to choose one or more part families to add to your parts list. You'll recognize these choices from the catalog HTML interface as well as from the Part Builder interface. Note that the preview image comes from the *partname*.bmp file located in the catalog folder.

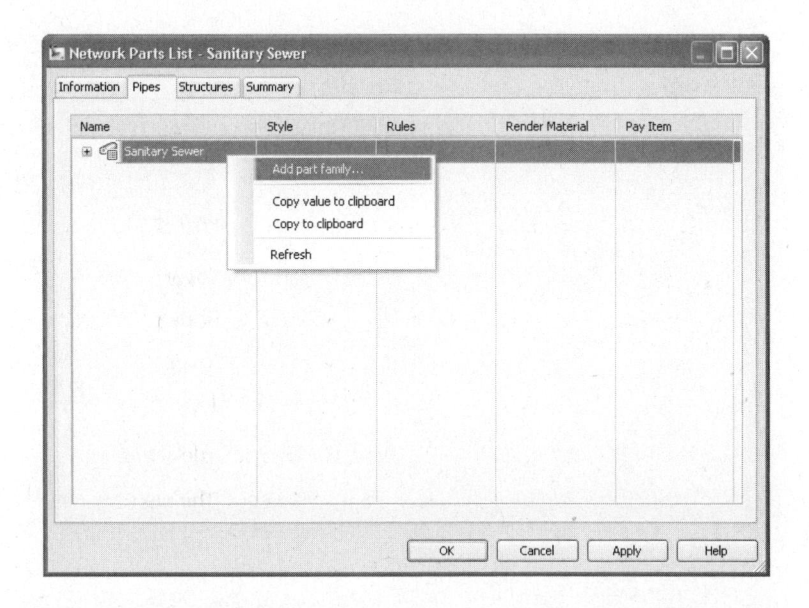

FIGURE 14.38
Adding a part family

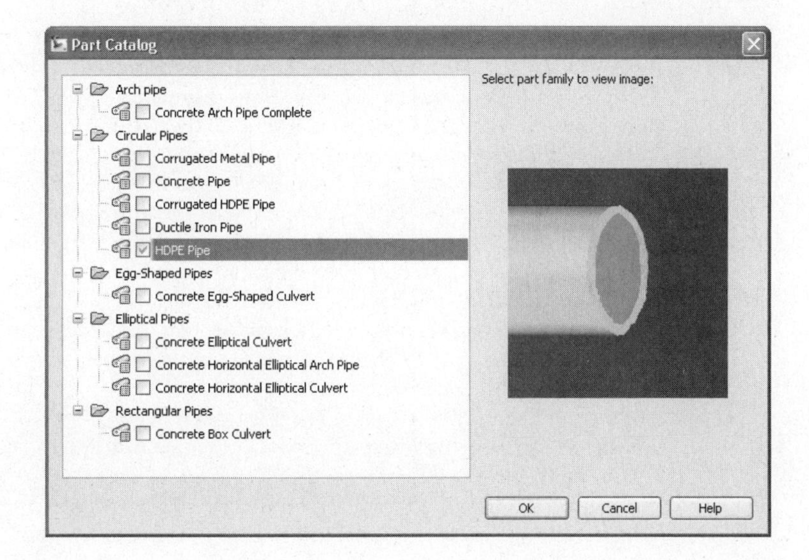

FIGURE 14.39
The Part Catalog dialog

Once the part family has been added to your Pipes tab, you must choose the appropriate sizes. Choose the Part Family entry, right-click, and choose Add Part Size.

Part sizes can be added individually, or you can select the Add All Sizes check box to add every part size available for that part family. Sometimes it's easier to add all the sizes and delete a few that you don't need rather than add many sizes individually.

Also note that in the Part Size Creator dialog (Figure 14.40), you can assign optional properties such as the Manning Coefficient and the Hazen Williams Coefficient. Such optional properties are applied any time this particular pipe is added to your network using this parts list.

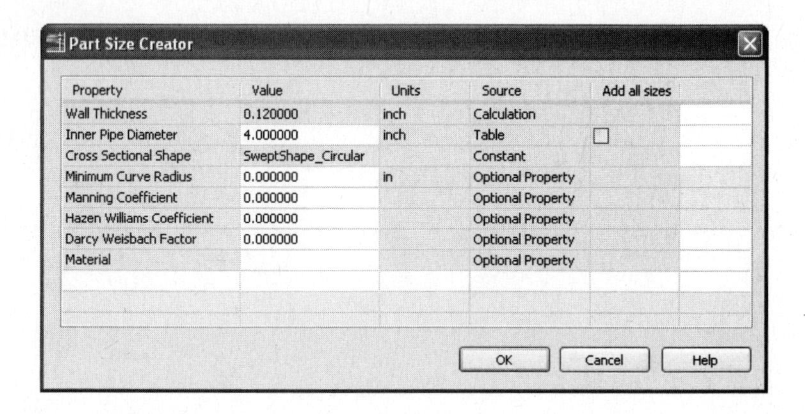

FIGURE 14.40
The Part Size Creator dialog

Once pipe sizes have been added to the Pipes tab, as shown in Figure 14.41, you can change their default descriptions by clicking in the first column. You can also delete an entry by selecting the size, right-clicking, and choosing Delete.

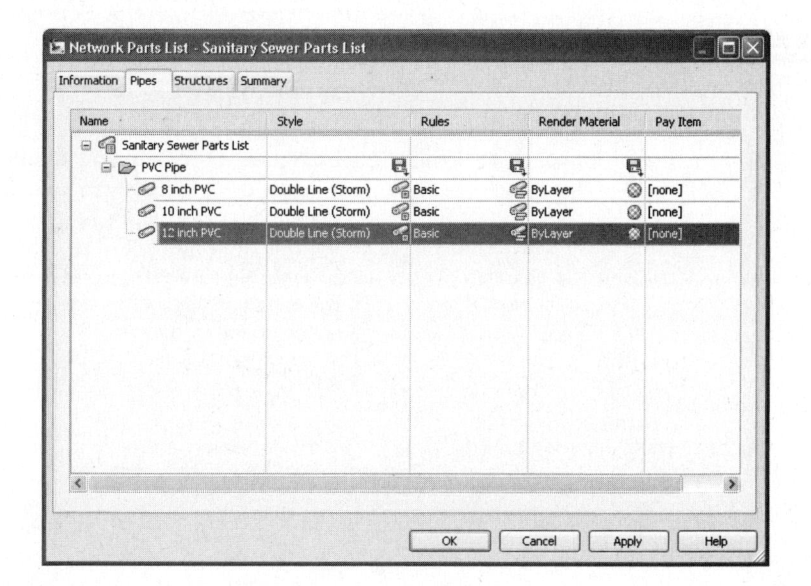

FIGURE 14.41
Pipe sizes added with descriptions edited. Note that styles and rules remain to be changed

MODIFYING STYLES AND RULES

Next to each pipe size are Style and Rules entries. If you have different styles and rules for each size of pipe, you can assign them individually. If you're applying the same style or rule set to all your sizes, click the button across from the Part Family entry (see Figure 14.42) to assign style or rules to all sizes in that part family. The Pay Items column shown here will be discussed further in Chapter 23, "Quantity Takeoff."

FIGURE 14.42
A completed Pipes tab for the hypothetical sanitary sewer project

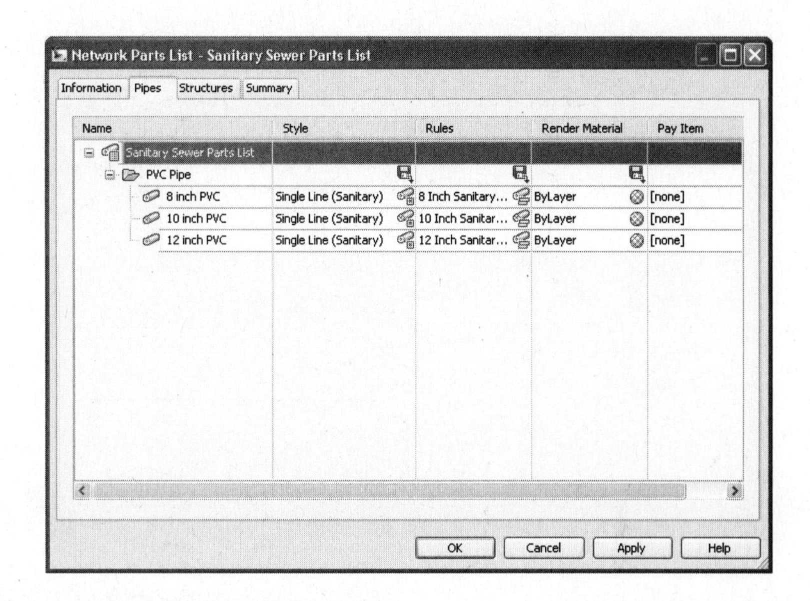

Adding Part Families on the Structures Tab

When you create a parts list, the Structures tab initially contains a null structure. You can add additional structure families by right-clicking the entry on the Structures tab (it may say New Parts List, or the name of your newly created parts list) and choosing Add Part Family.

In the Part Catalog dialog (see Figure 14.43), you're given the opportunity to choose one or more part families to add to your parts list. You'll recognize these choices from the catalog HTML interface as well as the Part Builder interface. Note that the preview image comes from the `partname.bmp` file located in the catalog folder.

FIGURE 14.43
Adding the Concentric Cylindrical Structure part family to your parts list

Part Catalog
Inlet-Outlets
CMP Rectangular End Section
Concrete Rectangular Box Culvert Headwall
Concrete Flared End Section
Concrete Rectangular Headwall Variable Height
Concrete Rectangular Headwall
Concrete Rectangular Winged Headwall
Junction Structures with Frames
Cylindrical Structure Slab Top Circular Frame
Rectangular Structure Slab Top Circular Frame
Rectangular Structure Slab Top Rectangular Frame
Eccentric Cylindrical 2-Tier Circular Frame
Rectangular Structure 2-Tier Circular Frame
Junction Structures without Frames
Cylindrical Junction Structure NF
Rectangular Junction Structure NF
Concentric Cylindrical Structure NF
Select part family to view image:
OK
Cancel
Help

Once the part family has been added to your Structures tab, you must choose the appropriate sizes. Choose the entry for the part family, right-click, and choose Add Part Size.

Part sizes can be added individually, or you can select the Add All Sizes check box to add every part size available for that part family. Sometimes it's easier to add all the sizes and delete a few that you don't need rather than add many sizes individually.

Also note that in the Part Size Creator dialog (see Figure 14.44), you can scroll down to assign optional properties such as Grate, Frame, and Cover Types. These properties will then be applied any time this particular structure is added to your network using this parts list.

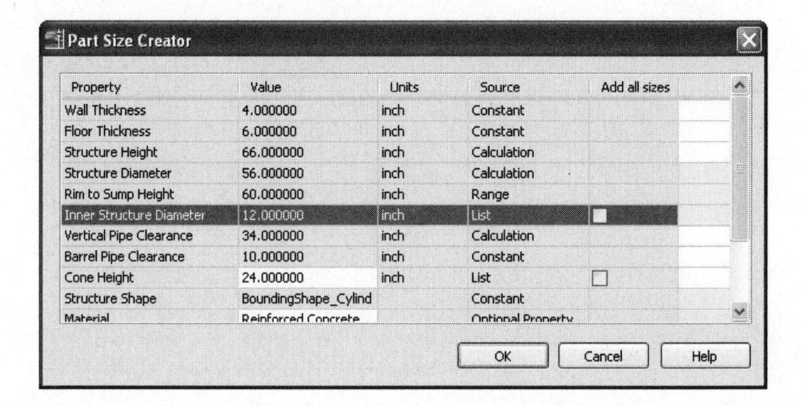

FIGURE 14.44
The Part Size Creator dialog with additional properties

Once structure sizes have been added to the Structures tab (see Figure 14.45), you can change their default descriptions by clicking in the first column. You can also delete an entry by selecting the size, right-clicking, and choosing Delete. The null structure can't be deleted; it serves as a placeholder between two pipes that are directly connected. In a gravity system, a null structure may never be necessary, but it must remain part of the Structures list in the case of two pipes being connected.

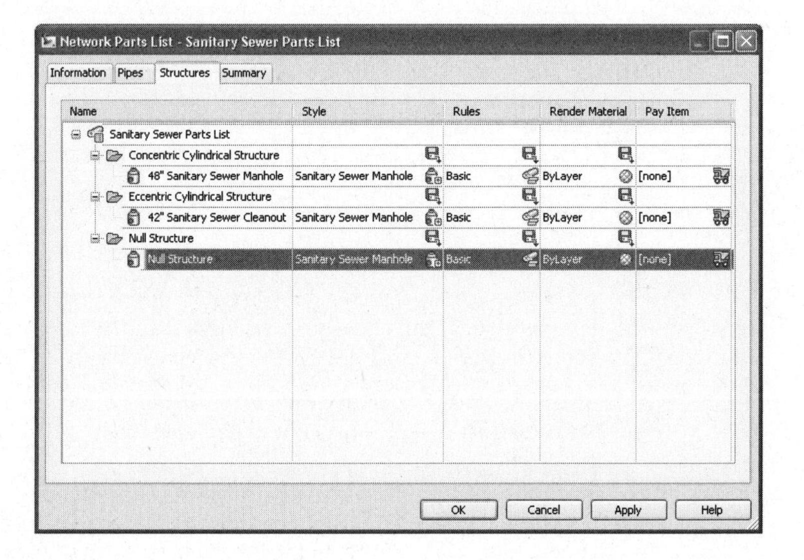

FIGURE 14.45
Structure sizes added with descriptions edited. Note that their styles and rules remain to be changed

MODIFYING STYLES AND RULES

Next to each structure entry are Style and Rules entries (see Figure 14.46). If you have different styles and rules for each structure, these can be assigned individually. If you're applying the same style or rule set to all your structures, click the Structure Style button in the Style column or the Structure Rule button in the Rules column on the same row as the Part Family entry to assign style or rules to all sizes in that part family.

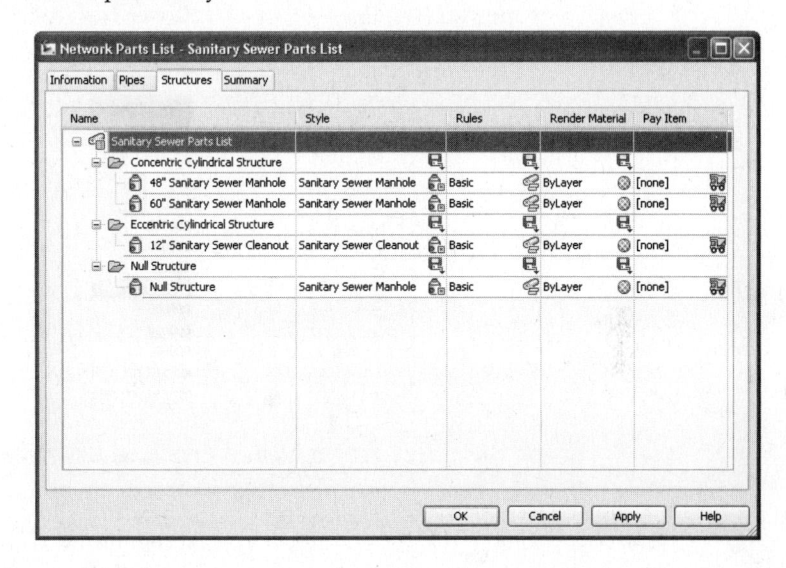

FIGURE 14.46
A completed Structures tab for the hypothetical sanitary sewer project

MODIFYING PAY ITEMS

Pay items can be assigned to objects or groups of objects in a drawing. You can manually tag pay items to individual pipe network objects or to groups of similar pipe network objects. After you have prepared codes, or selected pre-defined codes, you can automatically tag pipe networks with pay items. Figure 14.47 depicts a completed parts list with pay items assigned. Remember, the Pay Item column will be discussed in Chapter 23.

Creating a Parts List for a Sanitary Sewer

In this exercise, you'll combine the parts, styles, and rules created in the previous exercises:

1. Continue working in your drawing from the previous exercise.
2. Locate the Parts List entry on the Settings tab of Toolspace.
3. Locate the Sanitary Sewer parts list that is part of the default template. Select this parts list, right-click, and choose Delete. Parts lists must have unique names, and because you're about to create a list called Sanitary Sewer, you need to delete this one.
4. Select the Parts Lists entry under the Pipe Network tree, right-click, and choose Create Parts List. The Create Parts List dialog appears.
5. Enter **Sanitary Sewer** in the Name text box on the Information tab.
6. Switch to the Pipes tab. Right-click the New Parts List entry under the Name column, and choose Add Part Family. In the Part Catalog dialog, select the PVC Pipe check box, and click OK.

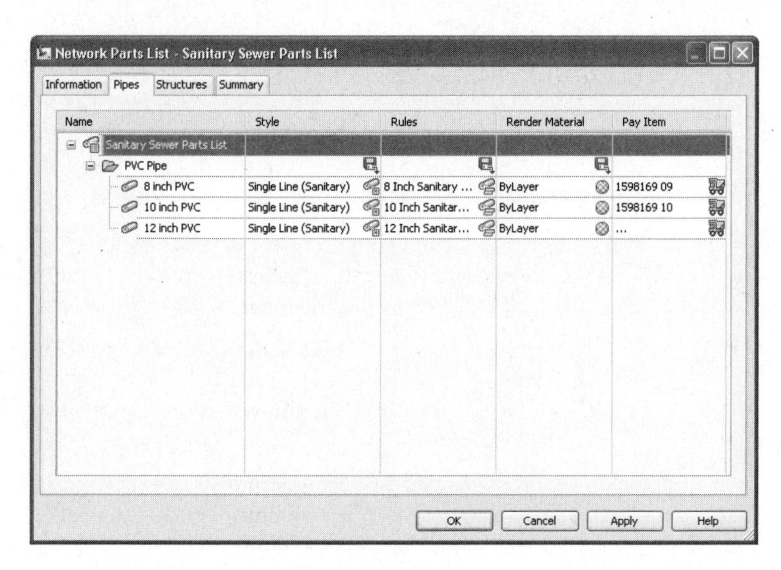

FIGURE 14.47
A completed Pipes tab for the hypothetical sanitary sewer project

7. Expand the Sanitary Sewer tree on the Network Parts List dialog to see the entry for PVC Pipe. Note that you've added the part family but you haven't yet added any pipe sizes.
8. Right-click the PVC Pipe entry, and choose Add Part Size.
9. In the Part Size Creator dialog, click in the Value field next to Inner Pipe Diameter to activate the drop-down list, and choose 8.00 (Figure 14.48). Click OK to add the pipe size. Repeat the process to add 10″ and 12″ pipes.

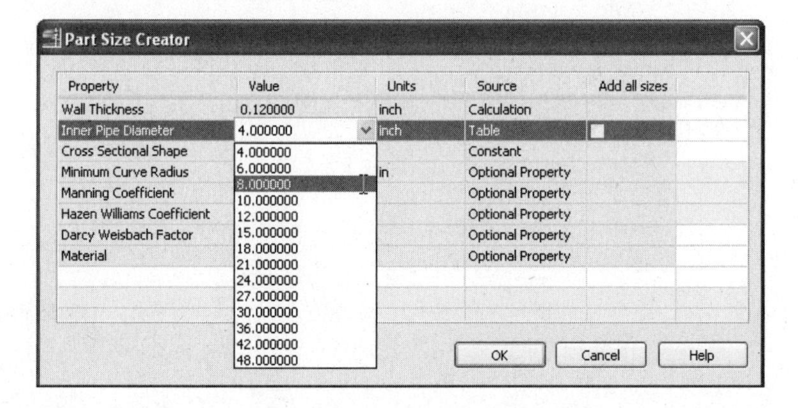

FIGURE 14.48
Choose a part size

10. Once you've returned to the Pipes tab, expand the PVC Pipe tree, click inside the description field for your 8″ pipe. If the pipe description says something like "8 inch PVC Pipe MCR_0.000000 ACMan_0.000000 ACHW_0.000000 ACDW_0.000000 Material_" by default, edit the description so that it reads **8 inch PVC Pipe**.
11. Repeat step 10 for the 10″ and 12″ pipes, if necessary.
12. Follow the procedures in the "Modifying Styles and Rules" subsection under "Adding Part Families on the Pipes Tab" earlier in this chapter to change the pipe style to Single Line (Sanitary) and the rules to the appropriate rule set, respectively, for each size pipe.

13. Switch to the Structures tab. Right-click the New Parts List entry in the Name field, and choose Add Part Family. In the Part Catalog dialog, select the Concentric Cylindrical Structure check box in the Junction Structures with Frames area, and select the Concentric Cylindrical Structure NF check box in the Junction Structures without Frames area. Click OK.
14. Right-click the Concentric Cylindrical entry in the Network Parts List dialog, and choose Add Part Size.
15. Following the same method used to add the different pipe sizes, add a 48″ and a 60″ structure size to the list using the Inner Structure Diameter Property.
16. Following the same method as in step 15, add a 12″ Concentric Cylindrical Junction Structure NF to the list. (Note that if you didn't do the Part Builder exercise in which you added the 12″ structure diameter, you won't have this choice. You may substitute any size in its place.)
17. Once you've returned to the Structures tab, click inside the Description field for your 48″ structure. It may say something like "Concentric Structure 48 dia 18 frame 24 cone 5 wall 6 floor Mat_Reinforced Concrete SF_Standard SG_Standard SC_Standard." Edit the description to read **48 inch Sanitary Sewer Manhole**.
18. Repeat step 17 for the other structure types, giving each an appropriate description. When adding the Cleanout structure, you'll want to rename it appropriately in addition to changing the description.
19. Follow the procedures found in the "Modifying Styles and Rules" subsection under "Adding Part Families on the Structures Tab" earlier to change the structure style to Sanitary Sewer Manhole and Sanitary Sewer Cleanout and the rules to Sanitary Structure Rules, respectively.
20. Check to make sure that the Pipes and Structures tabs look like what was shown previously in Figure 14.42 and Figure 14.46.

Real World Scenario

WATER NETWORK RULES AND PARTS LIST

The example in this chapter has been a gravity sanitary sewer network. Although a storm or other gravity network would have different parts, styles, and rules, the fundamental process is the same for all gravity pipe systems.

Pressure systems, however, need to take advantage of different parts, styles, and rules.

Parts In Civil 3D, pipes can't truly be connected without a structure between them. The structure acts like a glue that holds the pipes together. In the case of pipes-only networks, such as water, *null structures* are automatically placed wherever two or more pipes are directly connected. Null structures are part of every new parts list by default.

Styles When water pipes are joined together, there is often no visible structure in plan or profile view. Therefore, it's important to create a style for your null structures that is either invisible or No Plot. Some people prefer the cleaner look of an invisible style, knowing that they can always go into Prospector to locate and select the null structure. You can create a No Show style by turning off all the items on the Display tab of the Structure Style dialog.

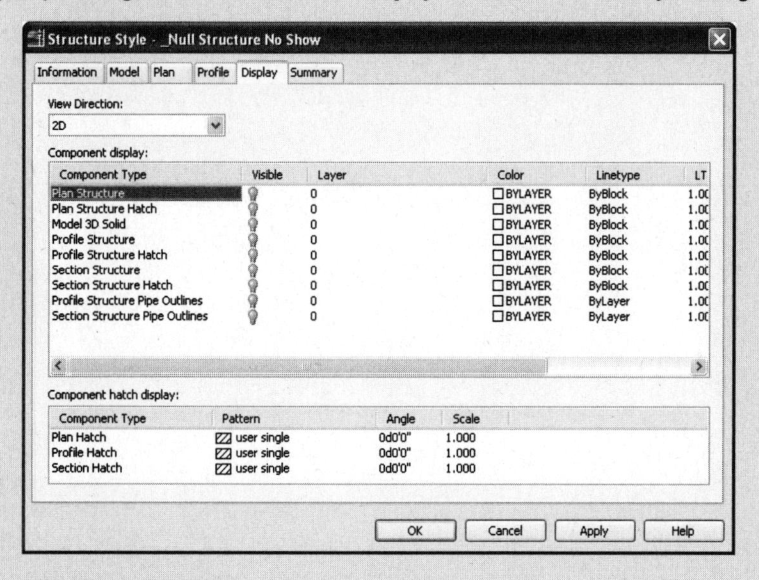

Once you apply this invisible style to your null structure, the pipe connection appears as if there is no structure present.

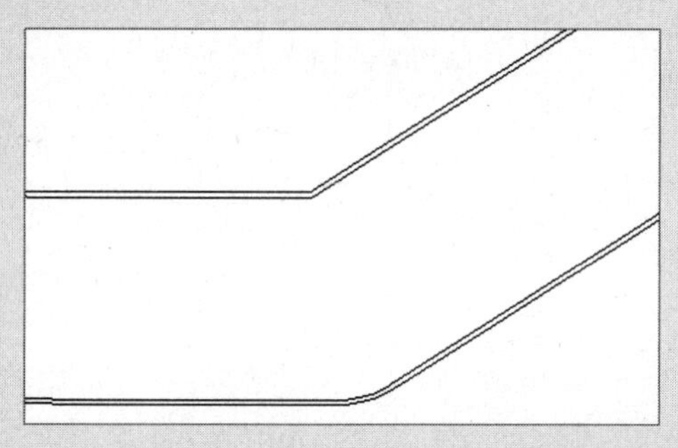

Some people prefer to use the No Plot style because it allows them to grab and grip-edit the location of the pipe connection; then, they can either ignore or freeze the No Plot layer for plan production.

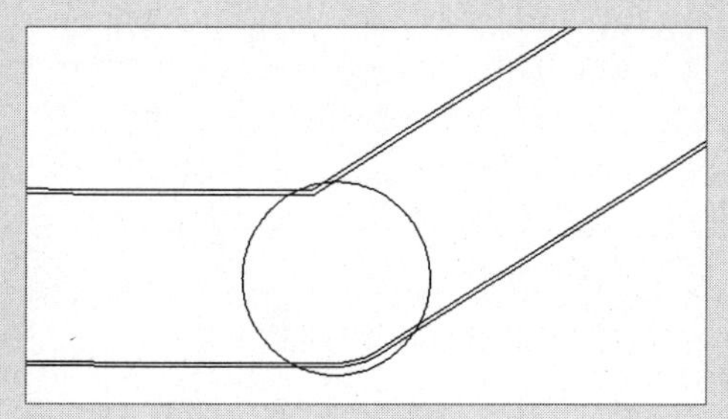

If inside or outside walls (or both) are shown when a pipe is drawn in plan view, there is an option that allows for cleanup at pipe connections.

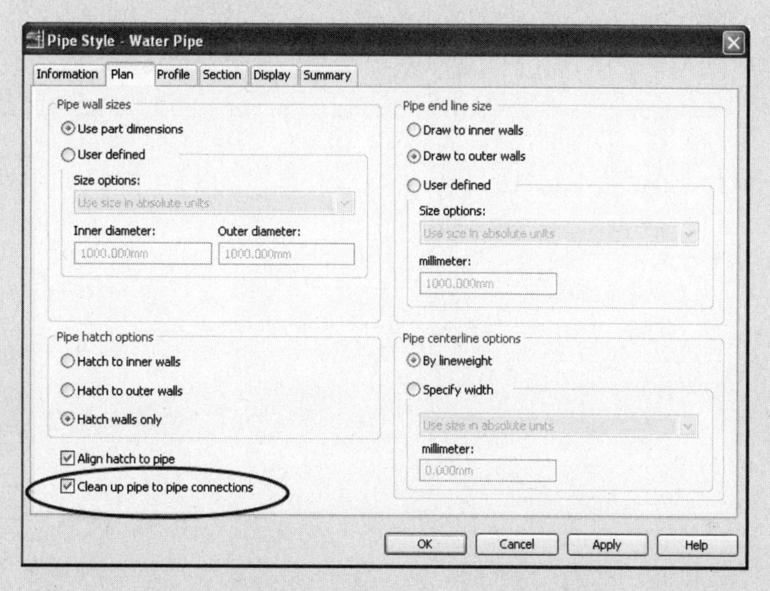

Rules In most water network design situations, the driving design factor is minimum cover. Often, there are no slope restrictions. In this case, you use the Cover Only rule instead of the Slope And Cover rule.

A second rule that may be added in a pipes-only situation is the Pipe to Pipe Match rule. This rule specifies what happens when two pipes are directly connected at a null structure.

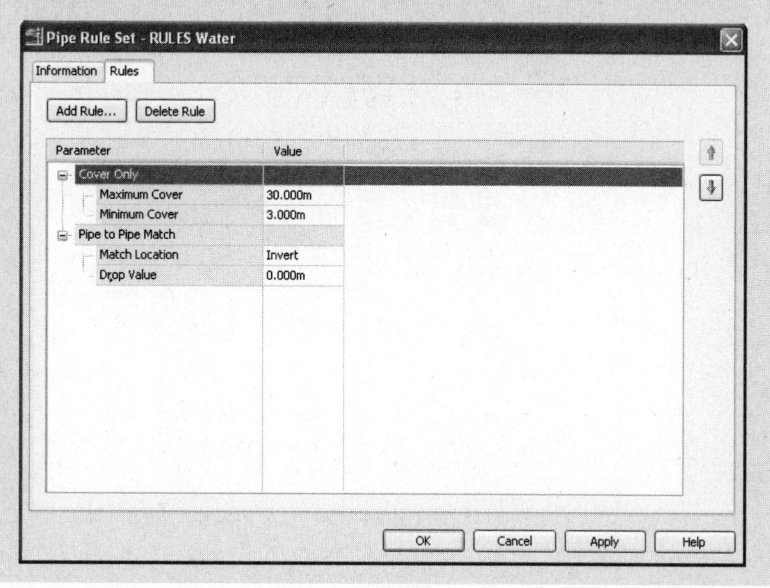

The process for creating a Water Network parts list is the same as creating a Sanitary parts list: identify all the parts you need to construct the network, identify how these parts will be represented in plan and profile views, and determine how these parts will behave.

Building your parts list using null structures, the appropriate style settings, and rules designed for a pipes-only network will ensure a smooth water network design.

The Bottom Line

Add pipe and structure families to a new parts list. Before you can begin to design your pipe network, you must ensure that you have the appropriate pipes and structures at your disposal.

Master It Create a new drawing from `_AutoCAD Civil 3D (Imperial) NCS Extended.dwt`. Create a new parts list called Typical Storm Drainage. (This template already includes a Storm Sewer parts list. Ignore it or delete it, but don't use it. Create your own for this exercise.)

Add 12″, 15″, 18″, 24″, 36″, and 48″ Concrete Circular Pipe to the parts list.

Add the following to the parts list: a Rectangular Structure Slab Top Rectangular Frame with an inner structure width and length of 15″, a Rectangular Structure Slab Top Rectangular Frame with an inner structure width of 15″ and an inner structure length of 18″, and a Concentric Cylindrical Structure with an inner diameter of 48″.

Create rule sets that apply to pipes and structures in a parts list. Municipal standards and engineering judgment determine how pipes will behave in design situations. Considerations include minimum and maximum slope, cover requirements, and length guidelines. For structures, there are regulations regarding sump depth and pipe drop. You can apply these considerations to pipe network parts by creating rule sets.

Master It Create a new structure rule set called Storm Drain Structure Rules. Add the following rules:

RULE	VALUES
Pipe Drop Across Structure	Drop Reference Location: **Invert** Drop Value: **0.01′** Maximum Drop Value: **2′**
Set Sump Depth	Sump Depth: **0.5′**

Create a new pipe rule set called Storm Drain Pipe Rules, and add the following rules:

RULE	VALUES
Cover And Slope	Maximum Cover: **15′** Maximum Slope: **5%** Minimum Cover: **4′** Minimum Slope: **.05%**
Length Check	Maximum Length: **300′** Minimum Length: **8′**

Apply these rules to all pipes and structures in your Typical Storm Drainage parts list.

Apply styles to pipes and structures in a parts list. The final drafted appearance of pipes and structures in a drawing is controlled by municipal standards and company CAD standards. Civil 3D object styles allow you to control and automate the symbology used to represent pipes and structures in plan, profile, and section views.

Master It Apply the following styles to your parts list:

- Single Line (Storm) to the 12″, 15″, and 18″ Concrete Circular Pipe
- Double Line (Storm) to the 24″, 36″, and 48″ Concrete Circular Pipe
- Storm Sewer Manhole to the Concentric Cylindrical Structure with an inner diameter of 48″
- Catch Basin to both sizes of the Rectangular Structure Slab Top Rectangular Frame

Chapter 15

Running Downhill: Pipe Networks

Once you understand the parts used to design and construct pipe networks, it's time to assemble those parts into a system or network.

In this chapter, you'll learn to:

- Create a pipe network by layout
- Create an alignment from network parts and draw parts in profile view
- Label a pipe network in plan and profile
- Create a dynamic pipe table

Exploring Pipe Networks

Parts in a pipe network have relationships that follow a network paradigm. A pipe network, such as the one in Figure 15.1a, can have many branches. In most cases, the pipes and structures in your network will be connected to each other; however, they don't necessarily have to be physically touching to be included in the same pipe network.

Land Desktop with civil design and some other civil engineering-design programs don't design piping systems using a network paradigm; instead, they use a branch-by-branch or "run" paradigm (see Figure 15.1b). Although it's possible to separate your branches into their own separate pipe networks in Civil 3D, your design will have the most power and flexibility if you change your thinking from a "run-by-run" to a network paradigm.

Pipe Network Object Types

Pipes are components of a pipe network that primarily represent pipes but can be used to represent any type of conduit such as culverts, gas lines, or utility cables. They can be straight or curved; and although primarily used to represent gravity systems, they can be adapted and customized to represent pressure and other types of systems such as water networks and force mains. The standard catalog has pipe shapes that are circular, elliptical, egg-shaped, and rectangular and are made of materials that include PVC, RCP, DI, and HDPE. You can use Part Builder (discussed in Chapter 14, "The Tool Chest: Parts Lists and Part Builder") to create your own shapes and materials if the default shapes and dimensions can't be adapted for your design.

Structures are components of a pipe network that represent manholes, catch basins, inlets, joints, and any other type of junction between two pipes. The standard catalog includes inlets, outlets, junction structures with frames (such as manholes with lids or catch basins with grates), and junction structures without frames (such as simple cylinders and rectangles). You can again take advantage of Part Builder to create your own shapes and materials if the default shapes and dimensions can't be adapted for your design.

FIGURE 15.1
(a) A typical Civil 3D pipe network, and
(b) a pipe network with a single-pipe run

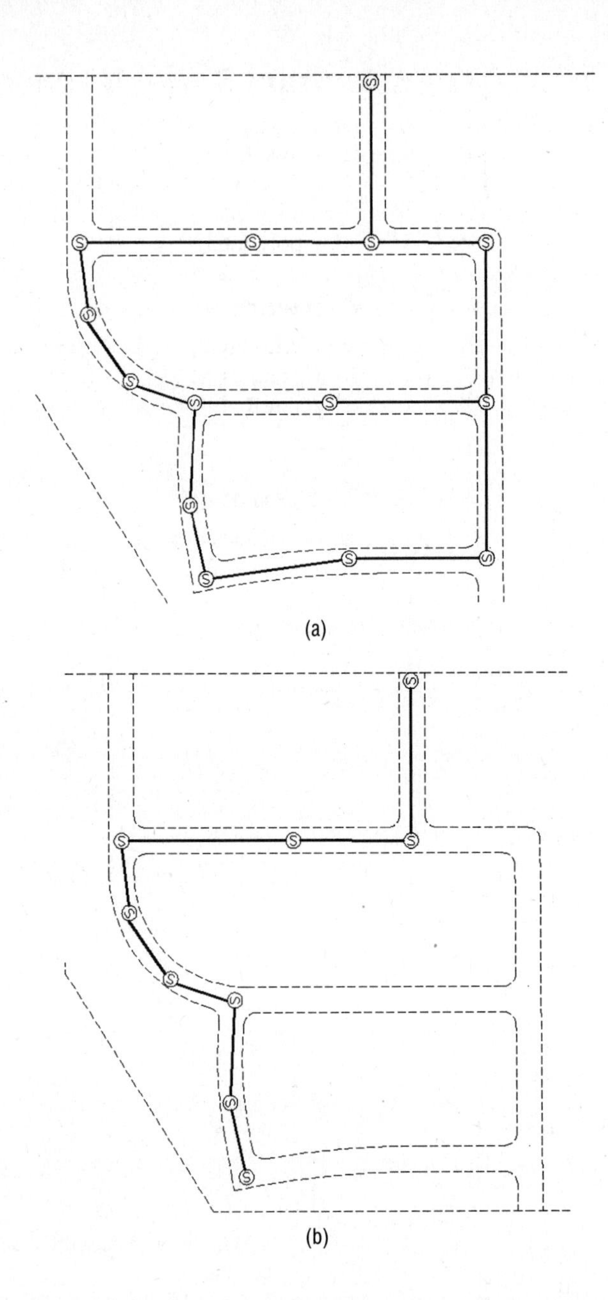

Null structures are created automatically when two pipes are joined together without a structure; they act as a placeholder for a pipe endpoint. They have special properties, such as allowing pipe cleanup at pipe intersections. Most of the time, you'll create a style for them that doesn't plot or is invisible for plotting purposes.

Creating a Sanitary Sewer Network

In Chapter 14, you prepared a parts list for a typical sanitary sewer network. This chapter will lead you through several methods for using that parts list to design, edit, and annotate a pipe network.

There are several ways to create pipe networks. You can do so using the Civil 3D pipe layout tools. Limited tools are also available for creating pipe networks from certain AutoCAD and Civil 3D objects, such as lines, polylines, alignments, and feature lines.

Creating a Pipe Network with Layout Tools

Creating a pipe network with layout tools is much like creating other Civil 3D objects, such as alignments. After naming and establishing the parameters for your pipe network, you're presented with a special toolbar that you can use to lay out pipes and structures in plan, which will also drive a vertical design.

Establishing Pipe Network Parameters

This section will give you an overview of establishing pipe network parameters. Use this section as a reference for the exercises in this chapter. When you're ready to create a pipe network, select Pipe Network ➢ Pipe Network Creation Tools from the Create Design panel on the Home tab. The Create Pipe Network dialog appears (see Figure 15.2), and you can establish your settings.

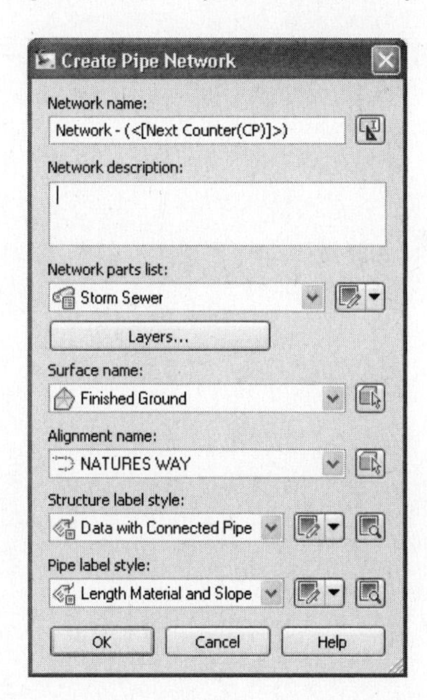

FIGURE 15.2
The Create Pipe Network dialog

Before you can create a pipe network, you must give your network a name; but more important, you need to assign a parts list for your network. As you saw in Chapter 14, the parts list provides a toolkit of pipes, structures, rules, and styles to automate the pipe-network design process. It's also important to select a reference surface in this interface. This surface will be used for rim elevations and rule application.

When creating a pipe network, you're prompted for the following options:

Network Name Choose a name for your network that is meaningful and that will help you identify it in Prospector and other locations.

Network Description The description of your pipe network is optional. You might make a note of the date, the type of network, and any special characteristics.

Network Parts List Choose the parts list that contains the parts, rules, and styles you want to use for this design (see Chapter 14).

Surface Name Choose the surface that will provide a basis for applying cover rules as well as provide an insertion elevation for your structures (in other words, rim elevations). You can change this surface later or for individual structures. For proposed pipe networks, this surface is usually a finished ground surface.

Alignment Name Choose an alignment that will provide station and offset information for your structures in Prospector as well as any labels that call for alignment stations and/or offset information. Because most pipe networks have several branches, it may not be meaningful for every structure in your network to reference the same alignment. Therefore, you may find it better to leave your Alignment option set to None in this dialog and set it for individual structures later using the layout tools or structure list in Prospector.

Using the Network Layout Creation Tools

After establishing your pipe network parameters in the Create Pipe Network dialog (shown previously in Figure 15.2), click OK; the Network Layout Tools toolbar appears (see Figure 15.3). No other command can be executed while the toolbar is active.

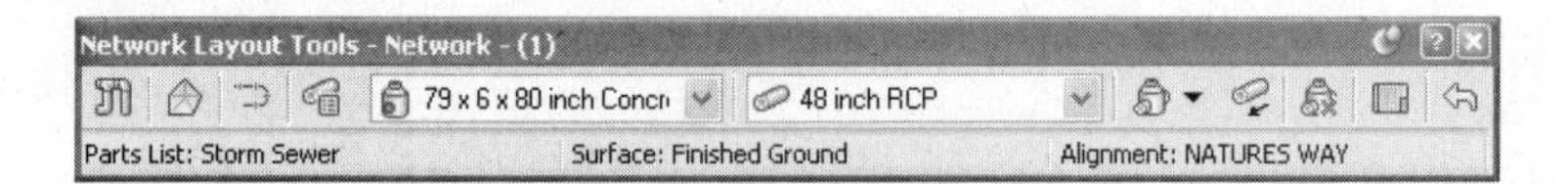

FIGURE 15.3
The Network Layout Tools toolbar

Clicking the Pipe Network Properties tool displays the Pipe Network Properties dialog, which contains the settings for the entire network. If you mistyped any of the parameters in the original Create Pipe Network dialog, you can change them here. In addition, you can set the default label styles for the pipes and structures in this pipe network.

The Pipe Network Properties dialog contains the following tabs:

Information On this tab, you can rename your network, provide a description, and choose whether you'd like to see network-specific tooltips.

Layout Settings Here you can change the default label styles, parts list, reference surface and alignment, master object layers for plan pipes and structures, as well as name templates for your pipes and structures (see Figure 15.4).

Profile On this tab, you can change the default label styles and master object layers for profile pipes and structures (see Figure 15.5).

Section Here you can change the master object layers for network parts in a section (see Figure 15.6).

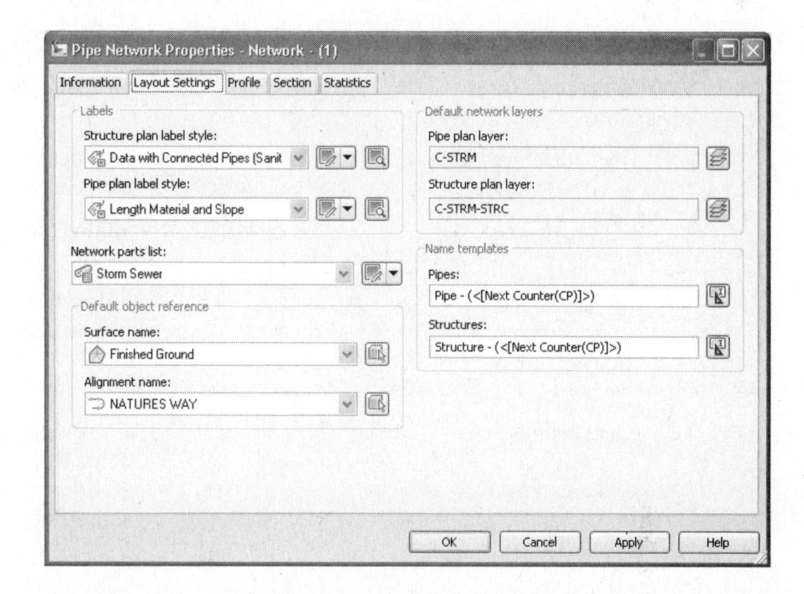

FIGURE 15.4
The Layout Settings tab of the Pipe Network Properties dialog

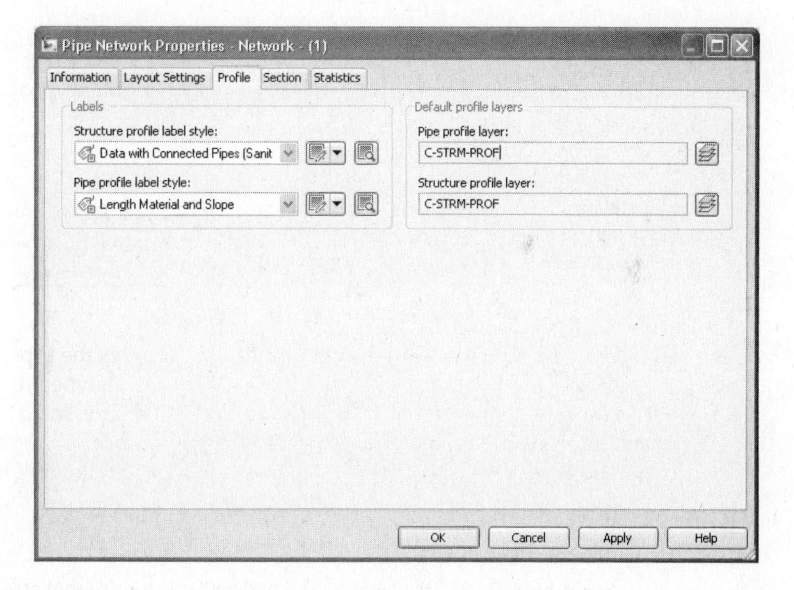

FIGURE 15.5
The Profile tab of the Pipe Network Properties dialog

Statistics This tab gives you a snapshot of your pipe network information, such as elevation information, pipe and structure quantities, and references in use (see Figure 15.7).

The Select Surface tool on the Network Layout Tools toolbar allows you to switch between reference surfaces while you're placing network parts. For example, if you're about to place a structure that needs to reference the existing ground surface, but your network surface was set to a proposed ground surface, you can click this tool to switch to the existing ground surface.

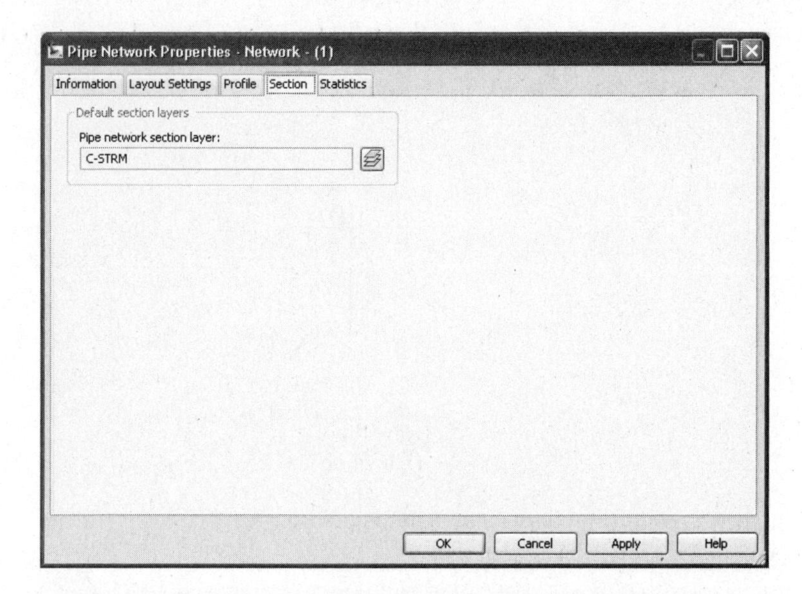

FIGURE 15.6
The Section tab of the Pipe Network Properties dialog

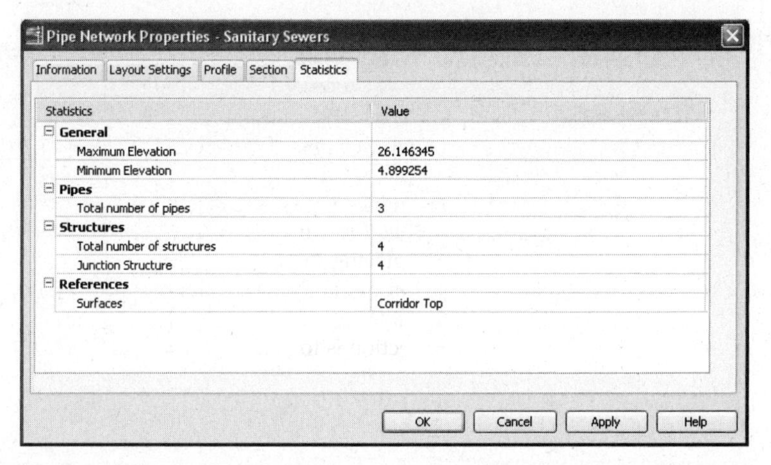

FIGURE 15.7
The Statistics tab of the Pipe Network Properties dialog

USING A COMPOSITE FINISHED GRADE SURFACE FOR YOUR PIPE NETWORK

It's cumbersome to constantly switch between a patchwork of different reference surfaces while designing your pipe network. You may want to consider creating a finished grade-composite surface that includes components of your road design, finished grade, and even existing ground. You can create this finished grade-composite surface by pasting surfaces together, so it's dynamic and changes as your design evolves.

The Select Alignment tool on the Network Layout Tools toolbar lets you switch between reference alignments while you're placing network parts, similar to the Select Surface tool.

The Parts List tool allows you to switch parts lists for the pipe network.

The Structure drop-down list (see Figure 15.8a) lets you choose which structure you'd like to place next, and the Pipes drop-down list (see Figure 15.8b) allows you to choose which pipe you'd like to place next. Your choices come from the network parts list.

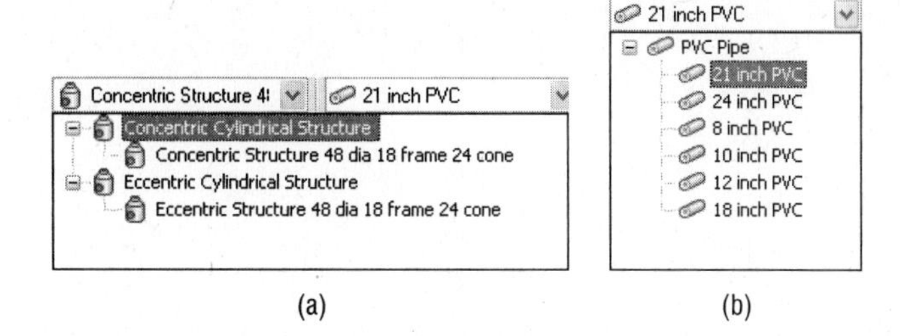

FIGURE 15.8
(a) The Structure drop-down list, and (b) the Pipes drop-down list

The options for the Draw Pipes and Structures category let you choose what type of parts you'd like to lay out next. You can choose Pipes and Structures, Pipes Only, or Structures Only.

PLACING PARTS IN A NETWORK

You place parts much as you do other Civil 3D objects or AutoCAD objects such as polylines. You can use your mouse, transparent commands, dynamic input, object snaps, and other drawing methods when laying out your pipe network.

If you choose Pipes and Structures, a structure is placed wherever you click, and the structures are joined by pipes. If you choose Structures Only, a structure is placed wherever you click, but the structures aren't joined. If you choose Pipes Only, you can connect previously placed structures. If you have Pipes Only selected and there is no structure where you click, a null structure is placed to connect your pipes.

While you're actively placing pipes and structures, you may want to connect to a previously placed part. For example, there may be a service or branch that connects into a structure along the main trunk. Begin placing the new branch. When you're ready to tie into a structure, you get a circular connection marker (shown at the top of Figure 15.9) as your cursor comes within connecting distance of that structure. If you click to place your pipe when this marker is visible, a structure-to-pipe connection is formed (shown at the bottom of Figure 15.9).

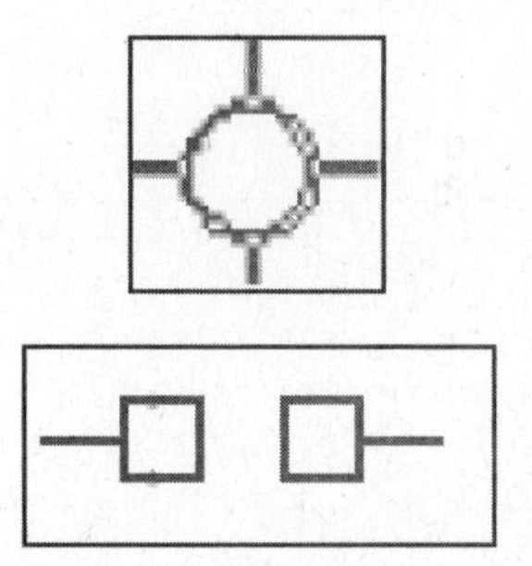

FIGURE 15.9
The Structure Connection marker (top), and the Pipe Connection marker (bottom)

If you're placing parts and you'd like to connect to a pipe, hover over the pipe you'd like to connect to until you see a connection marker that has two square shapes. Clicking to connect to the pipe breaks the pipe in two pieces and places a structure (or null structure) at the break point.

The Toggle Upslope/Downslope tool changes the flow direction of your pipes as they're placed. In Figure 15.10, Structure 9 was placed before Structure 10.

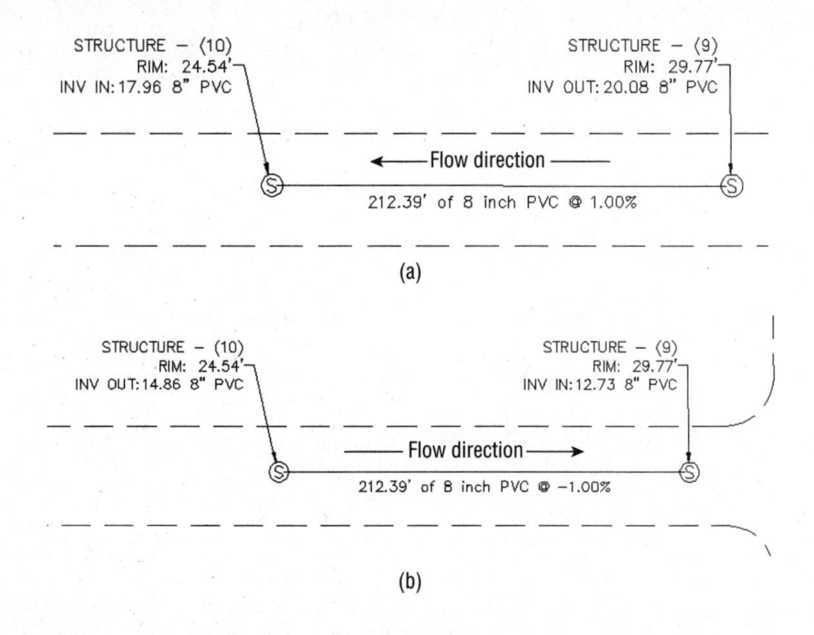

FIGURE 15.10 Using (a) the Downslope toggle and (b) the Upslope toggle to create a pipe network leg

OPTIMIZING THE COVER BY STARTING UPHILL

If you're using the Cover and Slope rule for your pipe network, you'll achieve better cover optimization if you begin your design at an upstream location and work your way down to the connection point.

The Cover and Slope rule prefers to hold minimum slope over optimal cover. In practice, this means that as long as minimum cover is satisfied, the pipe will remain at minimum slope. If you start your design from the upstream location, the pipe is forced to use a higher slope to achieve minimum cover. The following graphic shows a pipe run that was created starting from the upstream location (right to left):

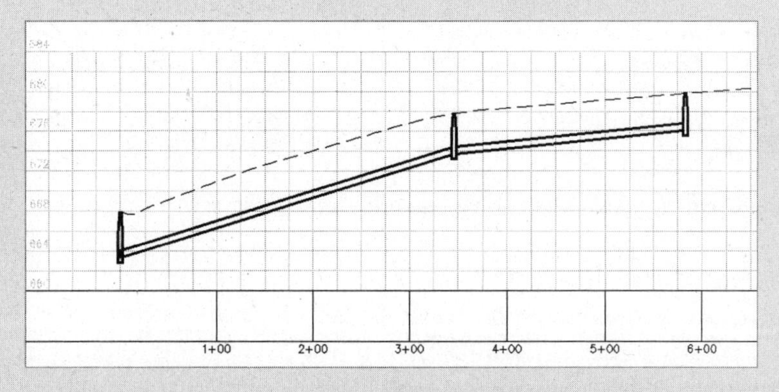

When you start from the downhill side of your project, the Minimum Slope Rule is applied as long as minimum cover is achieved. The following graphic shows a pipe run that was created starting from the downstream location (left to right):

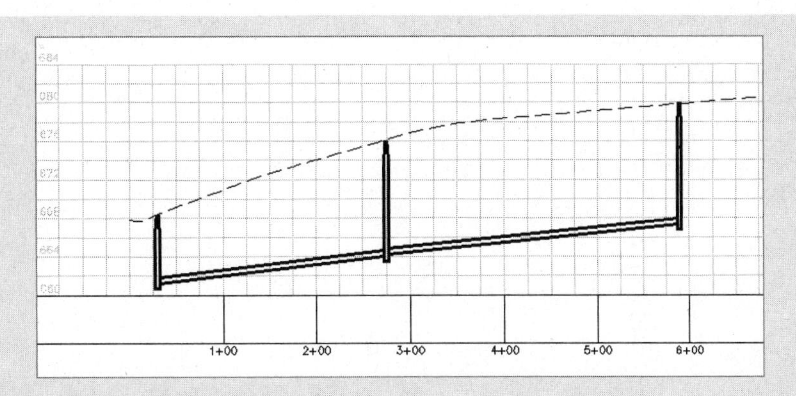

Notice how the slope remains constant even as the pipe cover increases. Maximum cover is a *violation only* rule, which means it never forces a pipe to increase slope to remain within tolerance; it only provides a warning that maximum cover has been violated.

Click Delete Pipe Network Object to delete pipes or structures of your choice. AutoCAD Erase can also delete network objects, but using it requires you to leave the Network Layout Tools toolbar.

Clicking Pipe Network Vistas brings up Panorama (see Figure 15.11), where you can make tabular edits to your pipe network while the Network Layout Tools toolbar is active.

FIGURE 15.11
Pipe Network Vistas via Panorama

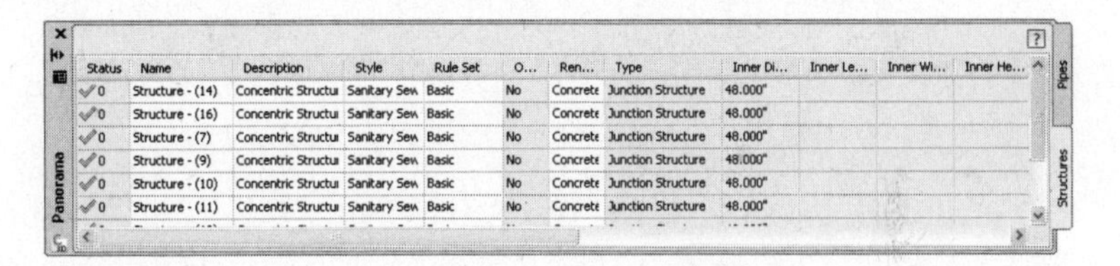

Status	Name	Description	Style	Rule Set	O...	Ren...	Type	Inner Di...	Inner Le...	Inner Wi...	Inner He...
0	Structure - (14)	Concentric Structu	Sanitary Sew	Basic	No	Concrete	Junction Structure	48.000"			
0	Structure - (16)	Concentric Structu	Sanitary Sew	Basic	No	Concrete	Junction Structure	48.000"			
0	Structure - (7)	Concentric Structu	Sanitary Sew	Basic	No	Concrete	Junction Structure	48.000"			
0	Structure - (9)	Concentric Structu	Sanitary Sew	Basic	No	Concrete	Junction Structure	48.000"			
0	Structure - (10)	Concentric Structu	Sanitary Sew	Basic	No	Concrete	Junction Structure	48.000"			
0	Structure - (11)	Concentric Structu	Sanitary Sew	Basic	No	Concrete	Junction Structure	48.000"			

The Pipe Network Vistas interface is similar to what you encounter in the Pipe Networks branch of Prospector. Many people think Pipe Network Vistas isn't as user friendly as the Prospector interface because it doesn't remember your preferred column order the way Prospector does. The advantage of using Pipe Network Vistas is that you can make tabular edits without leaving the Network Layout Tools toolbar. You can edit pipe properties, such as invert and slope, on the Pipes tab; and you can edit structure properties, such as rim and sump, on the Structure tab.

CREATING A SANITARY SEWER NETWORK

This exercise will apply the concepts taught in this section and give you hands-on experience using the Network Layout Tools toolbar:

1. Open `Pipes-Exercise 1.dwg`, which you can download from `www.sybex.com/go/masteringcivil3d2010`.
2. Expand the Surfaces branch in Prospector. This drawing has two surfaces: an existing ground and a finished ground, both of which have a _No Display style applied to simplify the drawing. Expand the Alignments and Centerline Alignments branches, and notice that there are several road alignments.

3. On the Home tab's Create Design panel, select Pipe Network ➤ Pipe Network Creation Tools.
4. In the Create Pipe Network dialog (shown previously in Figure 15.2), give your network the following information:
   - Network Name: **Sanitary Sewer Network**
   - Network Description: **Sanitary Sewer Network created by *your name* on *today's date***
   - Network Parts List: **Sanitary Sewer**
   - Surface Name: **Finished Ground**
   - Alignment Name: **Nature's Way**
   - Structure Label Style: **Data with Connected Pipes (Sanitary)**
   - Pipe Label Style: **Length Material and Slope**
5. Click OK. The Network Layout Tools toolbar appears.
6. Choose Concentric Structure 48 Dia 18 Frame 24 Cone from the drop-down list in the Structure menu and 8 Inch PVC from the Pipe list.
7. Click the Draw Pipes and Structures tool. Place two structures along Nature's Way somewhere between the road centerline and the right of way by selecting structure locations on your screen. (Doing so also places one pipe between the structures.) Note that the command line gives you additional options for placement. You can refer to the sections on pipe-network creation and editing for additional methods of placement.
8. Without exiting the command, go back to the Network Layout Tools toolbar and change the Pipe drop-down from 8 Inch PVC to 10 Inch PVC and then place another structure. Notice that the diameter of the pipe between your second and third structures is 10″.
9. Press ↵ to exit the command.
10. Next, you'll add a connecting length of pipe. Go back to the Network Layout Tools toolbar, and select 8 Inch PVC from the Pipe drop-down menu. Click the Draw Pipes and Structures tool button again. Add a structure off to one side of the road, and then connect to one of your structures within the road right-of-way. You'll know you're about to connect to a structure when you see the connection marker (a round, golden-colored glyph shaped like the one in Figure 15.12) appear next to your previously inserted structure.
11. Press ↵ to exit the command. Observe your pipe network, including the labeling that automatically appeared as you drew the network.
12. Expand the Pipe Networks branch in Prospector in Toolspace, and locate your sanitary sewer network. Click the Pipes branch; the list of pipes appears in the Preview pane. Click the Structures branch; the list of structures appears in the Preview pane.
13. Experiment with tabular and graphical edits, drawing parts in profile view, and other tasks described throughout this chapter.

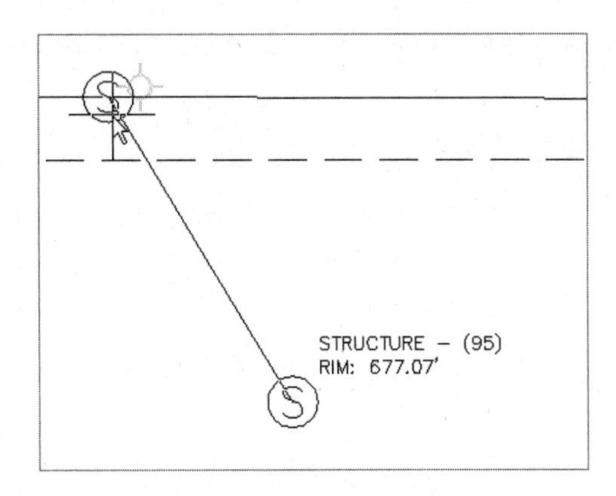

FIGURE 15.12
The connection marker appears when your cursor is near the existing structure

Creating a Storm Drainage Pipe Network from a Feature Line

If you already have an object in your drawing that represents a pipe network (such as a polyline, an alignment, or a feature line), you may be able to take advantage of the Create Pipe Network from Object command in the Pipe Network drop-down menu.

This option can be used for applications such as converting surveyed pipe runs into pipe networks and bringing forward legacy drawings that used AutoCAD linework to represent pipes. Because of some limitations described later in this section, it isn't a good idea to use this in lieu of Create Pipe Network by Layout for new designs.

It's often tempting in Civil 3D to rely on your former drawing habits and try to "convert" your AutoCAD objects into Civil 3D objects. You'll find that the effort you spend learning Create Pipe Network by Layout pays off quickly with a better-quality model and easier revisions.

The Create Pipe Network from Object option creates a pipe for every linear segment of your object and places a structure at every vertex of your object. For example, the polyline with four line segments and two arcs, shown in Figure 15.13, is converted into a pipe network containing four straight pipes, two curved pipes, and seven structures — one at the start, one at the end, and one at each vertex.

This option is most useful for creating pipe networks from long, single runs. It can't build branching networks or append objects to a pipe network. For example, if you create a pipe network from one feature line and then, a few days later, receive a second feature line to add to that pipe network, you'll have to use the pipe-network editing tools to trace your second feature line; no tool lets you add AutoCAD objects to an already-created pipe network.

Keep in mind that pipe networks can't be merged, and parts from one pipe network can't be connected to parts on another pipe network; so, it isn't typically to your advantage to create a separate pipe network for each object. Best practice is to use your longest object to start the pipe network and use the layout tools to trace and re-create the rest.

Creating a Storm Drainage Network from a Feature Line

This exercise will give you hands-on experience building a pipe network from a feature line with elevations:

1. Open the `Pipes-Exercise 2.dwg` file. (It's important to start with this drawing rather than use the drawing from an earlier exercise.)

FIGURE 15.13
(a) A polyline showing vertices, and (b) a pipe network created from the polyline

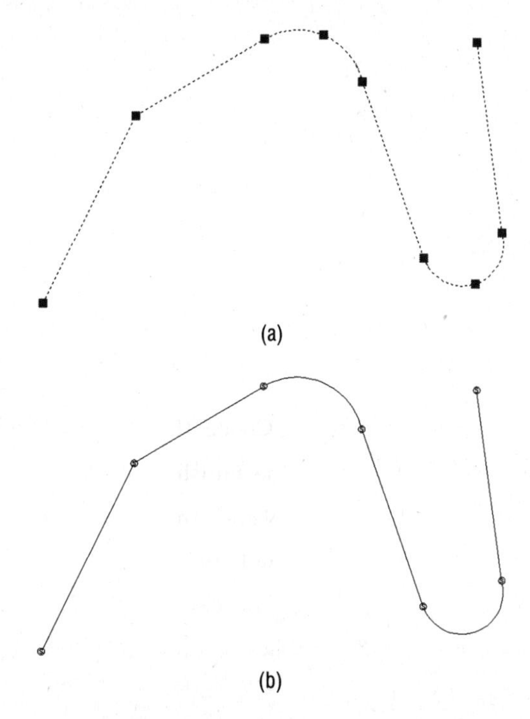

2. Expand the Surfaces branch in Prospector. This drawing has two surfaces: an existing ground and a finished ground, both of which have a _No Display style applied to simplify the drawing.
3. Expand the Alignments and Centerline Alignments branches, and notice that there are several road alignments. In the drawing, a yellow feature line runs a loop around the northern portion of the site. This feature line represents utility information for an existing storm-drainage line. The elevations of this feature line correspond with centerline elevations that you'll apply to your pipe network.
4. Choose Create Pipe Network from Object from the Pipe Network drop-down.
5. At the `Select Object or [Xref]:` prompt, select the yellow feature line. You're given a preview (see Figure 15.14) of the pipe-flow direction that is based on the direction in which the feature line was originally drawn.
6. At the `Flow Direction [OK/Reverse] <Ok>:` prompt, press ↵ to choose OK. The Create Pipe Network from Object dialog appears.
7. In the dialog, give your pipe network the following information:
   - Network Name: **Storm Network**
   - Network Description: **Storm Network created by *your name* on *today's date* from Feature Line**
   - Network Parts List: **Storm Network Parts List**
   - Pipe to Create: **15 Inch Concrete Pipe**

FIGURE 15.14
Flow-direction preview

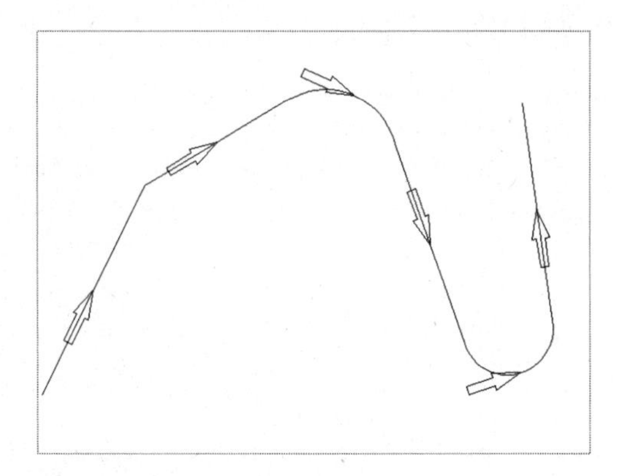

- Structure to Create: **15 x 15 Rect Structure 24 dia Frame**
- Surface Name: **Finished Ground**
- Alignment Name: **<none>**
- Erase Existing Entity check box: **selected**
- Use Vertex Elevations check box: **selected**

The two check boxes at the bottom of the dialog allow you to erase the existing entity and to apply the object's vertex elevations to the new pipe network.

If you select Use Vertex Elevations, the pipe rules for your chosen parts list will be ignored. The elevations from each vertex will be applied as a center elevation of each pipe endpoint and not the invert as you may expect. For example, if you had a feature line that you created from survey shots of existing pipe inverts and used this method, your newly created pipe network would have inverts that were off by an elevation equal to the inner diameter of your newly created pipe.

8. Click OK. A pipe network is created.

Changing Flow Direction

Choosing a pipe network object and right-clicking opens the Pipe Networks contextual tab. To change flow direction, select Change Flow Direction from the Modify panel. Change Flow Direction allows you to reverse the pipe's understanding of which direction it flows, which comes into play when you're using the Apply Rules command and when you're annotating flow direction with a pipe label-slope arrow.

Changing the flow direction of a pipe doesn't make any changes to the pipe's invert. By default, a pipe's flow direction depends on how the pipe was drawn and how the Toggle Upslope/Downslope tool was set when the pipe was drawn:

- If the toggle was set to Downslope, the pipe flow direction is set to Start to End, which means the first endpoint you placed is considered the start of flow and the second endpoint established as the end of flow.
- If the toggle was set to Upslope when the pipe was drawn, the pipe flow direction is set to End to Start, which means the first endpoint placed is considered the end for flow purposes and the second endpoint the start.

After pipes are drawn, you can set two additional flow options — Bi-directional and By Slope — in Pipe Properties:

Start to End A pipe label-flow arrow shows the pipe direction from the first pipe endpoint drawn to the second endpoint drawn, regardless of invert or slope.

End to Start A pipe label-flow arrow shows the pipe direction from the second pipe endpoint drawn to the first pipe endpoint drawn, regardless of invert or slope.

Bi-directional Typically, this is a pipe with zero slope that is used to connect two bodies that can drain into each other, such as two stormwater basins, septic tanks, or overflow vessels. The direction arrow is irrelevant in this case.

By Slope A pipe label-flow arrow shows the pipe direction as a function of pipe slope. For example, if End A has a higher invert than End B, the pipe flows from A to B. If B is edited to have a higher invert than A, the flow direction flips to be from B to A.

Editing a Pipe Network

You can edit pipe networks several ways:

- Using drawing layout edits such as grip, move, and rotate
- Grip-editing the pipe size
- Using vertical-movement edits using grips in profile (see the "Vertical Movement Edits Using Grips in Profile" section)
- Using tabular edits in the Pipe Networks branch in Prospector
- Right-clicking a network part to access tools such as Swap Part or Pipe/Structure Properties
- Returning to the Network Layout Tools toolbar by right-clicking the object and choosing Edit Network
- Selecting a network part to access the Pipe Networks contextual tab on the ribbon

With the exception of the last option, each of these methods is explored in the following sections.

Editing Your Network in Plan View

When selected, a structure has two types of grips, shown in Figure 15.15. The first is a square grip located at the structure-insertion point. You can use this grip to grab the structure and stretch/move it to a new location using the insertion point as a basepoint. Stretching a structure results in the movement of the structure as well as any connected pipes. You can also scroll through Stretch, Move, and Rotate by using your spacebar once you've grabbed the structure by this grip.

The second structure grip is a rotational grip that you can use to spin the structure about its insertion point. This is most useful for aligning eccentric structures, such as rectangular junction structures.

Also note that common AutoCAD Modify commands work with structures. You can execute the following commands normally (such as from a toolbar or keyboard macro): Move, Copy, Rotate, Align, and Mirror (Figure 15.16a). (Scale doesn't have an effect on structures.) Keep in mind that the Modify commands are applied to the structure model itself; depending on how you have your style established, it may not be clear that you've made a change. For example, if you execute the Mirror command, you can select the structures to see the results or use a 3D visual style, such as in Object Viewer, to see the modeled parts, as shown in Figure 15.16b.

FIGURE 15.15
Two types of structure grips

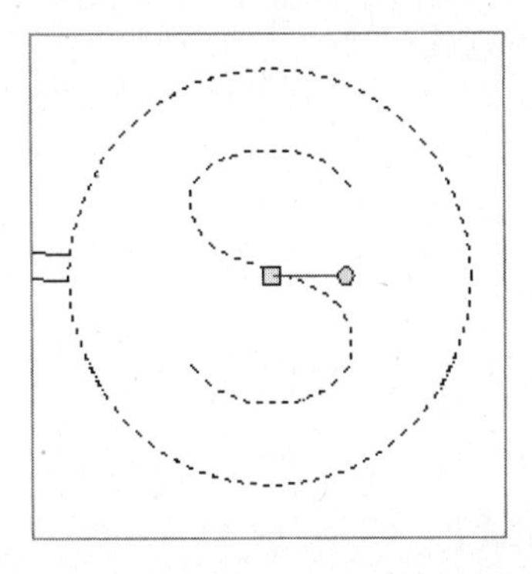

FIGURE 15.16
Mirrored structures seen (a) in plan view by their style and (b) in 3D wireframe visual style using Object Viewer

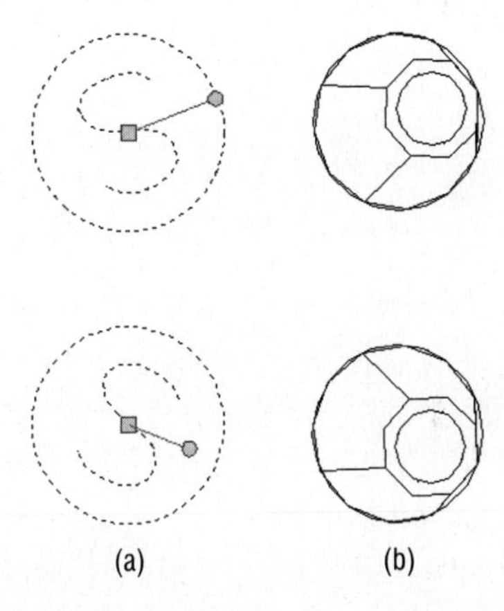

You can use the AutoCAD Erase command to erase network parts. Note that erasing a network part in plan completely removes that part from the network. Once erased, the part disappears from plan, profile view, Prospector, and so on.

When selected, a pipe end has two types of grips (see Figure 15.17). The first is a square Endpoint-Location grip. Using this grip, you can change the location of the pipe end without constraint. You can move it in any direction; make it longer or shorter; and take advantage of Stretch, Move, Rotate, and Scale by using your spacebar.

The second grip is a Pipe-Length grip. This grip lets you extend a pipe along its current bearing.

A pipe midpoint also has two types of grips (see Figure 15.18). The first is a square Location grip that lets you move the pipe using its midpoint as a basepoint. As before, you can take advantage of Stretch, Move, Rotate, and Scale by using your spacebar.

The second grip is a triangular-shaped Pipe-Diameter grip. Stretching this grip gives you a tooltip showing allowable diameters for that pipe, which are based on your parts list. Use this grip to make quick, visual changes to the pipe diameter.

Also note that common AutoCAD Modify commands work with pipes. You can execute the following commands normally (such as from a toolbar or keyboard macro): Move, Copy, Rotate,

Align, Scale, and Mirror. Remember that executing one of these commands often results in the modified pipe becoming disconnected from its structures. After the completion of a Modify command, be sure to right-click your pipe and choose Connect to Part to remedy any disconnects.

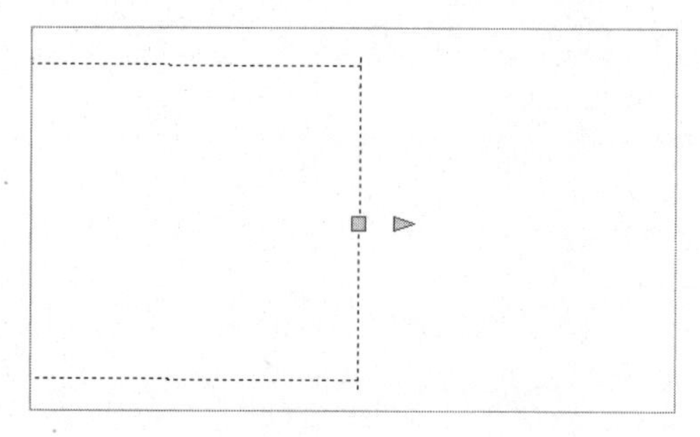

FIGURE 15.17
Two types of Pipe-End grips

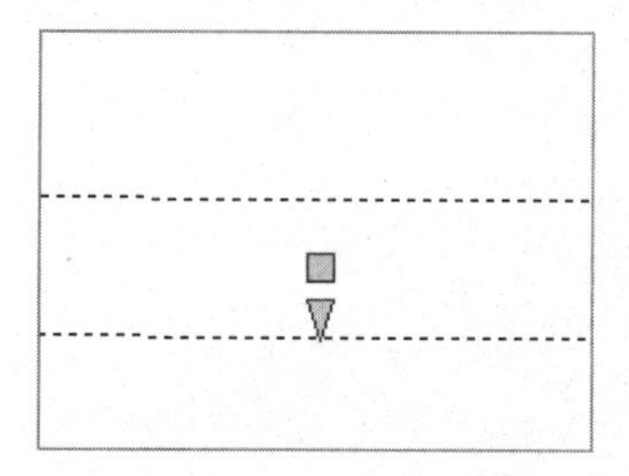

FIGURE 15.18
Two types of pipe Midpoint grips

You can use the AutoCAD Erase command to erase network parts. Note that erasing a network part in plan completely removes that part from the network. Once erased, the part disappears from plan view, profile view, Prospector, and so on.

DYNAMIC INPUT AND PIPE NETWORK EDITING

Dynamic Input (DYN) has been in AutoCAD-based products since the 2006 release, but many Civil 3D users aren't familiar with it. For some of the more command line–intensive Civil 3D tasks, DYN isn't always useful; but for pipe network edits, it provides a visual way to interactively edit your pipes and structures.

To turn DYN on or off at any time, click DYN at the bottom of your Civil 3D window.

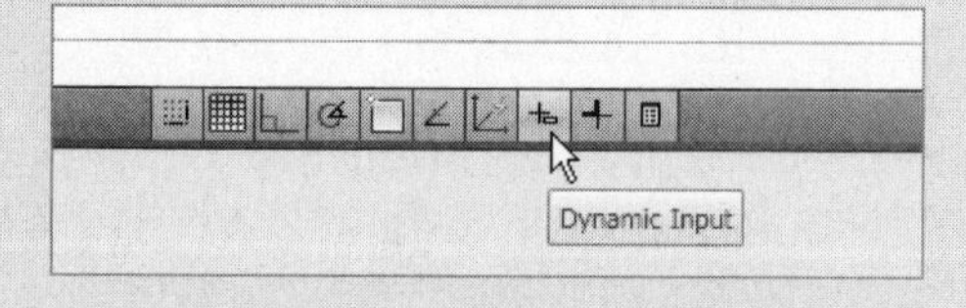

Structure Rotation While DYN is active, you get a tooltip that tracks rotation angle.

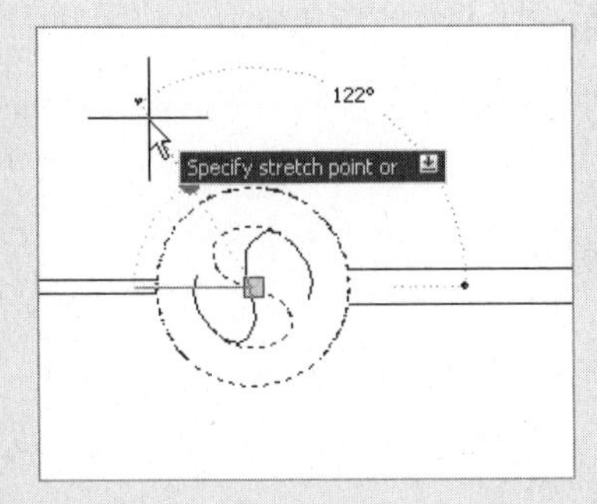

Press your down arrow key to get a shortcut menu that allows you to specify a basepoint followed by a rotation angle, as well as options for Copy and Undo. DYN combined with the Rotate command is beneficial when you're rotating eccentric structures for proper alignment.

Pipe Diameter When you're using the Pipe-Diameter grip-edit, DYN gives you a tooltip to assist you in choosing your desired diameter. Note that the tooltip depends on your drawing units; in this example, the tooltip diameter is shown as 1.5′.

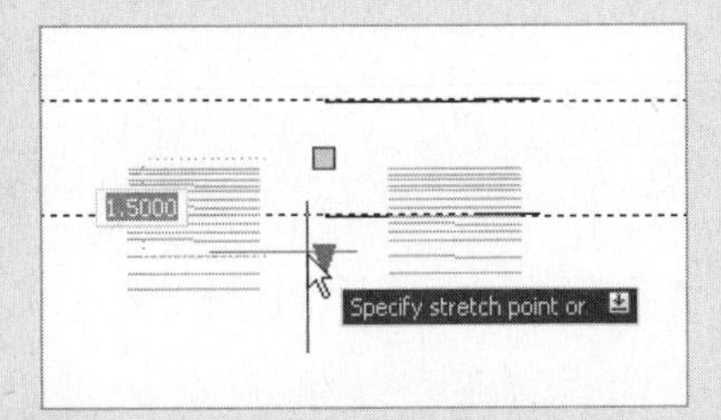

Pipe Length This is probably the most common reason to use DYN for pipe edits. Choosing the Pipe-Length grip when DYN is active shows tooltips for the pipe's current length and preview length, as well as fields for entering the desired pipe total length and pipe delta length. Use your Tab key to toggle between the total-length and delta-length fields. One of the benefits of using DYN in this interface is that even though you can't visually grip-edit a pipe to be shorter than its original length, you can enter a total length that is shorter than the original length. Note that the length shown and edited in the DYN interface is the 3D center-to-center length.

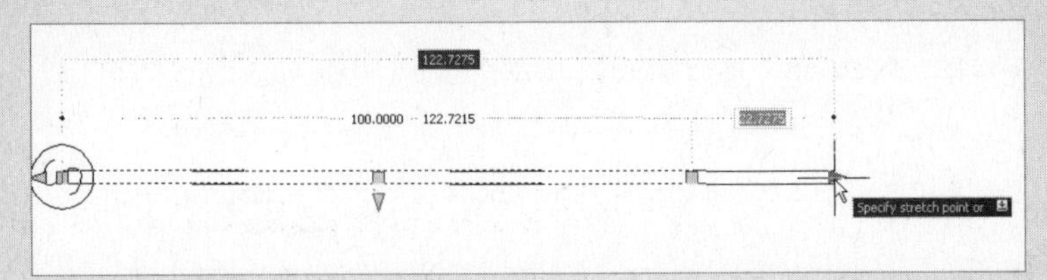

Pipe Endpoint Edits Similar to pipe length, pipe endpoint location edits can benefit from using DYN. The active fields give you an opportunity to input x- and y-coordinates.

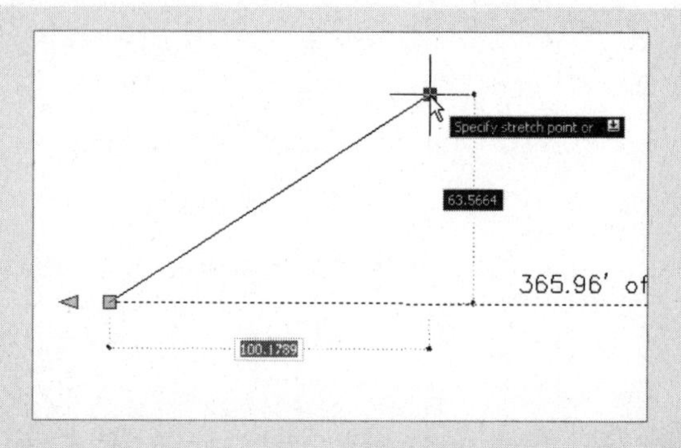

Pipe Vertical Grip Edits in Profile Using DYN in profile view lets you set exact invert, center-line, or top elevations without having to enter the Pipe Properties dialog or Prospector. Choose the appropriate grip, and note that the active DYN field is tracking profile elevation. Enter your desired elevation, and your pipe will move as you specify.

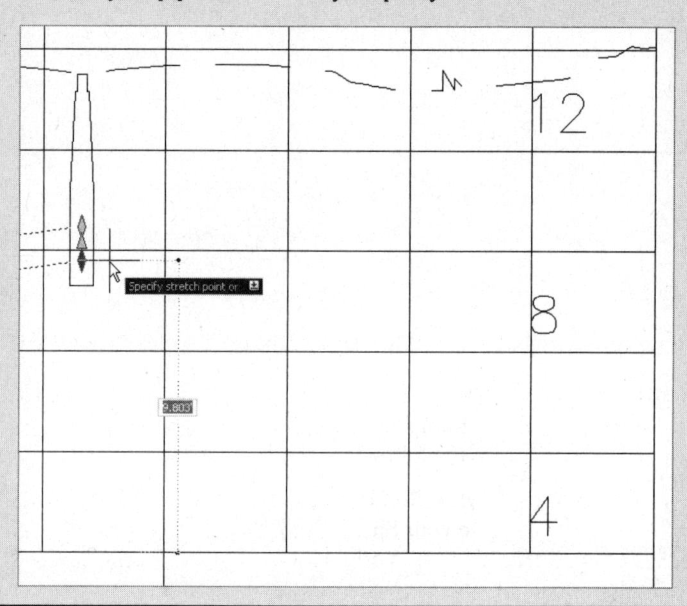

Making Tabular Edits to Your Pipe Network

Another method for editing pipe networks is in a tabular form using the Pipe Networks branch in Prospector (see Figure 15.19).

To edit pipes in Prospector, highlight the Pipes entry under the appropriate pipe network. For example, if you want to edit your sanitary sewer pipes, expand the Sanitary Sewers branch and select the Pipes entry. You should get a Preview pane that lists the names of your pipes and some additional information in a tabular form. The same procedure can be used to list the structures in the network.

White columns can be edited in this interface. Gray columns are considered calculated values and therefore can't be edited.

FIGURE 15.19
The Pipe Networks branch in Prospector

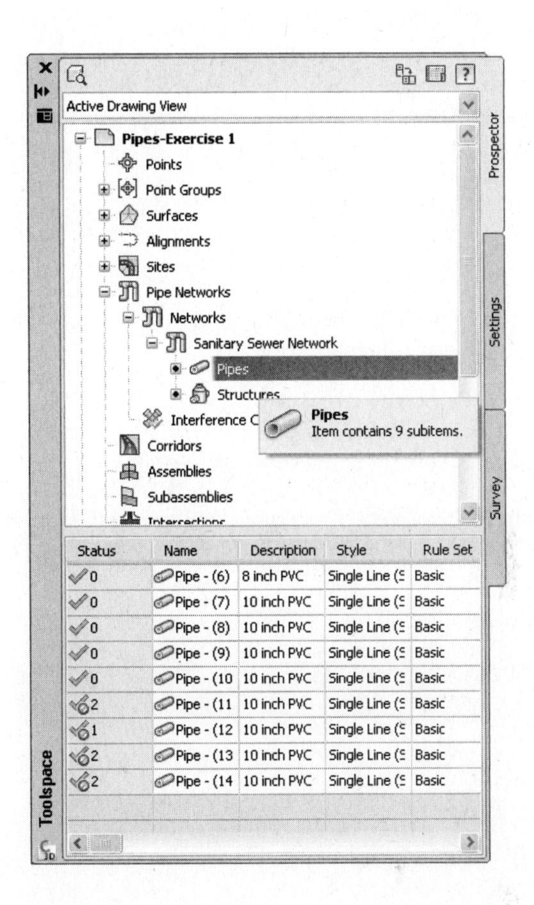

You can adjust many things in this interface, but you'll find it cumbersome for some tasks. The interface is best used for the following:

Batch Changes to Styles, Render Materials, Reference Surfaces, Reference Alignments, Rule Sets, and So On Use your Shift key to select the desired rows, and then right-click the column header of the property you'd like to change. Choose Edit, and then select the new value from the drop-down menu. If you find yourself doing this on every project for most network parts, confirm that you have the correct values set in your parts list and in the Pipe Network Properties dialog.

Batch Changes to Pipe Description Use your Shift key to select the desired rows, and then right-click the Description column header. Choose Edit, and then type in your new description. If you find yourself doing this on every project for most network parts, check your parts list. If a certain part will always have the same description, you can add it to your parts list and prevent the extra step of changing it here.

Changing Pipe or Structure Names You can change the name of a network part by typing in the Name field. If you find yourself doing this on every project for every part, check that you're taking advantage of the Name templates in your Pipe Network Properties dialog (which can be further enforced in your Pipe Network command settings).

You can Shift-select and copy the table to your Clipboard and insert it into Microsoft Excel for sorting and further study. (This is a static capture of information; your Excel sheet won't update along with changes to the pipe network.)

This interface can be useful for changing pipe inverts, crowns, and centerline information. It's not always useful for changing the part rotation, insertion point, start point, or endpoint. It isn't as useful as many people expect because the pipe inverts don't react to each other. If Pipe A and Pipe B are connected to the same structure, and Pipe A flows into Pipe B, changing the end invert of Pipe A does *not* affect the start invert of Pipe B automatically. If you're used to creating pipe-design spreadsheets in Excel using formulas that automatically drop connected pipes to ensure flow, this behavior can be frustrating.

Shortcut Menu Edits

You can perform many edits at the individual part level by using your right-click shortcut menu.

If you realize you placed the wrong part at a certain location — for example, if you placed a catch basin where you need a drainage manhole — use the Swap Part option on the shortcut menu (see Figure 15.20). You're given a list of all the parts from all the parts lists in your drawing.

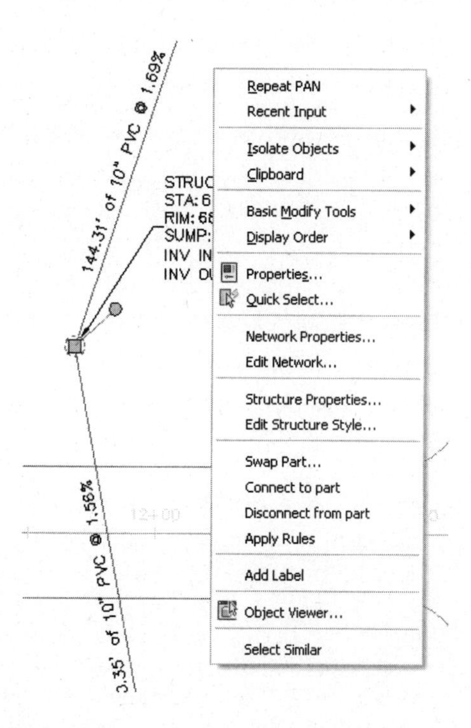

FIGURE 15.20
Right-clicking a network part brings up a shortcut menu with many options, including Swap Part

The same properties listed in Prospector can be accessed on an individual part level by using your right-click shortcut menu and choosing Pipe Properties or Structure Properties. A dialog like the Structure Properties dialog in Figure 15.21 opens, with several tabs that you can use to edit that particular part.

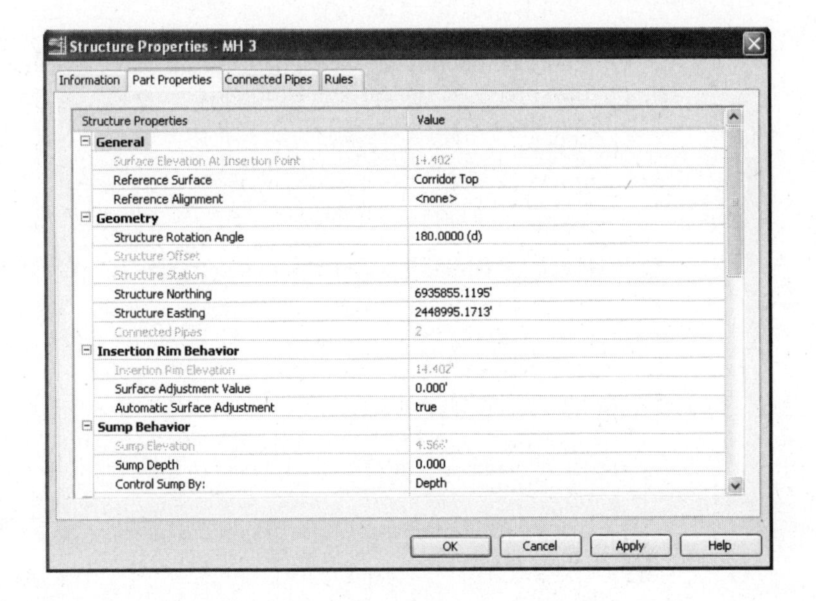

FIGURE 15.21
The Part Properties tab in the Structure Properties dialog gives you the opportunity to perform many edits and adjustments

Editing with the Network Layout Tools Toolbar

You can also edit your pipe network by retrieving the Network Layout Tools toolbar. This is accomplished by selecting a pipe network object, right-clicking, and choosing Edit Network, or by changing to the Modify tab and clicking Pipe Network on the Design panel.

Once the toolbar is up, you can continue working exactly the way you did when you originally laid out your pipe network.

This exercise will give you hands-on experience in making a variety of edits to a sanitary and storm-drainage pipe network:

1. Open the `Editing Pipes Plan.dwg` file. This drawing includes a sanitary sewer network and a storm drainage network as well as two surfaces and some alignments.
2. Select the structure STM STR 2 in the drawing. Right-click, and choose Swap Part. Select the 24 X 48 Rect Slab Top Structure from the Swap Part Size dialog. Click OK.
3. Select the newly placed catch basin so that you see the two Structure grips. Use the Rotational grip and your nearest osnap to align the catch basin with the pink property line.
4. Use the AutoCAD Erase command to erase Structure-(5).
5. Click DYN to turn it on, and select Pipe-(4). Use the triangular Endpoint grip and the DYN tooltip to lengthen Pipe-(4) to a total length of 310 feet. (Note that this is the 3D Center to Center Pipe Length.)
6. Select any pipe in the network. Right-click, and choose Edit Network.
7. Select Draw Structures Only from the drop-down menu in the Draw Pipes and Structures selection box. Place a concentric structure at the end of Pipe-(4).

8. Select Structure-(1) in the drawing. Right-click, and choose Structure Properties. Switch to the Part Properties tab. Scroll down to the Sump Depth field, and change the value to 0′. Click OK to exit the Structure Properties dialog.

9. Expand the Pipe Networks ➢ Networks ➢ Sanitary Sewer Network branches in Prospector in Toolspace, and select the Structures entry. Use the tabular interface in the Preview pane area of Prospector to change the names of Structures (1) through (4) to MH1 through MH4.

Creating an Alignment from Network Parts

On some occasions, certain legs of a pipe network require their own stationing. Perhaps most of your pipes are shown on a road profile, but the legs that run offsite or through open space require their own profiles. Whatever the reason, it's often necessary to create an alignment from network parts. Follow these steps:

1. Open the `Alignment from Network Parts.dwg` file.

2. Select Structure-(1).

3. Choose Alignment from Network on the Launch Pad panel.

4. The command line prompts you to `Select next Network Part or [Undo]`. Select Structure-(5).

5. Press ↵, and a dialog appears that is almost identical to the one you see when you create an alignment from the Alignments menu. Name and stylize your alignment as appropriate. Notice the Create Profile and Profile View check box on the last line of the dialog. Leave the box selected, and click OK.

6. The Create Profile from Surface dialog appears (see Figure 15.22). This dialog is identical to the one that appears when you create a profile from a surface. Choose both the Existing Ground and Finished Ground surfaces, and click Draw in Profile View. (See Chapter 8, "Cut to the Chase: Profiles," for further information about sampling profiles from surfaces.)

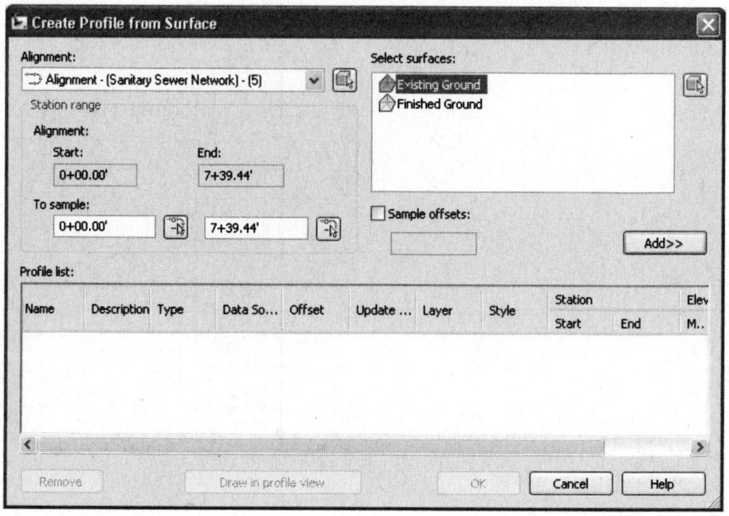

FIGURE 15.22
The Create Profile from Surface dialog

7. You see the Create Profile View Wizard (see Figure 15.23). Click the Next button in the Create Profile View Wizard until you reach the Pipe Network Display tab. You should see a list of pipes and structures in your drawing. Make sure Yes is selected for each pipe and structure in the Sanitary Sewer Network only. Click Create Profile View, and place the profile view to the right of the site plan.

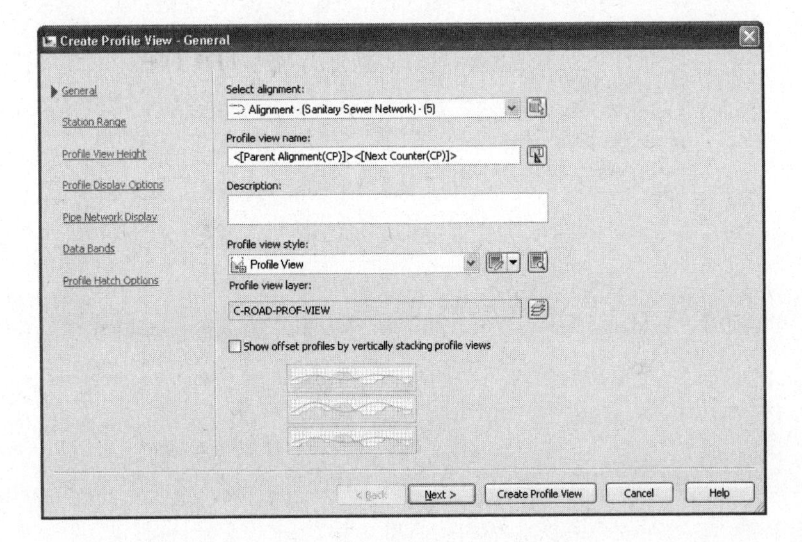

FIGURE 15.23
The Create Profile View Wizard

8. You see five structures and four pipes drawn in a profile view, which is based on the newly created alignment (see Figure 15.24).

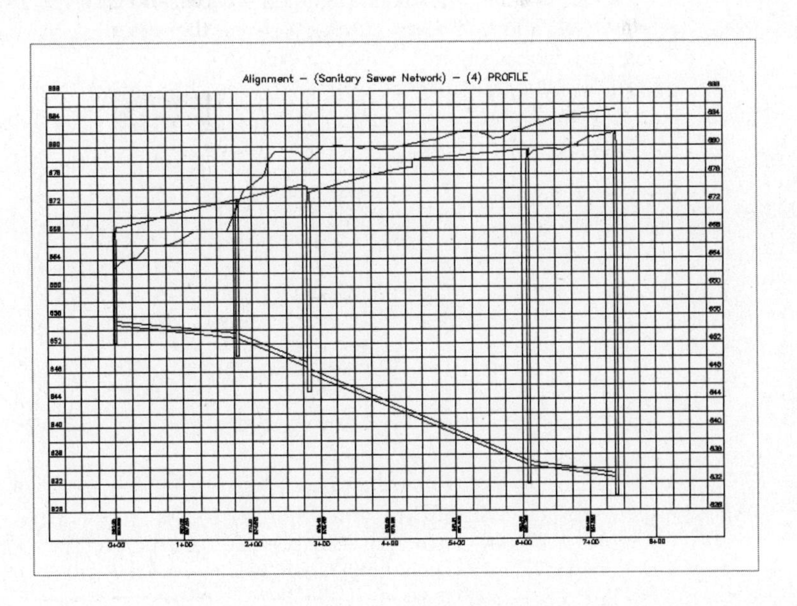

FIGURE 15.24
Creating a profile view

Drawing Parts in Profile View

If you've already created an alignment and profile view — for example, if you're going to show your pipes on the same profile view as your road design — select a network part, right-click, and choose Draw Parts in Profile from the Network Tools panel. When you're using this command, it's important to note that only selected parts are drawn in your chosen profile view.

If you neglect to choose specific parts that are meaningful to show in your profile view, you'll end up with a result like the one shown in Figure 15.25.

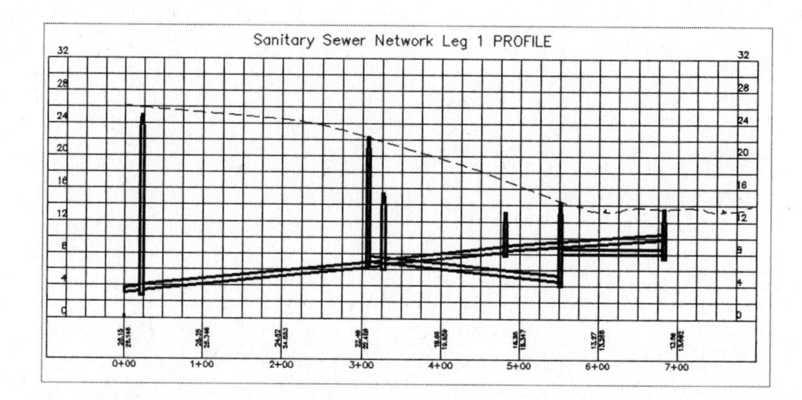

FIGURE 15.25
A profile view with inappropriate pipe network parts drawn on it

Also keep in mind when adding parts to your profile view that depending on the location of your alignment with respect to your pipes and structures, the labeled length from the model may not be the same as a pipe length you scale or measure from the profile view.

Profiles and profile views are always cut with respect to an alignment. Therefore, pipes are shown in the profile view on the basis of how they appear along that alignment or how they cross that alignment. Unless your alignment *exactly follows the centerline of your network parts*, your pipes will likely show some drafting distortion.

Let's look at Figure 15.26 as an example. This particular jurisdiction requires that all utilities be profiled along the road centerline. There's a curved road centerline, a sanitary network that jogs across the road to connect with an existing manhole in the middle, and a storm drain that crosses both the road centerline and your sanitary pipe.

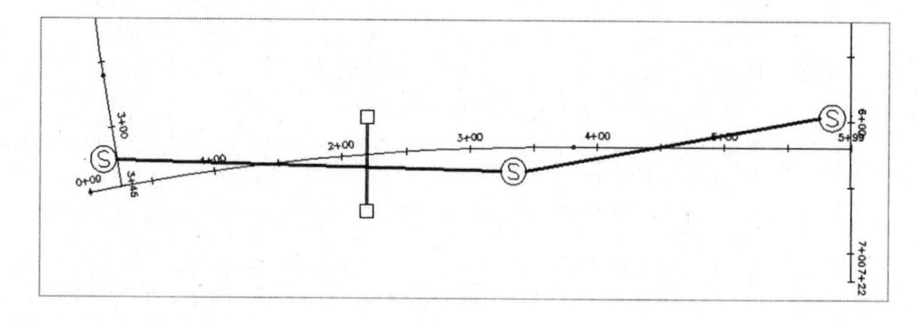

FIGURE 15.26
These pipe lengths will be distorted in profile view

At least two potentially confusing elements show up in your profile view. First, the distance between structures (2D Length – Center To Center) isn't the same in plan and profile (see Figure 15.27) because the sanitary pipe doesn't run parallel to the alignment. Because the labeling reflects the network model, all labeling is true to the 2D Length – Center To Center or any other length you specify in your label style.

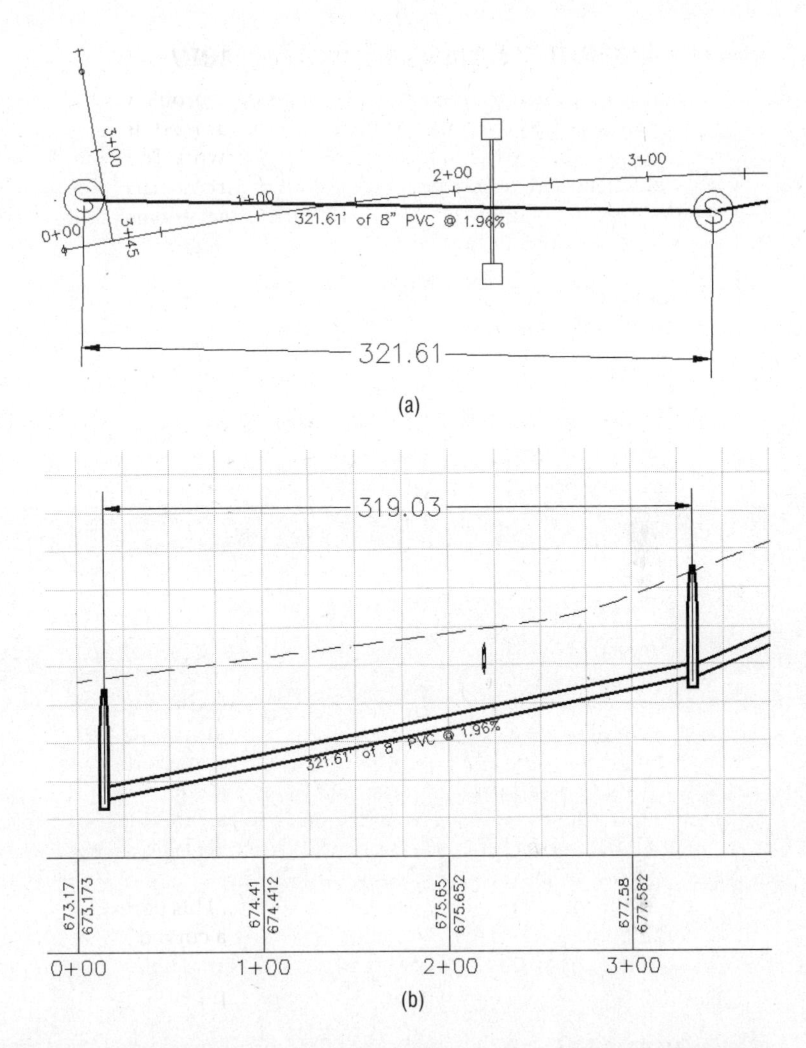

FIGURE 15.27
Pipe labels in (a) plan view and (b) profile view

The second potential issue is that the invert of your crossing storm pipe is shown at the point where the storm pipe *crosses the alignment* and not at the point where it crosses the sanitary pipe (see Figure 15.28).

If you're looking to ensure that your pipes have certain minimum crossing clearances, plotting this pipe in profile view won't give you the information you're seeking. It's best to use interference-checking for this application, as follows:

1. Open the `Draw Parts In Profile.dwg` file. The drawing has a sanitary pipe network, a storm drainage pipe network, and a profile view based on Nature's Way.
2. Select Structure-(3), Pipe-(3), Structure-(4), Pipe-(4), and Structure-(5).
3. From the Network Tools panel, choose Draw Parts in Profile.
4. The command line prompts you to `Select profile view`. Select the Nature's Way profile view. Three structures and two pipes appear in profile view.

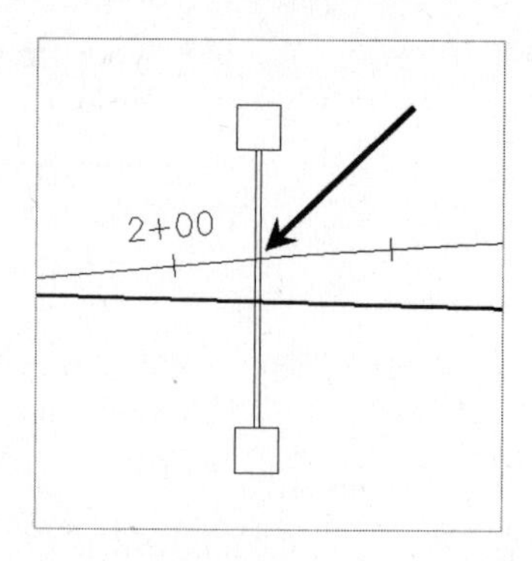

FIGURE 15.28
The invert of a crossing pipe is drawn at the location where it crosses the alignment

Vertical Movement Edits Using Grips in Profile

Although you can't make changes to certain part properties (such as pipe length) in profile view, pipes and structures both have special grips for changing their vertical properties in profile view.

When selected, a structure has two grips in profile view (see Figure 15.29). The first is a triangular-shaped grip representing a rim insertion point. This grip can be dragged up or down and affects the model structure-insertion point.

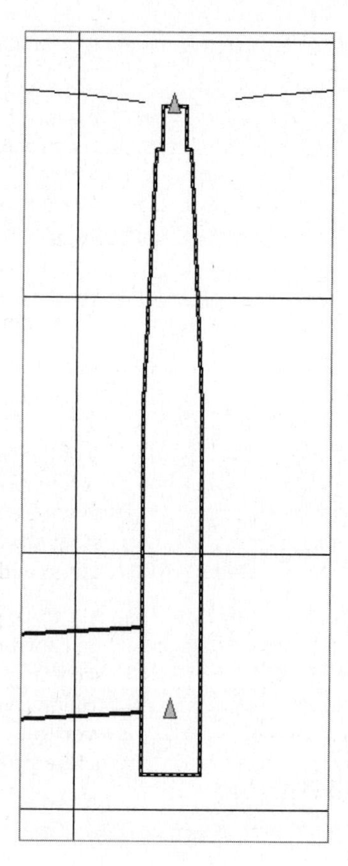

FIGURE 15.29
A structure has two grips in profile view

Moving this grip can affect your structure-insertion point two ways, depending on how your structure properties were established:

- If your structure has Automatic Surface Adjustment set to True, grip-editing this Rim Insertion-Point grip changes the surface adjustment value. If your reference surface changes, then your rim will change along with it, plus or minus that surface adjustment value.
- If your structure has the Automatic Surface Adjustment set to False, grip-editing this Rim grip modifies the insertion point of the rim. No matter what happens to your reference surface, the rim will stay locked in place.

Typically, you'll use the Rim Insertion-Point grip only in cases where you don't have a surface for your rims to target to or if you know there is a desired surface adjustment value. It's tempting to make a quick change instead of making the improvements to your surface that are fundamentally necessary to get the desired rim elevation. One quick change often grows in scope. Making the necessary design changes to your target surface will keep your model dynamic and, in the long run, will make editing your rim elevations easier.

The second grip is a triangular grip located at the sump depth. This grip doesn't represent structure invert. In Civil 3D, only pipes truly have invert elevation. The structure uses the connected pipe information to determine how deep it should be. When the sump has been set at a depth of zero, the sump elevation equals the invert of the deepest connected pipe.

This grip can be dragged up or down. It affects the modeled sump depth in one of two ways depending on how your structure properties are established:

- If your structure is set to control sump by depth, editing with the Sump grip changes the sump depth. The depth is measured from the structure insertion point. For example, if the original sump depth was zero, grip-editing the sump 0.5′ lower would be the equivalent of creating a new sump rule for a 0.5′ depth and applying the rule to this structure. This sump will react to hold the established depth if your reference surface changes, your connected pipe inverts change, or something else is modified that would affect the invert of the lowest connected pipe. This triangular grip is most useful in cases where most of your pipe network will follow the sump rule applied in your parts lists, but selected structures need special treatment.
- If your structure is set to control sump by elevation, adjusting the Sump grip changes that elevation. When sump is controlled by elevation, sump is treated as an absolute value that will hold regardless of the structure insertion point. For example, if you grip-edit your structure so its depth is 8.219′, the structure will remain at that depth regardless of what happens to the inverts of your connected pipes. The Control Sump By Elevation parameter is best used for existing structures that have surveyed information of absolute sump elevations that won't change with the addition of new connected pipes.

When selected, a pipe end has three grips in profile view (see Figure 15.30). You can grip-edit the invert, crown, and centerline elevations at the structure connection using these grips, resulting in the pipe slope changing to accommodate the new endpoint elevation.

When selected, a pipe in profile view has one grip at its midpoint (see Figure 15.31). You can use this grip to move the pipe vertically while holding the slope of the pipe constant.

You can access pipe or structure properties by choosing a part, right-clicking, and choosing Pipe Properties.

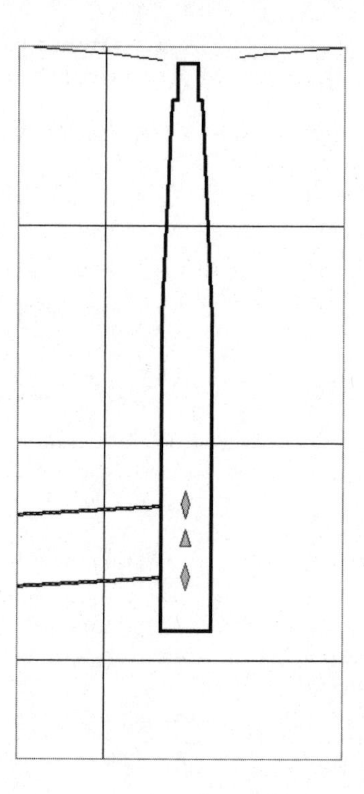

FIGURE 15.30
Three grips for a pipe end in profile view

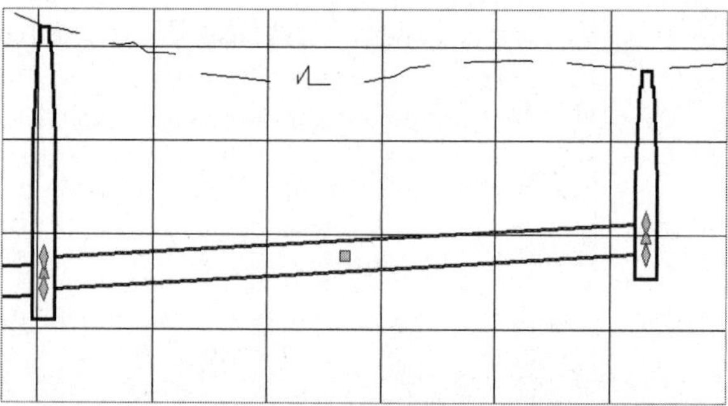

FIGURE 15.31
Use the Midpoint grip to move a pipe vertically

PART MATCHING

If you are running the Storm Sewer Extension, only stock parts can be matched up to Storm Sewer Parts.

Removing a Part from Profile View

If you have a part in profile view that you'd like to remove from the view but not delete from the pipe network entirely, you have a few options.

AutoCAD Erase can remove a part from profile view; however, that part is then removed from every profile view in which it appears. If you have only one profile view, or if you're trying to delete the pipe from every profile view, this is a good method to use.

A better way to remove parts from a particular profile view is through the profile view properties. You can access these properties by selecting the profile view, right-clicking, and choosing Profile View Properties.

The Pipe Networks tab of the Profile View Properties dialog (see Figure 15.32) provides a list of all pipes and structures that are shown in that profile view. You can deselect the check boxes next to parts you'd like to omit from this view. Also note that you can add parts to your view by deselecting the Show Only Parts Drawn in Profile View check box and making changes in the Draw column.

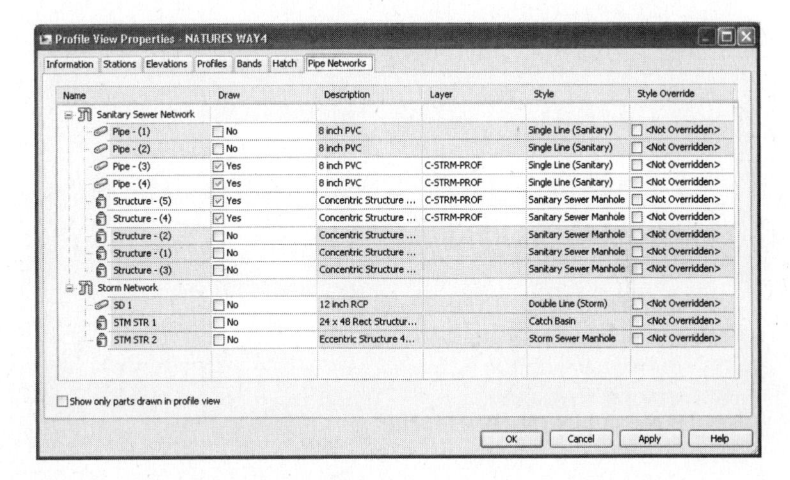

FIGURE 15.32
Deselect parts to omit them from a view

Showing Pipes That Cross the Profile View

If you have pipes that cross the parent alignment of your profile view, you can show them with a crossing style. A pipe must cross the parent alignment to be shown as a crossing in profile, and the vertical location of the pipe is shown where it crosses that alignment (see Figure 15.33).

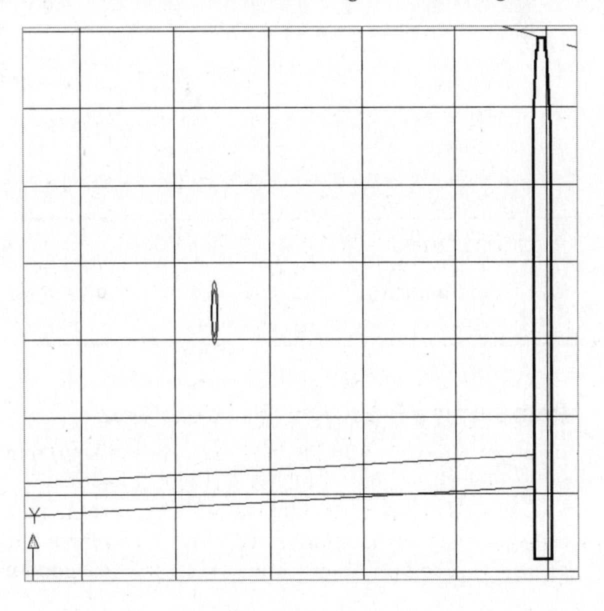

FIGURE 15.33
A pipe crossing a profile

For example, if you created an alignment directly from your network parts and then created a profile view for that alignment, any crossing pipes would be shown at the elevation where they cross the main run because your alignment and pipes coincide (see Figure 15.34).

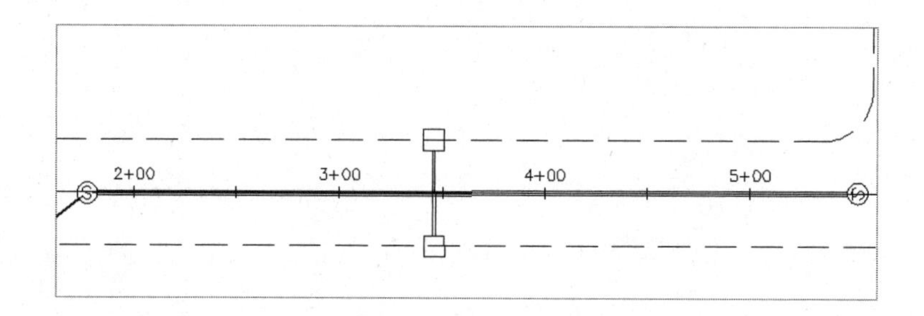

FIGURE 15.34
The invert of a storm crossing that runs along an alignment is shown in profile at the elevation where it crosses the sanitary line

If a leg of a pipe network runs offset from the road centerline, such as that shown in Figure 15.35, but you're showing those pipes in profile along the centerline alignment, any crossing pipes are shown where they *intersect the road centerline alignment*.

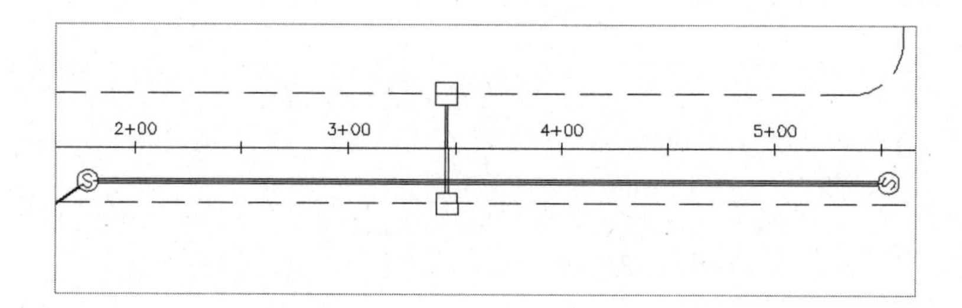

FIGURE 15.35
The invert of a storm crossing that is offset from the alignment is shown in profile at the elevation where it crosses the alignment

When pipes enter directly into profiled structures, they can be shown as ellipses through the Display tab of the Structure Style dialog (see Figure 15.36). See Chapter 14 for more information about creating structure styles.

The first step to display a pipe crossing in profile is to add the pipe that crosses your alignment to your profile view by either selecting the pipe, right-clicking, and selecting Draw Parts in Profile from the Network Tools panel, or by checking the appropriate boxes on the Pipe Network tab of the Profile View Properties dialog. When the pipe is added, it's distorted when it's projected onto your profile view — in other words, it's shown as if you wanted to see the entire length of pipe in profile (see Figure 15.37).

The next step is to override the pipe style *in this profile view only*. Changing the pipe style through pipe properties won't give you the desired result. You must override the style on the Pipe Networks tab of the Profile View Properties dialog (see Figure 15.38).

Locate the pipe you just added to your profile view, and scroll to the last column on the right (Style Override). Select the Style Override check box, and choose your pipe crossing style. Click OK. Your pipe should now appear as an ellipse.

If your pipe appears as an ellipse but suddenly seems to have disappeared in the plan and other profiles, chances are good that you didn't use the Style Override but accidentally changed the pipe style. Go back to the Profile View Properties dialog, and make the necessary adjustments; your pipes will appear as you expect.

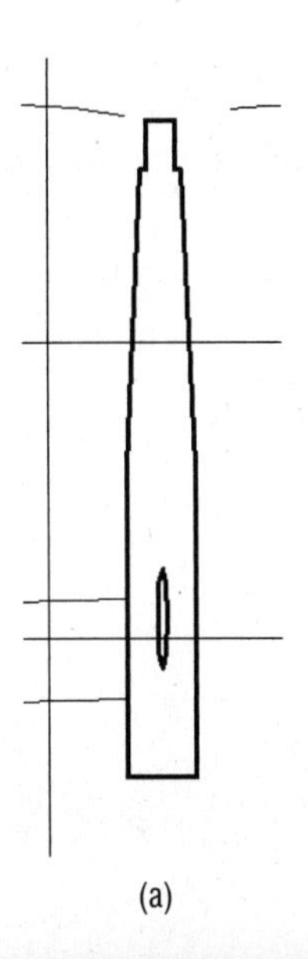

(a)

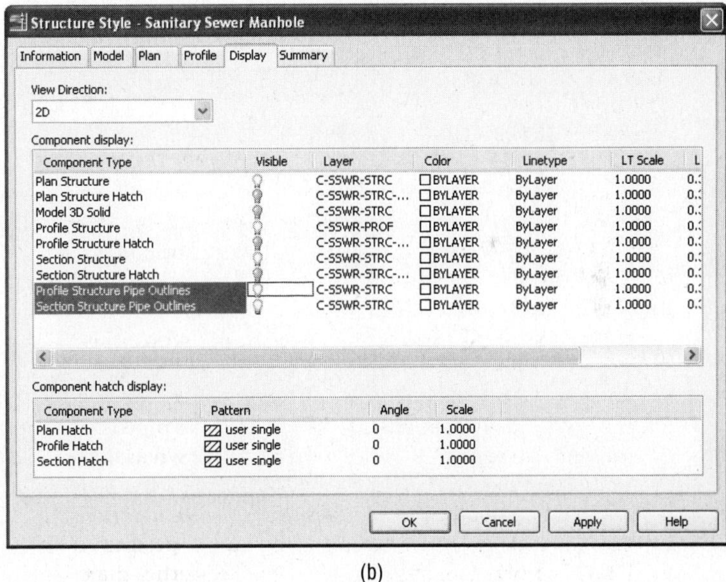

(b)

FIGURE 15.36
(a) Pipes that cross directly into a structure can be shown as part of the structure style (b) You can show crossing pipes by changing the Profile and Section Structure Pipe Outlines

Adding Pipe Network Labels

Once you've designed your network, it's important to annotate the design in a pleasing way. This section focuses on pipe network-specific label components in plan and profile views (see Figure 15.39).

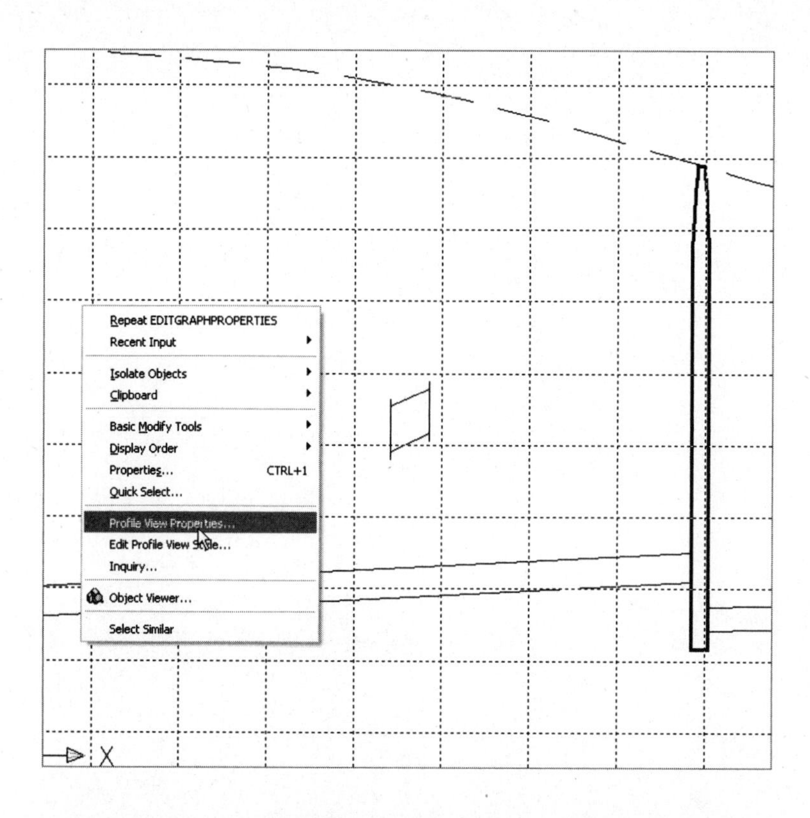

FIGURE 15.37
The pipe crossing is distorted

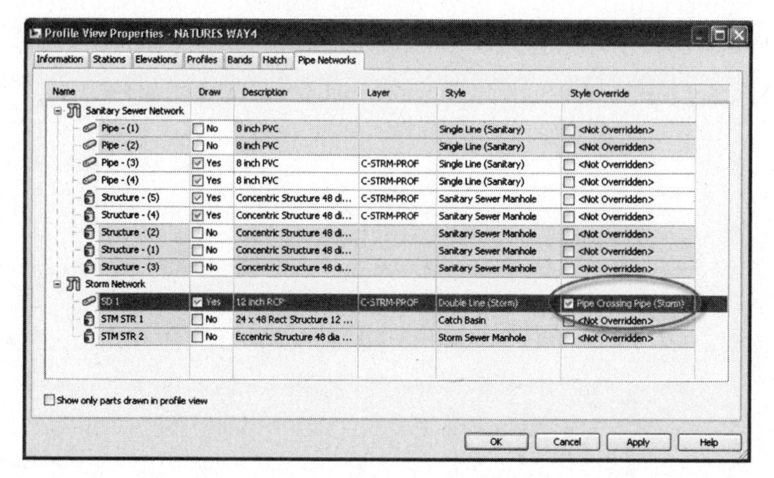

FIGURE 15.38
Correct the representation in the Profile View Properties dialog

Like all Civil 3D objects, the Pipe and Structure label styles can be found in the Pipe and Structure branches of the Settings tab in Toolspace.

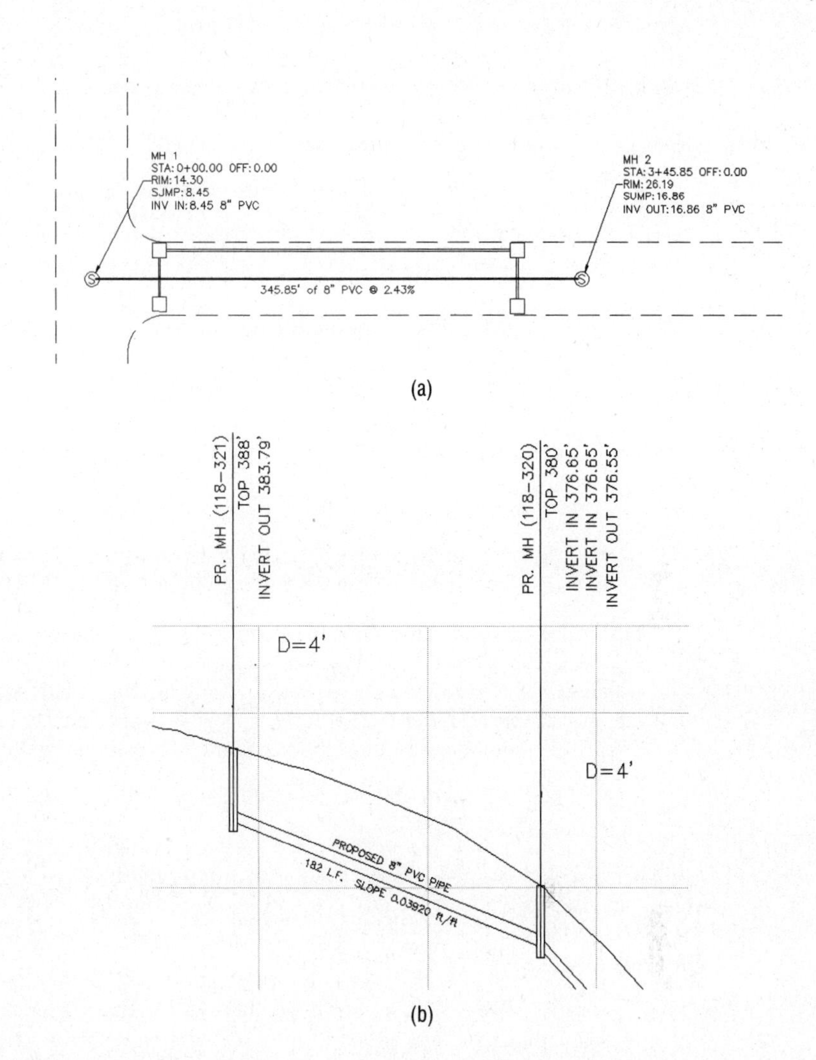

FIGURE 15.39
Typical pipe network labels (a) in plan view and (b) in profile view

Creating a Labeled Pipe Network Profile Including Crossings

This exercise will apply several of the concepts in this chapter to give you hands-on experience producing a pipe-network profile that includes pipes that cross the alignment:

1. Open the `Pipes-Exercise 3.dwg` file. (It's important to start with this drawing rather than use the drawing from an earlier exercise.)

2. Explore the drawing. It has two pipe networks: a sanitary network that doesn't follow the road centerline and a small storm network that crosses the road and the sanitary network.

3. From the Modify tab, choose Pipe Network to open the Pipe Networks contextual tab. Select Alignment from Network on the Launch Pad panel.

4. At the command-line prompt, first select Structure-5 and then select Structure-1 and press ↵ to create an alignment from the entire sanitary sewer network, with station zero being at the outlet point (Structure-5).
5. Choose the following options in the Create Alignment from Network Parts dialog:
   - Site: <none>
   - Name: **Sanitary Network Alignment**
   - Description: **Alignment from Structure 5 through Structure 1**
   - Alignment Style: Proposed
   - Alignment Label Set: All Labels
   - Create Profile and Profile View check box: selected

   Click OK.
6. Sample both the Existing Ground and the Finished Ground surfaces in the Create Profile from Surface dialog (shown previously in Figure 15.22).
7. Click Draw in Profile View to open the Create Profile View dialog. Click the Next button in the Create Profile View Wizard until you reach the Pipe Network Display tab. You should see a list of pipes and structures in your drawing. Make sure Yes is selected for each pipe and structure under the Sanitary Sewer Network. Click Create Profile View, and place the profile view to the right of the site plan.
8. Station 0+00 is the deepest end. Refer to previous sections for ideas on how to edit your pipes in profile view.
9. Select either a pipe or a structure to open the Pipe Network contextual tab, and on the Labels & Tables panel, select Add Labels ➤ Entire Network Profile.
10. The alignment information is missing from your structure labels because the alignment was created after the pipe network. Expand the Sanitary Sewer Network branch in Prospector in Toolspace to set your new alignment as the reference alignment for these structures. Click the Structures entry, and find the Structures list in the Preview pane.
11. Select all the structures in the Preview pane using Shift+click. Scroll to the right until you see the Reference Alignment column. Right-click the column header, and select Edit (see Figure 15.40). Choose Sanitary Network Alignment from the drop-down list.

    Once your reference alignment has been changed, you should see your structure labels update immediately.
12. Add the storm crossing by first choosing Pipe Network from the Design panel on the Modify tab. The Pipe Network contextual tab opens. From the Network Tools panel, select Draw Parts in Profile. Be careful to type **S** in the command line so you select just the crossing pipe and not the entire storm network. Press ↵, and then choose the profile view. Override the pipe style in this profile view only by following the procedure in the "Showing Pipes That Cross the Profile View" section earlier in this chapter.

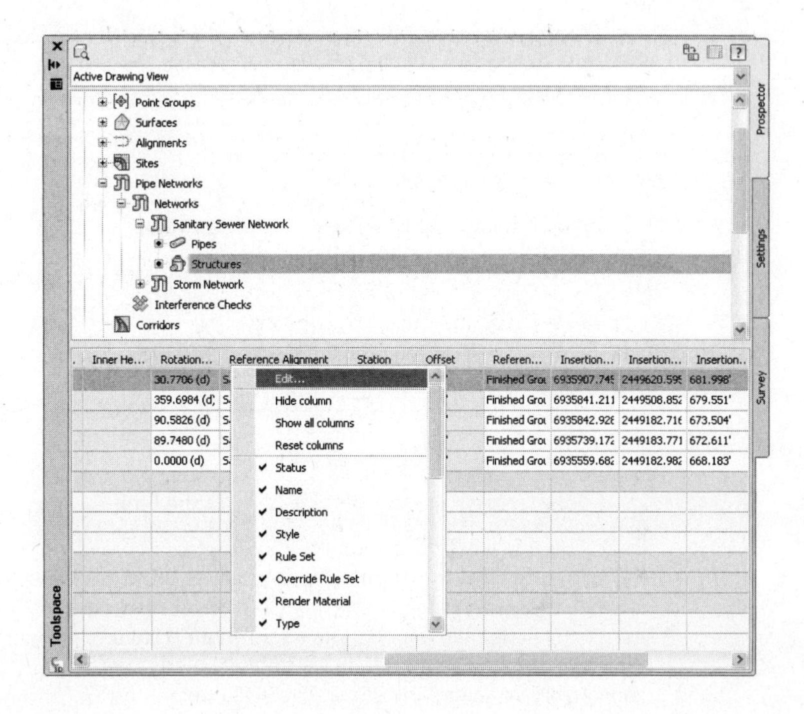

FIGURE 15.40
Change the alignment of all structures at once

Pipe Labels

Civil 3D makes no distinction between a plan-pipe label and a profile-pipe label. The same label can be used both places.

A pipe label is composed identically to most other labels in Civil 3D. In addition to text, line, and block, structure labels can also have a flow arrow and reference text, as shown in Figure 15.41.

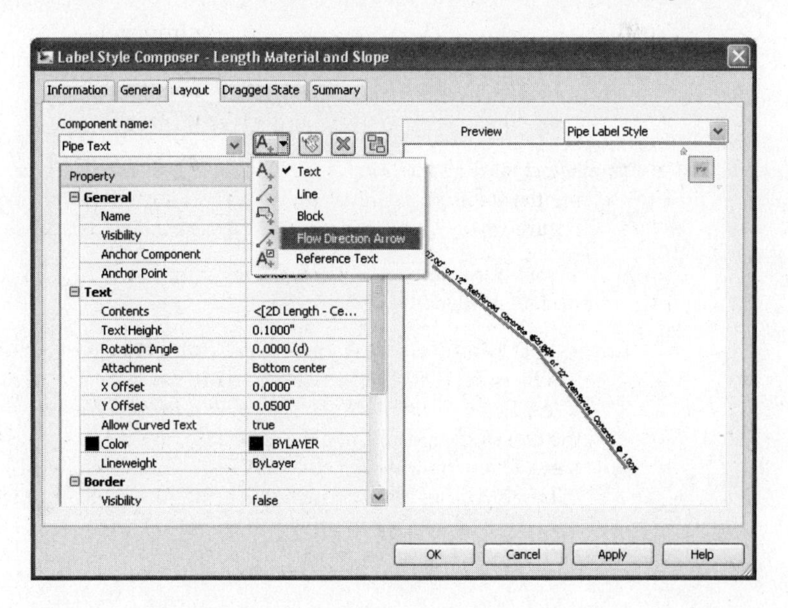

FIGURE 15.41
A flow-direction arrow and reference text are available in pipe and structure label styles

You can derive a large number of pipe properties from the model and use them in your label. If it's part of the design, it can probably be labeled. Spend some time in the Text Component Editor of your pipe label styles to fully explore all the information available.

SPANNING PIPE LABELS

In addition to single-part labels, pipes shown in either plan or profile view can be labeled using the Spanning Pipes option. This feature allows you to choose more than one pipe; the length that is reported in the label is the cumulative length of all pipes you choose.

Unlike Parcel spanning labels, no special label-style setting is required to use this tool. The Spanning Pipes option is on the Annotate tab. Simply select Add Labels ➤ Pipe Network ➤ Spanning Pipes Profile to access the command.

Structure Labels

As with pipes, Civil 3D makes no distinction between a plan-structure label and a profile-structure label. The same label can be used both places.

A structure label is composed identically to most other labels in Civil 3D. In addition to text, line, and block, structure labels can also have reference text and a component called Text for Each. Text for Each refers to text that pulls information from each connecting pipe. As mentioned earlier, structures don't have an understanding of invert. Therefore, if you want to label things such as connected pipe inverts, connected pipe diameter, and so forth, you need to add a Text for Each component to your label style.

Special Profile Attachment Points for Structure Labels

Although Civil 3D makes no distinction between plan labels and profile labels, structure labels have two special attachment points in profile view that you need to understand before you can harness their power.

Typical structure labels from the default template have limited flexibility, such as the label shown in Figure 15.42.

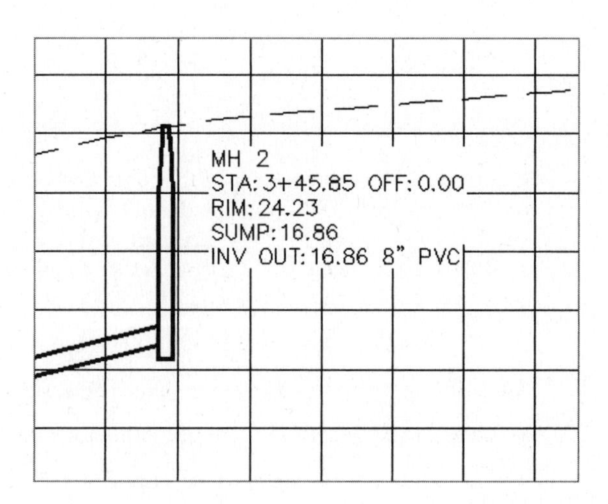

FIGURE 15.42
An example of a standard structure label from the default template

What if you wanted a profile-structure label with a line that could grow longer when you stretched it? If you go into the Label Composer for any structure label, you'll see two options for anchor points on the feature: Structure Dimension and Label Location. You can change and control the locations of both the Structure Dimension point and Label Location point.

Drilling into the settings (right-click Pipe Networks in the branch, and select Edit Feature Settings) shows you some options for customizing the locations of those two points (see Figure 15.43).

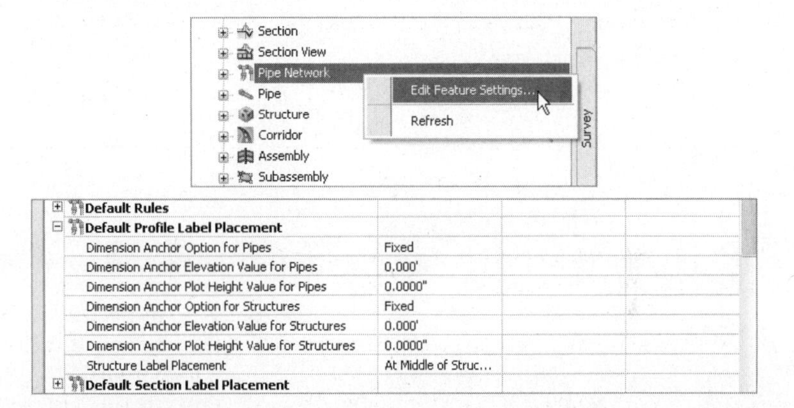

FIGURE 15.43
Accessing the Edit Feature Settings dialog

The Edit Feature Settings dialog has a cryptic batch of settings under Default Profile Label Placement. In a nutshell, these settings give your labels profile-view attachment points that you can customize. This section examines the structure labels closely. Note the default settings:

- Dimension Anchor Option for Structures: Fixed
- Dimension Anchor Elevation Value for Structures: 0′
- Dimension Anchor Plot Height Value for Structures: 0″
- Structure Label Placement: At Middle of Structure

If you highlight a structure label in profile, two cyan-colored grips appear. The lower grip appears at Profile View Elevation Zero. This is the structure dimension. Right now, it's fixed at elevation zero, which makes sense on the basis of the default settings. The location of the structure dimension is based on the profile view, and it can be grip-edited and stretched vertically after placement.

The second cyan-colored grip is the structure-label location. Right now, it's set to At Middle of Structure (see Figure 15.44) per the default settings. The structure-label location is always tied to the top, middle, or bottom of the structure and can't be grip-edited after placement.

If you go back into the Edit Feature Settings dialog (shown previously in Figure 15.43), you can change the values as follows:

- Dimension Anchor Option for Structures: Graph View Top
- Dimension Anchor Elevation Value for Structures: 0′
- Dimension Anchor Plot Height Value for Structures: 0″
- Structure Label Placement: At Top of Structure

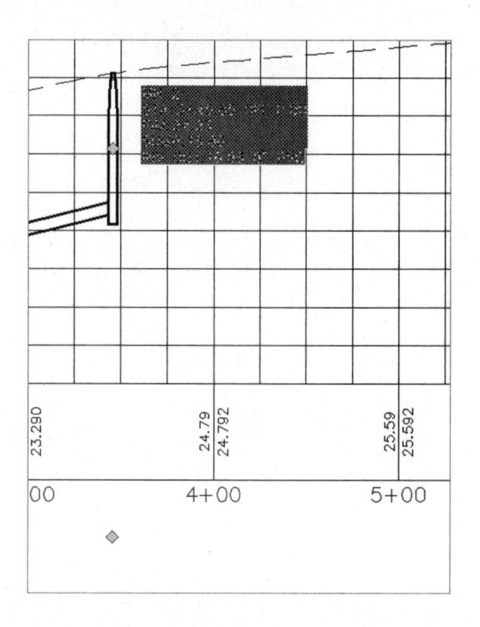

FIGURE 15.44
When a structure label is selected in profile, the structure-dimension location and the label location are marked with grips. This figure shows the structure dimension at elevation zero and the structure-label location set to At Middle of Structure

You must erase and replace your label to see the new attachment points. Once you do, it's clear that the structure dimension is now at Graph View Top, and the label location is set to At Top of Structure, as shown in Figure 15.45.

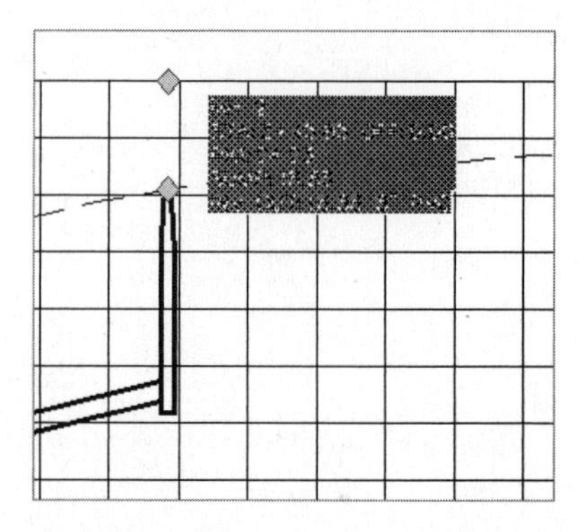

FIGURE 15.45
When a structure label is selected in profile, the structure-dimension location and the label location are marked with grips. This figure shows the structure dimension at Graph View Top and the structure-label location set to At Top of Structure

For label composition, adjusting the structure-label location and structure-dimension location is powerful because you can attach and orient lines, blocks, and text to these two points.

The foundation of the label shown in Figure 15.46 is the vertical line that runs from the top of the structure to Graph View Top.

Confirm that your feature settings have Structure Label Placement set to At Top of Structure and Dimension Anchor Option for Structures set to Graph View Top, and then make a new label style.

In the Label Style Composer, add a line that uses the label location as its start and the structure dimension as its end, as shown in Figure 15.47.

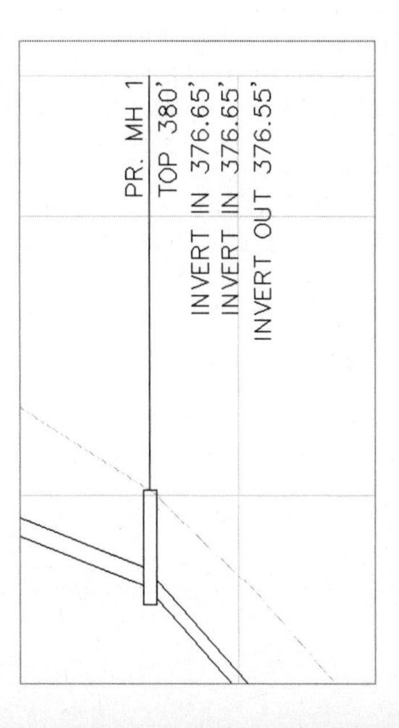

FIGURE 15.46
A structure label with a flexible line

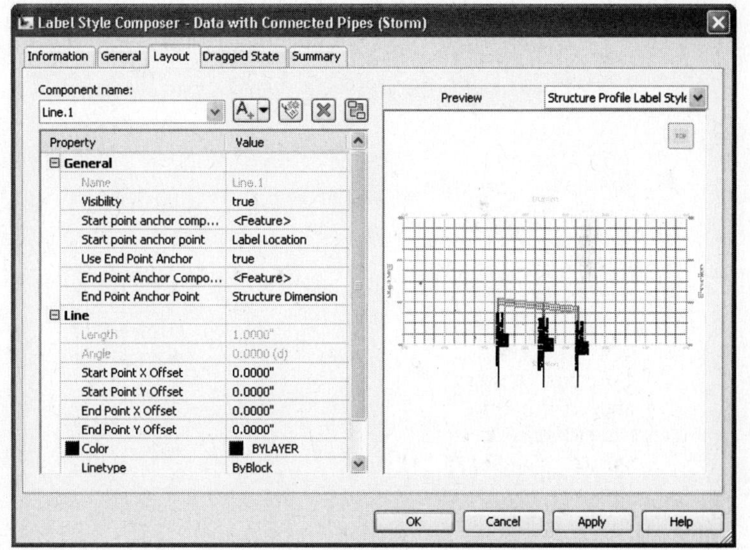

FIGURE 15.47
Add a line component anchored to Label Location and Structure Dimension

Add any other desired components, and examine your label. If you want to revise your label so the vertical line goes all the way down to the bottom of the structure, you don't have to change the label-style composition. In fact, editing the label style won't give you the desired results. The key is to change the structure-label location to At Bottom of Structure (see Figure 15.48).

Changing the feature settings requires you to reset the label, but once you replace the label you should see the line go all the way to the bottom of the structure.

You learned earlier that you can grip-edit the structure dimension to override the feature settings. If you grip-edit the structure dimension, you can drag the line vertically (see Figure 15.49).

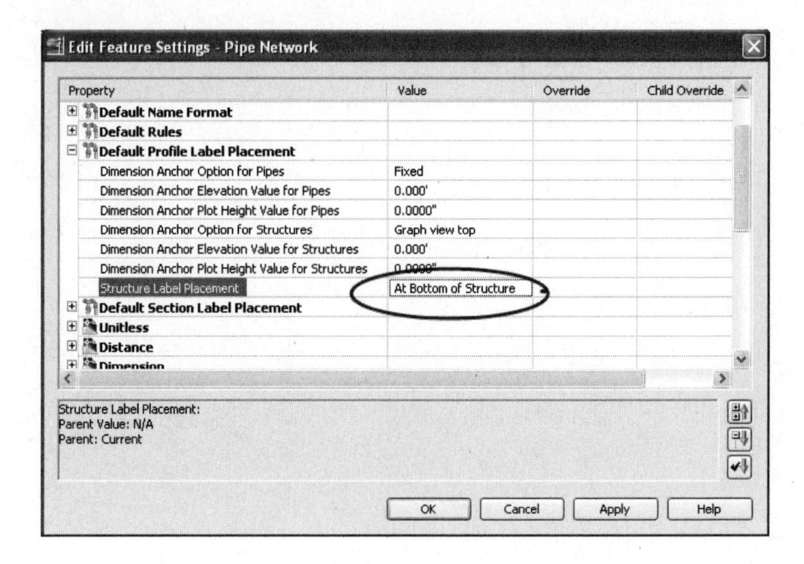

FIGURE 15.48
Change the Structure Label Placement option to achieve a line that connects to the bottom of structure

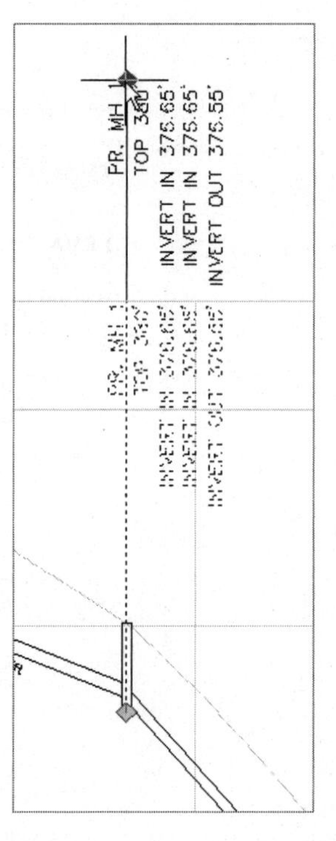

FIGURE 15.49
Use the grip to stretch the line and label

If you ever need to reset or customize the placement of this label, you can select the label, right-click, and choose Label Properties.

In the Properties dialog shown in Figure 15.50, you see Dimension Anchor Option from the feature settings and a Dimension Anchor Value, which is equal to the distance you stretched the grip. If you want to shorten the line so it touches the Graph View Top, change Dimension Anchor Value to 0.

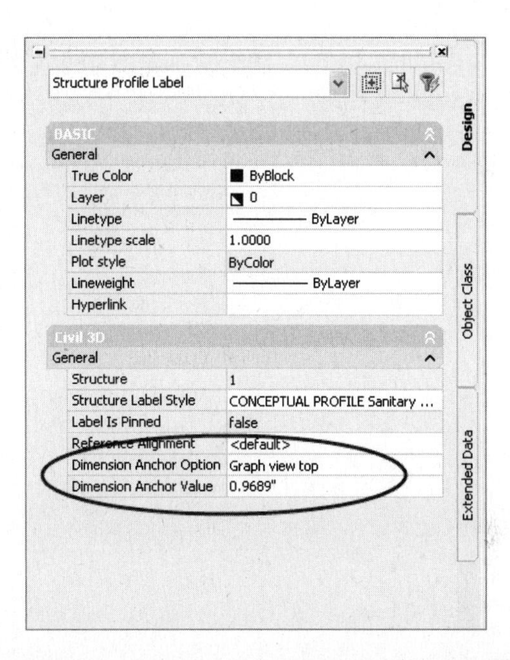

FIGURE 15.50
The AutoCAD Properties palette showing Dimension Anchor Option

Real World Scenario

ADDING EXISTING GROUND ELEVATION TO STRUCTURE LABELS

In design situations, it's often desirable to track not only the structure rim elevation at finished grade, but also the elevation at existing ground. This gives the designer an additional tool for optimizing the earthwork balance.

This exercise will lead you through creating a structure label that includes surface-reference text. It assumes you're familiar with Civil 3D label composition in general:

1. Open the `Pipes-Exercise 4.dwg` file, or continue working in any previous drawing.
2. Locate the Structure Label Styles branch on the Settings tab of Toolspace. (If Toolspace isn't visible, choose Toolspace on the palettes panel on the Home tab and make the Settings tab active.)
3. Right-click Label Styles, and select New. Refer to Chapter 1, "Getting Dirty: The Basics of Civil 3D," for more detailed information about label-style creation and settings. Set the options as follows:
   - On the Information tab, name your label.
   - On the General tab, you can choose your company text style and layer as well as change the label orientation and readability behavior.
   - On the Layout tab is a default text component called Structure Text. Click in the contents box to bring up the Text Component Editor.
   - Delete the "<[Description(CP)]>" text string, and then use the Properties drop-down menu to add the <Name> and RIM <Insertion Rim Elevation> (two decimal places).

4. Click OK to leave the Text Component Editor.
5. In the Label Style Composer, choose Reference Text from the Add Component drop-down menu.

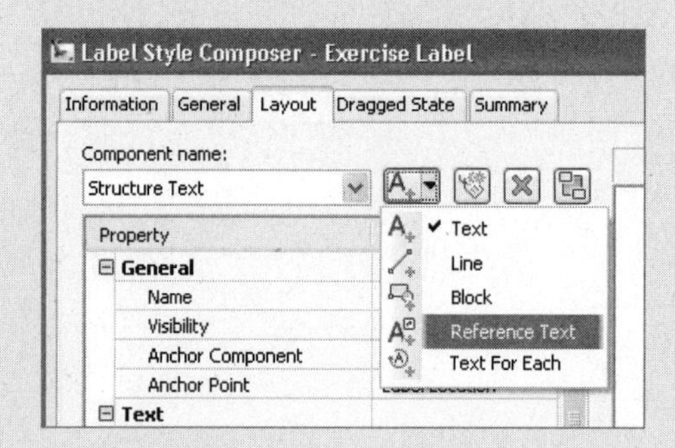

6. In the Select Type dialog, choose Surface.
7. Rename the component from Reference Text.1 to **Existing Ground**.
8. Click in the contents box to bring up the Text Component Editor.
9. Delete the "Label Text" text string, and then use the Properties drop-down menu to add the EG <Surface Elevation> (two decimal places).
10. Click OK to dismiss the Text Component Editor.
11. In the Label Style Composer, change Anchor Component for Existing Ground to Structure Text, change Anchor Point to Bottom Center, and change Attachment to Top Center.
12. Choose the Text for Each option from the Add Component drop-down menu.
13. In the Select Type dialog, choose Structure All Pipes.
14. Click in the Contents box to bring up the Text Component Editor.
15. Delete the "Label Text" string, and then use the Properties drop-down menu to add the INV <Connected Pipe Invert Elevation> (two decimal places).
16. Click OK to dismiss the Text Component Editor.
17. In the Label Style Composer, change Anchor Component for Text for Each.1 to Existing Ground and change Anchor Point to Bottom Center. Then, change Attachment to Top Center.
18. Click OK to dismiss the Label Style Composer.
19. Go into the drawing, and select one or more structure labels.
20. Right-click, and choose Properties. In the Properties dialog, choose your new label from the Structure Label Style drop-down menu.
21. Exit the Properties dialog, and clear your selections by pressing Esc. Look at the drawing; the Existing Ground part of the label is marked with question marks (???).
22. Choose one of the ??? labels, right-click, and select Label Properties. This time, you're given a Reference Text Objects drop-down menu where you can choose the Existing Ground surface.

If the existing ground surface doesn't appear on the drop-down menu, click the Entity Selection button, and then choose the existing ground surface by selecting it in the drawing.

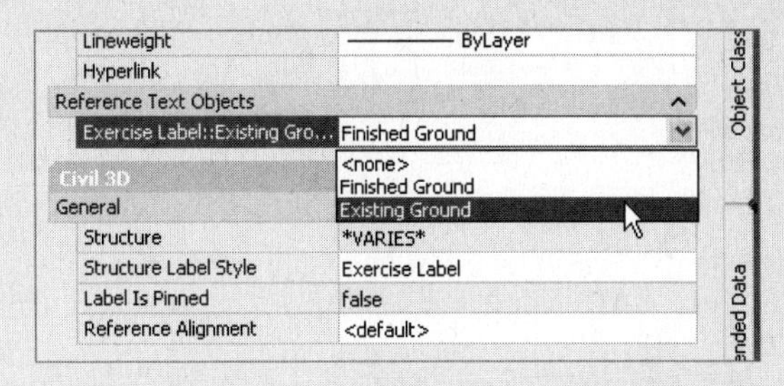

Look at the drawing; the Existing Ground part of the label is now populated.

Structure – (2)
RIM 672.61
EG 671.25
INV 652.87
INV 652.81

Explore label options to customize the layout and arrangement of your label using the tools mentioned in this and other chapters.

Creating an Interference Check between a Storm and Sanitary Pipe Network

In design, you must make sure pipes and structures have appropriate separation. You can perform some visual checks by rotating your modeling into 3D and plotting pipes in profile and section views (see Figure 15.51). Civil 3D also provides a tool called Interference Check that makes a 3D sweep of your pipe networks and lets you know if anything is too close for comfort.

The following exercise will lead you through creating a pipe network Interference Check to scan your design for potential pipe network conflicts:

1. Open the `Interference Start.dwg` file. The drawing includes a sanitary sewer pipe network and a storm drainage network.
2. Select a part from either network and choose Interference Check from the Analyze panel. You're prompted to `Select a part from the same network or different network`. Select a part from the network not chosen.
3. The Create Interferences dialog appears. Name the Interference Check **Exercise**, and confirm that Sanitary Sewer Network and Storm Network appear in the Network 1 and Network 2 boxes.

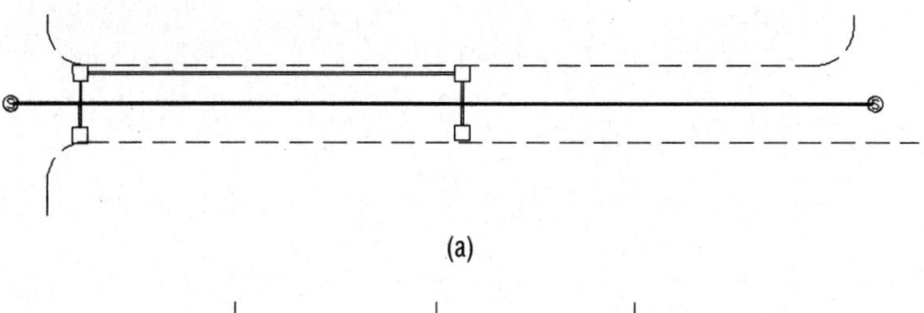

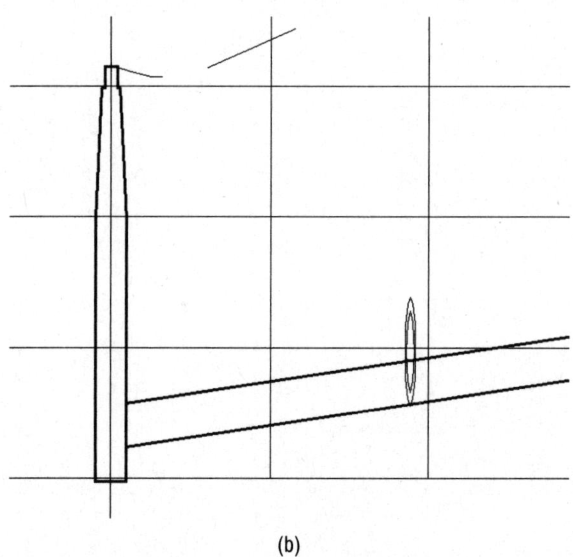

FIGURE 15.51
(a) Two pipe networks may interfere vertically where crossings occur. (b) Viewing your pipes in profile view can also help identify conflicts

4. Click 3D Proximity Check Criteria, and the Criteria dialog appears (see Figure 15.52).

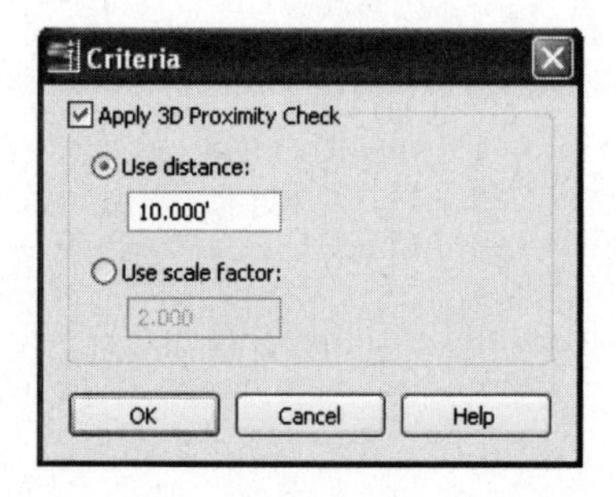

FIGURE 15.52
Criteria for the 3D proximity check

You're interested in finding all network parts that are within a certain tolerance of one another, so enter **3'** in the Use Distance box. This setting creates a buffer to help find parts in all directions that might interfere.

(If you were interested only in physical, direct collisions between parts in your networks, you'd leave the Apply 3D Proximity Check check box deselected and dismiss this dialog.)

AN INTERFERENCE CLOUD?

Think of the Interference Check as if each part is surrounded by a 3D "cloud." When that cloud touches another part, interference is flagged. If you specify distance, such as 2′, your cloud is created 2′ in all directions around each part. When you specify scale, such as 1.5, your cloud makes each part 1.5 times larger than its actual size.

5. Click OK to exit the Criteria dialog, and click OK to run the Interference Check. You see a dialog that alerts you to one interference. Click OK to dismiss this dialog.
6. Zoom in on the crossing of pipe SD-1 and Pipe-3. A small marker has appeared, as shown in Figure 15.53.

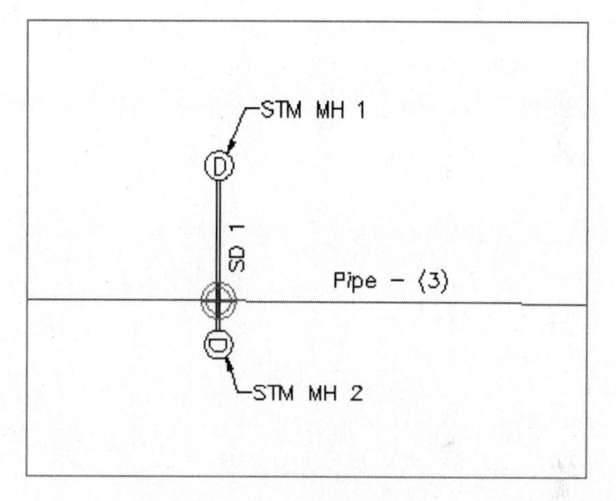

FIGURE 15.53
The interference marker in plan view

7. Change to the View tab and look at the Views panel. On the left is a list of preset 3D views, using the arrows in this box, scroll and then select SW Isometric. Zoom in on the crossing of pipe SD-1 and Pipe-3. The interference marker appears in 3D, as shown in Figure 15.54.

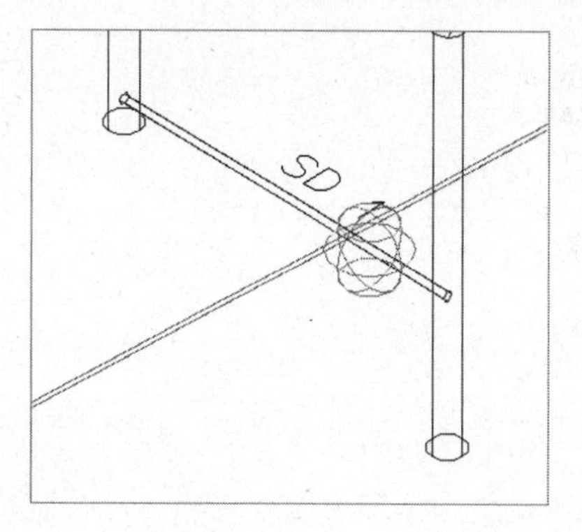

FIGURE 15.54
The interference marker in 3D

8. Using the same view navigation technique as you used in step 7, change back to a top-view orientation.

9. Drill into the Pipe Networks branch of Prospector, and you see an entry for Interference Checks. Note that each instance of interference is listed in the Preview pane for further study. Making edits to your pipe network flags the interference check as "out of date." You can rerun Interference Check by right-clicking Interference Check in Prospector. You can also edit your criteria in this right-click menu.

Creating Interference Styles

You can create your interference style to give you a visual cue in both plan and 3D views. Interference styles are located on the Settings tab under Pipe Network ➢ Interference Styles. The Interference Styles dialog has several options that control how you see the interference graphically. For example, the View Options tab includes the following:

- **Symbol Options** — Lets you select a marker style. You can choose one of the defaults or create your own marker style by expanding General ➢ Marker Styles on the Settings tab. Marker styles are similar to point styles and are used many places in Civil 3D.
- **Solid Options** — Allows you to choose between a true interference solid and a sphere:
  - A true interference solid assumes the shape of the exact overlap of the parts in question.

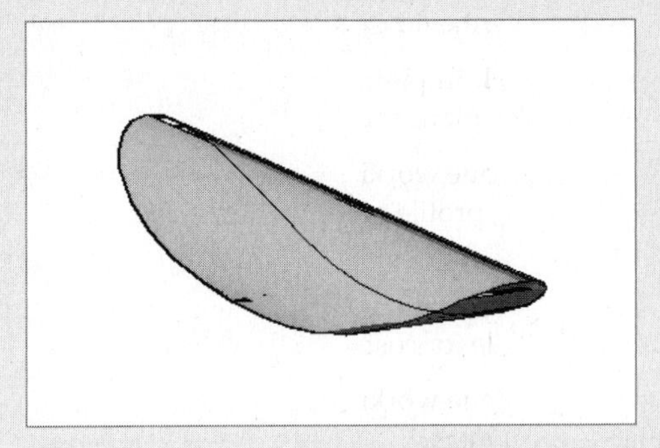

  - An interference sphere marks the location of the interference. You can control its size and behavior in this box. Choosing Diameter by True Solid Extents limits the size of the sphere to the overlap of the interference, similar to the true interference solid.

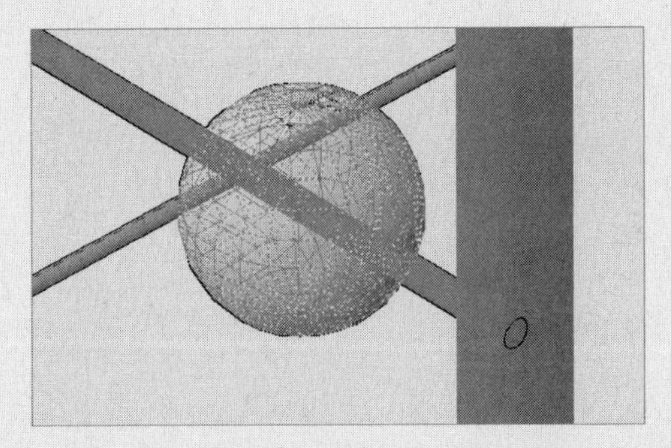

The Display tab lets you control visibility in plan and model views. Typically, you'll want to make sure the plan symbol is visible in 2D and the model symbol is visible in 3D, as shown in the Interference Check exercise.

The Bottom Line

Create a pipe network by layout. After you've created a parts list for your pipe network, the first step toward finalizing the design is to create a Pipe Network by Layout.

Master It Open the `Mastering Pipes.dwg` file. Choose Pipe Network Creation Tools from the Pipe Network drop-down of the Create Design panel on the Home tab to create a sanitary sewer pipe network.

Use the Finished Ground surface, and name only structure and pipe label styles. Don't choose an alignment at this time. Create 8″ PVC pipes and concentric manholes.

There are blocks in the drawing to assist you in placing manholes. Begin at the START HERE marker, and place a manhole at each marker location. You can erase the markers when you've finished.

Create an alignment from network parts, and draw parts in profile view. Once your pipe network has been created in plan view, you'll typically add the parts to a profile view on the basis of either the road centerline or the pipe centerline.

Master It Continue working in the `Mastering Pipes.dwg` file. Create an alignment from your pipes so that station zero is located at the START HERE structure. Create a profile view from this alignment, and draw the pipes on the profile view.

Label a pipe network in plan and profile. Designing your pipe network is only half of the process. Engineering plans must be properly annotated.

Master It Continue working in the `Mastering Pipes.dwg` file. Add the Length Material and Slope style to profile pipes and the Data with Connected Pipes (Sanitary) style to profile structures.

Create a dynamic pipe table. It's common for municipalities and contractors to request a pipe or structure table for cost estimates or to make it easier to understand a busy plan.

Master It Continue working in the `Mastering Pipes.dwg` file. Create a pipe table for all pipes in your network.

Chapter 16

Working the Land: Grading

Beyond creating streets and sewers, cul-de-sacs, and inlets, much of what happens to the ground as a site is being designed must still be determined. Describing the final plan for the earthwork of a site is a crucial part of bringing the project together. This chapter examines feature lines and grading groups, which are the two primary tools of site design. These two functions work in tandem to provide the site designer with tools for completely modeling the land.

By the end of this chapter, you'll learn to:

- Convert existing linework into feature lines
- Model a linear grading with a grading feature line
- Create custom grading criteria
- Model planar site features with grading groups

Working with Grading Feature Lines

There are two types of feature lines: corridor feature lines and grading feature lines. Corridor feature lines are discussed in Chapter 11, "Easy Does It: Basic Corridors," and grading feature lines are the focus of this chapter. It's important to note that grading feature lines can be extracted from corridor feature lines, and you can choose whether or not to dynamically link them to the host object.

As discussed in Chapter 5, "The Ground Up: Surfaces in Civil 3D," terrain modeling can be defined as the manipulation of triangles created by connecting points and vertices to achieve Delaunay Triangulation. In Land Desktop (LDT) and other software, this is often done with the use of native 3D polylines. In Civil 3D, the creation of the feature line object adds a level of control and complexity not available to 3D polylines. In this section, you look at the feature line, various methods of creating feature lines, some simple elevation edits, planar editing functionality, and labeling of the newly created feature lines.

Accessing Grading Feature Line Tools

The Feature Line creation tools can be accessed from the Home tab's Create Design panel as shown in Figure 16.1.

The Feature Line editing tools can be accessed via the Design panel on the Modify tab, or via the Feature Line context-sensitive tab that's available after you select an existing feature line (see Figure 16.2).

One thing to remember when working with feature lines is that they do belong in a site. Feature lines within the same site snap to each other in the vertical direction and can cause some confusion when you're trying to build surfaces. If you're experiencing some weird elevation data along your feature line, be sure to check out the parent site. If the concept of a parent site or sites in general doesn't make much sense to you, be sure and look at Chapter 6, "Don't Fence Me In: Parcels," on parcels before going too much further. Sites are a major part of the way feature lines interact with each other, and many users who have problems with grading completely ignore sites as part of the equation.

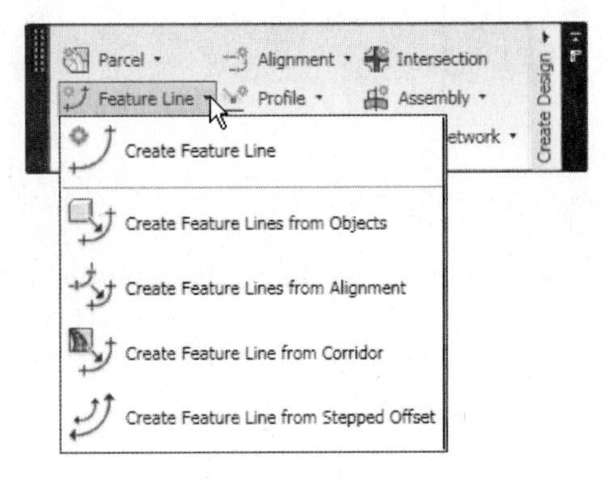

FIGURE 16.1
The Feature Line drop-down menu on the Create Design panel

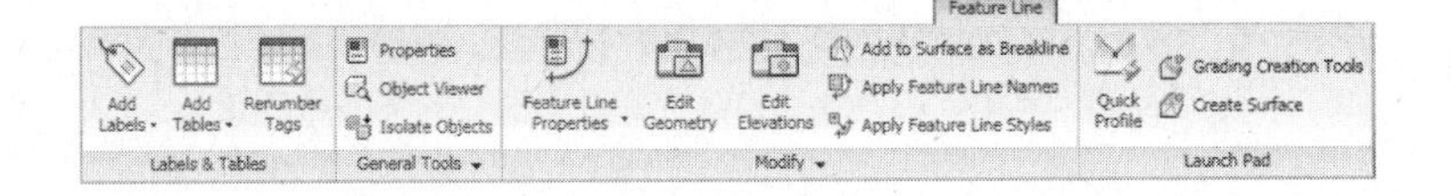

FIGURE 16.2
The Feature Line context-sensitive tab accessed by selecting an existing feature line

The next few sections break down the various tools in detail. You'll use almost all of them in this chapter, so in each section you'll spend some time getting familiar with the available tools and the basic concepts behind them.

Creating Grading Feature Lines

There are five primary methods for creating feature lines as shown in Figure 16.1. They generate similar results but have some key differences:

- The Create Feature Line tool allows you to create a feature line from scratch, assigning elevations as you go. These elevations can be based on direct data input at the command line, slope information, or surface elevations.
- The Create Feature Lines from Objects tool converts lines, arcs, polylines, and 3D polylines into feature lines. This process also allows elevations to be assigned from a surface or grading group.
- The Create Feature Lines from Alignment tool allows you to build a new feature line from an alignment, using a profile to assign elevations. This feature line can be dynamically tied to the alignment and the profile, making it easy to generate 3D design features based on horizontal and vertical controls.
- The Create Feature Line from Corridor command is used to export a grading feature line from a corridor feature line.
- The Create Feature Line from Stepped Offset is used to create a feature line from an offset and the difference in elevation from a feature line, survey figure, polyline, or 3D polyline.

You explore each of these methods in the next few exercises.

1. Open the `Creating Feature Lines.dwg` file. (Remember, all data can be downloaded from `www.sybex.com/go/masteringcivil3d2010`.) This drawing has the base line of a detention pond drawn and an overland swale already, along with a couple of critical points for drainage.

2. Change to the Home tab and select Feature Line ➤ Create Feature Line from the Create Design panel to display the Create Feature Lines dialog, as shown in Figure 16.3. You could assign a name to the feature line in this dialog, but you'll skip that process for now.

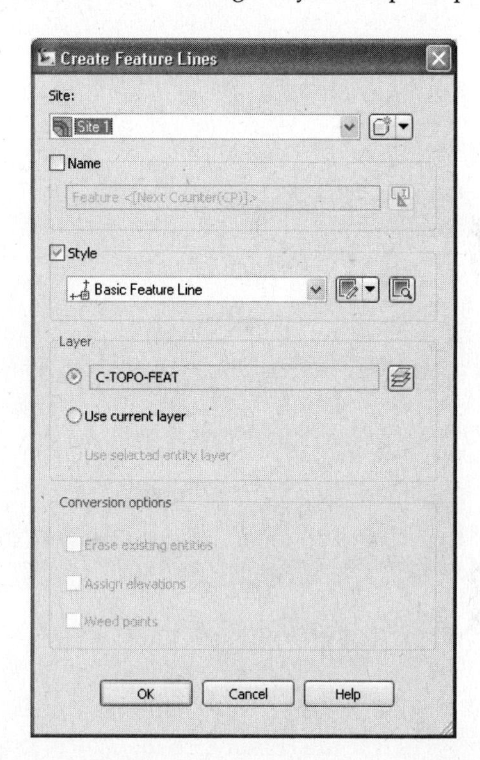

FIGURE 16.3
The Create Feature Lines dialog

3. Click OK. Note that a new site will be created. Grading objects and feature lines within the same site will react to each other. Poor site management early in the project lifecycle can lead to issues later.

4. Use a Center osnap to pick the small circle on the right. (The pick order is shown in Figure 16.4.)

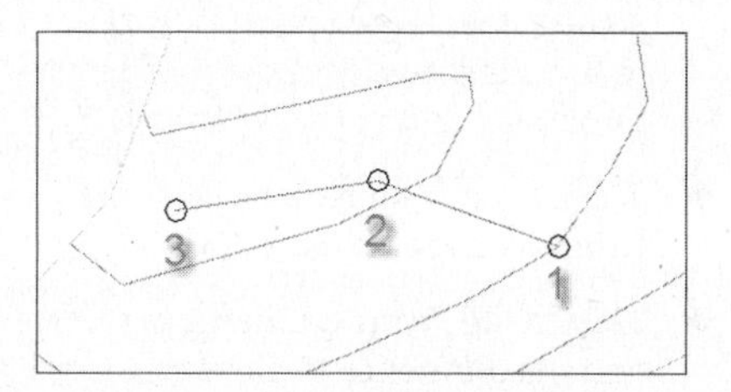

FIGURE 16.4
Completed breakline with pick order displayed

5. Enter **S**↵ at the command line to use a surface to set elevation information.
6. Press ↵ to accept the elevation offered.
7. Use a Center osnap to pick the center of the middle circle.
8. Enter **-2**↵ to set the grade between points.
9. Use a Center osnap to pick the center of the left circle.
10. Enter **SU**↵ at the command line to set the elevation from the surface.
11. Press ↵ to accept the elevation.
12. Press ↵ again to exit the command. Your screen should look like Figure 16.4.

Looking back at Figure 16.3, note that there is an unused option for assigning a Name value to each feature line. Using names does make it easier to pick feature line objects if you decide to use them for building a corridor object. The name option is available for each method of creating feature line objects, but will generally be ignored for this chapter except for one exercise covering the renaming tools available.

This method of creating a feature line connecting a few points seems pretty tedious to most users. In this next exercise, you convert an existing polyline to a feature line and set its elevations on the fly:

1. Open the `Creating Feature Lines.dwg` file if it isn't open already.
2. Change to the Home tab and choose Feature Line ➢ Create Feature Lines from Objects from the Create Design panel.
3. Pick the closed polyline representing the pond basin.
4. Right-click and select Enter, or press ↵. The Create Feature Lines dialog will appear. This dialog is similar to Figure 16.3; however, the Conversion Options near the bottom of the dialog are now active, so take a look at the options presented:
    - Erase Existing Entities removes the object onscreen and replaces it with a feature line object. This avoids the creation of duplicate linework, but could be harmful if you wanted your linework for planimetric purposes.
    - Assign Elevations lets you set the feature line PI (point of intersection) elevations from a surface or grading group, essentially draping the feature line on the selected object.
    - Weed Points decreases the number of nodes along the object. This is handy when you're converting digitized information into feature lines.
5. Check the Assign Elevations box. The Erase Existing Entities option is on by default, and you will not select the Weed Points option.
6. Click OK to display the Assign Elevations dialog. Here you can select a surface to pull elevation data from, or assign a single elevation to all PIs.
7. Check the Insert Intermediate Grade Break Points option. This inserts a PI at every point along the feature line where it crosses an underlying TIN line.

8. Click OK.

9. Pick the pond outline, and the grips will look like Figure 16.5.

> **ANOTHER POND?**
>
> Ponds are used frequently to demonstrate grading and feature line tools because they are suitable for demonstrating a myriad of tools. In some cases, a pond may be modeled as a corridor. Remember, there are typically several ways in which to approach grading challenges in Civil 3D.

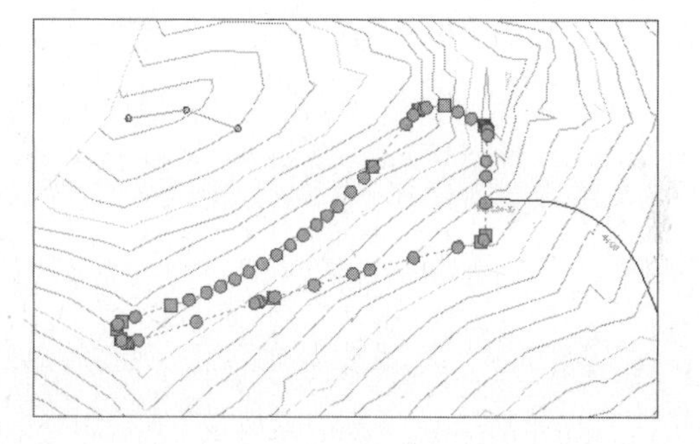

FIGURE 16.5
Conversion to a feature line object

Note that in Figure 16.5 there are two types of grips. Feature lines offer feedback via the grip shape. A square feature line grip indicates a full PI. This node can be moved in the *x*, *y*, and *z* directions, manipulating both the horizontal and vertical design. Circular grips are elevation points only. In this case, the elevation points are located at the intersection of the original polyline and the TIN lines existing in the underlying surface. Elevation points can only be slid along a given feature line segment, adjusting the vertical design, but cannot be moved in a horizontal plane. This combination of PIs and elevation points makes it easy to set up a long element with numerous changes in design grade that will maintain its linear design intent if the endpoints are moved.

Both of the methods used so far assume static elevation assignments for the feature line. They're editable, but are not physically related to other objects in the drawing. This is generally acceptable, but sometimes it's necessary to have a feature line that is dynamically related to an object. For grading purposes, it is often ideal to create a horizontal representation of a vertical profile along an alignment. Rather than build a corridor model as discussed in Chapters 11 and 12, a dynamic feature line can be extracted from a profile along an alignment, offset both horizontally and vertically, and used for grading. In the following example, a dynamic feature line is extracted from an alignment. Elevations for the vertices of the feature line are extracted from a profile, and finally, offset using the Create Feature Line from Stepped Offset tool to represent a swale.

1. Open the `Stepped Offset.dwg` file.

2. Change to the Home tab and select Feature Line ➢ Create Feature Lines from Alignment from the Create Design panel.

3. Select the alignment to the east of the pond to display the Create Feature Line from Alignment dialog shown in Figure 16.6. Note the Create Dynamic Link to the Alignment option near the bottom.

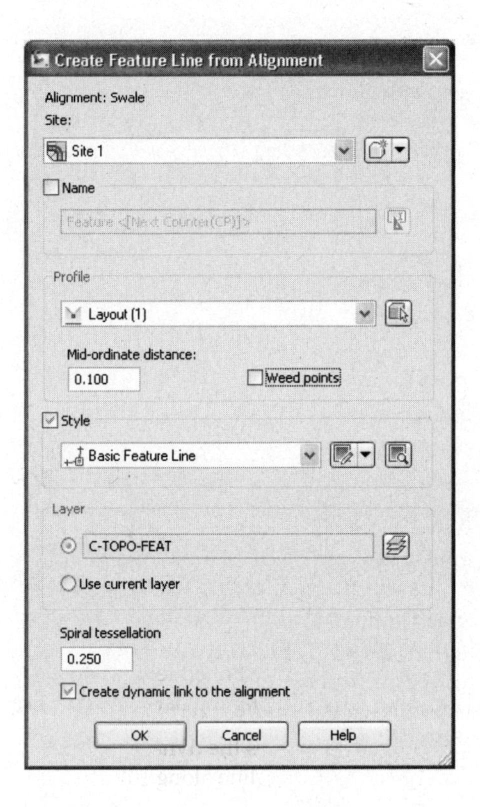

FIGURE 16.6
The Create Feature Line from Alignment dialog

4. Change the Profile drop-down list to Layout (1) and deselect the option to weed points as shown in Figure 16.6. Click OK.

5. Change to the Home tab and select Feature Line ➤ Create Feature Line from Stepped Offset.

6. Enter **10** at the command line and press ↵.

7. Select the dynamic feature line along the alignment.

8. Select a point north of the alignment when prompted to `Specify side to offset or [Multiple]:`.

9. The command prompt will offer you several choices for specifying elevations or grades to be used along the feature lines. If necessary, type the appropriate letters until the command prompt reads `Specify elevation difference or [Grade/Slope/Elevation/Variable]:`.

10. Enter **4** as the elevation difference at the command line and press ↵.

11. Repeat steps 8 through 10, but this time, pick a point south of the alignment. Your results should appear as shown in Figure 16.7. Leave this drawing open to use in the next exercise.

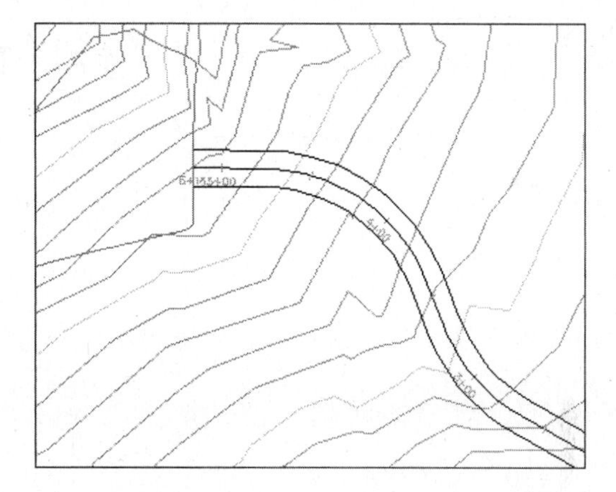

FIGURE 16.7
The completed alignment with offsets in place

If you click the alignment now, you will select the feature line, but you will not see any grips available. This is because the feature line is dynamically linked to the design profile along the alignment and can't be modified. If either the alignment or the profile changes, the dynamic feature line will automatically update. Simply repeat the preceding procedure to create new offsets if needed. These three feature lines can be included in a new surface definition as breaklines, as discussed in Chapter 5.

Because dynamically linked feature line objects are slightly different, you'll look at them in this next exercise before using the name and style buttons to update all the feature lines:

1. This exercise uses the dynamic feature line created in the previous exercise. Pan or zoom to view the feature line along the alignment to the east of the pond outline.
2. Select the feature line.
3. Choose Feature Line Properties from the Modify panel to open the Feature Line Properties dialog as shown in Figure 16.8. The information displayed on the Information tab is unique to the dynamic feature line, and you still have some level of control over the linking options.
4. Deselect the Dynamic Link option and click OK. Notice the grips appear. The dynamic relationship between the feature line and the alignment has been severed.
5. Select Feature Line Properties from the Modify panel and notice the Dynamic Link options have disappeared.
6. Select the two feature lines opposite the alignment.
7. Choose Apply Feature Line Styles from the Modify panel. The Apply Feature Line Style dialog will appear.
8. Select the Corridor Ditch style from the Style drop-down list and click OK to dismiss the dialog.

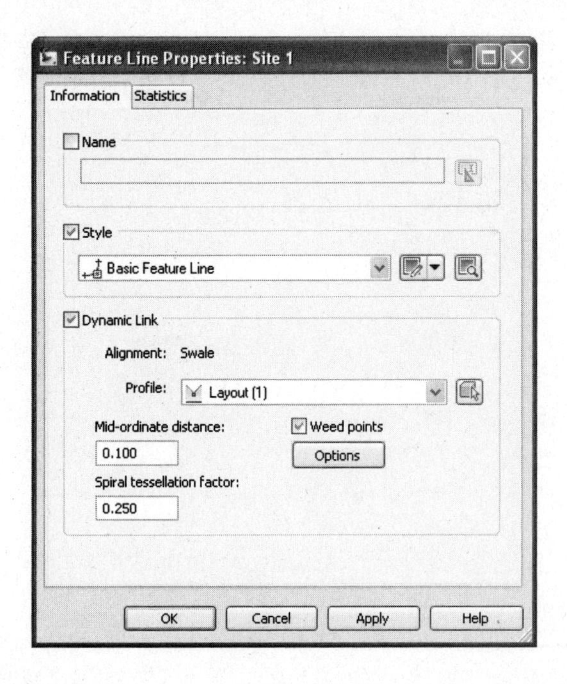

FIGURE 16.8
The Feature Line Properties dialog for an alignment-based feature line

Many people don't see much advantage in using styles with feature line objects, but there is one major advantage: linetypes. Zoom in on the feature lines on either side of the alignment and you'll see the Corridor Ditch style has a dashed linetype. 3D polylines do not display linetypes, but feature lines do. If you need to show the linetypes in your grading, feature line styles are your friend.

Within the Feature Line context-sensitive menu activated by selecting a feature line, several more commands can be found on the Modify panel, as shown in Figure 16.9, that are worth examining before we get into editing objects.

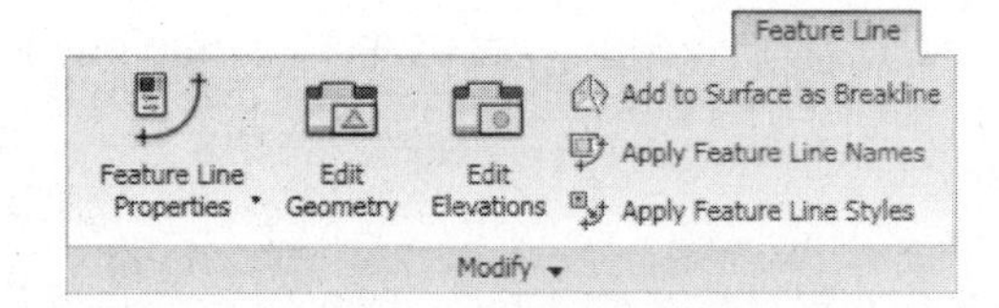

FIGURE 16.9
The Modify panel on the context-sensitive Feature Line tab

The Modify panel commands allow access to various properties of the feature line, the feature line style, and the feature line geometry as follows:

- The Feature Line Properties drop-down menu contains two commands. The first command, Feature Line Properties, is used to access various physical properties such as minimum and maximum grade. Only the name and style of the feature line can be modified on the Information tab as shown in Figure 16.10. The second command, Edit Feature Line Style, is used to access various display characteristics such as color and linetype.

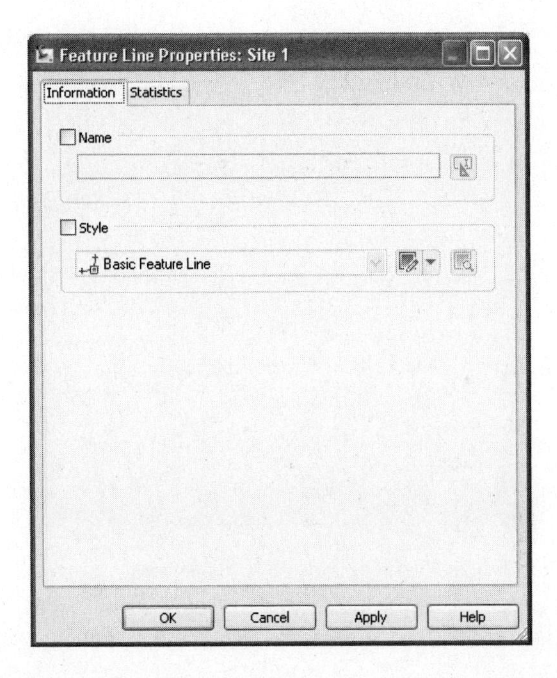

FIGURE 16.10
The Feature Line Properties dialog

- The Edit Geometry toggle opens the Edit Geometry panel on the Feature Line tab (see Figure 16.11). This panel will remain open until the Edit Geometry button is toggled off (it's highlighted when toggled on).

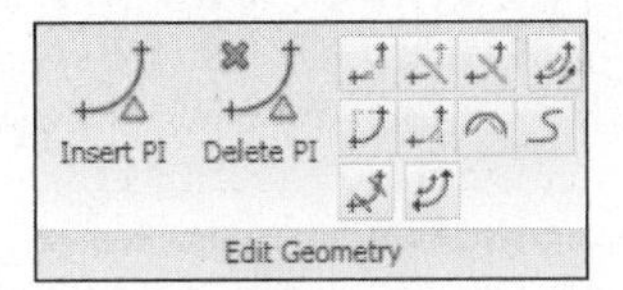

FIGURE 16.11
The Edit Geometry panel on the Feature Line tab

- The Edit Elevations toggle opens the Edit Elevations panel on the Feature Line tab (see Figure 16.12). This panel will remain open until the Edit Geometry button is toggled off (it's highlighted when toggled on).

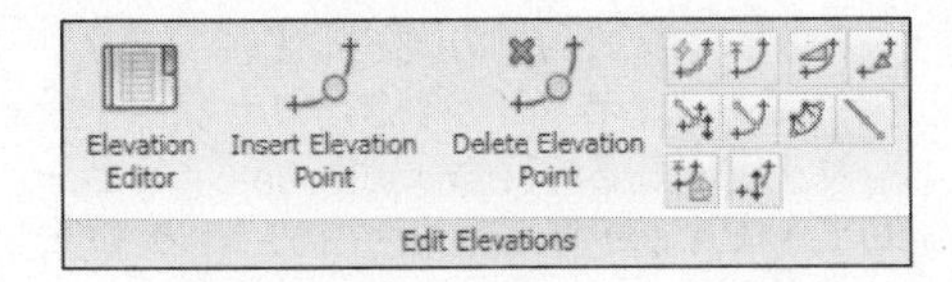

FIGURE 16.12
The Edit Elevations panel on the Feature Line tab

- The Add to Surface as Breakline tool allows you to select a feature line or feature lines to add to a surface as breaklines.
- The Apply Feature Line Names tool allows you to mass change a series of feature lines based on a new naming template. This can be helpful when you want to rename a group

or just assign names to feature line objects. This tool cannot be used on a feature line that is tied to an alignment and profile.

- The Apply Feature Line Styles tool allows for the mass change of feature line objects and their respective styles. Many users don't apply styles to their feature line objects because the feature lines are found in grading drawings and not meant to be seen in construction documents. But if you need to make a mass change, you can.

Once the Feature Line tab has been activated, the Quick Profile tool is available on the Launch Pad panel. The Quick Profile tool generates a temporary profile of the feature line based on user parameters found in the Create Quick Profiles dialog (Figure 16.13).

FIGURE 16.13
The Create Quick Profiles dialog

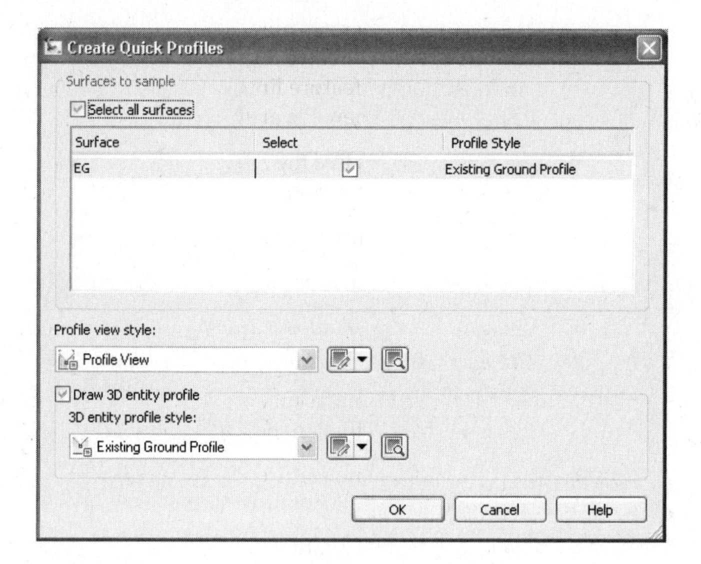

A few notes on this operation: Civil 3D creates a phantom alignment that will not display in the Prospector as the basis for a quick profile. A unique alignment number is assigned to this alignment, so your number might be different from that shown in Figure 16.13; a closed feature line will generate a parcel when the quick profile is executed; and finally, Panorama will display a message to tell you that a profile view has been generated. You can close Panorama or move the Panorama palette out of the way if necessary.

Now that you've created a couple of feature lines, you'll edit and manipulate them some more.

Editing Feature Line Horizontal Information

Creating feature lines is straightforward. The Edit Geometry and Edit Elevation tools make them considerably more powerful than a standard 3D polyline. The Edit Geometry and Edit Elevation tools can be accessed by changing to the Modify tab and choosing Feature Line, or simply by selecting an existing feature line. Both tools are found on the Modify panel of the Feature Line tab. The Edit Geometry functions are examined in this section, and the Edit Elevation functions are described in the next section.

Grading revisions often require adding PIs, breaking apart feature lines, trimming, and performing other planar operations without destroying the vertical information. To access the commands for editing feature line horizontal information, first change to the Modify tab and choose Feature Line from the Design panel. Then choose Edit Geometry from the Modify panel to open the Edit Geometry panel. The first two tools are designed to manipulate the PI points that make up a feature line:

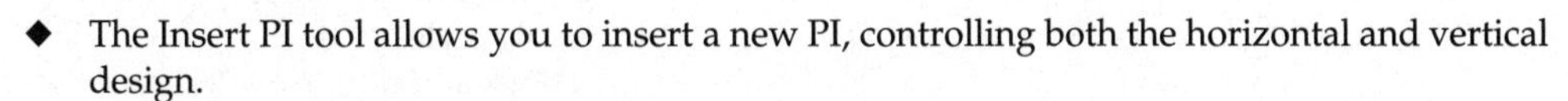

- The Insert PI tool allows you to insert a new PI, controlling both the horizontal and vertical design.

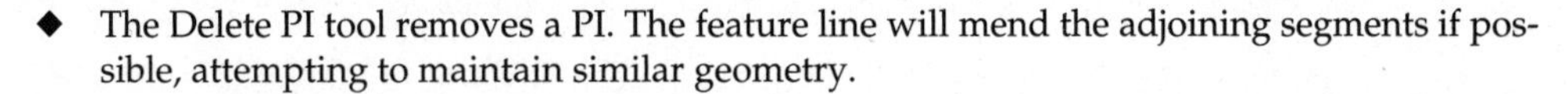

- The Delete PI tool removes a PI. The feature line will mend the adjoining segments if possible, attempting to maintain similar geometry.

The next few tools act like their AutoCAD counterparts, but understand that elevations are involved and add PIs accordingly:

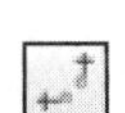

- The Break tool operates much like the AutoCAD Break command, allowing two objects or segments to be created from one. Additionally, if a feature line is part of a surface definition, both new feature lines are added to the surface definition to maintain integrity. Elevations at the new PIs are assigned on the basis of an interpolated elevation.

- The Trim tool acts like the AutoCAD Trim command, adding a new end PI on the basis of an interpolated elevation.

- The Join tool creates one feature line from two, making editing and control easier.

- The Reverse tool changes the direction of a feature line.

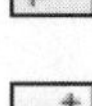

- The Edit Curve tool allows you to modify the radius that has been applied to a feature line object.

- The Fillet tool inserts a curve at PIs along a feature line and will connect feature lines sharing a common PI that are not actually connected.

The last few tools refine feature lines, making them easier to manipulate and use in surface building:

- The Fit Curve tool analyzes a number of elevation points and attempts to define a working arc through them all. This is often used when the corridor utilities are used to generate feature lines. These derived feature lines can have a large number of unnecessary PIs in curved areas.

- The Smooth tool turns a series of disjointed feature line segments and creates a best-fit curve. This tool is great for creating streamlines or other natural terrain features that are known to curve, but there's often not enough data to fully draw them that way.

- The Weed tool allows the user to remove elevation points and PIs on the basis of various criteria. This is great for cleaning up corridor-generated feature lines as well.

- As discussed in detail earlier, the Stepped Offset tool allows offsetting in a horizontal and vertical manner, making it easy to create stepped features such as stairs or curbs.

By using these controls, it's easier to manipulate the design elements of a typical site while still using feature lines for surface design. In this exercise, you manipulate a number of feature lines that were created by corridor operations:

1. Open the `Horizontal Feature Line Edits.dwg` file. A number of feature lines in the area of the detention pond have been brought into this drawing for editing and refinement.
2. Pick feature line A, as shown in Figure 16.14.

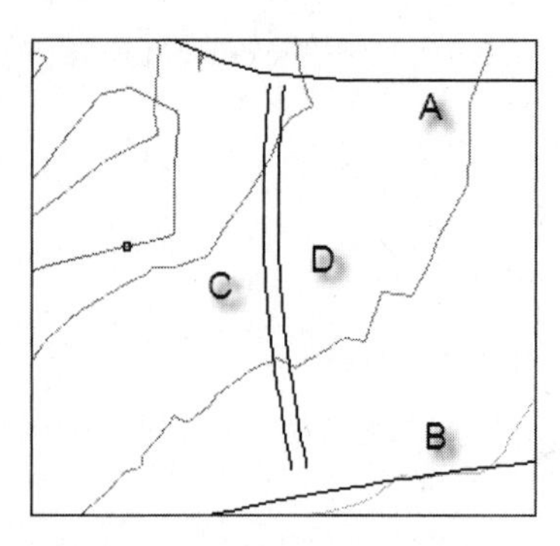

FIGURE 16.14
Feature lines A, B, C, and D

3. Choose Edit Geometry from the Modify panel to enable the Edit Geometry panel.
4. Click the Break tool.
5. Enter **F**↵ to pick the first point of the break. If this option is unavailable, pick feature line A again. When prompted to `Specify second break point or [First point]:`, enter **F**↵.
6. Using an Endpoint osnap, pick the endpoint on A just to the northwest of Feature Line C as shown in Figure 16.15.

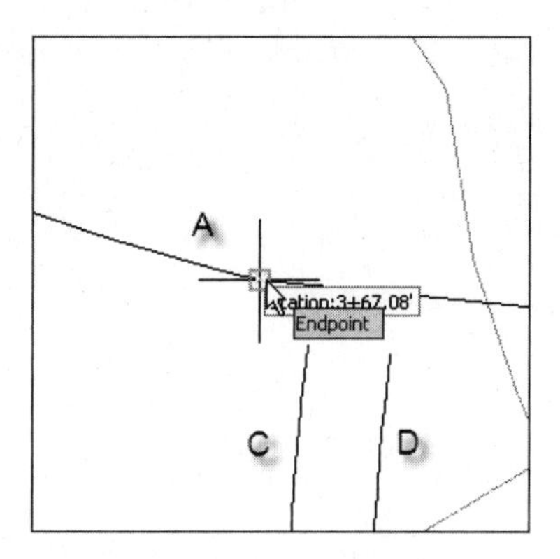

FIGURE 16.15
Using the Endpoint object snap to select a point on feature line A

7. Using an Endpoint osnap, pick the next endpoint on A to the east, leaving a gap, as shown in Figure 16.16.
8. Pick feature lines C and D to activate grips on both lines. Notice the large number of grips. Press Esc.
9. Pick feature line C and click the Weed tool on the Edit Geometry panel to display the Weed Vertices dialog.
10. Pick feature line C again when prompted to `Select a feature line, 3d polyline or [Multiple/Partial]:`. The Weed Vertices dialog will be displayed.
11. Enter **25′** for the Length, and then press Tab to register the value. The dialog changes to reflect the number of vertices that will be removed from the feature line. Additionally, the glyphs on the feature line itself will change from green to red to reflect nodes that will be

removed under the current setting. Click OK to complete the command and dismiss the Weed Vertices dialog.

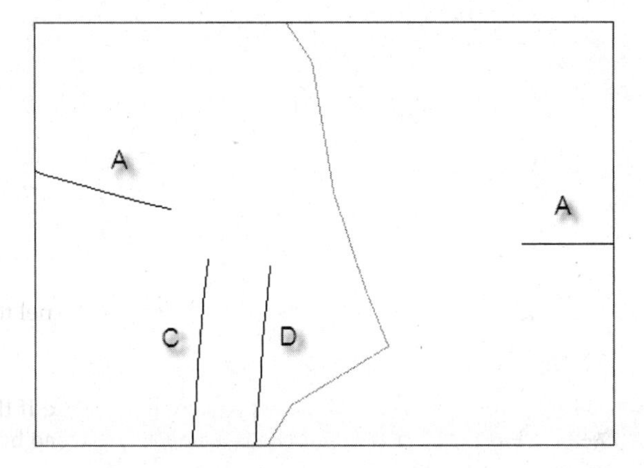

FIGURE 16.16
The feature line after executing the Break command

12. Pick feature line D so grips are active on both lines, as shown in Figure 16.17. Notice that the number of grips on feature line C has been reduced by comparison. Press Esc.

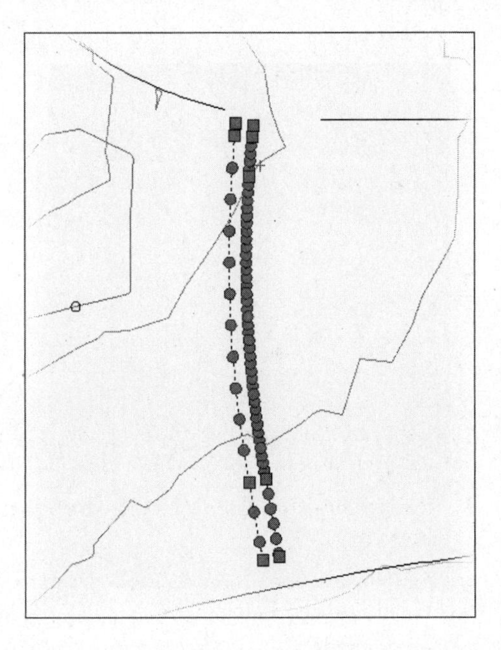

FIGURE 16.17
Feature lines after weeding (left) and before (right)

13. Using a standard AutoCAD Extend command, extend both feature lines C and D to feature line B.

14. Pick feature line C and click the Trim tool from the Edit Geometry panel. A standard AutoCAD trim will not work in this case. Pick feature lines C and D as the cutting edges

and press ↵. Choose feature line B between feature lines C and D when prompted to `Select objects to trim:`. Press ↵ to complete the command and review the results (see Figure 16.18). Press Esc.

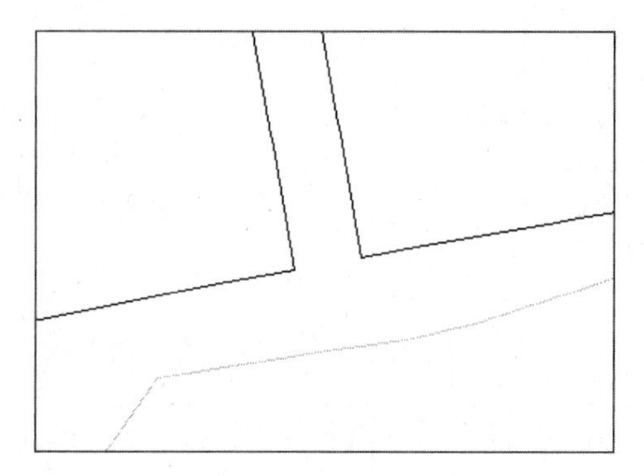

FIGURE 16.18
Feature line B trimmed

15. Select feature line D and click the Join tool on the Edit Geometry panel.
16. Select feature line B and press ↵. Notice the grips and that the two feature lines have been joined.
17. Select the Fillet tool. Enter **R**↵ at the command line to adjust the radius value. Enter **10**↵ to update the radius value.
18. Move your cursor toward the intersection of feature lines B and D until the glyph appears. Pick near the glyph to fillet the two feature lines.
19. Click the Edit Curve tool from the Edit Geometry panel and pick the curve when prompted to `Select feature line curve to edit or [Delete]:`. The Edit Feature Line Curve dialog opens as shown in Figure 16.19.

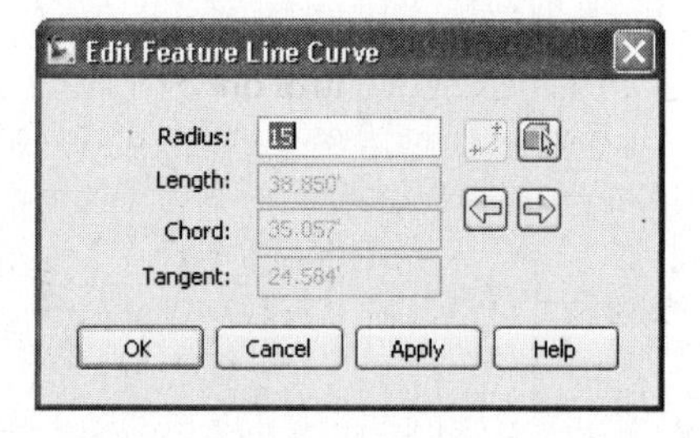

FIGURE 16.19
The Edit Feature Line Curve dialog

20. Enter a Radius value of **15** as shown, and click OK to close the dialog. Press ↵ to exit the command.
21. Select the feature line just modified and it will look similar to Figure 16.20.

When modifying the radius of a feature line curve, it's important to remember that you must have enough tangent feature line on either side of the curve segment to make a curve fit, or the program will not make the change. In that case, tweak the feature line on either side of the arc until there is a mathematical solution. Sometimes it is necessary to use the Weed Vertices tool to remove vertices and create enough room to fillet feature lines. In some cases, it may be necessary to plan ahead when creating feature lines to ensure that vertices will not be placed too closely together.

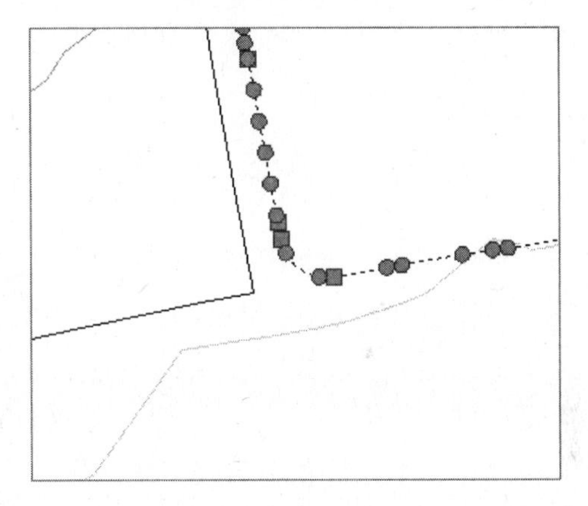

FIGURE 16.20
Filleted feature lines

Editing Feature Line Elevation Information

To access the commands for editing feature line elevation information, first change to the Modify tab and choose Feature Line from the Design panel. Then choose Edit Elevations from the Modify panel to open the Edit Elevations panel. Moving across the panel, you find the following tools for modifying or assigning elevations to feature lines:

- The Elevation Editor tool activates a palette in Panorama to display station, elevation, length, and grade information about the feature line selected. Feature lines, survey figures, and parcel lines can be edited using this tool.

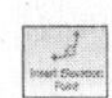

- The Insert Elevation Point tool inserts an elevation point at the point selected. Note that this point can control only elevation information; it does not act as a horizontal control point. Feature lines, survey figures, parcel lines, and 3D polylines can be edited using this tool.

- The Delete Elevation Point tool deletes the selected elevation point; the points on either side then become connected linearly on the basis of their current elevations. Feature lines, survey figures, parcel lines, and 3D polylines can be edited using this tool.

- The Quick Elevation Edit tool allows you to use onscreen cues to set elevations and slopes quickly between PIs on any feature lines or parcels in your drawing.

- The Edit Elevations tool steps through the selected feature line, much like working through a polyline edit at the command line. Feature lines, survey figures, parcel lines, and 3D polylines can be edited using this tool.

- The Set Grade/Slope Between Points tool sets a continuous slope along the feature line between selected points. Feature lines, survey figures, parcel lines, and 3D polylines can be edited using this tool.

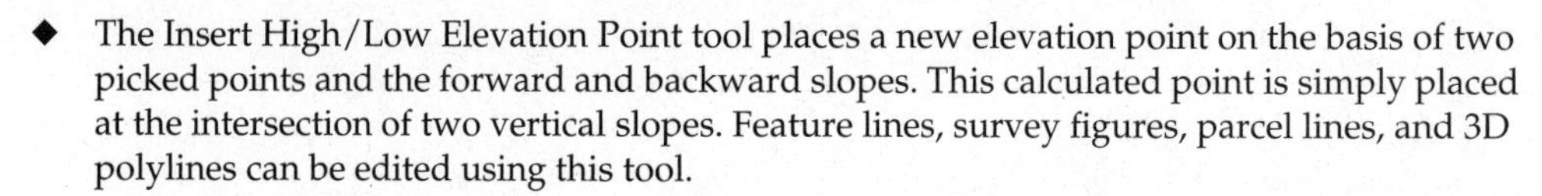

- The Insert High/Low Elevation Point tool places a new elevation point on the basis of two picked points and the forward and backward slopes. This calculated point is simply placed at the intersection of two vertical slopes. Feature lines, survey figures, parcel lines, and 3D polylines can be edited using this tool.

- The Raise/Lower tool simply moves the entire feature line in the *z* direction by an amount entered at the command line. Feature lines, survey figures, parcel lines, and 3D polylines can be edited using this tool.

- The Set Elevation by Reference tool sets the elevation of a selected point along the feature line by picking a reference point, and then establishing a relationship to the selected feature line point. This relationship isn't dynamic! This button also acts as a flyout for the next three tools. Feature lines, survey figures, parcel lines, and 3D polylines can be edited using this tool.

- The Adjacent Elevations by Reference tool allows you to adjust the elevation on a feature line by coming at a given slope or delta from another point or feature line point. Feature lines, survey figures, parcel lines, and 3D polylines can be edited using this tool.

- The Grade Extension by Reference tool allows you to apply the same grades to different feature lines across a gap. For example, you might use this tool along the back of curbs at locations such as driveways or intersections. Feature lines, survey figures, parcel lines, and 3D polylines can be edited using this tool.

- The Raise/Lower by Reference tool allows you to adjust a feature line elevation based on a given slope from another location. Feature lines, survey figures, parcel lines, and 3D polylines can be edited using this tool.

- The Elevations from Surface tool sets the elevation at each PI and elevation point on the basis of the selected surface. It will optionally add elevation points at any point where the feature line crosses a surface TIN line. Feature lines, survey figures, parcel lines, and 3D polylines can be edited using this tool.

Quite a few tools are available to modify and manipulate feature lines. You won't use all of the tools, but at least you have some concept of what is available. The next few exercises give you a look at a few of them. In this first exercise, you'll make manual edits to set the grade of a feature line segment:

1. Open the `Editing Feature Line Elevations.dwg` file.
2. Select the feature line to the northwest of the pond area and choose Elevation Editor from the Edit Elevations panel. The Grading Elevation Editor in Panorama will open.
3. Click in the Grade Ahead column for the first PI. It's hard to see in the images, but as a row is selected in the Grading Elevation Editor, the PI or the elevation point that was selected will be highlighted on the screen with a small glyph.
4. Change the value to **-1.5**, as shown in Figure 16.21.
5. Click the green check mark in the upper-right corner to dismiss the Panorama.

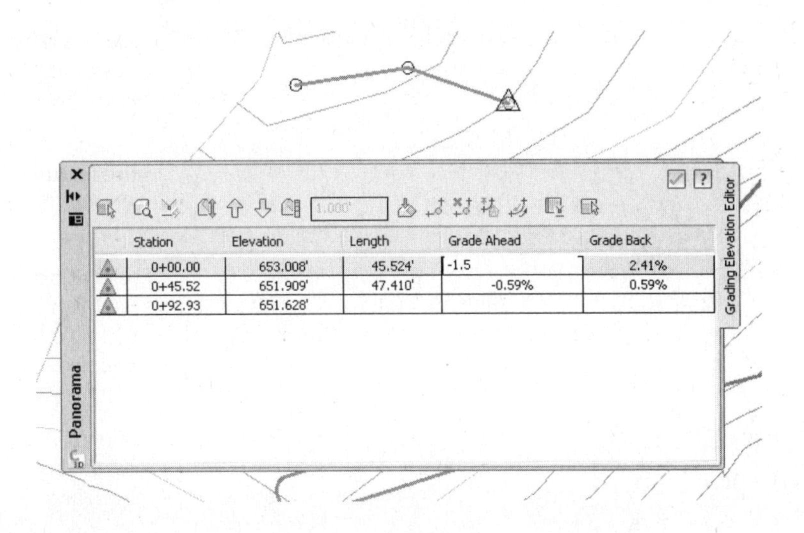

FIGURE 16.21
Manual edits in the Grading Elevation Editor

This is the most basic way to manipulate elevation information. You'll use it again to set the basic bottom of your pond, and then use the other tools to refine that part of your design. In this exercise, you remove some extraneous data points just as an example of the quick edits possible:

1. Zoom in on the northern edge of the pond, as shown in Figure 16.22. Pick the outline of the pond bottom.

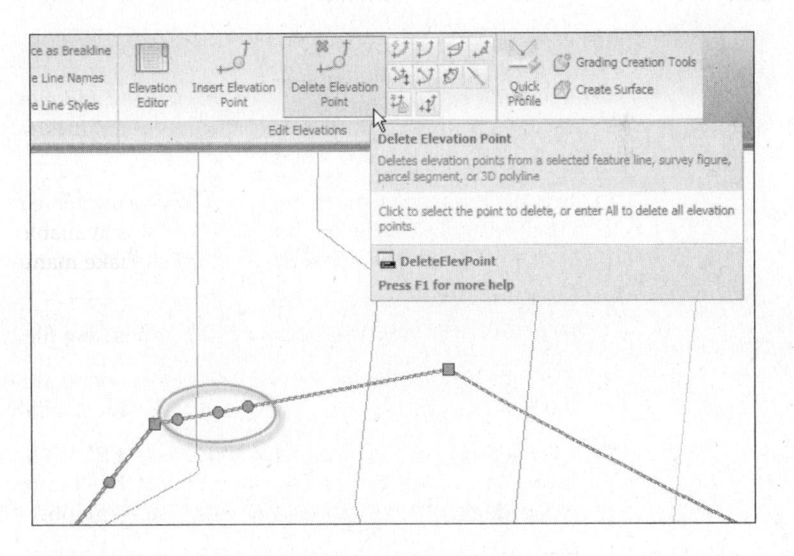

FIGURE 16.22
Elevation points to be deleted

2. Click the Delete Elevation Point tool on the Edit Elevations panel.

3. Grips are replaced with glyphs. Pick near one of the three circular glyphs on the feature line to remove the elevation point, and repeat for the other two.
4. Press ↵ twice to end the command and return to the `command:` prompt. You'll see something like Figure 16.23. Notice that the three circular grips are gone, leaving two PIs (the square grips) at the segment ends.

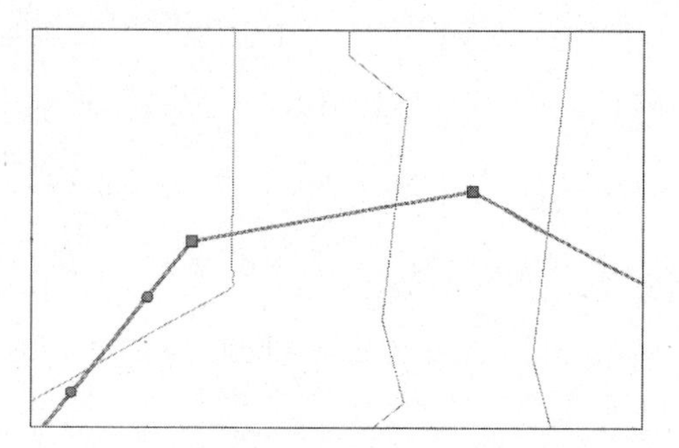

FIGURE 16.23
The feature line after deleting elevation points

Some of the relative elevation tools are a bit harder to understand, so you'll look at them in this next exercise and see how they function in some basic scenarios:

1. Open the `Editing Relative Feature Line Elevations.dwg` file. This drawing contains a sample layout with some curb and gutter work, along with some building pads.
2. Zoom to the ramp shown on the left-hand side of the intersection.
3. Select the feature line describing the ramp. Select Edit Elevations from the Modify panel to display the Edit Elevations panel if it isn't displayed already. Select Elevation Editor. The Panorama appears. Notice that the entire feature line is at elevation 0.000′. Click the green check mark in the upper right to close the Panorama. Press Esc to cancel the grips.
4. Select the feature line representing the flowline of the curb and gutter area.
5. Select the Adjacent Elevations by Reference tool. Civil 3D will prompt you to select the object to edit. You will edit the elevations along the feature line describing the ramp area.
6. Click the feature line describing the ramp area and Civil 3D will display a number of glyphs and lines to represent what points along the flowline it is using to establish elevations from, and will prompt you for the elevation difference (you could also use a grade or slope).
7. Enter **0.5** at the command line and press ↵ to update the ramp elevations. Press ↵ again to exit the command, and press Esc to cancel grips.
8. Select the feature line describing the ramp. Select Elevation Editor, and the Panorama appears. Notice that the PIs now each have an elevation as shown in Figure 16.24.
9. Click the green check mark in the upper right of the Panorama to dismiss it. Leave this drawing open to complete the next exercise.

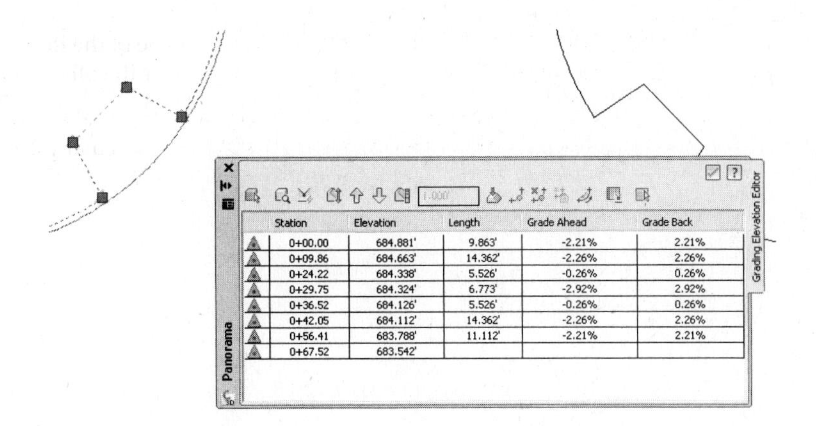

Station	Elevation	Length	Grade Ahead	Grade Back
0+00.00	684.881'	9.863'	-2.21%	2.21%
0+09.86	684.663'	14.362'	-2.26%	2.26%
0+24.22	684.338'	5.526'	-0.26%	0.26%
0+29.75	684.324'	6.773'	-2.92%	2.92%
0+36.52	684.126'	5.526'	-0.26%	0.26%
0+42.05	684.112'	14.362'	-2.26%	2.26%
0+56.41	683.788'	11.112'	-2.21%	2.21%
0+67.52	683.542'			

FIGURE 16.24
Completed editing of the curb ramp feature line

Next, you'll need to extend the grade along the line representing the flowline of the curb and gutter area on the left of the screen to determine the elevation to the south of your intersection on the right of the screen. You'll use the Grade Extension by Reference tool in the following exercise to accomplish this:

1. This exercise is a continuation of the previous exercise. Click the feature line representing the flowline of the curb and gutter area on the left side of the drawing.
2. Select the Grade Extension by Reference tool and select the flowline again when prompted to `Select reference segment` (see Figure 16.25). This tool will evaluate the feature line as if it were three separate components (two lines and an arc). Because you are extending the grade of the flowline to the east and across the intersection, it is important to select the tangent segment as shown in Figure 16.25.

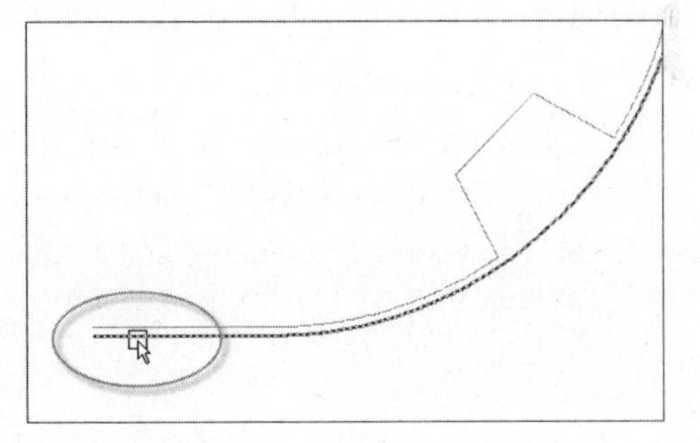

FIGURE 16.25
Selecting the flowline of the curb and gutter section

3. At the `Select object to edit` prompt, select the line representing the flowline of the curb and gutter area to the right of your screen as shown in 16.26.
4. At the `Specify point:` prompt, pick the PI at the left side of the tangent as shown in Figure 16.26.

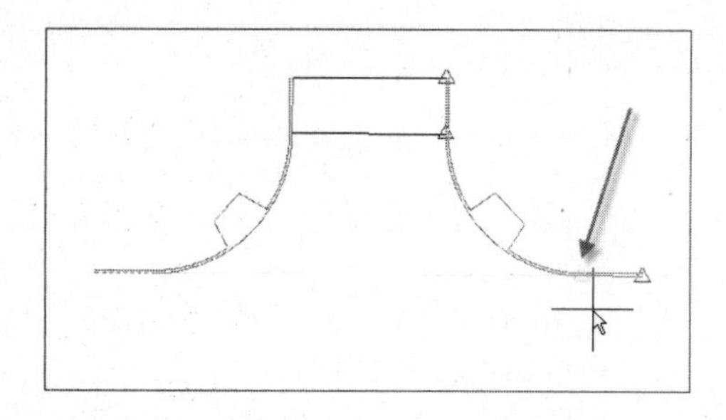

FIGURE 16.26
Selecting the PI at the left side of the tangent segment representing the flowline of the curb and gutter section

5. At the `Specify grade or [Slope/Elevation/Difference] <2.21>:` prompt, press ↵ to accept the default value of 2.21 (this is the grade of the flowline of the curb and gutter section to the left). Press ↵ to end the command, and press Esc to cancel grips.
6. Select the feature line representing the flowline of the curb and gutter section to the right of your screen to enable grips.
7. Move your cursor over the top of the PI as shown in Figure 16.27, but do not click. Your cursor will automatically snap to the grip, and the grip will change color. Notice that the elevation of the PI is displayed on the status bar as shown in Figure 16.27. This is a quick way to check elevations of vertices when modeling terrain. If your coordinates are not displayed, type the AutoCAD command **COORDS**, and set the value to **1** to display them. Leave this drawing open for the next exercise.

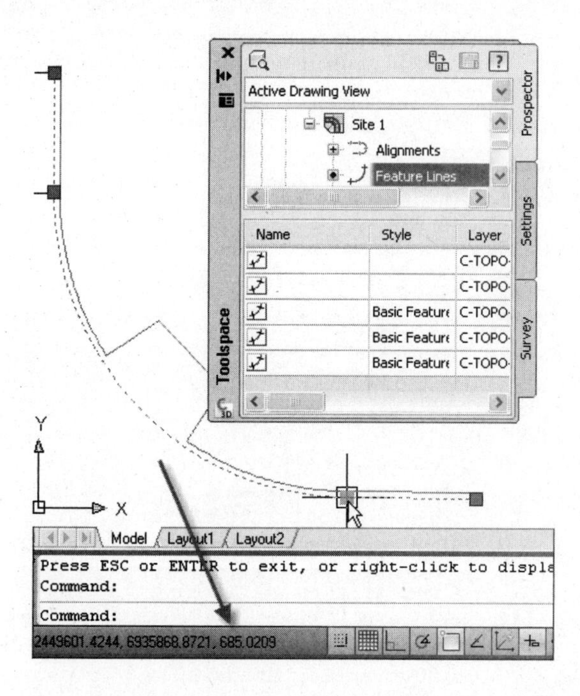

FIGURE 16.27
The x-, y-, and z-coordinates of the PI displayed on the status bar

With the elevation of a single point determined, the grade of the feature line representing the curb and gutter section can be modified to ensure positive water flow. In the following exercise, you'll use the Set Grade/Slope Between Points tool to modify the grade of the feature line:

1. This exercise is a continuation of the previous exercise. Select the feature line representing the flowline of the curb and gutter section to the right of your screen. Select the Set Grade/Slope Between Points tool from the Edit Elevations panel.
2. When prompted to Specify the start point, pick the PI as shown previously in Figure 16.26. The elevation of this point has been established and will be used as the basis for grading the entire feature line.
3. At the Specify elevation <685.021> prompt, press ↵ to accept the default value. This is the current elevation of the PI as established earlier.
4. When prompted to Specify the end point, pick the PI as shown in Figure 16.28.

FIGURE 16.28
Specifying the PI to establish the elevation at the flowline of the curb and gutter section

5. At the Specify grade or [Slope/Elevation/Difference] prompt, type **2** and press ↵ to set the grade between the points at 2 percent. Do not exit the command.
6. When prompted to Select object, select the feature line representing the flowline of the curb and gutter section to the right again.
7. When prompted to Specify the start point, pick the PI (shown previously in Figure 16.26) again.
8. At the Specify elevation <685.021> prompt, press ↵ to accept the default value. This is the current elevation of the PI as established earlier.
9. When prompted to Specify the end point, pick the PI at the far bottom right along the feature line currently being edited.
10. At the Specify grade or [Slope/Elevation/Difference] prompt, type **-2** (negative two) and press ↵ to set the grade between the points at negative 2 percent. Press ↵ to exit the command but do not cancel grips.
11. Select the Elevation Editor tool from the Edit Elevations panel to open Panorama. Notice the values in both the Grade Ahead and Grade Back columns as shown in Figure 16.29.

FIGURE 16.29
The grade of the feature line set to 2 percent

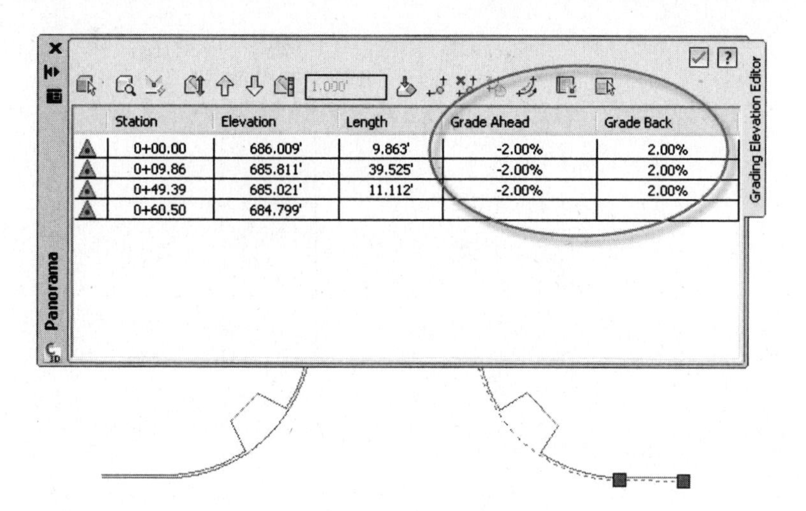

The possibilities are endless. In using feature lines to model proposed features, you are limited only by your creative approach. You've seen many of the tools in action, so you can now put a few more of them together and grade your pond.

 **Real World Scenario**

DRAINING THE POND

You need to use a combination of feature line tools and options to pull your pond together, and get the most flexibility should you need to update the bottom area or manipulate the pond's general shape. Here is the method that was used when this pond was designed:

1. Open the Pond `Drainage Design.dwg` file. The storm designer gave us a little bit more information about the pond design, as shown here:

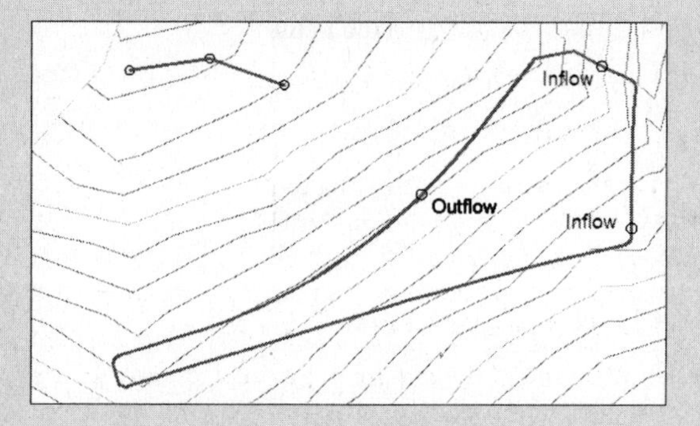

2. Click the pond basin outline and click the Insert Elevation Point tool.

3. Use the Center osnap to insert the elevation points in the center of each circle. Enter **655.5** as the elevation of each inflow, and accept the default elevation at the outflow. (You will change the elevation at the outflow in a moment.) Press ↵ to exit the command, and press Esc to cancel grips.
4. Draw a polyline, similar to the one shown here, from the southern inflow point across the bottom of the pond, through the outflow, and snap to the endpoint of the drainage channel feature line created earlier. Be sure to place a PI at the center of the circle designating the outflow point. This polyline will be the layout for the pilot channel.

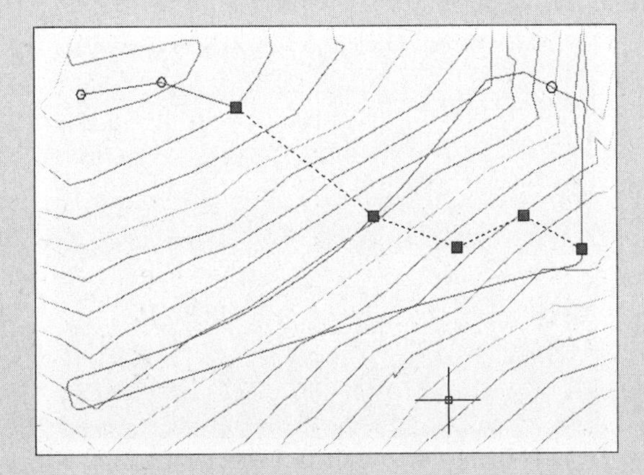

5. Change to the Home tab and choose Feature Line ➤ Create Feature Lines from Objects from the Create Design panel. Pick the polyline just drawn and press ↵. The Create Feature Lines dialog appears.
6. Click OK to complete the conversion without changing any options. Because this feature line was created in the same site as the bottom of the pond, the elevation of the PI at the southernmost inflow will reset to match the elevation of the endpoint of the pilot channel.
7. Select the feature line representing the pilot channel and pick the Fillet tool from the Edit Geometry panel.
8. Enter **R**↵, and enter **25**↵ for the radius.
9. Enter **A**↵ to fillet all PIs.
10. Press ↵ to exit the Fillet command.
11. Select the feature line representing to pilot channel. Click the Set Grade/Slope between Points tool on the Edit Elevations Panel.
12. Pick the PI at the inflow.
13. Enter **655.5**↵ to set this elevation.
14. Pick the other end of the feature line as shown here. All the PIs will highlight, and Civil 3D will display the total length, elevation difference, and average slope at the command line.

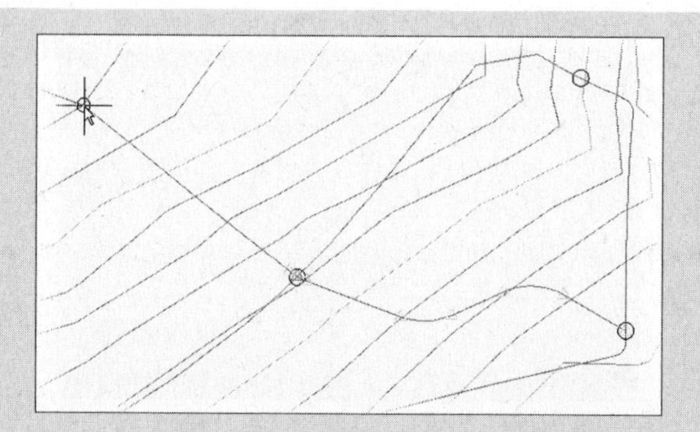

15. Press ↵ again to accept the grade as shown. This completes a linear slope from one end of the feature line to the other, ensuring drainage through the pond and outfall structure.
16. Press ↵ to exit the command. Press Esc to cancel grips.
17. Select the bottom of the pond and choose the Elevation Editor tool to open the Grading Elevation Editor in Panorama.
18. Click in the Station cells in the Grading Elevation Editor to highlight and ascertain the elevation at the outfall, as shown here. In this case, the elevation is 654.092′. Some minor variation might occur depending on your pick points and the length of your pilot channel. Notice also that the icon for this point in the Panorama display is a white triangle, indicating that this point is a point derived from a feature line intersection. This is called a *phantom PI*.

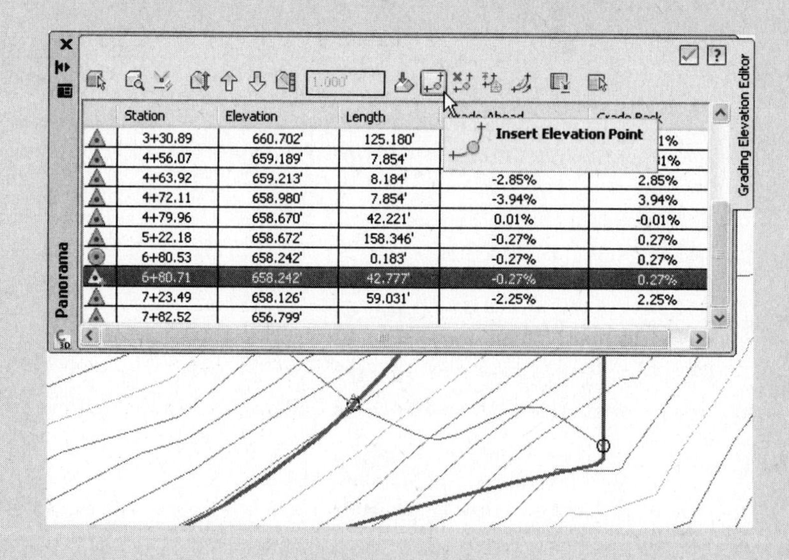

19. Click the Insert Elevation Point tool in the Panorama.

20. Using an intersection snap, set a new elevation point at the intersection of the pilot channel and the flowline. The Insert PVI dialog opens. Enter an elevation of **654.092′** (or the elevation you ascertained earlier) in the Elevation text box. This has to be added after the pilot channel has been created because the fillet process would tweak the location. Click OK to exit the Insert PVI dialog.
21. Close Panorama. A round grip is displayed at this new elevation point.
22. Select the pond bottom and click the Quick Elevation Edit tool on the Edit Elevations panel. Move your cursor over the southernmost PI at the southwest corner of the pond bottom. Pick the PI. Enter **657** and press ↵ twice to exit the command.
23. Click the Set Grade/Slope between Points tool on the Edit Elevations panel.
24. Move your cursor over the southernmost PI once again and pick. Press ↵ to accept the default elevation set previously as 657.
25. Pick the PI at the southern inflow. This should register an elevation of 655.50.
26. Press ↵ to accept the grade. This sets the elevations between the two selected PIs to all fall at the same grade.
27. Pick the same initial PI and press ↵ to accept the elevation of 657.
28. Moving your mouse clockwise to set the direction of the grading change you're about to make, use a center snap and pick the PI near the outflow shown here:

29. Press ↵ to exit the command. Select the Quick Elevation Editor tool from the Ribbon.
30. Hover your cursor over the outflow and wait for a moment for the tooltip to display. Be sure the elevation noted in the tooltip is 654.092, pick, and press ↵ to tie to the pilot channel elevation.
31. Pick the PI at the northern inflow, and press ↵ to accept the elevation (it should be 655.5).
32. Hover your cursor over the outflow and wait for a moment for the tooltip to display. Be sure the elevation noted in the tooltip is 654.092, click near the PI, and press ↵ to tie to the pilot channel elevation. Press ↵ to exit the command.

The entire outline of the pond bottom is graded except the area between the two inflows. Because you want to avoid a low spot, you'll now force a high point.

33. Click the Insert High/Low Elevation Point tool.
34. Pick the PI near the north inflow as the start point, and pick the PI near the southern inflow as the endpoint. Enter **0.5**↵ as the grade ahead.
35. Enter **0.5**↵ as the grade behind. A new elevation point will be created as shown here:

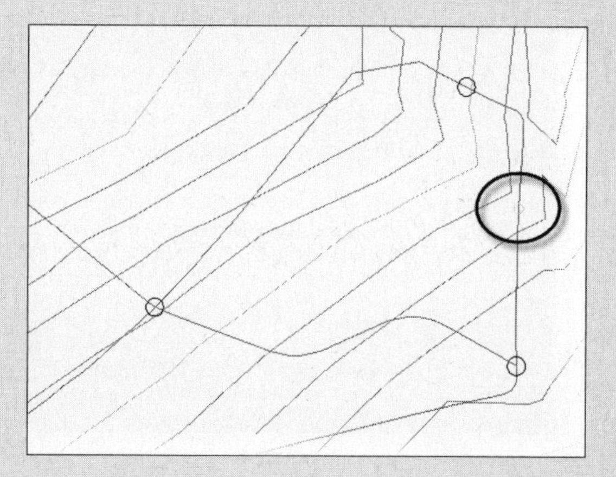

By using all the tools in the Feature Lines toolbar, you can quickly grade elements of your design and pull them together. If you have difficulty getting all of the elevations in this exercise to set as they should, slow down, and make sure you are moving your mouse in the right direction when setting the grades by slope. It's easy to get the calculation performed around the other direction; that is, clockwise versus counter-clockwise. This procedure seemed to involve a lot of steps, but it takes less than a minute in practice.

There are roughly 25 ways to modify feature lines using both the Edit Geometry and Edit Elevations panels. Take a few minutes and experiment with them to understand the options and tools available for these essential grading elements. By manipulating the various pieces of the feature line collection, it's easier than ever to create dynamic modeling tools that match the designer's intent.

Stylizing and Labeling Feature Lines

Though it's not common, feature lines can be stylized to reflect particular uses, and labels can be applied to help a reviewer understand the nature of the object being shown. In the next couple of exercises, you'll create a feature line style and then label a few critical points on your pond design to help you understand the drainage patterns better.

Feature Line Styles

Feature line styles are among the simplest in the Civil 3D program. Feature lines have options only for the display of the feature line itself, 2D and 3D, respectively. Many users never stylize their feature lines because there's so little to gain, but the following exercise gives you a look at the options:

1. Open the `Labeling Feature Lines.dwg` file. This is the completed Pond Pilot Channel exercise from the previous section.
2. Pick the Pilot Channel feature line and select Feature Line Properties from the Modify panel. The Feature Line Properties dialog opens.
3. On the Information tab, check the Style box and click the down arrow on the Style Options button to open the drop-down menu.
4. Select Create New to display the Feature Line Style dialog as shown in Figure 16.30.

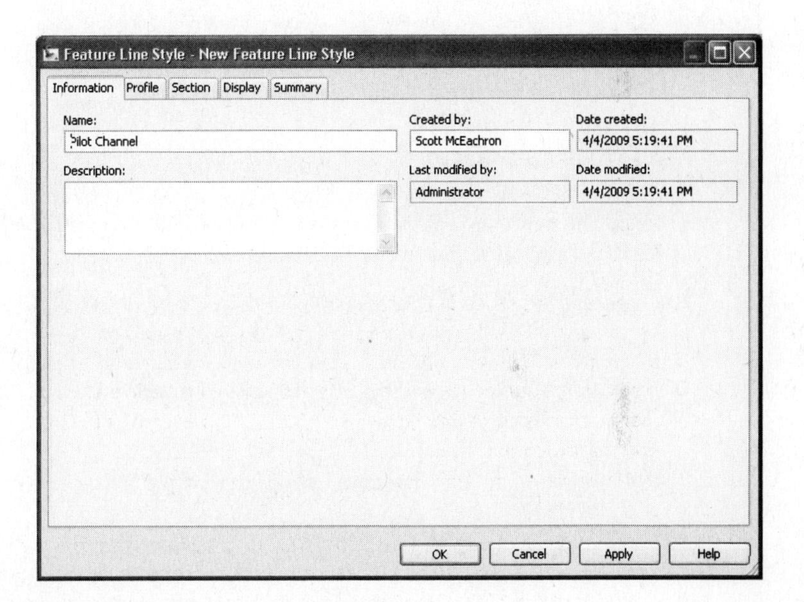

Figure 16.30
Creating a new feature line style

5. On the Information tab, enter **Pilot Channel** in the Name text box.
6. Switch to the Display tab.
7. Click the Linetype cell to open the Select Linetype dialog, and select Hidden. Click OK to dismiss the dialog.
8. Enter **3** in the LT Scale cell.
9. Click OK to dismiss the Feature Line Style dialog.

10. Click OK again to dismiss the Feature Line Properties dialog. Your screen should look like Figure 16.31.

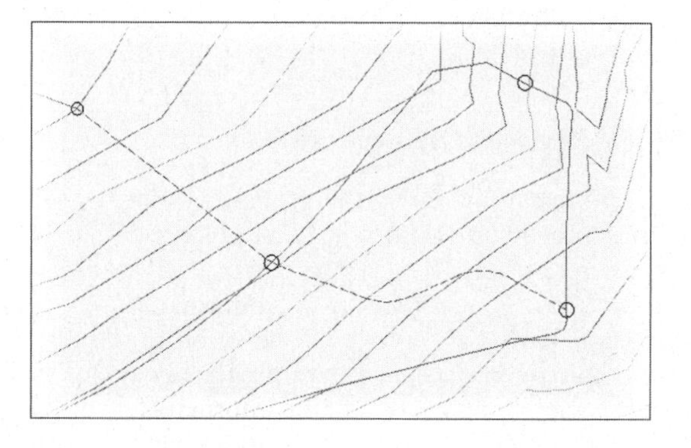

FIGURE 16.31
Pilot channel after styling the feature line

The style helps convey that this is a flow path for water, but you don't have much help in determining the actual slope or distances involved.

FEATURE LINE LABELS

Feature lines do not have their own unique label styles but instead share with general lines and arcs. The templates that ship with Civil 3D contain styles for labeling segment slopes, so you'll label the critical slopes in the following exercise:

1. Open the `Labeling Feature Lines.dwg` file if you closed it.
2. Change to the Annotate panel and choose Add Labels from the Labels & Tables panel.
3. Change the Line Label Style to Grade Only and Curve Label Style to Radius Only, as shown in Figure 16.32.

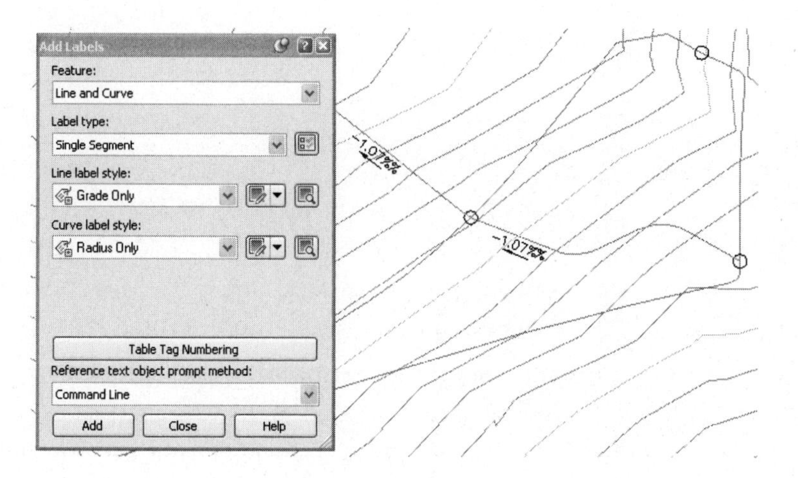

FIGURE 16.32
Adding feature line grade labels

4. Click Add.

5. Pick a few points along the Pilot Channel feature line tangents to create labels as shown in Figure 16.32.

Although it would be convenient to label the feature line elevations as well, there's no simple method for doing so. In practice, you would want to label the surface that contains the feature line as a component.

Grading Objects

Once a linear feature line is created, there are two main uses. One is to incorporate the feature line itself directly into a surface object as a breakline; the other is to create a grading object (referred to hereafter as simply a grading or gradings) using the feature line as a baseline. A grading consists of some baseline with elevation information, and a criteria set for projecting outward from that baseline based on distance, slope, or other criteria. These criteria sets can be defined and stored in grading criteria sets for ease of management. Finally, gradings can be stylized to reflect plan production practices or convey information such as cut or fill.

In this section, you'll start with defining a criteria set, use a number of different methods to create gradings, edit those gradings, stylize the grading, and finally convert the grading group into a surface.

Defining Criteria Sets

Criteria sets are merely collections of grading criteria organized for ease of use. These criteria might be something along the lines of "Grade at 3:1 to a Surface" or "Grade at 100% for 6" Elevation Change." These criteria can be built as mere guides, or they can act as straightforward commands offering little user interaction in the process. In this exercise, you'll create a new criteria set and add a few grading criteria:

1. Open the `Defining a Criteria Set.dwg` file. The typical pond section is shown in Figure 16.33. To achieve this section, create a grading criteria set for the various slopes, the ledges, and the surface tie-in.

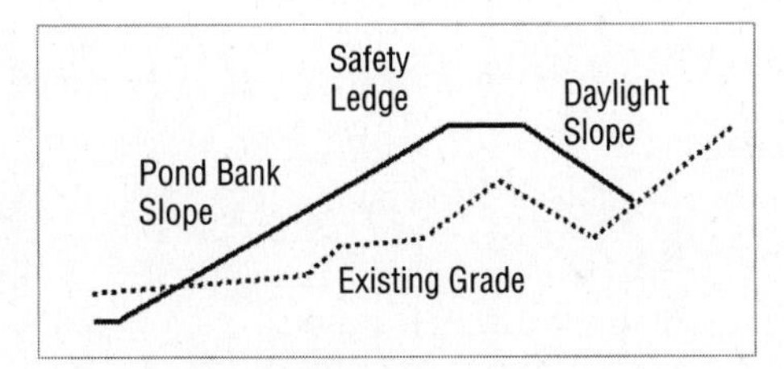

FIGURE 16.33
Typical pond section

2. Switch to the Settings tab.
3. Expand the Grading ➢ Grading Criteria Sets branches.
4. Right-click Grading Criteria Sets and select New. The Grading Criteria Set Properties dialog opens.
5. Enter **Pond Grading** in the Name text box and click OK to dismiss the dialog.
6. Expand the Grading Criteria Sets branch, and then right-click the Pond Grading grading criteria set and select New.

7. Enter **3:1 to Elevation** in the Name text box.
8. Switch to the Criteria tab.
9. Change the values as shown in Figure 16.34.

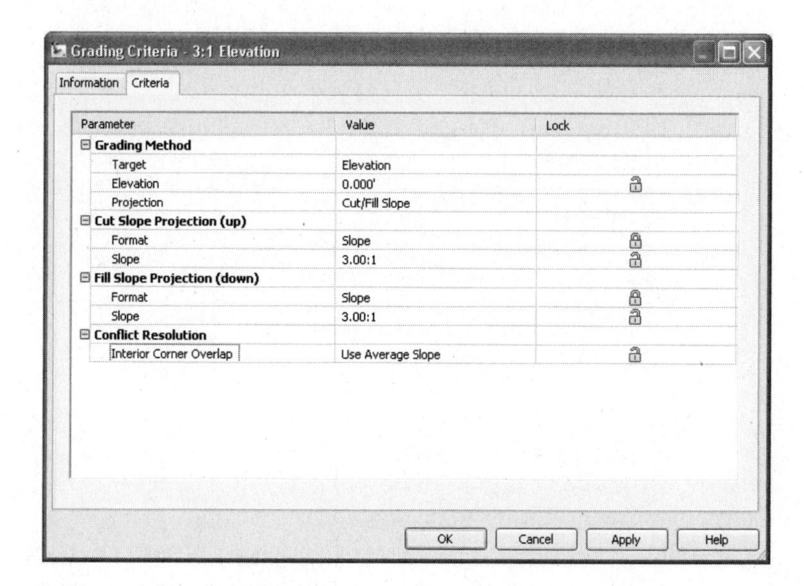

FIGURE 16.34
Creating a grading criteria

10. Click OK to close the Grading Criteria dialog.

Take a few moments to review the options available on the Criteria tab:

- Grading Method controls the projection method used. The Target cell lets you set the target type. Surface, Elevation, Relative Elevation, and Distance are the valid options, and as each is selected, options appear directly underneath.
- Slope Projection allows the selection of slopes (X:1) or grades (X%) as the format for specifying the grading projections.
- Conflict Resolution determines how Civil 3D will handle overlapping projection lines on interior corners. Use Average Slope is the typical answer, cleaning up the grading object most cleanly when conflicts arise.

Note that each of the rows has a padlock to the far right. If you lock any individual piece, the user will not be prompted for that value or be allowed to change it. In the case of Figure 16.34, the only user input required will be the Target Elevation for the grading. In this exercise, you'll complete the rest of the Grading criteria set for your pond design:

1. Expand the Pond Grading branch, right-click the 3:1 to Elevation branch, and select Copy, as shown in Figure 16.35.

FIGURE 16.35
Copying a Grading criteria set

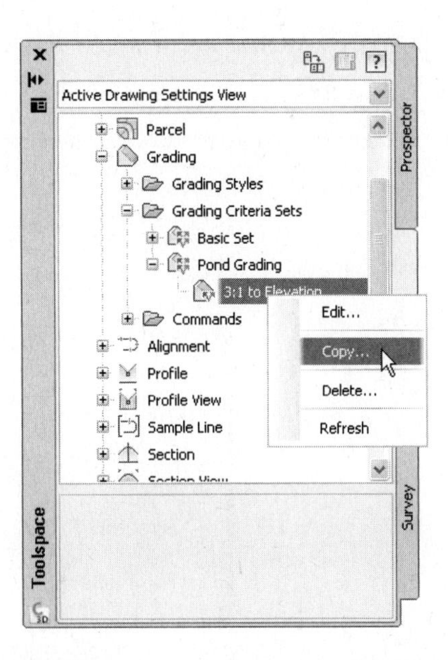

2. Enter **3:1 to Surface** in the Name text box.
3. Switch to the Criteria tab.
4. Select Surface from the drop-down list in the Target selection box, and unlock the Slope values for both Cut and Fill Slope Projections, as shown in Figure 16.36.

FIGURE 16.36
3:1 to Surface Grading Criteria settings

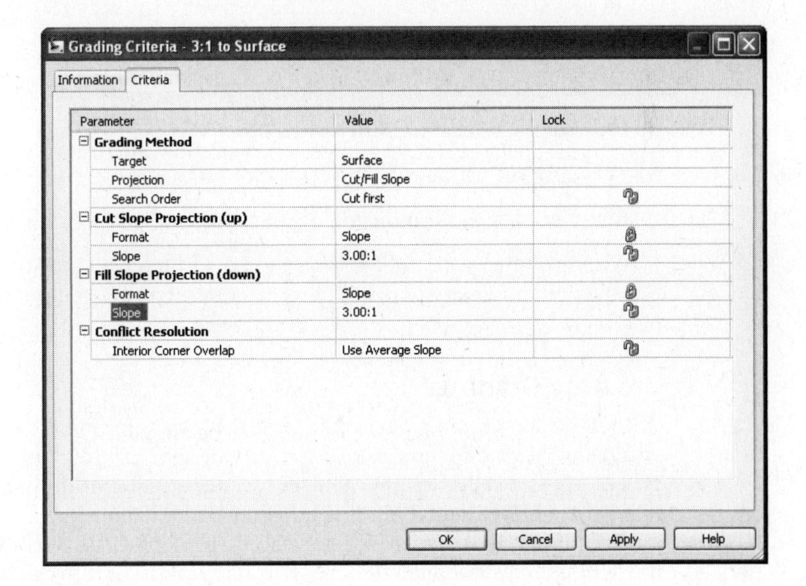

5. Click OK to close the dialog.
6. Right-click Pond Grading and select New.
7. Enter **Flat Ledge** in the Name text box.
8. Switch to the Criteria tab.
9. Select Grade from the drop-down list in the Slope Projection Format selection box, as shown in Figure 16.37.

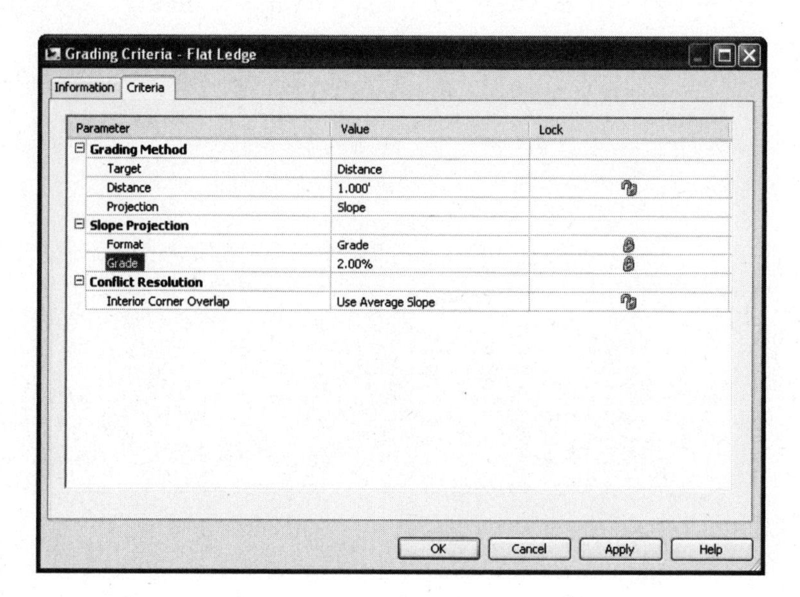

FIGURE 16.37
Flat ledge grading criteria settings

10. Enter **2%** for the Grade value.
11. Lock the Grade row.
12. Click OK to close the dialog and complete the exercise.

With this criteria set, you've locked many of the user inputs, making grading a fast and easy process with minimal opportunity for user input error. You put these grading criteria to use in the next section.

Creating Gradings

With the criteria set in place, you can now look at grading the pond as designed. In this section, you'll look at grading groups and then create the individual gradings within the group. Grading groups act as a collection mechanism for individual gradings and let Civil 3D understand the daisy chain of individual gradings that are related and act in sync with each other.

One thing to be careful of when working with gradings is that they are part of a site. Any feature line within that same site will react with the feature lines created by the grading. For that reason, the exercise drawing has a second site called Pond Grading to be used for just the pond grading.

1. Open the `Grading the Pond.dwg` file.
2. Pick the pond bottom.

3. Choose the Move to Site tool from the Modify drop-down panel. The Move to Site dialog appears.

4. Click OK to change the Pond feature line to the Pond Grading site. This will avoid interaction between the pond banks and the pilot channel you laid out earlier.

5. With the feature line still selected, choose Grading Creation Tools from the Launch Pad panel. The Grading Creation Tools toolbar, shown in Figure 16.38, appears. This toolbar is similar to the toolbar used in pipe networks. The left section is focused on settings, the middle on creation, and the right on editing.

FIGURE 16.38
The Grading Creation Tools toolbar

6. Click the Set the Grading Group tool to the far left of the toolbar to display the Site dialog. Choose the Pond site and click OK. The Create Grading Group dialog is displayed.

7. Enter **Pond** in the Name text box as shown in Figure 16.39 and click OK. You'll revisit the surface creation options in a bit.

FIGURE 16.39
The Create Grading Group dialog

8. Click the Select a Criteria Set tool to display the Select a Criteria Set dialog.

9. Select Pond Grading from the drop-down list and click OK to close the dialog.

10. Click the Create Grading tool on the Grading Creation Tools toolbar, or click the down arrow next to the Create Grading tool and select Create Grading, as shown in Figure 16.40.

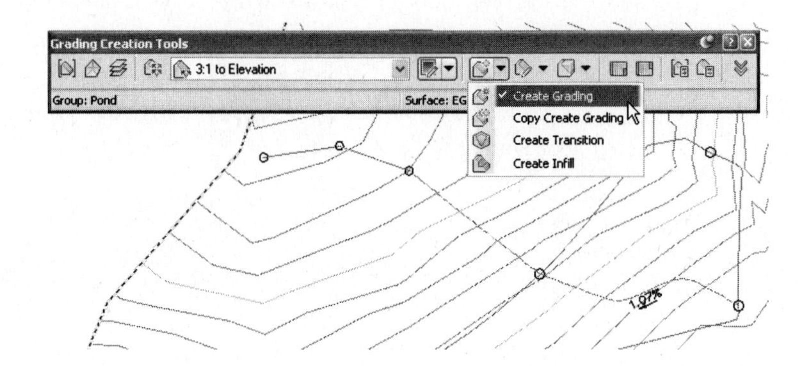

FIGURE 16.40
Creating a grading using the 3:1 to Elevation criteria

11. Pick the pond outline.
12. Pick a point on the outside of the pond to indicate the direction of the grading projections.
13. Press ↵ to apply the grading to the entire length of the pond outline.
14. Enter **664↵** at the command line as the target elevation. The first grading is complete. The lines onscreen are part of the Grading style.
15. In the Select a Grading Criteria drop-down list on the Grading Creation Tools toolbar, select Flat Ledge.
16. Pick the upper boundary of the grading made in step 15, as shown in Figure 16.41.

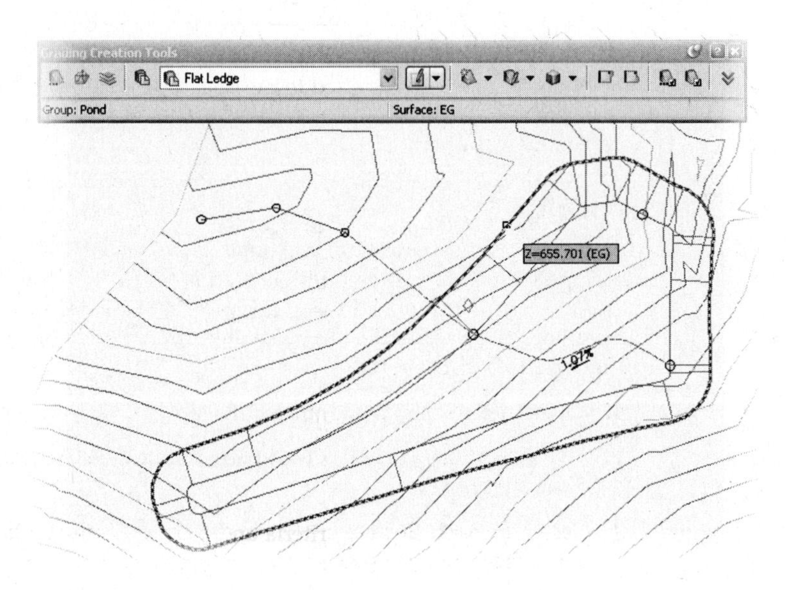

FIGURE 16.41
Creating a daisy chain of gradings

17. Press ↵ to apply to the whole length.
18. Enter **10**↵ for the target distance to build the safety ledge.
19. In the Select a Grading Criteria drop-down list on the Grading Creation Tools toolbar, select 3:1 to Surface.
20. Pick the outer edge of the safety ledge just created.
21. Press ↵ to apply to the whole length. Press ↵ to exit the command. Your drawing should look similar to Figure 16.42.

FIGURE 16.42
Complete pond interior grading

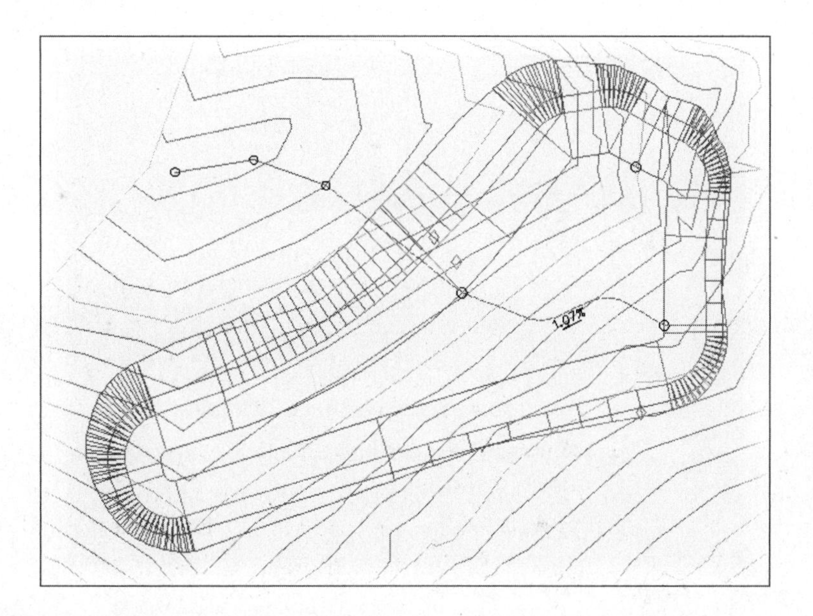

Each piece of this pond is tied to the next, creating a dynamic model of your pond design on the basis of the designer's intent. What if that intent changes? The next section describes editing the various gradings.

Editing Gradings

Once you've created a grading, you often need to make changes. A change can be as simple as changing the slope or changing the geometric layout. In this exercise, you'll make a simple change, but the concept applies to all the gradings you've created in your pond. Because the grading criteria were originally locked to make life easier, you'll now unlock them before modifying the daylight portion of the pond:

1. Open the `Editing Grading.dwg` file.
2. On the Settings tab of the Toolspace, expand the Grading ➢ Grading Criteria Sets ➢ Pond Grading branches.
3. Select the 3:1 to Surface criteria under Pond Grading, right-click, and select Edit. The Grading Criteria dialog opens.

4. On the Criteria tab, unlock the Slope values for both the Cut and Fill Slope Projections, as shown in Figure 16.43. Click OK.

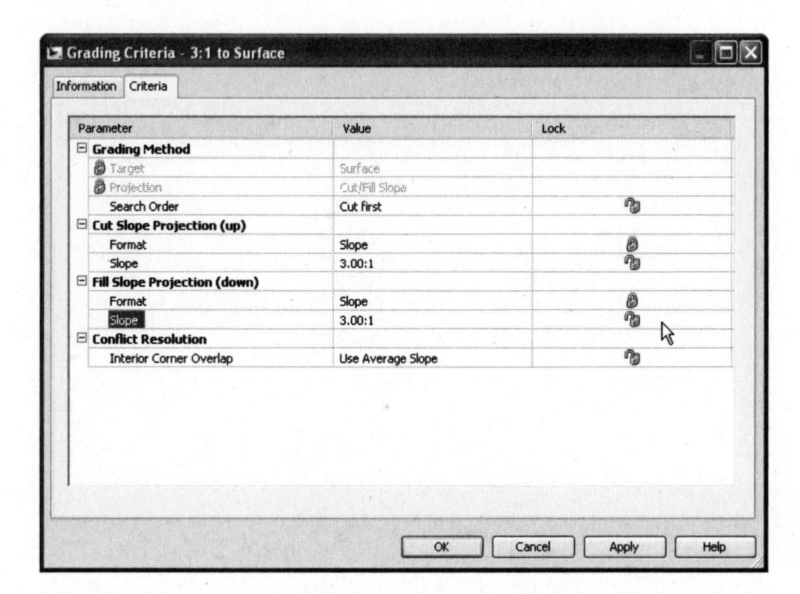

FIGURE 16.43
Modifying the 3:1 to Surface Grading criteria

5. Pick one of the projection lines or the small diamond on the outside of the pond.
6. Choose the Grading Editor tool from the Modify panel. The Grading Editor in Panorama appears.
7. Enter **4:1** for a Fill Slope Projection as shown in Figure 16.44.

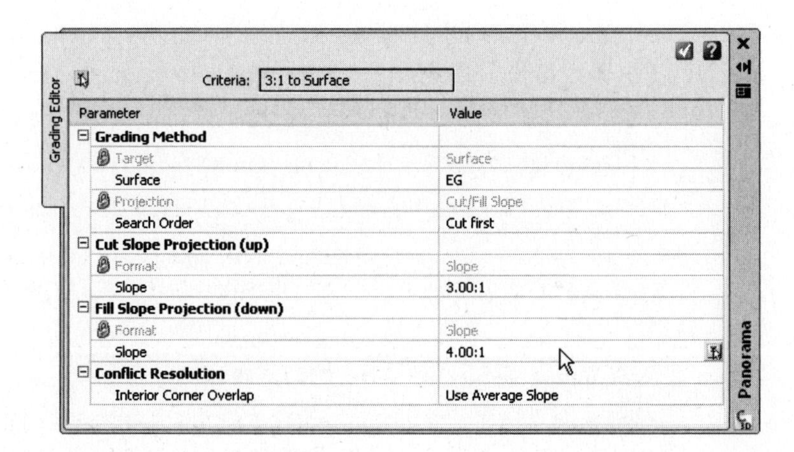

FIGURE 16.44
Editing the Fill Slope value

8. Close Panorama. Your display should look like Figure 16.45. (Compare this to Figure 16.42 if you'd like to see the difference.)

FIGURE 16.45
Completed grading edit

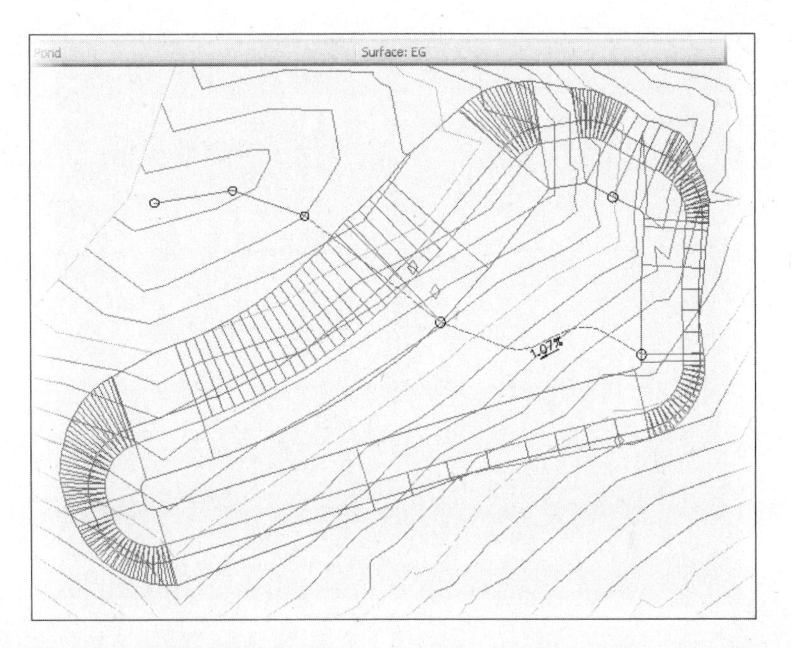

Editing any aspect of the grading will reflect instantly, and if other gradings within the group are dependent on the results of the modified grading, they will recalculate as well. Now that you have a working design, you can get a better view of the pond on the basis of stylizing the gradings, as described in the next section.

Grading Styles

The stock grading styles work fairly well, but you'd like to show the slopes using some pretty typical civil engineering symbology. This information gives immediate feedback about the gradings without you having to pull up a Properties dialog or inquire about slopes.

1. Open the `Grading Styles.dwg` file.
2. Expand the Grading ➤ Grading Styles branches on the Settings tab of the Toolspace.
3. Right-click Grading Styles and select New. The Grading Style dialog appears.
4. On the Information tab, enter **Pond Slopes** in the Name text box.
5. Switch to the Slope Patterns tab. Check the Slope Pattern box in the Options area.
6. Click the down arrow on the Style Options button to open the drop-down menu as shown in Figure 16.46, and select Create New. The Slope Pattern Style dialog appears.

YES, YOU'RE GOING AROUND THE LONG WAY

Slope pattern styles are contained in the Multipurpose styles collection under the General branch near the top of the Settings tree. Because you want to use this in direct correlation with a grading style, you're going through the Grading Style dialog to reach that same point. It's merely a reminder that there are a multitude of approaches to working within Civil 3D, and everyone will have his or her own favorite methods.

FIGURE 16.46
Creating a new slope pattern style

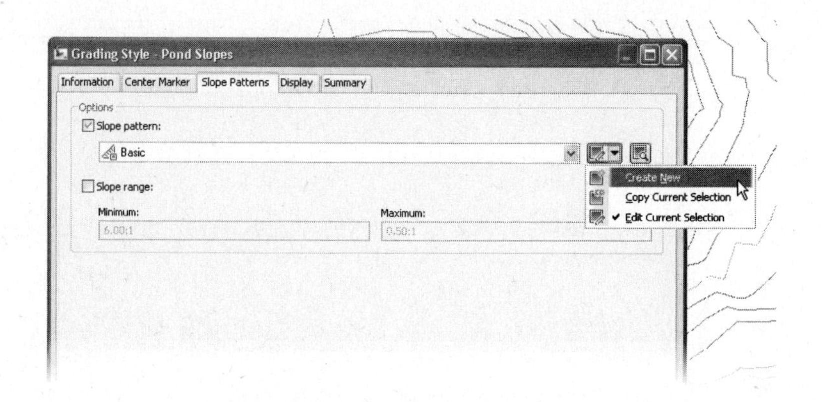

7. On the Information tab, enter **Full Slopes** in the Name text box.

8. Switch to the Layout tab. Change the values for Component 1, as shown in Figure 16.47. There are a large number of options for creating varied slope indicators, but this simple indicator works well for many cases.

FIGURE 16.47
Slope Pattern Style Layout tab

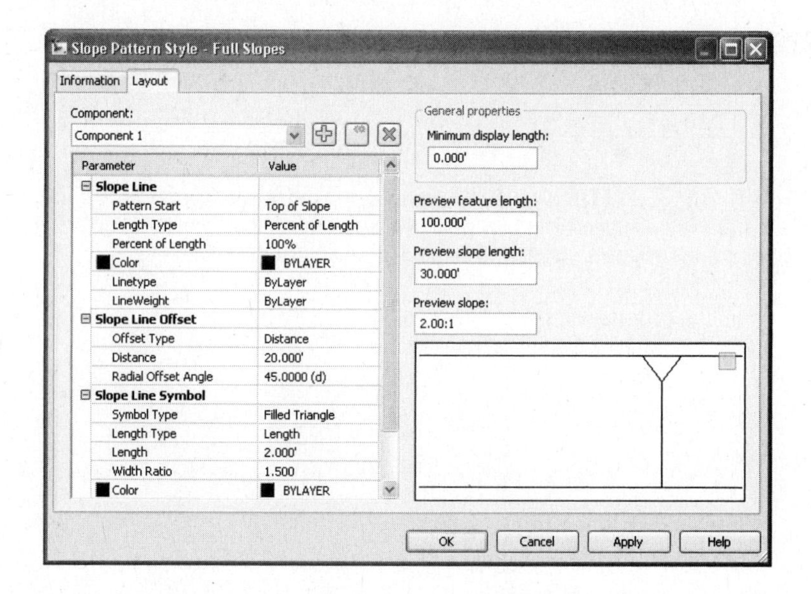

9. Click Apply.

10. Select Component 2 from the drop-down list in the Component selection box.

11. Click the red X button to delete Component 2, and click OK to dismiss the dialog.

12. Check the Slope Range box in the Options area. This option set limits for when the slope pattern will be applied. The default values will conveniently *not* put slope patterns on the berm grading, leaving a cleaner appearance. Leave the default values in place.

13. Switch to the Display tab. Turn the Projection Line component off in the Plan View Direction, and click OK to dismiss the dialog.
14. Pick one of the gradings by picking a projection line or the small diamond.
15. Select the Grading Properties tool from the Modify panel to display the Grading Properties dialog shown in Figure 16.48.

FIGURE 16.48
Changing grading styles

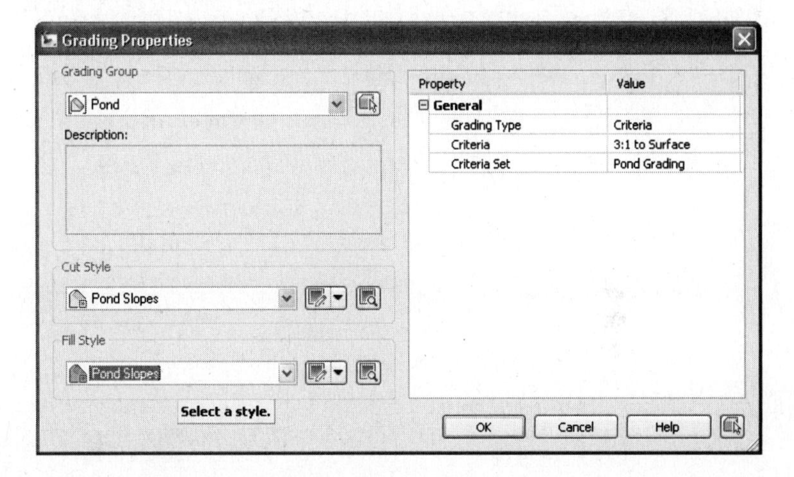

16. Select Pond Slopes from the drop-down lists in the Cut Style and Fill Style selection boxes, as shown in Figure 16.48. Click OK to dismiss the dialog.
17. Repeat this process for the other two gradings. The Ledges will only have one slope style to pick from, but you should still use the Pond Slopes option. Your completed work should look similar to Figure 16.49.

FIGURE 16.49
Completed grading style changes

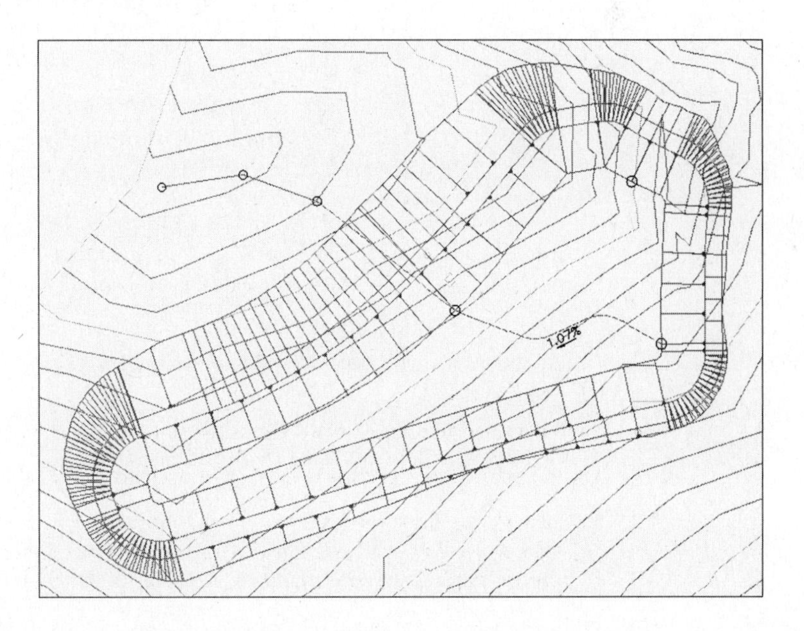

Although there are still some issues in the curve areas because of the way the gradings build feature lines, this view of your pond gives immediate feedback about the basic design. Note that the southeast corner contains slope arrows coming back to the pond area, indicating that the design is actually in a cut situation in this area. How much cut? You'll find out in the next section.

Creating Surfaces from Grading Groups

Grading groups work well for creating the model, but you have to use a TIN surface to go much further with them. In this section, you'll look at the conversion process, and then use the built-in tools to understand the impact of your grading group on site volumes.

1. Open the `Creating Grading Surfaces.dwg` file.
2. Pick one of the diamonds in the grading group.
3. Select the Grading Group Properties tool from the Modify panel to display the Grading Group Properties dialog.
4. Switch to the Information tab if needed. Check the box for Automatic Surface Creation.
5. In the Create Surface dialog that appears, enter **Pond Surface** in the Name field.
6. Click in the Style field, and then click the ellipsis button. The Select Surface Style dialog appears. Select Contours 1′ and 5′ from the drop-down list in the selection box, as shown in Figure 16.50. Click OK twice to return to the Grading Group Properties dialog.

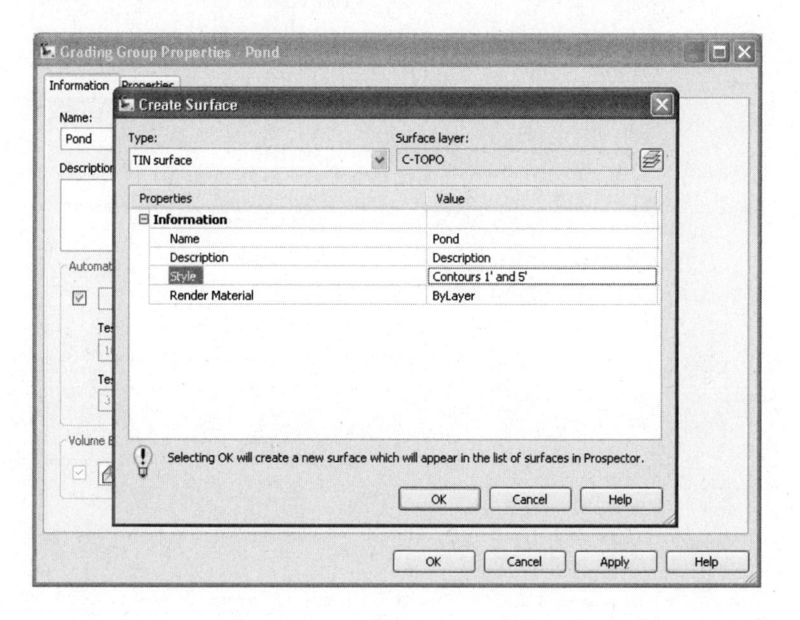

FIGURE 16.50
Creating a grading group surface

7. In the Grading Group Properties dialog, check the Volume Base Surface option to perform a volume calculation. Click OK to dismiss the dialog.

You're going through this process now because you didn't turn on the Automatic Surface Creation option when you created the grading group. If you're performing straightforward gradings, that option can be a bit faster and simpler. There are two options available when creating a surface from a grading group. They both control the creation of projection lines in curved area:

- The Tessellation Spacing value controls how frequently along an arced feature line TIN points are created and projection lines calculated. A TIN surface cannot contain any true curves the way a feature line can because it is built from triangles. The default values typically work for site mass grading, but might not be low enough to work with things such as parking lot islands where the 10′ value would result in too little detail.
- The Tessellation Angle value is the degree measured between outside corners in a feature line. Corners with no curve segment have to have a number of projections swung in a radial pattern to calculate the TIN lines in the surface. The tessellation angle is the angular distance between these radial projections. The typical values work most of the time, but in large grading surfaces, a larger value might be acceptable, lowering the amount of data to calculate without significantly altering the final surface created.

There is one small problem with this surface. If you examine the bottom of the pond, you'll notice there are no contours running through this area. If you mouse to the middle, you also won't have any Tooltip elevation as there is no surface data in the bottom of the pond. To fix that (and make the volumes accurate), you need a grading infill.

8. From the Modify panel, select Create Grading Infill. The Select Grading group dialog will appear.
9. Click OK twice to accept the default selection of the Pond group in the Pond site and Grading Style. Civil 3D will prompt you at the command line to select an infill area.
10. Hover your cursor over the middle of the pond and the pond feature line created earlier will be highlighted, indicating a valid area for infill.
11. Click once to create the infill, and press ↵ to apply. Civil 3D will calculate, and Panorama may appear. Dismiss it. You should now have some contours running through the pond base area as shown in Figure 16.51.
12. Zoom in if needed and pick one of the grading diamonds again to select one of the gradings. Make sure you grab one of the gradings and not the surface contours that are being drawn on top of them.
13. Select Grading Group Properties from the Modify panel to open the Grading Group Properties dialog.
14. Switch to the Properties tab to display the Volume information for the pond, as shown in Figure 16.52. This tab also allows you to review the criteria and styles being used in the grading group.

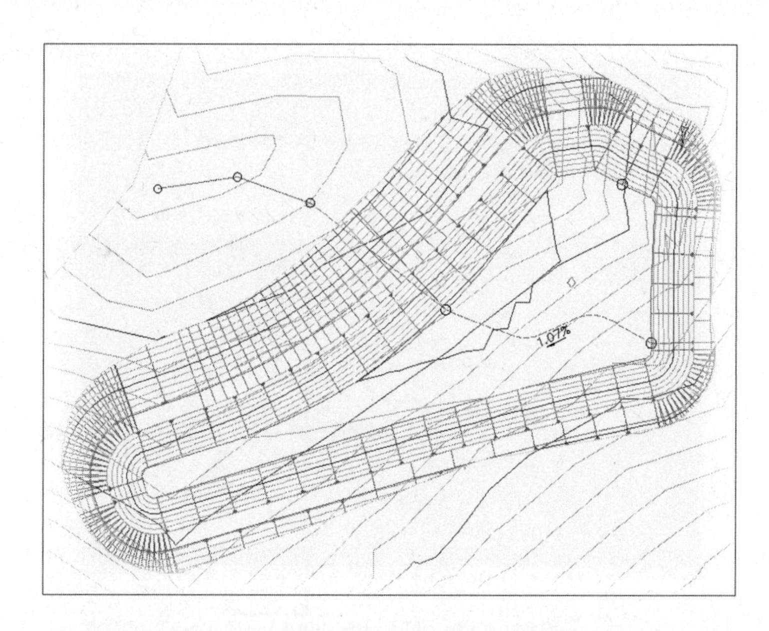

FIGURE 16.51
The pond after applying an infill grading

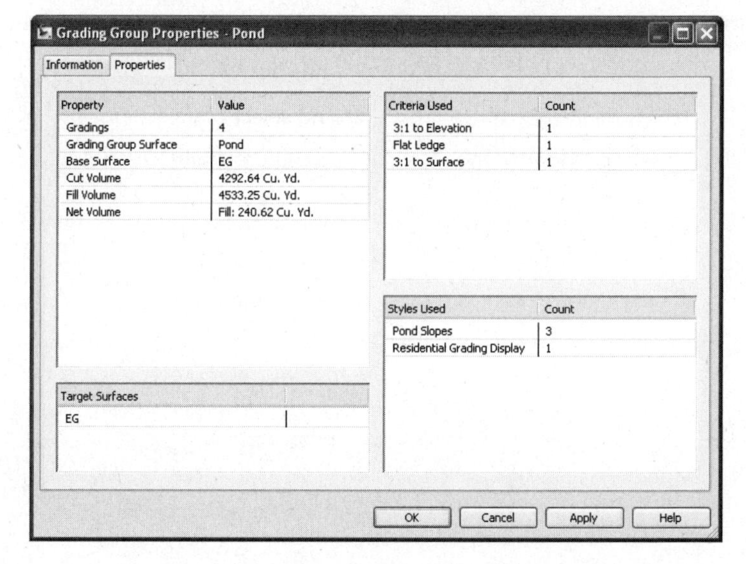

FIGURE 16.52
Reviewing the grading group volumes

This new surface is listed in Prospector and is based on the gradings created. A change to the gradings would affect the grading group, which would, in turn, affect the surface and these volumes. In the following exercise, you'll pull it all together:

1. Open the `Creating Composite Surfaces.dwg` file.
2. Right-click Surfaces in Prospector and select Create Surface.
3. In the Create Surface dialog, enter **Composite** in the Name text box. Click in the Style field, and then click the ellipsis to open the Select Surface Style dialog. Select the Contours 2′ and 10′ (Design) option from the drop-down list box, and click OK.

4. Click OK to dismiss the Create Surface dialog and create the surface in Prospector.
5. In Prospector, expand the Surfaces ➢ Composite ➢ Definition branches.
6. Right-click Edits and select Paste Surface. The Select Surface to Paste dialog in Figure 16.53 appears.

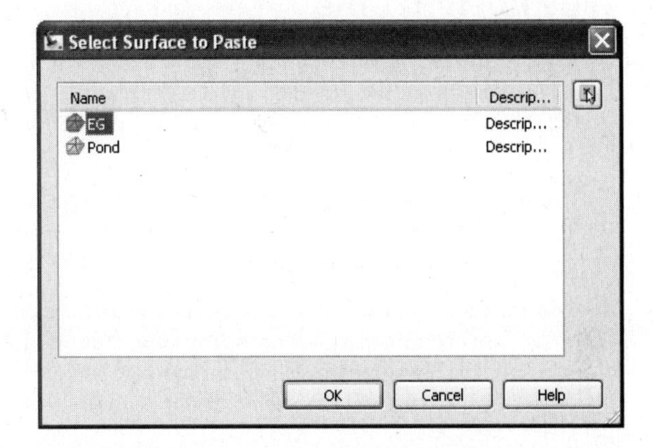

FIGURE 16.53
Pasting surfaces together

7. Select EG from the list and click OK. Dismiss Panorama if it appears.
8. Right-click Edits again and select Paste Surface one more time.
9. Select Pond Surface and click OK. The drawing should look like Figure 16.54.

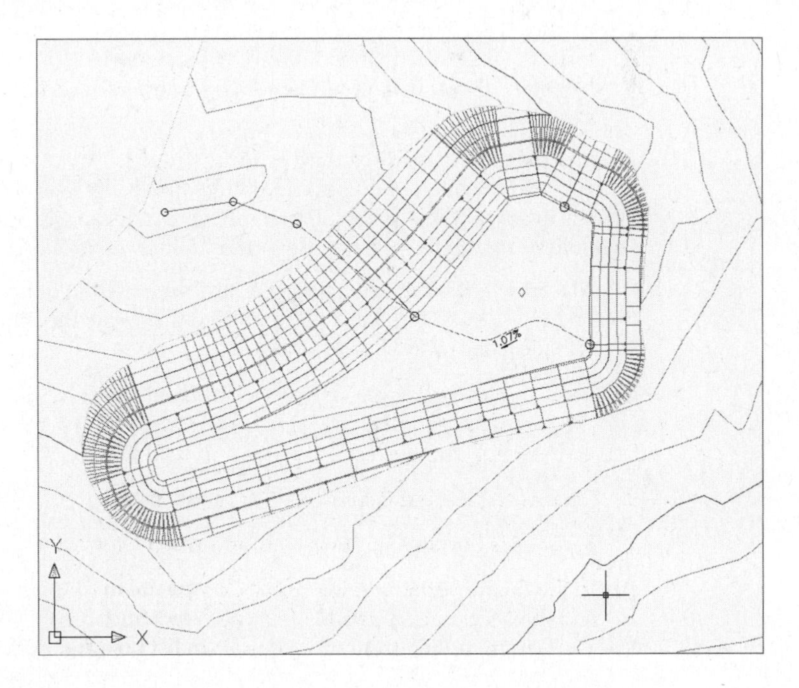

FIGURE 16.54
Completed composite surface

By creating a composite surface consisting of pasted together surfaces, the TIN triangulation cleans up any gaps in the data, making contours that are continuous from the original grade, through the pond, and out the other side. With the grading group still being dynamic and editable, this composite surface reflects a dynamic grading solution that will update with any changes.

The Bottom Line

Convert existing linework into feature lines. Many site features are drawn initially as simple linework for the 2D plan. By converting this linework to feature line information, you avoid a large amount of rework. Additionally, the conversion process offers the ability to drape feature lines along a surface, making further grading use easier.

Master It Open the `Mastering Grading.dwg` file from the data location. Convert the polyline describing a proposed temporary drain into a feature line and drape it across the EG surface to set elevations.

Model a simple linear grading with a feature line. Feature lines define linear slope connections. This can be the flow of a drainage channel, the outline of a building pad, or the back of a street curb. These linear relationships can help define grading in a model, or simply allow for better understanding of design intent.

Master It Add 200′ radius fillets on the feature line just created. Set the grade from the top of the hill to the circled point to 5 percent and the remainder to a constant slope to be determined in the drawing. Draw a temporary profile view to verify the channel is below grade for most of its length.

Create custom grading criteria. Using grading criteria to limit the amount of user input can help cut down on input errors and design mistakes. Defining a set to work with for a given design makes creating gradings a straightforward process.

Master It Continuing with the `Mastering Grading.dwg` file, create two new grading criteria: the first to be named Bottom, with a criteria of 5′ at 2 percent grade, and the second to be named Sides, with criteria of daylight to a surface at 6:1 slope in cut areas only; in fill, do not grade.

Model planar site features with grading groups. Once a feature line defines a linear feature, gradings collected in grading groups model the lateral projections from that line to other points in space. These projections combine to model a site much like a TIN surface, resulting in a dynamic design tool that works in the Civil 3D environment.

Master It Use the two grading criteria just used to define the pilot channel, with grading on both sides of the sketched centerline. Calculate the difference in volume between using 6:1 side slopes and 4:1 side slopes.

Chapter 17

Sharing the Model: Data Shortcuts

No man is an island, and it's the rare designer who gets to work alone. Even in a one-person design team, breaking the project into finite elements, such as grading, paving, and utilities, makes sense from a plan production and management standpoint. To do this effectively, your design software has to understand and have some method for sharing the design information behind the drawing objects. Civil 3D has two mechanisms for creating a project environment with project information sharing: data shortcuts and Vault. In this chapter, we look at data shortcuts as a project-management scheme.

By the end of this chapter, you'll be able to:

- List the project elements that are available for sharing through data shortcuts and those that are not
- Create new data shortcuts from Civil 3D objects
- Create data references in new drawings
- Restore a broken reference and repath shortcuts to new files

What Are Data Shortcuts?

Data shortcuts allow the cross-referencing of design data between drawings. To this end, the data is what is made available, and it's important to note that the appearance can be entirely different between host and any number of data shortcuts. We'll use the term *data shortcut* or, more simply, *shortcut* when we discuss selecting, modifying, or updating these links between files.

What About Vault?

This chapter focuses strictly on the creation and use of data shortcuts. In Chapter 19, "Teamwork: Vault Client and Civil 3D," you'll learn how using Vault compares to data shortcuts and the relative advantages of each method. You'll also look at a possible project-management methodology. The same techniques can generally be used with shortcuts. The two methods work in parallel, and similar but different material is covered in both chapters, so be sure to read it all!

There are two primary situations in which you need data or information across drawings, and they are addressed with external references (XRefs) or shortcuts. These two options are similar but not the same. Let's compare:

- **XRef functionality** — Used when the goal is to get a picture of the information in question. XRefs can be changed by using the layer control, XRef clips, and other drawing-element controls. Though they can be used for labeling across files in Civil 3D 2010, the fact that you have to bring the entire file to label one component is a disadvantage.

- **Shortcut functionality** — Brings over the design information, but generally ignores the display. Shortcuts only work with Civil 3D objects, so they will have their own styles applied. This may seem like duplicitous work because you have already assigned styles and labels in one drawing and have to do it again, but it offers the advantage in that you can have completely different views of the same data in different drawings.

As noted, only Civil 3D objects can be used with shortcuts, and even then some objects are not available through shortcuts. The following objects are available for use through data shortcuts:

- Alignments
- Surfaces
- Profile data
- Pipe networks
- View frame groups

You might expect that corridor, parcel, assembly, point, or point group information would be available through the shortcut mechanism, but they are not. Parcels and corridors can only be accessed via XRef, but once they've been XReferenced, you can use the normal labeling techniques and styles. Now that you've looked at what objects you can tackle with shortcuts, you'll learn how to create them in the next section.

A NOTE ABOUT THE EXERCISES IN THIS CHAPTER

Many of the exercises in this chapter — as well as in Chapter 18, "Behind the Scenes: Autodesk Data Management Server," and Chapter 19, "Teamwork: Vault Client and Civil 3D" — assume you've stepped through the full chapter. It's difficult to simulate the large number of variables that come into play in a live environment. To that end, many of these exercises build on the previous one. For the easiest workflow, don't close any files until the end of the chapter.

Additionally, you'll have to make some saves to data files throughout this chapter. Remember that you can always get the original file from the media that you downloaded from `www.sybex.com/go/masteringcivil3d2010`.

Publishing Data Shortcut Files

Shortcut files are simply XML files that contain pointers back to the drawing containing the object in question. These shortcuts are managed through Prospector and are stored in a working folder. Creating shortcuts is a matter of setting a working folder, creating a shortcut folder within that folder, and then creating the shortcut files. You look at all three steps in this section.

As a precursor to making your first project, you should establish what a typical folder structure looks like. Civil 3D includes a mechanism for copying a typical project folder structure into each new project. By creating a blank copy of the folder structure you'd like to have in place for your projects, you can use that as the starting point when a new project is created within Civil 3D.

1. Open Windows Explorer and navigate to `C:\Civil 3D Project Templates`.
2. Create a new folder titled **Mastering**.

3. Inside `Mastering`, create folders called **Survey**, **Engineering**, **Architecture**, **Word**, and **Con Docs**, as shown in Figure 17.1.

FIGURE 17.1
Creating a project template

This structure will show inside Civil 3D and in the working folder when a project is created. We've included a `Word` folder as an example of other, non–Civil 3D–related folders you might have in your project setup for users outside the CAD team, such as accountants or the project manager. If you have any files in these folders (such as a project checklist spreadsheet, a blank contract, or images), they will be checked into new projects as they are created.

The Working and Data Shortcuts Folders

The working folder concept was originally introduced with Vault, but for shortcuts, you can think of it as a project directory. The working folder will contain a number of projects, each in turn having a shortcuts folder where the shortcut files actually reside. Each time you create a new shortcut folder within the Prospector framework, you'll have the opportunity to create a full project structure. In this exercise, you'll set the working folder and create a new project:

1. Create a new blank drawing using any template.
2. Within Prospector, make sure the View drop-down list is set to Master View.
3. Right-click the Data Shortcuts branch and select Set Working Folder to display the Browse for Folder dialog shown in Figure 17.2.
4. Click Local Disk (C:) to highlight it, and then click the Make New Folder button.
5. Type **Mastering Shortcuts** as the folder name and click OK to dismiss the dialog.
6. Right-click the Data Shortcuts branch in Prospector, and select New Data Shortcuts Folder to display the New Data Shortcut Folder dialog shown in Figure 17.3.
7. Type **Niblo** for the folder name, and toggle on the Use Project Template option as shown in Figure 17.3.

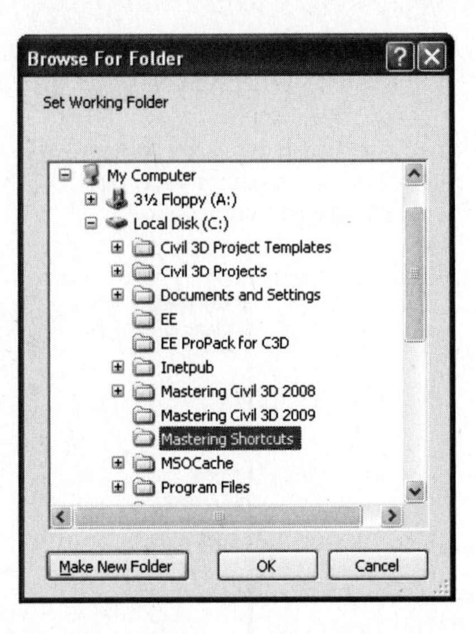

FIGURE 17.2
Creating a new working folder

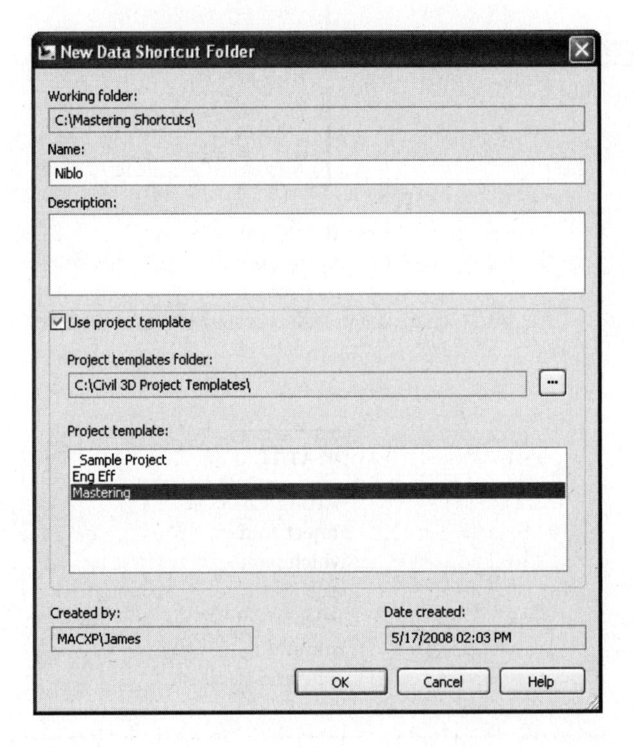

FIGURE 17.3
Creating a new shortcut folder (aka a project)

8. Select the `Mastering` folder from the list as shown, and click OK to dismiss the dialog. Congratulations, you've made a new Civil 3D project. Notice that the Data Shortcuts branch in Prospector now reflects the path of the Niblo project.

If you open Windows Explorer and navigate to `C:\Mastering Shortcuts\Niblo`, you'll see the folder from the Mastering project template plus a special folder named `_Shortcuts`, as shown in Figure 17.4. One common issue that arises is that you already have a project folder inside the working folder. This is typically done during some bidding or marketing work, or during contract preparation. If you already have a project folder established, it will not show up in Civil 3D unless there is a `_Shortcuts` folder inside it. To this end, you can manually make this folder. There's nothing special about it, except that it has to exist.

FIGURE 17.4
Your new project shown in Windows Explorer

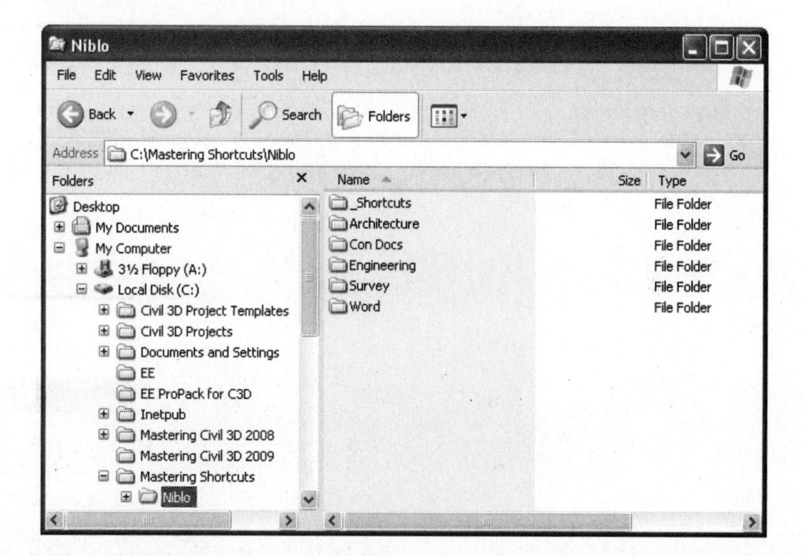

The *working folder* and *data shortcuts folder* are simply Civil 3D terms for a projects folder and individual projects. If you're familiar with Land Desktop, the working folder is similar to the project path, with various projects. Finally, remember that drawing files and working folders are not tied together. When you change drawings, the data shortcuts folder will not change automatically. You must right-click the Data Shortcuts branch and select a different data shortcuts folder.

ANOTHER WAY TO LOOK AT IT

One other option when setting up projects in Civil 3D with shortcuts is to set the working folder path to your particular project folder. Then you could have a CAD folder and within that would be the `_Shortcuts` folder, which would be the same for every project. This results in a folder structure like `H:\Projects\`*`Project Name`*`\CAD\_Shortcuts`, but some users may find this more useful and easier to manage, particularly if your company standards dictate that the top level of a project folder shouldn't include something like a `_Shortcuts` folder. Both methods are workable solutions, but you should decide on one and stick to it! We'll use the more conventional approach shown in the previous exercise throughout this text.

Creating Data Shortcuts

With a shortcut folder in place, it's time to use it. One difference between Vault and data shortcuts is that a drawing creating a shortcut for a given project can actually be stored anywhere. In spite of this, it's really best practice to keep the drawing files in the same rough location as your shortcut files, just to make things easier to manage. In this exercise, you publish data shortcuts for the alignments and layout profiles in your project:

1. Open the `Creating Shortcuts.dwg` file. This drawing contains samples of each shortcut-ready object as shown in Figure 17.5.

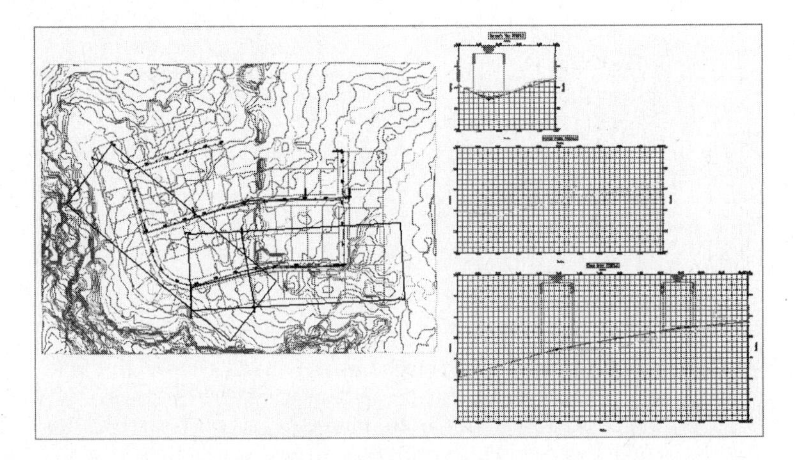

FIGURE 17.5
The Creating Shortcuts drawing file

2. In Prospector, right-click the Data Shortcuts branch, and select Create Data Shortcuts to display the Create Data Shortcuts dialog shown in Figure 17.6.

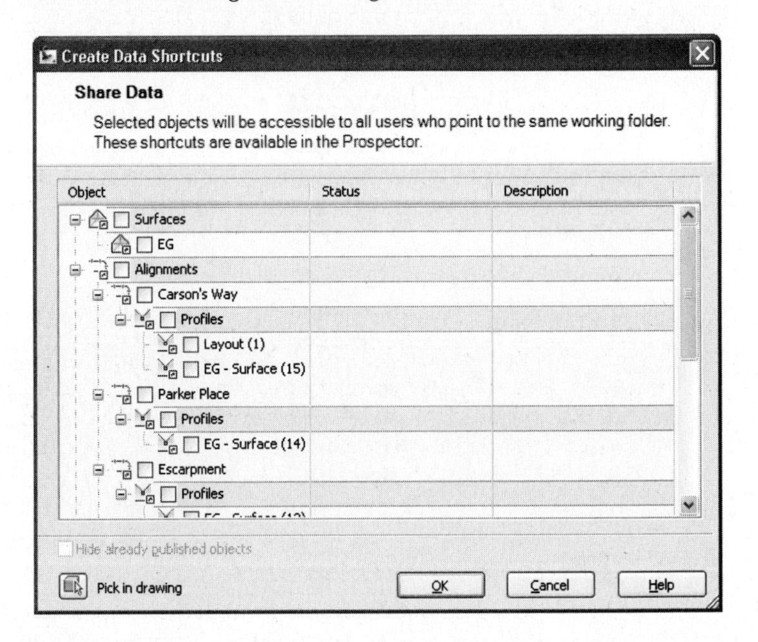

FIGURE 17.6
The Create Data Shortcuts dialog

3. Check the Surfaces and Alignments options, and the sub-items will all be selected. Leave the Pipe Networks and View Frame Groups unchecked. Note that the Profiles associated with each alignment are also being selected for publishing.

4. Click OK to dismiss the dialog and create the data shortcuts.
5. Expand the branches under the Data Shortcuts heading as shown in Figure 17.7, and you should see all of the relevant data listed. The listing here indicates that the object is ready to be referenced in another drawing file.

FIGURE 17.7
Data shortcuts listed in Prospector

The data shortcut mechanism changed greatly between the 2008 and 2009 releases, but it didn't change significantly in the 2010 release. In 2008, you had to manually manage the XML data reference files. In 2010, Civil 3D manages these files for you. They are stored in the magic `_Shortcuts` folder as individual XML files. You can review these XML files using XML Notepad, but it's worth noting that the first comment in the XML file is `PLEASE DO NOT EDIT THIS FILE!` In the past, some users found they needed to edit XML files to fix broken or lost references. This is no longer necessary with the addition of the Data Shortcuts Editor, which you look at later in this chapter.

Using Data Shortcuts

Now that you've created the shortcut XML files to act as pointers back to the original drawing, you'll use them in other ways and locations. Once a reference is in place, it's a simple matter to update the reference and see any changes in the original file reflected in the reference object. In this section, you look at creating and exploring these references, and then learn about updating or editing them.

MY SCREEN DOESN'T LOOK LIKE THAT!

Many of the screen captures and paths shown in this chapter reflect the author's working folders during the creation of the data. Your screen will be different depending on how you have installed the data, network permissions, and so on. Just focus on the content and don't let the different paths fool you — you're doing the right steps.

Creating Shortcut References

Shortcut references are made using the Data Shortcuts branch within Prospector. In this exercise, you'll create references to the objects you previously published to the Niblo project:

1. Open the `Creating References.dwg` file. This is an empty file built on the Extended template.

2. In Prospector, expand Data Shortcuts ➤ Alignments.
3. Right-click the Alexander Ave alignment and select Create Reference to display the Create Alignment Reference dialog shown in Figure 17.8.

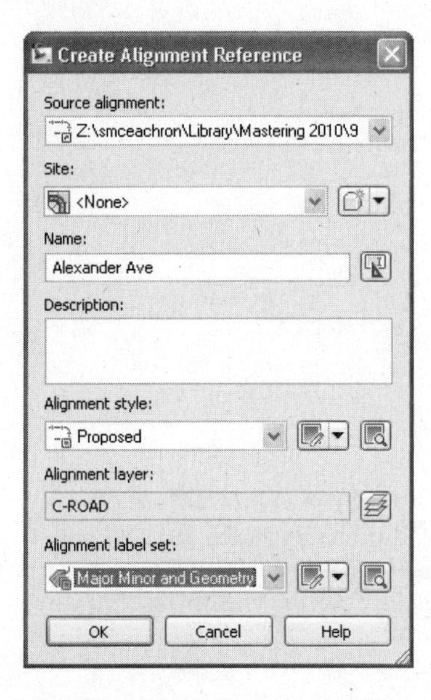

FIGURE 17.8
The Create Alignment Reference dialog

4. Set the Alignment Style to Proposed and the Alignment Label Set to Major Minor and Geometry Points as shown in Figure 17.8.
5. Click OK to close the dialog.
6. Perform a zoom extents to find this new alignment.
7. Repeat steps 3 through 5 for the Carson's Way, Parker Place, and Rose Drive alignments in the shortcuts list. When complete, your screen should look similar to Figure 17.9.

Each of these alignments is simply a pointer back to the original file. They can be stylized, stationed, or labeled, but the definition of the alignment cannot be changed. This is more clearly illustrated in a surface, so add a surface reference now:

1. Expand Data Shortcuts ➤ Surfaces.
2. Right-click the EG surface and select Create Reference to display the Create Surface Reference dialog.

WHAT'S IN A NAME?

The name in the XML file doesn't seem to have much effect on the creation of references; however, it does seem to have some effect on their maintenance. Some users have reported issues with broken references when they changed the name of an object during the Create Reference part of the process. No one seems to have a good feel for why it happens, but we don't recommend that you test it. Leave the name alone during the Create Reference step!

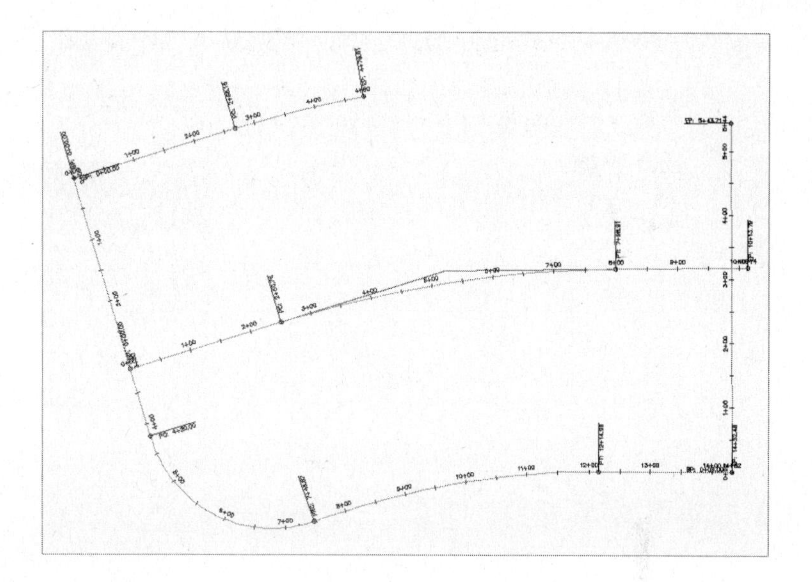

FIGURE 17.9
Completed creation of alignments references

3. Change the Surface Style to Contours 1 & 5 Background, and then click OK. Your screen should look like Figure 17.10.

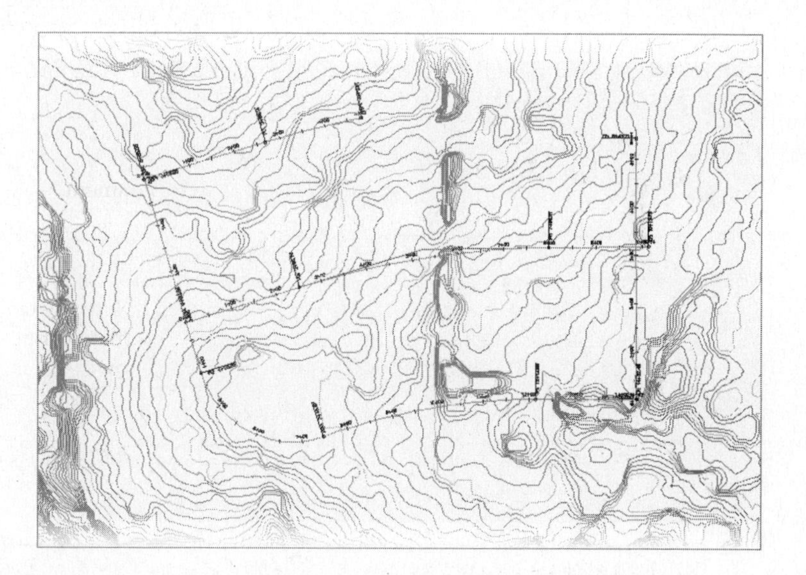

FIGURE 17.10
Data referenced surface and alignments

4. Expand the Surfaces branch of Prospector and the EG surface as shown in Figure 17.11.
5. Right-click the EG surface and select Surface Properties.

   Here are a couple of important things to note:

   - The small arrow next to the EG surface name indicates that the surface is a shortcut.

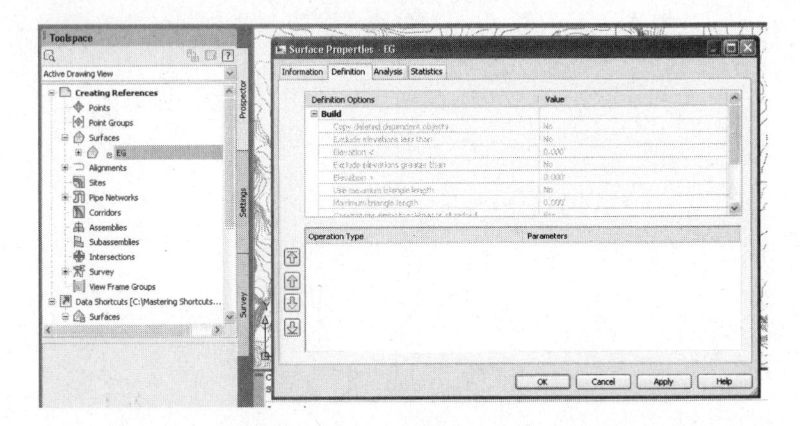

FIGURE 17.11
The EG surface in Prospector and the EG Surface Properties dialog

- There's no Definition branch under the EG surface. Additionally, in the Surface Properties dialog, the entire Definition tab is grayed out, making it impossible to change by using a shortcut.

6. Click OK to close the dialog.

To make a reference into a live object in the current drawing, simply right-click its name and select Promote. This breaks the link to the source information and creates an editable object in the current drawing.

Now that you've created a file with a group of references, you can look at how changes in the source drawing are reflected in this file.

Real World Scenario

Do I Need to Know This?

Yes, actually you do. Even if you are using Vault as your project-management scheme, there are times and situations where using data shortcuts is the only real solution. Let's look at a couple of cases where in spite of using Vault, we've used data shortcuts to pull the proverbial rabbit out of the hat.

When working with Vault, a drawing is typically attached to a particular project. Though this generally isn't a problem, it does limit your resources somewhat. When you have a multiphase project that has been divided in Vault, it is impossible to use the Vault mechanism to grab information from a drawing that is not part of the set project.

To get around this limitation, use a shortcut! Go into the file containing the desired object and export out a shortcut. Then, in the target file, you can import the data shortcut and reference the data accordingly. As the source file changes, the shortcut will update, keeping the two phases in synchronization.

Many firms just starting out with Civil 3D will try to put too much in one drawing. Since the program can handle it, why shouldn't they? This is fine until they run into a deadline at the end of the project. With no real sharing methodology in place, it becomes almost impossible to split up the work.

Even if you don't want to get into sharing data at that point in the job, you can use the data shortcut mechanism to break out individual pieces to other files. Once a shortcut has been created in the target file, right-click the object name in Prospector and select Promote as shown in the graphic.

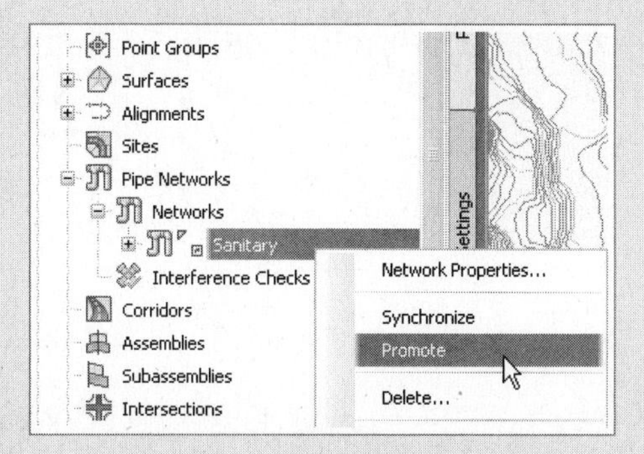

The act of promoting a shortcut breaks the link with the original data, and isn't a recommended practice, but sometimes the deadlines win. When it's time to get the project out the door, you do what works!

Updating and Managing References

As it is, if the reference were just a copy of the original data, you'd have done nothing more than cut and paste the object from one drawing to another. The benefit of using shortcuts is the same as with XRefs: when a change is made in the source, it's reflected in the reference drawing. In this section, you'll make a few changes and look at the updating process, and then you'll see how to add to the data shortcut listings in Prospector.

Updating the Source and Reference

When it's necessary to make a change, it can sometimes be confusing to remember which file you were in when you originally created a now-referenced object. Thankfully, you can use the tools in the Data Shortcut shortcut menu to very simply jump back to that file, make the changes, and refresh the reference:

1. In Prospector, expand Data Shortcuts ➤ Alignments.
2. Right-click Alexander Ave, and select Open Source Drawing. You can also do this by selecting the object in the drawing window and right-clicking to access the shortcut menu. The Open Source Drawing command is on the shortcut menu when a reference object is selected.
3. Make a grip-edit to Alexander Ave's northern end, dragging it up and further north, as shown in Figure 17.12.
4. Save the drawing.

   Once a change is made in the source drawing, Civil 3D will synchronize references the next time they are opened. In the current exercise, the reference drawing is already open. The following steps show you how the alert mechanism works in this situation.

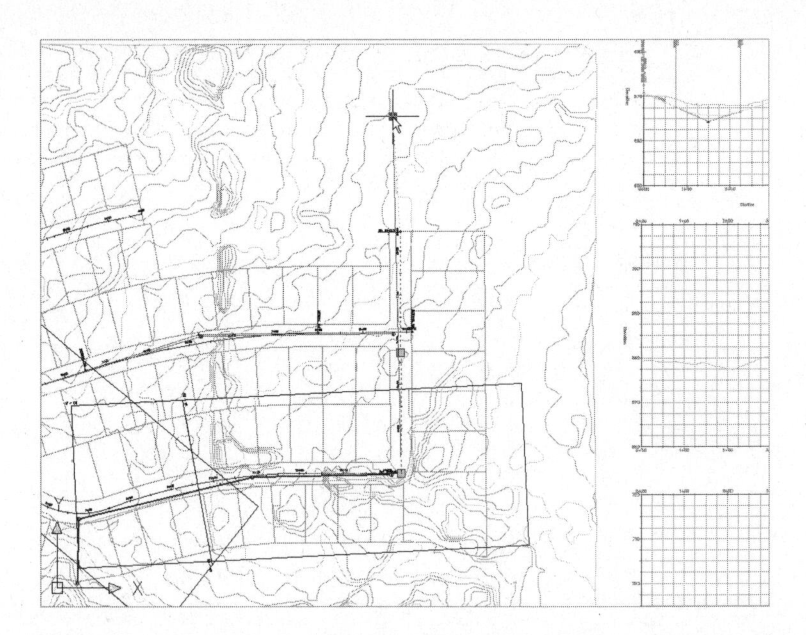

FIGURE 17.12
Grip-editing Alexander Ave

5. Use Ctrl+Tab on your keyboard to change to the `Creating References.dwg` file and make it active, or right-click `Creating References.dwg` in the Prospector and select Switch To. An alert bubble like the one shown in Figure 17.13 should appear. This may take a few minutes, but you can also expand Creating References ➤ Alignments ➤ Centerline Alignments in Prospector and you will see a series of warning chevrons.

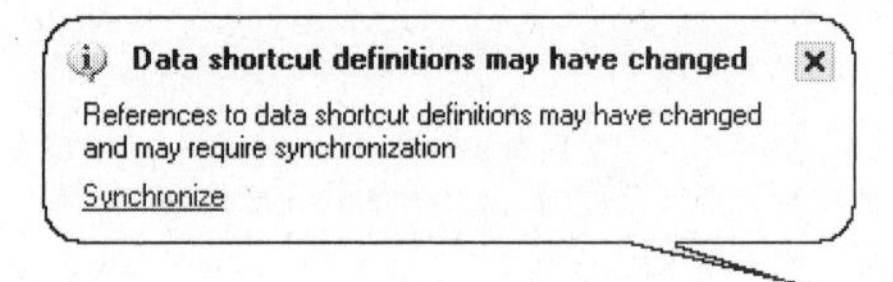

FIGURE 17.13
Data Reference Change alert bubble

6. Click Synchronize in the alert bubble to bring your drawing current with the design file and dismiss the Panorama window if you'd like. Your drawing will update, and the Alexander Ave alignment will reflect the change in the source. If you do not get the bubble, you can also select individual references within Prospector and right-click them to access the Synchronize command.

This change is simple enough and works well once file relationships are established. But suppose there is a change in the file structure of the source information and you need to make a change? You'll look at that in the next section.

Managing Changes in the Source Data

Designs change often — there's no question about that — and using shortcuts to keep all the members of the design team on the same page is a great idea. But in the scenario you're working with in this chapter, what happens if new, additional alignment data is added to the source file? You'll explore that in this exercise:

1. Return to the `Creating Shortcuts.dwg`.
2. Create a new alignment, named Carson Extension, extending from the end of Carson's Way to intersect with the now extended Alexander Ave. Your screen should look like Figure 17.14, with the new alignment circled.
3. Save the file.

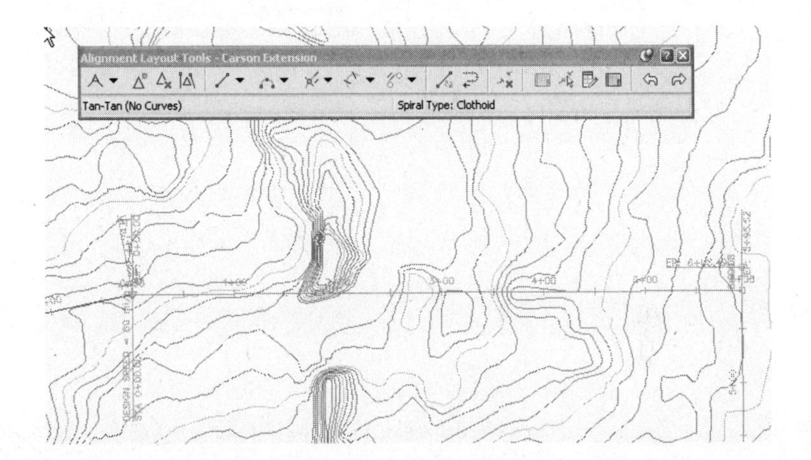

FIGURE 17.14
Carson Extension alignment drawn and completed

4. In Prospector, right-click Data Shortcuts and select Create Data Shortcuts to display the Create Data Shortcuts dialog.
5. Check the Hide Already Published Objects option in the lower left of the dialog to make finding the new object easier.
6. Check the Carson Extension alignment and click OK to dismiss the dialog.
7. Switch back to Creating References file, and add the Carson Extension alignment to your data shortcuts as in earlier examples.

By using shortcuts to handle and distribute design information, it's quite simple to keep adding information to the design as it progresses. It's important to remember that simply saving a file does not create new shortcut files for all of the Civil 3D objects contained; they have to be created from the Data Shortcuts branch.

FIXING BROKEN SHORTCUTS

One of the dangers of linking things together is that eventually you'll have to deal with broken links. As files get renamed, or moved from the `Preliminary` to the `Production` folder, the data shortcut files that point back simply get lost. Thankfully, Civil 3D includes a tool for handling broken or editing links: the Data Shortcuts Editor. You explore that tool in this exercise:

1. Open the file `Repairing References.dwg`. This file contains a number of references pointing to a file that no longer exists, and Panorama should appear to tell you that five problems were found. Close the Panorama window.
2. In Prospector, expand Repairing References ➤ Surfaces and you will see that EG has a warning chevron next to it.

3. Right-click EG and select Repair Broken References from the shortcut menu that appears to display the Choose the File Containing the Referenced Object dialog.

4. Navigate to the `Creating Shortcuts.dwg` file in the data set. Click OK to dismiss the dialog. An Additional Broken References dialog will appear as shown in Figure 17.15, offering you the option to repair all the references or cancel.

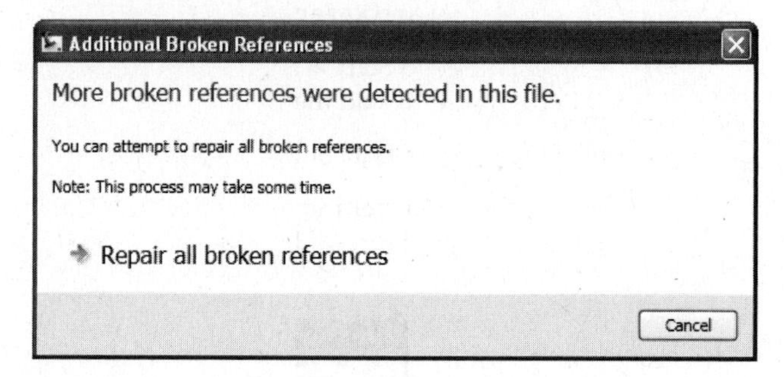

FIGURE 17.15
The Additional Broken References dialog

5. Click the Repair All Broken References button to close this dialog. Civil 3D will crawl through the file linked in step 4, and attempt to match the Civil 3D objects with broken references to objects in the selected drawing.

6. Perform a zoom extents, and your drawing should look like Figure 17.16, with a surface and four alignments.

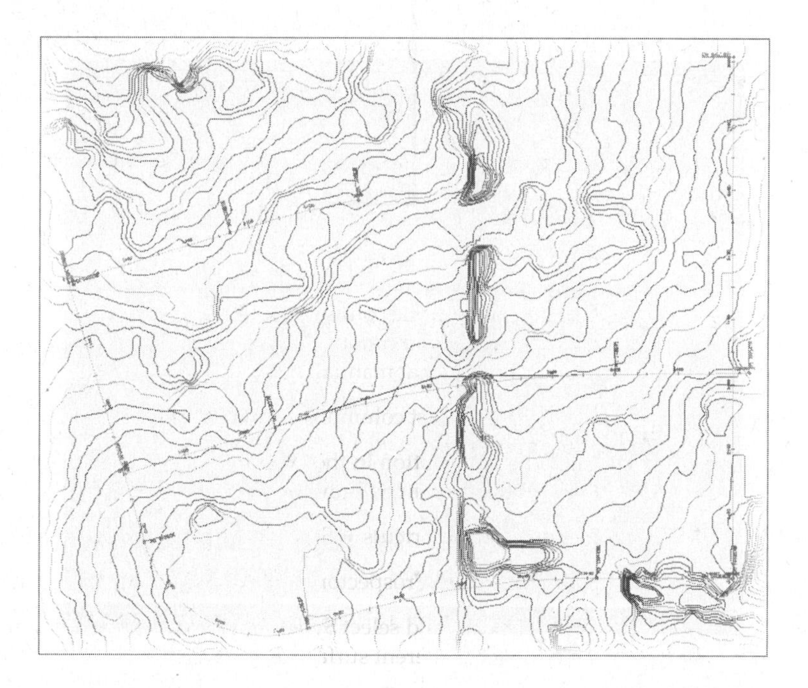

FIGURE 17.16
Repaired references within an older file

The ability to repair broken links helps make file management a bit easier, but there will be times when you need to completely change the path of a shortcut to point to a new file. To do this, you need to use the Data Shortcuts Editor.

THE DATA SHORTCUTS EDITOR

The Data Shortcuts Editor (DSE) is used to update or change the file to which a shortcut points. This can happen when an alternative design file is approved, or when you move from preliminary to final design. In the following exercise, you'll change the EG surface shortcut to point at a surveyed site surface instead of the Google Earth surface the shortcut file was previously using:

1. Make sure `Creating References.dwg` is still open.
2. In Windows, go to Start ➢ All Programs ➢ Autodesk ➢ Autodesk Civil 3D 2010 ➢ Data Shortcuts Editor to load the DSE.
3. Select File ➢ Open Data Shortcuts Folder to display the Browse for Folder dialog.
4. Navigate to `C:\Mastering Shortcuts\Niblo` and click OK to dismiss the dialog. Your DSE should look like Figure 17.17. (Your paths might be different from the author's.)

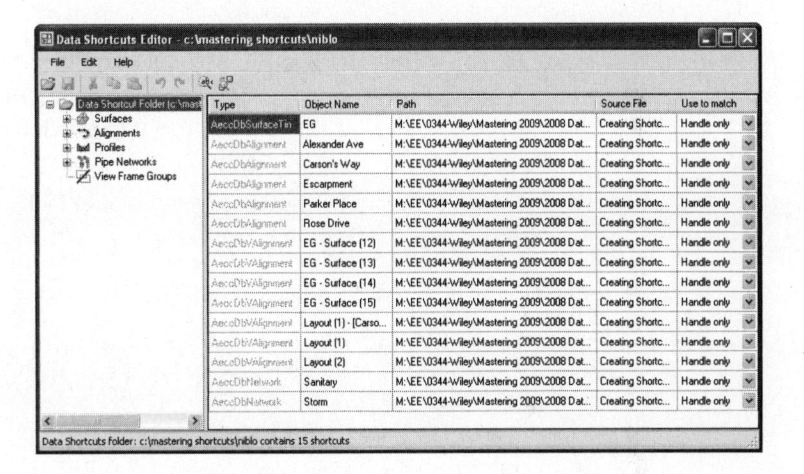

FIGURE 17.17
Editing the Niblo data shortcuts

5. On the left, select the Surfaces branch to isolate the surfaces.
6. Click the Source File column on the EG row, and type **XTOPO.dwg** as the Source file name. Unfortunately, you cannot browse. If you need to change the target file's path, you would need to change that manually as well.
7. Verify that the last column is set to Handle or Name.
8. Click the Save button in the DSE and switch back to Civil 3D. You should still be in the `Creating References` drawing file.
9. Expand Data Shortcuts ➢ Surfaces. Right-click EG and select Repair Broken Shortcut.
10. Scroll up within Prospector, and expand the Surfaces branch.
11. Right-click EG and select Synchronize. Your screen should look like Figure 17.18, showing a completely different surface as the EG surface.

The ability to modify and repoint the shortcut files to new and improved information during a project without losing style or label settings is invaluable. When you use this function though, be sure to validate and then synchronize.

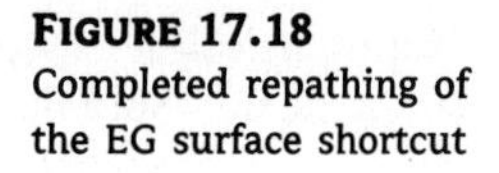

FIGURE 17.18
Completed repathing of the EG surface shortcut

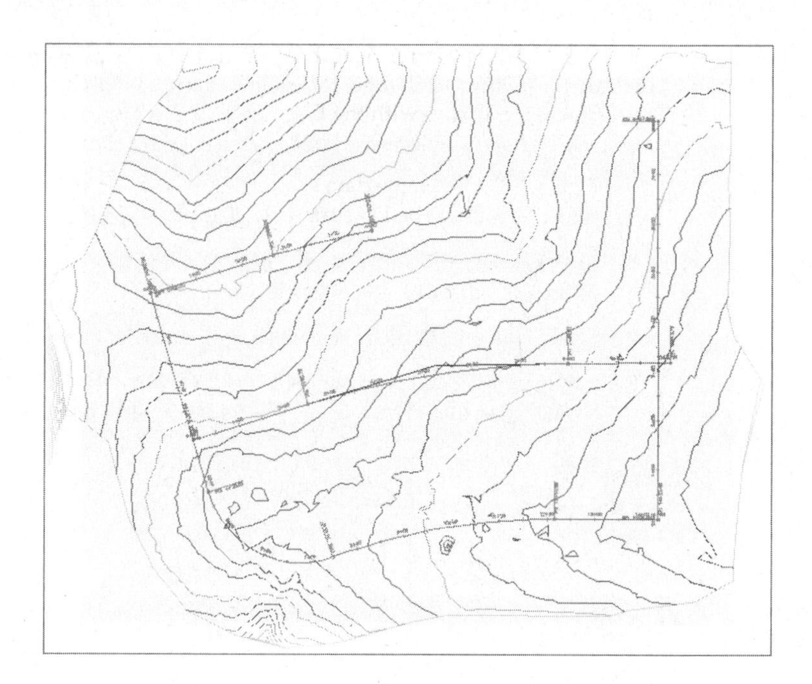

The Bottom Line

List the project elements that are available for sharing through data shortcuts and those that are not. The ability to load design information into a project environment is an important part of creating an efficient team. The main design elements of the project are available to the data shortcut mechanism, but some still are not.

Master It List the top-level branches in Prospector that cannot be shared through data shortcuts.

Create a data shortcut file from Civil 3D objects. Just creating a Civil 3D object isn't enough to share it using data shortcuts. You still have to go through the process of creating a shortcut file and exporting the linking information to allow other users to reference and analyze your data.

Master It Open the `Creating Shortcuts.dwg` file and create shortcut files for both of the pipe networks contained in that drawing.

Import a data shortcut and create references. Once a shortcut to a design element has been created, references can be made and managed. Team members should communicate that these files have been created and take appropriate steps so they are not overwritten with bad information. Creating and managing references is the most critical part of keeping project team members in sync with one another.

Master It Create a new drawing, and create references to both the EG surface and a pipe network.

Restore a broken reference and repath shortcuts to new files. During a design project, it's common for items to change, for design alternatives to be explored, or for new sources to be required. By working with the Data Shortcuts Editor, you have the power to directly manipulate what Civil 3D objects are being called by shortcuts and manipulate them as needed.

Master It Change the data shortcut files such that the file made in the previous Master It exercise points to the EG surface contained in the `Creating References` file instead of the XTOPO file.

Chapter 18

Behind the Scenes: Autodesk Data Management Server

Data sharing in Civil 3D can be accomplished with two methods: the data shortcuts reviewed in Chapter 17, ''Sharing the Model: Data Shortcuts,'' or via Vault and Autodesk Data Management Server (ADMS). Vault brings true project management into play, but it requires a fair amount of behind-the-scenes machinery. This chapter will look at the server component of this equation. ADMS handles much of the file interaction that data shortcuts require but brings its own set of criteria to the table.

In this chapter, you'll learn to:

- Recommend a basic ADMS and SQL installation methodology
- Create multiple vaults, users, and user groups
- Manage working folders

What Is Vault?

Vault is a document-management system initially created by Autodesk for the manufacturing industry. As Civil 3D evolved and a more robust project-management tool was required beyond the Data Shortcut mechanism in Civil 3D 2006, the Autodesk developers decided that Vault's concepts and mechanics could be applied across the board. The adoption of Vault as a platform-wide project-management system meant that each product team could take the platform tool and use it to make their product stronger. Autodesk created ADMS to be the solution for all the products. Let's first look at how ADMS operates at a high level and then explore some of the options in the creation of an ADMS server and the database on which it runs.

A SPECIAL NOTE ABOUT THIS CHAPTER AND THE NEXT

ADMS and Vault are specific and process oriented. In these chapters, each exercise will build on the previous one. Some figures show the use of a second machine to handle ADMS duties instead of installation to the local machine. The text will point out cases where this makes a difference in the images. Finally, the installation of ADMS requires Windows permissions and may require changes to your local Windows system configuration. The chapters attempt to note these areas; but in a corporate environment, please visit your systems administrator before making these changes.

ADMS and Vault

Civil 3D uses the Vault mechanism to handle standard document-management issues such as versioning and user rights, and also to handle creating and managing the links between project data items such as alignments, surfaces, and pipe networks. Vault is the front end to ADMS that most users see and use.

The basic breakdown is as shown in Figure 18.1. Data created on Vault clients is checked in to the project. This data is copied to the ADMS file store, and a record of the transaction is stored in the database. As another client requests that data, a lookup is made in the database, and the relevant information is pulled from the file store and checked out to the requesting client. ADMS handles the tracking of these transactions as well as security and version issues, and it acts as the management tool for both the file store and the database. It's important to note that in most installations, ADMS, the file store, and the database are all on one machine. They can be distributed to other machines, but that is an advanced installation and won't be covered in this text.

FIGURE 18.1
ADMS and Vault client schematic

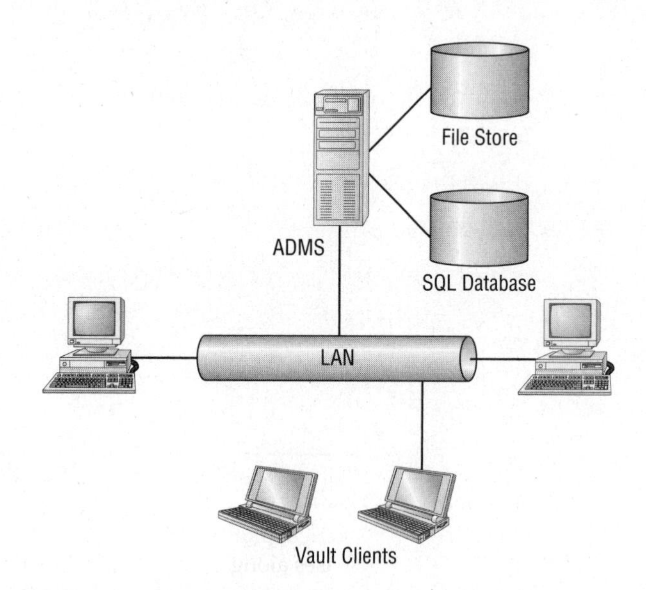

ADMS? VAULT SERVER? WHAT'S ALL THIS ABOUT?

If you're familiar with the 2007 release, you remember Vault Server. In 2008 versions, the naming convention changed, and Vault Server became ADMS. Now with 2010, you'll see both names thrown about. This book uses ADMS when talking about the server side and Vault when discussing the front end that Civil 3D uses to interact with the ADMS. (There's nothing like a straightforward naming convention.)

ADMS and SQL

The ADMS platform has two main options for installation. The first is based on Microsoft SQL Server Express 2005 (referred to as Express from now on) and is designed for individuals or small groups. This alternative is free and is included on the Civil 3D installation media. The second option is to use a full Microsoft SQL Server in one of its dedicated flavors: Workgroup, Standard, or Enterprise. Here are the main differences:

Express Limited to one processor and can access only 1GB of RAM. The database is limited to 4GB in size. Express allows a fair number of users, but the performance limits of the processor and RAM make using it beyond 30 to 40 users undesirable. Additionally, because Express is often installed on spare workstations or on a user's workstation, the Windows XP limit of 10 concurrent users comes into play. In spite of these limits, most small engineering firms find that using Express is a fine solution, especially when they're just getting into the process.

Workgroup, Standard, or Enterprise Edition Designed for larger firms. These incarnations typically sit on dedicated servers with multiple processors where the memory is limited only by the operating system (OS). The database size is unlimited, and the functional limits of the database depend more on hardware than software.

Because the nature of full SQL installations reaches beyond the scope of this text, this chapter and Chapter 19, "Teamwork: Vault Client and Civil 3D," use the default ADMS installation of Express. For more information on the various flavors of SQL Server, visit Microsoft's website as well as Autodesk's. Both offer great information pertaining to the SQL Server version selection.

How Much Server Do You Need?

The hardware requirements for ADMS are surprisingly low. Many firms are scared off of looking at Vault and ADMS because their trusted advisors in the hardware and software realm tell them they need to invest in huge dedicated servers built for launching the space shuttle. Don't buy it. SQL Express in a pilot-project environment or small workgroup can be run from an old AutoCAD workstation! The "second" machine that this chapter's ADMS is installed on is actually a virtual machine running XP; some engineering firms use virtual servers to run larger installations. You can also upgrade if you outgrow the Express edition. Don't let misinformation keep you from using Vault!

With the growth of the open software movement, one common complaint is that ADMS functions only on a Microsoft SQL platform. Although there have been numerous inquiries about using Oracle or MySQL databases along with ODBC connections, these aren't supported solutions. You *may* be able to get it to run, but you'll be in unsupported territory and will have no help available if something goes wrong. Stick with Microsoft on this one, even if you don't on other Enterprise solutions.

Installing ADMS

ADMS installation is included on the Civil 3D installation disc. The process is straightforward but requires that a few things be in place: Internet Information Services (IIS), .NET Framework 3.5 SP1, MSXML 6, and the Microsoft WSE 3.0 Runtime. Don't worry if none of these sound familiar. Most of these components are installed as part of the ADMS process, and if there are issues, you'll receive a warning during the preinstall system checks. In the following exercise, you'll perform a standard Vault install to familiarize yourself with the settings:

1. Load the Vault Server media, and launch `Setup.exe` if it does not launch automatically.
2. Click Install Products to move to the product-selection screen.

3. Select the Autodesk Vault 2010 (Server) option.
4. Read any informational notices and click Next as appropriate until you arrive at the License Agreement. Accept the License Agreement and click Next to open the Product and User Information dialog.
5. Enter the required personal information and select Next to open the Review – Configure – Install dialog.
6. Click the Configure button to display the Autodesk Vault 2010 (Server) installation options shown in Figure 18.2.

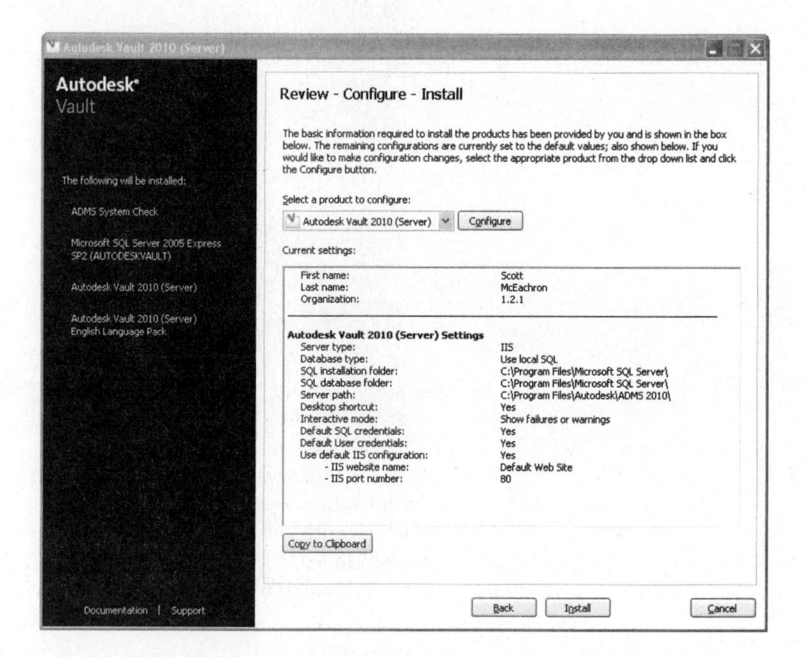

FIGURE 18.2
The ADMS Configuration screen

Because ADMS can be used in a variety of scenarios, ranging from the single-engineer office to the multinational design firm, a number of configuration options can be set during the installation. The configuration process allows for customization on the basis of your needs. In this case, you won't modify many settings because the default is for the small local user, but let's look at them. Figure 18.3 shows the first page of the ADMS installation options.

At installation, there are two pages of options and a confirmation page. The first page features the following major options:

Use Local Database or Remote Database SQL Remote SQL is an option only when you use it in conjunction with Autodesk Product Stream Replicator. This is a product that allows the distribution of project data across a WAN. Replicator is outside the scope of this text, but you can find more information at Autodesk's website or by contacting qualified consultants.

Specify Location of SQL Installation Folder; Specify Location of SQL Database; and Specify Location of Server Application You can change these locations, but doing so will make it more difficult for any support personnel to assist. Don't change these values unless you have a good reason.

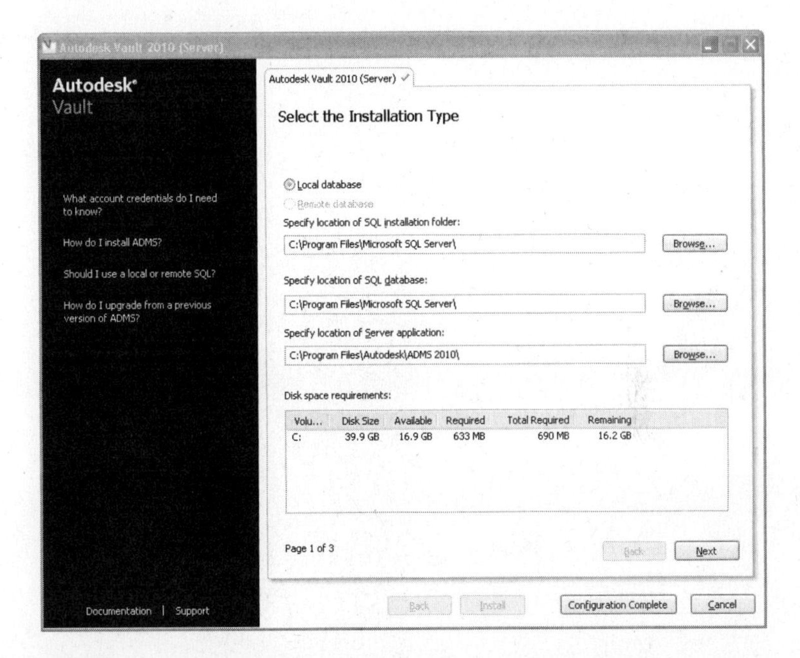

FIGURE 18.3
ADMS configuration options

Clicking Next takes you to the second page of the ADMS configuration (Figure 18.4), which deals with ADMS preferences and credentials. Within the preferences, the only option requiring explanation is Interactive Mode for 'Server System Check.' There are three options: always show interactive mode, show it only at failure, or show it only at failure and warnings. This setting determines the amount of feedback given to you during the installation process. As ADMS attempts to install, it offers a message to alert you if it finds issues or fails at any point. This is definitely worth watching.

If you're using another SQL installation or you're in a Replicator environment, you may need to sign in to the SQL database using something other than the default Autodesk security values. Enter this information here if needed. Again, don't change it unless you must.

Let's continue with the installation:

7. Change Interactive Mode for 'Server System Check' to Show Interactive Interface only for Failures or Warnings, and then click Next.

8. Click Configuration Complete.

9. Verify your settings on the screen, and click Install to begin copying the files to the local machine. The ADMS install process stops midway through to give the system readiness report shown in Figure 18.5.

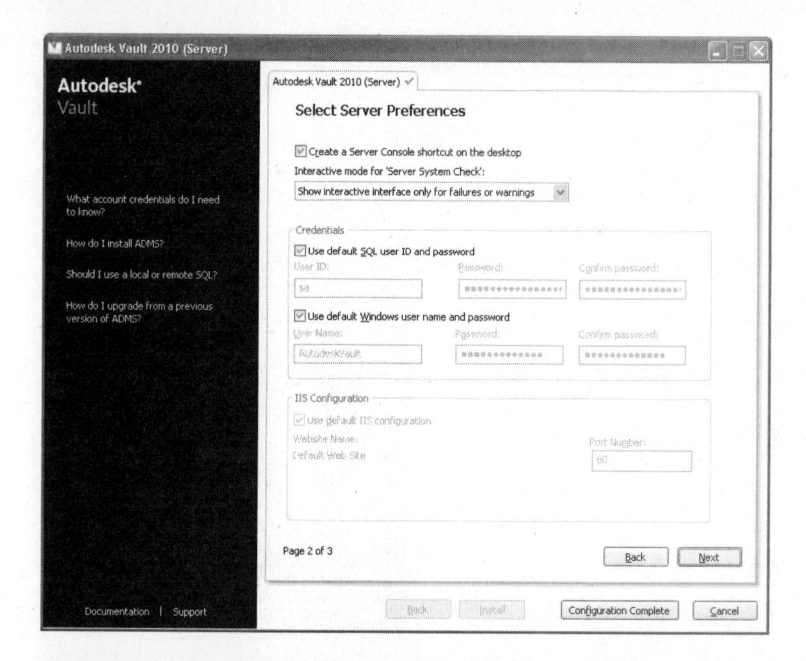

FIGURE 18.4
More ADMS configuration options

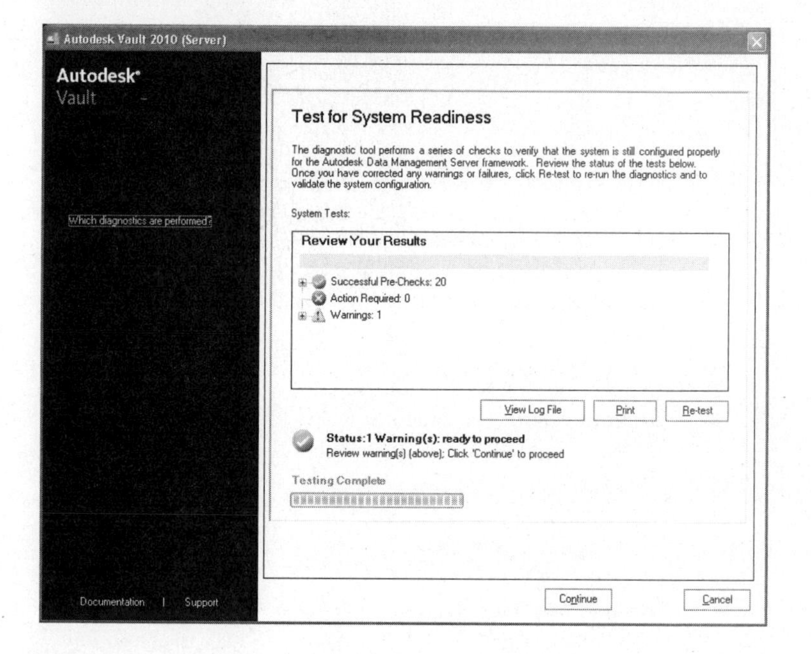

FIGURE 18.5
ADMS Installation and System Readiness results

10. Verify that no actions are required and that the warnings shown are acceptable. A large number of issues are possible here — be sure to read and understand them. The ADMS documentation is quite good; refer to it for more detailed troubleshooting.
11. Click Continue if the results are acceptable.
12. Get a cup of coffee. The installation can take a few minutes. You may have time to run to Starbucks, depending on your system speed.
13. When the install completes, the window shown in Figure 18.6 appears. Deselect the View the Autodesk Vault 2010 (Server) Readme check box, unless you'd like to read it now.

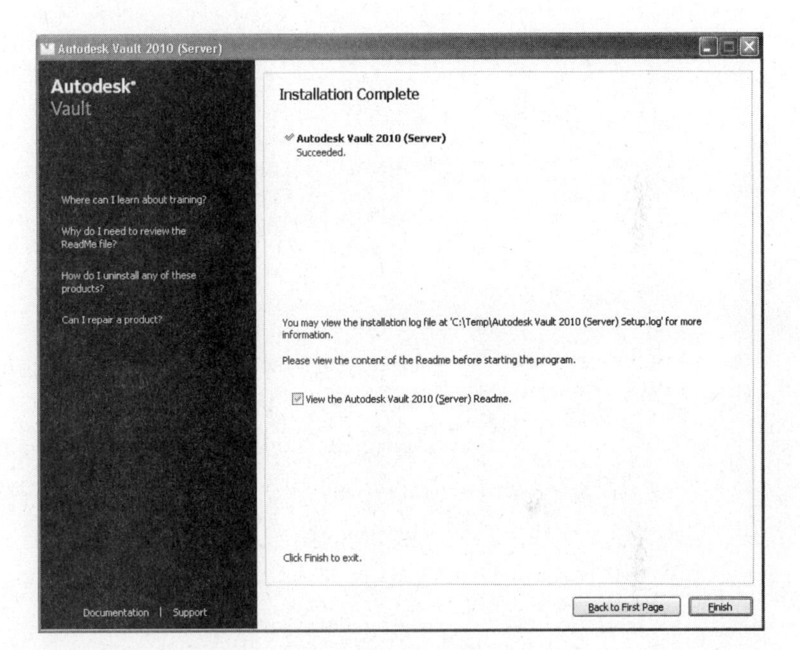

FIGURE 18.6
Completed ADMS installation

14. Click Finish to exit the installation program.

ADMS is now installed and ready to manage all of your Civil 3D projects.

OPENING THE DOORS TO ADMS

At its heart, ADMS uses standard Internet protocols to communicate, which is why IIS is a requirement and part of the system check. This change to your OS basically creates a web server that handles much of the communication between ADMS and the Vault clients. This is great when you're installing on a server; servers are typically set to perform this communication anyway. In the case of a small office or a test case like you're building here, however, XP is often used as the host OS, and the Windows XP Firewall that came with SP2 will probably get in the way. You'll take care of that now:

1. On the ADMS computer, select Start ➤ Control Panel ➤ Security Center to open the Windows Security Center dialog.

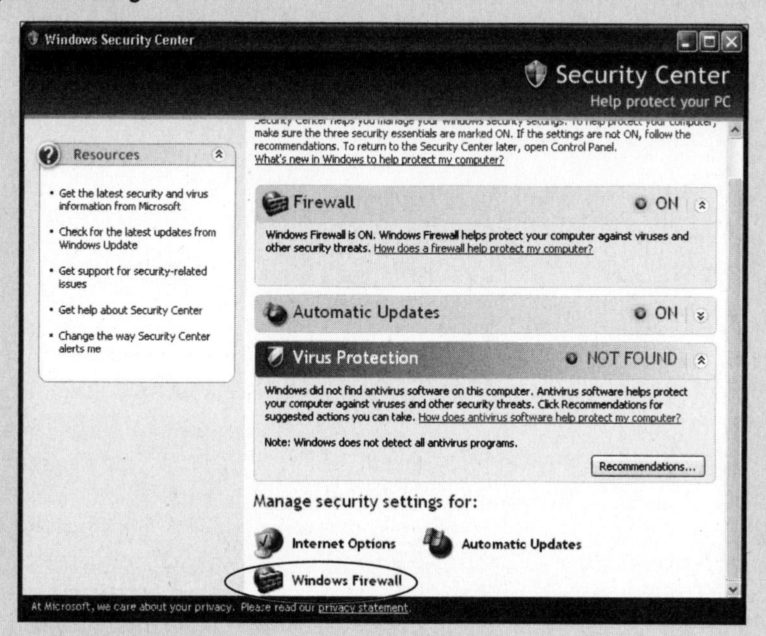

2. Click the Windows Firewall option to open the Windows Firewall dialog.
3. Switch to the Exceptions tab.
4. Click the Add Port button to display the Add a Port dialog. Enter **ADMS** in the Name text box and **80** in the Port Number text box.
5. Click OK once to close the Add a Port dialog and again to close the Windows Firewall dialog.

ADMS should now be accessible from other machines on your network. You'll verify that using Vault Explorer later in this chapter.

Managing ADMS

Two components that allow interaction with ADMS are ADMS Console and Vault Explorer. These programs focus on different management tools within the system, so the next few sections look at them independently.

ADMS Console

ADMS Console is only installed during the ADMS installation. This program manages the ADMS and starts and stops Windows services behind the scenes, creates vaults, and performs data auditing at the database level. ADMS Console is a complex program, and many of its functions fall outside the scope of this book. In this exercise, you'll look at the pieces involved in getting a Civil 3D vault and project up and running:

1. Launch ADMS by clicking the desktop icon or by selecting Start ➤ All Programs ➤ Autodesk ➤ Autodesk Data Management ➤ Autodesk Data Management Server Console 2010.
2. The program presents a login dialog when launched. The default settings are a username of Administrator and no password. You'll use these settings for this book, but you should change them before you go live with ADMS. Allowing normal users access to the Vault Administrator isn't a good security policy.

WINDOWS AUTHENTICATION?

In 2009, Autodesk added the ability to manage ADMS accounts using Active Directory authentication. To use this feature, you'll have to upgrade to a product called ProductStream. Talk with your reseller if you need to learn more about this advanced option.

3. Click OK to log in. You're presented with the screen of the active ADMS, as shown in Figure 18.7. Note that the text *BETABOX* near the top center refers to the name of the host computer. Your install will be different.

FIGURE 18.7
The ADMS 2010 Console screen

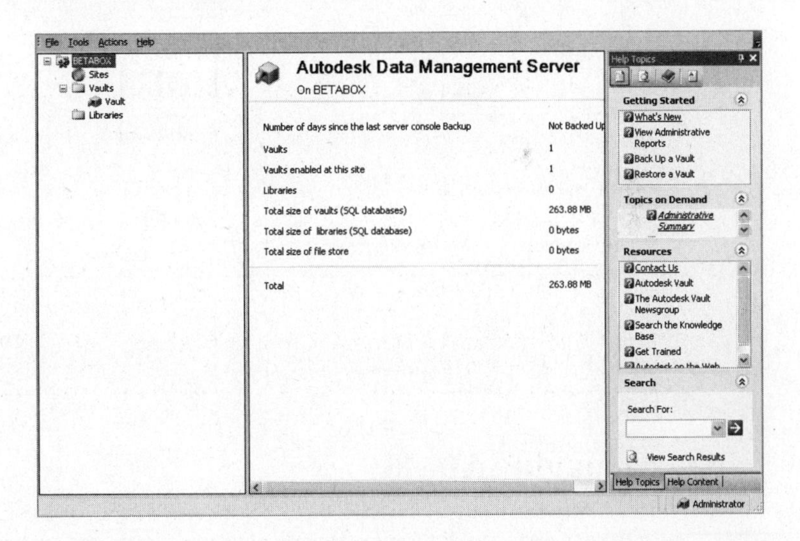

The main things to note on this screen are the Total Size of Vaults and Total Size of File Store values. When the total size of the vaults gets near 3GB, it's time to consider moving to a full SQL version instead of Express. When the server itself begins to complain about file storage or disk space, look at Total Size of File Store: this is the storage being used by record copies of all the files checked in to the ADMS.

When the total size grows to a level that begins to make the IT folks unhappy, it's time to purge the vault. You can do so manually or with scripts. The operation is too complex to cover here; refer to the excellent help and online resources available for dealing with ADMS and vault maintenance. In ADMS Console, search Help for *backup*, and you'll be headed in the right direction.

CREATING AND MANAGING VAULTS

In the left pane of Figure 18.7 is an Explorer-like interface to help you navigate through the various vaults on the ADMS machine. Sites are only involved with Replicator-based installations, and Civil 3D doesn't use libraries, so these exercises ignore them. During installation, a vault is created in the ADMS. This default vault is named Vault, which can lead to some confusion; so, in this exercise, you'll create another vault:

1. Select Vaults in the left pane.
2. Select Actions ➤ Create to display the Create Vault dialog, shown in Figure 18.8.

FIGURE 18.8
Creating a new vault

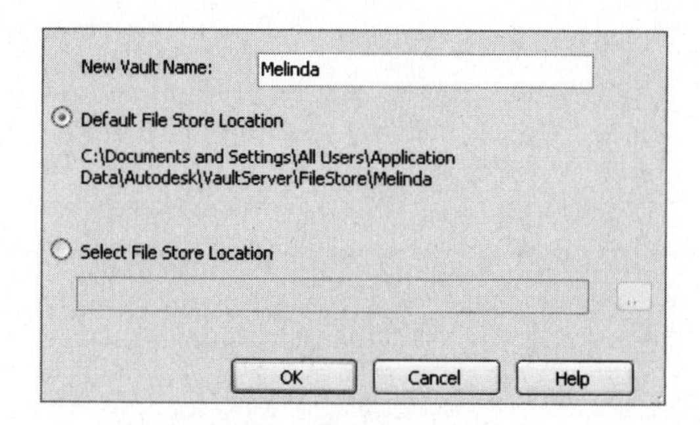

3. In the New Vault Name text box, enter the name **Melinda**.

MELINDA?

Yes, Melinda. When you're creating vaults for your office, you can name them anything you like. It doesn't matter much as long as you and the other users know which vault is which. Common practice involves naming vaults for office locations, clients, or years. It's entirely up to you.

---

4. Click OK, and wait for a confirmation message that says "Vault 'Melinda' was successfully created."
5. Click OK again to dismiss the message. Melinda now appears in the Vaults collection in the left pane.
6. Select Melinda to display the summary pane in the middle of the console, as shown in Figure 18.9.

The information displayed in the summary pane can help you determine the strategy for purging or defragmenting this vault. Purging is required only when the File Store Size value becomes unacceptable to the IT staff that allocates drive space. The need to defragment the vault database is indicated by a message at lower right. This area is circled in Figure 18.9, indicating that defragmentation isn't recommended at this time. This makes sense, because you just created the vault. Vault-specific tasks are available by right-clicking the vault name in the left pane and selecting from the context menu.

FIGURE 18.9
Summary of the newly created Melinda vault

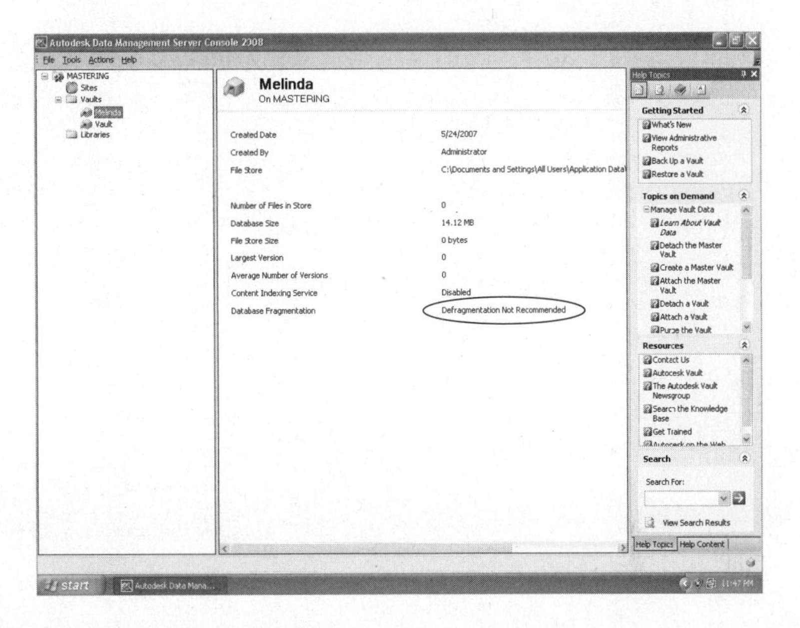

CREATING AND MANAGING USERS

One of the primary functions of implementing ADMS and Vault is to create a log of changes and version tracking within the Civil 3D environment. To perform these types of data audits, transactions must be tied to individual users. In this exercise, you'll make a series of users so that you can use them in other exercises and in Chapter 19, when you'll interact with the vault from within Civil 3D. Follow these steps:

1. From the ADMS Console, select Tools ➤ Administration to display the Administration dialog.
2. Select the Security tab, and click Users to display the User Management console.
3. Click the New User button to display the User Profile dialog, shown in Figure 18.10.
4. Enter the user information shown in Figure 18.10 for the new user Jolene. Enter **jolene**, all lowercase, as the password.
5. Click Roles to display the Add Roles dialog.
6. Check the Document Editor (Level 2). You want all Civil 3D users to be able to check files in and out of the vault, so they need to be document editors. Click OK.

LEVEL 1 VERSUS LEVEL 2

ADMS with 2010 provides two editor roles. Users assigned the role of Document Editor Level 1 can't delete files or folders, but users designated as Document Editor Level 2 can. This can be a handy differentiation if you have some users who are fond of cleaning up folders without thought; but you'll generally use the Document Editor Level 2 role.

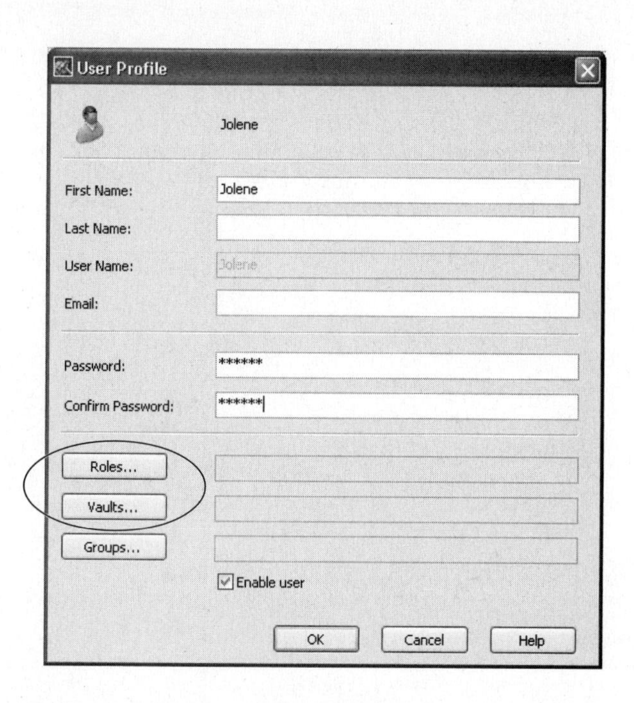

FIGURE 18.10 Creating a new vault user

7. Click Vaults, check the Melinda vault, and then click OK.
8. Click OK.
9. Repeat the process, and create a user named Jim as a Document Editor Level 2 in the Melinda vault.
10. In the User Management dialog, select View ➢ By Vault to display the User Management screen shown in Figure 18.11.

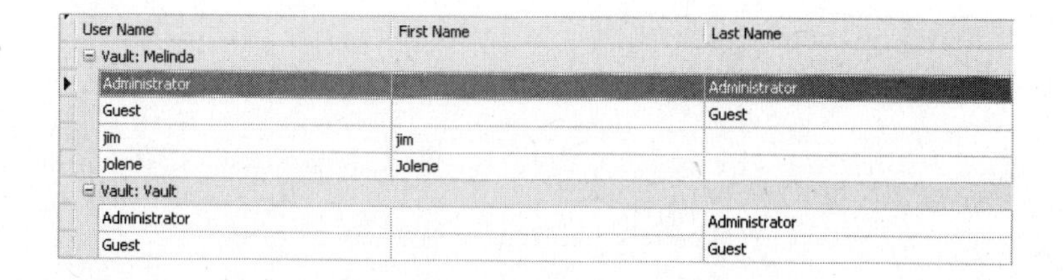

User Name	First Name	Last Name
Vault: Melinda		
Administrator		Administrator
Guest		Guest
jim	jim	
jolene	Jolene	
Vault: Vault		
Administrator		Administrator
Guest		Guest

FIGURE 18.11 Users created and assigned to roles and vaults

11. Select File ➢ Exit to close the User Management dialog and then Close to close the Administration dialog.

ADMS isn't the place to try to manage security. All too often, a user will leave the office for what's expected to be a short period and then be gone all day. In this situation, you should make it easy for one of their peers to check a drawing back in without involving the full IT staff in changing passwords or other things. For this reason, vault passwords are commonly simple. Windows or network passwords should be your security method.

CREATING GROUPS

Civil 3D can prevent certain users from accessing certain folders within the project. It's common for surveyors and engineers to limit the amount of access the other group has to their information. In the case of items such as plats, it can be a legal issue. Instead of controlling access by individual user (not bad with 10, inconceivable with 1,000), you can use groups to handle security levels. In the following exercise, you'll make a few groups to help manage your design teams:

1. Select Tools ➤ Administration to display the Administration dialog.
2. Click Groups to display the Groups dialog.
3. Click the New Group button to display the Group dialog, shown in Figure 18.12.

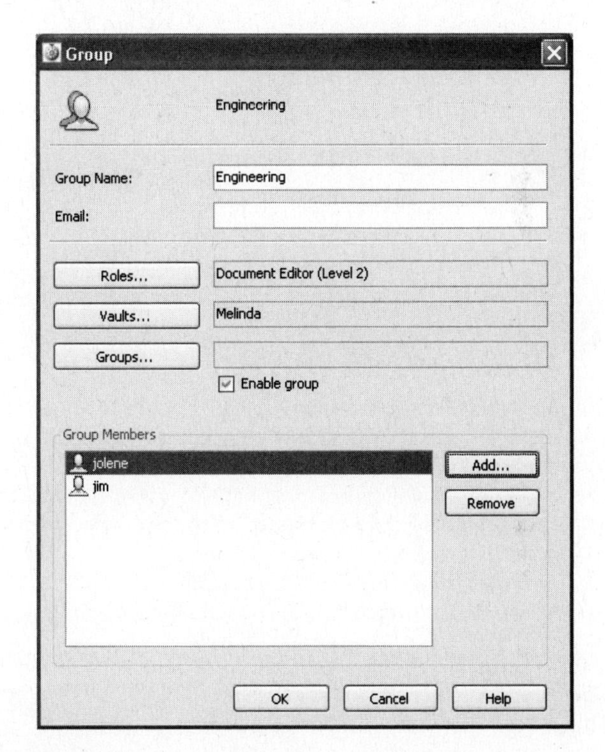

FIGURE 18.12 The completed Group dialog

4. Enter **Engineering** for a group name.
5. Click Roles, and select Vault Editor as you did with the users earlier.
6. Click Vaults, and select Melinda as you did with the users earlier.
7. Click Add to select members in the Add Members dialog, shown in Figure 18.13.
8. Select Jim, and click Add.
9. Select Jolene, and click Add. You can also use Ctrl+click to select multiple users at once.
10. Close the dialogs to return to the ADMS Console.

FIGURE 18.13
The Add Members dialog

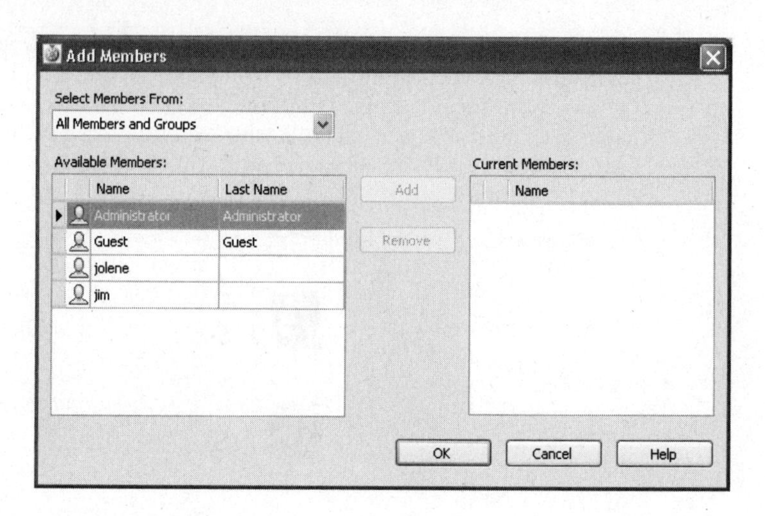

The creation of a single group will suffice for now. Most ADMS computers are managed by IT staff, and the CAD-management group may or may not have access to the console. For this reason, you'll finish building your User and Groups objects via the control that is installed with Civil 3D: Vault.

Accessing Vaults via Vault

Most user interaction with ADMS happens through the Vault program. In addition to controlling users and groups, Vault lets you perform other file-level management and review. This section looks first at user administration and then file management.

Vault is installed with Civil 3D, if you selected it during the initial installation. If not, you can add it by using Add/Remove Programs. Once Autodesk Vault is installed, you should have a desktop icon for Autodesk Data Management Server Console 2010.

Logging in to the vault is where the firewall issue can rear its head in a small-office installation, so you need to verify your connection before you go any further:

1. Double-click the Autodesk Vault 2010 icon on your desktop, or select Start ➤ All Programs ➤ Autodesk ➤ Autodesk Data Management ➤ Autodesk Vault 2010.
2. At Vault startup, you see a Welcome screen similar to Figure 18.14. The three large buttons on the left are short help files that point you in other directions. Click Log In at lower right to display the Log In dialog.
3. In the User Name text box, enter **Administrator**. The password is blank for now.
4. In the Server text box, enter the name of your ADMS computer, or your local host if you have installed it locally.
5. Click the ellipsis button circled in Figure 18.15 to bring up the Databases dialog. At this point, if all is well with the world, you'll see both Melinda and Vault listed as options.
6. Select Melinda, and click OK to return to the Log In dialog.
7. Click OK to log in.

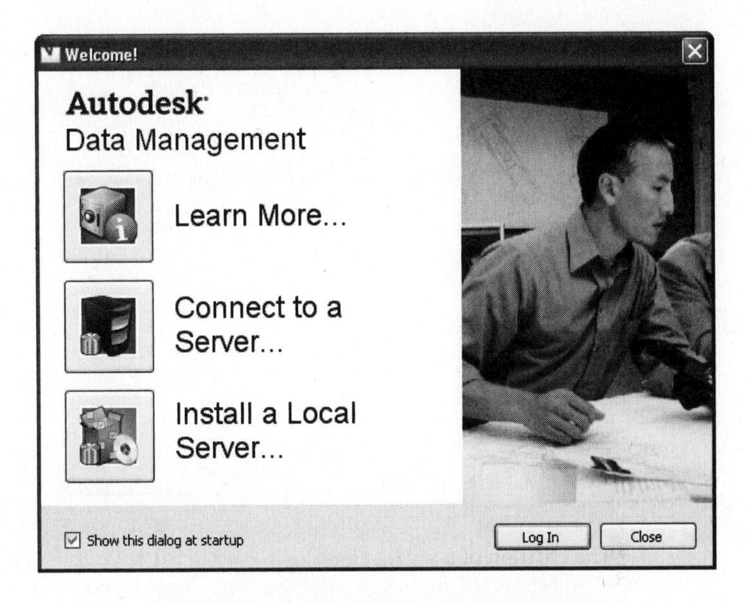

FIGURE 18.14
The Vault Welcome splash screen

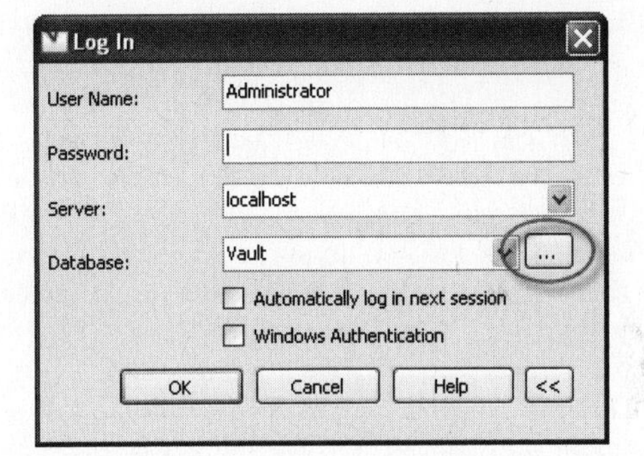

FIGURE 18.15
The Vault Log In dialog

LOGGING IN AUTOMATICALLY

Figure 18.15 shows an Automatically Log In Next Session option that you don't select. Because you're logging in as the vault administrator, this probably isn't an option you should leave on. It opens a security hole, because anyone with access to the computer can log into Vault with full privileges. For most day-to-day users, that isn't a problem: it's a good suggestion to have users select this option on their personal workstations so they don't have to go through the login process every time. The same option is available when logging in via the Civil 3D interface.

Vault looks similar to the console application, but because it's focused on one vault instead of the server, you can only see information about a single vault and the project it contains. Managing users and groups is similar to the methods employed in the console, so the steps aren't repeated here. The next section will look at vault management through the Vault interface.

Vault Management via Vault

Most of your daily interactions will be with vaults, not the ADMS. Using Vault from any station on the network, you can handle common management tasks such as managing users, project folders, and reviewing file versions. This section will examine a couple of important options in the Vault interface, explore how to reach the same user and group functions you did in the ADMS Console, and discuss Vault working folders.

Vault Options

Beyond the database users and privilege controls, a number of options are available that dictate how users interact with data in the vault. Vault options control many of these options, so let's look at what is available. Figure 18.16 shows the Options dialog, which you access by selecting Tools ➢ Options.

FIGURE 18.16
The Options dialog with two common options selected

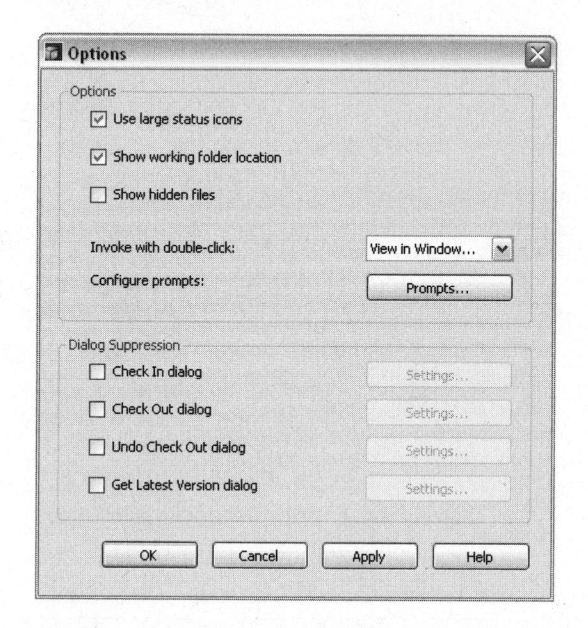

The only option you need to concern yourself with right now is Show Working Folder Location. Toggling this option displays the path of the working folder (discussed in the next section) in the head of the Vault pane. The other options for dialog suppression can be set to let you blow past some warnings and dialogs while working in a vault environment. Once you're comfortable with the Vault interface, you can consider shutting down some of these options, but it's not critical.

Vault Administration and Working Folders

The Administration panel in Vault is similar to the ADMS Console, but it's not exactly the same. Because Vault and ADMS are used with a variety of programs, some options are available that don't make complete sense in the Civil 3D market. Most users won't use the Enforce Unique File Names, Status, or Disable Check In of Design Files options. However, there is one option that is critical: Define Working Folder Options. Understanding the working folder is crucial to understanding how Civil 3D with Vault works.

Figure 18.17 shows that the working folder acts as an intermediary between the ADMS and the Vault client. When a client requests a file from Vault, a query is made to the SQL database to check the status of that file (available for checkout, protected, and so on). If the request is allowable, then the file is copied to the working folder from the ADMS file store. The file store records that the file is checked out to the requesting user. The user can work with that file in the working folder as long as they like — opening, closing, making changes, saving as needed, all the normal tasks. The working folder essentially acts as a local copy of project files. Changes aren't reflected in the project until the files are checked back into the Vault.

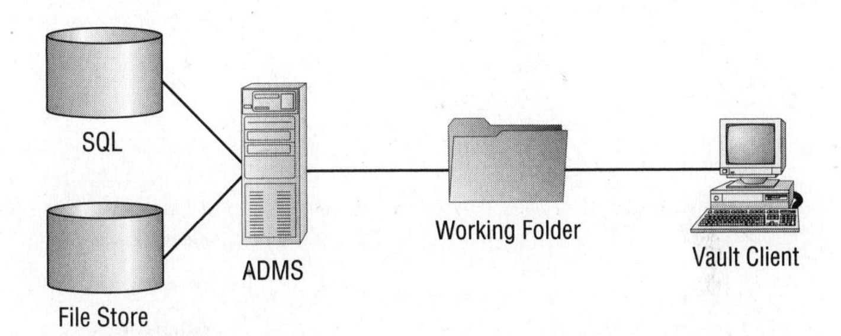

FIGURE 18.17
ADMS and working folder schematic

At the end of the day or when a critical part of the design has changed, the user checks the file back in. This check-in triggers the copying of the file in the working folder to the ADMS file store, and a new record is added to the SQL database. In this exercise, you'll set the options:

1. Launch and log into the Melinda vault.
2. Select Tools ➢ Administration to display the Administration dialog.
3. Switch to the Files tab (Figure 18.18).

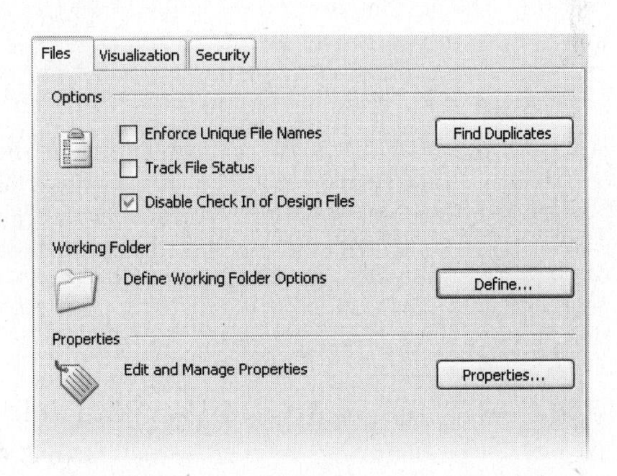

FIGURE 18.18
The Files tab in Vault's Administration dialog

4. Click Define in the Working Folder section to display the Working Folder Options dialog, shown in Figure 18.19.

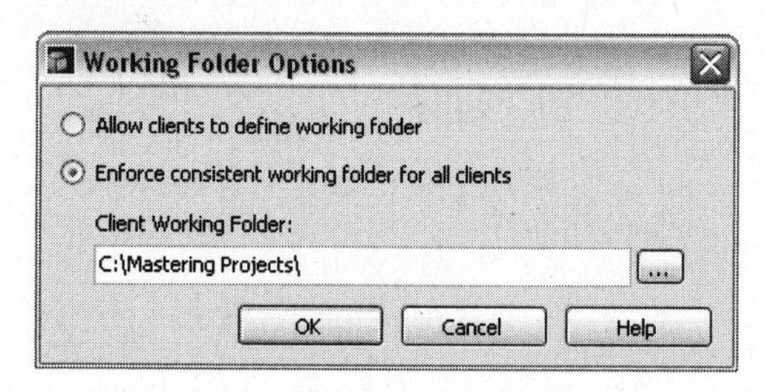

FIGURE 18.19
Setting the working folder

5. Select the Enforce Consistent Working Folder for All Clients radio button.
6. Set the path to `C:\Mastering Projects\`, as shown in Figure 18.19. This folder may not exist; typing it in will create it.
7. Click OK to close the dialog.
8. In the Administration dialog, change to the Security tab. Note that the options here are the same as they were on the console. Any user with a role of vault administrator can add users via this panel.
9. Close the Administration dialog.

There are two schools of thought about setting the working folder. The first is to set it as in this process, letting each user have a local path to aid in speedy operations. The other idea is to set all users to point to a shared working folder, creating a common project repository. This approach creates a project structure similar to what more firms have in place prior to adopting ADMS; this is the authors' preferred methodology.

It's important to note that a change in working-folder enforcement kicks in only when a user logs out and then logs in again. When a login occurs after the enforcement change, users receive an alert dialog that the path has been set.

Vault's administration panel is similar to ADMS because it's still dealing with the database side at heart. Chapter 19 looks at some other Vault Explorer functions.

VISUALIZATION? IT'S REALLY TALKING ABOUT DWF FILES.

Prior to the 2009 release, this functionality was handled as a DWF Attachment option. Many users like to turn off the creation of DWF files during the check-in of Civil 3D files. To do so, switch to the Visualization tab in the Administration dialog, click the Define button, and deselect the Enable Visualization File Attachment Options check box. DWF averted!

The Bottom Line

Recommend a basic ADMS and SQL installation methodology. The complexity of corporate networks makes selecting the right combination of ADMS and Microsoft SQL software part of the challenge. Understanding the various flavors of SQL Server 2005 at a basic level, along with some limitations of the ADMS, will aid in selecting the right combination for your own needs.

Master It For each of the following descriptions, make some general recommendations for a general setup for their ADMS and SQL needs — for example, a 30-person engineering firm with two offices that rarely share project data. The solution might be a pair of SQL Express machines, one for each office, running a flavor of Windows Server to get around any limits on concurrent connections.

1. A 100-person engineering firm with one office.
2. A 15-person firm with three offices equally split. They share data constantly.
3. A 5-person engineering firm with one office.

Create multiple vaults, users, and user groups. The enterprise nature of ADMS means it can handle multiple sets of data and user inputs at once. By creating multiple vaults to help manage data, you can make life easier on end users. Creating a login to the ADMS for each user lets you track file transactions, and creating groups makes security easier to handle.

Master It Create a new vault titled EE; add Mark, James, Dana, and Marc as users to this new vault. Additionally, make Marc a vault consumer, Mark and Dana vault editors, and James an administrator. Add an architecture group. Marc is the only architect; add the other new users to the engineering group.

Manage working folders. The working folder acts as your local desktop version of the project files. Drawings are saved and edited in the working folder until checked back into the Vault. Enforcing the working folder creates a constant path for XRef information and makes support easier.

Master It Set the working folder for the new EE vault to `C:\EE Projects\`. (Be sure to set this back to `C:\Mastering Projects` before you begin Chapter 19.)

Chapter 19

Teamwork: Vault Client and Civil 3D

Now it's time to pull it all together. Every chapter to this point has focused on individual tasks or design elements. This chapter focuses on using Vault to assemble the pieces into a complete set of plans. Although you won't create every single note on every sheet, you'll look at the various plans and how they come together. You'll learn about a potential workflow and then explore the mechanics of populating Vault with design data for sharing, handling changes and updates, and finally, tracking and restoring versions.

As you work through this chapter, remember that the puzzle has three pieces: Vault (Server), which was covered in the last chapter; the Autodesk Vault (aka Vault Explorer) application; and the vault object that stores all the data. This chapter uses the broader term *ADMS* for the server side and *Vault* for the application; if you get confused, look at the context, and things should work out.

In this chapter, you'll learn to:

- Describe the differences between using Vault and data shortcuts
- Insert Civil 3D data into a vault
- Create data references
- Restore a previous version of a design file

Vault and Project Theory

When users begin to look into using Vault as their project-management methodology, they may have a large number of questions. Chapter 18, "Behind the Scenes: Autodesk Data Management Server," discussed some of the infrastructure and setup concerns. This section focuses on the two big questions that remain: Why use Vault instead of data shortcuts? And, how does your team come together with Vault?

A SPECIAL NOTE ABOUT THIS AND THE PREVIOUS CHAPTER

ADMS and Vault are specific and process oriented. In these chapters, each exercise will build on the previous one. Some figures show the use of a second machine to handle ADMS duties instead of installation to the local machine. The text will point out cases where this makes a difference in the images. Finally, the installation of ADMS requires Windows permissions and may require changes to your local Windows system configuration. The chapters attempt to note these areas; but in a corporate environment, please visit your systems administrator before making these changes.

Vault versus Data Shortcuts

For many people, this choice is the sticking point. They look at the low overhead of data shortcuts, as discussed in Chapter 17, "Sharing the Model: Data Shortcuts," and see a simple solution that works. Prior to Civil 3D 2007, data shortcuts were the only option for sharing data, so many users got comfortable with that model and have stuck with it. However, as you can see in Table 19.1, some functions aren't available in shortcuts. Many users consider these differences trivial, but it's good to be aware of them. More differences in the use and operation of the data will become apparent once you're inside Civil 3D.

TABLE 19.1: Vault Compared with Data Shortcuts

Product-Sharing Mechanism	Vault	Data Shortcuts
Reference of Civil 3D objects	X	X
Synchronizing changes	X	X
Labeling references	X	X
Locking source objects	X	X
Multioffice support	X	
Managing non-CAD data	X	
User controls	X	
Archives and backups	X	

Project Timing

Data workflow tends to be the biggest hurdle for people when making the transition to Civil 3D. You have myriad options for pulling it all together. XRefs, data shortcuts, and Vault all come into play. At the end of the day, you're likely to use all three to create your complete project. This section will look at the general schematic of the project before the chapter gets into the mechanics of using data references.

But That's Not How We Do It!

Before you wade too far into this discussion, remember, this chapter talks about one workflow that has worked for a fair number of Civil 3D users. It may not mesh perfectly with your office or project requirements. Workflow and data management are the most complex and flexible parts of the program; for this reason, it's suggested that you get qualified assistance when you take on this task. The Civil 3D adoption period is a good time to evaluate the way your office *does* do it and see if it still makes sense. Civil 3D changes the game from a design side — is it any surprise that it changes the way engineering teams work together?

Most engineering firms operate in a team environment, dividing labor as it makes sense from a training and cost-perspective angle:

- The project manager operates with little knowledge of the CAD systems but understands design, submittal requirements, and general project management.
- Designers do the bulk of the heavy lifting during the design process and generally have some level of CAD knowledge, but they may not be comfortable using it as a design tool. These are typically highly experienced technicians who have moved to focus more on design or engineering.
- Technicians typically have less design knowledge but know how to put together a good set of plans. They're intimately familiar with the CAD system, creating most of the CAD work within the firm.

There are obviously exceptions to this broad generalization, but this breakdown of roles lends itself to an examination of the standard project timelines, conceptually illustrated in Figure 19.1.

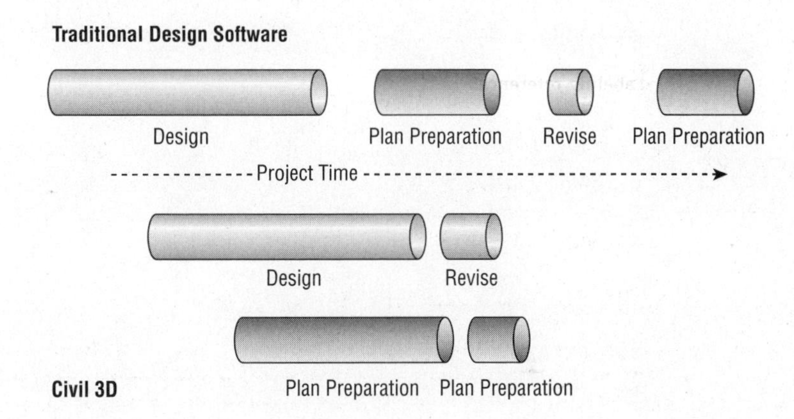

FIGURE 19.1 Comparing Civil 3D and other design software timelines

In the past, using typical software meant that design processes wound up in a linear state. Technicians didn't get involved until very late in the process because any change to the design would change so much of the labeling and other plan-level information that the technicians' time would be wasted. Project managers had to wait until the documentation process was moving along to get a feel for the project and make sure the team members were on track. With this sort of project timeline, nothing was more dreaded than a last-minute change, because it meant throwing out so much work. Lost work means lost time and lost profitability.

Working in the Civil 3D environment lets this timeline break apart into pieces that run in parallel. Because of the dynamic nature of the object model and labeling in Civil 3D, the technicians can begin documentation shortly after the designers start. Plans can be prepared for the project manager's review at a concept stage, with the understanding that the design is still in flux but looking for thoroughness of documentation and adherence to and review of agency standards. The entire project timeline can be reduced by days, weeks, or even months.

Project Workflow with Vault and Civil 3D

To shorten the project timeline as just discussed, the workflow has to change as well. Changing to a process where the tasks are running concurrently requires a different approach than prior software solutions. To make this process as efficient as possible, it's broken down more like Figure 19.2.

FIGURE 19.2
File and design information breakdown

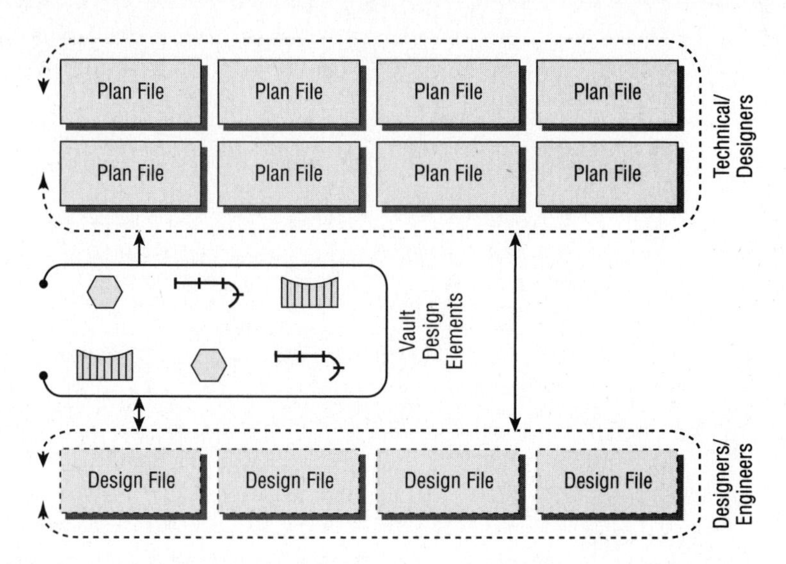

The concept is that the design staff creates and models the Civil 3D design elements in the respective files. When they push these files into the vault, design elements become available to project team members through Vault data references in addition to the standard AutoCAD XRef functionality. By using the Civil 3D objects to feed the creation of plan files, the technicians can begin creating and labeling plans as soon as the first iteration of the design has been checked in to the vault.

Feedback from the Vault

There's one last thing to look at before you start working with Vault in this chapter's exercises. Although you use Vault in a project, you'll often see icons next to drawing names. These icons indicate the status of the file in the working folder versus the ADMS file store. Table 19.2 shows the possible icons and explains their meaning.

TABLE 19.2: Civil 3D Vault Status Icons

STATUS ICON	IS FILE CHECKED OUT?	FILE STORE VS. WORKING FOLDER	NOTES
○	No	In sync	
(Green fill)	No	Working folder is newer	Usually means someone edited the file out of turn. Also known as a *dirty edit*.
(Red fill)	No	Master file is newer	Usually occurs when a previous version has been restored.

TABLE 19.2: Civil 3D Vault Status Icons *(CONTINUED)*

Status Icon	Is File Checked Out?	File Store vs. Working Folder	Notes
✔	Yes, to you	No working folder copy	
	Yes, to you	In sync	Occurs when you check out a file but haven't made a save yet to update the local copy.
(Green fill)	Yes, to you	Working folder is newer	Occurs when you perform a local save on a checked-out drawing.
(Red fill)	Yes, to you	File store is newer	Occurs when restoring a previous version of the file.
✖	Yes, to someone else	No working folder copy	
	Yes, to someone else	In sync	
(Green fill)	Yes, to someone else	Working folder is newer	Can occur when someone else performs a save on a checked-out drawing, and you're sharing a working folder on the network.
(Red fill)	Yes, to someone else	File store is newer	Someone else has checked out this file and then checked it back in. Your working folder is local, so you don't have the latest version in your working folder.

A similar table exists in the Civil 3D Help files. One good suggestion is to print that chart and place it next to your monitor until you're familiar with these icons and what they mean. Now, let's take all this theory and put it to work using Vault inside Civil 3D.

Working in Vault

Using Vault with Civil 3D requires some mental effort. It's a new process for moving and sharing data within the team. This section describes the mechanics of all this interaction, including setting up the project, checking data into the vault, and creating data references.

Preparing for Projects in Civil 3D

For most AutoCAD-based products, creating a folder in Vault Explorer is enough to generate a project available in the program. Not so with Civil 3D. Because of the project mechanism used to pass information from one drawing to another in Civil 3D, certain files must exist in a Vault folder for Civil 3D to recognize it as a valid project. First, you'll set up a project template that matches the desired workflow, and then you'll create the project itself.

ESTABLISHING A PROJECT TEMPLATE

Civil 3D 2010 includes a mechanism for creating a typical project folder structure when you're working with Vault. You can create a blank copy of the folder structure you'd like to have in place for typical projects and use that as the starting point when a new project is created in Civil 3D. Follow these steps:

1. Open Windows Explorer, and navigate to `C:\Civil 3D Project Templates`.
2. Create a new folder titled **Mastering**.
3. Inside `Mastering`, create folders called **Survey**, **Engineering**, **Architecture**, **Word**, and **Con Docs**, as shown in Figure 19.3.

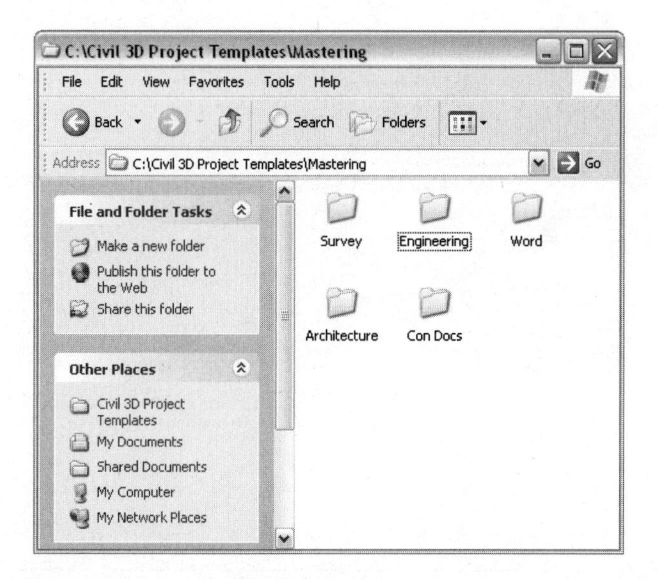

FIGURE 19.3
Creating a project template

This structure will appear in Civil 3D, in the working folder when a project is created, and in Vault Explorer. The `Word` folder is an example of non-Civil 3D–related folders you may have in your project setup for users outside the CAD team, such as accountants or the project manager. If these folders contain any files (such as a project checklist spreadsheet, a blank contract, or images), these additional files are added to new projects during the setup process.

CREATING A NEW CIVIL 3D PROJECT

With your project template in place, you can now use the Projects section of Prospector to create a project. Just a reminder: Projects created via Vault Explorer or any other clients won't appear in Prospector, because they're missing the "magic" files (which you'll examine in the last part of this exercise). Follow these steps:

1. Launch Civil 3D. In Prospector, choose Master View from the View drop-down menu at the top of the palette if it isn't already selected.

2. Right-click the Projects branch, and select Log In to Vault.
3. Select your server or local host, as appropriate, and select the Melinda database you created in Chapter 18. Click OK to return to Civil 3D.
4. Right-click Projects again, and notice the host of new options. Select New to bring up the New Project dialog shown in Figure 19.4.

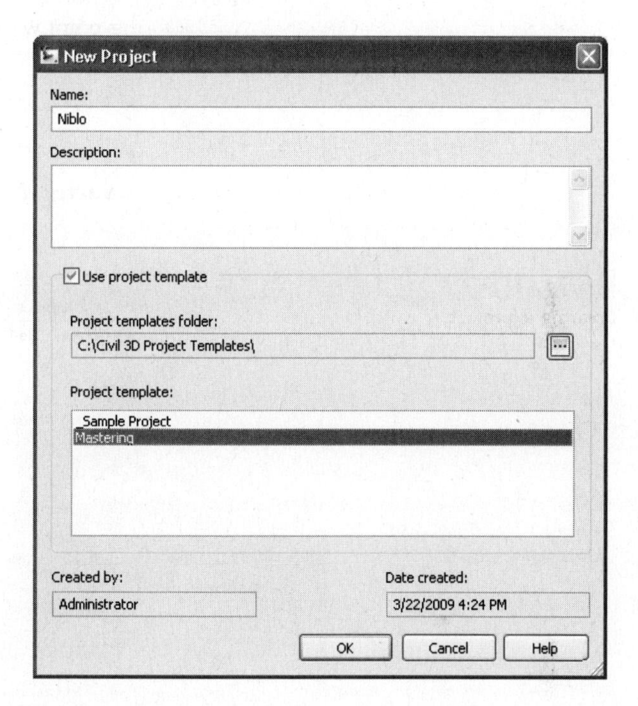

FIGURE 19.4
Creating the Niblo project

5. Enter **Niblo** in the Name text box, and select the Mastering project template as shown. Click OK to create the new Niblo project.
6. Expand Projects to reveal the project structure, as shown in Figure 19.5.
7. Launch Autodesk Vault, and log in to the Melinda vault.
8. Select the `Niblo` folder, as shown in Figure 19.6.

The files shown in the upper-right pane of Autodesk Vault are the critical files for Civil 3D. Each file manages part of the project information that Civil 3D requires to operate. These files are the reason you can't simply create folders in Vault Explorer and expect them to show up in the Civil 3D Projects collection. However, once a project is created and these files exist, you can add folders if needed to flesh out a project structure:

1. Right-click Niblo in Vault Explorer, and select New Folder.
2. Enter **Excel** for the name, and click OK.
3. In Civil 3D, right-click Niblo in Prospector, and select Refresh.

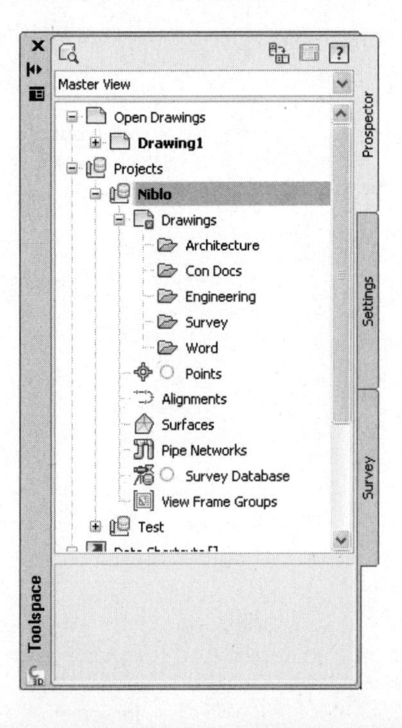

FIGURE 19.5 Prospector after creating the Niblo project from the Mastering project template

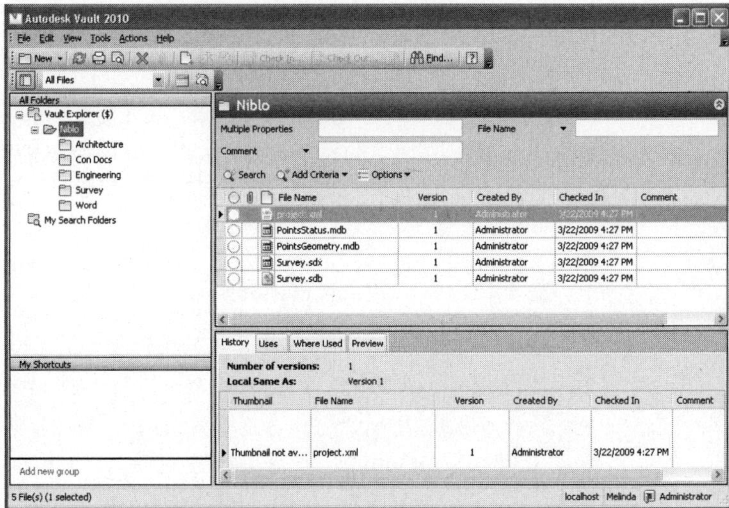

FIGURE 19.6 Niblo project viewed in Autodesk Vault

Note that the `Excel` folder you just made in Vault Explorer appears in the folder list in Prospector. The two programs are working from the ADMS, so they stay in sync. Now that you've created a project, let's pull in some information.

Populating Vault with Data

Projects are useless without design files. In this section, you'll learn about the Civil 3D elements available for sharing through Vault and insert some of these design elements into your project.

Vault Eligibility

Although more than a dozen Civil 3D–specific object types are used in modeling designs, only certain ones are available for sharing as data objects in Vault. The rest are typically used as components to other pieces of the model, but a few objects haven't yet been implemented as Vault data.

The following objects can be shared via Vault (*vaultable*):

- Point databases
- Alignments
- Surfaces
- Pipe networks
- Survey database
- View Frame groups

Parcels are an obvious candidate for project-level sharing but unfortunately aren't accessible at this time except via XRefs.

> **XRefs and Labels**
>
> Vault data references are frequently used to handle labeling requirements in different files. Although parcels aren't vaultable per se, with 2010, you can label XRef information using typical Civil 3D labels, so this concern isn't as major as before. Remember, you don't have to do things the same way you always have, and you probably shouldn't.

Loading the Vault

All drawings related to a project should be part of the vault. As drawings are added to the vault, four steps take place:

1. The drawing is copied to the working folder if it's not in there already.
2. A copy of the file is made to the ADMS file store.
3. The SQL database is updated to reflect the existence of the file within the scope of the project.
4. Depending on the options selected during this process, the file is checked in and closed or checked out and left open.

New drawings, existing internal drawings, and drawings coming from external sources are three common sources for data you can put into a vault. Let's look at getting each of them into the Vault project.

Loading New Drawings

New drawings are the simplest to add, but you should be aware of some nuances to avoid issues with duplicate data later. In this exercise, you'll look at the steps involved with adding a new drawing to the vault:

1. In Civil 3D, create a new file using the file `_AutoCAD Civil 3D (Imperial( NCS .dwt`.

2. Choose Save As from the application menu, and navigate to `C:\Civil 3D Projects\Niblo\Engineering`.
3. Name the file **XStorm.dwg**, and click Save. By saving the file into the working folder, you avoid creating a duplicate copy when the file is checked in to the vault.
4. Right-click the drawing name in Prospector, and select Add to Project, as shown in Figure 19.7.

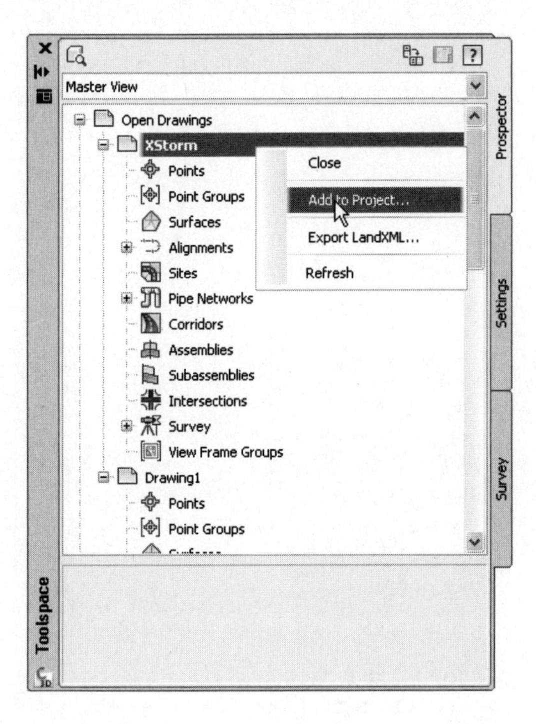

FIGURE 19.7
Adding a drawing to a project

5. Click Yes to close the warning about the drawing being saved. The Add to Project dialog appears.
6. Select the Niblo project, and click Next.
7. Verify that the `Engineering` folder is selected, and click Next. Notice that this tree matches your other views of the Vault project.
8. Deselect the Publish DWF check box, and click Finish. Prospector adds a project-status icon, as shown in Figure 19.8.

Some things to note about that last step: You removed the DWF option because the DWF created is often limited in its usefulness. It's used as a preview in Prospector, but it doesn't plot a DWF of the full Model or Paper space — only the current view at check-in. Additionally, many users aren't comfortable with DWF and are surprised to see a plot dialog appear when they thought they were saving their file. Chapter 18 discussed turning off this option, but this chapter includes the step in case you didn't match those steps or are using another ADMS instance. You didn't include a comment at this point, but you can use comments to indicate major milestones or indicate what was changed. These comments appear when you're browsing through the project files in Vault Explorer.

FIGURE 19.8
Drawing added to the Niblo project

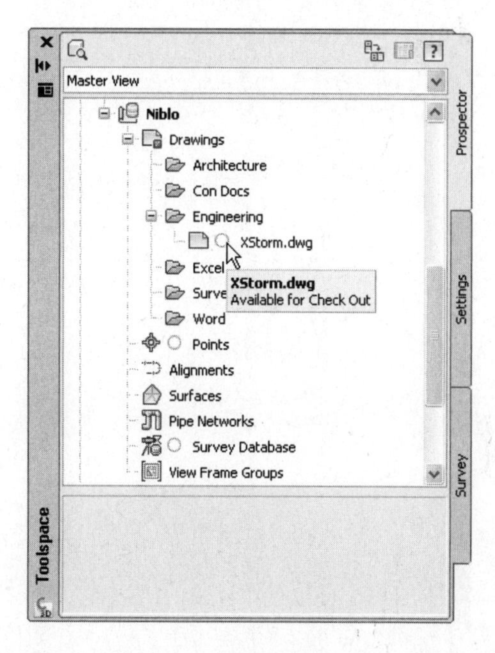

There are a couple of things to notice in Figure 19.8. First, the drawing name in Prospector now has a Vault status icon next to it. Second, the application window reflects the attachment of a project and gives feedback about the drawing's status. Blank drawings are easy; now, you'll add some existing data.

> **IT BEARS REPEATING**
>
> Be careful where you save those files when you load external data into the vault. It's easy to save a copy to the root of a project but check it in to a subfolder in the project. Doing so leaves a file sitting in the project that other users may find and assume is good data. If you make this mistake, erase *both* files, and then get the latest version from the vault. This ensures that the right file, in the right version, is in the right place.

Loading Existing Drawings

When project information exists in various file locations around the office, adding it to the project is an important part of data management. In this exercise, you'll add some of the files you've worked with in earlier chapters and look at the data being made available to the project team:

1. Open the `XTopo.dwg` file. Remember that all data can be downloaded from `www.sybex.com/go/masteringcivil3d2010`.
2. Right-click the drawing name, and select Add to Project as before.

3. Click Yes at the warning about the file being saved.
4. Select the Niblo project, and click Next.
5. Select the `Engineering` folder, and click Next.
6. Deselect the Keep Files Checked Out and Publish DWF check boxes, as shown in Figure 19.9, and click Next.

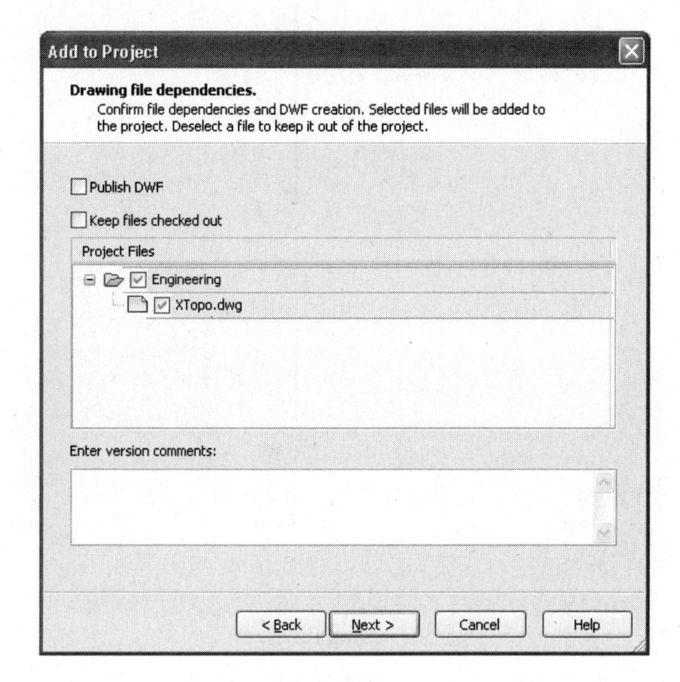

FIGURE 19.9
Setting options for DWF creation and file check-in

7. Select the Surfaces check box, as shown in Figure 19.10.
8. Click Finish. The drawing is added to the vault and then closed.

Let's look at a few things that were different in this process. You had to select a project folder this time because the drawing wasn't stored in the working folder before being added to the project. If this drawing came from an outside source, such as a consultant, it would be wise to save it in the working folder as part of the data-acceptance process. Doing so would help avoid duplicate files. As it is, you now have a copy of the drawing in the data location from which you opened it and in the working folder. This type of duplication can lead to errors later. It bears mentioning, remember the working folder location. By setting the proper working folder current, it is possible to add drawings and such to the vault without disturbing the original file locations.

The second major change was deselecting the Keep Files Checked Out check box. This option means that once the check-in is complete, you no longer want to keep that file open and reserved. You may find it helpful to think of this as a library where only one person can check out a book at a time.

Finally, the biggest change was the option to share the surface data. Because surfaces are among your vaultable objects, the check-in process offers the option of sharing this data. This screen displays each time a vaultable object is in the drawing and hasn't been shared previously.

Unfortunately, if you don't want to share something, you have to go through these options every time you check the drawing in to the vault.

FIGURE 19.10
Choosing to share the EG surface within the project

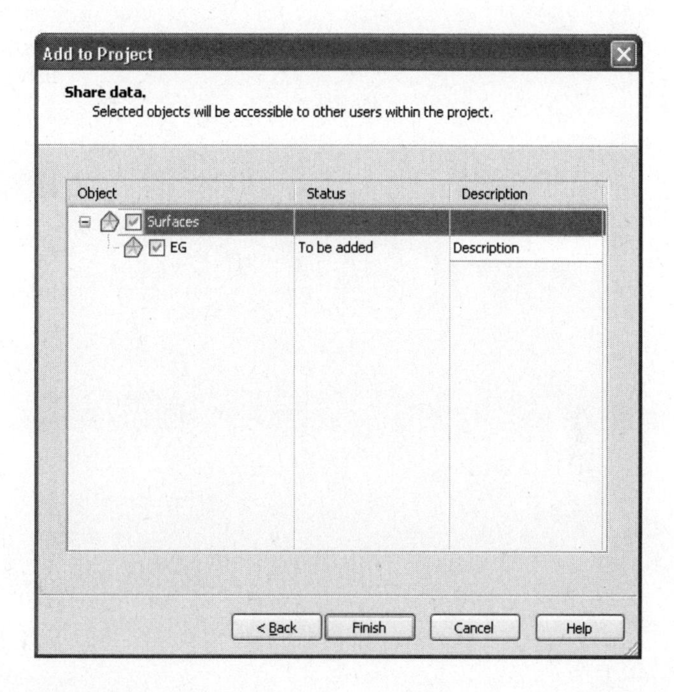

Next, you'll add one more drawing to repeat the process:

1. Open the `XRoad.dwg` file in Civil 3D.
2. Right-click, and select Add to Project.
3. Add the XRoad drawing file to the Niblo project in the `Engineering` folder. Don't keep it checked out, but do share all the alignments. Click Finish to complete the drawing.
4. In Prospector, expand Projects ➤ Niblo ➤ Drawings ➤ Engineering, Niblo ➤ Alignments, and Niblo ➤ Surfaces. Prospector should look like Figure 19.11.
5. Repeat the process and add the file `XPlat.dwg` to the `Survey` folder.

The blank drawing you made is still open and checked out to you. You'll use it to design and create data references in the next section.

Working with Vault Data References

Similar to XRefs or data shortcuts, data references allow you to access the *data* in other drawings while applying styles and labels as required for the current drawing. Using data references instead of XRefs brings into play all the design information, allowing you to not only paint a pretty picture of that data but also use it for design or inquiry. In this section, you'll learn how to create basic references for labeling or information and then make some changes to see how it all plays together.

FIGURE 19.11
The Niblo project shown in Prospector

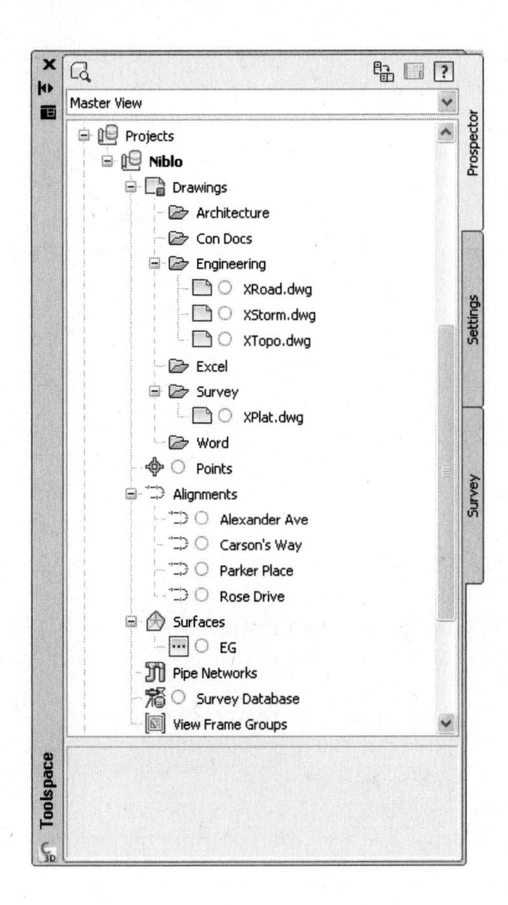

CREATING A DATA REFERENCE

Data references work by backtracking through the project interface to find the original design information in its host drawing, and then displaying that data on the basis of the current drawing's style requirements. This sounds more complicated than it is, so let's do a simple exercise:

1. Right-click on the `XStorm.dwg` file as shown in Prospector, and then select Check Out to display the Check Out Drawing dialog. Click OK.
2. In Prospector, expand Projects ➢ Niblo ➢ Surfaces.
3. Right-click the EG surface, and select Create Reference to display the Create Surface Reference dialog. Click OK.
4. Expand XStorm ➢ Surfaces ➢ EG, as shown in Figure 19.12.

   It's important to note the small shortcut arrow on the EG surface and that the Definition branch no longer appears under the EG surface as it did. Because EG in this drawing is a reference to another drawing, the definition can't be changed. But the cursor is giving feedback about the surface, indicating an elevation in the tooltip. The data is available, and reference objects can be stylized as needed.

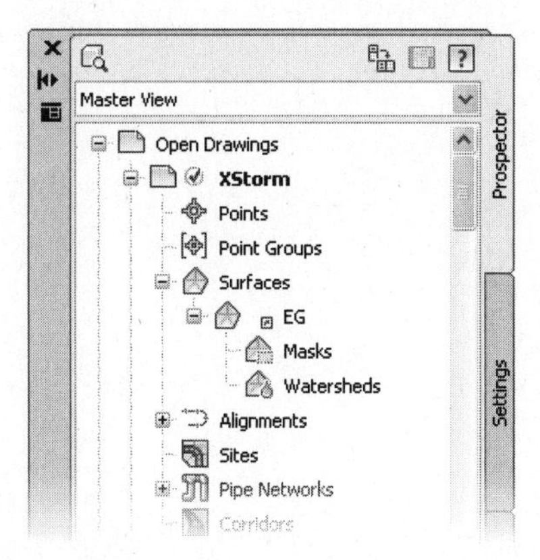

FIGURE 19.12
EG data reference created in the XStorm drawing

5. Change the EG surface style to Border Only.
6. Click Save.

SAVE? I THOUGHT THIS WAS VAULT!

It is. A save at this point saves to the working folder. The version in the ADMS file store hasn't changed, and it is still the blank drawing you added earlier in this chapter. You don't need to check in a drawing every time you make a change or are heading to lunch. Actually, it's a bad idea to do so. ADMS stores a *full* copy of the drawing every time a new version is checked in, so the file store can fill quickly if you check in your 10MB file every time you change a single piece of text or a line. Saving gives you a failsafe point, leaves the file checked out for editing to you, and doesn't cause file-store bloat.

UPDATING THE SOURCE

When design objects have been referenced, the original file is still available for checkout. Any changes that are then checked in to the vault are communicated to drawings that reference that data, and those drawings can be updated as needed. In this exercise, you'll modify a surface and update the vaulted version by checking it back in:

1. Expand Projects ➢ Niblo ➢ Surfaces.
2. Right-click EG, and select Check Out Source Drawing.
3. Click OK to get the latest version from the vault and place it in the working folder.
4. Draw a polyline like the one shown in Figure 19.13. Because you're dealing with only part of your site, you can place a boundary on the surface to limit the amount of data being shown.
5. Expand XTopo ➢ Surfaces ➢ Definition.
6. Right-click Boundaries, and select Add. Select the Non-Destructive Breakline check box, click OK, and select the polyline when prompted.
7. Right-click XTopo in Prospector, and select Check In.
8. Deselect the Keep Files Checked Out and Publish DWF check boxes, and click Finish.

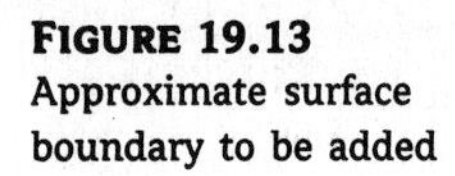

FIGURE 19.13
Approximate surface boundary to be added

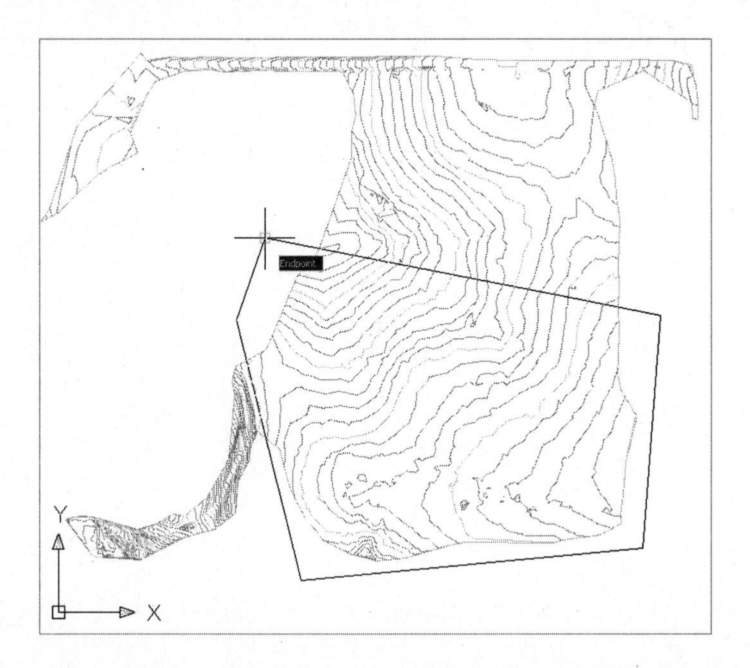

This process saved the file to the working folder and created a second full copy in the file store. Although the EG file used in this example is relatively small, you can see how the file store size could bloat quickly if you checked in a file every time you made even a minute change.

UPDATING THE REFERENCE

When source information changes, the `project.xml` file (one of those magic, Civil 3D–required files) is updated. As each session of Civil 3D that is accessing a particular project communicates with the vault, out-of-date items are flagged as needing to be updated. Now, you can update the surface reference that was modified in the last exercise:

1. Switch back to the XStorm drawing.
2. In Prospector, expand Surfaces, as shown in Figure 19.14.
3. Right-click EG, and select Synchronize. The surface updates, and the out-of-date warning is removed, as shown in Figure 19.15.
4. Right-click the XStorm drawing name in Prospector, and select Check In.
5. Select the Keep Checked Out check box, and then click Finish to complete the process.

UM, I DIDN'T KEEP IT CHECKED OUT . . .

If you miss the Keep Checked Out check box at any point along the way, expand Projects ➢ Niblo ➢ Drawings ➢ Engineering, right-click the needed file, and select Check Out. This is how you'll access files on a daily basis, so it's good to know.

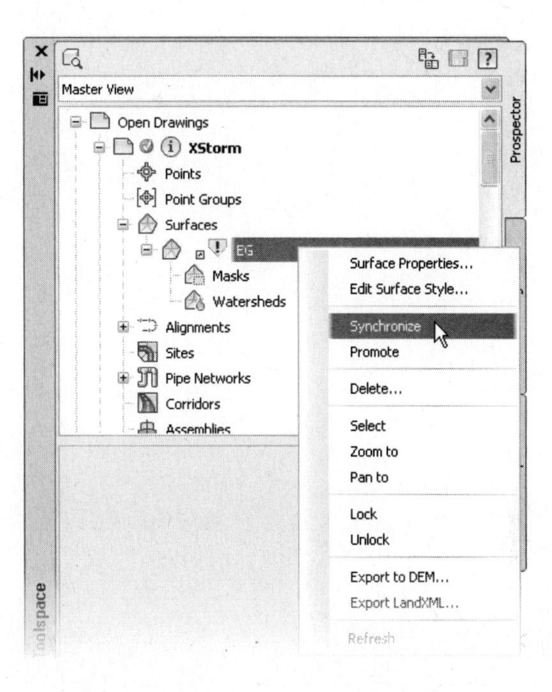

FIGURE 19.14
EG surface out-of-date warning and the Synchronize option

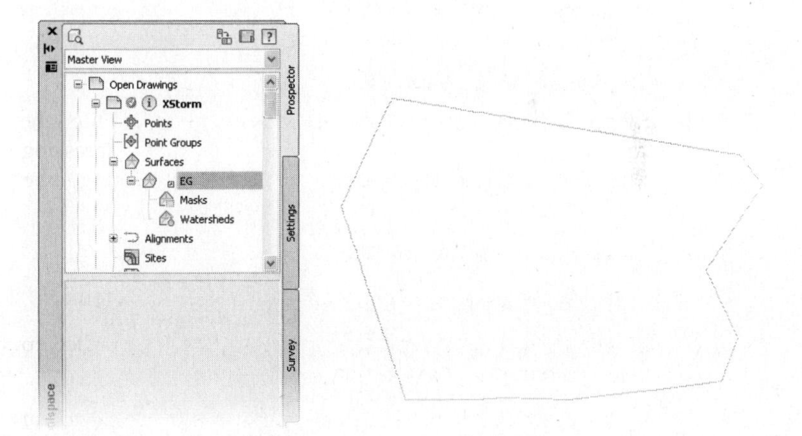

FIGURE 19.15
XStorm drawing after synchronizing project information

Taking this idea to the next level, the engineer can begin working with road-and-paving design and checking in the initial layout and profile information as soon as it's created. A technician can begin creating references, labeling things in the styles, and doing the formatting required to meet submittal specifications when the design is little more than a sketch. Using the preliminary design, sheets can be plotted for project-manager review, looking for the level of detail that suits your needs. As the design is updated, labels and entire sheets are updated with a simple synchronization and replotted.

Breaking the Reference

Occasionally, reference objects need to become part of the drawing they're being referenced in, with no connection to the original design. You can do this to consolidate data, to explore design alternatives, or to prepare a file for sharing with an outside source. Follow these steps:

1. Switch to the XStorm drawing if necessary.
2. Expand XStorm ➤ Surfaces.
3. Right-click EG, and select Promote.
4. Expand EG, and notice that the Definition branch is now available.
5. Check in `XStorm.dwg` again, but, this time, don't keep it checked out. On the Share Data page of the Check In Drawing dialog, you'll see a warning that EG can't be checked in. Vault won't allow duplicate object names in the same project, but you can check in the drawing without sharing that surface.

At this point, the EG in `XStorm.dwg` has nothing to do with the EG in the project. It's a copy of the data but won't reflect any changes to the source information. That's a dangerous situation to be in if you want to keep using this as a design file. A later section of this chapter will look at restoring previous versions.

Pulling It Together

Using a single reference is fairly straightforward; the complication comes when you try to understand how these references, XRefs, design files, and sheets all come together. This section examines how you can use all the tools in Civil 3D and Vault to create a plan and profile sheet for one road. This is a longer series of exercises, but it illustrates the full nature of the connections you can handle in Vault. Additionally, this section acts as a mini-exam over all the skills you've learned throughout the book.

Pulling the Needed References

To begin your design, you'll need a file that will be used to explore design ideas for creating a road corridor:

1. Create a new drawing using the NCS Extended template.
2. Save the file as **XCorridor.dwg** in `C:\Civil 3D Projects\Niblo\Engineering`.
3. Right-click the drawing name in Prospector, and select Add to Project.
4. Add the drawing to the Niblo project, but keep it checked out.
5. In Prospector, expand Projects ➤ Niblo ➤ Surfaces.
6. Right-click EG, and select Create Reference. Set the style to Border Only, and then click OK.
7. Expand Niblo ➤ Alignments.
8. Right-click Rose Drive, and select Create Reference to display the Create Alignment Reference dialog shown in Figure 19.16. Note the Source Alignment drop-down menu at the top of the figure. Occasionally, Vault gets lost and chooses the wrong alignment. You should always verify that the correct file and alignment are listed.

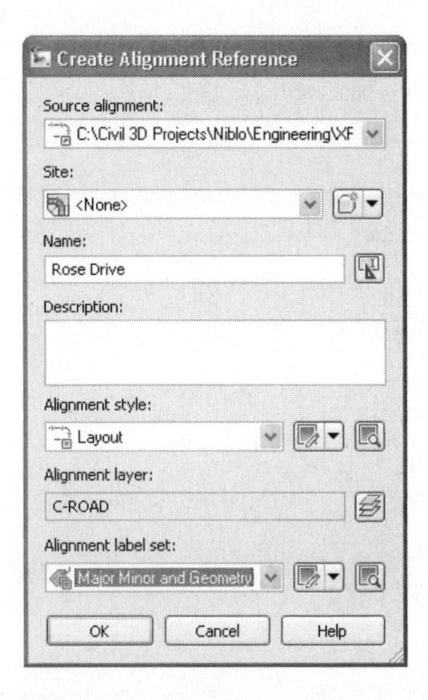

FIGURE 19.16
The Create Alignment Reference dialog

9. Set Alignment Style and Alignment Label Set as shown, and then click OK to create the alignment reference.

GOING VERTICAL WITH DESIGN

Now that you have a surface and the necessary horizontal alignments, you can pull together some vertical information:

1. Sample the EG surface using the Rose Drive alignment and create a profile.
2. Sample left and right offsets at 25′.
3. Set the profile styles as shown in Figure 19.17.

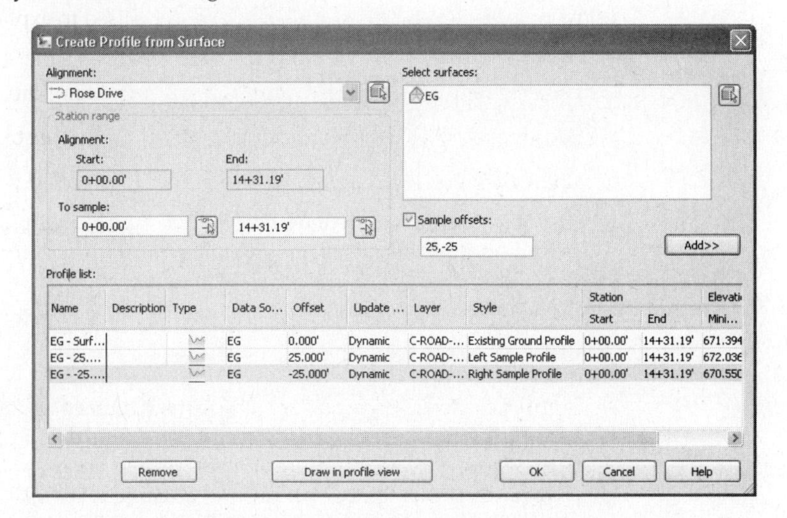

FIGURE 19.17
Creating profiles in the design file

4. Click Draw in Profile View to display the Create Profile View Wizard.
5. Set the Profile View style to Full Grid, and click Create Profile View.
6. Pick a point to the right of the surface boundary to place the profile view.

   At this point, you're going to cheat a bit. The data files include a profile design for Rose Drive.

7. Change to the Home tab and chose Profile ➢ Create Profile from File. In the dialog, select the `Rose Drive Prof.txt` file from the Chapter 19 folder. Change the Profile Name to **Rose Drive TC**, the Profile Style to Design Profile, and the Profile label set to Complete Label Set.
8. Click OK. Your drawing should look like Figure 19.18.

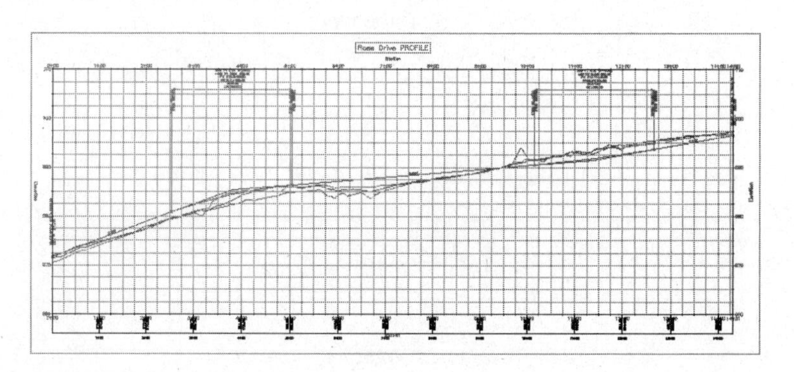

FIGURE 19.18 Sampled profiles from referenced alignment and surface data

BUILDING A ROAD MODEL

Now that you have horizontal and vertical elements for your design, you can build the road. To do this, you'll use another shortcut. You're going to use a prebuilt assembly that shipped with Civil 3D to create your road. Autodesk has already created a typical 27′ street assembly with a mountable curb and greenspace and placed it on the Assemblies – Imperial tool palette. Follow these steps:

1. Open the Tool Palettes by changing to the Home tab and choosing the Tool Palettes button from the Palettes panel. Find the Assemblies – Imperial palette.
2. Click the Through Road Main assembly on the palette.
3. Click somewhere on the screen to place the assembly, and then press Esc to exit the command.
4. Create a basic corridor using the Rose Drive alignment, the Rose Drive TC profile, and this assembly to arrive at the view in Figure 19.19. Don't forget to set a target surface for day-lighting.

PREPARING THE SHEETS

Now that you've completed a level of design, you can bring on another team member. This team member can begin working with the sheet files to create plans that match reviewer requirements. In this exercise, you'll break out some plan sheets and add them to the project:

1. Change to the Output tab and select Create View Frames from the Plan Production panel to display the Create View Frames Wizard.

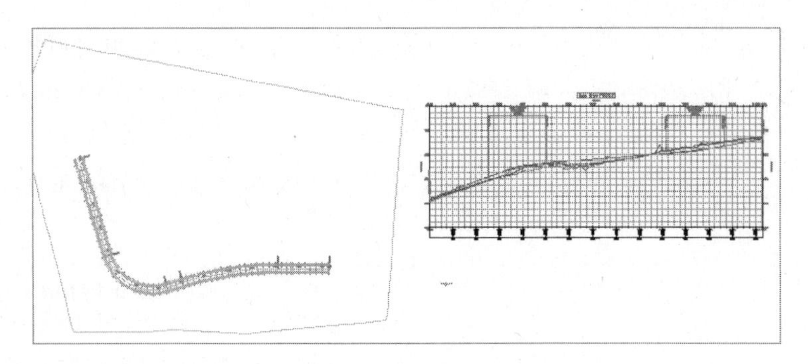

FIGURE 19.19
Completed corridor model in the design file

2. Click the Create View Frames button. (You're accepting the wizard's defaults. See Chapter 20, "Out the Door: Plan Production," for a full explanation of this new feature.) Your drawing should look like Figure 19.20.

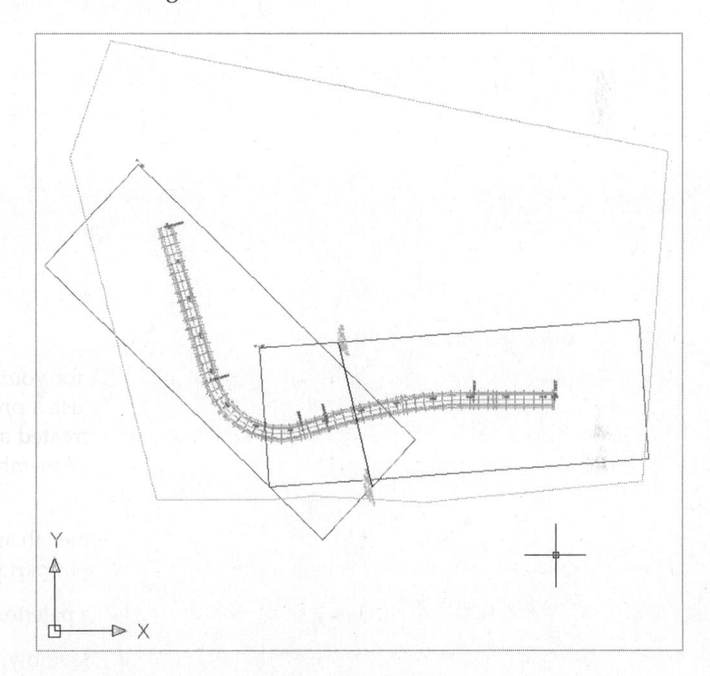

FIGURE 19.20
View frames created for Rose Drive

3. Choose Create Sheets from the Plan Production panel.
4. Select the All Layouts in One New Drawing radio button, as shown in Figure 19.21, and then click Next to move to the Sheet Set step.
5. Click the ellipsis button next to the Sheet Set File (`.DST`) Storage Location text box, shown in Figure 19.22. Select the `Con Docs` folder, and click OK. (There are two ellipsis buttons — you'll do this step twice.)
6. Click the other ellipsis button (next to the Sheet Files Storage Location text box) and select the `Con Docs` folder again. This sets the folder in the vault where the sheets will be stored.
7. Change the Sheet File Name to **ROAD**. The naming scheme seen previously uses X as a prefix to designate design files; no X means a sheet file.

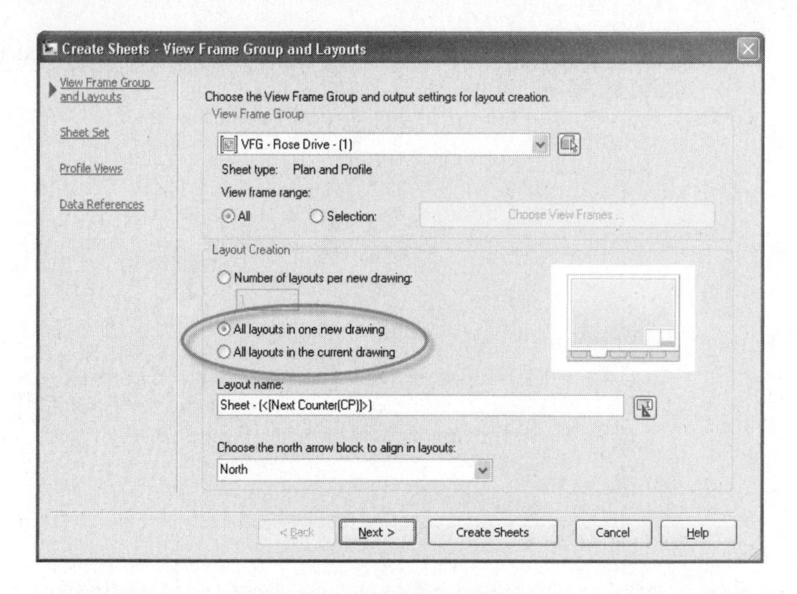

FIGURE 19.21
Controlling the placement of new sheet layouts

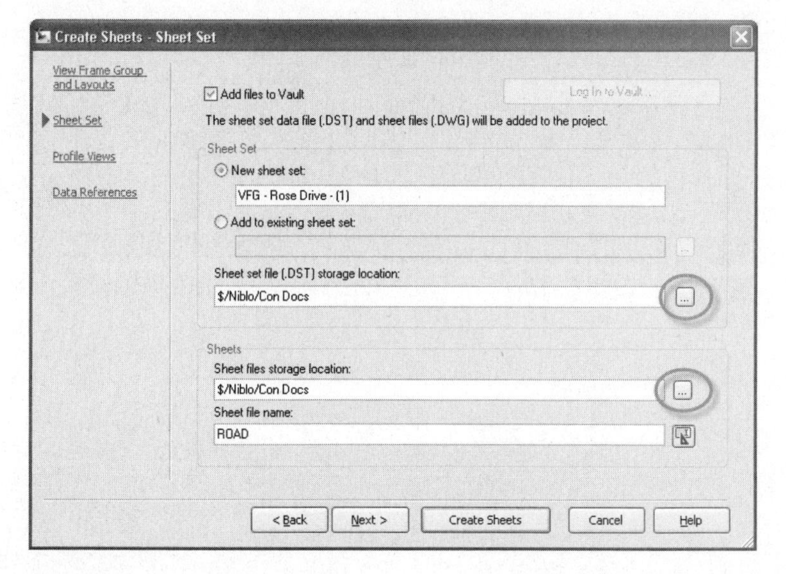

FIGURE 19.22
Setting paths for files and vault storage

8. Click Next to complete the Sheet Set step and move to the Profile Views step. Click Next to accept the settings in the Profile Views step and move to the Data References step.
9. Click the EG surface, as shown in Figure 19.23.
10. Click Create Sheets, and click OK at the warning that appears.
11. Pick a point to the right of the profile view already in the drawing to set an origin for the sheet profile views to be created.
12. Close the Sheet Set Manager palette that appears, and close Panorama if it's over your drawing.

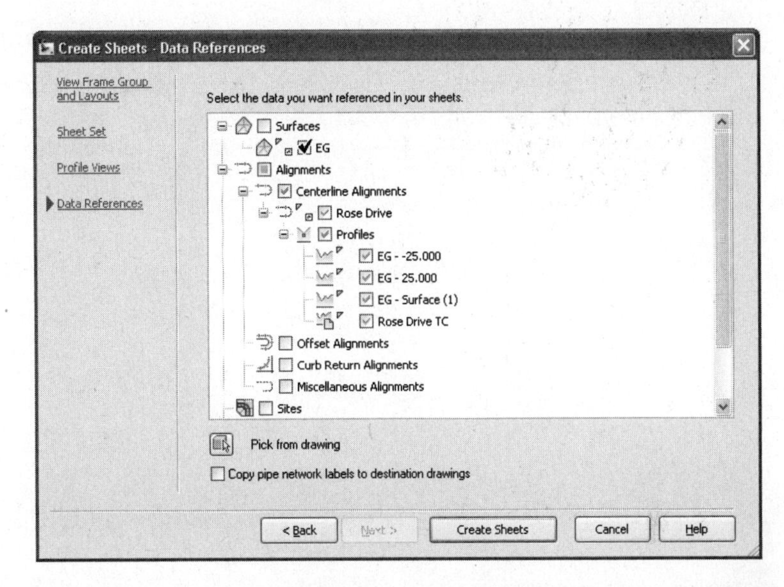

FIGURE 19.23
Selecting data references for the sheet file

13. In Prospector, expand Projects ➤ Niblo ➤ Drawings ➤ Con Docs. Right-click on Niblo and select Refresh. You should see the `Road.dwg` file (see Figure 19.24).

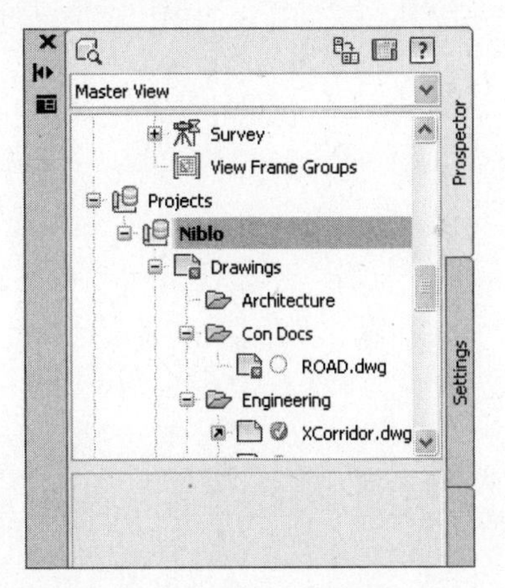

FIGURE 19.24
The sheet file is created, added to the project, and checked in

14. Check in your XCorridor drawing, and don't keep it checked out.

15. Select the View Frame Groups check box on the Share Data step, and click Finish.

At this point, you've effectively shortened the design- and plan-production process. Your technician can check out the Road drawing and begin to make any additional labels, titles, legends, notes, style changes, and so on as needed to complete the plans. A change in your design will be reflected via the updating mechanism, and the design and plan sheets will always stay in sync.

One last *very* cool thing to notice: The profile isn't in the same file as the alignment, allowing you to design the vertical in a different file from the horizontal. This means the surveyor who wants to lay out the street centerline to drive right-of-way calculation can share those alignments to keep the site designer in sync with the legal docs. That's one more step in the process that can be shared and split to make the project workflow more efficient.

Team Management in Vault

One of the common concerns with using Civil 3D is that the surveyor, engineer, technician, and project manager all have access to the design data. Many firms have taken this to an extreme in the past, creating separate projects for engineering versus surveying versus landscape architecture. This section will look at how you can use Vault to keep permissions straight, and what you can do when something does get changed by mistake.

Vault Folder Permission

ADMS and Vault bring a new level of functionality and security to the groups and users they manage. Previous versions basically depended on Vault roles to handle data-access control. Thankfully, this has changed, and now users and groups can manage access, similar to Windows Users and Groups.

Real World Scenario

KEEP THEM OUT OF MY DRAWINGS!

You've probably heard this a hundred times. The nature of CAD and engineering work means that sometimes you need to limit file access to certain groups or individuals. Now that you have files spread throughout your project, you can use these new functions to clearly delineate data access:

1. Launch Vault, and log in to the Melinda vault.
2. Right-click the `Engineering` folder, and select Properties to display the Properties for 'Engineering' dialog.

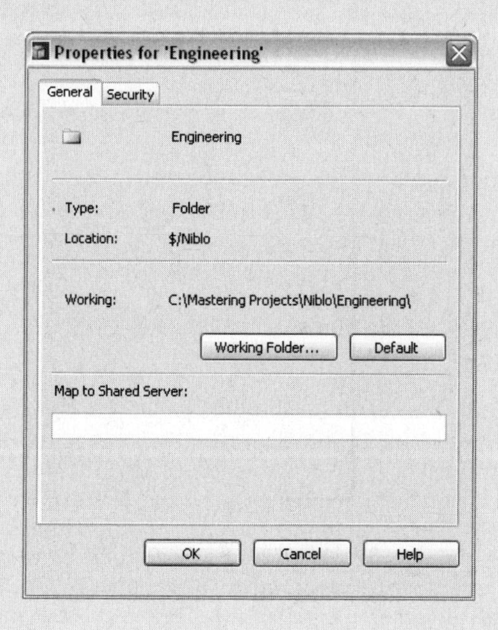

3. Switch to the Security tab, and click Add near the bottom to display the Add Members dialog.
4. Select Jolene, and click Add. Select Jim, and click Add again. Then, click OK to dismiss the dialog.

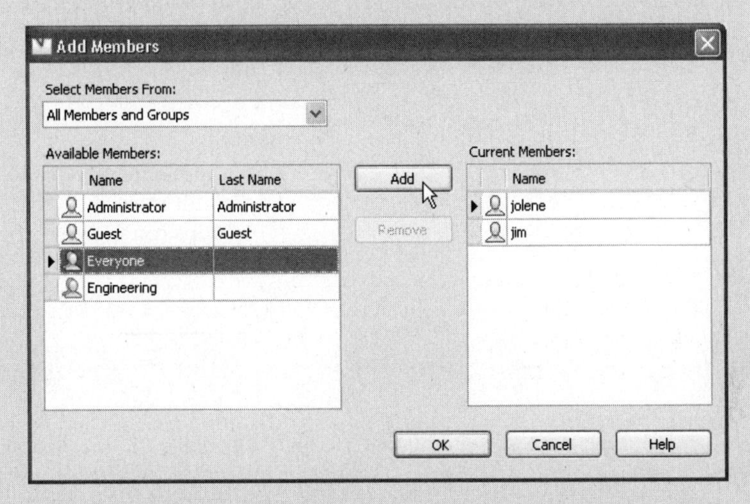

5. In the Properties for 'Engineering' dialog, notice the default permissions are Read Only. Click OK to dismiss the dialog.

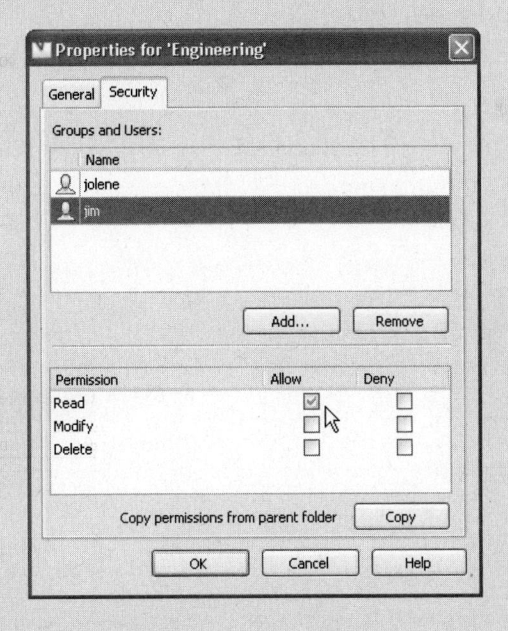

6. Switch back to Civil 3D. In Prospector, right-click Projects, and select Log Out.
7. Log back in as Jim (remember, the password is "jim").

8. Expand Projects ➤ Niblo ➤ Drawings ➤ Engineering, and select XCorridor.
9. Right-click, and select Check Out. Click OK to dismiss the Check Out Drawing dialog.
10. Vault does some quick checking and presents you with the following alert. Click Yes.

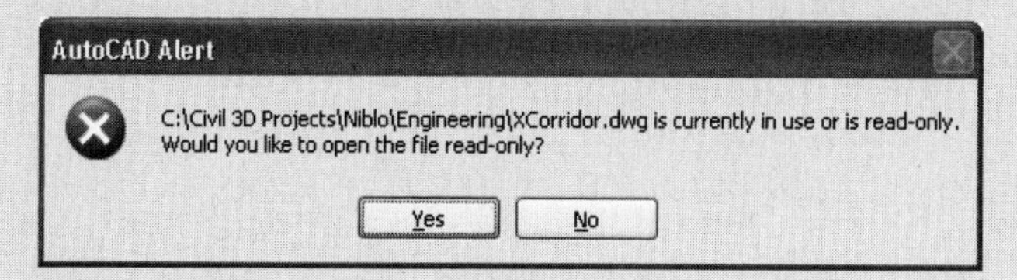

11. Right-click Projects, and log out as Jim.

Although this example uses individuals, a more corporate approach would be to use groups to manage permissions. This type of security should go a long way in reassuring surveyors, engineers, and technicians that they can coexist in harmony on design projects.

One common request with the integration of Civil 3D and ADMS and all its user management was the option to use Windows Active Directory to manage these settings. Some of this desired functionality has been added to the 2010 versions, but for full integration, you'll have to purchase Autodesk ProductStream.

Restoring Previous Versions

Every time a drawing is checked in to Vault, a full copy is made in the file store. Although this makes the bloat factor a real concern, it also means it's relatively simple to pull any version of a file that was ever used out of the file store and bring it back to current status. You'll try it in this exercise:

1. Log into the Melinda vault as Administrator, expand Niblo ➤ Engineering, and select `XTopo.dwg`, as shown in Figure 19.25.
2. Switch to the Where Used tab in the lower pane to review where this file is being referenced.
3. Right-click XTopo in the top pane, and select Get Previous Version to display the Get Previous Version of 'XTopo.dwg' dialog, as shown in Figure 19.26.
4. Select Version 1 from the Version drop-down menu (see Figure 19.26), and click OK to dismiss the dialog. The XTopo status icon changes in Vault. Remember, the working folder is merely a temporary holding location as files are edited. Overwriting the version in the working folder does nothing to what is in the vault and what is considered current.
5. Switch to Civil 3D.
6. Right-click Projects, and select Log In To Vault. Use the Administrator account for this example — you'll need to have read and write permissions.
7. Expand Project ➤ Niblo ➤ Engineering, and select XTopo.
8. Right-click, and select Check Out.

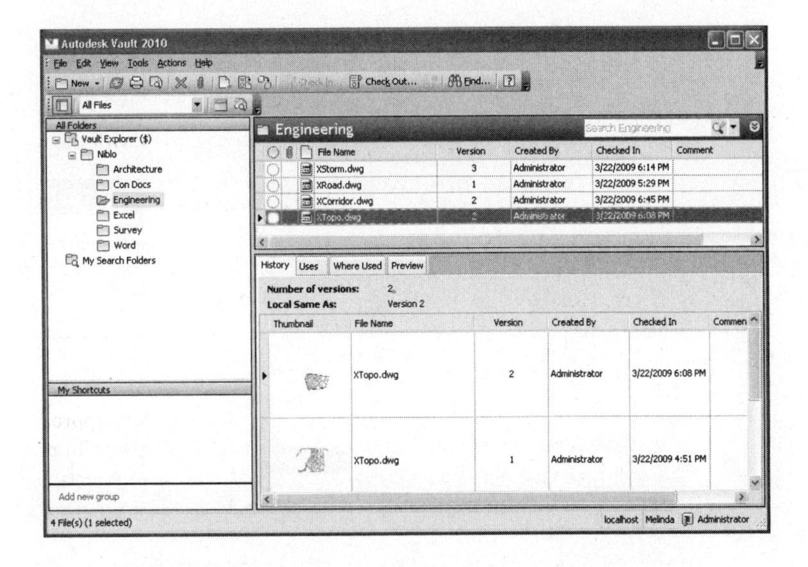

FIGURE 19.25
XTopo.dwg history in Vault

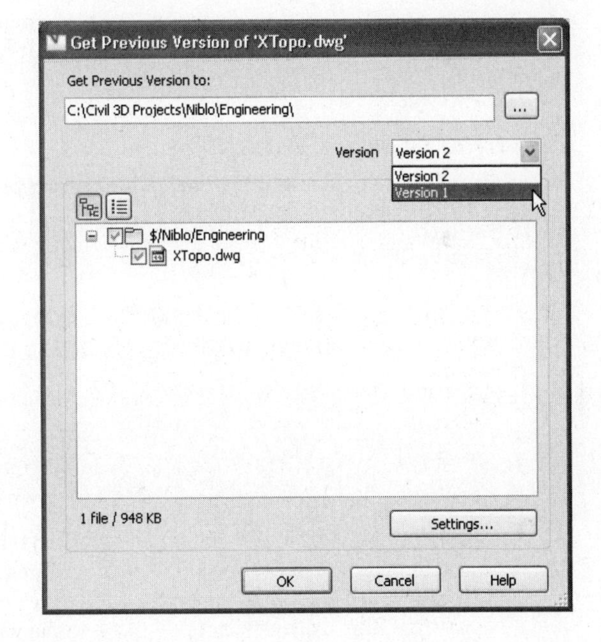

FIGURE 19.26
Use the Get Previous Version dialog to restore a version of a file

9. In the Check Out Drawing dialog, deselect the Get Latest Version check box, as shown in Figure 19.27. *This step is crucial!*
10. Click No To All in the Confirm File Replace dialog.
11. Click Yes to confirm that you will be opening the file with read-only permissions. Your screen regenerates, and the original version of the surface is displayed.

FIGURE 19.27
Don't get the latest version when you're attempting to restore a previous version

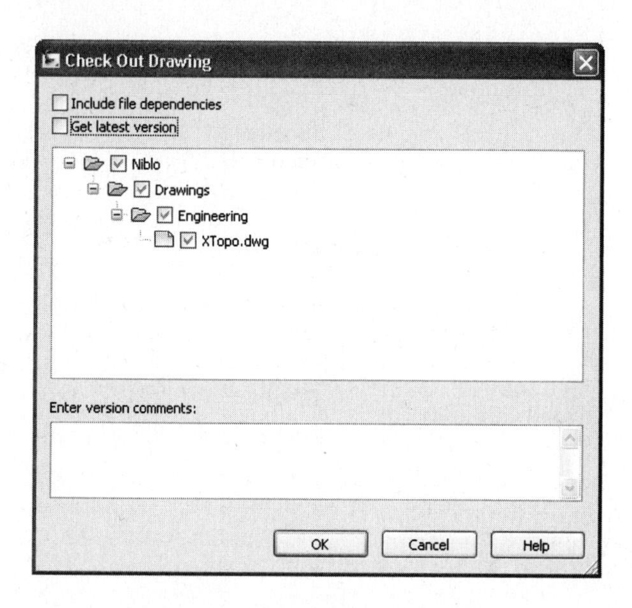

12. Right-click XTopo in Prospector, and select Check In. Deselect the Keep Files Checked Out check box in the Check In Drawing dialog if necessary.
13. Click No To All in the File Not Checked Out dialog. You do not want to check the file out. Click Finish.
14. Jump back to Vault, and press F5 to refresh the display.
15. Select XTopo one more time, and note that version 3 is now in the vault as shown in Figure 19.28.

FIGURE 19.28
The Administrator has checked in version 3

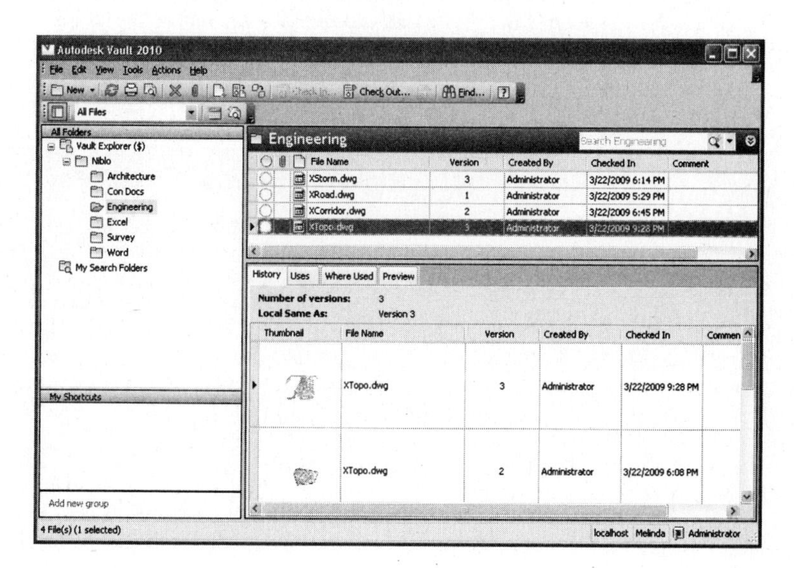

By placing an older version in the working folder, checking out without pulling the latest version from the file store to the working folder, but then checking back in the older version, you've effectively brought a file back from the past. In this case, it was a simple surface, but consider the sea of changes that sometimes occur when laying out a site. Being able to go back to the first conceptual design with almost no downtime is an enviable position to be in as a design firm.

The Bottom Line

Describe the differences between using Vault and data shortcuts. The two mechanisms in Civil 3D for sharing data are similar. Every firm has its nuances, and it's important to evaluate both methods of sharing data to see which one would be a better fit for the team's workflow. Recognizing the differences will allow you to make these recommendations with confidence.

Master It Describe three major differences between the two data-sharing methods.

Insert Civil 3D data into a vault. Creating a project in Vault requires some care to make sure all the pieces go in the right place. Using a project template to create organizational structures similar to what is already in place is a good way to start. Adding individual components to the vault as they fit into this structure makes for clean data organization and easy integration of new team members.

Master It Create a new project titled **Bailey** in the Melinda vault using the Mastering project template. Add the `Bailey's Run.dwg` file to the project in the `Engineering` folder, sharing all the data within it.

Create data shortcuts. Creating duplicate data is never good. By creating links to original data instead of working with copies, you maintain the integrity of the project information. Using data references instead of XRefs allows you to use the data behind the object instead of just the picture. This means styling and label changes can be part of the end file and allow team members to work on separate parts of the job at the same time.

Master It Create a new drawing called **Mastering Vault**, add it to the Bailey project in the `Engineering` folder, and create references to the FG surface and to the Natures Way and Old Settlers Way alignments. Use a Contours 2′ and 10′ (Design) style for the surfaces.

Restore a previous version of a design file. One of the major advantages of using a database-driven system is the ability to track file changes and restore them if necessary. Vault has a mechanism for moving data in and out of the file store. By using the restoration tools and checking in the proper files, it's easy to roll back to a prior version of a file.

Master It Erase the Founders Court alignment from the Bailey's Run file, and check it in. Then, use the Vault tools to restore the previous version containing the Founders Court alignment. What version number of the Bailey's Run drawing is listed when this is complete?

Chapter 20

Out the Door: Plan Production

So you've toiled for days, weeks, or maybe months creating your design in Civil 3D, and now it's time to share it with the world — or at least your corner of it. Even in this digital age, paper plan sets still play an important role. You generate these plans in Civil 3D using the Plan Production feature. This chapter takes you through the steps necessary to create a set of sheets, from initial setup, to framing and generating sheets, to data management and plotting.

In this chapter, you'll learn to:

- Create view frames
- Create sheets and use Sheet Set Manager
- Edit sheet templates and styles associated with plan production

Preparing for Plan Sets

Before you start generating all sorts of wonderful plan sets, you must address a few concepts and prerequisites. Civil 3D takes advantage of several features and components to build a plan set. Some of these components have existed in AutoCAD and Civil 3D for years (for example, layout tabs, drawing templates, alignments, and profiles). Others are new properties of existing features (such as Plan and Profile viewport types). Still others are entirely new objects, including view frames, match lines, and view-frame groups. Let's look at what you need to have in place before you can create your plotted masterpieces.

Prerequisite Components

The Plan Production feature draws on several components to create a plan set. Here is a list of these components and a brief explanation of each. Later, this chapter will explore these elements in greater detail:

Drawing Template Plan Production creates new layouts for each sheet in a plan set. To do this, the feature uses drawing templates with predefined viewports. These viewports have their Viewport Type property set to either Plan or Profile.

Object and Display Styles Like every other feature in Civil 3D, Plan Production uses objects. Specifically, these objects are view frames, view-frame groups, and match lines. Before creating plan sheets, you'll want to make sure you have styles set up for each of these objects.

Alignments and Profiles In Civil 3D 2010, the Plan Production feature is designed primarily for use in creating plan and profile views. Toward that end, your drawing must contain (or reference) at least one alignment. If you're creating sheets with both plan and profile views, a profile must also be present.

With these elements in place, you're ready to dive in and create some sheets. The general steps in creating a set of plans are as follows:

1. Meet the prerequisites listed previously.
2. Create view frames.
3. Create sheets.
4. Plot or publish (hardcopy or DWF).

The next section describes this process in detail and the tools used in plan production.

Using View Frames and Match Lines

When you create sheets using the Plan Production tool, Civil 3D first automatically helps you divide your alignment into sections that will fit on your plotted sheet and display at the desired scale. To do this, Civil 3D creates a series of rectangular frames placed end to end (or slightly overlapping) along the length of alignment, like those in Figure 20.1. These rectangles are referred to as *view frames* and are automatically sized and positioned to meet your plan sheet requirements. This collection of view frames is referred to as a *view-frame group*. Where the view frames abut one another, Civil 3D creates *match lines* that establish continuity from frame to frame by referring to the previous or next sheet in the completed plan set. View frames and match lines are created in Model space, using the prerequisite elements described in the previous section.

FIGURE 20.1
View frames and match lines

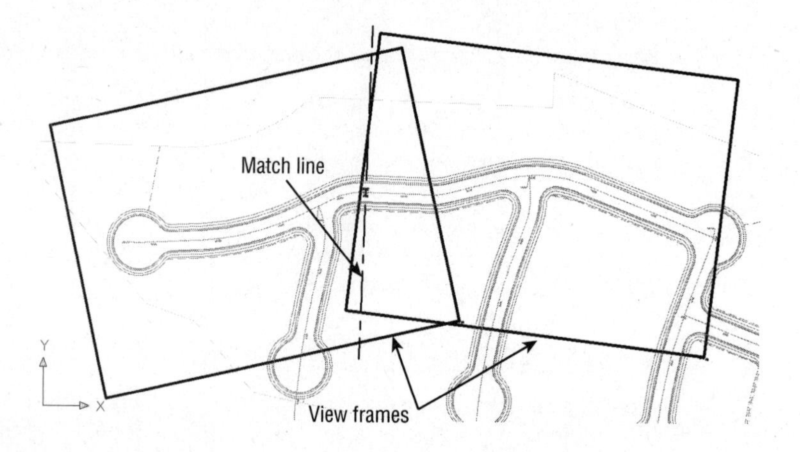

The Create View Frames Wizard

The first step in the process of creating plan sets is to generate view frames. Civil 3D provides an intuitive wizard that walks you through each step of the view frame–creation process. Let's look at the wizard and the various page options. After you've seen each page, you'll have a chance to put what you've learned into practice.

You launch the Create View Frames Wizard (Figure 20.2) by selecting the Create View Frames button on the Plan Production panel found on the Output tab of the ribbon. The wizard consists of several pages. A list of these pages is shown along the left side, and an arrow indicates which page you're currently viewing. You move among the pages using the Next and Back navigation buttons along the bottom of each page. Alternatively, you can jump directly to any page by clicking its name in the list on the left. The following sections walk through the pages of the wizard and explain their features.

FIGURE 20.2
The Create View Frames Wizard

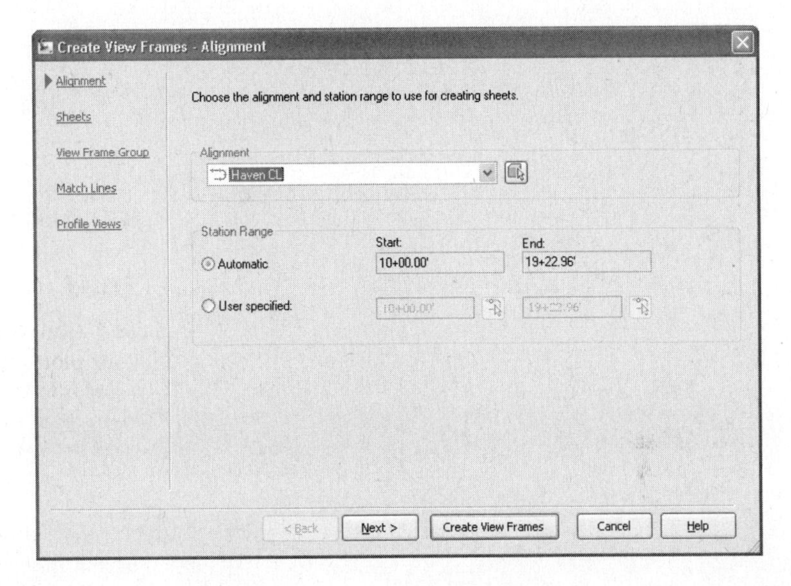

ALIGNMENT PAGE

You use the first page to select the alignment and station range along which the view frames will be created.

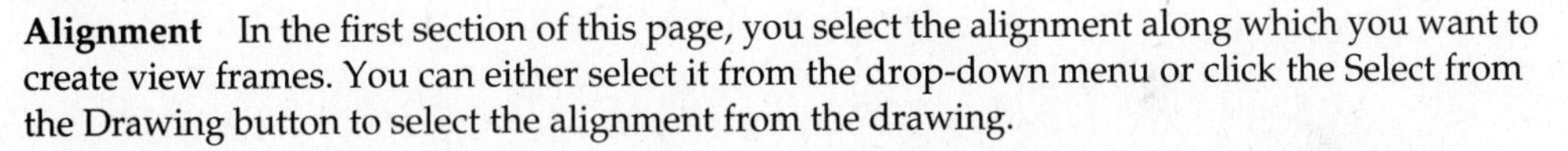

Alignment In the first section of this page, you select the alignment along which you want to create view frames. You can either select it from the drop-down menu or click the Select from the Drawing button to select the alignment from the drawing.

Station Range In the Station Range section, you define the station range over which the frames will be created. Selecting Automatic creates frames from the alignment Start to the alignment End. Selecting User Specified lets you define a custom range, by either keying in Start and End station values in the appropriate box or by clicking the button to the right of the station value fields and graphically selecting the station from the drawing.

SHEETS PAGE

You use the second page of the wizard (Figure 20.3) to establish the sheet type and the orientation of the view frames along the alignment. A plan production *sheet* is a layout tab in a drawing file. To create the sheets, Civil 3D references a predefined drawing template (`.dwt`) file. As mentioned earlier, the template must contain layout tabs, and in each tab the viewport's extended data properties must be set to either Plan or Profile. Later in this chapter, you'll learn about editing and modifying templates for use in plan production.

Sheet Settings In Civil 3D 2010, the Plan Production feature provides options for creating three types of sheets:

Plan and Profile This option generates a sheet with two viewports; one viewport shows a plan view and the other shows a profile view of the section of the selected alignment segment.

FIGURE 20.3
Create View Frames–Sheets

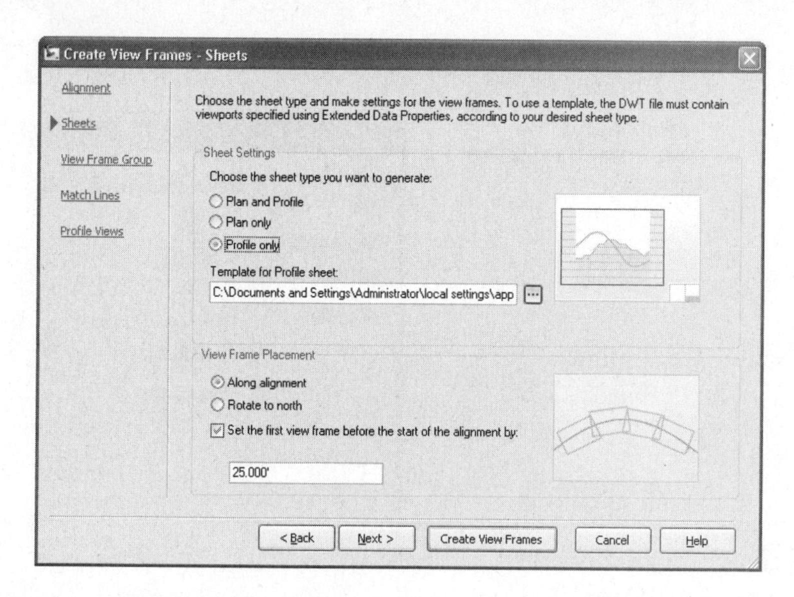

Plan Only As the name implies, this option creates a sheet with a single viewport showing only the plan view of the selected alignment segment.

Profile Only Similar to Plan Only, this option creates a sheet with a single viewport, showing only the profile view of the selected alignment segment.

GRAPHICS

Did you notice the nifty graphic to the right of the sheet-type options in Figure 20.3? This image changes depending on the type of sheet you've selected. It provides a schematic representation of the sheet layout to further assist you in selecting the appropriate sheet type. You'll see this type of graphic image throughout the Create View Frame Wizard and in other wizards used for plan production in Civil 3D.

After choosing the sheet type, you must define the template file and the layout tab within the selected template that Civil 3D will use to generate your sheets. Several predefined templates ship with Civil 3D and are part of the default installation.

Clicking the ellipsis button displays the Select Layout as Sheet Template dialog. This dialog provides the option to select the DWT file and the layout tab within the template. Clicking the ellipsis button in that dialog lets you browse to the desired template location. Typically, for Windows XP the default template location is

```
C:\Documents and Settings\<username>\Local Settings\Application
Data\Autodesk\C3D2010\enu\Template\Plan Production\
```

After you select the template you want to use, a list of the layouts contained in the DWT file appears in the Select Layout as Sheet Template dialog (Figure 20.4). Here you can choose the appropriate layout.

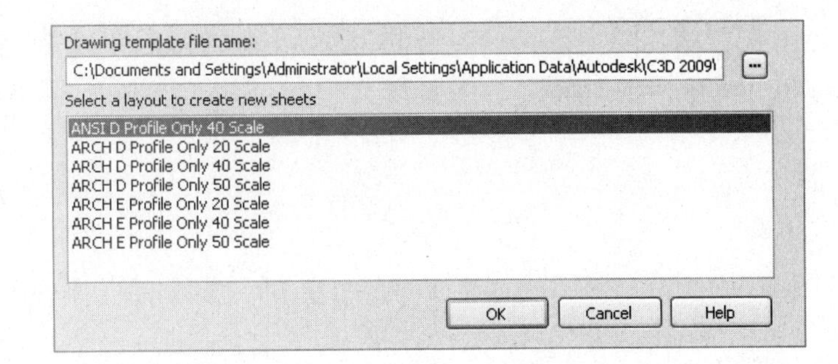

FIGURE 20.4
Use the Select Layout as Sheet Template dialog to choose which layout you would like to apply to your newly created sheets

View Frame Placement Your view frames can be placed in one of two ways: either along the axis or rotated north. Use the bottom section of the Sheets page of the wizard to establish the placement.

Along Alignment This option aligns the long axes of the view frames parallel to the alignment. Refer to the graphic to the right for a visual representation of this option.

Rotate to North As the name implies, this option aligns the view frames so they're rotated to the north directions (straight up), regardless of the changing rotation of the alignment centerline. *North* is defined by the orientation of the drawing. Again, refer to the graphic.

Set the First View Frame before the Start of the Alignment By Regardless of the view-frame alignment you choose, you have the option to place the first view frame some distance before the start of the alignment. This is useful if you want to show a portion of the site, such as an existing offsite road, in the plan view. When this option is selected, the text box becomes active, letting you enter the desired distance.

VIEW FRAME GROUP PAGE

You use the third page of the Create View Frames Wizard (Figure 20.5) to define creation parameters for your view frames and the view-frame group to which they'll belong. The page is divided into two sections: one for the view-frame group and the other for the view frames themselves.

View Frame Group Use these options to set the name and an optional description for the view-frame group. The name can consist of manually entered text, text automatically generated based on the Name Template settings (click the Edit View Frame Name button to open the Name Template dialog to adjust the name template), or a combination of both. In this example, the feature settings are such that the name will include manually defined prefix text (VFG-) followed by automatically generated text, which inserts the alignment name and a sequential counter number. For this example, this will result in a view-frame group name of VFG – Haven CL - 1.

The Name Template dialog isn't unique to the Plan Production feature of Civil 3D. However, the property fields available vary depending on the features to be named. If you need to reset the incremental number counter, use the options in the lower area of the Name Template dialog (Figure 20.6).

View Frame These options are used to set various parameters for the view frames, including the layer for the frames, view frame names, view frame object and label styles, and the label location. Each view frame can have a unique name, but the other parameters are the same for all view frames.

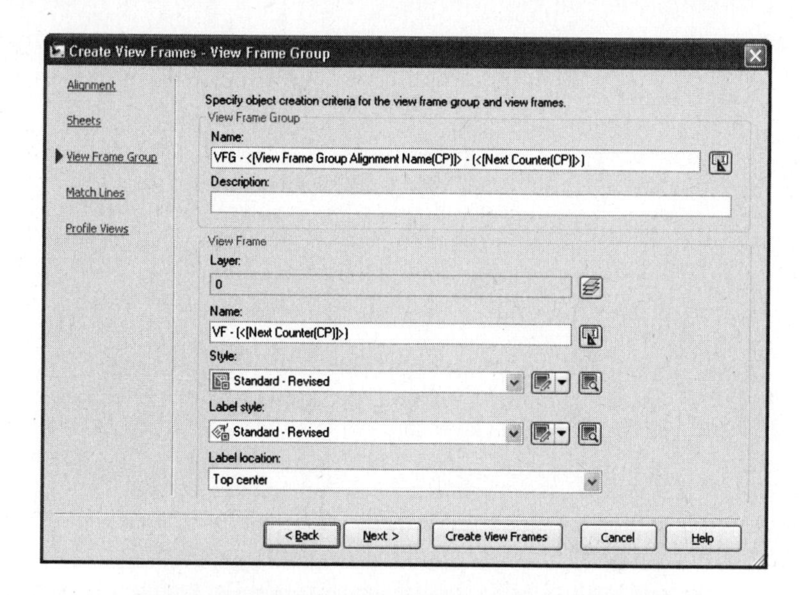

FIGURE 20.5
Create View Frames–View Frame Group

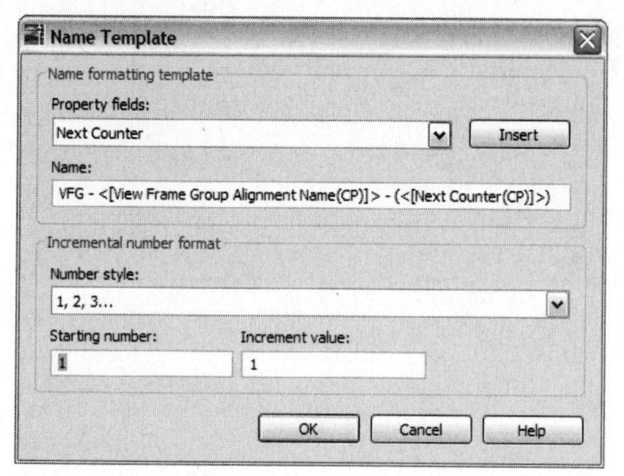

FIGURE 20.6
The Name Template dialog. Use the options in the Incremental Number Format section to adjust automatic numbering

Layer This option defines the layer on which the view frames are created. This layer is defined in the drawing settings, but you can override it by clicking the Layer button and selecting a different layer.

Name The Name setting is nearly identical in function to that for the view-frame group name discussed earlier. In this example, the default name results in VF-1, VF-2, and so on.

Style Like nearly all objects in Civil 3D, view frames have styles associated with them. The view frame style is simple, with only one component: the view frame border. You use the drop-down menu to select a predefined style.

Label Style Also like most other Civil 3D objects, view frames have label styles associated with them. And like other label styles, the view frame labels are created using the Label Style Composer and can contain a variety of components. The label style used in this example includes the frame name and station range placed at the top of the frame.

Label Location The last option on this page lets you set the label location. The default feature setting places the label at the Top Center of the view frame. Other options include Top Left, Top Right, Bottom Left, Bottom Right, and so on.

All view frame labels are placed at the top of the frame. However, the term *top* is relative to the frame's orientation. For alignments that run left to right across the page, the top of the frame points toward the top of the screen. For alignments that run right to left, the top of the frame points toward the bottom of the screen. You can make the view frame label display along the frame edge closest to the top of the screen by using a large Y-offset value when defining your view frame label style.

MATCH LINES PAGE

You use the next page of the Create View Frames Wizard (Figure 20.7) to establish settings for match lines. Match lines are used to maintain continuity from one sheet to the next. They're typically placed at or near the edge of a sheet, with instructions to "see sheet XX" for continuation. You have the option whether to automatically insert match lines. Match lines are used only for plan views, so if you're creating Plan and Profile or Profile Only sheets, the option is automatically selected and can't be deselected.

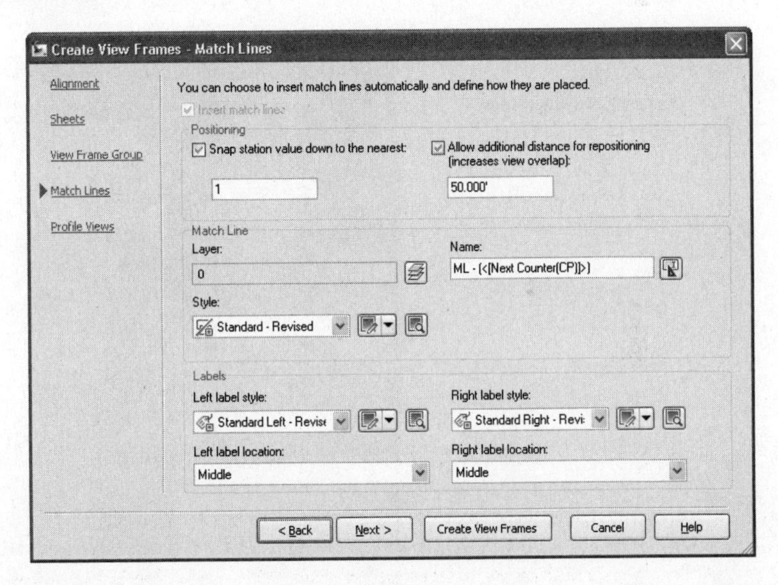

FIGURE 20.7
Create View Frames–Match Lines

Positioning The Positioning options are used to define the initial location of the match lines and provide the ability to later move or reposition the match lines.

Snap Station Value Down to the Nearest By selecting this option, you override the drawing station settings and define a rounding value specific to match-line placement. In this example, a value of 1 is entered, resulting in the match lines being placed at the nearest whole station. This feature always rounds down (snap station down as opposed to snap station up).

Allow Additional Distance for Repositioning Selecting this option activates the text box, allowing you to enter a distance by which the views on adjacent sheets will overlap and the maximum distance that you can move a match line from its original position.

Match Line The options for the match line are similar to those for view frames described on the previous page of the wizard. You can define the layer, the name format, and the match-line style.

Labels These options are also similar to those for view frames. Different label styles are used to annotate match lines located at the left and right side of a frame. This lets you define match-line label styles that reference either the previous or next station adjacent to the current frame. You can also set the location of each label independently using the Left and Right Label Location drop-down menus. You have options for placing the labels at the start, end, or middle, or at the point where the match line intersects the alignment.

PROFILE VIEWS PAGE

The final page of the Create View Frames Wizard is the Profile Views page (Figure 20.8). This page is optional and will be disabled and skipped if you chose to create Plan Only sheets on the second page of the wizard. The Plan Production feature needs to know what profile view and band set styles you intend to use for the profile views. This allows the correct positioning to be applied. Use the drop-down menus to select both the Profile View style and the Band Set style.

FIGURE 20.8
Create View Frames–Profile Views

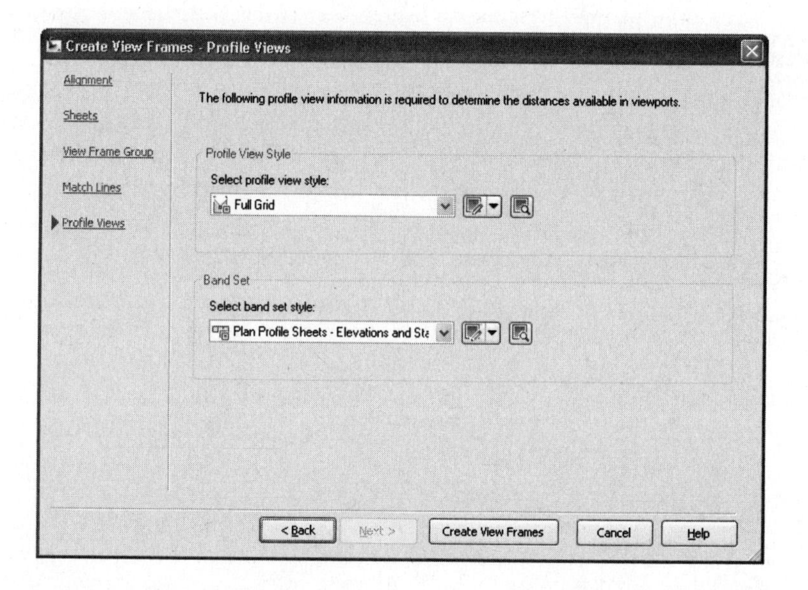

Civil 3D has difficulty determining the proper extents of profile views. If you find that your profile view isn't positioned correctly in the viewport (for example, the annotation along the sides or bottom is clipped), you may need to create *buffer* areas in the profile view band set style by modifying the text box width. The `_Autodesk Civil 3D NCS Extended.dwt` (both Imperial and Metric) file contains styles with these buffers created. The `readme.chm` file contained in `C:\program files\AutoCAD Civil 3D 2010\Setup\Docs` is a Windows Help file that has some additional information on the subject.

This last page of the wizard has no Next button. To complete the wizard, click the Create View Frames button.

Creating View Frames

Now that you understand the wizard pages and available options, you'll try them out in this exercise:

1. Open `View_Frame_Wizard.dwg`, which you can download from `www.sybex.com/go/masteringcivil3d2010`. This drawing contains several alignments and profiles as well as styles for view frames, view-frame groups, and match lines.
2. To launch the Create View Frames Wizard, select the Create View Frames button on the Plan Production panel found on the Output tab of the Ribbon.
3. On the Alignment page, select Haven CL from the Alignment drop-down menu. For Station Range, verify that Automatic is selected, and click Next to advance to the next page.
4. On the Sheet page, select the Plan and Profile option.
5.  Click the ellipsis button to display the Select Layout as Sheet Template dialog. Then, click the next ellipsis and browse to the Plan Production subfolder in the default template file location. Typically, for Windows XP, this is

```
C:\Documents and Settings\<username>\Local Settings\Application
Data\Autodesk\C3D 2010\enu\Template\Plan Production\
```

6. Select the template named `Civil 3D (Imperial) Plan and Profile.dwt`, and click Open.
7. A list of the layouts in the DWT file appears in the Select Layout as Sheet Template dialog. Select the layout named ARCH D Plan and Profile 20 Scale, and click OK.
8. In the View Frame Placement section, select the Along Alignment option. Then, select Set the First View Frame before the Start of the Alignment By. Note that the default value for this particular drawing is 25′. Click Next to advance to the next page.
9. On the View Frame Group page, confirm that all settings are as follows (these are the same settings shown previously in Figure 20.5), and then click Next to advance to the next page:

Setting	**Value**
View Frame Group Name	VFG - <[View Frame Group Alignment Name(CP)]> - (<[Next Counter(CP)]>)
View Frame Name	VF - (<[Next Counter(CP)]>)
Style	Standard - Revised
Label Style	Standard - Revised
Label Location	Top Center

10. On the Match Lines page, review the default settings and click Next to advance to the next page.

11. On the last page of the wizard, confirm that the settings are as follows (these are the same settings shown previously in Figure 20.8), and then click Create View Frames:

Setting	Value
Select Profile View Style	Full Grid
Select Band Set Style	Plan Profile Sheets - Elevations and Stations

The view frames and match lines are created and are displayed as a collection in Prospector, as shown in Figure 20.9.

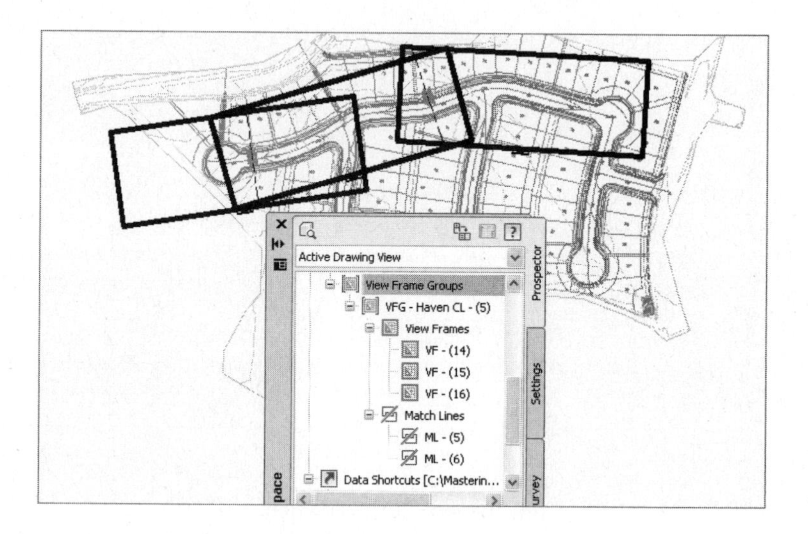

FIGURE 20.9
Finished view frames in the drawing (top) and in Prospector (bottom)

The numbering for your view frames, view-frame groups, and match lines may not identically match that shown in the images. This is due to the incremental counting Civil 3D performs in the background. Each time you create one of these objects, the counter increments. You can reset the counter by modifying the name template.

Editing View Frames and Match Lines

After you've created view frames and match lines, you may need to edit them. Edits to some view-frame and match-line properties can be made via the Prospector tab in the Toolspace palette. For both view frames and match lines, you can only change the object's name and/or style via the Information tab in the Properties dialog. All other information displayed on the other tabs is read-only.

You make changes to geometry and location graphically using special grip-edits (Figure 20.10). Like many other Civil 3D objects with special editing grips (such as profiles and Pipe Network objects), view frames and match lines have editing grips you use to modify the objects' location, rotation, and geometry. Let's look at each separately.

View frames can be graphically edited in three ways. You can move them, slide them along the alignment, and rotate them as follows:

- **To move a view frame:** The first grip is the standard square grip that is used for most typical edits, including moving the object.

FIGURE 20.10
View-frame and match-line grips

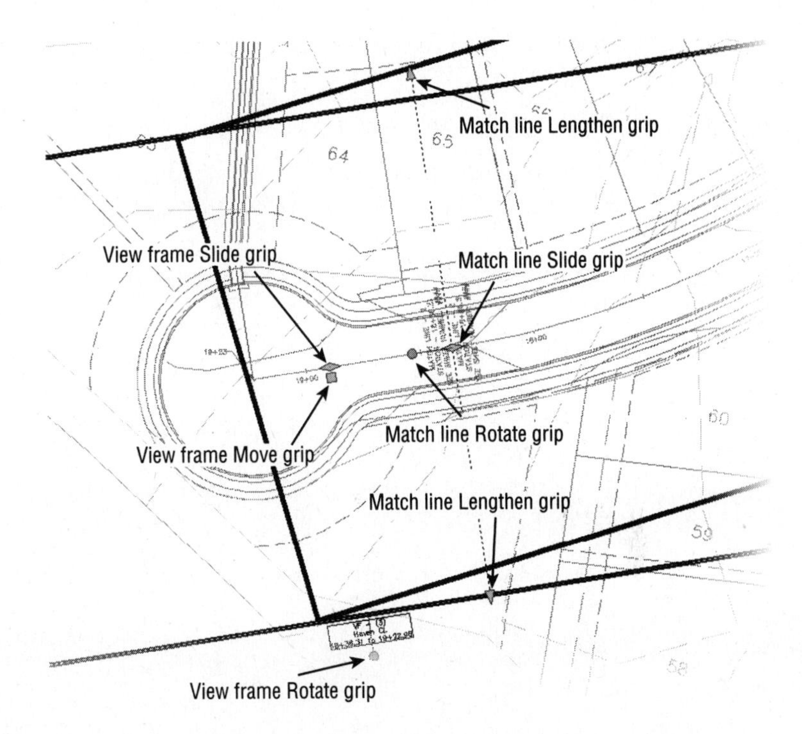

- **To slide a view frame:** Select the view frame to be edited, and then select the diamond-shaped grip at the center of the frame. This grip lets you move the view frame in either direction along the alignment while maintaining the orientation (Along Alignment or Rotated North) you originally established for the view frame when it was created.
- **To rotate a view frame:** Select the frame, and then select the circular handle grip. This grip works like the one on pipe-network structures. Using this grip, you can rotate the frame about its center.

 **Real World Scenario**

DON'T FORGET YOUR AUTOCAD FUNCTIONS!

While you're getting wrapped up in learning all about Civil 3D and its great design tools, it can be easy to forget you're sitting on an incredibly powerful CAD application.

AutoCAD features add functionality beyond what you can do with Civil 3D commands alone. First, make sure the DYN (Dynamic Input) option is turned on. This gives you additional functionality when you're moving a view frame. With DYN active, you can key in an exact station value to precisely locate the frame where you want it. Similar to view-frame edits, with DYN active, you can key in an exact rotation angle. Note that this rotation angle is relative to your drawing settings (for example, 0 degrees is to the left, 90 degrees is straight up, and so on). Also, selecting multiple objects and then selecting their grips while holding Shift makes each grip "hot" (usually a red color). This allows you to grip-edit a bunch of objects at once, like sliding a group of view frames along the alignment.

Whether you're learning Civil 3D yourself or training a group, it's a good idea to spend some time looking at the new AutoCAD features with every new release. You never know when you'll discover a nugget that cuts hours off your work day!

You can edit a match line's location and length using special grips. Like view frames, you can slide them along the alignment and rotate them. They can also be lengthened or shortened. Unlike view frames, they can't be moved to an arbitrary location.

- **To slide a match line:** Select the match line to be edited, and then select the diamond-shaped grip at the center of the match line. This grip lets you move the match line in either direction along the alignment while maintaining the orientation (Along Alignment or Rotated North) that you originally established for the view frame. Note that match line can only be moved in either direction a distance equal to or less than that entered on the Match Line page of the wizard at the time the view frames were created. For example, if you entered a value of 50′ for the Allow Additional Distance for Repositioning option, your view frames are overlapped 50′ to each side of the match line, and you can slide the match line only 50′ in either direction from its original location.
- **To rotate a match line:** Select the match line, and then selecting the circular handle grip. This grip works like the one on a view frame.
- **To change a match line's length:** When you select a match line, a triangular grip is displayed at each end. You can use these grips to increase or decrease the length of each half of the match line. For example, moving the grip on the top end of the match line changes the length of only the top half of the match line; the other half of the match line remains unchanged. See the sidebar "Don't Forget Your AutoCAD Functions!" for tips on using AutoCAD features.

The following exercise lets you put what you've learned into practice. Make sure Dynamic Input is active:

1. Continue working in the drawing from the previous exercise or open `Edit ViewFrames and MatchLines.dwg`. This drawing contains view frames and match lines. You'll change the location and rotation of a view frame and change the location and length of a match line.
2. Select the middle view frame, VF-2, and select its diamond-shaped sliding grip. Slide this grip to the left so the overlap with VF-1 isn't so large. Either graphically slide it to station 17+25 or enter **1725** in the Dynamic Input text box. Notice that the view-frame label is updated with the revised stations.
3. Select the circular rotation grip. Rotate the view frame slightly to better encompass the road. In the Dynamic Input text box, enter **282**. Then, press Esc to clear the grips from VF-2.
4. Now you'll adjust the match line's location and length. Select ML-1, and then select its diamond sliding grip. Either graphically slide it to station 14+70 or enter **1470** in the Dynamic Input text box. Notice that the match-line label is updated with the revised station.

5. Next, you'll lengthen the lower half of the match line so it extends the width of the view frame. Select the triangular lengthen grip at the lower end of ML-1, and either graphically lengthen it to 140′ or enter **140** in the Dynamic Input text box. Press Esc to exit the command.

Using Sheets

Civil 3D's Plan Production feature uses the concept of *sheets* to generate the pages that make up a set of plans. Simply put, *sheets* are layout tabs with viewports showing a given portion of your design model, based on the view frames previously created. The viewports have special properties set that define them as either Plan or Profile viewports. These viewports must be predefined in a template (DWT) file. You manage the sheets using the standard AutoCAD Sheet Set Manager feature; you can add sheet drawings and the drawing sheet-set file (DST) to your project in Vault. (See Chapter 19, ''Teamwork: Vault Client and Civil 3D,'' for more information on Vault.)

The Create Sheets Wizard

After you've created view frames and match lines, you can proceed to the next step of creating sheets. Like view frames, sheets are created using a wizard. Let's look at the wizard and the various page options. After you've walked through each page, you'll have a chance to put what you've learned into practice.

You launch the Create Sheets Wizard by switching to the Output tab and clicking the Create Sheets button on the Plan Production panel. A list of the wizard's pages is shown along the left side, and an arrow indicates which page you're currently viewing. You move among the pages using the Next and Back navigation buttons along the bottom of each page. Alternatively, you can jump directly to any page by clicking its name in the list on the left. Let's examine the wizard's pages and the features of each.

View Frame Group and Layouts Page

You use the first page of this wizard (Figure 20.11) to select the view-frame group for which the sheets will be created. It's also used to define how the layouts for these sheets will be generated.

View Frame Group In the first section of this page, you select the view-frame group either from the drop-down menu or by clicking the Select from the Drawing button to select the view-frame group from the drawing. After you've selected the group, you use the View Frame Range option to create sheets for all frames in the group or only for specific frames of your choosing.

All Select this option when you want sheets to be created for all view frames in the view-frame group.

Selection Selecting this option activates the Choose View Frames button. Click this button to select specific view frames from a list. You can select a range of view frames by using the standard Windows selection technique of clicking the first view frame in the range and then holding Shift while you select the last view frame in the range. You can also select individual view frames in nonsequential order by holding Ctrl while you make your view-frame selections. Figure 20.12 shows two of the three view frames selected in the Select View Frames dialog.

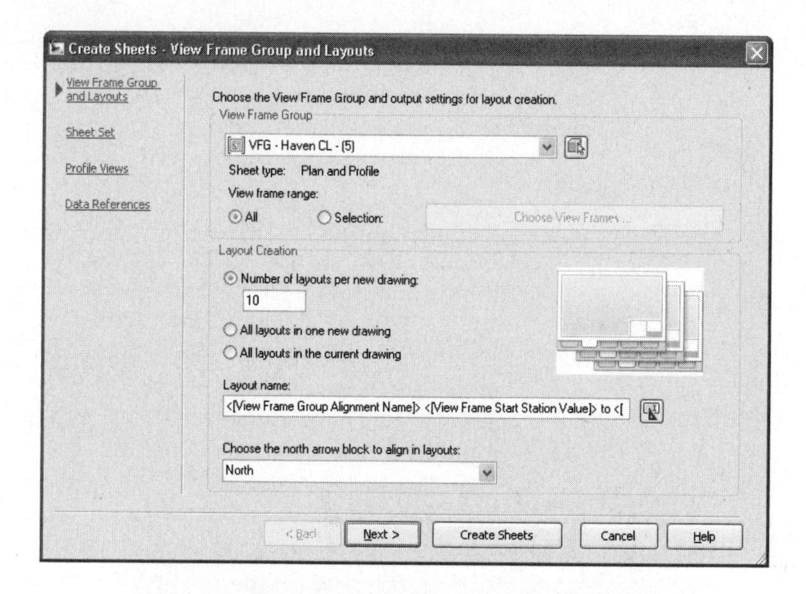

FIGURE 20.11
Create Sheets–View Frame Group and Layouts

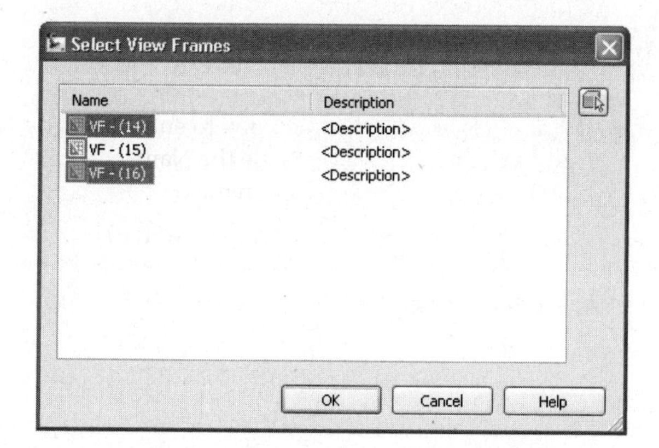

FIGURE 20.12
Select view frames by using standard Windows techniques

Layout Creation In this section, you define where and how the new layouts for each sheet are created as well as the name format for these sheets, and you specify information about the alignment of the north arrow block.

There are three options for creating layout sheets: all the layout tabs are created in the current drawing (the drawing you're in while executing the Create Sheets Wizard); all the new layouts are created in a new drawing file; or the layouts are created in multiple new drawing files, limiting the maximum number of layout sheets created in each file.

Number of Layouts Per New Drawing This option creates layouts in new drawing files and limits the maximum number of layouts per drawing file to the value you enter in the text box. For best performance, Autodesk recommends that a drawing file contains no more than 10 layouts. On the last page of this wizard, you're given the option to select the objects for which data references will be made. These data references are then created in the new drawings.

All Layouts in One New Drawing As the name implies, this option creates all layouts for each view frame in a single new drawing. Use this option if you have fewer than 10 view frames, to ensure best performance. If you have more than 10 view frames, use the previous option. On the last page of this wizard you're given the option to select the objects for which data references will be made. These data references are then created in the new drawings.

All Layouts in the Current Drawing When you choose this option, all layouts are created in the current drawing. You need to be aware of two scenarios when working with this option. (As explained later, you can share a view-frame group via Vault or Data Shortcuts and reference it into other drawings as a data reference.)

- When creating sheets, it's possible that your drawing references the view-frame group from another drawing or Vault (rather than having the original view-frame group in your current drawing). In this case, you're given the option to select the additional objects for which data references will be made (such as alignments, profiles, pipes, and so on). These data references are then created in the current drawing. You select these objects on the last page of the wizard.
- If you're working in a drawing in which the view frames were created (therefore, you're in the drawing in which the view-frame group exists), the last page of this wizard is disabled. This is because in order to create view frames (and view-frame groups), the alignment (and possibly the profile) must either exist in the current drawing or be referenced as a data reference (recall the prerequisites for creating view frames, mentioned earlier).

Layout Name Use this text box to enter a name for each new layout. As with other named objects in Civil 3D, you can use the Name template to create a name format that includes information about the object being named.

Align the North Arrow Block in Layouts If the template file you've selected contains a north arrow block, it can be aligned so that it points north on each layout sheet. The block must exist in the template. If there are multiple blocks, select the one you want to use from the drop-down menu.

WHERE AM I?

It's highly recommended that you set up the Name template for the layouts so that it includes the alignment name and the station range. This conforms to the way many organizations create sheets, helps automate the creation of a sheet index, and generally makes it easier to navigate a DWG file with several layout tabs.

CREATE SHEETS PAGE

You use the second page of the wizard (Figure 20.13) to determine whether a new or existing sheet set (`.dst`) is used and the location of the DST file. The sheet name and storage location are also defined here. Additionally, on this page you decide whether to add the sheet-set file (`.dst`) and the sheet files (`.dwg`) to the project vault.

Vault The first item on this page is the Add Files to Vault check box. This option is available only if you're logged in to Vault. If you aren't logged in, click the Log In to Vault button, and log in. Once you're logged in, the box is selected, and the drawing sheet set (`.dst`) and drawing sheets (`.dwg`) created by this wizard will be added to the project.

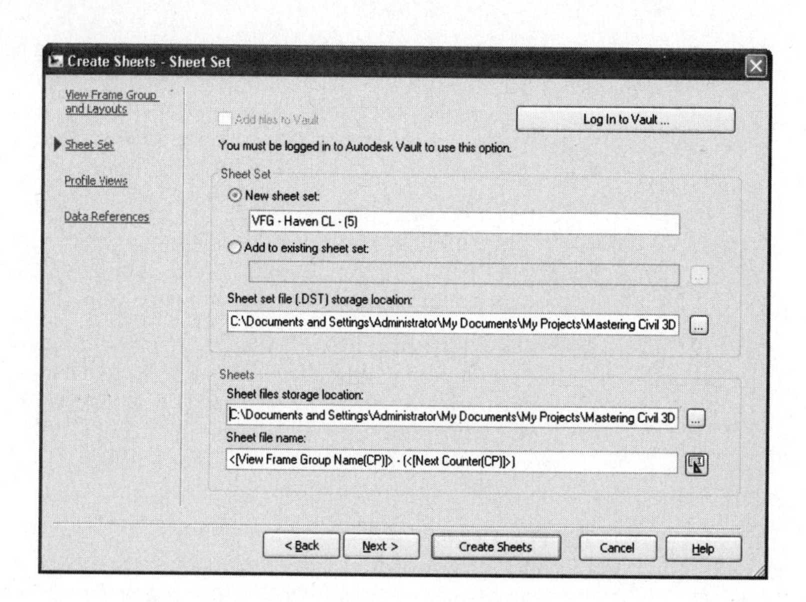

FIGURE 20.13
Create Sheets–Sheet Set

Sheet Set In this section of the page, you select whether to create a new DST file or add the sheets created by this wizard to an existing DST file.

New Sheet Set By selecting this option, you create a new sheet set. You must enter a name for the DST file and a storage location. By default, the sheet set is created in the same folder as the current drawing. You can change this by clicking the ellipsis and selecting a new location.

Add to Existing Sheet Set Selecting this option lets you select an existing sheet-set file to which the new sheets created by this wizard will be added. Click the ellipsis to browse to the existing DST file location.

Sheets You use the last section of the page to set the name and storage location for any new DWG files created by this wizard. On the previous page of the wizard, you had the choice of creating new files or creating the sheet layout in the current drawing. If you chose the latter option, the Sheets section on this page of the wizard is inactive. If you chose the former, here you enter the sheet file (DWG) name and the storage location.

WHAT IS A SHEET?

This page can be a little confusing due to the way the word *sheets* is used. In some places, *sheets* refers to layout tabs in a given drawing (.dwg) file. On this page, though, the word *sheets* is used in the context of Sheet Set Manager and refers to the DWG file itself.

PROFILE VIEWS PAGE

The next page of the wizard (Figure 20.14) lists the profile-view style and the band set selected in the Create View Frames Wizard. You can't change these selections. You can, however, make adjustments to other profile settings.

FIGURE 20.14
Create Sheets–Profile Views

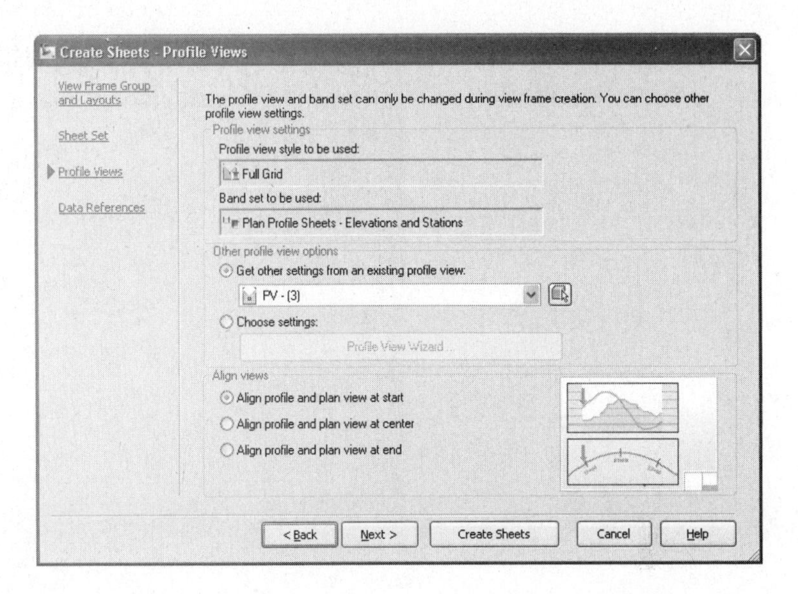

The Other Profile View Options section lets you modify certain profile-view options either by running the Profile View Wizard or by using an existing profile view in your drawing as an example. Regardless of what option you choose, the ''other options'' you can change are limited to the following:

- ◆ Split profile-view options from the Profile View Height page of the Profile View Wizard
- ◆ All options on the Profile Display page
- ◆ Most of the Data Band Page settings
- ◆ Profile Hatch Options
- ◆ All settings on the Multiple Plot Options page

See Chapter 8, ''Cut to the Chase: Profiles,'' for details of each of these settings.

DATA REFERENCES PAGE

The final page of the Create Sheets Wizard (Figure 20.15) is used to create data references in the drawing files that contain your layout sheets.

Based on the view-frame group used to create the sheets and the type of sheets (plan, profile, plan and profile), certain objects are selected by default. You have the option to select additional objects for which references will be made. You can either pick them from the list or click the Pick from Drawing button and select the objects from the drawing.

It's common to create references to pipe networks that are to be shown in plan and or profile views. If you choose to create references for pipe-network objects, you can also copy the labels for those network objects into the sheet's drawing file. This is convenient in that you won't need to relabel your networks.

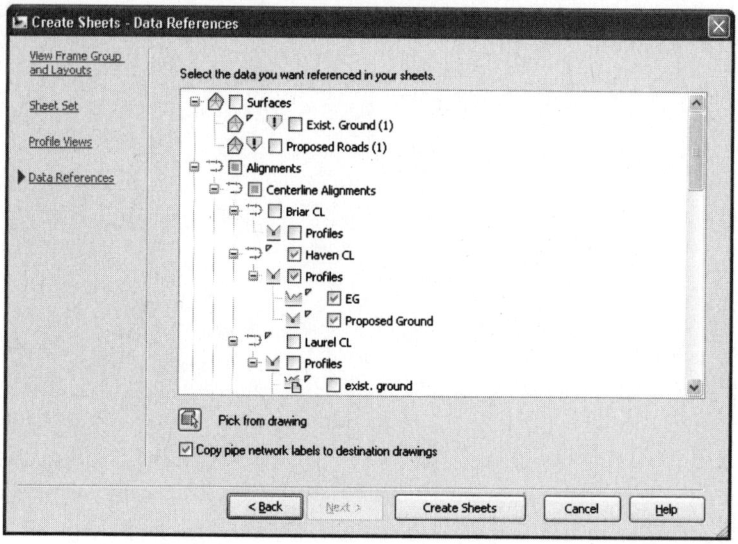

FIGURE 20.15
Select Create Sheets–Data References

Managing Sheets

After you've completed all pages of the wizard, you create the sheets by clicking the Create Sheets button. Doing so completes the wizard and starts the creation process. If you're creating sheets with profile views, you're prompted to select a profile-view origin. Civil 3D then displays several dialog boxes, indicating the process status for the various tasks such as creating the new sheet drawings and creating the DST file.

If the Sheet Set Manager (SSM) isn't currently open, it opens with the newly created DST file loaded. The sheets are listed, and the details of the drawing files for each sheet appear in the Details area (Figure 20.16).

If you double-click to open the new drawing file that contains the newly created sheets, you'll see layout-sheet tabs created for each of the view frames. The sheets are named using the Name template as defined in the Create Sheets Wizard. Figure 20.17 shows the names that result from the following template:

```
<[View Frame Group Alignment Name]> <[View Frame Start Station Value]>
 to <[View Frame End Station Value]>
```

To create the final sheets in this new drawing, Civil 3D externally references (XRefs) the drawing containing the view frames; creates data references (DRefs) for the alignments, profiles, and any additional objects you selected in the Create Sheets Wizard; and, if profile-sheet types were selected in the wizard, creates profile views in the final sheet drawing.

Finally, if you selected the Vault option, the drawing file containing the sheets and the DST file are checked into your project.

The following exercise pulls all these concepts together:

1. Open `Sheets_Wizard.dwg`, which you can download from `www.sybex.com/go/masteringcivil3d2010`. This drawing contains the view-frame group, alignment, and profile for Haven Road. Note that the drawing doesn't have profile views.

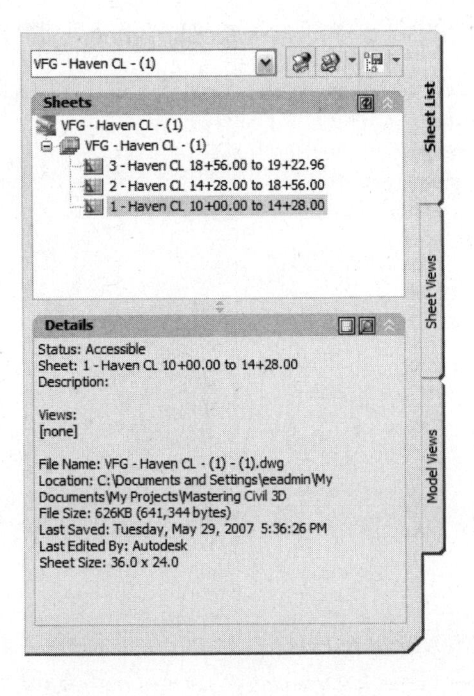

FIGURE 20.16
New sheets in the Sheet Set Manager

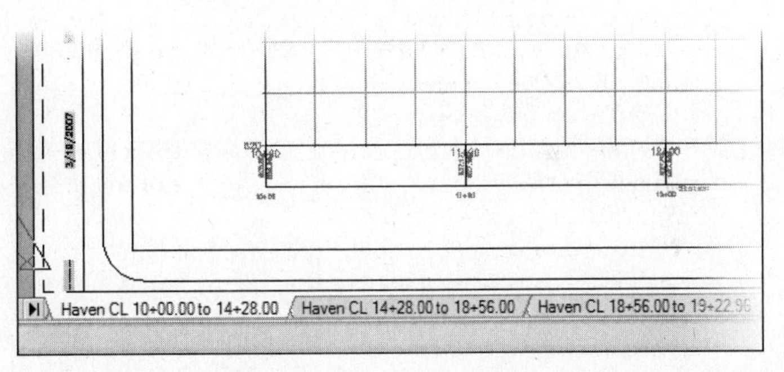

FIGURE 20.17
The template produces the Haven CL tab names shown here

2. To launch the Create Sheets Wizard, select the Create Sheets button on the Plan Production panel found on the Output tab of the ribbon.
3. On the View Frame Group and Layouts page, confirm that View Frame Range is set to All and that Number of Layouts Per New Drawing is set to 10. Click Next.
4. On the Sheet Set page, select the New Sheet Set option. For both Sheet Set File Storage Location and Sheet Files Storage Location, use the ellipsis to browse to `C:\Mastering Civil 3D 2010\CH 20\Final Sheets`, and click OK. Click Next.
5. On the Profile Views page, for Other Profile View Options, select Choose Settings and then click the Profile View Wizard button. The Create Multiple Profile Views dialog opens.
6. On the left side of the Create Multiple Profile Views dialog, click Profile Display Options to jump to that page.

For the Proposed Ground profile, scroll to the right and modify the Labels setting, changing it from Standard to Complete Label Set, as shown in Figure 20.18. After you select the label set, click OK; then, click Next to advance to the Data Bands page.

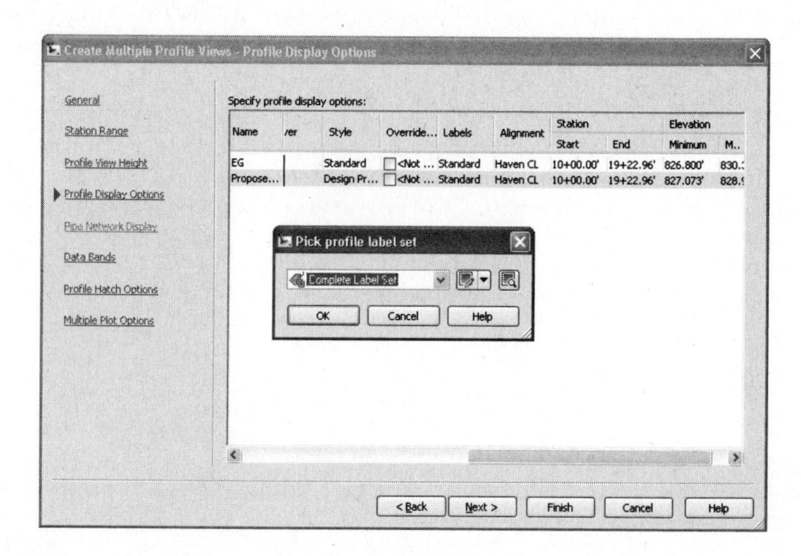

FIGURE 20.18
Change the labels for the Proposed Ground profile

7. On the Data Bands page, change Profile2 to Proposed Ground, as shown in Figure 20.19. Click Finish to return to the Create Sheets Wizard. Click Next to advance to the Data References page.

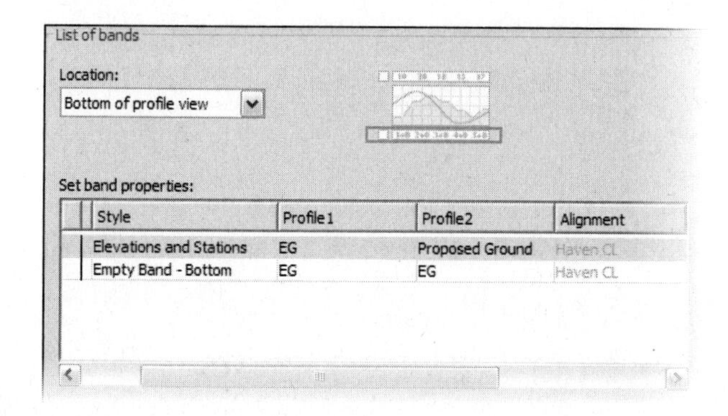

FIGURE 20.19
Set Profile2 to Proposed Ground

8. On the Data References page, confirm that Haven CL and both of its profiles are selected. Click Create Sheets to complete the wizard.

9. Before creating the sheets, Civil 3D must save your current drawing. Click OK when prompted. The drawing is saved, and you're prompted for an insertion point for the profile view. The location you pick represents the lower-left corner of the profile-view grid. Select an open area in the drawing, above the left side of the site plan. Civil 3D displays a progress dialog, and then Panorama is displayed with information about the results of the sheet-creation process. Close the Panorama window.

INVISIBLE PROFILE VIEWS

Note that the profile views are created in the current drawing only if you selected the option to create all layouts in the current drawing. Because you didn't do that in this exercise, the profile views aren't created in the current drawing. Rather, they're created in the sheet drawing files in Model space in a location relative to the point you selected in this step.

10. After the sheet-creation process is complete, the Sheet Set Manager window is open (Figure 20.20). Click sheet 1, named Haven CL 10+00.00 to 14+28.00. Notice that the name conforms to the Name template and includes the alignment name and the station range for the sheet. Review the details listed for the sheet. In particular, note the filename and storage location.

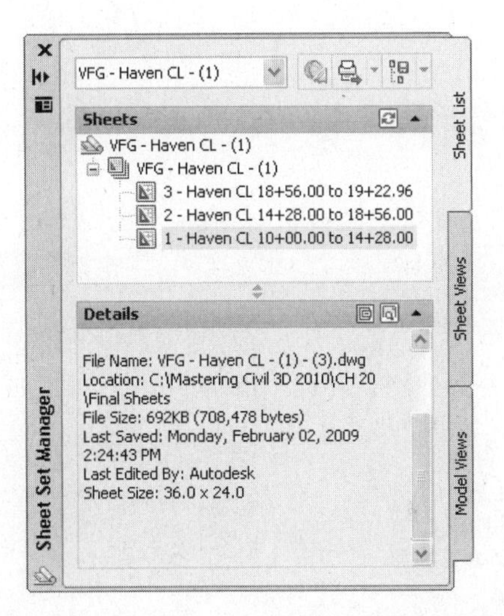

FIGURE 20.20
The Sheet Set Manager

11. Double-click this sheet to open the new sheets drawing and display the layout tab for Haven CL 10+00.00 to 14+28.00. Review the multiple tabs created in this drawing file. The template used also takes advantage of AutoCAD fields, some of which don't have values assigned.

Supporting Components

The beginning of this chapter mentioned that there are several prerequisites to using the Plan Production tools in Civil 3D. The list includes drawing templates (DWT) set up to work with the Plan Production feature and styles for the objects generated by this feature. In this section of the chapter, you'll look at preparing these items for use in creating your finished sheets.

Templates

Civil 3D ships with several predefined template files for various types of sheets that Plan Production can create. By default, these templates are installed in a subfolder called `Plan Production`, which is located in the standard `Template` folder. You can see the `Template` folder location by opening the Files tab of the Options dialog, as shown in Figure 20.21.

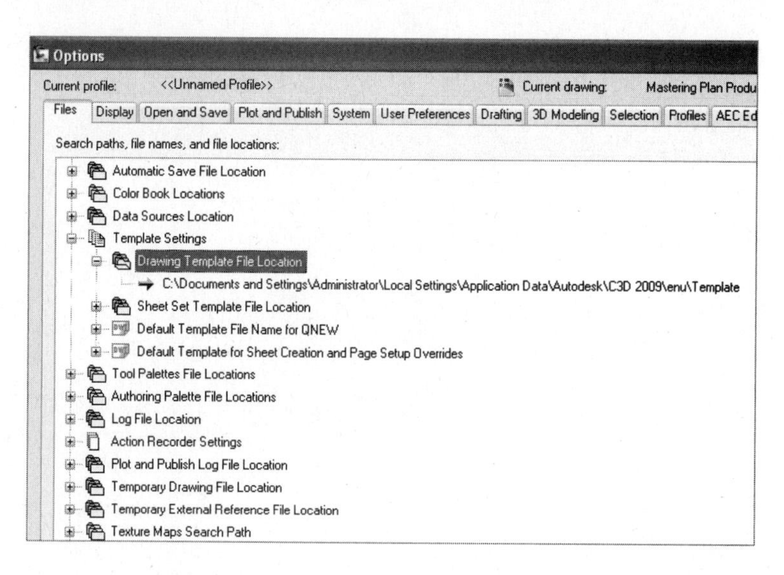

FIGURE 20.21
Template files location

Figure 20.22 shows the default contents of the `Plan Production` subfolder. Notice the templates for Plan, Profile, and Plan and Profile sheet types. There are metric and imperial versions of each.

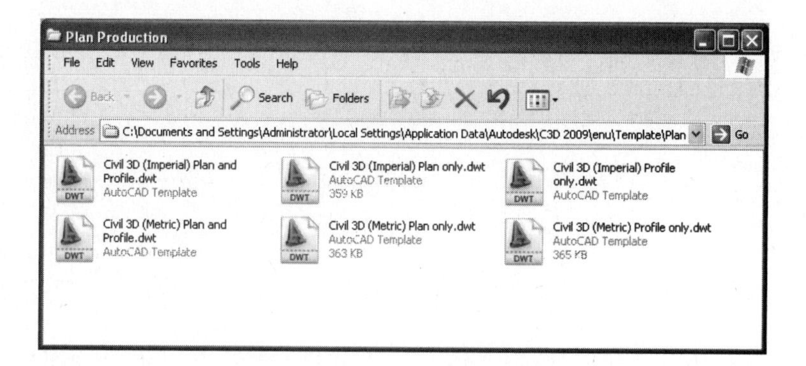

FIGURE 20.22
Plan Production DWT files

Each template contains layout tabs with pages set to various sheet sizes and plan scales. For example, the `Civil 3D (Imperial) Plan and Profile.dwt` template has layouts created at various ANSI and ARCH sheets sizes and scales, as shown in Figure 20.23.

The viewports in these templates must be rectangular in shape and must have Viewport Type set to either Plan or Profile, depending on the intended use. You set Viewport Type on the Design tab of the Properties dialog, as shown in Figure 20.24.

IRREGULAR VIEWPORT SHAPES

Just because the viewports must start out rectangular doesn't mean they have to stay that way. Experiment with creating viewports from rectangular polylines that have vertices at the midpoint of each side of the viewport (not just at the corners). After you've created your sheets using the Plan Production tool, you can stretch your viewport into irregular shapes.

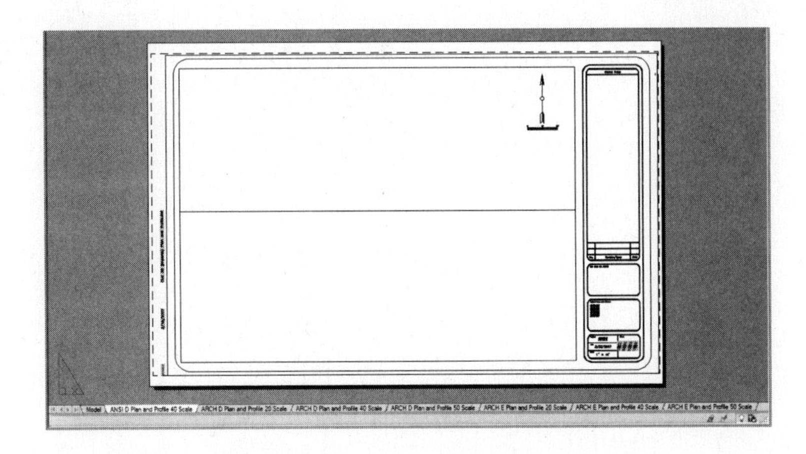

FIGURE 20.23
Various predefined layouts in standard DWT

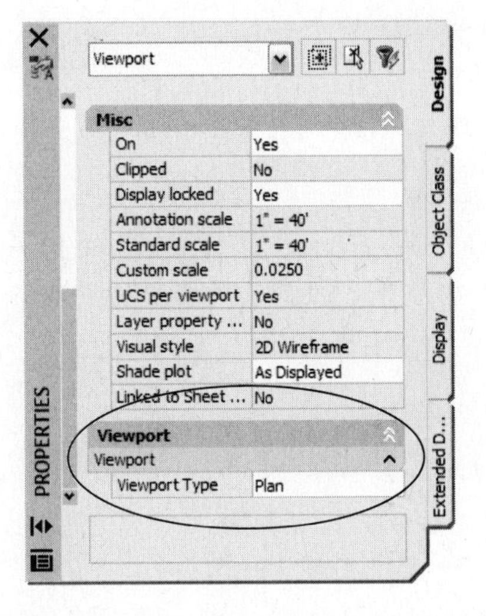

FIGURE 20.24
Viewport Properties–Viewport Type

Styles and Settings

If you've used previous versions of Civil 3D, you know it ships with several general template files that contain styles for all Civil 3D objects. These templates include object and label styles for the components used in plan production. They also include default feature and command settings for plan production. As this chapter has discussed, you use two main objects in plan production: view frames and match lines. Match lines are associated with view frames, which are collected in view-frame groups. Let's look at the object and label styles for each component in more detail:

View Frames The styles for view-frame objects and labels are straightforward. The object style has only a single component: View Frame Border. View-frame label styles are similar to other label styles used elsewhere in Civil 3D. You use the Label Style Composer to add Text, Line Block, and Reference Text components. Referenced Text can be used to include object information from an Alignment, COGO Point, Parcel, Profile, or Surface in your view-frame label.

Match Lines Like view-frame styles, match-line styles are pretty basic. The object style has two components: Lines and Match Line Mask. This second component, which controls the masking hatch that is displayed in the overlap area beyond the match line, warrants further explanation. Figures 20.25 and 20.26 show the Component Hatch Display settings for a match-line style and the resulting sheet.

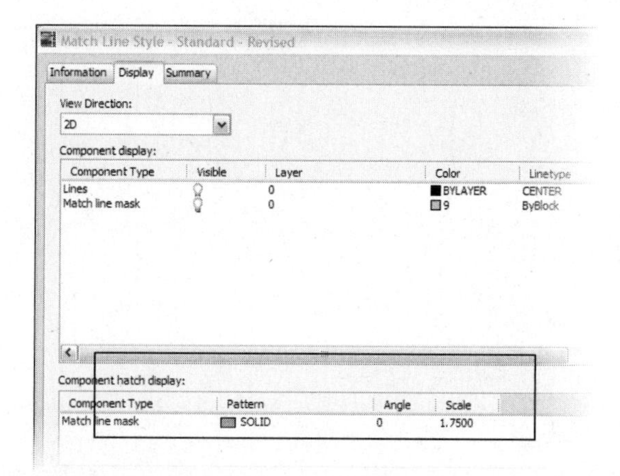

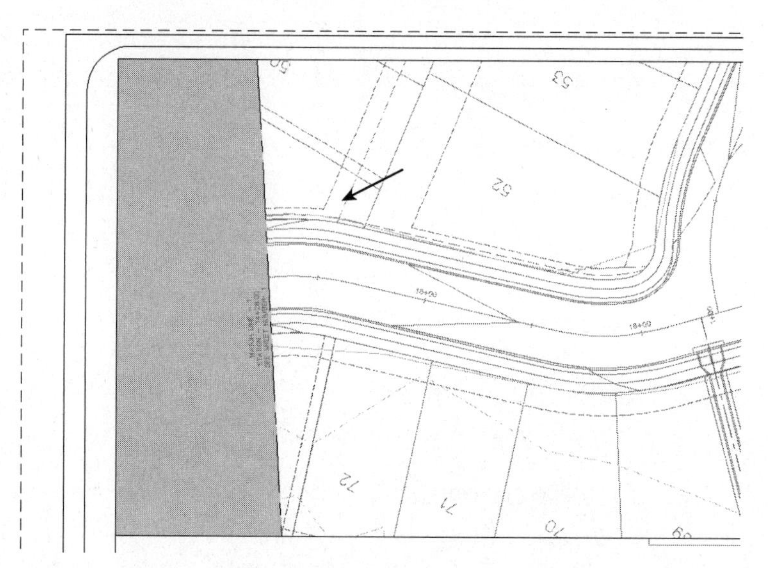

FIGURE 20.25 The display of the indicated area, beyond the match line, is controlled by the Component Hatch Display settings of the match-line object style

You define the match-line style in the drawing containing the view frames. Any changes to the style must be made in the view-frames drawing; the drawing must then be saved and the XRef updated in the sheets drawings. If you created the layout sheets in the drawing in which the view frames exist, the match-line style updates are automatically displayed on the layout sheets.

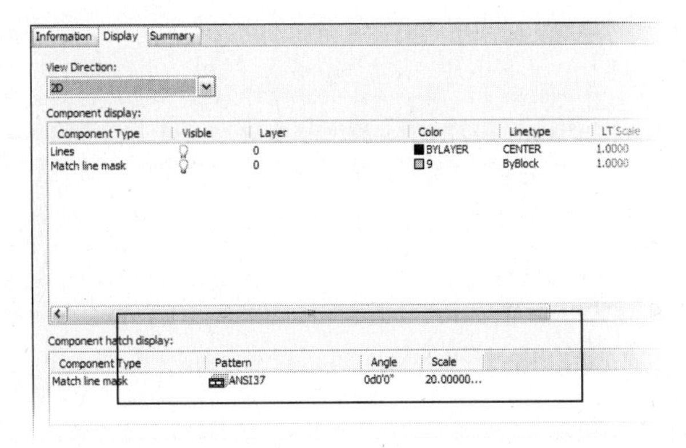

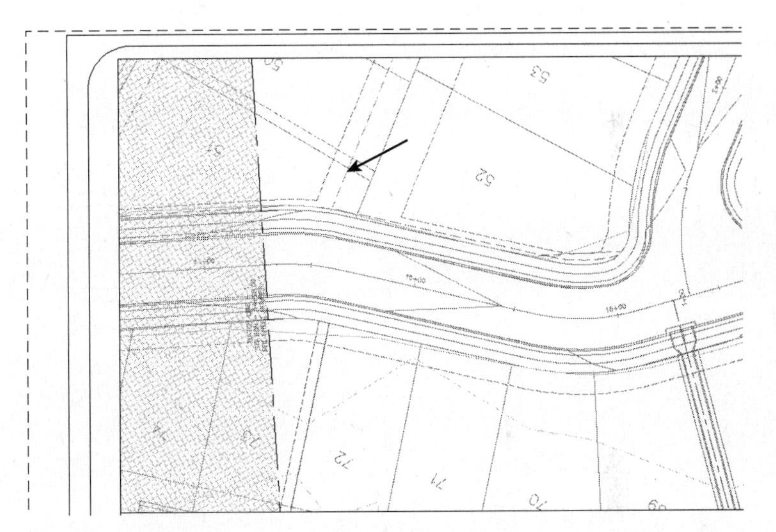

FIGURE 20.26 The hatching pattern has been updated to reflect the changes to the Component Hatch Display settings of the match-line object style

View Frame Groups The View Frame Group object has no styles — only feature and command settings (similar to subassemblies). The feature settings for this group are important because they're used to control the default settings for view frames and match lines. These default settings can be overridden by the command settings.

The Bottom Line

Create view frames. When you create view frames, you must select the template file that contains the layout tabs that will be used as the basis for your sheets. This template must contain predefined viewports. You can define these viewports with extra vertices so you can change their shape after the sheets have been created.

Master It Open the Mastering Plan Production drawing. Run the Create View Frames Wizard to create view frames for any alignment in the current drawing. (Accept the defaults for all other values.)

Create sheets and use Sheet Set Manager. You can create sheets in new drawing files or in the current drawing. Use the option to create sheets in the current drawing when a) you've referenced in the view-frame group, or b) you have a small project. The resulting sheets are based on the template you chose when you created the view frames. If the template contains customized viewports, you can modify the shape of the viewport to better fit your sheet needs.

Master It Continue working in the Mastering Plan Production drawing. Run the Create Sheets Wizard to create plan and profile sheets for Haven CL using the template `Mastering (Imperial) Plan and Profile.dwt`. This template has a custom-made viewport with extra vertices. (Accept the defaults for all other values.)

Edit sheet templates and styles associated with plan production. The Create View Frames command uses predefined styles for view frames and match lines. These styles must exist in the drawing in which you create the view-frame group. You can create match-line styles to meet your specific needs.

Master It Open the Mastering Plan Production 3 drawing, and modify the match-line component hatch display style setting to make the linework beyond the match line appear screened.

Chapter 21

Playing Nice with Others: LDT and LandXML

As great as it would be for the whole world to move to a Civil 3D model with real coordinates so everyone could share data without pause, it's not going to happen this year. In land development, you work with a large number of team partners on most projects. This can include architects, contractors, other civil engineers, planners, and landscape architects on a small job. On a large job, the number of people on this list may double or triple. An important part of being a good team member is sharing data cleanly. This chapter covers importing Land Desktop and LandXML data as well as exporting data in drawing and LandXML formats.

In this chapter, you'll learn to:

- Discern the design elements in a LandXML data file
- Import a Land Desktop project
- Import a LandXML file to create a surface, alignments, and profiles
- Create an AutoCAD drawing that can be read without special enablers

What Is LandXML?

For years, there was the humble text file. These files were the lingua franca of written documents and data input. Text files are still used for many tasks, in spite of the evolution of Microsoft Word DOC files, PDFs, and other common written formats. The problems occur when you try to send your Word document to someone running OpenOffice or WordPerfect. Formatting falls apart, font information is lost, and paragraphs don't justify as they should. Bring in the simple text file and stylize again! The idea is that by having a base-level format for the important part of the document in question, almost anyone can read and deal with the data as they wish.

LandXML is the text file of the land-development world. Created in January 2000 by an international consortium, LandXML is a nonproprietary data format that allows the exchange of land-development design elements between varying products. Members of the LandXML.org consortium include Autodesk Inc., Trimble Navigation, Bentley Systems Inc., and the U.S. Army Corps of Engineers. By working together to define a common language and then including LandXML tools in their own programs, these member groups have created a better translation mechanism for the land-development community at large.

Each version of this common language is known as a *schema* and contains important information about the format of the data that is included. It's important that you review and understand which schema you're attempting to read or write. Contact the software vendor or visit `landxml.org` for further information.

HANDY-DANDY NOTEPAD

The figures in this chapter show XML files in XML Notepad from Microsoft. XML Notepad is free and small, and it makes reading XML files much easier. It's highly recommend that you install it before beginning these exercises. Although you can view or edit XML files with the standard Windows Notepad, it's not something you want to do; it simply doesn't handle the data well, and it's very difficult to decipher.

An XML file is a specially formatted text file, similar to HTML or any of the modern web-based formats. Using an XML reader, you can read and edit these files. In this exercise, you'll look at a sample XML file and the sections it contains:

1. Open the `Niblo.xml` file. You can download this file from `www.sybex.com/go/masteringcivil3d2010`.
2. Expand the `LandXML` folder, as shown in Figure 21.1. This version is the LandXML schema that was used to create the file.

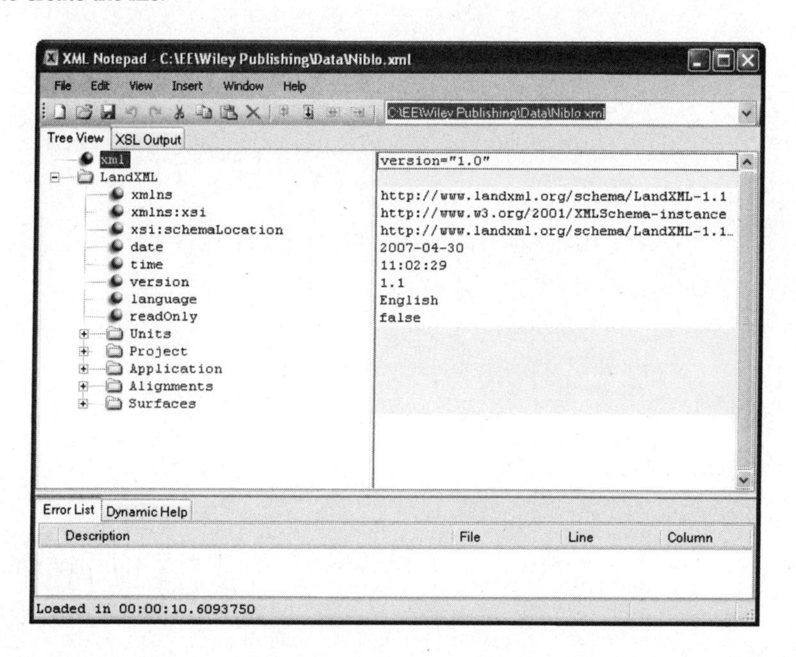

FIGURE 21.1
`Niblo.xml` file in XML Notepad

3. Expand the `Project` folder to view the name of the drawing from which the XML file was created.
4. Expand the `Application` folder to view the name of the program that created the file, the version of that program, and the creation timestamp on the XML file.
5. Expand the `Alignments` folder to explore the information associated with an alignment, as shown in Figure 21.2.
6. When you're done poking around, close the file without saving changes.

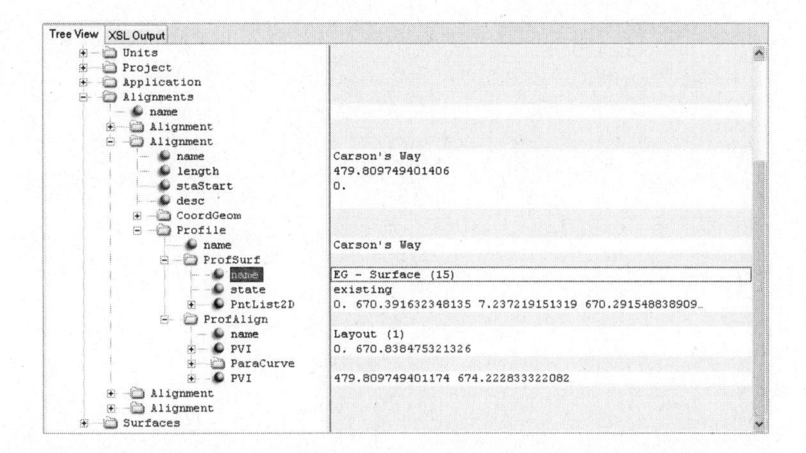

FIGURE 21.2
`Niblo.xml` file with the Carson's Way alignment section expanded

The schema defines how much information can be associated and defined in the XML file. In the case of alignments, Figure 21.2 shows both `ProfSurf` and `ProfAlign` folders under the `Profile` folder. These folders refer to a sampled-surface profile and a layout profile, respectively. By learning and understanding the names used in the XML file, you can discern the contained information before you import it to your Civil 3D model. The next section describes that step.

Handling Inbound Data

The data-conversion process takes place throughout a project. At the beginning of the job, bringing data into the Civil 3D object model is a crucial step. This data can be from existing Land Desktop (LDT) projects in the office, LDT data you've acquired from other consultants, or LandXML files from other packages. This section shows you how you can use both LDT and LandXML data to create a Civil 3D object model ready for design. You'll import Land Desktop project data and then some other data as a LandXML file.

Importing Land Desktop Data

Most of the firms moving to Civil 3D are coming from a Land Desktop background. In this environment, they had surfaces, alignments, profiles — the full gamut of project-design elements. By converting existing LDT data into a Civil 3D model, you ensure that these resources can continue to be used and pay dividends into the future. In this exercise, you'll import the LDT project that has provided much of the data in previous chapters:

1. Create a new Civil 3D drawing using the `AutoCAD Civil 3D (Imperial) NCS Extended` template file.
2. Change to the Insert tab and select Land Desktop from the Import panel.
3. Click the Browse button circled at upper right in Figure 21.3.
4. In the Browse For Folder dialog, expand the `LDT Import` folder, as shown in Figure 21.4, and click OK.
5. Click OK to dismiss the warning about pipes that appears. This warning is a reminder that you need a pipe part in the parts list to match the sizes described in the LDT project.

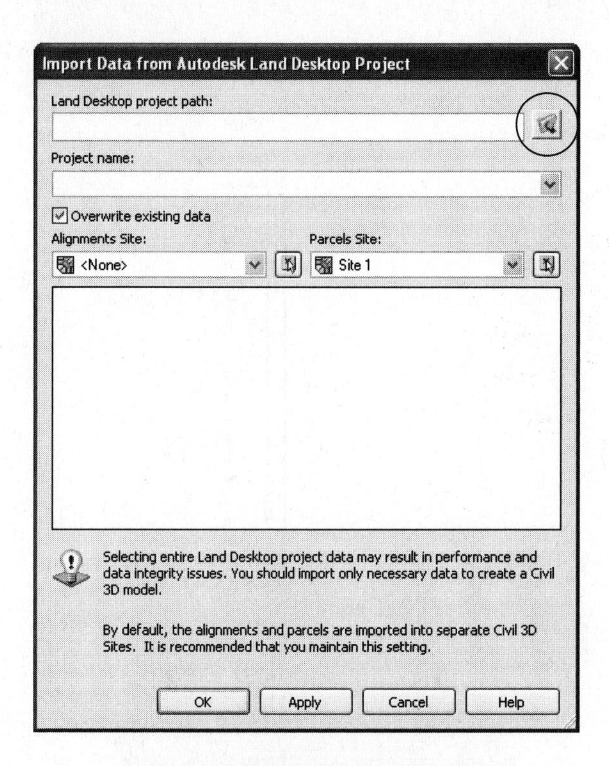

FIGURE 21.3
The Import Data from the Autodesk Land Desktop Project dialog

FIGURE 21.4
Selecting an LDT project path

6. Select Conklin.
7. Deselect the Pipe Runs portion of the data.
8. Deselect the Alignments heading, and then select the PROP RD alignment, as shown in Figure 21.5. Note that some branches have been collapsed for clarity.
9. Verify that Alignments Site is set to None.

FIGURE 21.5
The LDT project selected and ready for import

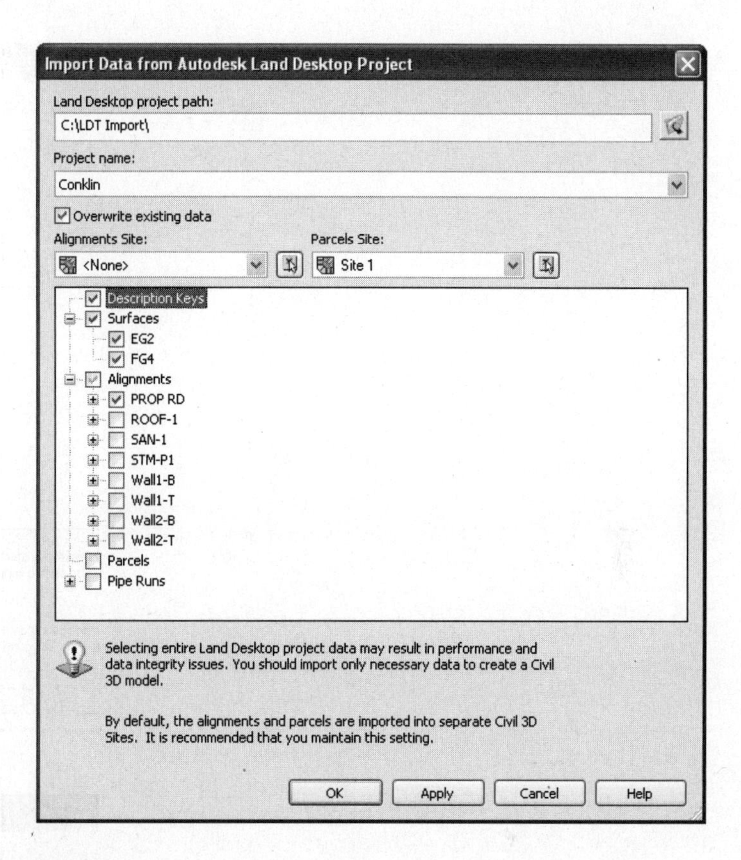

10. Click OK to import the LDT data.
11. Click OK to dismiss the Import Data from Autodesk Land Desktop Project dialog.
12. Perform a zoom extents to view the imported data, as shown in Figure 21.6.
13. Close the drawing without saving changes.

By working with existing LDT data, you extend its useful lifespan and make the transition to Civil 3D less of a traumatic change. Many firms have long histories of working with LDT data, and this ability continues to bring value to that history. The next section discusses importing LandXML, a methodology used frequently to communicate across firm boundaries.

WHAT ABOUT THE PARCELS?

In the example, you didn't attempt to use the Parcels import functions. Importing parcels in any format takes place as importing a series of closed objects. This means adjacent parcels often have duplicate lines. As you saw in Chapter 6, "Don't Fence Me In: Parcels," duplicate linework is one of the biggest causes of broken Civil 3D drawings. It's unfortunate, but it's generally recommended that you rework parcels to ensure a clean network topology.

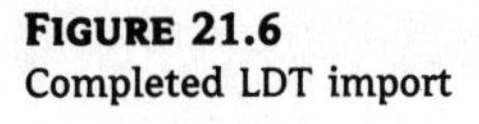

FIGURE 21.6
Completed LDT import

Importing LandXML Data

LandXML data comes from two major sources: land-development packages (such as Civil 3D, InRoads, and PowerCivil) and the fleet of design and analysis tools (such as StormCAD and IntelliSolve). You'll look first at a LandXML file created by Civil 3D and then at a LandXML file created by IntelliSolve.

LandXML data is more commonly sent from outside the organization, but it's sometimes used when you're trying to capture a snapshot between the LDT and Civil 3D data. Creating the XML file takes a picture at an instant in time so the source information in the XML file is always available for review. However, the LDT project may be in flux even after the import to Civil 3D takes place, so tracing the data back can be difficult.

In this exercise, you'll try the LandXML import process:

1. Create a new Civil 3D drawing using the `AutoCAD Civil 3D (Imperial) NCS Extended` template file.
2. Change to the Insert tab and select LandXML from the Import panel.
3. Navigate to the directory you have downloaded for this chapter, and select `Niblo.xml` to display the Import LandXML dialog.
4. Expand the Alignments Name and Surfaces branches, as shown in Figure 21.7, to review the data prior to importing the file.
5. Expand the Surfaces (1) ➤ Surface Desc branches to reveal the details in the XML file, as shown in Figure 21.8.

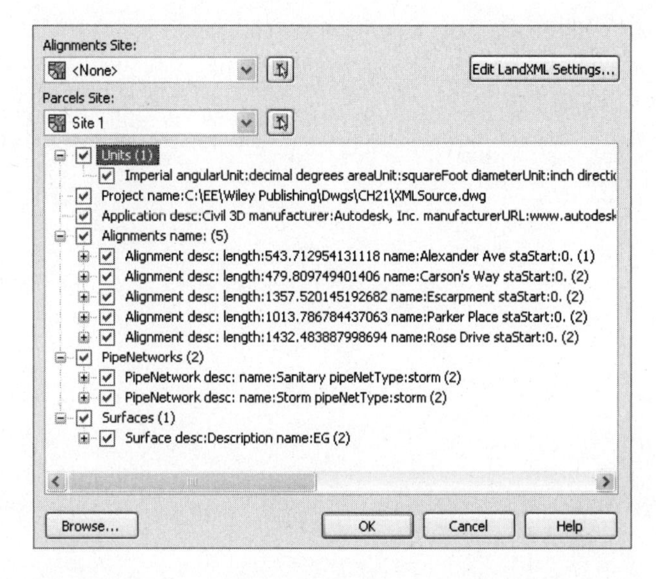

FIGURE 21.7
The LandXML file open for review

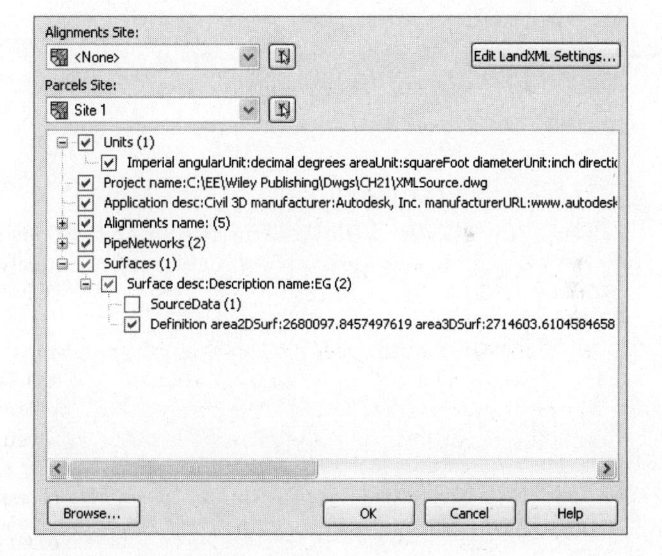

FIGURE 21.8
Expand the Surfaces branch in the Import LandXML dialog

6. Select the Source Data check box (see Figure 21.8) so the polylines from which the surface was created are *not* imported.

7. Click OK to import the data. Civil 3D imports the objects and zooms to the extents, as shown in Figure 21.9.

This data is ready for use, but let's look at some interesting things that come about during the LandXML process. The surface, alignments, and pipes are three major items of concern.

- Surfaces are created with a default style of Contours 2′ and 10′ (Background). The Surface Properties dialog in Figure 21.10 shows a more interesting option. Notice how the surface was built on the basis of the import of the XML file, but then a snapshot was automatically generated. Snapshots allow you to rebuild the surface from that point forward. What does this mean for you as a user? It means you could hypothetically send the drawing without the XML file, and the surface would still build. This method works well when you want to add a basic level of protection to your surface information. It's not perfect, but it can be interesting.

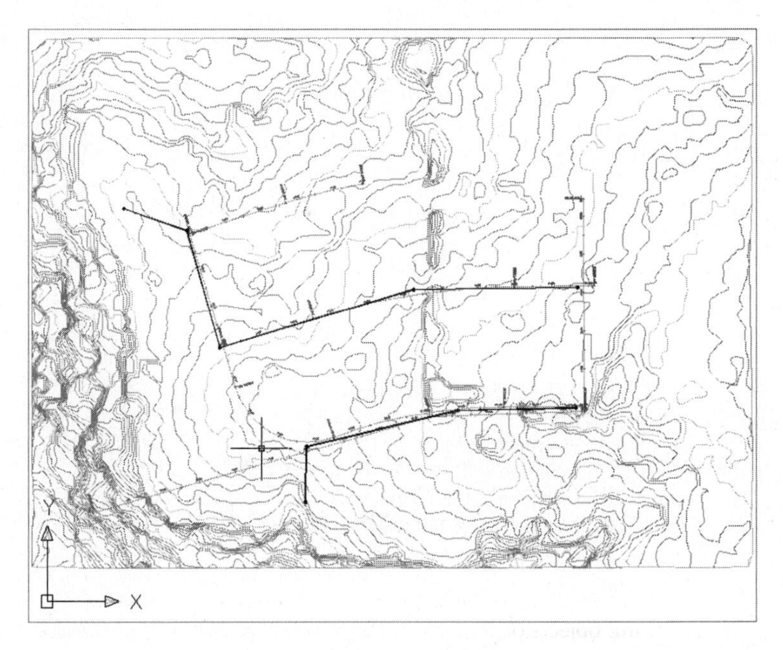

FIGURE 21.9
LandXML import completed

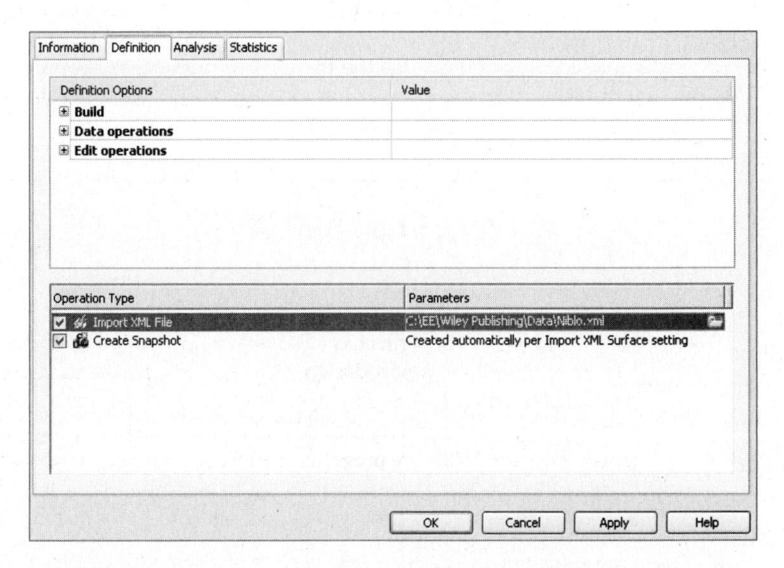

FIGURE 21.10
Surface Properties after LandXML import

- Alignments move to the siteless collection, but are considered Centerline Alignments as shown in Figure 21.11. Importing into the siteless Alignments collection means random parcels aren't created upon the XML import. Profiles are imported with alignments, but profile views aren't created. Review Chapter 9, "Slice and Dice: Profile Views in Civil 3D," for information on creating and manipulating profile views. Also, because LandXML schemas define alignments and other objects as hard-coded objects, the relationships aren't preserved. Each component of an alignment becomes a fixed entity, with no understanding of the component on either end of itself.

FIGURE 21.11
Imported alignments assigned with no site

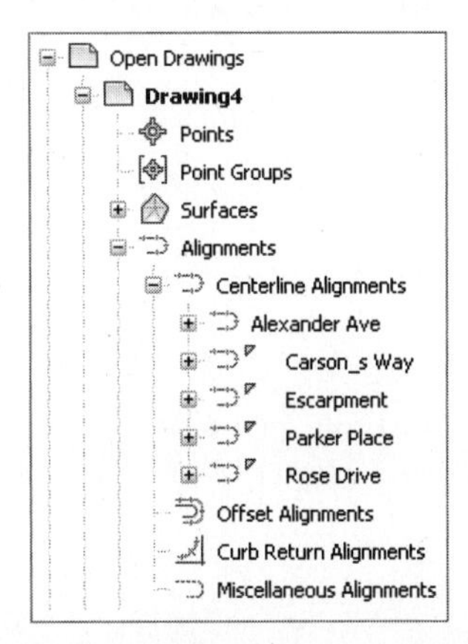

- Pipes must have a matching part size in both the target drawing and the source XML file. Your pipe network won't import if there is no valid pipe size in the current parts family as is specified in the XML file. See Chapter 14, "The Tool Chest: Parts Lists and Part Builder," for information about part lists and part families.

By using XML data from other sources, you eliminate the human error of importing by redrawing objects or the computational errors that come from importing points without breaklines in surfaces. LandXML allows for the easy sharing of critical data without the burden of the full drawing information.

By using LandXML as the translation system, even programs that have no direct hook into Civil 3D can be used as auxiliary design tools. Now that you've imported a number of different files and formats, let's look at the best way to get your Civil 3D model out to team members.

Real World Scenario

USING LANDXML IN ENGINEERING ANALYSIS

Most civil-engineering firms have at least one or two specialized analysis packages in the office. These can include tools for airport design, flow analysis, or soil mapping. Many of these packages are starting to include LandXML as an output format. To learn more about the various LandXML formats and how they have changed over time, visit `www.landxml.org`. In this exercise, you'll review an XML file created by the IntelliSolve program and import it into a blank Civil 3D drawing to see how the objects are converted:

1. Create a new Civil 3D drawing using the `AutoCAD Civil 3D (Imperial) NCS Extended` template file.
2. Change to the Insert tab and select LandXML from the Import panel.

3. Navigate to the `Data` directory, and select `Intellisolve.xml`. The Import LandXML dialog appears.

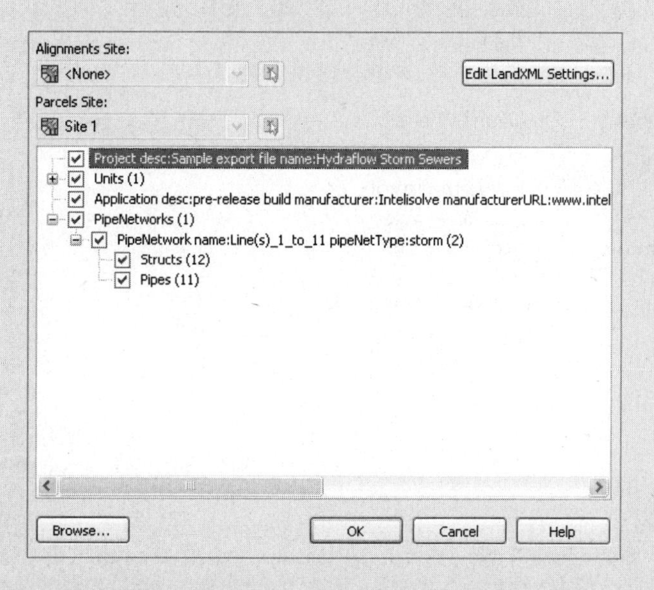

4. Click OK to finish the import. The drawing zooms to extents as in the other exercise and should look like this:

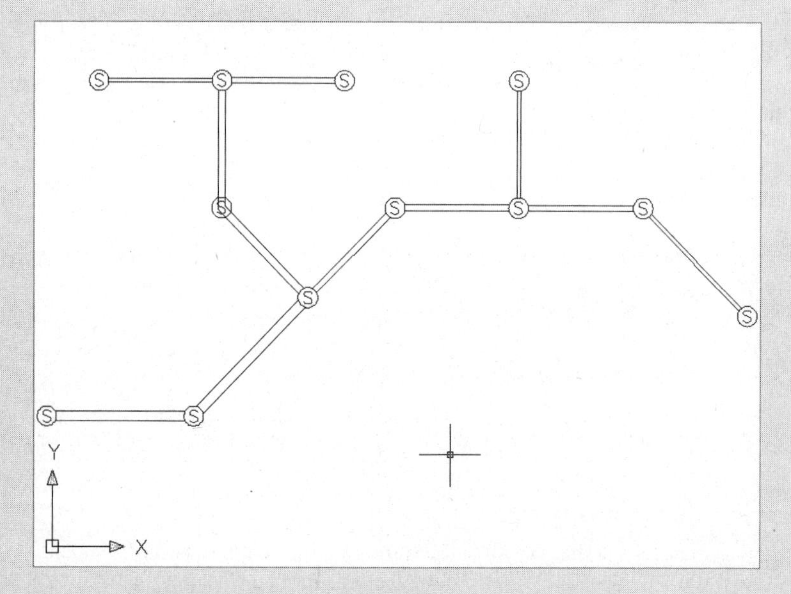

If you'd like to see more samples of LandXML from other design packages, check out `www.landxml.org` and click Samples. Sample files are available from many of the major consortium partners. You can use these files freely as long as you don't modify them.

Sharing the Model

Once you've created an active Civil 3D model, you'll inevitably have to share it. In many cases, this means printing plans. But your better team members will recognize the value in your design information and want something at a higher information level. This section shows you how to create XML files from Civil 3D as well as AutoCAD drawing files that can be shared.

Creating LandXML Files

Creating a LandXML file is simple — about three clicks total. The only hard part is dealing with the options and the schema involved on the other end. Because LandXML.org occasionally revises the schema to include new object information, going backward with LandXML isn't any easier than going backward with Civil 3D objects. In these exercises, you'll have to use the schema built into Civil 3D; but remember that your recipient may not have the latest schema. A little homework will go a long way in easing the frustrations.

In this exercise, you'll create a few XML files with various options and then analyze them in XML Notepad:

1. Open the `XMLSource.dwg` file.
2. In Prospector, right-click the drawing name, and select Export LandXML to display the dialog in Figure 21.12.

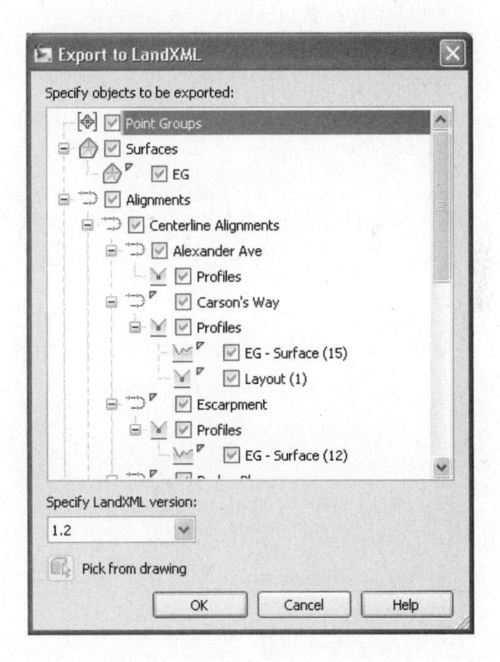

FIGURE 21.12
The Export to LandXML dialog

3. Click OK.
4. Navigate to the desktop (or some other suitable location), and click Save to create the XML file.

   That's it. Simple, right? Now, let's look at some of the options in the LandXML export:

5. Switch to the Settings tab in Toolspace.

6. Right-click the drawing name, and select Edit LandXML Settings.
7. Switch to the Export tab, as shown in Figure 21.13.

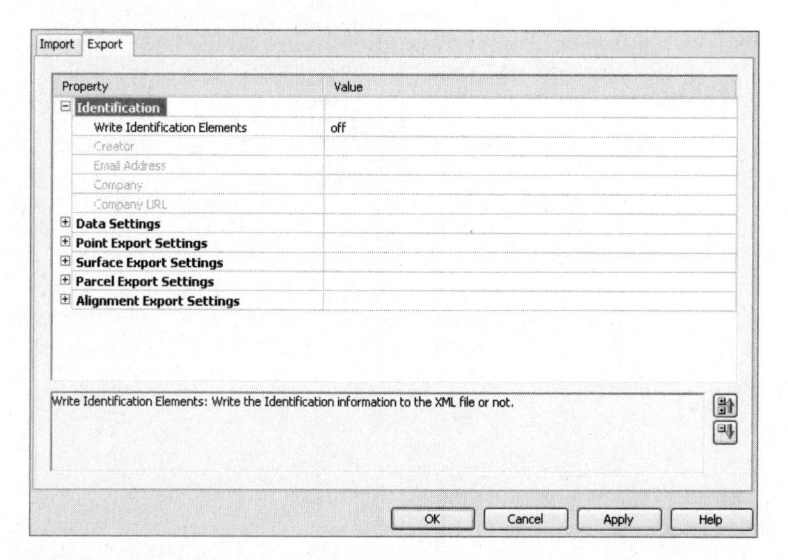

FIGURE 21.13
The Export tab in the LandXML Settings dialog

This dialog has the following options:

- **Identification** controls information written to the file to help identify the original source firm or person. This can include a name, phone number, e-mail address, or URL.
- **Data Settings** determine what foot standard is used, the rotation format, and the ability to create a read-only XML file. This read-only setting is applied both through the XML file and as a Windows read-only flag on the file.
- **Point Export Settings** control which point-data field is used as the code (or name field), the full description, whether to include point references (DRefs), the tolerance for points that tie to parcel COGO information, and description keys.
- **Surface Export Settings** control whether the surface is exported as points and faces or points, and whether watershed values are exported.
- **Parcel Export Settings** assign the direction of parcel creation: clockwise or counter-clockwise.
- **Alignment Export Settings** determine if cross-sectional data is exported with the alignments.

Programs may or may not use all these values, but any program built on the current schema should accept the file created with these values included. To see how these changes look in the output, complete the rest of this exercise.

8. Under Surface Export Settings, change the Surface Data option to Points Only, as shown in Figure 21.14.
9. Click OK to dismiss the dialog.

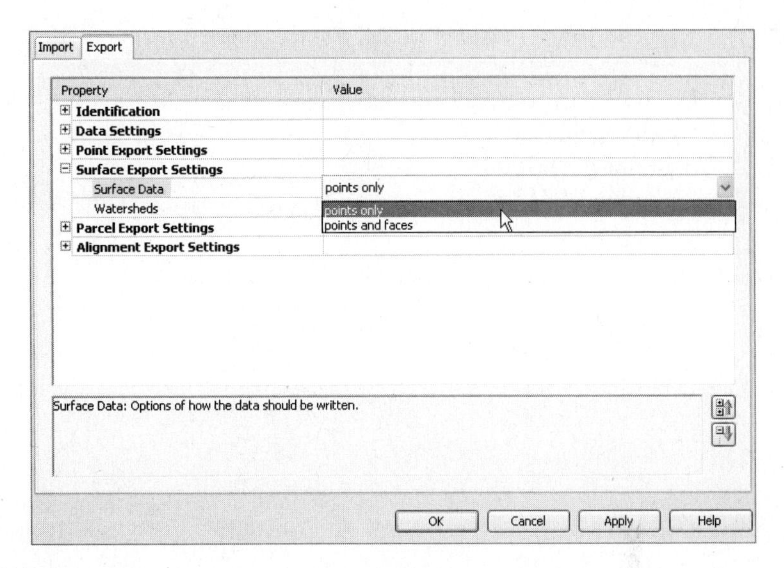

FIGURE 21.14
Changing the Surface Data option

10. Export a new LandXML file as you did in the steps 1 through 4 of this exercise. *Be sure not to overwrite the first file!*
11. Open the XML file from the first exercise in XML Notepad.
12. Scroll down in the tree view shown on the left, and expand the `Surfaces` folder, as shown in Figure 21.15.

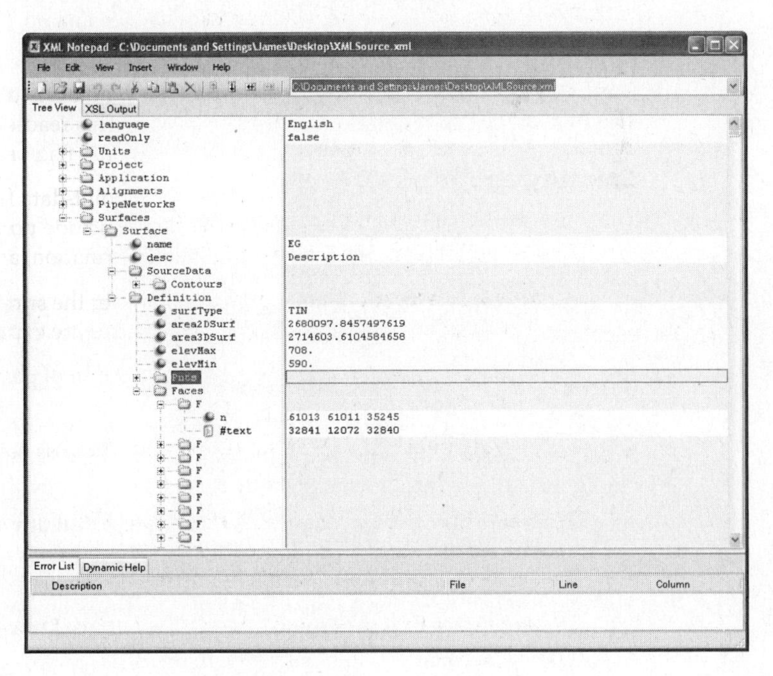

FIGURE 21.15
LandXML file with Surface set to Points and Faces

13. Choose Window ➤ New Window in XML Notepad.

14. Open the second XML file you created.

15. Expand the tree view as you did for the first file. Place the windows near each other to create a view similar to Figure 21.16.

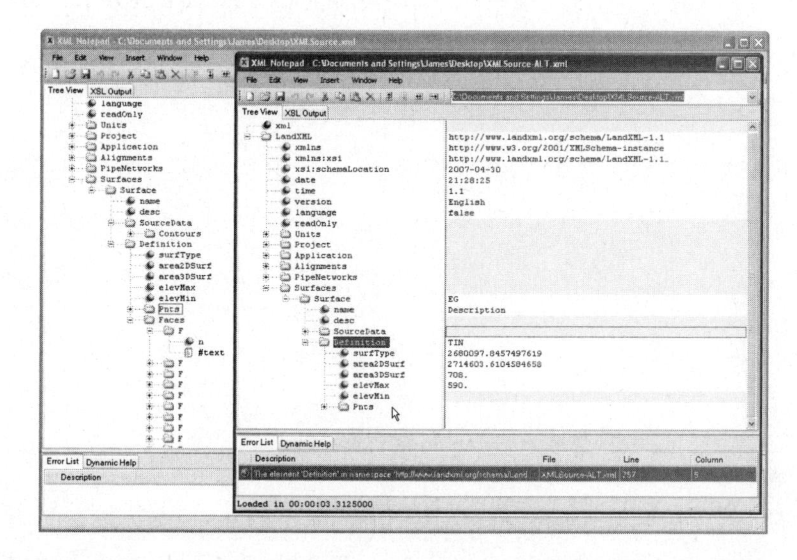

FIGURE 21.16
Comparing LandXML files with XML Notepad

Notice that the Surface definition in the second file (named XMLSource-ALT in the figure) lacks the Faces branch. By changing the settings, you changed the structure of the file created. In this case, you've created a lesser file, in that a surface created from the second LandXML file won't necessarily triangulate in the same way as the source file. Creating a LandXML file is straightforward, as you've seen, and unless you have a good reason, changing the settings generally isn't suggested.

Creating an AutoCAD Drawing

The unfortunate reality is that many of the partners in a land-development project team don't want your design data — they just want the linework. Because Civil 3D objects are so complex, just sending out the linework isn't as simple as it seems. This section discusses a couple of options for creating drawing files that may be suitable. The first option is to use proxy graphics or object enablers. The second is to ''dumb down'' your drawing to make it more palatable to other users.

Object Enablers and Proxy Graphics

For years, the main methods of dealing with custom objects (objects that don't exist in the core AutoCAD product) were to use proxy graphics or object enablers. Neither is an ideal situation, but if you can convince the recipient of your data to deal with one of these methods, it's better than the explosion of objects you'll look at in the next section.

Proxy graphics work by creating a simple linework version of the Civil 3D objects being drawn in your design file. These graphics are just that: graphics. They can't be used in any meaningful way for design, but they will display properly on the recipient's computer, assuming the drawing version is correct. The main problem with this approach is visible in Figure 21.17.

As you can see here, the file with proxy graphics is approximately 10 percent larger. This is for a simple Civil 3D file with a single surface, four alignments, and few labels. Because every label, profile view, alignment, contour, and so on in a drawing must be represented by a proxy graphic, this difference gets larger as the drawing gets more complex.

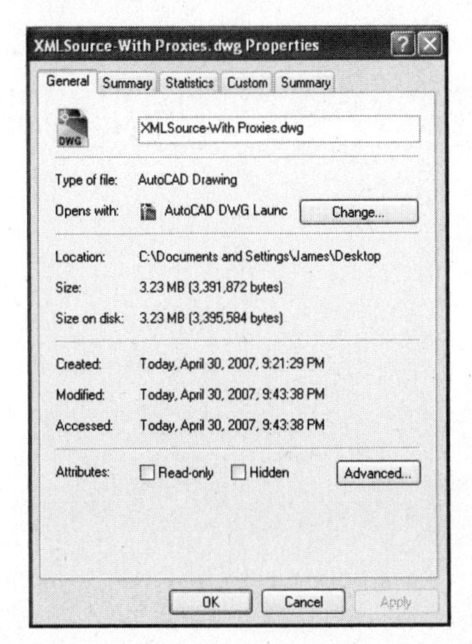

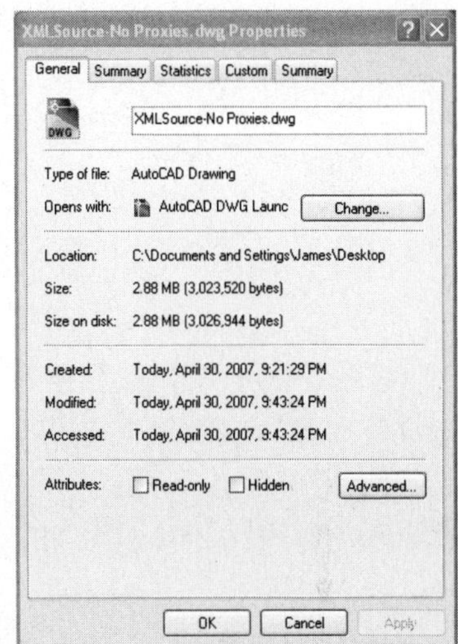

FIGURE 21.17
File Properties for a drawing with and without proxy graphics

PROXY GRAPHICS AND YOU

On the basis of feedback during the beta program for Civil 3D 2008, proxy graphics, turned on for many years, have been turned off by default in the shipping drawing templates. You'll have to turn them back on if you use these templates. To do this, type **proxygraphics** at the command line, and enter **1**.

The better answer (especially from your viewpoint!) is to use an object enabler on the recipient's end. These applications, written by Autodesk, provide the ability to draw Civil 3D objects without proxy graphics embedded in the file. They require an installation on the recipient's computer, but it's a one-time operation. Unfortunately, because these files have to be located, downloaded, and installed, many users won't do it. You can locate object enablers by searching the Autodesk site under the Utilities and Drivers page for the recipient's product.

Both options require that the recipient be on the same core AutoCAD version as the Civil 3D user. Because this isn't always the case, you'll learn how to export to other AutoCAD drawing file versions in the next section.

STRIPPING CUSTOM OBJECTS

The most drastic option for sharing data with non–Civil 3D users is to destroy the model and strip out the custom Civil 3D objects. This is a one-way trip, and the resulting file is made up of lines, arcs, polylines, and other native AutoCAD objects. In this exercise, you'll export a file with a variety of Civil 3D objects and then look at the resulting file:

1. Open the `XMLSource.dwg` file.
2. From the application menu (denoted with a large "C" in the upper-left corner of the ribbon), Choose Export ➤ AutoCAD DWG ➤ Export to AutoCAD 2007.

3. Navigate to the desktop or some convenient location, and click Save. The file is saved, but your original file is left untouched.
4. Open your exported file.
5. Zoom in on one of the "alignments," and pick any piece, as shown in Figure 21.18.

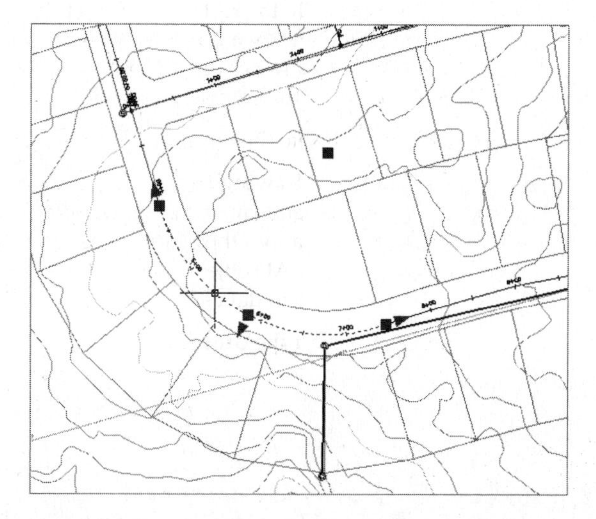

FIGURE 21.18
Alignment arc after export to AutoCAD

This file is considerably smaller than the original Civil 3D drawing but contains none of the original's intelligence. Re-creating the data would be possible, but because every piece is independent, it would be tedious.

WHAT VIEW DID YOU WANT?

One of the interesting facets of the Export to AutoCAD functionality is that it gives you exactly what you see on screen. Export a surface with a contour style in plan view, and you'll get flat polylines with no elevation. Turn to an isometric view before exporting, and you'll get the 3D representation of that surface. If your surface shows contours in the 3D view, you'll get polylines at elevation. A 3D TIN view will generate a whole host of 3D faces. It's all view dependent.

The Bottom Line

Discern the design elements in a LandXML data file. LandXML.org is an international consortium of software firms, equipment manufacturers, consultants, and other land-development professionals. By creating a lingua franca for the industry, they've made sharing design data between design and analysis packages a simple import/export operation.

Master It Download the `Campground.xml` file from the `landxml.org` website and list the creating program, along with the alignments and the surfaces it contains.

Import a Land Desktop project. Many Civil 3D firms have a huge repository of LDT projects. By using Civil 3D's Import Data from Land Desktop functionality, this data's lifespan can be extended, and the value of the information it contains can continue to grow.

Master It Import the STM-P3 Pipe Run from the Conklin LDT Project data source, and count the number of pipes and structures that successfully import.

Import a LandXML file to create a surface, alignments, and profiles. As more and more firms use Civil 3D, they'll want to share design data without giving up their full drawing files. Using LandXML to share import data allows the accurate transmission of data in a standardized format and gives the end recipient the ability to manipulate the entities as needed.

Master It Complete the import of the `Campground.xml` file into a new Civil 3D drawing.

Create an AutoCAD drawing that can be read without special enablers. Because many members of your design team won't be working with Civil 3D, it's important to deliver information to them in a manner they can use. Using the built-in tools to create drawings consisting of purely native AutoCAD objects strips these drawings of their design intelligence, but it provides a workable route to data sharing with almost any CAD user.

Master It Convert the drawing you just made from `Campground.xml` to an R2004 AutoCAD drawing.

Chapter 22

Get The Picture: Visualization

Because Civil 3D is built on AutoCAD, all the functionality of AutoCAD is available and adaptable to Civil 3D objects. Before the 2007 release, building 3D objects, 3D navigation, and rendering in the AutoCAD environment was possible but not very intuitive. In 2007, substantial changes were made to these tools in AutoCAD; those changes make creating rendered scenes simple. Also, several Civil 3D objects have built-in tools to make rendering straightforward.

This chapter is meant to be an introduction to visualization. For more information, see the AutoCAD Users Guide, which is built into the Help menu. Also, there are many robust AutoCAD rendering texts and learning resources.

By the end of this chapter, you'll be able to:

- Apply render materials to a corridor using a code set style
- Apply a Realistic visual style to a corridor model
- Create a 3D DWF from a corridor model

AutoCAD 3D Modeling Workspace

Civil 3D 2010 comes with several default workspaces. In previous chapters, we discussed the standard workspaces for Civil 3D–related tasks. One workspace is particularly suited for rendering work. The 3D Modeling workspace, shown in Figure 22.1, contains the View panel on the Home tab and the 3D Palettes panel on the View tab.

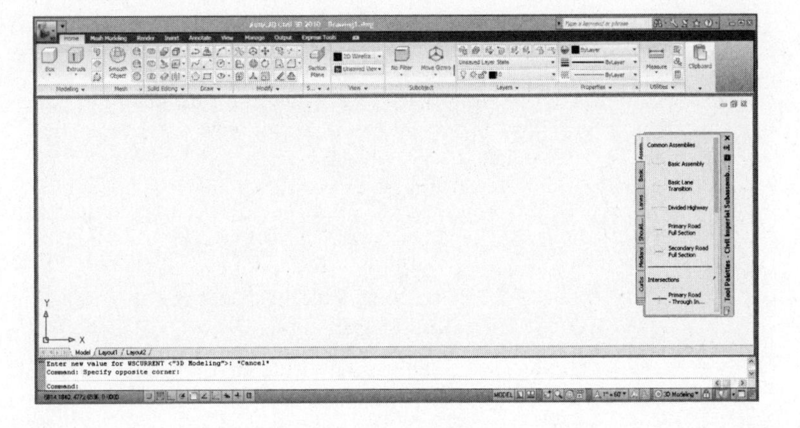

FIGURE 22.1
The Visualization and Rendering workspace

Applying Different Visual Styles

A visual style is a group of settings that stores your preferences for the display of faces, edges, shading, and texture. There are five default visual styles, and you can create your own custom visual styles as desired. The View panel on the Home tab (shown in Figure 22.2) is a panel you can add to any workspace.

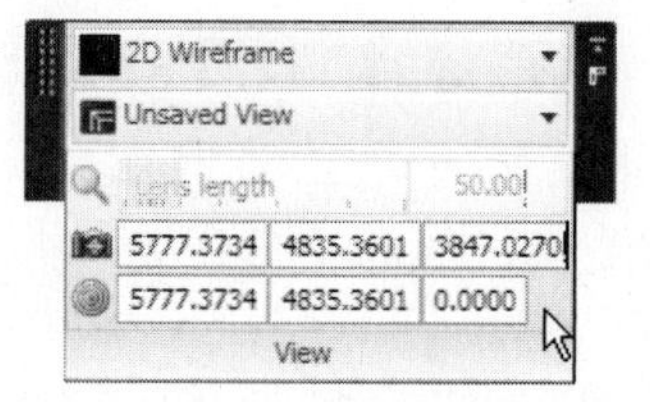

FIGURE 22.2
The View panel undocked

2D Wireframe When the visual style is set to 2D Wireframe, objects are shown as lines and curves. Most likely, you now do most of your work in 2D Wireframe. Raster Images, OLE objects, linetypes, and lineweights are shown in this visual style. Figure 22.3 shows a corridor in 2D Wireframe.

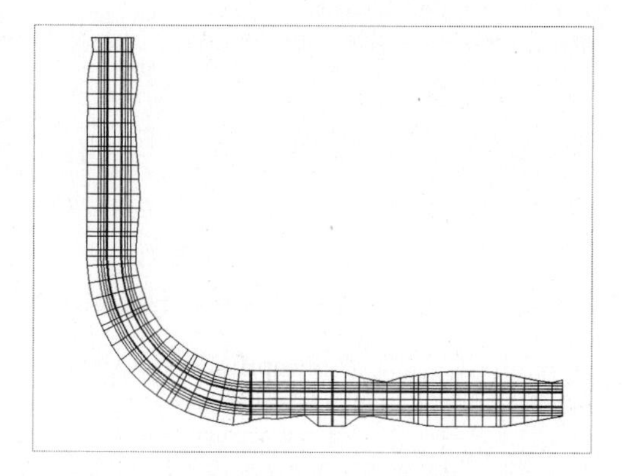

FIGURE 22.3
A corridor shown in top view with the 2D Wireframe visual style applied

SWITCHING TO 3D VIEWS

While working in the AutoCAD environment, you can navigate your model by using tools found in the 3D Modeling workspace on the Navigate panel and the Views panel found on the View tab. You can change from top view to an isometric view, use a constrained or unconstrained 3D orbit, or use any combination of these tools to see your model from a different perspective. This can be especially handy in terrain modeling, as discussed in Chapter 5, "The Ground Up: Surfaces in Civil 3D."

3D Wireframe Applying the 3D Wireframe visual style in plan view will look almost identical to 2D Wireframe. Objects are represented using lines and curves. A corridor in 3D Wireframe is shown in Figure 22.4.

3D Hidden The 3D Hidden visual style is similar to the 3D Wireframe visual style, as shown in Figure 22.5. The only difference is that any back faces are hidden.

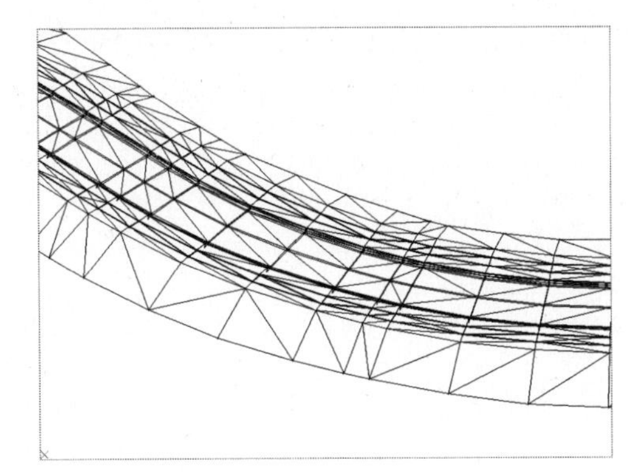

FIGURE 22.4
A corridor shown in isometric view with the 3D Wireframe visual style applied. Note that the 3D faces of the corridor are visible

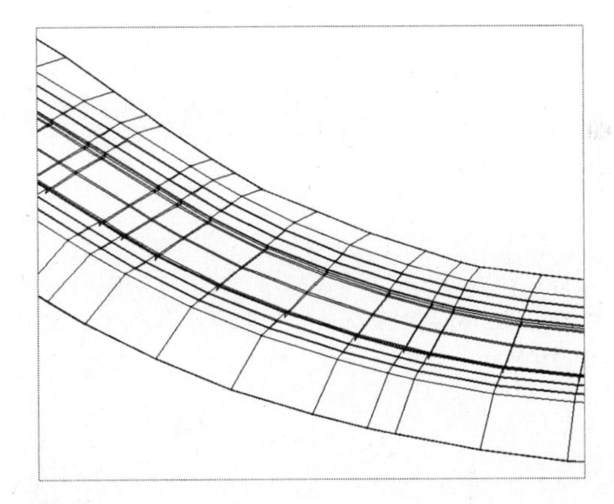

FIGURE 22.5
A corridor shown in isometric view with the 3D Hidden visual style applied. Note that the 3D faces of the corridor are not visible

Conceptual The Conceptual visual style was created to make your 3D objects appear smoother and perhaps easier to see without looking "real." It is similar to the 3D Hidden visual style with some warm and cool colors applied, as Figure 22.6 shows. This visual style is useful for giving your models a little more contrast without having to assign a render material.

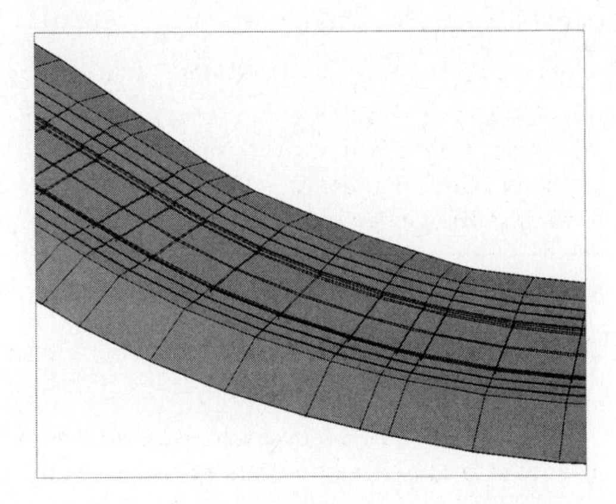

FIGURE 22.6
A corridor shown in isometric view with the Conceptual visual style applied. Note that the 3D faces are not visible

Realistic The Realistic visual style is similar to the 3D Wireframe visual style, as Figure 22.7 shows. In addition to showing the 3D faces of your model, this visual style smoothes the edges and shows any render materials you have assigned.

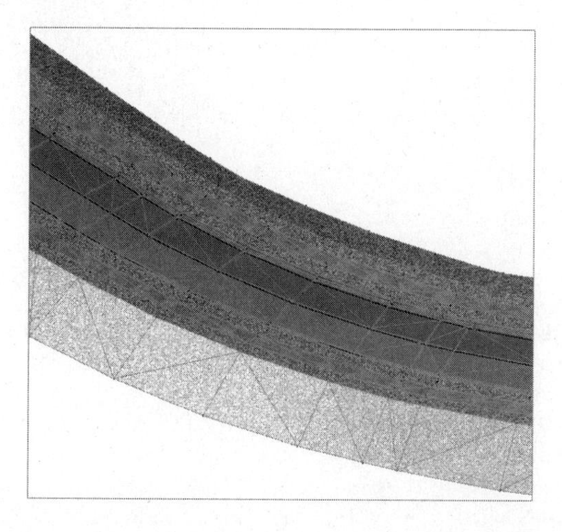

FIGURE 22.7 A corridor (with render materials applied) shown in isometric view with the Realistic visual style applied. Note that 3D faces are visible

The Visual Styles Manager The Visual Styles Manager provides an opportunity to customize the default visual styles and create new visual styles. Access the Visual Styles Manager by switching to the 3D Modeling workspace, changing to the View tab, and clicking Visual Styles from the 3D Palettes panel as shown in Figure 22.8.

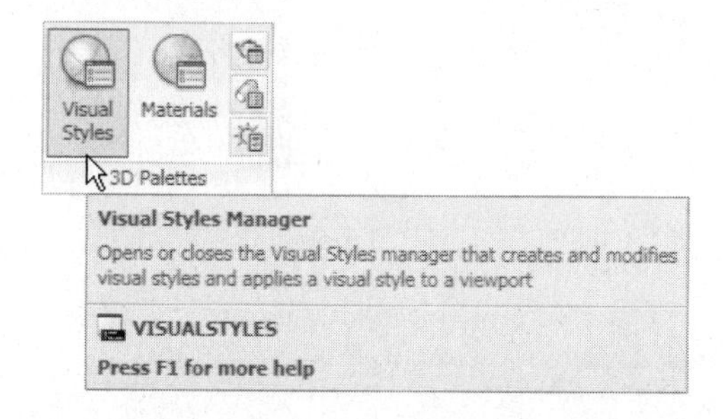

FIGURE 22.8 Accessing the Visual Styles Manager

In the Visual Styles Manager (shown in Figure 22.9), you can copy an existing visual style as a foundation for modification. All of the settings in the Visual Styles Manager are covered in detail in the AutoCAD Users Guide.

Render Materials

Render materials provide realistic colors, patterns, and textures to your drawing and Civil 3D objects. For example, you can assign a grass render material to an existing ground surface or a concrete render material to piping and sidewalks.

When you install Civil 3D, a materials library of approximately 100 render materials is installed by default. During installation, you will be given the choice of installing the full materials library that includes some 200 additional materials. If you neglect to toggle the button for this at install, you can always add the full materials library later by modifying your Civil 3D install under Add/Remove Programs.

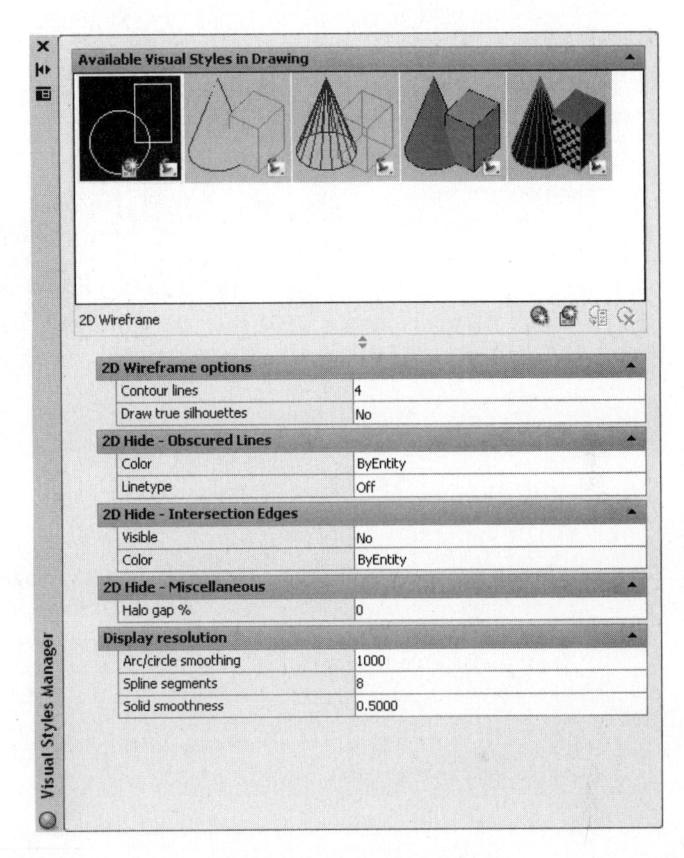

FIGURE 22.9
The Visual Styles Manager

In addition to the 300 materials that are provided by the materials library, you can create your own custom render materials from images, or edit those that are provided.

By default, many render materials have been built into the sample tool palettes, and additional material tool palettes can be created. Render materials can be dragged directly from the tool palette onto many objects. Figure 22.10 shows a simple box where an exposed concrete render material has been applied using the render material on the tool palette.

Civil 3D objects can be assigned render materials as part of their object properties or, in the case of corridors, their code set. You'll look into that in more detail in the next section.

Visualizing Civil 3D Objects

Several Civil 3D objects have rendering options built into their properties. When working on visualization of Civil 3D objects, it is probably best to create a new drawing for the purposes of visualization. One approach would be to create a new drawing and create data references of your desired objects in the visualization drawing, and then apply render materials and custom code sets. This will allow you to keep a dynamic model but prevent your main modeling drawings from getting bogged down with heavy materials and visual style applications.

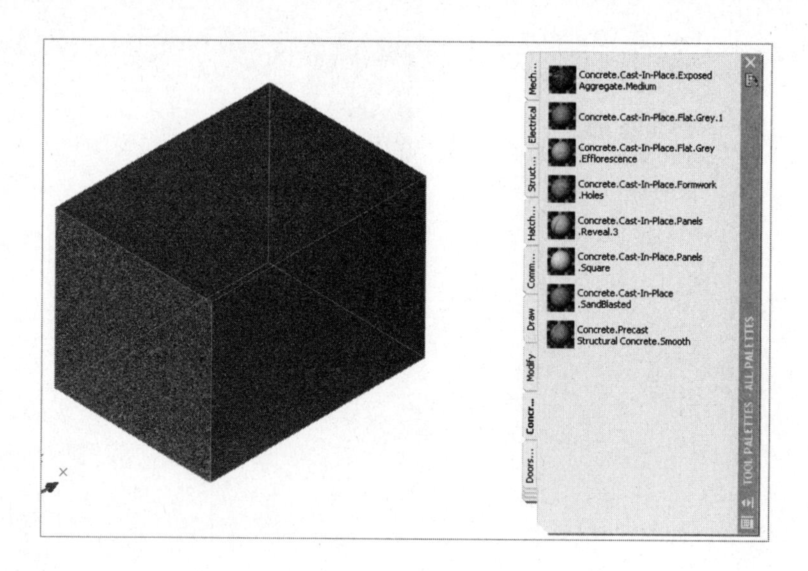

FIGURE 22.10
A simple box with a concrete render material applied from the tool palette

Applying a Visual Style

If you'd like to view your model in the AutoCAD environment to get an idea of what a true rendering would look like, you can simply apply a Realistic visual style to your drawing and switch to an isometric view.

> **DON'T SAVE "REALISTIC"**
>
> Do not save your drawing with the Realistic view applied. Before saving, switch to a Wireframe visual style. There have been occasions where saving the drawing with a Realistic visual style applied has caused drawing corruption problems.

The default Realistic visual style shows isolines. Because with Civil 3D objects isolines include TIN lines and 3D faces, they often cloud your model with lots of gray triangles. Also by default, textures are applied. Sometimes removing the texture application makes your model easier to comprehend in the AutoCAD environment.

To create a customized visual style, simply copy the default Realistic visual style by right-clicking it in the Visual Styles Manager and choosing Copy (Figure 22.11), and then right-click to paste the copy.

In your new visual style, change the Material Display toggle from Materials and Textures to Materials, and change the Edge Mode toggle from Isolines to None. You can rename your new visual style by right-clicking the style and choosing Edit Name and Description. Apply this customized visual style by right-clicking it and choosing Apply to Current Viewports. An example of this visual style is shown in Figure 22.12.

Visualizing a Surface

The first Civil 3D object that is meaningful to render is a surface. Every surface created in Civil 3D has a render material assigned upon creation, as shown in Figure 22.13.

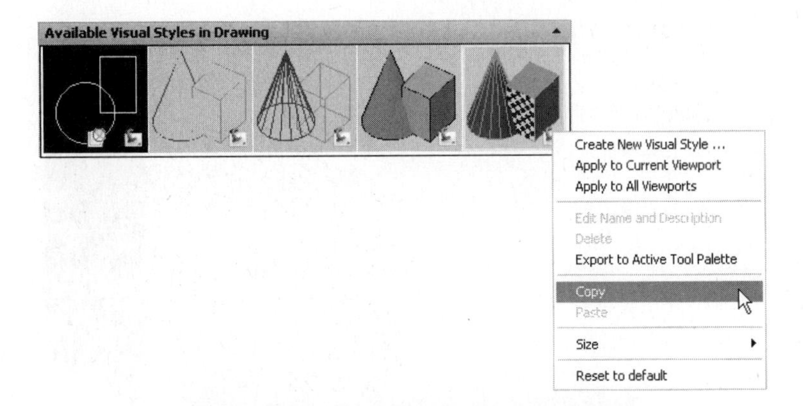

FIGURE 22.11
Make a copy of the default Realistic visual style

FIGURE 22.12
A corridor model and multiview blocks with a customized Realistic visual style applied

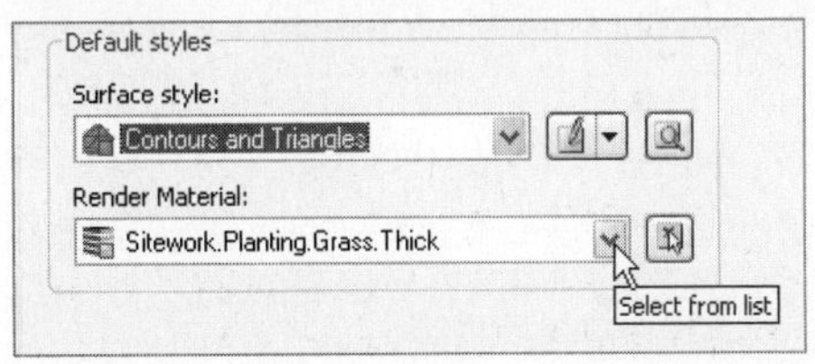

FIGURE 22.13
The Surface Properties box and Render Material drop-down button

In order for a render material to be visible when the Realistic visual style is applied, your current surface style must have triangles set to Visible, as shown in Figure 22.14. This applies to rendering in both 2D and 3D.

In this exercise, you'll learn how to apply a render material to a Civil 3D surface object, apply different visual styles, and customize a Realistic visual style:

1. Open the drawing file `Visualization-1.dwg`. (All of the data for this text can be downloaded from `www.sybex.com/go/masteringcivil3d2010`.) Note that an existing ground surface is shown in isometric view.
2. Switch your workspace to the 3D Modeling workspace.
3. Select the surface and select Surface Properties from the Modify panel.

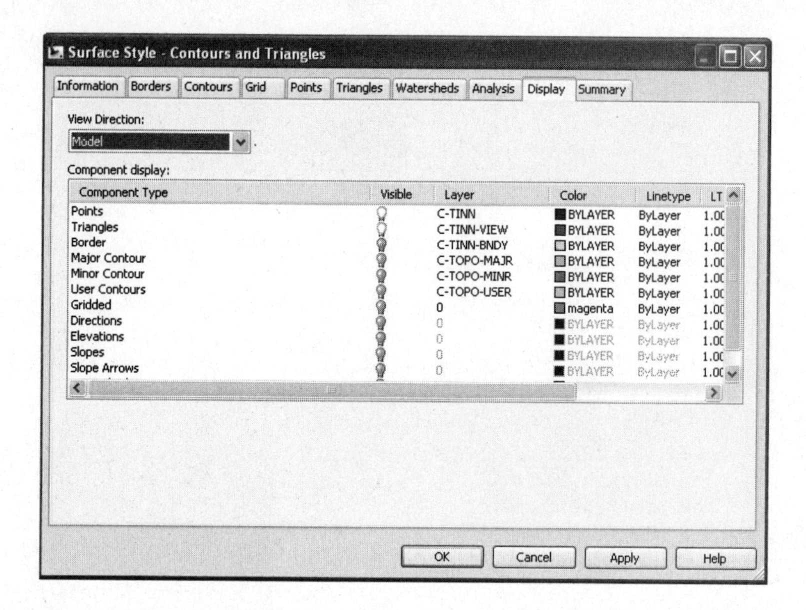

FIGURE 22.14
Triangles must be visible in 3D in order to see the render material applied

4. From the Render Material drop-down list, choose Sitework.Planting.Grass.Short. Click OK.
5. Use the View panel on the Home tab to change the visual style to Realistic.
6. Zoom in on the surface and note that the grass render material has been applied but that there are also still gray isolines visible from the TIN.
7. Change the visual style to Conceptual to see its effect. Note that warm and cool colors are applied to show a more "conceptual" effect on the surface.
8. Change the visual style to 3D Wireframe for ease of navigation.
9. Change to the View tab and choose Visual Styles from the 3D Palettes panel to access the Visual Styles Manager.
10. Make a copy of the Realistic visual style by choosing it, right-clicking, and choosing Copy from the shortcut menu. Then, right-click in the whitespace and click Paste. A new visual style should appear.
11. Select your visual style and right-click. Choose Edit Name and Description.
12. Change the name of the visual style to **Realistic-Civil 3D**. Delete the description and leave it blank. Click OK.
13. In the lower part of the Visual Styles Manager window, change the Edge mode from Isolines to None.
14. Right-click your new Visual Style icon in the top part of the Visual Styles Manager window and choose Apply to Current Viewport. Close the Visual Styles Manager palette.
15. Zoom in on your surface and notice that the gray isolines are gone.

16. Return to the Visual Styles Manager window. Select your Realistic Civil 3D visual style. In the bottom part of the Visual Styles Manager window, change the Material display from Materials and Textures to Materials.

17. Minimize or move your Visual Styles Manager out of the way so that you can see the effect of removing the textures from your visual style. Your surface should appear smooth and green.

Visualizing a Corridor

In Chapter 12, "The Road Ahead: Advanced Corridors," we explored the idea of links within the corridor object. For rendering, you can assign a certain render material based on the link code in a code set style. For example, the Paving code can be assigned an asphalt render material, and the Sidewalk code can be assigned a concrete render material. Additionally, you can assign preset styles to your corridor feature lines in the code set style.

Every time a code set style that was created for rendering is applied to a corridor, those links will automatically be assigned the correct render materials.

Creating Code Set Styles

Code set styles are located on the Settings tab of the Toolspace under the General ➢ Multipurpose Styles ➢ Code Set Styles.

The default template includes a code set that already has render materials assigned, as shown in Figure 22.15. You can also make your own custom code set by copying one of the existing sets.

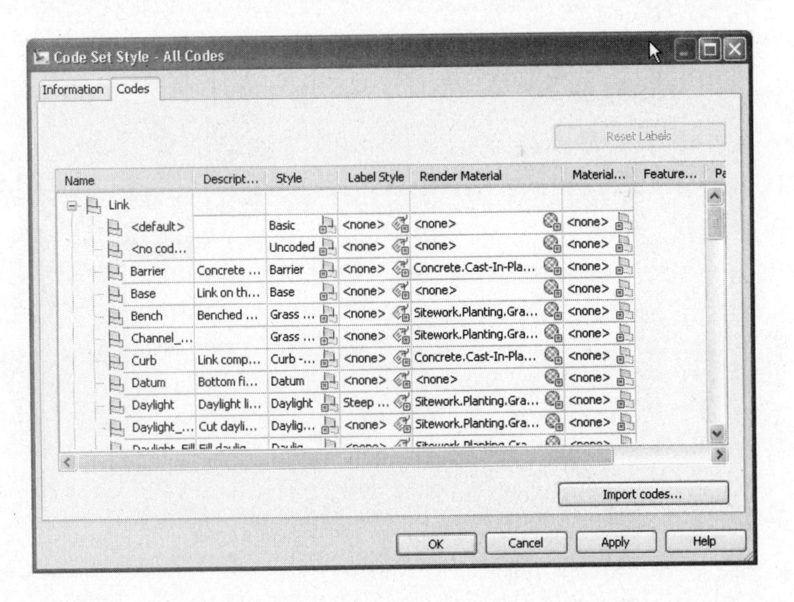

FIGURE 22.15
The Code Set Style dialog with the list of link code names on the left side

The Code Set Style dialog does resize, but you might find it easier to work if you scroll over to the right to locate the Render Material column. If it makes it easier, you can drag the Render Material column closer to the Link Name column as shown in Figure 22.16.

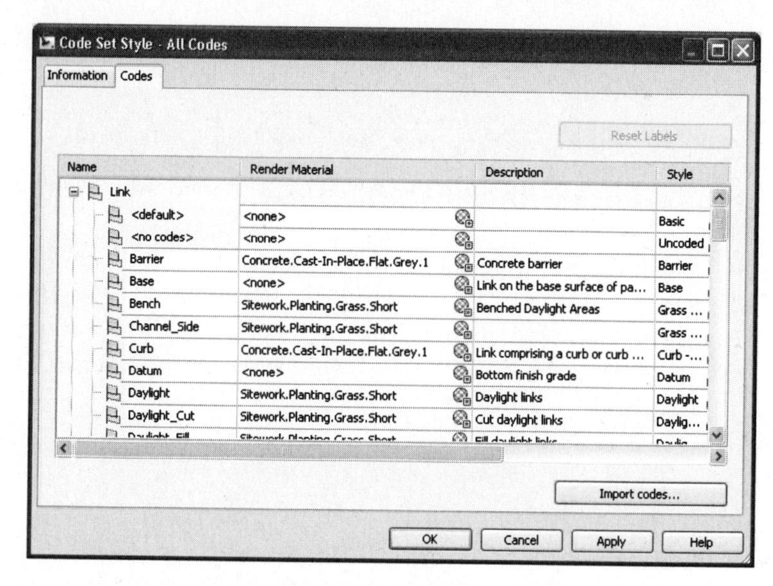

FIGURE 22.16
Move the Render Material column closer to the Link Name column

In each link row, you can assign an appropriate render material.

Once this code set is assigned in the Corridor Properties, you can see your rendering by changing from 2D Wireframe to a Realistic visual style (Figure 22.17).

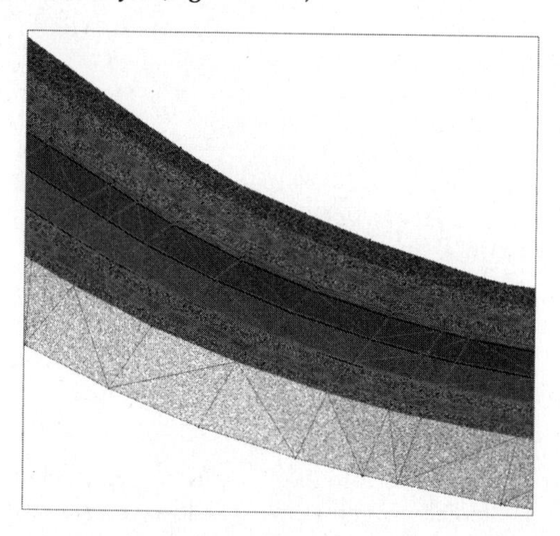

FIGURE 22.17
A corridor with render materials assigned through the code set style

In addition to adding render materials to your code set style, you may want to adjust the visibility of the links, points, shapes, and feature lines themselves.

CORRIDOR FEATURE LINES AND CODE SET STYLES

When you build your corridor, the default code set style in your Command settings is applied to links, points, shapes, render materials, material area fill styles, label styles, and feature lines. Once the corridor is built, changes to the code set style will update all of these items except the feature lines. Changes to feature lines once the corridor is built must be made in Corridor Properties.

The following exercise leads you through applying a code set style to a corridor model and applying a Realistic visual style:

1. Open the drawing file `Visualization-2.dwg`. Note that a corridor is shown. Change your workspace to 3D Modeling if it is not already current.
2. Select the corridor and click Corridor Properties from the Modify panel. Switch to the Codes tab.
3. Note that the Basic code set style is currently applied. In the Code Set Style drop-down menu, choose All Codes with Render Materials. Note that there are render materials listed in the Render Material column for this code set style. Click OK.
4. Switch the visual style from 2D Wireframe to Realistic. Note that materials and textures are applied.
5. Switch the visual style from Realistic to Realistic Civil 3D (the same visual style created in the previous exercise). Note that materials are applied, but isolines and textures are turned off.
6. Use the View panel to switch to the 3D Corridor View, as shown in Figure 22.18. Toggle between the different visual styles to see their effect in 3D.

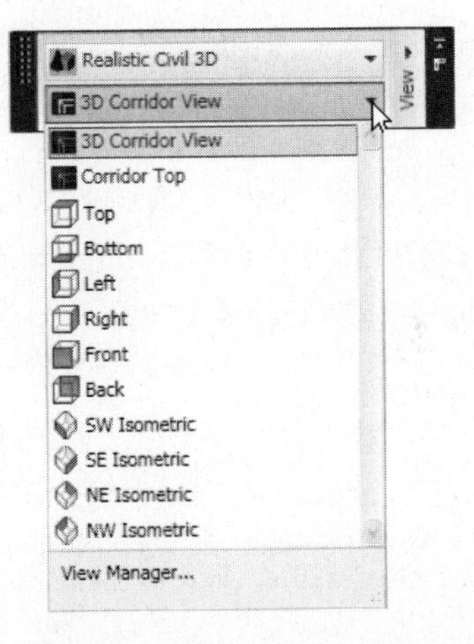

FIGURE 22.18
Using the View panel undocked to switch between views

7. Switch back to the Corridor top view using the Views drop-down list on the 3D Navigation toolbar. If you would like to save your drawing, be sure to refresh the 2D Wireframe visual style before saving.

Visualizing a Pipe Network

Civil 3D pipes and structures can be assigned appropriate render materials. There are items in the Standard materials library that mimic the look of ductile iron, concrete, plastic, or many others.

Pipes and structures can be assigned render materials using the individual pipe and structure properties once the pipe is in the drawing. A better method is to assign the render materials in your parts list before you begin designing your pipe network, as shown in Figure 22.19. This will ensure that every pipe created from this parts list already has the proper render material assigned.

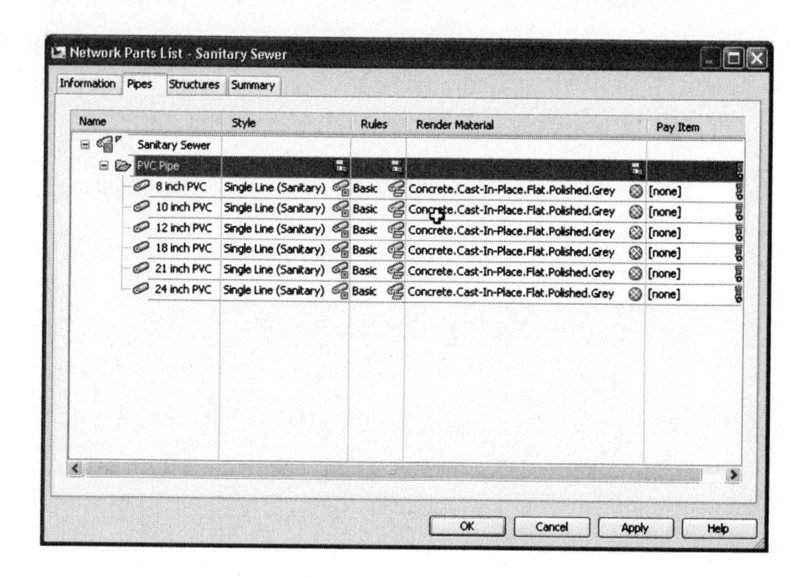

FIGURE 22.19
Assigning render materials for pipes in the parts list

The following exercise leads you through adding render materials to a drawing and applying them to two pipe networks:

1. Open the drawing file `Visualization-3.dwg`. Note that there are two pipe networks in the drawing: a concrete storm drainage network and a PVC sanitary sewer network. Change your workspace to 3D Modeling if it is not already current.
2. Bring in your tool palette by pressing Ctrl+3.
3. Right-click the spine of the tool palette and choose Materials Library (see Figure 22.20). This tool palette group should be present by default if you installed the Materials Library when installing Civil 3D.
4. Locate the Concrete – Materials Library tool palette. Select the Concrete.Cast-In-Place.Flat.Polished.Grey entry and hold down your left mouse button. Move your mouse over into the drawing area, and then release the left mouse button. This will "drag" the render material into the drawing and make it available to assign to objects.
5. Switch to the Wood and Plastics – Materials Library tool palette. Select the Woods-Plastics.Plastics.PVC.White entry and drag it into the drawing, as in step 4.
6. Dismiss the tool palette.
7. Select pipe SD-1 from the short storm network running north and south. Choose Pipe Properties from the Modify panel.
8. In the Render Material drop-down list, choose Concrete.Cast-In-Place.Flat.Polished.Grey. Click OK.
9. Select STM MH 1. Choose Structure Properties from the Modify panel.
10. In the Render Material drop-down list under the Information tab, choose Concrete.Cast-In-Place.Flat.Polished.Grey. Click OK.

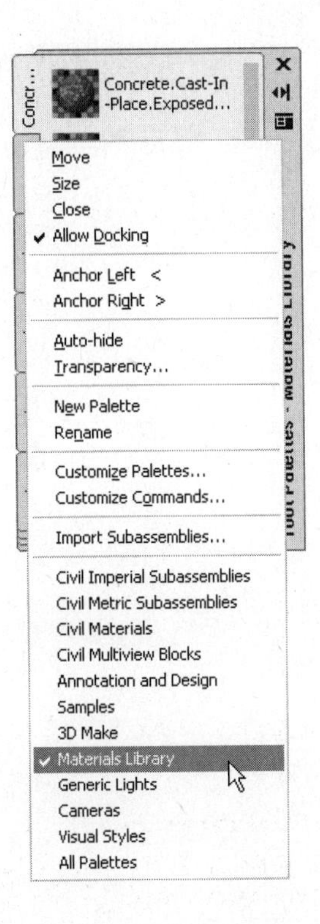

FIGURE 22.20
Switch to the Materials Library tool palette group

11. Locate the Sanitary Sewer Network under the Pipe Networks tree of the Prospector. Select the Pipes entry.
12. In the preview pane that shows the list of pipes, use your Shift key to select all pipes. Right-click the Render Material column heading and choose Edit.
13. In the Select Render Material dialog, choose Woods-Plastics.Plastics.PVC.White. Click OK. Your Prospector should look like Figure 22.21.

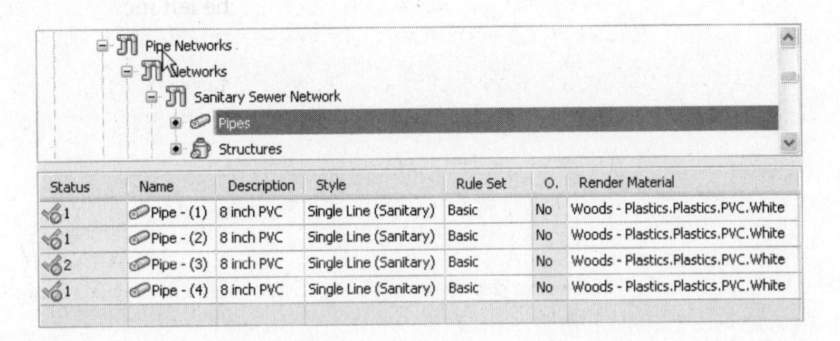

FIGURE 22.21
Assign a render material to a batch of pipes in Prospector

14. Select the Sanitary Sewer Network Structures tree. Using the same technique as you did in step 13, assign the Concrete.Cast-In-Place.Flat.Grey.1 render material to all of the sanitary structures.

15. Use the Views drop-down list on the 3D Navigation toolbar to switch to Southwest Isometric.

16. Switch to the Realistic visual style. Your pipes and structures should take on the appearance of concrete and PVC, as shown in Figure 22.22.

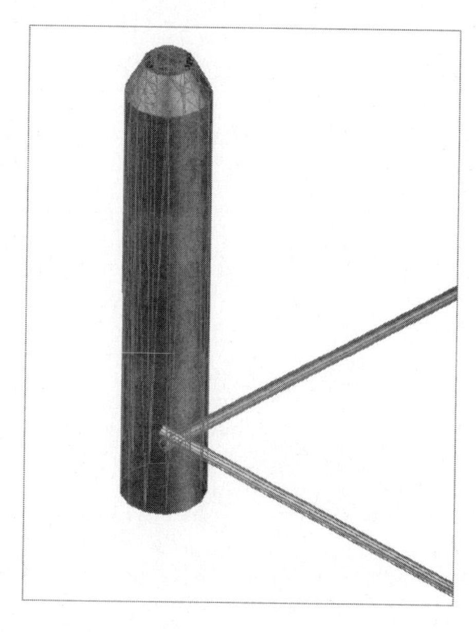

FIGURE 22.22
A close-up look at Structure-3 with the Realistic visual style applied

17. If desired, follow the steps from earlier exercises to create a Realistic visual style that turns off isolines and/or textures.

Visualizing AutoCAD Objects

Beyond applying rendering materials to Civil 3D objects, you can also use native AutoCAD objects to flesh out your visuals.

Solids

You may find AutoCAD solids to be useful as conceptual buildings or fixtures in your drawing. Explore these tools (shown in Figure 22.23) and see the AutoCAD User's Guide located in the Help menu for more information about these useful objects.

Blocks

There are many sources for 3D blocks that represent trees, benches, light posts, and other site development fixtures. With Civil 3D 2010, look for a Civil Multiview Blocks tool palette group that contains blocks for conceptual and realistic trees, signposts (shown in Figure 22.24), and more.

Additionally, you can obtain free samples of 3D blocks made by third parties through the DesignCenter Online. DesignCenter Online is accessed by opening the DesignCenter (Ctrl+2) and selecting the DC Online tab. You can also draw your own or buy an inexpensive CD of 3D blocks from many sources. There are many ways these blocks can be constructed, but typically they have a component suitable for plan view on one layer and a 3D component on another layer.

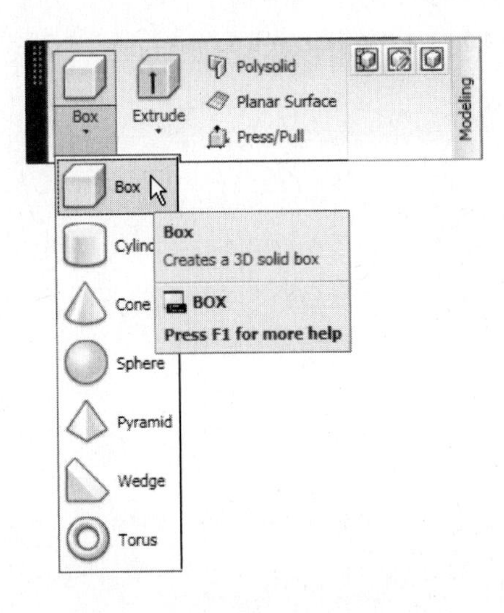

FIGURE 22.23
The solid creation tools on the AutoCAD Modeling panel shown undocked

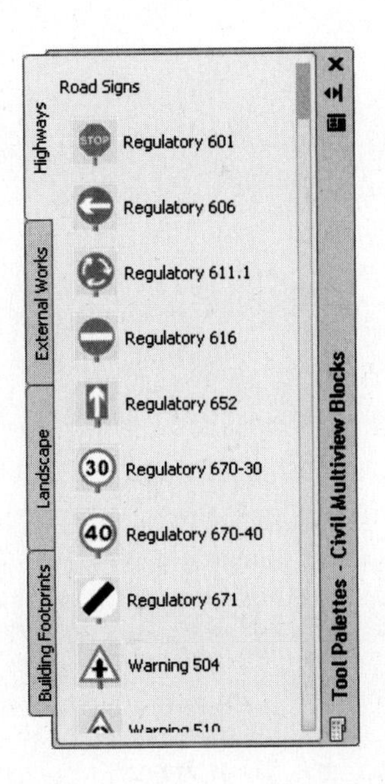

FIGURE 22.24
3D Signpost blocks from the Civil Multiview Blocks tool palette group

MULTIVIEW BLOCKS

Multiview blocks have been drawn to look a certain way in plan (Figure 22.25), profile, section, and 3D model views (Figure 22.26).

This eliminates the redundancy problem from typical 3D blocks. There is a collection of sample multiview blocks that install with Civil 3D. Blocks in this collection include trees, light standards, houses, buildings, and more. By default, the Multiview Block library is located at `C:\Documents and Settings\All Users\Application Data\Autodesk\C3D2010\enu\Data\Symbols\Mvblocks`.

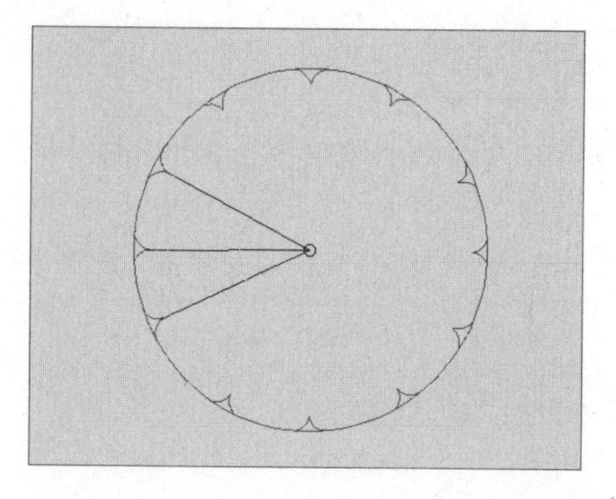

FIGURE 22.25
A tree multiview block in plan view

FIGURE 22.26
The same tree multiview block in 3D view

If you customize your installation, or you are using Windows Vista, you may have your Multiview Block library in a different location. You can find the Multiview Block library at `C:\ProgramData\Autodesk\C3D2010\enu\Data\Symbols\Mvblocks` for Windows Vista.

Multiview blocks can be inserted as traditional blocks, or can be used as part of a marker style or point style.

MOVING OBJECTS TO SURFACES AND EXTRACTING OBJECTS FROM SURFACES

In order to use items such as 3D blocks or multiview blocks, you need to move those blocks up to the appropriate elevation. In that case, you could use the Move Blocks to Surface command.

There may also be cases where you would like to extract a feature from a surface, such as a single contour line. The Extract Objects from Surface command is useful in that situation.

You can find these tools, and more, by selecting a surface in your drawing to open the Surface contextual tab.

You can also build 3D or multiview blocks into surface labels or point styles to achieve a similar effect. Trees, light standards, power poles, and similar blocks would be good choices for building into point styles.

The following exercise leads you through moving multiview tree blocks from elevation zero to surface elevation:

1. Open the drawing file `Visualization-4.dwg`. Note that there is a surface in the drawing, along with many multiview tree blocks inserted at elevation zero.
2. Switch to the Civil 3D workspace.
3. Change to the Modify tab and select Surface from the Ground Data panel to open the Surface contextual tab. Choose the Surface Tools drop-down panel and select Move Blocks to Surface.
4. In the Move Blocks to Surface dialog, choose both Apple and Colorado Blue Spruce blocks. Click OK.
5. Switch your workspace to 3D Modeling.
6. Using the View panel on the Home tab, switch the view to Southwest Isometric.
7. Zoom in on some trees. Your results should be similar to Figure 22.27.

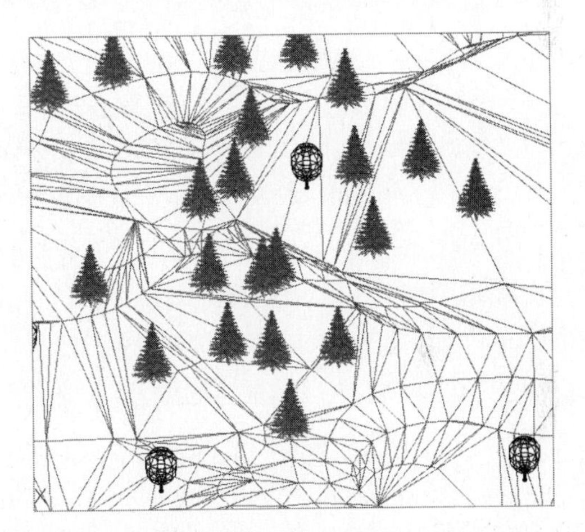

FIGURE 22.27
Multiview tree blocks moved to the surface

8. Switch to the Conceptual visual style to note its effects.

Creating a 3D DWF from a Corridor Model

Another simple way to visualize your model is through a 3D DWF. You may be familiar with standard DWFs for plan sheet transmittal, plotting, and archiving. DWF files can be shared with non-CAD users through DWF Design Review, a free application that can be downloaded from the Autodesk website. DWFs can also be embedded into web pages or Microsoft PowerPoint presentations, and they can be used in many other ways. A 3D DWF is the same idea, except it includes the model aspects of your drawing, including render materials, and allows for orbiting and other navigation tools.

To create a 3D DWF or DWFx (DWFx files can be opened natively by Windows Vista), you can use the Publish ➤ 3D DWF tool found in the application menu (the large "C" in the upper-left of the AutoCAD Civil 3D screen), or simply type **3DDWFPUBLISH** at the command line. Civil 3D will prompt you for a filename and location for your new DWF or DWFx file, and then it will ask if you would like to open the resulting file. The 3D DWF will open in Autodesk Design Review, as shown in Figure 22.28.

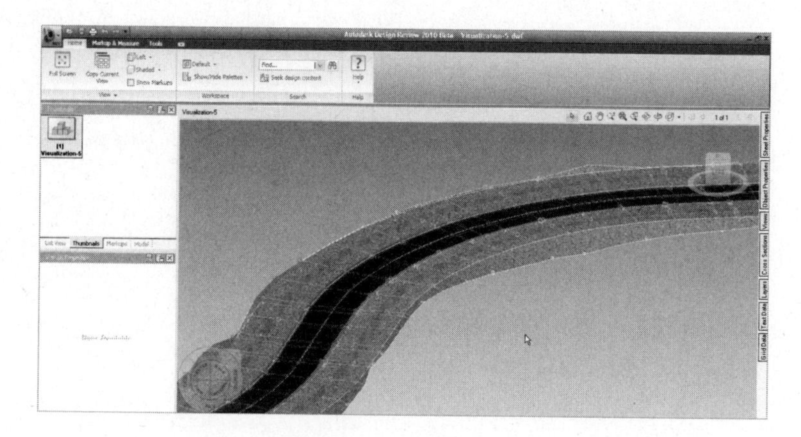

FIGURE 22.28
The Autodesk Design Review interface for viewing a 3D DWF

The navigation tools in the Design Review environment are identical to those in an AutoCAD-based interface, such as 3D orbit, pan, and zoom.

The following exercise leads you through creating a 3D DWF from a corridor model:

1. Open the drawing file `Visualization-5.dwg`. Note that there is a corridor in this drawing.
2. Switch to the 3D Modeling workspace if it is not already current.
3. Apply the Realistic Civil 3D visual style created earlier.
4. Type **3DDWFPUBLISH** at the command line. In the resulting dialog, name your DWF and choose a location to save the resulting file. Click OK.
5. A dialog will appear asking you if you would like to view the DWF. Click Yes.

Creating a Quick Rendering from a Corridor Model

A true rendering applies the render materials, lighting settings, shadows, and other advanced settings to an image or video clip. In this section, you prepare a more detailed image of a designed corridor.

A Quick Rendering

You can make a simple rendering with little preparation by simply navigating your model until your viewport/screen shows what you'd like to see rendered. Execute the Render command by switching to the 3D Modeling workspace and selecting Render from the Render panel, or by typing **-render** at the command line. The resulting image will look similar to Figure 22.29.

Adjusting Lighting

If you would like a more realistic-looking rendering, you can adjust the lighting settings to provide sunlight and shadow effects. One effect that improves the realism of your rendering is to change the geographic location of your drawing using both the Lights and Sun & Location panels as shown in Figure 22.30. You can type in a latitude and longitude, choose a city from a list, or pull a geographic location from Google Earth.

FIGURE 22.29
A quickly rendered scene at medium quality with no shadows, lighting, or other advanced settings

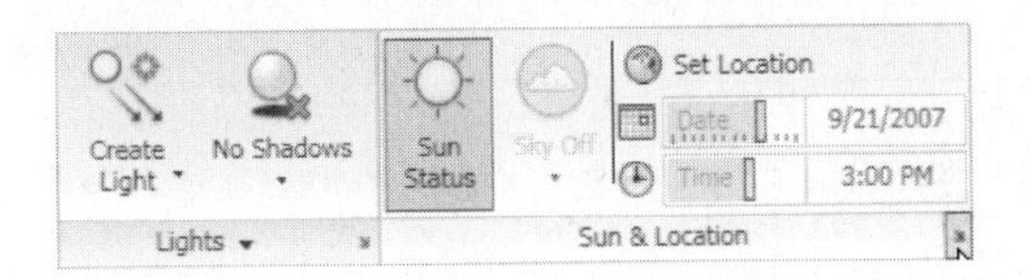

FIGURE 22.30
Make changes to geographic location, the sun, and other lighting settings

You can further adjust the sun by clicking the Sun Properties button on the Sun & Location panel to display the Sun Properties palette (shown in Figure 22.31), and then making the desired adjustments.

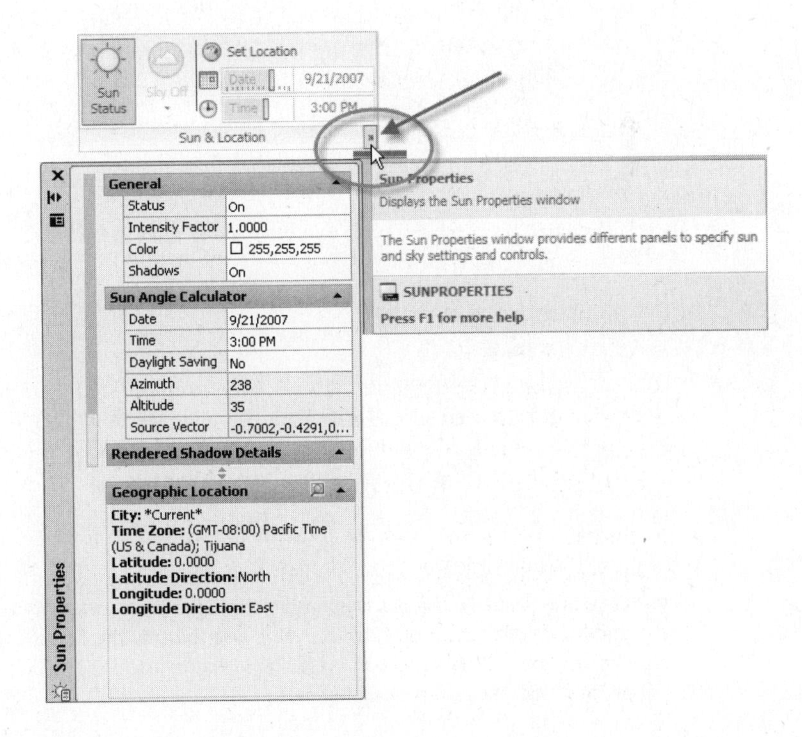

FIGURE 22.31
Adjust the sun settings to accommodate intensity details, different times of day, and shadow details

If you would like to further customize your scene, you can add custom lights, sky, and more. Take a look in the section of the AutoCAD Users Guide titled 'Creating Realistic Images and Graphics'.

The following exercise leads you through creating a simple rendered image from a corridor model:

1. Open the drawing file `Visualization-6.dwg`. Note that there is a corridor in this drawing.
2. Switch to the 3D Modeling workspace if it is not already current.
3. Change to the Render tab and locate the Render button on the Render panel. The Render button appears as a little gray teapot. Click the Render button.
4. A rendering window should appear and draw a quick rendered image of the scene. You can save this image if desired.

Real World Scenario

RENDERING A CORRIDOR AND BOULEVARD TREES

This exercise leads you through adding multiview blocks as points to a road design; then you'll create a custom visual style, create a simple rendering, and create a fly-through AVI.

1. Open the drawing `Visualization Corridor and Trees.dwg`.
2. Note that there is already a corridor in the drawing with the appropriate code style set applied, as well as a road surface built with a contour style applied. There is an existing ground surface in this drawing. If you select Existing Ground from the Prospector, right-click, and select Surface Properties, you will see that there is a grass render material applied (on the Information tab). Also note that there is an empty point group called Tree with a Tree point style applied.
3. Set some tree points on the corridor surface. From the Home tab's Create Ground Data panel, select Points ➢ Create Points - Surface ➢ Along Polyline/Contour. Use this command to set points on the corridor surface every 60′ along the thickened polylines 30′ to the right and left of the corridor baseline.
4. If your points do not immediately look like the Tree point style, right-click the Tree point group and choose Update. Your trees should appear as small trees along the polylines.
5. Erase the polylines.
6. Switch the drawing to isometric view.
7. Choose the corridor surface and change it to the _No Display style so that the triangles from the road surface do not get in the way of the corridor model.
8. Switch your workspace to the 3D Modeling workspace so you can access the Visual Styles Editor. Bring up the Visual Styles Editor.
9. Copy the Realistic visual style and create and apply a visual style that turns Edgemode to None and Material display to Materials Only. Use the Custom Realistic visual style that is already in the drawing as an example if needed.

10. Once you have applied your Custom Realistic visual style, you should see a black road, gray sidewalks, green daylight, and dark green existing ground. Toggle between other visual styles to observe their effect.
11. Switch to the 3D Modeling workspace if you haven't already.
12. Execute the Render command by changing to the Render tab and clicking the Teapot icon on the Render panel. A temporary render window will appear with a preview of your image. You can save this image, or go back to the drawing and change the view on your screen to be more zoomed in, a different angle, or something similar, and redo the rendering.
13. Type **3DDWFPUBLISH** at the command line. Name the DWF file and choose a location to save it. Once the file has been created, open the file and explore the Autodesk Design Review interface and the 3D model.
14. Change the visual style in the drawing to a wireframe style to make navigation faster.
15. Change back to top view and the Civil 3D workspace.

    Now you will create a motion path animation. You'll take a polyline from the road elevation and move it 15′ into the air to serve as the motion path.
16. Select the corridor and choose Polyline From Corridor from the Launch Pad drop-down panel to extract the crown feature line from the corridor.
17. Move this polyline 15′ in the positive *z* direction. The easiest way is to use the AutoCAD Move command to move the polyline from 0,0,0 to **0,0,15**.
18. Switch to the 3D Modeling workspace. If the Animations panel is not displayed on the Render tab, right-click the Render tab and click Show Panels ➢ Animations.
19. Change to the Render tab and choose Animation Motion Path from the Animations panel to open the Motion Path Animation dialog.
20. In the Motion Path Animation dialog, choose the following options to make a simple wireframe drive-through:
    - Camera: Choose Path, and then use the Select button to go into the drawing and choose your polyline. This is the path that the camera will follow.
    - Target: Leave the defaults as they are.
    - Animation Settings:
        - Frame Rate (FPS): **15**
        - Number of Frames: **300**
        - Duration (seconds): **20**
    - Visual Style: 3D Wireframe
    - Format: AVI
    - Resolution: 640 × 480
    - Corner Deceleration: Checked

- **When Previewing Show Camera Preview: Checked**

Your dialog should now look like what's shown here:

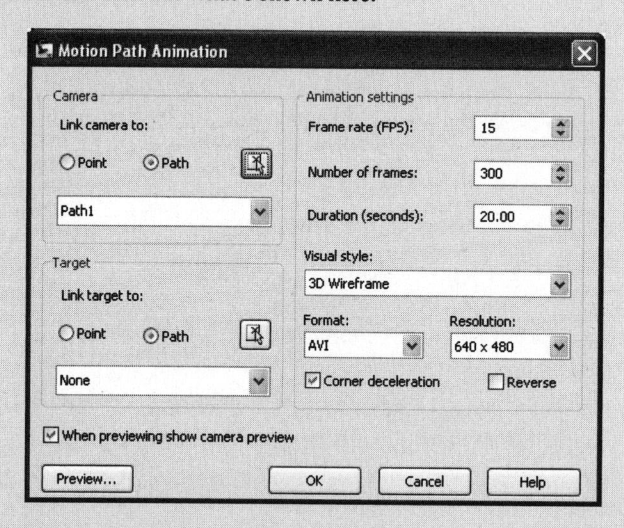

21. **Click OK to send the AVI to be created. Depending on your computer specifications, this could take a few minutes or longer. A preview will appear as shown here.**

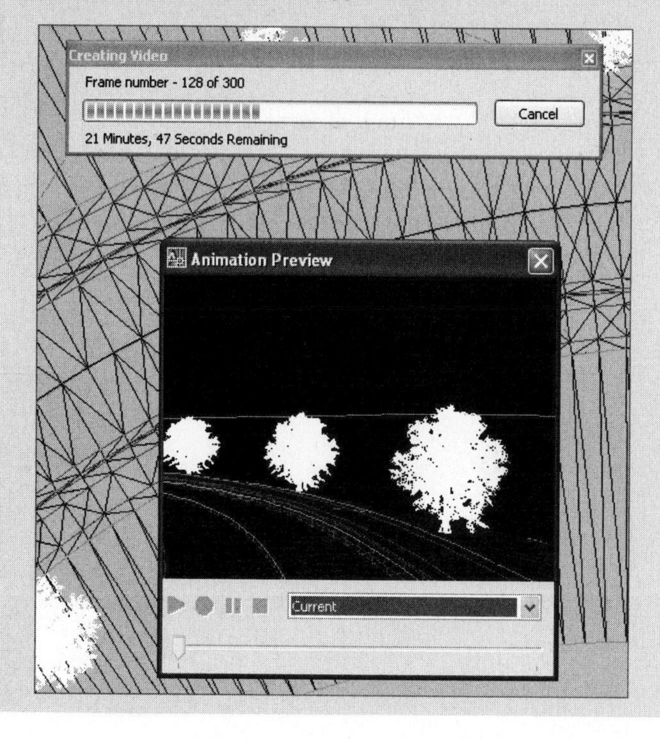

22. Once the preview window closes, locate the AVI file on your computer and play the video in your favorite media player.
23. Repeat the process using a different resolution or visual style. For a fully rendered drive-through, choose the Presentation option from the Render Presets drop-down menu. This may take several hours depending on your frame rate, resolution, and processor speed, but the results can be quite spectacular for a public hearing or similar event.

The Bottom Line

Apply render materials to a corridor using a code set style. Visualizing corridors is useful for creating complete site models for your own better understanding of the design, and for sharing with your clients and regulating agencies. The first step in visualizing a corridor is to apply a code set style that will automatically assign render materials to certain codes.

Master It Open the drawing file `Mastering Visualization.dwg`. Apply the All Codes code set style.

Apply a Realistic visual style to a corridor model. Once the appropriate code set style has been assigned, the corridor can be seen with its render materials with a Realistic visual style applied.

Master It Continue working in the file `Mastering Visualization.dwg`. Apply the Realistic visual style to the drawing.

Create a 3D DWF from a corridor model. One way you can share your visualized corridor is through 3D DWF. Autodesk Design Review is a free application that you can direct your clients to download from Autodesk.com. 3D DWFs can be opened, navigated, and analyzed using Design Review.

Master It Continue working in the file `Mastering Visualization.dwg`. Create a 3D DWF from the drawing and open it in Autodesk Design Review.

Chapter 23

Projecting the Cost: Quantity Takeoff

The goal of every project is eventual construction. Before the first bulldozer can be fired up and the first pile of dirt moved, the owner, city, or developer has to know how much all of this paving, pipe, and dirt are going to cost. Although contractors and construction managers are typically responsible for creating their own estimates for contracts, the engineers often perform an estimate of cost to help judge and award the eventual contract. To that end, many firms have entire departments that spend their days counting manholes, running planimeters around paving areas, and measuring street lengths to figure out how much striping will be required.

Civil 3D includes a Quantity Takeoff (QTO) feature to help relieve that tedious burden. You can use the model you've built as part of your design to measure and quantify the pieces needed to turn your project from paper to reality. You can export this data to a number of formats and even to other applications for further analysis.

By the time you complete this chapter, you'll learn to:

- Open and review a list of pay items along with their categorization
- Assign pay items to AutoCAD objects, pipe networks, and corridors
- Use QTO tools to review what items have been tagged for analysis
- Generate QTO output to a variety of formats for review or analysis

Inserting a Pay Item List and Categories

Before you can begin running any sort of analysis or quantity, you have to know what items matter. Various municipalities, states, and review agencies have their own lists of items and methods of breaking down the quantities involved in a typical development project. These lists are called Pay Item lists in Civil 3D.

When preparing quantities, different types of measurements are used based on the items being counted. Some are simple individual objects such as light posts, fire hydrants, or manholes. Only slightly more complicated are linear objects such as road striping, or area items such as grass cover. These measurements are also part of the Pay Item list.

In any project, there can be thousands of items to tabulate. To make this process easier, most organizations have built up pay categories, and Civil 3D makes use of this system in a Categorization file. This list can still be pretty confusing and overwhelming, so Civil 3D includes an option to create Favorite items that are used most frequently.

In this exercise, you'll look at how to open a Pay Item and Category list and add a few items to the Favorites list for later use:

1. Open the `QTO Paylists.dwg` file. Remember, all files can be downloaded from `www.sybex.com/go/masteringcivil3d2010`.
2. From the Analyze tab's QTO panel, select QTO Manager to display the QTO Manager palette shown in Figure 23.1.

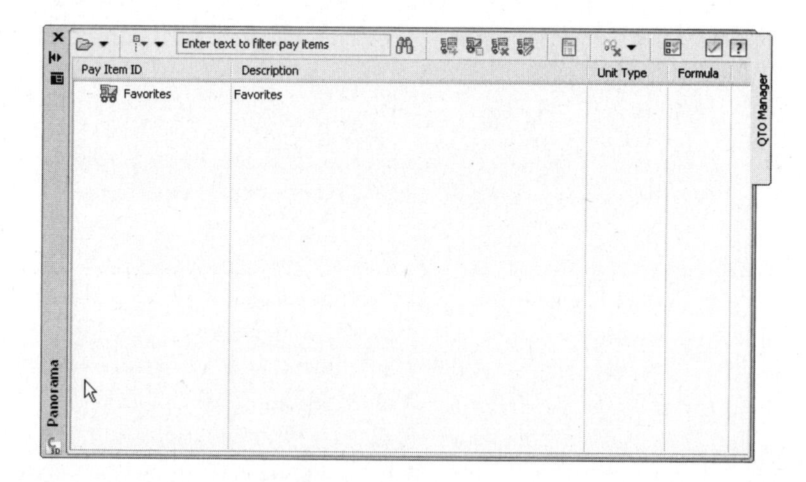

FIGURE 23.1
The empty QTO Manager Dialog palette

3. Click the Open button on the top left of the dialog to display the Open Pay Item File dialog.
4. From the Pay Item File Format drop-down list, select CSV (Comma Delimited). AASHTO TransXML and FDOT (Florida Department of Transportation) file formats may also be available, depending on your source data's format.
5. Click the Open (folder) button next to the Pay Item File text box to open the Open Pay Item File dialog.
6. Navigate to the Getting Started folder and select the `Getting Started.csv` file. In Windows XP, this file is found in `C:\Documents and Settings\All Users\Application Data\Autodesk\C3D 2010\enu\Data\Pay Item Data\Getting Started\`. In Vista, this file is found in `C:\ProgramData\Autodesk\C3D2010\enu\Data\Pay Item Data\Getting Started\`.
7. Click Open to select this Pay Item List.
8. Click the Open (folder) button next to the Pay Item Categorization file text box to display the Open Pay Item Categorization File dialog.
9. Navigate to the Getting Started folder and select the `Getting Started Categories.xml` file. Click OK to close the dialog. Your display should look similar to Figure 23.2.
10. Click OK to close the Open Pay Item File dialog, and Panorama will now be populated with a collection of "Divisions." These divisions came from the `Categories.xml` file.
11. Expand the Division 200 ➢ Group 201 ➢ Section 20101 branches, and select the item 20101-0000 CLEARING AND GRUBBING as shown in Figure 23.3.

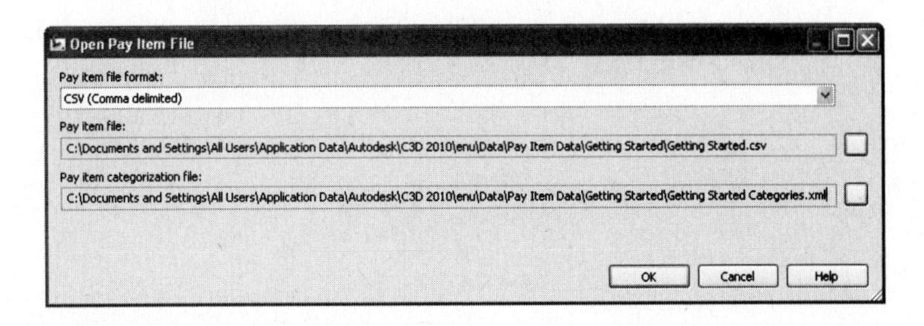

FIGURE 23.2 Completed file selection for the Open Pay Item dialog

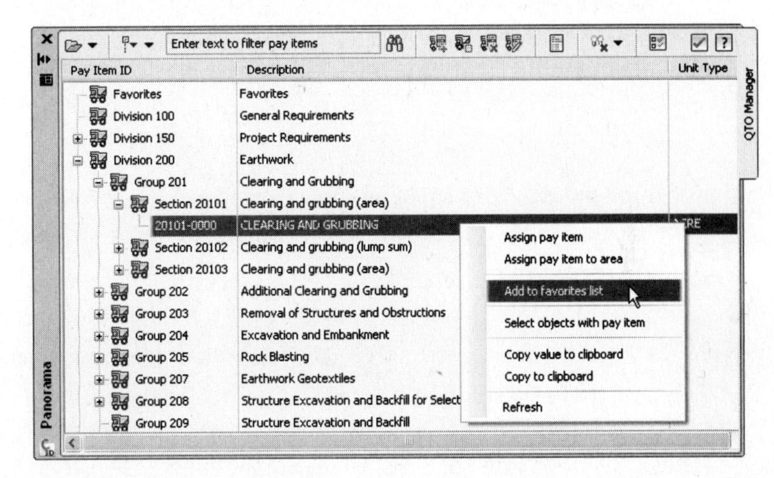

FIGURE 23.3 Selecting a Pay Item to add to Favorites

12. Right-click and select Add to Favorites List.
13. Expand a few other branches to familiarize yourself with this parts list, adding some to Favorites as you go. When complete, your QTO Manager should look similar to Figure 23.4.

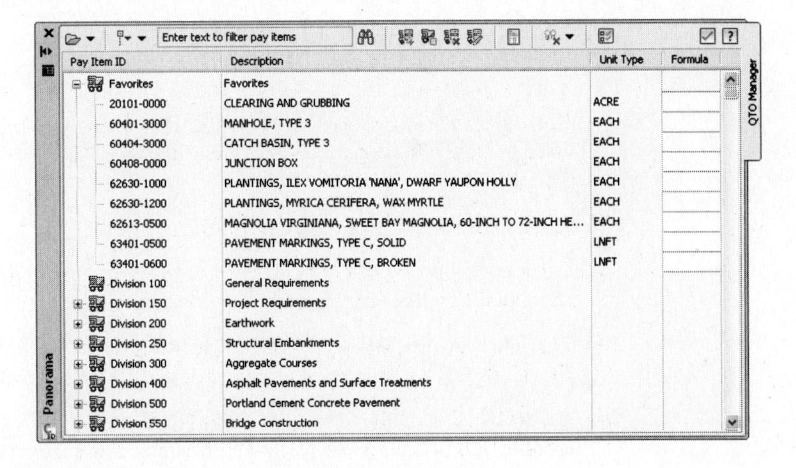

FIGURE 23.4 Completed creation of a Favorites list within the QTO Manager

Civil 3D ships with a number of pay item list categorization files, but only one actual pay item list. This avoids any issue with out-of-date data. Contact your reviewing agency for access to their pay item list and categorization files if they're not already part of the Civil 3D product. Once you have pay items to choose from, it's time to assign them to your model for analysis.

Real World Scenario

CREATING YOUR OWN CATEGORIES FILE

One of the challenges in automating any quantity takeoff analysis is getting the pay item list and categories to match up with your local requirements. Getting a pay item list is actually pretty straightforward — many reviewing agencies provide their own list for public use in order to keep all bidding on an equal footing.

Creating the category file can be slightly more difficult, and it will probably require some experimentation to get just right. In this example, you'll walk through creating a couple of categories to be used with a provided pay item list.

Note that modifying the QTO Manager palette affects all open drawings. You might want to finish the other exercises and come back to this when you need to make your own file in the real world!

1. Create a new drawing from the `_AutoCAD Civil 3D (Imperial)` template file.
2. From the Analyze tab's QTO panel, click the QTO Manager button to display the QTO Manager palette.
3. Click the Open Pay Item File button at top left to display the Open Pay Item File dialog.
4. Verify that the format is CSV (Comma Delimited).
5. Browse to open the `Mastering.csv` file as in the previous exercise. This file is in your downloaded *Mastering* files.
6. Click the Open (folder) button next to the Pay Item Categorization file text box to display the Open Pay Item Categorization File dialog.
7. Browse to open the `Mastering Categories.xml` file, and then click OK to dismiss the dialog. Your QTO Manager dialog should look like the following screen image.

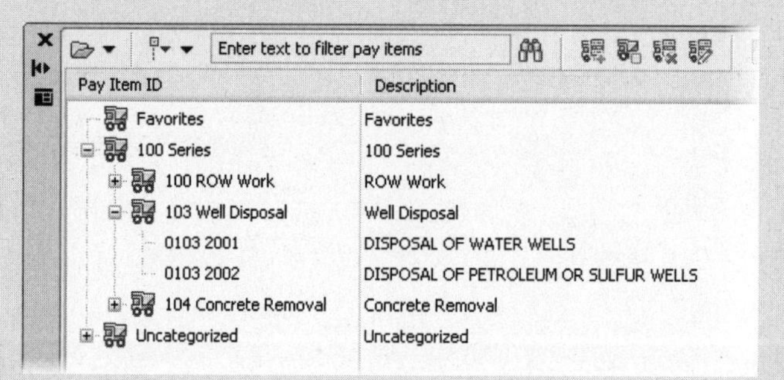

8. Launch XML Notepad. (You can download this from Microsoft as noted in Chapter 17, "Sharing the Model: Data Shortcuts.")
9. Browse and open the `Mastering Categories.xml` file from within XML Notepad.

10. Right-click the category branch as shown here and select Duplicate.

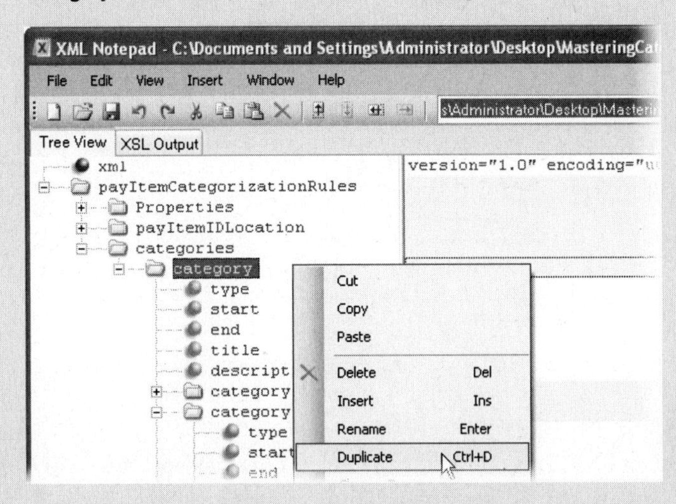

11. Modify the values as shown to match the following image. Be sure to delete the extra category branches under the new branch.

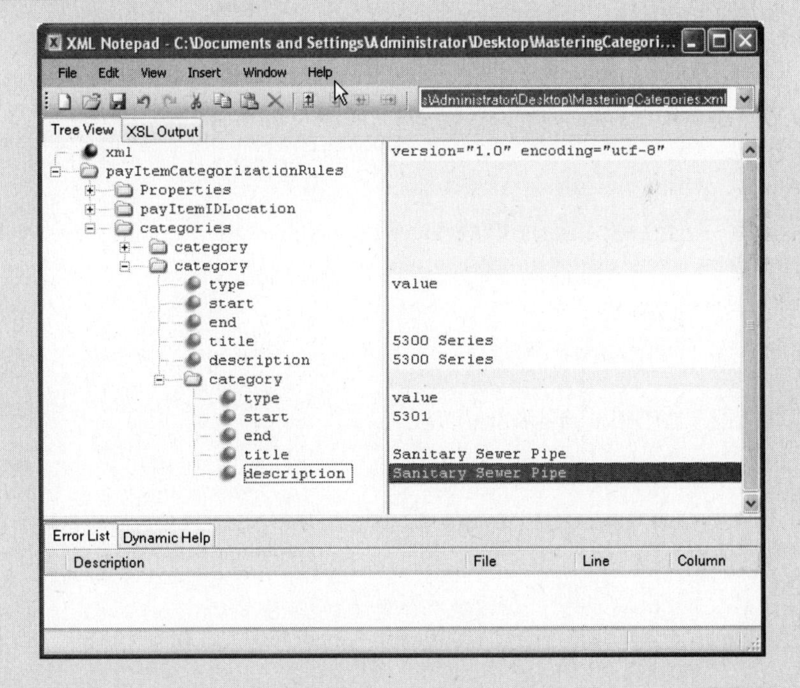

12. Save and close the modified `Mastering Categories.xml` file.

13. Switch back to Civil 3D's QTO Manager.

14. Click the Open drop-down menu in the top left of the QTO Manager, and select Open ➢ Categorization File.

15. Browse to the `Mastering Categories.xml` file again and open it. Once Civil 3D processes your XML file, your QTO Manager should look similar to the following screen image.

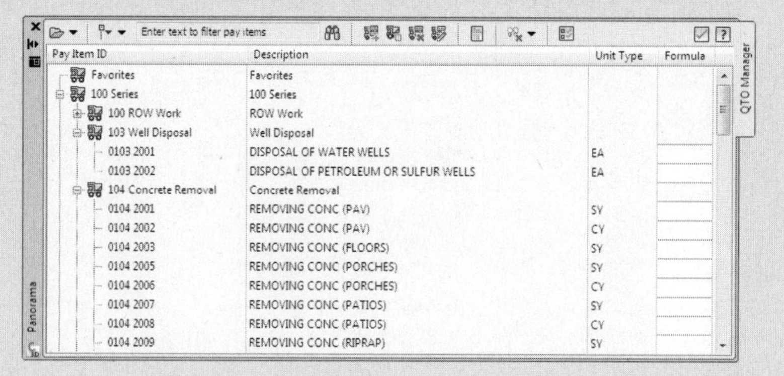

Creating a fully developed category list is a bit time-consuming, but once it's done, the list can be shared with your entire office so everyone has the same data to use.

Keeping Tabs on the Model

Once you have a list of items that should be accounted for within your project, you have to assign them to items in your drawing file. You can do this in any of the following ways:

- Assign pay items to simple items like blocks and lines.
- Assign pay items to Corridor components.
- Assign pay items to Pipe Network pipes and structures.

In the next few sections, you'll look at each of these methods, along with some formula tools that can be used to convert things such as linear items to individual quantity counts.

AutoCAD Objects as Pay Items

The most basic use of the QTO tools is to assign pay items to things like blocks and linework within your drawing file. The QTO tools can be used to quantify tree plantings, sign posts, or area items such as clearing and grubbing. In the following exercise, you'll assign pay items to blocks as well as to some closed polylines to see how areas can be quantified:

1. Open the `Acad Objects in QTO.dwg` file.
2. From the Analysis tab's QTO panel, click the QTO Manager button to display the QTO Manager palette.
3. Expand the Favorites branch, right-click on CLEARING AND GRUBBING, and select Assign Pay Item to Area as shown in Figure 23.5.
4. Enter **O** at the command line and press ↵ to activate the Object option for assignment.

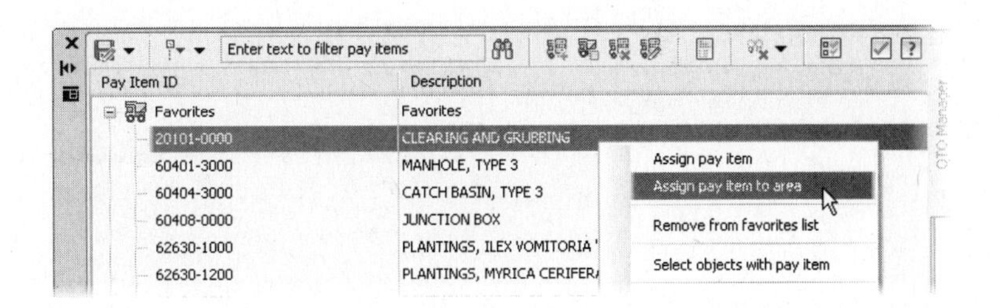

FIGURE 23.5
Assigning an area-based pay item

5. Click on the outer edge of the site as shown in Figure 23.6. Notice the entire polyline highlights to indicate what object is being picked! The command line should also echo, ''Pay item 20101-0000 assigned to object'' when you pick the polyline. Note that this will also create a hatch object reflecting the area being assigned to the pay item.

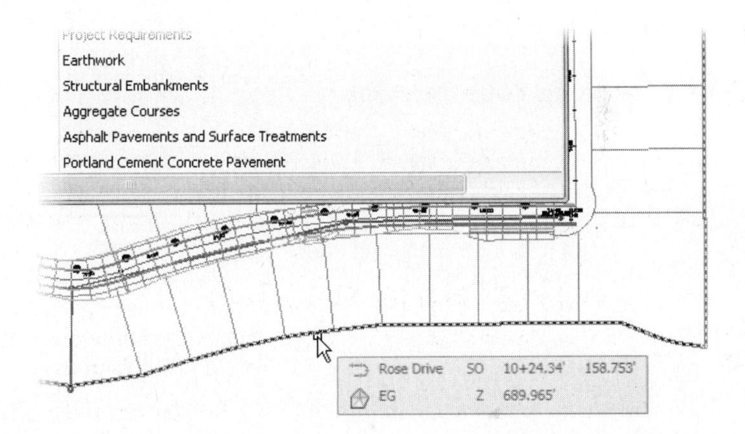

FIGURE 23.6
Selecting a closed polyline for an area-based quantity

6. Press ↵ again to complete the selection.
7. Move the QTO Manager palette to the side and zoom in on an area of the road where you can see two or more of the blocks representing trees.
8. Select one of the tree blocks. To select all of the tree blocks in the drawing, right-click and select Select Similar.
9. Back on the QTO Manager palette, under the Favorites branch, select PLANTINGS, MYRICA CERIFERA, WAX MYRTLE.
10. Near the top of the palette, click the Assign Pay Item button as shown in Figure 23.7. Notice the great tooltips on these buttons!
11. Press ↵ to complete the assigning.

The two assignment methods shown here are essentially interchangeable. Pay items can be assigned to any number of AutoCAD objects, meaning you don't have to redraw the planners' or landscape architects' work in Civil 3D to use the QTO tools.

Watch out for one quirk of the system! If you assign a pay item and then block that object out of your current drawing and into another, the pay item assignment goes along as well. You'll find out how to unassign pay items after you've looked through all the ways to assign them.

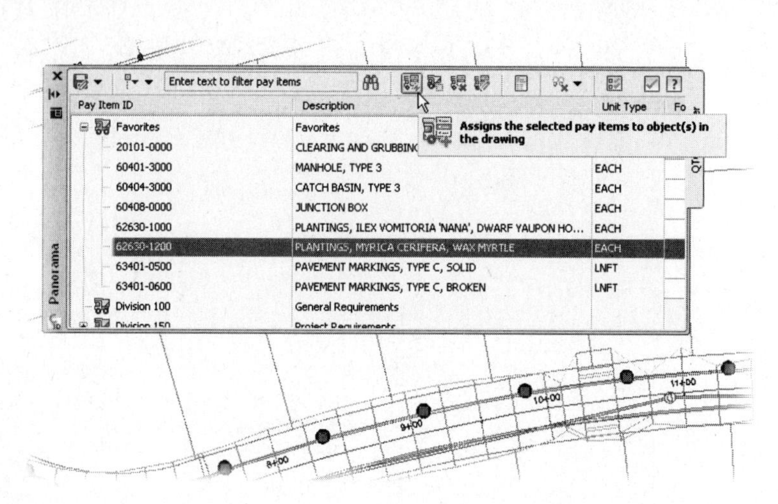

FIGURE 23.7
Assigning pay items to an existing selection set

Pricing Your Corridor

The corridor functionality of Civil 3D is invaluable. You can use it to model everything from streams to parking lots to mundane roads. With the QTO tools, you can also use the corridor object to quantify much of the project construction costs. Because a corridor is typically a linear object while many pay items are incremental counts, it would seem that using the corridor would be a poor tool for preparing quantities of individually counted items.

In this first example, you'll use the Pay Items list along with a formula to convert the linear curb measurement to an incremental count of light poles required for the project:

1. Open the `Corridor Object in QTO.dwg` file.
2. From the Analyze tab's QTO panel, select QTO Manager to display the palette.
3. In the text box near the top of the palette, type **light**. This will filter the pay item list for items that have "light" somewhere in the title or description.
4. Click on the drop-down list to the left of the text box and select Turn Off Categorization. The numerous category branches in a pay item list can hide filter results, so turning them off should give you something similar to Figure 23.8.

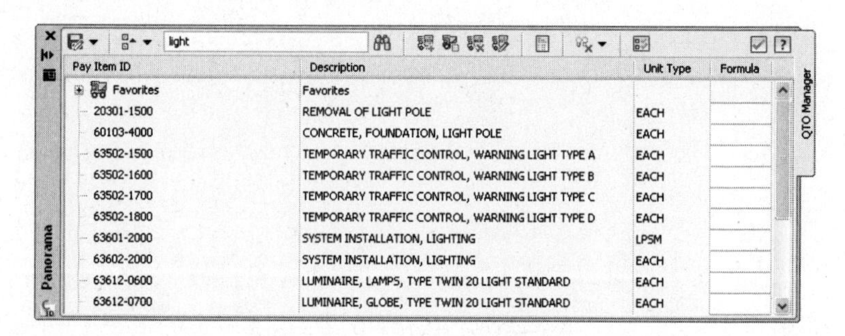

FIGURE 23.8
Filtering the pay item list for "light"

5. Right-click item number 63612-0600 and select Add to Favorites List.

6. Expand the Favorites branch to display your newly added Luminaire light fixture pay item.
7. Scroll to the right within the palette to reveal the column titled Formula and click in the empty cell on the Luminaire row. When you do, Civil 3D will display the alert box shown in Figure 23.9, which warns you that formulas must be stored to an external file.

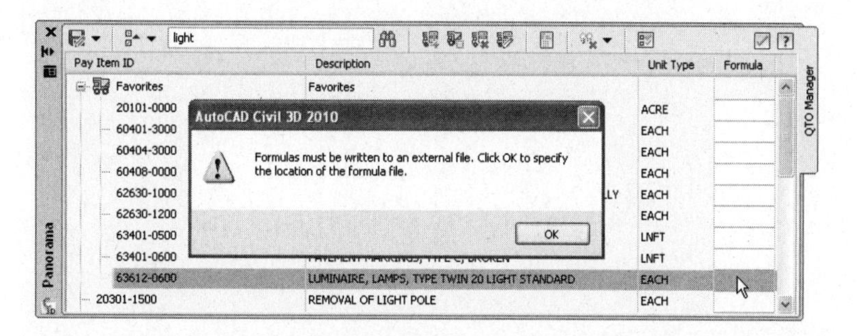

FIGURE 23.9
Click in the Formula cell to display this warning dialog.

8. Click OK to dismiss the warning, and Civil 3D will present the Select a Quantity Takeoff Formula File dialog. Navigate to your desktop, change the filename to **Mastering**, and click the Save button to dismiss this dialog.

   Civil 3D will display the Pay Item Formula: 63612-0600 dialog as shown in Figure 23.10. (The Expression will be empty, but you'll take care of that in the next steps.)

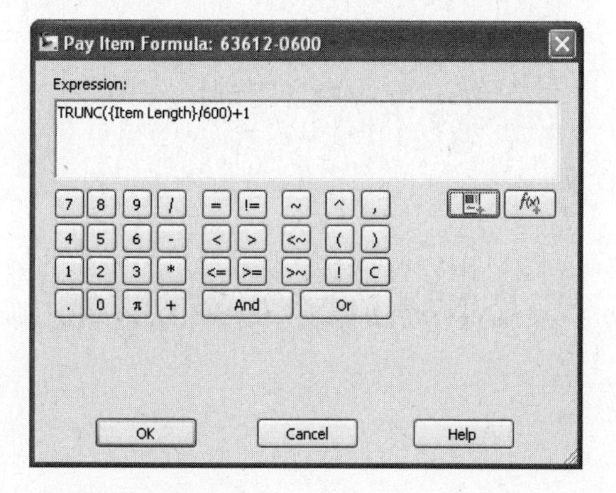

FIGURE 23.10
The completed pay item expression

Assume that you need a street light every 300 feet, but only on one side of the street. To do this, you will add up all the lengths of curb, and then divide by two because you want only one half of the street, and then divide by 300 because you are running lights in an interval. Finally, you'll truncate and add 1 to make the number conservative.

9. Enter the formula using the buttons in the dialog or typing it in to arrive at Figure 23.10, and then click OK to dismiss the dialog. Note that the pay item list now shows a small calculator icon on that row to indicate a formula is in use.

Now that you've modified the way the light poles will be quantified from your model, you can assign the pay items for light poles and road striping to your corridor object. This is done by modifying the code set as you'll see in this example:

1. Continue working with the `Corridor in QTO.dwg` file.
2. Within the Toolspace Settings tab, select General ➤ Mutlipurpose Styles ➤ Code Set Styles ➤ All Codes. Right-click and select Edit to display the Code Set Style – All Codes dialog.
3. Change to the Codes tab, scroll down to the Point section, and find the row for Crown as shown in Figure 23.11.

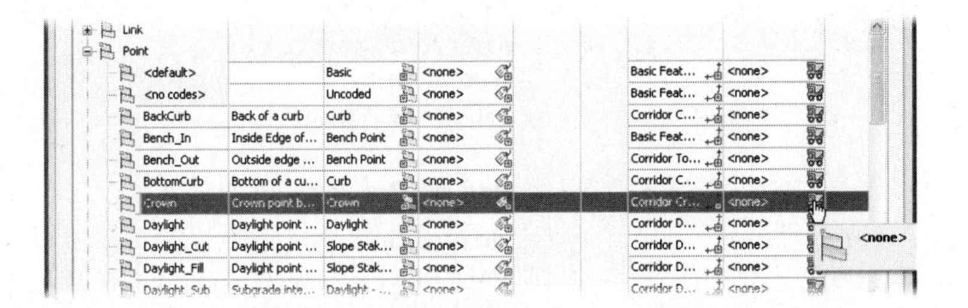

FIGURE 23.11 The Code Set Style dialog with the Point-Crown row selected

4. Click the truck icon in the Pay Item column as shown previously in Figure 23.8 to bring up the Pay Item List dialog.
5. Expand the Favorites branch and select PAVEMENT MARKINGS, TYPE C, SOLID. Click OK to return to the Code Set Style dialog. Your dialog should now reflect the pay item number of 63401-0500 as shown in Figure 23.12. Note that the tooltip also reflects this pay item.

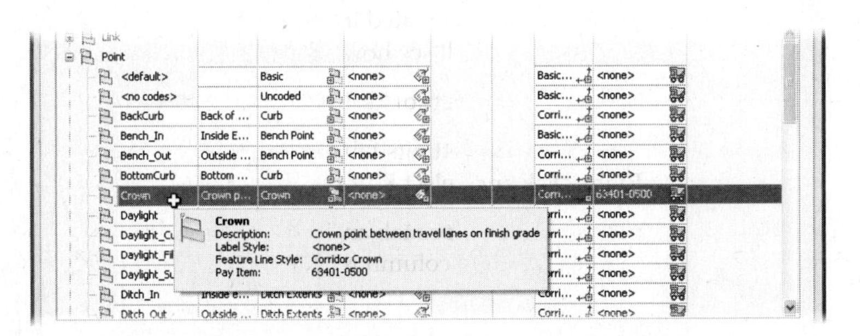

FIGURE 23.12 Crown point codes assigned a pavement marking pay item

6. Browse up to find the Back_Curb code row, and repeat steps 4 and 5 to assign 63612-0600 to the Back_Curb code. Remember, this is your luminaries' pay item with the formula from the previous exercise. Your dialog should look like Figure 23.13.
7. Switch to Prospector, and expand the Corridor branch. Right-click Corridor – (2) and select Properties to open the Corridor Properties dialog.
8. Change to the Codes tab and scroll down to the Point section. Notice that Back_Curb and Crown codes reflect Pay Items in the far-right column.
9. Click OK to close the dialog.

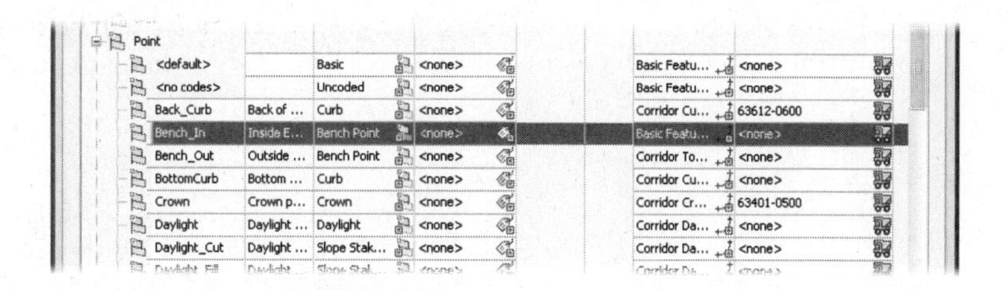

FIGURE 23.13
Completed Code Set editing for pay items

Corridors can be used to measure a large number of things. You've always been able to manage pure quantities of material, but now you can add to that the ability to measure linear and incremental items as well. Although we didn't explore every option, you can also use shape and link codes to assign pay items to your corridor models.

Now that you've looked at AutoCAD objects and corridors, it's time to examine the pipe network objects in Civil 3D as they relate to pay items.

Pipes and Structures as Pay Items

One of the easiest items to quantify in Civil 3D is the pipe networks. There are numerous reports that will generate pipe and structure quantities. This part of the model has always been fairly easy to account for; however, with the ability to include it in the overall QTO reports, it's important to understand how parts get pay items assigned. There are two methods: via the parts lists and via the part properties.

ASSIGNING PAY ITEMS IN THE PARTS LIST

Ideally, you'll build your model using standard Civil 3D parts lists that you've set up as part of your template. These parts lists contain information about pipe sizes, structure thicknesses, and so on. They can also contain pay item assignments. This means that the pay item property will be assigned as each part is created in the model, skipping the assignment step later.

In this exercise, you'll see how easy it is to modify parts lists to include pay items:

1. Open the `Pipe Networks in QTO.dwg` file.
2. Change to the Settings tab, and expand Pipe Network ➤ Parts Lists ➤ Storm Sewer. Right-click and select Edit to bring up the Network Parts List – Storm Sewer dialog.
3. Change to the Pipes tab and expand the Storm Sewer ➤ Concrete Pipe part family. Notice that the far-right column is the Pay Item assignment column, similar to the Code Sets in the previous section.
4. Click the truck icon in the 12 inch RCP row to display the Pay Item List.
5. Enter **12-Inch Pipe Culvert** in the textbox as shown in Figure 23.14 and press ↵ to filter the dialog. Remember that you can turn off categories with the button on the left of the text box.
6. Select the 12-INCH PIPE CULVERT item as shown, and click OK to assign this pay item to the 12 inch RCP pipe part.
7. Repeat steps 4 through 6 to assign pay items to 15, 18, 21, and 24 inch RCP pipe. Your dialog should look like Figure 23.15.
8. Change to the Structure tab of the Parts List dialog.

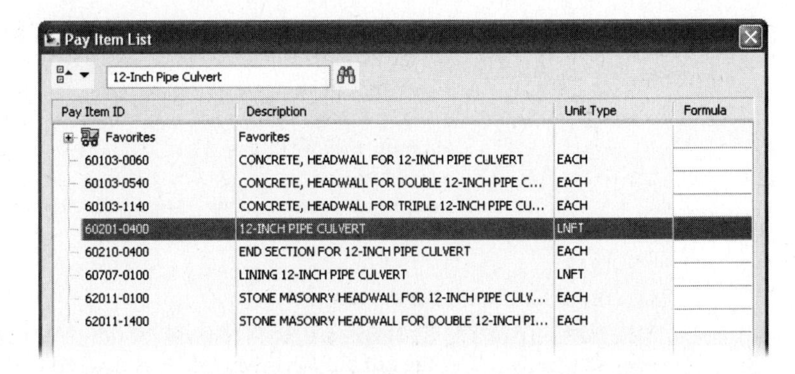

FIGURE 23.14 Filtering and selecting the 12-inch pipe culvert as a pay item

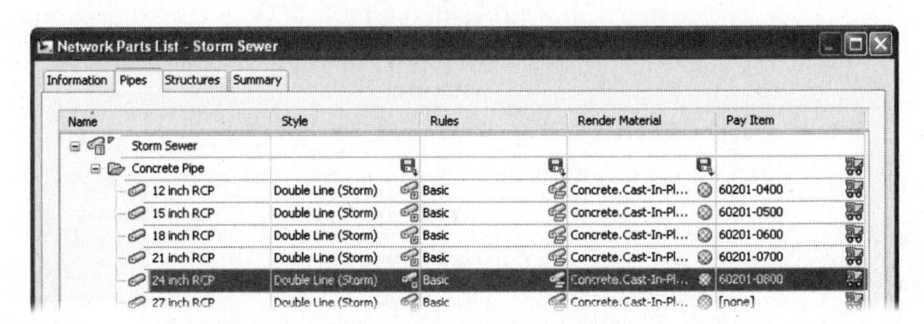

FIGURE 23.15 Completed pipe parts pay item assignment

9. Expand the Storm Sewer ➤ Concrete Rectangular Headwall branch.
10. Click the truck icon on the 105 × 6 × 102 Inch Concrete Rectangular Headwall row to display the Pay Item List.
11. Expand the Favorites branch and select CATCH BASIN, TYPE 3 as the pay item. Click OK to close the dialog.
12. Click the truck icon on the row of the Concentric Cylindrical Structure folder to display the Pay Item List. By picking at the level of the part family, you will be assigning the same pay item to *all* sizes of that part family.
13. Expand the Favorites branch and select MANHOLE, TYPE 3. Click OK to close the dialog. Your parts list should now look like Figure 23.16.

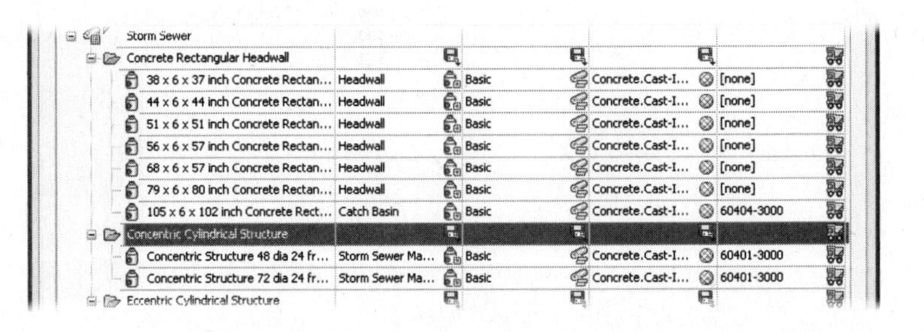

FIGURE 23.16 Completed structure parts pay item assignment

There is one major warning you should be aware of: pay items change when pipes and structures change only if the new part has a pay item assigned in the parts list. This means that if you build a pipe network with an 18-inch pipe (and the corresponding pay item assignment), but then change that pipe to a 21-inch diameter using the part properties or the Swap Part tool, the pay item still reflects an 18-inch pipe if the parts list doesn't have a pay item for the 21-inch pipe. Also, if you manually change a pipe property (such as diameter), this will not cause a change in the pay item assignment. You should use swap part to change sizes! This is not a major defect, but it's something you definitely want to keep an eye on. So, if you do need to change a pay item assignment to a part that's already in the network, how do you do it? You'll find that out in the next section.

Pay Items as Part Properties

If you have existing Civil 3D pipe networks that were built before your parts list had pay items assigned, or if you change out a part during your design, you need a way to review and modify the pay items associated with your network. Unfortunately, this isn't as simple as just telling Civil 3D to reprocess some data, but it's not too bad either. You simply remove the pay item association in place and then add new ones.

In this exercise, you'll add pay item assignments to a number of parts already in place in the drawing:

1. Open or continue working with the `Pipe Networks in QTO.dwg` file.
2. From the Analyze tab's QTO panel, select QTO Manager to display the QTO Manager palette.
3. Slide the QTO Manager to one side, and then select one of the manholes in the Storm Sewer Network (they have a D symbol on them).
4. Right-click and choose Select Similar from the contextual menu. Four manholes should highlight.
5. In QTO Manager, expand Favorites and select MANHOLE, TYPE 3 and then click the Assign Pay Item button as shown in Figure 23.17.

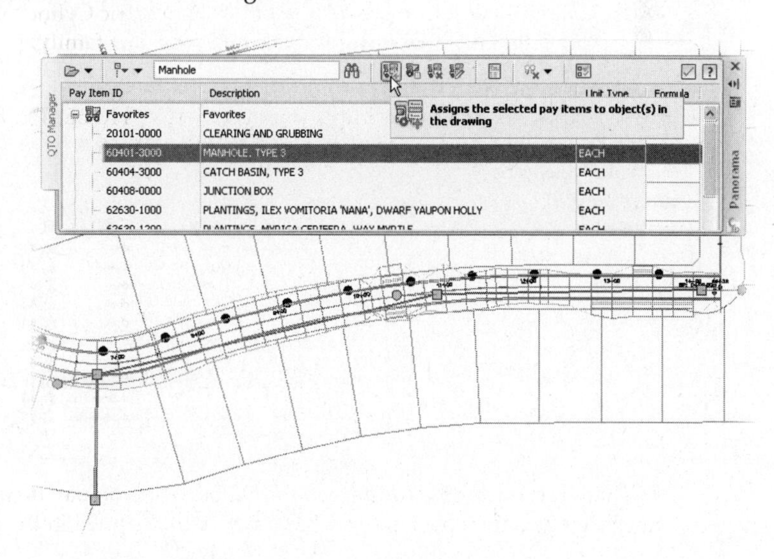

Figure 23.17
Assigning the MANHOLE pay item to the manhole structures

6. At the command line, press ↵ to complete the assignment. You can pause over one of the manholes, and the tooltip will reflect a pay item now, in addition to the typical information found on a manhole.

Assigning pay items to existing structures and pipes is very similar to adding data to standard AutoCAD objects. As mentioned before, the pay item assignments sometimes get confused in the process of changing parts and pipe properties, and they should be manually updated. To do this, you'll need to remove pay item data, and then add it back in as demonstrated in this exercise:

1. Pan to the northwest area of the site where the sanitary sewer network runs offsite.
2. Select the pipe as shown in Figure 23.18, right-click, and select the Swap Part as shown to display the Swap Part Size dialog.

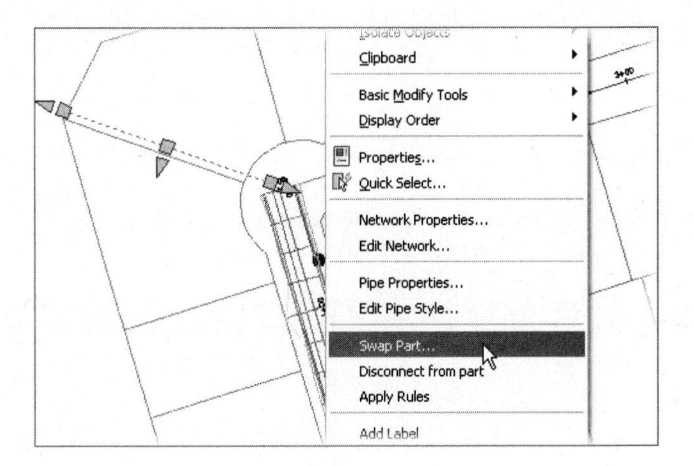

FIGURE 23.18
Swapping out the part size

3. Select 12 inch PVC from the list of sizes, and then click OK to dismiss the dialog.
4. Pause near the newly sized pipe and notice that the tooltip still reflects an 8-INCH PIPE CULVERT pay item as shown in Figure 23.19.

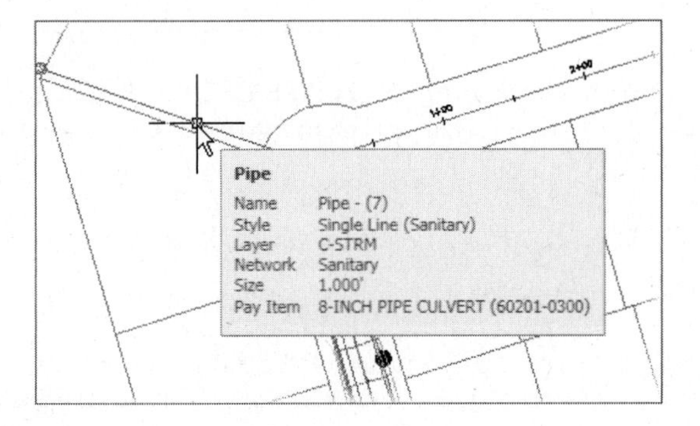

FIGURE 23.19
The tooltip indicates that pay items do not update!

5. Open the QTO Manager palette if it's not already open.

6. Enter **12-inch pipe** in the text box to filter the pay item list.
7. Select the 12-INCH PIPE CULVERT item in the list, right-click, and choose Assign Pay Item from the contextual menu.
8. Select the pipe you just modified to assign the pay item. The command line should echo, "Pay item 60201-0400 assigned to object."
9. Press ↩ to end the assignment command.
10. Click the Edit Pay Items on Specified Object button within the QTO Manager as shown in Figure 23.20.

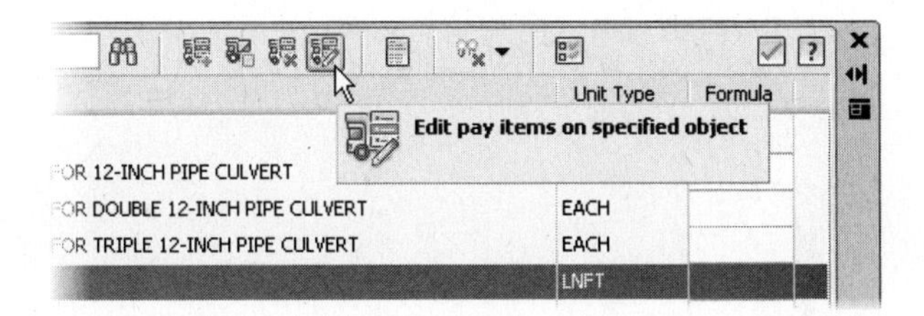

FIGURE 23.20
Editing pay item assignments

11. Pick the pipe one more time to display the Edit Pay Items dialog shown in Figure 23.21. Note that *both* pay item assignments are still on the pipe!

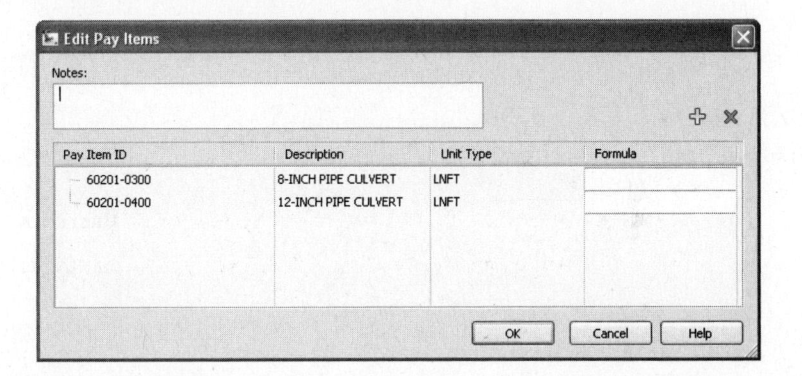

FIGURE 23.21
Editing pay items on a single object

12. Select the 8-INCH PIPE CULVERT row, and then click the red X in the upper right to remove this pay item from the pipe.
13. Click OK to dismiss the dialog.
14. Close the QTO Manager palette.

You might wonder why Civil 3D allows you to have multiple pay items on a single object. Linear feet of striping and tree counts can both be derived from street lengths; bedding and pipe material can both be calculated from pipe objects.

You've now built up a list of pay items, tagged your drawing a number of different ways, updated and modified pay item data, and looked at formulas in pay items. In the next section, you'll make a final check of your assignments before running reports.

Highlighting Pay Items

Before you run any reports, it's a good idea to make a cursory pass through your drawing and look at what items have had pay items assigned and what items have not. This review will allow you to hopefully catch missing items (such as hydrants added after the pay item assignment was done), as well as see any items that perhaps were blocked in with unnecessary pay items already assigned. In this exercise, you'll look at tools for highlighting objects with and without pay item assignments:

1. Open the `Highlighting Pay Items.dwg` file.
2. From the Analyze tab's QTO panel, select QTO Manager.
3. Within the QTO Manager, select the Highlight Objects with Pay Items as shown in Figure 23.22.

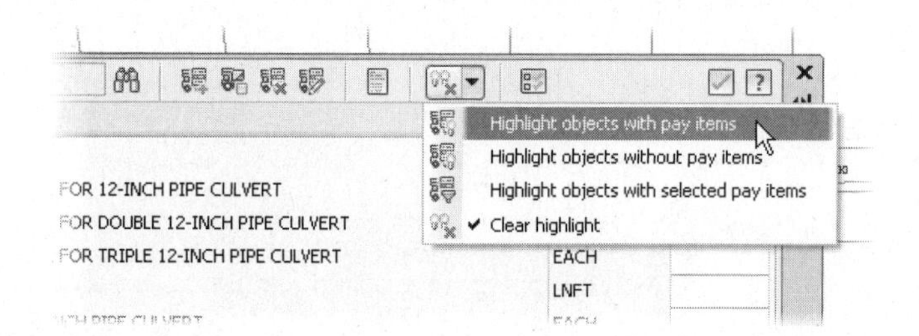

FIGURE 23.22
Turning on highlighting for items with pay items assigned

4. Pan to the middle of the drawing and note that some of the lot lines are still bright magenta instead of muted. This means they have a pay item assigned.
5. In QTO Manager, select Highlight Objects without Pay Items from the drop-down menu shown in Figure 23.22. Note that most of the alignments and lot lines are now highlighted.
6. In QTO Manager, select Clear Highlight to return the drawing view to normal.

While objects are highlighted, you can add and remove pay items as well as any other normal AutoCAD work you might perform. This makes it easier to correct any mistakes made during the assignment phase of the process. Finally, be sure to clear highlighting before exiting the drawing, or your peers might wind up awfully confused when they open the file!

Inventory Your Pay Items

At the end of the process, you need to generate some sort of report that shows the pay items in the model, the quantities of each item, and the units of measurement. This data can be used as part of the plan set in some cases, but it's often requested in other formats to make further analysis possible. In this exercise, you'll look at the Takeoff tool that works in conjunction with the QTO Manager to create reports:

1. Open the `QTO Reporting.dwg` file.
2. From the Analyze tab's QTO panel, select the Takeoff tool to display the Compute Quantity Takeoff dialog shown in Figure 23.23.

   Note that you can limit the area of inquiry by drawing or selection; by sheets if done from paper space; or by alignment station ranges. Most of the time, you'll want to run the full drawing. You can also report on only selected pay items if, for instance, you just wanted a table of pipe and structures.

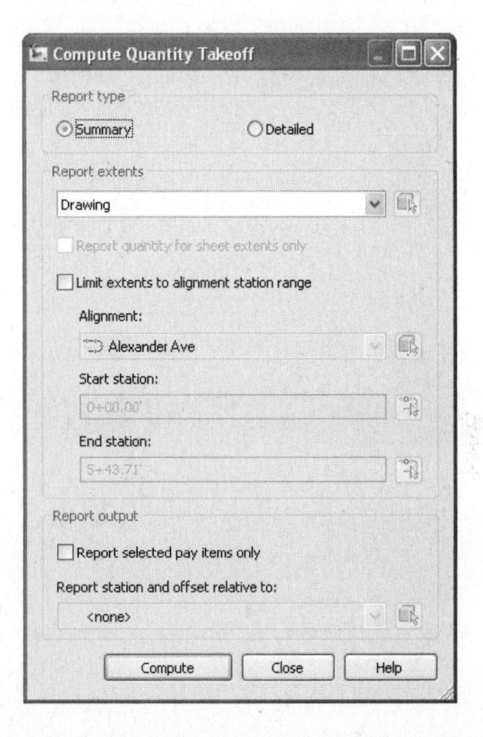

FIGURE 23.23
The Quantity Takeoff tool with default settings

3. Click Compute to open the Quantity Takeoff Report dialog as shown in Figure 23.24.

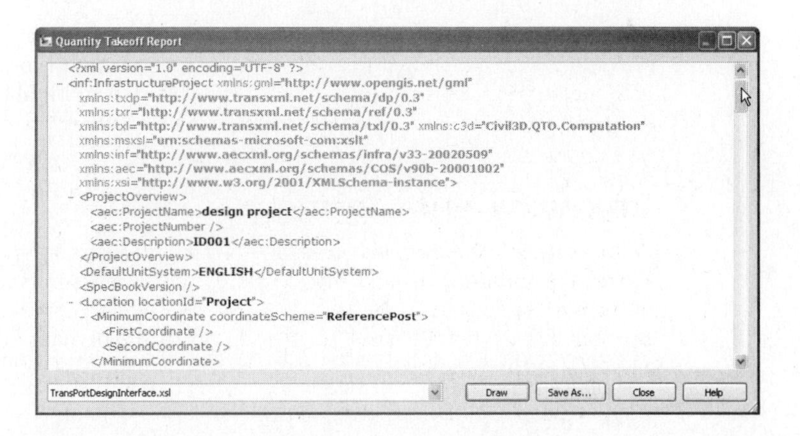

FIGURE 23.24
QTO reports in the default XSL format

4. From the drop-down menu on the lower-left of the dialog, select Summary (TXT).xsl to change the format to something more understandable. At this point, you can export this data out as a text file, but for the purposes of this exercise, you'll simply insert it into the drawing.

5. Click the Draw button at bottom of the dialog, and Civil 3D will prompt you to pick a point in the drawing.
6. Click near some clear space, and you'll be returned to the Quantity Takeoff Report dialog. Click Close to dismiss this dialog, and then click Close again to dismiss the Compute Quantity Takeoff dialog.
7. Zoom in where you picked in step 5, and you should see something like Figure 23.25.

FIGURE 23.25 Summary Takeoff data inserted into the drawing

Summary Takeoff Report

Pay Item ID	Description	Quantity	Unit
20101-0000	CLEARING AND GRUBBING	22.73	ACRE
60201-0300	8-INCH PIPE CULVERT	1300.788	LNFT
60201-0400	12-INCH PIPE CULVERT	180.814	LNFT
60201-0600	18-INCH PIPE CULVERT	735.083	LNFT
60201-0700	21-INCH PIPE CULVERT	145.535	LNFT
60401-3000	MANHOLE, TYPE 3	9	EACH
62630-1200	PLANTINGS, MYRICA CERIFERA, WAX MYRTLE	17	EACH
63401-0500	PAVEMENT MARKINGS, TYPE C, SOLID	1432.565	LNFT
63612-0600	LUMINAIRE, LAMPS, TYPE TWIN 20 LIGHT STANDARD	5	EACH

That's it! The hard work in preparing QTO data is in assigning the pay items. The reports can be saved to HTML, TXT, or XLS format for use in almost any analysis program.

The Bottom Line

Open and review a list of pay items along with their categorization. The pay item list is the cornerstone of QTO. You should download and review your pay item list versus the current reviewing agency list regularly to avoid any missed items.

Master It Add the 12-, 18-, and 21-Inch Pipe Culvert pay items to your Favorites in the QTO Manager.

Assign pay items to AutoCAD objects, pipe networks, and corridors. The majority of the work in preparing QTO is in assigning pay items accurately. By using the linework, blocks, and Civil 3D objects in your drawing as part of the process, you reduce the effort involved with generating accurate quantities.

Master It Open the `QTO Reporting.dwg` file and assign the Pavement Marking Type C Broken to the Alexander Ave., Parker Place, and Carson's Way alignments.

Use the QTO tools to review what items have been tagged for analysis. By using the built-in highlighting tools to verify pay item assignments, you can avoid costly errors when running your QTO reports.

Master It Verify that the three alignments in the previous exercise have been assigned with pay items.

Generate QTO output to a variety of formats for review or analysis. The Takeoff reports give you a quick and understandable understanding of what items have been tagged in the drawing, and they can generate text in the drawing or external reports for uses in other applications.

Master It Calculate and display the amount of Type C Broken markings in the `QTO Reporting.dwg` file.

Appendix

The Bottom Line

Chapter 1: Getting Dirty: The Basics of 3D

Find any Civil 3D object with just a few clicks. By using Prospector to view object data collections, you can minimize the panning and zooming that are part of working in a CAD program. When common subdivisions can have hundreds of parcels or a complex corridor can have dozens of alignments, jumping to the desired one nearly instantly shaves time off everyday tasks.

Master It Open `Sample Site.dwg` from the tutorials, and find parcel number five without using any AutoCAD commands.

Solution

1. In Prospector, expand Sites ➤ Rose Acres ➤ Parcels.
2. Right-click on Property :5 as shown and select Zoom To.

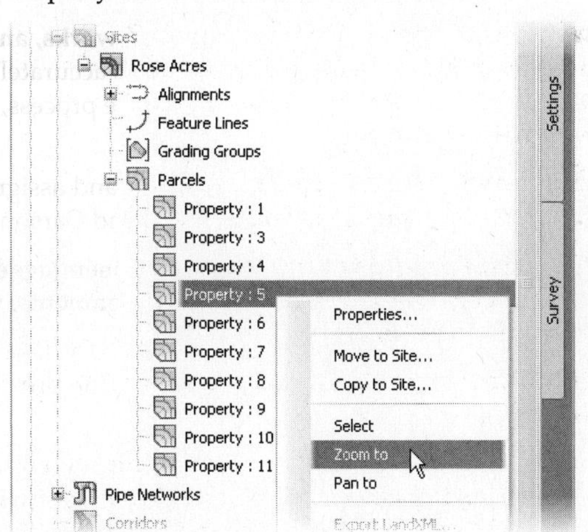

Modify the drawing scale and default object layers. Civil 3D understands that the end goal of most drawings is to create hard-copy construction documents. By setting a drawing scale and then setting many sizes in terms of plotted inches or millimeters, Civil 3D removes much of the mental gymnastics that other programs require when you're sizing text and symbols. By setting object layers at a drawing scale, Civil 3D makes uniformity of drawing files easier than ever to accomplish.

Master It Change `Sample Site.dwg` from a 200-scale drawing to a 40-scale drawing.

Solution Select 1″=40 from the Scale list in the lower right of the application window as shown here.

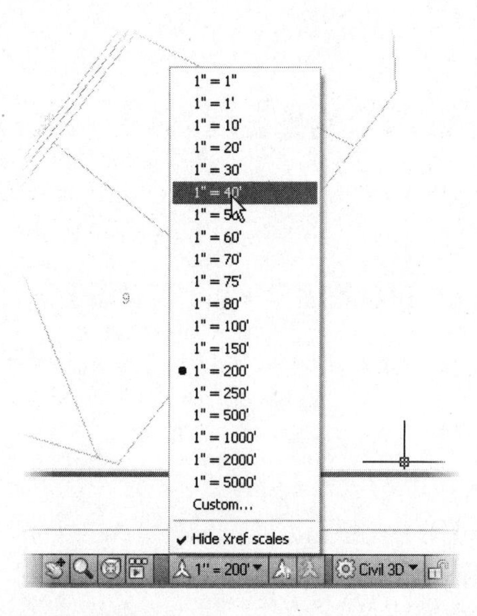

Modify the display of Civil 3D tooltips. The interactive display of object tooltips makes it easy to keep your focus on the drawing instead of an inquiry or report tools. When too many objects fill up a drawing, it can be information overload, so Civil 3D gives you granular control over the heads-up display tooltips.

Master It Within the same Sample Site drawing, turn off the tooltips for the Avery Drive alignment.

Solution Right-click the alignment in the drawing window or in Prospector, and bring up the Alignment properties. Deselect the box at lower left on the Information tab, as shown here.

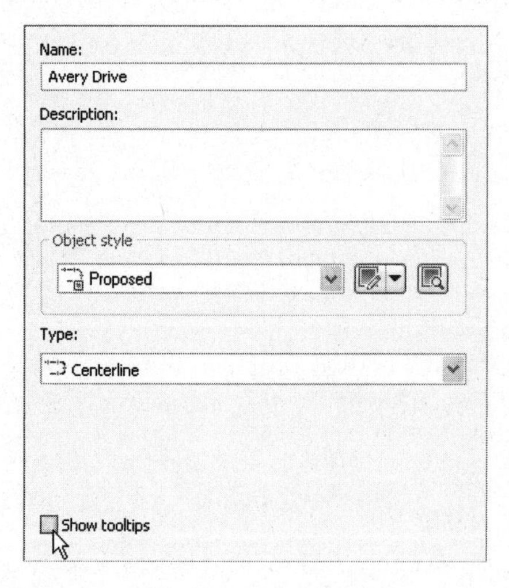

Add a new tool to the Toolbox. The Toolbox provides a convenient way to access macros and reports. Many third-party developers exploit this convenient interface as an easier way to add functionality without disturbing users' workspaces.

Master It Add the Sample Pipe macro from `C:\Program Files\Autocad Civil 3D 2010\Sample\Civil 3D API\COM\Vba\Pipe`, and select `PipeSample.dvb`.

Create a basic label style. Label styles determine the appearance of Civil 3D annotation. The creation of label styles will constitute a major part of the effort in making the transition to Civil 3D as a primary platform for plan production. Your skills will grow with the job requirements if you start with basic labels and then make more complicated labels as needed.

Master It Create a copy of the Elevation Only Point label style, name it Elevation With Border, and add a border to the text component.

Solution

1. Switch to the Settings tab in Toolspace. Expand the Point branch and then the Label Styles branch.
2. Right-click the Elevation Only label style, and select Copy.
3. Change the name to **Elevation with Border** on the Information tab.
4. On the Layout tab, set the Border visibility to True, and click OK.

Create a new object style. Object styles in Civil 3D let you quit managing display through layer modification and move to a more streamlined style-based control. Creating enough object styles to meet the demands of plan production work will be your other major task in preparing to move to Civil 3D.

Master It Create a new Surface style named Contours_Grid, and set it to show contours in plan views but a grid display in any 3D view.

Solution

1. In the Settings tab, right-click the Surface-Surface Styles folder, and select New to display the Surface Style dialog.
2. Rename this new surface style on the Information tab.
3. On the Display tab, with the View Direction set to 2D, turn off Border and turn on Contours.
4. Change the View Direction to 3D, turn off Triangles, and turn on Grid.
5. Click OK to close the dialog.

Navigate the ribbon's contextual tabs. As with AutoCAD, the ribbon is the primary interface for accessing Civil 3D commands and features. When you select an AutoCAD Civil 3D object, the ribbon displays commands and features related to that object. If several object types are selected, the Multiple contextual tab is displayed.

Master It Using the ribbon interface, access the Alignment Style Editor for the Proposed Alignment style. (Hint: it's used by the Avery Drive alignment.)

Solution

1. Select Avery Drive to display the Alignment ribbon
2. Select the Alignment Properties menu to display the Edit Alignment Style command as shown here.

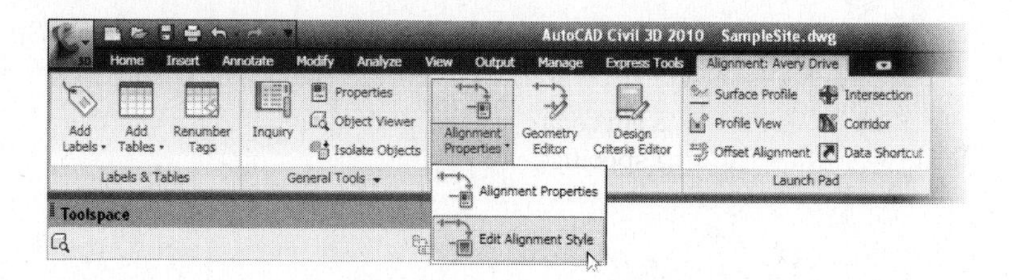

Chapter 2: Back to Basics: Lines and Curves

Create a series of lines by bearing and distance. By far the most commonly used command when re-creating a deed is Line by Bearing and Distance.

Master It Open the `Mastering Lines and Curves.dwg` file, which you can download from `www.sybex.com/masteringcivil3d2010`. Start from the Civil 3D point labeled START, and use any appropriate tool to create lines with the following bearings and distances (note that the Direction input format for this drawing has been set to DD.MMSSSS):

- N 57°06′56.75″ E; 135.441′
- S 41°57′03.67″ E; 118.754′
- S 27°44′41.63″ W; 112.426′
- N 50°55′57.00″ W; 181.333′

Solution

1. On the Draw panel, select Lines ➢ Create Lines ➢ Line by Bearing.
2. Use the Point Object transparent command to select the point labeled START.
3. On the command line, specify quadrant **1** ↵ **57.065675** ↵ **135.441** ↵.
4. Specify quadrant **2** ↵ **41.570367** ↵ **118.754** ↵.
5. Specify quadrant **3** ↵ **27.444163** ↵ **112.426** ↵.
6. Specify quadrant **4** ↵ **50.555700** ↵ **181.333** ↵.
7. Press ↵ again to leave the command.

Use the Inquiry commands to confirm that lines are drawn correctly.

Master It Continue working in your drawing. Use any appropriate Inquiry command to confirm that each line has been drawn correctly.

Solution

1. Change to the Analyze tab and click the Line and Arc Information button on the Inquiry panel.
2. Select each line drawn in the previous Master It exercise, and compare to the list at the beginning of the exercise.

Create a curve tangent to the end of a line. It's rare that a property stands alone. Often, you must create adjacent properties, easements, or alignments from their legal descriptions.

Master It Create a curve tangent to the end of the first line drawn in the first exercise that meets the following specifications:

- Radius: 200.00′
- Arc Length: 66.580′

Solution

1. Select Lines/Curves ➤ Create Curves ➤ Curve from End of Object.
2. Select the first line drawn (N 57° 06′ 56.75″ E; 135.441′).
3. On the command line, press ↵ to confirm the radius.
4. On the command line, type **200.00**, and then press ↵.
5. Type **L** to specify the length, and then press ↵.
6. Type **66.580**, and then press ↵.

Create a best-fit line for a series of Civil 3D points. Surveyed point data is rarely perfect. When you're creating drawing linework, it's often necessary to create a best-fit line to make up for irregular shots.

Master It Locate the Edge of Pavement (EOP) points in the drawing. Use the Create Line by Best Fit command to create a line using these points.

Solution

1. On the Draw panel, click Create Line by Best Fit.
2. In the dialog, choose From Civil 3D Points, and click OK.
3. Select the EOP point objects, and press ↵.
4. A Panorama window appears. Review the results, and then dismiss the window by clicking the green check mark. A best fit line is drawn.

Label lines and curves. Although converting linework to parcels or alignments offers you the most robust labeling and analysis options, basic line- and curve-labeling tools are available when conversion isn't appropriate.

Master It Add line and curve labels to each entity created in the exercises. Choose a label that specifies the bearing and distance for your lines and length, radius, and delta of your curve.

Solution

1. Change to the Annotate tab.
2. From the Labels & Tables panel, select Add Labels ➤ Line and Curve ➤ Add Multiple Segment Line/Curve Labels.
3. Choose each line and curve. The default label should be acceptable. If not, select a label, right-click, and choose Label Properties. In the resulting AutoCAD properties box, you can select an alternative label.

Chapter 3: Lay of the Land: Survey

Properly collect field data and import it into AutoCAD Civil 3D 2010. You learned best practices for collecting data, how the data is translated into a usable format for the survey database, and how to import that data into a survey database. You learned what commands draw linework in a raw data file, and how to include those commands into your data collection techniques so that the linework is created correctly when the field book is imported into the program.

Master It In this exercise, you'll create a new drawing and a new survey database and import the `Shopping_Center.fbk` file into the drawing.

Solution

1. Create a new drawing using a template of your choice.
2. On the Survey tab, create a new survey database.
3. Create a new network in the newly created survey database.
4. Import the `Shopping_Center.fbk` file and edit the options to insert both the figures and the points.

Set up styles that will correctly display your linework. You learned how to set up styles for figures. You also learned that figures can be set as breaklines for surface creation and lot lines and that they can go on their own layer and be displayed in many different ways.

Master It In this exercise, you'll use the `Mastering1.dwg` file and survey database from the previous exercise and create figure styles and a figure prefix database for the various figures in the database.

Solution

1. Open the drawing file.
2. Open the survey database created in the previous exercise if it's not already open.
3. Expand the Figures branch of the survey database.
4. Create and apply styles to the various figures in the database to display the information as you'd like.

Create and edit field book files. You learned how to create field book files using various data collection techniques and how to import the data into a survey database.

Master It In this exercise, you'll create a new drawing and survey database. Open the `field notes.pdf` file and use a data collector (data collector emulators can be downloaded from the websites of the data collector manufacturers) to input the field notes. Export the raw data, convert it into an FBK file, and import it into the new survey database.

Solution

1. Create a new drawing and survey database and import the `traverse.fbk` file into a network.
2. Create construction lines to determine your reference angle.
3. Using the Translate Survey Network command, rotate the network.
4. Export a new FBK file.
5. Create another new drawing and survey database and import the new FBK file. Your drawing should look like `mastering2.dwg`.

Manipulate your survey data. You learned how to use the traverse analysis and adjustments to create data with a higher precision.

Master It In this exercise, you'll use the survey database and network from the previous exercises in this chapter. You'll analyze and adjust the traverse using the following criteria:

- Use the Compass Rule for Horizontal Adjustment.
- Use the Length Weighted Distribution Method for Vertical Adjustment.
- Use a Horizontal Closure Limit of 1:25,000.
- Use a Vertical Closure Limit of 1:25,000.

Solution

1. Open a new drawing, create a new survey database and network, and import the `traverse.fbk` field book.
2. Create a new traverse from the four points.
3. Perform a traverse analysis on the newly created traverse, and apply the changes to the survey database.
4. You can find the results from the traverse analysis in the following files:
   - `batrav.trv.txt`
   - `fvtrav.trv.txt`
   - `antrav.trv.txt`
   - `trav.lso.txt`

Chapter 4: X Marks the Spot: Points

Import points from a text file using description-key matching. Most engineering offices receive text files containing point data at some point during a project. Description keys provide a way to automatically assign the appropriate styles, layers, and labels to newly imported points.

Master It Create a new drawing from _AutoCAD Civil 3D (Imperial) NCS.dwt. Revise the Civil 3D description key set to use the parameters listed below.

Civil 3D Description Key Set Parameters

Code	Point Style	Point Label Style	Format	Layer
GS	Standard	Elevation Only	Ground Shot	V-NODE
GUY	Guy Pole	Elevation and Description	Guy Pole	V-NODE
HYD	Hydrant (existing)	Elevation and Description	Existing Hydrant	V-NODE-WATR
TOP	Standard	Point#-Elevation-Description	Top of Curb	V-NODE
TREE	Tree	Elevation and Description	Existing Tree	V-NODE-TREE

Import the MasteringPointsPNEZDspace.txt file from the data location, and confirm that the description keys made the appropriate matches by looking at a handful of points of each type. Do the trees look like trees? Do the hydrants look like hydrants?

Solution

1. Select File ➢ New, and create a drawing from _AutoCAD Civil 3D (Imperial) NCS Extended.dwt.
2. Switch to the Settings tab of Toolspace, and locate the Description Key Set called Civil 3D. Right-click this set, and choose Edit Keys.
3. Delete any two keys in this set by right-clicking each one and choosing Delete.
4. Revise the remaining key to match the GS specifications listed under the "Master It" instructions for this exercise.
5. Right-click the GS key, and choose New. Create the four additional keys listed in the instructions. Exit the dialog.
6. On the Ground Data Panel, select Import Points. Select PNEZD Space Delimited from the drop-down menu in the Format selection box, and navigate out to the MasteringPointsPNEZDspace.txt file. Click OK.
7. Zoom in to see the points. Note that each description-key parameter (style, label, format, and layer) has been respected. Your hydrants should appear as hydrants on the correct layer, your trees should appear as trees on the correct layer, and so on.

Create a point group. Building a surface using a point group is a common task. Among other criteria, you may want to filter out any points with zero or negative elevations from your Topo point group.

Master It Create a new point group called Topo that includes all points *except* those with elevations of zero or less.

Solution

1. In Prospector, right-click Point Groups, and choose New.
2. On the Information tab, enter **Topo** as the name of the new point group.

3. Switch to the Exclude tab
4. Click the Elevation checkbox to turn it on and enter **=<0** in the field.
5. Click OK to close the dialog.

Export points to LandXML and ASCII format. It's often necessary to export a LandXML or an ASCII file of points for stakeout or data-sharing purposes. Unless you want to export every point from your drawing, it's best to create a point group that isolates the desired point collection.

Master It Create a new point group that includes all the points with a raw description of TOP. Export this point group via LandXML and to a PNEZD Comma Delimited text file.

Solution

1. In Prospector, right-click Point Groups, and choose New.
2. On the Information tab, enter **Top of Curb** as the name of the new point group.
3. Switch to the Include tab.
4. Select the With Raw Descriptions Matching check box, and type **TOP** in the field.
5. Click OK, and confirm in Prospector that all the points have the description TOP.
6. Right-click the Top of Curb point group, and choose Export to LandXML.
7. Click OK in the Export to LandXML dialog.
8. Choose a location to save your LandXML file, and then click Save.
9. Navigate out to the LandXML file to confirm it was created.
10. Right-click the Top of Curb point group, and choose Export Points.
11. Choose the PNEZD Comma Delimited Format and a destination file, and confirm that the Limit to Points in Point Group check box is selected. Click OK.
12. Navigate out to the ASCII file to confirm it was created.

Create a point table. Point tables provide an opportunity to list and study point properties. In addition to basic point tables that list number, elevation, description, and similar options, you can customize point-table formats to include user-defined property fields.

Master It Create a point table for the Topo point group using the PNEZD format table style.

Solution

1. Change to the Annorate tab, and select Add Tables ➤ Points.
2. Choose the PNEZD format for the table style.
3. Click Point Groups, and choose the Topo point group.
4. Click OK.
5. The command line prompts you to choose a location for the upper-left corner of the point table. Choose a location on your screen somewhere to the right of the project.
6. Zoom in, and confirm your point table.

Chapter 5: The Ground Up: Surfaces in Civil 3D

Create a preliminary surface using freely available data. Almost every land development project involves a surface at some point. During the planning stages, freely available data can give you a good feel for the lay of the land, allowing design exploration before money is spent on fieldwork or aerial topography. Imprecise at best, this free data should never be used as a replacement for final design topography, but it's a great starting point.

Master It Create a new drawing from the Civil 3D Extended template and bring in a Google Earth surface for your home or office location. Be sure to set a proper coordinate system to get this surface in the right place.

Solution

1. On the main menu, choose File ➢ New.
2. Select the `NCS Extended.dwt` file, and click Open.
3. Change to the Settings tab, and right-click the drawing name to open the Drawing Settings dialog. Select an appropriate coordinate system.
4. In Google Earth, locate your home or office using the search engine.
5. In Civil 3D, change to the Insert panel.
6. On the Import panel, select Google Earth ➢ Google Earth Surface

Modify and update a TIN surface. TIN surface creation is mathematically precise, but sometimes the assumptions behind the equations leave something to be desired. By using the editing tools built into Civil 3D, you can create a more realistic surface model.

Master It Modify your Google Earth surface to show only an area immediately around your home or office. Create an irregular shaped boundary and apply it to the Google Earth surface.

Solution

1. Draw a polyline that includes the desired area.
2. Expand the Google Earth Surface branch in Prospector.
3. Expand the Definition branch.
4. Right-click Boundaries and select the Add option.
5. Select the newly created polyline and click Add to complete the boundary addition.

Prepare a slope analysis. Surface analysis tools allow users to view more than contours and triangles in Civil 3D. Engineers working with nontechnical team members can create strong meaningful analysis displays to convey important site information using the built-in analysis methods in Civil 3D.

Master It Create an Elevation Banding analysis of your home or office surface and insert a legend to help clarify the image.

Solution

1. Right-click the surface and bring up the Surface Properties dialog.
2. Change the Surface Style field to Elevation Banding (2D).
3. Change to the Elevation tab and run an Elevation analysis.
4. Click OK to close the Surface Properties dialog.
5. Select the surface to display the Tin Surface tab on the Ribbon.
6. On the Labels and Tables panel, click Add Legend Table. Enter **E** and then **D** at the command line and pick a placement point on the screen to create a legend.

Label surface contours and spot elevations. Showing a stack of contours is useless without context. Using the automated labeling tools in Civil 3D, you can create dynamic labels that update and reflect changes to your surface as your design evolves.

Master It Label the contours on your Google Earth surface at 1′ and 5′ (Design).

Solution

1. Change the Surface Style to Contours 1′ and 5′ (Design).
2. Select the surface to display the Tin Surface tab.
3. On the Labels & Tables panel, select Add Labels ➤ Contour – Multiple.
4. Pick a point on one side of the site, and draw a contour label line across the entire site. Repeat this step as needed to label the site appropriately.

Chapter 6: Don't Fence Me In: Parcels

Create a boundary parcel from objects. The first step to any parceling project is to create an outer boundary for the site.

Master It Open the `Mastering Parcels.dwg` file, which you can download from `www.sybex.com/go/masteringcivil3d2010`. Convert the polyline in the drawing to a parcel.

Solution

1. From the Home tab's Create Design panel, select Parcel ➤ Create Parcel from Objects.
2. At the `Select lines, arcs, or polylines to convert into parcels or [Xref]:` prompt, pick the polyline that represents the site boundary. Press ↵.
3. The Create Parcels – From Objects dialog appears. Select Subdivision Lots; Property; and Name, Square Foot and Acres from the drop-down menus in the Site, Parcel Style, and Area Label Style selection boxes, respectively. Keep the default values for the remaining options. Click OK to dismiss the dialog.
4. The boundary polyline forms parcel segments that react with the alignment. Area labels are placed at the newly created parcel centroids.

Create a right-of-way parcel using the right-of-way tool. For many projects, the ROW parcel serves as frontage for subdivision parcels. For straightforward sites, the automatic Create ROW tool provides a quick way to create this parcel.

Master It Continue working in the `Mastering Parcels.dwg` file. Create a ROW parcel that is offset by 25′ on either side of the road centerline with 25′ fillets at the parcel boundary.

Solution

1. From the Home tab's Create Design panel, select Parcel ➤ Create Right of Way.
2. At the `Select parcels:` prompt, pick your newly created parcels on screen. Press ↵ to stop picking parcels. The Create Right Of Way dialog appears.
3. Expand the Create Parcel Right of Way parameter, and enter **25′** in the Offset from Alignment text box.
4. Expand the Cleanup at Parcel Boundaries parameter. Enter **25′** in the Fillet Radius at Parcel Boundary Intersections text box. Select Fillet from the drop-down menu in the Cleanup Method selection box.
5. Click OK to dismiss the dialog and create the ROW parcels.

Create subdivision lots automatically by layout. The biggest challenge when creating a subdivision plan is optimizing the number of lots. The precise sizing parcel tools provide a means to automate this process.

Master It Continue working in the `Mastering Parcels.dwg` file. Create a series of lots with a minimum of 10,000 square feet and 100′ frontage.

Solution

1. From the Home Tab's Create Design panel, select Parcel ➤ Parcel Creation Tools
2. Expand the Parcel Layout Tools toolbar.
3. Change the value of the following parameters by clicking in the Value column and typing in the new values:
   - Default Area: **10000 Sq. Ft.**
   - Minimum Frontage: **100′**
4. Change the following parameters by clicking in the Value column and selecting the appropriate option from the drop-down menu:
   - Automatic Mode: On
   - Remainder Distribution: Redistribute Remainder
5. Click the Slide Line – Create tool. The Create Parcels – Layout dialog appears.
6. Select Subdivision Lots, Single Family, and Name Square Foot & Acres from the drop-down menus in the Site, Parcel Style, and Area Label Style selection boxes,

respectively. Keep the default values for the rest of the options. Click OK to dismiss the dialog.

7. At the `Select Parcel to be subdivided:` prompt, pick the label of 90 for one of your property parcels.
8. At the `Select start point on frontage:` prompt, use your Endpoint osnap to pick the point of curvature along the ROW parcel segment.
9. The parcel jig appears. Move your cursor slowly along the ROW parcel segment, and notice that the parcel jig follows the segment. At the `Select end point on frontage:` prompt, use your Endpoint osnap to pick the point of curvature along the ROW parcel segment.
10. At the `Specify angle at frontage:` prompt, type **90**. Press ↵.
11. At the `Accept Result:` prompt, press ↵ to accept the lot layout.
12. Repeat steps 5 through 11 for the other property parcels, if desired.

Add multiple parcel segment labels. Every subdivision plat must be appropriately labeled. You can quickly label parcels with their bearings, distances, direction, and more using the segment labeling tools.

Master It Continue working in the `Mastering Parcels.dwg` file. Place Bearing over Distance labels on every parcel line segment and Delta over Length and Radius labels on every parcel curve segment using the Multiple Segment Labeling tool.

Solution

1. From the Annotate tab, select Add Labels ➤ Parcels ➤ Add Parcel Labels.
2. In the Add Labels dialog, select Multiple Segment, Bearing over Distance, and Delta over Length and Radius from the drop-down menus in the Label Type, Line Label Style, and Curve Label Style selection boxes, respectively.
3. Click Add.
4. At the `Select parcel to be labeled by clicking on area label or [CLockwise /COunterclockwise]<CLockwise>:` prompt, pick the area label for each of your single-family parcels.
5. Press ↵ to exit the command.

Chapter 7: Laying A Path: Alignments

Create an alignment from a polyline. Creating alignments based on polylines is a traditional method of building engineering models. With Civil 3D's built-in tools for conversion, correction, and alignment reversal, it's easy to use the linework prepared by others to start your design model. These alignments lack the intelligence of crafted alignments, however, and you should use them sparingly.

Master It Open the `Mastering Alignments.dwg` file, and create alignments from the linework found there.

Solution Use Create Alignments from Objects on the AutoCAD geometry in the file.

Create a reverse curve that never loses tangency. Using the alignment layout tools, you can build intelligence into the objects you design. One of the most common errors introduced to engineering designs is curves and lines that aren't tangent, requiring expensive revisions and resubmittals. The free, floating, and fixed components can make smart alignments in a large number of combinations available to solve almost any design problem.

Master It Open the `Mastering Alignments.dwg` file, and create an alignment from the linework on the right. Create a reverse curve with both radii equal to 200 and with a passthrough point in the center of the displayed circle.

Solution

1. Trace both lines with Fixed Segments.
2. Use the Floating Curve (From Entity with Passthrough Point) tool to draw an arc from the northern tangent segment with a passthrough point in the center of the circle.
3. Use the Free Curve (Between Two Entities Radius) tool to fillet the floating curve created in step 2 and the last fixed segment using a 200′ radius.

Replace a component of an alignment with another component type. One of the goals in using a dynamic modeling solution is to find better solutions, not just a solution. In the layout of alignments, this can mean changing components out along the design path, or changing the way they're defined. Civil 3D's ability to modify alignments' geometric construction without destroying the object or forcing a new definition lets you experiment without destroying the data already based on an alignment.

Master It Convert the arc indicated in the `Mastering Alignments.dwg` file to a free arc that is a function of the two adjoining segments. The curve radius is 150′.

Solution

1. Select the indicated alignment, and right-click to edit alignment geometry.
2. Delete the indicated arc segment using the Delete Subcomponent tool.
3. Select Free Curve Fillet (Between Two Entities, Radius).
4. Pick the two segments, and set the radius at the command line.

Create a new label set. Label sets let you determine the appearance of an alignment's labels and quickly standardize that appearance across all objects of the same nature. By creating sets that reflect their intended use, you can make it easy for a designer to quickly label alignments according to specifications with little understanding of the requirement.

Master It Within the `Mastering Alignments.dwg` file, create a new label set containing only major station labels, and apply it to all the alignments in that drawing.

Solution

1. On the Settings tab, expand the Alignments ➢ Label Styles ➢ Label Sets branch.
2. Right-click Major and Minor Only, and select Copy.
3. Change the name to **Major Only**.
4. Delete the Minor label on the Labels tab.
5. Right-click each alignment, and select Edit Alignment Labels.
6. Import the Major Only label set.
7. Repeat for each label.

Solutions may vary!

Override individual labels with other styles. In spite of the desire to have uniform labeling styles and appearances between alignments within a single drawing, project, or firm, there are always exceptions. Using AutoCAD's Ctrl+click method for element selection, you can access commands that let you modify your labels and even change their styles.

Master It Create a copy of the Perpendicular with Tick Major Station style called Major with Marker. Change the Tick Block Name to Marker Pnt. Replace some (but not all) of your major station labels with this new style.

Solution

1. On the Settings tab, expand the Alignment ➢ Label Styles ➢ Station ➢ Major Station branch.
2. Right-click Perpendicular with Tick, and select Copy.
3. Change the name to **Major with Marker**.
4. Change to the Layout tab.
5. Change to the Tick Component.
6. Change the AeccTickLine block selection to Market Pnt.
7. Click OK to close the dialog.
8. Open the AutoCAD properties dialog.
9. Ctrl+click a major station label.
10. Change the style to Major with Marker.

Chapter 8: Cut to the Chase: Profiles

Sample a surface profile with offset samples. Using surface data to create dynamic sampled profiles is an important advantage in working with a three-dimensional model. Quick viewing of various surface slices and grip-editing alignments makes for an effective preliminary planning tool. Combined with offset data to meet review agency requirements, profiles are robust design tools in Civil 3D.

Master It Open the `Mastering Profile.dwg` file and sample the ground surface along Alignment - (2), along with offset values at 15′ left and 25′ right of the alignment.

Solution

1. On the Home tab's Create Design panel, select Profiles ➤ Profile Creation Tools.
2. Set the values as shown here, and click OK:

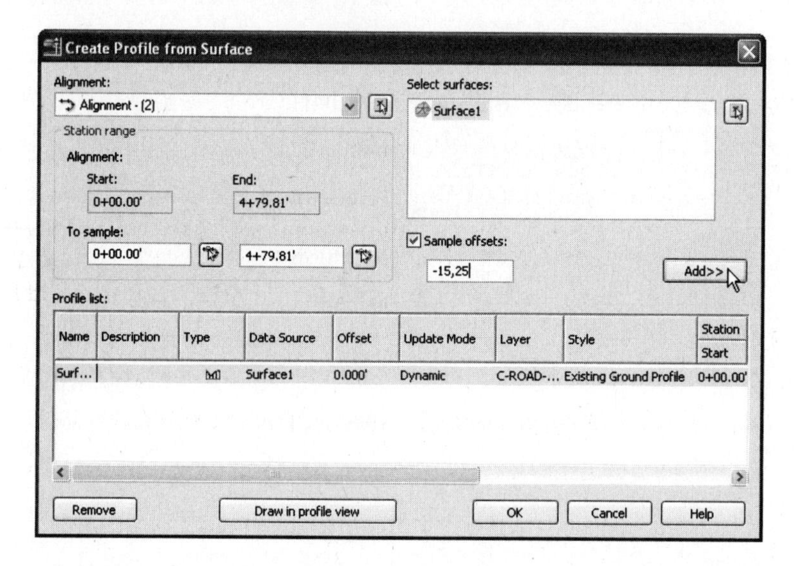

Lay out a design profile on the basis of a table of data. Many programs and designers work by creating pairs of station and elevation data. The tools built into Civil 3D let you input this data precisely and quickly.

Master It In the `Mastering Profiles.dwg` file, create a layout profile on Alignment (4) with the following information:

Station	PVI Elevation	Curve Length
0+00	694	
2+90	696.50	250′
5+43.16	688	

Solution

1. Pick the Alignment - (4) profile.
2. Set the L-value of the Curve settings to **250′**.
3. Use the Transparent Commands toolbar to enter station-elevation data.

Alternatively, you can import a text file.

Add and modify individual components in a design profile. The ability to delete, modify, and edit the individual components of a design profile while maintaining the relationships is an important concept in the 3D modeling world. Tweaking the design allows you to pursue a better solution, not just a working solution.

Master It In the `Mastering Profile.dwg` file, move the third PVI (currently at 9+65, 687) to 9+50, 690. Then, add a 175′ parabolic vertical curve at this point.

Solution

1. Grid-edit the PVI to the desired point, or use Panorama to set the value.
2. Use the Free Vertical Parabola (PVI-based) tool, and enter an L setting of **175**.

Apply a standard label set. Standardization of appearance is one of the major benefits of using Civil 3D styles in labeling. By applying label sets, you can quickly create plot-ready profile views that have the required information for review.

Master It In the `Mastering Profile.dwg` file, apply the Road Profiles label set to all layout profiles.

Solution

1. Pick the layout profile, right-click, and select the Edit Labels option.
2. Click Import Label Set, and select the Road Profile Labels set.
3. Repeat this procedure for all layout profiles.

Chapter 9: Slice and Dice: Profile Views in Civil 3D

Create a simple view as part of the sampling process. You will seldom want to sample a surface without creating a view of that data. By combining the steps into one quick process, you'll save time and effort as profile views are generated.

Master It Open the `Mastering Profile Views.dwg` file and create a view using the Full Grid profile view style for the Alexander Ave alignment. Display only the layout profile, the EG and offsets at 15′ left and 25′ right.

Solution

1. From the main menu, select Profiles ➢ Create Profile View.
2. Select the Alexander Ave alignment from the drop-down list, and change the Style to Full Grid. Then click Create Profile View.
3. Open the Profile View Properties dialog and modify it as shown here.

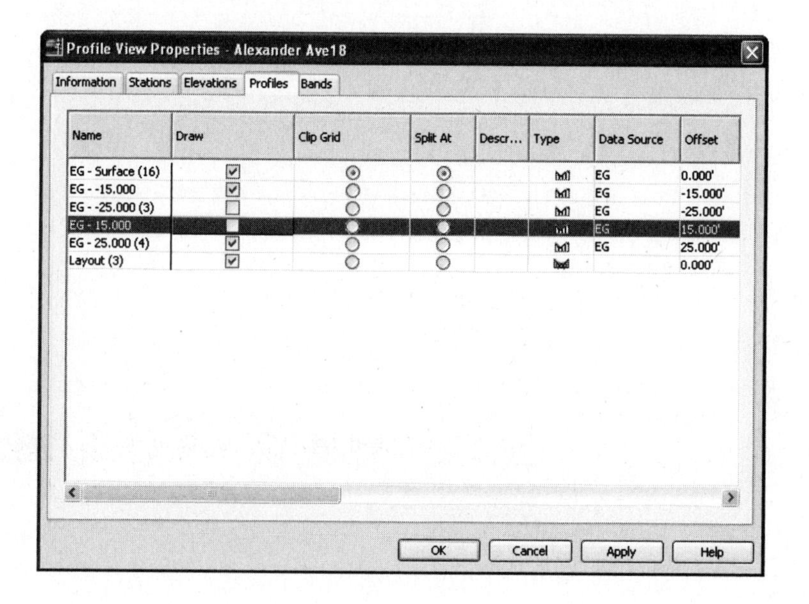

Change profile views and band sets as needed. Using profile view styles and band sets allows for the quick customization and standardization of profile data. Because it's easy to change the styles and bands, many users design using one style and then change the style as required for submission.

Master It Change the Alexander Ave profile view to the Mastering style and assign the EG+FG and Offsets band set. Assign appropriate profiles to the bands.

Solution

1. Select the Alexander Ave profile view and right-click to access the Profile View Properties dialog.
2. Change the Profile View Style to Mastering, and then switch to the Bands tab.
3. Click the Import Band Set button, and make the choices shown here.

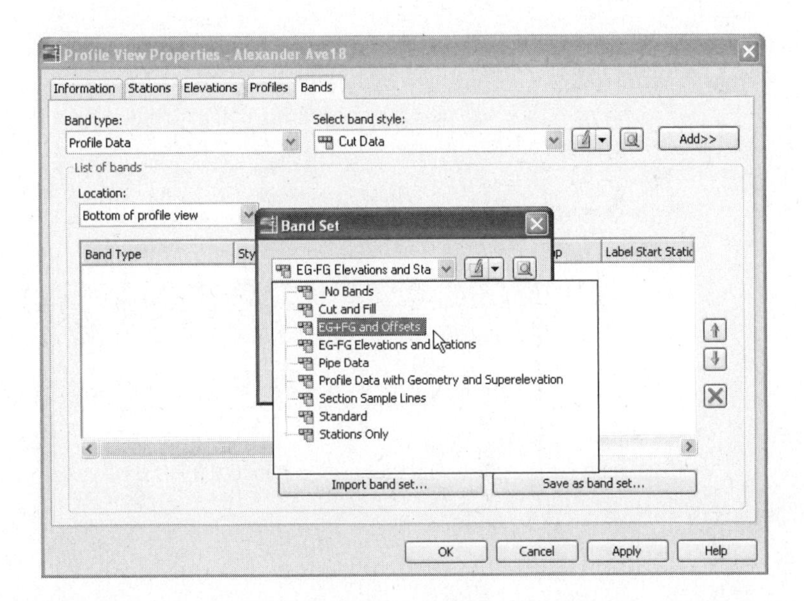

4. Return to the Bands tab and set the remaining options.

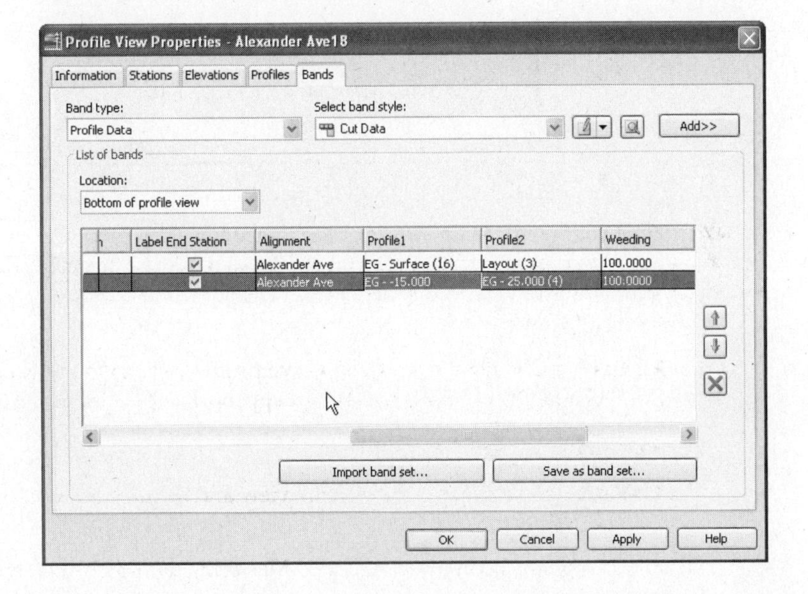

Split profile views into smaller views. Designing in one continuous profile view makes the designer's job easier, but plotting typically requires multiple views. Using the wizard or individual profile view properties makes it easy to split apart profile view information for presentation or submittal purposes.

Master It Create a new pair of profile views for the Parker Place alignment, each 600′ long. Assign the Mastering profile view style but no bands.

Solution From the Home tab's Profile & Section Views panel, select Profile View ➤ Create Multiple Profile Views and set the options as shown in the following images. Note: The profile views may need to have the elevations reassigned to Automatic in order to have a decent grid buffer.

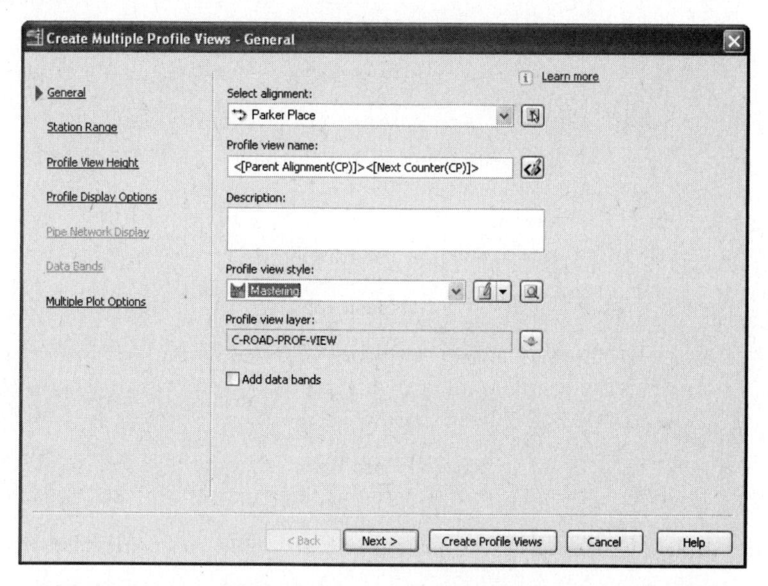

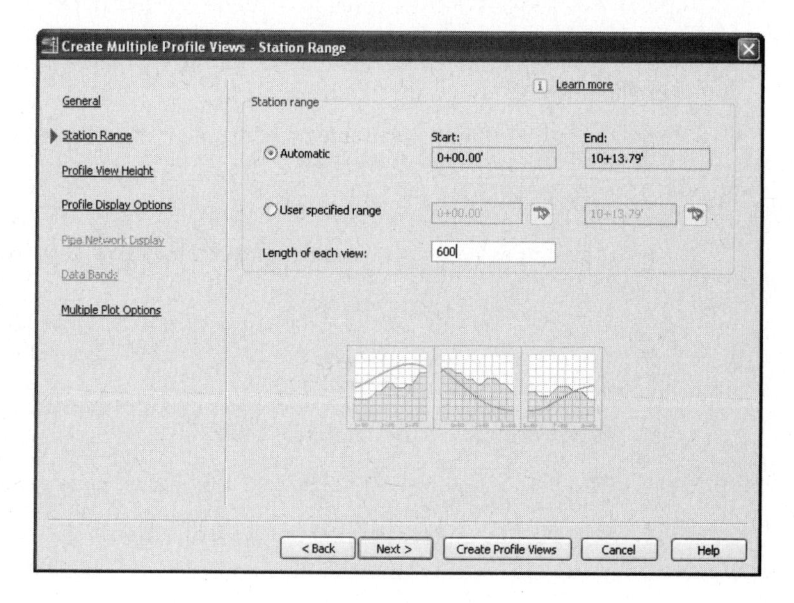

Chapter 10: Templates Plus: Assemblies and Subassemblies

Create a typical road assembly with lanes, curb, gutter, and sidewalk. Most corridors are built to model roads. The most common assembly used in these road corridors is some variation of a typical road section consisting of lanes, curb, gutter, and sidewalk.

Master It Create a new drawing from the `_AutoCAD Civil 3D (Imperial) NCS.dwt` template. Build a symmetrical assembly using BasicLane, BasicCurbandGutter, and BasicSidewalk. Use widths and slopes of your choosing.

Solution

1. Create a new drawing from the `_AutoCAD Civil 3D (Imperial) NCS Extended.dwt` template.
2. From the Home tab's Create Design Panel, select Assembly ➢ Create Assembly.
3. Name your assembly and set styles as appropriate.
4. Pick a location in your drawing for the assembly.
5. Locate the Imperial-Basic tab on the tool palette.
6. Click the BasicLane button on the tool palette. Use the AutoCAD Properties palette, and follow the command-line prompts to set the BasicLane on the left and right sides of your assembly.
7. Repeat the process with BasicCurbandGutter and BasicSidewalk. (Refer to the "Subassemblies" section of this chapter for additional information.)
8. Save the drawing for use in the next Master It exercise.

Edit an assembly. Once an assembly has been created, it can be easily edited to reflect a design change. Often, at the beginning of a project, you won't know the final lane width. You can build your assembly and corridor model with one lane width and then later change the width and rebuild the model immediately.

Master It Working in the same drawing, edit the width of each BasicLane to 14′, and change the cross slope of each BasicLane to −3.08%.

Solution

1. Pick your assembly, and right-click. Select the Assembly Properties option.
2. Switch to the Construction tab.
3. On the left side of the dialog, select the first BasicLane. On the right side, change the slope to −3.08% and the width to 14′.
4. Repeat the process for the second BasicLane.
5. Click OK, and view the results. Your lane should have an adjusted size and slope.
6. Save the drawing for use in the next Master It exercise.

Add daylighting to a typical road assembly. Often, the most difficult part of a designer's job is figuring out how to grade the area between the last hard engineered point in the cross section (such as the back of a sidewalk) and existing ground. An extensive catalog of daylighting subassemblies can assist you with this task.

Master It Working in the same drawing, add the DaylightMinWidth subassembly to both sides of your typical road assembly. Establish a minimum width of 10′.

Solution

1. Locate the Imperial-Daylight tab on the tool palette
2. Click the DaylightMinWidth button on the tool palette. Use the AutoCAD Properties palette, and follow the command-line prompts to set the DaylightMinWidth on the left and right sides of your assembly.

   You should now have daylighting subassemblies visible on both sides of your assembly.

Chapter 11: Easy Does It: Basic Corridors

Build a single baseline corridor from an alignment, profile, and assembly. Corridors are created from the combination of alignments, profiles, and assemblies. Although corridors can be used to model many things, most corridors are used for road design.

Master It Open the `Mastering Corridors.dwg` file. Build a corridor on the basis of the Project Road alignment, the Project Road Finished Ground profile, and the Project Typical Road Assembly.

Solution

1. From the Home tab's Create Design panel, select Corridors ➤ Create Simple Corridor.
2. Pick the Project Road alignment, the Project Road Finished Ground profile, and the Project Typical Road Assembly. Target the Existing Ground surface.

Create a corridor surface. The corridor model can be used to build a surface. This corridor surface can then be analyzed and annotated to produce finished road plans.

Master It Continue working in the `Mastering Corridors.dwg` file. Create a corridor surface from Top links.

Solution

1. Open the Corridor Properties dialog and switch to the Surfaces tab.
2. Click Create Surface and then the plus (+) sign to add Top links.
3. Click OK to dismiss the dialog. The corridor and surface will build.

Add an automatic boundary to a corridor surface. Surfaces can be improved with the addition of a boundary. Single baseline corridors can take advantage of automatic boundary creation.

Master It Continue working in the `Mastering Corridors.dwg` file. Use the Automatic Boundary Creation tool to add a boundary using the Daylight code.

Solution

1. Open the Corridor Properties dialog and switch to the Boundaries tab.
2. Right-click on the surface entry and click Add Automatically ➤ Daylight.
3. Click OK to dismiss the dialog. The corridor and surface will build.

Chapter 12: The Road Ahead: Advanced Corridors

Add a baseline to a corridor model for a cul-de-sac. Although for simple corridors you may think of a baseline as a road centerline, other elements of a road design can be used as a baseline. In the case of a cul-de-sac, the EOP, the top of curb, or any other appropriate feature can be converted to an alignment and profile and used as a baseline.

Master It Open the `Mastering Advanced Corridors.dwg` file, which you can download from `www.sybex.com/go/masteringcivil3d2010`. Add the cul-de-sac alignment and profile to the corridor as a baseline. Create a region under this baseline that applies the Typical Intersection assembly.

Solution

1. Select the corridor, right-click, and choose Corridor Properties. Switch to the Parameters tab.
2. Click Add Baseline. Choose Cul de Sac EOP in the Pick a Horizontal Alignment dialog.
3. In the Profile column, click inside the <click here . . . > box. Choose Cul de Sac EOP FG in the Select a Profile dialog.
4. Right-click the new baseline. Choose Add Region. Select Intersection Typical in the Select an Assembly dialog box.
5. Click OK to leave the Corridor Properties dialog and build the corridor.

Add alignment and profile targets to a region for a cul-de-sac. Adding a baseline isn't always enough. Some corridor models require the use of targets. In the case of a cul-de-sac, the lane elevations are often driven by the cul-de-sac centerline alignment and profile.

Master It Continue working in the `Mastering Advanced Corridors.dwg` file. Add the Second Road alignment and Second Road FG profile as targets to the cul-de-sac region. Adjust the Assembly Application Frequency to 5′, and make sure the corridor samples are profile PVIs.

Solution

1. Select the corridor, right-click, and choose Corridor Properties. Switch to the Parameters tab.
2. Click the Target Mapping button in the appropriate region.
3. In the Target Mapping dialog, assign Second Road as the transition alignment and Second Road FG profile as the transition profile. Click OK to leave the Target Mapping dialog.

4. Click the Frequency button in the appropriate region. Change the Along Curves value to 5′ and the At Profile High/Low Point value to Yes. Click OK to exit the Frequency to Apply to Assemblies dialog.
5. Click OK to leave the Corridor Properties dialog and build the corridor.

Use the Interactive Boundary tool to add a boundary to the corridor surface. Every good surface needs a boundary to prevent bad triangulation. Bad triangulation creates inaccurate and unsightly contours. Civil 3D provides several tools for creating corridor surface boundaries, including an Interactive Boundary tool.

Master It Continue working in the `Mastering Advanced Corridors.dwg` file. Create an interactive corridor surface boundary for the entire corridor model.

Solution

1. Select the corridor, right-click, and choose Corridor Properties. Switch to the Boundaries tab.
2. Select the corridor surface, right-click, and choose Add Interactively.
3. Follow the command-line prompts to add a feature line–based boundary all the way around the entire corridor.
4. Type **C** to close the boundary, and then press ↵ to end the command.
5. Click OK to leave the Corridor Properties dialog and build the corridor.

An example of the finished exercise can be found in `Mastering Advanced Corridors Finished.dwg`.

Chapter 13: Stacking Up: Cross Sections

Create sample lines. Before any section views can be displayed, sections must be created from sample lines.

Master It Open `sections1.dwg` and create sample lines along the alignment every 50′.

Solution Using the Elizabeth Lane alignment, sample all data except the datum surface. Create sample lines by station range and set your sample line distance to 50′.

Create section views. Just as profiles can only be shown in profile views, sections require section views to display. Section views can be plotted individually or all at once. You can even set them up to be broken up into sheets.

Master It In the previous drawing, you created sample lines. In that same drawing, create section views for all the sample lines.

Solution Using the Create Multiple Sections command, create section views for all sample lines. Use the Plot All option, and don't add any labels to the views.

Define materials. Materials are required to be defined before any quantities can be displayed. You learned that materials can be defined from surfaces or from corridor shapes. Corridors must exist for shape selection and surfaces must already be created for comparison in materials lists.

Master It Using `sections4.dwg`, create a materials list that compares EG with Elizabeth Lane Top Road Surface.

Solution To create this materials list, you will need to use Cut and Fill criteria. The EG surface will be Surface2, and the Datum will be Elizabeth Lane Top Road Surface.

Generate volume reports. Volume reports give you numbers that can be used for cost estimating on any given project. Typically, construction companies calculate their own quantities, but developers often want to know approximate volumes for budgeting purposes.

Master It Continue using `sections4.dwg`. Use the materials list created earlier to generate a volume report. Create an XML report and a table that can be displayed on the drawing.

Solution Use the Cut and Fill table style to display the volume calculations on the drawing. It should be set to dynamically update in the event that the quantities change (if the profile is adjusted or the alignment is moved).

Chapter 14: The Tool Chest: Parts List and Part Builder

Add pipe and structure families to a new parts list. Before you can begin to design your pipe network, you must ensure that you have the appropriate pipes and structures at your disposal.

Master It Create a new drawing from `_AutoCAD Civil 3D (Imperial) NCS Extended.dwt`. Create a new parts list called Typical Storm Drainage. (This template already includes a Storm Sewer parts list. Ignore it or delete it, but don't use it. Create your own for this exercise.)

Add 12″, 15″, 18″, 24″, 36″, and 48″ Concrete Circular Pipe to the parts list.

Add the following to the parts list: a Rectangular Structure Slab Top Rectangular Frame with an inner structure width and length of 15″, a Rectangular Structure Slab Top Rectangular Frame with an inner structure width of 15″ and an inner structure length of 18″, and a Concentric Cylindrical Structure with an inner diameter of 48″.

Solution

1. Choose File ➢ New, and create a drawing from `_AutoCAD Civil 3D (Imperial) NCS Extended.dwt`.
2. Locate the Parts Lists entry on the Settings Tab of Toolspace under Pipe Network. Right-click, and choose Create New Parts List.
3. On the Information tab, enter **Typical Storm Drainage** in the Name text box.
4. Switch to the Pipes tab, right-click the Parts List entry, and choose Add Part Family.
5. In the Part Catalog dialog, select the Concrete Pipe check box and click OK.
6. Expand the entries on the Pipes tab to find Concrete Pipe.
7. Right-click the Concrete Pipe entry, and choose Add Part Size.
8. Either add each size of pipe individually from the drop-down list in the Inner Pipe Diameter selection box, clicking OK each time and then returning to the Part Size Creator dialog and repeating the process — or select the Add All Sizes check box and delete any unwanted pipes from the list on the Pipes tab.

9. Switch to the Structures tab.
10. Right-click the Parts List entry, and choose Add Part Family.
11. In the Part Catalog dialog, select the check boxes next to Concentric Cylindrical Structure and Rectangular Structure Slab Top Rectangular Frame. Click OK.
12. Right-click the Concentric Cylindrical Structure entry, and choose Add Part Size.
13. In the Part Size Creator dialog, choose 48″ Inner Structure Diameter. Click OK.
14. Right-click the Structure Slab Top Rectangular Frame entry, and choose Add Part Size.
15. In the Part Size Creator dialog, change the Inner Structure Length parameter to 15″. Click OK. Repeat the process for an Inner Structure Length of 18″.
16. Click OK. Save your drawing.

Create rule sets that apply to pipes and structures in a parts list. Municipal standards and engineering judgment determine how pipes will behave in design situations. Considerations include minimum and maximum slope, cover requirements, and length guidelines. For structures, there are regulations regarding sump depth and pipe drop. You can apply these considerations to pipe network parts by creating rule sets.

Master It Create a new structure rule set called Storm Drain Structure Rules. Add the following rules:

Rule	Values
Pipe Drop Across Structure	Drop Reference Location: **Invert** Drop Value: **0.01′** Maximum Drop Value: **2′**
Set Sump Depth	Sump Depth: **0.5′**

Create a new pipe rule set called Storm Drain Pipe Rules, and add the following rules:

Rule	Values
Cover And Slope	Maximum Cover: **15′** Maximum Slope: **5%** Minimum Cover: **4′** Minimum Slope: **.05%**
Length Check	Maximum Length: **300′** Minimum Length: **8′**

Apply these rules to all pipes and structures in your Typical Storm Drainage parts list.

Solution

1. On the Settings tab, right-click Structure Rule Set, and choose New.
2. On the Information tab, enter **Structure Rule Set Storm Drain Structure Rules** in the Name text box.
3. Switch to the Rules tab. Click Add Rule.

4. In the Add Rule dialog, choose Pipe Drop Across Structure in the Rule Name drop-down list. Click OK. (You can't change the parameters.)
5. Change the parameters in the Structure Rule Set dialog to match the directions given earlier.
6. Click Add Rule.
7. In the Add Rule dialog, choose Set Sump Depth in the Rule Name drop-down list. Click OK. (You can't change the parameters in this box.)
8. Change the Sump Depth parameter in the Structure Rule Set dialog to match the directions, and click OK.
9. Right-click the Pipe Rule Sets on the Settings tab of Toolspace under the Pipe tree, and choose New.
10. On the Information tab, enter **Pipe Rule Set Storm Drain Pipe Rules** in the Name text box.
11. Switch to the Rules tab, and click Add Rule.
12. In the Add Rule dialog, choose Cover And Slope from the Rule Name drop-down list. Click OK. (You can't change the parameters.)
13. Modify the parameters to match the parameters listed in the directions, and click OK.
14. You should now have one structure rule set and one pipe rule set.
15. Select your Typical Storm Drainage parts list on the Settings tab. Right-click, and choose Edit.
16. On the Pipes tab, click the first button under the Rules column. Choose your Storm Drain Pipe Rules Set from the drop-down list. Your rule set should be applied to all concrete pipes.
17. Switch to the Structures tab. Use the same technique to apply your Storm Drain Structure rule set to each structure family.
18. Click OK. Save your drawing.

Apply styles to pipes and structures in a parts list. The final drafted appearance of pipes and structures in a drawing is controlled by municipal standards and company CAD standards. Civil 3D object styles allow you to control and automate the symbology used to represent pipes and structures in plan, profile, and section views.

Master It Apply the following styles to your parts list:

- Single Line (Storm) to the 12″, 15″, and 18″ Concrete Circular Pipe
- Double Line (Storm) to the 24″, 36″, and 48″ Concrete Circular Pipe
- Storm Sewer Manhole to the Concentric Cylindrical Structure with an inner diameter of 48″
- Catch Basin to both sizes of the Rectangular Structure Slab Top Rectangular Frame

Solution

1. Right-click your Typical Storm Drainage parts list on the Settings tab, and choose Edit.
2. On the Pipes tab, use the Style column to select the appropriate style for each size pipe.
3. On the Structures tab, use the Style column to select the appropriate style for each size of structure.
4. Click OK. Save your drawing.

Chapter 15: Running Downhill: Pipe Networks

Create a pipe network by layout. After you've created a parts list for your pipe network, the first step toward finalizing the design is to create a Pipe Network by Layout.

Master It Open the `Mastering Pipes.dwg` file. Choose Pipe Network Creation Tools from the Pipe Network drop-down of the Create Design panel on the Home tab to create a sanitary sewer pipe network.

Use the Finished Ground surface, and name only structure and pipe label styles. Don't choose an alignment at this time. Create 8″ PVC pipes and concentric manholes.

There are blocks in the drawing to assist you in placing manholes. Begin at the START HERE marker, and place a manhole at each marker location. You can erase the markers when you've finished.

Solution

1. From the Home tab's Create Design panel, select Pipe Network ➤ Create Pipe Network by Layout.
2. In the Create Pipe Network dialog, set the following parameters:
    1. Network Name: **Sanitary Sewer**
    2. Network Parts List: Sanitary Sewer
    3. Surface Name: Finished Ground
    4. Alignment Name: <none>
    5. Structure Label Style: Name Only (Sanitary)
    6. Pipe Label Style: Name Only
3. Click OK. The Pipe Layout Tools toolbar appears.
4. Set the structure to Concentric Manhole and the pipe to 8 Inch PVC. Click Draw Pipes and Structures, and use your Insertion osnap to place a structure at each marker location.
5. Press ↵ to exit the command.
6. Select a marker, right-click, and choose Select Similar. Click Delete.

Create an alignment from network parts, and draw parts in profile view. Once your pipe network has been created in plan view, you'll typically add the parts to a profile view on the basis of either the road centerline or the pipe centerline.

Master It Continue working in the `Mastering Pipes.dwg` file. Create an alignment from your pipes so that station zero is located at the START HERE structure. Create a profile view from this alignment, and draw the pipes on the profile view.

Solution

1. Select the structure labeled START HERE to display the Pipe Networks contextual tab and select Alignment from Network on the Launch Pad panel.
2. Select the last structure in the pipe run. Press ↵ to accept the selection.
3. In the Create Alignment dialog, make sure the Create Alignment and Profile View check box is selected. Accept the other defaults, and click OK.
4. In the Create Profile dialog, sample both the Existing Ground and Finished Ground surfaces for the profile. Click Draw in Profile View.
5. In the Create Profile View dialog, click Create Profile View and choose a location in the drawing for the profile view. A profile view showing your pipes appears.

Label a pipe network in plan and profile. Designing your pipe network is only half of the process. Engineering plans must be properly annotated.

Master It Continue working in the `Mastering Pipes.dwg` file. Add the Length Material and Slope style to profile pipes and the Data with Connected Pipes (Sanitary) style to profile structures.

Solution

1. Change to the Annotate tab and select Add Labels.
2. In the Add Labels dialog, change the Feature to Pipe Network, and then change the Label Type to Entire Network Profile. For pipe labels, choose Length Material and Slope; for structure labels, choose Data with Connected Pipes (Sanitary). Click Add, and choose any pipe or structure in your profile.
3. Drag or adjust any profile labels as desired.

Create a dynamic pipe table. It's common for municipalities and contractors to request a pipe or structure table for cost estimates or to make it easier to understand a busy plan.

Master It Continue working in the `Mastering Pipes.dwg` file. Create a pipe table for all pipes in your network.

Solution

1. Change to the Annotation tab, and select Add Labels ➤ Add Tables ➤ Add Pipe.
2. In the Pipe Table Creation dialog, make sure your pipe network is selected. Accept the other defaults, and click OK.
3. Place the table in your drawing.

Chapter 16: Working the Land: Grading

Convert existing linework into feature lines. Many site features are drawn initially as simple linework for the 2D plan. By converting this linework to feature line information, you avoid a large amount of rework. Additionally, the conversion process offers the ability to drape feature lines along a surface, making further grading use easier.

Master It Open the `Mastering Grading.dwg` file from the data location. Convert the polyline describing a proposed temporary drain into a feature line and drape it across the EG surface to set elevations.

Solution

1. From the Home tab's Create Design panel, select Feature Lines ➤ Create Feature Lines from Objects.
2. Pick the polyline.
3. Toggle the Assign Elevations check box.
4. Select the EG surface in the Assign Elevations dialog.
5. Click OK twice to close the dialogs and return to your model.

Model a simple linear grading with a feature line. Feature lines define linear slope connections. This can be the flow of a drainage channel, the outline of a building pad, or the back of a street curb. These linear relationships can help define grading in a model, or simply allow for better understanding of design intent.

Master It Add 200′ radius fillets on the feature line just created. Set the grade from the top of the hill to the circled point to 5 percent and the remainder to a constant slope to be determined in the drawing. Draw a temporary profile view to verify the channel is below grade for most of its length.

Solution

1. Display the Feature Line toolbar's Edit Geometry panel, and select the Fillet tool.
2. Pick the feature line.
3. Enter **R**↵ to change the radius.
4. Enter **200**↵ for the radius.
5. Enter **A**↵ to fillet all the points.
6. Toggle on the Edit Elevations panel and select Insert Elevation Point.
7. Use the Center osnap to pick the center of the circle. Press ↵ to accept the elevation.
8. From the Edit Elevations panel, select the Set Grade between Points tool.
9. Pick the feature line. Pick the PI near the top of the hill.
10. Press ↵ to accept the elevation. Pick the elevation point at the circle's center.
11. Enter **-5**↵ to set the grade.

12. Pick the feature line again.
13. Pick the elevation point in the circle. Press ↵ to accept the elevation.
14. Pick the PI at the downstream end of the channel. Press ↵ to accept the elevation.
15. Press ↵ to exit the command.
16. Pick the feature line. Right-click and select Quick Profile.
17. Select Layout from the drop-down list in the 3D Entity Profile Style selection box.
18. Click OK and pick a point on the screen to draw a profile view.

Create custom grading criteria. Using grading criteria to limit the amount of user input can help cut down on input errors and design mistakes. Defining a set to work with for a given design makes creating gradings a straightforward process.

Master It Continuing with the `Mastering Grading.dwg` file, create two new grading criteria: the first to be named Bottom, with a criteria of 5′ at 2 percent grade, and the second to be named Sides, with criteria of daylight to a surface at 6:1 slope in cut areas only; in fill, do not grade.

Solution

1. On the Settings tab of Toolspace, expand the Grading ➤ Grading Criteria Sets branches.
2. Right-click Basic Set and select New.
3. Enter **Bottom** in the Name text box.
4. Switch to the Criteria tab and set as shown here:

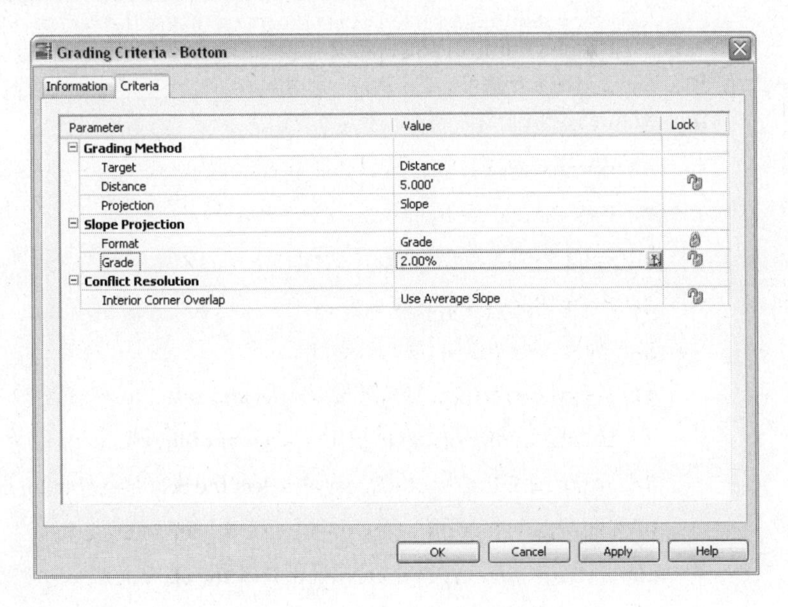

5. Click OK.

6. Right-click Basic Set and select New.
7. Enter **Sides** in the Name text box.
8. Switch to the Criteria tab and set as shown here:

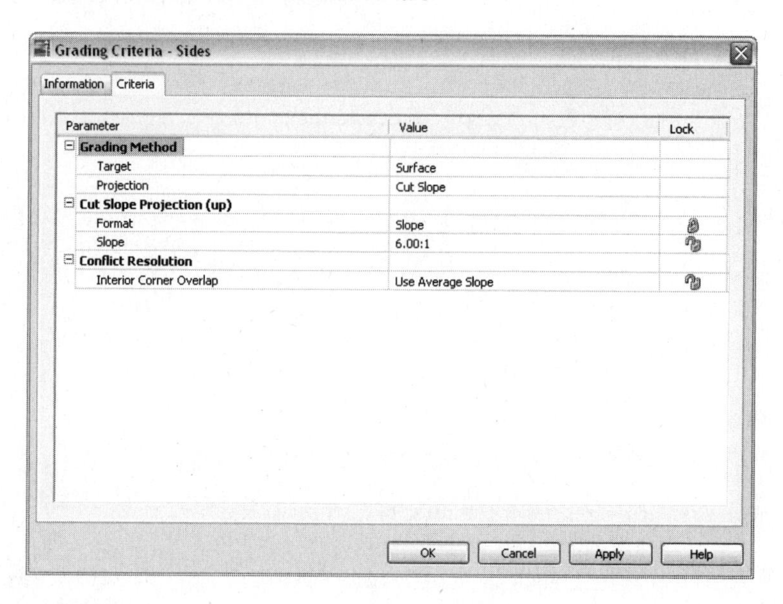

9. Click OK.

Model planar site features with grading groups. Once a feature line defines a linear feature, gradings collected in grading groups model the lateral projections from that line to other points in space. These projections combine to model a site much like a TIN surface, resulting in a dynamic design tool that works in the Civil 3D environment.

Master It Use the two grading criteria just used to define the pilot channel, with grading on both sides of the sketched centerline. Calculate the difference in volume between using 6:1 side slopes and 4:1 side slopes.

Solution

1. From the Home tab's Create Design panel, select Grading ➤ Create Grading to activate the Grading Creation Tools toolbar.
2. Click the Set The Grading Group tool.
3. Check the Automatic Surface Creation option.
4. Check the Volume Base Surface option, and click OK.
5. Click OK to accept the surface creation options.
6. Change the Grading Criteria to Bottom.
7. Click the Create Grading tool and pick the feature line.
8. Pick the left or right side. Press ↵ to model the full length.

9. Press ← to accept 5′ width.
10. Press ← to accept 2 percent grade.
11. Pick the main feature line again and grade the other side.
12. Change to the Sides Criteria and grade both left and right sides, accepting the default values.
13. Right-click to complete the gradings. The complete channel should look like what's shown here:

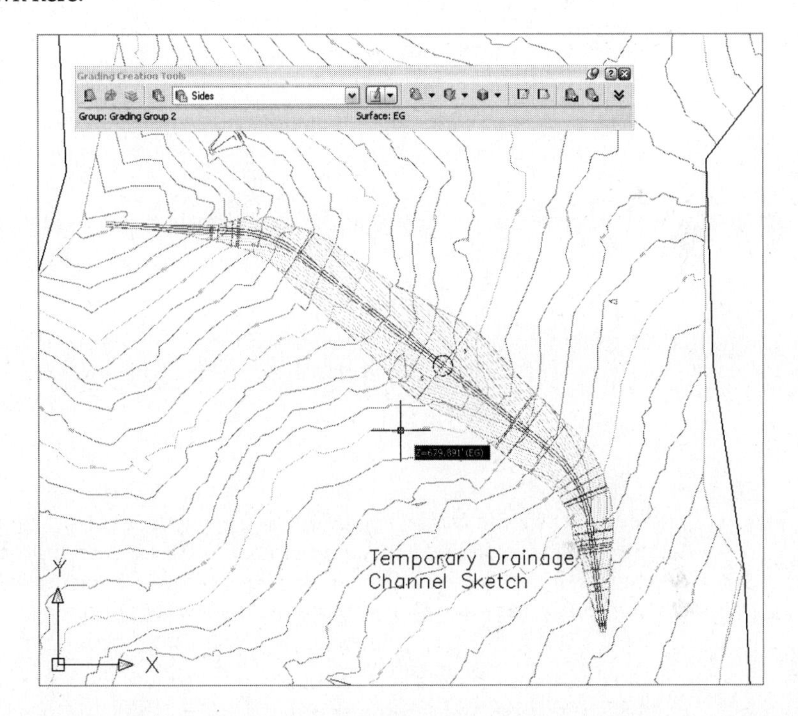

14. Pick one of the diamonds in the grading. Right-click and select Grading Group Properties.
15. Switch to the Properties tab and note the volume (approximately 17,800 Cu. Yd. Cut Net). Click OK.
16. Right-click the diamond representing the Sides grading.
17. Click the Grading Editor tool.
18. Change the Slope to 4:1. Click OK.
19. Repeat for the other side of the channel.
20. Pick the diamond again and select Grading Group Properties.

    The new net volume is approximately 12,500 Cu.Yd. The difference is approximately 5,300 Cu.Yd.

Chapter 17: Sharing the Model: Data Shortcuts

List the project elements that are available for sharing through data shortcuts and those that are not. The ability to load design information into a project environment is an important part of creating an efficient team. The main design elements of the project are available to the data shortcut mechanism, but some still are not.

Master It List the top-level branches in Prospector that cannot be shared through data shortcuts.

Solution Points, point groups, sites, corridors, assemblies, subassemblies, and survey pieces are all unavailable to shortcuts.

Create a data shortcut file from Civil 3D objects. Just creating a Civil 3D object isn't enough to share it using data shortcuts. You still have to go through the process of creating a shortcut file and exporting the linking information to allow other users to reference and analyze your data.

Master It Open the `Creating Shortcuts.dwg` file and create shortcut files for both of the pipe networks contained in that drawing.

Solution

1. Within Prospector, right-click Data Shortcuts and select General ➤ Data Shortcuts ➤ Edit Data Shortcuts to display the Data Shortcuts palette in Panorama.
2. Delete any shortcuts.
3. Click Create Data Shortcuts and pick the two pipe networks.
4. Click OK to close the dialog.

Import a data shortcut and create references. Once a shortcut to a design element has been created, references can be made and managed. Team members should communicate that these files have been created and take appropriate steps so they are not overwritten with bad information. Creating and managing references is the most critical part of keeping project team members in sync with one another.

Master It Create a new drawing, and create references to both the EG surface and a pipe network.

Solution

1. In Prospector, right-click on the Data Shortcuts branch and select Create Data Shortcuts.
2. Select the Pipe Networks check box to select all of the pipe networks in the drawing.
3. Click OK to create the shortcut files.

Restore a broken reference and repath shortcuts to new files. During a design project, it's common for items to change, for design alternatives to be explored, or for new sources to be required. By working with the Data Shortcuts Editor, you have the power to directly manipulate what Civil 3D objects are being called by shortcuts and manipulate them as needed.

Master It Change the data shortcut files such that the file made in the previous Master It exercise points to the EG surface contained in the `Creating References` file instead of the `XTOPO` file.

Solution

1. Launch DSE.
2. Select File ➤ Open Data Shortcuts Folder to display the Browse for Folder dialog.
3. Navigate to `C:\Mastering Shortcuts\Niblo` and click OK to dismiss the dialog.
4. On the left, select the Surfaces branch to isolate the surfaces.
5. Click in the Source File column on the EG row, and type **Creating References.dwg** as the Source file name.
6. Click the Save button in the DSE and switch back to Civil 3D. You should still be in the `Creating References` drawing file.
7. In Prospector, right-click the Data Shortcuts branch and select Validate Data Shortcuts. Prospector will alert you of a broken reference.
8. Expand Data Shortcuts ➤ Surfaces. Right-click EG and select Repair Broken Shortcut.
9. Scroll up within Prospector, and expand the Surfaces branch.
10. Right-click EG and select Synchronize. Your screen should now show a completely different surface as the EG surface.

Chapter 18: Behind the Scenes: Autodesk Data Management Server

Recommend a basic ADMS and SQL installation methodology. The complexity of corporate networks makes selecting the right combination of ADMS and Microsoft SQL software part of the challenge. Understanding the various flavors of SQL Server 2005 at a basic level, along with some limitations of the ADMS, will aid in selecting the right combination for your own needs.

Master It For each of the following descriptions, make some general recommendations for a general setup for their ADMS and SQL needs — for example, a 30-person engineering firm with two offices that rarely share project data. The solution might be a pair of SQL Express machines, one for each office, running a flavor of Windows Server to get around any limits on concurrent connections.

1. A 100-person engineering firm with one office.
2. A 15-person firm with three offices equally split. They share data constantly.
3. A 5-person engineering firm with one office.

Solution

1. They should use a single Windows Server machine with a full version of SQL (not Express).
2. Because they want to share project data, they should consider Replicator, which requires full versions of SQL and Windows Servers. ADMS can be run over a WAN, but network performance may not be acceptable. This one's the gray area of ADMS.
3. A single XP box running SQL Express will suffice if there's enough storage on the XP computer.

Create multiple vaults, users, and user groups. The enterprise nature of ADMS means it can handle multiple sets of data and user inputs at once. By creating multiple vaults to help manage data, you can make life easier on end users. Creating a login to the ADMS for each user lets you track file transactions, and creating groups makes security easier to handle.

Master It Create a new vault titled EE; add Mark, James, Dana, and Marc as users to this new vault. Additionally, make Marc a vault consumer, Mark and Dana vault editors, and James an administrator. Add an architecture group. Marc is the only architect; add the other new users to the engineering group.

Solution

1. Within ADMS Console, click the Vaults folder in the left pane, right-click, and select the Create option.
2. Name the vault **EE**, and click OK.
3. Select Tools Administration, and click Users. Click New User, and assign roles, vaults, and groups as described for these users.

Manage working folders. The working folder acts as your local desktop version of the project files. Drawings are saved and edited in the working folder until checked back into the Vault. Enforcing the working folder creates a constant path for XRef information and makes support easier.

Master It Set the working folder for the new EE vault to `C:\EE Projects\`. (Be sure to set this back to `C:\Mastering Projects` before you begin Chapter 19.)

Solution

1. Log out of Vault Explorer if necessary, and log back in, selecting the new EE vault.
2. Select Tools ➢ Administration. On the Files tab, click Define in the Working Folder area, and select the To Enforce Consistent Working Folder for All Clients radio button.
3. Type in the desired path, and click OK.
4. Return to Civil 3D, and log out. Then, log in to the EE vault to verify the change.

Chapter 19: Teamwork: Vault Client and Civil 3D

Describe the differences between using Vault and data shortcuts. The two mechanisms in Civil 3D for sharing data are similar. Every firm has its nuances, and it's important to evaluate both methods of sharing data to see which one would be a better fit for the team's workflow. Recognizing the differences will allow you to make these recommendations with confidence.

Master It Describe three major differences between the two data-sharing methods.

Solution

- Vault requires ADMS, which is a bit more complicated than the XML files of shortcuts.
- Vault allows for user-security levels on project folders.
- Vault allows for easy version tracking and restoration.

There are other differences between Vault and data shortcuts, but these are the big ones.

Insert Civil 3D data into a vault. Creating a project in Vault requires some care to make sure all the pieces go in the right place. Using a project template to create organizational structures similar to what is already in place is a good way to start. Adding individual components to the vault as they fit into this structure makes for clean data organization and easy integration of new team members.

Master It Create a new project titled **Bailey** in the Melinda vault using the Mastering project template. Add the `Bailey's Run.dwg` file to the project in the `Engineering` folder, sharing all the data within it.

Solution

1. Right-click Projects, and log in to the Melinda vault.
2. Right-click Projects, and select New.
3. Name the project **Bailey**.
4. Select the Mastering template.
5. Click OK.
6. Open the Bailey's Run drawing.
7. Right-click the drawing name in Prospector, and select Add to Project.
8. Select the `Engineering` folder.
9. Share all the data.

   Prospector should appear as shown here when complete.

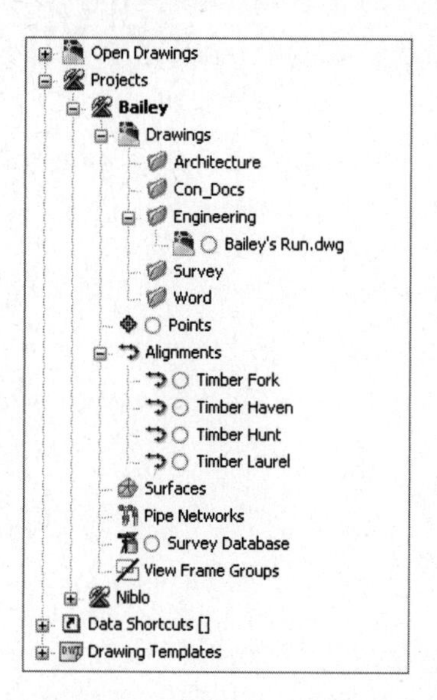

Create data shortcuts. Creating duplicate data is never good. By creating links to original data instead of working with copies, you maintain the integrity of the project information. Using data references instead of XRefs allows you to use the data behind the object instead of just the picture. This means styling and label changes can be part of the end file and allow team members to work on separate parts of the job at the same time.

Master It Create a new drawing called **Mastering Vault**, add it to the Bailey project in the `Engineering` folder, and create references to the FG surface and to the Natures Way and Old Settlers Way alignments. Use a Contours 2′ and 10′ (Design) style for the surfaces.

Solution

1. Create a new drawing from the NCS Extended template.
2. Save the drawing as **Mastering Vault** in `C:\Mastering Projects\Bailey\Engineering`.
3. Right-click the filename in Prospector, and select Add to Project.
4. Keep the drawing checked out, and step through the check-in process.
5. Expand Projects ➤ Bailey ➤ Surfaces, and select FG.
6. Right-click, and select Create Reference.
7. Click OK to make the reference.
8. Expand Projects ➤ Bailey ➤ Alignments, and select Natures Way.
9. Right-click, and select Create Reference.
10. Click OK to finish creating the reference.
11. Repeat this procedure for Old Settlers Way.

Restore a previous version of a design file. One of the major advantages of using a database-driven system is the ability to track file changes and restore them if necessary. Vault has a mechanism for moving data in and out of the file store. By using the restoration tools and checking in the proper files, it's easy to roll back to a prior version of a file.

Master It Erase the Founders Court alignment from the Bailey's Run file, and check it in. Then, use the Vault tools to restore the previous version containing the Founders Court alignment. What version number of the Bailey's Run drawing is listed when this is complete?

Solution

1. Expand Projects ➤ Bailey ➤ Alignments, and select Founders Court.
2. Right-click, and select Check Out Source Drawing.
3. Use Prospector to delete Founders Court.
4. Check Bailey's Run back in to the vault.
5. Expand the Alignments collection under the Bailey collection to confirm the erasure of the Founders Court alignment from the project.
6. In Vault, select the Bailey project's `Engineering` folder.
7. Select Bailey's Run in the upper-right pane.
8. Right-click, and select Get Previous Version.
9. Select Version 1, and click OK.
10. Return to Civil 3D.
11. In Prospector, check out Bailey's Run from Projects ➤ Bailey ➤ Drawings ➤ Engineering. Be sure to deselect Get Latest Version.
12. Right-click Bailey's Run, and check it back in.
13. Select Founders Court in the Share Data step as you go through the check-in process.
14. Click Finish.
15. Switch to Vault, and refresh the view of the `Engineering` folder. The version for Bailey's Run should be **3**.

Chapter 20: Out the Door: Plan Production

Create view frames. When you create view frames, you must select the template file that contains the layout tabs that will be used as the basis for your sheets. This template must contain predefined viewports. You can define these viewports with extra vertices so you can change their shape after the sheets have been created.

Master It Open the Mastering Plan Production drawing. Run the Create View Frames Wizard to create view frames for any alignment in the current drawing. (Accept the defaults for all other values.)

Solution

1. Change to the Output tab, and then select Plan Production Tools ➤ Create View Frames.
2. On the Alignment page, select Haven CL from the alignment drop-down menu. Click Next.
3. On the Sheets page, select the Plan and Profile option. Click the ellipsis button to display the Select Layout as Sheet Template dialog. In this dialog, click the ellipsis button, browse to `C:\Mastering Civil 3D 2009\CH 20\Data`, select the template named `Mastering (Imperial) Plan and Profile.dwt`, and click Open.
4. Select the layout named ARCH D Plan and Profile 20 Scale, and click OK.
5. Click Create View Frames.

Create sheets and use Sheet Set Manager. You can create sheets in new drawing files or in the current drawing. Use the option to create sheets in the current drawing when a) you've referenced in the view-frame group, or b) you have a small project. The resulting sheets are based on the template you chose when you created the view frames. If the template contains customized viewports, you can modify the shape of the viewport to better fit your sheet needs.

Master It Continue working in the Mastering Plan Production drawing. Run the Create Sheets Wizard to create plan and profile sheets for Haven CL using the template `Mastering (Imperial) Plan and Profile.dwt`. This template has a custom-made viewport with extra vertices. (Accept the defaults for all other values.)

Solution

1. Change to the Output tab, and then select Plan Production Tools ➤ Create Sheets.
2. On the View Frame Group and Layouts page, under the Layout Creation section, select All Layouts in the Current Drawing.
3. Click Create Sheets.
4. Click OK to save the drawing.
5. Click a location as the profile origin.
6. Dismiss the events Panorama.
7. In Sheet Set Manager, double-click the sheet named 1- Haven CL 10+00.00 to 14+28.00.
8. Select the top, Plan View viewport. Notice the extra vertices.
9. Grip-edit the shape of the viewport by stretching the various vertices to new locations to make an irregularly shaped viewport.

Edit sheet templates and styles associated with plan production. The Create View Frames command uses predefined styles for view frames and match lines. These styles must exist in the drawing in which you create the view-frame group. You can create match-line styles to meet your specific needs.

Master It Open the Mastering Plan Production 3 drawing, and modify the match-line component hatch display style setting to make the linework beyond the match line appear screened.

Solution

1. On the Settings tab of the Toolspace palette, expand the Match Line group, and then expand the Match Line Styles collection. Right-click Standard – Revised, and select Edit.
2. On the Display tab, in the Component Display section, for the Match Line Mask component, click the color to open the Select Color dialog.
3. On the True Color tab, select RGB from the Color Model drop-down menu.
4. For Red, Green, and Blue, enter **255**, and click OK. This tells Civil 3D to set the color of the hatch pattern to match the background color.
5. In the Component Hatch Display section, click the pattern name SOLID. In the box that pops up, change the type to Predefined, and select Dots in the Pattern Name box. Click OK to return to the previous screen. Change Angle to **22** degrees and Scale to **3**.
6. Click OK, and examine the linework beyond the match line. The dots of the hatch pattern obscure a portion of the linework, making it appear screened. Depending on the scale of your final plotted plans, you may need to experiment with the rotation and scale to get the screening effect you want.

Chapter 21: Playing Nice With Others: LDT and LandXML

Discern the design elements in a LandXML data file. `LandXML.org` is an international consortium of software firms, equipment manufacturers, consultants, and other land-development professionals. By creating a lingua franca for the industry, they've made sharing design data between design and analysis packages a simple import/export operation.

Master It Download the `Campground.xml` file from the `landxml.org` website and list the creating program, along with the alignments and the surfaces it contains.

Solution

Program	Eagle Point Civil/Survey 2002
Alignments	Lunar Loop, Short Street, Quiet Road, Sunset Strip, Shady Lane, Sunshine Street
Surfaces	Original Ground Model, Campsite Model

Import a Land Desktop project. Many Civil 3D firms have a huge repository of LDT projects. By using Civil 3D's Import Data from Land Desktop functionality, this data's lifespan can be extended, and the value of the information it contains can continue to grow.

Master It Import the STM-P3 Pipe Run from the Conklin LDT Project data source, and count the number of pipes and structures that successfully import.

Solution

1. Create a new Civil 3D drawing using the `_AutoCAD Civil 3D (Imperial) NCS Extended` template file.
2. On the Insert tab's Import panel, click the Land Desktop button.

3. Click the Browse button.
4. Navigate to the data directory, and select LDT Import.
5. A warning appears regarding pipes. Click OK to dismiss it. This warning is a reminder that you need a pipe part in the parts list to match the sizes described in the LDT project.
6. Select Conklin in the `Import` directory. (This is similar to selecting the Project path in LDT.)
7. Deselect all components except the STM-P3 pipe network, as shown here:

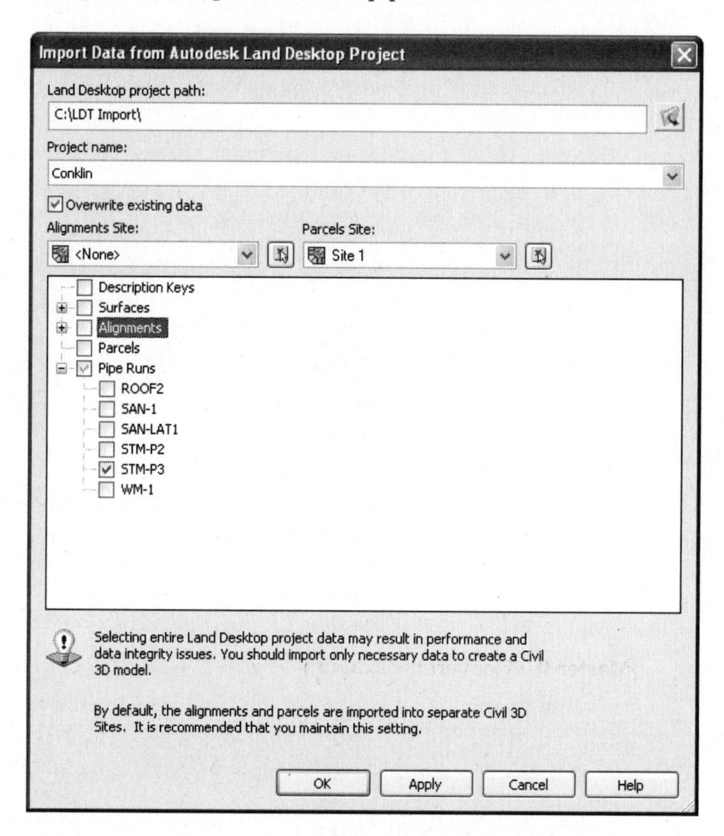

8. Click OK. Four pipes and five structures should successfully import.

Import a LandXML file to create a surface, alignments, and profiles. As more and more firms use Civil 3D, they'll want to share design data without giving up their full drawing files. Using LandXML to share import data allows the accurate transmission of data in a standardized format and gives the end recipient the ability to manipulate the entities as needed.

Master It Complete the import of the `Campground.xml` file into a new Civil 3D drawing.

Solution Create a new Civil 3D drawing, and from the Insert tab's Import panel, click the LandXML button. Navigate to `campground.xml`, and click OK to import. The drawing should look like this:

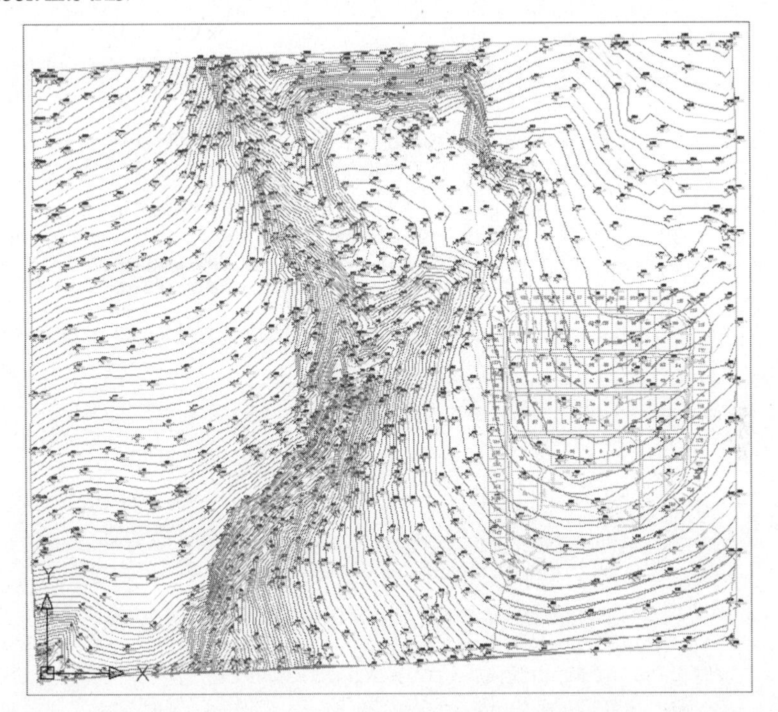

Create an AutoCAD drawing that can be read without special enablers. Because many members of your design team won't be working with Civil 3D, it's important to deliver information to them in a manner they can use. Using the built-in tools to create drawings consisting of purely native AutoCAD objects strips these drawings of their design intelligence, but it provides a workable route to data sharing with almost any CAD user.

Master It Convert the drawing you just made from `Campground.xml` to an R2004 AutoCAD drawing.

Solution Choose C:\ ➤ Export ➤ AutoCAD DWG ➤Export to AutoCAD R2004 Format. The resulting file should display this message at the command line upon opening:

```
Opening an AutoCAD 2004/LT 2004 format file.
Regenerating model.
```

Chapter 22: Get The Picture: Visualization

Apply render materials to a corridor using a code set style. Visualizing corridors is useful for creating complete site models for your own better understanding of the design, and for sharing with your clients and regulating agencies. The first step in visualizing a corridor is to apply a code set style that will automatically assign render materials to certain codes.

Master It Open the drawing file `Mastering Visualization.dwg`. Apply the All Codes code set style.

Solution

1. Select the corridor, right-click, and choose Corridor Properties. Switch to the Codes tab.
2. On the Code Set Style drop-down menu, choose All Codes.
3. Click OK.

Apply a Realistic visual style to a corridor model. Once the appropriate code set style has been assigned, the corridor can be seen with its render materials with a Realistic visual style applied.

Master It Continue working in the file `Mastering Visualization.dwg`. Apply the Realistic visual style to the drawing.

Solution

1. Continue working in `Mastering Visualization.dwg`.
2. Switch to the Visualization and Rendering workspace if it is not already current.
3. Using the Visual Styles control panel on the Dashboard, switch from the 2D Wireframe style to the Realistic visual style.

Create a 3D DWF from a corridor model. One way you can share your visualized corridor is through 3D DWF. Autodesk Design Review is a free application that you can direct your clients to download from Autodesk.com. 3D DWFs can be opened, navigated, and analyzed using Design Review.

Master It Continue working in the file `Mastering Visualization.dwg`. Create a 3D DWF from the drawing and open it in Autodesk Design Review.

Solution

1. Continue working in `Mastering Visualization.dwg`.
2. Type **3DDWFPUBLISH** at the command line.
3. Give your DWF a name and choose a location to save it. Click OK.
4. Click Yes to open the DWF and use the navigation tools to study the resulting DWF.

Chapter 23: Projecting the Cost: Quantity Takeoff

Open and review a list of pay items along with their categorization. The pay item list is the cornerstone of QTO. You should download and review your pay item list versus the current reviewing agency list regularly to avoid any missed items.

Master It Add the 12-, 18-, and 21-Inch Pipe Culvert pay items to your Favorites in the QTO Manager.

Solution

1. Open the QTO Manager palette.
2. Enter **12-Inch Pipe** in the text box to filter.
3. Right-click the 12-INCH PIPE CULVERT line and select Add to Favorites.
4. Repeat for the other sizes.

Assign pay items to AutoCAD objects, pipe networks, and corridors. The majority of the work in preparing QTO is in assigning pay items accurately. By using the linework, blocks, and Civil 3D objects in your drawing as part of the process, you reduce the effort involved with generating accurate quantities.

Master It Open the `QTO Reporting.dwg` file and assign the Pavement Marking Type C Broken to the Alexander Ave., Parker Place, and Carson's Way alignments.

Solution

1. Open the QTO Manager palette.
2. Expand the Favorites branch and select the PAVEMENT MARKINGS, TYPE C, BROKEN item.
3. Right-click and select Assign Pay Item.
4. Select the three alignments, and press ↵ to complete the selection.

Use the QTO tools to review what items have been tagged for analysis. By using the built-in highlighting tools to verify pay item assignment, you can avoid costly errors when running your QTO reports.

Master It Verify that the three alignments in the previous exercise have been assigned with pay items.

Solution

1. Turn on the Highlight Objects with Pay Items in the QTO Manager Palette.
2. Pan and ensure that all three alignments in question are highlighted and not shown in black, indicating no pay item assignment.

Generate QTO output to a variety of formats for review or analysis. The Takeoff reports give you a quick and understandable understanding of what items have been tagged in the drawing, and they can generate text in the drawing or external reports for uses in other applications.

Master It Calculate and display the amount of Type C Broken markings in the `QTO Reporting.dwg` file.

Solution

1. From the Analyze tab's QTO panel, select the Takeoff command, and click OK to run the report with default settings.
2. Change the report style to Summary (TXT).xsl. There should be 2037.309 LNFT of Type C Broken marking.

Index

Index

Note to the Reader: Throughout this index **boldfaced** page numbers indicate primary discussions of a topic. *Italicized* page numbers indicate illustrations.

D

E

G

H

I

J

K

L

S

T

U

W

X

Y

Z